Biology for Life:

A Guide to Our Living World

Melanie E. Peffer

Cengage

Australia • Brazil • Canada • Mexico • Singapore • United Kingdom • United States

Cengage

***Biology for Life: A Guide to Our Living World*, First Edition**
Melanie E. Peffer

EVP, General Manager, Cengage Academic: Nhaim Khoury

VP, Portfolio Management: Mark Santee

Portfolio Director: Maureen McLaughlin

Senior Portfolio Manager: Diana Baniak

Portfolio Assistant: Linh Lowry

Learning Designer: Kate Scheibel

Content Manager: Renee Schnee, Brendan Killion

Digital Project Manager: Yogalakshmi Mohana Krishnan

Development Editor: Mary Ann Murray

Product Marketing Manager: Valerie Hartman

Content Acquisition Analyst: Nichole Nalenz

Production Service: Lumina Datamatics Ltd.

Senior Graphic Designer: Chris Doughman

Cover Image Source: Bestproductsforsales1 /Dreamstime.LLC

Library of Congress Control Number: 2026901084

ISBN: 978-0-357-97598-5

Cengage
10650 Toebben Drive, Suite A
Independence, KY 41051
USA

Cengage is a leading provider of customized learning solutions. Our employees reside in nearly 40 different countries and serve digital learners in 165 countries around the world. Find your local representative at **www.cengage.com**.

To learn more about Cengage platforms and services, register or access your online learning solution, or purchase materials for your course, visit **www.cengage.com**.

Printed at CLDPC, USA, 2-26

Dedication

To my children—you are the light of my life. This book, and all my books, are always for you.

© Kevin Snyder Photography

Table of Contents

About the Author

© Kevin Snyder Photography

Dr. Melanie Peffer is a best-selling author, passionate biology educator, and accomplished science communicator whose work to help people rethink and find joy in biology has inspired thousands worldwide. In two decades of teaching introductory biology, students consistently praise her ability to bring the real world into the classroom in an engaging, creative manner.

Dr. Peffer is an in-demand public speaker and gave a TED talk (*How Rethinking Biology Can Positively Change Your Life*), developed a TED-ED lesson (*The Artist Who Won a Nobel Prize...in Medicine*), and regularly gives professional development workshops on various aspects of science communication, including how to write engaging science content for readers of various backgrounds and interest levels.

Her first book, *Biology Everywhere*, spent time on the Amazon.com best-seller list and won an EVVY gold award for educational text in 2022 from the Colorado Independent Publishers Association. After being selected as the 2021 High Plains Library District Writer in Residence, she wrote a follow-up children's spinoff of *Biology Everywhere* called *On the River*, which won EVVY gold in the children's category in 2022.

Dr. Peffer teaches introductory molecular biology courses at the University of Colorado and leads a learning sciences research laboratory. Her research focuses on how people learn, understand, and engage with biology content. She also uses textile arts like quilting to tell science stories and promote science engagement. Dr. Peffer's quilts have been displayed at county fairs and academic conferences alike where diverse audiences enjoy interacting with and telling their own creative science stories after viewing her quilts.

Outside of work, Dr. Peffer enjoys exploring all Colorado has to offer with her husband and their children. She also enjoys playing the flute and piccolo, reading, and sewing.

Acknowledgments

When *Biology Everywhere* was published, everyone asked me, "How long did it take you to write your book?" and my answer was always that the writing doesn't take long, it's the revising! Now with a textbook, we add to the list so many more steps, from art development to getting permission to use photos, to composing pages…and so much more! Although my name is on the front cover, I wouldn't be here without the massive team of people working behind the scenes. I am incredibly grateful to the dozens of talented people from around the world who answered my questions (and there were *a lot*!), gave feedback, checked my work, gave me pep talks, and contributed their own artistic skills to make *Biology for Life* a success.

First, to my family—Kevin, Felix, and Jasper. Thank you for all that you've sacrificed over the last few years as I've worked late time and again on this project. I appreciate your patience with me, your willingness to have your own stories included, and the never-ending supply of hugs.

Second, I want to thank my students. I have the privilege of teaching the absolute *best* students at University of Colorado Boulder. I always say that I learn just as much from my students as they learn from me—and that's never changed. I appreciate your support, excitement, and willingness to be guinea pigs on textbook material and provide feedback. I also want to acknowledge the students who answered random questions from me, ranging from Taylor Swift to checking if cultural references from me (a millennial) made sense to them. Sko Buffs!

I'd also like to thank Alison Williams for helping with the kids so much during the first year of the book project and all of her kindness and support with the long hours I've put into writing. To Elizabeth Warson and Megan Jett, I honestly don't know what I'd do without either of you—thank you for your support.

Now, moving onto the core members of the book team. To Diana Baniak—thank you for believing in me and all the many pep talks along the way. I appreciate your kindness, support, and leadership so much. To Renee Schnee, thank you for your organization and aid in staying on track and answering so many emails about what to prioritize on any given day. To Mary Ann Murray, authors are only as good as their editors—and I appreciate all your time and attention to detail to help turn my ideas into reality. A huge thank you to Craig Durant and Mike Demaray at Dragonfly Media Group for the fantastic art.

I also am grateful to the subject matter experts who contributed in various ways to the project, from answering my questions to writing end of chapter questions: Kate Scheibel, Holly Basta, and Jennifer Myka. I would also like to thank Hayley Chwazik-Gee for her help coordinating with the National Geographic Explorers featured in the text. I am enormously grateful to our team of reviewers for their helpful feedback and comments:

Adeyemi Adams, *PhD, Dallas College*
Aigerim Bijelic, *PhD, Harold Washington College*
Aiwei Borengasser, *PhD, University of Arkansas - Pulaski Technical College*
Alexandra Barnard, *PhD, National Park College*
Alex Landy, *The University of Tampa*
Amy Bekins, *Ed.D., University of Northern Colorado*
Andrea Green, *University of Montana*
Anica Lee, *PhD, Texas State University*
Ashley Spring, *PhD, Eastern Florida State College*
Becky Green-Marroquin, *PhD, Los Angeles Valley Community College*
Begona De Velasco, *PhD, California State University-Dominguez Hills*
Bridget Stuckey, *PhD, Fayetteville Technical Community College*
Cara Gormally, *PhD, Gallaudet University*
Cheryl Berry, *Saint Leo University*
Cheryl Chaney, *MS, Arkansas Tech University*
Donee Taylor, *Holmes Community College*
Erica Cain, *PhD, Northeast Lakeview College*
Eric Scott, *North Carolina Agricultural & Technical State University*
Fleur Bond, *MS, Community College of Denver*
Genevieve Chung, *EdD, Broward College*
Helene Engler, *PhD, University of Texas at Austin*
Ibrahim Ekaidi, *PhD, Southern University at New Orleans*
Jack Taha, *PhD, Oakton Community College*
Kara Reed, *Kansas City, Kansas Community College*
Kari Clifton, *PhD, University of West Florida*
Karry Kazial, *PhD, State University of New York at Fredonia*
Katie Rushing-Anderson, *Black Hawk College*
Katy Hinman, *PhD, American Association for the Advancement of Science*
Kevin Oley, *MS, Appalachian State University*
Kristy McClain, *PhD, Holmes Community College*
Lalitha Krishnan, *MS, Tyler Junior College*
Larry Baranowski, *PhD, East Tennessee State University*
Latasha Smith, *PhD, Champion Baptist College*
Lauren Weingarten, *Hinds Community College*
Leigh Delaney-Tucker, *MS, University of South Alabama*
Leslie Harrell, *Wilson Community College*
Le Su, *PhD, Jacksonville State University*
Lindsay Jackson, *MS, Purdue University Fort Wayne*
Lorena Harris, *PhD, SUNY Schenectady County Community College*
Marcia Pierce, *PhD, Eastern Kentucky University*
Marirose Ethington, *PhD, Genesee Community College*
Mary Dougherty, *Catawba Valley Community College*
May Coulson, *Oakton Community College*
Michael Rutledge, *PhD, Middle Tennessee State University*
Michele Glass, *Northeast State Community College*
Michelle Zurawski, *MS, Moraine Valley Community College*
Miranda Chen Musgrove, *PhD, East Tennessee State University*
Mitch Price, *NAEP*
Pamela Hobbs, *Doctor of Physical Therapy, DPT, Lee University*
Rosalyn Hunter, *Hill College*
Ryan Kelly, *PhD*
Samiksha Raut, *PhD, The University of Alabama at Birmingham*
Stephen Henry, *Houston Community College*
Stephen Page, *PhD, The Community College of Baltimore County*
Steven Merkley, *Cochise College*
Sukanya Subramanian, *PhD, Collin College*
Tiffany McFalls-Smith, *Elizabethtown Community and Technical College*
Tom Gehring, *PhD, Central Michigan University*
Winnifred Bryant, *PhD, University of Wisconsin-Eau Claire*
Yevgeniya Lapik, *PhD, Harold Washington College*

Finally, I want to thank some of my favorite authors: Diana Gabaldon, Deborah Harkness, Sarah J. Maas, Andy Weir, and Rebecca Yarros. Your words and style of writing have both added positively to my life and shaped my own writing. Thank you for sharing your stories—and inadvertently teaching me how to tell science-filled stories.

Mariposa is a quilt by the author that appears in Chapter 16 to represent hope and the belief that teaching biology will lead to a better world.

Preface

Dear student:

The very first biology class I took in college, I failed and had to re-take. So, despite my later success in biology, I know first hand what it's like to not do well in a biology class. I also know what it's like to try *really hard* to succeed in a class…only to fail. I know how frustrating and seemingly worthless textbooks can feel. By the time I graduated from college with my BS in molecular biology, I was convinced that there was no reason to use a textbook other than to complete homework assignments. This point hits even harder today—why purchase a textbook if you can easily Google your question or ask generative AI to explain it to you?

When I set out to write *Biology for Life*, I wrote the textbook that I needed as a student. While yes, of course, there's a lot of biology information in this text, it's presented largely through stories to make the content interesting and relevant. That's why the book is called *Biology **for** Life*. All the content you'll see here had to pass the test of relevancy—otherwise, it was cut. In addition, I made many cross-disciplinary connections so you can see the larger relevancy of biology. For more information on my approach, please see page XV.

Biology for Life is also designed using the science of learning. I carefully considered everything from the arrows used in figures (see Appendix A for the significance of the arrows used in this text) to the marginal elements (which include my avatar, pictured here) to the inclusion of coaching in every chapter, with suggestions for how you can study for success in your biology course.

My hope is that in addition to being the textbook that ***I*** needed as a student, this text meets its intended purpose—to be the textbook ***you*** need for success. As I wrote this book, the entire time I was thinking about what you, as a non–biology major student, need from a biology course. So. Here is a textbook that is not "dumbed down" biology, but rather a book written specifically for someone not majoring in biology.

I look forward to hearing about how *Biology for Life* helped you see how biology is part of your life. You can reach me at www.cengage.com/contact ATT: Melanie Peffer.

Melanie Peffer

Dear fellow biology educators:

Where have you experienced biology today?

Although it's perhaps simple for us to answer, so many of our students struggle with this question. Biology often feels inaccessible and abstract to non-major students, and most of them sign up for our classes simply because they need to check one more general education requirement off the list to graduate. But not engaging with biology isn't an option in the real world. We all need to make biologically informed decisions as members of society. These decisions include whether to get a flu shot, or if we'll purchase genetically modified foods, or how to vote on a measure when issues related to biology appear on the ballot.

Biology for Life: A Guide to Our Living World is more than just a textbook. It is biology told through our daily experiences. Helping students attain the needed knowledge of biology to make informed decisions in our ever-changing world starts with helping them to identify biology in their lives. When we see biology as not just understandable, but relatable, it's a logical next step is to find the material accessible and relevant to one's own lives.

Biology for Life is packed full of examples demonstrating how we experience biology as part of our daily lives. In addition, these examples are told in an innovative manner, such as using stories, cartoons, and high-quality artwork. In line with trends in biology education, the book also includes guidance on how to learn the material presented. There are also plenty of active learning exercises built into the various feature boxes throughout the text, challenging students with current biology topics and encouraging critical thinking about these real-world questions.

I look forward to hearing how *Biology for Life* will help you to teach biology, and help your students to learn. You can reach me at www.cengage.com/contact ATT: Melanie Peffer

Melanie Peffer

Text Philosophy

© Melanie Peffer

In 2024, I completed my first narrative quilt, called *Look Again* (pictured here). The *Look Again* quilt is a pictorial representation of my teaching philosophy for biology education—one that undergirded my TEDx talk in 2021 as well as my first book, *Biology Everywhere*, which is foundational to this text, *Biology for Life*. Many people, including non-major biology students, think that they dislike biology (or science in general) or lack the confidence to engage. Incidentally, the block on the black-and-white side of the quilt—the rightmost column, second row down—contains words like "hard" and "boring," which are direct quotes from my non-major students at the beginning of the semester in response to me asking about their thoughts on the semester ahead. The left side of the quilt represents these feelings—and the closed eye at the top represents the desire that many have to look away. Note the blocks with a rotational feel, to represent confusion, and the presence of stereotypes, like the mad, white, male scientist and jargon.

Now, consider the right side of the quilt, with the open eye—when we look again at biology, we can see that it's more colorful, interesting, and unexpected than one might believe. What does music have to do with biology, after all? Well, many things—from cues about our shared evolutionary history with other animals who sing to what is chemically happening in our brains when we hear or perform a lovely piece of music.

That is fundamentally the philosophy behind the text: When we look at biology as being integrated into our lives, it becomes accessible. When we hear stories about biology in our lives, it's relevant. When we look again, or in this case, help our students to look again, we can see biology not as boring or hard, but as an integral, interesting, and surprising part of our lives. As such, the text is full of personal touches and stories inspired by biology in my life—and offered to you to find ways to make connections to your own life.

Text Organization

The text is organized around six thematic units, following the order of a traditional one-semester biology course. Each of these thematic units connects the core content to the real world: Nutrition, Cancer, Human Reproduction, Tracing Life's Path, Climate Change, and Biology and You. Please note that the last three chapters of the book on conservation biology, human anatomy and physiology, and plant physiology are designed so that you can teach them like a typical chapter or take a modular approach. For example, depending on the political climate in your state, you may choose to skip Chapter 16 altogether, or skip the section of Chapter 16 that specifically discusses the impacts of climate change. Chapters 17 and 18 are designed to be cumulative, and so the content included in these chapters can support the coverage in chapters earlier in the semester, as desired.

Each chapter is structured around the study cycle as it is introduced in Chapter 1. Although there are many different versions of the study cycle, I chose a simplified version to unify the text and increase accessibility for students. Each chapter opens with a Preview It! activity, to get students to reflect on the material ahead and pique their curiosity. At the end of each chapter is a different Understand It! These features are a unique aspect of the text and are designed to give direct instruction to the student on how to effectively study the material. Where appropriate, they are tied directly to the content. This is particularly important for non-major students, who may, for example, be taking the biology course while majoring in the arts, where they would be assessed differently than in a science class. In addition, research[1] on student study habits suggests that students tend to not use evidence-based study strategies, instead requiring direct instruction[2] on how to learn to be successful. Finally, each chapter includes a Practice It! section, which includes a variety of multiple-choice and critical thinking questions to help prepare students for exams or other assessment modalities.

---

1 Carpenter, S. K., Pan, S. C., & Butler, A. C (2022). The science of effective learning with spacing and retrieval practice. Nat Rev Psychol 1, 496–511. https://doi.org/10.1038/s44159-022-00089-1.

2 Biwer, F., Oude Egbrink, M. G., Aalten, P., & De Bruin, A. B. (2020). Fostering effective learning strategies in higher education–a mixed-methods study. *Journal of Applied Research in Memory and Cognition*, 9(2), 186–203.

In addition, each chapter contains three kinds of feature boxes: Biology for Life, Process of Science, and Digging into Data. These feature boxes can be used in a variety of ways, but I particularly like having students read these feature boxes in class and discuss the questions using Think-Pair-Share.[3] Biology for Life boxes are ancillary to the core text and explore current events and issues related to the chapter content in deeper depth, such as looking at the de-extinction of animals. Process of Science boxes build from the descriptions of authentic science practices given in Chapter 1 to highlight science discoveries pertinent to the chapter content. This way, science literacy content extends across the text and is not limited to the first chapter. Finally, the *Digging into Data* feature boxes allow students to work with real-world data, analyzing and interpreting graphs and other visual displays of data, to gain familiarity and confidence for when they might encounter such data in the real world. For additional pedagogical support, a primer on data interpretation is included as Appendix D.

Best Practices for Learning

As a learning scientist and a person identifying with several disadvantaged groups, it was important to me to create a text that was informed by the science of learning while also being inclusive. Therefore, throughout the text, I took specific care not only to educate students on their own learning, but to ensure that the content was presented in a manner based on the most recent educational advances. For example, the marginal elements are designed to keep and/or capture students' attention, to keep them reading. In addition, I removed extraneous information to streamline the text and increase readability.

Significant care was also put into the art development process. Based on research regarding the confusion that textbook arrow styles cause students,[4] I worked with the development team to create an arrow style guide for the text, ensured consistency in arrow usage throughout, and included a detailed guide for students in Appendix A. Art was also designed with consideration of cognitive load[5] in mind, to clearly present a pedagogical point while eliminating extraneous details.

In addition, the text was designed using backward design,[6] starting with learning objectives and assessments and working backward from that starting point. Note that all items in the Practice It! sections are aligned with their respective learning objectives. The text also incorporates a variety of other evidence-based pedagogies, from comics to storytelling.

As a professor, my first goal is to always create a welcoming classroom where all students, regardless of their background, can succeed. I always tell my students on the first day of class that we must "Maslow before Bloom"[7]—meaning that their well-being (described by Maslow's hierarchy of needs)[8] comes before learning (referring to Bloom's taxonomy)[9]. One of the most critical ways that I've attempted to address this in *Biology for Life* is through attention to diversity, equity, and inclusion throughout the text. This includes everything from highlighting social justice issues, such as those we see related to racism and environmental justice, to highlighting scientists and stories from a wide variety of cultural backgrounds. I've tried to ensure that there is something in this text for everyone—if you don't see yourself or your students in these pages, or have a suggested change to make the text more inclusive, please reach out! You can reach me at www.cengage.com/contact ATT: Melanie Peffer.

---

[3] Guenther, A. R., & Abbott, C. M. (2024). Think-pair-share: Promoting equitable participation and in-depth discussion. *PRiMER: Peer-Reviewed Reports in Medical Education Research, 8*, 7.

[4] Wright, L. K., Cardenas, J. J., Liang, P., & Newman, D. L. (2017). Arrows in biology: lack of clarity and consistency points to confusion for learners. CBE—Life Sciences Education, 17(1), ar6.

[5] Skulmowski, A., Xu, K. M (2022). Understanding cognitive load in digital and online learning: a new perspective on extraneous cognitive load. Educ Psychol Rev 34, 171–196. https://doi.org/10.1007/s10648-021-09624-7.

[6] Reynolds, H. L., & Kearns, K. D. (2016). A planning tool for incorporating backward design, active learning, and authentic assessment in the college classroom. College Teaching, 65(1), 17–27. https://doi.org/10.1080/87567555.2016.1222575.

[7] Mutch, C., & Peung, S. (2021). 'Maslow before bloom': implementing a caring pedagogy during COVID-19. New Zealand Journal of Teachers' Work, 18(2), 69–90.

[8] Mathes, E. W. (1981). Maslow's hierarchy of needs as a guide for living. Journal of Humanistic Psychology, 21(4), 69–72.

[9] Krathwohl, D. R. (2002). A revision of bloom's taxonomy: an overview. Theory Into Practice, 41(4), 212–218. https://doi.org/10.1207/s15430421tip4104_2.

Book Features

Evidence-based Pedagogy

Dr. Peffer's 10+ years researching how students learn biology ensures that best practices from the educational research, including design elements and effective strategies for learning the material, facilitate student success.

Understand It! Mindset

Mindset is the beliefs we have about ourselves. These beliefs impact how we behave, our outlook on life, and our attitude, and can directly impact our classroom performance. A *fixed mindset* is the belief that your ability is static. You either have it, or you do not. Someone with a fixed mindset believes that we are born with certain skills and not others. As a result, people with fixed mindsets avoid challenges, give up easily, feel threatened or jealous of others, and are focused on performance and looking good to others.

Growth mindset is the belief that skills and abilities are developed. People with growth mindsets embrace and persevere through challenges. Failure is viewed as an essential part of acquiring new skills. People who have a growth mindset focus on improving, learning, and growing, and are inspired by the success of others.

People can have both a growth and a fixed mindset, depending on the situation. For example, one may have a fixed mindset in the biology classroom, but a growth mindset in their artistic pursuits. Remember that our brains are changeable! We can go from having a fixed to a growth mindset. As Carol Dweck, the psychologist who coined the terms fixed and growth mindset says, embrace the power of YET. I don't understand cellular respiration, YET. YET implies that one believes in their ability to learn and understand the material. I believe in your ability to succeed in this course!

1. What is something in your life that you have a fixed mindset about? How about a growth mindset?
2. Can you think of a time when you didn't know something, but later mastered it?
3. What is something from this chapter that you don't know YET but want to work on?

Current and Relevant

Each chapter details a scientific discovery, insight, or phenomenon focused on current data, while providing tools to distinguish real science from pseudoscience and misinformation.

Biology for Life

Social Media and the Rise of the Infodemic

The rise of social media has facilitated the rapid, worldwide dissemination of unreliable information. During the COVID-19 pandemic, the spread of false information about the virus was so pervasive that the World Health Organization (WHO) referred to it as an *infodemic*. Mark Twain accurately described the phenomenon of infodemics long before the existence of social media when he said "*A lie can travel halfway around the world while the truth is still putting on its shoes.*" Pseudoscientific and false information pervades the internet and falls into two categories, misinformation and disinformation. Misinformation is the unintentional or accidental spread of false information. The person spreading the information believes they are spreading accurate information but they are misinformed. In contrast, disinformation is the intentional and malicious spread of false information. Unfortunately, disinformation can be designed to manipulate people by triggering irrational thinking through tapping into our existing biases or emotions. Table 1.1 has some tips to help you spot mis/disinformation.

Table 1.1 Helpful Tips for Spotting Mis/disinformation on the Internet

How to spot mis/disinformation	Example
Claim is vague.	Essential oils boost your immune system.
Claim involves mysterious or supernatural phenomena.	The COVID-19 pandemic was caused by an alien from the planet Narp.
Lack of data, or questionable data.	Well, my aunt saw this, so it must be true.
Contrary evidence is inconsistent, ignored, dismissed, or denied.	We can't believe anything in the scientific literature because the peer reviewers were bought off by Big Pharma.

Now that you know how to spot mis/disinformation, what can you do about it? Do not re-share or interact with this content on social media. These engagements will cause the underlying social media algorithms to show this information to more people. Instead, use the "report" function to have the content flagged or taken down. Also, be cautious about engaging with anyone who promotes disinformation. A common tactic for those who spread disinformation is to present information in such a way that it undermines a critic's credibility. It also helps to be aware of our own biases and susceptibility to mis/disinformation. One of the most powerful things we can do to combat mis/disinformation is to amplify the positive voices in our community and promote good science. Even if you are new to biology, you can still help spread good information!

Discuss with a friend about where you have observed mis/disinformation. Have you fallen for it before? Why did you fall for it? What could you do to combat the infodemic?

Inspire Curiosity

Spark students' curiosity to learn more with interesting chapter openers focused on real world, thought-provoking questions relating to the unit theme and chapter.

What Can You Do with a Cell?

Most people typically do not think about their cells—especially when they are working as they should. Your body has about 30 trillion cells, all working together like a big beehive as you go about your day. Like bees in a beehive, these cells all have jobs, too. We have hair, skin, and heart cells. Routine medical care often involves an examination of our cells. A blood draw can reveal not only the levels of various molecules in the blood, but also provides a look at the number of cells in our blood. For example, red and white blood cells are two types of cells that are commonly assayed after a blood draw. If there are not enough healthy red blood cells, one may have anemia. Too many white blood cells could mean that you have an infection.

A single cell is powerful. One cell contains all your genetic information and therefore the blueprint for what makes you, well, you. One of your gametes, an egg or sperm (see the photo that begins this chapter), combined with another's complementary gamete, can make a whole new person. And a single cell, if left to grow unchecked, can lead to cancer (Chapter 6). What can we do with a cell? The cell is the smallest unit of life—and one can therefore do many things with a cell. But what it is about the cell that is so special? Why is a single cell so powerful?

Real-world Application

Written from the ground up to include vivid, relatable examples that connect complex concepts to everyday life; the real world is a central feature, not an aside.

Biology for Life

Comparing Antibiotics and Antifungal Medications

Why do we treat bacterial and fungal infections differently? And why is it so much harder to clear up a fungal infection? It comes down to the difference in the structures of the cells. The fungi that cause everything from athlete's foot to vaginal yeast infections are part of Domain Eukarya, the same as us. Bacteria, however, are prokaryotic and many have a structure that our human cells do not: a cell wall. The cell wall protects the cell and supports its shape (Figure 5.4). Penicillin-class antibiotics work by inhibiting the synthesis of a bacterial cell wall, thus killing the bacteria. Note that cell walls are not membranes and they do not consist of lipids. Therefore, taking an antibiotic has a direct negative effect on the bacteria only, not on our human cells.

The challenge with fungal infections is that our cells and the invading fungal cells are both eukaryotic, and so there are fewer differences between host and pathogen cells to take advantage of with a treatment strategy. As a result, clearing up a fungal infection typically requires longer and/or repeated treatments, and making environmental changes, like wearing cotton socks or undergarments that make our skin less hospitable for fungal growth.

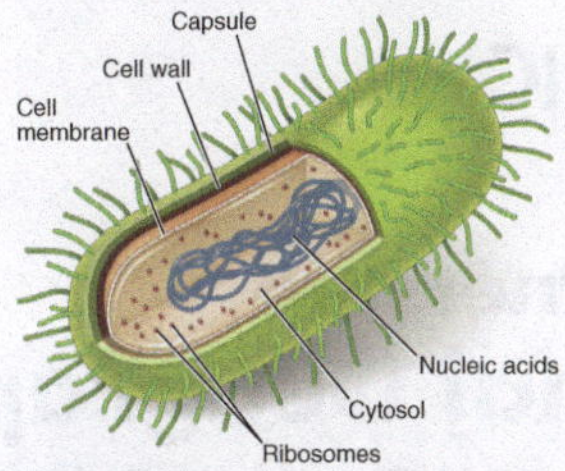

Figure 5.4 Generalized prokaryotic cell structure

Note the presence of the universal cell features, plus differentiating features such as the cell wall and capsule.

Topical (surface) treatments for fungal infections are typically preferred over pills. Can you think of a reason why?

Scientific Literacy

Help students understand that science is a process, not a set of facts, with the *Process of Science* feature, which highlights unexpected and exciting ways science works in the real world.

Process of Science

Indigenous Wisdom and Scientific Practices

Robin Wall Kimmerer is a member of the Citizen Potawatomi Nation. She is also a botanist, or a biologist who studies plants (Figure 4.13). In her book, *Braiding Sweetgrass*, Kimmerer recounts a story from her freshman year of college where, when asked why she wanted to study botany, she explained that it was because she wanted to know "why asters and goldenrod looked so beautiful together." She was told that her question was not scientific.

Or is it? As Kimmerer continued in her career, she explored the question further. It comes back to the pigments in plants and the wavelengths they reflect. Turns out, the purple of asters and the yellow of goldenrod complement one another such that when the colors are placed next to each other, they create a strong contrast. The striking contrast makes these flowers stand out not only to us but also to pollinators, insects like bees and butterflies that facilitate plant reproduction by moving pollen between plants.

So, if we rephrase the question slightly to "Do asters and goldenrods grow together because they are more likely to get pollinated?" we see what some would consider a hypothesis. This is an example of how Kimmerer's Indigenous perspective, which focuses on the beauty perspective to create a testable hypothesis, ultimately providing new insight into the world. Kimmerer's story is an example of how cultural backgrounds can influence scientific inquiry and of the importance of including diverse perspectives in science.

Figure 4.13 Dr. Robin Wall Kimmerer

Dr. Kimmerer combines her Indigenous background as a member of the Citizen Potawatomi Nation with her career as a botanist to ask questions that honor both.

What are some things that we can do to promote diverse perspectives in science?

Critical Thinking

Students are asked to problem-solve and draw conclusions with *Digging into Data*, a feature that provides a scaffolded opportunity to interpret scientific research.

Digging into Data

A Burning Concern

Your body contains about 9.5% carbon atoms by weight. Those carbon atoms were originally in CO_2, a gas in Earth's atmosphere. Plants use CO_2 to make sugars in photosynthesis. When you consume plants and other organisms, that carbon becomes part of your body. CO_2 is then released back into the atmosphere when sugars are used for energy during cellular respiration.

About 8,000 years ago, humans began burning forests for agriculture, and since the mid-1800s, have burned fossil fuels for heat and power. These activities have created new sources of CO_2 gas. Today, the atmospheric CO_2 level is higher than it has been for at least 800,000 years. How do we know? Scientists study CO_2 levels in gas bubbles trapped in ice cores drilled from thick sheets of ice in Greenland and Antarctica (Figure 4.17). The bubbles form when snow, which contains tiny pockets of air, is compacted during the formation of glacial ice. Scientists can measure the CO_2 levels of Earth's ancient atmosphere in these bubbles and compare them to current levels. (Figure 4.18). We'll return to the relationship between CO_2 and climate change in Chapter 16.

Figure 4.17 Example of an ice core

In this photo, a scientist from the Climate and Environmental Physics department at the University of Bern examines ice core samples that are more than 1.2 million years old. These ice samples contain samples of Earth's atmosphere trapped in the ice at the time the ice formed. The deeper in the ice the sample is from, the older the air in the bubbles.

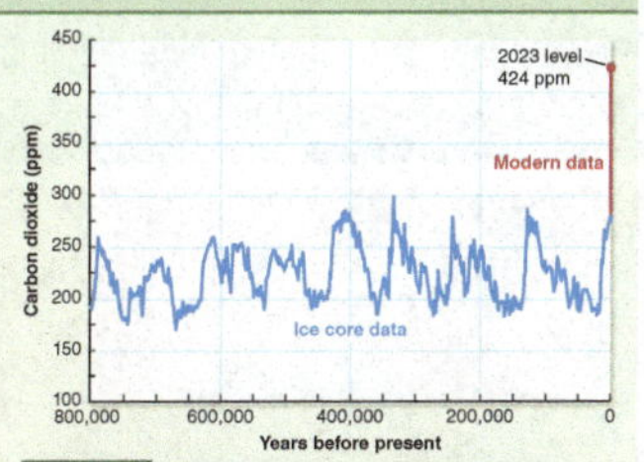

Figure 4.18 Atmospheric CO_2 over 800,000 years

The concentration of atmospheric CO_2 has fluctuated over the last 800,000 years but never rose above 300 ppm (parts per million) until the early 1900s. Note that the most recent data shown here are direct measurements from the air, not ice. In the last 150 years, human activities have released enough carbon dioxide to raise the atmospheric level above 400 ppm. Our burning of fossil fuels is by far the largest source of these emissions.

1. Based on the data in Figure 4.18, how much did the CO_2 content of the atmosphere change from 25,000 years ago to what it is today?
2. Before modern times ("0" on the *x*-axis), what was the highest concentration of CO_2 ever recorded? What was the lowest?
3. Based on the data provided, would you say that the current, high concentrations of CO_2 in the atmosphere fall within the normal fluctuations you see in the historic record?

Exclusive partnership with National Geographic

NATIONAL GEOGRAPHIC LEARNING

Exclusive partnership with National Geographic

Stunning visuals from National Geographic take center stage, captivating students' imaginations.

Connect your students with **National Geographic Explorers** from diverse backgrounds, whose interests in biological questions extend to the far reaches of the globe.

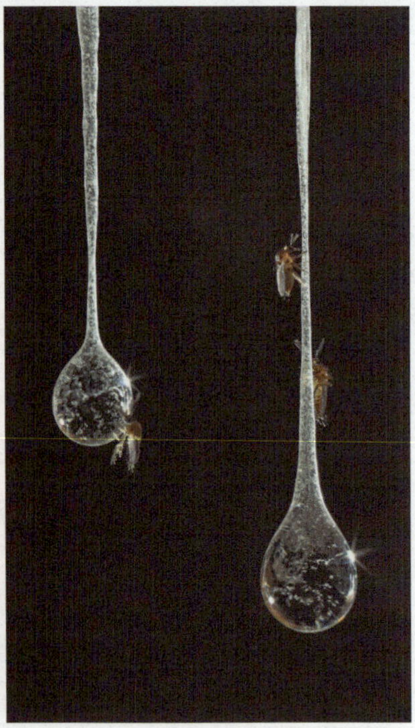

Gibbs Kuguru

Shark Scientist & Geneticist

A featured Explorer is Gibbs Kuguru, a Kenyan scientist and shark conservationist who uses genetic research to better understand the unique elements in DNA that shape shark populations.

Source: National Geographic Image Collection

National Geographic Explorers

Andrian Gajigan
Chapter 01

Dr. Aristide Takoukam Kamla
Chapter 16

Fernando Trujillo
Chapter 14

Gibbs Kuguru
Chapters 12 and 16

Heather Lynch
Chapter 14

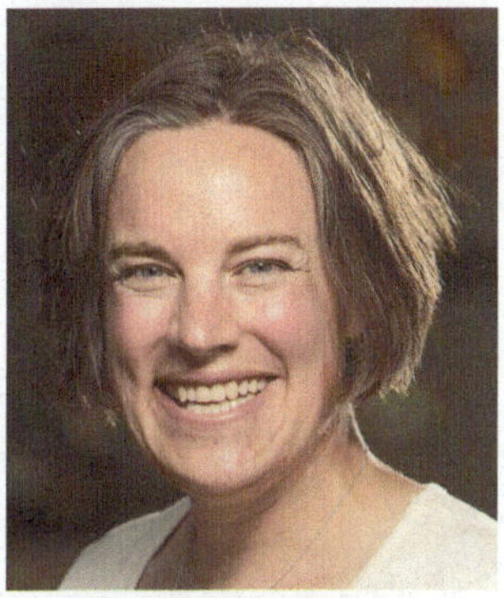

Dr. Isla Meyers-Smith
Chapter 14

Jodi Rowley
Chapter 13

Katlin Bowman
Chapter 15

Lee Berger
Chapter 06

Dr. Maria Albo
Chapter 11

Marina Arbetman
Chapter 18

Nalini Nadkarni
Chapter 18

Reuben Shipway
Chapter 16

Shane Campbell-Staton
Chapter 11

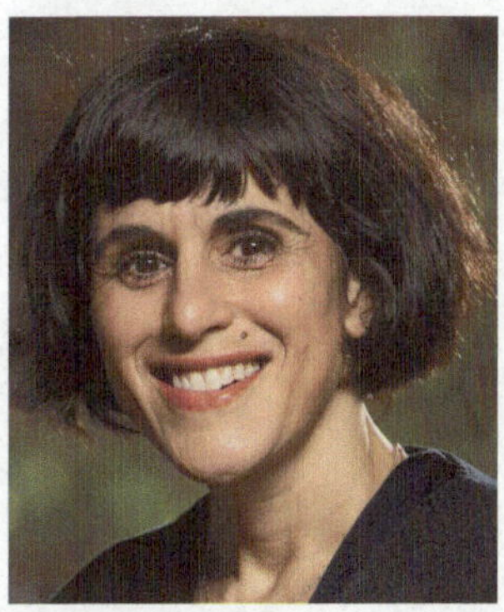

Theanne Schiros
Chapter 09 and 16

NATIONAL GEOGRAPHIC LEARNING

1

Invitation to Biology

Chapter Outline:

A Tale of Curiosity for the Natural World

The wind gently ruffled the young woman's hair as she considered the mushroom. It was an excellent example of *Agaricus campestris* or the humble field mushroom. She compared the specimen with her careful drawing, and then set her pencil down on her notebook and sighed. The sun was just beginning to set over the rolling Scottish countryside, and the end of summer was that much closer. Her annual summer holiday with her family in the Scottish Highlands was nearing an end.

The young woman's interest in mushrooms didn't end with simply illustrating them. As she grew older, she turned her attention fully to *mycology*, or the study of fungi. Her ability to carefully observe and artistically represent organisms provided critical insight into the natural world. Using a microscope, she discovered that fungi grew from spores, much like plants grew from seeds. Like any other scientist, she wrote up her findings to share with other scientists. But she couldn't share her findings directly with other scientists. Why? Because she was a woman in 19th-century England. In 1897 she submitted her paper to the Linnean Society of London…but the Linnean Society did not allow women until 1905. So, another scientist presented her paper. Although the paper was well received, the young woman declined to continue her scientific pursuits. It wasn't until much later that her contributions to science were recognized. Nevertheless, she became a household name in the early 1900s. Why? That was when she published her most famous work: *The Tale of Peter Rabbit*. Peter Rabbit would go on to become the oldest licensed fictional character and a household name for well over 100 years.

Who was the young woman? Beatrix Potter.

Visitors to the Scottish Highlands are often inspired by its rugged beauty. Located in the northern part of the United Kingdom, the Scottish Highlands are known for beautiful mountains, lakes (or lochs), and solitude.

Figure 1.1 **The author of Peter Rabbit was a mycologist**

A. This is a portrait of Beatrix Potter taken in 1913.

B. Potter included drawings of fungus in her children's stories. Here we see Squirrel Nutkin from *The Tale of Squirrel Nutkin* with mushrooms in the foreground.

C. This image shows one of Potter's scientific drawings demonstrating the reproductive structures of *Hygrocybe coccinea*.

Potter wasn't the first or last curious artist to make their mark on biology (**Figure 1.1**). In fact, the early biologists relied on their ability to draw. In the pre-photography era, drawings were the only way to record and share observations of the natural world. Scientific discoveries were also inspired by art. For instance, Santiago Ramón y Cajal, now known as the Father of Neuroscience, discovered that neurons were individual cells after repeatedly drawing magnified neural tissue (**Figure 1.2**). He later went on to win a Nobel Prize for his work.

Even now, with digital photography, art still inspires scientific discovery. Like Cajal, scientists today gain new insights into our natural world through art. For example, developmental biologists, who study growth from a single cell (the fertilized egg) into a fully functioning organism, use hand-drawn standard images of *model organisms* to understand development. These drawings are important for communicating scientific findings and also for comparing findings between different biologists.

Why study biology? Humans are curious, and this innate curiosity drives us to explore our natural world and the organisms in it. Arguably, the earliest biology experiment was poking something with a stick to see what would happen next. In this chapter, we will learn about modern scientific methodologies and biology, the study of life. We will close with a brief introduction to how to use the study cycle to succeed in your biology coursework.

Figure 1.2 **The Father of Neuroscience was an artist**

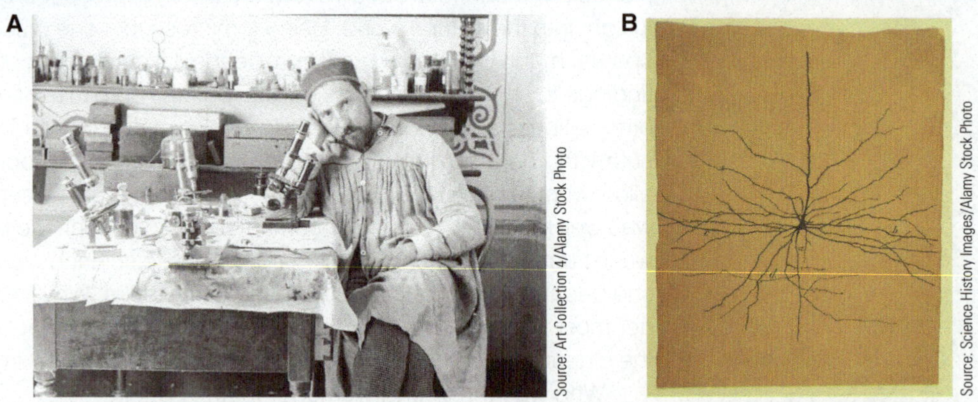

A. This is a self-portrait of Santiago Ramón y Cajal from around 1885.

B. Cajal's pen and paper drawings of the pyramidal neuron of the cerebral cortex show a central body with radiating branches. Through his repeated drawings of his microscopic observations of neural tissues, he developed the neuron doctrine, which maintains that the nervous system is comprised of individual cells.

Preview It!

Before you read further, pause and reflect on what interests you about biology. What do you hope to learn about the natural world in your biology class?

1.1 The Science of Nature and the Nature of Science

Learning Objectives

- **1-1** Explain how science works.
- **1-2** Compare and contrast hypotheses, scientific theories and scientific laws.
- **1-3** Explain what happens if data are produced that are inconsistent with an established theory.
- **1-4** Discuss some ways to identify pseudoscience.
- **1-5** Compare and contrast mis/disinformation.
- **1-6** Name strategies to protect yourself from mis/disinformation.
- **1-7** Differentiate scientific evidence from personal opinion.

Science Is a Verb

What do you think about when you hear *science*? Are you brought back to either pleasant or distressing memories of school science fairs? Maybe you think of massive textbooks that weigh as much as a human infant. Or do you imagine the stereotypical scientist—the White guy with crazy hair and a lab coat (Figure 1.3)? Or that one time in high school when you had to memorize a bunch of abstract, seemingly pointless *stuff*?

It's easy to think of science as a series of nouns—experiments, information, researchers. These nouns are an important part of science. In reality, you can think of science as a verb. Rather than being a collection of facts, science is a creative process for learning about our natural world. As a process, science answers questions such as how cells become cancerous, what features distinguish and unify life, and where we can find life. Keep in mind that science is limited. For example, the process of science can't answer all questions, such as the purpose of our existence. These questions belong to different inquiry-focused disciplines, such as philosophy. Importantly, although science is a useful process for generating new knowledge and insights, the scientific process does not reveal how to use scientific knowledge in society. Rather, we as humans assign value and purpose to that knowledge. In addition to facilitating learning about our natural world, science and related

Figure 1.3 **The stereotypical scientist**

Many people view scientists as old, male-conforming individuals with frazzled hair wearing lab coats, and goggles, and playing with glass vessels containing colored liquids.

Biology for Life

The Draw-a-Scientist Test

In the latter part of the 20th century, psychologists developed the Draw-a-Scientist Test. The Draw-a-Scientist Test is used to determine when children begin to develop stereotypical views of scientists. In this test, students are asked to draw a picture of what a scientist looks like. The researchers then score the drawings based on the presence of features like a lab coat.

In a 2018 *meta-analysis* of the data (that is, a scientific study that examines the combined results of many separate scientific studies), David Miller and colleagues noted that children draw an equal number of male-conforming and female-conforming scientists until elementary school. Starting around age 7-8, children begin to draw more male-conforming scientists and the percentage increases as students age. Miller noted that over the decades of data he examined, as more women entered scientific fields, the rate at which students drew male scientists in their drawings began to decrease.

This trend may also be influenced by the rise in cartoon characters like Ada Twist, who stars in Andrea Beaty's book *Ada Twist, Scientist*, illustrated by David Roberts (Figure 1.4). Ada Twist is a female-conforming child of color who is curious about the world around her, much to the chagrin of her parents, pet cat, and brother. In the story, we follow her journey to find out *why* and realize that there are many scientific questions we can ask in the confines of our own home. Although in the 2020s children still tend to draw more male scientists than female, children's book characters such as Ada may be changing children's stereotypical views of scientists.

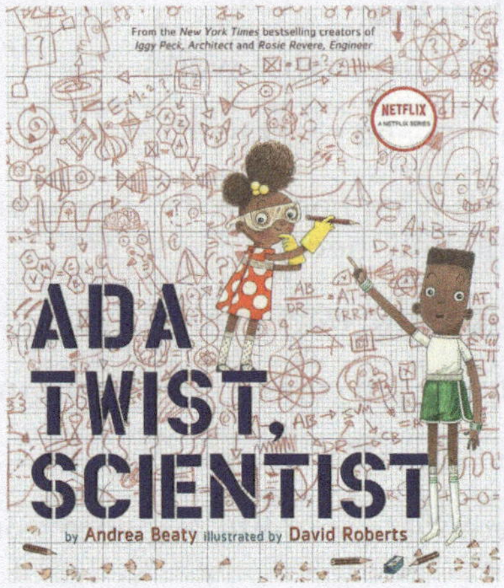

Source: Abrams Books for Young Readers

Figure 1.4 Ada Twist, Scientist®

The character Ada Twist is a modern view of who can be a scientist.

How would you draw a scientist? In your chosen career trajectory, what can you do to challenge stereotypical views of scientists?

disciplines, like engineering and medicine, are used to solve problems in society. For example, knowledge of biology might tell us how a modest pond dweller, the plant-like organism called algae, could serve as a source of biofuel and help combat the devasting impact of climate change.

Scientific Methodologies

Have you learned in other classes about *the* scientific method? That is, the idea that all science boils down to following a recipe-like series of steps to obtain facts? Good news! Science is *far* more interesting, exciting, and surprising. Good science employs a variety of approach and techniques, which we'll refer to as **scientific methodologies**. The appropriate method to use is the one that matches the question at hand. Science is iterative, meaning it builds on itself. Science also has a human aspect; it's a collaborative process, subject to cultural influences, and can be influenced by biases humans introduce to the process. Examination of how science really works in the real world is a tapestry of stories including everything from the weird, the wonderful, the biased, the strange, and the surprising. Throughout the book look for the Process of Science boxes to learn more about the process of science as it directly relates to the content in that chapter.

Scientific methods generate evidence. Imagine that you are walking through the woods and come across purple aliens singing show tunes (Figure 1.5). You recognize this as an important moment for science and realize you need to document the moment: the aliens, their behavior, and their choice of music. You note that they seem to be singing *Money* from the musical *Cabaret*. You ask yourself: Why are they here? Where did they come from? Why are they singing? Do they want money, or is the song a comment on human society and what makes the world go 'round?

What do you do next? This is a major scientific discovery! How can you use scientific methods to capture this momentous discovery?

Figure 1.5 **Finding yourself in a critical moment for science**

What would you do if walking through the woods one day you came across aliens singing show tunes in a barbershop quartet? What is an appropriate scientific methodology to use to capture this momentous occasion?

Although there is no universal scientific method, scientific methodologies all share a common goal: to generate evidence. Data serve as the basis for evidence. **Data** is factual information generated by both deductive and inductive scientific methods. **Deductive methods** start with a theory or hypothesis—that is, an idea or explanation for how something works—which is then tested by collecting observations or running experiments. In contrast, **inductive methods** start with data, often in the form of observations about the natural world, that are used to generate a hypothesis. Deductive and inductive methods go hand in hand (Figure 1.6).

Figure 1.6 **The winding path of how science works**

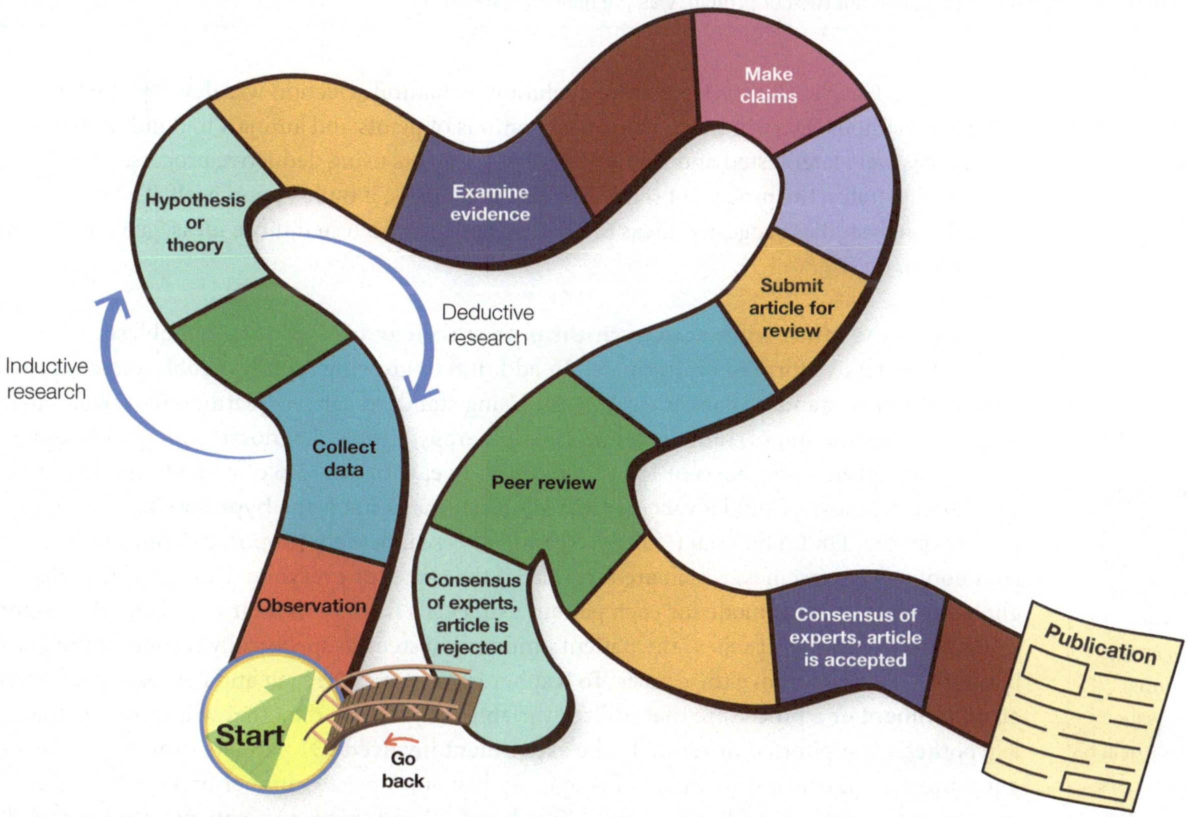

Rather than a single method, science is an evidence-generating process that involves vetting of claims by other experts to reach a scientific consensus. A scientific investigation starts with an observation, practical problem, or personal interest in a phenomenon. From there, an investigator tests or generates hypotheses and theories. Then evidence is examined, claims are made, and packaged into an article for peer review. Reviewers will agree that the claims are supported by evidence, and the paper will be accepted for publication. Or, reviewers will conclude that the claims are not supported by evidence and the paper will be rejected, sending the researcher back to the beginning. Note that although not shown here, there are many times a researcher might be forced to return to the beginning (e.g., evidence doesn't support a hypothesis), and oftentimes publishing a paper is the beginning of a new investigation.

Figure 1.7 Science is an iterative process that involves many different people separated by time and space

Although in this figure we see scientists who are all contemporaries of one another, the sharing of information and building on another's ideas can also occur over time. For example, the discovery of deoxyribonucleic acid (DNA) as the genetic material began in the mid-1800s with Gregor Mendel with his pea plant experiments. However, it was 100 years later when Alfred Hershey and Martha Chase demonstrated that DNA, and not RNA or protein, was the genetic material.

For example, Charles Darwin's theory of evolution by natural selection was developed using inductive methods. Darwin made many observations of plants and animals to build his theory. These ideas were later tested and refined by other scientists using deductive processes. As this example illustrates, the process of science is iterative, that is, it builds upon itself. The results from Darwin's studies suggested ideas to other scientists, who tested those ideas, generated more data, and so on (Figure 1.7).

Deductive methods in practice: finding new cancer treatments. Glioblastoma is the most aggressive form of brain cancer. In addition to growing rapidly, glioblastoma is also resistant to most standard cancer treatments. Using standard cancer treatments, such as surgery to remove the tumor, radiation, and chemotherapy, people diagnosed with glioblastoma are typically given a prognosis of 15–17 months to live, with only 5% of patients still living five years after diagnosis. Could a vaccine treat glioblastoma? That is the **hypothesis**, or testable prediction, that Dr. Linda Liau (Figure 1.8) and her research team proposed. Unlike other commonly given vaccines, which are formulated the same for everyone who receives it, the glioblastoma vaccine is made for each patient using cells from both their blood and the tumor. The resulting vaccine "teaches" the patient's immune system to specifically recognize the glioblastoma cells and remove the cancer. To test her hypothesis, Dr. Liau and her team performed an **experiment** or a procedure that utilizes variables to generate data that will show whether a hypothesis is supported or refuted. The experiment involved 331 patients from four different countries. Having a wide range of people, and an appropriate number of participants, is important to create generalizable results. Two hundred and thirty-two patients were assigned to the **treatment group**. The treatment or experimental group in any experiment is the group where the new factor or condition is tested. The treatment group is compared to a **control group**. The control group is used to provide a baseline for comparison. The control group in this study included 99 people who received a **placebo**. A placebo is designed to look like a real treatment but is not a treatment. All patients, regardless of which group they were assigned, received standard treatment for glioblastoma. Therefore, any differences in experimental

Which study participants are in the treatment group is a secret to both patients and doctors. This is because of the placebo effect, which arises when people who are treated are convinced that a certain outcome has occurred just because they were treated, even if they didn't receive the medicine. The placebo effect can mask real results.

results between the two groups were likely to be an effect of changing the variable. The research team collected data on how long people survived after diagnosis in both the treatment and control groups. Dr. Liau found that patients in the treatment group lived longer than those in the control group. After examining the data, Dr. Liau and her team concluded that the glioblastoma vaccine treatment helped patients live longer. In this case, we can say that the initial hypothesis was supported.

Figure 1.8 Dr. Linda Liau researches new treatments for brain cancer

Dr. Liau has an M.D. and Ph.D. and specializes in neurological surgery. Only 6% of all board-certified practicing neurosurgeons are women. In addition to her cutting-edge work on glioblastoma treatment, she is also a trailblazer for women in neurological surgery.

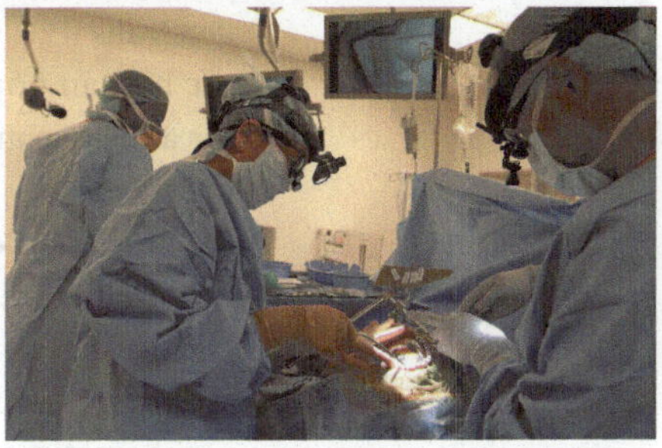

Source: LUCY NICHOLSON/Reuters

! Point of Confusion: A hypothesis cannot be "right" or "wrong." Rather, the current data either "support" or "refute" the hypothesis.

Inductive methods in practice: characterizing marine microorganisms. The ocean waters near the Philippines, where National Geographic Explorer Andrian Gajigan was born, are teeming with fish and other marine life. But what about microscopic life? Gajigan (Figure 1.9) and colleagues knew that microscopic life was important for life elsewhere in the ocean, but nobody knew what kinds of microscopic life existed in this particular region of the ocean and what role these life forms played in these waters. Like deductive methods, inductive methods also produce data. However, rather than using the data to support or refute a hypothesis, the data is instead describing the natural world and is used to generate new hypotheses. Although making observations and taking detailed notes doesn't fit into the mold of what we think of as an experiment, this is still a valid scientific endeavor. In this case, Gajigan and his research colleagues traveled to the Philippines and collected water samples from various water depths. To find out which organisms were present in the water, the research team collected and analyzed genetic material from the microorganisms. The team noted that there were about ten major groups of bacterial life present and that the microorganisms present varied depending on the depth. If we break the experiment down, we see that there is no control group or initial hypothesis. Which microorganisms are present and where the researchers found them is the data. Using this data, Gajigan or other researchers can go on to do similar studies and characterize other regions of the ocean. Or, researchers may develop a hypothesis on the role of specific microorganisms in the ocean and perform an experiment.

National Geographic Explorer and oceanographer
Andrian Gajigan

Source: National Geographic

Figure 1.9 Gajigan is a biologist interested in characterizing microscopic ocean life in the water surrounding his native Philippines.

Scientific claims are testable. After gathering scientific evidence through either a deductive or inductive method, scientists interpret the data to make **claims**. A claim is a data-informed statement that can be challenged through experimentation. Since testing is at the heart of good scientific methodology, scientific claims must be **falsifiable**. Falsifiable means that data can be sought that refutes the claim. In contrast, subjective claims, such as a personal opinion or belief (including supernatural beliefs), are not falsifiable. For example, one could claim that the COVID-19 pandemic was the result of a vengeful deity. The existence of a deity is a personal belief—you can't test it, let alone falsify it. And what a deity claims to do (or not do) is also not testable. We will talk more about unfalsifiable claims in our discussion of **pseudoscience**, or nonscientific processes, at the end of this section.

Statistical testing can help support claims. Once data is collected, how does one know if there was a real effect? Scientists look at **statistics**. Statistics is a branch of mathematics that includes mathematical tools (or tests) to assess if an observed phenomenon has an underlying connection, or if the observed difference is due to random chance. Statistical analyses can also tell us if enough individual samples were analyzed to make a meaningful claim. For example, a study with a small sample size (meaning, there were few participants) may not be generalizable. **Generalizable** means that the data can be used to make bigger statements across the human population. For example, let's say that a researcher was testing a new treatment for a disease and studied it in five people all from one location, and reports that the treatment worked. With five people all from one place, one can't rule out the possibility that the results had to do with something special about the five individuals or the single place. This is why human research, such as Dr. Liau's work, often involves people from different countries. Dr. Liau's team also involved independent statisticians to verify her findings and the use of an appropriate sample size.

Sometimes, events can appear related even when one has nothing to do with the other. Let's take a hypothetical example and say that the number of people owning cats and the number of people owning novelty socks in the United States have both increased over the last 100 years. You could look at the data and surmise that cat owners need socks to protect their toes from cat scratches. Or that they use the socks to play with their cats. This is an example of a *correlation*—that the increase in cat ownership appears to match the increase in the number of people who own novelty socks. Does one cause the other? Likely not; the average United States population has also increased in the last 100 years. So if this trend between cats and socks does exist, it is likely due to there being more people, rather than any relationship between cat ownership and purchasing novelty socks.

Correlation is a type of statistic that describes the relationship between two variables. A positive correlation means that the variables change in the same way: when one increases, the other increases, like our example with the cats. A negative correlation means the variables change in the opposite way: when one increases, the other decreases. Although correlations can be useful, correlation tells you nothing about causation. Correlation does not mean one thing causes the other (Figure 1.10).

How does one know if a relationship is correlation or causation? This is an opportunity to perform a scientific experiment and use deductive scientific methodology. The potential causative agent would be used in the treatment group and compared to a control group. If there is a difference between the two groups, we could infer that the phenomenon at hand is due to causation, not correlation.

Peer review is scientific quality control. Scientific methodologies have quality control features built in, many of which rely on **skepticism**. Skepticism in science is approaching any new scientific claim with an attitude of doubt. Therefore, many of the scientific quality control features rely on other skeptical scientists reviewing the claims of others in a peer-review process. **Peer review** is when other experts in the same field (that is, a scientist's "peers") review and critique a scientist's work. This can occur at different levels. For example, when a scientist has an idea they would like to test or explore, they first must request a grant of money or "grant" to do the experiments. These grant applications are extensively reviewed by a team of scientists and only the best designed and promising grant applications are funded.

Generally, scientists don't work alone. After enough experiments are finished and data collected, scientists write up their work in a **manuscript** and submit it for publication. The results are then peer-reviewed by another team of experts. The experts determine if the scientific methods chosen were appropriate and if the data presented supports the scientific claims. Reviewers often ask scientists to revise their work or perform more experiments before a manuscript is accepted for

Figure 1.10 A humorous view of correlation and causation

Death comes for us all, without regard to whether we confuse correlation and causation or understand the distinction completely. This example shows how confusing correlation and causation can needlessly scare people.

publication. This is similar to how you may turn in a term paper in an English class, and your professor returns it to you with changes. Note that not all work is deemed of sufficient quality for publication.

After publication, other scientists will often try to replicate the results for additional quality control. Controversial, unconventional, or surprising experimental results are especially subject to scrutiny by other scientists, many of whom will try to replicate the results. If other scientists cannot replicate the results, or evidence of data falsification comes to light, a manuscript may be retracted by the journal. For example, in 1998 Andrew Wakefield and 12 colleagues published an article in the *Lancet* suggesting that there was a link between the measles, mumps, and rubella (MMR) vaccine and the development of autism in children. Several scientists sought to replicate these results and were unable to do so. In 2010, the *Lancet* formally retracted the article. Claims about a link between the MMR vaccine and autism are no longer considered to be scientifically based.

Process of Science

Design for Bias-Free Experiments

Careful design of experiments is important to avoid introducing bias into the results. *Bias* is when one is predisposed to a belief, often in a way considered to be unfair or distorted. One of the problems with Andrew Wakefield and colleagues' 1998 paper in the *Lancet* was bias, both in the form of *conflict of interest*, which is when a person's interest or beliefs compromise their judgments, and a specific kind of bias called *selection bias*, or when subjects are intentionally chosen to support a claim. Today, many reputable scientific journals require that scientists who write papers disclose potential financial conflicts of interest to reviewers and readers to maintain full transparency.

There are many different types of bias. See if you can find another type of bias that could influence the design of an experiment or interpretation of results. Share what you find with a classmate.

Hypotheses, Theories, and Laws are Different Types of Scientific Knowledge

What if I told you I have a theory that chocolate chip cookies are the *best* cookies? Or that I have a theory that feeding chocolate chip cookies to students before their biology exam will improve their scores? Many people say "I have a theory..." when they mean to say they have a hypothesis, such as in my second example. Or, they use the word "theory" to express a personal opinion, like my thoughts on cookies. Our everyday use of the word theory is different from how scientists view a theory. Understanding the difference is important, and being able to distinguish between hypotheses, theories, and laws is necessary to distinguish scientific claims from pseudoscientific claims.

Hypotheses, *theories*, and *laws* each describe different types of scientific knowledge. A hypothesis doesn't turn into a theory, nor does a theory turn into a law. As we learned earlier in this section, a hypothesis is a testable prediction. For example, I could hypothesize that planting red flowers in my backyard will attract more pollinators, like hummingbirds, butterflies, and bees. I could then test this hypothesis by planting differently colored flowers and collecting data on which colored flowers receive the most visitors.

A **theory** differs from a hypothesis in that a theory has explanatory power. For example, evolutionary theory can explain where the diversity of life on Earth came from. Theories, like hypotheses, are also testable. Hypotheses can contribute to theories, but a single hypothesis does not "mature" into a theory. The best theories are those that explain most of the phenomena we see. Theories, like all scientific knowledge, are subject to revision as new evidence is discovered. For example, scientists used to believe that life could form from non-living matter. This theory is known as spontaneous generation. With the advent of technologies such as the microscope, scientists obtained new evidence and rejected spontaneous generation in favor of cell theory—that all cells arise from existing cells (Chapter 5).

Perhaps you heard someone say "Well, that's *just* a theory"? A scientific theory is not simply an idea that someone has; rather, a scientific theory is thoroughly evaluated and scrutinized. This is why saying something is "*just* a theory" is not fully accurate: theories are not immature ideas. Rather, a theory is supported by multiple lines of evidence. If the majority of scientists agree with the evidence at hand, the theory is "accepted," along with the possibility that new data inconsistent with the theory might be found, and therefore the theory may change. Theories that are accepted are those that have a consensus of experts agreeing that the current experimental data supports a given theory. If new data is found, the theory will be revised or rejected. New results do not invalidate previous results, but the interpretation of what the results mean can change. Thus, new data inconsistent with a theory may trigger its revision. For example, modern atomic theory has been modified many times since it was proposed by John Dalton in 1803. If someone ever discovers matter that does not consist of atoms and subatomic particles, the atomic theory would be revised to include the exception, something like "all matter consists of atoms and subatomic particles except…" If many exceptions accumulate, the theory will be rethought and rewritten so it better accounts for the discrepancies.

A **law** does not have explanatory power, nor is it a "grown-up" theory. Rather, scientific laws describe an unvarying relationship. It's easiest to think of laws in terms of mathematical equations. For example, Boyle's Law describes the relationship between the pressure (P) and volume (V) of an ideal gas at constant temperature as

$$P_1V_1 = P_2V_2$$

In other words, the law states that at a constant temperature, the product of the pressure and volume of a certain amount of gas will always be the same. That means that if you make the container holding the gas smaller, the pressure of the gas goes up. That makes sense, right? Now, Boyle's law doesn't *explain* why $P_1V_1 = P_2V_2$. Rather, it describes an unvarying relationship. Since a theory is used to explain a phenomenon and a law describes relationships, they are used together to create a fuller picture of how the world works.

Distinguishing Scientific from Pseudoscientific

Between 1849 and 1882, approximately 180,000 immigrants arrived in the United States from China to help build the transcontinental railroad. The Chinese immigrants brought traditional remedies from home with them, including snake oil to treat arthritis. The snake oil used in traditional Chinese medicine contained omega-3 fatty acid-rich oil from a Chinese water snake (**Figure 1.11A**) and actually does reduce inflammation and relieve the pain of arthritis. Reports of the therapeutic effects of snake oil were popularized when Clark Stanley, the "Rattlesnake King," sliced open a rattlesnake at the 1893 World's Exposition in Chicago, boiled it in front of the crowd, removed the fat, and claimed it to be "Stanley's Snake Oil" with all the same properties as the snake oil made from a Chinese water snake (**Figure 1.11B**).

There are a few concerning problems with this story. First, rattlesnakes and Chinese water snakes are different kinds of snakes. Rattlesnakes do not have nearly the necessary amount of omega-3 fatty acids to help with inflammation. Second, the "snake oil" Stanley peddled did not contain any snake-derived components. Analysis of shipments seized in 1917 (after the passage in 1906 of the Pure Food and Drug Act) indicated that the "snake oil" was mineral oil with added animal fat that was likely from cows, plus red pepper and turpentine. So began the use of the phrase *snake oil* to symbolize medical fraud. In Stanley's time, this fraud was typically perpetrated when a traveling salesman would pose as a "doctor" selling a medical remedy. An accomplice in the crowd of onlookers would be magically "cured" by the remedy—people would believe in the magical cure and purchase it—and then the con artists would move on to the next town (**Figure 1.11C**).

Because Stanley falsely represented his product as being the same as a different substance that had genuine curative properties and staged "evidence" demonstrating its supposed

Figure 1.11 **The origins of snake oil and medical mis/disinformation**

A B C

A. Shown here is a Chinese water snake. A traditional remedy for inflamed joints is the omega-3-rich oil from this snake.

B. Clark Stanley, the "Rattlesnake King," marketed his own snake oil he said was made from rattlesnakes and claimed it worked just like oil made from the Chinese water snake.

C. Artist's representation of the snake oil salesman. A salesman posing as a "doctor" would magically "cure" an accomplice he selected from the crowd to encourage people to buy the false medicine.

effectiveness, this is an example of pseudoscience, i.e., fake science. Is this historical example relegated to the past? Unfortunately, no. During the COVID-19 pandemic, for example, a wide variety of miracle cures appeared for sale on the internet. Unlike science, pseudoscience has no requirement for accuracy, truthfulness, or objectivity. In the worst of cases, pseudoscientific claims framed in rhetoric intended to persuade the general public can jeopardize our individual and collective welfare.

Some of the most common pseudoscientific claims we see are deliberately vague. For example, think of the last time you were on social media or in a grocery store: did you see an advertisement or a product that was advertised to "support your immune system"? Our immune systems are complicated and involve a wide array of biological processes. What processes are supported? What does *supported* even mean in this context? The claim is too vague to test.

Sometimes even a seemingly diligent search for information about a pseudoscientific claim will generate results that appear to validate the claim. This is due to **confirmation bias**. If one's personal beliefs or values are aligned with a pseudoscientific claim, then it is easy to deny, dismiss, or ignore evidence that is contrary to it, even unconsciously.

Biases arise from mental shortcuts used by our brains. Our brains process massive amounts of information daily, much of it subconsciously, and this process is imperfect. The key is to know your biases so you can avoid being trapped by pseudoscientific claims.

Biology for Life

The SIFT and CRAAP Tests

The internet is awash with "scientific" information. How do you know what sources to believe? Scientists have created strategies with easy-to-remember acronyms to help you evaluate sources. For example, Mike Caulfield, an information literacy scientist, developed the SIFT method. SIFT stands for Stop, Investigate the source, Find better coverage, and Trace claims, and is designed to help you fact-check information you find on the internet. The CRAAP test was developed by librarian Sarah Blakeslee and colleagues to provide an easy-to-remember method for evaluating claims. CRAAP stands for Currency (is the information recent?), Relevance, Authority, Accuracy, and Purpose. One common hallmark of questionable scientific claims is information presented to get you to buy something.

Working with a classmate, find two or three claims on the internet and analyze them with the SIFT and CRAAP tests. Are these valid claims? Why or why not?

Process of Science

How Do We Know That Vaccines Are Safe?

One of the most powerful aspects of the scientific process is that it is ongoing. Vaccines are extensively tested first in animals and later in humans before receiving approval by the U.S. Food and Drug Administration (FDA). After approval, scientists continue to monitor the safety and efficacy of vaccines. For example, in the United States, the FDA and Centers for Disease Control and Prevention (CDC) co-manage a system called the Vaccine Adverse Event Reporting System (VAERS). VAERS is a system used for ongoing monitoring of vaccine safety. Since new research continuously adds to our scientific knowledge (one of the most powerful benefits of science!), if new evidence came to light that a type of vaccine was not safe, it would be thoroughly investigated and, if warranted, approval revoked.

Visit the CDC website's page on Historical Vaccine Safety Concerns. What are some reasons that vaccines have been recalled or reassessed? What was the response of the scientific community?

Biology for Life

Social Media and the Rise of the Infodemic

The rise of social media has facilitated the rapid, worldwide dissemination of unreliable information. During the COVID-19 pandemic, the spread of false information about the virus was so pervasive that the World Health Organization (WHO) referred to it as an *infodemic*. Mark Twain accurately described the phenomenon of infodemics long before the existence of social media when he said "*A lie can travel halfway around the world while the truth is still putting on its shoes.*" Pseudoscientific and false information pervades the internet and falls into two categories, misinformation and disinformation. Misinformation is the unintentional or accidental spread of false information. The person spreading the information believes they are spreading accurate information but they are misinformed. In contrast, disinformation is the intentional and malicious spread of false information. Unfortunately, disinformation can be designed to manipulate people by triggering irrational thinking through tapping into our existing biases or emotions. Table 1.1 has some tips to help you spot mis/disinformation.

Table 1.1 Helpful Tips for Spotting Mis/disinformation on the Internet

How to spot mis/disinformation	Example
Claim is vague.	Essential oils boost your immune system.
Claim involves mysterious or supernatural phenomena.	The COVID-19 pandemic was caused by an alien from the planet Narp.
Lack of data, or questionable data.	Well, my aunt saw this, so it must be true.
Contrary evidence is inconsistent, ignored, dismissed, or denied.	We can't believe anything in the scientific literature because the peer reviewers were bought off by Big Pharma.

Now that you know how to spot mis/disinformation, what can you do about it? Do not re-share or interact with this content on social media. These engagements will cause the underlying social media algorithms to show this information to more people. Instead, use the "report" function to have the content flagged or taken down. Also, be cautious about engaging with anyone who promotes disinformation. A common tactic for those who spread disinformation is to present information in such a way that it undermines a critic's credibility. It also helps to be aware of our own biases and susceptibility to mis/disinformation. One of the most powerful things we can do to combat mis/disinformation is to amplify the positive voices in our community and promote good science. Even if you are new to biology, you can still help spread good information!

Discuss with a friend about where you have observed mis/disinformation. Have you fallen for it before? Why did you fall for it? What could you do to combat the infodemic?

Personal anecdotes are not a form of evidence. A personal anecdote is a short story about a person's experience. Many people feel that their personal anecdotes are valid evidence due to a phenomenon called **firsthand bias**. Firsthand bias is the tendency to believe something is more common than it is because it happened to us or someone close to us. However, firsthand bias only reflects one observation and a potentially biased opinion based on one's beliefs about a phenomenon. For example, according to the U.S. Department of Health and Human Services, life-threatening reactions to vaccines can occur in about one or two people per million doses given. That's a 0.0001% chance. However, if you happen to be that one-in-a-million person or know the one-in-a-million person, vaccines seem substantially more dangerous. Keep in mind, there are around 332 million people in the United States, so back of the napkin math would suggest you could find ~33 people who had a severe vaccine reaction.

Digging into Data

Evaluating the Source of Information to Decide on a Treatment Option

Treatment A: Alreadyusedicib

Dr. Young is a physicist the University of Yourtown who supports the use of the drug Alreadyusedicib. They read reports and found that Alreadyinusedicib is an effective treatment for similar hemorrhagic diseases and therefore, it should be an effective treatment for HID. They claim there is no need to study it further and it can be used immediately.

Treatment B: All Natural Treatment

MxNatural is a healthy-eating blogger who writes articles promoting the benefits of consuming only natural products. If it's natural, it's definitely better! MxNatural is promoting their All Natural Treatment on their website. MxNatural says it will cure HID and improve energy reserves, all without dangerous chemicals for the low price of $25 a bottle!

Treatment C: Newmedicib

Dr. Rose is an epidemiologist and president of the Scientific Federation for Public Health, an international organization of public health scientists. They are promoting the use of Newmedicib, a medicine that was tested in a dozen countries with thousands of people by different scientists. Dr. Rose represents the group's opinion that Newmedicib is the best treatment for HID.

Figure 1.12 **Three sources of medical advice and the treatments they propose**

For this exercise, imagine that the world has been hit by a new, mysterious pandemic. This time, the novel virus emerged when members of an illegally hunted animal species in Africa entered the black market. Human contact with the animal (the natural reservoir for the virus) caused the virus to "jump" to humans, causing a disease called Hemorrhagic Inflammation Disease (HID). HID causes extensive bleeding and has a mortality rate of 50%. The world goes into lockdown mode as public health officials scramble to understand the biology of the virus, devise and test new vaccines to prevent new infections, and uncover potential treatments. Everyone is sifting through information and wants to know what can be done to protect themselves and their families.

The latest news is on potential lifesaving treatments for people ill with the disease. Your best friend is sick with the disease. Knowing that you are a biology all-star, they ask for your opinion about what treatment option your friend should pursue. They are overwhelmed by the amount of information online and are unsure which treatment option to pick between the three candidates (Figure 1.12).

1. Review each of the treatment options and the person cheerleading each. Which option seems to be supported by science? How do you know?
2. Which individual(s) are promoting pseudoscientific claims? How do you know?
3. Which treatment would you recommend to your friend? Why?

1.1 Summary

- Science is an iterative process that generates evidence.
- Scientific claims are tested using scientific methodologies. All other sources, including opinions and personal anecdotes, may suggest avenues for scientific questioning but are not scientific.
- Pseudoscience often includes vague or non-falsifiable claims. Pseudoscience may only use information that justifies the claim while ignoring contradictory information.
- Misinformation is the unintentional spread of false information while disinformation is intentionally spread of false information to manipulate people.
- To protect yourself from mis/disinformation, always fact-check claims that seem too good to be true, help spread good scientific information, and report mis/disinformation.
- Hypotheses are testable explanations for phenomena. Scientific theories are supported by lots of data and have explanatory power.
- Scientific laws have a specific mathematical relationship and no explanatory power.
- Hypotheses do not become theories, nor do theories become laws. Rather, each describes a different form of scientific knowledge.
- When data inconsistent with an established scientific theory are found, the theory may be modified to include supported exceptions or revised to include the new data.

1.2 Biology Basics

Learning Objectives

1-8 Describe the field of biology.

1-9 Describe the criteria used to define life.

1-10 Explain the five big ideas in biology.

What Is Biology?

Biology is the scientific study of life. What, exactly, is "life"? We may never actually come up with a satisfying definition. Living organisms are too diverse, and they consist of the same basic components as nonliving things. When we try to define life, we end up with a long list of properties that differentiate living from nonliving things. Generally speaking, something alive has underlying organization, uses energy, maintains homeostasis, reproduces, and as a species, evolves (Figure 1.13).

Living organisms have an underlying organization. From the smallest molecules to biological tissues to organs and overall anatomy, how things are organized influences their function. We see this in non-biological contexts as well. For example, desks tend to have firm flat surfaces. This surface is essential for the desk's function of holding a computer, sewing machine, or sketchpad. Biologists view life in increasingly inclusive levels of organization. **Organization** refers to the underlying pattern of structures that make up an organism. Interacting components of one level compose larger, more complex structures and systems of the next. The interactions give rise to new properties that emerge at each level. For example, the **cell** (Chapter 5) is the functional unit of life. Cells make up tissue. **Tissue** is a group of similar cells arranged in a pattern. Tissues make up an organ, like the brain. An **organ** is a structure composed of tissues that carry out a specific task. In the case of the brain, it coordinates many bodily processes such as maintaining homeostasis, planning, and memory. The brain is part of a system called the nervous system. A **system** is a group of organs that all work together to perform a specific function. In the case of the nervous system, it helps you move, think about, and experience the world around you (Table 1.2).

Figure 1.13 **Biology is the study of life, and life has five key properties**

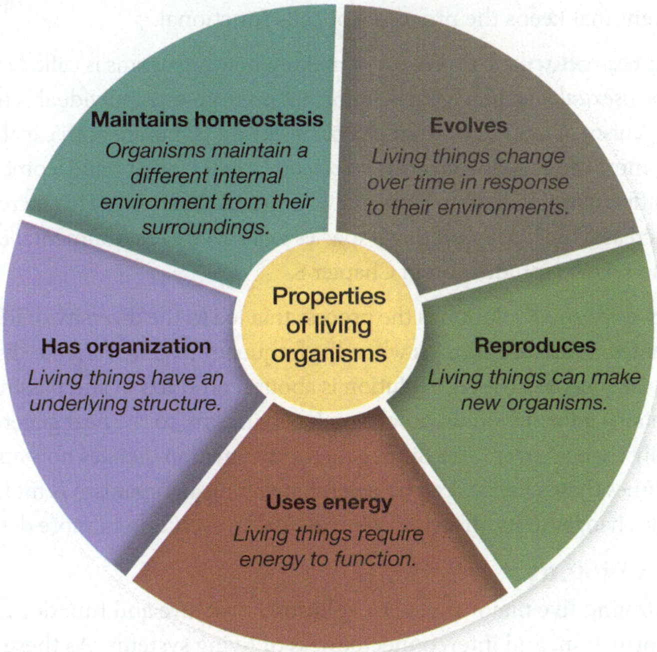

For something to be considered alive, it must use energy, have organization, be able to maintain homeostasis, evolve, and reproduce.

Table 1.2 Living Organisms Have an Underlying Structure

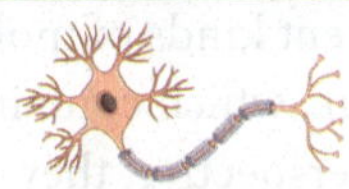 Cell	The smallest unit of life is the cell. Cells can be very different. This type of cell is called a neuron.	All living things are made up of cells. Organisms can be single-celled, like bacteria, or multicellular, like humans.
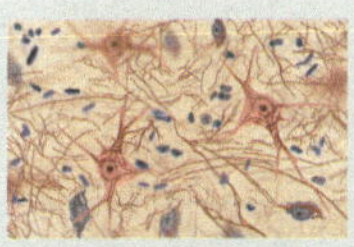 Tissue	Tissue is comprised of cells that are similar and have a shared function. One cell type found in brain tissue is the neuron. Neurons receive and transmit electrical signals.	In addition to nervous tissue, you also have other tissue types, like muscle and connective tissue in your body.
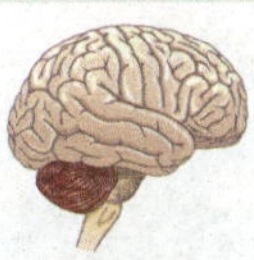 Organ	The brain is an organ made up of many kinds of cells, including neurons. The brain regulates many processes, including our wake/sleep cycles, learning, and emotions.	Other organs in your body include heart, skin, liver, and lungs.
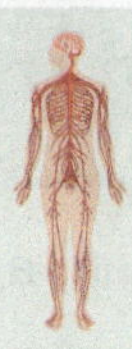 System	The brain is part of a larger system called the nervous system. As part of the nervous system, the brain communicates with the entire body, coordinating our responses to stimuli such as heat, cold, and pain.	The nervous system is only one of many systems in your body. Other examples of systems include the digestive, reproductive, and circulatory systems.

Living organisms use energy. Just like a car requires gasoline to run, life requires energy. **Energy** is the ability to do work (Chapter 4). Most life on Earth is dependent on the Sun for energy. Organisms like plants can convert solar energy from the Sun into a chemical form. This chemical energy can be consumed by animals. In Chapter 4, we will learn about energy transfer between autotrophs such as plants, and heterotrophs such as animals.

Living organisms maintain homeostasis. The process by which an organism maintains a constant internal environment is called **homeostasis**. This is why when you go outside, your body stays the same temperature instead of matching the outside temperature. Humans maintain homeostasis by sweating when the outside temperature is too high and shivering when the temperature is too low. This helps maintain the body's internal temperature, creating a constant environment that keeps the processes of cells functional.

Living organisms reproduce. The process of making new organisms is called **reproduction**. Reproduction can be *asexual*, meaning genetic information from one individual is used to create a second, identical organism or clone. For example, bacteria and plants are both capable of cloning themselves to make more individuals. *Sexual reproduction*, on the other hand, combines the genetic information from different individuals to come up with unique offspring. This increases variety in a population and increases the likelihood that a species will survive environmental changes. We'll return to asexual and sexual reproduction in Chapter 8.

Living organisms evolve. Evolution is the process that led to the diversity of life on Earth. **Evolution**, or descent with modification, is when the frequency of gene variants changes in a population (Unit 3). At the most basic level, evolution is about reproduction and passing on genetic material, which contains information on how to make organisms, to the next generation. Genetic information naturally changes over time. Since genetic information dictates how an organism looks and behaves, sometimes these changes can be helpful, meaning an organism is more likely to reproduce and pass on the changes to its offspring. We will discuss evolution in more detail in Chapter 10.

The Big Ideas in Biology

Biology is unified around five major themes: evolution, structure and function, information flow, energy transformation, and interconnectedness of living systems. As these are major themes in biology, we will return to these ideas throughout this book. Look for margin features with the symbol to see how what you are learning is connected to these big ideas.

Let's consider these themes in the context of a well-known virus: SARS-CoV-2. SARS-CoV-2 infection causes COVID-19. In 2019, SARS-CoV-2 sparked a worldwide pandemic that as of 2024 had killed upward of 7 million people. Viruses are made up of two different kinds of molecules: protein and nucleic acid. A protein wrapper encases the nucleic acid core, which contains the information on how to make a virus. Although simple from a structural perspective, they can cause illness, not only in humans and other animals but also in bacteria and plants. There are a wide variety of viral shapes (Figure 1.14). Note that while many viruses cause a variety of familiar human illnesses, others, such as bacteriophages, only infect bacterial cells.

? Did You Know?

Bacteriophages are used in the food industry to prevent bacterial growth. This helps minimize bacterial growth on processing equipment and extend the shelf life of food.

Figure 1.14 There are many types of viruses

Viruses cause diseases in humans, ranging from the common, such as adenovirus (colds) and influenza (flu), to the uncommon, such as the Ebola virus (Ebola hemorrhagic fever). SARS-CoV-2, a type of coronavirus, is responsible for COVID-19. Vaccines exist for humans for hepatitis B, influenza, and SARS-CoV-2. Bacteriophages are viruses that specifically infect bacteria and not humans.

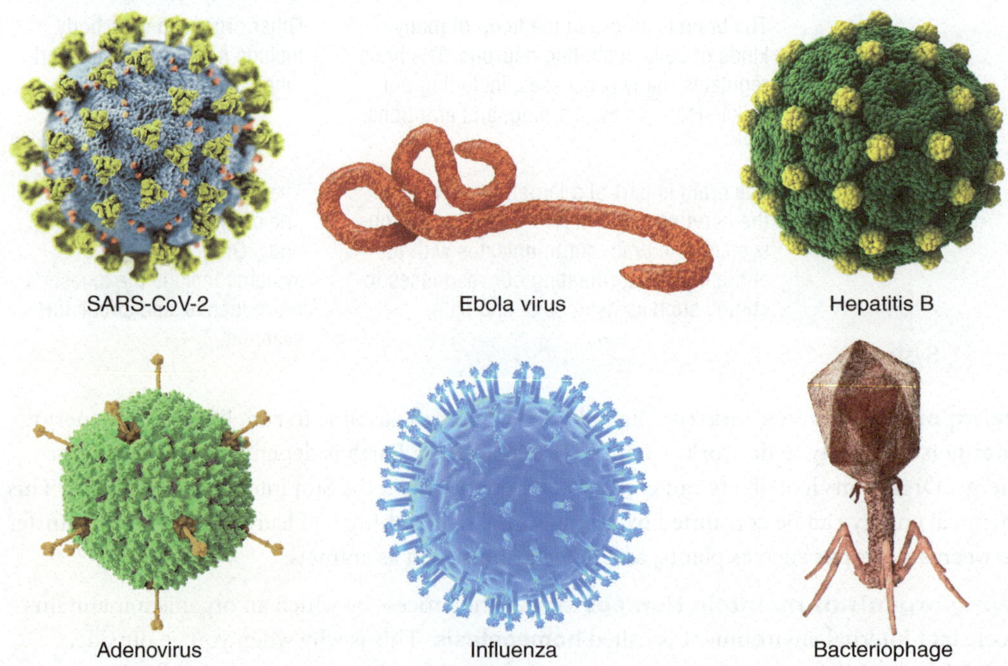

Big Idea #1: Evolution. If we look at SARS-CoV-2, every time the virus reproduced or made copies of itself, mistakes were introduced into the genetic material. A version of SARS-CoV-2 with several changes is called a variant of the original. For example, the first characterized version of SARS-CoV-2 is the Wuhan-Hu-1 strain. Wuhan is the name of the Chinese city where SARS-CoV-2 was first reported. In late 2020, the delta variant of SARS-CoV-2 appeared. Although the symptoms caused by the original and variant strains were the same, delta was 80-90% more contagious than the original strain. By 2021, the omicron variant appeared and was even more contagious than the delta variant, although there is evidence that it also resulted in less severe disease. We can expect that as more people are infected or reinfected with SARS-CoV-2, more variants will arise.

Big Idea #2: Structure and Function. SARS-CoV-2 is called a *coronavirus* because of a series of fingerlike projections that stick out from the surface. Since the central part of the virus is a ball, the "fingers" sticking out make it look like the virus is wearing a crown—and *corona* is the Latin word for crown (see Figure 1.14). These finger-like proteins are called the spike proteins. The virus uses the spike proteins to get inside human cells. Once in your cells, the virus hijacks the molecular machinery in your cells to make more copies of itself.

Big Idea #3: Information Flow. Genetic material stores information and this information is shared across individuals. For example, each new variant of SARS-CoV-2 is both similar and different from the previous variants. How people respond to a SARS-CoV-2 infection is also a matter of information flow. For example, a group of biologists were trying to determine why some people who develop COVID-19 become very sick and end up in the hospital. One potential reason is that in some people the immune system responds to SARS-CoV-2 infection by attacking itself rather than the virus. Since the immune system was not functioning properly, patients were more susceptible to infection and therefore were sicker. Another research group identified human gene variants that made patients' cells more susceptible to SARS-CoV-2 infection. After identifying these candidate genes, the biologists looked for medicines that interfere with SARS-CoV-2's ability to infect cells and spread throughout the body. This highlights how information flows and is exchanged (and blocked!) between the virus and its human host.

Big Idea #4: Energy Transformation. Most living things get energy from the Sun or by consuming other organisms. But where does a virus like SARS-CoV-2 get energy? It neither captures energy from the Sun nor gets energy by eating. Rather, viruses steal energy from their hosts. For example, the influenza virus uses about 1% of the host cell's energy reserves to produce about 6,000 copies of itself. In the case of SARS-CoV-2 infection, the virus redirects the function of the mitochondria, the part of the cell that makes energy. The virus causes the mitochondria to stop making energy for the cell and instead help make more viruses. Damage to the mitochondria continues even after a person no longer tests positive for COVID-19 and may explain why some people suffer long-term COVID-19 symptoms and organ dysfunction persists after a COVID-19 infection.

Big Idea #5: Interconnectedness of Living Systems. All living organisms on Earth are connected. This relates to our first big idea—that we are all related to one another and evolved from the same ancestral cells. Not only do we share basic genetic information, but organisms interact with one another. For example, SARS-CoV-2 interacts with our cells to make more copies of itself. The consequent infection causes a wide range of physiological responses in our bodies: we begin running a fever to kill off the invader, and our noses begin to run to flush out viral particles. As our bodies mount their immune responses to SARS-CoV-2, different versions of SARS-CoV-2 evolve that are better at evading these immune responses, leading to the emergence of new SARS-CoV-2 variants.

Biology for Life

Why Do I Need a Flu Shot Every Year?

Have you ever wondered why you need a new flu shot each year, or why you can get vaccinated against the flu and still get the flu? But for other viruses, like polio, you do not need to be vaccinated each year. When viruses reproduce, some are better than others at "proofreading." What do we mean by that? Just like you proofread your work before you turn it in, viruses can "check their work" too and see how well they copied themselves. A virus that is good at proofreading can not only detect errors but can also correct them, just like you can use backspace on your keyboard to correct an error. The flu virus isn't very good at proofreading and therefore doesn't recognize or fix its copying mistakes, so lots of errors are made when new flu virus copies are created. These errors in the virus's genetic material accumulate until eventually the virus is so different from the original that it is considered a new variant of the flu. Public health officials determine the most common types of flu virus present in the population each year to make recommendations on how to make the best vaccine. The good news is that even if you get the flu after you've been vaccinated, you are less likely to get very sick because the flu viruses are related. Therefore, the vaccine still gives you some protection.

The Centers for Disease Control and Prevention (CDC) maintains an interactive FluView application on their website. Using this tool, explore data on the current flu season. For example, which age group has had the most flu cases during the current season? Which strains of flu are making people sick? How does this season's statistics compare to last season? Five seasons ago? What questions do you have about the flu that you can answer with FluView?

With the understanding of biology you now have from reading this section, it should be easy to determine if something is alive, right? Is a strawberry alive? Was it "killed" when it was picked? It has seeds and could produce another strawberry plant, so is it living? How about artificial intelligence (AI)? Computer scientists were intrigued with the idea of artificial intelligence starting in 1950 when Alan Turing first proposed the "imitation game," now known as the Turing test, to identify the equivalence of human intelligence in a computer. Although no computer has ever passed the Turing test, today people routinely use generative AI for myriad purposes. Artificial intelligence has evolved, but not necessarily in a biological sense. Viruses, the main subject in this section, are also tricky to consider as living. On their own, viruses can neither use energy nor can they reproduce—they require a host for both. Remember, these questions are part of the *fun* of biology! Studying biology is so much more than simply memorizing facts, something we will explore more in the next section.

1.2 Summary

- Biology is the scientific study of life.
- Things that are alive have structure, use energy, maintain homeostasis, reproduce, and evolve.
- Biology is unified by five big ideas: evolution, structure and function, information flow, energy transformation, and interconnectedness.
- Evolution is when species change over time.
- Structure and function means that the shape of structures influences their function.
- Information flow can occur due to transmission of genetic material.
- Living organisms transform energy.
- Interconnectedness of living systems refers to the fact that all living organisms interact with one another.

1.3 Biology in Practice

Learning Objectives

1-11 Describe the places where biological research occurs.

1-12 Describe the kinds of people who do science.

1-13 Compare and contrast different disciplines of biology.

The Breadth of Biology as a Practice

What comes to mind when you think about what a biologist does? Is a biologist someone who dissects animals all day in a laboratory? Does a biologist always have to work in a laboratory? Do all biologists work with animals? Or maybe biologists just sit around all day and come up with new vocabulary words for unsuspecting students who just want to fulfill their general education requirement and graduate? In reality, biologists are diverse, and the places where they do biological research and the questions they ask are wide-ranging. Yes, some biologists work in laboratories. And some biologists work in nature or in "the field." There is even a subdiscipline of biology called taxonomy devoted to naming organisms, generating new words for you to know. But biology as a practice is far more than that. In the next section, we profile biologists with a variety of research interests, starting with a trip into outer space.

Biology in space. A *biofilm* forms when different kinds of microorganisms attach to a surface. You experience biofilms every day when you brush your teeth to remove plaque. Biofilms are not only on your teeth, though. They grow in places like the International Space Station, too. In the same way not regularly attending to the biofilms on your teeth can cause issues later on, biofilms in space can wreak havoc on equipment such as spacesuits or water recyclers. That's why scientists need to understand the microorganisms that grow in space and what they grow on.

The Space BioFilms team is a group of scientists and engineers from universities and space agencies who work together to answer these questions. One member of the team is microbiologist Dr. Pamela Flores Ayuso (Figure 1.15). *Microbiology* is a subdiscipline of biology concerned with the study of microscopic life. Because Flores works closely with engineers as well as biologists, her research is an example of **interdisciplinary research**, in which the study of a scientific question involves concepts from multiple disciplines.

Figure 1.15 Dr. Pamela Flores Ayuso, microbiologist

Dr. Flores studies how microorganisms in biofilms grow on surfaces in the International Space Station.

Source: Dr. Pamela Flores Ayuso

Point of Confusion: Although it sounds like "molecular" and "micro" biologists study the same thing, these are two different subfields. Molecular biologists are interested in what is occurring at the cellular and subcellular levels in the body. Microbiologists are interested in studying microscopic life, like viruses and bacteria.

Biology on tropical islands. Belize is a country and popular tourist destination located on the east coast of Central America. It also supports a population of crocodiles. Dr. Marisa Tellez (Figure 1.16) is an Indigenous and Latina scientist who earned her Ph.D. (Doctor of Philosophy, the highest degree awarded by a university) studying the interactions between crocodiles and parasites on Ambergris Caye, the largest island in Belize. She went on to co-found the Crocodile Research Coalition in Belize, a nonprofit organization that works to conserve crocodiles and their environments in Central America and the Caribbean through the involvement of local communities. Tellez's work spans biology subdisciplines including *ecology* (the study of relationships between living organisms and their environments) and *conservation biology* (the study of how to preserve nature and biodiversity by protecting species, plus habitats and ecosystems).

Figure 1.16 Dr. Marisa Tellez, Indigenous and Latina biologist

Dr. Tellez studies the ecology of tropical crocodiles and founded the Crocodile Research Coalition in Belize.

Source: crocodileresearchcoalition

Figure 1.17 **Dr. Lauren Esposito, evolutionary biologist**

Dr. Esposito studies scorpions and is the co-founder of 500 Queer Scientists, an organization dedicated to increasing the visibility of LGBTQI+ scientists.

Source: San Francisco Chronicle/Hearst Newspapers via Getty Images/Getty Images

Studying biology at a museum. Did you know that scorpions are related to spiders? Or that they obtain food by grabbing it with their claws and stinging it with their tail? Dr. Lauren Esposito (Figure 1.17) is an evolutionary biologist at the California Academy of Sciences, a natural history museum and research center in San Francisco, California. *Evolutionary biology* is a subdiscipline of biology concerned with the study of the evolutionary processes that led to the diversity of life on Earth. Esposito studies the evolution of scorpions and the venom that scorpions use to kill their prey. In addition to their work in evolutionary biology, Esposito is also the co-founder of 500 Queer Scientists. 500 Queer Scientists is a visibility campaign for LGBTQI+ people that provides LGBTQI+ role models and increases the visibility of LGBTQI+ scientists.

Studying biology to fight racism. Julian Lewis (Figure 1.18) was born in 1891 and the son of a man formerly enslaved. Lewis was the first African American scientist in the United States to earn both a Ph.D. and medical doctorate, or M.D. Lewis was interested in a subdiscipline of biology called *immunology*. Immunology is the study of the immune system, which is the system that protects our bodies from infection. In 1942, Lewis published a book called *The Biology of the Negro*, which detailed racial bias in medicine. In his book, Lewis provided evidence refuting racist notions as they concerned human biology. Lewis died in 1989 at the age of 97, but his legacy, not only as an African American scientist but also his work using science to fight racist ideas, continues today.

Figure 1.18 **Dr. Julian Lewis, early twentieth-century African American doctor**

Dr. Lewis was an associate professor at the University of Chicago and published in the field of immunology. His work helped expose racist medical beliefs.

Source: Scientist spotlight

1.3 Summary

- Biological research occurs in a variety of locations including inside space, in the field, in museums, and in laboratories or hospitals.
- Individuals from diverse backgrounds have contributed to science.
- There are many different disciplines of biology. For example, ecology focuses on the relationships between organisms and their environment, molecular biology focuses on living cells, microbiology studies microscopic organisms, and immunology is the study of the immune system.

1.4 How Do I Succeed in a Biology Class?

Learning Objectives

1-14 Identify the parts of the study cycle.

1-15 Match different study strategies to different parts of the study cycle.

Have you ever studied *really* hard for an exam, only to do poorly? Unfortunately, we have all been there! It can certainly feel frustrating and disheartening to work hard and not see the desired results. Science classes can be particularly difficult to study for, given the need to synthesize and apply abstract knowledge. In this next section, we will cover best study practices. There is also additional information on how to plan out your study in Appendix B. Throughout the textbook, we will return to these ideas and give specific advice in each chapter as part of the Understand It! section on how to apply evidence-based learning strategies in your biology class. Look for study tips in the margins and *Preview It!, Understand It!* And *Practice It!* sections in each chapter.

The Study Cycle

Learning new material in a classroom is no different than learning any other skill. Let's take learning to play an instrument as an example. First, you need some basic familiarity with musical terminology: notes, fingerings, keys, chords, embouchure, and parts of an instrument. This is equivalent to the first part of the study cycle, which we call *previewing*. Next, you learn how to play an instrument. For example, learning how to find middle C, play a C chord, and a C major scale on the piano. This is equivalent to the second part of the study cycle, *understanding*. Once we have this basic understanding of how to play an instrument, we can do the final part of learning: *practicing* And of course, once we've practiced how to play a song, we can perform it for another.

Now, what would happen if you skipped previewing and tried to play a C major scale on a piano? What's C? What's a major scale? It would be very overwhelming to navigate learning to play a C major scale without understanding first what we mean by a musical scale. Similarly, the class can feel very overwhelming if we don't have a basic understanding of the material before going in. What's next after understanding how to play a C major scale? Practice! This includes practicing the scale and applying the knowledge of the scale to playing music in C major. Only then can we perform a song. Understanding comes before practice, otherwise, we're trying to practice something we don't understand, which is inefficient at best and ineffective at worst. In addition, practice involves doing. Can we learn to play a song by studying the notes and not actually manipulating the keys on an instrument? No! We actually have to *do* it to learn the music—or to learn the new material you will encounter in this text.

How do these principles apply to help you learn? The **study cycle**, which is a framework for learning and includes previewing/reviewing, understanding, and practicing, is summarized in Figure 1.19. In the next section, we'll go over tips for helping you preview, understand, and practice in a biology class.

Figure 1.19 Use the study cycle to succeed in your studies

The study cycle has three parts: previewing/reviewing, understanding, and practicing. There are evidence-based strategies you can use at each part to succeed.

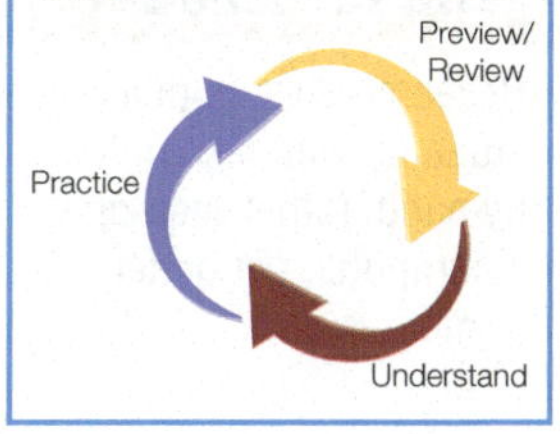

Previewing/reviewing. **Previewing** is the process of gaining some familiarity with the material before class. For example, you may want to learn a few basic vocabulary terms before coming to class. The review part is when you return to this phase later in the semester as needed to jog your memory about different terms, after completing the full study cycle.

Previewing is a particularly important activity to do in classes when you feel overwhelmed during a lecture. Having a baseline knowledge of the vocabulary will make the class go smoother. To preview, you may choose to review the textbook and/or posted class materials (like presentation slides) before class. For example, you could come to class a few minutes early and look through the material. You may also choose to schedule 30-60 minutes every weekend to preview course material for the week. At the end of your previewing activity of choice, write in your notebook "In class this week/today, I am learning about ... [make a list]." You could also write three things in your notes that you expect to learn in class that day.

Understanding. The second phase of the study cycle is understanding. **Understanding** is the process of storing information in long-term memory. The understanding phase occurs shortly after you have been introduced to new material, preferably within 24 hours of your class meeting. This is the period when you work to make sense of the information and move the information into long-term memory. Short-term memory holds information for a matter of seconds whereas memories stored in your long-term memory can persist for up to a lifetime.

There are several evidence-based study strategies to move information into your long-term memory. The first strategy is **elaboration**. Elaboration is the process of connecting new information we are learning to things that we already know. One way to do this is to look at a concept from class and explain it in your own words. For example, can you explain in your own words why structure-function relationship is a Big Idea in Biology? The key is to add your own spin on it, not repeat the text in your book or what your instructor told you. This is a similar process to the second strategy, **self-explanation**. One of the best ways to learn a new concept while also checking your understanding is by explaining it to someone else. Having a study group or partner is an excellent way to practice self-explanation for you and whomever you are studying with.

The third strategy is the use of **concrete ideas**. Given the way the brain works, it is far easier to understand concrete rather than abstract ideas. Science classes can have many abstract terms. Many instructors will use anecdotes, metaphors, and analogies to help illustrate scientific ideas. These concrete ideas are also found throughout this textbook. The key is to remember both the concrete idea and the abstract scientific principle behind them. For example, in Chapter 4 we refer to the molecules that "carry" energy to the next step in the pathway as "electron buckets" that can be "full" or "empty." This is useful to help remember the function of these molecules, but it is important to know the actual chemistry, too.

? Did You Know?

Research tells us that students who take notes by hand, rather than on a computer, get better grades.

Our fourth and final strategy for reaching understanding is called **dual coding**. Dual coding means that you use two different formats for representing information. For example, you could draw pictures *and* take notes. This way, one sees the information in visual and written format. Since there are two different kinds of information coming in, more areas of the brain are activated. This will help you remember and integrate the new information into your long-term memory.

How do you know if you were successful at understanding the material? I highly recommend doing self-checks of your understanding throughout your studying. You may also ask a tutor, friend, or teaching assistant to quiz you as well. It is MUCH better to find out you don't understand something *before* the exam when you can still do something about it, rather than have an Oh-No! moment during the exam! You may need to do some trial and error to figure out which strategy or strategies work best for you. If you are struggling to find something that works, the various learning insights presented in the Understand It! sections in each chapter are a great place to start.

Practicing. One of the biggest culprits behind "but I studied so hard and didn't do well on the test" is not practicing the material. What do I mean by practice? **Practicing** is the process of retrieving information from your long-term memory. Let's look at some non-academic examples. If you play sports, do you only go to games? No! You have practice. You do drills, exercise, and practice working within a team. How about a musician, do you show up and play a sonata from memory perfectly the first time? Nope. It takes years of practicing technique and scales, followed by dedicated practice sessions just on the music! The game or performance is like the test. To play hockey well, you must practice playing hockey. To make beautiful music, you must play the instrument. You don't, for example, study fingering charts and say it's sufficient the play an instrument.

If we bring this back to studying, you must practice pulling information out of your long-term memory and applying it (what you have to do on exams) *before* you take the exam. Just like looking at written instructions on how to play an instrument doesn't help you play an instrument, looking at your notes or memorizing terms with flashcards doesn't help you learn how to apply the information.

Several lines of evidence suggest that **retrieval practice**, or the process of pulling information from your long-term memory and using it, is one of the most effective study strategies. Using retrieval practice helps you develop not only durable memories (meaning, you are less likely to forget the information) but also accessible memories. Memories that are accessible mean they are easy for you to remember. Another way of thinking about having accessible memories is that "easy" test questions that you can answer quickly without thinking are linked to highly accessible memories. Using retrieval practice will not only help you in this course but will also help you remember this information and use it in other classes or as part of your life.

How do you use retrieval practice? Retrieval practice is any form of studying that causes you to remember and use the information you recall. For example, after studying for 30 minutes, close your book and write down everything that you remember on a sheet of paper. Practicing problems, doing practice quizzes or exams, or quizzing your friends (or having them quiz you) are all excellent ways of doing retrieval practice. Evidence suggests that students who do multiple practice sessions over time, called **spaced practice**, not only get higher scores on the exams but finish their exams faster. This brings us back to our opening example—the more you practice something, the faster you get. The more you practice scales, the faster you can play them. The more drills you do in hockey or soccer, the faster you get. The more you practice retrieving and using knowledge, the faster you can do it.

Remember, the study cycle is a *cycle*—you need to understand something before you can practice it. This is why it is important to ensure that you really understand the material before doing practice problems. You cannot practice using information that is not stored in your brain! If you are trying to solve problems and keep stopping to look things up, this is a sign you need to return to the understanding phase of the study cycle. The study cycle then completes by going back and reviewing what you've learned.

1.4 Summary

- There are three different phases of the study cycle: Preview/review, Understand, and Practice.
- Previewing is looking over the material before class.
- Understanding is moving information into your long-term memory shortly after class.
- Practice is recalling information stored in your long-term memory.
- Elaboration is connecting new information to things we already know.
- Self-explanation is explaining a new concept to someone else.
- Using concrete ideas help make sense of abstract terms.
- Dual coding means using two different formats to represent information, which helps the brain integrate the new information into long-term memory.

Understand It! Imposter Syndrome

Did you know that our emotions can influence how we learn? Emotions can both help and hurt us when we are trying to learn new material. The amygdala is the part of the brain that controls our emotions. The amygdala can override the cerebral cortex, which is the part of the brain that controls thinking. Humans evolved to have the amygdala override the rest of the brain as a safety mechanism. We do not need to be fully aware of a danger to feel fear and automatically jump away from a venomous snake underfoot.

So what do emotions have to do with forming memories and therefore learning in a biology class? Emotional memories tend to be very durable. Think back to a happy memory that you have—can you remember the details of that day? Now, compare it to a (likely) unemotional experience. What did you have for lunch last Wednesday? Can you remember it as clearly as the happy memory you picked?

The opposite can also be true. Sometimes strong emotions like anxiety or fear can interfere with our ability to form memories. Although biology is a fascinating subject, sometimes people feel anxiety about it or like they don't belong in a science classroom. This is a common psychological phenomenon called *imposter syndrome*. Imposter syndrome is when someone feels like a phony or that they don't belong. One of the best ways to address imposter syndrome and negate its impact on learning is by talking about it.

Now, thinking about the semester ahead, ask yourself these questions.

1. What content are you excited to learn about this semester?
2. What content are you nervous to learn this semester?

Chapter Review

1.1 The Science of Nature and the Nature of Science

Science is an iterative process that uses **scientific methodologies** to generate **data**. Scientific methodologies can be **deductive** or **inductive**. Regardless of the method, the ultimate goal of any scientific methodology is to generate **data** that is **generalizable**. Deductive methods rely on testing **hypotheses**, whereas inductive methods start with observations and gather data to generate new hypotheses. A **theory** is a well-supported explanation for a phenomenon, whereas a **law** describes an unvarying relationship, typically with a mathematical equation. **Pseudoscience, misinformation**, and **disinformation** are kinds of information not supported by scientific data. As a result, these claims are not supported by a consensus of science experts.

1.2 Biology Basics

Biology is the study of life. Living organisms have **organization**, meaning they have an underlying pattern. The **cell** is the smallest unit of life. Cells make up **tissue**, which is a collection of specialized cells that work together to perform a function. An **organ** is made up of tissue. Organs, are structures that perform a specific task. Different organs work together as part of a **system**. Living organism also use **energy** to maintain life, can maintain **homeostasis**, or create a stable internal environment can **reproduce** or produce offspring, and undergo **evolution** which can result in changes to the organism.

1.3 Biology in Practice

Biological research occurs in a wide variety of places and is done by diverse people. Subdisciplines of biology include molecular biology, ecology, and immunology. The defining factor is that all disciplines of biology study life using scientific methodologies. **Interdisciplinary research** is when scientists from different disciplines collaborate on one research area.

1.4 How Do I Succeed in a Biology Class?

Learning content in a classroom is the same process as learning any skill. It involves gaining a basic familiarity with the material, or **previewing**; moving that information into long-term memory, or **understanding**; and practicing using the material, or **practicing**. There are evidence-based study strategies associated with each phase of the **study cycle**.

Practice It!

Retrieval practice is the process of pulling information from your long-term memory and using it. Here are some questions to help with your retrieval practice.

Self-Quiz

1. Which of the following is an example of data? (LO 1-7)

a. My neighbors are Martians.

b. My neighbors leave for work between 8:00–8:15 a.m. on 82% of weekdays.

c. Someone said they saw my neighbors in their backyard at 4 a.m.

d. The lights go on and off at unusual times at the house next door.

2. A personal anecdote is not science because it is ______. (LO 1-7)

a. the unintentional spread of false information

b. generated using deductive scientific methods

c. a recollection of an experience of one person

d. based on mysterious phenomena that cannot be explained

3. In our example of the singing aliens, what kind of scientific methodology would be best to use to collect data to share with the scientific community? (LO 1-1)

a. Deductive methods; consult science fiction literature for a hypothesis to test

b. Inductive methods; document your observations

c. Deductive methods; use humans as a control group to see if aliens sing better

d. Inductive methods; use a scientific law to figure out why the aliens are singing

4. Which of the following is an opinion? (LO 1-7)

a. A chicken egg is 46% larger than a quail egg.

b. Quail eggs are better than chicken eggs.

c. Quail eggs contain twice the amount of iron contained in chicken eggs.

d. Five quail eggs have 71 calories versus 155 calories in one chicken egg.

5. Your light does not turn on when you flip the switch. Which of the following is a hypothesis? (LO 1-1)
   a. The light does not turn on when the switch is flipped.
   b. The lightbulb may be bad and needs to be replaced.
   c. After the lightbulb is replaced, the light is turned on.
   d. The lightbulb looks dark.
6. Which of the following is a scientific law and not a scientific theory? (LO 1-2)
   a. All living organisms are made of cells.
   b. The atom is the basic unit of matter.
   c. Evolution is the process of change over generations by natural selection.
   d. Buoyancy is determined by $F_B = w_{fl}$.
7. Which of the following is true about disinformation? Disinformation______. (LO 1-5)
   a. is based on misinterpretation of data through a lack of understanding
   b. is false information that is shared as truth but is wrong on purpose
   c. is the tendency to recognize only information that confirms what you think
   d. is based on whether the information matches your opinion
8. Which of the following is not a characteristic of life shared by all living organisms? (LO 1-9)
   a. reproduction
   b. movement
   c. homeostasis
   d. organization
9. The same metal can be used to make a hammer and a nail. The difference in ______ is what leads to the different ______ of a hammer and a nail. (LO 1-10)
   a. organization; evolution
   b. scientific method; organization
   c. energy transformation; structures
   d. structure; functions
10. A lawnmower is not alive. Which of the defining criteria for life are missing in a lawnmower? Select all that apply. (LO 1-9)
    a. organization
    b. use of energy
    c. evolves
    d. reproduces
11. The fact that proteins do not work properly if they are not folded into the correct shape relates to which big idea in biology? (LO 1-10)
    a. evolution
    b. structure and function
    c. energy transformation
    d. information flow
12. Which of the following would be studied by an ecologist? (LO 1-13)
    a. how the immune system recognizes a virus
    b. how trees interact with the soil and other trees
    c. how cells turn glucose into cellular energy
    d. how microscopic bacteria reproduce
13. What are the phases of the study cycle, in order? (LO 1-14)
    a. practice, preview, understand
    b. understand, preview, practice
    c. preview, understand, practice
    d. memorize, practice
14. Which part of the study cycle helps you move new information into long-term memory? (LO 1-14)
    a. previewing
    b. reviewing
    c. understanding
    d. practicing
15. Pulling information from your long-term memory and using it is a process called ______. (LO 1-15)
    a. previewing and reviewing
    b. retrieval practice
    c. memorization
    d. dual coding

Critical Thinking

1. How would you explain why a grasshopper is alive but a furnace is not? Which criteria of living organisms does that furnace exhibit? What is it missing? (LO 1-9)
2. A friend tells you that they have found the best supplement to take to have energy all day long. What would you want to know to determine if the reported effects of the supplement are scientifically supported or if your friend is being swayed by pseudoscience? (LO 1-4)
3. You are working on a hypothesis to use in your biology lab exercise and came up with *Ants prefer honey*. Your lab partner reminds you that your hypothesis must be falsifiable and testable. How could you reword your hypothesis to be falsifiable and testable? (LO 1-1, 1-2)
4. A news story reports that over the weekend there were four accidents involving drivers who were subsequently charged with a DUI (driving under the influence, i.e., while intoxicated). The drivers said they could not possibly be drunk because they were drinking alcohol "on the rocks," meaning their alcohol was served with ice. Why is the claim that adding ice to a drink inhibits intoxication and prevents drunk driving not supported by this data? (LO 1-4, 1-5)
5. How would you explain to your study partner that just looking at their notes will not help them do well on the upcoming exam? What are they missing by using only this strategy? (LO 1-15)

Source: d3sign/Getty Images

Unit

1

Nutrition

The Chemistry of Life

Today is Saturday, which means it is time for your weekly trip to the supermarket. As you grab your cart and head toward the dairy section, you're stopped from moving down aisle seven because someone has smashed a jar of pickles on the floor and the aisle is closed. So, you turn down aisle eight, the vitamins section. Halfway down the aisle you pause and ask yourself: *what is all this* ***stuff?*** Vitamins, amino acids, and supplements with unpronounceable names like ashwagandha. What is a vitamin or an amino acid? Aren't acids bad? And isn't zinc an element?

Shaking your head, you continue, leaving your questions behind in aisle eight. You reach the dairy section. It seems that your brain is feeling curious today. Scanning the shelves, you see a variety of milk options. There's the regular 2%, skim, and whole milk options. But you also see plant-based milk too. Almond milk, soy milk, coconut milk, rice milk. Why so many different varieties of milk? You have a friend who is lactose intolerant and drinks almond milk. What is lactose anyway, and what does it have to do with choosing almond over cow milk? You grab a bottle of chocolate milk, intent on escaping the store before more pesky biology and chemistry questions come to mind.

Visiting the supermarket is a basic task that everyone is familiar with. What may be new to you is how much biology and chemistry go into what we eat. To that end, our first unit discusses chemistry and energy through the lens of nutrition. We start in Chapter 2 with an overview of chemistry basics, learning about elements, atomic structure, and how atoms interact with one another to form bonds. We'll learn about the unique properties of water and briefly discuss acids and bases.

In Chapter 3, we'll learn about the molecules of life and discuss the organic compounds that make life possible. We'll identify the four macromolecules—carbohydrates, proteins, lipids (fats), and nucleic acids—and relate them to our daily lives. For example, we'll discuss the relationship between glucose (a carbohydrate) and diabetes. We'll also learn about amino acids, the building blocks of proteins, and why we tend to cook meat before we eat it.

In Chapter 4, we'll learn about energy. All living things require energy to live. Our bodies ingest food for energy, and we depend on organisms like plants that transform the Sun's energy through photosynthesis into a chemical form that we can use. We will also discuss how humans capitalize on these chemical processes to produce beer, bread, and yogurt.

Preview It!

Based on reading this unit opener, what questions do you have about the biology and chemistry of nutrition? What do you want to learn?

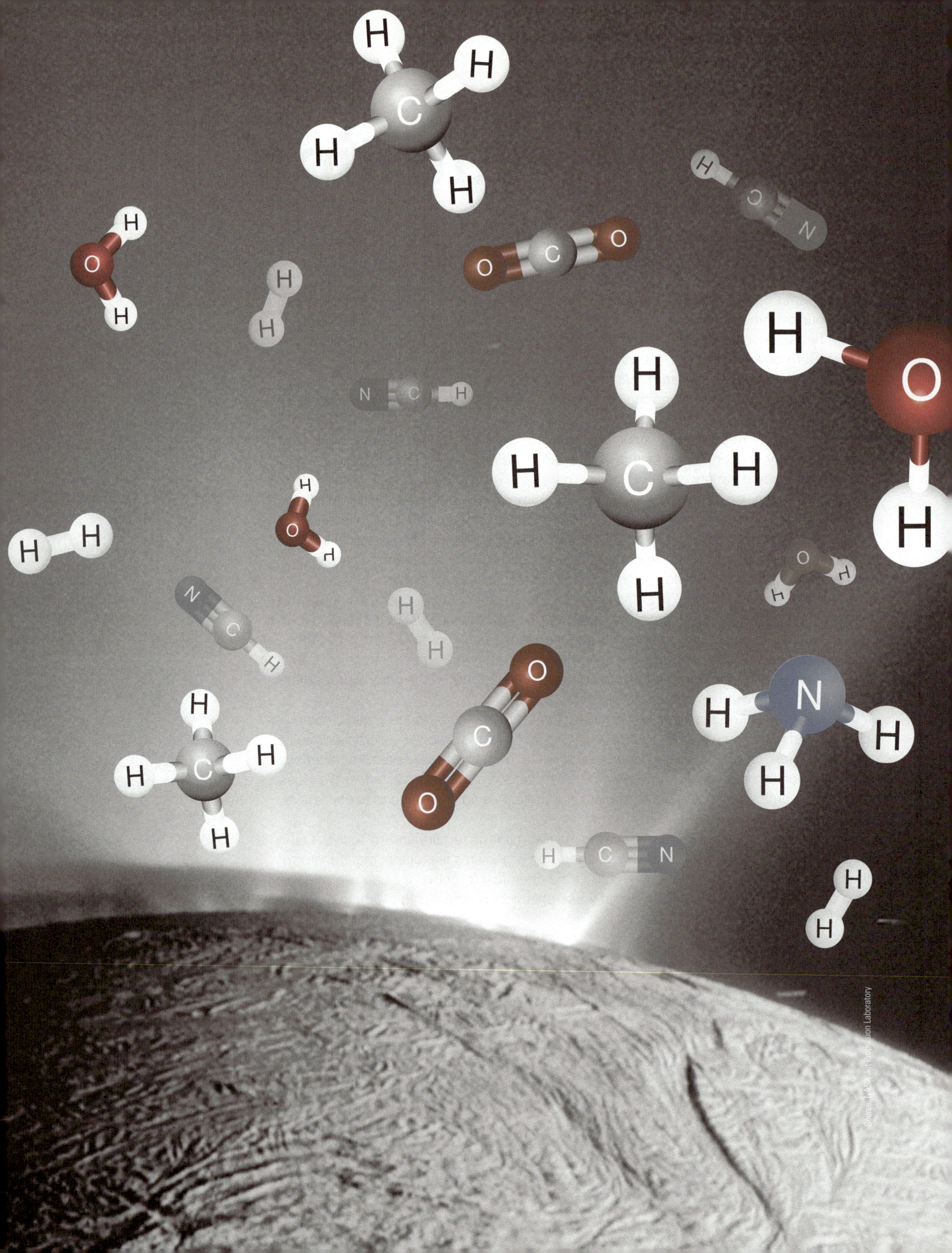

Source: NASA/Jet Propulsion Laboratory

2

The Chemistry of Biology

Chapter Outline:

Recipe for Life: Add Water

When it comes to the search for extraterrestrial life, biologists looking beyond Earth get very excited when they find two things: liquid water and high-energy molecules. Why? First, life as we know it requires water. This seems obvious, right? Humans cannot survive without a continuous supply of fresh water. We bathe in it, drink it, cook with it, play with it. You may even be sipping on water at this moment! Why are high-energy molecules important? If there is life out there, it must eat something. For example, alien life could be surviving on the energy from molecules it creates through a process like photosynthesis. Or, it could be living off the energy found in chemicals naturally occurring in the world where it exists. Discovering energy-containing molecules in the atmosphere or the rocks of an alien world means that food is available, should hungry life forms be wondering what's for dinner. Water, and high-energy molecules, are the ingredients that make life as we know it possible. But what does all of this have to do with chemistry? Why study chemistry in a biology class?

Frozen water and organic molecules spew from fissures in the icy crust of Enceladus, a moon of Saturn. The detection of water (H_2O) and organic molecules methane (CH_4) and hydrogen cyanide (HCN), plus molecules of ammonia (NH_3), hydrogen (H_2), and carbon dioxide (CO_2) in plumes ejected into space through fractures in Enceladus's crust, suggest conditions favorable to life exist in the moon's subsurface ocean. NASA scientist Jonah Peter says, "The discovery of hydrogen cyanide was particularly exciting because it's the starting point for most theories on the origin of life." Reactions with hydrogen cyanide produce many molecules important for life on Earth, so it could serve as a building block for life on Enceladus.

To truly understand life on Earth and elsewhere, we must study the unique chemistry of water and the chemical properties of matter. Life depends on chemistry, and in particular, water-based chemistry. Liquid water helps facilitate chemical reactions by dissolving substances. In this chapter, you will learn about the chemistry behind life.

Preview It!

Write down one thing that you already know about the chemistry of water. What questions do you have about chemistry? What would you like to learn about water's role in the chemistry of life?

2.1 Chemistry Is Key to Life

Learning Objectives

2-1 Describe the relationship between chemistry and biology.

2-2 Identify the parts of the atom.

2-3 Compare and contrast atoms and elements.

2-4 Describe what happens to electrons in a reduction–oxidation reaction.

You may be wondering why a biology textbook has an entire chapter devoted to chemistry. Chemistry is a scientific discipline that studies the properties and behavior of **matter**. Matter is anything that has mass and takes up space. If we think about this in the most simplistic terms, chemistry is the study of the "stuff" that makes up the planet and universe beyond. This includes living organisms.

As you sit reading this textbook, a wide range of chemical reactions are occurring. For example, if you happen to be eating food while you read, chemical reactions in your body work to break down the food in your mouth and stomach. Eventually, the nutrients will be absorbed in your small intestine and either chemically converted into energy or incorporated into the tissues and bones of your body. At the same time, inside your cells and tissues, a variety of chemical processes are keeping you alive. But it is not just inside our bodies where chemistry plays a role. We are aware of and respond to the outside environment—be it the touch of a kitty's fur or a cool breeze—through chemical interactions.

How do these chemical interactions work? To understand the relationship between chemistry and life, we must first understand basic chemical principles. In the following section, we will learn about atoms, the elements they make up, and how variability in the number of subatomic particles is important for life.

Atomic Structure

The **atom** is the smallest unit of matter. Atoms are made up of subatomic particles. There are three kinds of subatomic particles: **protons, neutrons,** and **electrons** (Figure 2.1). Protons have a positive electrical charge, electrons have a negative electrical charge, and neutrons have no charge (they are neutral). These electrical properties are important to understanding how atoms interact with one another. For example, oppositely charged particles attract one another, and particles with the same charge repel each other. We will return to the importance of charged particles throughout this chapter.

Protons and neutrons make up the core, or **nucleus**, of every atom. Protons and neutrons are about the same size, and the number of protons in the nucleus is generally equal to the number of neutrons. In some atoms, however, the number of neutrons varies slightly. We will discuss this more, later in this section.

Different atoms have different numbers of subatomic particles, but typically atoms have the same number of electrons as protons. The negative charge of an electron is the same

Figure 2.1 Carbon is an element important for life

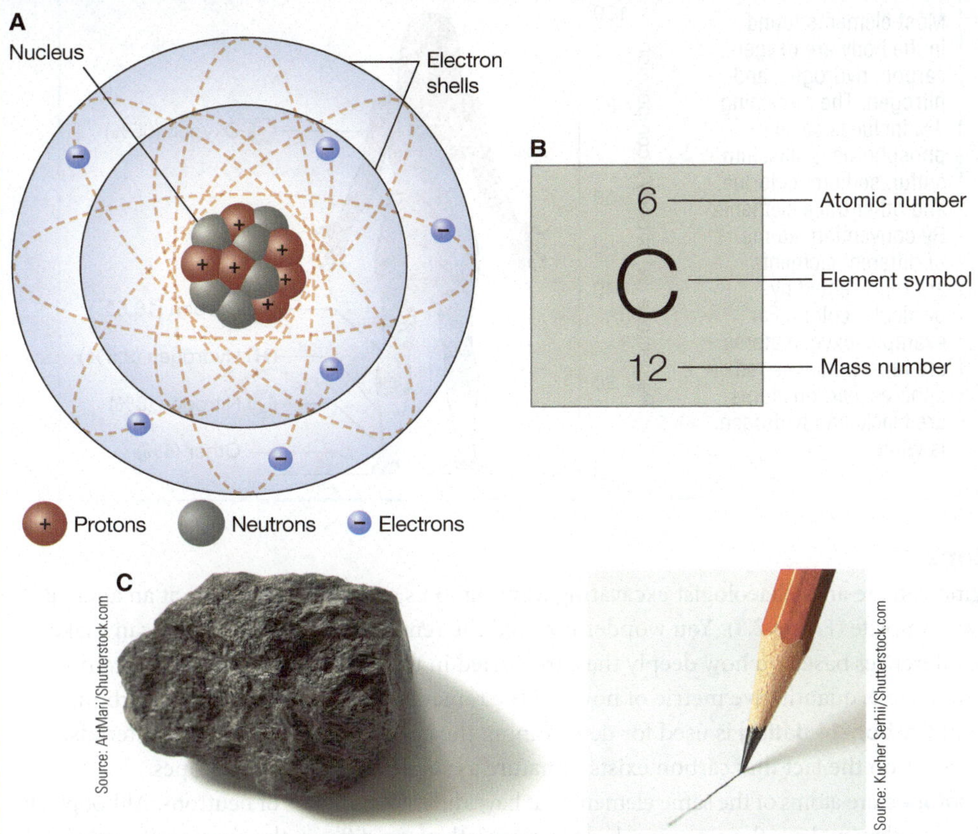

A. Atoms consist of electrons moving around a nucleus of protons and neutrons. Models such as this don't show the atom to scale because it can't—electrons move in defined, three-dimensional spaces about 10,000 times larger than the nucleus.

B. Simplified representation of carbon on the periodic table. Here, the mass number, 12, equals the number of neutrons and protons in the carbon nucleus. In reality, the average mass number of carbon is 12.01 because a very small fraction of carbon atoms has one or two extra neutrons.

C. Graphite is made of carbon and comprises the "lead" in a pencil.

magnitude as the positive charge of a proton, so an atom with the same number of electrons and protons carries no charge.

Elements

Elements are substances that cannot be broken down into other substances. For example, table sugar (sucrose) can be broken down into two other substances, fructose, and glucose, and so sugar is not an element. Conversely, carbon (Figure 2.1) is an element and when broken apart will not form another substance. The **atomic number** of an element refers to the number of protons in that element, and scientists use atomic numbers to define the elements. Carbon has an atomic number of 6, and all atoms with six protons in their nucleus are carbon atoms, no matter how many electrons or neutrons they have. Electrons have almost no mass and adding together the total number of protons plus neutrons in an atom gives you that atom's **mass number**.

> **? Did You Know?**
>
> Carbon's symbol, C, is from *carbo*, the Latin word for coal, which is mostly carbon.

Scientists know of 118 elements, and each is represented by a symbol that is an abbreviation of its name. These elements are arrayed in the periodic table. The **periodic table** organizes all elements based on their atomic number and chemical interactions (Appendix C). Note that the human body is mostly made up of just ten elements. Of those ten, our bodies are mainly oxygen, carbon, hydrogen, and nitrogen (**Figure 2.2**).

Figure 2.2 **Although there are 118 elements, only 10 are found in the human body**

Most elements found in the body are oxygen, carbon, hydrogen, and nitrogen. The remaining 4% includes calcium, phosphorus, potassium, sulfur, sodium, chlorine, and other trace elements. By convention, atoms of different elements are represented by particular colors. For example, oxygen atoms are represented as red spheres, carbon atoms are black, and hydrogen is white.

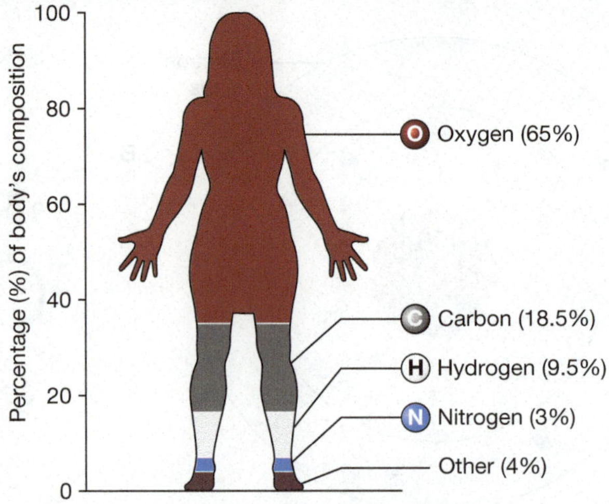

Figure 2.3 **^{14}C dating of prehistoric animals found at an ancient hunting site**

Carbon-14 levels are constant in a living organism, then decrease after death. Since carbon-14 decays at a consistent rate, called a half-life, biologists can measure the amount present in a sample to determine its age.

Sam Abell/National Geographic Image Collection

Isotopes

Imagine you are an archaeologist excavating hunting tools and animal remains at an ancient human campsite (Figure 2.3). You wonder how old the remains are. Although you can make some inferences based on how deeply they are buried in the ground, how can you obtain a more objective, quantitative metric of how old the remains are? By analysis of the carbon present! **Carbon-14 dating** is used for determining the age of various biological materials, and it relies on the fact that carbon exists in nature as stable and unstable isotopes.

Isotopes are atoms of the same element that have different numbers of neutrons. Although all atoms of an element have the same number of protons, they can differ in the number of neutrons. Most carbon atoms have six neutrons. However, some carbon atoms can have seven or even eight neutrons. As we learned previously, the total number of neutrons plus the number of protons in the nucleus of an isotope equals its mass number. Mass number is written as a superscript to the left of the element's symbol. The most common isotope of carbon has six protons and six neutrons (see Figure 2.1B), so it is ^{12}C, which is pronounced carbon-12 or C-12 for short. The other naturally occurring carbon isotopes are ^{13}C (6 protons, 7 neutrons), and ^{14}C (6 protons, 8 neutrons).

Some isotopes, like ^{14}C, are unstable. In an unstable isotope, the protons and neutrons that make up the nucleus can spontaneously transform into, respectively, neutrons and protons. During this process, the nucleus emits radiation in the form of new subatomic particles, energy, or both. This process is called **radioactive decay** (also called radioactivity). Radioactive decay occurs at predictable rates and forms predictable products. For example, when ^{14}C decays, one of its eight neutrons splits into a proton and an electron. The proton remains in the nucleus, and the electron is emitted as radiation.

! Point of Confusion: Radioactive decay is not the same thing as radiation. Radiation refers to the movement of energy as a wave, such as heat, light, or microwaves. Radioactive materials are potentially dangerous because radioactive decay *produces* a particular type of radiation that can damage our cells. One example of dangerous radiation produced by radioactive material is *ionizing radiation*, which is the release of high-energy alpha particles, beta particles, or gamma rays.

The nucleus has lost a neutron and gained a proton, so it now has seven of each. All atoms with seven protons are nitrogen atoms. Thus, an atom of ^{14}C (6 protons, 8 neutrons) decays into an atom of ^{14}N (7 protons, 7 neutrons). Scientists have measured the rate of this decay, so we know that half of the atoms in any sample of ^{14}C will be ^{14}N atoms after 5,730 years. Another way of saying this is that 5,730 years is the **half-life** of carbon-14.

How does this help you, an archeologist, determine the age of items you discover at your excavation site? Radioisotope decay is so predictable that researchers can estimate the age of a

Figure 2.4 **Overview of carbon-14 dating**

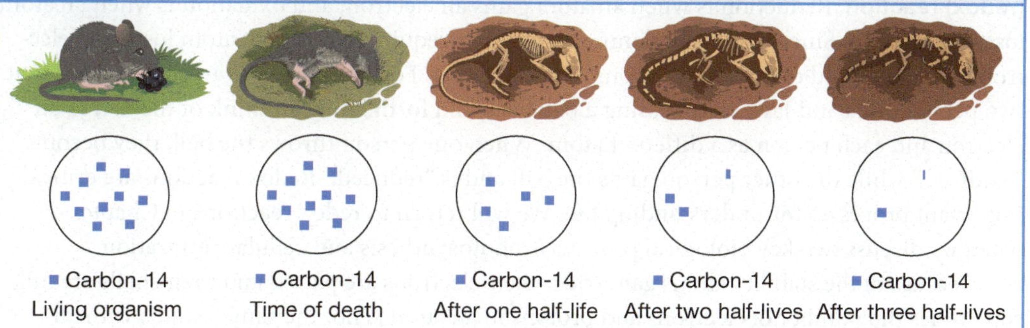

Carbon-14 levels are constant in a living organism, then decrease after death. Since carbon-14 decays at a constant rate, researchers measure the amount present in a sample to determine its age.

carbon-containing artifact by measuring its carbon isotope content. This is how it works: when an organism is alive, it takes in ^{14}C naturally present in the environment as part of its diet, or in the case of plants, ^{14}C is incorporated with CO_2 uptake during photosynthesis (Chapter 4). While the organism lives and grows, the carbon-14 in that organism remains constant. Once an organism dies, it is no longer ingesting gases from the atmosphere or food containing ^{14}C. Carbon-14 uptake stops, and the level of carbon-14 in the body starts to decline due to radioactive decay. Scientists can measure the amount of carbon-14 left in an organism's remains to estimate when the organism died (Figure 2.4). This technique is enormously useful and can be applied to a variety of samples including animal remains, carbon-containing ash from firepits, and the wooden artifacts and tools used by ancient people.

Dance of the Electrons, Act I

Just like the dancers moving across the stage is essential to a ballet, the movement of electrons is essential to chemical reactions. Electrons are tiny, compared to protons. They move at high speed around the nucleus and are found in certain regions of the atom called **shells**. Atoms have between one and seven shells. Atoms fill their innermost shells first with electrons, then move outward. Within these shells, electrons are found singly or paired. If all electrons are paired and the shells are full, the atoms tend to be unreactive and therefore stable. If there are unpaired electrons, it creates a vacancy in the shell (Figure 2.5A). Atoms "want" to fill these vacancies. As a result, atoms with vacancies are reactive and are more likely to interact with other atoms to fill the vacancies. Filling the vacancies results in new electron pairs forming, ultimately leading to the formation of bonds with other atoms (Figure 2.5B). We will discuss bonding in more detail in Section 2.2.

Atoms or molecules that contain atoms with unpaired electrons are called **free radicals**. Free radicals are unstable and consequently are highly reactive. As a result of this reactivity, they try to give away their unpaired electron to other atoms or they steal electrons from other atoms to create a stable pair. Such interactions damage organic molecules such as DNA and contribute to the development of cancer (Chapter 6). Consequently, free radicals can be dangerous to life.

? Did You Know?

An antioxidant is a chemical that neutralizes free radicals, preventing damage to cells or cellular components like DNA. Antioxidants can be found in foods such as blueberries and dark chocolate or drinks such as green tea.

Figure 2.5 **The shell model of the atom**

A. Each circle (shell) represents one energy level. Electrons fill shells from the innermost shell outwards until there are as many electrons as protons in the atom.

B. Electron vacancies dictate atom stability. Since hydrogen has a vacancy in its valence shell and helium does not, hydrogen is more reactive.

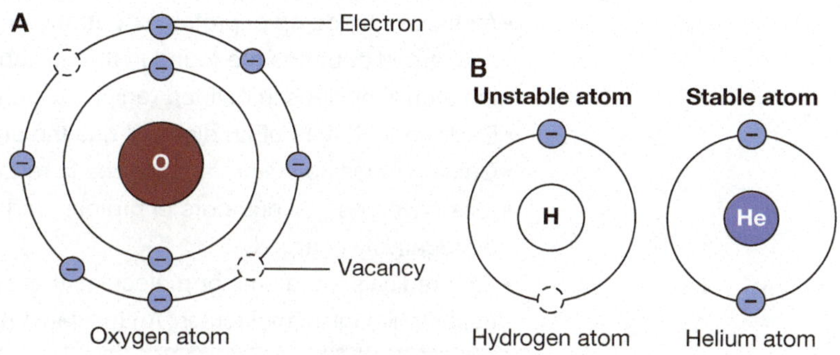

Figure 2.6 Considering redox reactions with sports

You can think of the ball as an electron and Alba and Javier as adjacent atoms. As Javier throws the ball, Javier becomes oxidized. Conversely, when Alba catches the ball, Alba is reduced.

Source: Petri Oeschger/Moment/Getty Images

The process by which an atom loses or gains an electron is called a **reduction–oxidation (redox) reaction**. Reduction is when an atom gains an electron, and oxidation is when an atom loses an electron. Since one atom gaining an electron requires a different atom losing an electron, reduction, and oxidation reactions occur together. For example, in Figure 2.6 you see that two people, Alba and Javier, are tossing a ball back and forth. You can think of the ball as an electron and each person as a different atom. When one person throws the ball, they become "oxidized," while the other person gains the ball and is "reduced." Redox reactions are critically important processes for understanding life. We will return to redox reactions in Chapter 4 when we discuss two key biological processes, photosynthesis and cellular respiration.

Because of the stability atoms gain when their electrons are paired and their shells are full, sometimes the number of electrons and protons in an atom is not the same. Atoms with an unequal number of protons and electrons are called **ions**. Since there is an uneven number of positively charged protons and negatively charged electrons, ions have a net (or overall) charge. For example, a sodium ion has one more proton than it has electrons, so it carries one positive charge. An ion's charge is indicated by a superscript to the right of the element symbol. In this case, a sodium ion, which has the two-letter abbreviation Na, is followed by a superscript + to indicate a single charge. Therefore, Na^+ is the designation for a sodium ion.

Ions are responsible for many biological processes. For example, zinc ions (Zn^{2+}) help the enzymes (which you'll learn about in Chapter 4) in your immune system work properly. This is why many over-the-counter medicines to fight the common cold contain zinc.

Biology for Life

Using Radioactivity to Detect Cancerous Cells

Neutrons have little effect on the chemistry of atoms, so all isotopes of an element have essentially the same chemical properties. This means that radioactive isotopes of an element work in the body the same as non-radioactive isotopes. Doctors use this property when studying biological processes or diseases, such as cancer. For example, positron emission tomography (PET) is used to visualize cellular activity. This is particularly useful for finding out if a person has cancer and/or determining how well a person with cancer is responding to treatment. In a PET scan, cancer cells are visualized as bright spots of light. This is because cancerous cells divide rapidly and absorb more of the tracer than the neighboring non-cancerous cells.

Radiation is useful not only for detecting cancer cells but also for treating cancer cells. Working with a partner, make a list of ways radiation is used to treat cancer. What kinds of cancer can be treated with radiation?

2.1 Summary

- Chemistry is the scientific study of matter and its interactions.
- Chemistry is important to biology because all living organisms and nonliving things are made of matter.
- All matter consists of atoms.
- Atoms are made up of protons, neutrons, and electrons. Positively charged protons and uncharged neutrons are found in the nucleus of an atom. Negatively charged electrons orbit the atom's nucleus in defined regions called shells.
- Each neutral atom of an element has the same number of protons, the atomic number.
- Gaining or losing electrons happens in redox reactions.
- Ions have unequal numbers of protons and electrons and are therefore positively or negatively charged.
- Free radicals are atoms or molecules that contain atoms with unpaired electrons and can destroy biological molecules. Antioxidants neutralize free radicals.

2.2 Chemical Interactions

Learning Objectives

2-5 Explain how electrons participate in bonding.

2-6 Compare ionic and covalent bonding.

2-7 Describe the characteristics of carbon that make it so versatile in building complex molecules.

2-8 Explain the various ways a chemical can be represented.

2-9 Distinguish between polar and nonpolar covalent bonds.

Dance of the Electrons, Act II

In the last section, we learned that electrons are negatively charged particles in an atom that move in the space beyond the nucleus. Electron behavior is important for understanding how atoms can interact with one another to form bonds. A **chemical bond** is a strong attractive force connecting two atoms that arises from the loss, gain, or sharing of electrons.

Recall from Section 2.1 that electrons fill shells around an atomic nucleus. The outermost shell is called the **valence shell**. Within the valence shell are the **valence electrons**, and these are the electrons that most readily participate in chemical reactions. Recall that atoms are most stable when their electrons are paired (see Figure 2.5B) and one way atoms ensure that their electrons are paired is by gaining or losing electrons, forming ions. Another way for an atom to fill vacancies in its valence shell is to participate in bonding. Two types of bonding relevant to biology are ionic and covalent bonds.

Ionic bonds. An **ionic bond** occurs when two charged molecules or atoms (called ions), stick together because of a strong mutual attraction. The mutual attraction arises because each ion participating in the bond has an opposite charge. For instance, a neutral sodium atom contains an unpaired electron in its valence shell, and a neutral chlorine atom is missing an electron. The unpaired electron in the sodium atom can be transferred to the chlorine atom, creating a sodium ion (Na^+) and a chlorine ion (Cl^-). Since Na^+ and Cl^- have opposite charges, there is also strong mutual attraction between the two. When sodium and chlorine combine to form an ionic bond, it results in sodium chloride (NaCl), more commonly known as table salt.

Covalent bonds. Some atoms can fill their vacancies by sharing electrons with other atoms. The sharing of electrons between atoms is called a **covalent bond** (Figure 2.7). Covalent bonds are stronger than ionic bonds. How many covalent bonds an atom can form depends on its

Figure 2.7 **Atoms can share, one, two, or three pairs of electrons**

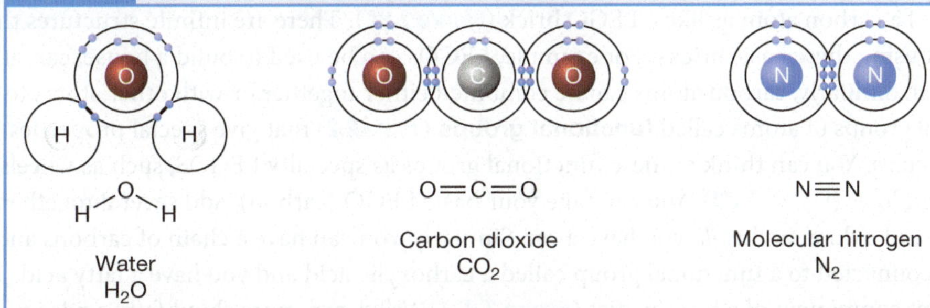

Water
H_2O

A. Single bond Sharing a single pair of electrons creates a single bond and is represented with a single line.

Carbon dioxide
CO_2

B. Double bond When two pairs of electrons are shared, a double bond is formed and is represented with two lines.

Molecular nitrogen
N_2

C. Triple bond Sharing three pairs of electrons creates a triple bond and is represented by three lines.

Table 2.1 Comparing Different Ways to Represent Molecules	
Familiar name	Water
Chemical name	Dihydrogen monoxide
Chemical formula	H_2O
Structural formula	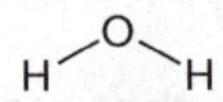
Ball-and-stick model	
Space-filling model	

valence electrons. The number of valence electrons dictates how many electron pairs can be shared and ultimately the number of bonds that can form. If one pair of electrons is shared, a single bond is formed. As shown in Figure 2.7, a single bond is shown as a single line between atoms. A double bond, or the sharing of two pairs of electrons, is represented as two lines. The sharing of three pairs of electrons is a triple bond and is represented as three lines.

Representing molecules. When two or more atoms covalently bond, a **molecule** is formed. Molecules can include the same or different elements covalently bonded together. For example, a molecule of oxygen gas, O_2, contains two atoms of oxygen covalently bonded together, whereas a molecule of water, H_2O, contains one atom of oxygen and two atoms of hydrogen covalently bonded together. A **compound** is a type of molecule that has two or more different elements bonded together. As you can see, water, H_2O is a compound.

Since the actual structure of organic molecules can be complex, we typically represent molecules in a simplified format (Table 2.1). First, there is the **structural formula**. A structural formula lists the atoms present using their shorthand letter abbreviations. Interactions (bonds) between molecules are shown as lines. Next, there is the **ball-and-stick model**. Rather than using only letters to represent atoms, colored balls are also used. Bonds are still represented as a stick. Ball-and-stick models are useful for showing what a molecule looks like in three dimensions. Finally, there is the **space-filling model**. The space-filling model is used to show a molecule's overall three-dimensional shape. A space-filling model is different from a ball-and-stick model in that the bonds are not drawn out.

Point of Confusion: Not all compounds are molecules. For example, sodium chloride consists of a sodium ion and a chloride ion ionically bound together and thus is an *ionic compound*. It is not a molecule because sodium and chloride do not share a covalent bond.

Building organic molecules. Did you know that how we build structures with the classic LEGO® bricks is a great way to think about how our bodies build organic molecules? Organic molecules all contain covalently bonded carbon and hydrogen atoms. The atomic structure of carbon makes it uniquely capable of forming complex and biologically active molecules. It has four valence electrons and can form up to four covalent bonds, with itself and with other atoms.

Think of a carbon atom as like a LEGO brick (Figure 2.8A). There are infinite structures that can be built using these basic bricks. For example, LEGOs can be used to build a house, car, or roller coaster. Similarly, carbon atoms can be combined either together or with other atoms to form special groups of atoms called **functional groups** (Table 2.2) that give special properties to the molecules. You can think of these functional groups as specialty LEGOS, such as wheels, doors, or windows (Figure 2.8B). You can take your basic LEGO (carbon), add several together, add wheels and a door, and *voila* you have a car. Similarly, you can have a chain of carbons and hydrogens connected to a functional group called a carboxylic acid and you have a fatty acid, an important component of a healthy diet (Figure 2.8C). We'll learn more about fatty acids in Chapter 3. In addition, different functional groups can also serve as molecular connectors for making new molecules. We'll come back to the idea of molecular connectors in Chapter 3.

Big Ideas in Biology

Structure and Function

Since carbon atoms have four valence electrons, they can form up to four covalent bonds. This structure allows carbon to form the backbone of many organic molecules.

Point of Confusion: To a chemist, organic molecules are those that contain carbon covalently bonded to hydrogen. Organic products at the grocery store refer to specific standards used in their production.

Table 2.2 Summary of Functional Groups

Name	Structure
Carboxylic acid	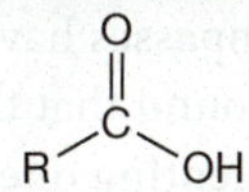
Amino	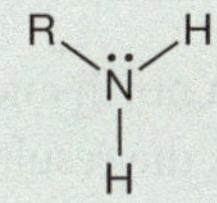
Hydroxyl	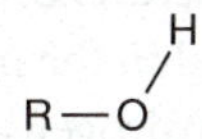
Aldehyde	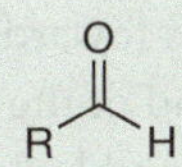
Ketone	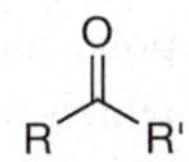
Sulfhydryl	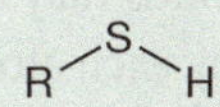
Phosphate	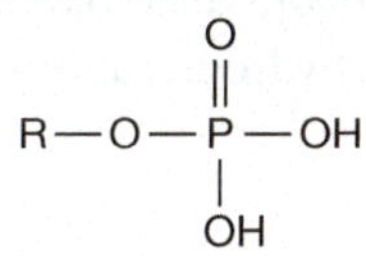

Figure 2.8 **Carbon atoms are the building blocks of organic molecules**

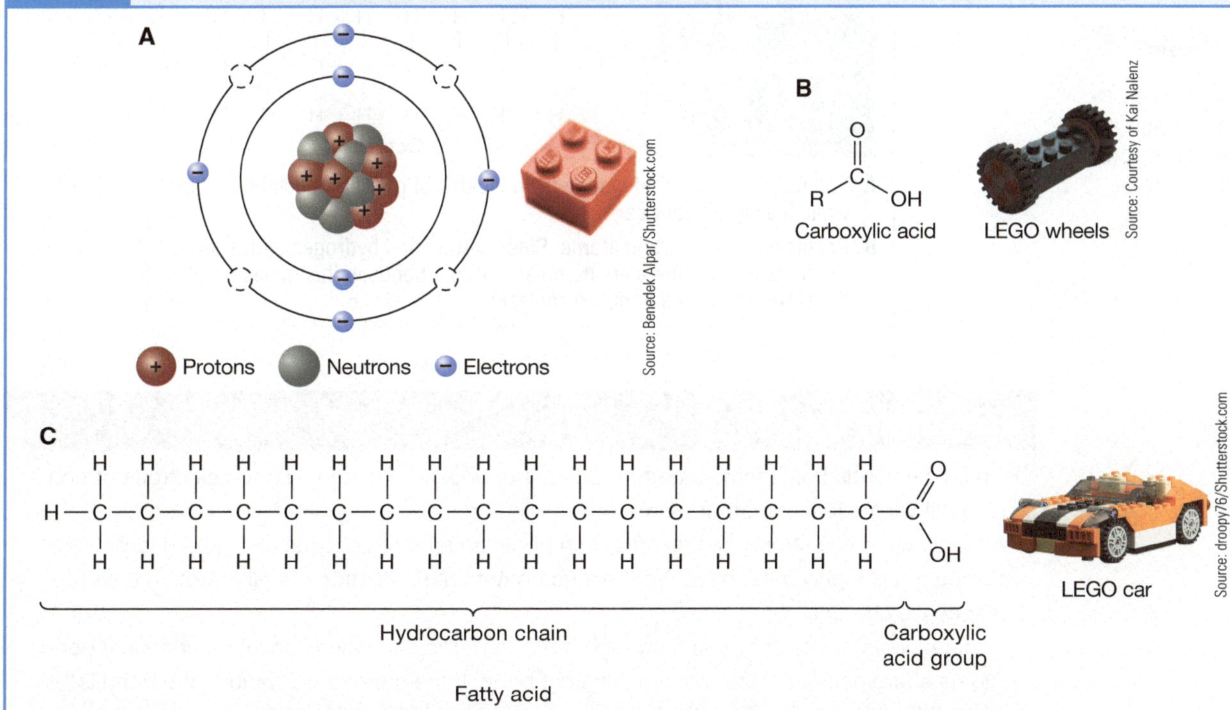

A. Carbon has four vacancies in its outermost electron shell and can therefore form up to four bonds with other molecules and generate a variety of structures. A carbon atom is like a LEGO® brick that can be used to generate a variety of structures.

B. In the same way that specialty LEGO components such as wheels bring new functionality to a LEGO model, functional groups bring new functionality to molecules.

C. Combining carbon atoms (the basic LEGO brick) with functional groups creates molecules. In this case, carbon atoms were combined to create a fatty acid, while the various LEGOs were combined to make a car.

Figure 2.9 **Water is a polar molecule**

Note the bent shape and negative charge around the oxygen atom and the positive charges at the hydrogen atoms.

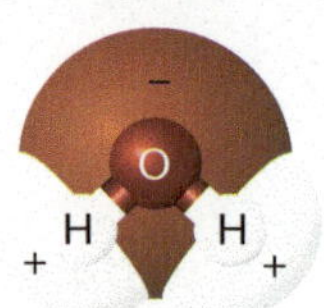

Bond Polarity

Before discussing bond polarity, let's first consider what we mean by "polarity" with a relatable example. As you know, a magnet has a north and south pole. So does Earth, and traditional compasses have metal needles that point toward the North Pole. For millennia, humans have found that the presence of a magnetic pole, and a tool for detecting that pole, is useful for orienting oneself. Similarly, we can use the "poles" of molecules to infer their chemistry.

Rather than north and south magnetic poles, molecules form positively and negatively charged poles as the result of differences in the distribution of positive and negative charges within their structure. Let's consider water (**Figure 2.9**). Water molecules have a V-shape. As a result of the bend in the molecule, negatively charged oxygen is on one end and the two positively charged hydrogens are on the opposite end. Since the oxygen atom hogs the negatively charged electrons toward itself and away from the hydrogen atoms, this creates a negative pole on the oxygen end of the molecule and two positive poles on the hydrogen end. The bonds between oxygen and hydrogen atoms are **polar covalent bonds**. A molecule with polar covalent bonds is called a polar molecule. Water is a polar molecule, and this property makes it essential for life on Earth, which we will learn more about in the next section.

Covalent bonds may also be non-polar. In a **non-polar covalent bond**, electrons are shared equally by each atom and thus no pole is formed. For example, octane (**Figure 2.10**), a hydrocarbon and major component of gasoline, contains only hydrogen and carbon atoms. Carbon and hydrogen atoms share their electrons equally and no poles are formed.

? Did You Know?

The difference between regular, midgrade, and premium gasoline is how much octane it contains.

Figure 2.10 **Octane is a hydrocarbon and a major component of gasoline**

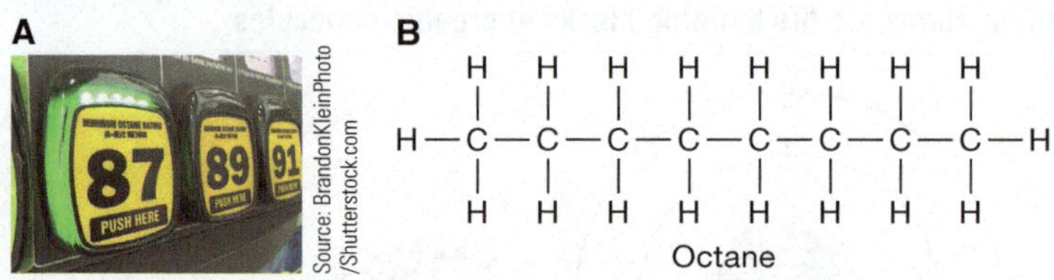

Source: BrandonKleinPhoto /Shutterstock.com

A. You may have heard of octane from looking at gasoline's octane ratings while fueling up your car.

B. Octane has eight carbon atoms. Since carbon and hydrogen atoms share electrons equally, there are no polar covalent bonds in this molecule, and therefore octane is a nonpolar molecule.

2.2 Summary

- Chemical bonds hold atoms together. Covalent bonds form when valence electrons of one atom fill vacancies in another atom.
- An ionic bond is a strong mutual attraction between ions of opposite charge. An ionic compound is highly polar because the ions do not share electrons, so they retain their full respective charges.
- A molecule consists of atoms held together in a particular way by covalent chemical bonds.
- Atoms share pairs of electrons in a covalent bond. If the sharing is unequal, the bond is polar. If the sharing is equal, the bond is nonpolar.
- Uneven sharing of electrons causes molecules to have slight positive and negative regions. These partial charges can be used to attract other charged molecules.
- For chemists, organic molecules are those that contain carbon–hydrogen bonds.

2.3 Special Properties of Water

Learning Objectives

2-10 Describe hydrophobic/hydrophilic interactions.

2-11 Explain why molecules must be polar to hydrogen bond.

2-12 Differentiate between acids and bases.

2-13 Explain what pH is and why pH is important in biological systems.

Life Relies on Water

All cellular organisms known to science are mostly water. Many organisms spend all or part of their lives in water. When considering the chemistry of life, all chemical reactions are carried out in water-based fluids. What is it about water that makes it so fundamentally important for life as we know it?

It comes back to bonding and polarity. As we learned in the last section, water is a polar molecule and has two polar covalent bonds. The V-shape of the molecule is important for separating the positive and negative charges (Figure 2.9). This separation is important because opposite charges attract one another, and like charges repel. Therefore, the negative end of water can easily attract the positively charged ends of other polar molecules—including other water molecules—as well as positively charged ions (Figure 2.11). Alternatively, the positive end of water can attract negatively charged ions or the negative end of other polar molecules. As a result of these interactions, water molecules can pull apart ionic compounds and surround the ions. This causes ionic compounds like sodium chloride to dissolve in water. Since water can pull apart other compounds, it is an excellent solvent. A **solvent** is a substance that can dissolve something else.

One reason water can so easily participate in chemical reactions is that it can spontaneously break apart and form positively charged hydrogen ions (H^+) and negatively charged hydroxide ions (OH^-). Note that hydrogen ions are simply protons—a hydrogen atom missing its electron. Although it would be tempting to overlook protons as too simple to be important, that is not the case, and scientists keep count of the number of protons in a solution relative to the number of hydroxide atoms by measuring the solution's pH. **pH** is a measure of the number of protons in solution. If a fluid has an equal number of H^+ and OH^- ions, its pH is 7 (or neutral). Pure water is neutral. If there are more H^+ ions than OH^- ions, the pH is below 7 (or acidic). Conversely, if there are fewer H^+ ions than OH^- ions, the pH is above 7 (or basic). A one-unit difference in pH corresponds to a tenfold difference in hydrogen ion concentration (Figure 2.12).

Figure 2.11 **Water interacts with ions in solution**

Note that the positive hydrogen atoms orient themselves toward the negative chloride ion (Cl^-) and the negative oxygen atoms orient themselves toward the positive sodium ion (Na^+).

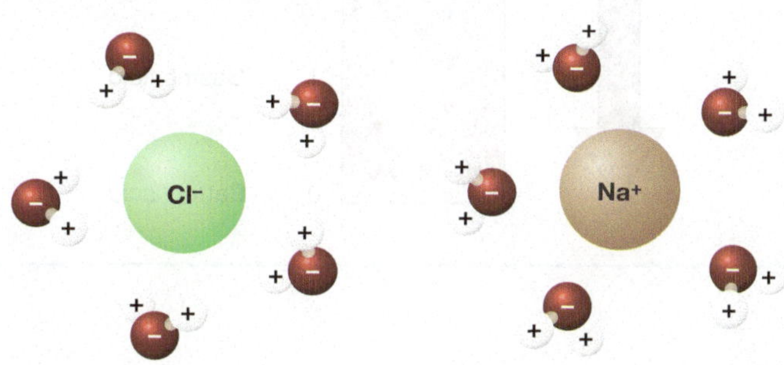

Figure 2.12 The pH scale

This pH scale ranges from 0 (most acidic) to 14 (most basic). A change of one unit on the scale corresponds to a tenfold change in the hydrogen ion concentration. Red dots signify hydrogen ions (H^+) and blue dots signify hydroxide ions (OH^-). Also shown are the approximate pH values for some common solutions. Note that the more protons, the lower the pH, and the more hydroxide ions, the higher the pH.

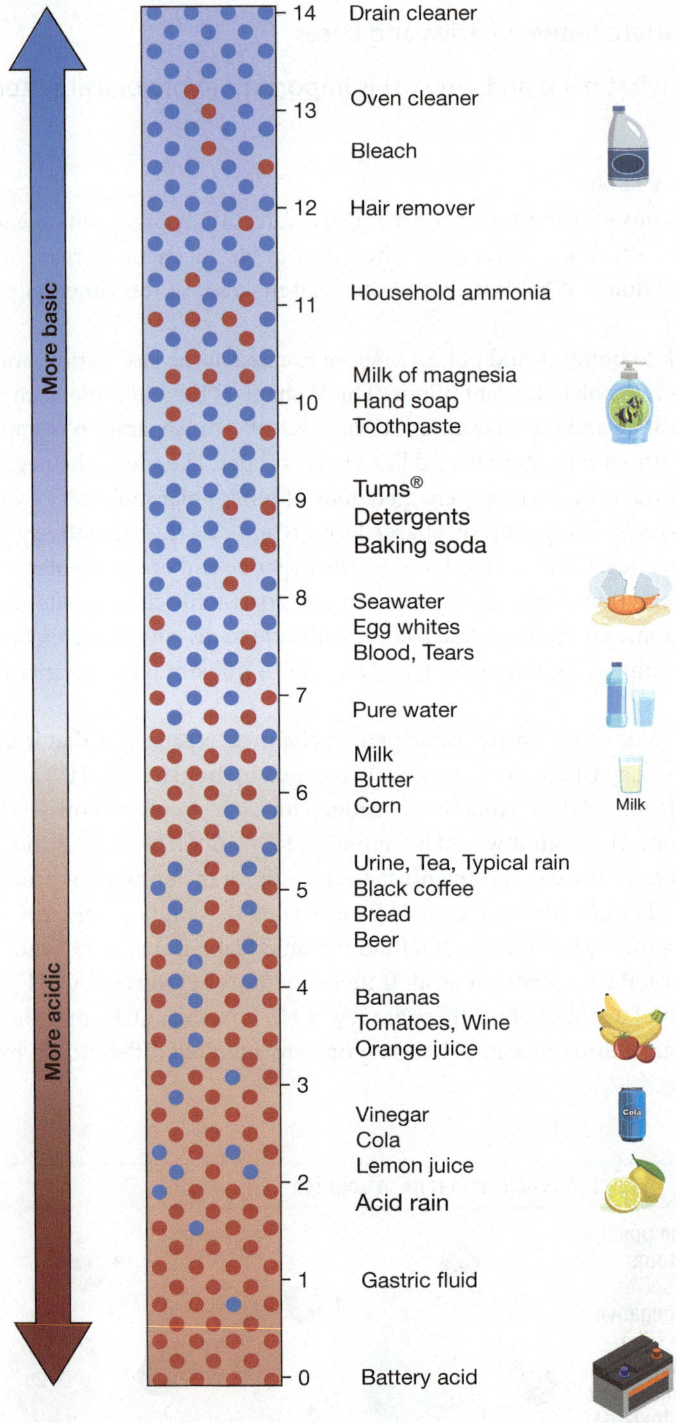

Sources: Ayman alakhras/Shutterstock.com; Sheila Fitzgerald/Shutterstock.com; Steve Cukrov/Shutterstock.com; Jr images/Shutterstock.com

? Did You Know?

Sour Lake in Yellowstone National Park has a pH of 2.5, the same pH as acid rain. The low pH comes from *Sulfolobus*, a single-celled microorganism that belongs to a group of organisms called *Archaea. Sulfolobus* converts sulfur-containing compounds into sulfuric acid.

Process of Science

Life Gets into Hot Water

When most people consider Yellowstone National Park, they immediately think about its most famous geyser, Old Faithful, and its variety of large mammals, such as wolves and bison. But did you know that Yellowstone is also famous for its remarkable microbial life forms? Geologists visiting Yellowstone in the 1870s were the first to document the wide variety of colors (Figure 2.13) surrounding the various thermal features. In 1889, geologist Walter Weed first proposed that the colors represented different kinds of microbial life. But how could life survive in water that would burn a human?

It is easy to take a view of life based on humans' tolerance for heat and imagine that it's impossible for life to live in near-boiling water. However, the discovery of life in the hot springs of Yellowstone changed our understanding of heat tolerance in organisms and where life is possible. Life, we now know, can thrive in a variety of environments, including some very hot water! Interestingly, *Phormidium* bacteria, which are found in the Grand Prismatic Spring, form rounded structures called stromatolites. Fossilized stromatolites represent the earliest signs of life on Earth and suggest that the first organisms may have evolved in high-temperature waters. Scientists study life forms in these extreme environments to better understand the origins of life on Earth and gain insights into where life may be possible elsewhere in the universe.

Source: © Tom Murphy/National Geographic Image Collection

Figure 2.13 Grand Prismatic Spring in Yellowstone National Park

The Grand Prismatic Spring is not only the largest hot spring in Yellowstone National Park but also in all North America. Its temperature ranges from 63–87 °C (145–188 °F). The center of the hot spring (the clear blue area) is the hottest part, and the runoff areas (yellows and oranges) contain cooler water. The colors of each of these areas derive from the presence of different microbial life. As the water cools to approximately 74 °C (165 °F), we begin to see green *Synechococcus*, a photosynthetic cyanobacteria. As the water temperatures drop to 57 °C (135 °F), we see *Phormidium* bacteria. In the coolest areas, around 30 °C (86 °F), we see dark brown *Calothrix* bacteria.

Research other types of extreme environments where we can find life. What conditions surprise you the most? What examples of non-microbial life can you find living in extreme environments?

In a chemical reaction, water can act as an acid or a base. An **acid** is a substance that releases hydrogen ions. The addition of acid to a solution can lower its pH. In contrast, a **base** is a substance that picks up hydrogen ions, so it can raise pH. Metabolic processes in our cells continuously produce acids and bases, but the pH of fluids in our bodies and cells typically does not fluctuate much. Why not? Buffers stabilize their pH. A **buffer** is a chemical that keeps pH stable by reacting with substances that affect pH. Maintaining a stable pH is important because most biological molecules only function properly in a narrow pH range. For example, changing the pH can alter the charge on proteins. This can affect the folding of proteins and impact their function. We will return to this idea in Chapter 3.

We preserve foods such as pickles in acidic environments not only because people like the sour taste of acids, but also because it prevents the growth of dangerous microorganisms.

Digging into Data

How Can an Acid-Loving Bacteria Be Used to Clean Polluted Water?

In 1986, Jim Pisarowicz, a park ranger and avid caver, discovered "snottites" in a cave during an expedition to Mexico (**Figure 2.14**). Snottites (sometimes called "snoticles") are made by bacteria that live inside the thick droplets. Since snottite-producing bacteria live in the dark, the microbes use high-energy sulfur compounds for energy instead of relying on light energy from the Sun. These adaptations feed an entire community, including the tiny flies called midges you can see clinging to the snottites in the photo.

One of the by-products of the sulfur compounds consumed by snottite-producing bacteria is sulfuric acid, a very strong acid. That means that the bacteria live in highly acidic conditions that would kill most of the life we are accustomed to seeing every day. Pisarowicz suggested that if life can survive in such dark, hostile conditions on Earth, maybe we could find life below the surface on Mars, where life would exist without sunlight but at the same time be protected from harsh conditions on Mars' surface.

Life in extreme environments not only gives us insight into what life could look like elsewhere in the solar system but also has practical uses here on Earth. Some species of snottite bacteria can remove heavy metals from wastewater. This is important because heavy metals such as lead pose a significant environmental health hazard, especially to pregnant women and children. Scientists are studying one type of snottite bacteria, *Acidithiobacillus thiooxidans*, to better understand how it grows in a laboratory. If we understand the conditions under which it grows, it will be easier to use *Acidithiobacillus thiooxidans* to remove heavy metals from polluted water.

1. Snottite bacteria produce sulfuric acid from the sulfur compounds they consume. Looking at **Figure 2.15**, describe how the bacteria change the pH over time.
2. If these bacteria were used at a water treatment plant to remove heavy metals, suggest problems the bacteria could create that would make the water unhealthy to drink. What do you think people at the water treatment plant could do to correct these problems?

Source: © Mark Thiessen/National Geographic Image Collection

Figure 2.14 **Drops of "snottite" found in the Cave of the Lighted House (Cueva de Villa Luz) in Mexico**

The "snot" consists of a bacteria-rich biofilm hanging from the walls of the cave. The bacteria live in perpetual darkness and "eat" sulfur-containing compounds, and in turn provide food for other organisms in the cave, such as these midges seen snacking on the biofilm.

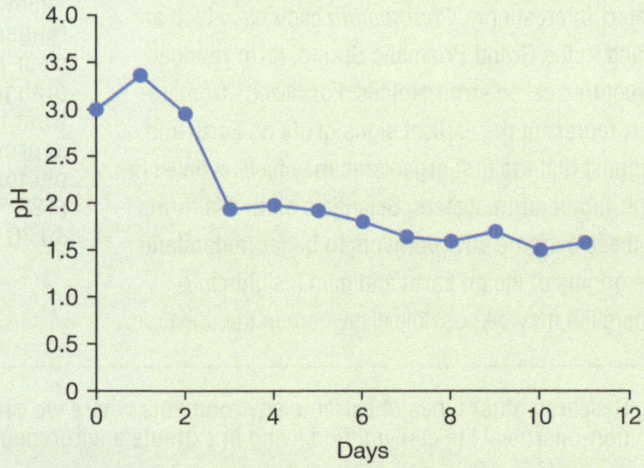

Figure 2.15 **pH level of an *Acidithiobacillus thiooxidans* culture over time**

Scientists grew the bacterium in a laboratory and measured the pH of the solution containing the microbe every day.

For more information on how to interpret graphs, see Appendix D.

Graph adapted from: Ibáñez, A., Barreiro, C., Diez-Galán, A., Cobos, R., Calvo-Peña, C., & Coque, J. J. R. (2023). Molecular Identification and Acid Stress Response of an Acidithiobacillus thiooxidans Strain Isolated from Rio Tinto (Spain). *International Journal of Molecular Sciences, 24*(17), 13391.

Figure 2.16 **Preparing vinaigrette-based salad dressing requires mixing the hydrophilic (watery) and hydrophobic (oily) layers**

Before mixing, the oil layer is floating on top of the water and vinegar layer (left). Immediately after mixing, the oil and water layers are fully mixed (right). Over time, the layers will reseparate.

Splashy Molecular Interactions

Have you ever wondered why a bottle of vinaigrette-based salad dressing naturally separates into two layers (Figure 2.16)? The top layer is oily and the lower layer has water and vinegar. Since the oil will not mix with water on its own, to mix the layers you shake the bottle, in effect adding energy to the system. However, if we stop shaking the bottle, the layers of the salad dressing will reseparate. Why? And why is this important to life?

How atoms are bonded together to form molecules dictates how molecules can interact with one another. How and which molecules can interact with one another explains a wide variety of biological processes, from how proteins fold (Chapter 3) to the formation of the cell membrane (Chapter 5). In the following sections, we will explore how molecules interact with one another. Given the importance of water to life, many of these interactions are described in relation to the properties of water.

Big Ideas in Biology

Interconnectedness of Living Systems

Biologists have observed that all living systems are organized around the same principles of energy use and utilize the same basic chemicals, like water. That is why they believe studying life on Earth provides insights into life on other worlds.

Hydrophilic interactions. The word **hydrophilic** can be broken down into two parts, *hydro* (water) and *phillic*. *Phillic* is derived from the Greek word *philia*, which means love or friendship. Molecules that are hydrophilic interact with both water and other hydrophilic molecules. Since water is polar, it can dissolve molecules that are also polar. You can also think of this as "like dissolves like"—a polar substance will dissolve and interact with other polar substances.

How do you know if something is hydrophilic? One method is to look for functional groups (see Table 2.2). Many of the functional groups contain electron-greedy atoms and polar covalent bonds. The presence of these functional groups increases the ability of a carbon-containing molecule to dissolve in water.

Hydrophobic interactions. What does it mean if someone has a phobia? A phobia is an extreme fear or aversion to something else. A *hydrophobic* molecule is so named because it has a "fear" (*phobia*) of water (*hydro*). **Hydrophobic** molecules are non-polar and get their reputation because they do not interact with water, which is polar. Remember, a non-polar molecule doesn't have polar covalent bonds or charged ends that would readily interact with the positive or negative sides of another molecule. Common non-polar molecules include oils and fats. Going back to our salad dressing example, the oil layer separates from the water layer because it is hydrophobic.

Hydrophobic and hydrophilic interactions explain the structure and function of the cell membrane, how molecules enter and exit the cell, and how proteins spontaneously fold.

! Point of Confusion: When we think about solvents, likely the first thing that comes to mind is water. However, non-polar liquids such as oil and gasoline can also act as solvents, and they are good for dissolving non-polar molecules, such as grease and other oils.

Figure 2.17 Water hydrogen bonds with itself

A

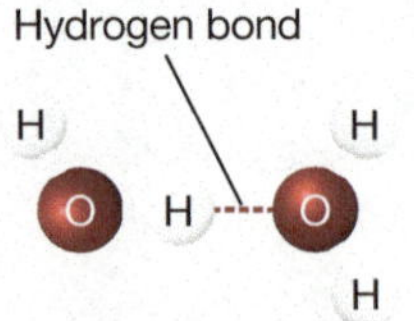

The positively charged hydrogen of one water molecule interacts with the negatively charged oxygen atom on an adjacent water molecule.

B

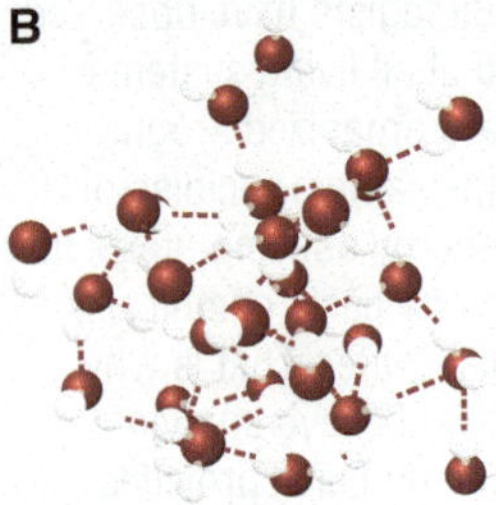

In water, many water molecules hydrogen bond with one another.

Hydrogen Bonding

The polarity of individual water molecules attracts them to one another: The slight positive charge on a hydrogen atom in one water molecule is drawn to the slight negative charge on an oxygen atom in another. This type of interaction is called a **hydrogen bond** (Figure 2.17A). Rather than atoms interacting by sharing electrons or via the attraction between ions, the slight opposite charges on the hydrogen and oxygen atoms of adjacent polar molecules attract the polar molecules to each other. Hydrogen bonds can form between hydrogen atoms and electron-greedy atoms such as nitrogen, oxygen, and fluorine. The interaction that underlies hydrogen bonding is the same phenomenon as rubbing a balloon on your head and then using the balloon to make your hair stand up. The balloon picks up electrons from your hair and your hair gives up electrons to the balloon. The attraction between the now negatively charged balloon and the positively charged strands of hair occurs because of the opposing charges.

Hydrogen bonds are relatively weak interactions. Since they are weak, they form and break easily, much more so than covalent or ionic bonds. This property is important, as we shall see when we look at the interactions between enzymes and their substrates (Chapter 4). Like Velcro loops and hooks, even though the individual hydrogen bonds are weak, many together are very strong. Hydrogen bonds form in tremendous numbers among water molecules (Figure 2.17B), and they also stabilize the characteristic structures of biological molecules such as proteins and nucleic acids (Chapter 3).

> **! Point of Confusion:** Although called hydrogen bonds, electrons are not shared between molecules, like in covalent bonds. Rather, hydrogen bonds are the result of weak electrostatic interactions.

Hydrogen Bonding and Life

Water's ability to participate in hydrogen bonding has consequences that go beyond what happens in the tiny world of molecules. Hydrogen bonding also makes possible water's properties of cohesion and adhesion, which can affect how an organism moves, how tall a tree can grow, and what a droplet looks like on a leaf.

Water's ability to stick to itself is why doing a belly flop hurts! The painful slap you feel is from water's reluctance to part when your body hits the surface.

Cohesion. In water, hydrogen bonds collectively exert a continuous pull on individual water molecules and keep them together, a phenomenon called **cohesion**. You can see cohesion as *surface tension*, a property of liquid water that allows the surface of liquid water to behave a bit like a sheet of elastic. This allows insects like water striders (Figure 2.18A) to walk on water.

Water's cohesion plays a role in many processes that sustain multicellular life. Consider how sweating helps keep your body cool during hot, dry weather. Sweat, which is about 99 percent water, cools the skin as it evaporates. Why? **Evaporation** is the process in which molecules escape from the surface of a liquid and become vapor. The evaporation of water is resisted by hydrogen bonding among individual water molecules. In other words, overcoming cohesion takes energy. When we sweat, the heat from our bodies is absorbed by the water molecules and this energy allows the water molecules to convert to water vapor. As the water evaporates, heat is carried away and we feel cooler.

Cohesion works inside organisms, too. Consider how plants absorb water from soil and move it to their leaves, where photosynthesis occurs (Chapter 4). Because water molecules stick together, when water molecules evaporate from leaves, it is replaced by water pulled upward from roots. Cohesion is one of the forces that makes it possible for columns of liquid water to rise from roots to leaves inside narrow pipelines in the stems and trunks of plants. In some trees, these pipelines extend hundreds of feet above the soil. Think about it: a giant Sequoia has no "heart" to pump water to its highest leaves. Water travels to the tops of trees because of cohesion and adhesion.

Figure 2.18 Water can stick to itself and other surfaces through cohesion and adhesion

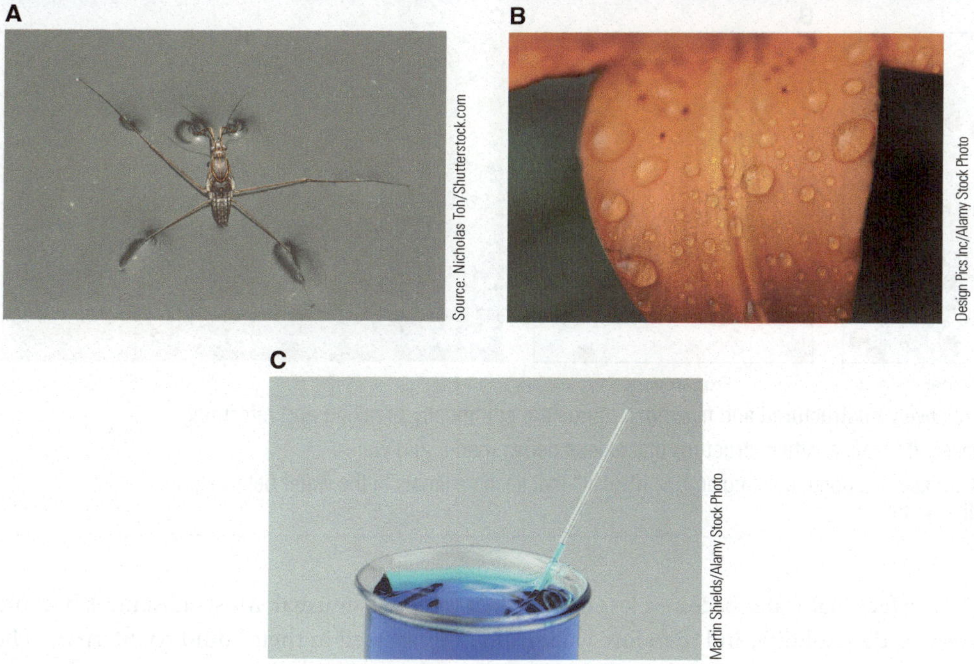

Source: Nicholas Toh/Shutterstock.com

Design Pics Inc/Alamy Stock Photo

Martin Shields/Alamy Stock Photo

A. Water's property of cohesion creates surface tension, which allows animals such as this water strider to walk on water.

B. Water's ability to stick to itself is why water will form a rounded, bead-like structure on a flower petal. Note that the water is also sticking to the flower petal via adhesion.

C. Water's ability to stick to itself and other substances explains how it can "climb" up a straw.

Study Tip

To keep cohesion and adhesion straight, think of a **co**-worker as someone you work with (**co**hesion is water molecules sticking to each other). Adhesion is water sticking to other substances, so think of **ad**hesives, which stick things together. Water **adheres** to other substances.

Adhesion. The ability of water to stick to other surfaces is called **adhesion**. It is the property that allows water to stick to a surface, such as water beads sitting on a flower petal (Figure 2.18B). The water sticks to itself through cohesion and forms a ball, and sticks to the petal via adhesion. Adhesion is also the reason why water will "climb." For example, if you put a straw in a glass of water, the water will begin to move up the straw (Figure 2.18C). Going back to our earlier example of how water can rise through trees, water can move through the vascular tissue of a tree not only because it is sticking to itself, but also because it is sticking to the vascular tissue of the tree.

Water stabilizes temperature. Water's ability to hydrogen bond also explains its ability to stabilize temperature. All molecules, including water, move. We measure the energy of this motion as **temperature**. Adding energy (in the form of heat, for example) speeds up the movement, so the temperature rises. Hydrogen bonds help resist movement. For example, for water to boil it requires breaking hydrogen bonds. As a result, it takes more heat to raise the temperature of water and for water to boil, compared with other liquids that do not hydrogen bond. Water's resistance to temperature change helps our bodies maintain a consistent internal temperature. This is important for life because most of the molecules of life function properly only within a certain range of temperature.

Coastal regions tend to have milder summers and warmer winters, compared to inland regions. This is because they are adjacent to large bodies of water, which resist temperature changes.

Now consider ice (frozen water). When water freezes, the molecules lock together in a lattice structure in which the individual molecules are slightly further apart than they are in liquid water (Figure 2.19A, B). This makes ice less dense than liquid water and thus it floats.

Figure 2.19 **Water is less dense as a solid**

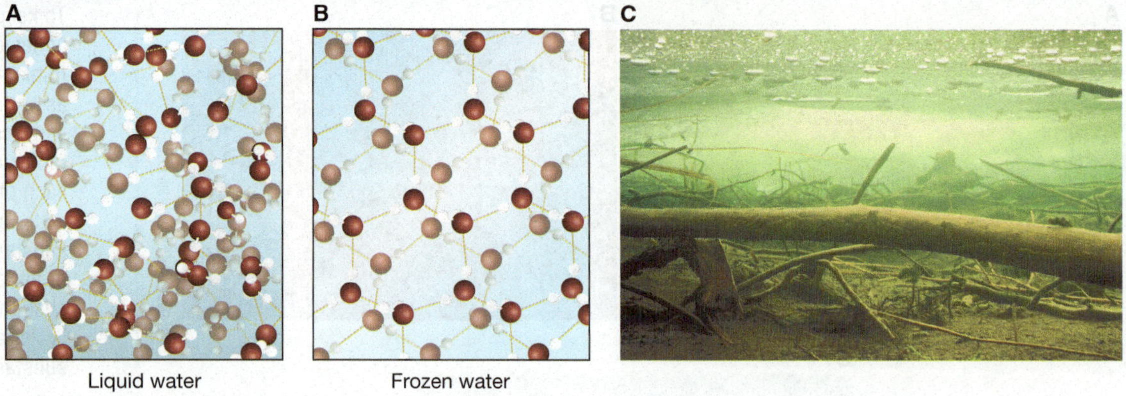

A. Liquid water is relatively unstructured and hydrogen bonds are continually breaking and reforming.

B. When water freezes, it forms a lattice structure that is less dense than liquid water.

C. Ice floats on the surface of a pond, creating an "ice blanket" that keeps animals in the water below from freezing during the winter.

The fact that water becomes less dense when frozen is unusual. Most substances become denser as they solidify, and therefore would sink if immersed in their liquid counterpart. The fact that ice floats, combined with its ability to resist temperature changes, explains why animals can live in frozen bodies of water during the winter. In subfreezing air temperatures, sheets of ice form on the surface of bodies of water such as lakes and seas. This "ice blanket" insulates the underlying water and can keep it from freezing (**Figure 2.19C**). If ice sank, on the other hand, then the liquid water remaining at the surface would be exposed to freezing temperatures, it would freeze and sink, and so on, until the entire body of water froze solid, along with the organisms living in it.

2.3 Summary

- Water is essential for life as we know it and can dissolve many substances.
- pH is a measure of the number of protons (hydrogen ions, H^+) in a solution. A neutral pH is 7 and the number of H^+ and OH^- ions are equal. If there are more H^+ ions than OH^- ions, the pH is lower and the solution is acidic. Basic solutions have a pH higher than 7 and have fewer H^+ ions.
- Most biological molecules function properly only within a narrow range of pH. Buffers keep pH within these ranges by accepting and releasing hydrogen ions as needed.
- Water is a polar molecule. It interacts with itself and other polar molecules through hydrophilic interactions.
- Uncharged, non-polar molecules do not interact with water. Hydrophobic interactions occur between non-polar molecules and each other.
- Cohesion refers to water's ability to stick to itself. Adhesion is water's ability to stick to surfaces. These properties explain how trees get water from the ground to their leaves.
- A hydrogen bond is an electrostatic interaction between two molecules, one of which possesses a hydrogen atom with a slight positive charge and the other possessing an atom with a slight negative charge, usually oxygen, nitrogen, or fluorine. Hydrogen bonds can also form between different atoms in the same molecule.
- Hydrogen bonds are individually weak but collectively strong.
- Extensive hydrogen bonding among water molecules arises from the polarity of the individual molecules, and it gives rise to special properties of water that make life as we know it possible.

Understand It! Concrete Examples

Our brains evolved to be to remember concrete information more easily than abstract information. What does concrete mean? Concrete information is something that we can easily see. Therefore, a concrete idea is a specific example used to illustrate something we cannot easily see, like a molecular or chemical process.

Usually, these ideas are tied to something we already know. For example, in this chapter, we used salad dressing to illustrate the concepts of hydrophobic and hydrophilic. Salad dressing is a concrete idea. People often remember seeing how the layers separate. We can then "connect" these memories of salad dressing to new knowledge on the abstract principles behind why salad dressing separates. In the case of salad dressing, the abstract idea is that water molecules are polar and oil molecules are nonpolar. Polar and nonpolar molecules do not interact with one another and will therefore separate.

Many chemistry topics, like hydrophobic and hydrophilic, are abstract. So given the way our brains work, it helps to learn abstract material using concrete examples. Remember, it is far easier to remember the concrete example, just don't forget the abstract idea underlying the concrete example. Here are some examples of ways you can study this chapter's materials using concrete ideas.

1. Using material such as clay or putty, build your own model of a water molecule. With your model, show how water can interact with itself or other molecules, such as ions.
2. Give an example of another substance around the house that will separate into oil and water layers.
3. What other concrete examples can you think of that fit this chapter's material? Share your ideas with a friend.

Chapter Review

2.1 Chemistry is Key to Life

Atoms are the smallest unit of **matter**. Atoms contain three subatomic particles: positively charged **protons**, negatively charged **electrons**, and uncharged **neutrons**. Each **element** has a different **atomic number** which indicates the number of protons in its nucleus. Atoms of the same element with different numbers of neutrons are called **isotopes**, and the **mass number** indicates the number of protons plus neutrons in an atom. **Radioactive decay**, or the emission of subatomic particles and/or energy, is the process behind **carbon-14 dating**. When atoms gain or lose electrons, charged ions are formed. **Reduction–oxidation** or **redox** reactions are when molecules pass electrons between each other.

2.2 Chemical Interactions

Molecules are formed when atoms join together in **chemical bonds**. Differently charged ions are held together through **ionic bonds**. When atoms share electrons, **covalent bonds** are formed. If electrons are shared equally between atoms, **nonpolar covalent bonds** result. Unequal sharing of electrons in **polar covalent bonds** leads to regions in molecules with slightly positive and slightly negative charges and creates polar molecules. We can represent chemical structures in a variety of forms, including **structural formulas**, **ball-and-stick models**, and **space-filling models**. Carbon atoms can form up to four covalent bonds with other atoms. This makes carbon atoms an excellent building block for making organic molecules. Organic molecules can also contain **functional groups**, which are specific groups of atoms that give the molecule particular molecular characteristics.

2.3 Special Properties of Water

Water is a polar molecule. Due to its V-shaped molecular structure, the slight negative and positive charges on the molecule are separated. This allows water to be an excellent **solvent**. Water can also act as an **acid** or **base**. Acids release protons while bases accept protons. The higher the concentration of H^+ ions, the lower the pH, with 7 being a neutral pH. **Buffers** maintain pH in biological settings. Due to the importance of water-based chemistry for life, other chemicals are described based on their ability to interact with water. Molecules that are polar and can interact with water are **hydrophilic**. Molecules that are non-polar and do not chemically interact with water are **hydrophobic**. Polar molecules interact with one another through **hydrogen bonding**. Water is particularly good at hydrogen bonding, and this explains why water is so good at sticking to itself (**cohesion**) and to other objects (**adhesion**).

Practice It!

Retrieval practice is the process of pulling information from your long-term memory and using it. Here are some questions to help with your retrieval practice.

Self-Quiz

1. Which of the following is made of matter? (LO 2-1)
   a. the air you breathe
   b. a puppy
   c. a piece of broccoli
   d. a rock
   e. All of the above are made of matter.

2. Which of these correctly lists particles in the nucleus of an atom? (LO 2-2)
   a. protons, neutrons, and electrons
   b. protons and neutrons
   c. protons and electrons
   d. neutrons and electrons

3. The negatively charged particle found in an atom is called a(n) ______. (LO 2-2)
   a. proton
   b. neutron
   c. electron
   d. None of these are negatively charged.

4. Which substance can help prevent the destructive activity of free radicals? (LO 2-4)
   a. isotope c. atom
   b. antioxidant d. ion

5. How many neutrons are in an atom that has an atomic number of 5 and a mass number of 11? (LO 2-2)
   a. 5 c. 10
   b. 6 d. 11

6. If an oxygen atom has a total of 6 electrons in its valence shell, how many open spaces for bonding does it have? (LO 2-5, LO 2-6)
   a. 1 c. 3
   b. 2 d. 4

7. If one atom shares an electron with another atom, what type of bond is formed? (LO 2-6)
   a. covalent c. weak
   b. ionic d. hydrogen

8. In a polar covalent bond, ______. (LO 2-9)
   a. protons are shared evenly
   b. protons are shared unevenly
   c. electrons are shared unevenly
   d. electrons are shared evenly

9. A molecule in which all electrons are shared equally would have what type of bond? (LO 2-9)
   a. nonpolar covalent c. ionic
   b. polar covalent d. hydrogen

10. Which of the following properties explains why water is a good solvent? (LO 2-10)
    a. Its buffering ability c. Its polarity
    b. Its ability to stick to itself d. Its ability to stabilize temperature

11. Which of the following cannot be involved in hydrogen bonding? (LO 2-11)
    a. water c. polar molecules
    b. non-polar molecules d. ions

12. Acids ______ hydrogen ions and make the pH ______. (LO 2-12)
    a. release, lower c. release, higher
    b. accept, lower d. accept, higher

13. In an acid base pair, bases have ______ hydrogen ions than acids. (LO 2-12)
    a. more
    b. fewer
    c. equal
    d. None of these are correct.

14. Buffers ______. (LO 2-12, LO 2-13)
    a. stop hydrogen bonding
    b. cause cohesion of water molecules
    c. cause water to be a solvent
    d. help maintain a constant pH

Critical Thinking

1. Water is an unusual molecule because it is less dense as a solid (ice) than as a liquid (water). What would happen if ice were more dense than liquid water, so it sank instead of floating? How might this affect aquatic life? (LO 2-10)

2. Joshua caught a stomach bug at school. He ended up with a severe case of diarrhea. After two days, he was rushed to the hospital. He was diagnosed with hyperchloremic acidosis caused by losing too much sodium bicarbonate with his severe diarrhea. Sodium bicarbonate is known to be a buffer. Why would losing too much buffer cause problems? (LO 2-12, LO 2-13)

3. Zeke poured rubbing alcohol (isopropanol) into a glass. He noticed that as soon as the alcohol reached the brim, it started to spill over. He then poured water into the same type of glass. He noticed that the water was able to rise above the edge of the glass and form a slight dome. What properties of water allowed this to happen? Since the alcohol spilled over the sides of the glass, what property of water might alcohol be missing? (LO 2-10)

4. Living organisms have a constant amount of carbon-14 in their tissues, equal to the amount of carbon-14 in their environment. When the organism dies, the amount of carbon-14 is set and then declines over time due to radioactive decay. Carbon-14 decays to nitrogen-14 with a half-life of 5,730 years. An archeologist determined that the amount of carbon-14 in a bone is one-fourth of the amount in a living organism. How old is the bone? (LO 2-7)

5. Vitamins can be fat-soluble or water-soluble. Why might consuming too much of a fat-soluble vitamin be more dangerous than too much of a water-soluble vitamin? *Hint*: you may want to check out how the body handles wastes and stores nutrients. (LO 2-10)

Source: iStock.com/PhotographerJohnce

3

Molecules of Life

Chapter Outline:

A Lunch Date with Biology

Imagine that you are walking into your favorite restaurant with a friend. You get settled into a sunny corner booth and the server gives each of you a menu. You order the apple poppyseed spring green salad because it sounds good, and salads are healthy, right? Your friend pulls out their phone and tells you they are tracking their macros and want to see where they stand on the day before ordering. Huh? Macros? What are those?

You ask your friend what it means to track your macros. They turn their phone around and show you a pie chart. They explain that they are working with a personal trainer because they want to increase their muscle mass. To do so, their trainer recommended that about 30% of their calories come from protein. Your friend reflects that they have not eaten enough protein so far that day and so plan to get a salad with chicken to correct the deficit. You look over the menu again. Is your chicken-less salad as healthy as you initially thought?

Many people track their macros. "Macros" is short for *macronutrients*—one way of referring to the "big" molecules we ingest to keep healthy. Tracking macros can help with weight loss, ensuring the consumption of a balanced diet. The three macronutrients people track are *carbohydrates*, *protein*, and *fats* (i.e., *lipids*). In biology, these macronutrients correspond to three classes of *macromolecules* that have essential roles in cellular function. *Nucleic acids* are a fourth class of macromolecule, and while we consume them too, they are less important from a dietary standpoint. How these macromolecules interact or bond with one another is important for understanding how they function in the body. In this chapter, you will learn about macromolecules—the molecules of life.

Do you track your macros? "Macros" refers to *macronutrients*: the carbohydrates, proteins, and lipids that are critical components of a healthy diet. These macronutrients, along with nucleic acids, are the molecules of life we will be talking about in this chapter.

Preview It!

Write down three things you know about macromolecules already. What is one question you have about macromolecules?

3.1 Life Runs on Energy

Learning Objectives

3-1 Describe the concept of energy.

3-2 Describe how we know that energy cannot be created or destroyed.

3-3 Explain how cells store and retrieve energy in organic molecules.

Just like a car requires gasoline to run, organisms require energy in the form of food. In our bodies, macromolecules are the raw materials our cells use to make energy (Chapter 4). **Energy** is the capacity to do work. Said otherwise, energy is not a material, but a measurement of how much "stuff" or "work" can be done using that energy. As part of doing work, energy is transferred.

For example, let's consider pushing a box across the floor. In this case, a body (you) transfers energy to another body (the box) to make it move (the work). Similarly, a plant cell works to make sugars. Inside the cell, one set of molecules harvests the energy of light and then transfers it to another set of molecules. The second set of molecules uses the energy to build sugars from carbon dioxide and water. This kind of energy transfer involves the conversion of light energy to chemical energy. Most other types of cellular work occur by the transfer of chemical energy from one molecule to another. We will come back to chemical pathways that transfer energy in Chapter 4.

Energy can be classified as kinetic energy and potential energy. Kinetic is an adjective used to describe something that is moving. In the context of a biological system, **kinetic energy** refers to the movement of atoms and molecules. **Potential energy** is a form of "stored" energy that arises from an object's position or other property of matter (**Figure 3.1**). In the context of biology, potential energy is thought of as the energy stored in chemical bonds.

Thermodynamics

To begin our discussion of thermodynamics, we will look at a real-world example. To raise a single 1,000-pound cow requires around 10,000 pounds of food. What happens to the other 9,000 pounds? Nine thousand pounds is roughly the weight of an ambulance—not a trivial amount! About half of the food ingested gets processed and uh, left behind, by the animal. Ultimately, 15% of the initial input of food is used to build the animal's body mass—and the rest is lost as heat.

Figure 3.1 Illustration of potential and kinetic energy

Water molecules at the top of the waterfall have more potential energy than water molecules at the bottom of the waterfall. As a drop falls, potential energy is converted to kinetic energy. When the water reaches the bottom of the fall and strikes the rocks, its kinetic energy is converted to mechanical energy, which breaks up the rocks and produces sound. This example also illustrates the first two laws of thermodynamics. Energy is not lost as the water falls, it is just converted into different forms (first law of thermodynamics). The outcome—splashing water, noise, and tumbled rocks—is chaotic and disordered (second law of thermodynamics).

Why does this occur? Energy transfers, such as the conversion of the macromolecules in our food into energy (Chapter 4), are inefficient. In this section, we will briefly explore the formal study of heat and other forms of energy or thermodynamics (*thermo*– means "heat"; *dynam* means "power"). By making careful measurements, scientists discovered that the total amount of energy in a system before and after every conversion is always the same. In other words, energy cannot be created or destroyed—a phenomenon called the **first law of thermodynamics**.

When we consider kinetic and potential energy in biology in the context of the first law of thermodynamics, we see that when a chemical bond between atoms is formed or broken, the energy to make or break that bond is transferred, not lost. Kinetic energy is stored in the bond as potential energy. If we break the bond between two atoms, the potential energy is converted back to kinetic energy. The amount of energy required to break a bond is equivalent to the amount of energy released when the bond forms. How much energy is stored in that bond depends on the atoms in the molecule. For instance, a bond between a hydrogen and oxygen atom holds more energy than a bond between two hydrogen atoms.

Figure 3.2 The universe tends toward disorder

Since the universe tends towards disorder, clothes taken out of the dryer or thrown in your hamper form a disordered mess. Folding your clothes and creating order requires an input of energy.

Source: Mike Focus/Shutterstock.com

The **second law of thermodynamics** refers to the universe's tendency toward disorder. Let's take your laundry as an example. If you took your laundry hamper and dumped it on the floor, would your clothes fall out neatly folded? No. They would tumble out into a mess on the floor. Folding laundry and making it ordered takes energy. Therefore, unless someone puts energy into the system, your clothes will form the disordered Mount Laundry on the floor (**Figure 3.2**).

Chemical Reactions

A chemical reaction is the process of changing a starting set of molecules (called reactants) into new molecules (products). Intermediate molecules may form between reactants and products. We show a chemical reaction as an equation in which an arrow points from reactants to products (**Figure 3.3**). A number before a chemical formula in such equations indicates the number of molecules; a subscript indicates the number of atoms of that element per molecule. Notice that atoms shuffle around in a reaction, but they never disappear: The same number of atoms that enter a chemical reaction remain at the reaction's end.

Storing and retrieving energy. Our cells use chemical reactions to store and retrieve energy. For example, cells store potential energy in molecular bonds. During a chemical reaction, bonds are both formed and broken. If the end products have less energy than the reactants, energy is released during the reaction. This is called a favorable or **spontaneous chemical reaction**. Said otherwise, it will happen on its own without any outside help. In contrast, if the products have more energy than the reactants, it is a **nonspontaneous chemical reaction**, that is, not favorable. For a nonspontaneous chemical reaction to occur, an energy input is required (**Figure 3.4**).

In cells, spontaneous chemical reactions occur with the help of *enzymes*, which are specialized molecules, typically proteins, that speed up chemical reactions (Chapter 4). For example, during photosynthesis, enzymes build organic molecules like glucose. Cells retrieve energy stored in organic molecules through reactions that disassemble the high-energy molecules. For

Figure 3.3 Chemical bookkeeping

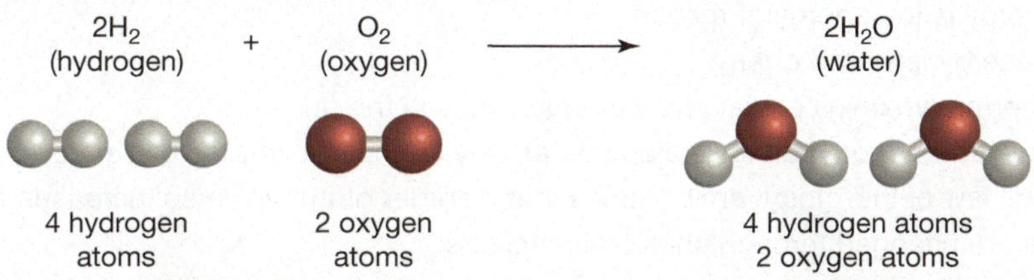

In equations that represent chemical reactions, reactants are written to the left of an arrow that points to the products. A number before a chemical formula indicates the number of molecules. Atoms may shuffle around, but the same number of atoms that enter a chemical reaction remain at its end.

Figure 3.4 Cells obtain the energy stored in organic molecules through chemical reactions

Energy is put into reactions to create macromolecules. When these macromolecules are broken down, energy is released.

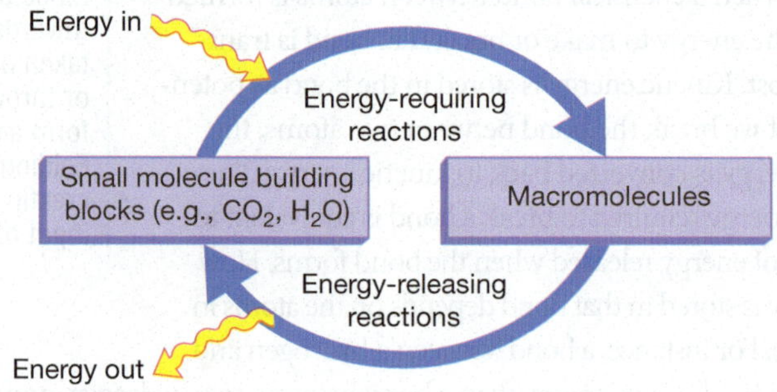

example, the reactions of aerobic respiration collectively release the energy of glucose by breaking bonds between its carbon atoms. As the bonds break, energy is released and used by the cell. We will discuss the process of assembling organic molecules through photosynthesis and breaking them down through cellular respiration in Chapter 4.

Digging into Data

Feed Conversion Ratios and the First Law of Thermodynamics

Animals raised for human consumption can be fed in a variety of ways, but the majority are fed grains like corn to increase weight gain. The amount of energy used for weight gain is calculated using a feed conversion ratio (FCR). The FCR is the ratio of the amount of food needed (energy in) compared to the amount of weight gained, given in kilograms. The higher the ratio, the more food or energy is needed for that weight gain. Another way to say this is that the higher the FCR, the more resources are needed to produce the meat. In Figure 3.5, you can see the FCR for three animals commonly consumed in the United States: beef cattle, pigs, and chicken.

1. What are the ranges of FCR for each of these food animals?
2. Which of these meats is the least resource-intensive? The most resource-intensive?
3. Does this information change your thoughts on eating any or all of these meats?

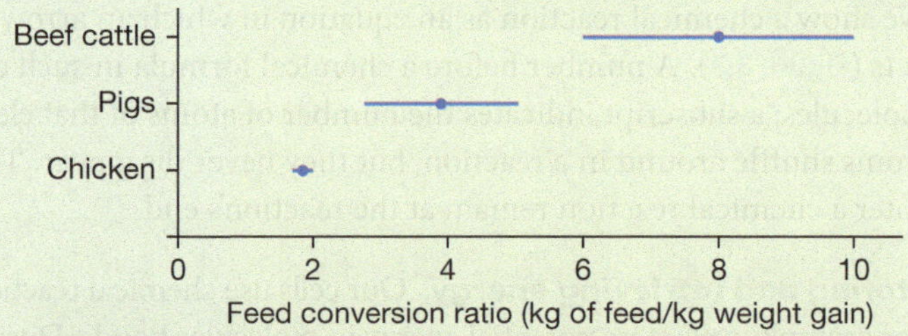

Figure 3.5 Feed conversion ratio for beef cattle, pigs, and chickens

Feed conversion ratios are the kilograms of feed required for an animal to gain one kilogram of weight. Some animals are more efficient at gaining weight and require less feed.

Source: J.P. Fry et al. (2018) Feed conversion efficiency in aquaculture: do we measure it correctly? Environ. Res. Lett. 13.

3.1 Summary

- Energy is a measurement of how much work can be done.
- Energy can be either kinetic or potential.
- Kinetic energy is the energy of motion.
- Potential energy is stored energy.
- Laws of thermodynamics explain how energy works in the universe.
- The first law of thermodynamics states that energy cannot be created or destroyed.
- The second law of thermodynamics states that disorder of the universe increases over time.
- Molecules are changed through chemical reactions.
- Chemical reactions are accompanied by energy transfers.
- Cells constantly build and break down organic molecules. When molecules are built, energy is stored in the chemical bonds. When molecules are broken down, stored energy is released.

3.2 The Building Blocks of Life

Learning Objectives

3-4 Identify the four macromolecules.

3-5 Describe the distinguishing characteristics of the four macromolecules.

3-6 Explain the difference between dehydration synthesis and hydrolysis.

3-7 Describe how monomers form polymers using dehydration synthesis.

In the chapter opener, we learned that macronutrients (carbohydrates, proteins, and lipids) are an essential part of our diet. Together with nucleic acids, these four groups of molecules make up the **macromolecules**. Each macromolecule has distinguishing characteristics (Table 3.1). We will talk about each molecule in depth later in the chapter.

In the last section, we learned how the building and breaking of molecules is how cells store and retrieve energy. Now, we will begin putting these pieces together: what are the specific chemical reactions that cells use to break down the four macromolecules we consume? Once broken down, how can our bodies reform the basic building blocks into new macromolecules?

Big Ideas in Biology

Evolution

All biological systems are based on the same macromolecules, a similarity that is one of many legacies of life's common origins.

Polymers Are Built of Monomers

Generally, macromolecules each have a monomer and polymer form (lipids are the exception). A **monomer** is a single subunit. Let's consider the classic children's toy, pop beads (Figure 3.6A). Think of a single pop bead as the monomer. One can connect the pop bead to another to create a chain. This chain of monomers is called a **polymer**. Carbohydrates, proteins, and nucleic acids each have their own monomer and polymer form. We will discuss each of these in depth in the following sections.

How do cells build polymers from monomers? In the same way pop beads have two different ends that facilitate connecting to join a polymer, monomers have different ends as well. These ends are different functional groups (Chapter 2). The process of joining monomers together to form a polymer is called **dehydration synthesis** (Figure 3.6B). If we break down the phrase "dehydration synthesis", it literally means joining (synthesis) while losing water (dehydration). And that's what happens. The —H and —OH combine to form water as a byproduct, leaving the monomers

Figure 3.6 Polymers are made up of individual monomers

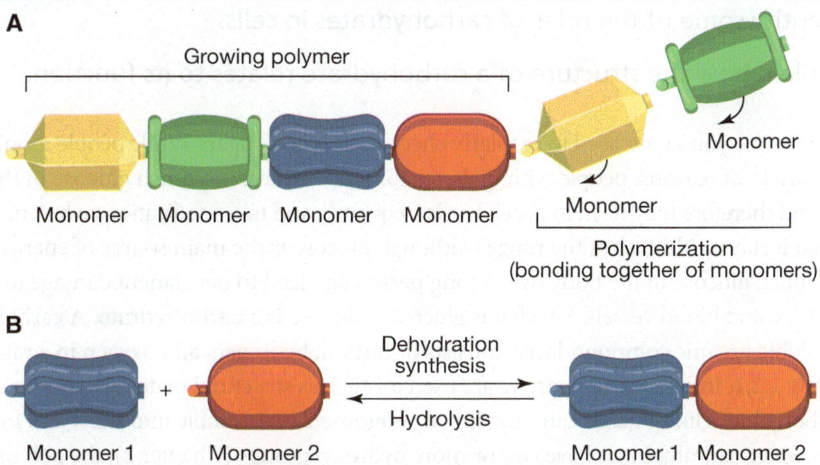

A. Like the children's toy "pop beads," different units snap together to form longer chains.

B. Polymers are made via dehydration synthesis and are broken down using hydrolysis. Different macromolecules use different functional group ends to form polymers.

Table 3.1 Four Classes of Macromolecules

Macromolecule	Distinguishing characteristics
Carbohydrates	Monomer is a monosaccharide. Carbohydrates have many polymer forms. Carbohydrates function as energy sources, building blocks, and in cell identification.
Proteins	Monomer is an amino acid. Proteins have hierarchical structure with primary, secondary, tertiary, and quaternary structure. Proteins must fold properly to function.
Nucleic acids	Monomer is a nucleotide. There are two types of nucleic acids, deoxyribonucleic acid (DNA) and ribonucleic acid (RNA). Many nucleic acids store information in the sequence of monomers.
Lipids	Lipids are not built from monomers. Lipids all have a hydrophobic part of their structure. Lipids make up the cell membrane, are a source of energy, and facilitate cells talking to one another.

covalently attached to one another. This process—the formation of covalent bonds between molecules and the removal of water—continues until a polymer of the correct length is formed.

Study Tip

Use the root words to help you remember these processes.

The opposite of dehydration synthesis is called *hydrolysis*. **Hydrolysis** means breaking (to lyse) with water (hydro). The addition of water between the bonded monomers causes the polymer to split into two separate molecules again. Note that water is broken apart in the reaction to recreate the ends of the monomer.

3.2 Summary

- There are four macromolecules: carbohydrates, proteins, nucleic acids, and lipids.
- Carbohydrates, proteins, and nucleic acids have monomer and polymer forms.
- Monomers are joined together to form polymers through dehydration synthesis. A molecule of water is created in this process.
- Polymers are broken apart using hydrolysis. Water is used up in this process.
- The functional groups at the end of the monomers facilitate these connections.

3.3 Carbohydrates

Learning Objectives

3-8 Describe the structure of carbohydrates.

3-9 Identify some of the roles of carbohydrates in cells.

3-10 Explain how the structure of a carbohydrate relates to its function.

Figure 3.7 Glucose is an example of a monosaccharide

Glucose molecules are an important source of energy. Glucose is also the building block of many polysaccharides.

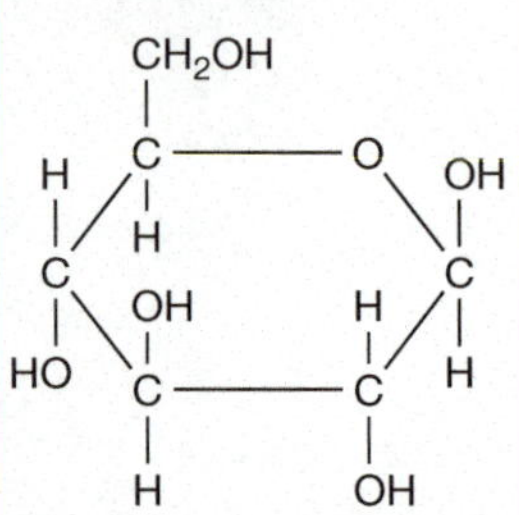

Do you or someone you know need to regularly check their blood sugar? While people monitor blood sugar for a variety of reasons, people with diabetes sometimes have too much glucose in their blood (Figure 3.7) and therefore they need to check levels frequently and take medications when necessary to ensure that it stays within a healthy range. Although glucose is the main source of energy in our bodies, too much glucose in the body over a long period can lead to permanent damage to the eyes, kidneys, nerves, and blood vessels. So what is glucose? Glucose is a **carbohydrate**. A carbohydrate is a water-soluble organic compound that consists of carbon, hydrogen, and oxygen in a ratio of approximately 1:2:1. In cells, carbohydrates are used for fuel, as structural materials, and for storing energy. Carbohydrates like glucose can function as monomers or assemble into short and long chains of polymers. Since carbohydrates have two or more hydroxyl groups (Chapter 2), the polymer form can have many branches.

Point of Confusion: Carbohydrates are also called carbs, sugars, and/or saccharides. Saccharide means sugar.

Biology for Life

Carbohydrates Are Molecular Name Tags

Monosaccharides on the outside of cells function as molecular nametags. For example, having Type A or B blood refers to having A or B type sugars on the outside of the cell. Someone with AB blood has both A and B sugars, and someone with Type O blood has neither A nor B type sugars. Since the immune system uses these sugars to recognize "self," someone with Type O blood can donate to anyone (the universal donor). However, someone with Type O blood can only receive Type O blood—otherwise, their immune system would recognize the sugars as foreign and attack the donated red blood cells. Keep in mind that the + or – in a blood type refers to the presence or absence of a protein called the Rh factor.

What blood type do you have?

Carbohydrate Monomers Are Simple Sugars

A single carbohydrate monomer is called a **monosaccharide** or sugar. Sometimes the monomer is referred to as a *simple sugar* because they are the simplest carbohydrates, and many of them have a sweet taste. Simple sugars are burned more readily in the body than complex sugars. As a result, people are encouraged to limit consumption of simple sugars, as they can contribute to blood sugar spikes.

Monosaccharides have a variety of extremely important biological roles. For example, one way that cells store energy is as long chains of glucose. Ribose and deoxyribose sugars are components of the nucleotide monomers of RNA and DNA, respectively. Cells also use monosaccharides as structural materials to build larger molecules. We will talk more about the polymer forms of carbohydrates next.

Carbohydrate chains. Carbohydrates can function as chains of two monomers (**disaccharide**), a short chain of approximately three to six monomers (**oligosaccharide**), or a long chain containing hundreds or thousands of monomers (**polysaccharide**).

Disaccharides include lactose, a component in milk, and sucrose or table sugar. People who are lactose intolerant cannot break down the monomers that make up lactose. As a result, they suffer indigestion after consuming milk. People who are lactose intolerant use tablets that contain lactase, an enzyme that breaks down lactose, before eating to avoid discomfort.

One common example of oligosaccharide that you may have encountered is raffinose. Raffinose contains three monomers and is common in beans. Humans cannot digest raffinose … but bacteria can. The bacteria in your gut feast on the undigested raffinose and produce gas as a byproduct. Discreetly or not, you then release the gas to the delight of those around you.

Polysaccharides are also called "complex" carbohydrates. The most common polysaccharides—cellulose, starch, and glycogen—all consist only of glucose monomers. These polysaccharides vary in how these monomers are connected. We will discuss each of these polysaccharides and their functions next.

> **Big Ideas in Biology**
>
> **Structure and Function**
>
> The differences in polysaccharide structure formed from various ways of connecting glucose monomers relate to the function of each polysaccharide.

Cellulose. The most abundant organic molecule on Earth is cellulose. **Cellulose** is a structural polysaccharide made by plants. In cellulose, hydrogen bonds cross-link long, straight chains of covalently bonded glucose monomers (Figure 3.8A). As a result, the stacks of interconnected carbohydrates create a sturdy structure that resists stress. A 200-foot redwood tree is a towering example of the strength of cellulose, the structural component of wood.

Due to the strength of cellulose, it is not easily broken down. Some bacteria and fungi make enzymes that can break it apart into its component sugars, but humans and other mammals do not. Bacteria that live in the guts of termites and grazers such as cattle and sheep help these animals digest the cellulose in plants.

In our diets, we refer to cellulose as fiber. Since we cannot digest fiber (meaning, we cannot break cellulose down into glucose monomers), it has many uses as part of a healthy diet. For example, fiber can help someone feel fuller while eating, making them less likely to overeat. Consumption of fiber also helps bulk up stool, keeps your microbiome healthy, and promotes regularity.

> **Did You Know?**
>
> Humans capitalize on the strength of cellulose. Many movie classics were recorded on cellulose nitrate film—and 100+year-old films are still playable!

Figure 3.8 **Cellulose, starch, and glycogen are all examples of carbohydrate polymers**

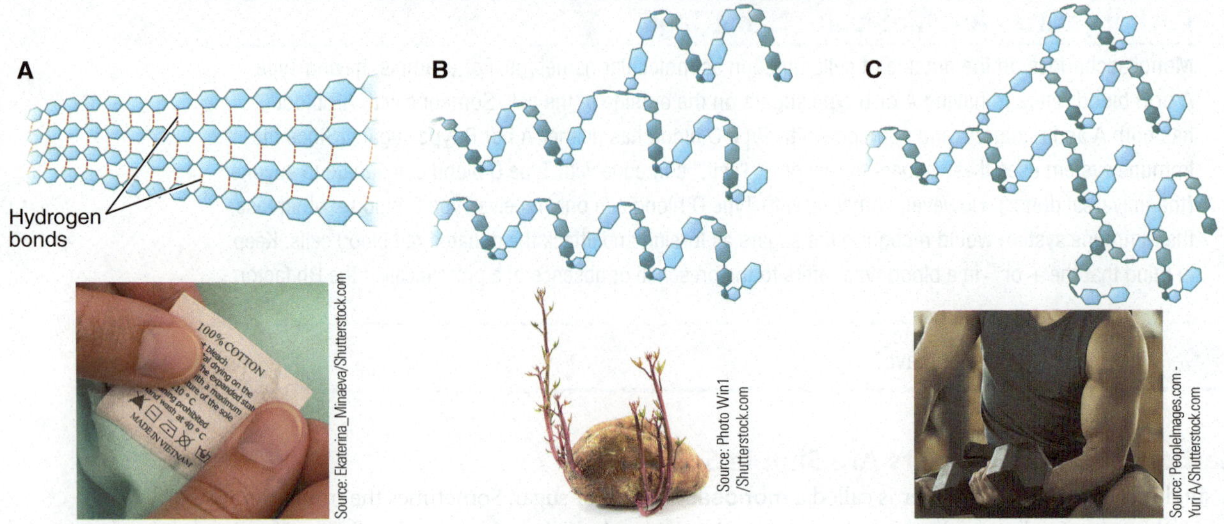

Cellulose is formed of linear polymers. These polymers align into regular arrays with hydrogen bonds in between. Consequently, cellulose forms strong structures, such the cotton in a T-shirt.

Starch forms a helical polymer. Plants, such as this potato plant, store their energy in the form of starch. Starch is the reason why potatoes will sprout.

Glycogen forms a highly branched helical structure. Muscle and liver cells stockpile glycogen to provide energy when the body's energy demands are high, such as during exercise.

Has a potato ever sprouted in your kitchen? The new growth is fueled by starch.

Starch. In addition to cellulose, plants also produce starch. **Starch** is a polysaccharide in which the bonding pattern between glucose monomers creates a chain that is highly branched and/or coils up into a spiral (Figure 3.8B). Starch does not dissolve easily in water, but it is easier to break down than cellulose. These properties make the molecule ideal for storing sugars in the watery interior of plant cells. For example, sugars are moved into developing seeds and then converted to starch for storage. A mature seed may rest for a long time before it sprouts. Then, enzymes break the bonds between the starch's glucose monomers. The released glucose is used to fuel the growth of the young plant until it can make its own food. Humans also have enzymes that break down starch, so this carbohydrate is an important component of our food.

Glycogen. While plants use starch to store energy, animals store energy as glycogen. Like starch, **glycogen** is a polysaccharide that consists of highly branched chains of glucose monomers (Figure 3.8C). Muscle and liver cells contain most of the glycogen in an animal's body. When the blood sugar level falls, enzymes in liver cells break down the glycogen, and the released glucose subunits enter the blood. During a marathon, runners can run out of glycogen. This is called "hitting a wall" while running. To avoid "hitting a wall" many runners will fuel with simple sugars before or during their run.

Process of Science

Human Milk Oligosaccharides: A Decades-Old Mystery

Human milk contains carbohydrates called human milk oligosaccharides (HMOs) that the baby cannot digest. Although scientists have known about HMOs since the 1930s, it wasn't until 2018 that scientists knew the purpose of these carbohydrates. HMOs feed bacteria in the gut, promoting the development of a healthy gut microbiome. The gut microbiome helps create a healthy immune system, and ultimately a healthy child.

Can you think of another example where scientists knew about something for decades but did not know what it was for?

Digging into Data

Uncovering the Impact of Macromolecule Consumption on Blood Sugar Levels

In Section 3.3 we learned about glucose, a monosaccharide, and the relationship between blood glucose and diabetes. Diabetic individuals regularly monitor their blood sugar to ensure that it stays within a healthy range. Although people traditionally monitored their glucose levels by pricking the finger with a needle to obtain a drop of blood, now more people are using continuous glucose monitors (CGMs). For example, CGMs available in the early 2020s are roughly the size of a penny and send glucose readings to a smartphone.

In 2018, Nishino and colleagues examined how the order in which people consume macromolecules during a meal influenced blood sugar levels. The study participants consumed rice (high in starch), vegetables (high in cellulose or fiber), and meat (high in protein) in different orders.

The authors then assessed glucose levels in the research subjects before eating (time = 0 minutes), and at 30, 45, 60, 90, and 120 minutes after eating. You can see their results in Figure 3.9.

1. Which group had the largest blood glucose spikes and which group had the smallest after eating?
2. Does the order in which we eat different macromolecules matter? Why or why not?
3. What will you do differently, if anything, about the order in which you consume foods?

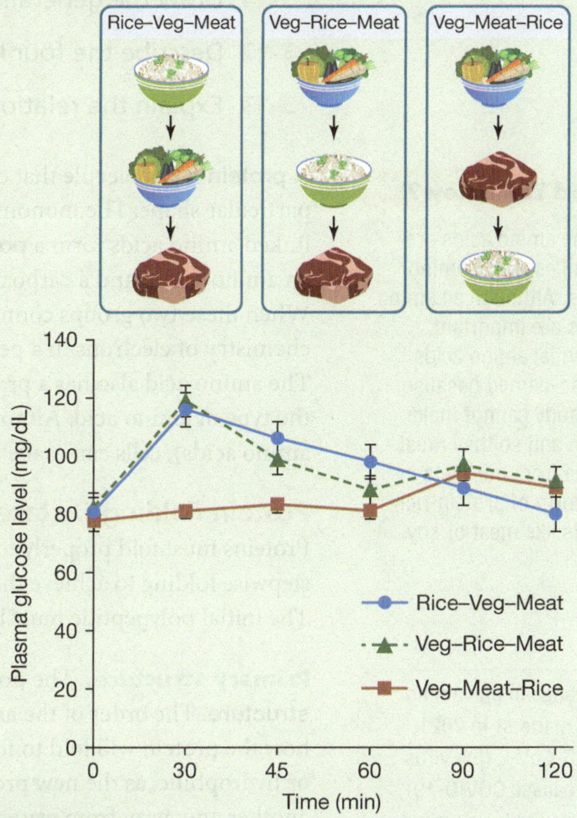

Figure 3.9 How does the order in which food is consumed impact blood sugar?

A. Nishino and colleagues' participants ate the same food, but in different orders.

B. Blood sugar results in each of the three conditions.

Figure based on: Nishino, K., Sakurai, M., Takeshita, Y., & Takamura, T. (2018). Consuming carbohydrates after meat or vegetables lowers postprandial excursions of glucose and insulin in nondiabetic subjects. *Journal of nutritional science and vitaminology, 64*(5), 316–320.

3.3 Summary

- Carbohydrates are water-soluble molecules containing carbon, hydrogen, and oxygen.
- Carbohydrate monomers are called monosaccharides, or simple sugars. Carbohydrates also exist as disaccharides, oligosaccharides, and polysaccharides.
- Carbohydrates have roles as sources of energy, structural function, and cellular identity.
- Long straight chains of hydrocarbons make cellulose an ideal building material for plants. Polysaccharide structure in starch and glycogen make these molecules ideal for energy storage.

3.4 Proteins

Learning Objectives

3-11 Draw the generalized structure of an amino acid.

3-12 Describe the four levels of protein structure.

3-13 Explain the relationship between protein folding and protein function.

? Did You Know?

Some amino acids are called essential amino acids. Although all amino acids are important, essential amino acids are so-named because the body cannot make them and so they must come from our diets in the form of protein-rich foods like meat or soy.

A **protein** is a molecule that contains amino acids linked together in a chain that folds into a particular shape. The monomer form of protein is called an **amino acid** (Figure 3.10A), and linked amino acids form a **polypeptide**. Amino acids have two functional groups (Chapter 2): an amino group and a carboxylic acid group. These groups are used for making the polypeptide. When these two groups connect, they form a **peptide bond** (Figure 3.10B). Due to the specific chemistry of electrons in a peptide bond, they are much stronger than other covalent bonds. The amino acid also has a proton and an R group. R groups can vary and are used for defining the type of amino acid. Although there are 20 different R groups (and consequently 20 different amino acids), cells can make thousands of different proteins from these 20 basic building blocks.

Protein Folding Dictates Function

Proteins must fold properly to be able to function. For example, making an origami boat requires stepwise folding to achieve the final shape and function—to float on water. Proteins are similar. The initial polypeptide must be folded properly to achieve a final functional form (Figure 3.11).

A change in one amino acid matters! In 2021, SARS-CoV-2 (the virus that causes COVID-19) mutated to have a single amino acid change, creating the delta variant. The delta mutation was 50% more contagious than the original strain.

Primary structure. The polypeptide, or chain of linked amino acids, is called the **primary structure**. The order of the amino acids in the series defines the type of protein and influences how the protein will fold to form the protein's final shape. Since amino acids can be hydrophobic or hydrophilic, as the new protein is made, hydrophobic amino acids will move towards one another and away from water, causing the protein to begin to fold.

Figure 3.10 Generalized structure of an amino acid

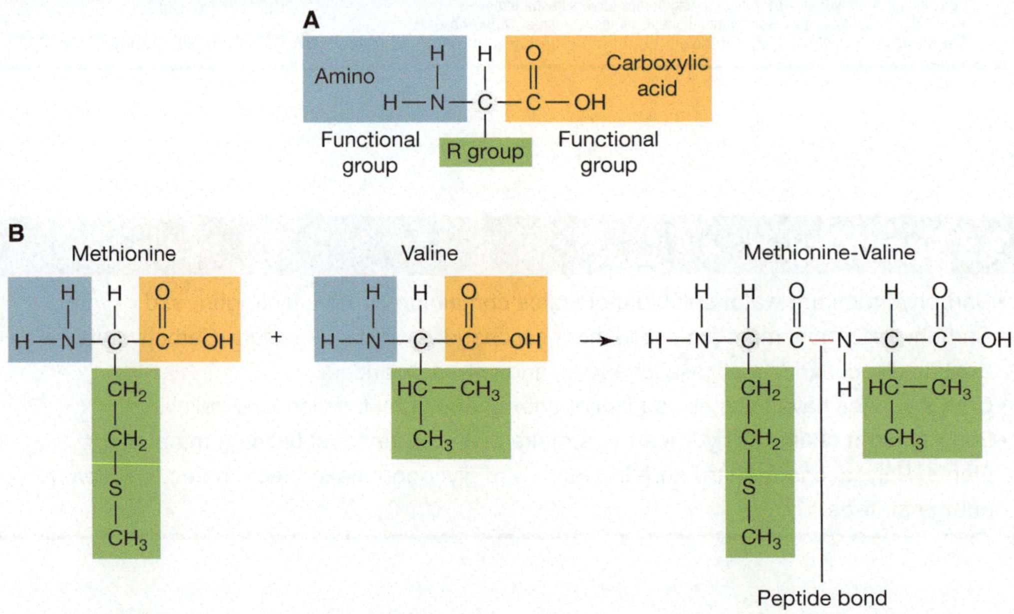

A. Each amino acid consists of an amine functional group (blue), a carboxylic acid functional group (yellow), and an R group (green) bonded to the same carbon atom. The R group determines the type of amino acid.

B. Amino acids connect via a peptide bond. The carboxylic acid and amino groups serve as the molecular connectors to form a peptide bond.

Figure 3.11 **The hierarchical structure of proteins**

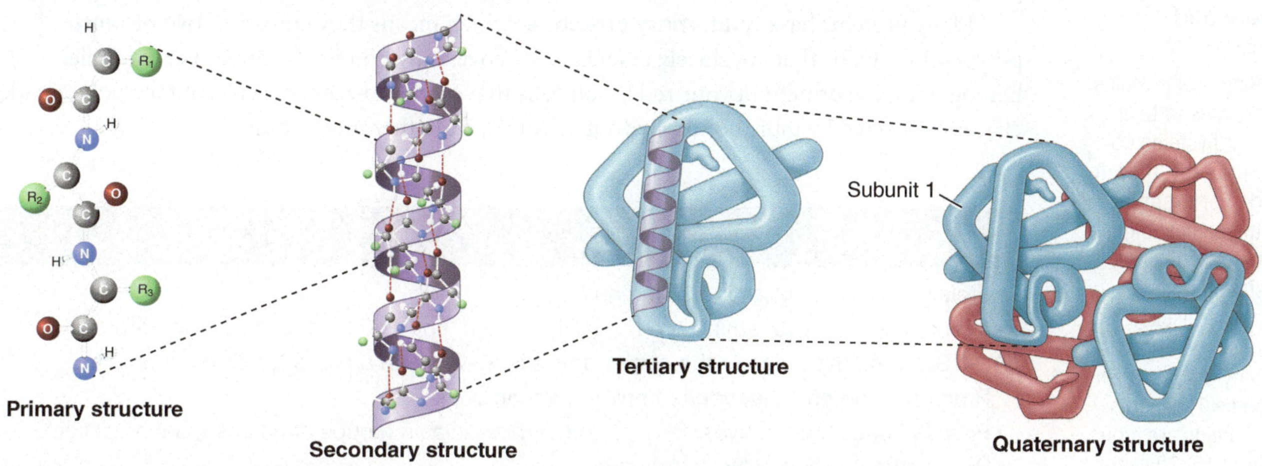

The first level of protein structure is the primary structure, which is simply the polypeptide chain. Hydrogen bonding between the peptide backbone leads to two types of secondary structures, the α helix and the β pleated sheet, shown here is an α helix. Interactions between R groups lead to tertiary structure. If multiple polypeptides interact, a quaternary structure forms.

Secondary structure. As the new polypeptide is made, hydrogen bonds form between the oxygen and hydrogen atoms in the peptide bonds along its backbone. As a result of hydrogen bonding, the structure begins to twist the molecule into two types of shapes, and **secondary structure** forms. These characteristic shapes are called alpha (α) helices and beta (β) pleated sheets. Each type of protein has a unique primary structure, but almost all proteins have helices, sheets, or both.

Study Tip

Grab a pipe cleaner and fold it like shown in primary, secondary, and tertiary structures. You can grab a second pipe cleaner and repeat these steps, bringing the two pipe cleaners together to represent a quaternary structure.

Tertiary structure. When the single polypeptide begins to fold in on its self to create the final 3-dimensional shape, **tertiary structure** forms. The secondary structure brings the R groups closer together, and they begin to interact with one another to form hydrogen, covalent, and ionic bonds. Some proteins are fully functional once they fold into tertiary structure. Proteins can form into a variety of shapes. For example, barrel-shaped proteins can function as tunnels to allow molecules to pass through cell membranes.

Biology for Life

The Protein Biochemistry behind Texture and Color Changes While Cooking Meat

Why does meat change color when you cook it (Figure 3.12)? Heat causes proteins to unfold or denature. When we heat meat, the proteins unfold and we can visualize this as a color and texture change. This makes the proteins easier to eat and denatures the proteins of any microorganisms living on the meat. This kills the microorganisms and prevents illness.

Figure 3.12 **Heating proteinaceous foods causes many chemical changes, including the unfolding of proteins**

Note that cooked bacon has a darker color and firmer texture.

Can you think of another example from your kitchen of how proteins look different when cooked?

Big Ideas in Biology

Structure and Function

The structure of proteins directly relates to their function. If proteins do not fold properly, they do not work. For example, one of the earliest signs of Alzheimer's disease, a devastating neurodegenerative disease, is the presence of misfolded amyloid β protein in the brain. These misfolded proteins are typically found 15–20 years before any clinical symptoms arise.

Quaternary structure. While some polypeptides can fold and function on their own, other polypeptides need to link up with different polypeptides to be functional.

Many proteins have **quaternary structure**, which means they consist of two or more polypeptide chains that are closely associated or covalently bonded together. For example, hemoglobin, the protein in your red blood cells that carries oxygen, consists of four polypeptide chains connected by ionic bonds, hydrogen bonds, and other interactions.

3.4 Summary

- Proteins are made of amino acid monomers.
- Amino acids have an amino group ($-NH_2$) on one end and a carboxylic acid group ($-COOH$) on the other. In the center, there is a $-CH$ with a place for an R group. The R group is the part that varies between amino acids.
- Peptide bonds form between amino and carboxylic acid groups of adjacent amino acids.
- Proteins have a hierarchical structure.
- Primary structure is the sequence of linked amino acids in a polypeptide.
- Hydrogen bonds in the peptide backbone make secondary structure (alpha helices, beta pleated sheets).
- Interactions between R groups makes tertiary structure.
- Multiple polypeptides interacting together form quaternary structures.
- Functional proteins are properly folded, like a paper folded into a floating origami boat.
- Proteins can unfold in the presence of heat. Unfolded proteins do not function.

3.5 Nucleic Acids

Learning Objectives

3-14 Describe the parts of a nucleotide.

3-15 Differentiate ribo- and deoxyribonucleic acids.

3-16 Explain how the chain of nucleotides relates to the function of a nucleic acid.

Nucleic acids can transmit information through their sequence of monomers. There are five different kinds of monomers, or **nucleotides**: adenine (A), guanine (G), cytosine (C), thymine (T), and uracil (U). Combining these "letters" in different sequences can represent information. You can make an analogy with the English language. Your text is written using 26 letters, combined in a variety of sequences. Your other texts are written using the same 26 letters. By combining these letters differently, new information is transmitted. In terms of biology, the information to create a new organism is transmitted via nucleic acids from each of the biological parent(s).

Study Tip

Think **de**oxyribose like losing an oxygen atom. Ribose sugar has a hydroxyl at the second carbon and deoxyribose sugar does not.

Nucleotides Are Made of a Sugar, a Nitrogenous Base, and Phosphate Functional Groups

Nucleotides have three parts (Figure 3.13A). In the middle is a five-carbon sugar. We look at this sugar to distinguish the two types of nucleic acids. **Ribonucleic acid (RNA)** has sugars made of ribose. Ribose has two hydroxyl groups at the second and third carbons. In **deoxyribonucleic acid (DNA)**, nucleotides instead have deoxyribose sugar. Deoxyribose sugar does not have an oxygen atom at the second position.

Connected to the sugar are a phosphate functional group (Chapter 2) and a nitrogenous base. As you see in Figure 3.13B, both DNA and RNA can have nitrogenous bases

Figure 3.13 Nucleic acids are made up of different nucleotides

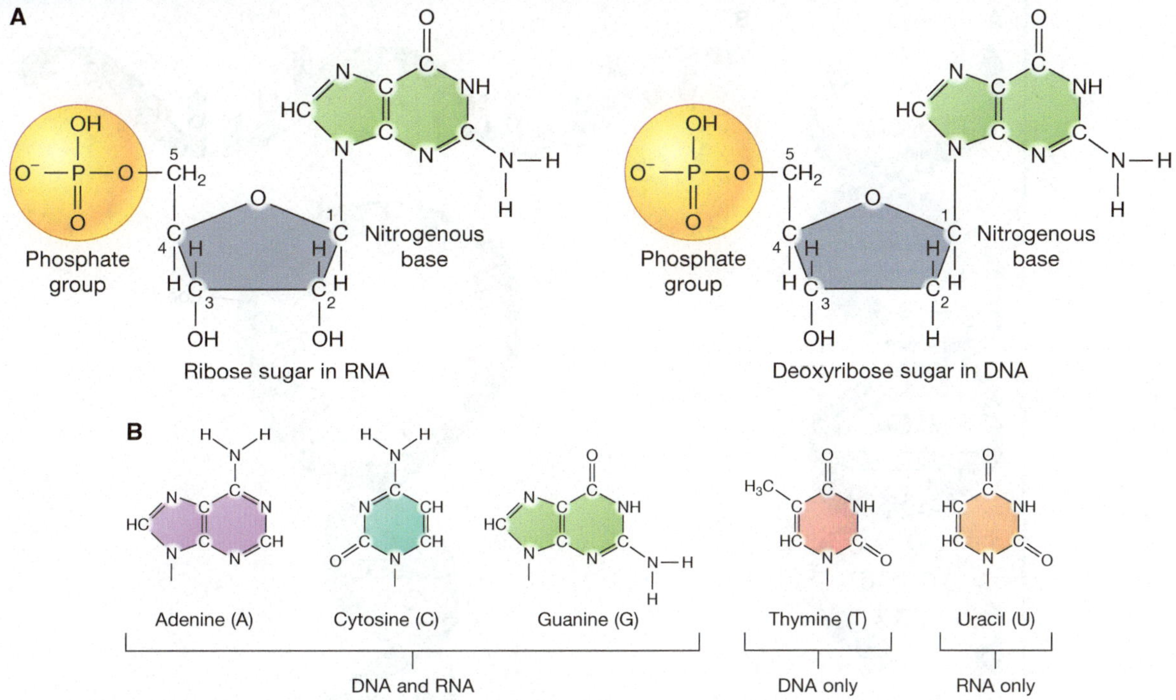

A. The nucleotide has three components, the phosphate group, sugar, and nitrogenous base. Note that at the C_2 carbon, there is a hydroxyl group in RNA, making the sugar a ribose sugar, and a hydrogen in DNA, making the sugar deoxyribose.

B. DNA and RNA have adenine, cytosine, and guanine nitrogenous bases. Only DNA has thymine and only RNA has uracil.

corresponding to adenine (A), cytosine (C), and guanine (G). Only DNA has the base thymine (T), and RNA substitutes thymine for uracil (U). Nucleotides form nucleic acids through dehydration synthesis using the phosphate groups at the carbon five position and a hydroxyl group at the carbon two position.

> **! Point of confusion:** In addition to storing information, nucleotides can also be used to store energy. For example, the nucleotide **ATP** (adenosine triphosphate) serves an especially important role as an energy carrier in cells.

DNA. A DNA molecule consists of two strands of nucleotides twisted into a double helix (Figure 3.14A). The order of nucleotides composing the DNA represents information about the organism. You can think of DNA as like a blueprint. Rather than containing instructions for how to build a house, it instead contains information on how to build an organism. Each cell starts life with DNA inherited from a parent cell. That DNA contains all of the information necessary to build a new cell and, in the case of multicellular organisms, an entire individual.

Due to its helical structure, DNA is extremely stable. In fact, scientists are using ancient DNA to try and bring extinct organisms, like mammoths, back to life!

RNA. Unlike DNA, most RNA molecules are single-stranded. RNA plays important roles in a variety of biological processes, such as protein synthesis and gene expression. RNA can also act like a protein and fold in a particular shape to function (Figure 3.14B). Since RNA has protein-like characteristics in addition to its information storage function, molecular biologists have hypothesized that RNA may have existed first.

Figure 3.14 DNA and RNA are polymers of nucleotides that have three dimensional structures

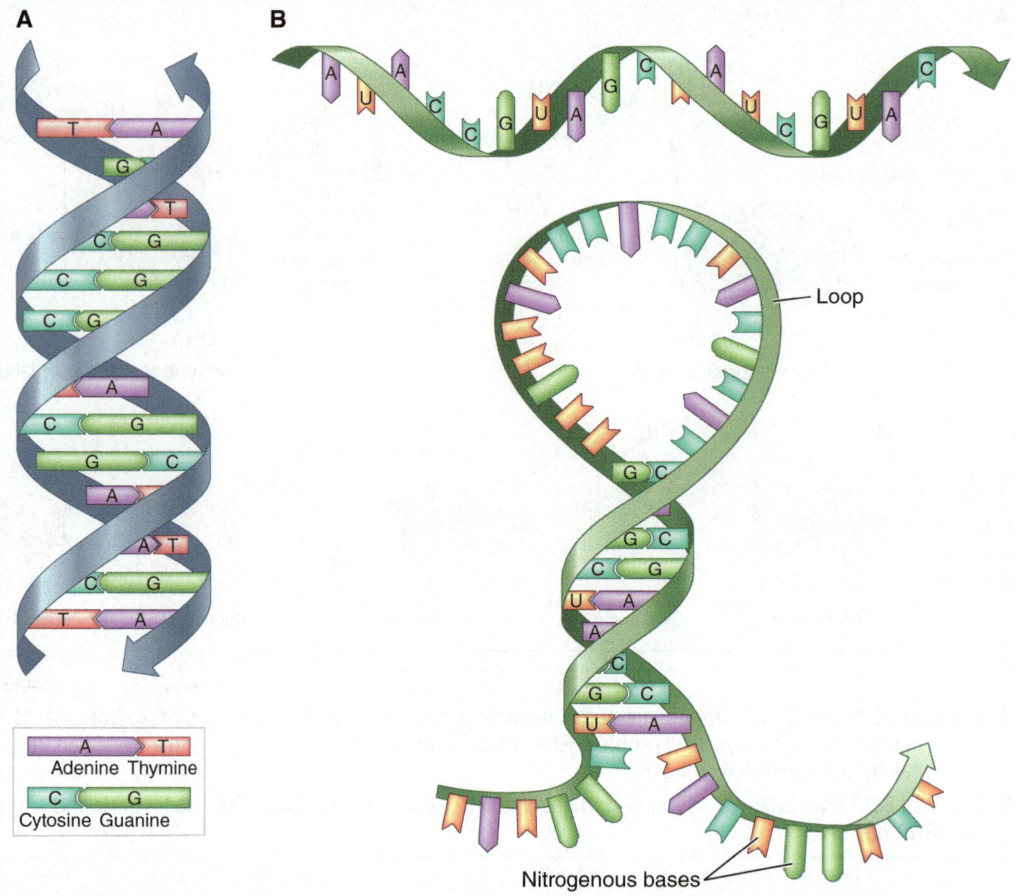

A. DNA is a double-stranded nucleic acid. Note that adenine (A) bonds with thymine (T) and guanine (G) bonds with cytosine (C).

B. RNA is a single-stranded nucleic acid that, like proteins, can fold into a three-dimensional shape, such as a loop. Notice that in RNA, the thymine (T) in DNA is replaced by uracil (U).

3.5 Summary

- Nucleic acids are made up of monomers called nucleotides.
- Nucleotides are made of a sugar, a nitrogenous base, and a phosphate group.
- There are two types of nucleic acids, DNA and RNA.
- Ribonucleic acid (RNA) has a ribose sugar with two –OH groups. Deoxyribonucleic acid (DNA) has a deoxyribose sugar with only one –OH group.
- Both DNA and RNA use adenine, cytosine, and guanine nitrogenous bases (A, C, G). Only RNA uses uracil (U) and only DNA uses thymine (T).
- Nucleic acids store information in the sequence of bases and are important for passing information along to the next generation.

3.6 Lipids

Learning Objectives

3-17 Identify different types of lipids.

3-18 Explain the difference between saturated, unsaturated, and trans fatty acids.

3-19 Explain how a phospholipid is both hydrophilic and hydrophobic.

Unlike carbohydrates, proteins, and nucleic acids, lipids are not built from monomers. Rather, a lipid is a any molecule that does not dissolve in water. Lipids can be either completely hydrophobic or amphiphillic. An **amphiphilic** molecule has both hydrophilic and hydrophobic characteristics. Amphiphilic molecules are very important to the structure and function of the cell membrane. We will come back to the structure of the cell membrane in Chapter 5. For the remainder of this section, we will explore some common types of lipids that you may hear about as part of your life.

> **Study Tip**
> An amphibian is an animal that lives in and out of water. An amphiphilic molecule has parts that interact with water and other parts that do not.

Fatty Acids

According to the Centers for Disease Control and Prevention (CDC), it is better to wash your hands with soap and water than to use hand sanitizer. What is so special about washing your hands with soap and water? Hand sanitizer is great for killing germs on your skin, but washing your hands with soap and water physically removes all germs (including those not easily killed by hand sanitizer!). How? Soap is predominantly made up of a type of lipid called a **fatty acid.** A **fatty acid** has a long hydrocarbon "tail" with a carboxylic acid functional group (Chapter 2) "head." The tail is hydrophobic (fatty); the carboxyl group makes the head hydrophilic (and acidic, hence the name fatty *acid*). As a result, fatty acids are amphiphilic. When you wash your hands, the hydrophobic tails attract dirt along with the membranes surrounding microorganisms, and the hydrophilic heads help dissolve the dirt in water (Figure 3.15). When you wash the water away, the dirt and microorganisms, wash away with the water.

Fatty acids can be saturated, unsaturated, or trans. A **saturated fatty acid** has only single bonds linking the carbons in their tails (Figure 3.16A). An **unsaturated fatty acid** contains at least one double bond between carbons making up its tail (Figure 3.16B). These double bonds are *cis* or *trans*, depending on their geometry (Figure 3.16C).

> **Study Tip**
> You can keep saturated and unsaturated fats straight by remembering that saturated fats are completely *saturated* with hydrogens.

Figure 3.15 Soap is made up of amphiphilic fatty acids

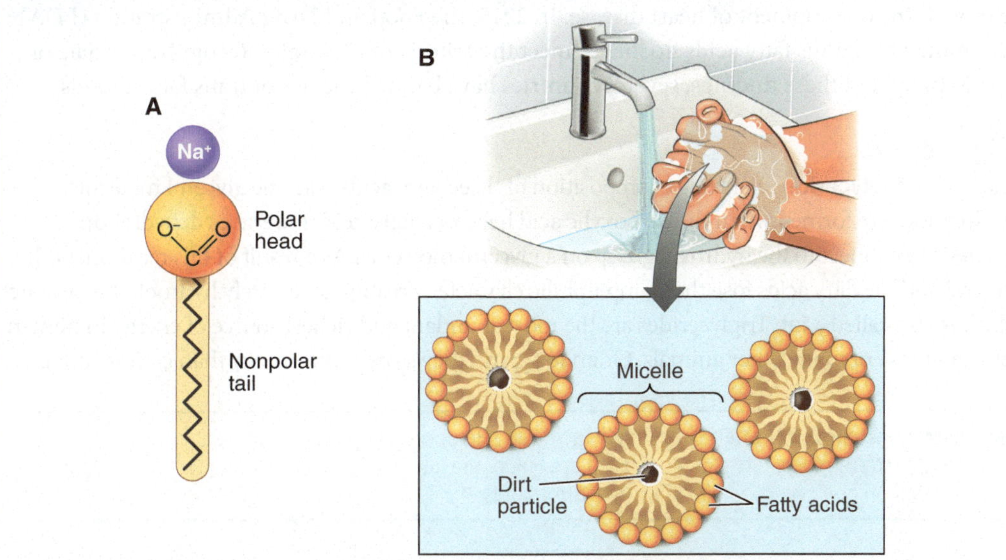

A. General structure of soap. Note the polar head, with a negative charge, and the non-polar tail. The negatively charged head forms an ionic bond with a positively charged sodium ion (Na^+), forming soap.

B. When you wash your hands with soap, the hydrophobic tails stick to dirt or microorganisms on your skin. When you rinse with water, the hydrophilic heads interact with the water molecules, taking the soap and the attached dirt away.

Figure 3.16 **Saturated, unsaturated, and trans fatty acids have different structures**

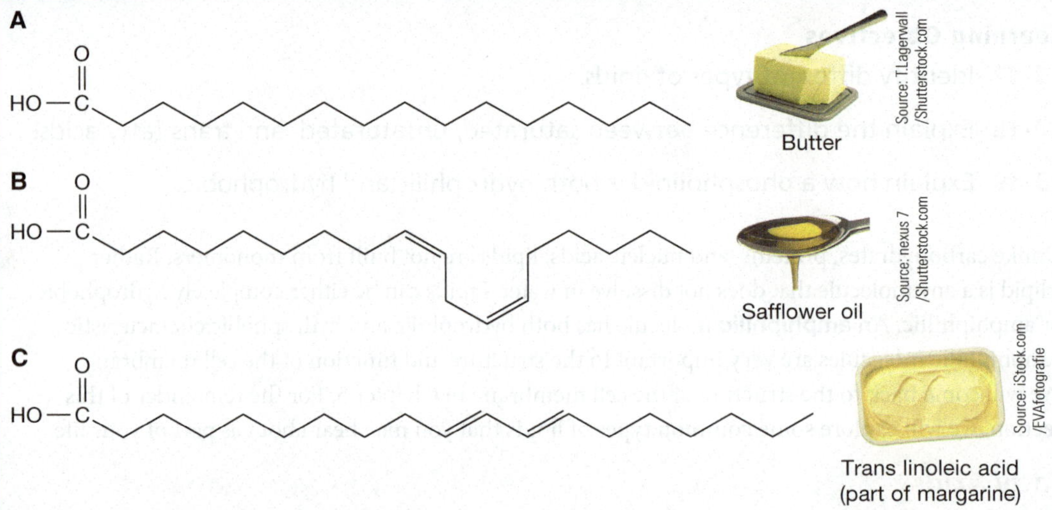

A. Saturated fats contain only single bonds.
B. Unsaturated fats contain at least one double bond.
C. Trans fats are artificially produced by adding hydrogens to unsaturated fats to increase shelf life.

Trans fats are often found in processed foods because these fats have a longer shelf life. Since they are detrimental to your health, the Food and Drug Administration requires that they be listed on nutrition labels.

Since saturated fatty acids have long straight tails, they stack together neatly at the molecular level. This leads to fats that are solid at room temperature, such as butter. Unsaturated fatty acids, such as safflower, olive, and fish oils, with their kinky tails do not stack neatly and are liquid at room temperature. **Trans fatty acids** come from artificially adding hydrogen to vegetable oil, an unsaturated fatty acid, to make the vegetable oil solid.

Unsaturated fatty acids such as omega-3 fatty acids are important building blocks for our bodies and have a variety of health benefits. The Centers for Disease Control (CDC) recommends that people consume predominantly unsaturated fatty acids to decrease their risk of heart disease. While saturated fatty acids are less good for your health, trans fatty acids are the most detrimental to health. In the early 1990s, scientists first linked the consumption of trans fatty acids with the development of heart disease. In 2015, the Food and Drug Administration (FDA) determined that trans fatty acids no longer meet the criteria of Generally Recognized as Safe or GRAS. In light of these findings, several countries have banned the use of trans fats in foods.

Triglycerides

The "tri" in **triglyceride** refers to its composition of three fatty acids and one glycerol molecule. Triglycerides are formed when the carboxylic acid head of a fatty acid undergoes dehydration synthesis to connect to the hydroxyl group on a glycerol molecule. As a result of this reaction, both glycerol and the fatty acids lose their amphiphilic character, creating an entirely hydrophobic product, which is also called a **fat**. Triglycerides are the most abundant and richest source of energy in human cells and those of many other animals. Gram for gram, they store more energy than carbohydrates.

Blood tests can reveal the levels of triglycerides in your blood. High levels of triglycerides in your blood can lead to an increased risk of heart attack or stroke.

! Point of Confusion: We often refer to fatty acids as fats in our daily lives. Triglycerides are also commonly referred to as fats. Although both are lipids, the key to distinguishing the two is that fatty acids are amphiphilic and triglycerides are hydrophobic.

Phospholipids

A **phospholipid** is a type of lipid with two long hydrocarbon tails and a head with a phosphate group (Figure 3.17A). The tails are hydrophobic, and the polar phosphate group makes the head

Figure 3.17 **Phosphilipids make up the cell membrane**

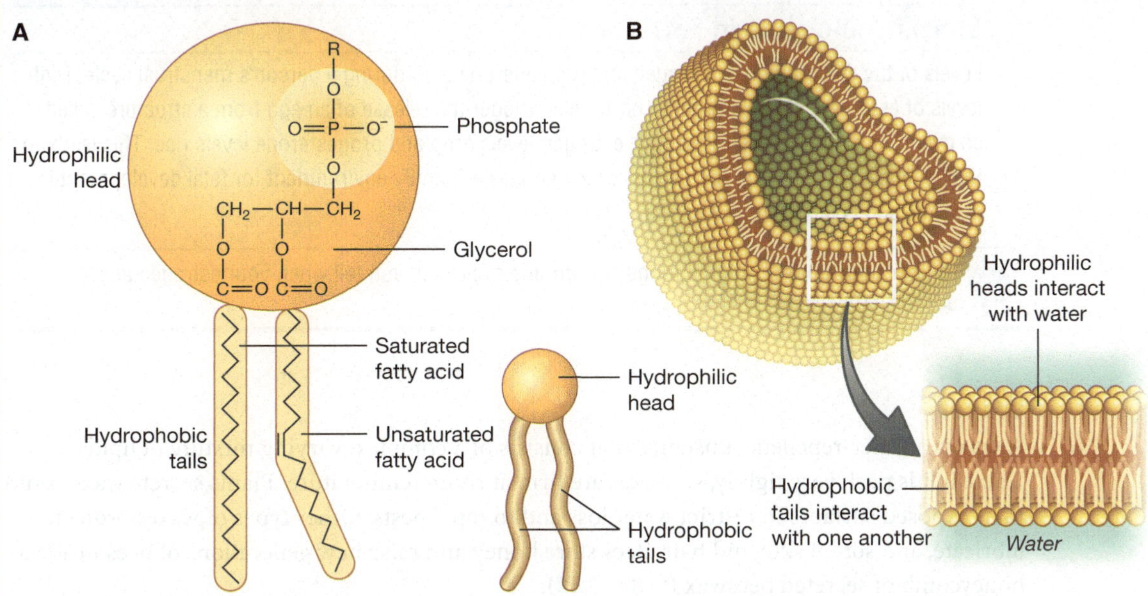

A. The structural formula of a phospholipid (left) and the cartoon representation (right).

B. Phospholipids make a layer with hydrophilic heads pointing towards the outside or inside of the cell, and hydrophobic tails pointing inward.

hydrophilic. Therefore, phospholipids are amphiphilic. The amphiphilic nature of phospholipids allows them to naturally assemble into a **lipid bilayer** when in water. Think of the lipid bilayer as a two-layered sheet of phospholipids. The hydrophilic heads of one layer face the cell's watery interior, and the hydrophilic heads of the other layer face the cell's watery exterior. The hydrophobic tails of all the phospholipids are sandwiched between the heads, so the interior of a lipid bilayer is highly hydrophobic. The lipid bilayer is the basic structure of all cell membranes (Figure 3.17B).

Big Ideas in Biology

Structure and Function

Knowing the structure of the lipid bilayer helps us understand how molecules get in and out of the cell.

Steroids

Steroids are lipids with no fatty acid tails. Rather, they have a rigid backbone with a characteristic pattern of four carbon rings. The types of functional groups attached to the rings define the type of steroid. Steroids have varied and important physiological functions in plants and animals (Table 3.2). **Cholesterol** (Figure 3.18), the most common steroid in animal tissue, is used to make other molecules such as vitamin D (required to keep teeth and bones strong) and steroid hormones, such as testosterone, estrogen, and progesterone. Steroid hormones are important for the development of secondary sexual characteristics, such as the development of facial hair (linked to testosterone) or breast development (linked to estrogen).

Table 3.2 Steroids and their Functions

Steroid	Function
Cholesterol	Supports the cell membrane, starting material for making other steroids
Estrogen Progesterone	Main female sex hormones
Testosterone	Main male sex hormone

Figure 3.18 **Cholesterol**

Cholesterol is an important type of steroid. Note the characteristic structure of four interlocking rings.

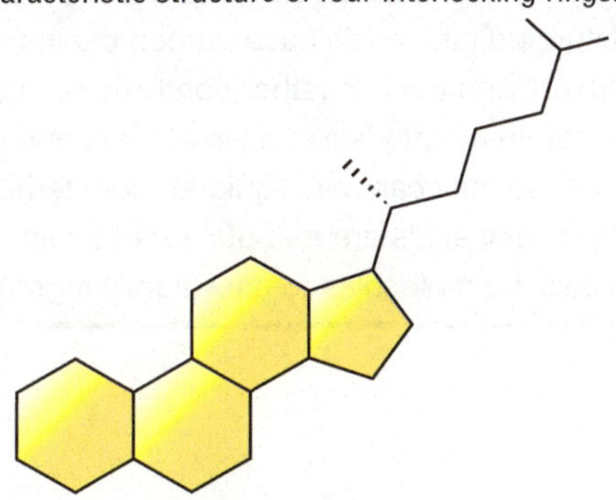

Biology for Life

Steroids and Human Fertility

Levels of the sex hormones estrogen and progesterone shift during a person's menstrual cycle. High levels of estrogen, along with other hormones, trigger the release of an egg from a structure called an ovary. After the release of the egg, estrogen levels drop and progesterone levels rise. This leads to the buildup of the lining of the uterus, which creates a friendly environment for fetal development in case the egg is fertilized by a sperm.

If your body goes through a monthly hormone cycle, can you tell when your estrogen and progesterone levels change?

Waxes

A **wax** is a water-repellent substance that consists of a complex, varying mixture of lipids. These lipids pack very tightly, so waxes are firm at room temperature. Plants secrete waxes onto their exposed surfaces to restrict water loss and to repel pests. Other types of waxes protect, lubricate, and soften skin and hair. Bees store honey and raise new generations of bees inside a honeycomb of secreted beeswax (Figure 3.19).

Figure 3.19 Beeswax is a type of lipid

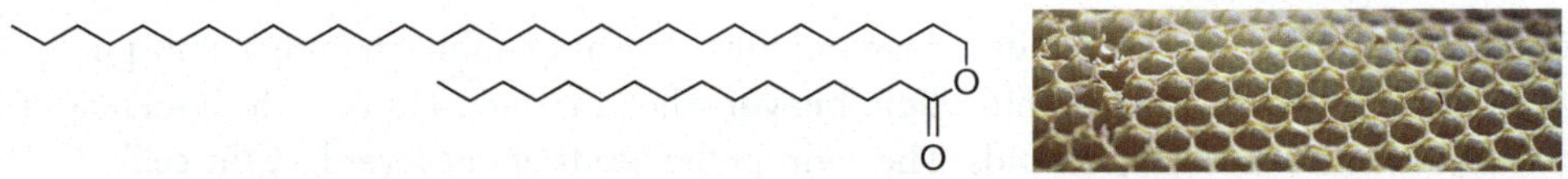

Source: Catherine Leblanc/Getty Images

Left is the chemical structure of beeswax. Right is the honeycomb structure created by bees. People have used beeswax to make candles since the 1500s.

3.6 Summary

- Lipids are not built from a set of monomers.
- Lipids all have at least one hydrophobic part of their structure.
- Some lipids are amphiphilic, meaning they have hydrophilic and hydrophobic portions.
- Fatty acids have a carboxylic acid head and fatty acid tail.
- Triglycerides are formed by three fatty acids plus one glycerol and are entirely hydrophobic.
- Phospholipids have two hydrophobic tails and a hydrophilic head and form lipid bilayers.
- Steroids like cholesterol and steroid hormones have a four-carbon ring backbone.
- Waxes repel water by packing together very tightly.
- Saturated fatty acids have carbon chains that are filled with hydrogen atoms so they are straight and pack together tightly (e.g., butter).
- Unsaturated fatty acids have at least one double bond which leads to a bend or kink in the chain so they can stay liquid at room temperature (e.g., vegetable oil).
- Trans fatty acids are manufactured by adding hydrogen to unsaturated fatty acids. This allows the molecules to pack tightly together and remain solid at room temperature.

Understand It! Elaboration

What if you were talking to someone and they asked you to elaborate on what you meant when you said last week's party was interesting? Perhaps you would give more details about how the stray dog showed up at the door right when the pizza was delivered, and how the dog jumped onto the delivery driver, who then dropped the pizza on the ground.

With those details, the concept of "interesting" is now expanded to include details about a dog and pizza. When we learn new material, one of the best ways to store it in our long-term memory is to add details to it, a process called *elaboration*. To elaborate on what you are learning in class means to put it in your own words while also connecting it with other information you already know. The more details you can add, the easier it will be to remember it later.

One way to elaborate on your knowledge is to make a concept map and connect new ideas with old ones. Another way is to look at a concept from class and explain it in your own words. Here are some sample questions to get you started:

1. Using the second law of thermodynamics, explain why your laundry does not come out of the dryer folded.
2. Explain why protein-based foods like beef change color when you cook them.
3. Explain the relationship between the nucleotide sequence and the information storage function of DNA.

Chapter Review

3.1 Life Runs on Energy

Energy is the ability to do work. **Kinetic energy** is the energy of motion. **Potential energy** is stored energy, such as the chemical energy stored in **covalent bonds**. The first two **laws of thermodynamics** state that 1. energy is neither created nor destroyed; and 2. over time, the universe tends to become more disordered, respectively. Chemical reactions can be **spontaneous** or **non-spontaneous**. Chemical reactions result in transfers of energy that change molecules.

3.2 The Building Blocks of Life

There are four kinds of **macromolecules: carbohydrates, proteins, nucleic acids**, and **lipids**. Except for lipids, macromolecules are built from **monomers** strung together to form **polymers**. Monomers join together through a reaction called **dehydration synthesis** and are broken apart by a reaction called **hydrolysis**. A water molecule is removed during the dehydration reaction, and a water molecule is added back during hydrolysis.

3.3 Carbohydrates

Carbohydrates are made of monomers called **monosaccharides**. These monosaccharides are also called simple sugars. Carbohydrates can exist as single monomers, two connected monomers (**disaccharide**), a short chain of monomers (**oligosaccharide**), or long chains of monomers (**polysaccharide**). Carbohydrates store energy, provide structure, and serve as cell identification markers. **Cellulose, starch**, and **glycogen**, are all example of energy storage polysaccharides.

3.4 Proteins

Proteins are made up of **amino acid** monomers with differing R groups. The chain of amino acids, called a **polypeptide**, is built through **peptide bonds**. Hydrogen bonds, ionic bonds, and other interactions between the amino acids cause the polypeptide chain to fold, and the resulting 3-D shape determines the function of the protein. There are four levels of protein structure: **primary, secondary, tertiary**, and **quaternary**. While the first three levels occur within one polypeptide chain, quaternary structure involves multiple different polypeptides. When proteins are unfolded, they are no longer functional.

3.5 Nucleic Acids

Ribonucleic acid (RNA) and **deoxyribonucleic acid (DNA)** are nucleic acids made up of **nucleotide** monomers. RNA monomers contain ribose sugar, which has one more hydroxyl group than DNA monomers, which contain deoxyribose sugar. Nucleic acids store information in the sequence of nucleotides, which in DNA are guanine (G), adenine (A), cytosine (C), and thymine (T). In RNA, thymine is replaced by uracil (U).

3.6 Lipids

Lipids are **hydrophobic** or **amphiphilic** molecules that are used in the body for energy (like **triglycerides**), as part of **lipid bilayers** in cell membranes (**phospholipids**), as **steroids** (like **cholesterol**), and as water-repellant **waxes**. Fatty acids contain a long hydrocarbon tail with a carboxylic acid group. Depending on the structure of the tail, a fatty acid can be **saturated** or **unsaturated. Trans fatty acids** are a type of artificially produced unsaturated fatty acid. Unsaturated fatty acids are the healthiest, followed by saturated fatty acids, and finally trans fatty acids.

Practice It!

Retrieval practice is the process of pulling information from your long-term memory and using it. Here are some questions to help with your retrieval practice.

Self-Quiz

1. When a food scientist determines the energy content of food, they put the food into a closed chamber and break it down completely by burning the food. They always find that the amount of energy calculated for the food is the same as the energy they detect being released through the burning process. This is an example of which law of thermodynamics? (LO 3-2)

a. The First Law of Thermodynamics: Energy cannot be created or destroyed

b. The First Law of Thermodynamics: The universe tends towards disorder

c. The Second Law of Thermodynamics: Energy cannot be created or destroyed

d. The Second Law of Thermodynamics: The universe tends towards disorder

2. You are walking to class and drop your notebook. Your papers fly everywhere. You notice that your papers scattered instead of landing in a neat pile. Which property of energy explains this result? (LO 3-1)

a. Kinetic energy is the energy of motion.

b. There is a tendency toward disorder.

c. Potential energy is stored energy.

d. Energy is the ability to do work.

3. Cells use dehydration synthesis to ______. (LO 3-6, 3-7)

a. break apart polymers into monomers

b. destroy polymers by adding water

c. build monomers

d. link monomers together to make polymers

4. Which of the following is NOT a macromolecule? (LO 3-4)

a. lipid
c. water
b. protein
d. carbohydrate

5. When the hydrolysis reaction breaks apart polymers to release monomers, how is water involved? (LO 3-7)

a. A hydrogen atom and an –OH are released from the polymer and form water.

b. A hydrogen atom and an –OH from water add to the polymer, breaking the bond that joins the monomers.

c. Hydrophilic interactions break apart polymers that are attracted to water.

d. Hydrophobic interactions keep water away during hydrolysis.

6. In order to be a carbohydrate, which elements must be present? (LO 3-5, 3-8)

a. carbon and oxygen

b. carbon, hydrogen, and oxygen

c. oxygen and hydrogen

d. None of these are correct.

7. Which of the following is NOT a function of carbohydrates? (LO 3-9)

a. energy storage
c. cell identification
b. information storage
d. cell structure

8. Table sugar (or sucrose) is a carbohydrate consisting of a monomer of glucose and a monomer of fructose. Therefore, table sugar is a(n) ______. (LO 3-8)

a. monosaccharide
c. oligosaccharide
b. disaccharide
d. polysaccharide

9. Blood sugar levels in humans are determined by measuring the blood levels of ______. (LO 3-9)

a. cellulose
c. glycogen
b. starch
d. glucose

10. What monomer makes up a protein? (LO 3-11)

a. monosaccharide
c. nucleotide
b. fatty acid
d. amino acid

11. What part of an amino acid varies? (LO 3-11)

a. amino group
c. R group
b. carboxylic acid group
d. carbon and hydrogen group

12. Which level of protein structure includes more than one polypeptide chain? (LO 3-12)

a. primary
c. tertiary
b. secondary
d. quaternary

13. Which level of protein structure includes hydrogen-bonded alpha helices? (LO 3-12)

a. primary
c. tertiary
b. secondary
d. quaternary

14. An unfolded protein is ______. (LO 3-12, 3-13)

a. in tertiary structure

b. nonfunctional

c. looks like a tunnel

d. functional

15. Which of the following lists the correct parts of an RNA nucleotide? (LO 3-14, 3-15)
    a. nitrogenous base, phosphate group, ribose sugar
    b. nitrogenous base, ribose sugar
    c. phosphate group, deoxyribose
    d. nitrogenous base, phosphate group, deoxyribose sugar

16. How do nucleic acids store genetic information? (LO 3-16)
    a. They can be broken down for lots of energy.
    b. They can be used to build important cellular structures.
    c. The order of their nucleotides can be used as a code.
    d. The types of atoms are different from other organic molecules.

17. Which of the following nucleic acid sequences is an RNA molecule? (LO 3-15)
    a. GATCG
    b. CTGAA
    c. TACAG
    d. AUGAC

18. Molecules that have both hydrophobic and hydrophilic properties are called ______. (LO 3-19)
    a. nucleic acids
    b. cellulose
    c. acidic
    d. amphiphilic

19. Which type of lipid is completely hydrophobic? (LO 3-18, 3-19)
    a. fatty acids
    b. triglycerides
    c. steroids
    d. phospholipids

Critical Thinking

1. Marcia is studying to be a hairstylist and is surprised to find out that biological macromolecules are involved in her field. Can you explain how proteins, nucleic acids, and lipids might be involved in growing and maintaining hair? (LO 3-4, 3-5)

2. All four macromolecules are found in food, but nutrition labels only include three of them. Which one is not included and why do you think it doesn't get mentioned? (LO 3-5, 3-9, 3-16, 3-17)

3. You're invited to a study session with your classmates. As you reach for a snack in your backpack, you notice the nutrition label. Your snack has protein, carbohydrates, and fat, all compounds you are learning about in class. For this activity, find a snack in your bag, dorm room, or kitchen, and determine the following. (LO 3-5, 3-9, 3-16, 3-17)
    a. What is the most abundant macromolecule in your snack in terms of mass? In terms of the percent daily value of the nutrient?
    b. Does your snack contain any trans fatty acids?
    c. Do you think your snack is healthy? Why or why not?

4. You have been camping with no access to a warm shower. Once you arrive home and are lathering up your hair, you recall something your professor said about shampoo containing lipid molecules. This seems confusing now because you know lipids don't mix with water. What is the property of some lipids that makes this possible? (LO 3-17, 3-19)

4

Energy Capture and Transformation

Chapter Outline:

Fueling Life in the Deep Ocean

Have you ever stood beside the ocean and wondered what was below the surface? How many fish were within a few feet of you? The world's oceans are mysterious places. As of the 2020s, about 75% of the ocean remains unexplored. Why is that? For one thing, the ocean is big. More than two-thirds of the planet is covered with water, at an average depth of about 3,700 m (2.3 miles). Sunlight only penetrates the top 200 m (656 feet), so most of the ocean is in complete darkness.

It was truly surprising then when, in 1977, scientists exploring the ocean floor discovered that instead of the rare crab they expected to see, they found a community of never-discovered organisms around hydrothermal vents. These organisms were not dependent on photosynthesis for energy, but instead on high-energy chemicals like hydrogen sulfide (H_2S) spewing from the vents. Scientists discovered that in the same way plants use sunlight to create energy-rich molecules, microorganisms in these communities use molecules such as H_2S to fix carbon into food that can be eaten by other organisms. This process is called *chemosynthesis*, and the hypothesis that all life on Earth depended on the Sun was thus refuted with this discovery. Why was it so monumental to discover life that doesn't rely on the sun? What does the sun have to do with the energy we need for our bodies to function?

Giant tube worms (*Riftia pachyptila*) live at the bottom of the ocean. Chemosynthetic bacteria found around hydrothermal vents can turn high-energy sulfur-containing chemicals into a food source not only for themselves, but for other organisms like these giant tube worms.

As we learned in Chapter 1, life requires energy. But how do we go from energy in the form of solar radiation to the sandwich you eat for lunch? The goal of this chapter is to describe how energy is captured by organisms, either on land or at the bottom of the sea, and how that energy is converted into a form that can then be used by other organisms.

We'll begin by discussing enzymes and their role in three key biochemical pathways covered in this chapter: photosynthesis, cellular respiration, and fermentation. At the end of the chapter, we will discuss how we obtain energy from the foods we ingest.

Preview It!

List three things you know about photosynthesis, cellular respiration, and fermentation already. Do you have a favorite fermented food or drink? What do you want to learn in this chapter?

4.1 Enzymes and Metabolic Pathways

Learning Objectives

4-1 Describe why cells need enzymes.

4-2 Explain the function of the active site.

4-3 Explain how an enzyme's environment impacts its function.

4-4 Describe the role of ATP in a cell's energy economy.

4-5 Describe the basic features of a metabolic pathway.

What's So Special about Enzymes?

The stress of your leaving for college was too much for your dear cat. Your cat, being a cat, now pees all over your clothes when you come home for a visit. The silver lining is that there's no need to schlep your clothes to the campus laundromat. Instead, you can use the washing machine at your parents' house. But the smell doesn't come out! So you rewash your clothes.

Back in your dorm room, you notice that your clothes still hold onto the strong ammoniacal smell characteristic of cat pee. What now? As you vent to your biology nerd roommate about your stinky clothes, they say you need to use a laundry detergent with enzymes. So you venture to the supermarket and study all the laundry detergent bottles until you find one that lists "amylase enzyme" and "protease enzyme" on the ingredients label. Lo and behold, the smell disappears after using the enzyme-containing detergent. What's so special about enzymes?

The need for speed. An **enzyme** is a molecule that can speed up, or **catalyze**, a chemical reaction but is not consumed during that reaction. Said otherwise, an enzyme can be used repeatedly to speed up reactions. Although enzymes are usually proteins, there are some cases where RNAs can act as enzymes. Why do we need to speed up chemical reactions? And what does this all have to do with why enzymes can help get the cat pee smell out of your clothes?

Without enzymes, the chemical reactions that sustain life would occur too slowly. For example, in the presence of oxygen, glucose spontaneously combusts to form CO_2 gas, water, and energy. Now, before you panic and remove the candy bar from your backpack, spontaneous does not mean fast. Without enzymes, this chemical reaction proceeds too slowly to be of concern. However, we need energy to sustain our bodily processes *now*—waiting years for even spontaneous reactions to convert food to energy takes too long. Too long for life to exist. Enzymes speed the process up.

Point of Confusion: In chemical terms, a "spontaneous" reaction is one that proceeds without the addition of energy. It doesn't mean it is fast. An example of a spontaneous chemical reaction that proceeds slowly is a rusting car. Cars rust spontaneously, but over many years.

Figure 4.1 **The active site of an enzyme**

An enzyme converts substrates into products by squeezing or pulling the molecules as needed to help form or break bonds. Since the enzyme is unchanged, it can repeat the process.

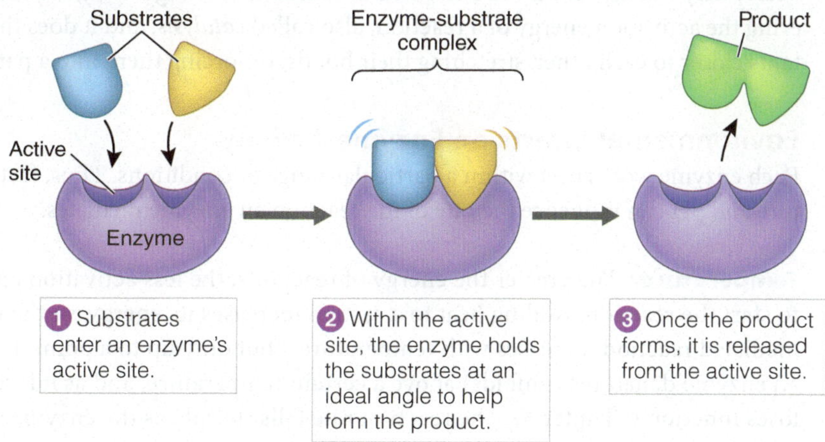

Enzymes also perform a wide variety of other functions. In the case of enzyme-containing laundry detergents, amylase is an enzyme that breaks down carbohydrates and proteases are enzymes that break down proteins. These enzymes break down the stinky chemicals found in cat pee to eliminate the smell.

How does an enzyme speed up chemical reactions? An enzyme makes it easier for products to form by bringing the starting materials, or **substrates**, closer together and holding them at the perfect angle in the **active site** of the enzyme (Figure 4.1). The active site is a location where substrates bind, forming an enzyme-substrate complex, and also where the reaction occurs. After the reaction, the active site releases the product(s), so more substrates can bind. Substrates are complementary in shape, size, polarity, and charge to an active site. That fit is the reason why each enzyme acts only on specific reactants.

Pharmaceutical companies take advantage of this specificity to design drugs. For example, a **competitive inhibitor** is a compound that also binds to the active site of an enzyme. The inhibitor and the substrate compete for the spot in the active site, just like children playing musical chairs. If there is too much inhibitor present, the active site stays occupied and the substrate is not converted to the product. For example, the chemotherapeutic drug methotrexate is a competitive inhibitor used to kill rapidly dividing cancer cells by binding to an enzyme used to make DNA.

In biochemical terms, we say that an enzyme decreases the **activation energy** of a reaction (Figure 4.2). Activation energy is the minimum amount of energy required for a chemical

Study Tip:

You can think of the active site as a lock and the substrate as the key. The key fits a specific lock, just like the substrate fits into the active site.

Big Ideas in Biology

Structure and Function

Certain medicines can bind to another region of an enzyme and cause the shape of the active site to change. If the active site is changed, it can no longer bind to the substrates and therefore no products are produced.

Figure 4.2 **Enzymes lower activation energy**

Without an enzyme, the amount of activation energy needed to convert reactants to products is much greater (taller double headed arrow). With an enzyme, the same reaction can proceed with less activation energy necessary (shorter double headed arrow). This results in a faster chemical reaction. Note that although activation energy changes, the amount of energy in the products and reactants does not change.

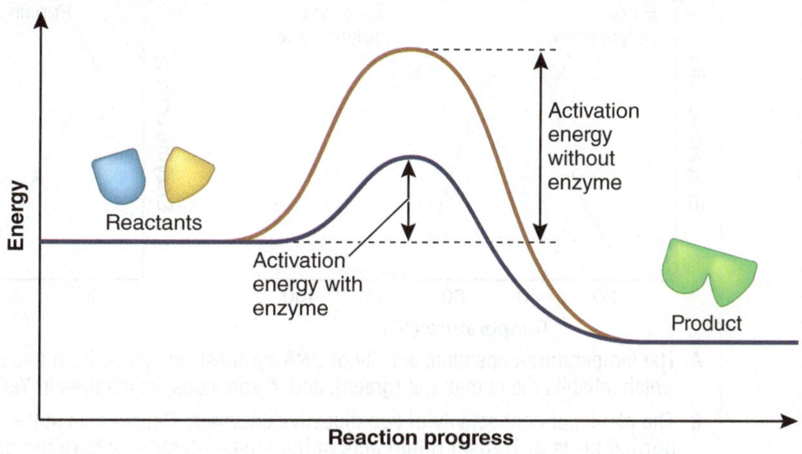

reaction to occur. You can think of activation energy as the "hill" that reactants need to "climb" before they can roll down the other side as reactants (see Figure 4.2). An enzyme works by lowering the activation energy of a reaction, also called *catalysis*, and it does this by bringing reactants closer to each other, stretching their bonds, or forcing them into a particular shape.

Environmental Effects on Enzyme Activity

Each enzyme works best within a particular range of conditions. Thus, factors such as temperature and pH can influence the rate of the reaction an enzyme mediates.

Temperature. The greater the energy of reactants, the less activation energy is required to start the reaction. Adding heat to a system increases its energy, so the rate of an enzyme-mediated reaction increases with temperature—but only up to a point. Like other proteins, an enzyme denatures (unfolds) above a certain temperature, and as it loses shape, it also loses function (Chapter 3). The reaction rate falls sharply as the enzyme stops working (Figure 4.3A).

pH. Most enzymes function within a specific range of pH (Chapter 2), and that range depends on the enzyme. Consider two enzymes that play a role in human digestion (Figure 4.3B). The enzyme pepsin, which works best at low pH, begins the process of protein breakdown in the acidic environment of the stomach (pH 2). During digestion, the stomach's contents pass into the small intestine, where the pH rises to about 7.5. Pepsin denatures above pH 5.5, so this enzyme stops working in the small intestine. Protein breakdown continues in the small intestine with the assistance of trypsin, an enzyme that functions well at a higher pH.

Enzymes Sometimes Require Helper Molecules

Most enzymes cannot function properly without assistance from metal ions or small organic molecules. **Cofactors** are inorganic ions, like metals, and are often derived from dietary vitamins and minerals. **Coenzymes** are organic molecules that are needed by an enzyme to function and can be modified during a reaction. They are regenerated in separate reactions, however, so like enzymes, they are not "used up" in a reaction and are instead recycled. These coenzymes carry chemical groups, atoms, or electrons from one reaction to another, and often into or out of

Figure 4.3 **Enzyme function is impacted by temperature and pH**

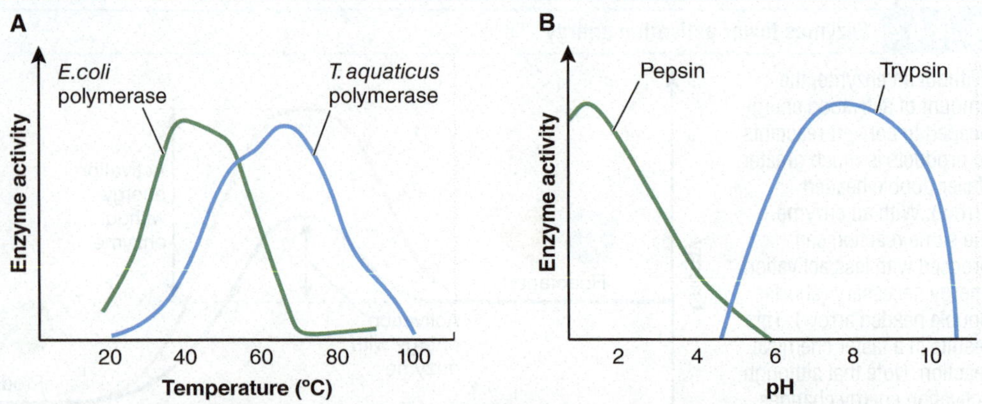

A. The temperature-dependent activity of DNA synthesis enzymes from two species of bacteria—*E. coli*, which inhabits the human gut (green), and *T. aquaticus*, which lives in Yellowstone's hot springs (blue).

B. The pH-dependent activity of two digestive enzymes. Pepsin acts in the stomach (green), where the normal pH is 2. Trypsin (blue) acts in the small intestine, where the pH is normally around 7.5.

organelles. For example, later in this chapter, we will discuss how one coenzyme, NAD^+ (nicotinamide adenine dinucleotide) is used by hundreds of different enzymes, including those in cellular respiration.

> **Big Ideas in Biology**
>
> **Evolution**
>
> ATP and the enzymes used to make ATP are found across all life on Earth, from humans to bacteria.

ATP: Energy for Enzymes

Have you ever watched the blinking of fireflies on a summer evening? Where does the energy come from, that fuels a firefly's light (Figure 4.4A)? From ATP! **Adenosine triphosphate**, or **ATP**, is a nucleotide that functions as an energy source in many reactions by donating and accepting phosphate functional groups (Chapter 2). Bonds between phosphate groups hold lots of potential energy—and ATP has two of them holding its three phosphate groups together (Figure 4.4B). When a phosphate group is transferred to or from a nucleotide, this bond energy is transferred along with it. Thus, by forming and breaking phosphate bonds, a nucleotide can pick up and store energy produced in one reaction, and it can donate energy in a different reaction that requires energy. ATP is like a universal currency for all energy transactions in living organisms. Note that throughout the art in this chapter, ATP will be represented as a gold coin 

.

A reaction in which an enzyme facilitates the attachment of a phosphate to a molecule is called **phosphorylation**. When an enzyme transfers a phosphate group from ATP to another molecule, ATP becomes ADP, or **adenosine diphosphate**. Note that it's called adenosine *di*phosphate because the molecule has only two (di-) phosphates.

Cells constantly break apart ATP to supply energy-requiring reactions (Figure 4.4C). Thus, they must constantly replenish their stockpile of ATP, and they do so by running *energy-releasing* reactions that provide energy to phosphorylate ADP. This cycle of using and replenishing ATP couples energy-requiring reactions with energy-releasing ones. ATP is a major part of a cell's energy economy.

Enzyme Assembly Lines

Much like how a car is constructed in a stepwise fashion along an assembly line, building, rearranging, or breaking down an organic molecule also occurs in a stepwise manner. A **metabolic pathway** is a process that takes a starting material and ends with a product using a series of enzymatic reactions. Some metabolic pathways are linear, meaning that the reactions

Figure 4.4 The nucleotide ATP serves as energy currency for a cell

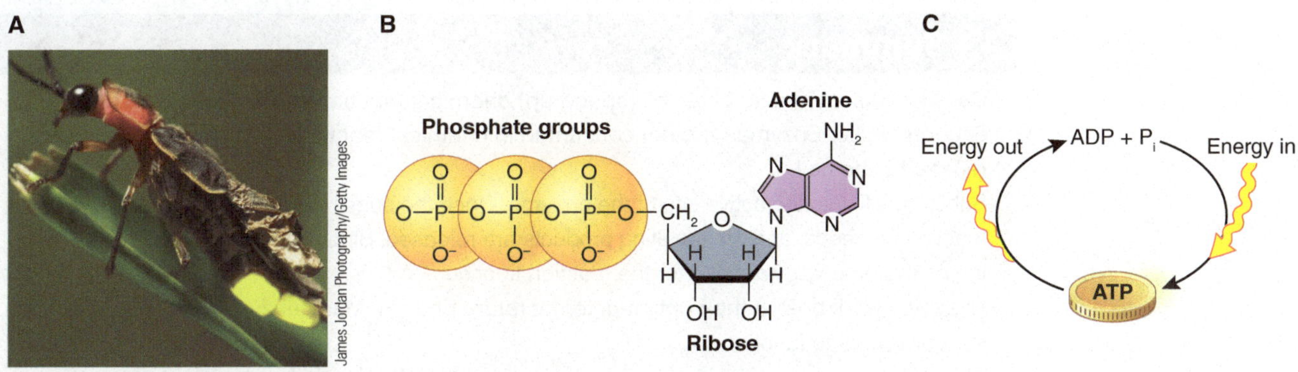

A. The bright light emitted by a firefly is used to attract mates. Light-producing reactions in these beetles run on energy provided by ATP.

B. The structure of ATP includes an adenine nucleotide, ribose sugar, and three phosphate groups. Bonds between the phosphate functional groups hold considerable energy.

C. ATP works like a rechargeable battery. When ATP loses one of its phosphate functional groups (indicated in the figure as P_i), it forms ADP. ADP is like a dead battery, and adding back energy in the form of a phosphate group recreates ATP.

Figure 4.5 Linear and cyclic metabolic pathways

A. Some metabolic pathways have a straight path from reactant to product.

B. Metabolic pathways can also occur in cycles, where the last step of the pathway regenerates the first reactant.

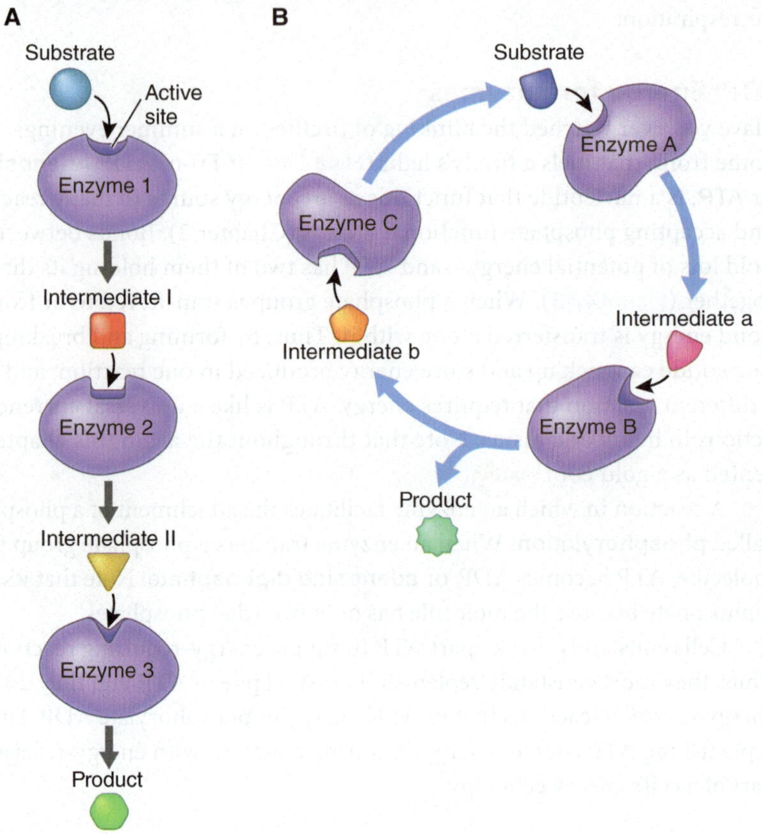

run straight from reactant to product (Figure 4.5A). Others are cyclic. In a cyclic pathway, the last step regenerates a reactant used in the first step (Figure 4.5B). Note that metabolic pathways can interact with one another through **feedback inhibition**, where the product of one pathway controls the activity of another.

4.1 Summary

- Cells need enzymes to catalyze (speed up) chemical reactions so they occur fast enough for life. Enzymes are not consumed in chemical reactions and can therefore be reused by the cell.
- Substrates fit into the active site of the enzyme. Once the substrate is bound, the chemical reaction proceeds, and the resulting products are released. Binding to an active site reduces the activation energy needed for the reaction to occur.
- Enzymes work best at their optimum temperature and pH. Many enzymes require cofactors or coenzymes to function.
- ATP stores energy by accepting a phosphate group and releasing that energy when the phosphate group is lost. In this way, ATP supplies energy for many different metabolic pathways in the cell.
- Metabolic pathways are a series of enzymatic reactions that transform an initial substrate into a final product.

4.2 Connecting Photosynthesis and Cellular Respiration

Learning Objectives

4-6 Describe the relationship between autotrophs and heterotrophs.

4-7 Explain the relationship between photosynthesis and cellular respiration.

It's study time and alas you are hungry. What do you do? Do you head over to the kitchen to grab a snack? Root around in your bag for a protein bar? Suppose you were a plant or a giant tube worm on the bottom of the ocean. Neither plants nor tube worms can move around and find something to eat. Where does their food come from?

Plants have it best because they are autotrophs. An **autotroph** is an organism that can take a non-nutritional inorganic substance, such as CO_2, and convert it into nutritious organic molecules, such as sugars. Autotrophs are different from heterotrophs. A **heterotroph**, such as you or any other animal, can only get energy from organic molecules. A heterotroph can't convert inorganic molecules into organic molecules, so heterotrophs must eat autotrophs or parts of autotrophs, like fruit, or other heterotrophs, such as other animals, to get the energy locked up in the organic molecules (Figure 4.6). Those hungry tube worms we mentioned? They are heterotrophs, and they rely on organic compounds produced by bacteria and archaea through chemosynthesis. **Chemosynthesis** is when organic molecules are produced by living organisms using an energy source other than the sun.

The Energy of Life: A Play in Two Parts

But where does the energy that autotrophs require to produce organic compounds come from? For plants, it's the Sun! Plants (and some other organisms) can make energy from sunlight through a process called photosynthesis. **Photosynthesis** harnesses the energy of sunlight to drive the assembly of sugars from CO_2 and water. Photosynthesis includes two different processes: an energy capture phase and a sugar building phase (Section 4.4). Sugars, in turn, are broken down to release energy held in their bonds, a process called **cellular respiration**. Cellular respiration has different pathways, depending on whether an organism is burning sugar for energy in the presence of oxygen (Section 4.5) or in the absence of oxygen (Section 4.6).

The relationship between photosynthesis and cellular respiration can be demonstrated by looking at the reactions that link them. In photosynthesis, organisms use the Sun's energy to join carbon atoms from CO_2 molecules to make sugar and produce oxygen gas. In cellular respiration, sugar is broken down to produce energy (ATP) for the cell, plus CO_2 and H_2O.

$$6CO_2 + 6H_2O \underset{\text{cellular respiration}}{\overset{\text{photosynthesis}}{\rightleftarrows}} C_6H_{12}O_6\ (\text{glucose}) + 6O_2$$

Figure 4.6 Photosynthesis sustains life

Heterotrophs, like a moose, get their energy from the organic molecules produced by autotrophs, such as the plants the moose is eating. Plants sustain themselves by carrying out photosynthesis.

Study Tip

The "photo" in *photosynthesis* means light and "synthesis" is to make. Photosynthesis is to make (something) with light. Glucose is made during photosynthesis.

Big Ideas in Biology

Interconnectedness of Living Systems

Autotrophs such as plants and heterotrophs such as animals need each other to survive. Autotrophs can create food for themselves and others. In return, heterotrophs give resources back to autotrophs. For example, all organisms leave behind detritus (dead tissue and/or waste like feces) that is recycled back into inorganic nutrients by decomposers for autotrophs to use.

Together, photosynthesis and cellular respiration are two of the most important metabolic pathways. Without autotrophs harnessing energy and creating organic compounds, there would be no life on Earth. A summary of the linked processes of photosynthesis and cellular respiration is shown in (Figure 4.7).

To help you follow the carbon atoms and energy transfers that occur in these processes, here are two tour guides for you: Carla, who represents carbon atoms, and Eddie, who represents energy. Look for Carla and Eddie at the bottom of each section for a summary of what happened to carbon atoms and energy in each phase.

Figure 4.7 **Photosynthesis and cellular respiration are linked processes essential for life**

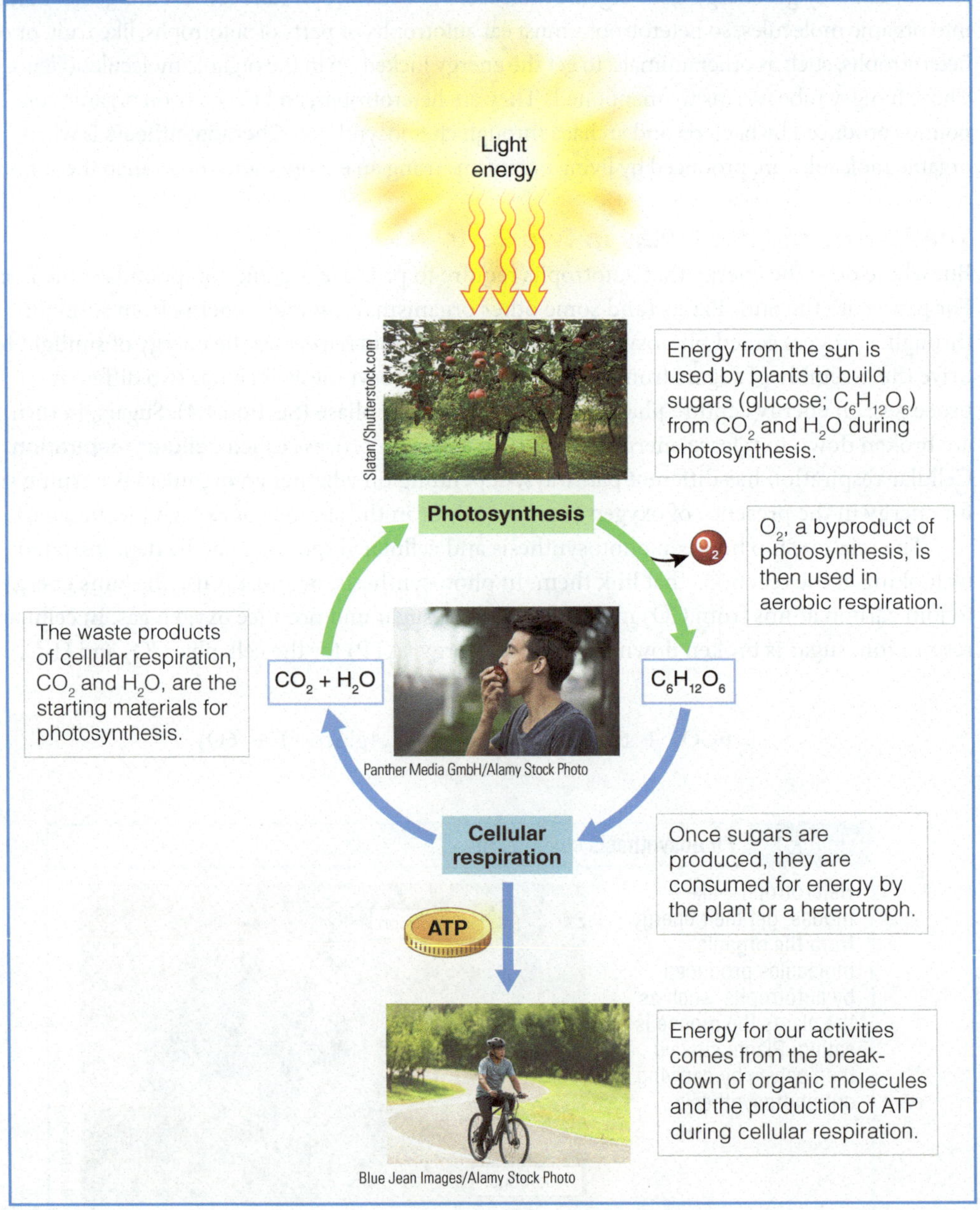

4.2 Summary

- Autotrophs can make their own food from inorganic molecules like CO_2 using light or chemical energy. Plants are photosynthetic autotrophs. Heterotrophs cannot synthesize their own food so they must ingest other organisms for their nutrition. Animals, including humans, are heterotrophs.
- Photosynthesis converts sunlight energy into chemical energy stored in sugars. Cellular respiration harvests energy from glucose and stores it as ATP. The substrates and products of photosynthesis and cellular respiration link these two processes. Photosynthesis takes CO_2 and H_2O and releases O_2 and sugars. Cellular respiration breaks down sugars using O_2 and releases CO_2 and H_2O.

4.3 Setting the Stage: Where Photosynthesis and Cellular Respiration Occur

Learning Objectives

4-8 Identify the parts of the chloroplast where photosynthesis occurs.

4-9 Identify the cellular locations where cellular respiration occurs.

4-10 Describe the features that make the leaf ideal for catching light.

Before talking about how energy is transferred and carbon atoms are moved around during photosynthesis and cellular respiration, we first need to discuss the cellular locations where photosynthesis and cellular respiration occur.

The Leaf as a Molecular Baseball Glove

Much like how a baseball glove has a large surface for catching baseballs, the leaf has a similarly wide surface for catching light (Figure 4.8). Within the leaf are the cellular structures necessary to conduct photosynthesis. For example, most land plants have leaves that are covered in a thin, waterproof layer called a **cuticle**. This cuticle limits evaporative water loss from their leaves.

Gases cannot diffuse across the cuticle. How does gas exchange occur during photosynthesis? To solve this problem, the surfaces of leaves (mainly the underside) are studded with tiny, closable

Figure 4.8 **The leaf is the part of the plant that catches the sunlight**

A. Just like a baseball glove has a wide surface for catching a baseball, the leaf has a wide surface for catching sunlight.

B. These tiny holes are stomata on the underside of a leaf. These pores close to conserve water and open to facilitate the gas exchange needed for photosynthesis.

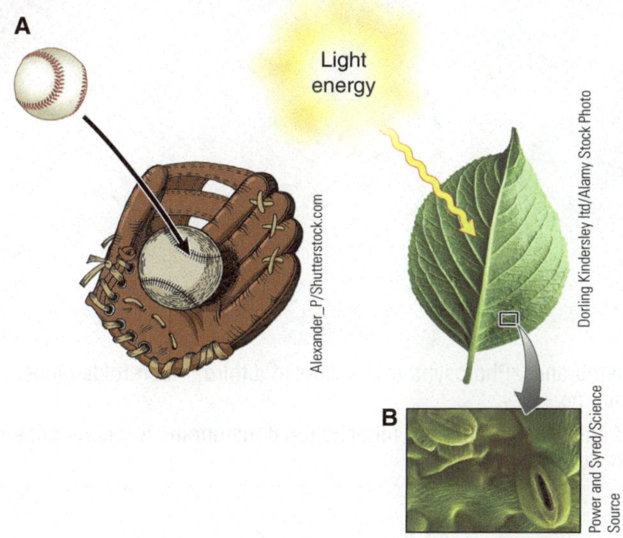

pores called **stomata** (singular, stoma). Stomata close to conserve water on hot, dry days, but are otherwise typically open during the day to allow CO_2 from the air to diffuse into the plant's tissues and O_2 to escape back into the air. Thus, stomata balance a plant's need for the gas exchange required for photosynthesis with its minimizing water loss.

Cells Contain Functional Compartments

Within the leaf's cells are two important intracellular compartments, or **organelles**. But why does a cell need organelles? Let's start by drawing an analogy between the organization of the cell and the setup of your home.

Regardless of whether you live in a dorm room, apartment, or house, your home is divided into areas based on activities done in each of those areas. For example, you likely store pots and pans in the kitchen. Your textbooks, computer, and notebooks probably live near your desk. Having an underlying organization facilitates function. When you want to cook something, it's easiest to have your pots and pans handy in the kitchen, rather than needing to retrieve them from your desk or bathtub.

Similarly, cells also have an underlying organization. Cells store enzymes and other molecules in particular places to use later. In Chapter 5, we will go into more depth exploring organelles and their specific functions. However, to understand photosynthesis and cellular respiration requires introducing two organelles—chloroplasts and mitochondria.

You can think of chloroplasts and mitochondria as like specific rooms in a house, each with a different purpose. Surrounding each of these organelles is a watery substance called the **cytosol**. Note that the **cytoplasm** includes the cytosol and anything suspended within. The cytosol is enclosed by a membrane made mainly of phospholipids (Chapter 3). We will discuss the membrane in detail in Chapter 5.

The **chloroplast** is an organelle, present in plant and algal cells plus some single-celled life, that is specialized for photosynthesis. Plant chloroplasts are oval or disk-shaped (**Figure 4.9A**). Each has two outer membranes enclosing a semifluid interior called **stroma**, which contains

Figure 4.9 **The chloroplast and mitochondrion are cellular organelles used for generating and harvesting energy**

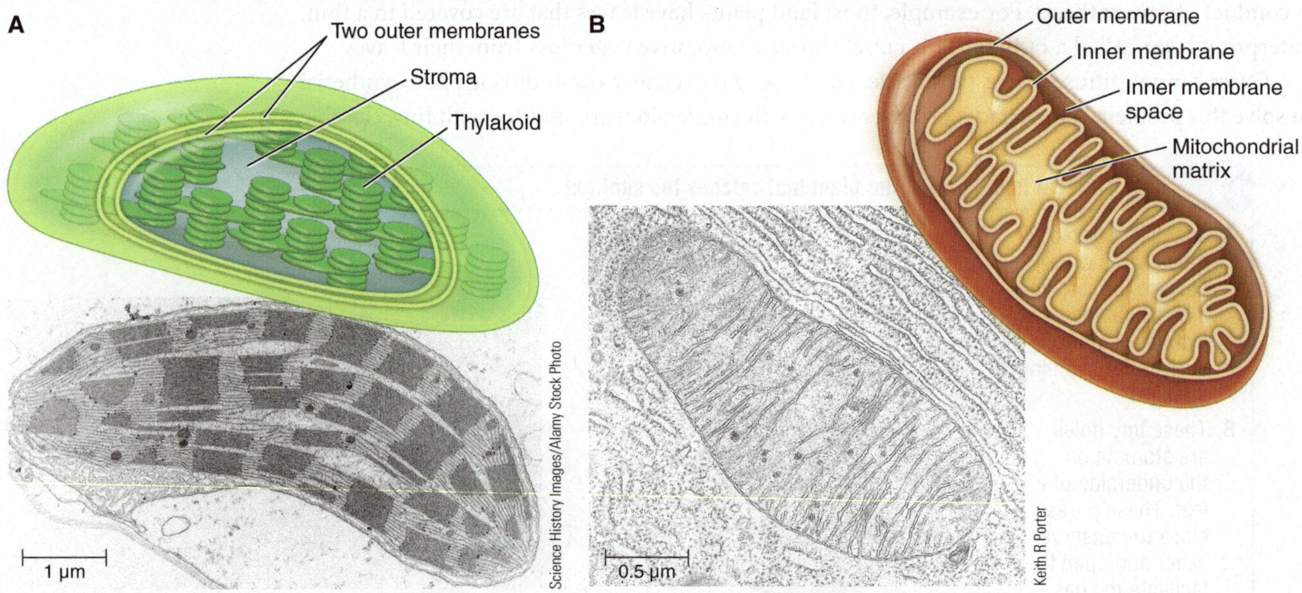

A. Plant chloroplasts have two outer membranes. Photosynthesis occurs at a third, much-folded inner membrane called the thylakoid membrane.

B. The mitochondrion has an inner and outer membrane. The mitochondrial membrane is where some of the reactions of cellular respiration occur.

enzymes and the chloroplast's DNA. In the stroma are stacks of disk-shaped structures called **thylakoids**. Since the thylakoids contain the photosynthetic pigments used for capturing light, this is also the location of photosynthesis.

Also within the leaf cell is a second organelle called the **mitochondrion** (plural, **mitochondria**). Almost all cells, except for bacterial cells and archaea, contain mitochondria. The mitochondrion is an organelle where two parts of cellular respiration occur (the rest occurs in the cytoplasm). Like a chloroplast, the mitochondrion is also roughly bean shaped, contains an inner and outer membrane, and has its own DNA (Figure 4.9B). In addition, the mitochondrion has an inner membrane space between its two membranes. The region within the inner membrane is called the mitochondrial matrix. Since mitochondria are where the majority of the cell's ATP is produced, you may have heard mitochondria referred to as the "powerhouse of the cell." Cells with a very high demand for energy, such as your muscle cells, typically have many mitochondria.

4.3 Summary

- Plant leaves have large surface areas to catch sunlight for photosynthesis. Their cuticles are waterproof to limit water loss. Leaves contain stomata, pores that allow CO_2 to enter and O_2 to leave.
- Chloroplasts are photosynthetic organelles.
- Cellular respiration occurs in two locations within the cell—the mitochondria and cytosol.

4.4 Conversion of Light into Chemical Energy

Learning Objectives

4-11 Explain the relationship between light wavelength and energy.

4-12 Explain why organisms use different pigments for photosynthesis.

4-13 Describe the conversion of light energy to chemical energy during photosynthesis.

4-14 Explain why oxygen forms during light-dependent reactions.

4-15 Identify the reactants and products of the light-dependent reactions.

4-16 Explain the process of carbon fixation in photosynthesis.

4-17 Explain how coenzymes connect the light-dependent reactions with the Calvin cycle.

4-18 Identify the reactants and products of the Calvin cycle.

4-19 Relate photosynthesis to climate change.

Have you ever wondered why an orange is orange? Or why a violet is purple? The different colors derive from different pigments found in the fruit of the orange and the flowers of a violet. Pigments are necessary for photosynthesis. To understand the variety of colors in plants and the connection to photosynthesis, we must start by considering the properties of light.

Light is electromagnetic radiation, a type of energy that moves through space in waves, like how waves move across an ocean. The distance between the crests (or troughs) of two successive

Figure 4.10 Visible light is a spectrum of color

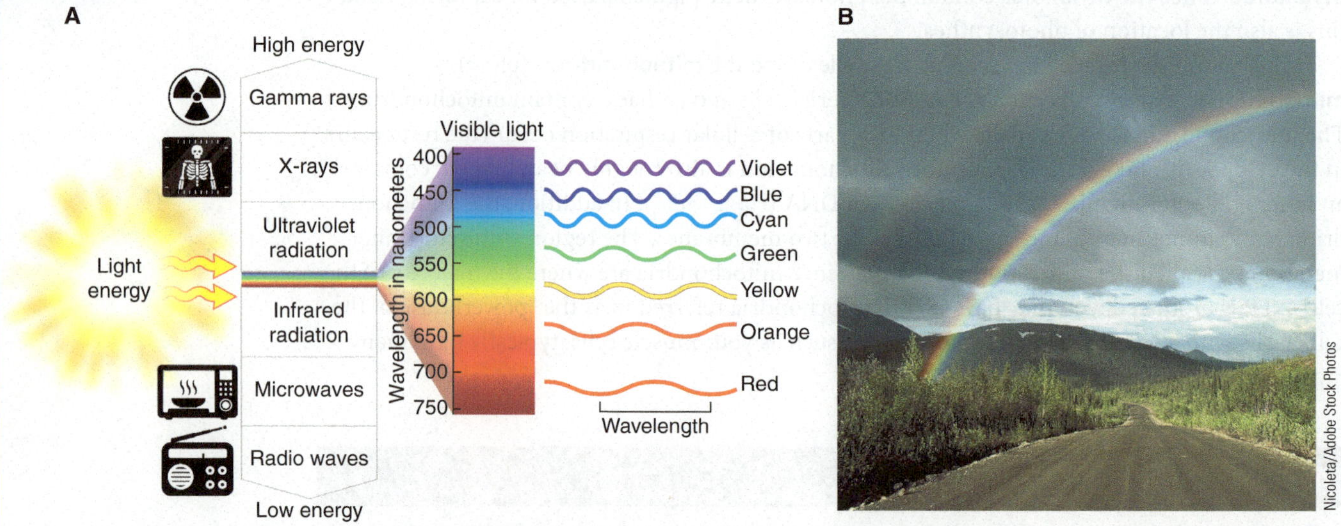

A. Electromagnetic radiation moves through space in waves that we measure in nanometers (nm). Visible light makes up a small part.

B. Raindrops can separate the different wavelengths of visible light, which we see as the different colors in a rainbow.

waves is called a **wavelength**, and it is measured in nanometers (nm), which is 10^{-9} meters (Figure 4.10A). Wavelength and energy are related: The shorter the wavelength, the higher the energy. All light of the same wavelength has the same amount of energy.

Light that is visible to the human eye is a small part of the spectrum of electromagnetic radiation emitted by the Sun. Visible light travels in wavelengths between 380 and 750 nm, and those are the main wavelengths of the Sun's energy that drive photosynthesis. Our eyes perceive wavelengths in this range as different colors; a combination of all these wavelengths appears white. White light separates into its component colors when it passes through a glass prism. A prism works by bending longer wavelengths more than it bends shorter ones, so a rainbow of colors forms when the light exits the prism. Raindrops can also act as tiny prisms, which is how rainbows form (Figure 4.10B).

? Did You Know?

Other animals can see more of the electromagnetic spectrum than we can. Some insects can see ultraviolet light, and snakes can see infrared radiation—light invisible to our eyes. The mantis shrimp can see light ranging from 300–740nm!

The Biology behind Fall Colors

If you live in a part of the world full of deciduous trees, you will be familiar with the transition every fall from an abundance of green leaves to the presence of yellow, orange, and red leaves. Where does the green color in leaves come from? Why do leaves change colors? And what does all of this have to do with photosynthesis?

The color we see comes from pigments. **Pigments** are organic molecules that selectively absorb light of specific wavelengths, a bit like antennas specialized for receiving light of certain colors. The light-trapping region of a pigment molecule has a special structure that allows electrons to move freely between its atoms. Only light with exactly enough energy to boost one of these electrons to a higher energy level (shell) is absorbed, which is why a pigment absorbs light of only certain wavelengths.

Whatever wavelengths of light that are not absorbed by a pigment are instead reflected. It's reflected light that gives each pigment its characteristic color. The main photosynthetic pigment in plants, algae, and photosynthetic microorganisms is **chlorophyll *a***. Chlorophyll *a* absorbs all wavelengths of visible light except green, which is reflected back so that this pigment appears

? Did You Know?

Humans have a pigment called melanin in their skin—and for dark-eyed individuals, also in their eyes. Melanin acts as a natural sunscreen. The more time you spend in the sun, the more melanin your skin produces—leading to a suntan.

Figure 4.11 **The diversity of plant colors derives from pigments**

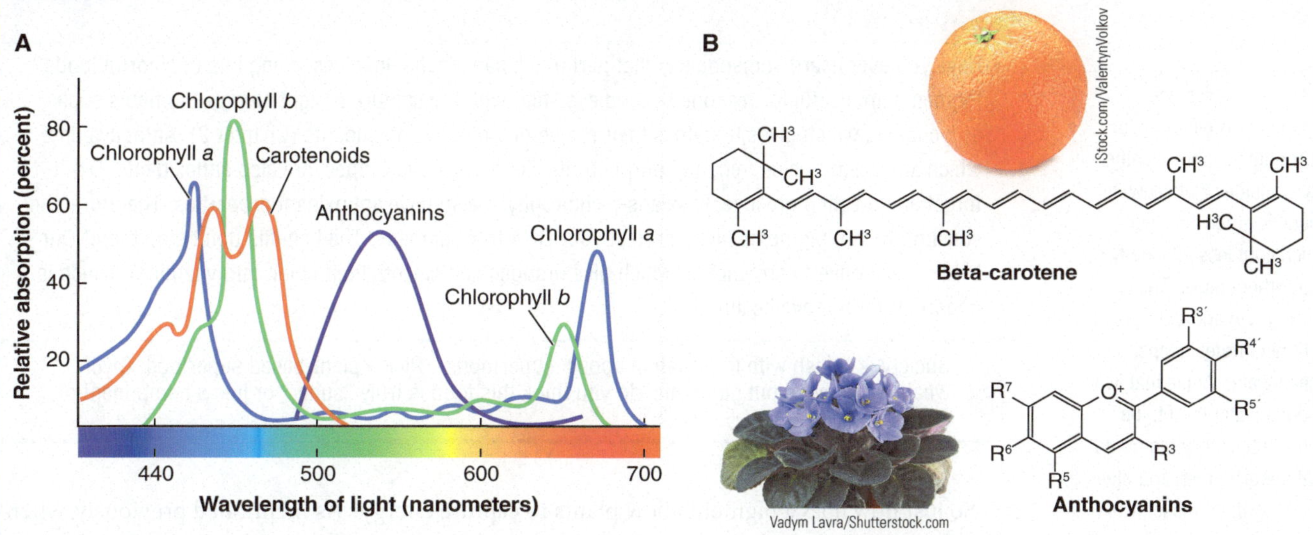

A. The curves in this graph show the efficiency with which a few photosynthetic pigments absorb different wavelengths of visible light. Using a combination of pigments maximizes the range of wavelengths that an organism can capture for photosynthesis.

B. Plants use pigments for reasons beyond photosynthesis. The bright colors of pigments such as beta-carotene (a carotenoid) and anthocyanins (shown here is the general structure of anthocyanins), can help attract pollinators to flowers and animals to ripening fruit, both of which facilitate plant reproduction.

green to us. This is why photosynthetic organisms and specific structures like leaves are green. In addition to chlorophyll *a*, most photosynthetic organisms have accessory pigments, including chlorophyll *b*, carotenoids, and anthocyanin, a class of pigments that range in color from red to purple to pink. (**Figure 4.11**).

Using accessory pigments expands the range of wavelengths organisms can capture for photosynthesis. These accessory proteins serve a variety of other functions. For example, anthocyanin, the pigment that contributes to a violet's blue color, serves as sunscreen for a plant. Appealing colors also attract pollinators to flowers and animals to ripening fruits, which facilitates plant reproduction.

In most plants, chlorophylls are usually so abundant that their green color masks the colors of the other pigments. Plants that change color during autumn are preparing for a period of dormancy; they conserve resources by moving nutrients from tender parts that would be damaged by winter cold (such as leaves) to protected parts (such as roots). Chlorophylls are not needed during dormancy, so they are disassembled and their components recycled. Yellow and orange accessory pigments are also recycled, but not as quickly as chlorophylls. Their colors begin to show as the chlorophyll content declines in leaves (**Figure 4.12**). In some plants, anthocyanin synthesis also increases just before dormancy, and this pigment adds red and purple tones to turning leaf colors.

Figure 4.12 **Accessory pigments become visible in fall**

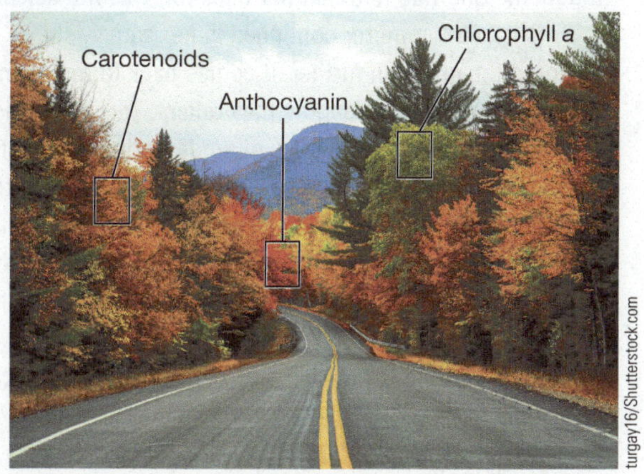

Chlorophyll *a*, the main photosynthetic pigment in most organisms, is green. Accessory pigments become visible in many plants as they prepare for dormancy in the fall.

Biology for Life

Eating in Color

Have you ever heard someone say that part of a balanced diet involves eating lots of colorful foods? Pigments are useful for reasons beyond assisting with the capture of light energy. Pigments such as lycopene, which gives tomatoes their red color, are also antioxidants (Chapter 2). Anthocyanins, which are found in blue, red, and purple fruits like berries and grapes, are also antioxidants. Don't forget about eating green leafy salads—chlorophylls also have antioxidant properties. Therefore, the consumption of pigments helps protect your DNA from damage. The benefits don't stop there! Our bodies can convert carotenoids, which give oranges and carrots their color, into vitamin A, which is important for our eye health.

Society is awash with information about "superfoods." Pick a plant-based superfood. Given what you know about pigments, do you think this food is truly "super" or just a component of a balanced diet?

Different species use different combinations of pigments for photosynthesis. Consider that seawater absorbs green and blue-green light less efficiently than other colors. Thus, more green and blue-green light penetrates deeper ocean water. Algae that live in deep water tend to be rich in accessory pigments that absorb green and blue-green light, so these organisms appear red to us.

So just how *does* a pigment allow plants to capture energy? As mentioned previously, when a pigment absorbs light energy, one of its electrons jumps to a higher energy level. The electron quickly drops back down to a lower shell by emitting its extra energy (Chapter 2). In the thylakoid membrane, energy emitted by a pigment's electron is not lost to the environment. In this special membrane, photosynthetic pigments occur in clusters held together by proteins. These clusters hold on to the emitted energy and are part of the *photosystems* that are essential for the first phase of photosynthesis, the light-dependent reactions.

Process of Science

Indigenous Wisdom and Scientific Practices

Robin Wall Kimmerer is a member of the Citizen Potawatomi Nation. She is also a botanist, or a biologist who studies plants (**Figure 4.13**). In her book, *Braiding Sweetgrass*, Kimmerer recounts a story from her freshman year of college where, when asked why she wanted to study botany, she explained that it was because she wanted to know "why asters and goldenrod looked so beautiful together." She was told that her question was not scientific.

Or is it? As Kimmerer continued in her career, she explored the question further. It comes back to the pigments in plants and the wavelengths they reflect. Turns out, the purple of asters and the yellow of goldenrod complement one another such that when the colors are placed next to each other, they create a strong contrast. The striking contrast makes these flowers stand out not only to us but also to pollinators, insects like bees and butterflies that facilitate plant reproduction by moving pollen between plants.

So, if we rephrase the question slightly to "Do asters and goldenrods grow together because they are more likely to get pollinated?" we see what some would consider a hypothesis. This is an example of how Kimmerer's Indigenous perspective, which focuses on the beauty perspective to create a testable hypothesis, ultimately providing new insight into the world. Kimmerer's story is an example of how cultural backgrounds can influence scientific inquiry and of the importance of including diverse perspectives in science.

Photo Courtesy of John D. and Catherine T. MacArthur Foundation

Figure 4.13 **Dr. Robin Wall Kimmerer**

Dr. Kimmerer combines her Indigenous background as a member of the Citizen Potawatomi Nation with her career as a botanist to ask questions that honor both.

What are some things that we can do to promote diverse perspectives in science?

Photosynthesis Act 1: The Light-dependent Reactions

Now we've set the scene for photosynthesis. We have a setting: the chloroplast. And we have part of our scenery: the pigment. Recall from the introduction to this section that photosynthesis has two parts: the light-dependent reactions and the Calvin cycle. We can think of photosynthesis occurring in two acts. In the first act, energy is harvested, and in the second act, the energy is stored as sugars (Figure 4.14).

Figure 4.14 Overview of photosynthesis

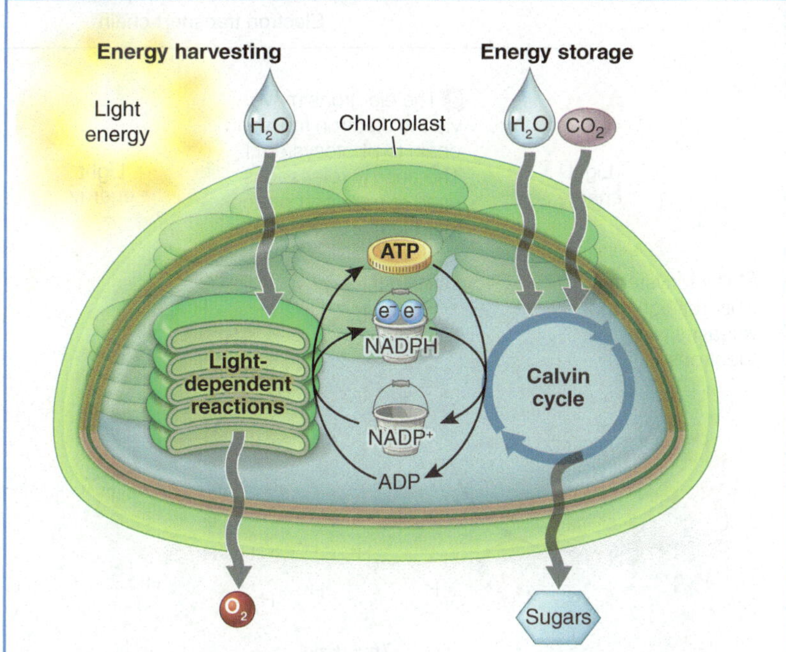

The entire process of photosynthesis has two parts, the light-dependent reactions and the Calvin cycle.

> **Point of Confusion:** The Calvin cycle is also referred to as the Calvin-Benson cycle, the dark reactions, the light-independent reactions, and the carbon reactions. Throughout the text, we will use the Calvin cycle.

Photosynthesis starts with the **light-dependent reactions**, which is when light energy is captured using a pigment molecule. What happens to the energy next? Coenzymes called nicotinamide adenine dinucleotide phosphate (NADPH) and nicotinamide adenine dinucleotide (NADH) are used to transport energy during these reactions. Later, in cellular respiration, there is an additional coenzyme called flavin adenine dinucleotide ($FADH_2$) that transports energy.

How do these molecules transport energy? Through redox reactions (Chapter 2)! The electron carriers—NADPH, NADH, and $FADH_2$—have two forms: a high-energy reduced form and a low-energy oxidized form. For example, $NADP^+$ is the oxidized form and NADPH is the reduced form. The reduced form is "full" of energy because it contains extra electrons. Like a bucket can be filled with water, these molecules fill with energy when they gain electrons and protons (hydrogen ions). Losing energy is like pouring water out of a bucket. Throughout the art in this chapter look for the  to represent energy-rich, reduced forms NADPH, NADH, and $FADH_2$ and the  to represent the energy-lacking, oxidized forms ($NADP^+$, NAD^+, FAD).

Essentially, the light-dependent reactions capture energy from the Sun and use it to make ATP, and reduce, or "fill up," $NADP^+$ with energy so that it becomes NADPH. How does this happen?

Part 1: Sunlight excites electrons in photosystem II. Within the thylakoid membrane are two photosystems, type I and type II (named in order of their discovery). Each **photosystem**, or complex of proteins that serve as the functional unit for photosynthesis, contains pigments for catching light energy (green dots, Figure 4.15). First, **photosystem II** absorbs light energy (1), then releases electrons that travel through an **electron transport chain** to photosystem I (3). An electron transport chain is where a series of redox reactions (Chapter 2) occur. Within the electron transport chain are various proteins that pass energy to one another, much like people could pass buckets to one another to transport water or sand. The first electron transport chain uses the released energy to move protons into the thylakoid space. These protons will be used to fuel **ATP synthase**, an enzyme that produces ATP. But, now we are missing electrons in photosystem II—how are they replaced? By breaking apart water (2). This replaces the missing electrons in photosystem II and produces oxygen.

> **Study Tip**
>
> Oxidation is the *loss* of electrons and protons, while *reduction* is the gain of electrons and protons. You can use a mnemonic like OIL RIG to help you remember. **O**xidation **I**s **L**oss and **R**eduction **I**s **G**ain.

Part 2: Sunlight excites electrons in photosystem I. Next, in **photosystem I**, light energy is absorbed and excites electrons (4). These high-energy electrons are also passed down

Figure 4.15 The light-dependent reactions of photosynthesis

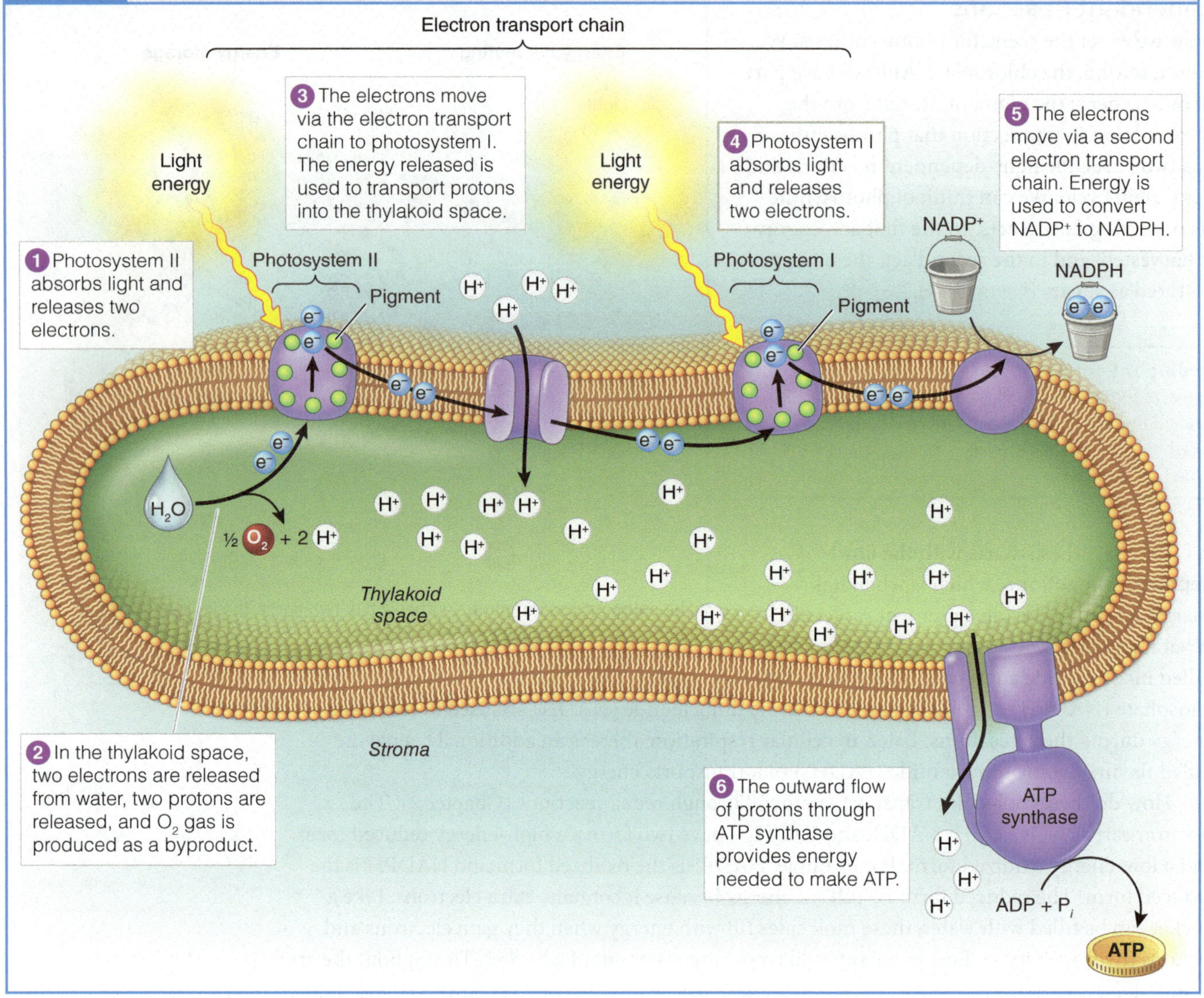

an electron transport chain 5. However, instead of fueling the movement of protons into the thylakoid space, the energy of these electrons is used to convert $NADP^+$ to NADPH, refilling this electron carrier with electrons as the final step of the light reactions—electrons that can then be used in photosystem II, starting the process over again.

The chemistry behind how solar panels generate electricity is similar to how energy is harvested during the light reactions. Light hits the panel causing electrons to move, which creates an electric current.

Remember the protons that were moved into the thylakoid space during the first electron transport chain? Since there are more protons on one side of the membrane than the other, the gradient created is a source of potential energy. As these protons flow back out through ATP synthase, an enzyme that makes ATP, it uses that energy to create ATP 6.

No need for Carla just yet! They'll make their debut in the Calvin cycle.

Carla

During the light reactions, energy from sunlight is captured by pigments and then stored in two high-energy molecules: NADPH and ATP. This crucial step is how energy from the Sun is captured for use by life on Earth.

Eddie

Photosynthesis Act 2: Making Sugars in the Calvin Cycle

In the last section, we learned that the first stage of photosynthesis is harvesting energy from the Sun through two different electron transport chains. As a result, high-energy molecules are produced. In the second stage of photosynthesis, the **Calvin cycle**, the energy in ATP and NADPH is converted and stored as glucose (Figure 4.16).

Since these reactions do not require sunlight, glucose can be produced night and day. This series of reactions occurs as a continuous cycle, where ATP, NADPH, and carbon atoms from CO_2 are used to build the carbon backbones of sugar molecules. Extracting carbon atoms from an inorganic source (such as CO_2) and incorporating them into an organic molecule is called **carbon fixation**. This entire process takes place in the stroma of chloroplasts.

Part 3: Carbon fixation. The Calvin cycle begins when the enzyme **rubisco** attaches CO_2 to a five-carbon organic compound called ribulose bisphosphate, or RuBP (Figure 4.16), 1. After rubisco combines three CO_2 molecules with three RuBP molecules, three six-carbon intermediates are formed.

Part 4: Energy storage. Each of the six-carbon molecules formed by rubisco is unstable and immediately splits into two three-carbon molecules called PGA 2. After the unstable intermediate splits, there will be a total of 6 PGA molecules. The NADPH and ATP generated in the light-dependent reactions are used to convert the PGA molecules to PGAL, a three-carbon sugar with an attached phosphate functional group. At this point, one molecule of PGAL exits the cycle 3. Other enzymatic processes are used to combine multiple molecules of PGAL to make glucose.

? Did You Know?

Some herbicides kill plants by blocking the electron transport chain. Without the ability to capture energy from the sun, the plant dies.

? Did You Know?

Rubisco is the most abundant enzyme on Earth. A good thing too—without the ability to fix carbon, there would be no life on Earth.

Figure 4.16 The Calvin cycle is how carbon fixation occurs

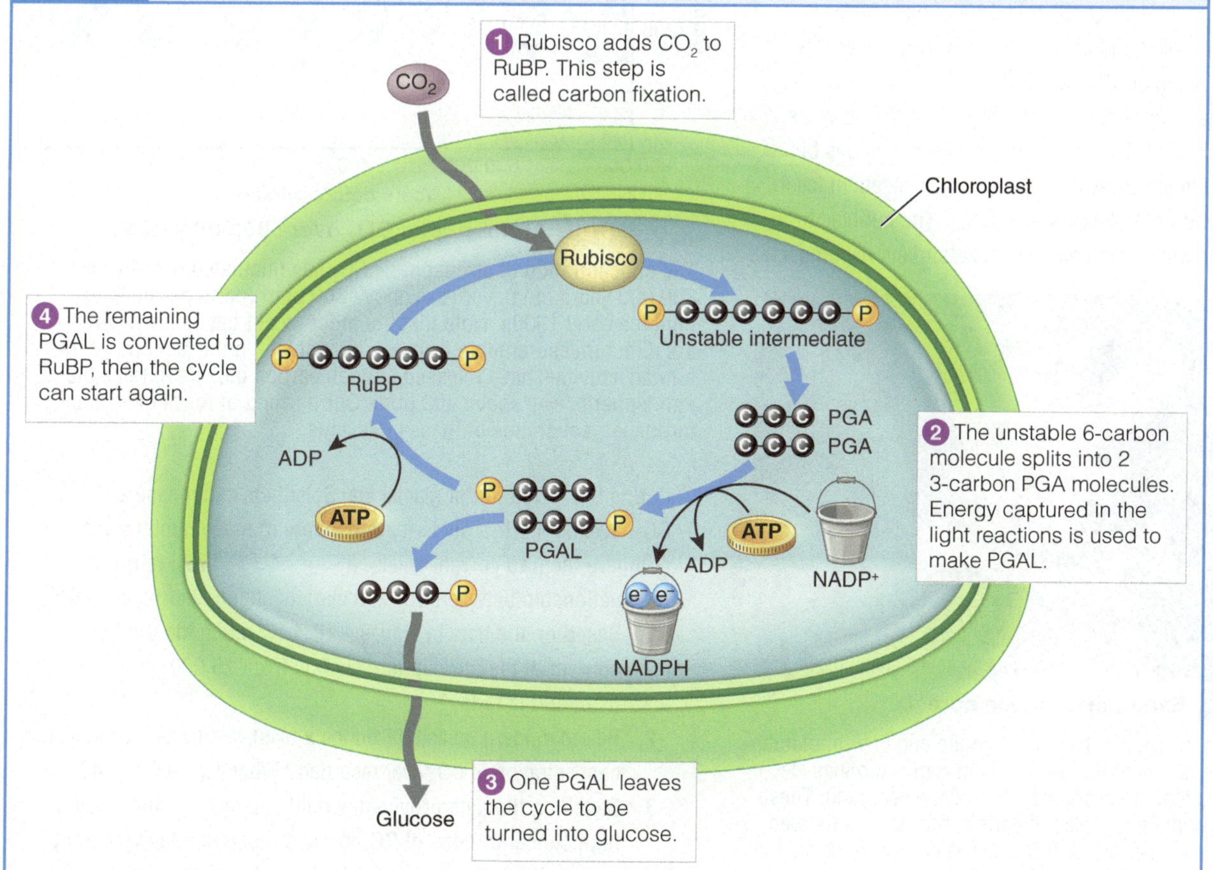

Part 5: Cycle completion.

Only one molecule of PGAL exits the cycle, leaving five molecules behind. Using ATP, the cell then regenerates the RuBP (4), and the cycle is ready to repeat.

Welcome, Carla! Carla entered the cycle as CO_2 gas. Now, Carla's carbon atom is on its way to getting turned into glucose.

Carla

High energy molecules from the light reactions are used in this step to make PGAL, which leaves the cycle to become glucose or is used to regenerate RuBP.

Eddie

Digging into Data

A Burning Concern

Your body contains about 9.5% carbon atoms by weight. Those carbon atoms were originally in CO_2, a gas in Earth's atmosphere. Plants use CO_2 to make sugars in photosynthesis. When you consume plants and other organisms, that carbon becomes part of your body. CO_2 is then released back into the atmosphere when sugars are used for energy during cellular respiration.

About 8,000 years ago, humans began burning forests for agriculture, and since the mid-1800s, have burned fossil fuels for heat and power. These activities have created new sources of CO_2 gas. Today, the atmospheric CO_2 level is higher than it has been for at least 800,000 years. How do we know? Scientists study CO_2 levels in gas bubbles trapped in ice cores drilled from thick sheets of ice in Greenland and Antarctica (Figure 4.17). The bubbles form when snow, which contains tiny pockets of air, is compacted during the formation of glacial ice. Scientists can measure the CO_2 levels of Earth's ancient atmosphere in these bubbles and compare them to current levels. (Figure 4.18). We'll return to the relationship between CO_2 and climate change in Chapter 16.

Figure 4.17 Example of an ice core

In this photo, a scientist from the Climate and Environmental Physics department at the University of Bern examines ice core samples that are more than 1.2 million years old. These ice samples contain samples of Earth's atmosphere trapped in the ice at the time the ice formed. The deeper in the ice the sample is from, the older the air in the bubbles.

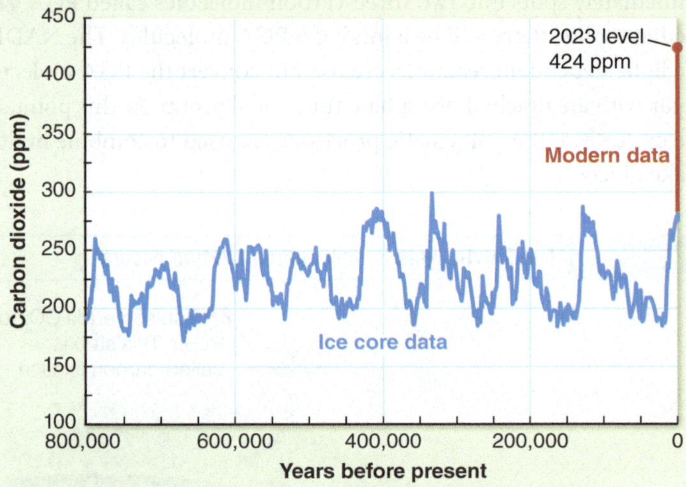

Figure 4.18 Atmospheric CO_2 over 800,000 years

The concentration of atmospheric CO_2 has fluctuated over the last 800,000 years but never rose above 300 ppm (parts per million) until the early 1900s. Note that the most recent data shown here are direct measurements from the air, not ice. In the last 150 years, human activities have released enough carbon dioxide to raise the atmospheric level above 400 ppm. Our burning of fossil fuels is by far the largest source of these emissions.

1. Based on the data in Figure 4.18, how much did the CO_2 content of the atmosphere change from 25,000 years ago to what it is today?
2. Before modern times ("0" on the *x*-axis), what was the highest concentration of CO_2 ever recorded? What was the lowest?
3. Based on the data provided, would you say that the current, high concentrations of CO_2 in the atmosphere fall within the normal fluctuations you see in the historic record?

The Big Picture: Photosynthesis and Climate Change

What's the big deal with CO_2? CO_2 is a greenhouse gas. A **greenhouse gas** is a gas that, as a component of the atmosphere, absorbs or radiates heat, causing Earth to warm up. Recall that rubisco is an enzyme that pulls CO_2 out of the atmosphere and attaches it to other molecules so that it can eventually be converted to other forms. Therefore, the mass of a plant comes not from nutrients in the soil, but from carbon captured from the atmosphere. Since plants actively remove CO_2 from the atmosphere, they help mitigate climate change.

Cutting down trees and burning them is particularly damaging to CO_2 levels in the atmosphere not only because of the loss of organisms that are removing CO_2 from the atmosphere, but also because the carbon stored in the trees is released into the atmosphere. Excessive CO_2 levels not only contribute to a warming planet but also have other environmental impacts, which we will return to in Unit 5.

Although there are many ways to mitigate climate change, our new understanding of photosynthesis highlights the importance of plants. By prioritizing natural areas and planting more trees, we can remove more CO_2 from the atmosphere and therefore help to reduce the effects of climate change.

? Did You Know?

Fossil fuels get their name because they are generated from the remains of plants and animals that lived millions of years ago. Burning these fuels releases carbon originally stored in these organisms as CO_2 gas.

4.4 Summary

- Light is electromagnetic radiation that travels as waves. The shorter the wavelength, the higher the energy of the light.
- Pigments selectively absorb specific wavelengths of light. Photosynthetic organisms use different pigments to absorb light of different wavelengths, allowing them to absorb energy from a larger part of the spectrum.
- Plants appear green because they contain Chlorophyll a, which reflects green light.
- The light-dependent reactions convert light energy into chemical energy. Photosystems I and II absorb light energy, releasing electrons. The electrons move through the electron transport chain, combining with $NADP^+$ and H^+ to produce NADPH. ATP is produced by ATP synthase as protons flow through the membrane.
- Water is broken apart to replace the electrons released from photosystem II. The remaining oxygen atoms combine to form O_2.
- Water, $NADP^+$, and ADP are the reactants of the light-dependent reactions. O_2, NADPH, and ATP are the products.
- Carbon fixation occurs when carbon is incorporated from an inorganic source into an organic molecule. In the Calvin cycle of photosynthesis, CO_2 supplies the carbon used to build sugars, starting with rubisco attaching CO_2 to RuBP.
- NADPH is the coenzyme that connects the light-dependent reactions with the Calvin cycle. $NADP^+$ collects electrons, forming NADPH, which, with ATP, is used to drive sugar production in the light-independent reactions.
- CO_2, ATP, and NADPH enter the Calvin cycle as reactants. ADP, $NADP^+$, and three-carbon sugars are the products.
- CO_2 is a greenhouse gas. Since plants actively remove CO_2 from the atmosphere, maintaining green spaces and prioritizing forests is one strategy for reducing the effects of climate change.

4.5 Aerobic Respiration: Burning Sugars in the Presence of Oxygen

Learning Objectives

4-20 Explain why aerobic respiration requires O_2 and releases CO_2.

4-21 Identify the reactants and products of glycolysis.

4-22 Identify the reactants and products of the citric acid cycle.

4-23 Identify the reactants and products of the electron transfer chain.

4-24 Describe the movement of energy during acetyl-CoA formation and the citric acid cycle.

4-25 Describe how ATP forms in the last stage of aerobic respiration.

4-26 Explain how oxidizing a macromolecule can fuel ATP production.

In the previous section, we learned how autotrophs like plants can take inorganic molecules, such as CO_2 and H_2O, and energy from the Sun to generate glucose, a sugar. The next stage is to break down the sugar to get the stored energy. The process of breaking down glucose to generate energy is called **cellular respiration**, which mainly occurs in mitochondria (Figure 4.19). Energy can be made in the presence of oxygen through a process called **aerobic respiration** (this section), or when there is no oxygen, through **fermentation** (Section 4.6).

Cellular Refueling

Study Tip

Aerobics are a type of exercise that gets you out of breath—you need oxygen! So remember: aerobic respiration requires oxygen.

Aerobic respiration and fermentation start with the same set of reactions, called **glycolysis**. Glycolysis (*glyco* = sugar and *lysis* = break) occurs in the cytoplasm and is the process of breaking apart glucose into two three-carbon molecules called **pyruvate**. From here, aerobic respiration and fermentation differ. In aerobic respiration, pyruvate moves into the mitochondria, where it is converted to acetyl-CoA before entering the citric acid cycle and then the electron transfer chain. In contrast, during fermentation, which occurs when no oxygen is present, pyruvate stays in the cytoplasm. Ultimately, both cellular respiration and fermentation produce ATP, although far more ATP is produced during cellular respiration.

Figure 4.19 Overview of cellular respiration

Both cellular respiration and fermentation start in the cytoplasm with glycolysis. In the presence of oxygen, aerobic respiration continues in the mitochondria with the citric acid cycle and the electron transport chain. If there is no oxygen, fermentation proceeds in the cytoplasm.

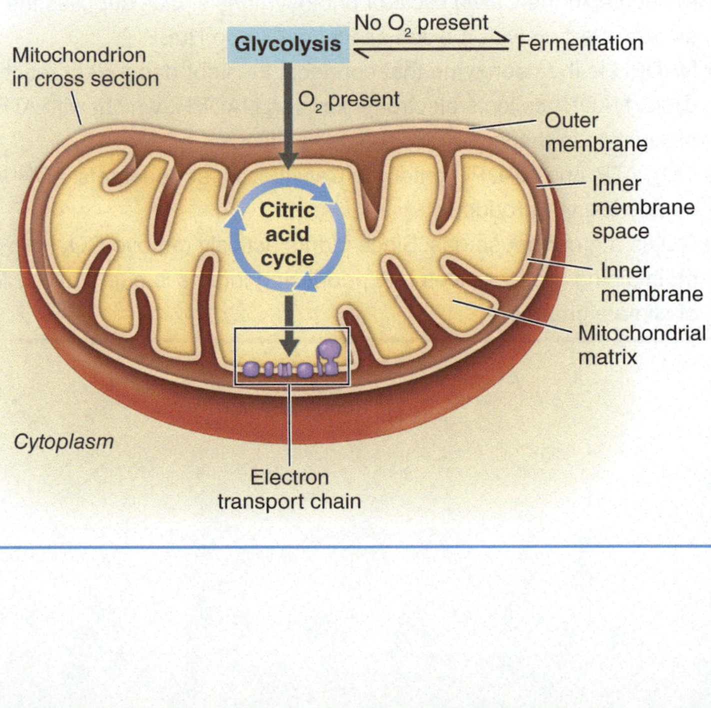

! Point of Confusion: Although we talk about how plants (autotrophs) get their energy through photosynthesis and how we (heterotrophs) get our energy through cellular respiration, plants also have mitochondria and get ATP via cellular respiration. For example, potato plants access the glucose stored in starch molecules (Chapter 3) for energy.

Part 1: Breaking down glucose. Glycolysis requires many steps and enzymes, but for our purposes we will only highlight a few key steps (Figure 4.20). Note that we can divide glycolysis into two phases. The first phase has two reactions that use ATP. This phase is sometimes called the energy-requiring phase. The second phase involves creating high-energy products, including NADH and ATP. This phase is sometimes called the energy-harvesting phase.

Figure 4.20 Glycolysis is the process of breaking down glucose into pyruvate

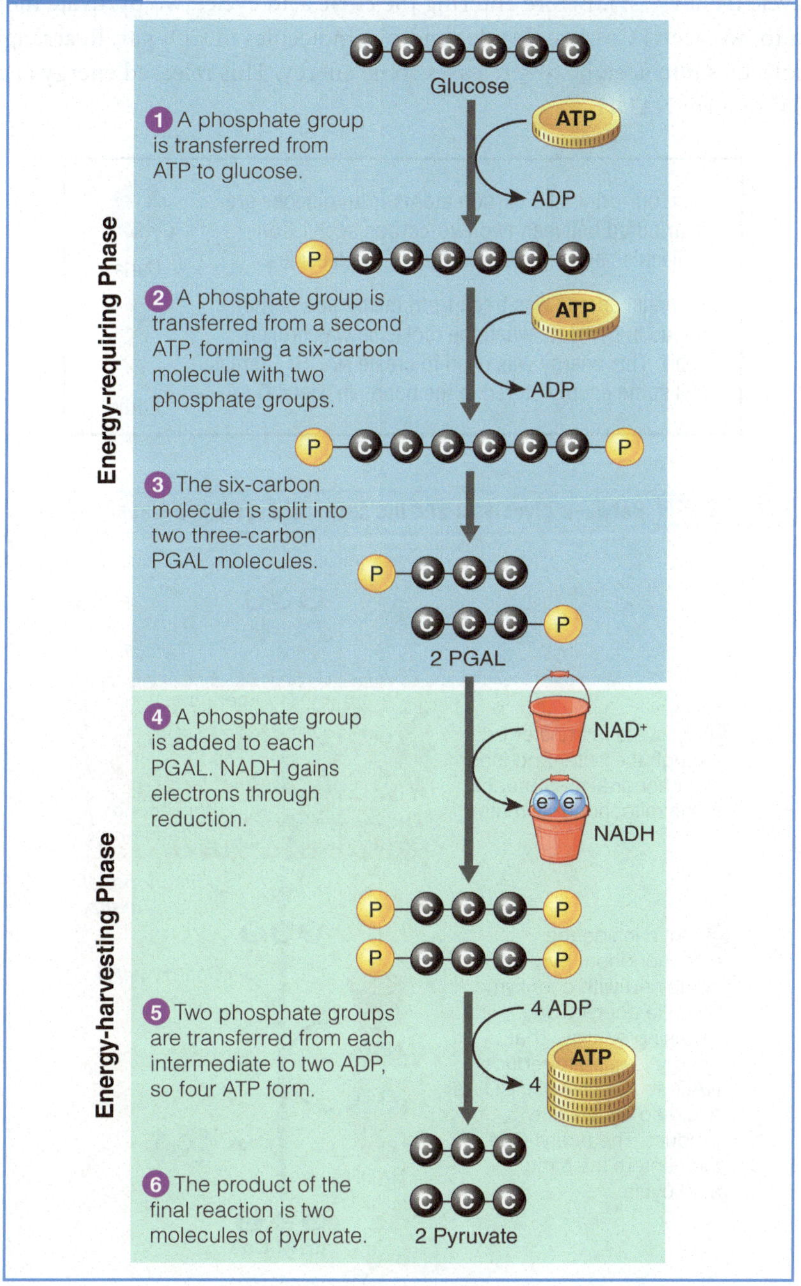

The six carbon atoms in glucose are now rearranged into two pyruvates, each with three carbons.

Carla

Glucose is full of potential energy. Only a small amount of this potential energy is collected. The rest of the energy is stored in the two pyruvate molecules. These pyruvate molecules will either enter the mitochondria and continue with cellular respiration or stay in the cytoplasm for fermentation (Section 4.6).

Eddie

Part 2: Transition between glycolysis and citric acid cycle. The transition from glycolysis to the citric acid cycle is not only a change to a different process, but also a change in cellular location—from the cytoplasm for glycolysis and into the mitochondrion for the citric acid cycle (Figure 4.21). Before entering the citric acid cycle, two pyruvate molecules are converted into two acetyl Co-A molecules and two molecules of CO_2 gas. Rearranging the pyruvate molecules into acetyl Co-A releases some energy. This released energy is used to reduce NAD^+ to form NADH.

You exhale CO_2 molecules each time you breathe. During photosynthesis, these same CO_2 molecules are removed from the air by plants and recycled to make sugar.

Now our original 6 carbon atoms from glucose are distributed between two two-carbon acetyl-CoA molecules and two one-carbon CO_2 molecules.

Carla

Potential energy was freed from the carbon-carbon bonds in pyruvate when the molecule was broken apart. This energy was used to create NADH. There is still some energy stored in the bonds in acetyl Co-A.

Eddie

Figure 4.21 **Between glycolysis and the citric acid cycle is a transition step**

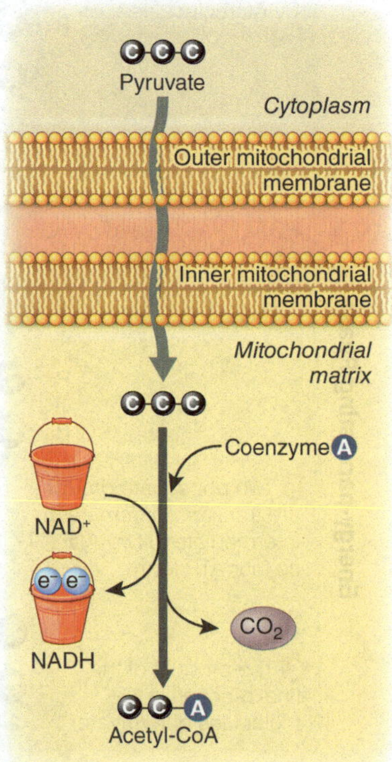

1 Pyruvate is transferred through the outer and inner mitochondrial membranes to the mitochondrial matrix.

2 Once inside the mitochondrion, pyruvate is combined with coenzyme A to make acetyl-CoA, releasing energy. That energy is used to reduce NAD^+ to form NADH; CO_2 is released as a waste product. The acetyl-CoA then enters the citric acid cycle.

Figure 4.22 **In the citric acid cycle, the conversion of molecules results in the release of carbon dioxide gas and the generation of high-energy molecules**

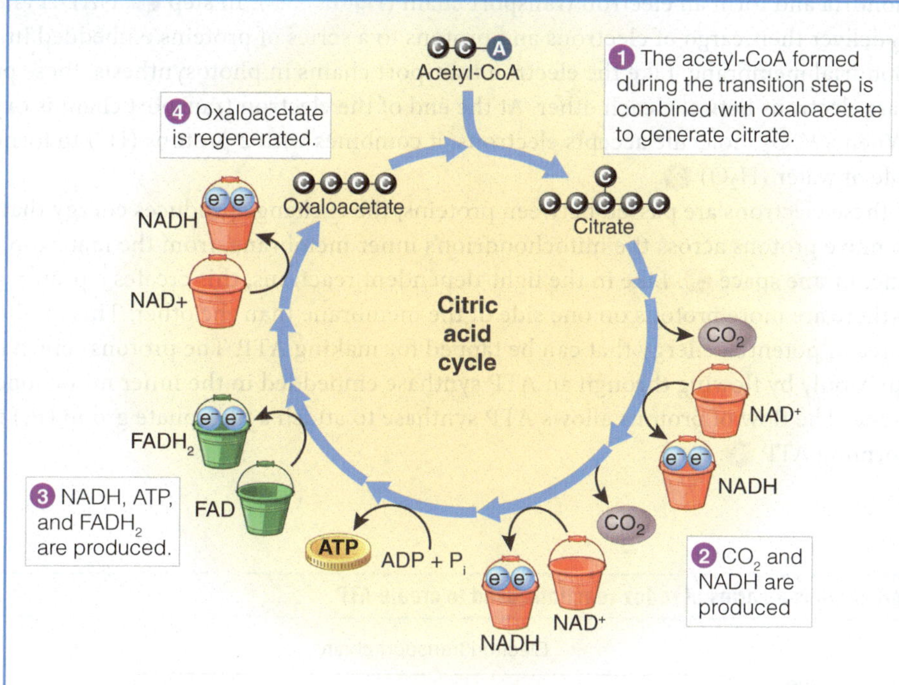

Part 3: The citric acid cycle dance. You may be familiar with traditional dances like the hora in which people hold hands, form a circle, and move together. Other individuals can enter or exit the circle by holding or releasing each other's hands. Similarly, the **citric acid cycle** also proceeds in a circle, with molecules entering and exiting. The citric acid cycle (Figure 4.22) is a cyclic pathway that releases energy by breaking apart acetyl-CoA. One molecule of acetyl-CoA enters the cycle at a time. Going back to our original glucose molecule, we can say that the cycle turns twice for each glucose molecule (one glucose molecule generates two acetyl Co-A). Like in glycolysis, the citric acid cycle involves many enzymes and intermediates. Figure 4.22 gives an overview and highlights the most important molecules for you to know.

? Did You Know?

You can buy NADH tablets at the supermarket. Although the research is limited, some scientists have found evidence that NADH can mitigate the effects of jet lag.

! Point of Confusion: The citric acid cycle is also called the TCA cycle for short, as well as the Krebs cycle.

As in all cyclic pathways, the first step in the cycle takes the starting material from the last turn of the cycle (oxaloacetate) and combines it with a new molecule entering the cycle. In this case, one four-carbon oxaloacetate molecule combines with one two-carbon acetyl Co-A molecule to form a six-carbon molecule called citrate 1. Two carbons are then plucked from citrate, one at a time. The leftover carbon eventually forms a molecule of CO_2 2. The energy released when these carbons are removed from citrate is used to reduce NAD^+, forming NADH. The next two steps 3 and 4 lead to the regeneration of oxaloacetate and the generation of high-energy molecules including $FADH_2$, NADH, and ATP.

Of the four carbon atoms that enter the cycle as acetyl Co-A, all will eventually be converted to CO_2 and ultimately exhaled from the body. From here, we say goodbye to Carla!

Carla

Potential energy was freed from the carbon-carbon bonds in citrate when the molecule was broken apart to form two molecules of CO_2. This energy was used to create NADH. During other chemical conversions in this cycle, energy is stored as ATP, $FADH_2$, and NADH. Overall, the citric acid cycle converts the energy from one molecule of glucose into 2 ATP, 6 NADH, and 2 $FADH_2$.

Eddie

In Chicago in 1982, seven people died after consuming cyanide-laced Tylenol®. Cyanide blocks oxygen from accepting electrons at the end of the electron transport chain, effectively shutting down the process and preventing ATP from forming. These murders led to changes in how medications are packaged to prevent tampering.

Part 4: Bouncing electrons and the electron transport chain. The last stage of aerobic respiration is a series of redox reactions that occur within the inner membrane of mitochondria and form an electron transport chain (Figure 4.23). In step 1, NADH and $FADH_2$ deliver their cargo of electrons and protons to a series of proteins embedded in the mitochondrial membrane. Like the electron transport chains in photosynthesis, these proteins pass electrons between each other. At the end of the electron transport chain is oxygen (O_2). When a ½ O_2 molecule accepts electrons, it combines with 2 protons (H^+) to form 1 molecule of water (H_2O) 2.

As these electrons are passed between proteins, the exchange produces energy that is used to move protons across the mitochondrion's inner membrane, from the matrix to the inner membrane space 3. Like in the light-dependent reactions, this creates a proton gradient—there are more protons on one side of the membrane than the other. This gradient is a source of potential energy that can be tapped for making ATP. The protons return to the matrix only by flowing through an ATP synthase embedded in the inner mitochondrial membrane. The flow of protons allows ATP synthase to attach a phosphate group (P_i) to ADP, forming ATP 4.

Figure 4.23 The electron transport chain is a series of redox reactions used to create ATP

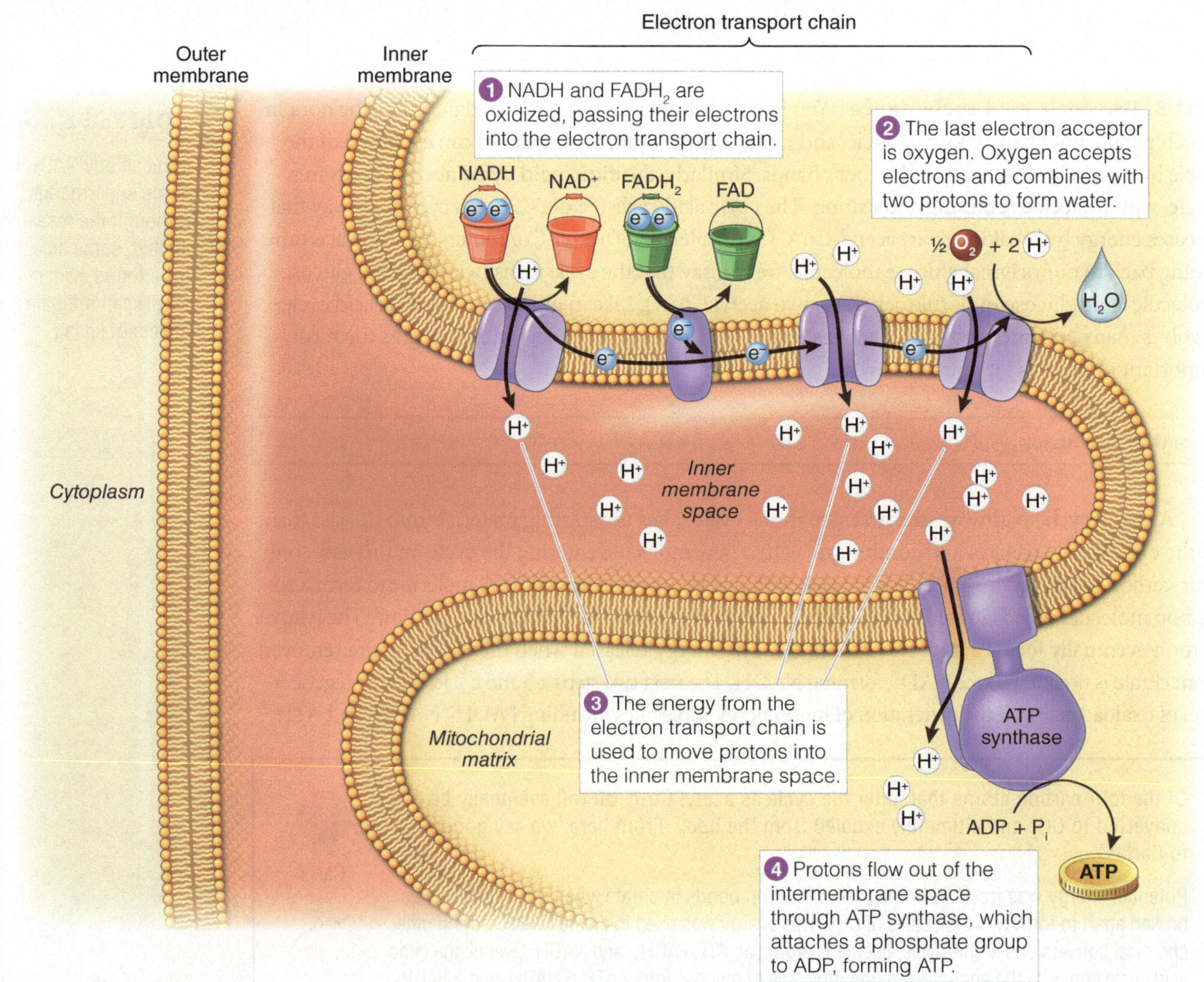

Recall that the inner mitochondrial membrane is folded. Represented in this image is one of the folds relative to the outer membrane.

From the NADH and $FADH_2$ produced in the citric acid cycle, we now have 38 ATP molecules.

Eddie

! Point of Confusion: Thirty-eight or 36 molecules of ATP? Note that NADH is used for transporting molecules. The amount of NADH used is equivalent to approximately 2 molecules of ATP; therefore, our theoretical grand total of energy produced from one molecule of glucose is 36 ATP.

? Did You Know?

Energy is transferred many times during cellular respiration, and with each transfer some energy is lost as heat, creating what is called "metabolic heat." Interestingly, metabolic heat generated in plant mitochondria allows the eastern skunk cabbage, which blooms in February in North America, to maintain a temperature as much as 15–35°C (27–63°F) above air temperature.

Making Connections: Obtaining Energy from Diet

In Chapter 3 we learned about macromolecules—the proteins, carbohydrates, and lipids (fats)—that make up our diet. In this chapter, we learn how sugars are turned into energy during cellular respiration, and how complex carbohydrates such as starch are broken down and utilized during fermentation.

But what about the lipids and proteins that we eat? How are they used to create the ATP our bodies need? In the same way that our cells use glucose with aerobic respiration, cells can break down macromolecules into molecules that can enter the reactions of aerobic respiration at various points (Figure 4.24). Remember the circular dance analogy? Proteins and lipids can be converted to new dance "partners" that can join various metabolic pathways at different points, all leading to the production of ATP.

Breakdown of proteins. Enzymes in the digestive system split dietary proteins into their amino acid subunits, which are absorbed into the bloodstream. Cells use these free amino acids to build proteins or other molecules. When you consume more protein than your body needs, the excess amino acids are broken down. The amino (—NH_2) group is removed, and the carbon backbone is split into fragments. The removed amino group is converted to urea, a waste product voided in urine. The carbon-containing fragments are converted to pyruvate, acetyl-CoA, or an intermediate of the citric acid cycle. These molecules enter aerobic respiration at the acetyl-CoA formation stage or the citric acid cycle.

Breakdown of carbohydrates. The digestive systems of humans and other mammals break down many polysaccharides into monosaccharides. Sucrose, or common table sugar, is a disaccharide that consists of glucose and fructose monomers. A digestive enzyme breaks the bond between the two monomers, releasing the monosaccharides. Glucose and fructose both enter glycolysis at different steps for conversion to pyruvate.

Breakdown of lipids. A triglyceride molecule has three fatty acid tails attached to a glycerol head (Chapter 3). Cells dismantle triglycerides by first breaking the bonds that connect the tails to the head. Fatty acids are converted to acetyl-CoA, which can enter the citric acid cycle. Enzymes in liver cells convert the glycerol to PGAL, which is then fed into glycolysis.

Figure 4.24 Proteins, carbohydrates, and lipids can be broken down into components that can enter directly into glycolysis or the citric acid cycle, or be converted to acetyl-CoA.

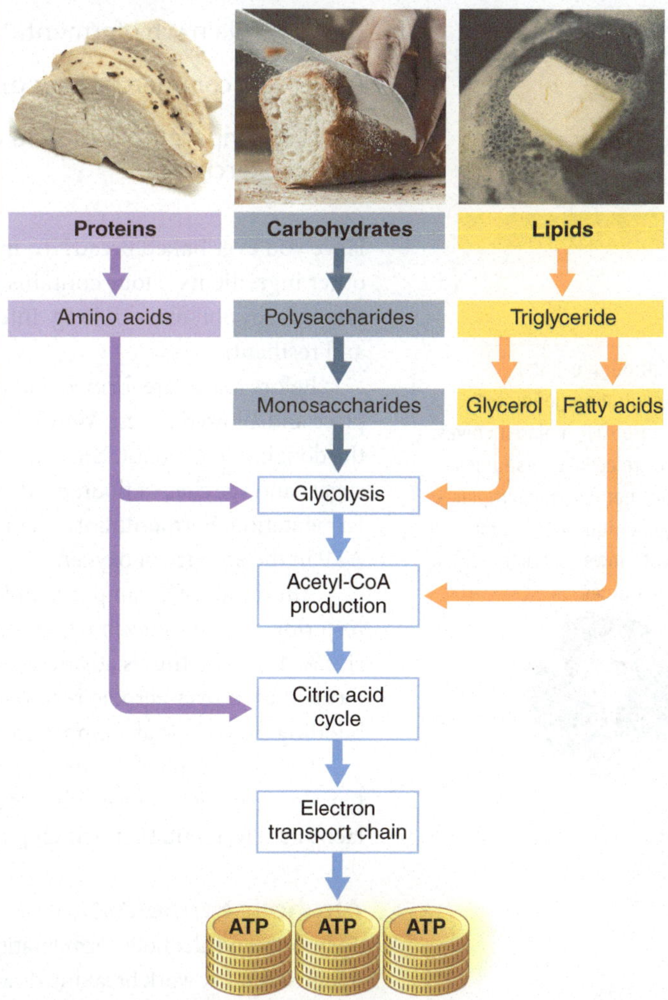

Note that each macromolecule can be broken into different smaller molecules that can enter into cellular respiration at various points in the process.

Media Whale Stock/Shutterstock.com; Master1305/Shutterstock.com; john shepherd/Getty Images

4.5 Summary

- Aerobic respiration occurs in the presence of oxygen, breaking down glucose to form pyruvate. By the end of the process, six CO_2 molecules are released, one for each carbon in glucose.
- Glucose, 2 NAD^+, and 2 ADP molecules enter glycolysis as reactants. 2 ATP, 2 NADH, and 2 three-carbon pyruvates exit as products.
- Two acetyl-CoA molecules, 3 NAD^+, 1 $FADH^+$, and ADP enter the citric acid cycle as reactants. ATP, 3 NADH, 1 $FADH_2$, and 2 CO_2 leave the citric acid cycle as products.
- High energy molecules from glycolysis and the citric acid cycle are used in the electron transport chain to fuel production of ATP.
- Cells break down fats, proteins, and carbohydrates into molecules that can be used in various parts of aerobic respiration.

4.6 Fermentation: Burning Sugars in the Absence of Oxygen

Learning Objectives

4-27 Describe how ATP is generated during fermentation.

4-28 Explain why fermentation requires no oxygen.

4-29 List commercial uses of alcoholic and lactic acid fermentation.

4-30 Compare how muscle cells use aerobic respiration and fermentation during exercise.

Have you ever baked bread? To make bread, flour is kneaded with water, yeast, and sometimes other ingredients. Flour contains starch and a protein called gluten. Kneading causes the gluten to form polymers in long, interconnected strands that make the resulting dough stretchy and resilient.

Before loaves are formed and put into the oven, the stretchy dough is put aside in a warm place and allowed to rest. Why? Resting gives the dough time to puff up and "rise." What causes the dough to rise? Yeast. Yeast are rapidly dividing single-celled microorganisms that digest the sugars and starches in flour, producing the CO_2 gas that causes bread to rise in a process called fermentation. **Fermentation** occurs when high-energy molecules are broken down to produce ATP in the absence of oxygen.

Almost all cells can perform fermentation. Fermentation is anaerobic, but aerobically respiring cells use it even when oxygen is present. Fermentation's fast, less resource intensive ATP production is advantageous in some circumstances. For example, by prioritizing fermentation over aerobic respiration, fast-growing cells can devote more of their resources to building proteins that support growth.

Fermentation occurs in the cytoplasm and always begins with glycolysis. There are two types of fermentation we will discuss: alcoholic fermentation, which produces ethanol; and lactic acid fermentation, which produces lactic acid.

Alcoholic Fermentation

To understand alcoholic fermentation, let's revisit our resting bread dough. Recall that the yeast in the dough is hard at work breaking down starch into digestible bits, which in this case are starch's glucose subunits. These glucose molecules are the starting material for alcoholic fermentation (Figure 4.25).

Figure 4.25 **Alcoholic fermentation**

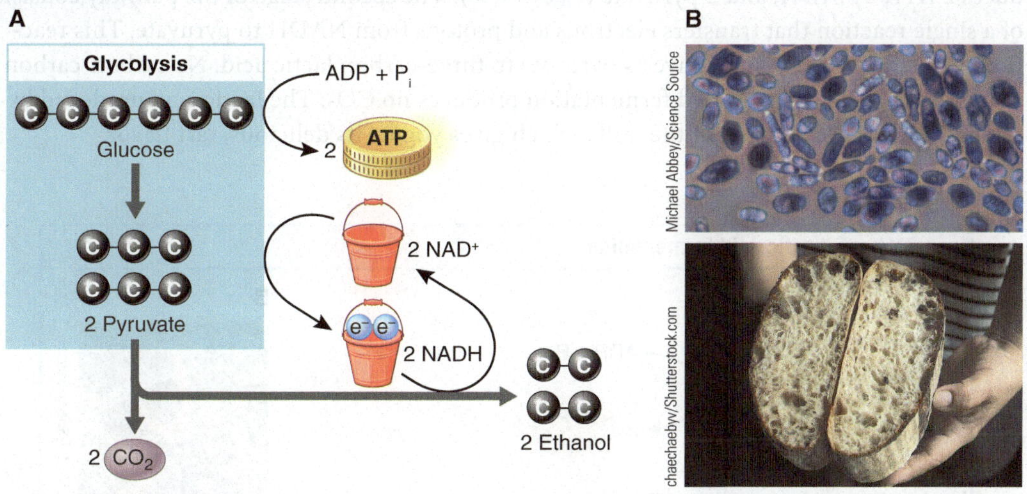

A. Alcoholic fermentation begins with glycolysis and produces carbon dioxide and ethanol. The overall energy yield is 2 ATP.

B. Yeast (top) is a microorganism that utilizes alcoholic fermentation and is used to make food like bread (bottom). Note that the holes in bread are produced by trapped CO_2 gas.

Fermentation is also used to produce alcoholic beverages like beer and wine. As the yeast cells make ATP for themselves, they also produce ethanol (which makes the beer alcoholic) and CO_2 (which makes it bubbly).

In **alcoholic fermentation**, glycolysis occurs first, and like in aerobic respiration, produces 2 ATP, 2 NADH, and 2 molecules of pyruvate. The second stage of the pathway begins with a reaction that breaks one carbon-carbon bond of pyruvate. This reaction produces a two-carbon intermediate and a molecule of CO_2. In our example, the CO_2 produced accumulates in bubbles that are trapped by the mesh of gluten, a type of protein found in some grains. As the bubbles expand, they cause the dough to rise. In the final reaction, electrons and hydrogen ions are transferred from NADH to the intermediate, regenerating NAD^+ and producing a two-carbon alcohol, ethanol. The ethanol produced during fermentation evaporates during baking.

Now our original 6 carbon atoms from glucose are distributed between two molecules of CO_2 and two molecules of ethanol (each molecule of ethanol contains two carbons).

Carla

Two ATP are produced during glycolysis using the energy liberated when NADH is converted to NAD^+.

Eddie

Lactic Acid Fermentation

Alcoholic fermentation is not the only way we can use fermentation to make tasty foods. For instance, growing bacteria such as *Lactobacillus bulgaricus* and *Streptococcus thermophilus* in milk produces yogurt. The bacterial cells first break down lactose (the sugar in milk) into its monosaccharide subunits, and then they use these sugars for lactic acid fermentation.

Now our original 6 carbon atoms from glucose are distributed between two lactic acid molecules.

Carla

Two ATP are produced during glycolysis using the energy liberated when NADH is converted to NAD^+.

Eddie

Lactic acid fermentation sustains animals such as freshwater turtles during hibernation. During the winter, freshwater turtles are unable to breathe normally because they are buried in mud or trapped under ice. A turtle's normal metabolic rate drops and very little ATP is required, so lactic acid fermentation is sufficient.

Like alcoholic fermentation, **lactic acid fermentation** begins with glycolysis and produces 2 ATP, 2 NADH, and 2 pyruvate (Figure 4.26). The second stage of the pathway consists of a single reaction that transfers electrons and protons from NADH to pyruvate. This reaction regenerates NAD^+ and converts pyruvate to three-carbon lactic acid. No carbon-carbon bonds are broken, so lactic acid fermentation produces no CO_2. The lactic acid produced by the bacteria reduces the pH of the milk, which gives yogurt its delicious, tart flavor.

Figure 4.26 **Lactic acid fermentation**

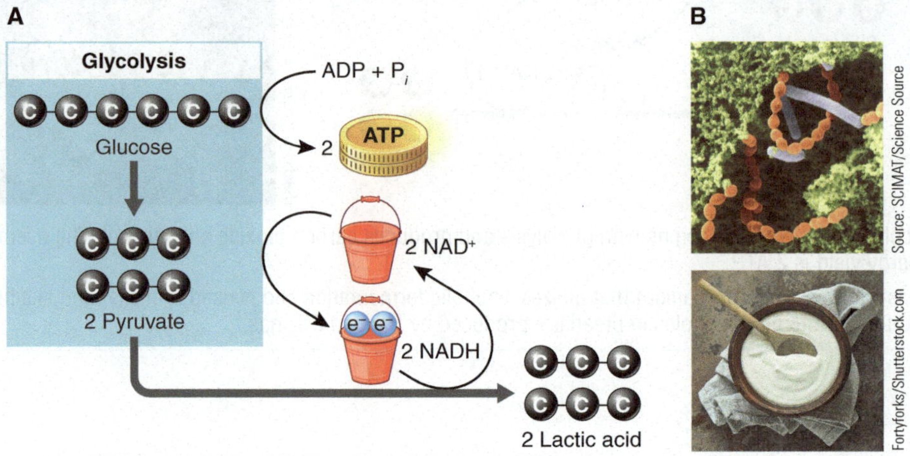

A. Like alcoholic fermentation, lactic acid fermentation begins with glycolysis. However, the final product is lactic acid. The overall energy yield is 2 ATP.

B. Bacteria such as *Lactobacillus bulgaricus* (orange beads) and *Streptococcus thermophilis* (light blue rods) (top image) are microorganisms that undergo lactic acid fermentation and are used to make food like yogurt (bottom).

Biology for Life

Bursts of Energy

Cells in animal skeletal muscles can carry out both aerobic respiration and lactic acid fermentation. In these cells, aerobic respiration, with its greater efficiency, predominates under most circumstances. However, there are times when fermentation is required; for example, when intense exercise depletes oxygen in muscles faster than it can be replenished. Under the resulting anaerobic conditions, muscle cells produce ATP mainly by lactic acid fermentation. This pathway makes ATP quickly, so it is useful for strenuous bursts of activity, but the low ATP yield does not support prolonged exertion (Figure 4.27).

Contrary to popular opinion, the buildup of lactic acid in muscles after exercise does not cause muscle soreness the following day. Lactic acid very quickly leaves muscles and enters the bloodstream, where cells that are not oxygen depleted take it up. These cells convert the lactic acid back to pyruvate for use in aerobic respiration. Instead, delayed muscle soreness is likely due to microscopic tears in muscle fibers that require repair following strenuous exercise.

Figure 4.27 **Lactic acid fermentation occurs in muscles while running**

Intense activity such as sprinting quickly depletes oxygen in muscles. Under anaerobic conditions, ATP is produced mainly by lactate fermentation. Fermentation does not make enough ATP to sustain strenuous activity for long.

Why would we or any other animal need to move so fast that our oxygen intake cannot keep up with our body's demand for energy?

4.6 Summary

- ATP forms during fermentation through glycolysis.
- While aerobic respiration requires oxygen at the end of the electron transport chain, fermentation only uses glycolysis, so no oxygen is required. Instead, NADH is used to form ethanol or lactic acid, and NAD^+ is regenerated for use in glycolysis.
- Alcoholic fermentation is used to make bread and alcoholic beverages. Lactic acid fermentation is used to make yogurt.
- Muscle cells use aerobic respiration when oxygen is available. However, when exercise depletes oxygen, lactic acid fermentation is used so that muscles can still contract.

Understand It! Mindset

Mindset is the beliefs we have about ourselves. These beliefs impact how we behave, our outlook on life, and our attitude, and can directly impact our classroom performance. A *fixed mindset* is the belief that your ability is static. You either have it, or you do not. Someone with a fixed mindset believes that we are born with certain skills and not others. As a result, people with fixed mindsets avoid challenges, give up easily, feel threatened or jealous of others, and are focused on performance and looking good to others.

Growth mindset is the belief that skills and abilities are developed. People with growth mindsets embrace and persevere through challenges. Failure is viewed as an essential part of acquiring new skills. People who have a growth mindset focus on improving, learning, and growing, and are inspired by the success of others.

People can have both a growth and a fixed mindset, depending on the situation. For example, one may have a fixed mindset in the biology classroom, but a growth mindset in their artistic pursuits. Remember that our brains are changeable! We can go from having a fixed to a growth mindset. As Carol Dweck, the psychologist who coined the terms fixed and growth mindset says, embrace the power of YET. I don't understand cellular respiration, YET. YET implies that one believes in their ability to learn and understand the material. I believe in your ability to succeed in this course!

1. What is something in your life that you have a fixed mindset about? How about a growth mindset?
2. Can you think of a time when you didn't know something, but later mastered it?
3. What is something from this chapter that you don't know YET but want to work on?

Chapter Review

4.1 Enzymes and Metabolic Pathways

Enzymes are molecules that speed up chemical reactions without being consumed. **Substrates** bind at the enzyme's **active site**, where they react to form **products**, which are then released. Enzymes decrease **activation energy**, the minimum amount of energy needed to start a chemical reaction. Enzymes are affected by factors such as **temperature** and **pH**. Some enzymes require **cofactors** or **coenzymes** to perform their activity. **ATP** is used to transfer energy through the **phosphorylation** of other molecules. A **metabolic pathway** is a series of enzymatic reactions.

4.2 Connecting Photosynthesis and Cellular Respiration

Autotrophs can build organic molecules from inorganic substances, whereas **heterotrophs** must obtain organic molecules through their diets. **Photosynthesis** is the assembly of sugars from CO_2 and water using energy from the Sun, whereas **chemosynthesis** uses high-energy chemicals to produce organic molecules. **Cellular respiration** is the process of breaking down organic molecules to release stored energy.

4.3 Setting the Stage: Where Photosynthesis Occurs in Plants

Leaves capture sunlight. Leaves have **stomata**, pores that can be closed to conserve water on hot, dry days. Cellular **organelles** are intracellular compartments where specific jobs occur. **Chloroplasts** perform photosynthesis while **mitochondria** are where the most energy from ATP is produced in eukaryotes.

4.4 Conversion of Light into Chemical Energy

Light is **electromagnetic radiation** composed of different **wavelengths**. **Pigments** are organic molecules that can selectively absorb light of specific wavelengths. **Chlorophyll *a*** is an important photosynthetic pigment that absorbs all wavelengths of visible light but green, which it reflects.

The **light-dependent reactions** of photosynthesis produce ATP, O_2, and NADPH through the excitation of electrons in **photosystem II**, which passes electrons through the **electron transport chain** to **photosystem I**. The movement of protons through **ATP synthase** fuels the generation of ATP. Water is broken down to replace the electrons lost in photosystem II. The electrons that left photosystem II replace missing electrons excited in photosystem I. When sunlight excites the electrons in photosystem I, the electrons travel along another electron transport chain and provide energy to convert $NADP^+$ back to NADPH.

The **Calvin cycle** takes the energy in ATP and NADPH to make PGAL, which will eventually form glucose. Therefore, this process stores the energy from photosynthesis. **Carbon fixation** is the process of extracting carbon from CO_2 and storing it in organic molecules. The **rubisco** enzyme fixes CO_2 onto RuBP, forming six-carbon intermediates that are split into three-carbon PGA molecules. NADPH and ATP from the light-dependent reactions convert PGA into PGAL.

4.5 Aerobic Respiration: Burning Sugars in the Presence of Oxygen

Cells obtain energy using **cellular respiration** to break down glucose and other organic molecules. When oxygen is used in the process, it is called **aerobic respiration**. Aerobic respiration begins with **glycolysis**, where glucose is broken into two three-carbon molecules (**pyruvate**). Glycolysis proceeds in two steps: an energy-requiring step that uses ATP, and an energy-harvesting step. The energy-harvesting step generates NADH and ATP. The transition step turns pyruvate molecules into acetyl-CoA molecules and CO_2 gas, and stores energy in NADH. The **citric acid cycle** joins two-carbon acetyl-CoA with a four-carbon molecule to form a six-carbon molecule. By the end of the cycle, two carbon atoms have been lost as carbon dioxide for each acetyl-CoA. This regenerates oxaloacetate which can then join with the next acetyl-CoA to continue the process. Additionally, NADH, FADH2, and ATP are produced.

The **electron transport chain** harvests the electron energy stored in the NADH and $FADH_2$. First, the electrons are passed from protein to protein in the mitochondrial membrane, where the energy released is used to fuel the movement of protons across the membrane, thus creating a proton gradient. Next, **ATP synthase** uses the potential energy stored in the proton gradient to create ATP from ADP. The final electron acceptor is **oxygen**, which combines with hydrogen ions to form water.

4.6 Fermentation: Burning Sugars in the Absence of Oxygen

Fermentation breaks down glucose to produce ATP without oxygen. **Alcoholic fermentation** produces ethanol and CO_2 gas, while **lactic acid fermentation** produces lactic acid. Fermentation regenerates NAD^+ by removing the electrons from NADH. Glycolysis can continue without the oxygen and aerobic respiration steps.

Practice It!

Retrieval practice is the process of pulling information from your long-term memory and using it. Here are some questions to help with your retrieval practice.

Self-Quiz

1. ______ are proteins that speed up chemical reactions. (LO 4-1)

a. Substrates
b. ATP
c. Feedback inhibition
d. Enzymes

2. A series of chemical reactions that together produce important molecules for the cell is called ______. (LO 4-5)

a. feedback inhibition
b. an activation energy pathway
c. phosphorylation
d. a metabolic pathway

3. A cat eats a bird. The bird ate a caterpillar. The caterpillar chewed on a plant leaf. The ______ is an autotroph. (LO 4-6)

a. cat
b. bird
c. caterpillar
d. plant

4. Plants use ______ as an energy source to drive photosynthesis. (LO 4-7)

a. sunlight
b. sugars
c. O_2
d. CO_2

5. Cellular respiration uses which products of photosynthesis? (LO 4-7)

a. glucose and CO_2 c. water and CO_2
b. glucose and O_2 d. water and O_2

6. Which cellular organelles produce sugars by capturing energy from sunlight? (LO 4-8)

a. lipid bilayer c. stomata
b. mitochondria d. chloroplasts

7. Which feature of leaves makes them perfect for catching sunlight? (LO 4-10)

a. stomata that can open and close
b. waterproof cuticles
c. large surface areas
d. strong stems

8. When conditions are hot and dry, plant leaves close their ______ to conserve water. (LO 4-10)

a. chloroplasts
b. mitochondria
c. cytosol
d. stomata

9. Where is the majority of ATP produced during cellular respiration? (LO 4-9)

a. mitochondria c. cytosol
b. chloroplasts d. thylakoid membrane

10. In the light-dependent reactions, ______ accumulate(s) in the thylakoid space of chloroplasts. (LO 4-15)

a. sugars c. O_2
b. protons d. CO_2

11. The carbon that land plants use for photosynthesis comes from ______. (LO 4-16)

a. glucose. c. water.
b. the atmosphere. d. soil.

12. Glycolysis occurs in the ______ of all cells. (LO 4-21)

a. nucleus
b. mitochondria
c. cytoplasm
d. membrane

13. From one glucose molecule, ______ molecules of pyruvate are formed, eventually resulting in six carbon atoms exiting the cell as CO_2. (LO 4-21, 4-22)

a. one c. three
b. two d. six

14. For each molecule of glucose oxidized in aerobic respiration, the citric acid cycle turns ______ time(s). (LO 4-22)

a. one c. three
b. two d. six

15. Which molecule accepts electrons at the end of the electron transport chain in aerobic respiration? (LO 4-23)

a. water c. O_2
b. H^+ d. NADH

16. Triglycerides are broken down into free fatty acids and glycerol, which enter aerobic respiration at which two steps? (LO 4-26)

a. glycolysis and citric acid cycle
b. glycolysis and acetyl-CoA transition step
c. acetyl-CoA transition step and citric acid cycle
d. citric acid cycle and electron transport chain

17. Fermentation is needed to regenerate ______ so that glycolysis can continue removing electrons from glucose without oxygen. (LO 4-27)

a. NAD^+ c. CO_2
b. ATP d. pyruvate

18. Fermentation does not need oxygen because it only uses which of the following steps? (LO 4-28)

a. glycolysis
b. the transition step
c. the citric acid cycle
d. the electron transport chain

19. If a person is engaging in strenuous exercise, in addition to cellular respiration, their cells are also undergoing: (LO 4-30)

a. alcoholic fermentation c. lactic acid fermentation
b. aerobic respiration d. photosynthesis

Critical Thinking

1. Plants and other photosynthesizers need CO_2 to produce sugars in the Calvin cycle. Hypothetically, how might the current rise in atmospheric CO_2 affect photosynthesis? Keep in mind that the real world is much more complex, as there are interactions between atmospheric CO_2 and other factors, such as temperature, drought, and rainfall. (LO 4-19)

2. About 200 years ago, Jan Baptista van Helmont wanted to know where growing plants obtain the materials necessary for them to increase in size. He planted a tree seedling weighing 2.2 kilograms in a barrel filled with 90 kilograms of soil and then watered the tree regularly. After five years, the tree had gained almost 75 kilograms and the soil's weight

was unchanged. He incorrectly concluded that the tree had gained all its additional weight by absorbing water. How did the tree really gain most of its weight? (LO 4-16)

3. While looking at an aquarium, you notice that one of the plants is releasing tiny bubbles. Knowing what you know about photosynthesis, what are the bubbles? (LO 4-14)

4. Explain why considerably more ATP is produced during the electron transport chain than during glycolysis. (LO 4-25)

5. Cyanide is a poison that blocks the end of the electron transport chain so that oxygen cannot accept the electrons and form water. Without the electron transport chain working, cells cannot generate the large amount of ATP usually produced from aerobic respiration. If you think carefully about what goes into and out of the electron transport chain, what molecule would build up in the mitochondria if this process is stopped? (LO 4-23)

6. Human babies, bears, and some rodents have specialized fat cells called "brown fat," which allow them to generate heat instead of ATP during the electron transport chain. A mitochondrial carrier protein in brown fat called thermogenin acts like a "hole" in the mitochondrial membrane, with the result that proton concentrations are not able to build up to run ATP synthase. Some people have tried using a molecule called DNP (dinitrophenol) to artificially bypass ATP synthase to lose weight. What can go wrong with this method of weight loss? *(Note: DNP is not approved for medical use because of this potential danger!)* (LO 4-25)

Unit 1 Review

In the opening of this unit, we visited a grocery store and considered the biochemistry of the foods, beverages, and supplements that crowd the grocery shelves. Let's now conclude our trip to the supermarket and "check out" what we've learned.

In Chapter 2, we learned about the basic chemistry that's important to life. For example, we learned about the atom and how atoms connect to form various molecules. We also learned about the structure of water and why water is so important to life. We learned that acids are not bad, and some acids, like acetic acid or vinegar, are tasty and important for food preservation.

In Chapter 3, we learned about macromolecules. When we go to the store, the food we buy is comprised of macromolecules. Crackers, bread, and chips are full of carbohydrates. Protein is a major component of meat and tofu. Butter and oils are lipids. Although we also ingest nucleic acids when we eat plant and meat products, this fourth class of macromolecule is typically not something that finds its way onto our shopping list. We also learned how proteins fold and we can visualize how they unfold when we cook meat.

Why do we need to eat? To get energy! In Chapter 4, we learned that plants are autotrophs and capture energy from the sun, converting it into a chemical form that they (or we!) can burn for energy. To gain this energy, we eat plants or consume animals that eat plants. Molecules from our diets feed into metabolic pathways such as cellular respiration to create energy. Some of our favorite foods, like bread, are made using a metabolic pathway called fermentation, which also creates usable energy.

Next time you visit the store, look at what you are buying and ask why those items are in your shopping cart. What new connections can you make? Can you relate the amino acid supplement to proteins? What macromolecule is predominant in your shopping cart? Why are pickles packed in vinegar and not water? The questions are endless!

Credit: LIVINGRELICPRODUCTIONS.COM

Unit 2

Cancer

Cell Biology

Did you know that dinosaurs could develop cancer? In fact, the earliest recorded evidence of cancer was found in a 240-million-year-old shell-less turtle *Pappochely rosinae*, such as the animals shown here by a Triassic lake. The ancient turtle's hind leg showed evidence of osteosarcoma, a type of bone cancer that can also affect humans. Was bone cancer the only type of cancer that ancient creatures such as dinosaurs could suffer from? Likely not, although we don't see it in the fossil record. Other types of tissues affected by cancer are too soft and delicate to be preserved as fossils.

Although scientists found evidence of osteosarcoma in ancient humans (more on that in Chapter 6), the first written record of human cancer was recorded by the ancient Egyptians around 3000 BCE. We call it "cancer" because the ancient Greek physician Hippocrates thought that the finger-like projections from tumors looked like crab legs—so he called them *carcinos*, the Greek word for crab. The name *cancer* that we use today is the Latin word for crab. Carcinogens, the term for cancer-causing chemicals, can trace its roots to *carcinos*.

Cancer, unfortunately, touches too many of our lives in a negative manner. We know cancer is a devastating disease and has something to do with cells—but what is cancer exactly? Where does cancer come from? How can cancer be treated? In this unit, we'll explore cell biology and relate basic cell biology to cancer. In Chapter 5, we will learn the basics of the cell, including the parts of the cell, the different kinds of cells, the bustling activity within them, and how we're able to visualize cells using microscopes. Once we have an understanding of what cells are, we'll talk about different cellular behaviors, such as how cells give rise to new cells by mitosis and how cells can be rid by the body through a process called apoptosis. We'll relate how misregulation of either of these processes can contribute to cancer and how cancer treatments can target these processes. Finally, we'll discuss strategies for protecting our cells and preventing cancer.

Preview It!

What do you want to learn about cancer in this unit?

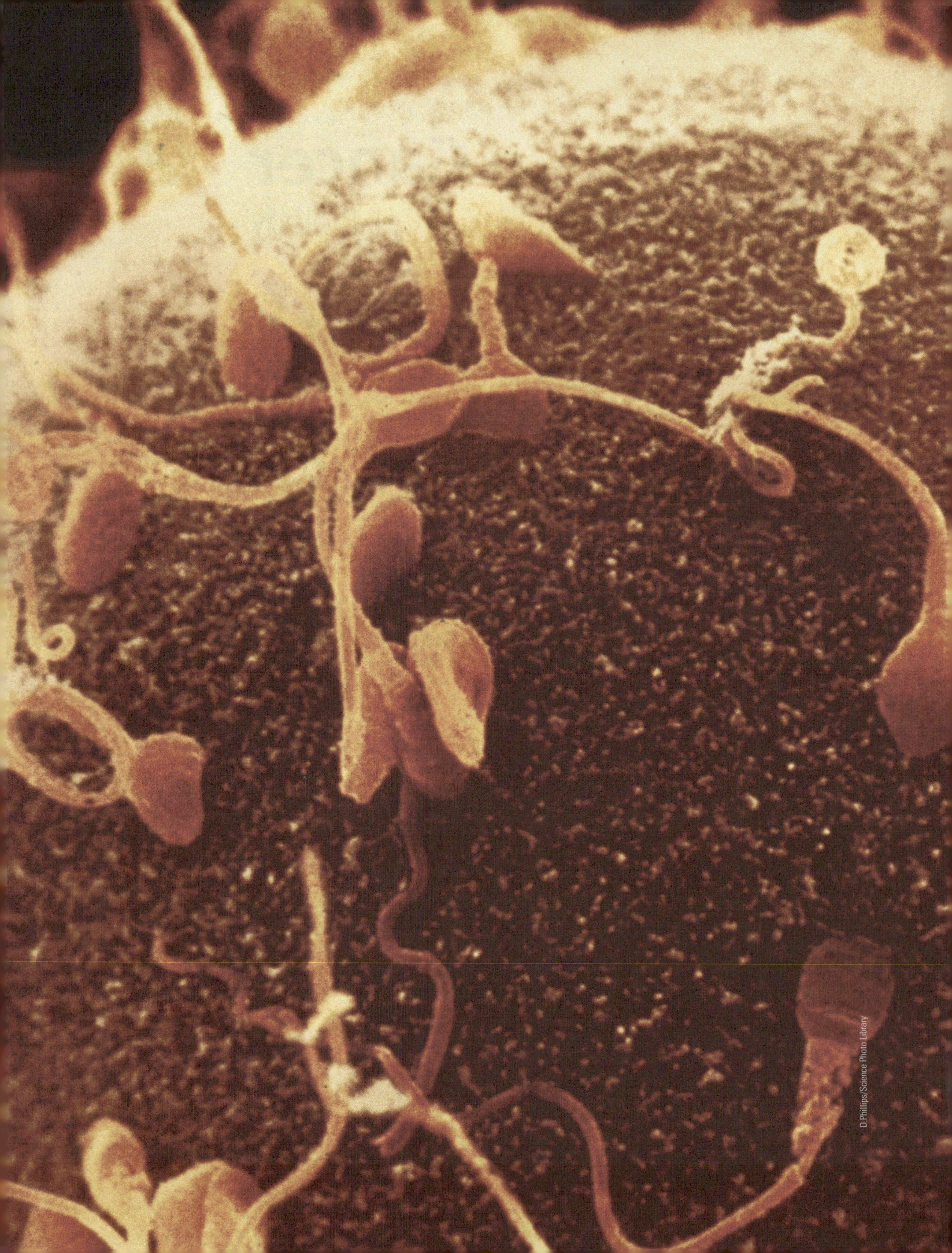
D.Phillips/Science Photo Library

5

The Secret Life of the Cell

What Can You Do with a Cell?

What Can You Do with a Cell?

Most people typically do not think about their cells—especially when they are working as they should. Your body has about 30 trillion cells, all working together like a big beehive as you go about your day. Like bees in a beehive, these cells all have jobs, too. We have hair, skin, and heart cells. Routine medical care often involves an examination of our cells. A blood draw can reveal not only the levels of various molecules in the blood, but also provides a look at the number of cells in our blood. For example, red and white blood cells are two types of cells that are commonly assayed after a blood draw. If there are not enough healthy red blood cells, one may have anemia. Too many white blood cells could mean that you have an infection.

A single cell is powerful. One cell contains all your genetic information and therefore the blueprint for what makes you, well, you. One of your gametes, an egg or sperm (see the photo that begins this chapter), combined with another's complementary gamete, can make a whole new person. And a single cell, if left to grow unchecked, can lead to cancer (Chapter 6). What can we do with a cell? The cell is the smallest unit of life—and one can therefore do many things with a cell. But what it is about the cell that is so special? Why is a single cell so powerful?

Through a single cell, one can reproduce Shown here is a colorized scanning electron micrograph (SEM) of an egg surrounded by sperm. Notice how much larger the egg is than the sperm. Although there are many sperm, only one sperm cell will successfully penetrate the egg to fertilize the cell.

In this chapter, we'll learn about the inner workings of the cell. First, we'll discuss the nature of the cell and compare the cells of bacteria, archaea, and eukaryotes. Note that the majority of the chapter will focus on eukaryotic cells. We'll discuss the boundary of the cell—its outer membrane—including the structure of the cell membrane and how cells transport molecules across the membrane. Next, we'll discuss the cellular organelles. We first learned about organelles in Chapter 4 when we discussed the chloroplast and mitochondria and their respective roles in photosynthesis and cellular respiration. Here, we'll cover the remainder of cellular organelles. Finally, after learning about individual cells, we'll discuss how cells are connected through physical structures and biochemical communication.

Preview It!

Write down three things you know about cells. What do you want to learn about cells in this chapter?

5.1 What Is a Cell?

Learning Objectives

- **5-1** Explain the principles of cell theory.
- **5-2** Describe the four components that all cells have.
- **5-3** Explain why the surface-to-volume ratio limits cell size.
- **5-4** Compare and contrast bacterial, archaea, and eukaryotic cells.
- **5-5** Compare different types of microscopy.

Some of the earliest evidence for the cell came about in 1683 when Dutch scientist Antoine van Leeuwenhoek scraped plaque from his teeth and examined it under a rudimentary microscope. He described his plaque as containing "many very little living animalcules, very prettily a-moving." A contemporary of van Leeuwenhoek, Robert Hooke is credited with devising the name "cell" after examining cork (bark from the cork tree) under a microscope. He noted that it was made of many small structures that he likened to the rooms that monks lived in, called cells. The work of van Leeuwenhoek, Hooke, and others ultimately led to the generation of cell theory. Recall from Chapter 1 that theories are well supported by data and have explanatory power. **Cell theory** is one of the foundations of modern biology and has three tenets: (1) all living organisms are made up of one or more cells; (2) the cell is the basic structural and functional unit of life; and (3) cells come from preexisting cells.

All living things—even the largest, most complex organisms—are composed of these small units of life. How does a free-living cell like a bacterium differ from a skin cell in your arm? And how do trillions of individual cells function as a collective to form your body? To understand the answers to these questions, we must learn about the structure of the cell, the diversity of cells, and how cells work together as part of a tissue.

Why Are Cells Small?

Although there is some variety in the size of an individual cell—for instance, a frog egg is a single cell and can be seen without the aid of a microscope—most cells tend to be small and require a microscope to see.

Why so small? The primary reason is that a living cell must exchange substances with its environment. Nutrients must enter the cell fast enough to sustain its metabolism, and wastes must leave the cell before reaching toxic levels. These exchanges occur across the cell's membrane (Section 5.2), which can handle only so many exchanges at a time. The rate of exchange across a membrane depends on its surface area: The greater the surface area, the

Table 5.1 Surface-to-Volume Ratio of Three Spheres

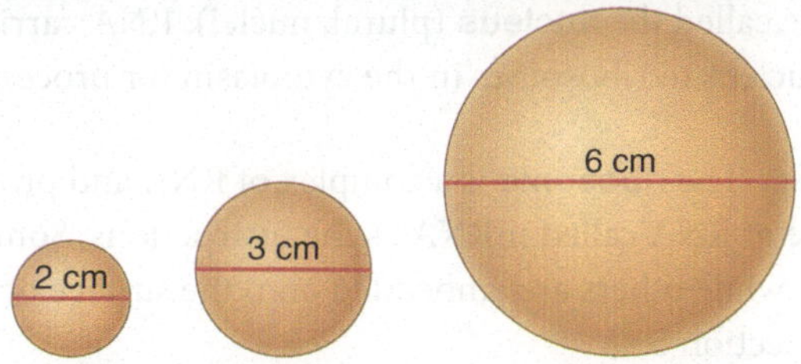

Diameter (cm)	2 cm	3 cm	6 cm
Surface area (cm^2)	12.6	28.3	113.1
Volume (cm^3)	4.2	14.2	113.1
Surface-to-volume ratio	3:1	2:1	1:1

Some nerve and muscle cells are up to 1 meter (3 feet) long. Since these cells are very thin and thread-like, they still have an acceptable surface-to-volume ratio and can therefore still exchange substances efficiently with fluids in the surrounding tissue.

more substances that can cross it during a given interval. Thus, cell size is limited by a physical relationship called the **surface-to-volume ratio**.

When a cell expands in diameter, its volume increases faster than its surface area. Using three spheres as an example, the smallest of the spheres shown in Table 5.1 has the largest surface-to-volume ratio (3:1), with three times as much surface area as volume compared with the largest sphere (1:1), and therefore would be the most effective at exchanging nutrients and wastes with its environment. The larger the cell, the smaller the surface-to-volume ratio and the less efficient the exchange. If the cell gets too big, the inward flow of nutrients and the outward flow of wastes across that membrane would not be fast enough to keep the cell alive.

Universal Cell Features

The diversity of cell types and functions is breathtaking. Despite this variety, cells have many universal features. Features shared by all cells include an outer cell membrane, cytosol, nucleic acids, and ribosomes (Figure 5.1).

The membrane. A cell's outer **membrane** separates its contents from the external environment. With little variation, the basic structure of a cell's outer membrane (and any other cell membrane) is the lipid bilayer (Chapter 3). In Section 5.2, we will discuss the organization of the cell's membrane, including the various kinds of proteins embedded in the lipid bilayer or attached to one of its surfaces that carry out membrane functions. Only certain materials can cross cell membranes, so the membrane controls the exchange of materials between the cell's internal and external environments.

Cytosol. The membrane encloses a jellylike mixture of water, sugars, ions, and proteins called the **cytosol**. A major part of a cell's metabolism occurs in the cytosol, and the cell's other internal components, including **organelles** or sub-cellular components, are suspended in it. Note that although the cytosol refers to the liquid surrounding the organelles, the term **cytoplasm** is used to describe everything inside of the cell—the cytosol plus whatever is suspended in it, like organelles or nucleic acids.

Nucleic acids. All cells contain nucleic acids, which include DNA and RNA (Chapter 3). Note that some cells, such as red blood cells, lose their DNA after maturing. DNA contains the blueprint, or instructions, to make that cell. DNA is stored in the cell in different

Figure 5.1 Universal cell features

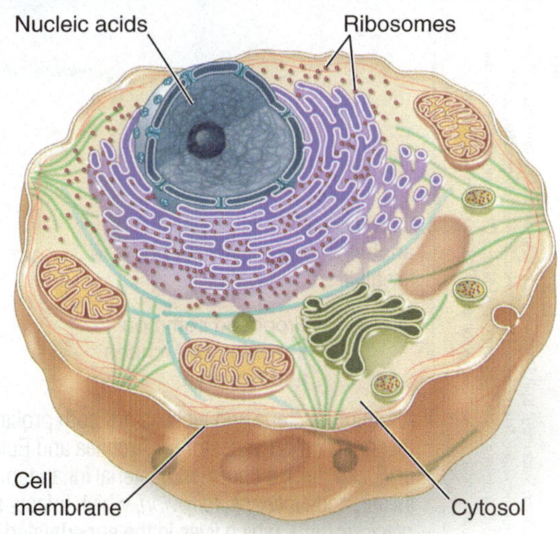

This illustration of a generalized cell highlights the features that all cells have: cytosol, nucleic acids, ribosomes, and an outer cell membrane.

ways, depending on the type of cell. In prokaryotes (bacteria and archaea), DNA is suspended directly in the cytosol. Eukaryotic cells (plant and animal cells, for example) store their DNA in an organelle called the **nucleus** (plural, nuclei). RNA carries instructions contained in the DNA from the nucleus to ribosomes in the cytoplasm for processing (Chapter 7).

Big Ideas in Biology

Structure and Function

Cells come in a wider variety of shapes, sizes, and textures than standard classroom models would lead you to believe. These different structural elements are important for understanding how cells function. For example, skin cells are flattened to help create a barrier between your body and the outside world.

Ribosomes. The **ribosome** is a complex of RNA and protein used for building new proteins, using a type of RNA called mRNA as the instructions. Some ribosomes are found floating in the cytosol, while others are embedded onto the surface of an organelle called the *endoplasmic reticulum* (Section 5.4).

Three Classes of Cells and the Three Domains of Life

In 1990, biologists proposed organizing life into three domains based on the three cell types that exist on Earth. The three domain system is divided into Bacteria, Archaea, and Eukarya (Figure 5.2). Domain Bacteria and Domain Archaea include prokaryotic cells. **Prokaryotes** are single-celled organisms that do not have a nucleus or membrane-bound organelles. However, individual cells of many prokaryotic species cluster in filaments or colonies, such as the plaque Leeuwenhoek scraped from his teeth. Within these colonies, individual cells can work for the collective good of the colony, which is a multicellular behavior, but the individual cells themselves are considered unicellular. **Bacteria** are the largest and most well known prokaryotes, and are the culprits behind many common infections, like strep throat.

Outwardly, cells of the two groups appear so similar that archaea were once presumed to be an unusual group of bacteria. Both bacteria and archaea lack a nucleus, so they were named prokaryotes, a word that means "before the nucleus." By 1977, it had become clear that **archaea** are genetically more closely related to eukaryotes than to bacteria, and contain unique features such as a different membrane composition, so they were given their own separate domain. The term "prokaryote" is now an informal designation only.

Figure 5.2 **The three-domain model of life divides life based on three distinct types of cells**

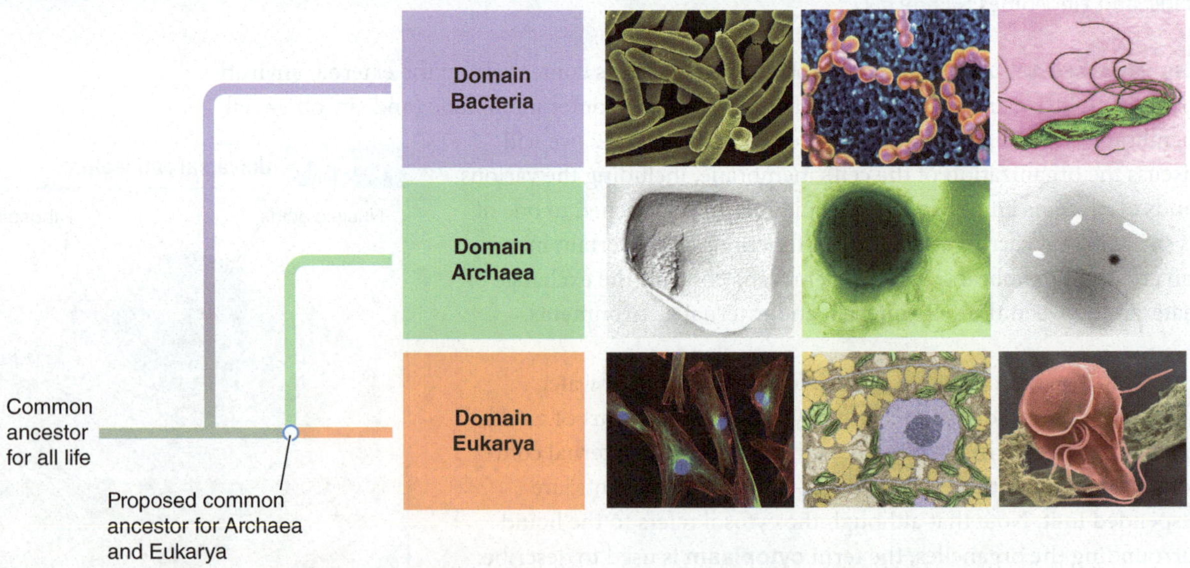

Although bacteria and archaea are both prokaryotic, archaea cells are more like eukaryotic cells, suggesting a more recent common ancestor for Archaea and Eukarya. At the top are three examples of bacteria, including *Escherichia coli*, or *E. coli* (left), a common bacterial inhabitant of the human gut; *Streptococcus pyogenes* (middle), which causes strep throat; and *Helicobacter pylori*, which infects the stomach lining and can cause ulcers. Archaea cells include *Ferroglobus placidus* (left), which lives in the superheated water spewing from the ocean floor; *Thermococcus gammatolerans* (middle), which is the most radiation-resistant organism ever discovered; and *Haloquadratum walsbyi*, which lives in brine pools saltier than soy sauce. Eukaryotic cell examples include a human skin cell (left), a plant cell (middle), and an example of a single-celled eukaryotic organism, *Giardia lamblia*. Giardia infection is the most common cause of hiker's diarrhea.

Biology for Life

The Microbiome: A Menagerie of Microscopic Life

Did you know that you are home to a menagerie of microscopic life? About 1–3% of your total body mass isn't "you" but the collection of bacteria, archaea, viruses, and eukaryotic life living on and in your body. Before you run for the nearest shower, realize that this microbial life is essential for your health, producing vitamins, helping with digestion, and protecting your intestines from colonization by harmful types of bacteria.

The most microbe-rich portion of your body is within your gastrointestinal system (Figure 5.3). The microbiota-gut-brain axis describes the bidirectional relationship between your gut and its resident microoganisms and the brain. Some research suggests that psychiatric symptoms such as anxiety and depression are linked to changes in the gut microbiome, and new treatments for these conditions might be possible based on increasing the levels of certain bacteria in the gut through the consumption of particular probiotics.

Our gut microbiome is largely established by exposure to vaginal fluids during birth. However, babies born via a Cesarean section (C-section) are born without exposure to their mother's vaginal microbiome. Doctors are investigating using vaginal seeding as a microbiome intervention in C-section newborns. In this procedure, the mother's vagina is swabbed and then used to swab the baby's mouth.

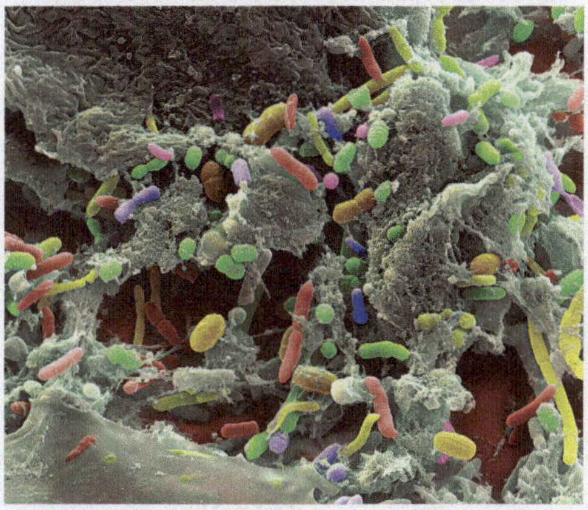

EYE OF SCIENCE/Science Source

Figure 5.3 The microbiome within our gastrointestinal system is critical to our health

In this false-color electron micrograph of human fecal matter, we see a small portion of the 30,000 species of bacteria that normally inhabit our large intestine.

What are the potential benefits of vaginal seeding? Can you think of any risks? Are these risks the same, less, or different than what a baby would be exposed to during birth?

Domain Eukarya includes all organisms that are eukaryotes. **Eukaryotes** have cells that are characterized by membrane-bound organelles (Section 5.4). Note that eukaryotic life can be single eukaryotic cells, like yeast, or may join to make up a multicellular organism. Although traditional classroom models suggest that eukaryotic cells have a typical structure—with a round animal cell and a square or rectangular plant cell—in truth, there is a wide range of shapes and sizes of eukaryotic cells (see Figure 5.2).

? Did You Know?

People often get stomach upset when taking antibiotics because the "good" bacteria that live in our gastrointestinal tract are also killed by the antibiotic.

Visualizing Cells

Sometimes, scientific discoveries are contingent on the emergence of new technologies. Perhaps surprisingly, cell theory and the discovery of cells would not have been possible without the development of the telescope. Galileo Galilei, who is best known for his work in astronomy and physics, also built the first microscope, adapting his telescope to see tiny objects. Van Leeuwenhoek built on Galileo's original design to create better lenses for microscopes and is now considered the father of microscopy. Two hundred years after Van Leeuwenhoek's experiments with his dental plaque, Rudolf Virchow used a microscope to discover that cancer cells derive from normal, non-cancerous cells. Microscopes are important to this day for everything from medical examination of tissue samples to scientific research. Today, we have a variety of microscopes and techniques that allow us to see different aspects of cell structures, including light microscopy, fluorescence microscopy, and electron microscopy.

Why do you brush your teeth to get them clean, instead of just swishing with water? Bacteria in your mouth have a sticky outer layer called a capsule that allows them to stick to your teeth and each other. This forms plaque, which isn't good for your teeth. The only way to remove plaque is physical force via brushing.

Biology for Life

Comparing Antibiotics and Antifungal Medications

Why do we treat bacterial and fungal infections differently? And why is it so much harder to clear up a fungal infection? It comes down to the difference in the structures of the cells. The fungi that cause everything from athlete's foot to vaginal yeast infections are part of Domain Eukarya, the same as us. Bacteria, however, are prokaryotic and many have a structure that our human cells do not: a cell wall. The cell wall protects the cell and supports its shape (Figure 5.4). Penicillin-class antibiotics work by inhibiting the synthesis of a bacterial cell wall, thus killing the bacteria. Note that cell walls are not membranes and they do not consist of lipids. Therefore, taking an antibiotic has a direct negative effect on the bacteria only, not on our human cells.

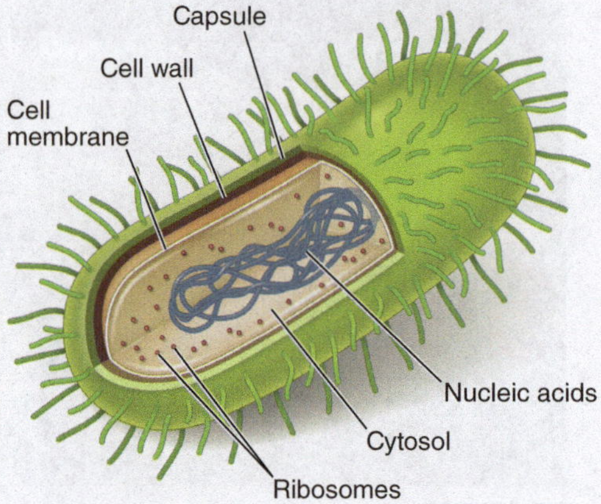

Figure 5.4 Generalized prokaryotic cell structure

Note the presence of the universal cell features, plus differentiating features such as the cell wall and capsule.

The challenge with fungal infections is that our cells and the invading fungal cells are both eukaryotic, and so there are fewer differences between host and pathogen cells to take advantage of with a treatment strategy. As a result, clearing up a fungal infection typically requires longer and/or repeated treatments, and making environmental changes, like wearing cotton socks or undergarments that make our skin less hospitable for fungal growth.

Topical (surface) treatments for fungal infections are typically preferred over pills. Can you think of a reason why?

Light microscopy. A **light microscope** uses visible light to illuminate a sample. Inside the instrument, curved lenses focus light that either passes through or is reflected from a specimen, to form a magnified image. The image of *H. pylori* in Figure 5.2 was taken with a light microscope and then colorized to distinguish its structural features.

Fluorescence microscopy. Microscopy that uses fluorescent dyes that absorb light of a particular color and then emit light of a different color is called **fluorescence microscopy**. These dyes are often used as tracers in microscopy to pinpoint the location of a molecule or structure of interest in a cell. The image of human cells in Figure 5.2 was taken using fluorescence microscopy.

Electron microscopy. Structures smaller than about 200 nanometers in diameter appear blurry under light microscopes. To observe objects of this size range clearly, an electron microscope is used, but instead of using glass lenses to focus light, **electron microscopy** uses magnetic fields as lenses to focus a beam of electrons onto a sample. Several of the images in Figure 5.2 (e.g., *E. coli*, *G. lamblia*) were taken with an electron microscope and colorized.

There are two types of electron microscope. A transmission electron microscope (TEM) directs electrons through a thin specimen, and the specimen's internal details appear as shadows in the resulting image. A scanning electron microscope (SEM) directs a beam of electrons across the surface of a specimen coated with a thin layer of metal. Electrons are reflected from the specimen and are focused into an image of the surface.

5.1 Summary

- All organisms are made up of one or more cells. The cell is the basic unit of life, and all cells arise from preexisting cells.
- All cells have an outer membrane, cytosol, nucleic acids, and ribosomes.
- Surface-to-volume ratio limits cell size because a cell with too much volume and not enough surface area cannot transport materials in and out fast enough to keep the cell alive.
- Prokaryotic cells are unicellular, although some cluster into filaments or colonies. Bacteria and archaea cells do not have a nucleus or organelles, but some have cell walls. Prokaryotic cells store their DNA in the cytosol.
- Eukaryotic cells store DNA in a nucleus and have a variety of organelles that carry out cellular functions. All multicellular organisms are made up of eukaryotic cells, but some eukaryotes are unicellular.
- Most cells can only be seen with a microscope. Light, fluorescence, and electron microscopy are all used to visualize cells.

5.2 The Cell Membrane

Learning Objectives

5-6 Describe the structure of the lipid bilayer.

5-7 Explain the fluid mosaic model of the cell membrane.

5-8 Describe the roles of proteins in the membrane.

Did you know that in ancient Rome, food was a popular theme for mosaics (Figure 5.5)? A mosaic is a picture created using colored pieces of material such as stone, tile, or glass arranged in a pattern. In the same way a mosaic contains different materials arranged in space, the cell membrane contains many different molecules—such as cholesterol and proteins—embedded in or attached to the lipid bilayer of a cell membrane (Chapter 3).

Fluid Mosaic Model of the Cell Membrane

Recall that the head of a phospholipid is polar and hydrophilic, so it interacts with water molecules. The two long hydrocarbon tails are nonpolar and hydrophobic, so they do not interact with water molecules. As a result of these opposing properties, phospholipids swirled into water will spontaneously organize themselves into a **lipid bilayer**, with the hydrophobic tails facing each other at the interior of the membrane and hydrophilic heads facing the water surrounding on the outside of the membrane (Figure 5.6).

We say that the membrane is fluid because phospholipids in the bilayer are not chemically bonded to one another; they stay organized in a watery environment because of hydrophobic and hydrophilic interactions. Thus, individual phospholipids in the bilayer drift sideways and spin around their long axis, and their tails wiggle. Similarly, the proteins embedded in the bilayer are not chemically bonded to the membrane and so can also move throughout the membrane.

The **fluid mosaic model** describes the cell membrane as a two-dimensional fluid of mixed composition. The "mosaic" part of the name comes from the many different types of molecules that make up the membrane, and the word "fluid" describes how phospholipids and proteins move relative to one another, where the proteins float like rubber ducks bobbing in the bathtub.

Figure 5.5 **A Roman mosaic**

Mosaics are comprised of many small pieces of glass or other materials arranged to make a shape, such as this ancient fruit basket.

Figure 5.6 **The fluid mosaic model**

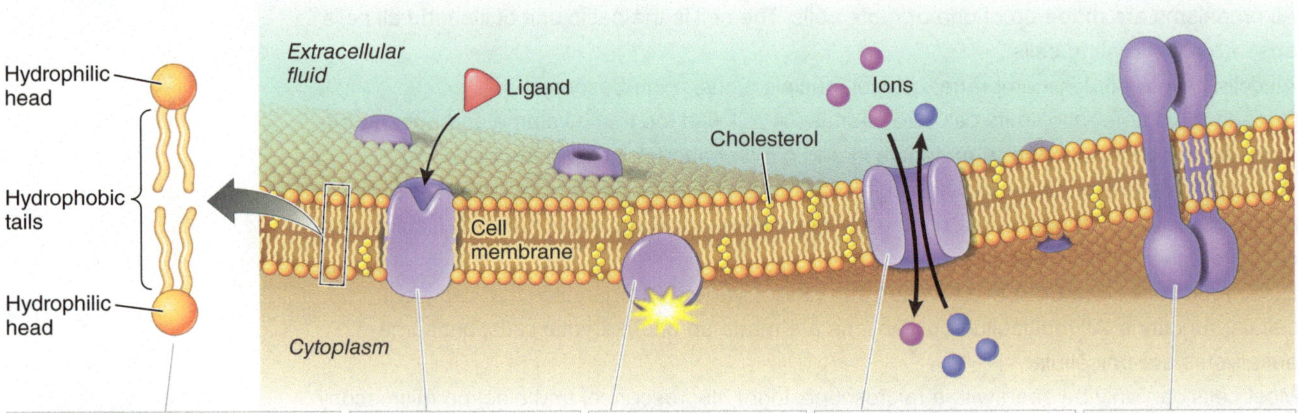

Lipid bilayer
The lipid bilayer forms the framework of a typical cell membrane. In water, phospholipids organize into two layers, with the hydrophobic tails pointing towards each other and the hydrophilic heads facing outward.

Receptors
These proteins transmit a signal from outside to inside the cell, often by binding a specific protein, such as a ligand.

Enzymes
Enzymes catalyze reactions that occur near the membrane.

Transporters
Transporters can physically move molecules, such as ions, from one side of the membrane to the other.

Adhesion proteins
These proteins link cells together.

Like a mosaic, the cell membrane is also comprised of diverse individual pieces—in this case, phospholipids, cholesterol, and various kinds of proteins, each with its own specialized function. We say the membrane is fluid because many of the components of the cell membrane can move around.

Archaea phospholipids are different from bacteria and eukaryote phospholipids. The structure of archaeal phospholipids gives rise to a dense, durable membrane that is stable even at extreme temperatures. This unique membrane composition was one of the clues that led scientists to place these organisms in their own domain.

Proteins Add Function to the Membrane

A cell membrane physically separates the external environment from the cell's internal one. However, that is not its only task. Many types of proteins are embedded in a cell membrane, and each kind of protein imparts a specific function to it. Thus, different cell membranes can have different tasks depending on which proteins are present. For example, cells stay organized in animal tissues because **adhesion proteins** in their outer membranes fasten them together and hold them in place (see Figure 5.6). This arrangement strengthens tissue and can constrain certain membrane proteins to an upper, lower, or side surface of the cell.

Both the membrane surrounding the cell and some internal membranes incorporate **receptor proteins**, which trigger a change in the cell's activities in response to a signal molecule that binds to a receptor. When a receptor is bound by a signal, depending on the type of receptor and the signal, it can result in many different cellular responses, including movement (Section 5.5), cell division (Chapter 6), cell death (Chapter 6), and changes in gene expression (Chapter 7).

The outer cell membrane always has **enzymes** (proteins that catalyze reactions; Chapter 4). Some of these enzymes act on other proteins or lipids that are part of the membrane; others use the membrane as a scaffold. Membranes also typically have **transport proteins**, which move specific substances across the bilayer. There are many kinds of transport proteins, including the proton pumps used in photosynthesis and cellular respiration (Chapter 4), and channels, which allow substances like ions to move through the membrane. Transport proteins are important because the hydrophobic core of a lipid bilayer is impermeable to most hydrophilic substances, including the ions and polar molecules that cells must take in and expel regularly, which we will discuss in the next section.

5.2 Summary

- The lipid bilayer consists of two layers of phospholipids with the hydrophilic heads facing the watery exterior outside and inside the cell and the hydrophobic tails sandwiched in the middle.
- The cell membrane is described as a fluid mosaic, with mixed composition containing proteins and lipids. Phospholipids and other molecules can move laterally in the membrane.
- Proteins in the cell membrane each have a specific function, such as in adhesion (holding cells together); acting as receptors for cells to receive messages from other cells; providing transport for substances across the membrane; or enzyme activity.

5.3 Transport across the Cell Membrane

Learning Objectives

5-9 Name three factors that influence the rate and direction of diffusion.

5-10 Explain selective permeability in terms of lipid bilayers.

5-11 Describe how osmosis works.

5-12 Explain why a cell requires transport proteins.

5-13 Explain why active transport requires energy and passive transport does not.

5-14 Describe the mechanisms by which cells take in and expel materials in bulk.

Last night, your roommate was out eating bean-laden burritos. And, as we learned in Chapter 3, beans contain raffinose, which ultimately leads to gas production in the small intestine and discomfort if the gas isn't released. But why does *your* half of the room stink if your roommate is merrily passing gas on the other side of the room? The reason is—diffusion!

Diffusion is the spontaneous spreading of atoms or molecules through a fluid or gas. Diffusion is one mechanism that allows molecules to move across membranes and through cells. Since an atom or molecule is never still, random movement causes it to bounce off nearby objects, including other atoms or molecules. Rebounds from such collisions propel molecules dissolved in a medium, called **solutes**, to disperse throughout. The result is a gradual and complete mixing, in much the same way that tea diffuses through hot water, or how smelly hydrogen sulfides spread through a room. Note that whatever medium the solute is dissolved in is called a **solvent**. Recall from Chapter 2 that water is a universal solvent.

Diffusion of Solutes

The rate at which molecules diffuse and where they diffuse depends on temperature, concentration, and charge. **Temperature** is a measure of molecular motion. Since atoms and molecules move faster at higher temperatures, they collide more frequently. If they collide more often, then they diffuse faster. **Concentration** refers to the amount of a solute present in a certain area. Concentration also impacts how fast molecules will diffuse.

Let's draw an analogy to a crowded elevator. Which would you be keener to do: shove your way *on* to a crowded elevator or escape *out* of the crowded elevator? More likely than not, you'd rather escape out!

When there is a lot of solute in one place compared to another, we say there is a **concentration gradient**. Like your desire to escape from a crowded space, solutes tend to diffuse "down" their concentration gradient, from a region of higher concentration to one of lower concentration. Why? Consider that moving objects (such as molecules) collide with each other more often when they are crowded together in one spot, just like you are more likely to bump into people on a crowded elevator. Once bumped, molecules will move in a different direction.

Figure 5.7 Selective permeability of lipid bilayers

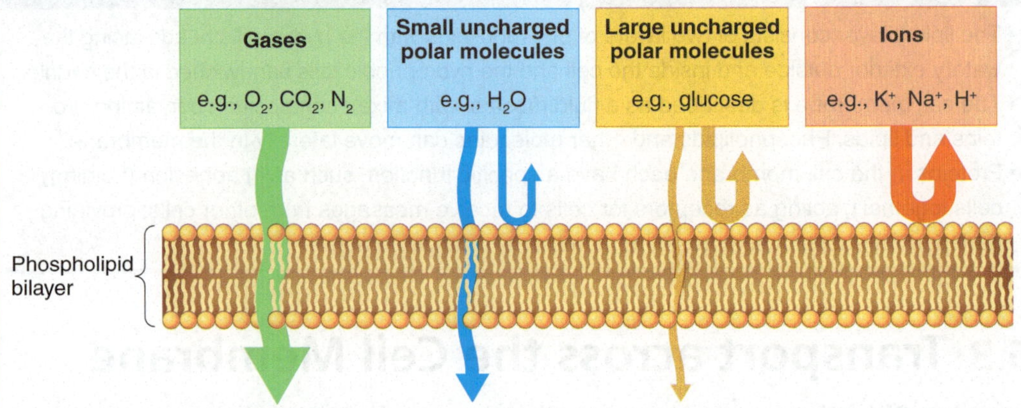

Water and other small polar molecules can diffuse through a lipid bilayer, but less freely than hydrophobic molecules and gases. Some large uncharged polar molecules can pass through on their own. A lipid bilayer is completely impermeable to ions and large polar molecules.

Charge (Chapter 2) can also influence diffusion. Each ion or charged molecule in a fluid contributes to the fluid's overall electric charge. A difference in charge between two regions of fluid (a charge gradient) can affect a solute's diffusion between them. For example, positively charged substances such as sodium ions (Na^+) will tend to diffuse toward a region with an overall negative charge, and negatively charged ions such as chloride (Cl^-) will tend to diffuse toward a region with an overall positive charge.

Selective Permeability of Lipid Bilayers

The lipid bilayer serves not only to define the boundaries of a cell, but also as a gatekeeper for what can and cannot penetrate the cell. Since some substances can readily diffuse through while others cannot, we say that the lipid bilayer is **selectively permeable** (Figure 5.7). The long, nonpolar tails of phospholipids make the core of a lipid bilayer hydrophobic. Gases and hydrophobic molecules (such as fats and steroids) diffuse freely through this core, but ions and large polar molecules (like sugars) cannot. Small polar molecules such as water diffuse slowly through a lipid bilayer by slipping between the lipid tails.

? Did You Know?

Plants get their water via osmosis—water moves into the more solute-rich root cells. Salting roads in the winter causes the soil to become more solute-dense than the root cells, resulting in the outflow of water, ultimately killing the plant.

Movement of Water

Have you ever wondered why drinking only saltwater will kill you? It comes down to **osmosis**, the diffusion of water through a semipermeable membrane.

The direction and rate of water's movement depends on the relative solute concentration of the two fluids across a membrane. For example, if a cell interior contains more solute than to the exterior of the cell, the outside solution is said to be **hypotonic** (*hypo–*, under) to the cell interior (Figure 5.8). This would cause a net movement of water via osmosis toward the interior of the cell, causing it to swell or even burst. If, on the other hand, there were a higher solute concentration to the exterior of the cell, that solution would be **hypertonic** (*hyper–*, over) to the cell interior, resulting in a net flow of water out of the cell, causing the cell to shrink. The diffusion will continue until the two fluids are **isotonic**, meaning that they have the same overall solute concentration.

The saline solutions given intravenously to someone in the hospital are always isotonic to the fluid in the blood to prevent damage to the blood cells.

! Point of Confusion: Osmosis is the movement of water down its concentration gradient. This means that water will move to an area where there is a high solute concentration and low water concentration. One way to remember this is to think that water follows the solute.

So, if an organism consumes saltwater, the concentration of salt increases in the fluid outside of cells but not inside the cells. The fluid outside of the cells would then be hypertonic relative to the fluid inside the cells, and osmosis will cause water to rush out of the cells. With less water inside the cell, the solute concentration of the cytosol would change. If it changes enough, the cell's enzymes would stop working (Chapter 4), with lethal results.

Figure 5.8 **Movement of water into and out of red blood cells can cause the cell's shape to change**

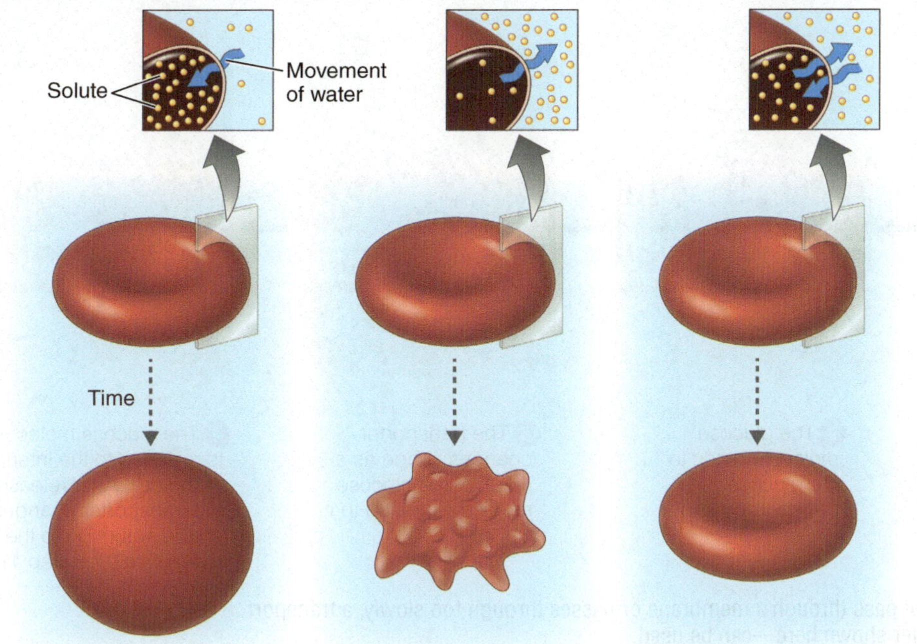

Hypotonic solution
The cell interior contains more solute than the solution outside of it; the net flow of water is into the cell, causing it to swell.

Hypertonic solution
The solution outside the cell contains more solute than inside the cell; the net flow of water is out of the cell, causing it to shrink.

Isotonic solution
The solute concentration is the same between the cell interior and the solution outside of it; there is no net flow of water into or out of the cell.

Study Tip

Remember that we determine hypertonic and hypotonic relative to one another, and that water diffuses from a hypotonic solution into a hypertonic one. Sometimes thinking of these words in relation to one another helps to keep them straight!

Passive and Active Transport

The lipid bilayer is selectively permeable, and so only certain molecules can freely pass through the membrane. How does the cell get the necessary molecules that can't diffuse through the membrane to the inside of the cell? They need a helper!

Transport proteins help molecules pass through the cell membrane, specific to the type of substance they move. For example, calcium pumps only move calcium ions; glucose transporters move only glucose; and so on.

Transporters can move molecules down or up their concentration gradient. Let's return to how concentration impacts diffusion. Molecules naturally move away from areas of high concentration to areas with low concentration, down a concentration gradient. Additionally, charged molecules will move away from areas of similar charge and towards areas where the charge is different. Since these are energetically favorable processes (Chapter 4), transport by these mechanisms requires no energy and is called **passive transport**.

There are two general kinds of passive transport: diffusion and facilitated diffusion. Diffusion, as we discussed earlier in this section, means that the molecules or ions can move down their concentration gradients without any help, from an area of higher concentration to an area of lower concentration (Figure 5.9).

Figure 5.9 **Passive transport is the movement of molecules or ions down their concentration gradient**

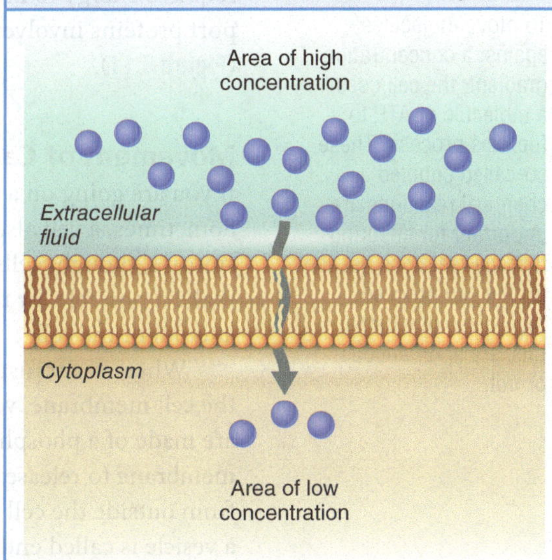

With simple diffusion, a molecule can move in space or across a membrane without any input of energy.

Figure 5.10 **Facilitated diffusion is passive transport that requires a protein**

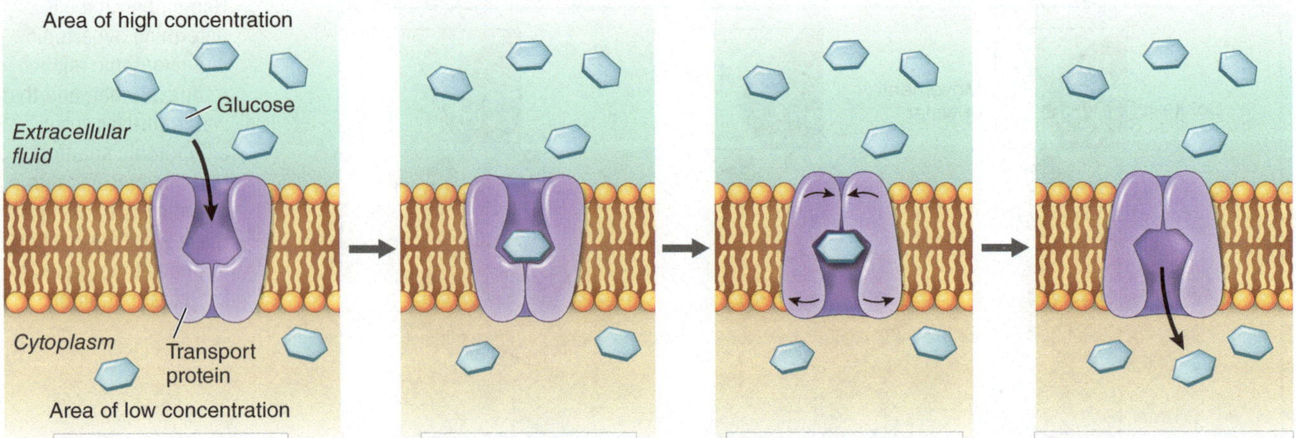

1 A glucose molecule enters a glucose transport protein.

2 The glucose molecule binds to the glucose transport protein.

3 The transporter changes shape as a result of the glucose molecule binding to it.

4 The glucose molecule is released to the interior of the cell. After release, the transporter changes shape, returning to the reset state (see step 1).

In cases where a molecule either cannot pass through a membrane or passes through too slowly, a transport protein—such as the glucose transporter shown here—can be used.

In some situations, however, such as when a molecule or ion needs to cross a membrane and can't diffuse on its own, a transport protein is needed. This process is called **facilitated diffusion**. For example, a transport protein may provide a pathway or physical channel through the membrane for ions to flow through, down their concentration gradients, such as the ion channels in our neurons (Chapter 17) that control the movement of electrical signals in the body. Other transporters will physically move a molecule, such as the movement of glucose through a glucose transporter (Figure 5.10).

Another type of transport protein, called a **proton pump**, moves protons against their concentration gradient. This is a form of **active transport**, which is when a transport protein requires energy to move a solute across a cell membrane. The function of many active transport proteins involves a change in shape triggered by a phosphate-group transfer from ATP (Figure 5.11).

> **Big Ideas in Biology**
>
> **Energy Transfer**
>
> To move molecules against a concentration gradient, the cell uses a molecule of ATP to fuel this process. These so-called coupled chemical reactions are important for ensuring that the reactions needed for life can proceed, regardless of whether they are spontaneous or not.

Movement of Cargo in Bulk

If you are going on a trip, you pack your clothes and other essentials into a suitcase. Sometimes, a cell also needs to move important materials around. Rather than using a suitcase, though, cells pack materials inside "bubbles" of membrane called **vesicles**. Vesicles make it possible for the cell to move materials in bulk, including inside the cell as well as into and out of the cell.

Whereas a transport protein moves one or a few molecules or ions at a time across the cell membrane, vesicles enable the transport of many molecules at once. Since vesicles are made of a phospholipid bilayer like the cell membrane, a vesicle can fuse with the cell membrane to release its cargo to the outside of the cell, or alternatively, collect molecules from outside the cell to bring them in. The process of importing materials into the cell using a vesicle is called endocytosis, and the process of exporting materials out of the cell using a vesicle is called exocytosis.

Endocytosis. In one type of endocytic pathway, receptors in the cell membrane concentrate molecules near one area of the membrane that will fold up to form a vesicle that balloons into the cytoplasm, a process called **endocytosis**. (Figure 5.12A). When the balloon pinches off in the cytoplasm, it becomes a vesicle that encloses a droplet of extracellular fluid, along with any suspended solutes and particles. This type of endocytosis is specific about what it brings into the cell. Endocytosis is one way that viruses enter our cells and make us sick.

Exocytosis. Cells expel materials in bulk by **exocytosis**, where a vesicle in the cytoplasm moves to the cell's surface and its membrane fuses with the cell's outer membrane. As the fusion occurs, the contents of the vesicle are released to the surrounding fluid. (Figure 5.12B)

Figure 5.11 Active transport requires energy to move molecules against their concentration gradients

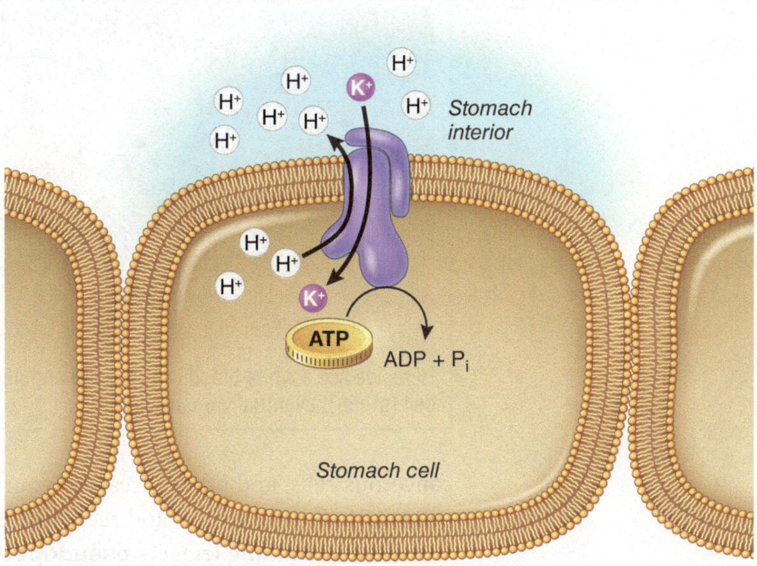

In the example shown here, a proton pump moves protons from a stomach cell to the stomach interior using active transport, where ATP is used as the energy source. A potassium ion moves to the cell interior in the process.

Figure 5.12 Endocytosis and exocytosis are processes used for moving materials into and out of the cell

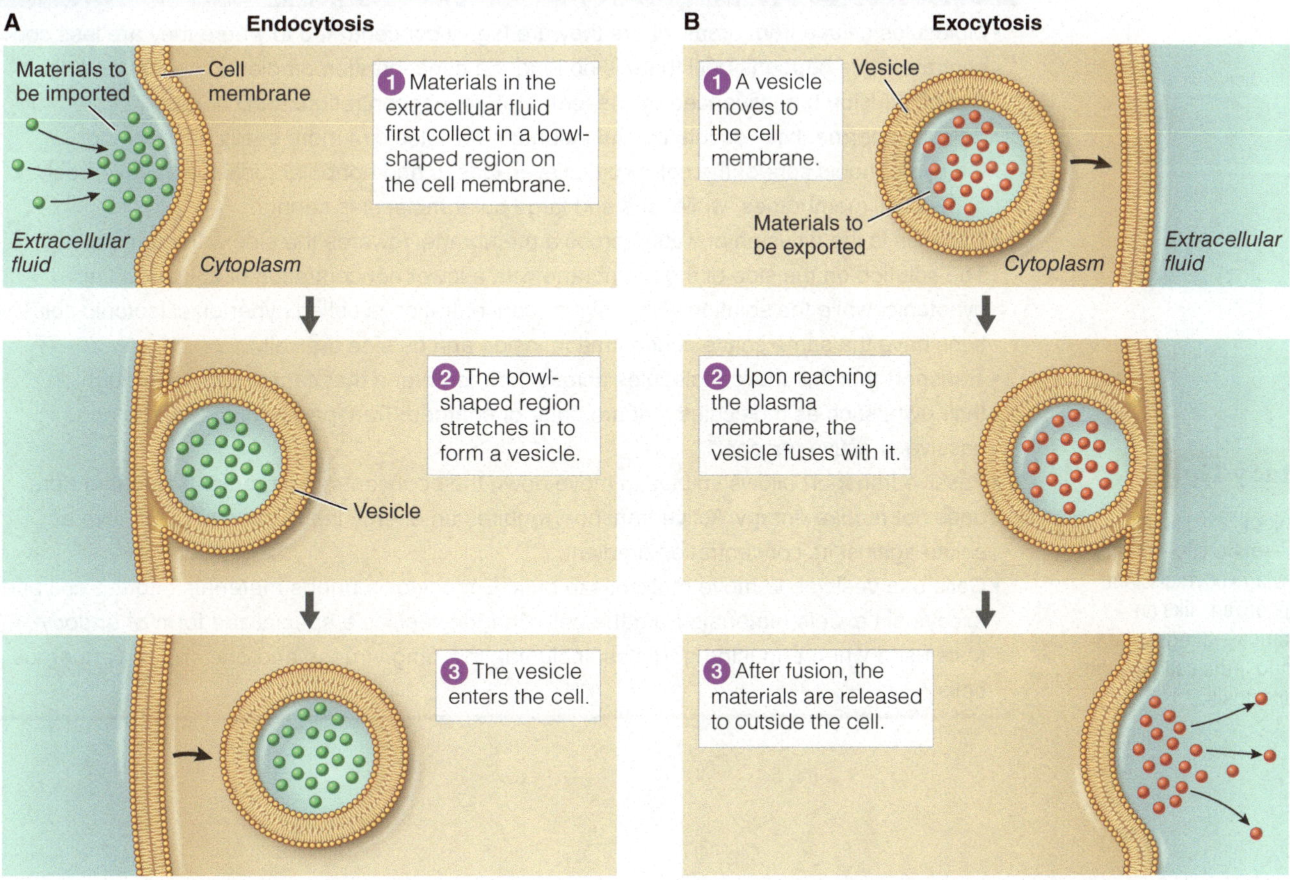

Figure 5.13 White blood cells in your immune system remove pathogens from your body via phagocytosis

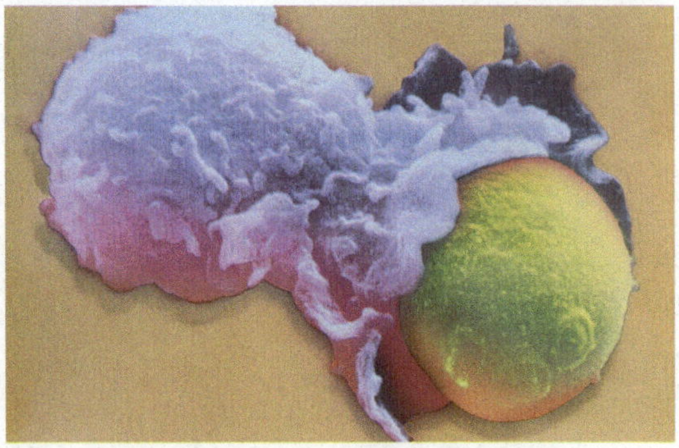

Here, we see a white blood cell (purple) engulfing a yeast cell (green), eventually destroying it.

Phagocytosis. Have you ever played the classic arcade game Pac-Man, where the cheerful yellow character races around to eat the dots while simultaneously avoiding the ghosts? It turns out, cells do something similar! **Phagocytosis** is when a cell engulfs a large particle such as a microorganism or cellular debris (Figure 5.13). Phagocytosis means "cell eating," and many single-celled protists such as amoebas feed by this special endocytic pathway. Some of your white blood cells use phagocytosis to engulf viruses and bacteria, cancerous cells, and other threats to your health (Chapter 17). Phagocytosis begins when a cell encounters a particle, and receptors in its membrane bind to molecules on the particle's surface. The binding triggers **pseudopodia** (so-called "false feet") to extend from the membrane and surround the particle. As the pseudopodia merge around the particle, it becomes trapped inside a vesicle that sinks into the cytosol.

5.3 Summary

- Molecules diffuse from areas where they are highly concentrated to where they are less concentrated; this happens until there is no longer a concentration gradient. The rate and direction of diffusion are influenced by the temperature, concentration, and charge.
- Selective permeability occurs because some molecules can more easily pass through the hydrophobic core of the cell's lipid bilayer. Gases and nonpolar molecules move freely across cell membranes, while ions and large polar molecules cannot.
- Osmosis is the diffusion of water across a membrane, towards the side with more solute. The solution on the side of the membrane with a lower concentration of solutes is called hypotonic, while the solution with a higher concentration is called hypertonic. Isotonic solutions have the same solute concentration inside and outside the cell.
- Transport proteins move molecules across the membrane that cannot diffuse through on their own, such as molecules that are polar or charged. Transport proteins can be used in passive or active transport.
- Passive transport allows solutes to move down their concentration gradients and therefore does not require energy. Active transport requires an energy input, often ATP, to move a solute against its concentration gradient.
- Cells use vesicles to move materials in bulk. Endocytosis brings materials into the cell and exocytosis expels materials from the cell. Phagocytosis is a specialized form of endocytosis used when engulfing particles, including invading viruses and bacteria or cancerous cells.

Study Tip

Think of "ex" like an exit—materials *exit* the cell via *exo*cytosis. And think of "en" like an entrance—things *enter* ("endo") the cell through *endo*cytosis.

5.4 Eukaryotic Organelles

Learning Objectives

5-15 Describe the relationship between the structure and function of the nucleus.

5-16 Describe the path of a new protein moving through the endomembrane system.

5-17 Explain the role of the lysosomes in the cell.

Earlier in this chapter, we discussed the cell membrane. In Chapter 4, we talked about how cells increase their efficiency through compartmentalization of different tasks into organelles. This compartmentalization is like how we organize our living spaces into distinct functional areas. Each of these organelles is enclosed in a membrane made up of phospholipids, similar to the membrane that encases the entire cell. Because each organelle is surrounded by a membrane, it can regulate the types and amounts of substances that enter and exit.

Through this control, the organelle maintains a specific internal environment that allows it to carry out a particular function—for example, isolating toxic or sensitive substances from the rest of the cell or transporting substances through the cytoplasm. Membrane-enclosed organelles—which include the nucleus, endoplasmic reticulum (ER), Golgi apparatus, chloroplasts, and mitochondria (see Chapter 4 for details about these last two organelles)—are generally characteristic of eukaryotes (Figure 5.14).

The Nucleus Is the Cellular Control Center

The nucleus serves two important functions. First, it keeps the cell's DNA—its only copy of the genetic material needed for the functioning of the organism—away from metabolic processes and reactions that might damage it. Isolated in its own compartment, the DNA stays

Figure 5.14 An idealized animal cell showing the various organelles

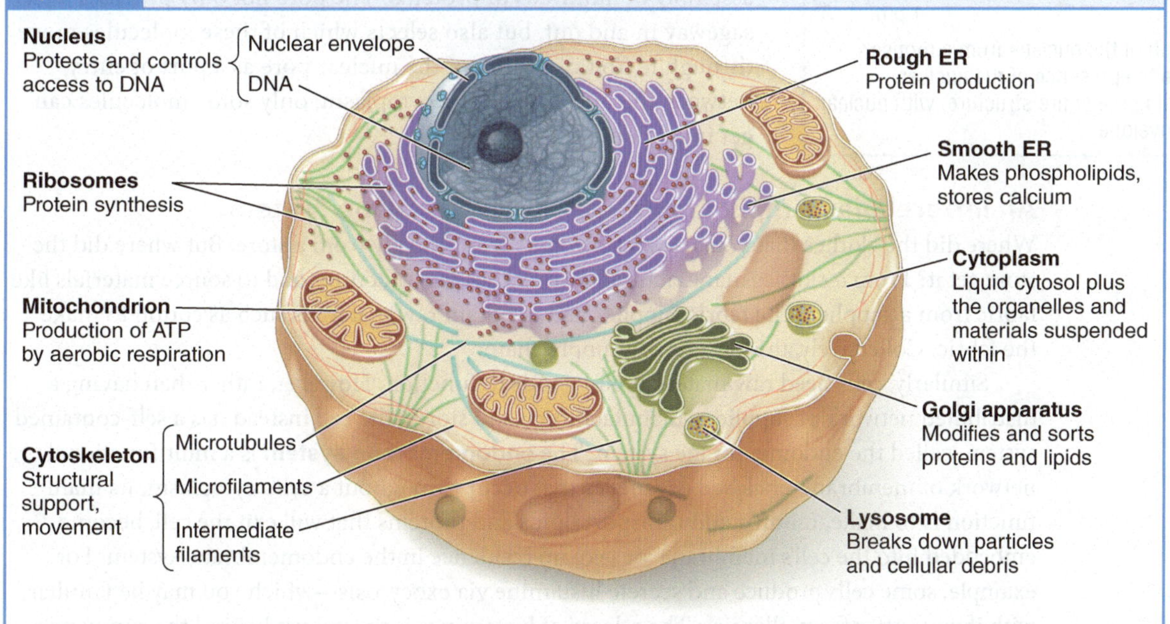

Biology for Life

Connecting Nuclear Pores to Alzheimer's Disease

Alzheimer's disease is a devasting neurogenerative condition affecting millions of Americans. The brain of a patient with Alzheimer's has two characteristics: groups of misfolded proteins stuck together, called amyloid plaques, that are found between brain cells; and tau tangles, found inside brain cells.

Tau tangles are fibrous clumps of misfolded proteins; they consist of a protein (called tau) that is produced incorrectly when subcellular structures called amyloid fibrils are present. Defective tau proteins interact abnormally with nuclear pore proteins, causing the tau proteins to aggregate into tangles, altering nuclear pores in a way that makes them leaky. Among other effects, proteins that should stay in the cytoplasm instead accumulate in the nucleus, and RNA molecules that should stay in the nucleus accumulate in the cytoplasm. These outcomes disrupt the function of the cell, eventually leading to cell death.

Can you find other human diseases that are caused by defects in the nuclear pores?

Figure 5.15 The structure of the nucleus

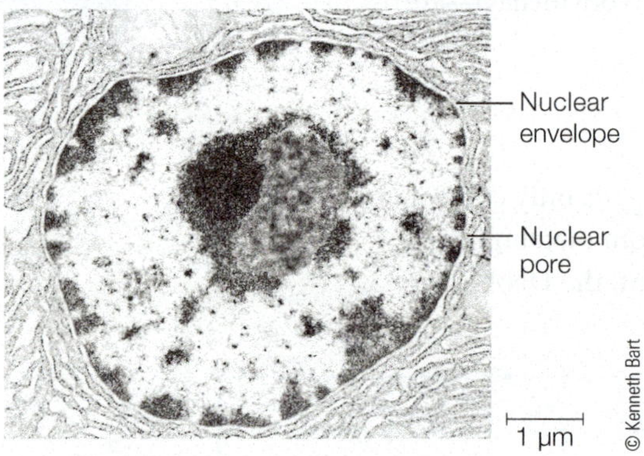

Electron micrograph of the nucleus from a mouse's pancreas cell. Note the presence of the nuclear envelope surrounding the entire structure, with nuclear pores within the envelope.

separated from the bustling activity of the cytoplasm. Second, the nucleus controls the passage of certain molecules across its membrane. The nucleus has a special membrane, the **nuclear envelope**, that protects the DNA while also allowing the passage of molecules into and out of the nucleus (Figure 5.15).

Like you might wrap a precious vase up in multiple layers to protect it during transit, the nuclear envelope wraps the DNA up in not one but two layers of phospholipid bilayer. Only certain molecules—such as mRNA and proteins—can enter or exit the nucleus through holes in the nuclear envelope called **nuclear pores**. Thousands of nuclear pores span both bilayers, each a complex assembly of hundreds of proteins. The pore not only provides a passageway in and out, but also selects which of these molecules move through it. You can think of the nuclear pore as a port of entry between the nucleus and the cytoplasm: only some molecules can get in and out.

An Intracellular Supply Chain: The Endomembrane System

Where did the clothes that you are wearing come from? Likely from a store. But where did the store get it? From a clothes manufacturer. The clothes manufacturer had to source materials like fabric from a supplier. The fabric supplier had to obtain raw materials such as cotton to make the fabric. Collectively, this describes a supply chain.

Similarly, cells need raw materials and goods to function. However, rather than having a distributed network of suppliers, manufacturers, and stores, the cell instead has a self-contained system, called the endomembrane system. The **endomembrane system** is a multifunctional network of membrane-enclosed organelles that occur throughout a cell's cytoplasm. Its main function is to make, modify, and transport lipids and proteins that will exit the cell, become embedded into the cell's membrane, or take up residence in the endomembrane system. For example, some cells produce and secrete histamine via exocytosis—which you may be familiar with if you suffer from allergies. The release of histamines is the culprit behind the runny nose, sneezing, and teary eyes you may experience in response to exposure to allergens like pollen or cat hair.

In addition to its role in manufacturing, the endomembrane system also destroys toxins, recycles wastes, and has other special functions. Components of the system vary among different types of cells, but its main components are the endoplasmic reticulum and the Golgi apparatus (Figure 5.16). Note that in the same way vesicles transport materials into and out of cells, vesicles throughout the endomembrane system are also used to transfer materials between organelles. Let's explore the journey of histamines as they are produced and move through the endomembrane system before finally being secreted from the cell.

Endoplasmic reticulum. The **endoplasmic reticulum (ER)** is an extensive system of sacs and tubes that enclose a single, continuous compartment. The ER makes lipids and proteins for the cell's membranes as well as proteins for secretion, such as histamine. Because the endoplasmic reticulum is an extension of the outer lipid bilayer of the nuclear envelope, the nuclear envelope is sometimes considered to be part of the endomembrane system.

> **Big Ideas in Biology**
>
> **Structure and Function**
>
> The vesicles that transport material throughout the endomembrane system are made of a lipid bilayer, just like the membrane surrounding the organelles and the entire cell. This allows the vesicle to seamlessly fuse with these structures.

Figure 5.16 The endomembrane system is an intracellular supply chain

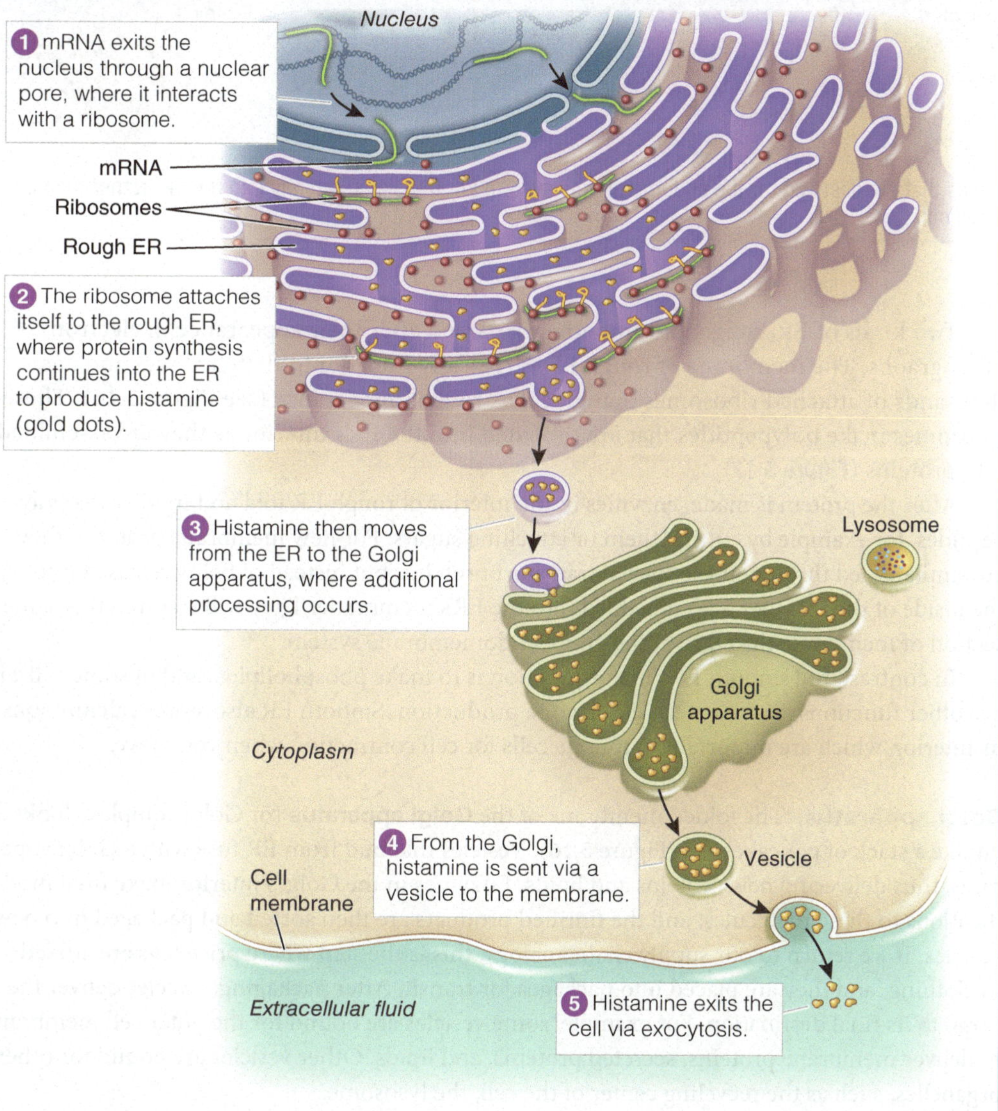

The production of a secreted molecule begins in the nucleus and ends when the molecule—in this case, histamine—is secreted at the cell membrane.

Figure 5.17 **Entry of proteins into the ER**

1. mRNAs that encode proteins destined for the endomembrane system also encode a tag-like signal sequence (red) that tells the ribosome to take it to the ER.
2. The tag gives access to a special transport protein in the ER membrane, and protein synthesis resumes as the histamine polypeptide is threaded through the transport protein.
3. The histamine polypeptide is released into the rough ER, where enzymes fold and modify the polypeptide into the finished protein.

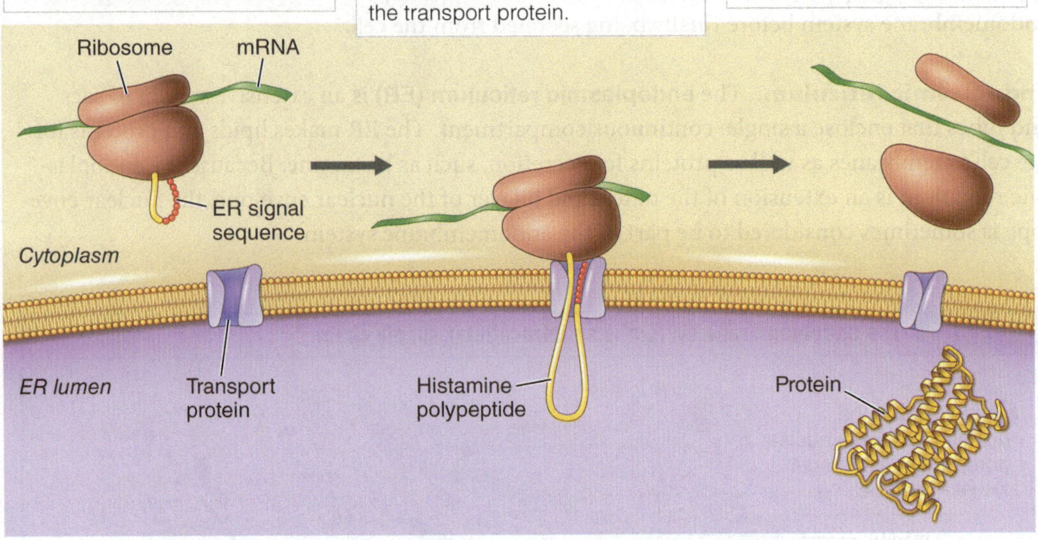

Proteins that are destined for the endomembrane system are translated directly into the ER, rather than in the cytosol.

Two kinds of ER, rough and smooth, are named after their appearance in electron micrographs. The membrane of rough ER is typically folded into flattened sacs and has thousands of attached ribosomes that give it a "rough" appearance (see Figure 5.16). These ribosomes make polypeptides that are threaded into the ER's interior as they are assembled into proteins (**Figure 5.17**).

After the protein is made, enzymes in the interior of rough ER fold and modify the polypeptides, for example by cutting them or attaching sugars. For new membrane proteins, they are similarly fed through these special transport proteins, but instead of being released into the inside of the ER, they are embedded into the ER's membrane. They will travel in this same section of membrane throughout the entire endomembrane system.

In contrast, the smooth ER's major function is to make phospholipids, and in some cells, it has other functions, such as steroid hormone production. Smooth ER also stores calcium ions in its interior, which are important in muscle cells for cell contraction when you move.

Golgi apparatus. The folded membrane of the **Golgi apparatus** (or Golgi complex) looks a bit like a stack of pancakes (see Figure 5.16). Vesicles that bud from ER fuse with a Golgi apparatus, thus delivering new proteins and lipids. Enzymes in the Golgi's interior make final modifications to these molecules, and the finished products are then sorted and packaged into new vesicles. If we return to our supply chain analogy, this is the step where price tags are affixed to clothing, and they are placed into packages for transit. After packaging, vesicles deliver the cargo to its final destination. For example, some vesicles are bound for the outer cell membrane to deliver membrane proteins, secreted proteins, and lipids. Other vesicles are bound for other organelles, such as the recycling center of the cell, the lysosome.

Process of Science

Camillo Golgi and the Neuron

The Golgi apparatus is named after its discoverer, Camillo Golgi. Other than his namesake organelle, Golgi is most well-known for his discovery of a staining technique called the black reaction. Using this stain, scientists were able to visualize neurons for the first time. Santiago Ramón y Cajal (whom we met in Chapter 1) used this technique when making his famous drawings that led to the proposition that the neuron is the functional unit of the brain, called the neuron doctrine. Due to their respective contributions to the understanding of the neuron, Golgi and Cajal shared the Nobel Prize in Physiology or Medicine in 1906.

Although they *shared* the prize, they gave contradictory acceptance speeches. Golgi was convinced of reticular theory, the idea that the brain was a single contiguous unit. Cajal, on the other hand, promoted the neuron doctrine—that the nervous system was made up of single cells. Both were convinced they were right. Sometimes, scientists can interpret the same results differently and it takes additional work by other scientists to provide new evidence and break a stalemate. In the case of the nervous system, later experimental results supported Cajal and not Golgi. Now reticular theory is considered obsolete and today the neuron doctrine is the predominant theory.

Relate this story to what we learned in Chapter 1. What non-scientific factors could have been at play that caused Golgi and Cajal to reach different conclusions, while looking at the same data?

Digging into Data

The Chloride Chronicles: A Cellular Water Park Crisis in Cystic Fibrosis

Imagine your body's cells are tiny water parks. In this scenario, a membrane transport protein called the cystic fibrosis transmembrane conductance regulator (CFTR) acts like a water slide operator, controlling the amount of water coming down the slide. As a membrane transporter, the CFTR protein is responsible for moving chloride ions through the cell's membrane. Water follows the chloride ions, creating a thin film of mucus that protects and lubricates the cavities and ducts of the lungs, liver, pancreas, intestines, and reproductive system.

In people with cystic fibrosis (CF), the water park is in crisis. Defects in the CFTR gene mean that there aren't enough CFTR protein slide operators in their water parks. Not enough chloride ions leave the cells, so not enough water leaves them either. Patients with CF have trouble breathing because too few CFTR proteins result in thick, dry mucus that clogs the airways to the lungs and other passages. These clogged airways can make breathing a real challenge and can lead to chronic lung infections.

To understand why people with CF do not have functional CFTR protein, researchers tracked the location of the CFTR protein in cells, in people with and without CF. They looked at the endoplasmic reticulum (where CFTR protein is made), vesicles (which transport CFTR protein between the ER and the Golgi apparatus), and the Golgi (where CFTR protein is packaged) (Figure 5.18).

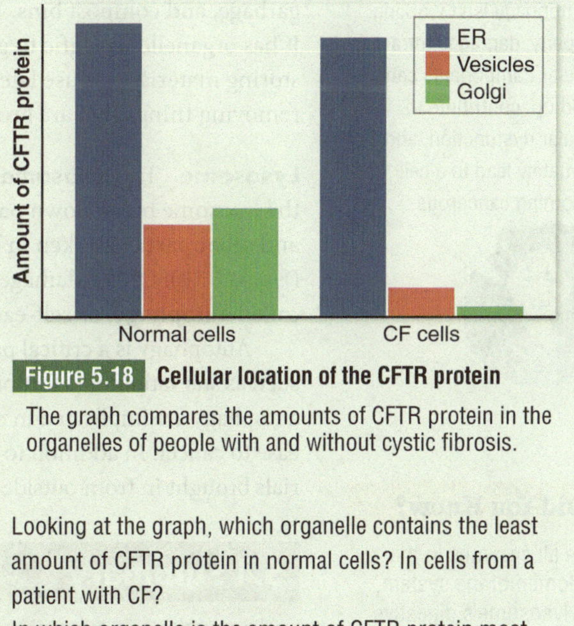

Figure 5.18 Cellular location of the CFTR protein

The graph compares the amounts of CFTR protein in the organelles of people with and without cystic fibrosis.

1. Looking at the graph, which organelle contains the least amount of CFTR protein in normal cells? In cells from a patient with CF?
2. In which organelle is the amount of CFTR protein most similar in both types of cells?
3. Using the data from the graph, where do you think the CFTR protein is getting held up in the cells of people who have CF?

Figure 5.19 **The lysosome is the recycling center of the cell**

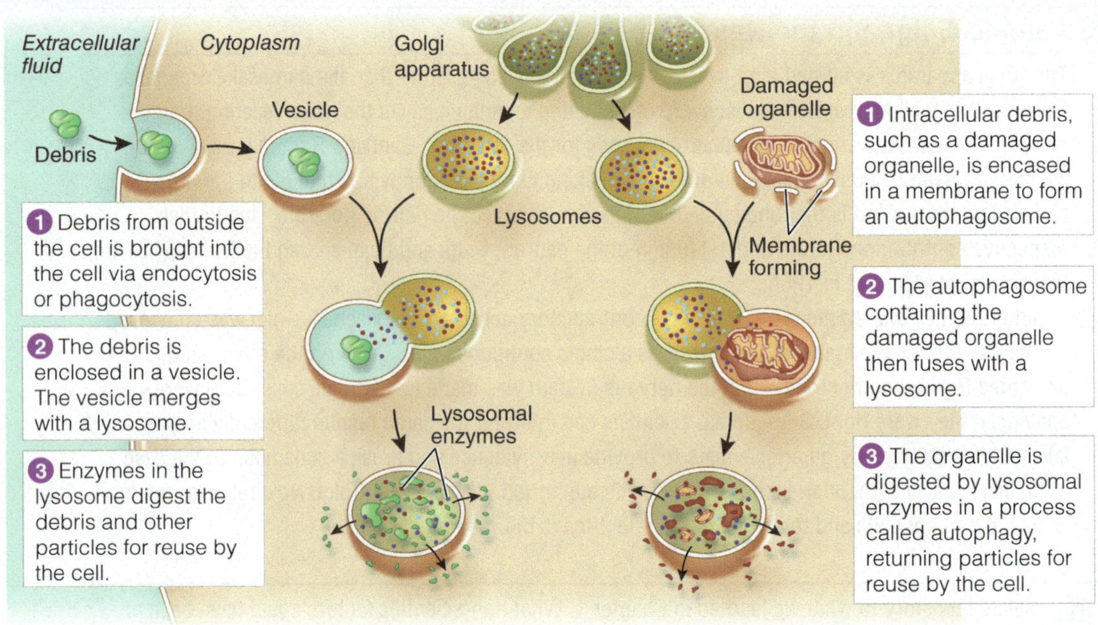

The lysosome can digest extracellular materials (left), or intracellular materials through autophagy (right).

Defects in autophagy are linked to the formation of malignant tumors. Biologists hypothesize that if autophagy isn't working properly, damaged intracellular components can build up, contribute to cellular dysfunction, and ultimately lead to a cell becoming cancerous.

Recycling and Trash Centers: The Lysosome

Wherever you live—whether it's a dorm room, apartment building, or house—you likely have a special place for things that you no longer need and want to get rid of. Perhaps trash disposal involves walking your stuff to a dumpster or sorting household waste into recycling, garbage, and compost bins. You know what kind of waste goes where. The cell is no different: It has organelles specific to getting rid of unwanted materials, recycling macromolecules, and storing materials for use later. The primary organelle responsible for keeping the cell neat, and removing things that are broken or unwanted, is the lysosome.

Lysosome. The **lysosome** is an organelle that acts a bit like the cell's stomach. Enzymes in the lysosome break down particles such as cellular debris and bits of waste. Bacteria, cell parts, and other particles taken in by the cell are delivered to lysosomes for disassembly and reuse (Figure 5.19). Old or damaged organelles are also digested in the lysosome as part of a process called **autophagy,** or self-eating.

Autophagy is a critical part of cell biology. For example, if the cell is under stressful conditions, such as starvation or infection, it can consume parts of itself to produce more energy. Defects in autophagy are implicated in diseases ranging from neurodegenerative diseases like Parkinson's disease to cancer. In addition to its intracellular recycling function, the lysosome is often where materials brought in from outside the cell, either from endocytosis or phagocytosis, are digested.

? Did You Know?

Like all enzymes in the endomembrane system, the lysosome's digestive enzymes traverse the ER and the Golgi apparatus. Why don't the enzymes eat the ER, the Golgi apparatus, or vesicles? Lysosomal digestive enzymes only work at the low pH found inside the lysosome.

5.4 Summary

- The nucleus is a membrane-enclosed organelle, present in eukaryotic cells only, that protects and controls access to the cell's DNA. Nuclear pores allow certain proteins and RNA molecules to exit and enter the nucleus through the nuclear envelope.
- A newly formed protein moves through the endomembrane system by first being built by a ribosome and threaded into the rough ER, moving via a vesicle to the Golgi apparatus for processing, and then being transported in a vesicle to its final destination—either another organelle or outside the cell.
- Lysosomes contain digestive enzymes that help the cell recycle and dismantle unwanted molecules.

5.5 Elements of Connection

Learning Objectives

5-18 Describe the function of each type of cytoskeletal elements in eukaryotes.

5-19 Compare the roles of cytoskeletal elements in cell motility.

5-20 Explain the different functions of extracellular matrix (ECM).

5-21 Describe the three major types of cell junctions in animals.

5-22 Explain how cells communicate with one another through gap junctions.

Just like you have a skeleton that supports your body, the cell also has an underlying supportive structure called the **cytoskeleton** (Figure 5.20). The cytoskeleton is a network of proteins. These proteins not only support the cell but also physically move the cell or provide highways for other proteins to traverse the cytoplasm. Some cytoskeletal elements are permanent; others form only at certain times. Here, we discuss three types of cytoskeletal elements that occur in eukaryotic cells: microfilaments, intermediate filaments, and microtubules.

Figure 5.20 **The cytoskeleton**

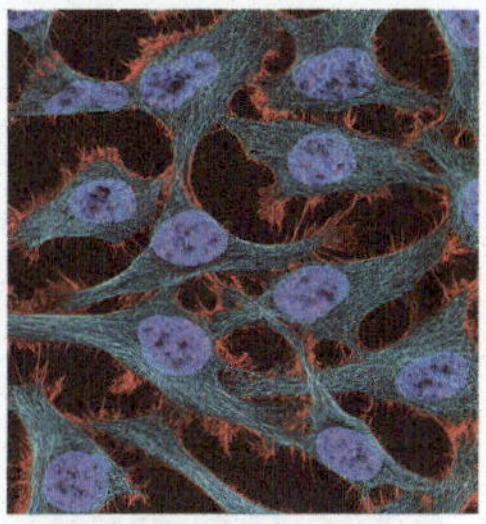

This fluorescent microscopy image of human cervical cells shows microfilaments (red) and microtubules (light blue). The dark blue circles are nuclei.

National Institutes of Health/Stocktrek Images/Getty Images

Cytoskeletal Elements

Microfilaments. Fine fibers called **microfilaments** consist primarily of actin subunits (Figure 5.21A). In many cells, a mesh of microfilaments is part of the cell cortex, a region of cytoplasm inside the membrane. The mesh, which connects to and supports the membrane, forms a scaffolding for proteins that function in cellular movement, contraction, shape changes, and migration. For example, the process of phagocytosis (Section 5.3) relies on actin in order to engulf particles.

Intermediate filaments. In addition to microfilaments, animal cells also contain intermediate filaments (Figure 5.21B). **Intermediate filaments** are made of protein subunits and form a stable framework that lends structure and resilience to cells and tissues. Several types of intermediate filaments are assembled from different fibrous proteins. For example, cells in your hair follicles assemble keratin, a structural protein, into intermediate filaments that make up your hair.

Microtubules. Long, hollow cylinders called **microtubules** are cytoskeletal elements that function in movement (Figure 5.21C). They consist of tubulin subunits and can rapidly assemble when they are needed and disassemble when they are not. For example, during the process of cell division (Chapter 6), microtubules assemble, separate the cell's chromosomes, and then disassemble. Some eukaryotic cells, including the amoeba and white blood cells (see Figure 5.13) extend their microtubules to form pseudopodia. As these temporary, irregular lobes bulge outward, they can move the entire cell or, in the case of single-celled amoeba, engulf a target such as prey.

Microtubules also function as highways throughout the cell. The "cars" that move along these highways are called **motor proteins**. Remember the vesicles that move between organelles and the membrane in the endomembrane

Figure 5.21 **Microfilaments, intermediate filaments, and microtubules are all part of the cytoskeleton**

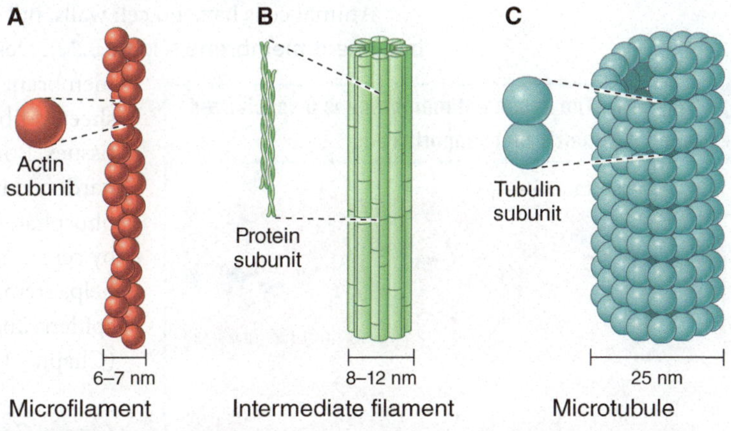

A. Microfilaments are the smallest part of the cytoskeleton and are made up of actin subunits. Microfilaments reinforce cell membranes and facilitate muscle contraction.

B. Intermediate filaments are made of protein subunits that provide structural support for cell membranes and tissues.

C. Microtubules are made up of tubulin subunits and are involved in both moving cargo throughout the cell and moving the entire cell.

Figure 5.22 **Microtubules facilitate a wide variety of movement**

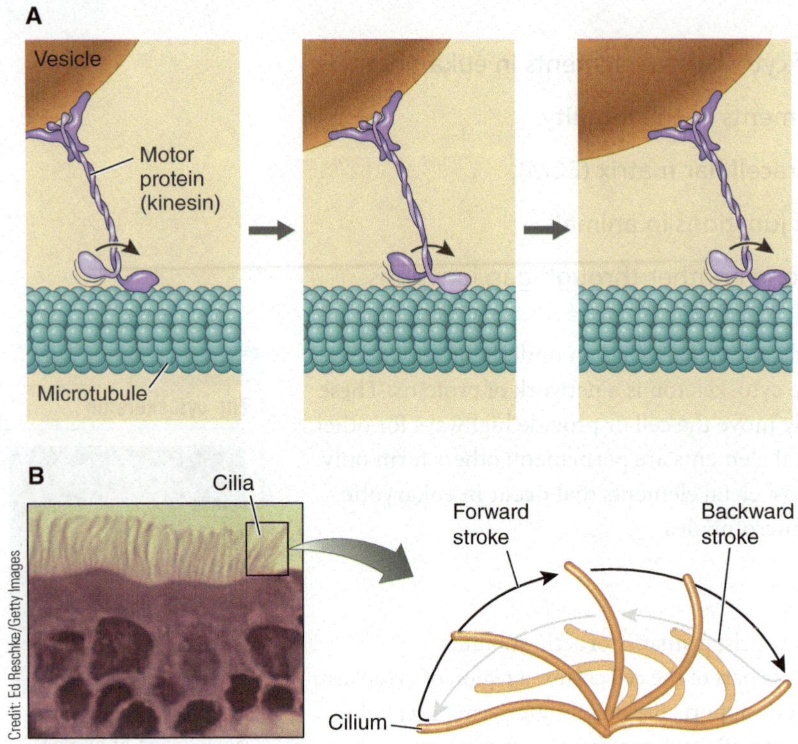

A. Kinesin motor proteins "walk" along microtubules carrying cargo such as vesicles.

B. Cilia in the airway move mucous out of your airways by spinning rhythmically in a circle.

system? Motor proteins like kinesin have molecular "feet" that walk along the microtubule while "carrying" the vesicle to its final destination (Figure 5.22A).

Another type of motor protein called dynein is used to move structures such as the flagella on a sperm cell (see the chapter opener). In the case of sperm, the dynein makes the flagella whip back and forth, propelling the cell forward. Dynein can also interact with microtubes to facilitate the rhythmic waving motion of short, hairlike **cilia** (singular, cilium) that project from the membrane of some eukaryotic cells (Figure 5.22B).

Cilia often occur in clumps that beat in unison. Their coordinated movement can propel a cell through fluid and stir fluid around a stationary cell. For example, in the trachea, cilia help move mucous and the dirt trapped within the mucous away from your lungs. The action of the cilia is why you cough up dirty mucous after sitting around a campfire—the dirt you see from the campfire smoke was trapped by the mucous and removed by the cilia before entering your lungs.

Extracellular Matrix

Many cells secrete extracellular matrix onto their surfaces. **Extracellular matrix (ECM)** is a complex mixture of molecules that structurally supports and physically protects the cells that secrete it. A cell wall is an example of ECM. Most prokaryotes, all plants, and some fungi and protists have cell walls. The composition of the cell wall differs among these groups, although they also have similarities. Like a prokaryotic cell wall, a eukaryotic cell wall is porous: water and solutes easily cross it on the way to and from the membrane.

Animal cells have no cell walls, but some types secrete an extracellular matrix called a **basement membrane** (Figure 5.23). Despite the name, the basement membrane is not a cell membrane as it does not consist of a lipid bilayer. Rather, it is a sheet of fibrous material that structurally supports and organizes tissues. Bone cells secrete an ECM of collagen, a fibrous protein, hardened into a mineral by bonding with calcium, magnesium, and phosphate ions. A covering called **cuticle** is a type of ECM secreted by cells at a tissue surface. In plants, a cuticle of waxes and proteins helps stems and leaves retain water and fend off insects. Crabs, spiders, and other arthropods have cuticles called *exoskeletons* (Chapter 13) that consistent primarily of chitin.

Figure 5.23 **The basement membrane is a specialized extracellular matrix that supports cells**

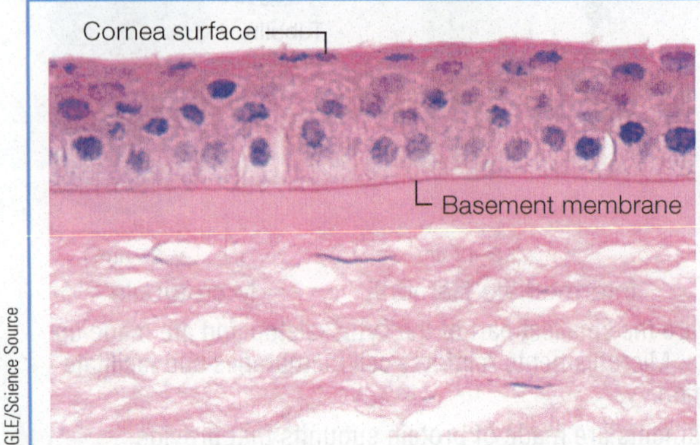

In this section of tissue from a human cornea (part of the eye), the basement membrane is visible as a dark line underneath the cells.

Close Connections between Cells

The milk mothers produce within the first 72 hours of a baby's birth is special. This thick, golden liquid is called colostrum. Colostrum is packed full of Mom's immune cells, proteins called antibodies that help prevent infections, and anti-microbial factors, among other beneficial components. Normally, these macromolecules cannot be absorbed in the intestine—except during the first few days of life, when cells in the newborn's gastrointestinal system lack a specific kind of structure called a tight junction (Figure 5.24A).

Tight junctions are proteins that join the membranes of adjacent cells together, much like quilting connects two pieces of fabric to make a blanket. Tight junctions create a barrier to absorption in the intestine. At birth, babies do not have tight junctions and so the larger goodies in colostrum can sneak in between the cells. As the baby's tight junctions form, the mother begins to make mature milk, which has a different composition.

Tight junctions are useful in any bodily tissue where it is necessary to prevent body fluids from seeping between the cells. For example, the lining of the stomach is leakproof because tight junctions seal its cells together. These junctions keep gastric fluid, which is acidic and contains digestive enzymes, safely inside the stomach. Some bacteria that infect the stomach's lining damage tight junctions, allowing gastric fluid to leak into and injure the underlying tissue, causing painful peptic ulcers.

Tight junctions are not the only way cells physically connect to one another. Cells also connect through **desmosomes**, which consist of anchor proteins that work like a rivet or button to connect two cells (Figure 5.24B). These junctions strengthen a tissue because they connect to cytoskeletal elements inside each cell. Contractile tissues (such as heart muscle) have many desmosomes, as do tissues subject to abrasion or stretching (such as skin).

? Did You Know?

Mother's milk also helps promote the formation of a colony of healthy microbes in the infant's gastrointestinal system.

Figure 5.24 Cells physically join with tight junctions, desmosomes, and gap junctions

A. Tight junctions form when two adjacent plasma membranes are quilted together to create a watertight barrier.

B. Desmosomes act like rivets between two cells, physically connecting the intermediate filaments of adjacent cells.

C. Gap junctions not only hold adjacent cells together, but also act as a portal between cells where molecules can move through, allowing neighboring cells to communicate with one another.

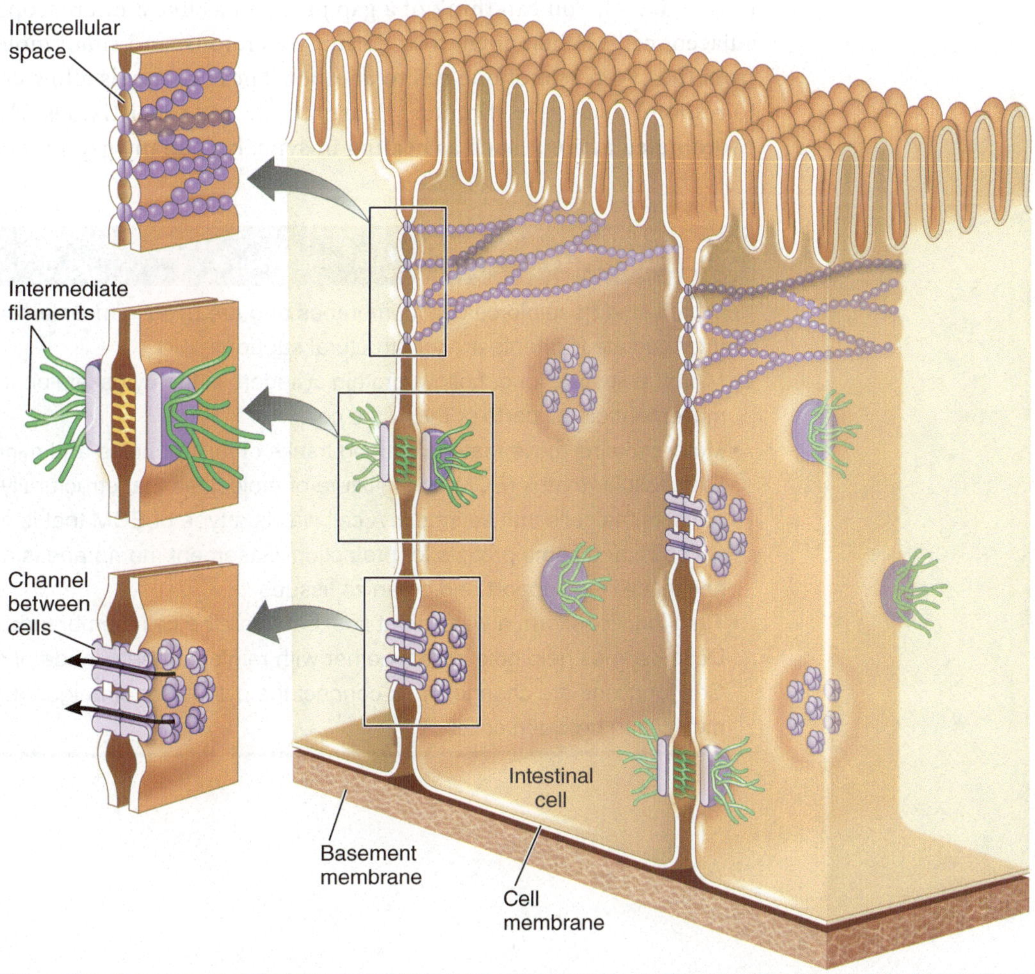

Biology for Life

The Cytoskeleton, Extracellular Matrix, and Cancer Metastasis

One of the hallmarks of a malignant cancerous cell is its ability to travel from one location of the body to another. This is called *metastasis*. Note that if a person has breast cancer that has moved to the brain and the lung, they do not have three kinds of cancer (breast, brain, and lung). Rather, they have metastatic breast cancer that moved to these other sites in the body.

How does this happen? The extracellular matrix that surrounds cells provides structural and biochemical support, and typically helps maintain biological controls on cells. However, when a cell becomes cancerous, the cells change the surrounding extracellular matrix to create an ideal environment for tumors to grow and cancerous cells to escape and invade the rest of the body. The cancer can move throughout the body by changing its cytoskeleton. Given the importance of the cytoskeleton and the extracellular matrix in cancer development and progression, biologists are studying ways to promote normal extracellular matrix, prevent the uncontrolled movement of cells, and prevent the formation of pro-cancerous extracellular matrix.

Healthy cells are connected to the ECM. How does it "help" a cancer cell to move throughout the body if it no longer "hears" the signals from the ECM?

The third way that cells connect also enables them to communicate with each other. Note that we'll discuss cell-cell communication more in Chapter 6. **Gap junctions** are channels that open and close and connect the cytoplasm of adjoining animal cells (**Figure 5.24C**). You can think of a gap junction as like a microscopic tube connecting two adjacent cells. When open, they permit water, ions, and small molecules to pass directly from the cytoplasm of one cell to another. These channels allow entire regions of cells to respond to a single stimulus. Heart muscle and other tissues where cells perform coordinated actions, such as beating in sync, have many gap junctions.

5.5 Summary

- Microfilaments reinforce cell membranes and are used in muscle contraction.
- Intermediate filaments are for structural support.
- Microtubules are long, hollow, protein cylinders in the cytoskeleton that are used for the movement of cell parts or the entire cell.
- Motor proteins move materials along tracks of microtubules and microfilaments.
- Extracellular matrix (ECM) is a mixture of molecules that structurally supports and physically protects the cells that secrete it. A cell wall is a type of ECM that is made by plant cells, fungal cells, and some protists for protection. Basement membrane is a form of ECM used by animal cells to support and organize tissues.
- Tight junctions form a waterproof seal between the cell membranes of adjacent cells. Desmosomes help hold cells together with reinforcements made of cytoskeletal elements. Gap junctions are channels that connect the cytoplasm of neighboring cells, facilitating communication between cells.

Understand It! Dual Coding

Have you ever found it easier to learn something if you see a picture with words? This is because of dual coding. When we receive information in two different ways, for example as a picture and in words, it is easier to remember the information. Visual information is processed differently by our brains than auditory information. Therefore, receiving information in two ways, such as looking at a picture while talking about it, activates more parts of the brain at once, which fosters the formation of durable memories.

Psychological research tells us that dual coding is one of the most effective ways of moving information into our long-term memories. It's also an opportunity to have fun while studying! For example, you can make a graphic organizer, drawings, cartoons, comics, or infographics.

Here are some ways you can study this chapter's material using dual coding.

1. Create a comic strip illustrating an intercellular process, such as how proteins move through the endomembrane system.
2. Draw an organelle and see if your study buddy can guess what you drew.
3. Create an infographic comparing organelles.

Chapter Review

5.1 What Is a Cell?

Cell theory is a major biological theory and has three components: That living organisms are made of at least one cell, the cell is the basic unit of life, and cells come from other cells. All cells contain an outer membrane, nucleic acids, cytosol, and ribosomes. The outer cell **membrane** separates a cell from its environment, and controls which materials enter and exit it. The membrane encloses the jellylike **cytosol** where the cell's other components (including **ribosomes**, which are involved in protein synthesis) are suspended. **Surface-to-volume ratio** limits cell size and influences cell shape. A **eukaryotic** cell's DNA is contained within a membrane-enclosed **organelle** called the **nucleus**. DNA is suspended in the cytoplasm of prokaryotic cells. **Prokaryotic** cells are always unicellular (although they can live in colonies with other similar cells), and they include **bacteria** and **archaea** cells. Almost all cells are too small to see with the naked eye, so we use **light, fluorescence**, and **electron microscopy** to observe them.

5.2 The Cell Membrane

The **lipid bilayer** is the basic structure of almost all cell membranes and has many other molecules attached to or embedded in it. A cell membrane can be described as a **fluid mosaic** of lipids, proteins, and other molecules. The properties of a cell membrane depend on the molecules that compose it. Cell membranes typically have **enzymes, transport proteins, receptor proteins,** and **adhesion proteins.**

5.3 Transport Across the Cell Membrane

Diffusion is the spontaneous dispersal of molecules down their **concentration gradient** (i.e., from higher toward lower concentration). The rate and direction of diffusion are influenced by **temperature** and regional differences in **concentration** and charge. A lipid bilayer is **selectively permeable.** Gases and nonpolar molecules move freely through it; water and some other small polar molecules can slowly diffuse through. Lipid bilayers are impermeable to ions and large polar molecules.

Osmosis is the diffusion of water through a semipermeable membrane. Water diffuses across the membrane, from the **hypotonic** to the **hypertonic** fluid. Water moves at an equivalent rate between two **isotonic** solutions.

Transport proteins move specific ions or molecules across membranes. In **facilitated diffusion**, a **solute** binds to a transport protein that releases it to the opposite side of the membrane. This is a type of **passive transport** (no energy input is required). With **active transport**, a transport protein uses energy (often in the form ATP) to move a solute across a membrane against its concentration gradient.

Exocytosis and endocytosis are vesicle-based transport mechanisms that move substances in bulk across membranes. Cells expel substances by **exocytosis**, and they bring substances into the cytoplasm by **endocytosis**. Some cells use **phagocytosis** to engulf large particles such as other cells.

5.4 Eukaryotic Organelles

Membrane-enclosed organelles maximize cellular efficiency by compartmentalizing tasks and substances. Eukaryotic cells have a **nucleus**, which protects and controls access to the cell's DNA. The outer boundary of the nucleus is a **nuclear envelope**, a double lipid bilayer with **nuclear pores** that allow select molecules, such as RNA and protein molecules, to pass into and out of the nucleus.

The **endomembrane system** is a series of interacting, membrane-enclosed organelles that produce lipids and proteins for the cell's membranes, as well as proteins for secretion. The **endoplasmic reticulum (ER)** is a continuous system of sacs and tubes extending from the nuclear envelope. Polypeptides made by ribosomes on rough ER are folded and modified in its interior. Smooth ER makes phospholipids, stores calcium ions, and has additional functions in some cells.

Vesicles that bud from ER deliver proteins and lipids to the **Golgi apparatus**, where they are given any needed final modifications and sorted into new vesicles. Some vesicles that bud from the Golgi apparatus deliver proteins and lipids to the membrane; others move to the **lysosome**, which is the organelle that digests cellular debris and other particles.

5.5 Elements of Connection

Cells maintain their shape and internal structure due to their cytoskeleton of protein filaments, including **microfilaments** and **intermediate filaments**. The cytoskeleton is also responsible for cell movement. In eukaryotes, ATP-driven **motor proteins** interact with **microtubules** to move cargo around the cell. Microtubules also make up **cilia**.

Many cells secrete **extracellular matrix**, or **ECM**, onto their surfaces. The composition and function of ECM varies depending on cell type. In animals, an ECM called basement membrane supports and organizes cells in tissues. A **cell wall** is another example of ECM. Cells of plants, fungi, and some protists have cell walls, but animal cells do not.

Cell junctions connect cells in tissues. **Tight junctions** form a waterproof seal between cells. **Desmosomes** connect animal cells to one another and the basement membrane using cytoskeletal elements. In animals, **gap junctions** are closable channels that connect adjacent cells and facilitate communication.

Practice It!

Retrieval practice is the process of pulling information from your long-term memory and using it. Here are some questions to help with your retrieval practice.

Self-Quiz

1. All cells have these four things in common: (LO 5-2)
- a. cytosol, a cell wall, nucleic acids, and organelles with membranes.
- b. a membrane, ribosomes, nucleic acids, and a nucleus.
- c. cytosol, nucleic acids, ribosomes, and a membrane.
- d. a cell wall, cytosol, ribosomes, and nucleic acids.

2. The surface-to-volume ratio ______. (LO 5-3)
- a. does not apply to prokaryotic cells.
- b. is part of cell theory.
- c. constrains cell size.
- d. only applies to archaea cells.

3. Cell membranes consist mainly of ______ and ______. (LO 5-6)
- a. phospholipids; carbohydrates
- b. phospholipids; proteins
- c. phospholipids; ECM
- d. phospholipids; ECM

4. In a lipid bilayer, the ______ of the phospholipids are sandwiched between the ______. (LO 5-6)
- a. hydrophilic tails; hydrophobic heads
- b. hydrophilic heads; hydrophilic tails
- c. hydrophobic tails; hydrophilic heads
- d. hydrophobic heads; hydrophilic tails

5. A solute will diffuse from a region where it is ______ concentrated to another where it is ______ concentrated. (LO 5-9)
- a. more; less
- b. less; more
- c. equally; less
- d. equally; more

6. ______ cannot diffuse directly through a lipid bilayer. (LO 5-10)
- a. Steroids
- b. Gases
- c. Ions
- d. Small nonpolar molecules

7. If you immerse a human red blood cell in a hypotonic solution, water will ______ the cell and the cell will ______. (LO 5-11)

a. move into; swell
b. move out of; swell
c. move into; shrivel
d. move out of; stay the same

8. A transport protein requires ATP to pump sodium ions across a membrane. This is an example of ______. (LO 5-13)

a. passive transport
b. active transport
c. facilitated diffusion
d. osmosis

9. Put the following structures in order according to the pathway of a secreted protein, starting with its production:

1. outer cell membrane 2. post-Golgi vesicles 3. Golgi apparatus 4. endoplasmic reticulum (LO 5-16)

a. 1, 2, 3, 4
b. 3, 2, 4, 1
c. 4, 3, 2, 1
d. 4, 1, 3, 2

10. The ______ is a waxy protective layer in plants that is part of their ECM. (LO 5-20)

a. cuticle
b. bark
c. root
d. stem

11. If a person has very stretchy skin that is unusually susceptible to tearing with less force than typical, which of the following cell connection structures might be weakened? (LO 5-20)

a. basement membrane
b. actin filaments in the cytoskeleton
c. motor proteins moving vesicles inside cells
d. microtubules in cilia

12. ______ keep waterproof connections between intestinal cells so that bacterial invaders cannot slip between them when brought in with food. (LO 5-21)

a. Gap junctions
b. Tight junctions
c. Plasmodesmata
d. Desmosomes

13. ______ allow animal cells to quickly share molecular signals through their cytoplasm. (LO 5-22)

a. Gap junctions
b. Tight junctions
c. Plasmodesmata
d. Desmosomes

Critical Thinking

1. Cholesterol can be a part of a lipid bilayer that adds stability and helps the membrane resist temperature changes. Consider animals such as elk in the frozen arctic tundra. They have warm bodies and thin legs with hooves to walk on the snow and ice. Recall that the cell membrane on different cells can have different compositions. If you were to compare the cholesterol composition of a cell from the elk's belly versus a cell from just above the hoof, which portion would you predict would have more cholesterol? Why? (LO 5-6, 5-7)

2. A person is declared to be dead upon the irreversible ceasing of spontaneous body functions: brain activity, or blood circulation and respiration. However, only about 1 percent of a person's cells must die for all these things to happen. How can someone be dead when 99% of their cells are still alive? (LO 5-1)

3. In a classic episode of *Star Trek*, a gigantic amoeba floating in space engulfs an entire starship. Mr. Spock blows the cell to bits before it has a chance to reproduce. Think of at least one inaccuracy that a biologist would identify in this scenario. (LO 5-3)

4. Giving a wilted plant distilled water helps it recover, but giving the same water to a person through an I.V. can lead to a life-threatening emergency. Explain the difference in these two outcomes. (LO 5-11)

5. Consider the role of the cilia in the respiratory system. Why is it significant that smoking cigarettes paralyzes the cilia in the trachea (windpipe) going to our lungs? (LO 5-18)

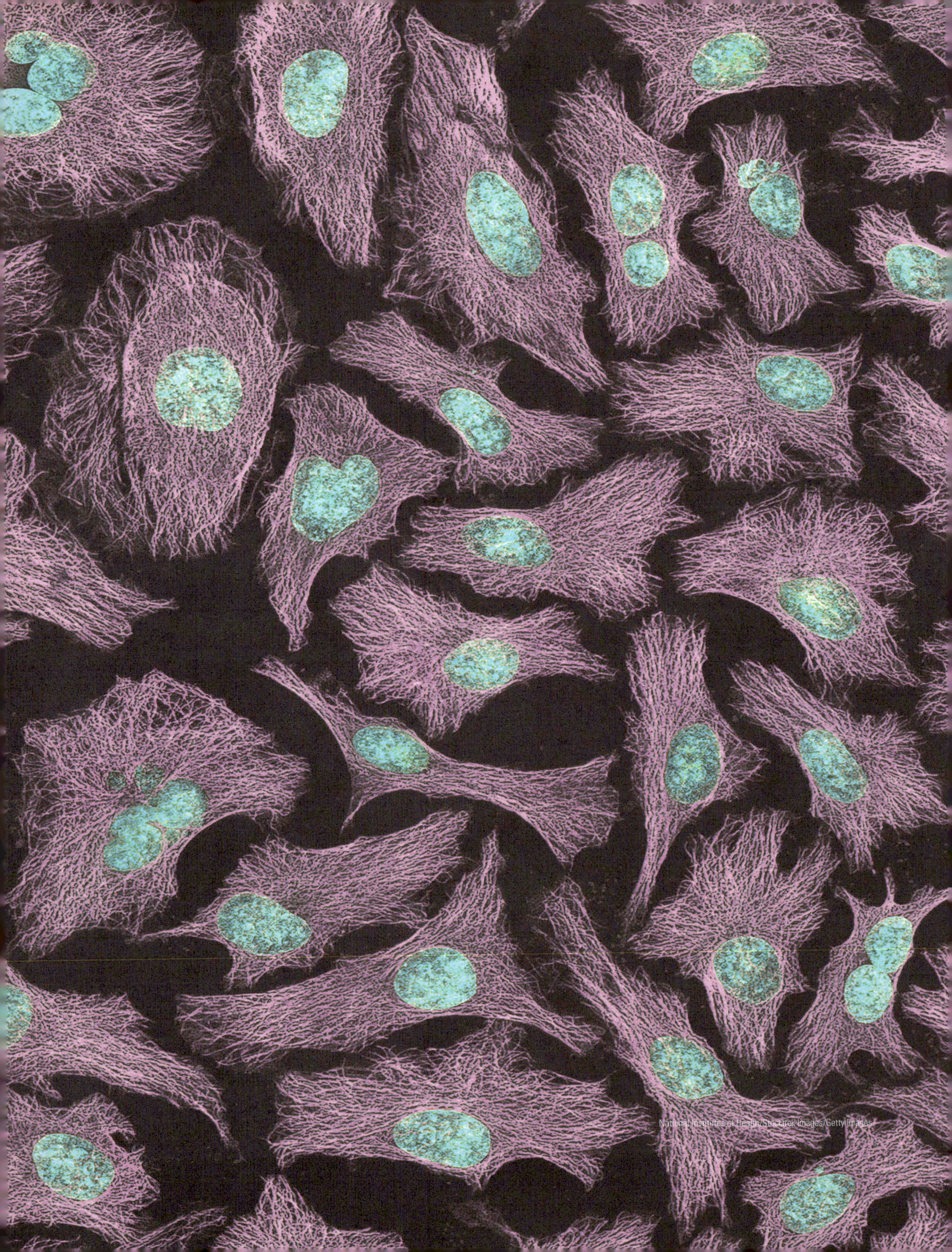
National Institutes of Health/Stocktrek Images/Getty Images

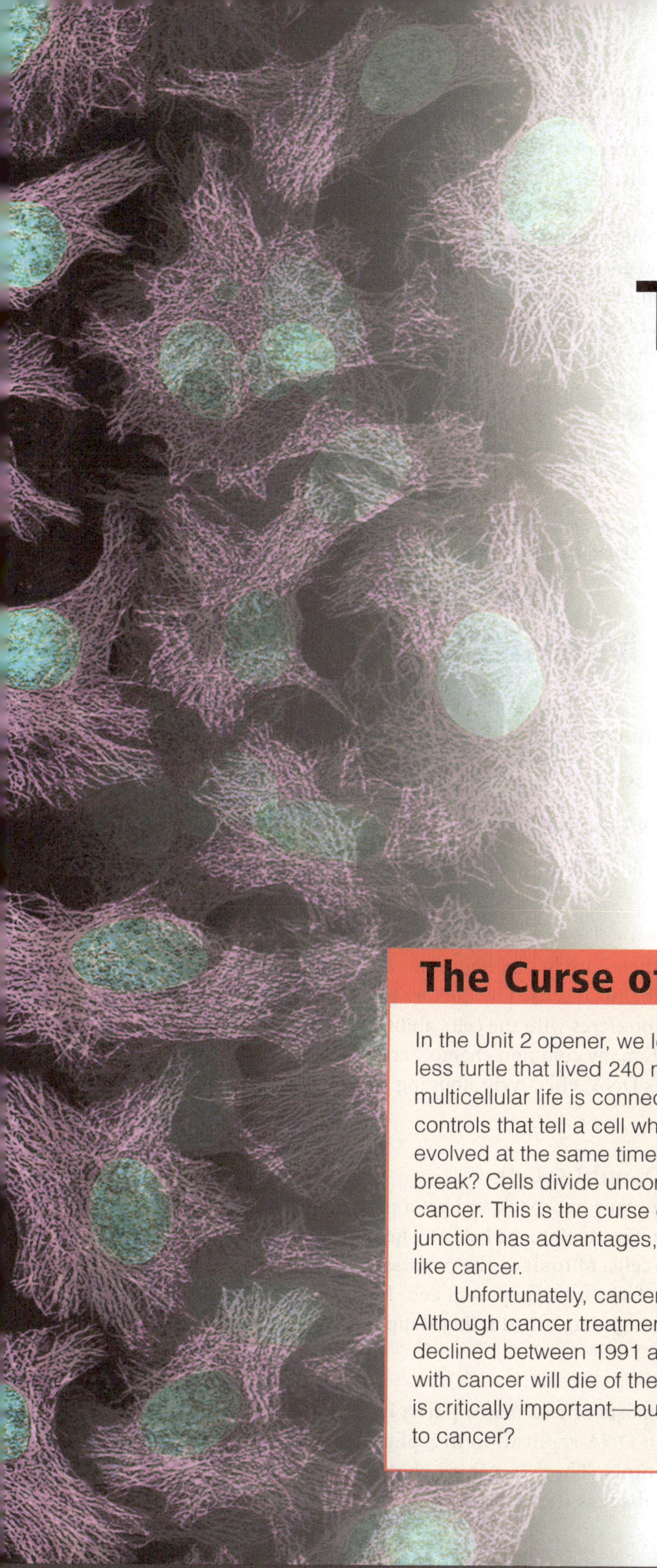

6

The Cell Cycle and Cancer

Chapter Outline:

- **6.1** Preparing for Mitosis: DNA Replication
- **6.2** Packing DNA for Mitosis
- **6.3** Mitosis and Cytoplasmic Division
- **6.4** Cell Cycle Control
- **6.5** Mutations and Routes to Cancer

The Curse of Multicellularity

In the Unit 2 opener, we learned that the earliest evidence of cancer is from a shell-less turtle that lived 240 million years ago. It turns out that the evolution of the first multicellular life is connected to the origin of cancer. That is because the biological controls that tell a cell when to divide and how to function as part of a tissue likely evolved at the same time life became multicellular. What happens if these controls break? Cells divide uncontrollably, invade different parts of the body, and cause cancer. This is the curse of multicellularity: While having many cells working in conjunction has advantages, there are more chances for disruption, leading to diseases like cancer.

Unfortunately, cancer is a disease that many of us are all too familiar with. Although cancer treatments have improved and the overall death rate from cancer declined between 1991 and 2020, as of 2024, about a third of people diagnosed with cancer will die of their disease. Understanding cancer and where it comes from is critically important—but what is cancer? What is going on within a cell that leads to cancer?

Deceptively beautiful, these cervical cancer cells eventually led to the death of the woman they were isolated from. We will learn more about Henrietta Lacks and her cervical cancer cells later in this chapter.

In the classic children's story, *Goldilocks and the Three Bears*, Goldilocks breaks into the bears' house and tries out their porridge, chairs, and beds until she finds the one that is "just right" for her. In the context of our cells and how many cells we have, we want the "just right" number of cells. Cells are individually alive while also part of a whole living organism. Under the right conditions, cells can grow (increase in size) or go through *mitosis* and make new cells.

We need to generate new cells through mitosis to replace cells that are damaged, such as when we get a paper cut, or as a person grows from a baby to an adult. If we have too many cells, or if a cell is not quite right, we can remove cells through a process called *apoptosis*.

In this chapter, we'll learn how our bodies make new cells through mitosis and remove cells through apoptosis. We'll also discuss the biological controls that regulate these processes, both within and outside a cell. Finally, we'll relate mutations in DNA to biological controls and how this can lead to the development of cancer.

Preview It!

Write down three things that you already know about mitosis. What kinds of cancers have you heard about before? What do you want to learn about mitosis, apoptosis, and cancer in this chapter?

6.1 Preparing for Mitosis: DNA Replication

Learning Objectives

6-1 Describe why the body requires mitosis.

6-2 Explain the process of mitosis.

6-3 Describe the process of DNA replication.

6-4 Identify and explain the roles of various proteins in DNA replication.

Figure 6.1 The origins of cancer treatments in nature

Peter K. Ziminski/Shutterstock.com

The bark of the Pacific yew tree contains a chemical called paclitaxel, which inhibits the cell cycle.

Look at the Pacific yew tree in Figure 6.1. These trees are native to the Pacific Northwest region of the United States. Within their bark lies an important chemical called paclitaxel. Under the brand name Taxol®, this chemical is used by doctors to treat cancers—a procedure called **chemotherapy**. Why? Taxol interferes with the cell's ability to successfully split its DNA into two daughter cells. In fact, many common chemotherapies interfere with the cell's ability to either divide or replicate its DNA, effectively stopping the cell's ability to make new cells through mitosis.

Mitosis Generates Identical Daughter Cells

Imagine you are shuffling papers in your backpack and get a paper cut. The cut damaged your skin cells. How can your body fix the damaged skin? The body stimulates healthy cells to begin mitosis to form new cells. **Mitosis** is the process of generating two new, genetically identical daughter cells from a single parent cell. Mitosis is different from *meiosis* (Chapter 8), which produces genetically distinct daughter cells for the purposes of sexual reproduction.

The cell cycle. The cell cycle consists of two major parts: interphase and mitosis (Figure 6.2). In interphase, prior to DNA replication in S phase and again before mitosis are two quality control checkpoints. There is a third checkpoint during mitosis prior to the cell dividing. We will discuss the cell cycle, including the checkpoints, in this chapter.

? Did You Know?

Chemotherapeutics kill not only cancer cells but also normally dividing cells. This is why it's common to see chemotherapy side effects like hair loss, because your hair is produced by rapidly dividing cells.

Figure 6.2 **An overview of the cell cycle**

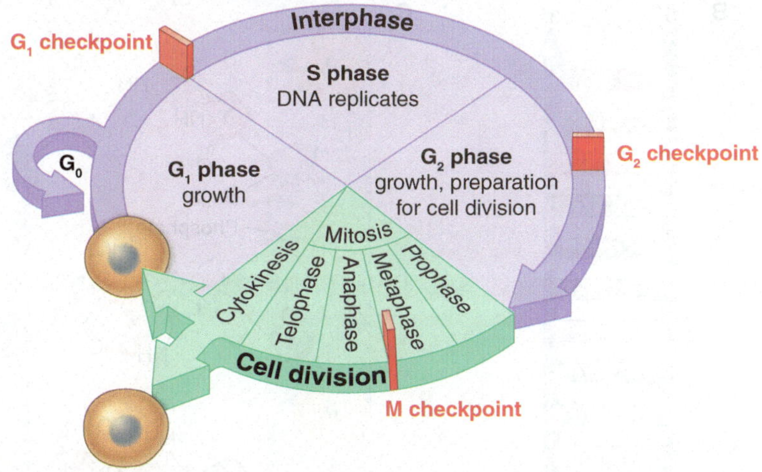

Cells spend most of their lives in interphase, only entering mitosis to divide. There are three quality control checkpoints during the cell cycle: G_1, G_2, and M. At any of these checkpoints, the cell may proceed with mitosis or stop. Some cells can exit the cell cycle entirely into a stage called G_0.

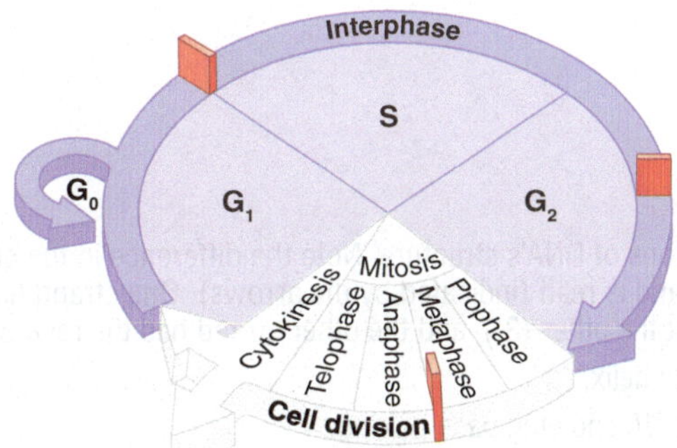

Producing genetically identical cells is a multistep process that includes copying DNA. A typical cell spends most of its life in a state called **interphase**. Interphase is the time between cell divisions when the cell carries out basic cellular functions, such as growth, copying DNA, and making proteins. Interphase has three major stages: G_1, S, and G_2. A cell starts out life in **G_1**. During this phase, the cell grows and produces the molecules needed for DNA replication. Cells in G_1 may temporarily or permanently exit the cell cycle by entering a nondividing state called **G_0**.

A cell that is going to divide continues in the cell cycle and moves from G_1 into **S phase**. S phase is when DNA replication occurs. After DNA replication, the cell enters **G_2**, where it makes the proteins and other cellular components needed for cell division. The remainder of the cell cycle consists of mitosis, which is when the cell divides (Section 6.3). Because each cell arises by the division of a preexisting cell (Chapter 5), the cell cycle begins and ends with cell division. We will discuss what happens at the interphase-specific checkpoints later in this chapter.

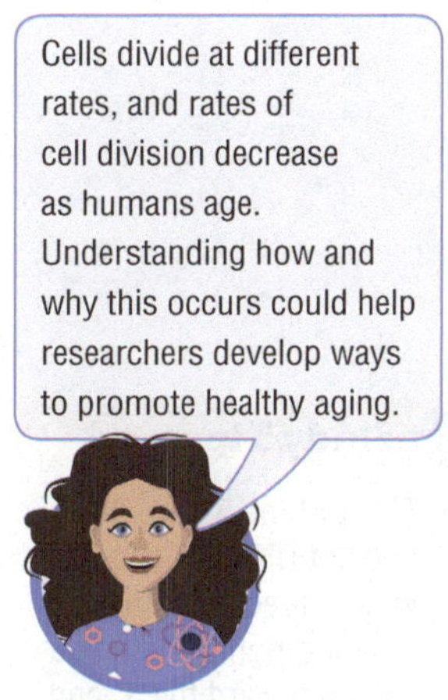

The First Step of Mitosis is DNA Replication

If mitosis produces genetically identical daughter cells, then each descendant cell must inherit a complete copy of the parent cell's genetic information. Thus, in preparation for division, a cell must copy all its DNA so that there are two sets: one for each of its daughter cells. The process of copying DNA is called **DNA replication**. Each molecule of DNA is copied in its entirety, and the two molecules of DNA produced are identical to the parent molecule.

Knowing the structure of DNA is important for understanding how it is copied. Since each nucleotide only binds to another specific nucleotide, the cell can use a single strand of DNA

Figure 6.3 **DNA has a three-dimensional structure and serves as a template for copying itself**

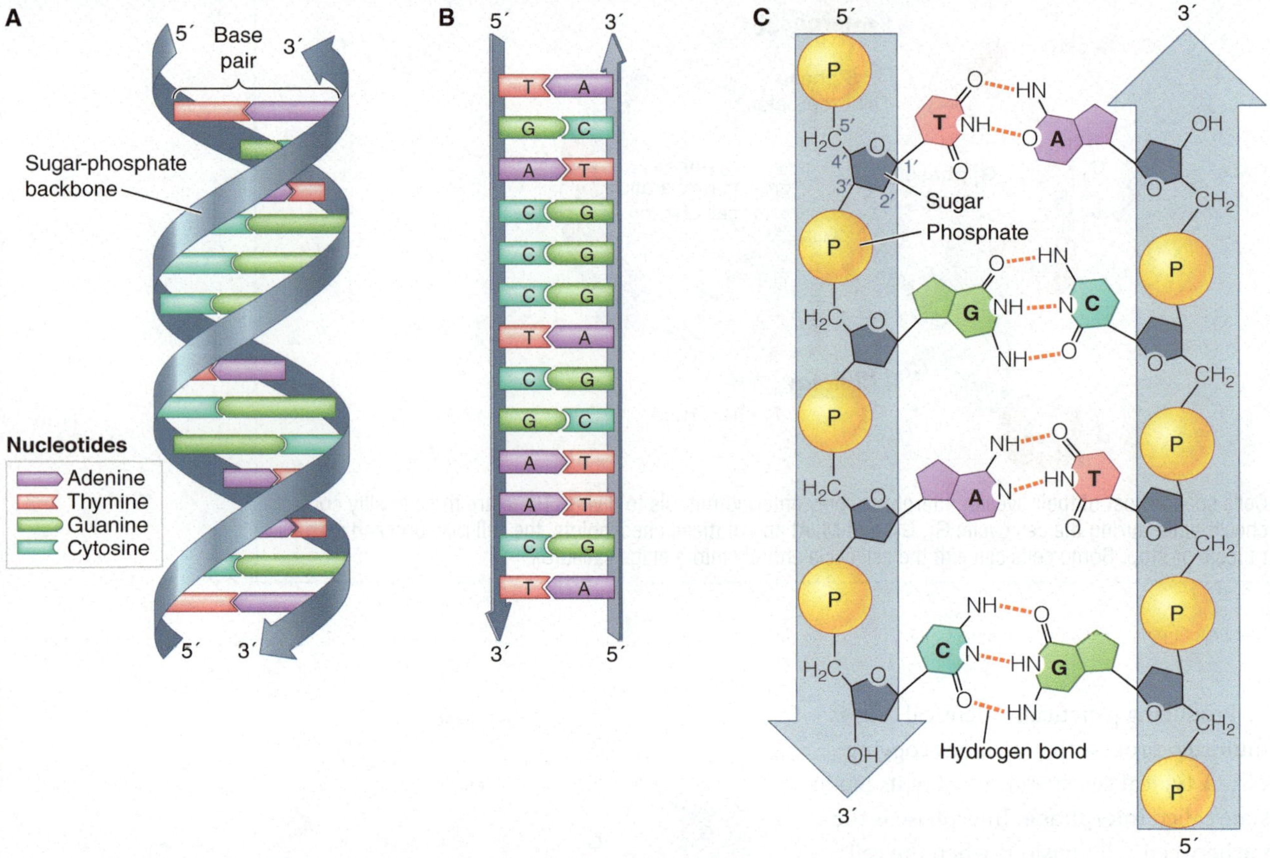

Shown are three representations of DNA's structure. Note the difference in the ends of the two strands and the direction that the strand is read (indicated by the arrows). One strand has a phosphate group at its end (5′) and hydroxyl at the other (3′), and the other strand has the reverse orientation.

A. DNA wound up as a double helix.

B. Ladder representation of DNA showing base pairing.

C. Detailed image showing the different functional groups at each end of DNA, plus the hydrogen bonds between base pairs.

as a template for making a new double-stranded piece of DNA. As we learned in Chapter 3, the structure of DNA is a double helix, and the nucleotides that compose the double strands of DNA are monomers of nucleic acids. These "bases" are adenine (abbreviated A), guanine (G), cytosine (C), and thymine (T).

Bases are complementary, which means that to create the double helix, a base on one DNA strand will form a hydrogen bond only with its "partner" on the opposite strand, where A binds only to T, and G binds only to C (and vice versa). DNA strands are also **antiparallel** to one another, which means one strand has a phosphate group ($-PO_4^{3-}$) at one end and a hydroxyl group (—OH) at the other (said to be running from 5′ to 3′). The other strand runs in the opposite direction, with the hydroxyl group and the phosphate group on opposite ends (said to run from 3′ to 5′) (Figure 6.3). We use 5′ and 3′ to orient ourselves on the DNA strand and track where different proteins are moving during the process of copying DNA.

Study Tip

Think of the 5′ and 3′ ends of DNA like you would directions. Instead of going right or left, you can go toward the 3′ end or the 5′ end.

Process of Science

Rosalind Franklin: The Mother of DNA

The discovery of the structure of DNA was one of the biggest discoveries in molecular biology—and highlights how non-scientific factors like culture can influence the process of science. By 1950, scientists knew about base pairing, that A paired with T and G with C. But the rest of the structure of DNA was a mystery. Multiple groups of scientists were working to solve the structure of DNA at the time, including a young woman named Rosalind Franklin (Figure 6.4A). Franklin was using a technique called X-ray crystallography to capture images of the structure of DNA.

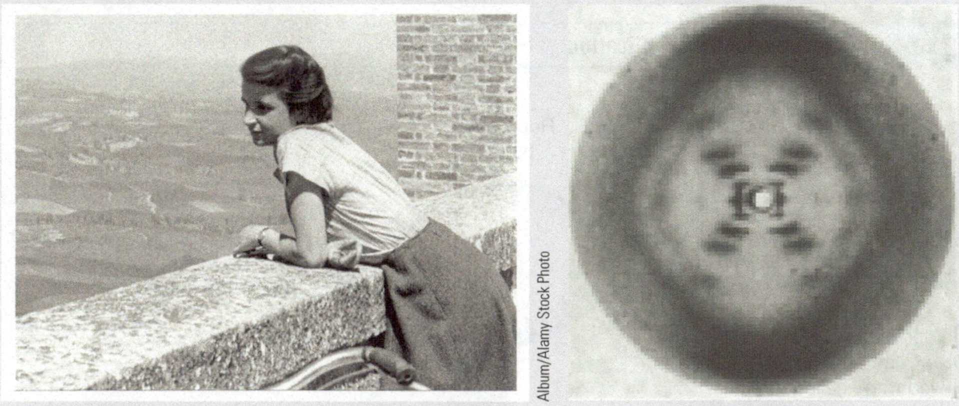

Album/Alamy Stock Photo

National Library of Medicine

Figure 6.4 The unsung mother of DNA: Rosalind Franklin

A. Image of Rosalind Franklin. Franklin did not fit into the mold of what was expected of women in 1950s England.

B. Franklin's famous Photo 51, which gave biologists their first look at the double helical structure of DNA.

Franklin's meticulous work, alongside that of her graduate student Raymond Gosling, yielded the first clear X-ray diffraction image of DNA (Figure 6.4B). Using this image, she determined that the structure of DNA was a double helix. Unbeknownst to her, Franklin's colleague, Maurice Wilkins, shared the photo with two other scientists also trying to solve the structure of DNA—James Watson and Francis Crick. This critical information inspired Watson and Crick's subsequent breakthrough. However, they did not give Franklin credit. Franklin and Gosling's work was published alongside Watson and Crick's landmark paper, but was published as the third paper in a series of three (a paper by Wilkins was the second paper on the topic). Both the Watson and Crick paper and the Wilkins paper downplayed the fact that Franklin and Gosling had provided the missing piece essential to solving the structure of DNA.

Today, many people remember Rosalind Franklin not for her talent as a scientist, but because of Watson's sexist characterization of her in his autobiography, *The Double Helix*, and the lack of credit for her contribution to the discovery of DNA's double helical structure. Some have gone so far as to say that if Franklin had received proper credit for her contribution, she would be less known to science today. Franklin's niece encourages people to remember Franklin today as not an "iconic feminist representative," but rather as "just who women should be." And just how should women be? Taking her achievements into account, Franklin's story is of a woman persevering and clearing the road for other female scientists to follow in her footsteps.

1. What are some ways we can ensure that the discoveries of marginalized people, both past and present, are appropriately credited?
2. How do you think cultural attitudes influence science today?
3. Can you name other women who made major discoveries in science who aren't well known? What did they do?

Study Tip

Sometimes it's helpful to think of complicated molecular processes like a musical. There are actors (enzymes) doing tasks, props (materials) needed for the process, and scenery (where the action is happening). Consider making a table where you list the actors, props, and scenery of DNA replication. You can use the same strategy for other processes throughout the book.

DNA Replication

The process of DNA replication involves a wide variety of enzymes, proteins, nucleic acids, and other materials all working together in synchrony. In the next section, we will go through each step of DNA replication. You may find it useful to follow along with Figure 6.5.

Step 1: DNA strands separate. Before DNA can be copied, the two strands must first be pulled apart by an enzyme called **helicase** ①. This opening in the DNA is called a **replication bubble** ②. As helicase pulls the two strands apart, another enzyme called **topoisomerase** keeps knots from forming in the DNA ③. Think of it like a tangled necklace.

Figure 6.5 The process of DNA replication

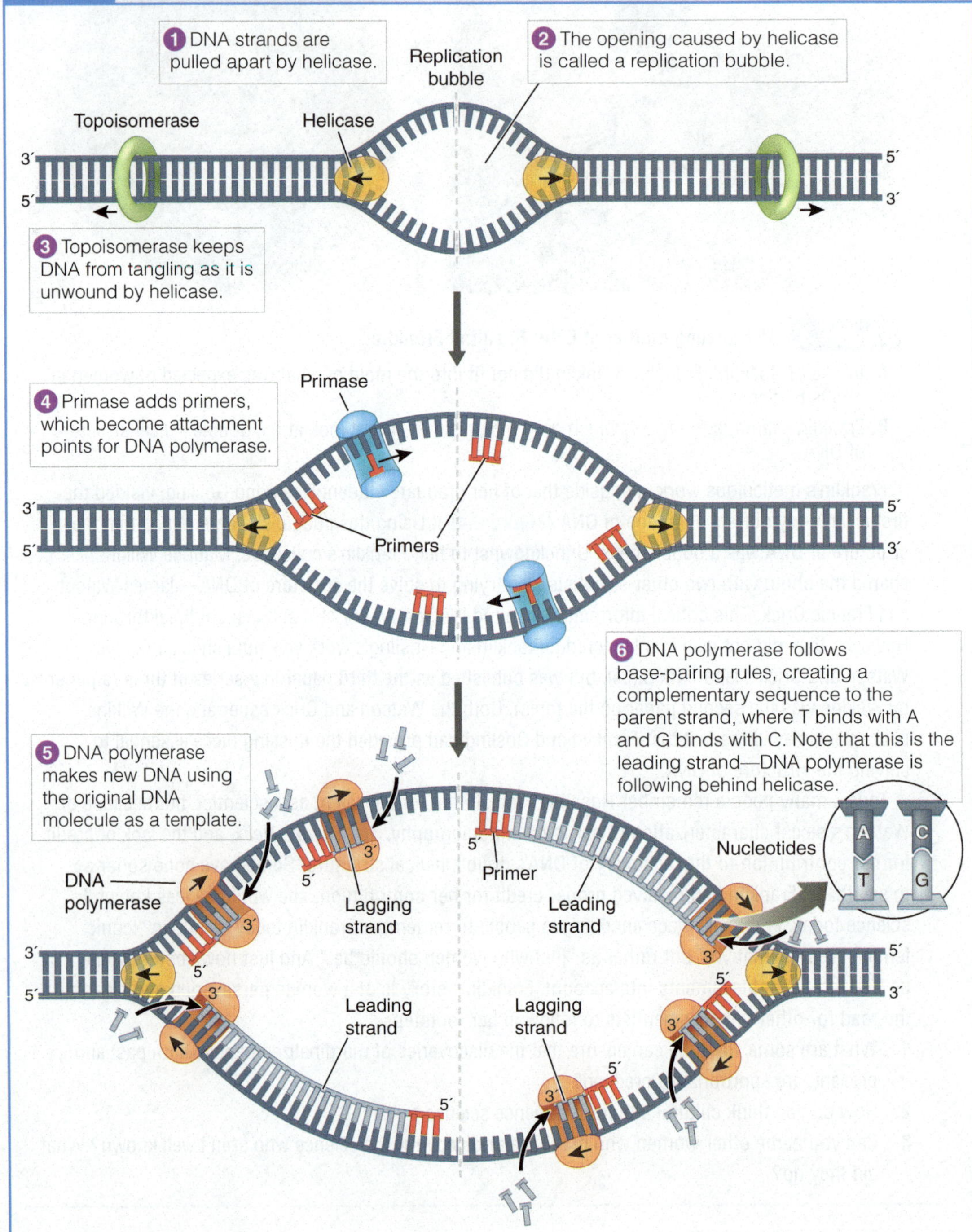

When you are trying to untangle a necklace and you pull, a knot can form. The same thing can happen with DNA—and topoisomerase keeps those knots from forming as helicase pulls the double helix apart.

Step 2: Primers attach to DNA. The enzyme that carries out DNA synthesis is called **DNA polymerase**. DNA polymerase uses an existing strand of DNA as a template to make a copy. Before that can happen, an enzyme called **primase** comes in to add primers ④. A **primer** is a short, single strand of RNA that serves as an attachment point for DNA polymerase.

Step 3: DNA polymerase makes new DNA. Using the original DNA molecule as a template, DNA polymerase "reads" the DNA and adds in new complementary bases ⑤. DNA polymerase follows base-pairing rules: When it reaches an A in the template strand, it adds a T to the new DNA strand; when it reaches a G, it adds a C; and so on. Thus, the DNA sequence of each new strand is complementary to the parental strand ⑥.

DNA polymerase can only make DNA in one direction: 5′ to 3′. If one strand is 5′ to 3′ and the other strand is 3′ to 5′, then how can DNA polymerase copy both strands of DNA? Think of the two strands of DNA as like traffic going in two directions on a highway, with a southbound lane and a northbound lane. The cars in each lane are moving in opposite directions relative to one another. DNA polymerase does the same thing.

As illustrated in ⑥, on the **leading strand**, DNA polymerase follows behind helicase, like two cars would follow each other going the same direction on the highway, copying DNA just as fast as the helix is opened. The **lagging strand** is going in the opposite direction that the helicase is moving. This means that primers need to be put in place as the helix continues to open and DNA polymerase moves in the opposite direction of helicase (that is, per our analogy, one is moving southbound on the highway, while the other heads northbound). Since DNA polymerase must continually jump backward toward helicase, it copies DNA more slowly (i.e., it *lags* behind). This creates lots of pieces of DNA that an enzyme called **ligase** then seals together to make one piece of DNA. Our result? Two identical DNA molecules, ready to make their journey to the new daughter cells.

Study Tip

Think of these primers like red carpets laid out in preparation for the star of the show (DNA polymerase).

Did You Know?

DNA polymerase can "proofread" as it makes DNA, just like you proofread and correct errors as you type! This is important for preventing errors in DNA that could ultimately lead to cancer.

! Point of Confusion: DNA replication in humans occurs to the left and right, on both strands at the same time. That's why we draw a line through the replication bubble, like the dashed line in Figure 6.5, so that we can look at the left and right sides separately. This means that which strand is lagging and which is leading is different, depending on if we are looking at the left or right side of the replication bubble. If helicase and DNA polymerase are moving in the same direction, this portion of DNA is the leading strand. If helicase and DNA polymerase are moving in opposite directions, this portion of DNA is the lagging strand.

6.1 Summary

- The body uses mitosis to make new cells for growth and repair.
- DNA replication makes identical copies of DNA. Therefore, daughter cells each have their own identical copy.
- In DNA, guanine (G) always pairs with cytosine (C), while adenine (A) always pairs with thymine (T). This complementary base pairing allows DNA to be replicated accurately by using the parental strand as a template.
- DNA helicase separates the two DNA strands to produce the replication bubble. Topoisomerase prevents tangling as DNA is pulled apart. Primase adds primers for DNA polymerase to attach to and adds complementary nucleotides to the growing DNA strand.

6.2 Packing DNA for Mitosis

Learning Objectives

6-5 Describe how DNA is organized in a chromosome.

6-6 Explain how karyotypes are used to analyze chromosomes.

In one of your cells, there are 6 billion base pairs, which equates to 6.5 feet (2 meters) of DNA. How is it possible to fit so much material into a single cell? And how do we successfully sort DNA into new cells? The same way we move anything big—by packing it up!

DNA Packaging

Have you ever put holiday lights away? Are the lights easier to get out of the box the following year if you throw them in the box and call it a day, or if you instead spend time winding them onto a spool? Similarly, newly replicated strands of DNA wind around protein "spools" called **histones**. DNA loops around histones approximately two times. These histones interact, forming a structure that looks like an old-fashioned curlicue telephone cable of loosely packed chromatin. These coils twist up around a scaffolding protein, finally forming the **chromosome**, which is a structure containing DNA and associated proteins (Figure 6.6).

The DNA of a eukaryotic cell is divided among chromosomes that differ in length. The number of chromosomes in a cell of a given species is called the **chromosome number**. For example, the chromosome number of humans is 46. Each of our body cells contains 23 types

Figure 6.6 **DNA packing in eukaryotic chromosomes**

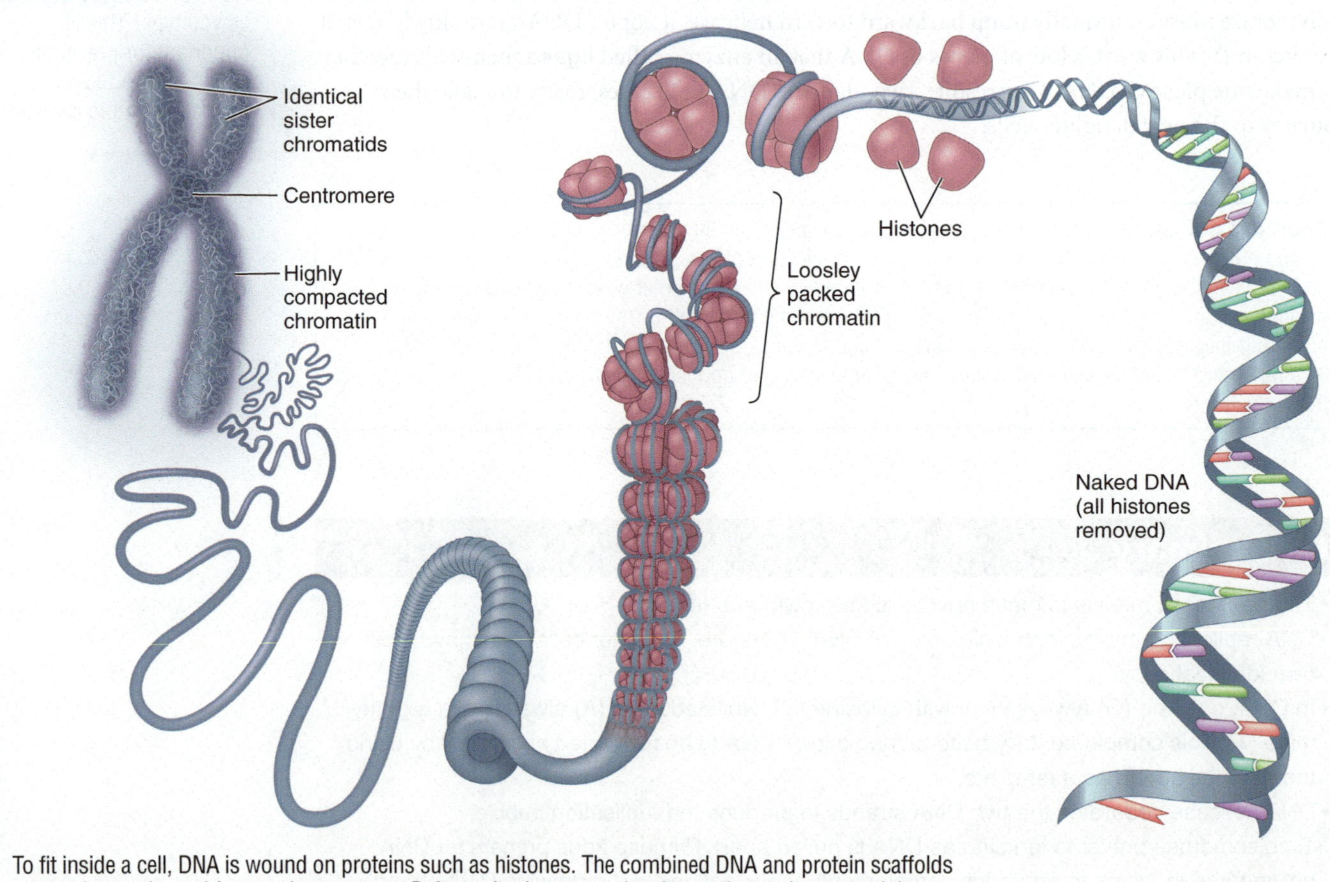

To fit inside a cell, DNA is wound on proteins such as histones. The combined DNA and protein scaffolds create chromatin, and form a chromosome. Before mitosis occurs, the chromatin condenses and the chromosome is visible using a light microscope.

of chromosomes. These chromosomes are of different lengths. We have two copies of each chromosome (one from our biological mother, and one from our biological father, which we will discuss more in Chapter 8) for a total of 46 chromosomes. Cells with two sets of chromosomes are **diploid**, or 2*n* (*n* stands for number).

Since one chromosome came from each biological parent, although they are the same type of chromosome, they are not genetically identical. These are called **homologous chromosomes**. After DNA replication occurs, we have identical copies of each of the 46 chromosomes for 92 chromosomes total. Identical chromosomes are called **sister chromatids**. Sister chromatids are attached to one another at a region called the **centromere**. Just before the cell divides, its DNA packs into an even more highly condensed form to facilitate transport and decrease the chance of damage. This form looks like an "X" under a microscope. We will discuss homologous and sister chromosomes in more detail in Chapter 8.

Figure 6.7 A human karyotype

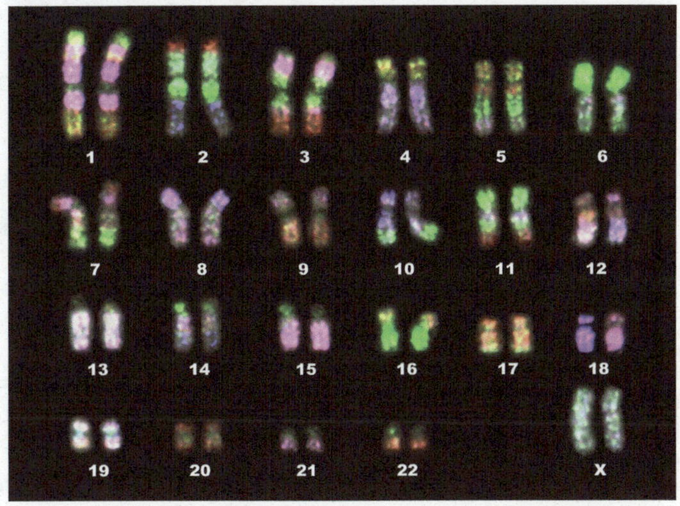

Karyotypes are used for examining chromosomes; for example, to ensure that someone has an expected number of chromosomes.

DEPT. OF CLINICAL CYTOGENETICS, ADDENBROOKES HOSPITAL/Science Photo Library

Chromosome number and structure are revealed in a **karyotype**, an image of an individual cell's chromosomes (Figure 6.7). To create a karyotype, cells are treated to make the chromosomes condense, and then stained so the chromosomes can be distinguished under a microscope. The image is digitally rearranged so that the chromosomes are arranged according to size, shape, and length. Note that there are 23 pairs of chromosomes shown, but the 23rd pair is different: these are the sex chromosomes. In this case, the 23rd pair is XX, which indicates that this individual, at least at the chromosomal level, is female. We will discuss sex determination and sex chromosomes in Chapter 8.

One hallmark of cancerous cells is that they have abnormal numbers of chromosomes. A cancerous cell can divide even if it hasn't replicated its chromosomes correctly or if the chromosomes aren't positioned properly in preparation for separation into the two daughter cells.

6.2 Summary

- Chromosomes form with the progressive winding of DNA, first around histone proteins, then around successive protein scaffolds.
- A karyotype is an arrangement of condensed chromosomes used to find abnormal chromosome structure or number.

6.3 Mitosis and Cytoplasmic Division

Learning Objectives

6-7 List the phases of mitosis.

6-8 Describe what is occurring during each phase of mitosis.

6-9 Describe the role of microtubules in mitosis.

6-10 Explain the cell division limit.

6-11 Describe the role of apoptosis in maintaining the appropriate number of cells.

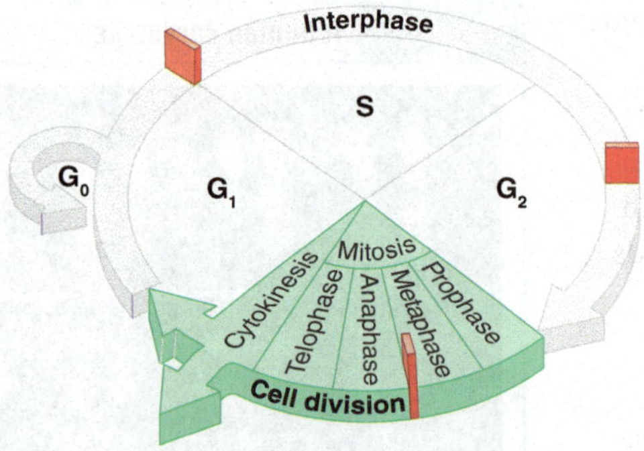

When the cell is ready to divide, it passes through the phases of mitosis. The events of mitosis proceed in four main stages: prophase, metaphase, anaphase, and telophase. Across each of these stages, the duplicated chromosomes condense further, move to the center of the cell, and then are pulled apart. The final part of cell division is **cytokinesis**, when the cytoplasm splits into two cells. In this section, we will discuss each of these phases in detail.

Study Tip

Use an acronym like PMAT to help you remember the steps of mitosis.

The Phases of Mitosis

We will now go through each of the phases of mitosis in order. You may wish to refer to Figure 6.8 as you read the descriptions that follow.

Prophase. Mitosis begins with **prophase**. During prophase, chromosomes pack into their most compact forms. Tight packing keeps the chromosomes from getting tangled and breaking as the nucleus divides. The nuclear envelope breaks up as the **spindle**, which contains the microtubules used for moving chromosomes during mitosis, lengthens from two regions on opposite sides of the cell and attach to the chromosomes at their centromeres. The microtubules form a spindle, a temporary structure that moves chromosomes during nuclear division. The areas where the spindle originates are called **spindle poles**.

Big Ideas in Biology

Structure and Function

Recall from Chapter 5 that microtubules are made of tubulin subunits. These subunits are added and subtracted to make the microtubules different lengths, thereby moving the chromosomes.

Metaphase. By the end of prophase, one sister chromatid from each chromosome is attached to microtubules extending from one spindle pole. The other sister chromatid is attached to microtubules extending from the opposite spindle pole. To line up chromosomes in the center, the microtubules pull against one another, like a game of tug of war, until the chromosomes are in the center. This alignment marks **metaphase**.

Anaphase. During **anaphase**, sister chromatids detach from one another and move toward opposite ends of the cell. When sister chromatids separate, each becomes an individual, unduplicated chromosome.

Telophase. **Telophase** begins when one complete set of 46 chromosomes (in humans) arrives at each spindle pole and forms a cluster. At this stage, the spindle disassembles, the chromosomes decondense, and a new nuclear envelope forms around the chromosomes as the chromosomal packing loosens. The result is two identical cells, and mitosis is complete.

Cytokinesis. Mitosis is followed by division of the cell's cytoplasm. In animals and single-celled eukaryotes, the cytoskeleton reorganizes to form a band of microfilaments and motor proteins that wrap around and squeeze until the cell is cut in half, forming two daughter cells.

! Point of Confusion: Multiplication by division? Mitosis is the process of making new cells by splitting one original cell into two daughters. This is not the same as cell growth, which refers to the size of an individual cell growing larger.

When Do Cells Stop Dividing?

Have you ever replaced bald tires on a car? As we drive, the tread on the tires slowly wears off until the tread is worn out—rather than having a textured tread, the tires are smooth. Without sufficient tread, a car has less traction, increasing the chance of an accident, particularly in slippery conditions.

Figure 6.8 Mitosis is the part of the cell cycle where a cell divides to produce two daughter cells

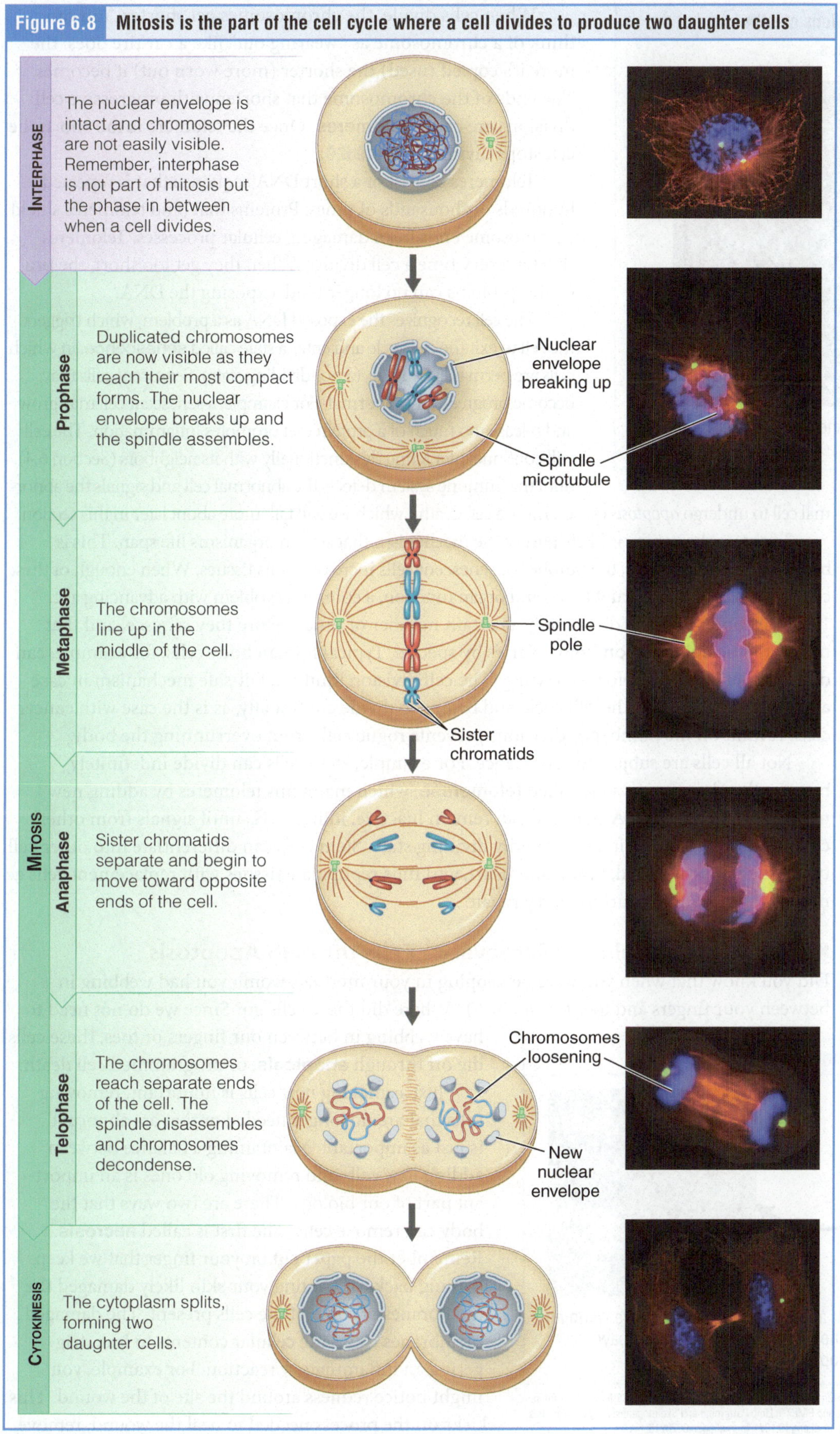

Source: DR. JUAN F. GIMENEZ-ABIAN/SCIENCE PHOTO LIBRARY

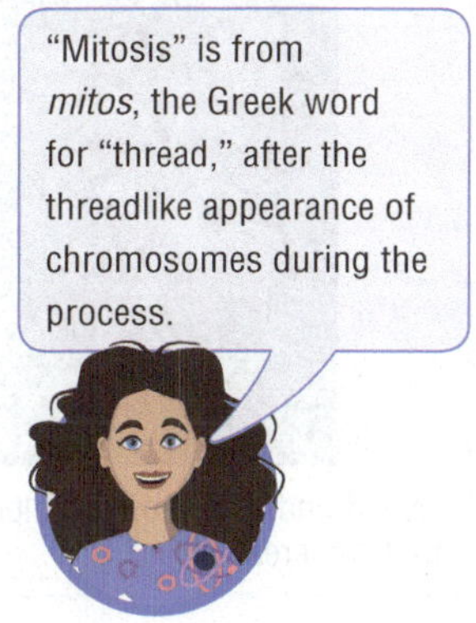

Figure 6.9 **Telomeres are regions of DNA found at the ends of eukaryotic chromosomes**

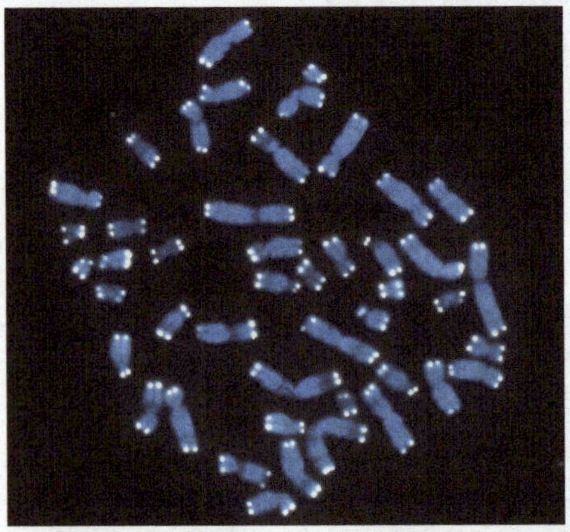

DNA strands are shown in blue, and the bright dots at the ends are telomeres.

BSIP SA/Alamy Stock Photo

When cells divide, the chromosomes get shorter. You can think of a chromosome as "wearing out" like a car tire does: the more it's copied (used) the shorter (more worn out) it becomes. The ends of the chromosome that shorten with progressive cell divisions are called **telomeres**. Once the telomere is too short, the cell stops dividing (Figure 6.9).

Telomeres consist of a short DNA sequence that is repeated hundreds or thousands of times. Proteins that bind telomeres shield chromosome ends from damaging cellular processes. Telomeres shorten every time a cell divides. When they get too short, the protective proteins can no longer bind, exposing the DNA.

The cell recognizes the exposed DNA as a problem, which triggers the cell to exit the cell cycle and enter a state called **senescence**, in which cells are permanently in G_0 (a nondividing state). Senescent cells may become progressively abnormal. For example, a senescent cell may grow and release too much of a protein that promotes inflammation. The cell will continue interacting dysfunctionally with its neighbors (Section 6.4) until the immune system detects the abnormal cell and signals the abnormal cell to undergo *apoptosis* (programmed cell death), which we will talk more about later in this section.

Cancer cells produce telomerase, which is why they can divide indefinitely. Some cancer researchers are developing treatments that target telomerase, with the goal of inducing senescence in cancer cells.

Telomere shortening is likely part of the mechanism that sets an organism's life span. This is because as an animal ages, the number of senescent cells increases in its tissues. When enough of these cells accumulate, they disrupt tissue and organ function, a common problem with advancing age.

Most body cells can divide only a certain number of times before they senesce, and that number—the **cell division limit**—varies by species. Typical human body cells, for example, can divide about 50 times before senescing. The cell division limit is a fail-safe mechanism in case a cell loses control over the cell cycle and begins to divide continually, as is the case with cancer cells. A limit on the number of divisions prevents rogue cells from overrunning the body.

Not all cells are subject to senescence. For example, stem cells can divide indefinitely because they have an enzyme called **telomerase**, which maintains telomeres by adding new nucleotides to the ends. A stem cell can remain inactive, idling in G_0 until signals from other cells (Section 6.4) move it into a rapidly dividing state. Stem cells can differentiate into other cell types, such as skin, blood, brain, and heart, and thus repopulate tissues with replacement cells as needed—for example, during injury repair.

Keeping Cells in Balance: Removal of Cells through Apoptosis

Did you know that when you were developing in your mother's womb you had webbing in between your fingers and toes (Figure 6.10)? Where did these cells go? Since we do not need to have webbing in between our fingers or toes, these cells die off through **apoptosis**, or programmed cell death.

Figure 6.10 **Apoptosis is a normal part of our biology**

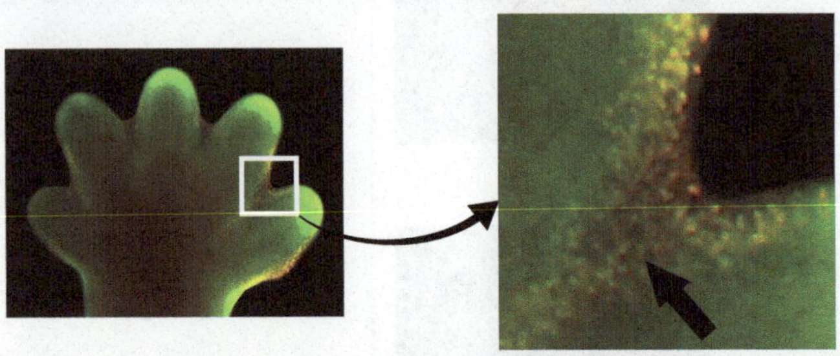

During embryonic development, apoptosis is used to remove webbing in between our fingers. Shown here is a developing mouse paw. Orange cells are undergoing apoptosis.

Source: Bandyopadhyay, Amitabha, Kunikazu Tsuji, Karen Ann Cox, Brian D Harfe, Vicki Rosen, and Clifford James Tabin. 2006. "Genetic Analysis of the Roles of BMP2, BMP4, and BMP7 in Limb Patterning and Skeletogenesis." PLOS Genetics 2 (12). Public Library of Science: e216–e216. doi:https://doi.org/10.1371/journal.pgen.0020216.

Just as making new cells is important, removing cells that are no longer needed or that are abnormal is just as important. Maintaining a balance between adding new cells and removing old ones is an important part of our biology. There are two ways that the body can remove cells. The first is called **necrosis**. Remember the paper cut on your finger that we keep coming back to? Cutting your skin likely damaged the membranes of some of the cells present. The damaged membranes cause the cellular contents to leak, triggering an inflammatory reaction. For example, you might notice redness around the site of the wound. This kicks off the process needed to heal the wound, remove

Biology for Life

Henrietta Lacks' Immortal Cells

As early as the mid-1800s, researchers were trying to grow human cells in the lab, which would allow them to study human diseases (and potential cures) without experimenting directly on people. Efforts were unsuccessful until one day in 1951, when a doctor collected cells from a cervical tumor in a young Black woman named Henrietta Lacks (**Figure 6.11A**). The doctor shared the cells (**Figure 6.11B**) with a research team who named the cells HeLa, representing the first two letters of Lacks' first and last names. As was standard at the time, Lacks was not informed of the researchers' use of her cells.

The HeLa cells vigorously divided in the lab. Sadly, cancer cells in Lacks' body were dividing just as fast. Only eight months after her diagnosis with cervical cancer, Lacks was dead. However, her cells lived on and were used in many scientific breakthroughs, such as the development of the polio vaccine, which has saved millions of lives. HeLa cells are still used today to study cancer and many other biological processes. HeLa cells helped several researchers win Nobel Prizes and have even traveled into space.

The Lacks family, however, did not receive any compensation for the use of Henrietta Lacks' cells until a landmark settlement in 2023 with Thermo Fisher Scientific, who made millions selling the cells. The family was unaware that HeLa cells existed until 1975, 24 years after Lacks died. Lacks' story came to light to the public in 2010 with the publication of the book *The Immortal Life of Henrietta Lacks* by Rebecca Skloot, and again in 2017, with an HBO documentary with the same title. Since then, Lacks' story, and HeLa cells, are synonymous with discussions about privacy, patient rights—especially in marginalized groups—and who owns and says how biological information and materials should be used.

A

Bridgeman Images

B

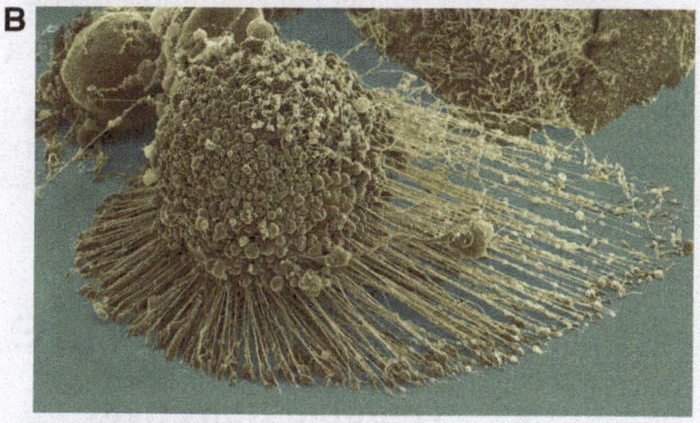

Courtesy of Tom Deerinck/National Institute of General Medical Sciences (NIGMS)/NIH

Figure 6.11 HeLa cells are a legacy of Henrietta Lacks, a person with an aggressive cancer

A. Henrietta Lacks died of cancer in 1951, but her cells live on in laboratories all over the world.

B. A micrograph of HeLa cells reveals membrane extensions that allow cancer cells to invade tissues.

1. In the 1950s, patients did not have rights regarding how their biological materials were used. Thus, the physician treating Henrietta Lacks didn't ask permission before sharing a sample of her cancerous cervix with the researchers. These days, patients' tissues can be used for research only with their signed consent. Would you permit your tissues to be used in research? Why or why not?
2. Cancer cells cultured from a person's tissue sample may divide indefinitely when maintained properly in a laboratory, producing generation after generation of descendant cells. Do you think the descendant cells belong to the person from whom the tissue was taken or to the laboratory that maintains them?
3. Since Lacks' family shares genetic material with HeLa cells, what rights does the family have in saying what or how the cells are used? Do these rights extend only to direct descendants or do extended family members have a say as well?
4. Why do you think researchers use HeLa cells instead of cells that could be obtained using informed consent?

cellular debris, and remove any invading bacteria. Necrosis is a messy version of cell death—but a necessary mess for getting new cells to take the place of the damaged ones.

Apoptosis, on the other hand, is a clean and tidy way of destroying a cell that starts from signals either inside the cell or from a neighboring cell (Section 6.4). For example, in the case of the webbing in between fingers, these cells are no longer needed. Signals from within the cell cause it to self-destruct, and then the cell breaks apart into self-contained pieces that are cleaned up by the immune system, much like you would vacuum up a mess at home. If a cell is identified as abnormal, such as at a cell cycle checkpoint (Section 6.4), an internal chain of events can begin to remove the offending cell. This is an important biological protection against cancer, as we will learn more about in Section 6.5.

6.3 Summary

- Mitosis occurs after (and before) interphase. The phases of mitosis are prophase, metaphase, anaphase, and telophase. The splitting of the cytoplasm (cytokinesis) occurs after telophase.
- During prophase, DNA condenses into chromosomes, and each sister chromatid is attached to a spindle pole by microtubules.
- During metaphase, the sister chromatids are lined up at the cell's middle.
- During anaphase, sister chromatids separate and are pulled to opposite ends of the cell.
- During telophase, each set of chromosomes clusters together, a new nuclear envelope forms, and chromosomes unpack into the newly formed nucleus.
- After mitosis is complete, cytokinesis splits the cytoplasm and organelles into two daughter cells.
- Most cells can only divide a set number of times before their telomeres get too short and they go into senescence, a stage where they can no longer divide.
- Cells can be removed from the body through necrosis or apoptosis. Apoptosis is a form of programmed cell death and is used to remove unwanted or abnormal cells.

6.4 Cell Cycle Control

Learning Objectives

6-12 Describe the role of checkpoints in the cell cycle.

6-13 Compare paracrine and endocrine signaling.

6-14 Relate cell signaling to the cell cycle.

6-15 Describe how cell signaling breakdowns can lead to cancer.

Have you ever traveled internationally? Depending on the country you are visiting, there are many different checkpoints you must clear before you can even board a plane! For example, one must typically have their documents checked with the ticket agent, at security, and during boarding. At one's destination, there is typically a port of entry with required security checks.

Why so many checks? To keep people safe! Like international air travel, cells also have many checkpoints to clear before they can divide. **Cell cycle checkpoints** exist to keep the body safe from cell overgrowth, which is one of the drivers of cancer.

Cell Cycle Checkpoints

The timing of cell division is determined by many different molecules, each working at one of three main checkpoints. The **G_1 checkpoint** is first and occurs between the G_1 and S phase. The **G_2 checkpoint** occurs prior to mitosis, and the **M checkpoint** occurs during mitosis.

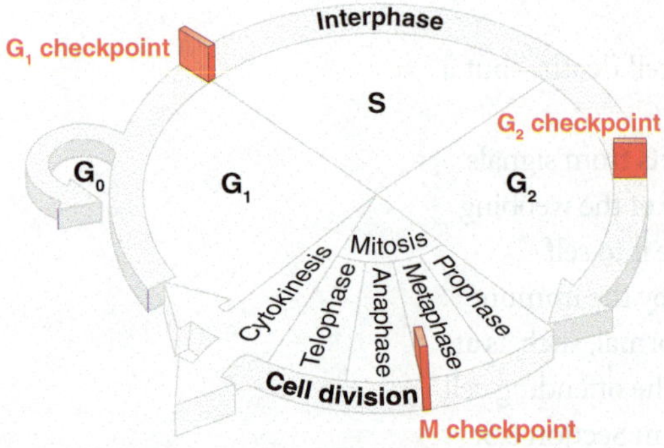

G_1 checkpoint. At this checkpoint, various proteins check to see if the cell can successfully divide; if the cell should exit the cell cycle and enter G_0; or if the cell is damaged and should undergo apoptosis. For example, if the DNA in a cell is damaged, a protein called p53 (which we will talk more about in Section 6.5) can trigger apoptosis. Other quality control checks at this phase include seeing if the cell is large enough to divide and if it has adequate nutrients to provide the energy needed to divide.

G_2 checkpoint. The second checkpoint occurs before mitosis begins. At this point, the cell's DNA is checked again, to ensure that the DNA was replicated properly. If the cell detects DNA damage, it will direct DNA repair enzymes (Section 6.5) to repair the damage. If the DNA cannot be repaired, the cell may undergo apoptosis.

M checkpoint. At this checkpoint, DNA is checked one final time. Various proteins inspect the chromosomes to ensure they are properly attached to the microtubules making up the spindle. If the microtubules are not attached properly, they will not split properly at anaphase and the daughter cells may receive the wrong numbers of chromosomes.

Intra- and Extracellular Chatter

In the previous section, we learned about how proteins within a cell can communicate with one another. This internal signal, such as the recognition of DNA damage, triggers a cascade of events within a cell. This cascade is like an array of dominoes triggered to fall: One domino falling onto another domino triggers the next to fall, and so on. The signal is passed between various proteins resulting in a final change in the cell's behavior—such as a signal to wait, go ahead and enter mitosis, or self-destruct with apoptosis. **Cell signaling** is the process by which cells talk to themselves and to each other.

Like we communicate in different ways (for example, face-to-face conversation, text message, or a phone or video call), cells also have different ways of communicating with one

Digging into Data

Finding a Needle in a Haystack: Using Knowledge of the Cell Cycle in the Fight against Cancer

In the hunt for new anticancer drugs, researchers often test a multitude of compounds with the hope of finding the proverbial "needle in a haystack" that works. One possible class of compounds to test is called cyclin-dependent kinase (CDK) inhibitors. CDK inhibitors block the action of proteins called CDKs.

In normal cells, CDKs facilitate the cell's transition to the next phase of the cell cycle; that transition can be delayed by CDK inhibitors. This form of cell cycle regulation helps to ensure that cells move through the cell cycle only when more cells are needed. Since cancer cells often bypass cell cycle controls, one strategy for treating cancer is to target CDKs. Therefore, biologists are developing a class of chemical CDK inhibitors to determine their impact on the cell cycle of breast cancer cells.

Using this strategy, researchers treat cancer cells with a promising compound and then record the number of cells at each stage of the cell cycle (G_1, S, and G_2/M). At any given time, most cells are in G_1, meaning they are not dividing. Once cells begin mitosis, they must create two new cells—or else undergo apoptosis. Typically, mitosis proceeds quickly. However, if cells seem to be stalling, it could be due to a CDK inhibitor. Therefore, a CDK inhibitor-based treatment could help prevent the generation of more breast cancer cells.

The graph in Figure 6.12 compares the percentage of breast cancer cells at G_1, S, and G_2/M, with and without treatment with a CDK inhibitor.

Graph based on: Fry, D. W., Harvey, P. J., Keller, P. R., Elliott, W. L., Meade, M., Trachet, E., Albassam, M., Zheng, X., Leopold, W. R., Pryer, N. K., & Toogood, P. L. (2004). Specific inhibition of cyclin-dependent kinase 4/6 by PD 0332991 and associated antitumor activity in human tumor xenografts. Molecular Cancer Therapeutics, 3(11), 1427–1438. https://pubmed.ncbi.nlm.nih.gov/15542782/

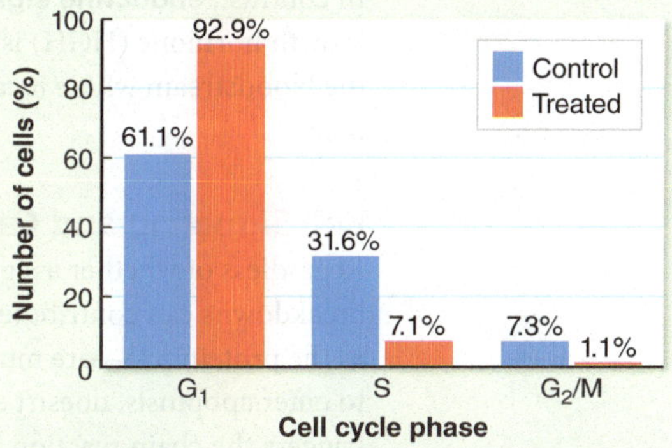

Figure 6.12 Percentage of breast cancer cells at three stages of the cell cycle, with and without treatment

The graph compares the percentage of human breast cancer cells in G_1, S, or G_2/M cell cycle stages in untreated (control) cells and treated cells.

1. Based on the graph, what happens to cell cycle progression in treated versus untreated (control) cells? Which phase of the cell cycle is affected in treated cells?
2. If the compound stops breast cancer cells at a certain point in the cell cycle, how would this affect tumor growth? Based on this data, would you recommend further study of this CDK inhibitor?
3. How might blocking a cell cycle checkpoint affect normal cells in a patient? How does this explain common side effects of chemotherapy, such as loss of hair?

Figure 6.13 **Cells use different communication strategies depending on target cell proximity**

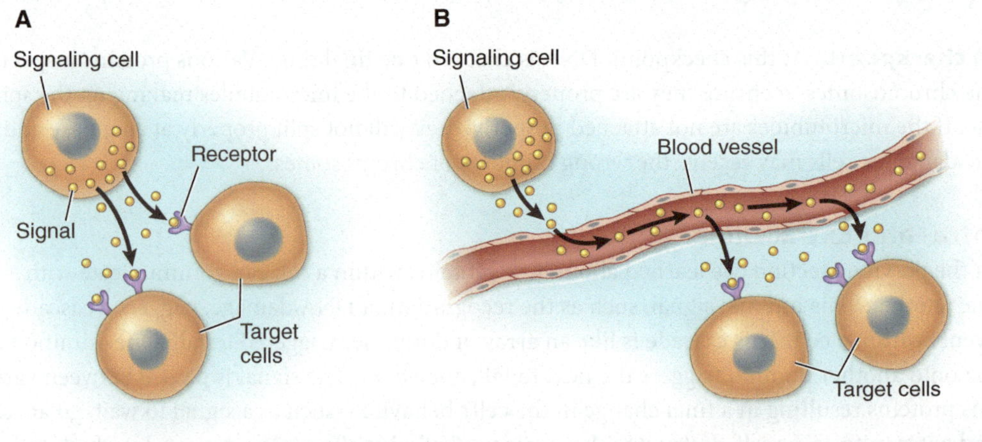

A. Paracrine signaling is when adjacent cells communicate with one another and requires no other bodily structures.

B. Endocrine signaling occurs over long distances in the body and requires other structures, such as blood vessels, to carry messages between the signaling and target cells.

another, depending on if they are close or far away (Figure 6.13). We already learned one way that cells near one another communicate in Chapter 5 via gap junctions. **Paracrine signaling** is when one cell produces a signal that impacts the cells immediately surrounding it. This signal diffuses into the nearby environment, where it binds to signal receptors on neighboring cells. In contrast, **endocrine signaling** occurs over long distances in your body. For example, human growth hormone (HGH) is produced in the brain during childhood. From the brain, it enters the bloodstream where it causes tissues, such as bones, to grow resulting in children growing taller.

Cell Signaling and Cancer

Regardless of whether a signaling cascade originates within or outside of a cell, communication breakdowns can contribute to cancer. For example, if proteins that detect DNA damage—such as the protein p53—are mutated and no longer work, the signal to stop the cell cycle, or for it to enter apoptosis, doesn't exist. If we go back to our dominos analogy, the initiating signal that triggers the chain reaction is lost—the cell divides even when it should not.

Conversely, cells can be stuck always dividing, even without signals to divide from their neighbors. In addition, neighboring cells may recognize that something is wrong with a cell and signal for it to enter apoptosis. Cancerous cells can "ignore" such signals and continue to divide. We will discuss how cell signaling breakdowns can contribute to cancer development in the next section.

6.4 Summary

- Cell cycle checkpoints allow the cell to stop cell division, repair DNA, or undergo apoptosis.
- The G_1 checkpoint confirms that the cell has enough resources for cell division.
- The G_2 checkpoint confirms that the DNA is properly replicated.
- The M phase checkpoint confirms that the chromosomes are properly attached to the spindle fibers.
- Cells rely on cell signaling to confirm that they are functioning properly and to know when to prepare for cell division. Paracrine and endocrine signaling are ways for cells to communicate with one another.
- Breakdowns in cell signaling can lead to cancer if the target cells fail to respond appropriately to cell signals to divide and/or undergo apoptosis.

6.5 Mutations and Routes to Cancer

Learning Objectives

6-16 Explain how mutations arise.

6-17 Describe cellular mechanisms that prevent mutations from occurring.

6-18 Explain different sources of DNA mutations.

6-19 Explain why mutations can give rise to tumors.

6-20 Compare and contrast the roles of tumor suppressors and proto-oncogenes in the cell cycle.

6-21 Describe how malfunctions in tumor suppressors can lead to cancer.

6-22 Describe how proto-oncogene to oncogene transformation can lead to cancer.

6-23 Compare and contrast malignant and benign cells.

Consider for a moment where you have heard about mutations in pop culture. Maybe you enjoy the X-Men franchise or recall the genetically engineered "muttations" in *The Hunger Games* books and movies. In both cases, mutations result in some kind of fantastical change, sometimes leading to superhuman powers or other abilities. What do mutations have to do with cancer? Where do mutations come from? And are mutations necessarily a bad thing? We will explore these questions, and more, in this section.

How Do Mutations Occur?

Recall that DNA holds information in its sequence of bases. If the base sequence of a cell's DNA changes, the information it encodes can change as well. A permanent change in the DNA sequence of bases is called a **mutation**. The cell has systems (such as cell cycle checkpoints, Section 6.4) for monitoring DNA quality, as well as mechanisms to repair damaged DNA. These mechanisms help cells protect the integrity of their genetic information.

Even so, the system is not perfect. Major alterations in DNA typically cause the cell to die. Cancer, for example, begins with mutations, as do many genetic disorders. In some cases, mutations have no effect or are beneficial to the organism, something we will return to in Chapters 7 and 11. Mutations that do not disrupt the cell's function enough to prevent mitosis are passed on to the cell's descendants. Where do these mutations come from?

> **Big Ideas in Biology**
>
> **Evolution**
>
> Mutations aren't necessarily a bad thing! Mutations that lead to beneficial traits can drive evolution. We will return to the connection between mutations and evolutionary advantages in Chapter 11.

Replication errors. Sometimes, newly replicated DNA is not complementary to its parent strand. A nucleotide may get deleted during DNA replication, for example, or an extra one might be inserted (Chapter 7). Occasionally, the wrong nucleotide is added. Although DNA polymerase can proofread and correct its work, it is not failproof, and mistakes can and do occur.

Errors missed during DNA polymerase's proofreading process can be corrected later. For example, mismatched bases can cause the double helix to be misshapen. The mistake can then be recognized by enzymes and repaired through a process called mismatch repair. In **mismatch repair,** the strand is cut, and the section containing the mismatch is removed. DNA polymerase then uses the remaining parental strand as a template to replace the missing nucleotides, and DNA ligase seals any gaps in the new strand.

Sometimes, proofreading and mismatch repair fail to correct an error. If the error is not caught before the next round of DNA replication, the error is copied and the two strands can base pair properly. Consequently, the error is no longer recognizable to either DNA polymerase

Figure 6.14 **Replication errors can lead to DNA mutations**

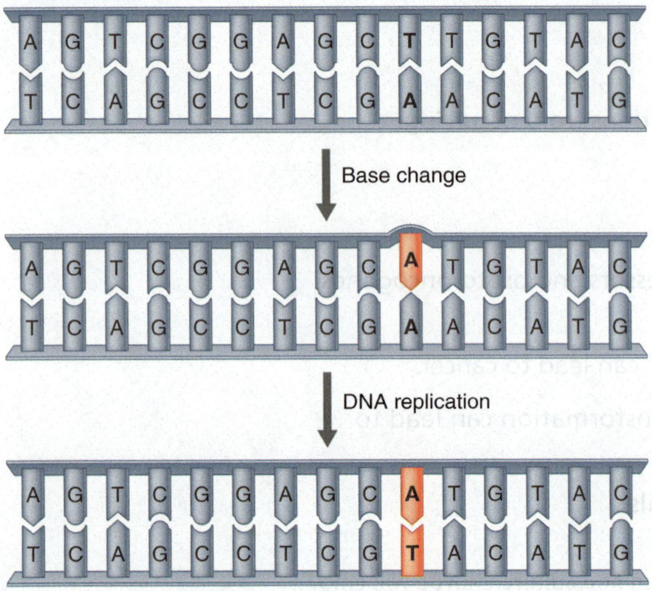

Repair enzymes can detect a mismatched base (red) and correct it unless a second round of DNA replication has occurred. After the additional round of replication, both strands will base-pair properly, but with a different sequence than the original DNA.

or repair enzymes, and the change becomes permanent (Figure 6.14). When the cell divides, its descendants will inherit the error as a mutation.

Mutations accumulate over the course of one's lifetime, which explains why older individuals are more likely to develop cancer. Since mitosis produces genetically identical daughter cells, all mutations are passed onto the cell's descendants. Accordingly, if a mutation leads to a cell growing when it should not, the initial cell's daughter cells also divide when they should not, and so on. This is why it is important to take steps to protect your DNA from mutations, such as by not inhaling cigarette smoke or polluted air, and by avoiding excessive exposure to ultraviolet (UV) light, all of which may increase the risk of developing cancer. These precautions are particularly important for protecting young children, who will carry mutations throughout their lives.

Exposure to mutagens. There are many different **mutagens**, or agents that cause DNA damage. In this section, we will learn about two common mutagens: electromagnetic radiation and **carcinogens**, or cancer-causing chemicals.

Exposure to UV light causes DNA bases to stick together inappropriately. The resulting shape makes a kink in the DNA double helix. This malformed DNA causes mutations because polymerases can't copy kinked DNA correctly. UV exposure can also cause chromosomal breaks. Usually, chromosomal breaks cause cell death. Cells that survive often have mutations at the breakage sites. This is due to the highly inaccurate nature of repairing double-strand breaks in DNA. The poor repair process can result in mutations and ultimately contribute to developing cancer.

? Did You Know?

For every second a skin cell is exposed to UV light, 50 to 100 malformations occur in its DNA. This is why unprotected exposure to UV light can lead to skin cancer.

The extent to which DNA is damaged by X-rays increases with the duration of exposure. Your routine dental X-rays are considered safe.

UV light is not the only form of electromagnetic radiation that can cause DNA damage. X-rays have enough energy to knock electrons off atoms, forming free radicals (atoms or molecules with unpaired electrons; see Chapter 2) that can damage DNA. Exposure to X-rays can break chromosomes or cause chromosomes to stick together, thereby preventing DNA replication. Cells can do some repair, but the process is complicated and does not always work well. Other forms of electromagnetic energy, such as radiation, can cause extreme changes (Figure 6.15).

Some chemicals, such as those in cigarette smoke, are carcinogenic, capable of generating mutations that could lead to cancer. According to the Centers for Disease Control, cigarette smoking is the leading cause of preventable death in the United States. Keep in mind that using other tobacco products often marketed as a "safe alternative"—such as smokeless tobacco, snuff, and chewing tobacco—is also dangerous. These products contain the same mutation-causing chemicals as tobacco smoke.

The Road to Cancer: Connecting Mutations with Cell Cycle Control

In Section 6.4, we learned about cell cycle checkpoints and how various proteins within and outside the cell work together to let only healthy cells divide, thus protecting the body from cancer. One of the major drivers of cancer is when mutations occur in either the DNA that codes for the proteins involved in cell cycle checkpoints or in the signaling pathways that tell a cell when to divide. In humans and other multicellular life, a mutation that disrupts a cell cycle checkpoint can result in a **tumor**—a mass of abnormally dividing cells in a tissue. Tumors can disrupt body tissues, both structurally and functionally.

Tumors can be classified as benign or malignant. A **benign** tumor (root word from the Latin *benignus,* meaning kind or good) is an abnormal collection of cells, but the tumor grows slowly and stays put. For example, the moles on your skin are benign skin growths. However, some benign cells (including your moles) can become **malignant**. Malignant cells are dangerous because they can invade other tissues. For example, the proteins that hold a cancer cell in place may be defective or missing. This allows cells to slip easily into and out of vessels of the circulatory and lymphatic systems, migrating to other parts of the body and establishing new tumors (Figure 6.16).

The process by which malignant cells break loose from the primary tumor in their home tissue and invade other parts of the body is called **metastasis**. These new growths are called secondary tumors. Unless chemotherapy, surgery, or another procedure eliminates cancerous cells from the body, their presence can result in death.

Figure 6.15 Mutated flowers from Chernobyl

In 1986, a catastrophic accident at the Chernobyl nuclear power plant in Ukraine released a massive amount of radiation. These *Ranunculus* flowers were collected in the Exclusion Zone, an area of about 1,000 square miles that is still dangerously contaminated. All but one (circled) are abnormal—the result of DNA mutations.

Taavi Tuulik

! Point of Confusion: If cells metastasize and form tumors in another part of the body, this does not mean the person has multiple forms of cancer. Cancers are named based on their tissue of origin. For example, breast cancer tends to metastasize to the brain, bones, lungs, and liver. This does not mean the person has those types of cancer. Instead, we say they have breast cancer that spread to those parts of the body.

Proto-oncogenes and Tumor Suppressors

Generally, multiple mutations are required to transform a normal cell into a cancerous one. Mutations in the genes that encode proteins (Chapter 7) involved in cell cycle regulation are linked to many forms of cancer. Additionally, some mutations may be inherited (which we will discuss more in Chapter 8), but usually, mutations arise due to DNA damage.

Figure 6.16 Metastasis is the spread of cancer cells to other parts of the body

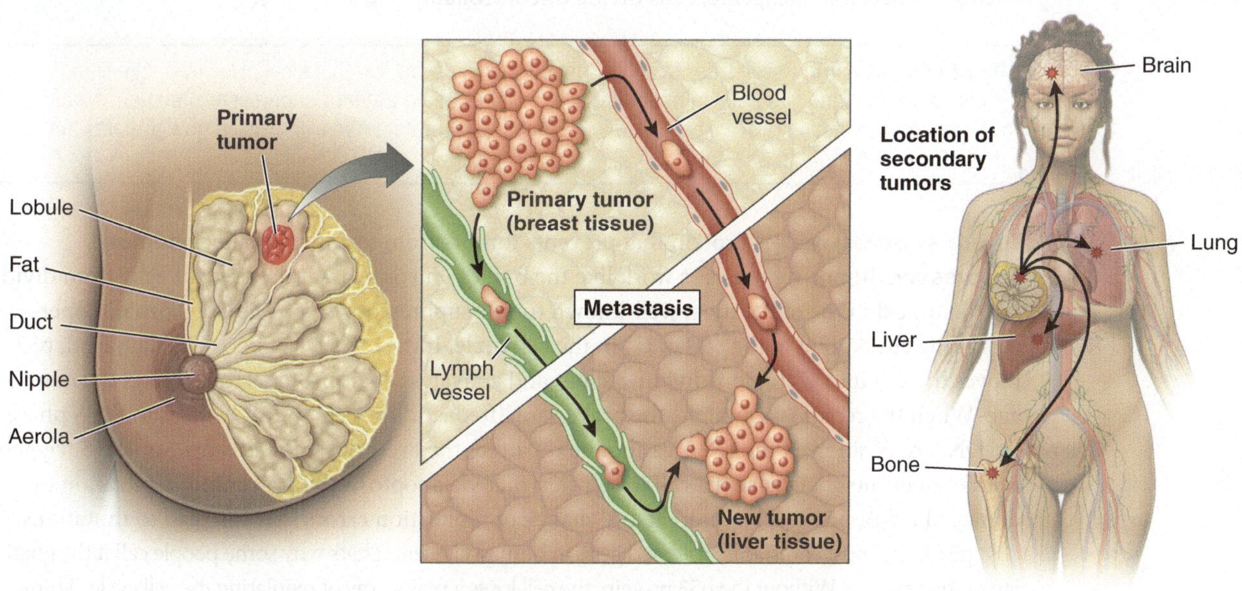

This person has a primary tumor in their breast that migrated through lymph and blood vessels to form new, secondary tumors in other parts of the body (the bone, liver, brain, and lung).

Biology for Life

Is a Universal "Cure" for Cancer Possible or Is Prevention Key?

Although various kinds of cancer share similar hallmarks, such as cells growing out of control and spreading to other parts of the body, "cancer" is not one thing. People talk about finding a cure for cancer, with the idea that it would be a universal cure. However, some oncologists (doctors who specialize in treating cancer) would argue that it's not possible to actually "cure cancer"—because how and why people get cancer, and the way that individual cancers behave, is vastly different.

For example, some cancers like cervical cancer are linked to infection by certain strains of the human papillomavirus (HPV). Since there is a vaccine that protects against HPV, we can prevent many cases of cervical cancer. The "cure" for cervical cancer may therefore be found in prevention. Similarly, melanoma is a deadly form of skin cancer that is almost entirely preventable by limiting exposure to UV light. This emphasizes the importance of education and regular examination of one's skin.

Other forms of cancer, like certain forms of breast, colon, and prostate cancer, have a genetic basis that increases risk. However, increased risk does not guarantee that one will develop cancer. Factors that individuals can control, such as environment and diet, can go a long way toward preventing cancer. Education regarding the early signs of cancer and regular screening can also help. New, personalized treatments for cancer are also revolutionizing how we treat cancer. We will return to personalized medicine and bioengineering cancer treatments in Chapter 9.

1. Can you think of ways to reduce your risk of cancer by making changes in your daily life?
2. Despite education regarding cigarette smoke and cancer, many people still smoke. What do you think would be an effective strategy for helping people make choices to prevent cancer?

Next, we'll discuss two classes of proteins, present in all cells, that control when your cells will divide: tumor suppressors and proto-oncogenes. As an analogy, Figure 6.17 represents tumor suppressors and proto-oncogenes/oncogenes as factory workers. Since tumor suppressors keep damaged cells from copying themselves (Figure 6.17A), their loss results in a buildup of damaged cells (Figure 6.17B).

Proto-oncogenes tell the cell when to divide and are therefore under tight control about when to cause the cell to divide (represented by the "Go" button in Figure 6.17C). If proto-oncogenes become oncogenic, cells divide uncontrollably (Figure 6.17D).

Point of Confusion: Recall from Chapter 3 that proteins do work within a cell. However, genes, not proteins, can be mutated. Keep in mind that when we talk about a tumor suppressor mutation, we are referring to a change in DNA that leads to a malfunctioning protein. We will discuss how proteins are made from DNA more in Chapter 7.

Tumor suppressors. A protein that prevents tumors from forming is called a **tumor suppressor**. Tumor suppressors work like the brakes on your car—they prevent cells from dividing too quickly. One of the most well-known tumor suppressors is p53. The p53 protein operates at the checkpoint at the end of G_1 in the cell cycle. If the cell's DNA becomes damaged, p53 moves the cell into G_0 (a non-dividing state) and activates repair proteins that can fix the damage. When the repair is complete, the brakes are lifted, and the cell cycle advances to the S phase for DNA replication. If the DNA cannot be repaired, the cycle does not advance, and p53 triggers a series of events that cause the cell to self-destruct via apoptosis. By preventing replication of damaged DNA, p53 helps minimize the chance of replication errors that can lead to mutations.

p53 is the most commonly mutated gene in human cancers. That's why some people call it the guardian of the genome. Without the p53 protein, the cell loses a major way of regulating the cell cycle. Think of it like losing the brakes on your car. Without your brakes, you cannot slow down your car, and without p53 or other tumor suppressors, cells divide when they are not supposed to, which can lead to cancer.

Another example of a mutated gene that may lead to cancer is *BRCA1*—the gene implicated in familial breast cancer. The BRCA1 protein acts as a tumor suppressor and has many roles in the cell. For example, BRCA1 can trigger DNA repair, preventing the formation of abnormal

Figure 6.17 A cartoon analogy of the genetic basis of cancer

A Our first cartoon example focuses on the role of tumor suppressors in making new cells. Although not a perfect biological representation, the cartoon captures the heart of what tumor suppressors do—they prevent damaged cells from persisting in the body. Healthy cells proceed to the end and can go through mitosis. Unhealthy cells are either repaired or removed via apoptosis.

B If a tumor suppressor gene is mutated, resulting in a non-functional protein (shown here as sleeping workers), then the important quality control checkpoints are lost. Cells that are damaged persist in the body instead of being removed. These defective cells can make copies of themselves, causing more potentially cancerous cells to form.

C Proto-oncogenes are a normal part of our cell biology. Their job is to tell a cell when to divide, usually in response to signals from outside the cell (shown here as the loudspeaker saying "Go!").

D If a proto-oncogene mutates, it becomes an oncogene. Oncogenes make proteins that are always active. In this illustration, we show the oncogene as not listening to signals from outside the cell to stop making new cells, instead repeatedly hitting the "Go" button. The body becomes overwhelmed by the influx of cells, and other quality control measures, like the action of tumor suppressors, is overwhelmed.

Biology for Life

Cancer over Human History

We've all heard of the various trappings of modern life that "cause" cancer—pollution, synthetic chemicals, consumption of ultra-processed foods, and so forth. While arguably our longer lifespans and modern existence result in more people with cancer, our human-like ancestors contracted cancer as well.

National Geographic Explorer Lee Berger (**Figure 6.18A**) was part of the team that discovered a foot bone with osteosarcoma (the same kind of bone cancer that afflicted the Triassic-era shell-less turtle in the Unit 2 opener) in a human ancestor who died approximately 1.7 million years ago (**Figure 6.18B**). One of Berger's coauthors, Edward Odes, stated about the discovery that "You can opt for the paleo diet, you can have as clean a living environment as you want, but the capacity for these diseases is ancient." This brings us back to the chapter's opening story: the curse of multicellularity. Humans have had a relationship with cancer long before we (*Homo sapiens*) evolved (Chapter 12). Some cancers, such as osteosarcoma, happen due to defects in our biology—and in this case, disproportionately affect the young.

The good news is, unlike our ancient ancestors, we can look toward a future with better cancer treatments and prevention strategies, armed with a modern understanding of biology and cutting-edge technology.

A B

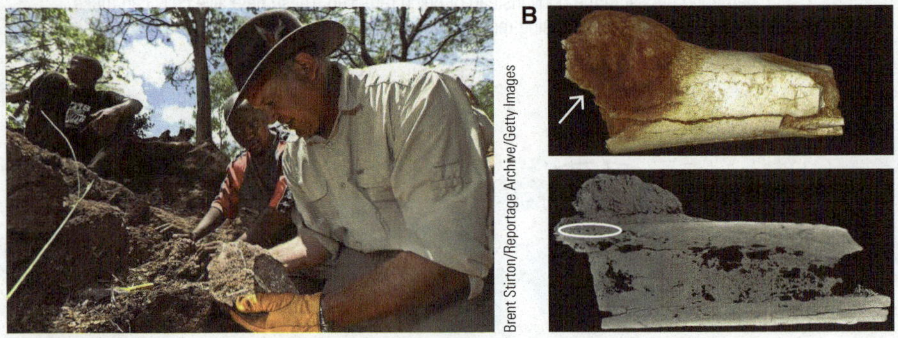

Brent Stirton/Reportage Archive/Getty Images

Odes, E. J., Randolph-Quinney, P. S., Steyn, M., Throckmorton, Z., Smilg, J. S., Zipfel, B., Augustine, T. N., de Beer, F., Hoffman, J. W., Franklin, R. D., & Berger, L. R. (2016). Earliest hominin cancer: 1.7-million-year-old osteosarcoma from Swartkrans Cave, South Africa. *South African Journal of Science*, 112(7/8), 5. https://doi.org/10.17159/sajs.2016/20150471

Figure 6.18 The discovery of cancer in an ancient human ancestor

A. National Geographic Explorer Lee Berger and a team member on site in South Africa, near the Malapa cave where an ancient toe bone with cancer was discovered.

B. The large brown mass (arrow) shows the tumor growing on the toe bone of a human relative who lived 1.7 million years ago. The lower image is a cross-section of the above image. The circled region indicates the cauliflower-like appearance of the mass, which is consistent with osteosarcoma.

1. Using the Internet, research how different rates of cancers have changed over time. What kinds of cancers are more common today? Less common today?
2. What factors can you find that influence cancer diagnoses? What about deaths from cancer?
3. How did the COVID-19 pandemic impact cancer diagnosis? Cancer survival rates? What does this tell you about current approaches to detecting and managing cancer?

cells. If a cell's *BRCA1* gene is mutated, it can result in a protein product that doesn't work properly, where DNA damage can then accumulate in the cell, leading to cancer.

Proto-oncogenes. The second class of cell cycle regulators is called **proto-oncogenes**. Proto-oncogenes are genes that encode proteins that tell your cells when to divide. Remember, cell division at the right time and place is a good thing! How else would you be able to heal that paper cut?

Proteins made from proto-oncogenes are under very strict control by other molecules in the cell. When a proto-oncogene is mutated, it creates a protein that stops listening to signals from outside and inside the cell telling it when it can divide. The mutated, rogue form of a proto-oncogene is called an **oncogene**. Oncogenes result in cells that divide continually, driving the formation of cancer. Going back to our car analogy, instead of losing our brakes, now our accelerator pedal is stuck to the floor, and we cannot release it.

6.5 Summary

- A wrong nucleotide added during DNA replication or the action of mutagens can cause mutations.
- Cellular mechanisms that prevent mutations include proofreading by DNA polymerase to repair mistakes made during DNA replication, and mismatch repair, which fixes mis-paired bases.
- Mutations that lead to extra cells can eventually result in tumors. Additional mutations that allow cells to spread to other areas of the body create malignant tumors capable of metastasis.
- Tumor suppressor proteins stop the cell cycle when there are problems. Proto-oncogene proteins tell the cell when to divide.
- Tumor suppressors act like "brakes" on cell division. When they malfunction, cells can divide out of control.
- When proto-oncogenes mutate to form an oncogene, cells divide uncontrollably. Oncogenes act like a stuck "gas pedal" and cells divide out of control.
- Malignant tumors can divide out of control and migrate to new areas of the body. When cancer cells start growing in new tissues, we say the cancer has metastasized. Benign tumors are abnormal growths that do not pose a danger to the body.

Understand It! Pomodoro Method

What if I told you that one of the secrets to learning biology had to do with a tomato—and not just any tomato, but a Pomodoro tomato? In the 1980s, Francesco Cirillo developed a study strategy inspired by his Pomodoro tomato-shaped kitchen timer. This study technique, called the Pomodoro Method, is used to break up your study section into 25-minute intervals. During the 25 minutes, you study the class material using any of the techniques introduced in the Understand It! sections. Then you take a 5-minute break.

The science behind it comes down to how our brains work. Cognitive psychologists found that taking regular short breaks results in an increased ability for people to stay on task longer. Psychologists think that taking short breaks works because our brains focus on changes in our environment.

If you want to try this out for yourself, first decide how long of a work period versus a break period you want to take. Most people use 25 minutes of work to 5 minutes of break time, but this ratio can change depending on the task. For example, some people prefer to do a 50:10 ratio when writing a paper. Then, set a goal for the work interval, such as explaining the material in Section 6.1 in your own words, and set a timer—you can use a tomato-shaped kitchen timer if you want, or the timer on your phone, or any of a large variety of free timer apps. Then reward yourself with a break! After about four Pomodoro Technique sessions, it's time for a longer break.

1. What do you think is your ideal Pomodoro Method interval for studying biology?
2. Doing something after a short break is not the only way you can change your environment to improve your productivity. Can you think of other changes you could make that will help you focus?

Chapter Review

Section 6.1 Preparing for Mitosis: DNA Replication

Multicellular organisms need to keep their cell numbers in balance to maintain homeostasis and function. Eukaryotic cells make genetically identical copies of themselves through a process called **mitosis.** DNA replication ensures each daughter cell has a complete copy of DNA. **Chemotherapy** is a cancer treatment that uses chemicals to block mitosis. Some chemotherapeutic drugs target **DNA replication.**

Interphase is the time between cell divisions and is comprised of G_1, S, and G_2. The **G_1 phase** is when cells grow bigger and make the molecules needed for DNA replication. The **S phase** is when DNA is synthesized. The **G_2 phase** is when the cell makes additional proteins and cellular organelles for the new cells to come. After everything is ready, the cell enters mitosis and begins the process of division.

The DNA double helix is **antiparallel,** meaning it consists of two strands, each with a 5′ end and a 3′ end, running in opposite directions. The cell uses one strand as a template to make new DNA.

DNA replication starts with **helicase** separating the two strands, forming the **replication bubble**, while **topoisomerase** prevents knots from forming as helicase pulls the strands apart. Next, **primers** bind to the DNA so **DNA polymerase** can attach. DNA polymerase synthesizes the new DNA strand by adding **complementary bases** according to the parental strand sequence. Guanine (G) pairs with cytosine (C), and adenine (A) pairs with thymine (T).

DNA polymerase only adds new bases onto the 3′ end, so all DNA is made in one direction (5′ to 3′). This means that polymerase must bind to the 3′ end of the parent strand of DNA. **Leading strand** synthesis is faster because it adds new DNA in the same direction as helicase separates the two strands. **Lagging strand** synthesis is slower because the DNA polymerase has to "leapfrog" to make new DNA in pieces, since it is moving in the opposite direction of helicase. **Ligase** seals together pieces to make one long piece of DNA.

Section 6.2 Packing DNA for Mitosis

DNA is packaged by first wrapping around **histones**, then onto a protein scaffold, and finally into a fully condensed **chromosome**. During S phase, DNA replicates, resulting in two identical molecules of DNA in each chromosome, called **sister chromatids**. Chromatids are attached by the **centromere**.

Each species has a characteristic **chromosome number**. Humans are diploid, having 23 pairs of chromosomes for a chromosome number of 46. **Karyotypes** are used to analyze chromosome number and structure.

Section 6.3 Mitosis and Cytoplasmic Division

After the DNA is replicated and interphase is complete, cells enter mitosis. In **prophase**, chromosomes condense, and sister chromatids are attached at their centromeres to the **spindle**. In **metaphase**, the chromosomes align in the center of the cell. In **anaphase**, the sister chromatids separate to opposite poles in the cell. Finally, in **telophase**, a new nucleus forms around each group of chromosomes, which then decondense. Following mitosis, **cytokinesis** divides the cytoplasm and organelles into the two daughter cells.

Section 6.4 Cell Cycle Control

Cell cycle checkpoints occur at the **G_1**, **G_2**, and **M** phases. If a problem is detected at a checkpoint, cell division will pause until the problem is repaired or the cell undergoes apoptosis. The **cell division limit** is the number of times a cell can divide before it enters **G_0**, or **senescence**. **Telomeres** protect the ends of chromosomes. They shorten with each cell division, and once worn away, a cell will senesce. **Stem cells** and some cancer cells can undergo unlimited cell division because they have **telomerase** to repair their telomeres. **Necrosis** and **apoptosis** are two ways that cells are removed from the body.

Cells signal other cells in the body for a variety of reasons, including cell cycle control. **Paracrine** signals diffuse between cells that are close to each other while **endocrine** signaling occurs between cells in different parts of the body. Cancer can result if cells do not respond properly to cell signaling.

Section 6.5 Mutations and Routes to Cancer

Mutations are changes in the nucleotide sequence in the DNA of a cell. Mutations can be caused by DNA replication errors or through damage caused by **mutagens**. Cancer-causing mutations may lead to **tumors**, or abnormal growths. Some tumors are benign and often harmless. **Malignant** tumors that can spread, or **metastasize**, to new areas of the body are dangerous to one's health. **Tumor suppressors** are proteins that keep cells from dividing too often, and if they malfunction, cell division can proceed unchecked, which can lead to cancer. **Proto-oncogenes** encode proteins that tell cells when to divide, and if they transform into **oncogenes**, cells may divide out of control, which can also lead to cancer.

Practice It!

Retrieval practice is the process of pulling information from your long-term memory and using it. Here are some questions to help with your retrieval practice.

Self-Quiz

1. Interphase is the part of the cell cycle when ______. (LO 6-2)
   a. a cell ceases to function
   b. the spindle forms prior to nuclear division
   c. a cell grows and replicates its DNA
   d. telomeres are replicated

2. In DNA replication, ______ forms the replication bubble by separating the two DNA strands of the double helix. (LO 6-3, 6-4)
   a. DNA polymerase
   b. DNA ligase
   c. helicase
   d. primase

3. In DNA replication, ______ adds primers so that DNA polymerase can bind and start adding complementary nucleotides to form the new DNA strand. (LO 6-3, 6-4)
   a. DNA polymerase c. helicase
   b. DNA ligase d. primase

4. In DNA replication, ______ reads the original strand and adds the complementary base pairs to form the newly replicated strand. (LO 6-3, 6-4)
   a. DNA polymerase c. helicase
   b. DNA ligase d. primase

5. Which of the following is (are) not required for DNA replication to occur? (LO 6-3, 6-4)
   a. DNA polymerase c. primers
   b. nucleotides d. all are required

6. In eukaryotic chromosomes, DNA wraps around ______. (LO 6-5)
   a. histone proteins
   b. sister chromatids
   c. centromeres
   d. nucleotides

7. Karyotypes analyze ______. (LO 6-6)
   a. the DNA sequence of nucleotides
   b. the number and structure of chromosomes
   c. mutations affecting one nucleotide
   d. the amount of histone proteins in a chromosome

8. During mitosis, the spindle fibers are used to anchor and move the chromosomes. These spindle fibers are composed of what type of cytoskeletal structure? (LO 6-8)
   a. mitochondria
   b. microtubules
   c. nuclear membrane
   d. vesicles

9. After interphase, what is the correct order of steps in cell division? (LO 6-7)
   a. prophase, anaphase, telophase, metaphase
   b. anaphase, telophase, metaphase, prophase
   c. prophase, metaphase, anaphase, telophase
   d. metaphase, anaphase, prophase, telophase

10. Sister chromatids line up in the middle of the cell during ______ of mitosis. (LO 6-8)
   a. anaphase
   b. prophase
   c. telophase
   d. metaphase

11. Cell contents, including the new nuclei, are separated into two daughter cells during ______ in mitosis. (LO 6-8)
   a. anaphase c. telophase
   b. prophase d. metaphase

12. After mitosis, the chromosome number of a descendant cell is ______ the parent cell. (LO 6-8)
   a. the same as
   b. one-half of
   c. rearranged compared to
   d. doubled compared to

13. The ______ checkpoint ensures that the cell has all the resources it needs to replicate its DNA and enter mitosis. (LO 6-12)
   a. G_1 c. M
   b. G_2 d. S

14. ______ are short, repeated DNA sequences that wear down with each cell division. (LO 6-10)
   a. Primers c. Promoters
   b. Telomeres d. Enhancers

15. Cell signaling between cells that are in close proximity is called ______. (LO 6-13)
   a. endocrine signaling
   b. paracrine signaling
   c. amplification signaling
   d. autocrine signaling

16. Cancer is characterized by cells that ______. (LO 6-19, 6-23)
   a. divide out of control
   b. break loose from their home tissue
   c. invade and grow in other parts of the body
   d. all of the above

17. Normal proteins that signal the cell to start the process of mitosis are called ______. (LO 6-20)
   a. proto-oncogenes
   b. oncogenes
   c. tumor suppressors
   d. mutations

18. Normal proteins that stop cells from entering mitosis if there are problems are called ______. (LO 6-20)
   a. proto-oncogenes
   b. oncogenes
   c. tumor suppressors
   d. mutations

Critical Thinking

1. Why is it important for DNA proofreading and mismatch repair that DNA molecules are double-stranded? (LO 6-17)

2. Xeroderma pigmentosum is an inherited disorder characterized by the rapid formation of skin sores. All forms of radiation trigger these symptoms, including fluorescent light, which contains UV light. In most affected individuals, a certain class of proteins is missing or defective. What is the collective function of these proteins? (LO 6-17)

3. Mice lacking telomerase age prematurely, with a life expectancy that is half that of normal mice. However, if you give telomerase back to mice lacking it at the end of their lives, they become youthful again. Worn-out tissue in the brain and other organs repairs itself, and the once-geriatric individuals begin to reproduce again. Therefore, increasing the activity of telomerase holds therapeutic promise for rejuvenating aged tissues. However, this approach can be dangerous. Why? (LO 6-10)

4. A friend of yours shares that their aunt had breast cancer. He mentions that while the doctors said they treated the breast cancer, his aunt later developed lung and liver cancer. How could you explain to your friend that it is not very likely that his aunt had three distinct cancers arise independently? *Hint*: Consider metastasis in your explanation. (LO 6-19)

5. You are talking with a friend about reducing cancer risk. You mention that you've heard that lifestyle changes can help to prevent cancer formation. What kinds of lifestyle choices could help prevent cancer and why? (LO 6-17, 6-19)

Unit 2 Review

Although we think of cancer as a modern disease that has appeared only in recent years—as human life spans have increased, along with potential exposure to carcinogens—we can see from fossil evidence that both ancient ancestral humans and even dinosaurs could get cancer. Cancer is a disease that is fundamentally about our cells—and in Chapter 5 and 6, we learned about the biology of our cells.

In Chapter 5, we covered the basics of the cell, including different kinds of cells and their subcellular components. We learned about the structure of the cell's outer membrane, the proteins within it, and how cells transport cargo across the membrane. We also covered the cellular organelles and the functions of each, and how the organelles work together to benefit cells. We closed Chapter 5 with a discussion of how cells are connected to one another.

In Chapter 6, we learned about cellular behaviors, and how changes in cellular behaviors can contribute to making too many cells, ultimately resulting in the development of cancer. We discussed how new cells are generated through mitosis, including how DNA is replicated during S phase. We talked about how a cell can transmit messages intracellularly and receive messages from other cells, such as when to divide or be removed via apoptosis. Finally, we discussed how DNA mutations can impact cell cycle control, ultimately leading to cancer.

Understanding the nature of cells is important when we consider better treatments and cures for cancer. Cancer cells can get around cell–cell controls that we learned about in Chapters 5 and 6. Catching cancerous cells early and removing them surgically, before they have a chance to spread throughout the body, is a traditional form of cancer treatment. Newer treatments involve supercharging a patient's own cells to become cancer-fighting machines—something we'll return to in Chapter 9.

Although it takes changes in an individual cell to get cancer, cells can fight back. And with your new knowledge of the cell, you can think about ways to make healthy decisions to benefit your own cells, and body.

Kevin Snyder Photography

Unit

3

Human Reproduction

Genetics

We all have a family. Family can mean different things to different people, and there is definitely something to be said for family being what you make, rather than who you share your DNA with. In this unit, we're going to be focusing on genetics—and so the version of family we'll be covering here is individuals with whom we share our DNA.

I'm fortunate to have two biological children, pictured here. Their genes and traits were passed down from my husband and me, our parents before us, our grandparents, and so on. One child has brown eyes, and the other blue. One is tall for their age, and the other is average height. Both love to go-go-go and have blonde hair, despite my husband and I having brown hair. If my two kids have the same biological parents, why are they so different from each other? How is it possible for two brown-haired parents to have blonde children?

These are questions that can be answered through genetics—the study of the passage of genes and traits from one generation to the next. Unit 3 focuses on genetics, with an emphasis on human reproduction, including what genes are and how they are expressed; the passage of genes and traits from parents to offspring, including challenges with human reproduction; and finally, biotechnology.

In Chapter 7, we'll start by learning what genes are as well as exploring the process of gene expression, which is how information encoded in our DNA leads to our characteristics: height, response to medications, risk for developing a disease, and so forth. In Chapter 8, we will talk about how traits are passed down or are inherited in families, and how the expression of that DNA is also inherited. We'll also discuss problems with human reproduction that can lead to pregnancy loss. In Chapter 9, we'll discuss biotechnology and all the ways in which genetic technology is changing our lives today, from genetically modified foods to fertility treatments.

Before we get started, though, a brief note on the terms "male" and "female" as they are used in the context of genetics. Gender is culturally derived, exists on a spectrum, and is not the same as biological sex. Biological sex also exists on a spectrum and includes a wide variety of traits, including but not limited to intersex anatomy, differences in hormone levels, and differences in chromosomes. For the purposes of this text, "male" refers to a person whose body makes sperm cells, and "female" refers to a person whose body makes (or made) egg cells.

Preview It!

List three things you know already about genetics. What is something you want to learn in this unit? What questions do you have about genetic technologies like genetic testing, genetically modified organisms, and personalized medicine?

National Geographic Creative

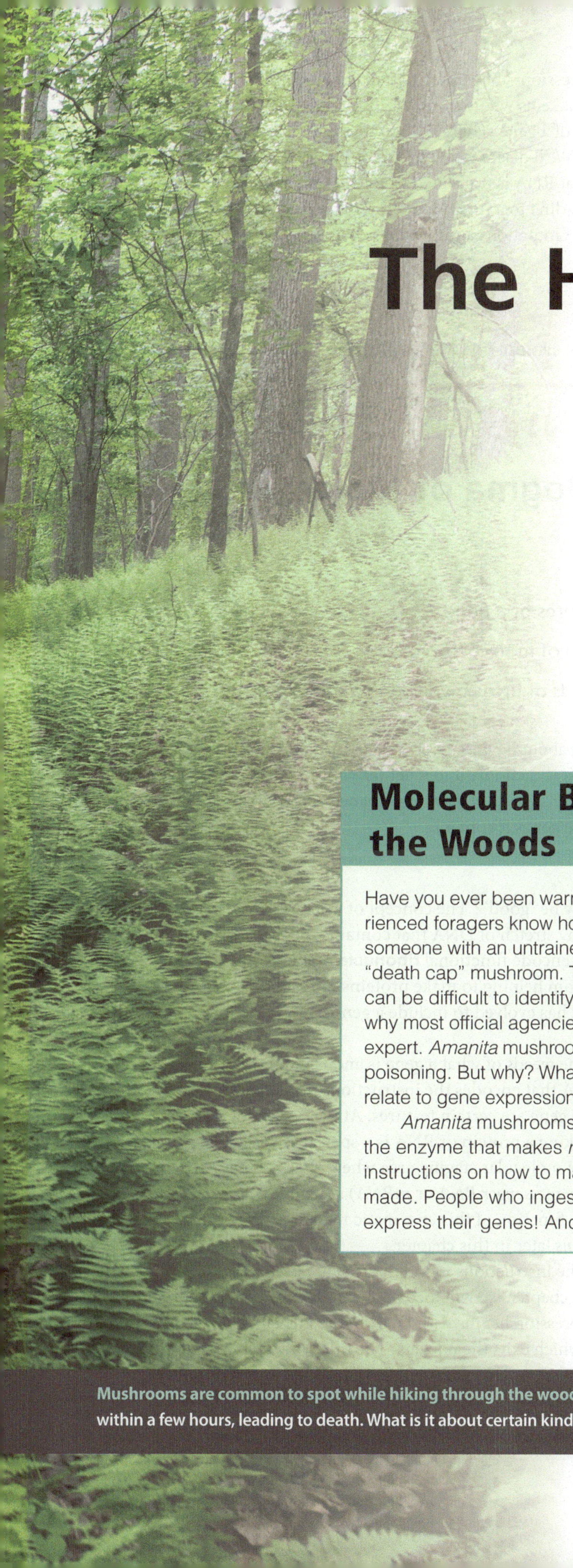

7

The Heart of Gene Expression

Chapter Outline:

Molecular Biology in a Walk Through the Woods

Have you ever been warned about picking and eating wild mushrooms? While experienced foragers know how to find and identify a variety of mushroom delicacies, someone with an untrained eye could accidentally pick up an *Amanita phalloides*, or a "death cap" mushroom. This unassuming white mushroom with a slightly greenish cap can be difficult to identify—unless you know exactly what you're looking for—which is why most official agencies recommend not eating wild mushrooms unless you are an expert. *Amanita* mushrooms are responsible for almost all fatalities from mushroom poisoning. But why? What makes these mushrooms so deadly? And how does this relate to gene expression, which is the subject of this chapter?

Amanita mushrooms contain chemicals called amatoxins. These toxins inhibit the enzyme that makes *messenger RNA* (mRNA). Without mRNA, there are no instructions on how to make proteins. No gene expression means no proteins are made. People who ingest *Amanita* mushrooms die because they can no longer express their genes! And the inability to express our genes is a death sentence.

Mushrooms are common to spot while hiking through the woods. While some are delicious, others can cause permanent organ damage within a few hours, leading to death. What is it about certain kinds of mushrooms that can cause fatal poisonings?

In this chapter, we'll explore gene expression—how information encoded in our DNA is used to make the proteins that ultimately make us who we are. We'll start with an overview of gene expression and the basic organization of a gene. Then we'll discuss the central dogma of molecular biology: DNA → RNA → protein. We'll discuss the process of transcription, making an RNA copy of DNA, and translation, how that RNA is translated from the language of nucleic acids to the language of proteins, ultimately leading to a functional protein product. In closing, we will discuss the functional consequences of mutations and the control of gene expression.

Preview It!

How would you illustrate the relationship between DNA, RNA, and protein? What is one question you have about gene expression?

7.1 The Central Dogma of Molecular Biology

Learning Objectives

7-1 Describe the general features of a gene.

7-2 Explain the central dogma of molecular biology.

7-3 Describe the different kinds of RNAs involved in gene expression.

Have you ever heard someone talking about having "good genes"? Or considered why you look the way that you do? While the clothes you choose and whether your hair is its natural color are decided by choices that you make, other aspects of your appearance are outside of your control. It all comes back to your DNA and the genes it contains.

Genes and Gene Expression

So, what is a gene? And can they really be "good"? The concept of a gene evolved over time. One outdated definition is that a gene is any stretch of DNA that contains the instructions to make a protein. Now we know that DNA can encode functional **ribonucleic acids (RNAs)**—a type of nucleic acid that can do everything from helping to make proteins to regulating gene expression. Therefore, the definition of *gene* has evolved to include a gene being a section of DNA that encodes a protein *or* RNA.

Without getting too much into the weeds of what a gene is and is not, for our purposes, we will define a **gene** as a unit of DNA that encodes the instructions needed to make a specific protein or RNA molecule. Within the gene are certain features. At the beginning of a gene, right next to the actual instructions for making a protein or RNA is a special area called a **promoter** (**Figure 7.1**, inset). A promoter is like a molecular staging area where the enzymatic machinery for making an RNA copy of DNA (the *messenger RNA* or *mRNA*) assembles. It's also an area where other proteins that regulate the function of this enzymatic machinery congregate. We will return to features like *introns* and *exons* later in this chapter.

Gene expression occurs when the instructions found in DNA are used to make protein or RNA in a multistep process. In this chapter, we will focus only on the expression of protein-coding genes. The process of gene expression begins with DNA in the nucleus. The DNA is used as a template to make an RNA copy, which exits the nucleus. The mRNA is then used to make a protein. The process of going from DNA → RNA → protein is called the **central dogma of molecular biology**, or simply the central dogma. The central dogma states that genes are expressed starting with DNA in the nucleus, where an RNA copy of the information in DNA is made through a process called *transcription*, and where the RNA copy then exits the nucleus and is used in conjunction with the ribosome to create a protein in a process called *translation* (Figure 7.1).

Figure 7.1 An overview of the central dogma

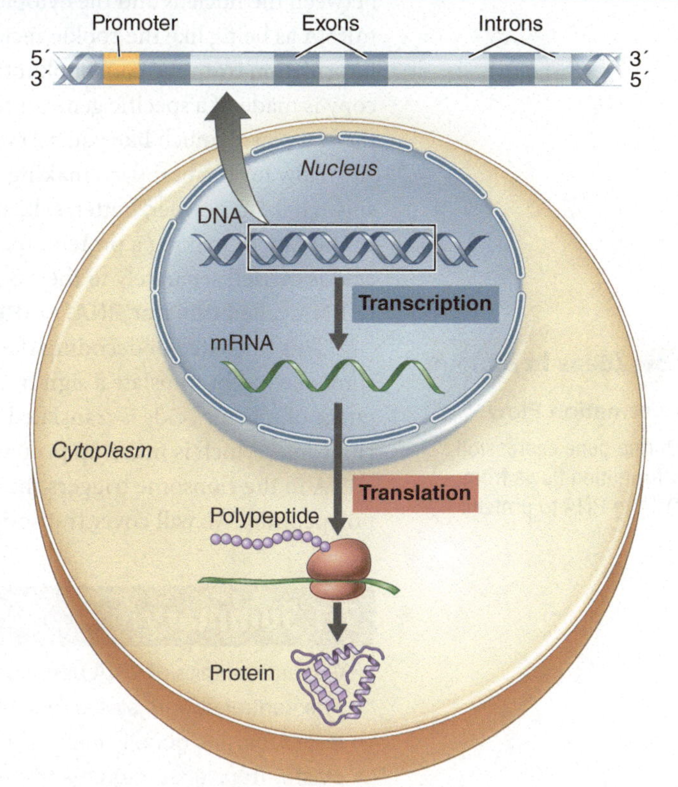

Gene expression begins in the nucleus, when an mRNA copy is made of the DNA. The mRNA copy is then exported to the cytoplasm, where the protein-making machinery is located. Note that the polypeptide that forms from the ribosome is primary protein structure and eventually folds into a functional protein (Chapter 3).

Cooking up Genes

A common way of thinking about the central dogma is like making cookies (Figure 7.2). Let's say that you are going to a friend's house to bake cookies. You need the recipe from your cookbook. Let's say your cookbook is a 5-pound behemoth complete with a lifetime's worth of cooking notes courtesy of your dear grandmother. If you need one recipe, do you need to bring the entire heavy, irreplaceable cookbook? Nope! In addition, if you are carrying your big cookbook around, you run the risk of it getting lost, damaged, or stolen. So instead, you make a copy of the recipe and then take it from your house to your friend's house, where you can then go about baking cookies.

Think of the cookbook as like your DNA. It's both large and comprehensive, with 20,000 to 25,000 genes (recipes) in it. And as we learned in Chapter 6, if your DNA gets damaged, it could cause a mutation that may contribute to the development of cancer. So, your DNA stays in the nucleus where it is safe and protected. But the ribosome, the protein-building machinery of the cell, is in the cytoplasm. How does the information in DNA make it from the nucleus to the ribosome?

Figure 7.2 Drawing parallels between the central dogma and cookie baking

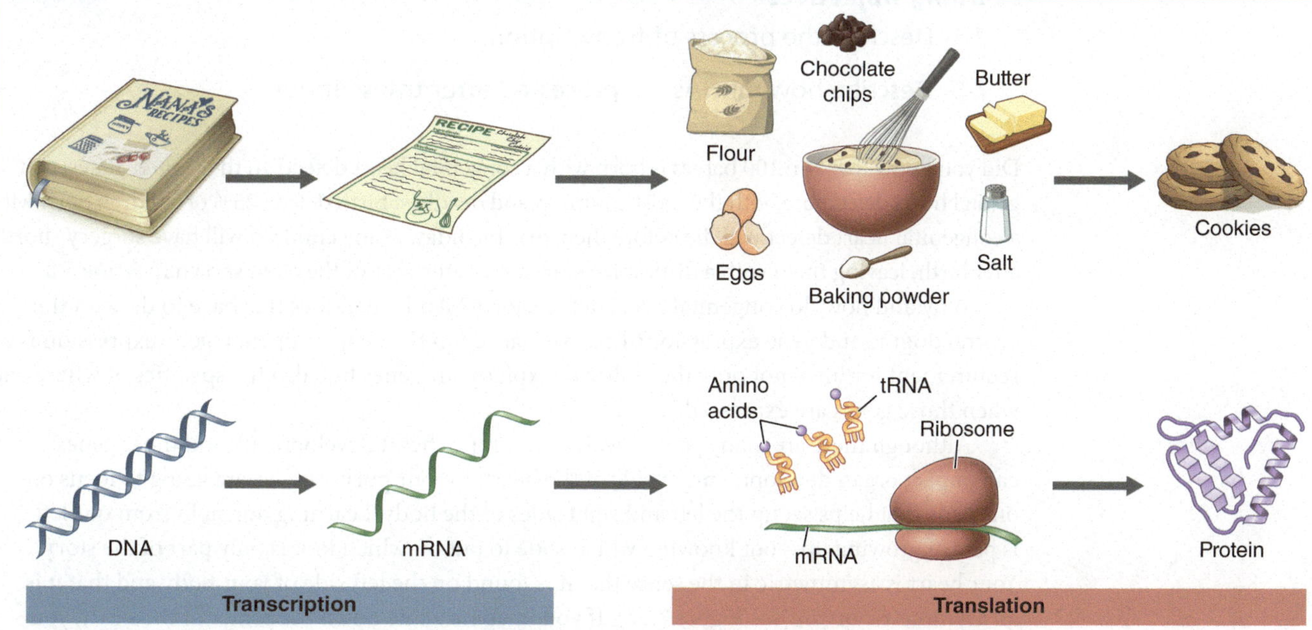

The central dogma describes the process of going from DNA (the cookbook) to RNA (the recipe) and finally to protein (the cookies). The molecular process includes two key phases: transcription and translation.

Enter: messenger RNA. **Messenger RNA (mRNA)** was named for its role as the "messenger" between the nucleus and the cytoplasm. Using our cookie baking analogy, you can think of mRNA as being like the cookie recipe that you copied from the cookbook. You *transcribe* the information from the cookbook into a portable form. Similarly, during transcription, an RNA copy is made of a specific gene on DNA. This mRNA carries the message from the nucleus into the cytoplasm, much like you carry the copied recipe from your house to your friend's house.

Now for the best part: making your cookies! First, you need your raw ingredients—flour, sugar, baking powder, butter, salt, and chocolate chips—which are combined in a mixing bowl. In the case of making a protein, remember that proteins are made of amino acids. Each amino acid is carried separately to the ribosome (the protein-constructing machinery) by another kind of RNA called **transfer RNA**, or **tRNA**.

Big Ideas in Biology

Information Flow

During gene expression, information flows from DNA to RNA to protein.

This final step of decoding the mRNA to form a protein is called translation. Just like someone might translate a sign from German into English, proteins are made when the language of nucleic acids is translated into the language of proteins. The site of translation is the ribosome, which is made up protein and another type of RNA called **ribosomal RNA**, or **rRNA**. rRNA in the ribosome triggers the chemical reaction that links amino acids together, forming a polypeptide. We will cover transcription and translation in more detail later in this chapter.

7.1 Summary

- Genes are units of DNA used to make an RNA or protein product.
- The central dogma states that information flows from DNA to RNA to protein.
- mRNA carries genetic information from the nucleus to the protein machinery in the cytoplasm. Translation requires the use of other types of RNA—tRNA and rRNA. tRNAs carry amino acids to the ribosome, and rRNAs trigger the reaction to connect these amino acids together into the growing polypeptide chain.

7.2 Transcription: DNA to RNA

Learning Objectives

7-4 Describe the process of transcription.

7-5 Describe how mRNAs are processed after transcription.

Did you know that 1 in 100 babies is born with a congenital heart defect? In the United States, congenital heart defects are both the most common and deadliest birth defect: 25% of children born with a congenital heart defect will die before their first birthday. Many children will have surgery shortly after birth, leaving them with a distinctive scar, such as the scar of the child shown in **Figure 7.3**.

Why and how do congenital heart defects form? And what does this have to do with the central dogma and gene expression? Like we learned in the chapter opener, gene expression is a requirement for life—not only the ability to express our genes but also the specifics of where and when those genes are expressed.

Although there are many genes involved in proper heart development, and those genes can impact organ development outside of the heart, for our purposes, we are going to focus on one gene that helps set up the left and right sides of the body. Learning our right from our left is part of growing up—but knowing which shoe to put on which foot is only part of the story. Your heart is asymmetric in the sense that it is found on the left side of your body and that it is an asymmetric organ (see Figure 7.12). If you imagine a line down the center of the heart, you can see that there are different features, like certain blood vessels, that are on the left side of the heart but not on the right.

When your heart is first forming, the body needs to know which side is left and which side is right so that the heart and other internal organs develop properly—for example, making sure the left ventricle (one of the chambers of the heart) in its correct position. But how does the body know which is left and which is right? It comes down to gene expression—and in particular, gene expression on the left side of the body only.

For our purposes, we will consider *lefty-1*, a gene that acts almost like a molecular fence differentiating the left and right sides of the body. We will return to the role of the Lefty-1 protein and congenital heart defects after discussing gene expression.

Figure 7.3 The scar of a heart warrior

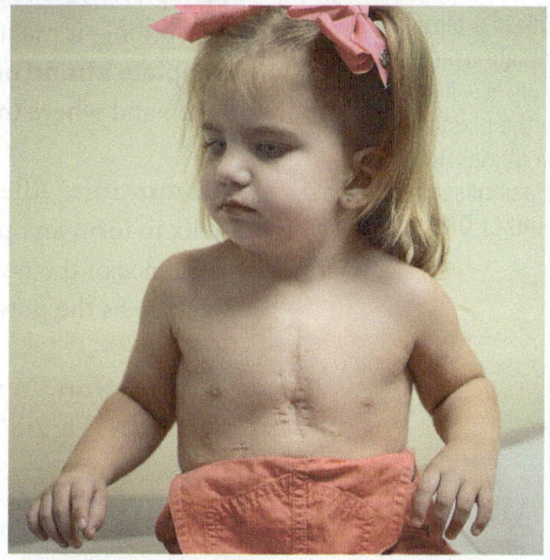

Daniel Tardif/Photodisc/Getty Images

Babies born with congenital heart defects often have open heart surgery shortly after birth and may go on to have other corrective heart surgeries.

Making mRNA from a DNA Template: The Process of Transcription

Recall from the previous section that transcription is the process of making a portable mRNA copy of a gene. We can describe this process as having three phases: initiation (getting started), elongation (making mRNA), and termination (wrapping up). After the mRNA is made, it needs to be processed before it is ready for use. We will go through these processes one at a time, starting with transcription (Figure 7.4).

Figure 7.4 Transcription

1 Initiation
During initiation, RNA polymerase binds to the promotor.

2 Elongation
RNA polymerase moves along the gene, unzipping the helix to form a transcription bubble. As it moves, the enzyme assembles a strand of RNA. The DNA re-winds behind the RNA polymerase.

3 Termination
RNA polymerase reaches the end of the gene and detaches. The mRNA transcript is released.

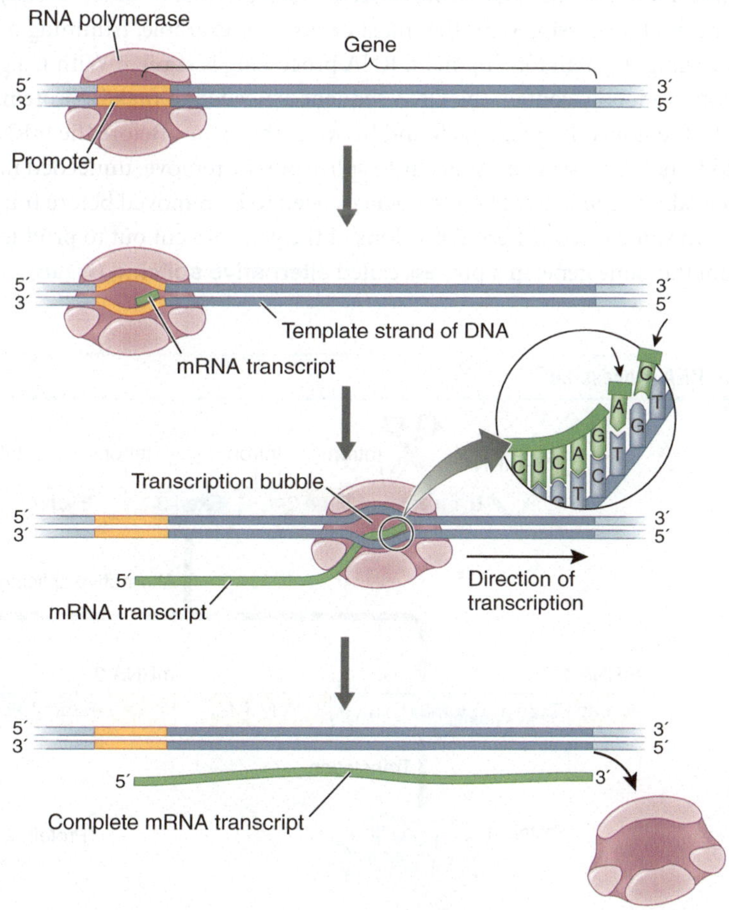

Biologists use lowercase letters and italics when referring to a gene, such as *lefty-1*, and non-italicized words beginning with a capital letter when referring to a protein, such as Lefty-1.

Remember the mushroom toxin from the chapter opener? It inhibits RNA polymerase. Therefore, if no mRNA can be produced, no proteins will be made and the organism will die.

Study Tip

Think of a template like a template used for tracing a picture—just like you use the plastic template as a guide, RNA polymerase uses the template strand of DNA as a guide.

Initiation. mRNA is made using an enzyme called **RNA polymerase**. Like DNA polymerase (Chapter 6), RNA polymerase reads the DNA to make a complementary copy, with one key exception—it makes RNA instead of DNA. RNA polymerase binds to a special region on the **template strand** of DNA called a promoter. A promoter contains special sequences that control when and where the gene is expressed (Figure 7.4, step 1).

Elongation. After binding, the polymerase starts moving along the gene, unwinding the double helix to form an opening called a **transcription bubble**. As the enzyme moves, it "reads" the base sequence of the noncoding DNA strand, assembling a strand of RNA. The double helix winds back up as the polymerase passes (Figure 7.4, step 2).

Termination. When the polymerase reaches the end of the gene, it releases the DNA and the new RNA (Figure 7.4, step 3). Now we have an immature mRNA that is ready for processing.

! Point of Confusion: Transcription and DNA replication have some similarities. For example, base-pairing rules are followed. However, with transcription, adenine (A) pairs with uracil (U) instead of thymine (T). Transcription and DNA replication also use a strand of DNA as a template to make something new. However, transcription copies only a portion of one strand, whereas DNA replication copies the entirety of both strands. Therefore, DNA replication produces two double-stranded DNA molecules, and transcription makes one single-stranded RNA molecule.

In the case of *lefty-1*, there are four exons and three introns. The introns are spliced out prior to translation.

Post-transcriptional RNA Processing

Think about any manufacturing process: Before a final product is ready to be shipped, there are some final processing steps that must occur—for example, trimming off excess material and packaging the item for shipment. RNA processing is similar. Within a gene are introns and exons. **Exons** generally contain the DNA sequences needed to make a protein, whereas **introns** are typically the noncoding regions found between the exons. Before the mRNA can leave the nucleus, additional processing must occur to splice out, or remove, unneeded material—typically, the noncoding sequences. These sequences need to be removed before translation can begin.

In some cases, different portions of the gene are cut out to produce different proteins, all from the same gene, in a process called **alternative splicing** (Figure 7.5). Alternative splicing

Figure 7.5 Post-transcriptional RNA processing

Alternative splicing is the removal of unneeded sequences (typically introns) and the splicing together of exons from the mRNA. Since one gene can encode multiple protein products (in this image, there are two possible mature mRNAs), alternative splicing represents another exception to the central dogma. Note that processing also includes the addition of a 5′ cap and 3′ poly-A tail.

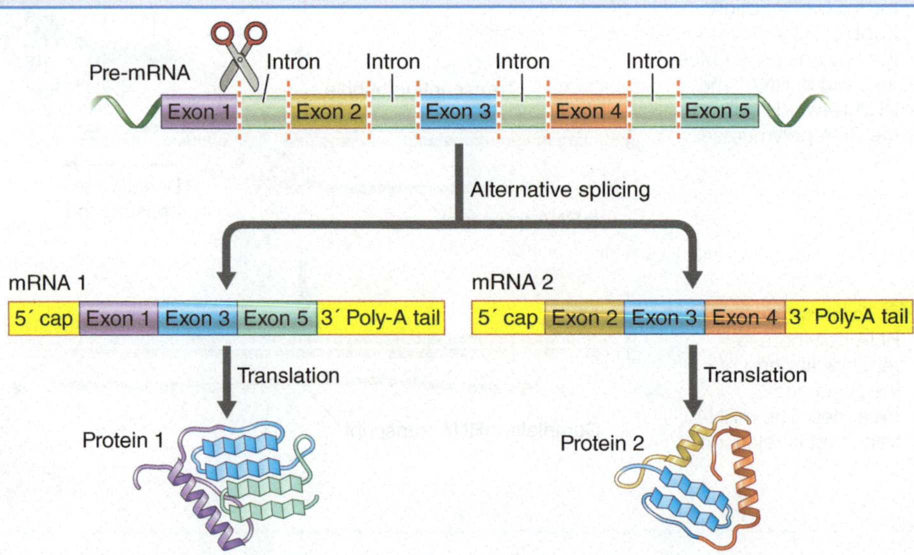

allows one gene to encode multiple versions of a protein. Problems with alternative splicing are implicated in a variety of diseases, including cancer, neurological disorders, autoimmune disorders, and many others.

Before an mRNA can leave the nucleus, it needs special modifications, including the addition of a **5′ cap** and a **3′ poly-A tail** (see Figure 7.5). The 5′ cap is a modified nucleotide attached at the 5′ end of the mRNA molecule. The 5′ cap facilitates translation (which we'll cover in the next section), keeping the mRNA stable and allowing its export from the nucleus. The 3′ poly-A tail is formed when a special enzyme adds a string of adenines to the 3′ end of the mRNA. The 3′ poly-A tail regulates the timing and duration of the mRNA's translation, and in eukaryotes, it is also a signal that allows the molecule to be exported from the nucleus.

? Did You Know?

About 95% of human protein-coding genes are alternatively spliced. This allows for more proteins to be expressed with less DNA.

Big Ideas in Biology

Information Flow

Alternative splicing is another exception to the central dogma. Rather than one gene encoding one protein, one gene can encode many proteins.

7.2 Summary

- Transcription uses a DNA template to copy a gene into mRNA form.
- Transcription consists of three steps: initiation, elongation, and termination. During initiation, RNA polymerase binds to the promoter of a gene. In elongation, RNA polymerase adds nucleotides to the growing mRNA. During termination, the RNA polymerase stops building mRNA and falls off the DNA.
- RNA processing is the creation of a mature RNA. This involves the splicing out of introns (and sometimes exons) and the addition of a 5′ cap and a 3′ poly A tail to the new mRNA, which help keep the mRNA stable.

Study Tip

Think of the 5′ cap and 3′ poly-A tail like the plastic piece (aglet) at the end of a shoelace that keeps the laces from unraveling.

7.3 Translation: RNA to Protein

Learning Objectives

7-6 Describe codons and their function.

7-7 Relate the function of a tRNA to its structure.

7-8 Describe the process of translation.

7-9 Explain how an mRNA specifies the order of amino acids in a polypeptide.

Have you ever exchanged secret coded messages with a friend? Or had a secret decoder ring? The basic premise is that you have a message that makes no sense without a way to decode it. You must translate between the code and the alphabet to make sense of the message. Translation is no different—the language of nucleic acids (nucleotides) is translated into the language of proteins (amino acids). Before we get to the process of translating the mRNA resulting from transcription, and the process of folding the protein at the end of the process, we first need to discuss the genetic code.

The Genetic Code

The protein-building information carried by an mRNA occurs in three-nucleotide units called **codons**. Each codon provides a specific instruction during translation: either "start," "stop," or "add a particular amino acid here." The order of the three nucleotides in a codon determines the instruction it specifies. With four possible nucleotides—U (which is used in RNA instead of the T that is used in DNA; see Chapter 6), C, A, or G—in each of the three positions of a codon, there are a total of 64 mRNA codons making up the **genetic code** (Figure 7.6).

Note as you look at Figure 7.6 that although there are 64 codons, there are only 20 amino acids. This is a way of protecting an organism from harmful mutations. Generally, a mutation has functional consequences only if it results in a change in protein structure. Therefore, a single mutation may not lead to a change in amino acid because there is redundancy in the genetic

Figure 7.6 The codon table represents each codon and its corresponding amino acid or a stop codon that will end translation

First base in codon	Second base in codon: U	C	A	G	Last base in codon
U	UUU Phe	UCU Ser	UAU Try	UGU Cys	U
U	UUC Phe	UCC Ser	UAC Try	UGC Cys	C
U	UUA Leu	UCA Ser	UAA STOP	UGA STOP	A
U	UUG Leu	UCG Ser	UAG STOP	UGG Trp	G
C	CUU Leu	CCU Pro	CAU His	CGU Arg	U
C	CUC Leu	CCC Pro	CAC His	CGC Arg	C
C	CUA Leu	CCA Pro	CAA Gln	CGA Arg	A
C	CUG Leu	CCG Pro	CAG Gln	CGA Arg	G
A	AUU Ile	ACU Thr	AAU Asn	AGU Ser	U
A	AUC Ile	ACC Thr	AAC Asn	AGC Ser	C
A	AUA Ile	ACA Thr	AAA Lys	AGA Arg	A
A	AUG Met (START)	ACG Thr	AAG Lys	AGG Arg	G
G	GUU Val	GCU Ala	GAU Asp	GGU Gly	U
G	GUC Val	GCC Ala	GAC Asp	GGC Gly	C
G	GUA Val	GCA Ala	GAA Glu	GGA Gly	A
G	GUG Val	GCG Ala	GAG Glu	GGG Gly	G

Big Ideas in Biology

Evolution

The genetic code is highly conserved, which means that all organisms use essentially the same code.

code. For example, if the third U in UCU is replaced with a C in an RNA molecule, the protein product will still be serine (Ser), so the appropriate protein will be produced. We will return to mutations in Section 7.4.

Keep in mind that most codons specify amino acids. For example, the codon UUU specifies the amino acid phenylalanine (Phe), and UUA specifies leucine (Leu). One notable exception is stop codons—these do not code for an amino acid; rather, they signal for a release factor to bind to the ribosome, thus ending translation.

Figure 7.7 Transmission of information from DNA to RNA to protein

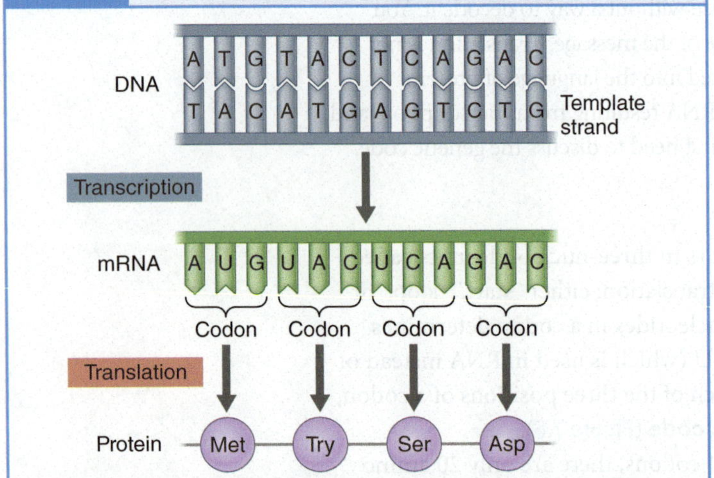

The gene (in DNA) is transcribed to mRNA. The mRNA is subdivided into codons based on the position of the start codon. Note that only the beginning of a gene is shown here.

The first AUG in a typical mRNA is a signal to start translation, so it is called a **start codon.** AUG is also the codon for the amino acid methionine. Therefore, methionine is always the first amino acid in a polypeptide. Codons occur one after another along the length of an mRNA. When the mRNA is translated, the order of its codons determines the order of amino acids in the resulting polypeptide (Figure 7.7).

Making Proteins: The Process of Translation

Translation is the process of building a new protein using an mRNA template. Translation always begins in the cell's cytoplasm and will either complete in the cytoplasm or move into the endoplasmic reticulum (Chapter 5). Like transcription, it has three general phases: initiation, elongation, and termination (Figure 7.8).

Figure 7.8 Translation

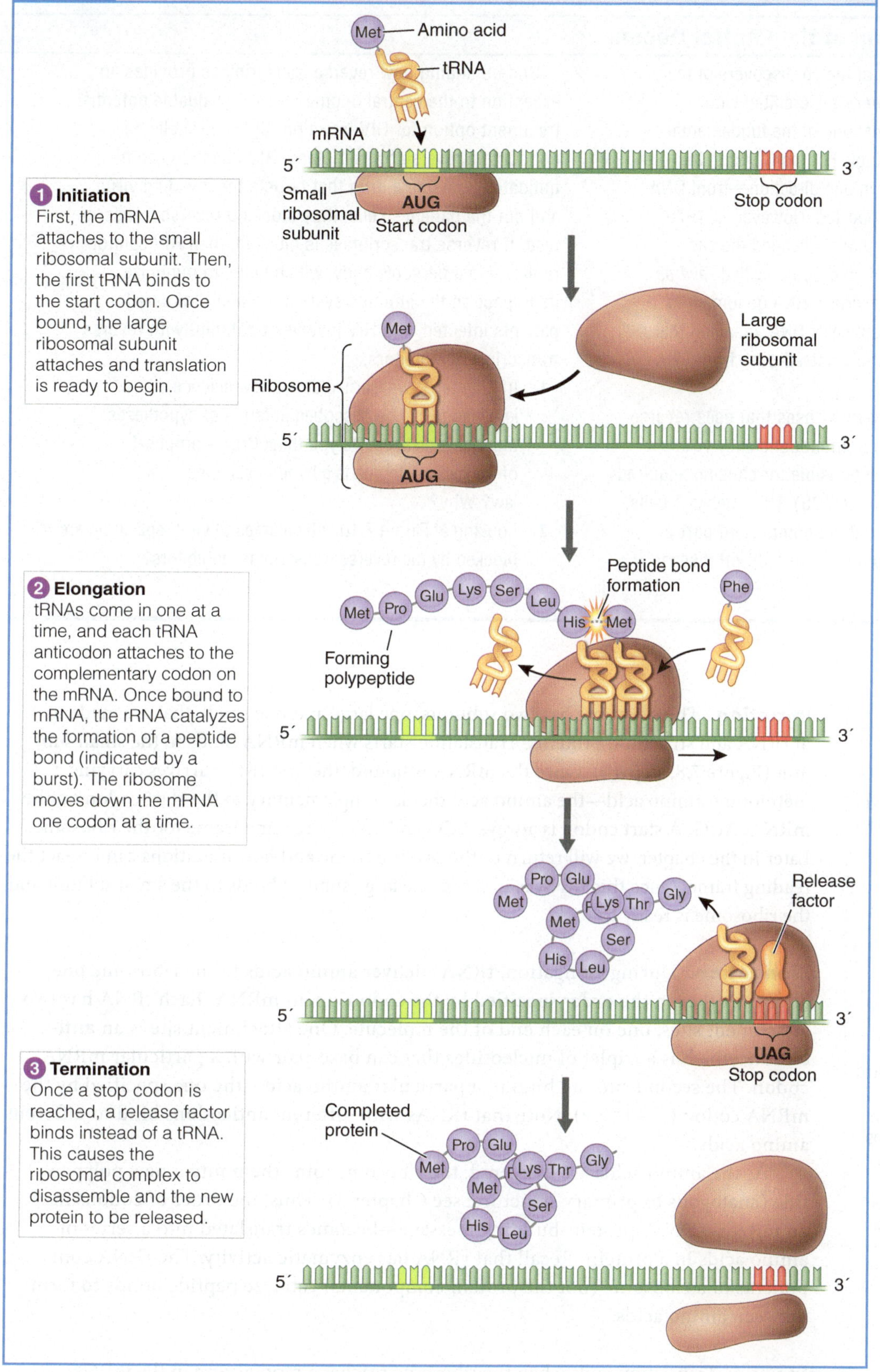

Process of Science

HIV and the Evolution of the Central Dogma

In 1958, Francis Crick (one of the co-discoverers of the structure of DNA; see Chapter 6) is credited with proposing the central dogma, one of the fundamental principles of molecular biology. For many years, the idea that information flows in one direction—from DNA to RNA to protein—was supported. However, in 1970, a new discovery was made that challenged the central dogma: the discovery of an enzyme, called *reverse transcriptase*, that turns the virus's RNA genome into DNA that inserts into the host cell's DNA, meaning that the flow of genetic information instead goes from RNA to DNA.

One of the most well-known viruses that uses reverse transcriptase is the human immunodeficiency virus (HIV), which is the virus responsible for causing acquired immunodeficiency syndrome (AIDS). HIV attacks T cells, a type of white blood cell that is an important part of your immune system, helping you to fight off disease (**Figure 7.10**).

Understanding how reverse transcriptase provides an exception to the central dogma reveals a valuable potential treatment option for HIV infection. Since our cells do not contain reverse transcriptase, HIV infection can be mitigated by using a drug that targets the invading virus, and not the patient's cells, thus blocking reverse transcriptase. If reverse transcriptase is blocked, then HIV cannot replicate in a person's body, which helps to minimize its impact on the immune system. Standard of care for patients infected with HIV involves treatment with reverse transcriptase inhibitors.

1. In Chapter 1, we learned about how science knowledge can be described in terms of hypotheses, theories, and laws. Do you think Crick's proposal of the central dogma is a hypothesis, theory, or law? Why?
2. Looking at Figure 7.10, which steps of viral replication are blocked by the reverse transcriptase inhibitors?

Figure 7.9 tRNA

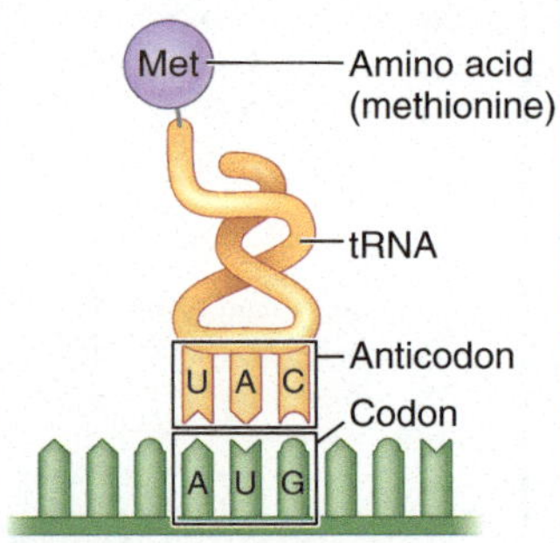

A tRNA contains both nucleic acid and protein components. The complementary anticodon binds to the mRNA codon during translation, where the tRNA donates the amino acid it is carrying (in this case, Met, or methionine) to the growing polypeptide chain.

Initiation. The ribosome has two subunits, one large and one small, that are made up of rRNA and structural proteins. Translation starts when mRNA binds to the small subunit (Figure 7.8, step 1). Once the mRNA is bound, the first tRNA arrives carrying a methionine amino acid—the amino acid that is complementary to the start codon on the mRNA, AUG. A start codon is always AUG and sets the reading frame for the ribosome. Later in the chapter, we will return to the reading frame and how mutations can impact the reading frame. Once this tRNA is in place, the large subunit binds to the small subunit and the ribosome is ready to rock!

Elongation. During elongation, tRNAs deliver amino acids to the ribosome one after the other in the order specified by the codons in an mRNA. Each tRNA has two attachment sites, one on each end of the molecule. One attachment site is an **anticodon**, which is a triplet of nucleotides that can base-pair with a particular mRNA codon. The second site can bind to a particular amino acid—the one specified by the mRNA codon (**Figure 7.9**). Note that tRNAs with different anticodons bind to different amino acids.

As the amino acids are delivered, the ribosome joins them into a new polypeptide (analogous to primary structure; see Chapter 3). Thus, the order of codons in an mRNA—DNA's protein-building message—becomes translated into a series of amino acids in a protein. Recall that rRNA has enzymatic activity. The rRNA components of a ribosome (not the protein components) catalyze peptide bonds to form between amino acids.

Termination. Translation ends when the ribosome reaches a **stop codon** in the mRNA. Rather than a tRNA binding, a **release factor** binds to the ribosome where a tRNA would typically bind. The ribosome then breaks apart into its small and large subunits, releasing the mRNA and the newly formed polypeptide.

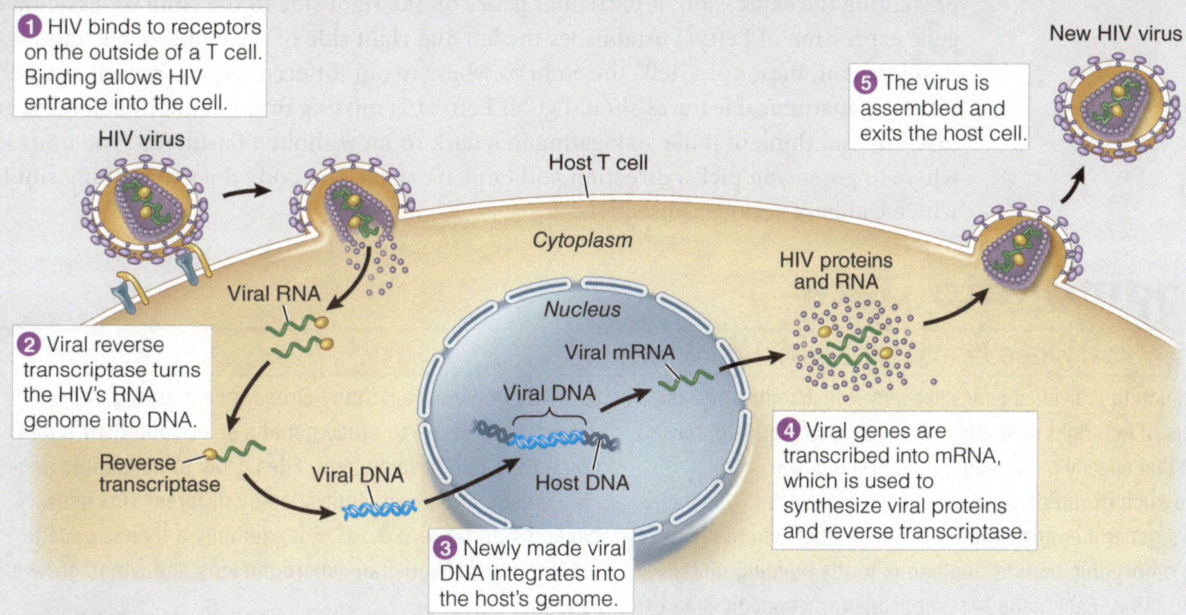

Figure 7.10 **Overview of HIV infection** HIV uses reverse transcriptase to infect a host's cells.

Folding a Protein Is Like Origami

Recall from Chapter 3 that proteins must fold into their final shapes to be functional, just like folding paper in origami (Figure 7.11). Different amino acids have different chemistries—some are hydrophobic ("water-fearing"), and others are hydrophilic ("water-loving"). Hydrophobic amino acids tend to be on the interior of a protein, to avoid the watery cytosol, whereas hydrophilic amino acids are toward the outside. Proteins thus begin to fold during translation, with hydrophilic amino acids pointing outward and hydrophobic amino acids pointing toward the interior of the new protein, eventually folding into their functional shape. If a protein has lots of hydrophobic amino acids, a special class of helper proteins called chaperones cover these hydrophobic amino acids and help with folding.

> **? Did You Know?**
>
> Many neurodegenerative diseases, like Parkinson's disease, are linked to protein misfolding.

Figure 7.11 **The Lefty-1 polypeptide folds to make a functional protein**

Note that only part of the Lefty-1 polypeptide is shown here. In the folded functional Lefty-1 protein (which helps differentiate the left and right sides of the embryo), you can see the two types of secondary structure—alpha helices (spirals) and beta-pleated sheets (represented as arrows).

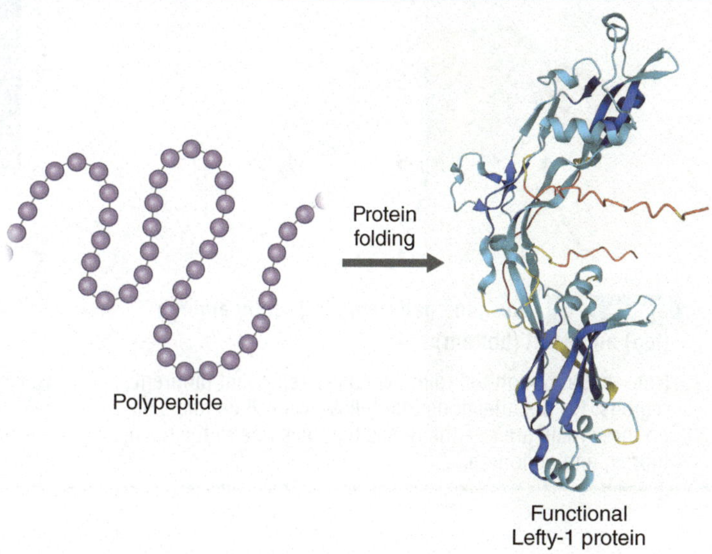

Putting the Pieces Together: Lefty-1 and Congenital Heart Defects

Now that we have covered normal gene expression, let's have a look at what happens when the process goes awry. At the beginning of Section 7.2, we discussed the relationship between gene expression and congenital heart defects. Now, it's time to bring our story full circle. What does Lefty-1 protein do, and how does it relate to heart development?

Lefty-1 acts like a molecular fence down the middle of the developing embryo, preventing the expression of particular genes on the right side of the embryo. Essentially, gene expression of Lefty-1 establishes the left and right side of the embryo. During development, these cues "tell" the embryo where to put different organs as well as where different anatomical features should go. If Lefty-1 is missing or non-functional, these cues are lost. You can think of it like navigating in a dark room without a flashlight--you don't know where to go, so you pick a direction and hope it's right. The body does something similar, which can have serious consequences.

Digging into Data

Body Maps: When Plans (and Genes) go Wrong

In addition to individual genes like *lefty-1* that control specific aspects of left–right patterning, there are also groups of related genes that specify patterning during development.

One such group of genes are called *Hox* genes. *Hox* genes act like general contractors on building sites, where the developing embryonic body is analogous to the building under construction. Each gene is responsible for a specific area of the body, thus ensuring that everything grows in the right place. For example, you would expect the bathtub to be placed in the bathroom in your house and that your general contractor would ensure these instructions are followed. Similarly, there are complete sets of directions that must be executed together to grow complicated parts of our bodies in the right places.

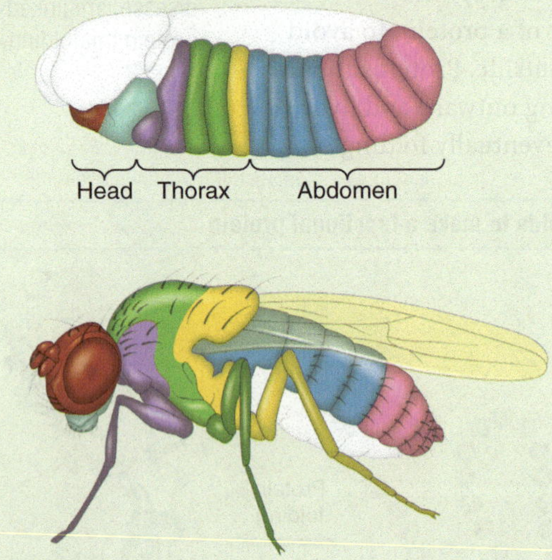

Figure 7.14 ***Hox* gene patterning in fruit fly embryo (top) and adult (bottom)**

Note that although the rainbow colors represent different regions, for the questions that follow, we will be focusing on three main areas—the genes that give rise to the head, thorax, and abdomen.

Researchers often study *Drosophila melanogaster* (the common fruit fly) to study genetic errors and their impact on embryonic development. Flies grow to have three general body segments, each controlled by different *Hox* gene expression (Figure 7.14). For example, a fly has genetic instructions that dictate where the legs and wings grow on the thorax.

There can be functional consequences when *Hox* gene expression goes awry, as shown in Figure 7.15.

1. Compare the normal and abnormal fruit flies in Figure 7.15. What differences do you see? How do you think geneticists created these mutants?
2. Using your knowledge of the central dogma, how do you think changing the gene expression of a *Hox* gene could lead to changes in the fruit fly body plan?
3. What do you think would happen to a fruit fly if head-specific *Hox* genes were expressed on the abdomen instead of the head?

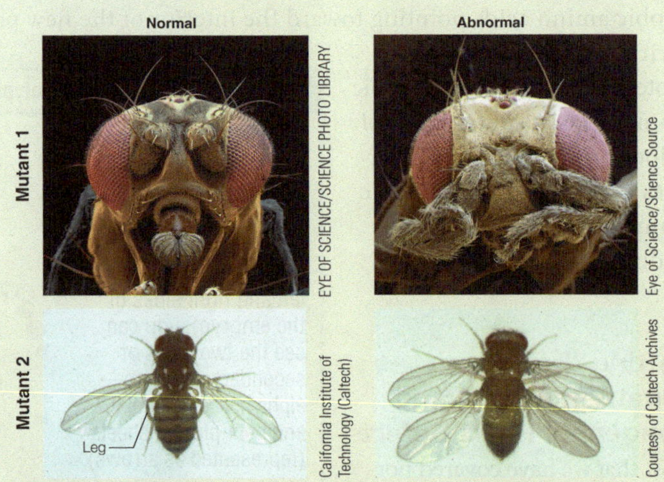

Figure 7.15 **Normal and abnormal *Hox* gene expression**

A normal fruit fly head and body (left) and an abnormal fruit fly head and body.

Expression of Lefty-1 begins with the directional flow of cilia moving a ligand toward receptors (Chapter 5) on one side of the developing embryo. Without Lefty-1, it is like trying to navigate in a new place and not knowing which direction is north: Essentially, one must pick a direction and hope for the best. As the heart develops, the signals saying "Put this vessel here" or "Put this chamber here" are lost, so the heart can develop with a wide range of defects.

One such condition related to the loss of left-right signals that can affect heart development is called heterotaxy (*heteros* is Greek for "other than," and *taxis* means "arrangement"), which is involved in 3% of all congenital heart defects (**Figure 7.12**). In the case of the diagram of the heart shown here, the patient (Levi) had right atrial isomerism, meaning that both sides of his heart had the features of the right side. It wasn't just his heart that was off, either—he also had two right lungs and a right-sided stomach. We can infer that Levi's right-side-predominant anatomy was likely due to a rare genetic mutation in gene(s) involved with setting up the left–right patterning of the body.

Unfortunately, Levi's condition was not compatible with life, and he passed away at 4 months of age. Levi's rare condition can be traced back to inappropriate gene expression and highlights the importance of the central dogma to our bodily function.

If left-sided genes are expressed on the right side of the body instead of the left, the internal organs all develop correctly, but in a mirror image to what you'd expect. This rare condition is called *situs inversus*.

Figure 7.12 Comparison of a typical heart and Levi's heterotaxy heart

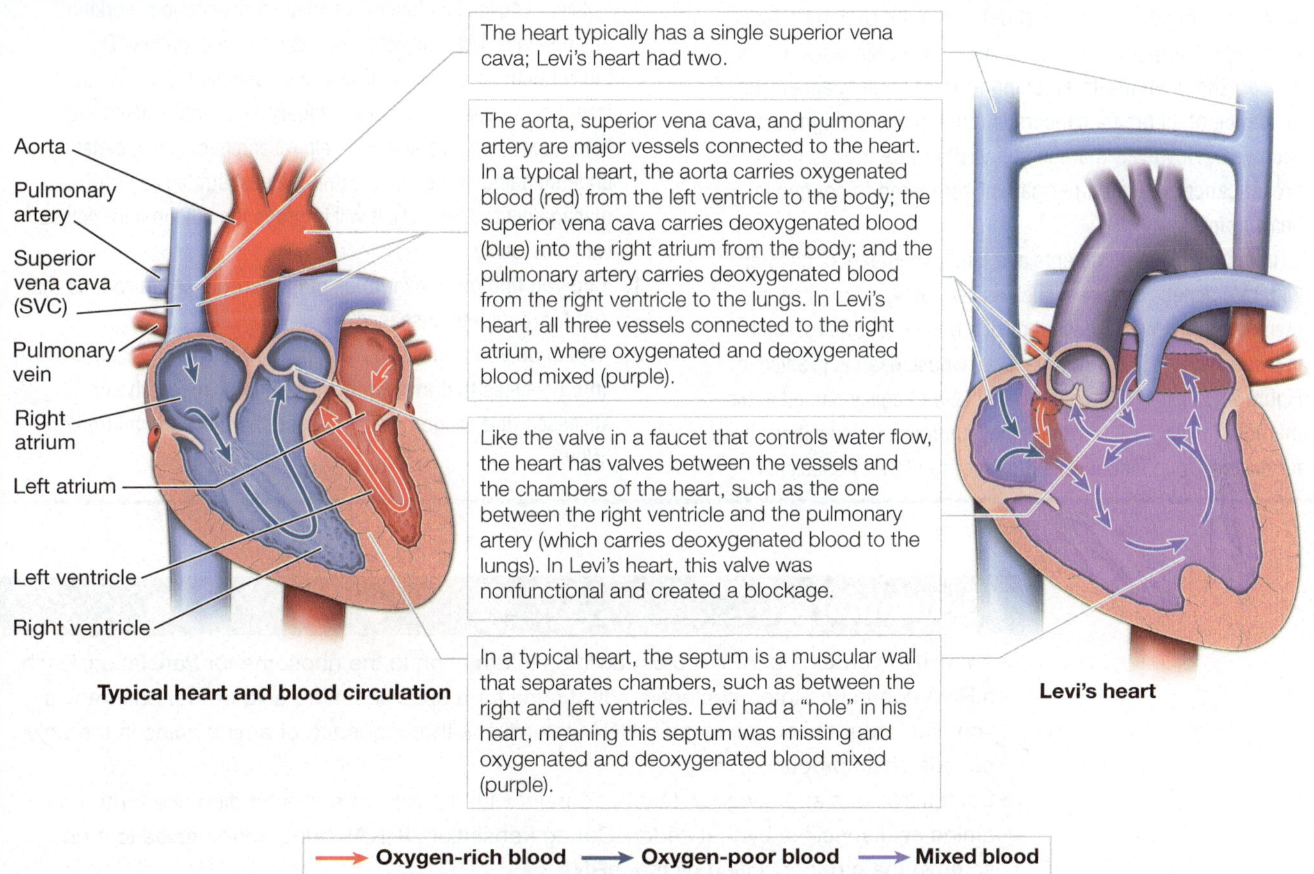

In a typical heart, the deoxygenated blood in the right side of the heart never mixes with oxygenated blood in the left, but in Levi's heart, the blood mixed together. This made it difficult to deliver oxygenated blood to his tissues.

Process of Science

How Money Flows Influence Scientific Research

Throughout this book, we have learned about different examples of how nonscientific factors can influence the process of science. Money, unfortunately (or not), makes the world go around—and money flow directly impacts what gets researched. Some research areas are better funded than others, and diseases that impact White Americans are disproportionately funded compared to diseases that impact people of color. For example, more research dollars are spent on cystic fibrosis than sickle cell anemia, a disease that predominantly affects people of African, Mediterranean, and Middle Eastern descent.

In 2024, the American Heart Association awarded $600,000 in research grants for the study of congenital heart defects. That may seem like a lot of money, unless you look at it in comparison to funds to support breast cancer research, which totaled $74.75 million for 2025–2026 from a *single* foundation, The Breast Cancer Research Foundation. Granted, the scale is different—there are an estimated 40,000 incidences of congenital heart defects in the United States each year, compared to 310,720 new case of invasive breast cancer (the most life-threatening form of breast cancer; the incidence of all breast cancers is much higher) diagnosed each year. However, there is still vastly more money put into breast cancer research per patient compared to congenital heart defects.

In response, some parents and family members of children with rare genetic diseases, such as heterotaxy, found their own nonprofits with a goal of raising money to fund research in a specific area. For example, Levi, whose heart is profiled in Figure 7.12, has a foundation called Levi's Legacy named in his memory (Figure 7.13). Others may fundraise for specific causes for existing foundations, like the American Heart Association.

Figure 7.13 The Levi's Legacy logo

Levi's Legacy was formed by his parents in his memory to both support congenital heart defect patients and their families, and to fundraise money for research grants to study congenital heart defects such as heterotaxy.

Levi's Legacy has funded a variety of researchers working on congenital heart defects. For example, one grantee, Dr. Emily Buchholz at Boston Children's Hospital, received a grant from Levi's Legacy to build a database featuring patients with heterotaxy. This database will help researchers gain a better understanding of the disease and the best strategies for disease management. Ultimately, it will improve the outcome for babies with heterotaxy.

1. Can you find other examples of grassroots organizations that fund rare genetic diseases?
2. Which do you think is more important to fund—research into diseases that impact more people or research into diseases that predominantly impact babies and children? Why?

7.3 Summary

- An mRNA carries a gene's protein-building information to the ribosome for translation. Each mRNA codon has three nucleotides that code for a specific amino acid or that halt translation. The order of codons in the mRNA determines the sequence of amino acids in the polypeptide chain produced.
- Each tRNA has an anticodon that base-pairs with the mRNA and a binding site for the amino acid specified by that codon. During translation, tRNAs bring amino acids to a ribosome in the order specified by the mRNA.
- During initiation, the first tRNA binds and the ribosomal subunits assemble. During elongation, tRNAs deliver amino acids to the ribosome in the order specified by mRNA. Peptide bonds link the amino acids together in a polypeptide chain. Termination occurs when the ribosome reaches a stop codon and releases the newly formed polypeptide.

7.4 Mutations and Translation

Learning Objectives

7-10 Describe the types of mutations related to translation.

7-11 Explain how each type of mutation affects protein structure.

7-12 Explain why some mutations are not harmful to an organism.

As we learned in the Unit 2 opener, two cell types are required in human reproduction—a sperm and an egg cell. Egg cells are safely hidden deep within the body—far beyond the reach of the penis, which deposits sperm in the reproductive tract during sexual intercourse. How, then, does a sperm reach an egg cell? By swimming! Conception in humans requires sperm to swim from the vagina through the cervix into the uterus and, finally, into the fallopian tubes. Once in the fallopian tubes, sperm will sit and wait until they receive signals from the egg. The egg acts like a homing beacon for the sperm, which puts the sperm on the move yet again. Perhaps not surprisingly, a leading cause of infertility in those whose bodies make sperm is sperm swimming poorly, or not at all.

What can cause sperm to swim poorly? Although many different factors can contribute to poor sperm movement, biologists have discovered different frameshift mutations that can lead to reduced sperm movement, which we'll explore next.

Losing and Gaining Letters: Frameshift Mutations

A **frameshift mutation** occurs when new DNA bases are added or deleted in an interval other than three. Why three? Think of a codon—it's three base pairs long. If there is an insertion or deletion of three, it is likely that a functional protein will result—the rest of the sequence is intact. Let's look at this sentence as an example of two 3-letter words being added in.

The fat cat ate the sad rat.

The fat cat ***sat and*** *ate the sad rat.*

The sentence is different, but still conveys meaning. However, an insertion or deletion of one or two nucleotides throws off how the ribosome reads the mRNA, leading to a completely different protein product. A frameshift deletion usually has drastic consequences because it garbles the message, just as deleting the first "T" in the following sentence leads to an incorrect grouping of the remaining letters, losing the meaning of the words in a sentence:

The fat cat ate the sad rat.

- hef atc ata tet hes adr at

Adding in nucleotides can have a similar effect. Let's look at our sentence about the fat cat again, and notice the misplaced "a" in "fat":

The fat cat ate the sad rat.

The faa tca tat eth esa dra t

Base-Pair Substitutions

Changing one single base pair can also impact the protein produced. A mutation that changes a base pair but not the amino acid is called a **silent mutation**. Recall from Section 7.3 that the genetic code is redundant—multiple codons can code for the same amino acid. This can help protect the body from potentially harmful mutations. Since the protein is the functional product, the underlying base-pair sequence doesn't matter as long as the same protein is produced.

Figure 7.16 The consequences of missense mutations

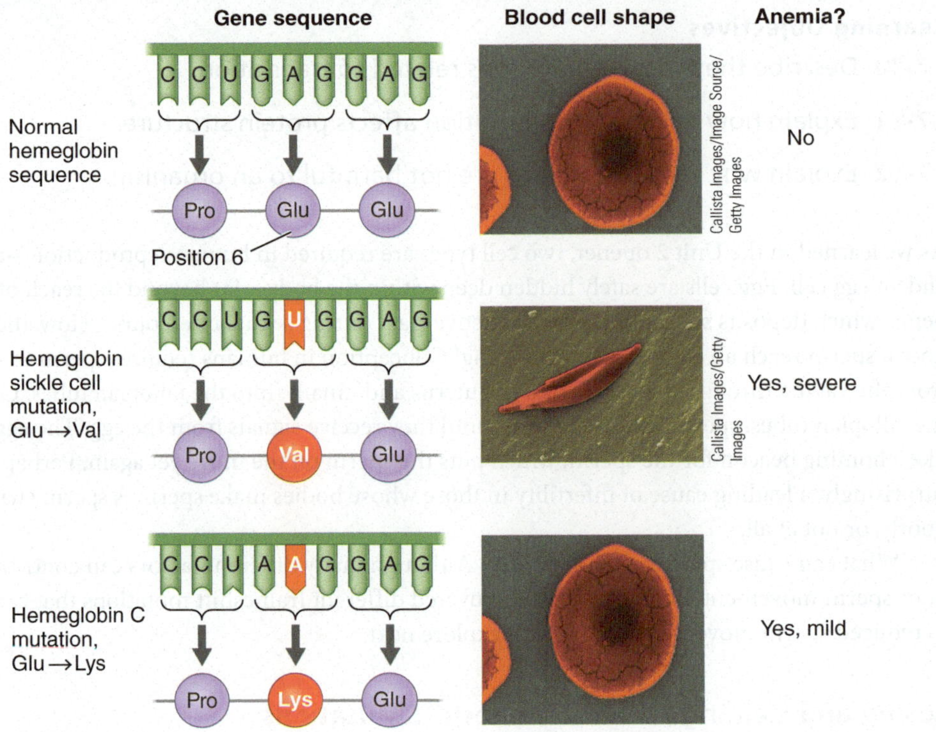

A single base-pair substitution that causes a missense mutation in the same amino acid of a polypeptide can cause sickle cell anemia (middle), which can significantly impact one's life; or mild symptoms (bottom), depending on the change. Note that only a subset of amino acids is shown here—the polypeptide is much longer.

? Did You Know?

People with sickle cell anemia are resistant to malaria.

A **missense mutation** occurs when a single base-pair change leads to a different amino acid. If we go back to our sentence analogy:

The fat cat ate the sad rat.

The ***mad*** *cat ate the sad rat.*

The sentence is still the same length and still conveys meaning, but the meaning is slightly different. With a missense mutation, the type of amino acid is changed, and whether the new amino acid is like the original one will impact the proper functioning of the protein.

For example, in the case of sickle cell disease, a valine is included at position 6 instead of glutamic acid in the hemoglobin protein (Figure 7.16). Glutamic acid carries a negative charge, but valine carries no charge. As a result of that one base-pair substitution, a tiny patch of the beta globin polypeptide, which is part of hemoglobin, that is normally hydrophilic, becomes hydrophobic. This change slightly alters the behavior of hemoglobin, and is enough to cause the structure of a blood cell to change from a round disk into a crescent-like sickle shape. Sickled cells clog tiny blood vessels, disrupting blood circulation throughout the body. Over time, repeated episodes of sickling can damage organs and eventually cause death.

In contrast, a base-pair substitution that changes the glutamic acid at the same position to a lysine forms hemoglobin C. Glutamic acid and lysine are both charged amino acids, and so the protein is less affected and hemoglobin C does not clump or distort red blood cells. It can cause mild anemia, but most people who carry the mutation have no symptoms at all.

The final type of mutation is called a **nonsense mutation**. A nonsense mutation occurs when a single base pair change causes a codon to switch from encoding an amino acid to a stop codon. This results in a short nonfunctional protein. Let's look at our sentence again for an example.

The fat cat ate the sad rat.

The fat cat ate the.

Ate the.…what? The sentence does not make sense—which is why these are called *nonsense* mutations.

Big Ideas in Biology

Evolution

Mutations are one of the forces that drive evolution. Not all mutations lead to a trait that is detrimental to an organism. In fact, some can be beneficial, and what is helpful, not helpful, or neutral is dictated by the environment. We will return to the relationship between mutations and evolution in Chapter 11.

Biology for Life

What Happen to the Cat's Hair?

Remember alternative splicing? The machinery that cuts mRNA is very sensitive to single base-pair changes at intron–exon boundaries. The mutation that causes the hairless appearance of sphynx cats occurs due to a base substitution at an intron–exon boundary, in a gene called *keratin*.

Keratin is a fibrous protein that makes up hair. The substitution alters a series of bases that normally mark an intron–exon boundary. Molecules that splice new RNAs don't recognize the altered boundary, so they don't remove the intron. The intron happens to contain a stop codon in the middle. The protein translated from this mRNA is too short and cannot properly assemble into hair. Cats that have the mutation still make hair, but it falls out before it gets very long.

Elena Medoks/Shutterstock.com

Figure 7.17 Sphynx cats are hairless due to a frameshift mutation

A single nucleotide change gives rise to the hairless appearance of a sphynx cat. The change alters the instructions in DNA for making keratin.

1. What does this mutation tell you about the nature of mutations? Are mutations necessarily "bad"?
2. Blue eyes are one example of a genetic mutation that we typically don't think of as "bad." Can you think of other examples?
3. What would be the disadvantage if we were able to eliminate all mutations? Can you think of a mutation that provides a benefit?

Study Tip

Think of nonsense like no—no protein forms, and ***mis***sense like ***mis***-spelled. The amino acid sequence is misspelled. Some misspellings don't impact meaning—one can infer the nature of the creature eating the rat regardless of if it's a fat cat or a fat kat. A silent room means it's quiet, so the modification isn't "heard."

7.4 Summary

- Frameshift mutations result from insertions or deletions in multiples of one or two bases.
- Silent mutations change one base pair in the DNA but don't result in a different amino acid or final protein.
- Missense mutations change a codon such that a different amino acid is specified. This may or may not have a significant consequence on the protein, depending on whether the new amino acid is like the original one.
- Nonsense mutations change an amino acid codon to a stop codon, resulting in a shortened protein that is typically nonfunctional.

7.5 Control of Gene Expression

Learning Objectives

7-13 Explain the basic principles of gene expression control.

7-14 Identify the reasons why gene expression control is necessary.

7-15 Describe how DNA methylation is a form of gene expression control.

7-16 Explain how genes can be regulated before, during, or after transcription.

Regulating gene expression is an important part of homeostasis, because it allows control over the kinds and amounts of substances present in a cell at any given time. By adjusting expression of particular genes, cells respond appropriately to changes in their internal and external environments. For example, bacteria alter gene expression based on changes in the availability of nutrients in their environment. When a bacterium encounters a specific nutrient, it begins transcribing genes for enzymes that break it down. When the nutrient is no longer available, transcription of those genes stops. Thus, the cell does not waste energy and resources producing gene products that are not needed.

Adjustments to gene expression also affect form and function in multicellular organisms. A typical differentiated or specialized body cell uses only about 10% of its genes at any given time. Some of the expressed genes affect structural features and metabolic functions common to all cells; others are used only by certain cells. For example, all your body cells express genes for glycolysis enzymes, but only immature red blood cells express globin genes. Such differences begin early in embryonic development.

Gene Expression Is Regulated at Many Levels

All steps of gene expression are regulated. Expressing the correct genes at the right time and place is important not only for development but also to save cellular energy and resources, so that only needed proteins are made. Although there are many levels of gene regulation, we will be focusing on three key areas of gene regulation: pre-transcriptional regulation, transcriptional regulation, and post-transcriptional regulation (Figure 7.18).

Figure 7.18 **Overview of levels of transcriptional regulation**

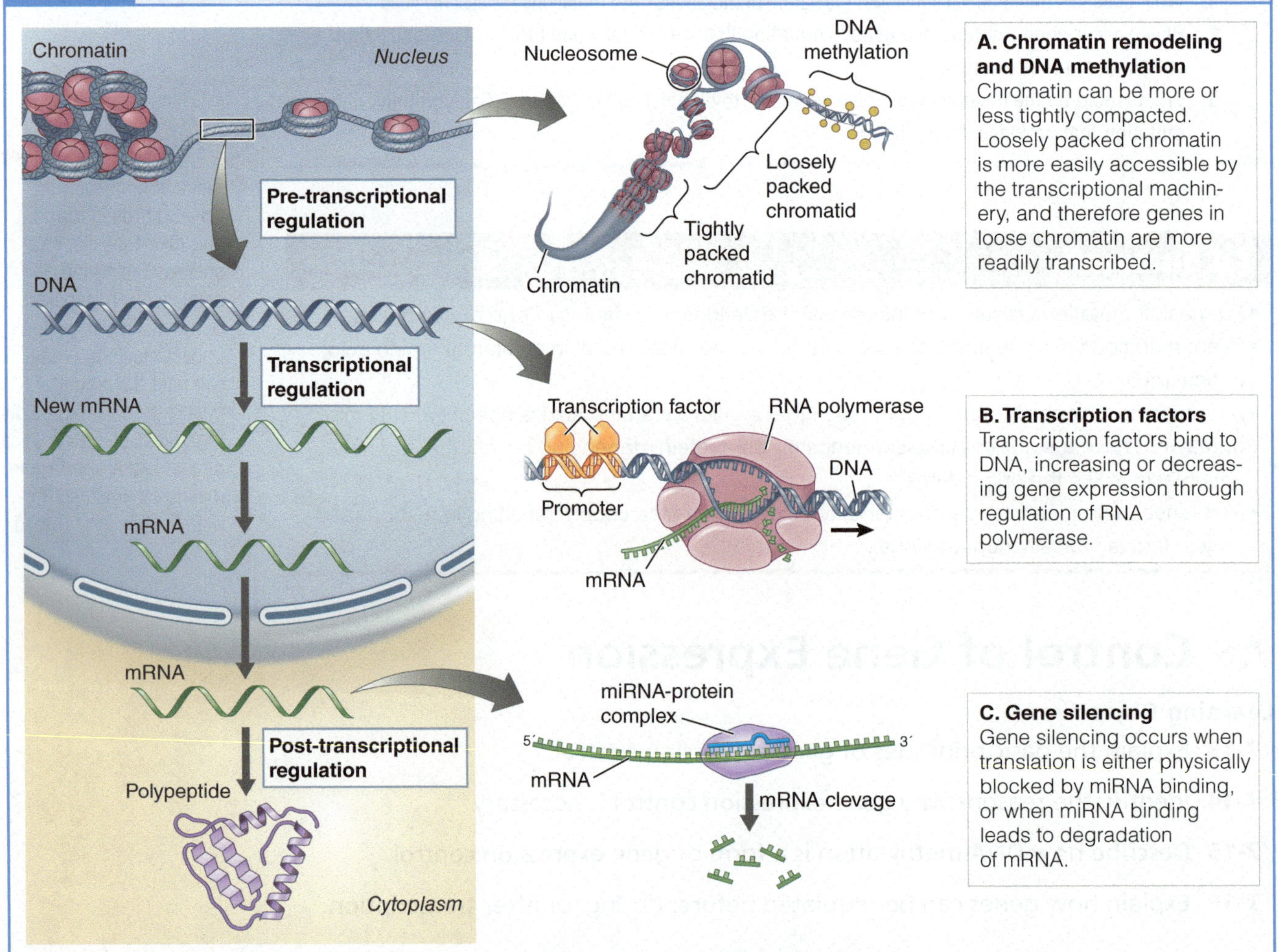

Pre-transcriptional regulation. One of the main ways transcription is controlled is via **chromatin remodeling** (Figure 7.18A). Recall from Chapter 6 that DNA is wound around proteins called histones so that it can fit into a cell. The DNA–protein complex is called **chromatin**. Think about chromatin and transcription like your belongings packed in a suitcase. Let's say you need to find your underwear in the suitcase. Is it easier to find in a tightly packed suitcase, or after you've upended your suitcase over the hotel bed? Once you've unpacked it, of course!

We see a similar phenomenon with gene expression. Regions of DNA that are loosely packaged are more accessible to proteins like RNA polymerase, making it easier to transcribe genes. Conversely, regions of DNA that are tightly packaged are inaccessible to RNA polymerase, making it too difficult to transcribe genes in these regions. In special cases, an organism can turn off an entire chromosome through a process called X-inactivation. We will return to X-inactivation in Chapter 8.

DNA methylation, or the addition of methyl groups to cytosines in DNA (Figure 7.18A), can cause the chromatin to become more condensed, thus decreasing gene expression. Between 3% and 6% of DNA is methylated in normal, differentiated body cells, but which sites are methylated varies by the individual. This is because methylation is influenced by environmental factors encountered during an individual's lifetime. For example, exposure to cigarette smoke alters DNA methylation in a person's cells. Chemicals in the smoke change the methylation of certain promoters in a pattern that also occurs in cancer cells. Synthetic compounds such as bisphenol A (BPA) and phthalates (both found in plastics; Chapter 17) alter methylation patterns of several genes. Scientists think these changes in gene expression contribute to the adverse health effects associated with exposure to these chemicals. Factors that influence DNA methylation patterns can have multigenerational effects—which we will return to in Chapter 8.

? Did You Know?

DNA methylation is a critical part of embryonic development. Genes actively expressed in the cells of an embryo become silenced as their promoters get methylated, and this silencing is part of the changes in gene expression that drive differentiation.

Transcriptional level. Regulating gene expression does not involve changes to DNA sequence; rather, cells use molecules that act as "switches" to turn genes on or off. These molecules can increase or decrease the rate of individual steps of gene expression. For example, proteins called **transcription factors** affect whether and how much a gene is transcribed by binding directly to regulatory sites in DNA (Figure 7.18B). Some of these proteins enhance transcription by helping RNA polymerase bind to promoters; others inhibit transcription by preventing RNA polymerase from binding to a promoter, or by blocking its progress.

Post-transcriptional level. Prolactin is a hormone that stimulates milk production in humans. When prolactin is present, there is more of a milk protein called casein produced. Why? Prolactin makes the casein mRNA more stable. The longer mRNA is around, the more protein can be made and consequently the higher the gene expression.

This is not the only way that the cell can regulate stability of an mRNA transcript to impact gene expression. Some species of small RNAs (~20 nucleotides) can impact gene transcription. For example, a complex of RNA and associated proteins called microRNAs (miRNAs) can bind to the 3′ end of mRNA and block translation, cause the mRNA to degrade, or cut the mRNA into pieces (mRNA cleavage) (Figure 7.18C). This process is called **gene silencing**. miRNAs have varied roles in our biology, including controlling the cell cycle, the differentiation of stem cells, and the development of diseases like cancer and rheumatoid arthritis.

Another form of post-transcriptional regulation is one we learned about earlier in the chapter—alternative splicing (see Figure 7.5). mRNA transcripts can be spliced in many ways, leading to different proteins and ultimately differences in the cell.

? Did You Know?

Researchers are studying ways to use gene silencing in clinical applications. The first gene silencing therapy was approved by the Food and Drug Administration in 2018 to treat nerve damage caused by a genetic condition.

Biology for Life

Transcription Factors and Embryonic Development

An animal body starts out as a tiny cluster of identical cells, all expressing the same genes. These cells divide repeatedly, and their descendants begin to differentiate as they start expressing different subsets of genes. As the cells diverge in form and function, they give rise to tissues, organs, and other body parts.

Embryonic development is orchestrated by transcription factors that are produced in cascades of gene expression. In these cascades, the transcription factor product of one gene affects the expression of other transcription factor genes, whose products in turn affect the expression of others, and so on. Some of these genes are called master regulators. Expression of a master regulator begins a gene expression cascade that ultimately changes cells in a lineage from one type to other, more differentiated type—a bit like a master switch turns on a whole system with a single flip.

1. Why is it important that cells regulate their gene expression? Wouldn't it be easier to just transcribe and translate all genes all the time?
2. You've learned that bacteria can alter their gene expression when they encounter new nutrients. Can you think of how your body might change gene expression in a way that is visible, either in response to an environmental stimulus or through aging?

7.5 Summary

- Regulating gene expression allows individual cells to respond appropriately to internal and external change. By controlling gene expression and transcribing only the necessary genes, cells save energy.
- Gene expression is regulated via chromatin remodeling and DNA methylation at the pre-transcriptional level, the presence of certain transcription factors at the transcriptional level, and regulation of mRNA stability by miRNAs and alternative splicing at the post-transcriptional level.

Understand It! Self-Explanation

When I was in graduate school and was learning new laboratory techniques, the more senior students would take a "watch one, do one, teach one" approach. First, I would watch someone else do it, then do it myself, and finally, I would either teach it to someone else or explain back to the student training me how to do it. Perhaps you have experienced something similar in your own life—the best way to see if you really understand something is to try and explain it to someone else.

This is called self-explanation. Rather that recalling something that was memorized, you have to explain how it works or what to do. This helps us to understand and remember the information because we have to retrieve it from our brain which, as we learned in Chapter 1, is one of the most effective study strategies. It also helps us to make sense of the topic in order to explain it, and then through the process of explaining it to someone else, we can identify gaps in our understanding.

Here are some suggestions while you are studying to learn this chapter's material:

1. After studying for 20 minutes, close your notebook and either write an explanation of what you've learned or teach it to someone else—be it your dog, roommate, parent, or teddy bear.
2. Working with a study partner, take turns explaining the concepts to the group.
3. Create a poster or other visual media illustrating the central dogma—something that could teach this concept to people waiting at a bus stop.

Chapter Review

7.1 The Central Dogma of Molecular Biology

A **gene** is a unit of DNA containing instructions for making a specific RNA or protein. Genes contain features such as **promoters**, **exons**, and **introns**. The **central dogma** specifies that information flows from DNA to RNA to protein.

During **transcription**, DNA is used as a template to make mRNA. During translation, that **messenger RNA (mRNA)** is used as a template to make a protein. In **translation**, **transfer RNAs (tRNAs)** bring amino acids to the ribosome to be added to the new protein, while **ribosomal RNAs (rRNAs)** are a part of ribosome structure that joins the amino acids together.

7.2 Transcription: DNA to RNA

Transcription is the process of making an mRNA copy of a gene, which takes place in the nucleus of a eukaryotic cell. Transcription starts with **initiation**, where **RNA polymerase** binds to the **promoter** of a gene on the **template strand** of DNA. **Elongation** is when **RNA polymerase** adds new RNA nucleotides. At the end of the gene, **termination** occurs, and the DNA, immature mRNA, and RNA polymerase separate.

RNA processing is used to splice the immature mRNA into a mature mRNA. A **5′ cap** and **3′ poly-A tail** are added to the 5′ and 3′ ends, respectively, to help with mRNA stability and nuclear export. The mRNA is also spliced, typically to remove introns. Alternative splicing can also occur.

7.3 Translation: RNA to Protein

Each **codon** of an mRNA consists of three nucleotides that code for a specific amino acid, or to stop translation. The **genetic code** is made up of 64 codons that code for 20 different amino acids.
Each **tRNA** molecule has an **anticodon** that base-pairs with a codon in the **mRNA** and a binding site for the amino acid specified by that mRNA codon. This ensures the tRNA hands off the correct amino acid to the newly forming polypeptide.

Translation is the process of making a protein from an mRNA template. Translation begins with **initiation** in the cytoplasm, when the ribosome assembles in a complex with a single tRNA carrying methionine bound to the **start codon** on an mRNA. During **elongation**, amino acids are delivered to the ribosome by **tRNAs** and sequentially added using the order dictated by the mRNA. Each tRNA's anticodon is matched to the mRNA codon to ensure the correct amino acid is added per the **genetic code**. The amino acids are linked by peptide bonds. **Termination** occurs when the ribosome reaches a **stop codon** in the mRNA. The newly released polypeptide folds into its three-dimensional functional shape.

7.4 Mutations and Translation

Frameshift mutations occur when DNA bases are added or deleted in multiples of one or two, changing the reading frame. Frameshift mutations generally result in nonfunctional proteins.

Silent mutations change one base pair in the DNA without changing the amino acid added to the protein, and therefore, do not affect the resulting protein. **Missense mutations** change one base pair in the DNA, eventually resulting in proteins with single amino acid changes that can be functional or detrimental. **Nonsense mutations** occur when a single base mutation in DNA results in an early stop codon, resulting in a short protein that is typically nonfunctional.

Not all mutations are dangerous. Silent mutations change the DNA but not the final protein and so are not harmful. Depending on the location of a mutation, the protein may still function properly, or sometimes even better, than the original protein, having no overall negative impact on the organism. Whether or not the change is beneficial, harmful, or neutral depends on the environment.

7.5 Control of Gene Expression

Gene expression is regulated at many steps. At the **pre-transcriptional level**, **chromatin remodeling** leads to chromatin that is more or less packed, altering access to specific genes. **DNA methylation** can also regulate transcription by altering access to DNA. **Transcription factors** affect how much a gene is transcribed by altering the action of RNA polymerase. **Post-transcriptional** control affects gene expression at the mRNA level, such as by regulating its stability or through alternative splicing.

Gene expression control is necessary for cells to respond to their environment and react to signals from other cells. The cell transcribes only genes needed at any given time, saving energy and adding flexibility to its responses.

Practice It!

Retrieval practice is the process of pulling information from your long-term memory and using it. Here are some questions to help with your retrieval practice.

Self-Quiz

1. A chromosome has many units called ______ that contain directions for building RNAs and proteins. (LO 7-1)

a. genes
b. double helix
c. ribosomes
d. chaperone proteins

2. RNAs form by ______; proteins form by ______. (LO 7-2)

a. replication; translation
b. translation; transcription
c. transcription; translation
d. replication; transcription

3. The central dogma of molecular biology states that information flows from ______. (LO 7-2)

a. RNA to DNA to RNA
b. RNA to protein to DNA
c. DNA to protein to RNA
d. DNA to RNA to protein

4. What would happen to gene expression if mRNA were blocked from exiting the nucleus? (LO 7-3)

a. Nothing; genes can be expressed in the nucleus.
b. No proteins would be expressed, and the organism would die.
c. No proteins would be expressed, and the organism would have to eat more protein to survive.
d. rRNA would not be able to catalyze the formation of a peptide bond.

5. Gene expression uses three types of RNA. The tRNA ______. (LO 7-3)

a. contains the directions for the ribosome
b. brings amino acids to the ribosome
c. makes up much of the structure of the ribosome
d. alters the nucleotide sequence at the ribosome

6. Transcription uses ______ as a template to copy a gene into ______. (LO 7-4)

a. DNA; mRNA
b. DNA; protein
c. mRNA; DNA
d. mRNA; protein

7. During initiation of transcription, RNA polymerase binds to which area of a gene? (LO 7-4)

a. stop codon
b. poly-A tail
c. promoter
d. template

8. Cells can make more than one protein using a single gene by ______. (LO 7-5)

a. adding a poly-A tail
b. chaperone proteins
c. DNA replication
d. alternative splicing

9. To ensure that the correct amino acid is added to a growing polypeptide, the tRNA has a(n) ______ that can base-pair to the mRNA. (LO 7-6, LO 7-7, LO 7-9)

a. anticodon
b. codon
c. polypeptide
d. amino acid

10. Biologists are studying ways to treat cancer cells with ribosomal inhibitors. Why would this be a valid treatment option? (LO 7-8)

a. Ribosomal inhibitors will cause the cell to undergo apoptosis.
b. Ribosomal inhibitors will cause transcription to halt, and the cell will die from lack of proteins.
c. Ribosomal inhibitors will cause translation to halt, and the cell will die from lack of proteins.
d. Both (b) and (c)

11. A mutation that changes an amino acid codon to a stop codon is a ______ mutation. (LO 7-10, LO 7-11)

a. frameshift
b. silent
c. nonsense
d. missense

12. A mutation that changes one amino acid in a protein is a ______ mutation. (LO 7-10, LO 7-11)

a. frameshift
b. silent
c. nonsense
d. missense

13. Muscle cells differ from bone cells because they ______. (LO 7-13)

a. have different genes
b. express different genes
c. are eukaryotic
d. Both (a) and (b)

14. Put the following processes in the order of their occurrence during gene expression: 1. mRNA processing; 2. translation; 3. transcription; 4; RNA leaves nucleus. LO 7-4, LO 7-5, LO 7-8)

a. 1, 2, 3, 4
b. 4, 1, 3, 2
c. 2, 3, 4, 1
d. 3, 1, 4, 2

Critical Thinking

1. Toxic ribosome-inactivating proteins (RIPs) kill cells because they stop a process of gene expression. Many antibiotics used to treat bacterial infections in humans also target processes of gene expression. For example, rifampicin binds to RNA polymerase and physically blocks the lengthening of a new RNA. Why do these antibiotics kill bacteria, but not humans? (LO 7-16)

2. An anticodon on a tRNA has the sequence GCG. What amino acid does this tRNA carry? What would be the effect of a mutation that changed the C of the anticodon to a G? (LO 7-4, LO 7-6, LO 7-11)

3. Use the genetic code to translate the following sequence of bases in an mRNA into an amino acid sequence, starting at the first base. (LO 7-4, LO 7-6)

   GGUGAAAAUGAGACCAUUUGUAGU

4. How would your previous answer change if the first G in the first codon GGU were deleted? What kind of mutation is this? (LO 7-4 to LO 7-11)

5. Each position of a codon can be occupied by one of four nucleotides—U, G, A, C. There are 20 naturally occurring amino acids but 64 mRNA codons. Why is it beneficial to have "extra" codons for some amino acids? (LO 7-8)

6. You've learned that genes can have both introns and exons and that the introns are typically removed during RNA processing. What could happen if an intron were accidentally left in an mRNA? (LO 7-5, LO 7-13)

Credit: Lopolo/Shutterstock.com

8

Sex, Genes, and Inheritance

Chapter Outline:

A Double Take on Inheritance

If you have a biological sibling, did you know that you share 38–61% of your DNA with them? This means you may be more (or less!) like your sibling than your biological parents, with whom you share approximately 50% of your DNA. The exception is identical twins—if you have an identical twin, you share 100% of your DNA.

But are twins truly identical? Have you ever noticed that twins look very similar when children, but can look less identical over time? It's not just a result of changing identities or getting away from parents who *insist* on dressing their kids in identical or complimentary clothing. The epigenome controls how our genes are expressed (such as methylation; see Chapter 7), can change over time in response to our environments.

When people think about learning genetics, they usually associate it with thoughts of Gregor Mendel, pea plants, and making predictions using a tool called a *Punnett square*. In actuality, how we inherit not only our genes from our biological parents but also our epigenome is far more interesting.

Why is it that how much DNA we share with our siblings can vary, but the amount we share with our parents does not? What is it about the epigenome that causes identical twins to become less identical over time?

Identical twins share 100% of their DNA with each other. Why is it that one twin can develop a disease while the other does not?

Genetics is the subdiscipline of biology concerned with genes, genetic variation, and how genes are inherited (heredity). Understanding genetics is essential to understanding how evolution works—one of the big ideas in biology we learned about in Chapter 1 and the subject of Unit 4. In this chapter, we will briefly touch on Mendel and basic Mendelian inheritance (which explains less than 10% of our genetics), starting with how gametes are formed through *meiosis* and the relationship between meiosis and Mendel's laws. For the remainder of the chapter, we will focus on the myriad of ways in which traits are inherited, including non-Mendelian inheritance, complex traits, and epigenetics. We'll also talk about how geneticists and genetic counselors can track the inheritance of traits through families using tools called *pedigrees*.

But before we get started, I want to touch on the terms "male" and "female" as they are used in the context of genetics. Gender is culturally derived, exists on a spectrum, and is not the same as biological sex. Biological sex also exists on a spectrum and includes a wide variety of traits, such as intersex anatomy, differences in hormone levels, and differences in chromosomes. For the purposes of this chapter (and text), "male" or "dad/father" refers to a person whose body makes sperm, and "female" or "mom/mother" refers to a person whose body makes (or made) eggs.

Preview It!

List three things you already know about genetics. What do you want to learn about genetics in this chapter?

8.1 Let's Talk about Sex

Learning Objectives

8-1 Describe the evolutionary advantages of sexual reproduction.

8-2 Differentiate homologous chromosomes and sister chromatids.

8-3 Differentiate autosomes from sex chromosomes.

8-4 Explain why sexual reproduction requires meiosis.

8-5 Explain how meiosis creates gametes.

8-6 Compare mitosis and meiosis.

Looking around the room at a family reunion, what features do people have in common? Maybe there's a characteristic height or nose shape. In both the Unit 3 opener and this chapter's opener, we considered the question of why siblings can be more (or less) like each other than their biological parents. Why so much diversity? It comes down to the fact that humans reproduce sexually.

Most, but not all, eukaryotes reproduce sexually. **Sexual reproduction** (Figure 8.1A) is the process in which offspring arise from two parents and inherit genes from both. In contrast, some organisms reproduce through **asexual reproduction**, where the offspring is identical to the one parent (Figure 8.1B). If the function of reproduction is the perpetuation of one's genes, then an asexual reproducer would seem to win the evolutionary race, passing all its genes to every one of its offspring, whereas only about half of a sexual reproducer's genes are passed to each of its offspring. In addition, asexual reproduction tends to be faster than sexual reproduction because it requires no partner. So why is sexual reproduction so widespread?

? Did You Know?

The type of banana found at the supermarket was bred to be unable to reproduce on its own. Rather, clones of the same plant are propagated for planting. The lack of genetic diversity among banana plants makes them vulnerable to devastation by disease.

Offspring of Asexual Reproducers Are Clones

Consider how all offspring of an asexual reproducer are clones: In the absence of new mutations, they have the same alleles as one another (and their one parent). This consistency is a good evolutionary strategy in a favorable, unchanging environment, because traits that help an organism survive and reproduce do the same for its offspring. However, most environments are constantly changing, and change is not always favorable. Individuals that are genetically identical are equally vulnerable to environmental challenges.

Offspring of Sexual Reproducers Are Diverse

In changing environments, sexual reproducers have the evolutionary edge because they give different alleles to their offspring. Sexual reproduction randomly mixes up the genetic information of two parents who have different alleles (and different forms of traits associated with those alleles). Offspring inherit new combinations of the alleles—and new combinations of the traits—so they differ from one another and from their parents. An environmental challenge may be unfavorable for some of these individuals, but others may be perfectly suited to the change. Thus, as a group, they have a better chance of surviving environmental challenges than clones do.

How is it possible for two humans—each of whom is **diploid (2*n*)**, meaning they have two copies of each chromosome in their cells —to make another diploid organism? Let's think about this in terms of coinage. Think about a diploid body cell as being like a $1 coin (Figure 8.2, top).

Figure 8.1 Eukaryotes can reproduce sexually or asexually

A. Offspring of sexual reproducers, such as these dogs, differ from one another and their parents.
B. This plant is reproducing asexually. Each of the offspring (the small plants hanging below the main plant) is genetically identical to the parent and one another.

Credits: cynoclub/Shutterstock.com; DEA/G.CIGOLINI/Getty Images

Big Ideas in Biology

Evolution

The influence of the environment in shaping which traits are favorable is a driving force behind evolution.

Figure 8.2 A coin analogy for human ploidy levels

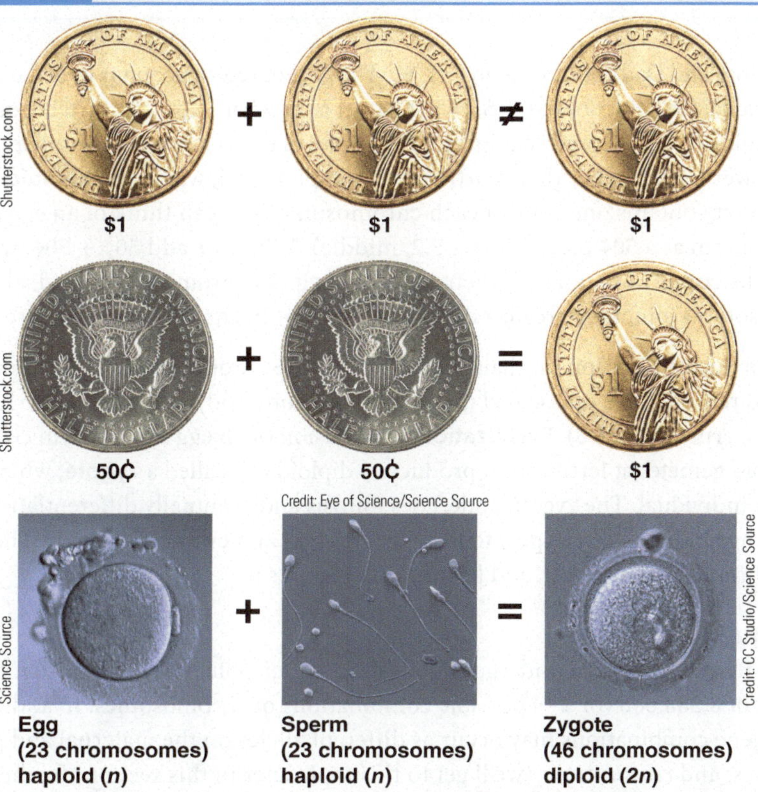

Credit: Sascha Burkard/Shutterstock.com
Credit: Steve Collender/Shutterstock.com
Credit: Eye of Science/Science Source
Credit: Pascal Goetgheluck/Science Source
Credit: CC Studio/Science Source

Think of a haploid gamete like a half dollar coin—it contains half the genetic material. Creating a dollar (a diploid organism) requires using two half dollars, not two dollars.

Figure 8.3 **Fertilization restores chromosome number**

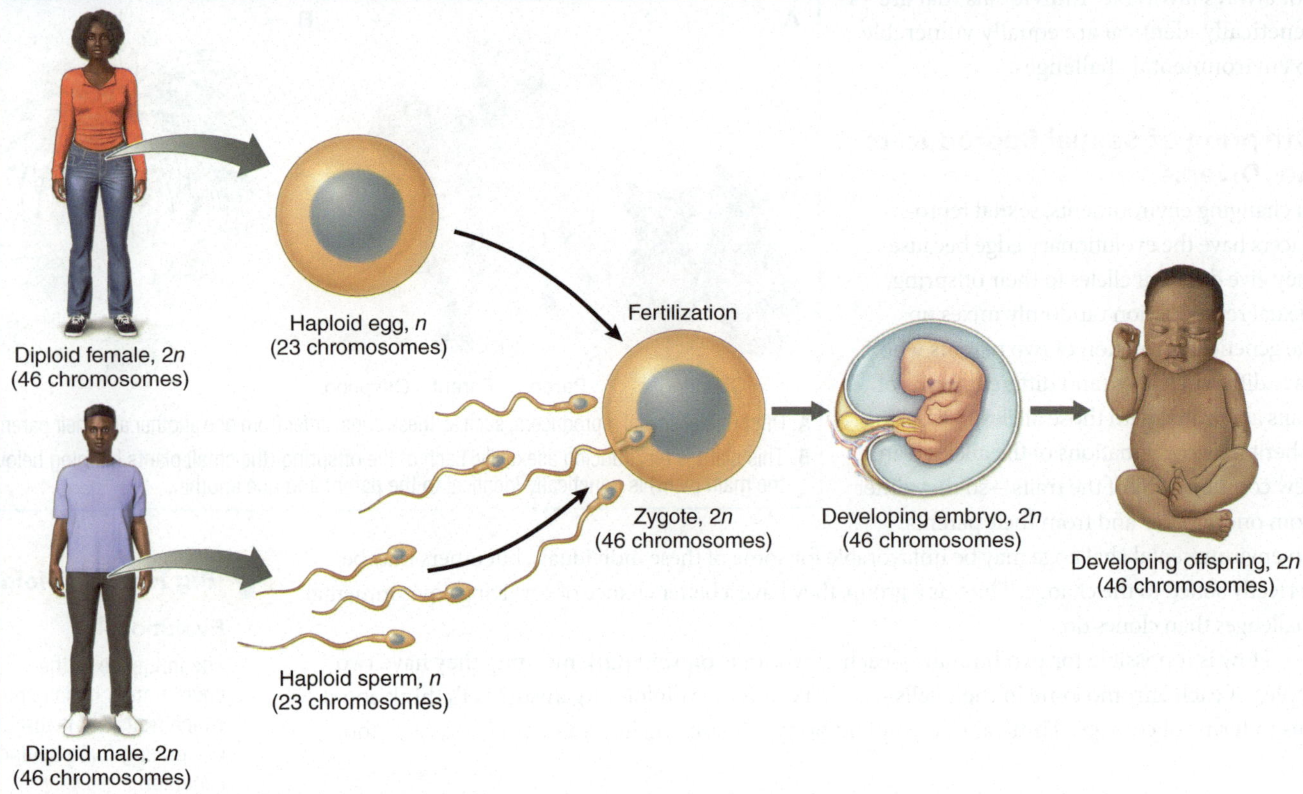

Two different individuals each donate one type of gamete, either an egg or a sperm cell. Fusion of the gametes produces a new diploid cell, the zygote. This single cell eventually forms an embryo and later, a fetus.

> **? Did You Know?**
>
> Achoo! You may be painfully familiar with plant sexual reproduction. All that pollen that makes you sneeze in the springtime? Pollen is wind-borne male haploid spores for plant reproduction produced by seed plants.

So, each of Mom and Dad's body cells are like $1. The goal of reproduction is to make up another diploid individual ($1). But does $1 + $1 = $1? Nope! Now what? How can we take something from Mom, something from Dad, and end up with two copies of each chromosome, rather than four?

The answer is that rather than starting with a diploid cell, we need a **haploid (*n*)** reproductive cell that contains one copy of each chromosome. You can think of an egg as a 50¢ piece and a sperm as a 50¢ piece (Figure 8.2, middle). When we add 50¢ + 50¢, we end up with $1. In the context of humans, when a haploid egg (23 chromosomes) and a haploid sperm (23 chromosomes) join, they create a diploid organism (46 chromosomes; Figure 8.2, bottom).

Meiosis is a nuclear division mechanism that halves the chromosome number and is used to form haploid reproductive cells called **gametes**. A person's body makes only one kind of gamete—egg or sperm (Figure 8.3). **Fertilization** is the fusion of an egg and a sperm cell. The fusion of two haploid gametes at fertilization produces a diploid cell called a **zygote**, which is the first cell of a new individual. The zygote undergoes mitosis and eventually differentiation (when cells begin to change into different types) to form an embryo, and eventually a baby. Thus, meiosis halves the chromosome number, and fertilization restores it.

Alleles and Diversity

Each time a human germ cell undergoes meiosis, the four potential gametes that form end up with one of 8,388,608 (or 2^{23}) possible combinations of chromosomes. In addition, any number of gene combinations may occur as different alleles on the maternal and paternal chromosomes, and *crossing over* (we'll get to this topic later in this section) further contributes to genetic diversity. Then, out of all the male and female gametes that form, the two that join at fertilization is a matter of chance. For instance, which sperm cell penetrates the egg first will vary. Depending on which genes you or your sibling inherit from the multiple possible combinations of genes from your parents, you may be more or less similar to one another.

Alleles are gene variants. Recall from Chapter 6 that each of your cells contains homologous chromosomes, which are similar but not identical to one another (Figure 8.4). The two chromosomes of each pair have the same genes, but their DNA sequences are not identical. This is because you inherited your chromosomes from two parents who differ genetically; unique mutations accumulated in their separate lines of descent over time. Thus, the DNA sequence of any of your genes may differ a bit from a corresponding gene on a homologous chromosome. Different forms of the same gene are called **alleles**. Recall that once DNA replication occurs, we have sister chromatids, which are identical chromosomes that are joined together until the cell divides.

New alleles arise by mutation. They may encode slightly different forms of a gene's product, and such differences influence the details of traits. Members of a species have the same traits because they have the same genes, but almost every shared trait varies a bit among individuals of a sexually reproducing species. We will return to alleles and inheritance later in this chapter.

Figure 8.4 Comparing homologous chromosomes and sister chromatids

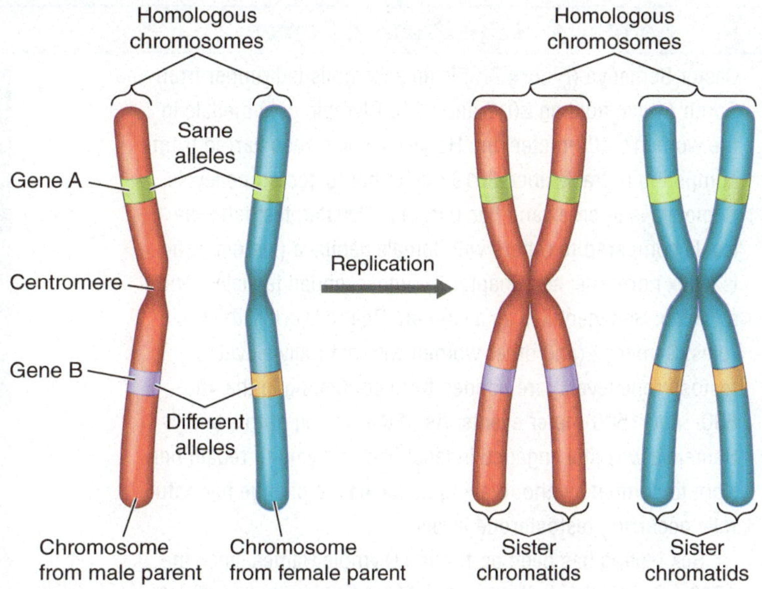

Homologous chromosomes are similar—they have the same genes—but are not identical, meaning there could be different alleles on each chromosome. Sister chromatids are identical copies.

Note that human cells contain two different kinds of chromosomes—autosomes and sex chromosomes. **Autosomes** make up most of your chromosomes (pairs 1–22). The 23rd pair of chromosomes is called the **sex chromosomes**. In humans, there are two kinds of sex chromosomes, X and Y. Sex chromosomes are one method of sex determination. In humans, the default developmental program is for female anatomy (e.g., ovaries, vagina). Therefore, someone with two X chromosomes typically has female anatomy. If a Y chromosome is present, a gene found on the Y chromosome called *SRY* turns on during embryonic development to cause the embryo to develop male anatomy (e.g., penis, testes). Therefore, people with an X and Y chromosome typically have male anatomy. Note that people can also be intersex, with chromosome patterns, anatomy, or other biological characteristics of both sexes.

Misunderstandings about the difference between biological sex and gender identity can contribute to prejudices. Gender identity is how a person describes their personal experience. Biological sex refers to measurable biological differences between people.

How is it possible for some people to have a single X chromosome and for others to have multiple X chromosomes? Typically, a developing embryo with too many or too few chromosomes will not survive. One notable exception is Down syndrome, which we will return to in Section 8.2. In the case of the X chromosome, one X chromosome can be turned off through a process called **X-inactivation**. The deactivated chromosome condenses into a tiny structure called a **Barr body** (Figure 8.5). Which X chromosome is turned off is random. Therefore, gene expression between different cells can be different depending on which X chromosome is turned off.

Diseases related to genes on the X chromosome are called **X-linked disorders** (see Section 8.6). In the case of X-linked ichthyosis, a condition that causes dry, itchy skin, a person with a single X chromosome containing the mutated form of the *KRT1* gene (one type of gene that can cause ichthyosis) would expect to have dry, itchy skin across their entire body. However, if a person has two X chromosomes, one with the mutated *KRT1* gene and one with the non-mutated, or wild-type, *KRT1* gene (said to be heterozygous, which we will return to in Section 8.3),

Figure 8.5 Barr bodies form in XX, but not XY, cells

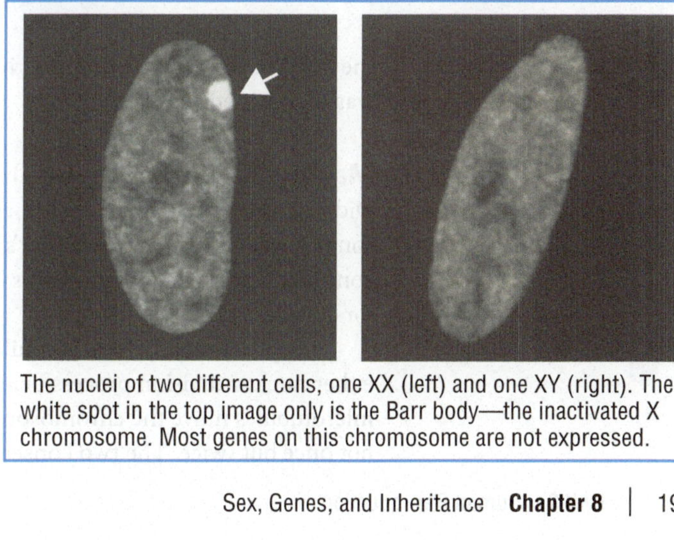

The nuclei of two different cells, one XX (left) and one XY (right). The white spot in the top image only is the Barr body—the inactivated X chromosome. Most genes on this chromosome are not expressed.

Biology for Life

Sex, Gender, and the Olympic Games

Caster Semenya (Figure 8.6) is an accomplished runner from South Africa, holding 2012 and 2016 Olympic gold medals in the women's 800-meter run. However, she's been barred from competing in track since 2019 due to her testosterone levels. Semenya is intersex and has naturally elevated testosterone levels compared to others with female genitalia (testosterone is a sex hormone; see Chapter 3), undescended testicles, and a vagina. She identifies as a woman. Due to World Athletics rules, Semenya (and other women with naturally elevated testosterone levels) are banned from competing in the 400-, 800-, and 1500-meter events. As of the writing of this text, Semenya was still engaged in legal battles trying to regain her right to compete without taking medicines to change her naturally occurring testosterone levels.

Sex testing has been part of the Olympic Games since the 1940s. Sex testing is the process of confirming that an athlete is competing in men's or women's sports according to the sex they were assigned at birth. Initially, sex testing was limited to a visual inspection of one's external genitalia. As time went on, sex testing involved more detailed physical examination, chromosome testing, and an examination of hormone levels, such as testosterone. Although some claim that sex testing is important to keep the games fair, especially in women's sports, others claim that the practice is discriminatory to those who are intersex as well as transgendered individuals.

Forcing individuals into a binary ignores that biological sex, like many traits, exists on a spectrum. For instance, sex-based characteristics include hormone levels and the presence or absence of certain chromosomes, external genitalia, or gonads (ovaries or testes). Being intersex is not a disorder or disease, but a normal part of biological variation—one that impacts 2 in 100 people. For example, an intersex person could have a penis and be assigned male at birth, but have high levels of the hormone estrogen, which is typically associated with females.

Figure 8.6 Caster Semenya

Semenya is an accomplished intersex runner fighting for her right to compete as a woman in the 400-, 800-, and 1500-meter track events.

1. Given the biological spectrum of sex characteristics, do you think it's fair to classify sports based on a sex binary? What about other physical characteristics that could give someone an advantage, such as height, in sports like swimming?
2. The idea of men's and women's sports intersects with another aspect of inclusion in athletics—the fact that women's sports are often seen as "less than" men's sports. For example, male athletes tend to make far more money than female athletes, and major athletic events in the United States, such as the Super Bowl and the Stanley Cup playoffs, have all-male teams. Do you think that changing rules around how people compete in sports could lead to more equity for all?

? Did You Know?

Swyer syndrome occurs when a genetically male individual (XY) has female anatomy due to a mutation in the *SRY* gene.

they will have patches of itchy skin and normal skin, depending on which X chromosome was inactivated (Figure 8.7).

Making Gametes: Spermatogenesis and Oogenesis

Did you know that females are born with a finite number of eggs? These eggs form when a person is developing in their mother's womb and set the limit on their fertility. Unlike males, who continually make new sperm cells over the course of their lifetime, egg cells pause in meiosis for decades.

The process of meiosis is similar to mitosis in several ways. A cell replicates its DNA before either nuclear division process begins (during interphase). As in mitosis, a spindle forms, and its microtubules move the chromosomes. However, meiosis sorts the chromosomes into new nuclei not once but twice. The two consecutive nuclear divisions are called *meiosis I* and *meiosis II*.

Figure 8.7 **People with more than one X chromosome have differential gene expression**

Since which X chromosome is turned off is random (see the Barr bodies; the paternally derived chromosome is blue and the maternally derived chromosome is red), there can be different gene expression throughout the body. In the case of X-linked ichthyosis, some skin cells will be normal (beige) and others affected (brownish red) depending on if they contain the mutated or wild-type *KRT1* gene.

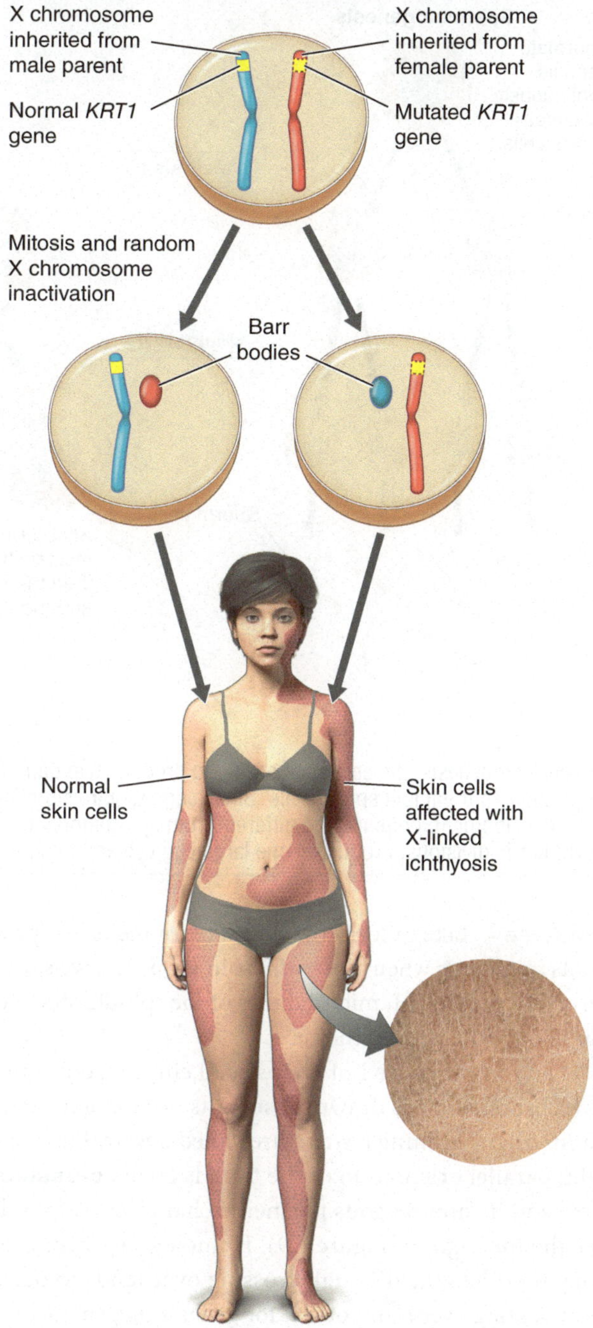

? Did You Know?

Calico cats get their characteristic orange and black patchy patterns due to X-inactivation. This is why calico cats are almost always XX.

Meiosis is different between males and females. While the process of making sperm in males, called **spermatogenesis**, results in four equally sized sperm cells, **oogenesis**, the process of making egg cells in females, results in a single large egg cell (Figure 8.8). Note that during our discussion of meiosis, we will be specifically focusing on spermatogenesis, which produces four gametes. In the margin of this text, the avatar will highlight differences with oogenesis, which produces a single gamete and two **polar bodies**—haploid cells that cannot be fertilized and undergo apoptosis. Follow along with Figure 8.9 as you read the following section.

Big Ideas in Biology

Information Flow

Gametes transmit genetic information between generations.

Meiosis I

Meiosis I in females begins in utero, whereas in males it occurs throughout the lifetime. The steps of meiosis are the same in both and are similar to mitosis.

Remember the role of scientific consensus that you learned about in Chapter 1, and that as new evidence comes to light, that consensus might change? As of the writing of this text, the scientific consensus was that females are born with a finite number of eggs. However, new evidence suggests that stem cells in the ovaries may lead to postnatal oogenesis. You can look up the latest research on your own!

Figure 8.8 **Sperm and egg production vary**

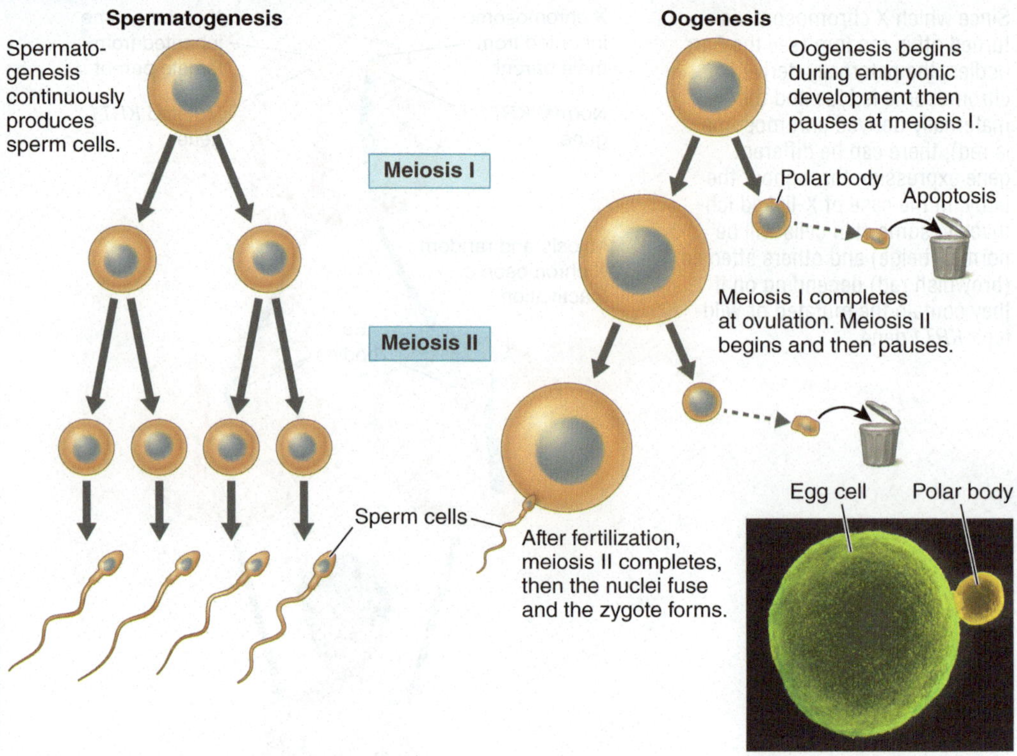

In spermatogenesis, the cell splits evenly to produce two daughter cells, which in turn split evenly to produce four haploid sperm cells. Since eggs require significantly more resources to support early development, oogenesis is different from spermatogenesis. Therefore, the egg undergoes two asymmetric divisions, producing one large egg cell and two polar bodies that undergo apoptosis.

Credit: Prof. P.M. Motta/Univ. ""La Sapienza"", Rome/Science Source

Oogenesis pauses during crossing over. The developing egg will remain in suspended animation until ovulation. Ovulation is the time in a person's menstrual cycle when an egg is released in preparation for fertilization. During ovulation, the egg "wakes up" and restarts the process of meiosis with metaphase I.

Prophase I. Like mitosis, the first stage of meiosis is prophase. Since this is the first of two points in meiosis when a cell will be in prophase, we say this is **prophase I** (see Figure 8.9). By the end of prophase I, microtubules of the spindle attach one chromosome of each homologous pair to opposite ends of the cell.

Early in prophase I of meiosis, all chromosomes in the cell pack tightly. When they do, each chromosome is drawn close to its homologous partner, so their corresponding chromatids align, forming a structure called a **tetrad** (since there are four chromatids total). This tight, parallel orientation of the tetrads favors **crossing over**, a process by which a chromosome and its homologous partner exchange corresponding pieces of DNA during meiosis (see the top-right of Figure 8.9). Homologous chromosomes may swap any segment of DNA along their length, although crossing over tends to occur more frequently in certain regions.

Crossing over is important for generating genetic diversity, because a completely new chromosome, called a **recombinant chromosome**, forms. This new chromosome is a fusion of what was originally two different chromosomes. Thus, crossing over introduces novel combinations of alleles—and new combinations of traits—among offspring.

Metaphase I. During **metaphase I**, microtubules grow and shrink, pushing and pulling the chromosomes until the chromosomes are aligned midway between the spindle poles.

Anaphase I. During **anaphase I**, the homologous chromosomes of each pair move away from one another and toward opposite ends of the cell.

Telophase I. During **telophase I**, the two sets of chromosomes reach the spindle poles, and new nuclear envelope re-forms around each set as the chromosomal packing loosens. The new nuclei have one complete set of chromosomes, the sister chromatids. Meiosis I finishes with cytokinesis, when the cytoplasm divides, producing two cells.

Figure 8.9 An overview of meiosis to produce sperm: meiosis I

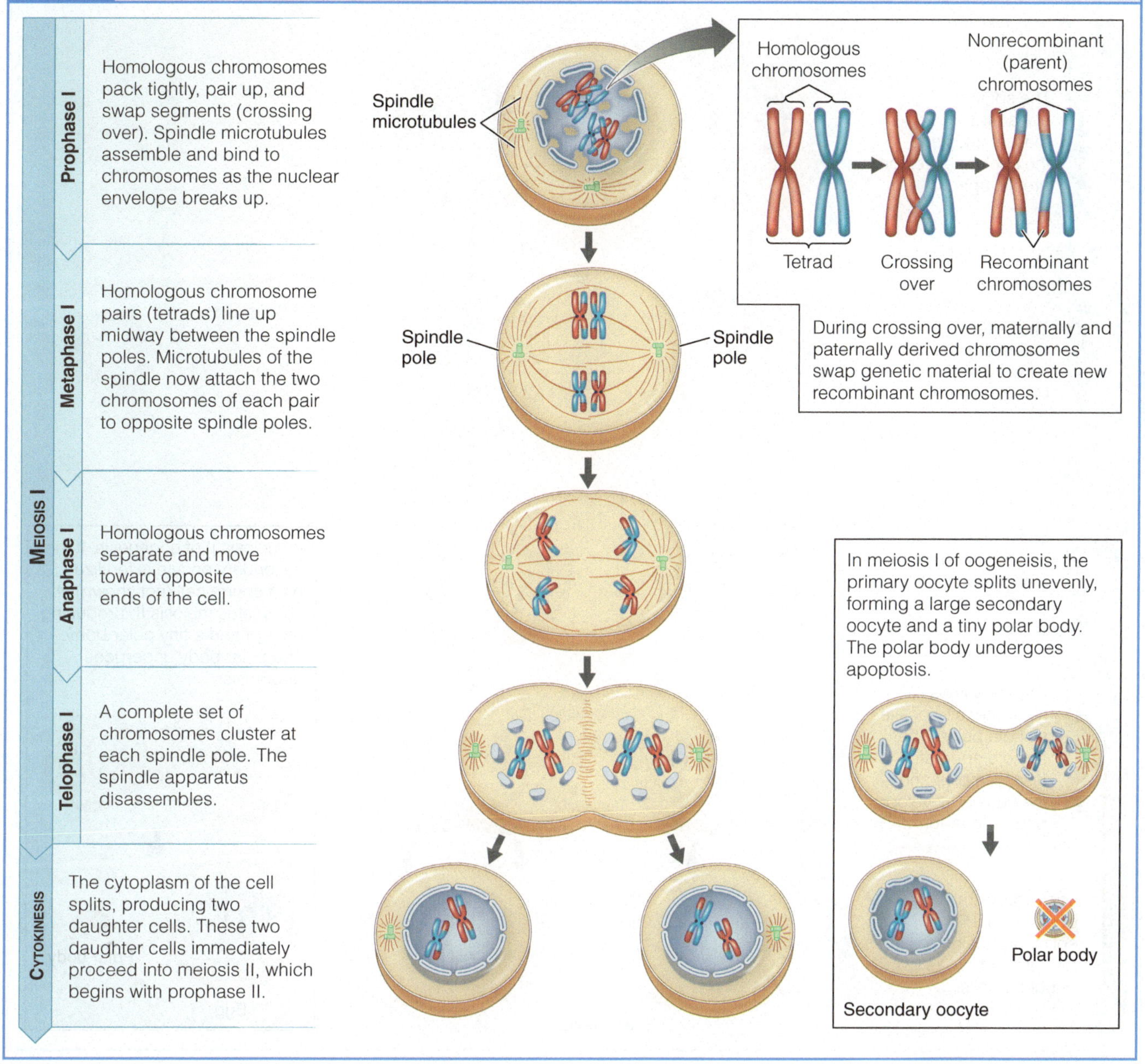

Point of Confusion: Meiosis I or meiosis II? In meiosis I, *homologous chromosomes* line up at the metaphase plate; this is the only time you will see two pairs of chromosomes lined up prior to separating from each other. In meiosis II, *sister chromatids* line up and are split from each other.

Did You Know?

In humans, between 46 and 95 crossovers occur per round of meiosis, so each pair of homologous chromosomes crosses over two or three times, on average.

Meiosis II

During meiosis II, sister chromatids separate and the final gamete(s) form. Meiosis II begins with the products of meiosis I. In egg cell production, meiosis II does not proceed until after a sperm penetrates the egg. In contrast, sperm are continually made.

Prophase II. During **prophase II**, the chromosomes pack tightly, and the nuclear envelope breaks up as a new spindle forms. By the end of prophase II, spindle microtubules attach to the sister chromatids of each chromosome.

Metaphase II. During **metaphase II**, the microtubules push and pull the chromosomes, aligning them midway between spindle poles.

Figure 8.9 An overview of meiosis to produce sperm: meiosis II, *continued*

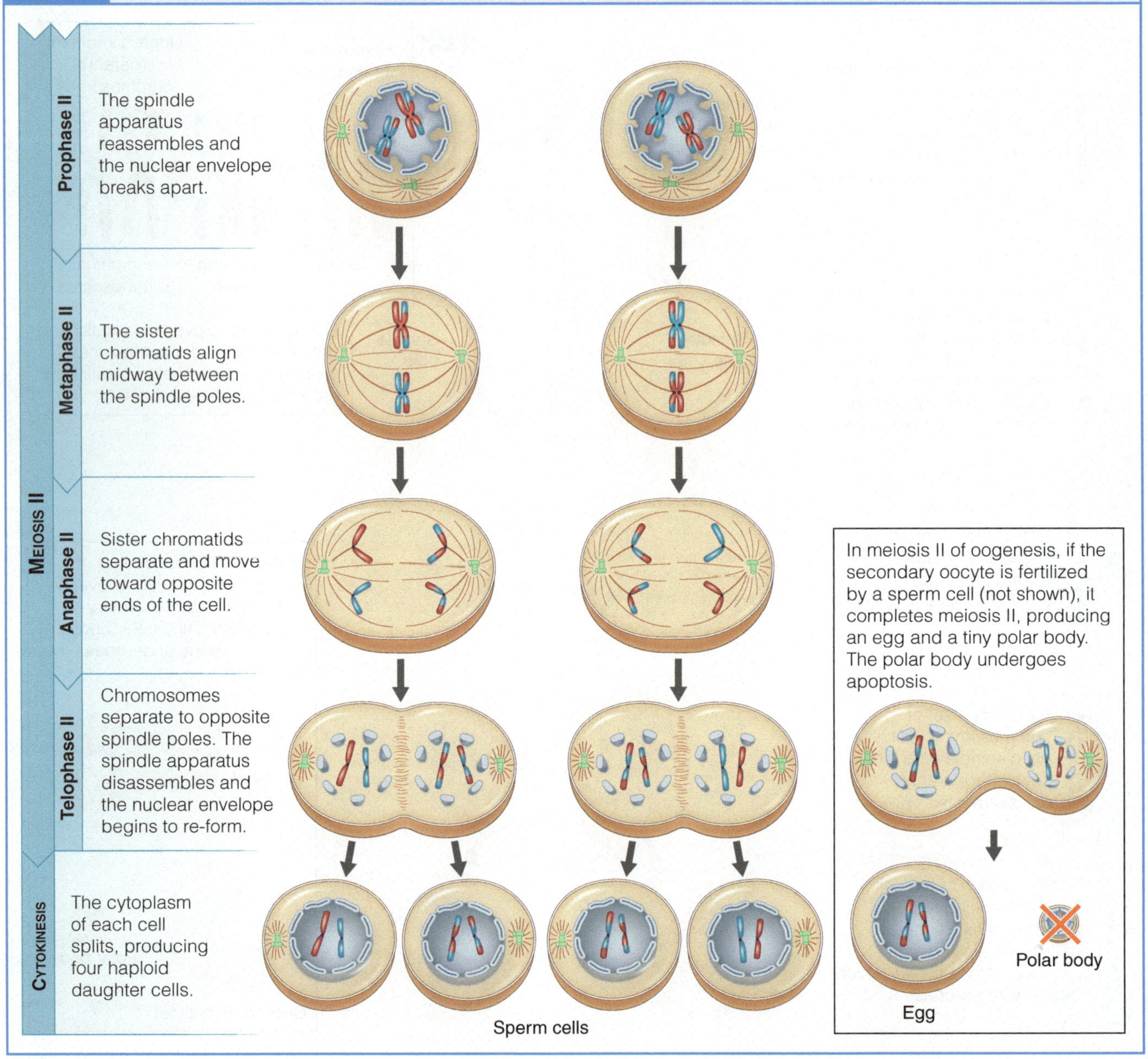

After ovulation, oogenesis pauses at metaphase II. If a sperm cell penetrates the egg, meiosis will complete. Then, the two haploid nuclei will fuse, and fertilization will be complete.

Anaphase II. During **anaphase II**, the sister chromatids separate and move toward opposite spindle poles, thus becoming individual chromosomes.

Telophase II. During **telophase II**, the chromosomes reach the spindle poles. New nuclear envelope forms around the clusters of chromosomes as their packing loosens. The new nuclei are haploid (n), with one complete set of chromosomes. Once the cytoplasm splits during cytokinesis, there are four genetically distinct haploid cells (for spermatogenesis).

Comparing Mitosis and Meiosis

Mitosis and meiosis are similar processes for making new cells but have several important distinguishing characteristics (Figure 8.10). In both processes, DNA replication occurs once to produce identical chromosomes (sister chromatids). Mitosis is similar to meiosis II in that

Figure 8.10 A comparison of mitosis and meiosis

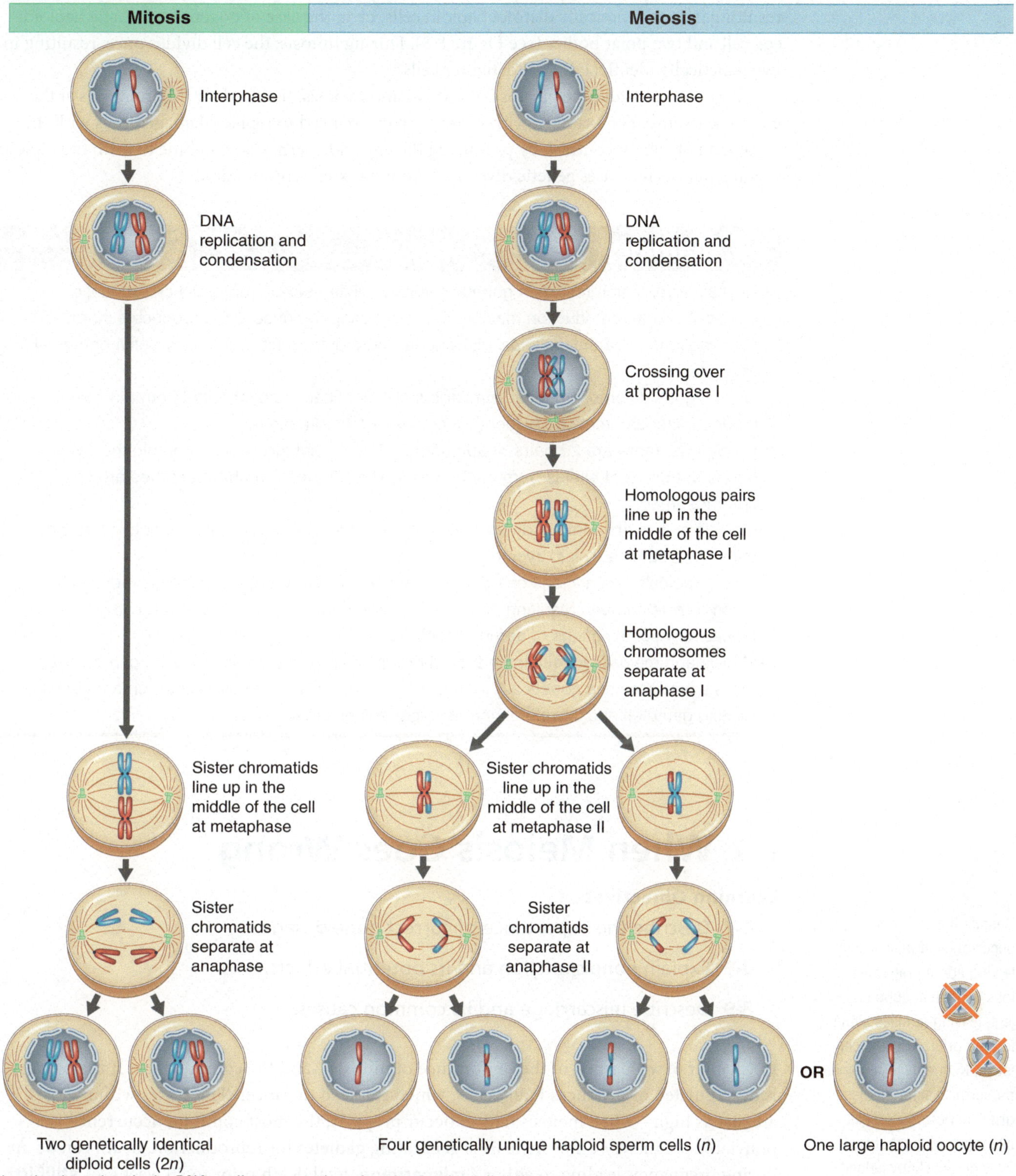

In mitosis and meiosis, DNA replicates one time. In mitosis, the cell splits once to produce genetically identical daughter cells. In contrast, during meiosis, the cell splits twice, producing four genetically distinct haploid daughter cells (spermatogenesis) or a single haploid egg and two polar bodies (oogenesis); the two polar bodies undergo apoptosis.

sister chromatids are separated from each other. However, in meiosis, recombination occurs via crossing over to produce genetically distinct chromosomes. Meiosis then splits the cell twice, resulting in four genetically distinct haploid cells, or in the case of oogenesis, a single haploid egg cell and two polar bodies (see Figure 8.8). During mitosis, the cell divides once, resulting in two genetically identical diploid daughter cells.

When we consider the purpose of mitosis and meiosis, the need for different cells at the end makes sense. For example, with wound repair, we need to replace damaged cells with the same kind of cell. In contrast, by producing the egg and sperm that combine to form the diploid zygote, meiosis increases genetic diversity through sexual reproduction.

8.1 Summary

- Sexual reproduction results in genetic diversity, while asexual reproduction produces clones. Asexual reproduction may be faster, but offspring of sexual reproducers inherit new combinations of parental alleles, allowing for more opportunities to survive environmental changes.
- Homologous chromosomes are similar but not identical, each inherited from a different parent. Sister chromatids are identical copies of a chromosome.
- In humans, there are 22 pairs of autosomes. The homologous pairs contain the same genes but different alleles from each parent. The 23rd pair of chromosomes are sex chromosomes.
- Meiosis halves the chromosome number to produce gametes. The fusion of two haploid gametes creates a diploid zygote.
- During meiosis, the nucleus of a diploid ($2n$) cell divides twice. This process reduces the chromosome number, resulting in haploid (n) gametes. Crossing over during prophase I makes new combinations of parental alleles.
- Mitosis and meiosis are similar in that DNA replication occurs only once in each process. Mitosis produces two genetically identical daughter cells, whereas meiosis creates unique haploid gametes and therefore increases genetic diversity.

8.2 When Meiosis Goes Wrong

Learning Objectives

8-7 Describe the importance of correct gene dosage.

8-8 Explain nondisjunction and its potential effects.

8-9 Describe miscarriage and its common causes.

Despite the common occurrence of miscarriage, talking about pregnancy loss is often a taboo subject, leaving families alone in their grief. Grief specialist Dr. Kenneth Doka coined the term "disenfranchised grief" to describe losses like miscarriage that typically go unacknowledged by society.

In Chapter 6, we learned of the importance of mitosis—and the consequences if the process is dysregulated or produces faulty cells, which can lead to cancer. The stakes with meiosis are just as high—when meiosis fails to occur properly, the most common occurrence is a nonviable pregnancy. Since meiosis is generating gametes for reproduction, death can occur during pregnancy, leading to either a **miscarriage**, fetal death prior to 20 weeks, or **stillbirth**, fetal death after 20 weeks.

Miscarriage is a common occurrence—and one that can result in trauma and grief for those who experience this kind of loss. According to the March of Dimes, greater than 30% of pregnancies end in miscarriage. Why do miscarriages happen? Although there are a wide variety of factors that can cause or contribute to miscarriage, in this section, we will focus on one common cause of miscarriages: chromosomal problems.

Relating Gene Dosage and Miscarriage

Healthy individuals require a complete set of DNA. This ensures that a person has the correct **gene dosage**, which is two copies of each autosome and consequently two copies of each gene. **Aneuploidy** is when there are too many or too few copies of a particular chromosome, which can lead to miscarriage. Aneuploidy is usually an outcome of **nondisjunction**, the failure of chromosomes to separate properly during meiosis. Nondisjunction of meiosis I occurs when homologous chromosomes fail to separate properly. In contrast, nondisjunction of meiosis II occurs when sister chromatids fail to separate properly.

Figure 8.11 **Down syndrome is caused by trisomy 21**

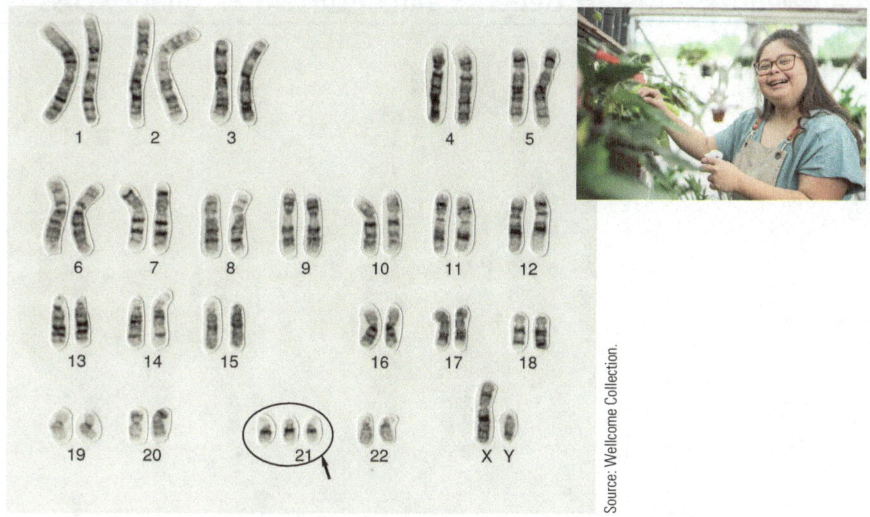

A. Karyotype of an XX individual with Down syndrome. Note the three copies of chromosome 21.
B. Person with Down syndrome working in a gardening center.

Nondisjunction during meiosis produces gametes with an abnormal number of chromosomes. Suppose a normal gamete fuses with a gamete that has an extra chromosome. The resulting zygote will have three copies of one type of chromosome and two of every other type, an aneuploid condition called **trisomy**. If a normal gamete fuses with a gamete missing a chromosome, the new individual will have one copy of one chromosome and two of every other type, a condition called **monosomy**. For example, Turner syndrome, or monosomy X, occurs when a person has a single X chromosome. People with Turner syndrome often struggle with fertility issues.

Turner syndrome is one example of aneuploidy that is not fatal. In addition, a person born with three copies of chromosome 21 (trisomy 21; Figure 8.11) will have Down syndrome and a high likelihood of surviving infancy. People with Down syndrome may have mild to moderate mental impairment and health problems such as heart disease; physical markers such as a somewhat flattened facial profile; and skeletal differences such as misaligned bones of the fingers, toes, and hips. Many people with Down syndrome live fulfilling lives.

? Did You Know?

Alzheimer's disease is associated with Down syndrome due to extra copies of a gene located on chromosome 21. Too many copies of this gene result in an abundance of the proteins that interfere with brain function, leading to Alzheimer's.

Biology for Life

Delayed Childbearing and Miscarriage Risk

Although many people think that rates of nondisjunction are linked solely to a female's age, sperm and egg quality both decrease over a life span. Sperm from older males typically have higher rates of DNA fragmentation, which can double the risk of a miscarriage. In females, because meiosis pauses at meiosis I—leaving eggs in suspended animation, sometimes for decades until ovulation—there is additional risk for DNA damage.

More and more people are delaying having children until they are older, beyond their peak reproductive years, so that they can focus on school or starting a career. Is there a way to preserve fertility and decrease the chance of miscarriage? Some biologists are examining the role of antioxidants, such as CoQ10, and their ability to preserve gamete quality.

Producing eggs and sperm requires a large amount of energy. For example, eggs have more mitochondria (Chapter 4) than almost all other cell types. CoQ10 naturally occurs in the mitochondrial membrane, where it participates in the electron transport chain. CoQ10 helps keep our cells healthy by minimizing oxidative damage (Chapter 2). However, CoQ10 levels decrease as we age, which impacts gamete quality. Therefore, CoQ10 supplementation may help preserve egg quality. The benefits aren't just for women, either. Researchers found that men who took daily CoQ10 produced not only more sperm, but more healthy sperm.

1. Do you think more guidance should be provided to young people who want to have kids someday on how to preserve their gamete quality until they are ready to have children? Why or why not?
2. If you are planning to have children, what do you plan to do regarding continuing your education and personal development during your peak reproductive years?

Digging into Data

Can We Risk It Again? Genetics and Recurrent Pregnancy Loss

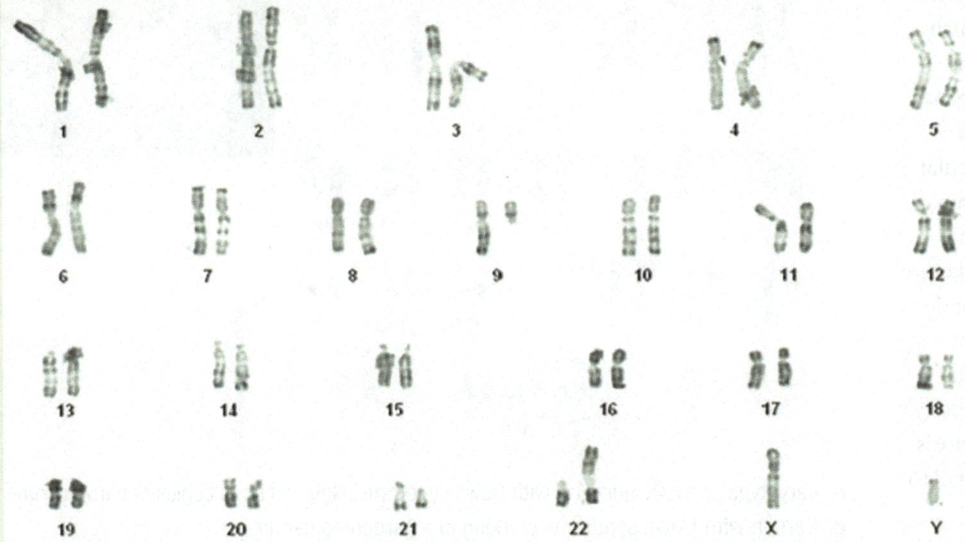

Figure 8.12 Scott's karyotype

Source: Hammami, W., Kilani, O., Khelifa, M. B., Ayed, W., Bouzouita, A., Zhioua, F., ... & Amouri, A. (2015). Genetic diagnosis in non-obstructive azoospermic Tunisian men. Austin J Reprod Med Infertil 2(2), 1012.

After several years of marriage, Scott and Celeste welcomed a baby girl, Mariposa, into their lives. A few years later, they decided to add to their family. Instead of another baby, however, they endured three miscarriages in a row. The physical and emotional toll was exhausting.

While a single miscarriage is common, it is unusual to have back-to-back miscarriages. Seeking answers, Scott and Celeste turned to Dr. Debra Valery, a fertility specialist. After weeks of testing for a variety of possible causes, Scott was diagnosed with a condition called *balanced translocation*, which occurs when part of one chromosome completely breaks off and reattaches to another chromosome (Figure 8.12). Scott is healthy because he has both portions of the broken chromosome. However, unless the fertilizing sperm cell contains the correct gene dosage from Scott, the resulting zygote will either not have the complete set of genetic instructions or will have too many copies of some genes. Therefore, a miscarriage will occur.

Dr. Valery explained that there were different options to pursue to have another baby while minimizing the chances of another miscarriage. For example, they could do artificial insemination with donor sperm or *in vitro* fertilization (IVF) with embryos that have had their chromosomes tested to ensure proper gene dosing. Scott and Celeste carefully weighed these options and made their decision. A year later, they welcomed Felicity, a healthy baby girl, to their family.

1. Approximately one-third of all pregnancies end in miscarriage. Do some research on miscarriages. What is the medical definition of a miscarriage? In your answer, differentiate between different kinds of miscarriage. Compare a chemical pregnancy, ectopic pregnancy, missed miscarriage, and molar pregnancy.
2. How were Scott and Celeste able to have a living child, without an intervention?
3. Look at Scott's karyotype (Figure 8.12). Recall that in karyotypes, chromosomes are paired by size. Between which two chromosomes did a balanced translocation occur? How do you know?

8.2 Summary

- Gene dosage, or having the proper number of gene copies, is important for the health of an organism. Without proper gene dosage, miscarriage or stillbirth can occur.
- Nondisjunction is when either homologous chromosomes or sister chromatids do not separate properly. Nondisjunction leads to aneuploidy, where there is either too many or too few chromosomes.
- Sex chromosome aneuploidy typically has milder effects on health than autosomal aneuploidy. Autosomal aneuploidy often leads to a miscarriage or stillbirth. Trisomy 21, which causes Down syndrome, is an exception.

8.3 Mendelian Inheritance

Learning Objectives

8-10 Explain the relationship between genotype and phenotype.

8-11 Compare dominant and recessive alleles.

8-12 Compare homozygous and a heterozygous genotypes.

8-13 Describe a single trait cross and how it can reveal a dominant–recessive relationship between alleles.

8-14 Explain the relationship between Mendel's law of segregation and Punnett squares.

8-15 Explain how Mendel's law of independent assortment contributes to genetic diversity.

Did you know that in the nineteenth century, people thought that hereditary material must be some type of fluid, with fluids from both parents blending at fertilization like milk into coffee? However, the idea of "blending inheritance" failed to explain what people could see with their own eyes. Children sometimes have traits that do not appear in either parent. For example, a child, but neither parent, could have freckles.

At the time, no one knew that hereditary information is divided into discrete units (genes), an insight that is critical to understanding how traits are inherited. Around 1850, Gregor Mendel, an Austrian monk, began an extended series of experiments breeding pea plants, which vary in traits such as flower color, height, and so on.

Mendel's meticulous work breeding pea plants and tracking their traits led him to conclude (correctly) that hereditary information passes from one generation to the next in discrete units. Little did he know that his work with pea plants would be the starting point for modern genetics.

Fundamental Principles of Inheritance

Even without knowing what a gene or DNA was, Mendel was able to figure out several important genetic principles. For example, he was able to infer the relationship of genotype and phenotype. **Genotype** refers to an organism's genetic makeup, or the unique set of alleles that they posses. Genotype is the basis of **phenotype**, which refers to the individual's observable traits. Note that gene expression (Chapter 7) is the process that connects genotype to phenotype. Genes are expressed into proteins, and these proteins contribute to an observed phenotype.

An organism's alleles can be dominant or recessive. An allele is **dominant** when its effect masks that of a recessive allele. **Recessive** alleles control traits that are observed only if a person has two copies of the recessive allele. A dominant allele is represented by an uppercase italic capital letter such as *A*. In contrast, recessive alleles are indicated by a lowercase italic letter such as *a*. If an individual has the same allele on both homologous chromosomes, they are **homozygous** for the allele. By contrast, an individual with different alleles of a gene is **heterozygous** for the allele.

> **Study Tip**
>
> Look at the root words to help you remember the difference between homozygous and heterozygous—*homo* means "the same," and *hetero* means "different."

In Mendel's peas, "white-flowered" and "purple-flowered" are examples of pea plant phenotypes that arise from differences in genotype. In these plants, the allele that specifies purple flowers (let's call it *P*) is dominant over the allele that specifies white flowers (*p*). Thus, a pea plant homozygous for the dominant allele (*PP*) has the dominant trait: purple flowers. A plant homozygous for the recessive allele (*pp*) has the recessive trait: white flowers. A heterozygous plant—one with both the dominant and the recessive allele (*Pp*)—has the dominant trait, which is purple flowers (Figure 8.13).

Figure 8.13 **The relationship between genotype and phenotype**

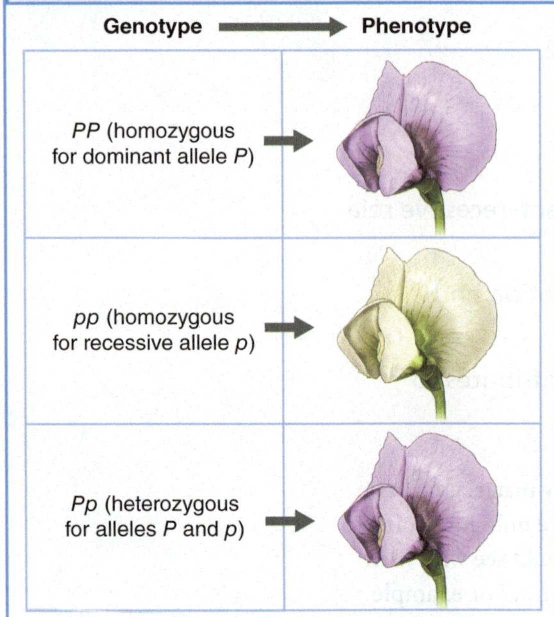

In this example, the dominant *P* allele specifies purple flowers, and the recessive *p* allele specifies white flowers.

Mendelian Inheritance

In the Unit 3 introduction, you learned that my children have different eye colors—brown and blue. How is it possible for my children to have different eye colors? Whether or not one has pigment in the iris of their eye, resulting in dark eyes, is a rare example of a Mendelian trait, or a trait that is purely dominant or recessive. Please note that there are many genetic factors influencing the precise eye color a person has (light brown, dark brown, hazel, blue, green, gray), but for the purposes of this example, we will stick with having dark- or light-colored eyes.

Dark eye color results if pigment is expressed in one's iris (the colored part of the eye). People with light-colored eyes (which includes blue) do not express pigment in the eye. Rather, the blue color that we see is due to the scattering of light within the eyeball—the same phenomenon that explains why large bodies of water appear blue.

The allele that results in pigment in the eye is dominant to the allele that does not result in pigment in the eye. If you have the dominant allele, you will have pigment in your eye, which prevents the scattering of light. Therefore, to have light eye color, one must be homozygous for the recessive allele.

We can describe the inheritance of the pigment-producing genes with a Punnett square. **Punnett squares** are a graphical representation of the potential gametes that a parent can produce and the potential genotypes of the offspring. Punnett squares are also useful for making predictions about how many offspring will have a particular genotype or phenotype.

Let's look at a sample Punnett square for my family, where we will examine the outcomes in a **single-trait cross**, a way of tracking the inheritance of one gene. In this case, between myself and my husband (Figure 8.14). Together, my husband and I make up the **parental (P)**

Figure 8.14 **Punnett squares are useful tools for predicting inheritance**

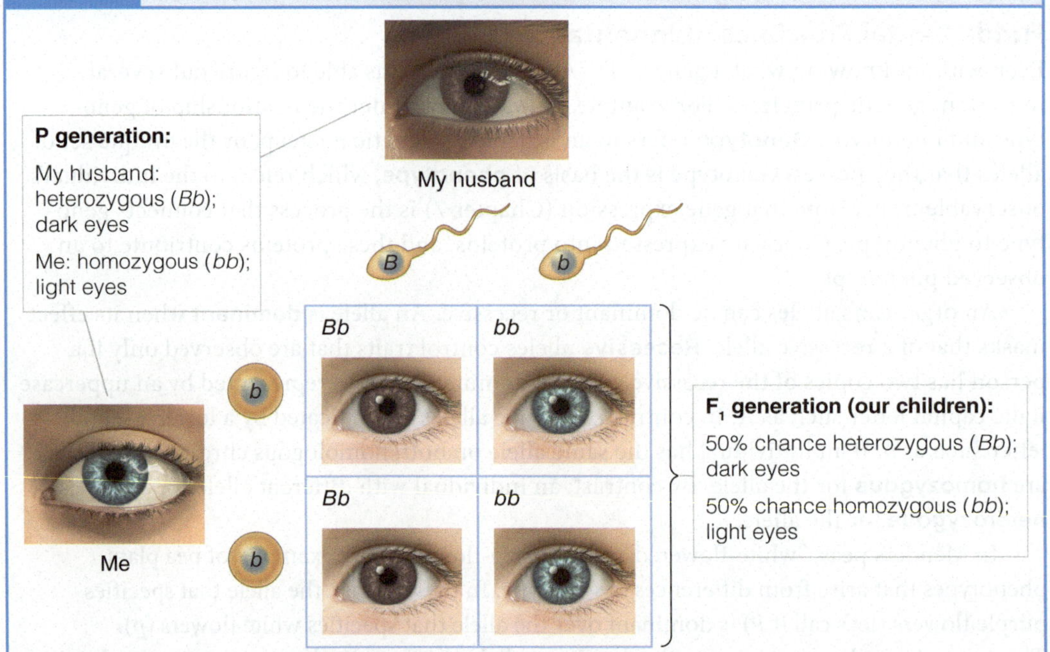

In this example, we are looking at the inheritance of light or dark eye color in my family. I have light eyes and can produce only *b* gametes. My husband has dark eyes and is heterozygous *Bb*, so his body can produce either *B* or *b* gametes. Looking at the Punnett square, there is a 50% chance of our children having light-colored eyes.

generation, and our children make up the **F_1 generation**. Since I have light eyes, I must have the *bb* allele combination, and because my husband has dark eyes, he is either *Bb* or *BB*. Now, how do we know if my husband is *Bb* or *BB*?

The most efficient way to figure this out is to look at my husband's parents' eye color, where his Mom had light eyes, and therefore, the *bb* allele combination. Since she could pass on only a *b* allele and my husband has dark eyes, that means he inherited the *B* allele from his father, and consequently, must be *Bb*. So, what are the possible genotypes and associated phenotypes of our children?

If we complete the Punnett square, we see that there is a 50% (2 of 4) chance of our offspring having light-colored eyes. When we look at the real-world data, among our two kids, the ratio holds true: One child has light eyes and the other dark.

Mendel's Laws

In addition to his observations pertaining to dominant and recessive traits and their inheritance, Mendel also devised three laws, two of which we will discuss. First, Mendel's **law of segregation** states that alleles on different chromosomes are inherited separately from one another. This makes sense when we consider meiosis—which chromosomes go where has no influence on other chromosomes, and each gamete gets only a single allele of each gene (Figure 8.15). At fertilization, which gametes combine is random, leading to different possibilities. Mendel's law of segregation also explains why we can use a Punnett square to trace genes and make predictions about inheritance.

Figure 8.15 The segregation of genes on homologous chromosomes into gametes

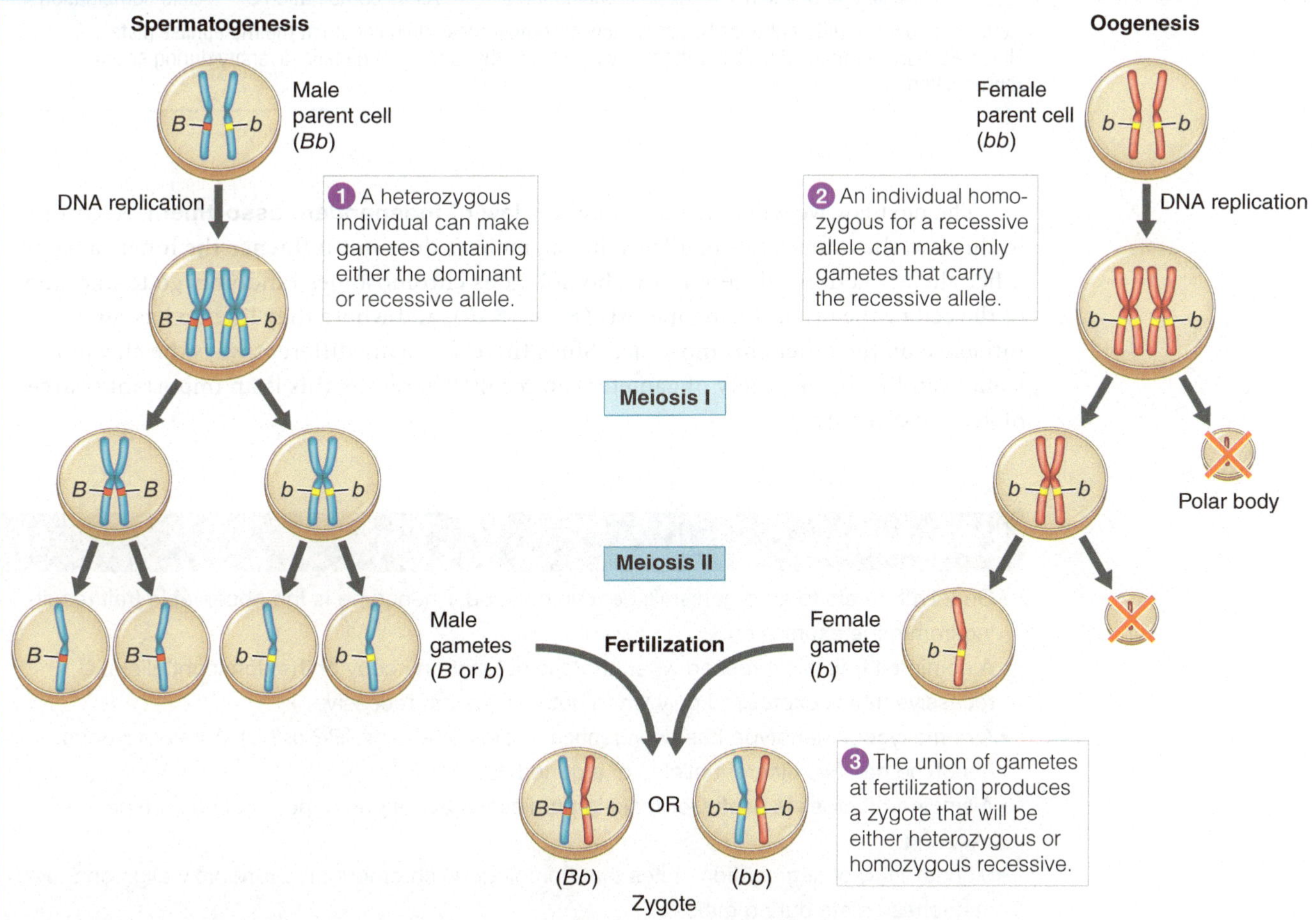

Homologous chromosomes separate during meiosis I, so the genes they carry separate, too. Each of the resulting gametes carries one of the two members of the gene pair. Notice that for oogenesis, one chromosome in the pair after meiosis I and II is lost when the polar bodies undergo apoptosis. Note also that this figure traces the movement of genes predicted in the Punnett square in Figure 8.14.

Figure 8.16 **Mendel's law of independent assortment describes the random arrangement of homologous chromosomes at meiosis I**

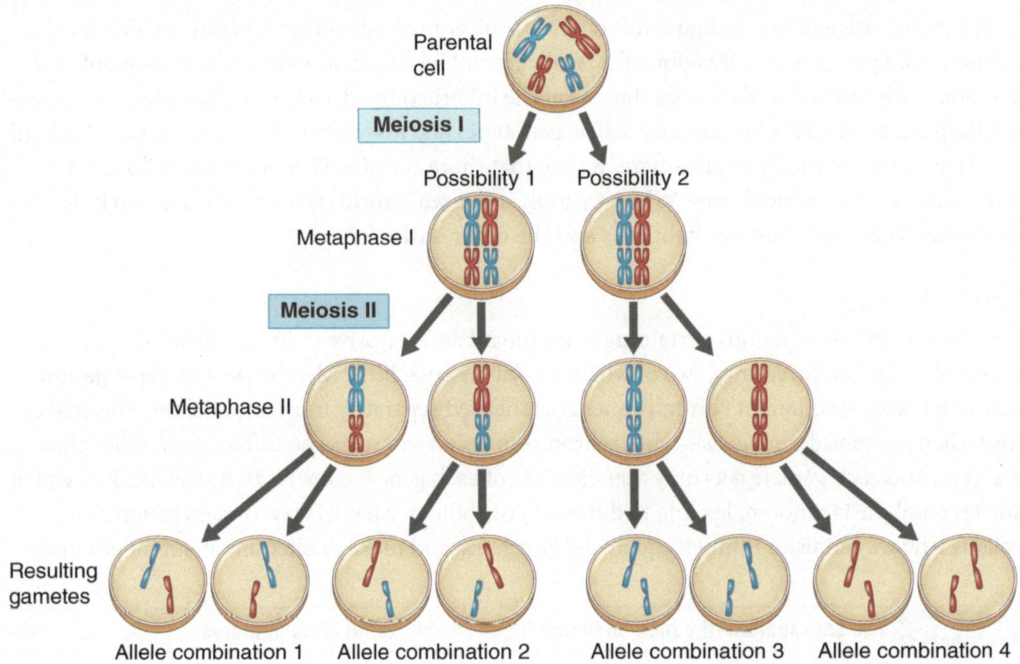

Here, we show two different possibilities of how chromosomes could line up at the metaphase plate. However, there are many different options—which is another source of genetic diversity during sexual reproduction.

The next law we will discuss is Mendel's **law of independent assortment**. According to this law, the inheritance of alleles during meiosis does not influence the inheritance of other alleles. Said in a different way, homologous chromosomes randomly go to one side of the cell or the other during meiosis (Figure 8.16), and where they line up has no influence on the other chromosomes. Since there are many different ways the chromosomes can line up, a variety of gametes can result. Therefore, this is an important source of genetic diversity.

8.3 Summary

- Genotype refers to an organism's genetic makeup. Phenotype is the observable trait resulting from gene expression.
- A dominant trait is expressed when there is at least one copy of the dominant allele. A recessive trait is expressed only when both alleles are recessive.
- A homozygous genotype has two identical alleles of a gene (*PP* or *pp*). A heterozygous genotype has two different alleles of a gene (*Pp*).
- A single-trait cross is useful for making predictions about genotypes and phenotypes in the offspring.
- Mendel's law of segregation states that homologous chromosomes randomly align on the metaphase plate during meiosis I.
- Mendel's law of independent assortment states that the inheritance of alleles on one chromosome does not affect the movement of alleles on different chromosomes.

8.4 Non-Mendelian Inheritance

Learning Objectives

8-16 Compare incomplete dominance and codominance.

8-17 Explain pleiotropy, or how one gene may influence multiple traits.

8-18 Describe polygenic traits and how they differ from other traits.

Scan any biology text and you'll find information about Gregor Mendel. Perhaps you even remember learning about Gregor Mendel and Mendelian inheritance in high school, or even earlier in your educational career. Although Mendel managed to figure out a truly remarkable amount of information about genetics with his pea plants, altogether, Mendelian inheritance explains a very small portion of our biology.

We've already seen one example of non-Mendelian inheritance in this chapter—X-inactivation (Section 8.1). With X-inactivation, one can have a gene but not express it—even if it's a dominant gene. If that gene happens to be on the X chromosome that is inactivated, the gene won't be expressed in that cell. In this section, we will learn about some other exceptions to the one gene:one phenotype paradigm of Mendelian inheritance. Other, more complex relationships between alleles and traits are more common.

Variations in Gene Dominance

Some genes are inherited in a Mendelian manner (Section 8.3), but the phenotypes produced are not what we'd expect to see. Rather, either one allele doesn't completely mask the other (incomplete dominance) or we see phenotypic contributions from both alleles (codominance). We'll examine both phenomena next.

Incomplete dominance. In our eye color example in Section 8.3, if a person expresses pigment in their eye, it completely masks the light eye phenotype. This is an example of complete dominance. **Incomplete dominance** is when one allele is not fully dominant over the other. Therefore, a heterozygous individual will have a phenotype that is an intermediate blend of the two homozygous phenotypes. In this case, the offspring do not look like either parent. For example, consider labradoodle dogs. If a straight-haired dog is bred with a curly haired dog, the resulting offspring will have wavy hair—an intermediate between curly and straight hair.

Codominance. If incomplete dominance means neither allele is completely dominant to the other, then **codominance** means the two alleles equally contribute to the phenotype. Another way of thinking about this is that neither allele is dominant nor recessive. For a heterozygous individual, one can see the phenotypic contributions of both alleles.

Let's look at a common example—human blood types. People can have one of four blood types: A, B, AB, or O. These blood types reflect the presence or absence of A or B sugars on the blood cells (Chapter 3). The ABO gene encodes an enzyme that modifies a carbohydrate on the surface of human red blood cells. Two alleles of the gene, *A* and *B*, encode slightly different versions of the enzyme, which in turn modify the carbohydrate differently. A third allele, *O*, has a frameshift mutation (Chapter 7), resulting in a nonfunctional protein and unmodified carbohydrate.

The alleles of the *ABO* gene determine the form of the carbohydrate on blood cells and are the basis of blood type (Figure 8.17). The *A* and *B* alleles are codominant. Let's say your genotype is *AB*. That means there are both A and B sugars on your blood cells and your blood type is AB. The *O* allele is recessive when paired with either the *A* or *B* allele. If your genotype is *AA* or *AO*, your blood type is A. If your genotype is *BB* or *BO*, it is type B. If you are *OO*, it is type O.

Study Tip

Think of ***in***complete dominance as leading to an ***in***termediate phenotype, and ***co***dominance as being like two ***co***workers working on the same project.

Figure 8.17 **Human blood typing and codominance**

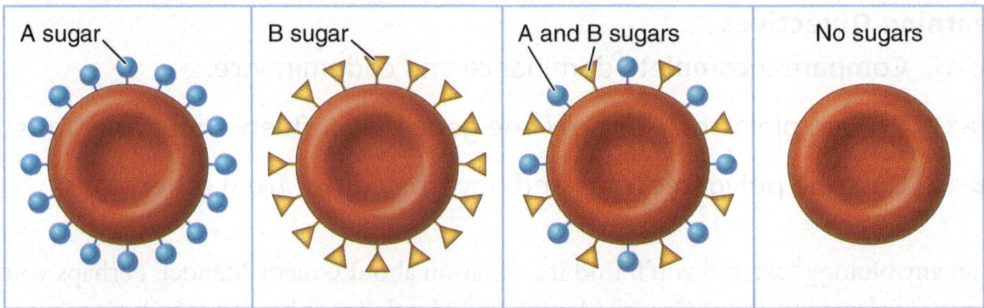

Genotype	*AA* or *AO*	*BB* or *BO*	*AB*	*OO*
Phenotype	A type blood	B type blood	AB type blood	O type blood

Combinations of different alleles (genotype) lead to different blood types (phenotype).

A Single Gene and Multiple Traits

Marfan syndrome occurs in about 1 in 5,000 people. People with Marfan syndrome are usually tall and thin, and are at high risk for aortic rupture during exercise. Why? Marfan syndrome is traced back to a mutation in the gene for fibrillin. Fibrillin is a critical part of the elastic tissues that make up the heart, skin, blood vessels, tendons, and other body parts. People with

Process of Science

Racial Inequality and Misinterpreting Inheritance

Mendel's work is the foundation of modern genetics—and until the last several decades, it encompassed our understanding of genetics. As a result, study of the central dogma (Chapter 7), Mendelian genetics, plus a few of the exceptions encompass most genetics information that students learn about.

The problem, however, is that Mendel's work with peas suggested that inheritance is straightforward. Unfortunately, this straightforward view of inheritance resulted in some groups of people misappropriating Mendel's work and using it to justify racial inequality. This phenomenon, called *genetic essentialism*, is a bias in humans that leads to believing that social categories are the result of genetics. In other words, people of one race are more genetically like one another than they are to another race.

The truth is, using modern genetic tools like genomics (Chapter 9) that allow biologists to analyze the entire genome with computers, we see that humans are 99.9% identical, which is one of several lines of evidence that there is no genetic basis for race. We'll discuss the evolution of human skin color and the importance of race as a cultural concept more in Chapter 12. Differences, for example, in health between racial groups is not the result of fundamental differences in genetics, but a reflection of systemic inequalities leading to, for example, different access to healthcare.

As our knowledge of genetics has evolved over time—from the one gene:one protein view of genetics to understanding complex traits and the influence of environment on gene expression—we can see how to use our understanding of genetics to push back against prejudiced beliefs based on misunderstandings about the genetics and race.

1. Now that you know that many easily observable traits are polygenic, can you identify times you might have accidentally oversimplified complex traits?
2. Why is it important to understand that race is not a genetic concept?
3. How might oversimplifying genetic traits contribute to stereotypes or biased thinking?

How can humans be 99.9% identical and yet in the chapter opener, we learned that we could share 38–61% of our DNA with our siblings? The variance between our siblings is in that 0.1% of DNA—which is still approximately 3 million base pairs. Within those 3 million bases, you share 38–61% with a sibling, plus the other 99.9%.

Figure 8.18 Comparing pleiotropy and polygenic traits

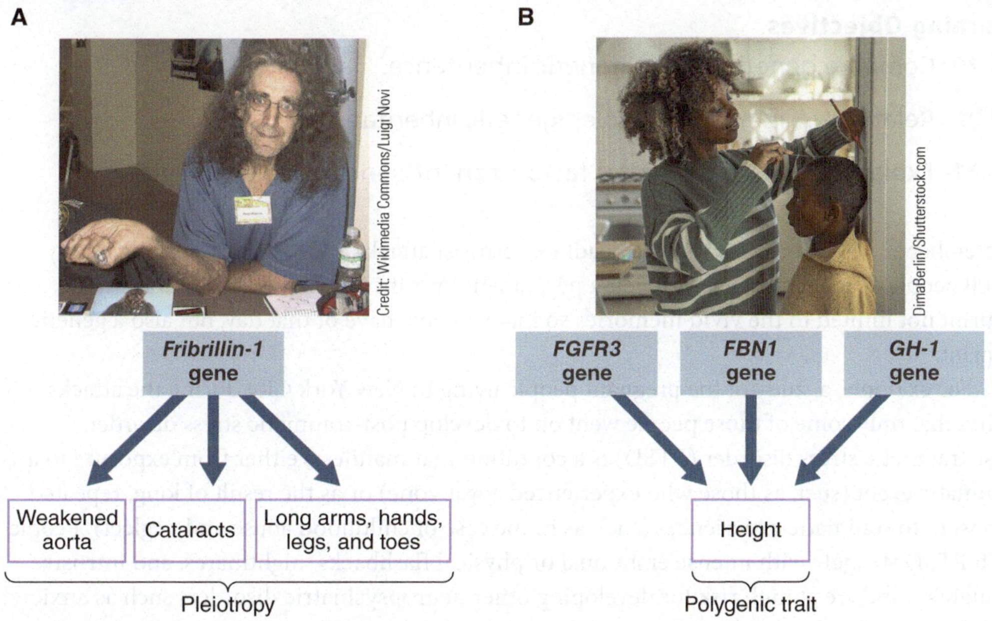

A. Peter Mayhew, who played Chewbacca in the original *Star Wars* movies, had Marfan syndrome. Marfan syndrome is an example of pleiotropy, or when one gene results in many different phenotypes.

B. With polygenic traits, many genes influence one trait. In the case of height, although only three examples are included here, there are many genes and thousands of slight variations of those genes that contribute to height.

Marfan syndrome have weakened aortas because their aorta is not as elastic as it should be, and it eventually stretches and becomes leaky. This can cause the aorta to rupture during exercise, causing immediate death.

Marfan syndrome is an example of a disorder resulting from pleiotropy. In **pleiotropy**, a single gene influences more than one trait, often because expression of that gene in different body tissues leads to more than one observable trait (Figure 8.18A).

> **Big Ideas in Biology**
>
> **Evolution**
>
> Pleiotropy gives biologists insight into how our genomes evolved. It may be that genes evolved to have different functions over evolutionary time—this would explain why mutations in a single gene can have an impact on many different body systems.

Multiple Genes and a Single Trait

The majority of phenotypes that we see are the result of multiple genes interacting together, or **polygenic inheritance**. Hundreds or even thousands of genes may be involved, with each making a small contribution to the phenotype. Many of the easily observable traits we see in ourselves and each other are polygenic. For example, height is a polygenic trait (Figure 8.18B). One study of human height estimates that up to 90% of human height variation is due to genetic factors—with different gene variants numbering in the thousands. Environmental influences, such as diet and socioeconomic status, also influence height. We'll return to the influence of environment on genetics in Section 8.4.

8.4 Summary

- For incompletely dominant traits, the heterozygous phenotype is a blend of the homozygous phenotypes.
- In codominance, both alleles contribute to the heterozygous phenotype.
- In pleiotropy, one gene affects multiple traits.
- Polygenic traits, like height, are the result of the expression of many different genes.

8.5 Epigenetic Inheritance

Learning Objectives

8-19 Compare genetic and epigenetic inheritance.

8-20 Relate DNA methylation to epigenetic inheritance.

8-21 Explain how environmental factors can influence gene expression.

Researchers also looked at the epigenetic inheritance of stress on the descendants of Holocaust survivors. Children of Holocaust survivors are more likely to experience neuropsychiatric disorders, such as PTSD, anxiety, and depression. These changes are linked to different methylation patterns on stress-related genes.

September 11, 2001, or 9/11, was the deadliest terrorist attack in U.S. history, leaving nearly 3,000 people dead. The attacks also left a permanent imprint across the United States—an imprint not limited to the vivid memories so many people have of that day, but also a genetic imprint.

For example, a study of the pregnant people living in New York City during the attacks found that only some of those people went on to develop post-traumatic stress disorder. Post-traumatic stress disorder (PTSD) is a condition that manifests either from exposure to a traumatic event (such as those who experienced a war zone) or as the result of long, repeated exposure to traumatic experiences (such as in the case of childhood abuse and neglect). People with PTSD struggle with intense emotional or physical flashbacks, nightmares, and intrusive thoughts—and are at high risk for developing other neuropsychiatric disorders such as anxiety and depression.

Why did only some pregnant people at or near the World Trade Center during the attack develop PTSD and not others? And why were the authors of the study particularly interested in looking at those who were pregnant during the attacks? It turns out we inherit from our parents not just genes, but also the way those genes are expressed. Recall from Chapter 7 that adding methyl groups to a gene's promoter can shut down transcription of the gene. These types of modifications that affect gene expression without changing the DNA sequence are said to be epigenetic. **Epigenetic inheritance** is when DNA modifications that do not change the DNA sequence itself are passed on to offspring, altering gene expression patterns in the offspring. Epigenetic inheritance can prepare offspring to adapt to an environmental challenge much more quickly than evolution could (Chapter 11), and the changes can be reversed much more quickly after an environmental challenge has faded.

Compared to genetic inheritance, which refers to the passage of DNA to an offspring, epigenetic inheritance refers to inheritance of gene expression. Notably, although we can't change the genes we receive from our parents, gene expression can be changed over the course of one's life.

Now, let's come full circle between epigenetic inheritance and the development of PTSD. People who were pregnant and near the World Trade Center during the attacks, and who went on to develop PTSD after the attacks, had lower cortisol levels. Cortisol is a hormone associated with the stress response, and people with lower cortisol levels are more at risk for developing PTSD. What's interesting is not only did the people who were pregnant and developed PTSD have lower cortisol levels, but so did their children. And the children of people living in the same area who did not go on to develop PTSD? They had normal cortisol levels.

The infants with the lowest cortisol levels were those whose parent was in the third trimester of pregnancy during the attacks. This suggests that a pregnant person's traumatic experience can have an impact on the development of a fetus. Therefore, the parent's experience and response to stress can "prime" the offspring to be more stressed out as well, thus increasing the likelihood of their survival. For instance, if one is more anxious and aware of their surroundings, they may be more likely to protect themselves from threats in their immediate environment, and therefore survive.

? Did You Know

In thale cress (*Arabidopsos thaliana*), a plant used for genetics research, geneticists found that the plant does not have random mutations throughout its genome. Rather, epigenetic changes in certain regions of the genome lead to better DNA repair and fewer mutations, whereas other locations without these epigenetic markers have more mutations.

Genes and the Environment

We will now explore the interactions between our genes, the environment, and gene expression. One such example is **DNA methylation**, the attachment of methyl groups to DNA. Methylation occurs in our cells specifically at promotor regions called CpG islands (Chapter 7), which are rich in cytosines and guanines. Which genes are methylated varies both by cell type and an individual's personal history. Methylation can vary between individuals because it is influenced by environmental factors or by exposure to a traumatic event.

Methylation patterns also change depending on nutrition, starting at conception. For example, children conceived during a period of famine acquire a particular methylation pattern that differs dramatically from that of children conceived during a period of abundance. The resulting differences in gene expression give rise to differences in glucose and fatty acid metabolism in the adults.

The phrase "nature versus nurture" refers to a centuries-old debate about whether human behavioral traits arise from one's genetics (nature) or from environmental factors (nurture). Today, we know that both play a substantial role. The environment affects the expression of many genes, which in turn affects phenotype—including behavioral traits.

Figure 8.19 Different environments result in twins becoming less identical over time

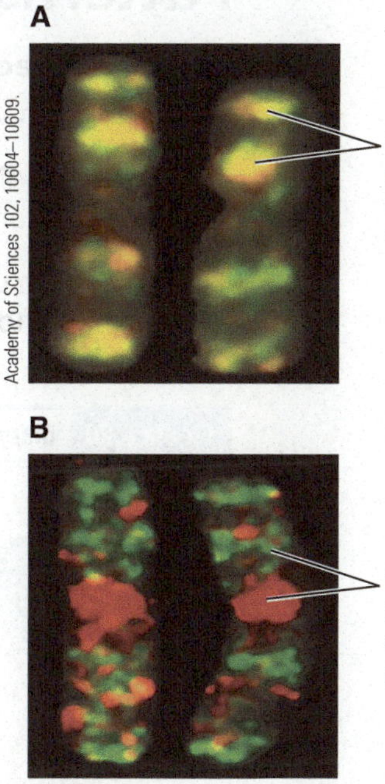

Credit: M. F. Fraga, et al., (2005), Epigenetic differences arise during the lifetime of monozygotic twins, Proceedings of the National Academy of Sciences 102, 10604–10609.

A. 3-year-old twins have more yellow, and therefore more similar gene expression.

B. 50-year-old twins have more red and green, and therefore less similar gene expression.

One striking example of the impact of the environment on gene expression comes from studies comparing the epigenome of identical twins. Recall from the introduction that twins share 100% of their genetic material. Early in life, when twins are exposed to similar environments, their gene expression is nearly identical (Figure 8.19A). However, as twins age and their environments change, their epigenomes change as well (Figure 8.19B).

Unlike our genes, which we cannot change, our epigenome is malleable. The fact that our epigenomes are malleable is both good and bad. If you make healthy decisions, your epigenome will change such that genes associated with metabolic syndrome—a series of conditions associated with increased risk of heart disease, stroke, and type 2 diabetes—are methylated and silenced. This also means that for those of us who have parents or grandparents who experienced trauma, we are not doomed to suffer from anxiety and depression.

However, the reverse is also true. Making less healthy lifestyle choices can lead to changes in gene expression that will promote the development of diseases like cancer. As an example, let's look at alcohol use disorder, a condition where people become emotionally and physically dependent on alcohol and cannot control their drinking. Alcohol use disorder has both genetic and epigenetic roots. Regarding the epigenetic roots, consumption of alcohol results in changes to the epigenome. These changes in gene expression in the brain lead to a growing dependence on alcohol, causing someone to drink more alcohol, which maintains these changes within the brain. This underscores the importance of awareness of disorders as not something we are "fated" to have based on our genetics, but something that can be to some extent controlled by our choices.

? Did You Know?

Seasonal changes in temperature and the length of day affect production of pigments that color fur. Hormonal signals triggered by seasonal changes cause fur to be shed, and new fur grows back with different types and amounts of pigments. The resulting change in phenotype provides these animals with seasonally appropriate camouflage from predators.

Given the role of the epigenome in the development of alcohol use disorder, using medicines to target the epigenome could be a potential treatment.

8.5 Summary

- Genetic inheritance is the inheritance of genes. Epigenetic inheritance refers to the inheritance of how those genes are expressed via heritable DNA modifications.
- One epigenetic modification is DNA methylation, which modifies gene expression without changing the DNA sequence.
- Environmental factors can trigger heritable changes in DNA methylation.

8.6 Genetic Counseling and Inheritance Patterns

Learning Objectives

8-22 Explain why pedigrees are used to study human inheritance patterns.

8-23 Use pedigrees to explain autosomal dominant and autosomal recessive inheritance patterns.

8-24 Explain why X-linked recessive disorders are more common in XY individuals than those who are XX.

Figure 8.20 Mary-Claire King

Credit: Andrew Toth/Getty Images

Mary-Claire King is a geneticist and the first person to show that a person has an elevated risk of breast cancer if they have a mutation in the gene *BRCA1*.

For many years, people have known that breast cancer arises more frequently in some families than in others. In the 1970s, Mary-Claire King proposed that the elevated risk of breast cancer in these families was due to inherited mutations (Figure 8.20). Today, the link between genetics and cancer is well accepted, but King's idea was radical at a time when cancer was assumed to be caused solely by external factors, like viruses.

King tested her hypothesis by developing mathematical models that predicted the theoretical incidence of cancer from genetic and nongenetic causes. By comparing these predictions against a large set of patient data collected by the National Cancer Institute, King's group discovered that a "breast cancer trait" arising from a dominant allele on an autosome could explain familial early-onset breast cancer.

This striking result set the stage for a major shift in the focus of cancer biology research toward genetic causes. In 1990, King's lab succeeded in identifying *BRCA1*, a gene encoding a tumor suppressor (Chapter 6). Mutations in *BRCA1* result in an increased susceptibility to breast cancer. Genetic testing (Chapter 9) allows people to test to see if they have the mutation and opt for increased cancer screenings and/or preventative treatments.

Some expectant parents want to know the risk of their offspring developing cancer or another disorder with a genetic component, such as cystic fibrosis. If parents know of the presence of genetic disorders within their families, they may choose to meet with a genetic counselor. A **genetic counselor** is a medical professional who supports and educates prospective parents about the risks of passing on disease-causing genes to their offspring. They may also give guidance regarding whether genetic testing is advisable. As we learned in the last two sections, since most genes are not inherited in a Mendelian manner and the environment influences phenotype, making predictions about inheritance patterns presents a challenge. Despite these limitations,

One limitation of pedigrees is that they cannot account for random mutations, such as those that may happen during meiosis.

there are some predictable patterns that genetic counselors can use to help patients—which we will turn to next.

Tracking Traits with Pedigrees

Because of the challenges involved in studying human inheritance, genetic counselors often use historical records to track traits through past generations of a family. They make and use **pedigrees**, which are charts that mark the appearance of a trait among generations of family members.

A pedigree is constructed in a standard way, with individuals represented as shapes and relationships between them as lines (Figure 8.21). Males are represented as squares; females, as circles. A filled-in shape indicates an individual with the trait being studied. A union (mating) is signified by a line between individuals, with offspring of the match connected to this line. Every generation appears in a row designated by a Roman numeral (I, II, III, and so on), and each individual in a generation is assigned a number.

Figure 8.21 A sample pedigree

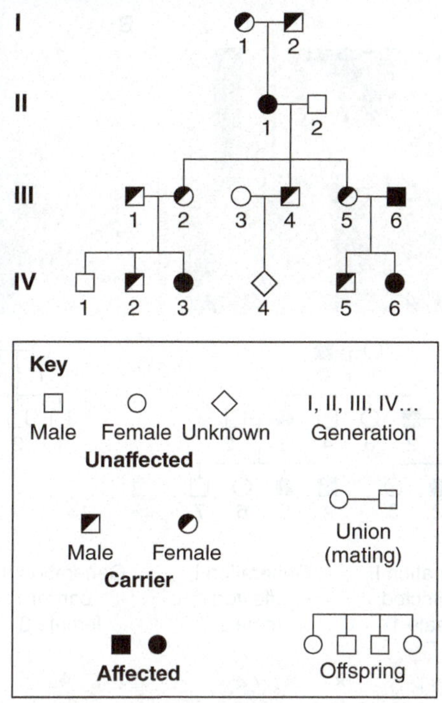

Use the pedigree's key to interpret information about the four generations of individuals shown here.

A pedigree allows genetic counselors to estimate the probability that a phenotype will reappear in future generations. Human genetic disorders inherited in a Mendelian pattern are typically categorized by the chromosome of origin (autosome or sex chromosome) and whether alleles associated with them are dominant or recessive.

Autosomal dominant inheritance pattern. A trait associated with a dominant allele on an autosome appears in heterozygous people as well as those who are homozygous. For example, the pedigree shown in Figure 8.22A represents achondroplasia or dwarfism. Achondroplasia is inherited in an **autosomal dominant pattern**, which means such traits can appear in every generation of a family, and they occur with equal frequency in both sexes. When one parent is heterozygous for a dominant allele, and the other is homozygous for the recessive allele, each of their children has a 50% chance of inheriting the dominant allele and having the associated trait (see the Punnett square in Figure 8.22A).

Autosomal recessive inheritance pattern. A trait associated with a recessive allele on an autosome creates an **autosomal recessive pattern**. In this pattern, the phenotype appears only in homozygous individuals. Heterozygous individuals are called carriers because they have the allele but not the trait. Conditions such as albinism, where people have low levels of skin pigmentation, are inherited in an autosomal recessive manner. Such traits appear in both sexes at equal frequency, and they tend to skip

Figure 8.22 **Some representative human inheritance patterns and associated Punnett squares**

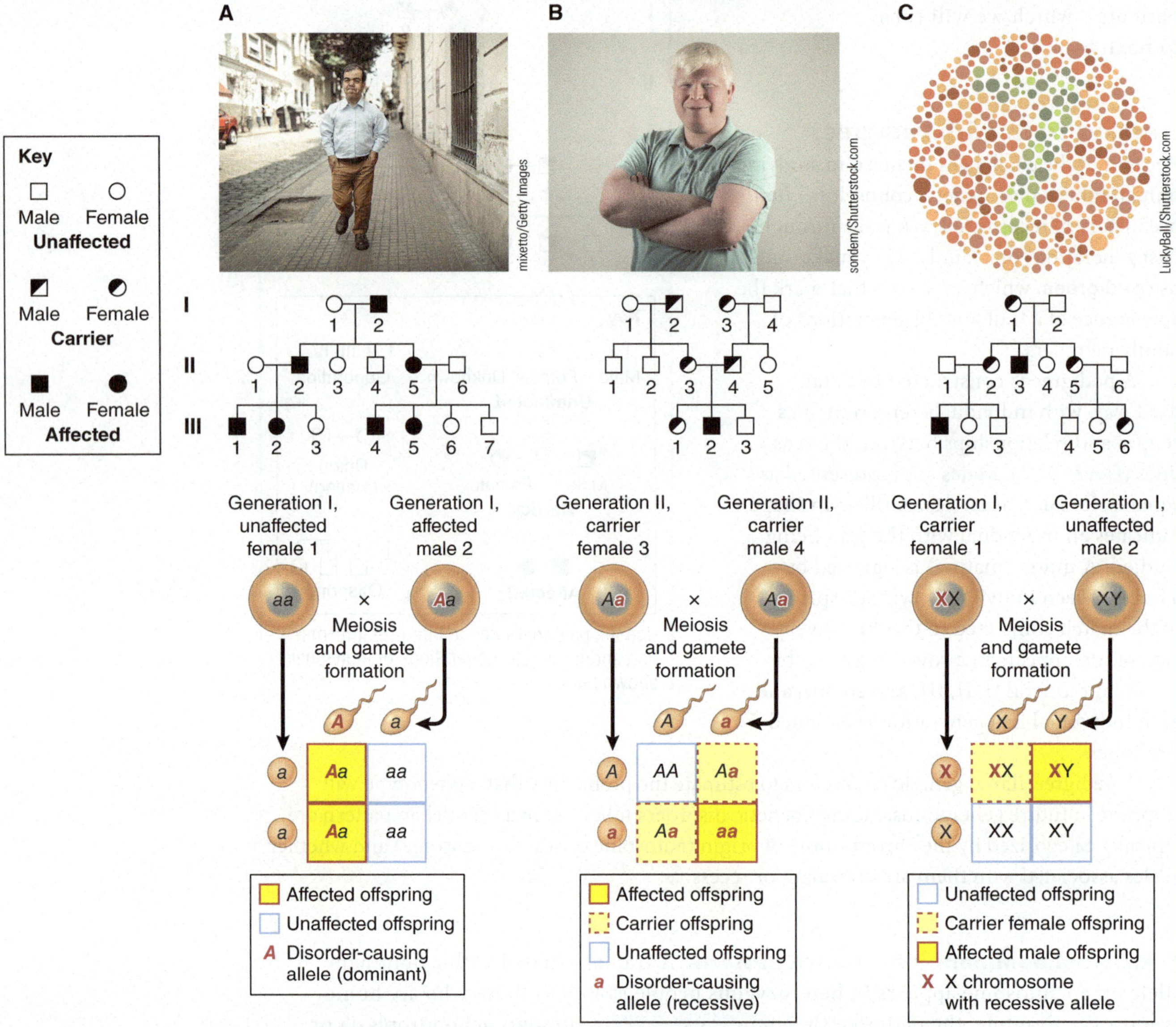

A. Achondroplasia is inherited in an autosomal dominant manner. Note that affected individuals appear in each generation.

B. Albinism is inherited in an autosomal recessive manner. Although there are carriers in each generation, only when two carriers reproduce do we see an affected individual.

C. Can you see a number here? Red-green colorblindness is inherited in a X-linked recessive manner. Note that in this pedigree, only males are affected.

> **? Did You Know?**
>
> Hemophilia, a clotting disorder that is inherited in an X-linked recessive manner, occurred quite frequently among European and Russian royals due to inbreeding among the royal families.

generations. Any child of two carriers has a 25% chance of inheriting the allele from both parents—and developing the trait (Figure 8.22B).

X-linked inheritance pattern. Traits inherited in an **X-linked pattern** (X-linked traits) arise from genes on the X chromosome and are typically seen in people who have a single X chromosome. One example of a trait inherited in an X-linked manner is red-green colorblindness. Red-green colorblindness is the most common form of colorblindness in humans and occurs when people perceive reds and greens differently. Most X chromosome alleles that cause genetic disorders are recessive, and these leave two inheritance clues. First, an affected father never passes one of these alleles to a son, because all children who inherit their father's X chromosome are female (Figure 8.22C). Second, the disorder appears more often in males than females. Having only one X chromosome, a male must inherit only one allele to be affected by the disorder; a female would need to inherit two, and inheriting two disorder-causing alleles is statistically less likely than inheriting one. Note that we're only discussing X-linked recessive inheritance here—there are also X-linked dominant traits as well.

Have you heard of King Henry VIII of England, his six wives, and his quest to have a male heir? The pattern of pregnancy and infant loss, plus his later mental decline, suggests that the King may have had McLeod syndrome, an X-linked recessive genetic disorder.

8.6 Summary

- Inheritance patterns are studied by tracking genetic traits in families. A pedigree is a chart that tracks the inheritance of a trait through a family.
- A trait on an autosome associated with a dominant allele may appear in every generation. Recessive traits on autosomes tend to produce disorders in every other generation.
- X-linked recessive disorders are more common in XY individuals than XX individuals because the former have only a single X chromosome.

Understand It! Metacognition

The genetics chapter in a text often has a different feel than other chapters in an introductory biology course due to the nature of solving genetics problems. For some, examining pedigrees and calculating probabilities feels more reminiscent of a math class.

Whether you are solving problems in a math class or a biology class, incorporating metacognition into your study practice is instrumental to your success. Metacognition is the awareness of your own thinking. There are many ways that metacognition can be useful as part of your study. For example, your instructor may assign a metacognitive activity after your exams to help you think about your approach to the exams and what you could do differently the next time. One such activity is included in Appendix B.

Research studies (including those done by your author!) show that students who use metacognitive processes—such as planning how to solve a problem and checking their work—do better than their peers who do not use metacognitive practices in genetics classes. In other words, the more you can think through your approach to solving a problem, and reflect on your success, the better you are likely to do at exam time.

Here are some ways to practice metacognition as you study the material in this chapter:

1. Make a list of steps you need to follow when solving genetics problems.
2. When solving a problem, can you identify what information you are missing?
3. How will you check your work after solving a problem?

Chapter Review

8.1 Let's Talk about Sex

In **asexual reproduction**, the daughter cells are identical clones of the parent. **Sexual reproduction** produces genetically distinct offspring. Offspring produced by sexual reproduction inherit new combinations of parental alleles, leading to genetic diversity.

During **meiosis**, a **diploid (2*n*)** cell divides twice to create a **haploid (*n*)** cell. The resulting reproductive cells are called **gametes**. The fusion of two haploid gametes at **fertilization** results in a diploid **zygote**, the first cell of a new individual.

Alleles are different forms of the same gene on different chromosomes (homologous pairs). Most chromosomes are **autosomes**. **Sex chromosomes** differ from each other, such as the human X and Y. In humans, when there are two X chromosomes, one is condensed into a **Barr body** by **X-inactivation**. Diseases related to genes on the X chromosome are **X-linked**.

Meiosis differs depending on if the final product is sperm or egg. **Spermatogenesis** produces four genetically distinct sperm cells. In contrast, oogenesis produces a single large egg and two **polar bodies**. Spermatogenesis occurs throughout a person's life. In contrast, **oogenesis** begins *in utero* and pauses at meiosis I for decades. Meiosis in egg cells does not complete until after fertilization.

Meiosis I includes **prophase I, metaphase I, anaphase I,** and **telophase I. Crossing over** during prophase I, when the two pairs of homologous chromosomes (a **tetrad**) exchange pieces of DNA, produces **recombinant chromosomes**. Meiosis I is when the homologous chromosomes split. **Meiosis II** includes **prophase II, metaphase II, anaphase II**, and **telophase II**. Meiosis II is when sister chromatids separate.

DNA replication occurs only once in both mitosis and meiosis. Mitosis results in two genetically identical daughter cells. In meiosis, recombination and crossing over lead to genetically distinct chromosomes.

8.2 When Meiosis Goes Wrong

Gene dosage is having the proper number of copies for each gene. **Aneuploidy** is too many or too few copies of a chromosome. The failure of chromosomes to separate properly during meiosis is called **nondisjunction**. Nondisjunction results in gametes with an abnormal number of chromosomes. **Trisomy** is when a zygote has three copies of one type of chromosome, while **monosomy** is having one copy. A **balanced translocation** can also impact gene dosage. Inappropriate gene dosage can contribute to **miscarriage** (a pregnancy lost prior to 20 weeks of gestation) or **stillbirth** (pregnancy loss after 20 weeks of gestation).

8.3 Mendelian Inheritance

The set of alleles in an individual is the **genotype**. If the same allele of a gene is found on both homologous chromosomes, the individual is said to be **homozygous** for that allele. **Heterozygous** individuals have nonidentical alleles of the same gene on each chromosome. The **phenotype** of an individual describes their observable traits. A **dominant** allele masks the effect of a **recessive** allele when paired with it in a heterozygous individual. Recessive traits are expressed only when the individual is homozygous for the gene.

A **Punnett square** is used to predict the outcome of crosses in the **parental**, or **P, generation**; their offspring are the **F_1 generation**. A **single-trait cross** is used to look at the inheritance of one gene.

Gregor Mendel's **law of independent assortment** states that alleles of one gene are distributed into gametes independently of alleles of other genes during meiosis. This law explains why we can use Punnett squares to make predictions. The **law of segregation** states that alleles on different chromosomes are inherited separately from each other during meiosis.

8.4 Non-Mendelian Inheritance

Incomplete dominance occurs when one allele is not fully dominant, so the heterozygous phenotype is a blend of phenotypes. The phenotypes resulting from homozygous dominant, heterozygous, and homozygous recessive genotypes are distinct from each other. **Codominance** occurs when there are multiple dominant alleles expressed in the heterozygous phenotype.

In **pleiotropy**, a single gene influences more than one trait, often because expression of that gene in different body tissues leads to more than one observable trait. **Polygenic inheritance** occurs when two or more genes affect a single trait. Most phenotypes are polygenic.

8.5 Epigenetic Inheritance

Genetic inheritance is the passage of DNA sequences from one generation to the next. **Epigenetic inheritance** is when DNA modifications that do not change the DNA sequence itself are passed on to offspring, resulting in the inheritance of gene expression patterns. For instance, **DNA methylation** patterns are heritable. Environmental factors such as diet, stress, and exposure to toxins can trigger DNA methylation.

8.6 Genetic Counseling and Inheritance Patterns

Genetic counselors are medical professionals who support and educate people about their own genetic risks of inherited disease and the possibility of genetic diseases in their children. **Pedigrees** are charts of families that track the inheritance of a particular trait over multiple generations.

The **autosomal dominant pattern** of inheritance will show a trait appearing in every generation of a family with equal frequency in both sexes. Alleles with an **autosomal recessive pattern** appear in both sexes with equal frequency and tend to skip generations. Alleles inherited in an **X-linked pattern** appear more frequently in XY individuals than XX individuals because there is only one X chromosome.

Practice It!

Retrieval practice is the process of pulling information from your long-term memory and using it. Here are some questions to help with your retrieval practice.

Self-Quiz

1. A cell with two of each type of chromosome is ______. (LO 8-5)

a. diploid
b. haploid
c. tetraploid
d. abnormal

2. Homologous chromosomes are ______. (LO 8-2)

a. inherited from both parents
b. sister chromatids
c. different in size and length
d. identical in DNA sequence

3. One evolutionary advantage of sexual reproduction over asexual reproduction is that it produces ______. (LO 8-1)

a. more offspring per individual
b. more variation among offspring
c. less harmful mutations
d. less adaptations

4. Alternative forms of the same gene are ______. (LO 8-2)

a. gametes
b. homologous
c. alleles
d. oncogenes

5. Nondisjunction at meiosis can result in ______. (LO 8-8)

a. base-pair substitutions
b. aneuploidy
c. crossing over
d. pleiotropy

6. An individual with one copy of a particular chromosome has ______. (LO 8-8)

a. polyploidy
b. monosomy
c. diplosomy
d. trisomy

7. When a pregnancy is lost before 20 weeks of gestation in humans, this is called a ______. (LO 8-9)

a. base-pair substitution
b. meiotic mistake
c. polygenic inheritance
d. miscarriage

8. The set of alleles carried by an organism make up its ______. (LO 8-10)

a. phenotype
b. variation
c. genotype
d. pedigree

9. The law of segregation states that alleles ______. (LO 8-14)

a. on different chromosomes are inherited separately from one another
b. of one gene on different chromosomes are distributed into gametes independently of alleles of other genes
c. are the same in any given genotype
d. are different in any given genotype

10. The law of independent assortment states that alleles ______. (LO 8-15)

a. on different chromosomes are inherited separately from one another
b. of one gene on different chromosomes are distributed into gametes independently of alleles of other genes
c. are the same in any given genotype
d. are different in any given genotype

11. Which of the following describes a heterozygous genotype? (LO 8-12)

a. *bb*
b. *AaBb* × *AaBb*
c. *Aa*
d. *Aa* × *Aa*

12. The offspring of the cross *Aa* × *aa* are ______. (LO 8-13)

a. all *AA*
b. all *aa*
c. all *Aa*
d. half *Aa* and half *aa*

13. If one parent is heterozygous for a dominant allele on an autosome and the other parent does not carry the allele, any child of theirs has a ______ chance of being heterozygous. (LO 8-13)

a. 25%
b. 50%
c. 75%
d. 100%

14. A plant that has red flowers is crossed with one that has white flowers. The resulting offspring all have pink flowers. This is an example of ______. (LO 8-16)

a. polygenic inheritance
b. codominance
c. pleiotropy
d. incomplete dominance

15. A person with type A blood and a person with type O blood have a child. If the parents are both homozygous for their blood types, which blood type do their children have? (LO 8-16)

a. A
b. B
c. O
d. AB

16. A trait influenced by several genes is an example of ______. (LO 8-18)

a. polygenic inheritance
b. codominance
c. pleiotropy
d. incomplete dominance

17. Traits like sickle cell anemia and phenylketonuria cause symptoms in unrelated organ systems, where people can have different symptoms. These traits are examples of ______. (LO 8-17)

a. codominance
b. incomplete dominance
c. pleiotropy
d. polygenic inheritance

18. When an environmental challenge genetically affects a developing fetus without changes in their DNA sequences, this is an example of ______. (LO 8-19)

a. nondisjunction
b. mutation
c. epigenetic inheritance
d. dominant/recessive inheritance

19. What sex chromosome does a genetically male child inherit from each of his parents? (LO 8-3)

a. X from female parent and X from male parent
b. Y from female parent and Y from male parent
c. X from female parent and Y from male parent
d. Y from female parent and X from male parent

20. When a pedigree analysis shows inheritance from parents to offspring that skips generations in a family, the gene in question is likely to be ______. (LO 8-23)

a. dominant
b. recessive
c. codominant
d. incompletely dominant

Critical Thinking

1. An error during meiosis gives rise to Down syndrome, a genetic disorder in which a person inherits three copies of chromosome 21 rather than two copies. What errors of meiosis could result in offspring inheriting the incorrect number of chromosomes, and in what phases of meiosis would they occur? (LO 8-8)

2. Marfan syndrome is inherited in an autosomal dominant pattern. What is the percent chance that a child will have the syndrome if one parent does not carry the associated allele and the other is heterozygous for it? (LO 8-13)

3. Which pattern of inheritance best describes each of the following scenarios? (LO 8-13, LO 8-16)

a. In mice, alleles of the *agouti* gene determine the distribution of pigment in the individual's fur. The dominant allele *A* of this gene gives rise to banded fur; the recessive allele *a* gives rise to solid fur. Heterozygous animals have banded fur.
b. Cats with a dominant allele of the *KIT* gene have white fur and blue eyes, and they are deaf. Homozygous recessive cats have brown fur, green eyes, and can hear.

c. In Andalusian chickens, a mating between an individual with black feathers and an individual with white feathers produces offspring with gray feathers.

d. In cattle, a mating between an individual with red hair and an individual with white hair produces offspring with roan coloration (an even distribution of red hair and white hair).

4. A famine in the Netherlands during the winter of 1944–45, referred to as the Dutch Hunger Winter, led to significant nutritional deprivation. Later studies of people who were *in utero* during that time showed that they had increased risk of metabolic and cardiovascular disease. What type of inheritance could cause these increased risks for disease? What process would you expect to be involved? (LO 8-20, LO 8-21)

5. Unicorn is a rare calico cat who came into a Colorado animal shelter with a secret—he's male. Only 1 in 3,000 calico cats are male. Based on what you learned in this unit, explain why it's unusual for a calico cat to be male. What does this suggest about Unicorn's sex chromosomes, and how do they differ from those of typical male cats? (LO 8-3)

Credit: Emily Gillis

9

Society and Genetic Technologies

Chapter Outline:

The Little Embryo That Could

As young adults, we're often bombarded with cautionary tales of unplanned pregnancies. The subtle message is that getting pregnant is easy and takes one sexual encounter. For some people, especially people in their teens and early 20s, this is true: Getting pregnant is often easy. However, the startling reality is that many people struggle to get pregnant (primary infertility), or they can conceive but not stay pregnant (secondary infertility). For example, recurrent miscarriages are a form of secondary infertility (Chapter 8).

According to the World Health Organization, 1 in 6 people are impacted by infertility. Infertility isn't a condition limited to impacting one's reproductive health, but also impacts the mental health of millions of couples grieving their inability to have biological children. Although generations of people have struggled with infertility, in the 1970s, everything changed. On July 25, 1978, the very first person conceived from *in vitro* fertilization (IVF), or fertilization outside of the body, was born. IVF is a form of assisted reproductive technology that opened the door for infertile and same-sex couples to have biological children. But how do assisted reproductive technologies relate to genetics? What other genetic technologies are influencing society today?

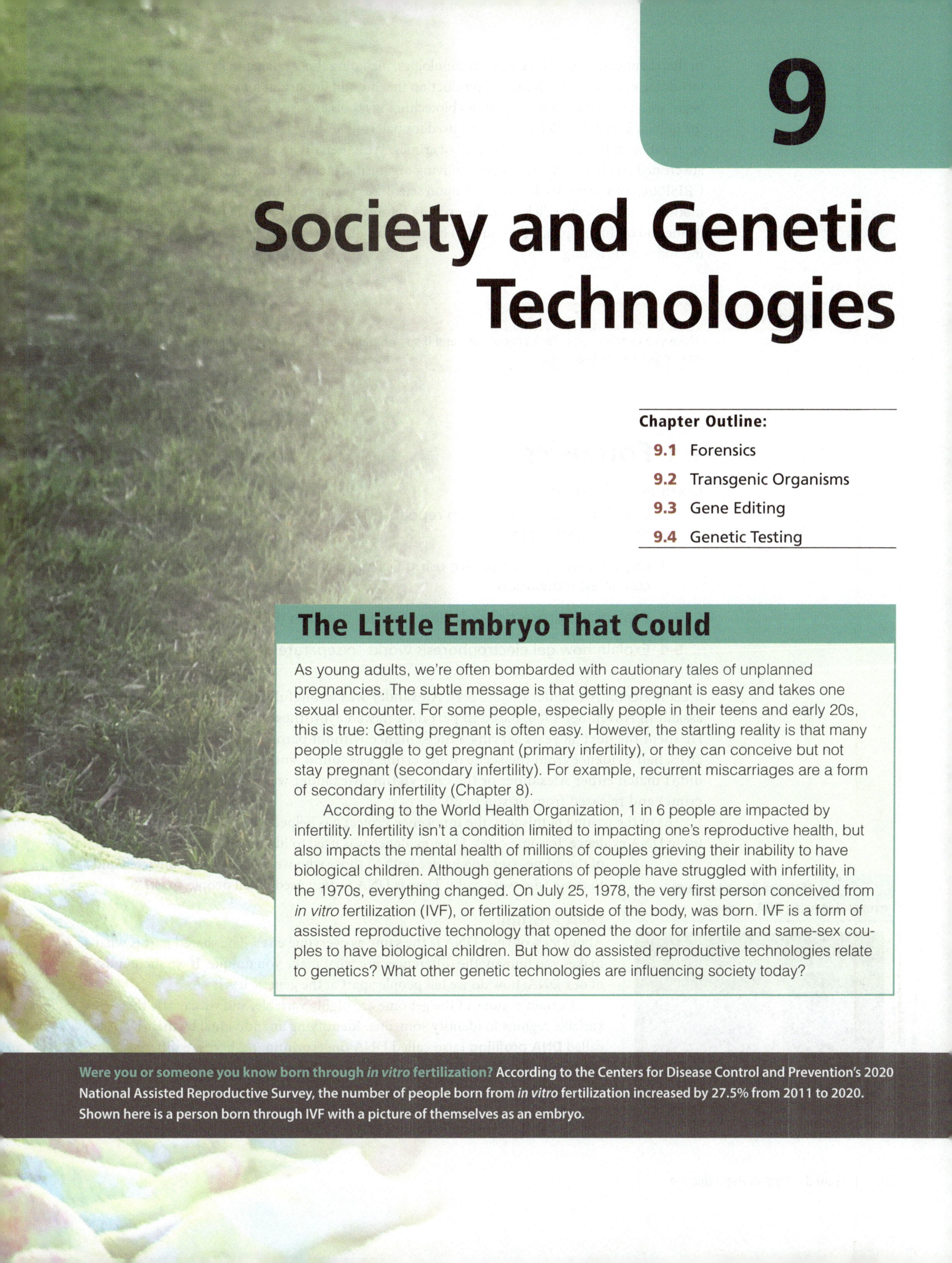

Were you or someone you know born through *in vitro* fertilization? According to the Centers for Disease Control and Prevention's 2020 National Assisted Reproductive Survey, the number of people born from *in vitro* fertilization increased by 27.5% from 2011 to 2020. Shown here is a person born through IVF with a picture of themselves as an embryo.

In this chapter, we'll explore genetic technologies, including those related to IVF, as well as technologies unrelated to human reproduction that are still important for you to know about. We'll begin with forensics and how common biotechnology tools are used to detect small variations in the human genome, which can then be used for paternity testing, forensic investigation, and the exoneration of those wrongly convicted of crimes. We'll then discuss how transgenic organisms are created and how DNA is changed in living organisms, including the editing of genes with the CRISPR/Cas9 system. We'll also touch upon scientific (and nonscientific) issues around genetically modified foods. Next, we'll look at the exciting future of gene editing and personalized medicine. At the end of this chapter, we'll return to human reproduction with a discussion of gene sequencing and genetic testing.

Preview It!

Have you ever had your DNA sequenced, and if so, what did you learn? What other genetic technologies would you like to learn about?

9.1 Forensics

Learning Objectives

9-1 Describe how short tandem repeats (STRs) are used to distinguish between different individuals.

9-2 Explain how geneticists are using DNA profiling to exonerate wrongfully convicted individuals.

9-3 Explain the process of polymerase chain reaction (PCR) and its uses.

9-4 Explain how gel electrophoresis works to separate DNA samples.

Meet Malcolm Alexander (**Figure 9.1**). Alexander was wrongfully convicted of sexual assault in November 1980, based in part on a flawed eyewitness identification process. He spent the next 38 years serving a life sentence without parole. What happened then? In 2013, hair evidence from the scene of the crime was found, and the DNA profile of the hair didn't match either Alexander or the rape victim. Alexander was exonerated (cleared of the crime) and released from prison in 2018.

As of the writing of this text, the Innocence Project has helped exonerate over 200 people using DNA evidence. In some cases, DNA evidence was used to identify the actual perpetrator of the crime—serving justice both to those who were wrongfully convicted and to those who were guilty. How is DNA testing used to tell people apart?

Figure 9.1 Malcolm Alexander was exonerated using DNA evidence

Credit: © Malcolm Alexander. Taken from malcolm-alexander/innocenceproject

DNA Profiling

About 99% of your DNA is the same as everyone else's. The shared part is what makes you human; the differences are what make you unique. If we share most of our DNA, how do we tell people apart at the genetic level?

Certain regions of the genome are highly variable, and we can use these variable regions to identify someone. Identifying an individual by their DNA is called **DNA profiling** (also called DNA fingerprinting). There are different ways of creating a person's DNA profile. One way is by using genotyping to examine single-nucleotide polymorphisms (Section 9.4). The other way, which is also more precise, is by looking at the number of short tandem repeats in a person's genome. **Short tandem repeats (STRs)**, or microsatellites, are sequences of

Figure 9.2 **Examining short tandem repeats during paternity testing**

A

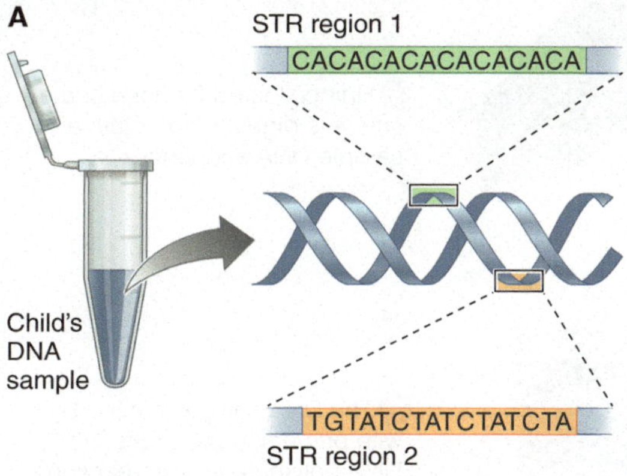

B

STR region #	# Repeats in child's sample	# Repeats possible father #1	# Repeats possible father #2	# Repeats possible father #3
1	8	5	8	6
2	4	18	4	4
3	11	8	11	16
4	15	13	15	18
5	10	12	10	17

A. STRs are found in DNA, such as the DNA from a child whose father is unknown. Here, in STR region 1, the sequence CA is repeated eight times and is highlighted in green; in STR region 2, the sequence TCTA is repeated four times and is highlighted in orange.

B. Investigators can determine the number of repeats for any given STR. In this case, by comparing the number of repeats between the child's sample with those of the three possible fathers (see the table), only one of the individuals (#2) matches the DNA of the child. Note that for simplicity, results for only a single chromosome are shown here.

DNA where a series of 2 to 6 nucleotides is repeated many times in a row. The number of times a sequence is repeated differs among individuals. For example, in an STR that consists of the repeated nucleotide sequence TCTATA, one person may have 18 repeats of this sequence, while another may have 4.

DNA profiles based on STR analysis are routinely used for paternity testing (Figure 9.2) and as evidence in criminal cases. Unless two people are identical twins, the chance that they would have an identical number of repeats at each of the 21 different STR locations typically assessed during a paternity test is extremely unlikely. Thus, an individual's array of short tandem repeats is, for all practical purposes, unique.

Let's circle back to Malcolm Alexander's case. How was STR analysis used in his exoneration, and what other tools were involved? To generate someone's profile, a biologist will use polymerase chain reaction to amplify 13 to 20 variable regions. **Polymerase chain reaction (PCR)** is a laboratory technique used to make millions of copies of a specific gene sequence—in this case, to make many copies of the STR region for analysis.

Essentially, PCR is DNA replication (Chapter 6) in a test tube. PCR can quickly transform a needle in a haystack—that one-in-a-million fragment of DNA—into a huge stack of needles with a little hay. The starting material for PCR is any sample that contains at least one molecule of DNA. It could be extracted from sperm, a hair left at a crime scene, or even an ancient mummy! In Alexander's case, hair from the crime scene was used in his exoneration (Figure 9.3).

PCR occurs in a machine called a **thermocycler**. The thermocycler goes through different temperatures in a set pattern, and each of these patterns is called a cycle. Performing PCR requires two synthetic **primers** designed to base-pair at opposite ends of a targeted section of DNA. During STR analysis, multiple PCR reactions will occur concurrently. For example, each person's DNA sample will be tested for each STR region using a different set of primers.

First, forensics experts collect genetic material from a crime scene, such as a hair sample (1). Researchers mix STR-specific primers with the DNA to be amplified, nucleotides, and heat-stable DNA polymerase (2). Then, the mixture is loaded into the thermocycler. The initial high temperature breaks the hydrogen bonds that hold the two strands of a DNA double helix together (Chapter 3). Consequently, every molecule of DNA unwinds into two separate strands (3). As the temperature of the reaction mixture is lowered, the primers bind to the DNA target sequence (4). Unlike DNA replication, these primers are specific to a gene of interest and

? Did You Know?

By 2024, the database of DNA fingerprints maintained by the Federal Bureau of Investigation (FBI) contained the STR profiles of more than 17 million convicted offenders.

The DNA polymerases of most organisms denature at the high temperature required to separate DNA strands. PCR uses *Taq* polymerase, an enzyme isolated from *Thermus aquaticus* bacteria, which live in Yellowstone National Park's hot springs. Since the polymerase is from bacteria that evolved to live in hot springs, it can withstand the high temperatures of PCR.

Figure 9.3 Overview of polymerase chain reaction (PCR)

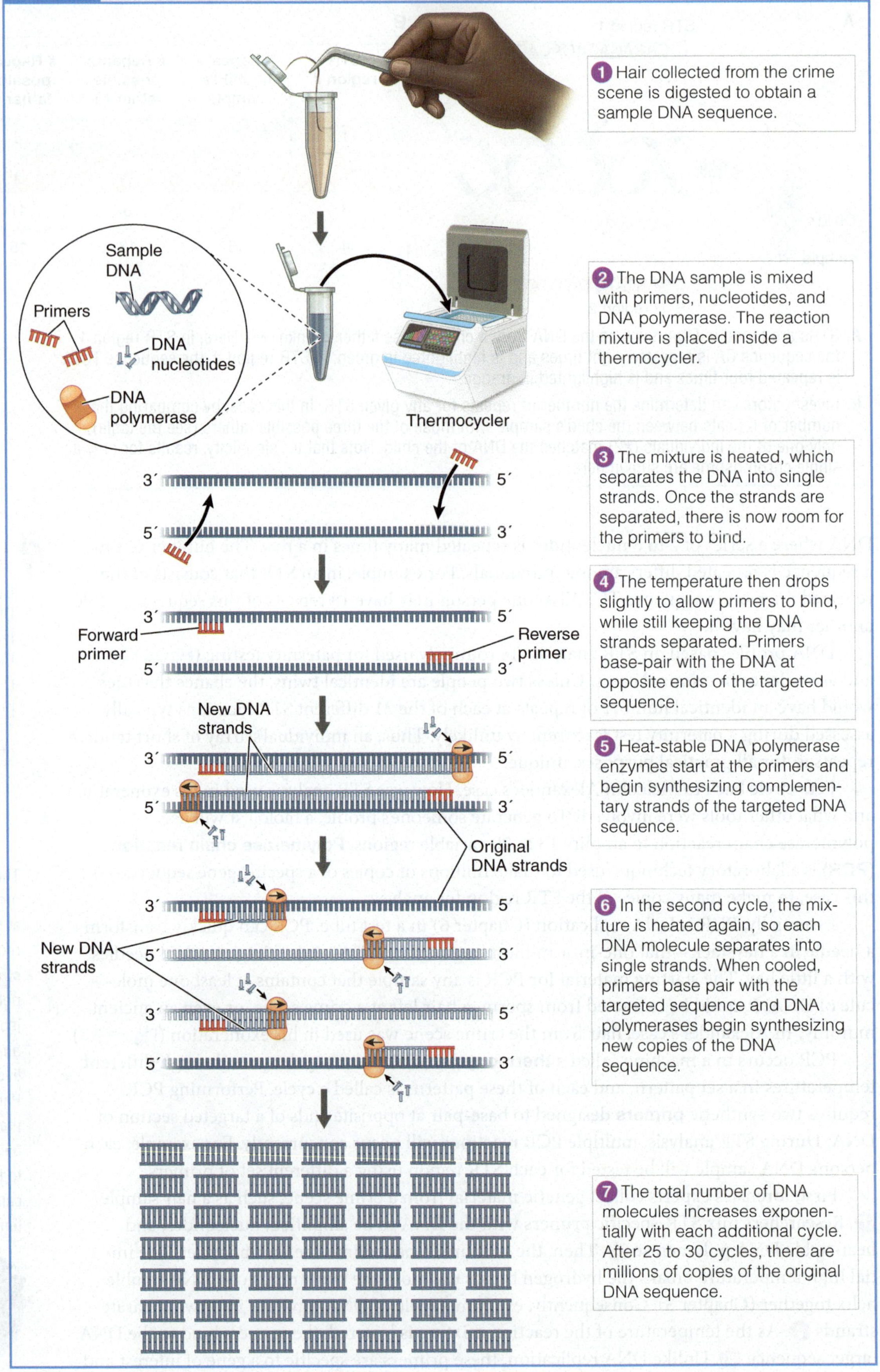

mark both sides of a gene—one a forward and one a reverse primer. Once the primers attach, DNA polymerase makes a copy of the DNA in the region indicated by the primers. Recall from Chapter 6 that the DNA polymerase reads the original DNA strands in the 3′→ 5′ direction, while making new DNA in the 5′→ 3′ direction (5). The reaction proceeds until the second round of PCR, when the temperature rises again, and the DNA (both the original and new strands) separates into single strands. When the mixture is cooled, primers hybridize (stick to the DNA), and DNA synthesis begins again (6). Each cycle of heating and cooling takes only a few minutes but doubles the number of copies of the targeted section of DNA. Eventually, after 20 to 30 cycles, there are millions of copies of DNA available (7), enough for further analysis.

? Did You Know?

Kary Mullis invented PCR in 1985. He went on to win the Nobel Prize in chemistry for his invention in 1993.

Since STRs come in different lengths depending on the number of repeats, PCR will produce DNA fragments of different lengths. Biologists can determine the length of the different DNA pieces using gel electrophoresis. **Gel electrophoresis** is a process for separating different-sized nucleic acids using an electric field to pull the fragments through a semisolid agarose gel (Figure 9.4).

Let's continue with use the DNA amplified with PCR in Figure 9.3 as the example. First, a researcher loads a solution containing DNA fragments whose sizes are known, called a ladder, into a small indentation in the gel called a well (1). The PCR mix containing the amplified DNA of the hair from the crime scene, as well samples from three suspects, is then loaded into the remaining wells. Once loaded, the researcher runs an electrical current through the gel, where—since DNA is negatively charged due to the phosphate backbone (Chapter 3)—it will move toward the positive end, or "bottom," of the gel (2).

Note that for simplicity's sake, we're showing only the workflow for a single hypothetical STR. For convictions or exonerations, multiple STR locations would be analyzed, making the gel much larger.

As the DNA fragments move through the gel, they migrate at different rates based on their size. The shorter a fragment is, the faster it will travel to the bottom of the gel. This is because shorter fragments slip through the tangled weblike molecules that make up the gel faster than

Figure 9.4 Gel electrophoresis is a procedure for separating amplified DNA by size

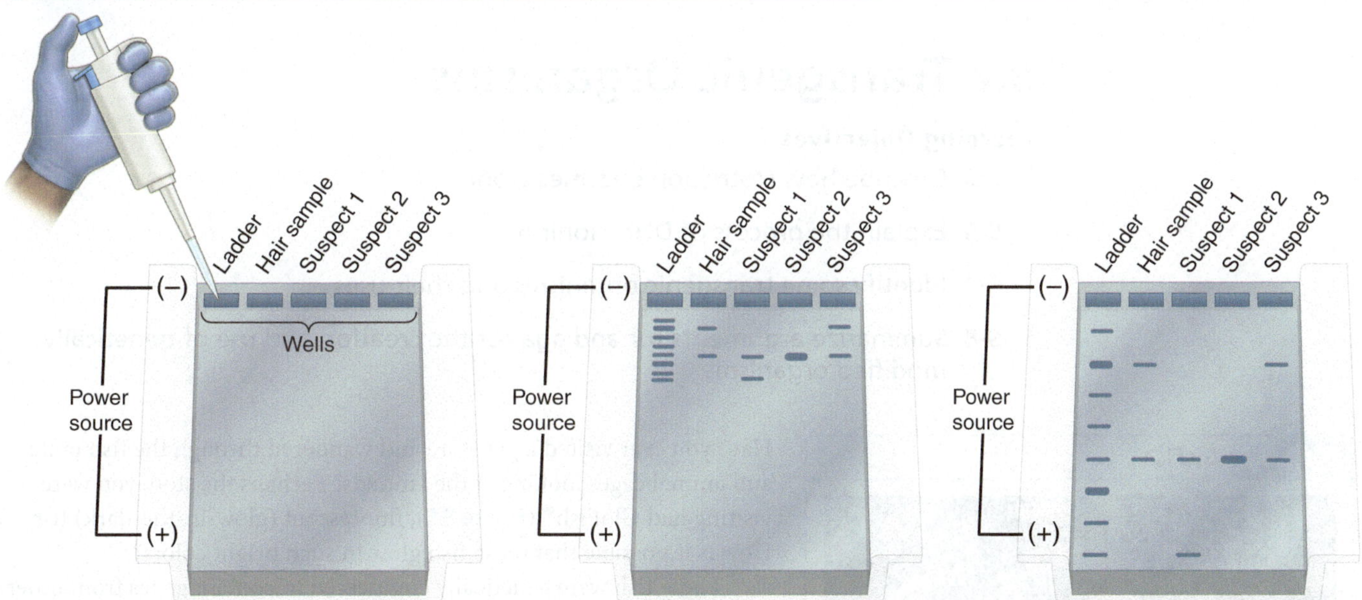

(1) A researcher loads a ladder—a solution with known DNA fragment sizes—as well as DNA samples from a crime scene and three suspects into wells of an agarose gel.

(2) An electrical current is then applied, where, due to the negative charge on the DNA backbone, the DNA fragments move down the gel toward the positive end. Smaller fragments move faster through the gel than larger fragments. Note that some bands are thicker than others—this is because there is a greater amount of that sized DNA present.

(3) After the bands are well separated, the researcher can then compare the positions of the DNA bands from the crime scene hair sample to those of the suspects. In this case, Suspect 3 is a match.

For simplicity, only the results from one STR region are shown here. Note that individuals have either two bands (are heterozygous for the STR) or a single large band (are homozygous for the STR).

the longer fragments do. Think of it like a small child moving through a playground—they can sneak around playground equipment much faster than their adult supervisor can. Therefore, once the procedure is completed, DNA fragments of STRs with fewer repeats will appear farther down in the gel than DNA fragments of STRs with more repeats 3.

The gel also contains chemicals that stick to the DNA and fluoresce in the presence of UV light. After the DNA separates, an investigator can expose the gel to UV light to visualize the bands.

! Point of Confusion: When looking at a gel, one can infer the length of DNA as well as its relative amount. Bands towards the bottom of the gel are smaller (shorter number of repeats). If two bands are at the same place on a gel, but one is thicker than the other, this means that there is more DNA in this band compared to the thinner band.

9.1 Summary

- DNA profiling identifies individuals by comparing specific highly variable regions in the human genome, such as short tandem repeats (STRs).
- STR analysis of DNA evidence from a crime scene can be used to convict or exonerate individuals by comparing DNA profiles.
- Polymerase chain reaction (PCR) replicates DNA by combining DNA polymerase, nucleotides, and primers with a template DNA sample in a thermocycler. PCR is used to quickly make enough copies of a specific DNA sequence for analysis.
- Gel electrophoresis separates the DNA fragments by size by running pieces of DNA through a gel-like material using an electrical current.

9.2 Transgenic Organisms

Learning Objectives

9-5 Describe how restriction enzymes work.

9-6 Explain the process of DNA cloning.

9-7 Identify some transgenic organisms and their uses.

9-8 Summarize arguments for and against the creation and use of genetically modified organisms.

Figure 9.5 GloFish

GloFish are genetically engineered fish that contain genes from other bioluminescent animals, which explains their ability to also glow in the dark.

Credit: Varvara Voinarovska/Shutterstock.com

Have you ever visited a pet store and wandered through the fish tanks and animal cages looking at the animals? Perhaps the store you were visiting had GloFish® (Figure 9.5), fluorescent (glow-in-the-dark) fish. How is it possible that these fish glow in such bright colors?

These fish were genetically engineered to contain genes from other animals that *do* glow in the dark, such as corals and other marine life. Since different animals produce different colors, the geneticists behind GloFish could create a rainbow of different colored fish. Any organism whose DNA has been altered to include DNA from a different organism—be it bacteria, plant, or animal—is called **transgenic**. In the case of GloFish, these fish contain a gene for fluorescence isolated from a jellyfish or coral, depending on the specific color.

Beyond their novelty uses, many biologists use transgenic organisms in their research. For example, using a mouse that expresses a human protein associated with a particular disease allows biologists to study how the mutated form of the protein acts within the context of the body. This allows researchers to test potential treatments in mice before using them in humans—an important first step in bringing new medicines to market.

Scientists currently use transgenic bacteria to make human insulin for people who are diabetic, where bacteria are given the human gene for insulin and then multiply, making thousands of copies of themselves while also churning out human insulin. These transgenic bacteria are made using recombinant DNA that consists of bacterial genetic material and the human gene that codes for insulin. When genes are thusly combined from two or more sources, the resulting DNA molecule is called **recombinant DNA**.

In this section, we will learn how recombinant DNA is generated using a process called *DNA cloning*. Once produced, recombinant DNA is useful for many genetic engineering applications, including the production of transgenic organisms.

> **! Point of Confusion:** Although both are colloquially called cloning, DNA cloning is not the same as animal cloning (somatic cell nuclear transfer), which we'll discuss in Chapter 11.

We'll also explore some of the promise of and controversy around genetically modified organisms. **Genetically modified organism (GMO)** is a catch-all term that includes any organism that was genetically engineered, whether this occurred through gene editing (Section 9.3) or the use of recombinant DNA. Unlike *artificial selection*, which is a form of selective breeding (Chapter 11), where an individual will breed an animal or a crop for desired traits, genetic engineering allows scientists to quickly and precisely select for certain traits and/or introduce genes from another organism for a desired trait.

There is no evidence that consumption of genetically modified foods is bad for one's health. In fact, 80% of processed foods contain genetically modified ingredients. As we'll discuss in Chapter 18, genetically modified foods could be a critical tool for protecting our food supply from the threats of climate change. That said, we'll also discuss legitimate concerns surrounding GMOs, such as passage of recombinant DNA into the wild or herbicide-resistant genetically modified plants. Note that the goal with this section is to present the science of GMOs as well as considerations for their use, so that you feel empowered to make informed decisions regarding which foods you purchase at the store and the policies you support.

> **! Point of Confusion:** Although people tend to group organisms that are transgenic or have edited genes together under the umbrella term "GMO," these two technologies are different. Gene editing may or may not introduce foreign DNA from another organism into the modified organism's DNA. If gene editing focuses solely on manipulating the genes that an animal already has, or is using genetic material from the same species or a closely related species, it is not considered transgenic even if it is genetically modified. Transgenic animals, like GloFish, *have* genetic material from another organism inserted into their DNA, and because they cannot be created through traditional selective breeding methods, they are often the subject of further scrutiny by regulatory agencies and the public.

Creating a Transgenic Organism Starts with Recombinant DNA

As described earlier in this chapter, transgenic organisms contain DNA from different sources (such as humans and bacteria), creating either an animal model of a human disease or an organism that is used to produce a valuable biological material, such as insulin. The first step in making a transgenic organism is to create recombinant DNA using DNA cloning. **DNA cloning** is a technique used for making lots of copies of engineered (recombinant) DNA using a cell (usually a bacterial cell).

The first step in making recombinant DNA is to insert a gene of interest into a cloning vector. A **cloning vector** is a piece of DNA, typically a circular piece of DNA from a bacterium called a **plasmid** (Figure 9.6). The plasmid contains special sequences of DNA. One of these sequences is called the **multiple cloning site**. The multiple cloning site contains restriction enzyme sites, which are specific sequences of DNA that are recognized by restriction enzymes. A **restriction enzyme** is a type of enzyme that cuts DNA to create a double-stranded break

Figure 9.6 A cloning vector

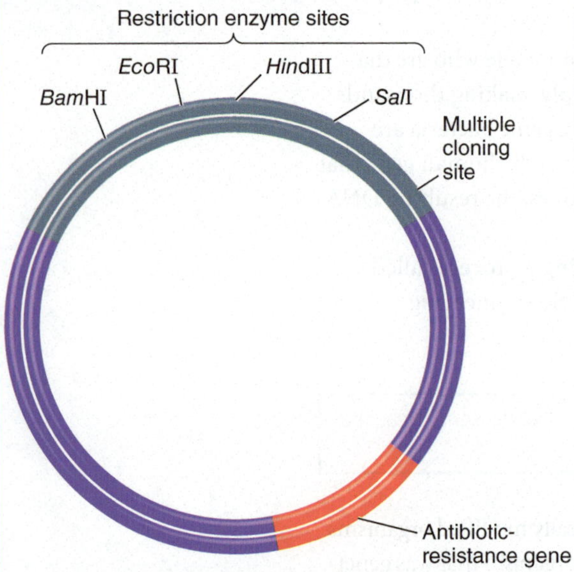

Cloning vectors, or plasmids, are circular bacterial chromosomes engineered to contain specialized sequences, like a multiple cloning site and an antibiotic-resistance gene, which are used to create recombinant DNA. Some common restriction enzyme sites are noted at the top.

only at a specific nucleotide sequence. Some common restriction enzymes, such as *Bam*HI, are noted in Figure 9.6. The foreign DNA will be inserted into the plasmid using at least one of these sites. Another important part of the plasmid is a gene for antibiotic resistance. This gene will be used for identifying bacteria that carry recombinant DNA, as you'll see in Figure 9.8.

To insert a gene of interest into a plasmid first requires both pieces of DNA to be cut so that they fit together like two pieces of a jigsaw puzzle. Restriction enzymes will cut the plasmid and the sample DNA to create complementary ends that will base-pair together, called **sticky ends** (Figure 9.7, 1). After cutting with the restriction enzyme, the DNA fragment and plasmid are mixed together 2. The sticky ends base-pair with one another, and DNA ligase seals the pieces together, forming the recombinant plasmid vector 3.

After a recombinant plasmid vector is produced, a researcher can begin the process of inserting the plasmid in a host cell, such as a bacterium (Figure 9.8). First, the recombinant plasmid vector is added to a vial containing special bacteria called competent cells 1. **Competent cells** are often *Escherichia coli* (abbreviated as *E. coli*) that are chemically treated to be able to take up new DNA. Exposure to heat causes some of the cells to take up the plasmid 2. The cells are then grown in the presence of an antibiotic 3, where only the cells that contain the plasmid (which has the antibiotic-resistance marker) will grow. From

Figure 9.7 Generation of recombinant DNA

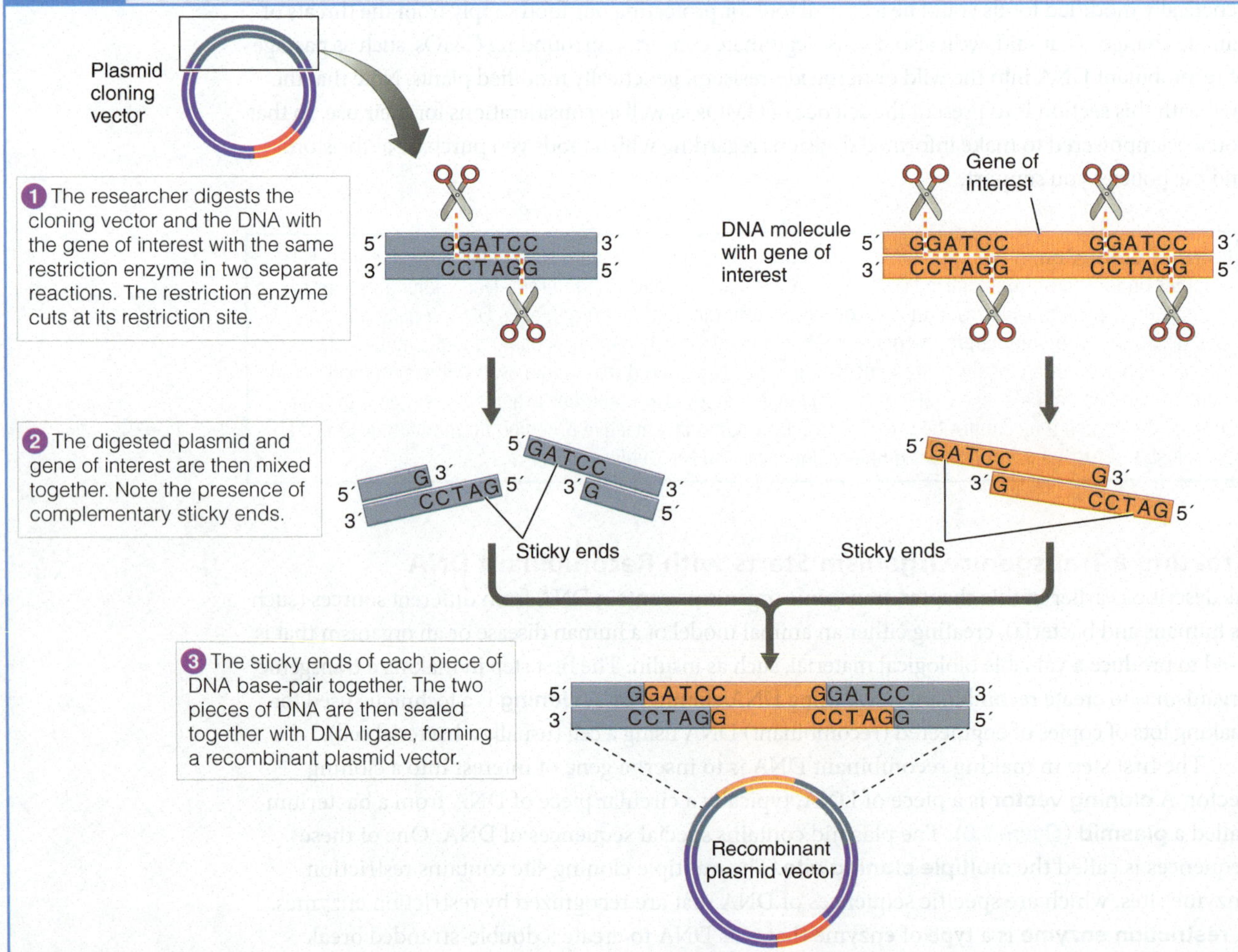

Figure 9.8 **Overview of inserting recombinant DNA into a host cell**

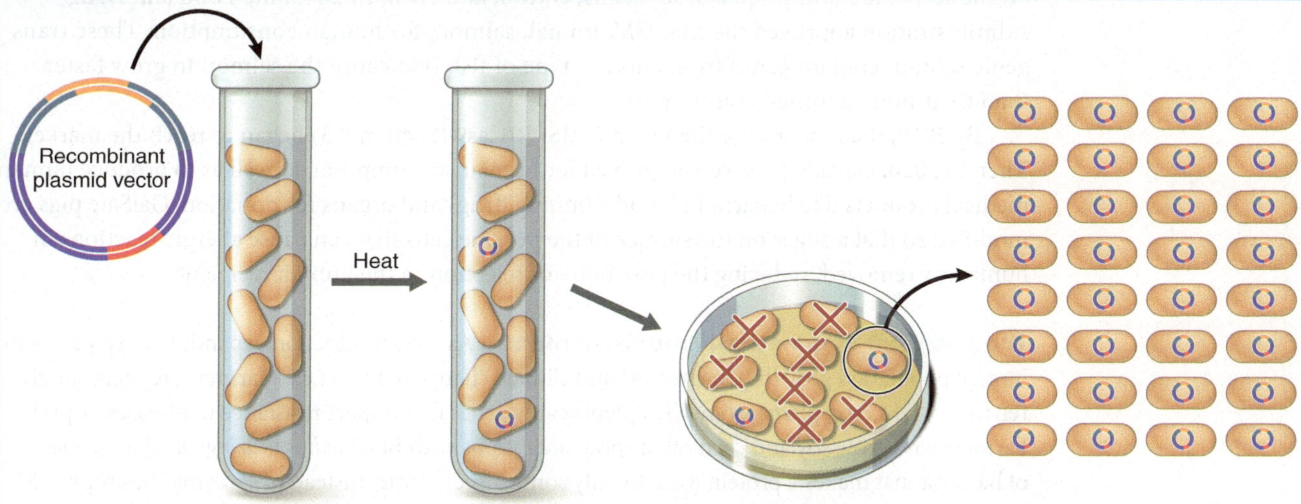

1 The recombinant plasmid vector is added to a vial containing competent bacterial cells.

2 The bacterial cells are exposed to high temperatures for a short period of time. This heat shock causes some of the cells to acquire the plasmid.

3 Bacterial cells are grown on an agar plate containing an antibiotic. The antibiotic corresponds to the antibiotic-resistance gene in the plasmid. Only the cells with the engineered plasmid will grow on the antibiotic-containing plate.

4 The researcher can select a single colony of antibiotic-resistance cells and grow more bacteria, all of which contain the engineered plasmid.

here, the investigator can choose one of the colonies of antibiotic-resistant bacteria to continue to grow and consequently make more copies of the recombinant plasmid 4.

Once recombinant DNA is generated, the plasmids can be used to store pieces of DNA long term, can be placed into a cell to make copies, or can be used to make a transgenic animal. For example, zebrafish are common laboratory model organisms (Chapter 13). A biologist can insert a plasmid-containing recombinant DNA in 1- to 4-cell zebrafish embryos and produce a transgenic animal.

The Potential and Perils of Genetically Modified Foods

DNA cloning and the generation of transgenic bacteria is a ubiquitous molecular biology technique. Creating transgenic organisms is important for a variety of applications, from making biofuels to combat climate change to creating transgenic organisms to study human diseases—as well as to produce useful compounds, such as insulin for people with diabetes. Genetically modified foods are common in our food supply—and important for creating a resilient food supply. Despite all of these benefits, many people are reluctant to eat genetically engineered foods due to concerns about potential health and environmental risks.

A brief history of genetically modified (GM) foods. Humans have selectively bred crops and animals to produce desirable traits for thousands of years (we will return to this process in Chapter 11). For example, wild strawberries are much smaller and tarter than their selectively bred cousins that you find in the grocery store (Figure 9.9). Although traditional breeding can be effective, it also can take a long time, and it may be difficult to achieve the desired results.

The process of cloning DNA was developed in 1973, and the first policy governing GM food was developed in 1986. The first GM produce hit the market in 1994—a tomato with increased

Figure 9.9 **Comparison of cultivated and wild strawberries**

Here, a cultivated strawberry of a variety typically found at the supermarket (left) is lying next to a vine of wild strawberries. Note the difference in size between the two.

Credit: Aleksandar Milutinovic/Alamy Stock Photo

shelf-life and resistance to fungal infections. Throughout the 1990s, more GM crops came on the scene, including squash, soybeans, cotton, and corn. In 2015, the Food and Drug Administration approved the first GM animal, salmon, for human consumption. These transgenic salmon contain genes from another type of fish that cause the salmon to grow faster than their non-modified counterparts.

By 2019, the first crops edited with CRISPR/Cas9 (Section 9.3) began to reach the market. Then in 2020, GalSafe pigs were approved for human consumption, as well as to produce human medical products like heparin (a blood thinning drug) and organs for donation. GalSafe pigs are modified so that a sugar on the surface of the porcine cells that can cause allergic reactions in humans is removed, reducing the possibility of rejection in the human recipient.

The potential of genetically modified plants. As crop production expands to keep pace with human population growth (Chapter 14) and climate change (Chapter 18), farmers are increasingly relying on GM crop plants. Some GM plants carry genes that impart resistance to diseases or pests. Farmers who grow soy and corn often spray their crops with Bt (*Bacillus thuringiensis*), a species of bacteria that makes a protein toxic to only some insect larvae. Instead of spraying the crops, GM plants instead have the *Bt* gene and produce the Bt protein. Larvae that feed on these engineered plants die shortly after eating the Bt-containing food. Consequently, farmers can use much less pesticide on their crops, since the modified crops make their own resistance to the pest.

Researchers are also developing transgenic crops for impoverished regions of the world. Scientists have already introduced genes that confer drought tolerance and enhanced nutritional value into plants such as corn, rice, beans, sugarcane, and wheat. For example, rice plants were engineered to overproduce β-carotene, an orange photosynthetic pigment that is remodeled into vitamin A by cells of the small intestine. These rice plants carry two genes in the β-carotene synthesis pathway: one from corn, the other from bacteria. One cup of their seeds—grains of Golden Rice (Figure 9.10)—has enough β-carotene to satisfy a child's daily need for vitamin A.

Controversies around Roundup® Ready crops. Generally speaking, scientists tend to have few concerns regarding the consumption of genetically modified foods. It's also important to note that the heavily processed foods derived from GM crops can have health risks associated with their consumption—irrespective of whether or not they contain GM ingredients. For example, processed foods tend to have lots of sugar, trans fats, and salt that can lead to increased risk of heart disease or certain kinds of cancer. There are, however, some legitimate concerns specifically related to the potential ecological impact of GM crops that are resistant to the herbicide RoundUp®.

The most common form of GM crop in the United States is the RoundUp Ready variety. RoundUp Ready crops contain a bacterial protein that leaves the plants unaffected by treatment with the herbicide glyphosate, the active ingredient in RoundUp. RoundUp Ready crops make it easier for farmers to grow food because RoundUp can be sprayed on the crops, killing the weeds that compete with the crops for resources like water while not killing the crops themselves.

The scientific consensus is that RoundUp Ready crops are as nutritious as other crops. However, their impact on wild plants and the environment is less clear. Genes that confer glyphosate resistance can appear in other wild plants or unmodified crops—which means that recombinant DNA can escape into the environment. Glyphosate-resistance genes can be transferred from GM plants to non-GM ones via pollen carried by wind or insects. Additionally, the rise of glyphosate-resistant super weeds means farmers are

Figure 9.10 Golden rice is an example of a genetically modified crop

Credit: IRRI Images

Plants that make Golden Rice (top) carry two genes that cause enhanced production of β-carotene, a pigment that is converted to vitamin A in the body, within the edible seeds (rice grains).

Biology for Life

Bioengineering and the Future of Fashion

Credit: Jon Aaron Brown

Figure 9.11 National Geographic Explorer Theanne Schiros

Here, Schiros is holding a sample of a biodegradable textile with color produced by genetically modified bacteria.

Credit: Vojce/Shutterstock.com

Figure 9.12 *Discosoma* coral

This coral contains a gene for red fluorescent protein, which gives the coral its bright red appearance.

National Geographic Explorer Theanne Schiros (Figure 9.11) is interested in how we can use tools like genetic engineering to harness nature's genius. Her research focuses on how we can create products for use that are compatible with nature. One area of her work is how we can generate next-generation clothing and textiles through genetic engineering.

Each year, the fashion industry produces around 92 million tons of waste. The waste ranges from the packaging used to wrap clothing to the waste chemicals left after dyeing fabrics to microplastics (Chapter 15) to returned clothes that, instead of going out for resale, go straight into a landfill. Due to the transport of raw materials and finished goods, the fashion industry alone is responsible for up to 10% of global carbon dioxide emissions (a greenhouse gas which contributes to climate change, Unit 5).

When we think back to Chapter 1, one of the purposes of science is to serve communities and solve society's problems. Schiros's work and that of her collaborators involves using protein engineering to produce biodegradable textiles with built in color. The color comes not from polluting synthetic dyes, but from *Discosoma* coral (Figure 9.12). This coral species has a gene that produces a brilliant red protein called red fluorescent protein (RFP), which gives *Discosoma* coral its vivid color and UV protection.

Members of Schiros's team inserted the RFP gene into bacteria using DNA cloning. Then, the engineered protein was harvested and purified. The resulting engineered protein was mixed with a plant-based protein, such as soy, and then pushed through a nozzle with tiny holes to form red fibers. Once processed in this way, these fibers can be knit or woven together to make textiles, like the one Schiros is holding in Figure 9.11.

Using sources from nature to derive color is not the only application. For example, this same process can be applied to make clothing waterproof without relying on the use of plastics. As Schiros says, rather than contributing to a growing waste problem, their work "takes only knowledge from nature" to make the Earth and our lives better.

1. The genetically engineered bacteria that are used to make textiles like Werewool remain in the lab, and only the textile itself leaves. Based on what you know about genetic engineering, what are the potential benefits of this approach compared to the use of genetically modified crops in agriculture?
2. The fashion industry creates a lot of pollution and waste, including the release of microplastics, chemical dyes, and CO_2 into the environment. How does using genetically engineered bacteria to produce biodegradable textiles like Werewool help to solve these problems?

turning to herbicides that are less safe than glyphosate, rather than returning to more environmentally friendly methods of managing weeds.

The safety of these crops is unclear, with different governmental agencies arriving at different conclusions. As of 2025, the U.S. Environmental Protection Agency stated that there are no risks to human health from the consumption of crops that contain glyphosate residue. This contrasts with the stance held by the World Health Organization's International Agency for Research on Cancer, which concluded in 2015 that glyphosate can directly cause cancer. To add additional color, a meta-analysis (when data from many studies on the same topic are assessed together; Chapter 1) of reports on glyphosate residue suggested methodological flaws with many of the studies used by regulatory agencies. The authors of the study cautioned that glyphosate residues may exist on food at much higher levels than previously reported.

The controversy with GM crops doesn't stop there, either. Additional concerns about business practices, factory farming practices, the need to feed people, and the ethics of genetically modifying animals all muddy the waters regarding the best response to GM crops or livestock. Ultimately, the question isn't about whether GM crops are themselves safe for human consumption, but about whether the use of GM technologies such as adding glyphosate-resistance genes to crops can have unintended harmful consequences. For example, these consequences include the aforementioned overuse of pesticides that are potentially harmful to humans, as well as to pollinators and soil. Understanding the science behind these crops is the first step toward forming your own educated opinions. We will return to genetically modified crops and whether they are superior (nutritionally or otherwise) to organic crops in Chapter 18.

? Did You Know?

Since all living organisms contain nucleic acids, you eat nucleic acids every day. In the case of transgenic crops, you are eating recombinant DNA.

9.2 Summary

- Restriction enzymes cut DNA at specific sequences of nucleotides and are useful for making recombinant DNA.
- In DNA cloning, restriction enzymes cut DNA to insert foreign DNA into a plasmid, which is then introduced into an organism.
- Genetically modified organisms (GMOs) include everything from bacteria to animals. GMOs can help improve human health by increasing the nutritional value of foods, such as Golden Rice, and creating crops that are more resistant to climate change. Concerns about GMOs include the risk of engineered DNA entering wild populations, and the potential health effects on humans and other organisms from any associated chemical use, such as the herbicide glyphosate used on RoundUp Ready crops.

9.3 Gene Editing

Learning Objectives

9-9 Describe applications of gene editing.

9-10 Explain how *ex vivo* and *in vivo* gene therapy work.

9-11 Describe how the CRISPR/Cas9 system works.

9-12 Compare how CRISPR/Cas9 generates gene knock-outs and knock-ins.

9-13 Explain how CAR-T cell therapy works.

In May 2020, Tom and Kari Whitehead received the heartbreaking news that their 5-year-old daughter Emily had acute lymphoblastic leukemia. After Emily relapsed twice and it was clear that her cancer was resistant to treatment, her medical team recommended that Emily start hospice care, saying that she only had weeks to live.

With nothing else to lose, Emily's parents decided to pursue an experimental treatment called *CAR-T cell therapy*. Emily was the first pediatric patient to receive CAR-T cell therapy. The experimental treatment worked, and from facing almost certain death, Emily's cancer was cured. In 2022, Emily celebrated 10 years cancer-free (**Figure 9.13**). As of the writing of this text, Emily is still cancer-free and attending college.

According to St. Jude Children's Research Hospital, CAR-T cell therapy in children with leukemia resulted in a 99% remission rate. As of 2025, there are seven different Food and Drug Administration-approved CAR-T cell therapies for different types of blood cancers, with hundreds more in development. What is so special about CAR-T cell therapy? And why is it so effective for treating cancer?

Figure 9.13 A decade cancer-free

Emily Whitehead was given weeks to live as a young child as a result of treatment-resistance leukemia. Subsequently, she received an experimental treatment, CAR-T cell therapy, that cured her cancer.

Credit: Emily Whitehead

The success of CAR-T cell therapy comes down to its use of gene editing to personalize the cancer treatment specific to the patient. **Gene editing** is a form of genetic engineering where an organism's DNA is changed. These changes could include inserting a new gene or modifying a gene; replacing a gene with a different version, called a **gene knock-in**; or removing a gene's function, called a **gene knock-out**.

In this section, we are going to learn about gene editing and the technologies that make CAR-T cell therapy possible, starting with gene therapy and then discussing precise genome editing with CRISPR/Cas9. We'll return to CAR-T cell therapy and how it works at the end of this section.

Replacing Missing or Broken Genes: Gene Therapy

Drugs and other treatments can minimize the symptoms of some genetic disorders, but the only possible cure is through correcting the underlying genetic problem. **Gene therapy** is an approach to curing a genetic disorder by replacing a faulty gene with a functioning one. For gene therapy to work, the disorder must be linked to a defect in a single gene (which, as we learned in Chapter 8, is rare), the sequence of the functioning gene must be known, and there must be a way to insert the functioning gene into the target tissue and control its expression. The last part is one of the most difficult aspects to achieve—as you've learned throughout Unit 2 and this unit, gene expression must be at the right time, at the right place, and to the right extent, or there can be catastrophic biological results.

> **Study Tip**
>
> The word *vivo* is from the Latin *vivus*, meaning "alive or living," whereas *ex* means "out of." Therefore, *ex vivo* means "away from the body," and *in vivo* means "in the body."

There are two ways of performing gene therapy—*ex vivo* and *in vivo*. ***Ex vivo* gene therapy** is when cells are removed from a person and genetically engineered viruses are used to introduce the normal gene into the cells. The cells are then re-introduced into the patient (**Figure 9.14A**).

> **Point of Confusion:** The viruses used for *ex vivo* gene therapy are called viral vectors. These are not the same as the plasmid vectors you learned about in Section 9.2.

Early *ex vivo* gene therapy treatments for an X-linked genetic disorder (Chapter 8) called SCID-X1—which causes recurrent infections in those affected due to a lack of immune cells (Chapter 17)—had mixed results. Some children were cured, while others developed leukemia due to the treatment. The cause of death was the recombinant viral vector inserting the gene at a site near a proto-oncogene, causing the gene to become an

Figure 9.14 ***Ex vivo* and *in vivo* gene therapy**

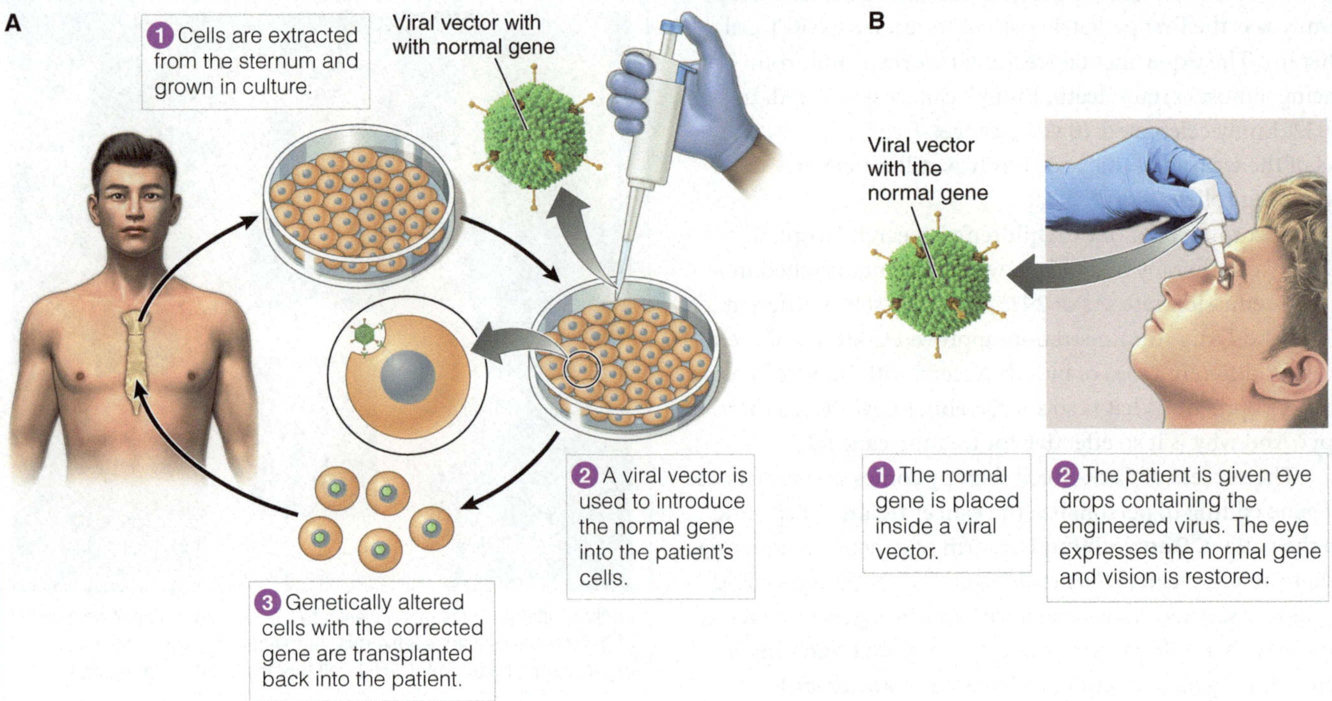

A. *Ex vivo* gene therapy is used to treat disorders caused by cells deep within the body.

B. *In vivo* gene therapy is used on parts of the body, such as the eye, that are easy to directly treat with the virus.

oncogene (Chapter 6), leading to cancer. Researchers have since developed viral vectors that are less likely to cause cancer, and some are instead opting to use technologies such as CRISPR/Cas9, which has more precise targeting and therefore less risk.

***In vivo* gene therapy** directs a genetically engineered virus into a person's body to deliver the normal copy of the gene to the impacted cells (Figure 9.14B). This requires the affected cells to be easily accessible to a virus. For example, in the case of Corey Haas, who was born with a rare genetic condition called Leber congenital amaurosis type 2 that causes blindness, *in vivo* gene therapy was life changing. Prior to treatment, the then 8-year-old was legally blind. Four days after receiving gene therapy, he started screaming during a trip to the zoo because his eyes hurt. Why did his eyes hurt? Because he was seeing sunlight for the first time. Although questions remain as to how long the benefits of gene therapy related to vision last, Corey's story is a touching and exciting reminder of the power of gene therapy.

Using the Knowledge of Nature: Editing Genes with the CRISPR /Cas9 System

> **Big Ideas in Biology**
>
> **Information Flow**
>
> Information stored in a bacterium's CRISPR array creates a bacterial "immune system" that acts like our own adaptive immune system (Chapter 17), allowing the bacterium to recognize and destroy foreign invaders.

In 1993, a team of biologists noticed that bacteria had a series of short repeating sequences, called repeats, separated by short pieces of DNA called spacers. The spacer sequencers in between these repeats were unique sequences of DNA with no similarity to one another. These spacer sequences remained a mystery until 2005, when researchers discovered that the sequences were not bacterial DNA at all…but viral DNA. What were viral DNA sequences doing in bacterial DNA?

A special type of virus called a **bacteriophage** (Chapter 1) infects bacteria by injecting its DNA into the host cell. The bacterial cell can prevent damage from the virus by degrading the viral DNA. After the infection is over, a piece of the degraded viral DNA can be saved in the bacterial genome. These pieces of viral DNA act like a record book, allowing the bacterium to recognize the invading virus in subsequent infections. This "record book" is called a **CRISPR array**. The **Cas9** protein, an enzyme that cuts DNA at specific sequences, uses the genetic information in the array to identify invading viral DNA and destroy it. Together, these form the **CRISPR/Cas9 system**, which acts like a defense system to protect the bacterium from future infections. Note that CRISPR stands for clustered regularly interspersed short palindromic repeat, and Cas9 stands for CRISPR-associated protein 9.

Researchers were able to repurpose this naturally occurring system to edit genes. In a bacterium, the Cas9 enzyme is guided to the invading viral DNA using information stored in the CRISPR array. In an experimental setting, researchers provided Cas9 with a short sequence of RNA called **guide RNA**, which directs it to a target gene, such as an allele that causes a disease (Figure 9.15 ①).

Once the Cas9 complex reaches the target gene, it will cut the DNA, causing a double-stranded break ②. From here, there are two outcomes. If no donor DNA is present, the cell will repair the double-stranded break caused by the Cas9 enzyme by haphazardly putting the two pieces of DNA together. This causes bases to be inserted or removed at random, resulting in a frameshift mutation (Chapter 7), essentially deactivating, or "knocking out," the gene ③.

Figure 9.15 Overview of CRISPR/Cas9 system for gene editing

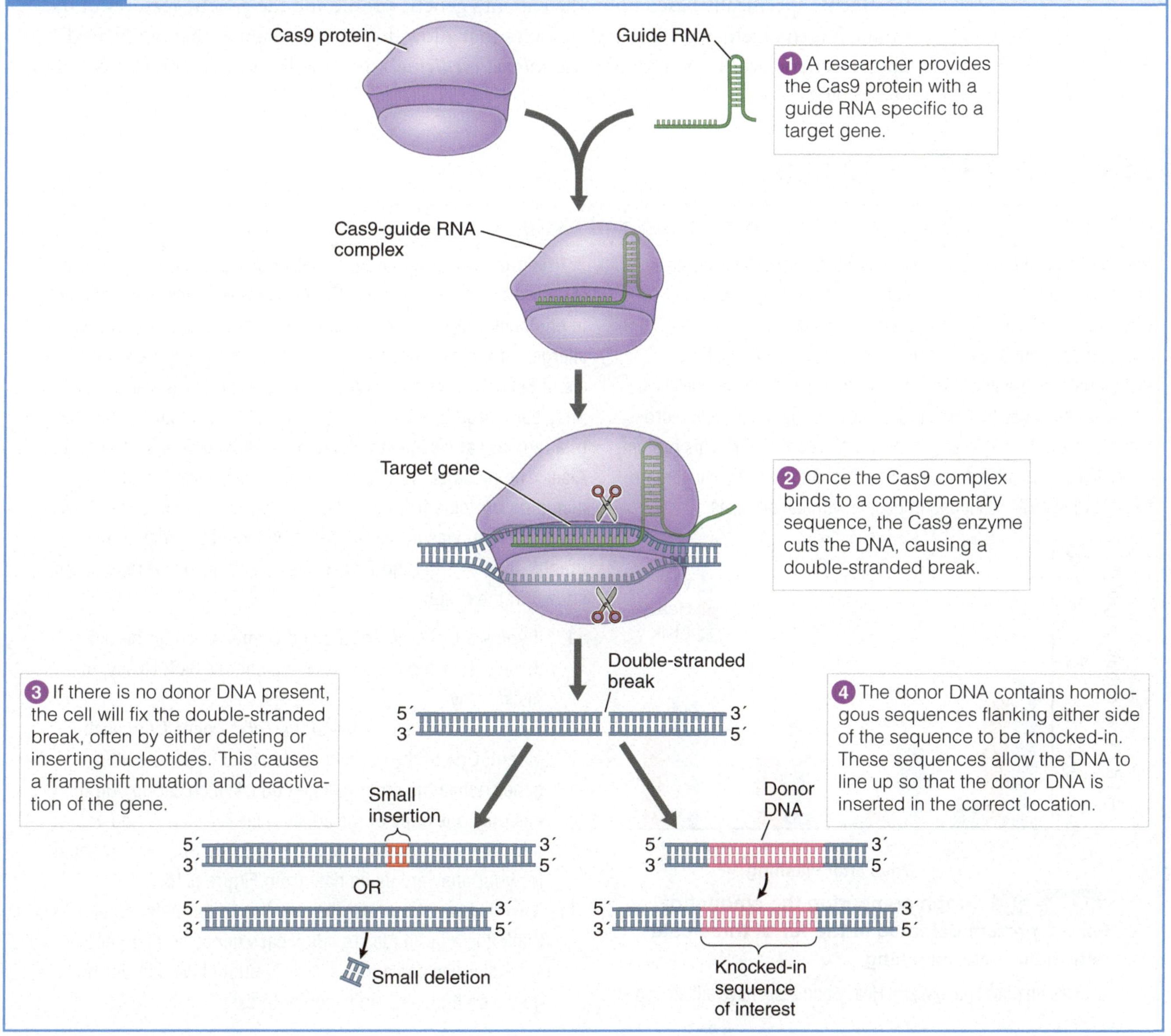

> **? Did You Know?**
>
> In 2023, CASGEVY™ was approved to treat people 12 years and older with sickle cell disease or beta thalassemia. CRISPR/Cas9 is used to edit the patient's blood-forming cells to produce healthy red blood cells instead of sickled cells (Chapter 7).

This is why deactivating a gene is called a knock-out. One application of generating knock-outs with CRISPR/Cas9 is to remove the genes in peanuts that produce the proteins that cause peanut allergies. Therefore, this technology could be a life-changer for the approximately 3 million Americans with a peanut allergy.

Alternatively, if the researcher provides donor DNA, it can be inserted into the region with the double-stranded break (4). Donor DNA is used if a researcher wants to, for example, replace a faulty disorder-causing gene with a normal copy, where the corrected gene is inserted into the DNA sequence, creating a gene knock-in.

The Future of Cancer Treatment: CAR-T Cell Therapy

For many years, approaches to cancer treatment involved procedures that interfere with the cell cycle, such as chemotherapy (Chapter 6) and radiation therapy to kill cancerous cells. While surgical removal of cancerous cells and radiation are more targeted approaches, they can miss cancer cells, and chemotherapy often has significant side effects, because it impacts *all* dividing cells, not just the cancerous cells.

As we learned at the beginning of this section, CAR-T cell therapy is revolutionizing how we treat cancers. In addition, genetic testing of tumors allows doctors to begin to tailor cancer treatments specifically based upon the patient's genetic profile and the genetic features of the tumor. Taken together, genetic technologies that allow doctors to design a treatment based upon a genetic profile are part of *personalized medicine*, which we'll discuss in greater detail in Section 9.4

Digging into Data

Allergy-Free Cats? A Look into Gene Editing and Asthma

Are you or is anyone you know allergic to cats? A common misconception is that it's the cat's fur that makes you sneeze, when in fact, it's a protein in the cat's saliva.

If you've spent any time around cats, you will recall how fastidious they are about the state of their fur. Cats routinely groom themselves by licking their fur. Grooming spreads a protein found on their skin and in their saliva to their fur. This protein, called Fel d 1, is often the culprit behind cat allergies, causing up to 90% of allergic reactions in people with cat allergies.

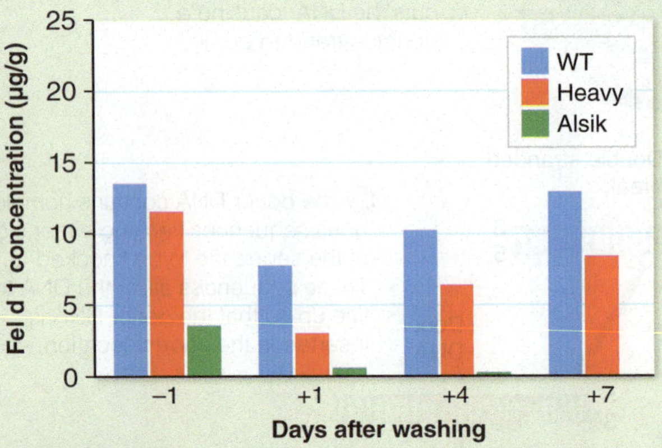

Figure based on: Lee, S. R., Lee, K. L., Song, S. H., Joo, M. D., Lee, S. H., Kang, J. S., ... & Kong, I. K. (2024). Generation of Fel d 1 chain 2 genome-edited cats by CRISPR-Cas9 system. *Scientific Reports, 14*(1), 4987.

Figure 9.16 Graph comparing the amount of Fel d 1 protein detected in the fur of three cats before and after washing

WT has unedited genes, and Heavy and Alsik have edited genes.

One group of researchers studied how to use CRISPR/Cas9 to remove the *Fel d 1* gene. Their goal was to use gene editing to engineer a hypoallergenic cat to reduce the amount of this allergen in the cat. In the experiment, researchers measured levels of Fel d 1 in the fur of cats after gene editing (Figure 9.16). They compared the concentration of Fel d 1 protein on the coat of a regular cat (WT) with two genome-edited cats, Heavy and Alsik. The researchers washed the cats to remove dander—including proteins from the skin and saliva—and checked how much allergen was in the fur before washing (−1 day in the graph), then at 1, 4, and 7 days after washing. Their results are shown in the graph.

1. If you were allergic to cats and wanted to adopt one of these cats, which one would you choose: WT, Heavy, or Alsik? Why?
2. Every cat receives a copy of the *Fel d 1* gene from each parent. One of the genome-edited cats has two edited genes, while the other has just one. Which cat do you think has only one copy of the edited gene? Which cat do you think has two copies of the edited gene? Provide a rationale for your answers using data from Figure 9.16.
3. Now that we can adjust the genetics of our pets, do you think it is a good idea to edit a cat's genes so that people with allergies can adopt a cat? Or should we edit our own genes so that we aren't allergic to cats?

Biology for Life

Inheritance of Genes Edited with CRISPR/Cas9

Recall that your body has two kinds of cells—diploid somatic cells, or body cells, and haploid gametes, which are used for reproduction. If you use gene editing to change your body cells, these changes are not passed down to your offspring. However, CRISPR/Cas9 could potentially be used to edit gametes, and these changes, called germline modifications, would be passed down to future offspring.

The practice is illegal in many countries, but in 2019, He Jiankui made headlines when he created the first gene-edited babies. Jiankui's goal was to produce HIV-resistant children. HIV enters cells by binding to a particular receptor (Chapter 7), and CRISPR/Cas9 was used to change the structure of the receptor so the virus could no longer bind to it. The researcher claimed that two baby girls, children of a man who is HIV positive, were born in 2018 with this edit. As of the writing of this text, the effect of the edit on the children is unknown. It may be that the children are less susceptible to contracting HIV, but off-target effects of the editing could result in other problems down the road, such as cancer.

Many scientists and ethicists condemned this experiment as a monstrous ethical violation, not only because it is unknown if the children will have negative consequences in the future, but also because the changes would be passed on to any offspring and future generations. This story also illustrates an important point about all genetic technology: Our ability to tinker with genetics may have surpassed our ability to understand the impact of the tinkering.

1. What are some possible risks and benefits when scientists use gene editing on humans, especially when making changes that could be passed on to future generations? What could happen if scientists avoid taking these risks?
2. Should society be able to decide or limit what scientists do with gene editing, especially when the changes might affect not only the person treated but also their future children? Why or why not?
3. CRISPR/Cas9 technology can be used to edit genes in adults with diseases like sickle cell anemia or in embryos before birth. How are these two uses different, and why might editing embryos raise more ethical questions?

CRISPR/Cas9 and principles from *ex vivo* gene therapy are leading to a new type of personalized treatment: CAR-T cell therapy. **CAR-T cell therapy** is a process by which T cells (immune system cells that target and destroy cancer cells; Chapter 17) are removed from a person's body, edited with CRISPR/Cas9 to become experts at fighting the patient's specific cancer, and then infused back into the patient's body (Figure 9.17).

Figure 9.17 Overview of CAR-T cell therapy

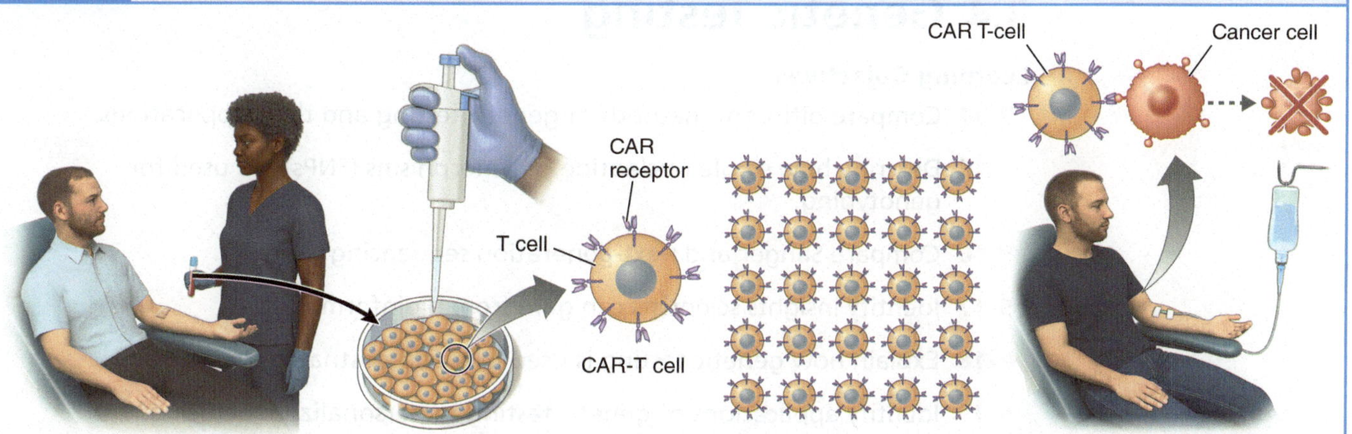

1 First, a sample of the patient's blood is removed.

2 The patient's T cells are grown in a lab and edited using the CRISPR/Cas9 system to contain a personalized chimeric antigen receptor (CAR). This produces CAR-T cells.

3 Millions of the patient's genetically engineered T cells (CAR-T cells) are grown in culture.

4 The engineered T cells are infused back into the patient, where they bind specifically to the cancer cells, killing them.

Early versions of CAR-T cell therapy used viruses to modify the patient's cells. Today, however, the CRISPR/Cas9 system is used due to its enhanced accuracy in gene editing compared to using genetically engineered viruses.

CAR stands for **c**himeric **a**ntigen **r**eceptor—where this receptor can be edited to target specific features on the surface of cancer cells and then added to a patient's T cells. Since each person's cancer is unique, CRISPR/Cas9 is used to edit the CAR to be specific to the patient's tumor cells. CAR-T cell experts like Renier Brentjens at Memorial Sloan Kettering Cancer Center in New York describe CAR-T cell therapy as a "living drug." Once this personalized living drug is created, the supercharged T cells are infused back into the patient, where they identify and kill cancer cells.

As mentioned at the beginning of this section, CAR-T cell therapy has had remarkable results, in both pediatric and adult patients. For example, patients with B-acute lymphoblastic leukemia treated using traditional therapies typically lived an additional 7 months after diagnosis. With CAR-T cell therapy, more than half of patients were still alive a year later.

9.3 Summary

- Gene editing can interfere with a gene's function, replace or fix a faulty gene, or add a gene for a desired trait.
- In *ex vivo* gene therapy, the gene is replaced in cells outside of the body and the engineered cells are infused back into the patient. In contrast, *in vivo* gene therapy directs a genetically engineered virus into a person's body to deliver the normal copy of the gene to the impacted cells.
- The CRISPR/Cas9 system acts like an adaptive immune system for a bacterium, allowing it to recognize and kill an invading virus in subsequent infections. Researchers repurposed this naturally occurring system to edit genes, using a guide RNA to bind a target gene so the Cas9 enzyme can cut the DNA at a precise location.
- Once the DNA is cut, the cell can repair the break on its own, causing a frameshift mutation and deactivation of a gene (gene knock-out), or swapping out DNA at the site of the break (gene knock-in).
- CAR-T cell therapy takes a patient's T cells, edits them with CRISPR/Cas9 to specifically recognize and attack the patient's cancer cells, and then infuses them back into the body.

9.4 Genetic Testing

Learning Objectives

9-14 Compare different methods of genetic testing and their applications.

9-15 Describe how single nucleotide polymorphisms (SNPs) are used for genotyping.

9-16 Compare Sanger and next-generation sequencing.

9-17 Identify insights scientists can gain from bioinformatics.

9-18 Explain how genetic testing is used pre- and postnatally.

9-19 Identify applications of genetic testing to personalized medicine.

In Chapter 7, we learned about how the misexpression of a single gene can have catastrophic effects, such as a right-sided heart. In Chapter 8, we learned about how defects in meiosis can lead to chromosomal problems, and that certain conditions like a balanced translocation can lead to a heightened risk of miscarriage. We also learned how genetic testing and pedigrees are useful for making predictions about disease inheritance.

Genetic testing can help expectant parents make informed decisions about family planning. These decisions are wide ranging, and some have ethical considerations, such as decisions regarding pregnancy termination (abortion). In addition, if a family finds out their child will have a genetic disorder, they may choose to deliver in a hospital well suited for taking care of high-needs babies. Other testing yields insight into which embryos to transfer during *in vitro* fertilization (called *preimplantation genetic diagnosis,* or PGD, discussed below).

In addition, genetic testing can provide adults with important health information, which clinicians can then use to assess a patient's risks for developing disease and/or to provide the best treatment options given the genetic makeup of, for example, a tumor.

In this section, we'll discuss genetic testing for screening and/or diagnosis of genetic conditions. We'll examine two different kinds of genetic testing used in medical applications: genotyping and gene sequencing. Note that we already talked about one type of genetic testing in Section 9.1—STR analysis. Here, instead of attempting to specifically tell individuals apart at the genetic level, our focus will be on the clinical uses of genetic testing.

SNPs are a built-in genetic record of family history. People who are related have similar patterns of these DNA sequence variations, with closer relatives having more similarities than distant relatives do. Therefore, genetic testing can help you understand where your ancestors hailed from.

Genotyping and At-Home Genetic Testing

At-home genetic testing services, such as 23andMe®, are useful for helping people to understand their genetic makeup. Other services, such as MyRisk® and Genomind®, provide testing that can be clinically useful. Although there are some genetic testing technologies that use Sanger or next-generation sequencing (discussed later in this section), most of these kits rely on genotyping. **Genotyping** is the process of determining one's genetic makeup. There are two main ways of doing this (outside of sequencing the full gene). One is to look at single-nucleotide polymorphisms, or SNPs (pronounced "snips"), and the other uses short tandem repeats, or STRs (Section 9.1).

Alleles of most genes differ by single nucleotides. These slight differences, or **single nucleotide polymorphisms (SNPs)**, are the basis of the variation in human traits and what make us unique—the chance that two people who are not identical twins would have the same set of these SNPs is less than one in 10^{21}! To determine the SNPs that a person has, first a DNA sample is collected, such as from saliva. The DNA sample is then loaded onto DNA microarray chips, which are tiny glass plates with thousands of microscopic spots of DNA stamped on them. Each spot contains a short, synthetic single strand of DNA with a unique SNP sequence of interest. When an individual's DNA is washed over a chip, it hybridizes with (sticks to) spots that have a complementary DNA containing the SNP. Fluorescent tracers reveal which spots the sample attached to in the microarray—and therefore, which SNPs are in that individual's DNA (Figure 9.18).

? Did You Know?

Compared to their parents, newborns have an average of 60 new SNPs. These changes are the result of naturally occurring gene mutations. The older the father (but not the mother) is at the time of conception, the more new SNPs in the offspring.

Figure 9.18 **A DNA microarray**

Each colored spot is a region where an individual's DNA has hybridized with a SNP.

Credit: Andre Nantel/Shutterstock.com

Decoding DNA through Sequencing

American author Kurt Vonnegut is known for his science fiction works that are often filled with satire and dark humor. In the short story *Unready to Wear*, humans live in a world where they can shed their bodies at will and live in either a disembodied form or in the body of another. In the story, people often lament the need to return to their bodies, and the inconvenience of even existing in a body. In real life, nowhere is the inconvenience of our bodies more apparent than during a norovirus infection. Noroviruses are highly contagious viruses that are the leading cause of gastroenteritis and/or food poisoning in the United States. The symptoms of norovirus are unmistakable, with the rapid onset of spectacular bouts of vomiting,

Biology for Life

Surprises from at Home Genetic Testing

In late 2007, 23andMe was the first company to offer DNA testing directly to consumers. In the years since, a variety of new insights into family trees have appeared. For instance, stories abound about people finding out about siblings they didn't know they had…and in some cases, evidence of extra-marital affairs. Genetic genealogy, or the practice of creating pedigrees (Chapter 8), is now used by criminal investigators to find criminals based on genetic evidence left at a crime scene.

Overall, the unexpected news from genetic testing can be positive or negative, depending on the situation. Although laws such as the Genetic Information Privacy Act of 2019 exist, there are still incidental findings through at-home genetic testing that could be construed as privacy violations.

In addition, when 23andMe declared bankruptcy in 2025, more concerns came to light regarding the privacy of the data collected from over 15 million users. Since 23andMe is a direct-to-consumer service, there are no protections from laws such as the Health Insurance Portability and Accountability Act (HIPAA). Additionally, 23andMe had provided deidentified data (data without names, birthdates, or other potentially identifying features) to pharmaceutical companies for research purposes—which is acceptable under federal human research laws. Users were not without power, though—one could delete their account and request that retained saliva samples be destroyed.

1. What are some possible concerns with making personal genetic information available through services like 23andMe?
2. Should companies like 23andMe provide information about close relatives? Why or why not?
3. Given what you know now about genotyping with SNPs or STRs, do you think deidentified genetic information should be sharable between companies? Why or why not?

? Did You Know?

Common food sources of norovirus include shellfish, particularly oysters, which are eaten raw. The other source is when tiny amounts of an infected person's feces or vomit are ingested, usually from not washing one's hands before eating.

diarrhea, and stomach cramping. Most of us would agree that a hiatus from our body during the infection would be ideal!

However, there are plenty of other reasons people can have vomiting or diarrhea, ranging from pregnancy to food intolerances. Knowing if a norovirus outbreak is occurring is important so that public health officials can work to stop transmission, such as closing a restaurant or recalling food. How do public health officials know if a norovirus outbreak is occurring?

A scientist can identify and track a norovirus outbreak using DNA sequencing. **DNA sequencing** is the process of determining the exact order of nucleotide bases in a DNA molecule, also known as its sequence. Fecal samples from infected people can be used to determine whether norovirus DNA is present and potentially find out the source of the infection. Although there are many different types of DNA sequencing, we will focus on two kinds: Sanger sequencing and one form of next-generation sequencing.

Study Tip

Sanger sequencing is called chain termination sequencing because the addition of a ddNTP means that another "link" (a nucleotide) can't be added to the "chain" (the newly synthesized DNA), causing the reaction to terminate (end).

Sanger sequencing. First developed in the 1970s (and resulting in a Nobel Prize in 1980), **Sanger sequencing**, also called chain termination sequencing, is used for decoding short sequences of DNA (~500–1,000 base pairs). Sanger sequencing uses specialized DNA nucleotides called dideoxynucleosides, regular DNA nucleotides (also called deoxynucleotide triphosphates, or dNTPs), a primer specific to the sequence or gene of interest, and capillary gel electrophoresis. **Dideoxynucleoside triphosphates (ddNTPs)** are DNA bases with their 3′—OH group removed (Figure 9.19A). These ddNTPs contain a fluorescent dye added for detection, and each base (A, T, C, G) is associated with a different color (Figure 9.19B). Without the 3′—OH group, DNA polymerase cannot add another base to the growing chain (Chapter 6), so the addition of these bases will stop the sequencing reaction.

The sequencing reaction begins by mixing ddNTPs, dNTPs, the gene-specific primer, the DNA sample, and DNA polymerase together (Figure 9.20 ①). As the sequencing reaction proceeds, a ddNTP will randomly be incorporated during the reaction. Once the ddNTP is added, no additional nucleotides can be added to the molecule. The ultimate result is DNA molecules of different lengths, each with a fluorescently labeled ddNTP at the end ②. When these molecules are separated by size using capillary gel electrophoresis—a process that is similar to the gel electrophoresis process shown in Figure 9.4, but where the matrix is in a

Figure 9.19 **Sanger sequencing uses regular DNA nucleotides and specialized ddNTPs**

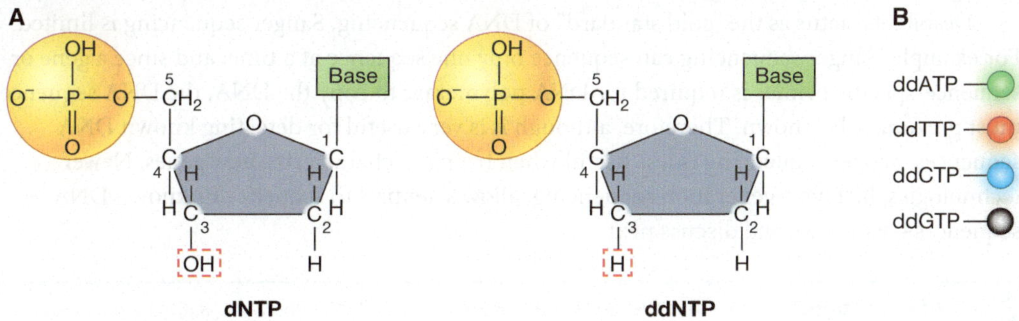

A. Recall from Chapter 6 that for DNA polymerization to occur, there must be a 3′—OH group present on the nucleotide. Therefore, incorporation of a ddNTP (which lacks the 3′ —OH) instead of a DNA nucleotide onto the growing chain will cause the synthesis reaction to halt.

B. Since the ddNTP molecules are tagged with specific colors depending on the base added (notice the ATCG following "dd" in each tag), the identity of the base can then be determined.

Figure 9.20 **An overview of Sanger sequencing**

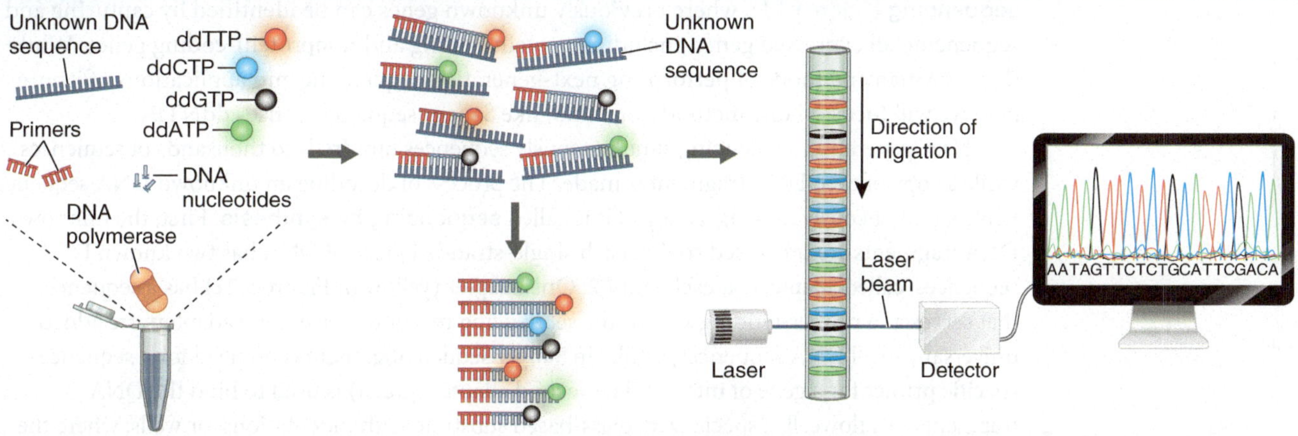

1 The Sanger sequencing reaction mix is similar to a PCR reaction mix—with DNA polymerase, DNA nucleotides, a sequence-specific primer, and the DNA sequence of interest. However, there is the addition of specialized ddNTPs, which interfere with DNA polymerase's ability to make DNA, stopping the reaction.

2 DNA polymerase randomly incorporates a ddNTP instead of a DNA nucleotide, resulting in differently sized fragments of DNA. Each fragment ends with a different fluorescently labeled nucleotide, per the base the sequence it ends on.

3 Capillary gel electrophoresis is used to separate differently sized pieces of DNA by size. A laser passes through the tube to a detector, which determines the color present at the end of each DNA molecule.

4 The detector is connected to a computer, which interprets the color to determine the nucleotide present at the end of the sequence. The curves represent the color detected. Putting all of these individual pieces together reveals the DNA sequence.

capillary tube instead of a tray—the ddNTP found at the end of each DNA molecule is identified using a laser and a detector 3, where the nucleotide at the end of each tagged sequence is read by a computer. When the computer analyzes the colors detected at the end of each DNA fragment, it generates an output called a **chromatograph**, where each color is presented as a curve for each base in the sequence 4. The chromatograph can then be converted into a list of bases, and ultimately, reveal the DNA sequence.

The benefit of Sanger sequencing is that it is a cheap and highly accurate method for sequencing DNA. As such, it is still a commonly used method for sequencing DNA, with wide applicability in public health and medicine, such as in our opening example of a public health official who wants to know if an outbreak of food poisoning is caused by norovirus. Sanger sequencing is also used to confirm results from newer technologies, like next-generation sequencing.

Despite its status as the "gold standard" of DNA sequencing, Sanger sequencing is limited. For example, Sanger sequencing can sequence only one sequence at a time, and since a gene or sequence-specific primer is required for DNA polymerase to copy the DNA, the DNA sequence must previously be known. Therefore, although it is very useful for detecting known DNA sequences, Sanger sequencing is less useful when trying to characterize new genes. Newer technologies, like next-generation sequencing, allow scientists to sequence unknown DNA sequences—which we will discuss next.

Big Ideas in Biology

Structure and Function

Recall from Chapter 2 that —OH is a functional group. In Chapter 4, we learned about the importance of the active site in an enzyme. The —OH group is required for the nucleotide to fit into the DNA polymerase enzyme's active site. Replacing the —OH with a —H means that the nucleotide no longer fits properly into the enzyme's active site, and polymerization cannot occur.

Point of Confusion: Some people refer to Sanger sequencing as "first-generation sequencing," since it was the first sequencing method developed, and to contrast it with newer sequencing technologies, like next-generation sequencing or single-molecule sequencing (sometimes called third-generation sequencing). Although Sanger sequencing was first, it was not replaced by newer sequencing technologies and is still routinely used today.

Next-generation sequencing. Sanger sequencing works with known sequences. How do researchers then find and analyze unknown sequences? That is the benefit of **next-generation sequencing** (Figure 9.21), where previously unknown genes can be identified by capturing and sequencing all expressed genes, including protein-coding and nonprotein-coding genes. While there are many methods of performing next-generation sequencing and applications of its use, here we will focus on one methodology that, like Sanger sequencing, uses ddNTPs.

Next-generation sequencing simultaneously sequences hundreds to thousands of sequences while a copy of each DNA fragment is made. The process of decoding an unknown DNA sequence while simultaneously making a copy of it is called **sequencing by synthesis**. First, the unknown DNA fragments are processed so that each single-stranded piece of DNA has two known DNA sequences, called adapters, at each end (1). One adapter (yellow in Figure 9.21) has a sequence that serves as a primer-binding site for the sequencing reaction. Since these adapters are added universally to all DNA sequences, unlike in Sanger sequencing, there is no need for a sequence-specific primer for a gene of interest. The second adapter (green) is used to bind the DNA fragments to a flowcell, a specialized, glass-based substrate with indentations, or wells, where the sequencing reactions occur. The flowcell is then loaded into the Illumina® sequencing machine.

Figure 9.21 An overview of next-generation sequencing using sequencing by synthesis

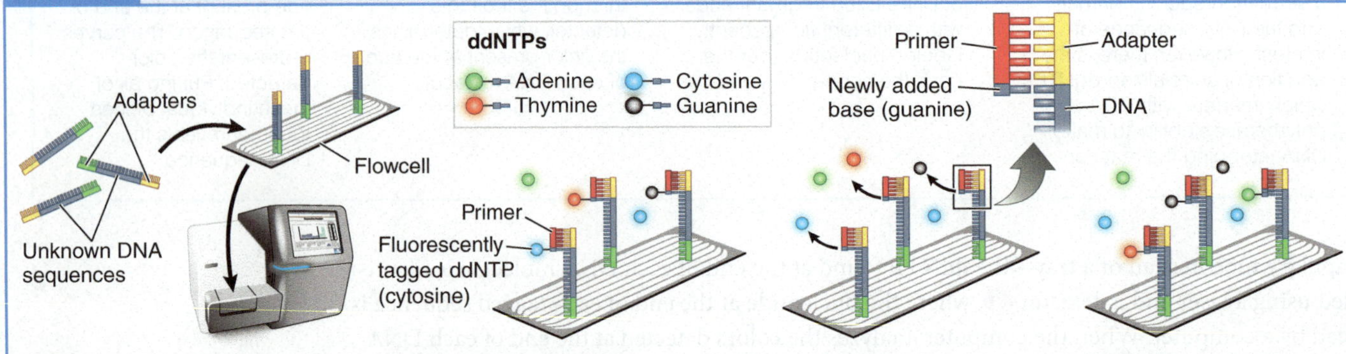

(1) Short segments of known DNA sequences, called adapters, are added to unknown DNA sequences. The yellow adapter is the primer-binding site, and the green adapter is used to bind the DNA sequence to a flowcell. The flowcell is then loaded into the Illumina® sequencer.

(2) In the sequencer, primers bind to the yellow adapters, initiating DNA amplification. DNA polymerase adds a ddNTP with a fluorescent tag to the first base in the unknown DNA sequence. An image is taken to record the color, and thus, the base at that position.

(3) The fluorescent tag is then removed, and the base at that position is chemically converted to a regular DNA nucleotide (dNTP).

(4) DNA polymerase moves to the next base in the unknown sequence, adding a fluorescently tagged ddNTP to that position. An image is taken to record the color. Steps 2 through 4 are repeated until all of the bases in the unknown sequence are determined.

In the sequencer, the DNA is then amplified via PCR (see Section 9.1) using DNA polymerase, a primer specific to the yellow adapter, and fluorescently tagged ddNTPs. Once the primer attaches to the yellow adapter, DNA polymerase adds a ddNTP with a fluorescent tag to the first base in the unknown DNA sequence (2). Recall from Sanger sequencing that once a ddNTP is added, no additional nucleotides can be added to the molecule. Therefore, the sequencing reaction stops once the base is added, and at this point, an image is taken to capture the color of the fluorescent tag that is present, and thus, the base at that position. The fluorescent label is then removed and the base is chemically converted from a ddNTP to a regular DNA nucleotide (a dNTP; (3)). Once the base is added, DNA polymerase can add a ddNTP to the next base in the unknown DNA sequence (4), and the process repeats until the entire DNA fragment is sequenced.

The latest sequencing technologies capture not only which genes are expressed in a cell, but where the genes are expressed. For example, within a single organ, individual cells might exhibit differences in gene expression. This provides important insight into how our organs work at the molecular level and could lead to new insight into the cellular basis of disease.

Bioinformatics and Handling Sequencing Data

A single experiment using next-generation sequencing can produce massive amounts of data—enough so that researchers needed to develop new methods of organizing and analyzing sequencing data. **Bioinformatics** is a relatively new field that creates methods or tools for managing, analyzing, and understanding large biological data sets. Bioinformatics is an interdisciplinary field that requires knowledge not only of biology, but also of computer science, programming, mathematics, and statistics. Bioinformatics professionals can analyze DNA, RNA, and protein data, as well as how these different macromolecules interact with one another.

Process of Science

The Influence of Technology on Scientific Progress

Ten years after the development of Sanger sequencing, scientists began to consider sequencing the entire human genome—all 3 billion nucleotides. Proponents of the idea said it could provide huge payoffs for medicine and research. Opponents said this daunting task would divert attention and funding from more urgent research. With the sequencing methods available in the 1980s, sequencing the 3 billion bases that make up a person's genome would have taken at least 50 years.

However, methods continued to improve rapidly, such as the advent of robotic technologies and PCR. As the technology improved, sequencing became faster. In 2003, 50 years after the discovery of the structure of DNA, the sequence of the human genome was officially completed. From start to finish, the Human Genome Project required about 13 years and approximately $2.7 billion, but it spurred development of much more efficient technologies for sequencing and data analysis. Today, an entire genome can be sequenced in a day for only a few hundred dollars.

Getting the price of genome sequencing under $1,000 was a goal set by the National Human Genome Research Institute in 2004. Improvements in technology advance science—and in the case of DNA sequencing, better technology made the process faster and cheaper (Figure 9.22). By working to make sequencing more accessible, researchers could use the technology to spur advances in biomedical research and healthcare. Today, gene sequencing is part of routine medical treatment. Overall, this story highlights the connection between scientific progress and advances in technology.

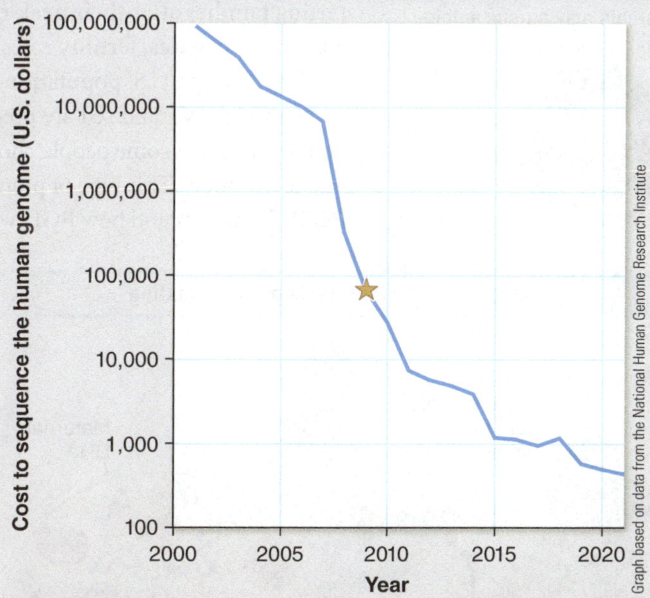

Figure 9.22 The rapid decline of genome sequencing costs

Between 2000 and 2020, the cost to sequence a human genome dropped dramatically as the technology improved. Note the star, which represents the steep decline in costs that occurred once next-generation sequencing technologies came onto the market.

1. How have technological improvements in DNA sequencing changed the feasibility and impact of large projects like the Human Genome Project?
2. Government funding was essential to the success of the Human Genome Project. Do you think its $2.7 billion price tag was worth the results? Should large-scale projects like this one be prioritized in the future?

? Did You Know?

Microchimerism extends beyond pregnancy. For instance, DNA from the baby can be detected in the mother's body for decades after giving birth.

NIPT largely replaced other more risky and invasive procedures, such as amniocentesis and chorionic villus sampling (CVS) for screening purposes. In both procedures, a needle is used to sample fetal cells from either the amniotic fluid or chorion. Since NIPT is a screening tool, and amniocentesis and CVS are diagnostic tests, many people will start with NIPT and if a heightened risk for aneuploidy is detected, the parents may pursue amniocentesis or CVS.

With so much information, how do biologists begin to make sense of the data? After obtaining DNA sequences using computer programs, there are many directions that a researcher could take, depending on their specific research question. For example, researchers might examine gene expression (Chapter 7) in two different cell types to gain insight into how cells respond differently to a treatment with a new medicine.

Medical Applications of Genetic Testing

There are a variety of applications for the use of genetic testing in medical settings. We'll begin by reviewing various forms of pre- and postnatal genetic testing, before ending with a discussion of personalized medicine.

Prenatal testing. In Greek mythology, a Chimera is a hybrid creature with the body of a lion, a goat's head, and a snake's tail, and possibly dragon wings to complete its fanciful look. Geneticists often use "chimera" in a different way, where **microchimerism** describes someone who carries the DNA of multiple people. The most common natural cause of microchimerism is pregnancy. When a person is pregnant, cells and DNA are exchanged across the placenta—an organ that forms inside the womb to provide oxygen and nutrients to the fetus. As a result, one can sample the mother's blood to test the fetus's DNA to identify potential chromosomal abnormalities—a process called **noninvasive prenatal testing (NIPT)** (Figure 9.23).

Preimplantation genetic diagnosis. Between 2012 and 2021, the number of babies conceived through assisted reproductive techniques doubled, including *in vitro* fertilization (IVF) and intrauterine insemination (IUI). Research suggests that the drivers behind this increase come from improved educational and career opportunities for women, leading to their starting families after their peak reproductive years. In fact, the birth rate has declined such that in the United States, fertility rates are below population replacement levels, meaning that absent immigration, the U.S. population is will eventually begin to decline.

Although IVF and IUI are predominantly utilized by couples experiencing infertility and by same-sex couples, some people chose IVF with genetic testing to avoid miscarriage and/or to identify genetic disorders. This type of prenatal genetic testing is called **preimplantation genetic diagnosis (PGD)**. To understand how PGD works as part of IVF, we will start by discussing how IVF works.

Figure 9.23 Overview of noninvasive prenatal testing

A

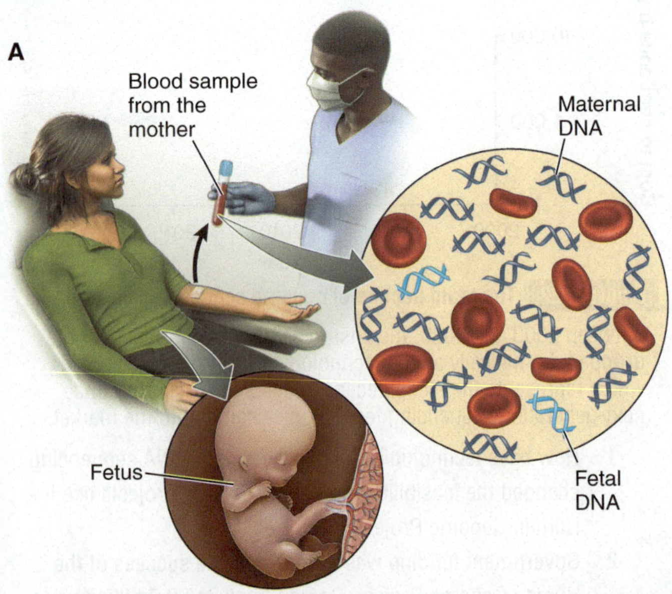

B

Content	Result
FETAL SEX	Consistent with Male
AUTOSOMAL ANEUPLOIDIES	
Trisomy 21 (Down syndrome)	Negative
Trisomy 18 (Edwards syndrome)	Negative
Trisomy 13 (Patau syndrome)	Negative
SEX CHROMOSOME ANEUPLOIDIES	
Monosomy X (Turner syndrome)	Not Detected
XYY (Jacob's syndrome)	Not Detected
XXY (Klinefelter syndrome)	Not Detected
XXX (Triple X syndrome)	Not Detected

A. In NIPT testing, blood is drawn from the mother, and the fetal DNA within her blood is analyzed to examine the chromosomes of the fetus using next-generation sequencing.

B. A sample NIPT output. Note that this fetus is XY and no aneuploidies were detected.

Figure 9.24 **Overview of *in vitro* fertilization with PGD testing**

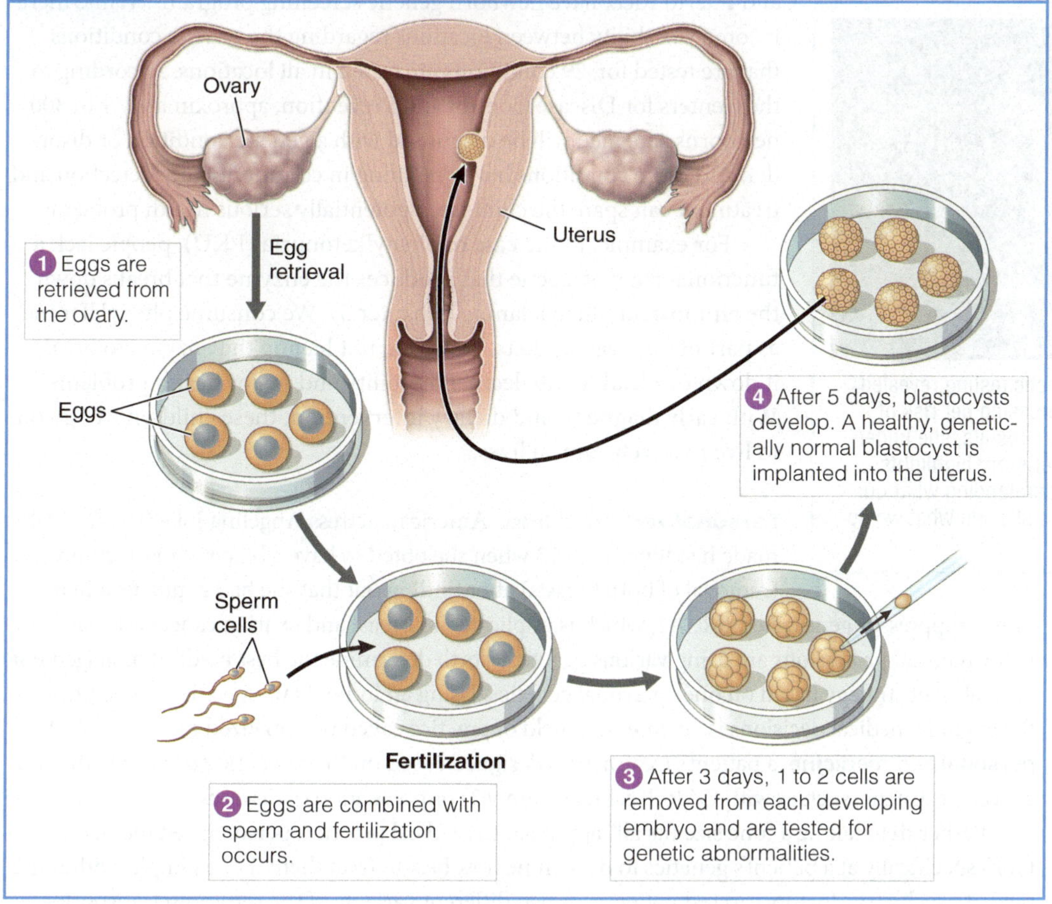

***In vitro* fertilization (IVF)** is an assisted reproductive technique that combines male and female gametes outside the body. First, gametes are isolated from a male and a female. Sometimes the gametes will both come from one couple, while other times a couple will use gametes donated by a sperm or egg donor. The isolated sperm are used to fertilize eggs in a laboratory (Figure 9.24). If fertilization is successful, the zygote (Chapter 8) will form and begin to divide and make more cells. Technicians can safely remove 1 to 2 cells from the embryo after 3 days of growth. The DNA from these cells undergoes genetic testing to identify whether any abnormalities exist. The type of genetic testing can vary depending on the specific test used, but may include next-generation sequencing to assess a specific gene of interest. After 5 to 6 days, a hollow ball of 200 to 300 cells called a **blastocyst** develops. After genetic testing, a healthy blastocyst is chosen for transfer into the mother. Other healthy blastocysts can remain frozen in storage and can be thawed out if the parents plan to have other children, or they can be donated to another individual or couple.

! Point of Confusion: Scientists consider blastocysts to be an embryo. The cells inside the ball (which are embryonic stem cells) will eventually become the embryo, where the outside cells will eventually become the placenta.

One of the ethical concerns with IVF relates to the fact that the number of healthy embryos produced is variable. Sometimes there is only a single healthy embryo, and other parents may have more healthy embryos than they can reasonably raise into adulthood. There is no way to predict how many healthy embryos will result from IVF, so there is concern as to what happens with extra embryos. Other ethical considerations include what to do with the embryos that have diagnosed genetic disorders, as well as the donation of embryos for use in embryonic stem cell research, which results in the destruction of the embryo.

Gattaca was a 1997 science fiction film depicting a dystopian future where preimplantation genetic testing was used to create genetically "perfect" people, and those who were not tested constituted an underclass. The questions raised about designer babies in 1997 are still pertinent today.

Figure 9.25 **American actress Angelina Jolie**

Credit: Kent Nishimura/Getty Images

Jolie made headlines when genetic testing revealed she had a *BRCA1* mutation, increasing her risk of breast cancer. After receiving the results, she underwent a bilateral mastectomy. Her story highlights the potential concerns with understanding what our genes tell us about our genetic risks and what we do to mitigate those risks.

Newborn genetic testing. All 50 states, plus the District of Columbia and Puerto Rico, have newborn genetic screening programs. While there is some variability between locations regarding the genetic conditions that are tested for, 29 conditions are tested in all locations. According to the Centers for Disease Control and Prevention, approximately 1 in 300 newborns screened will be diagnosed with a genetic condition or disorder, and these conditions have one thing in common: Early detection and treatment can spare the child from potentially serious health problems.

For example, in the case of phenylketonuria (PKU), people lack a functional copy of a gene that produces the enzyme that breaks down the amino acid phenylalanine (Chapter 3). We consume phenylalanine as part of our regular diets. However, not having the enzyme to break it down can lead to intellectual disability and major health problems. With early diagnosis and dietary intervention, these children can go on to live relatively typical lives.

Personalized medicine. American actress Angelina Jolie (Figure 9.25) made headlines in 2013 when she opted to have a bilateral mastectomy (removal of both breasts) after finding out that she had a mutation in a tumor suppressor gene called *BRCA1*, which is implicated in breast and ovarian cancers. Today, many medical professions are using various genetic tests to determine the best medical management or treatment strategy based on one's personal genetics. Using an individual's specific genetic profile to guide medical decisions is an emerging field of practice called personalized medicine. With **personalized medicine**, a patient's treatment is designed based on their genetic profile, whether for cancer treatment (such as with CAR-T cells; Section 9.3) or prevention of diseases.

? Did You Know?

Triple-negative breast cancer, a particularly aggressive form of cancer, is "triple negative" because it doesn't express three genes that are common targets for breast cancer treatments. Since there are fewer treatment options, triple-negative breast cancer has a worse prognosis than other types of breast cancer.

Rather than taking a "one size fits all" approach to medical practice, personalized medicine looks specifically at a patient's genetics to determine how best to treat them. For example, within the field of psychiatry, doctors can test patients for two different variants of the serotonin transporter gene, *SLC6A4*. Some forms of *SLC6A4* contain a deletion in the promotor region (Chapter 7). This variant is called the short or "s" form. People who have either one or two copies (Chapter 8) of the short *SLC6A4* form are more likely to experience adverse side effects when treated with a selective serotonin reuptake inhibitor (SSRI, which includes most frontline anti-anxiety and antidepressant medications). Therefore, with this information in hand, psychiatrists can avoid prescribing SSRIs to people with this specific genetic profile and instead choose different medications.

Many people are now offered the option to undergo hereditary cancer screening as part of their regular wellness checks. For example, Myriad Genetics has a panel called myRisk® which uses next-generation sequencing to detect genetic variants associated with 11 different types of cancer. These tests can provide insight for people and their medical providers regarding risks for developing different kinds of familial cancers. This can impact decision-making within families. For example, if your parent has one of the genes associated with cancer risk, you will have a 50% chance of having it as well (Chapter 8) and may therefore choose to be tested.

Knowing that you have a particular gene variant could drive whether you have cancer screenings more frequently or starting at a younger age. For example, some forms of hereditary colon cancer can develop before the recommended age for colonoscopies, which are routine screenings for colon cancer. Therefore, someone who has these gene variants may choose to start getting colonoscopies earlier in life.

Using Genetic Information

One of the challenges with genetic testing is that having a specific genetic mutation does not guarantee someone will get cancer, only that their risk is greater. As we learned in Chapter 8, epigenetics plays a large role in understanding our gene expression and biology. Since epigenetics is influenced by our environment, an individual's environment can have a nontrivial impact on whether they will develop cancer.

Cancer can also develop due to random genetic mutations. Therefore, someone without these known markers should not assume that they don't need to follow through with routine cancer screenings like colonoscopies and mammograms. Alternatively, if someone finds out they have genes that predispose them to cancer, this can have a negative impact on their mental health. Considering their genetic testing results, people may feel hopeless about their health prospects. Genetic counselors (Chapter 8) are important members of a medical team who can help patients not only understand their risks of passing potentially disease-causing genes to their offspring but also make lifestyle choices.

As genetic testing becomes cheaper and more readily available, pursuit of extreme measures such as surgical removal of body parts in favor of preventing cancer development could be a direction people wish to go in. However, surgery is no guarantee of avoiding a cancer diagnosis. Most genes do not function in isolation, and many genes plus environmental factors play into whether someone will develop cancer or other diseases. Overall, personal genetic testing can be a wonderful tool to help people be the healthiest they can be—or it can cause stress and familial strife.

9.4 Summary

- Genetic testing is use to ascertain a person's genetic makeup and includes technologies like genotyping and gene sequencing.
- Single nucleotide polymorphisms (SNPs) are alleles that differ due to variations in single DNA nucleotides. Many at-home genetic testing kits identify one's genotype by examining SNPs.
- Sanger sequencing is used for single genes and known DNA sequences. Next-generation sequencing can be used to identify unknown genes. Both rely on using ddNTPs, but have different ways of decoding the sequence.
- Bioinformatics is a discipline involved with managing, analyzing, and understanding large genomic data sets. This can provide insights into understanding, for example, the genetic basis of certain diseases.
- Genetic testing is used prenatally to identify whether an embryo or fetus has certain genetic anomalies, such as aneuploidy. This information can be used to make healthcare decisions, such as if an abortion may be warranted, if delivery in a hospital with a sophisticated neonatal care is necessary, or even which embryos to implant during IVF. Postnatal genetic testing is useful for identifying conditions where early treatment is essential for preventing serious health problems.
- Personalized medicine uses an individual's genetic profile to guide treatment. For example, cancer treatments could be chosen based on a tumor's specific genetic makeup.

Understand It! Affect

You may have heard people talk about grammatical rules regarding "affect" and "effect." "Affect" is typically used as a verb (as in, "I was affected by that movie!"), whereas "effect" is a noun ("One effect of watching that movie was that I couldn't sleep!"). In psychology, "affect" means something else entirely, referring to one's emotional state. The Understand It! features in the next few chapters will examine the role of emotions in learning, including Test Anxiety (Chapter 10), Motivation (Chapter 11), and Procrastination (Chapter 12). What's the big deal with emotions and learning?

Fundamentally, humans are emotional beings. Having an awareness of one's emotions is not only an important part of personal wellness and essential to forming healthy relationships, but is also an important aspect of learning. For example, *hot cognition* occurs when we associate a strong emotion with a topic—and discussion of various genetic technologies can definitely trigger strong emotions!

Oftentimes, people can remember things better if there are emotions involved. This is why people can recall joyful memories clearly many years after the event. When we learn information in class that triggers emotions (such as excitement over new genetic technologies, or anger over bad actors in the field), it can help us to remember the information better. Associating our classrooms or subjects with good emotions can make learning easier.

However, the opposite is also true. Emotions can interfere with learning. Negative associations with a topic or class can make it difficult to learn. Emotions like anxiety, as you'll learn about in the Chapter 10 Understand It!, can also interfere with our ability to form and recall memories. As we learned in Chapter 1, triggering people to think emotionally, rather than rationally, is a key way that disinformation spreads.

Emotions are important for learning and can be either helpful or harmful. That's why it's important to pay attention to our emotional states so that we know when our emotions are helping us to succeed in the classroom, or if they are getting in the way. Here are some questions for you to consider regarding your emotional states and learning.

1. When you think about your experiences in your biology class this semester, what emotions come up? Were they the same or different at the beginning of the semester as they are now?
2. What emotions came up for you as you were learning the material in this chapter? Do you think these emotions will help you remember this material or interfere?

Chapter Review

9.1 Forensics

DNA profiling identifies individuals by comparing **short tandem repeats (STRs)**, repeating sequences of 2–6 nucleotides. Since DNA profiling conclusively identifies individuals, it can be used to prove guilt or to exonerate wrongfully convicted people, or for paternity testing.

Polymerase chain reaction (PCR) uses heat-tolerant DNA polymerase and complementary **primers** to copy template DNA using a **thermocycler**, which varies the temperature in cycles to produce millions of copies of DNA. **Gel electrophoresis** separates DNA molecules by size using an electric field, with shorter fragments moving faster through a semisolid gel.

9.2 Transgenic Organisms

Genetic modifications are used to produce desired traits in an organism. Genes are edited to generate **transgenic** organisms, which are organisms that contain DNA from another organism. All transgenic organisms are **genetically modified organisms (GMOs)**, but not all GMOs are transgenic. Rather, some GMOs are edited to remove undesirable traits, and therefore do not contain foreign DNA.

DNA cloning is a technique for making and copying **recombinant DNA**. Recombinant DNA is made from two or more different sources. During the cloning process, a DNA fragment is inserted into a **plasmid. Restriction enzymes** cut the DNA at specific locations called **restriction enzyme sites**. These sites in a plasmid are located in a region called the **multiple cloning site**. Restriction enzymes cut the plasmid DNA and the DNA with the gene of interest at the same location, creating **sticky ends** that facilitate complementary base-pairing between the two. The engineered plasmid is then placed into **competent cells**, which are chemically treated to be able to take up new DNA. The cells are grown in culture to make more copies of the recombinant plasmid.

9.3 Gene Editing

Scientists use **gene editing** to change, remove, or replace genes to treat genetic conditions or to add desired traits. If a gene is replaced, it is a **gene knock-in**; when a gene is removed or inactivated, it is a **gene knock-out**.

Gene therapy replaces a faulty gene with a functional one to cure a genetic disease. In **ex vivo gene therapy**, gene modification occurs outside of the body, while **in *vivo* gene therapy** takes place within a person's body.

A special type of virus called a **bacteriophage** infects bacteria by injecting its DNA into the host cell. A bacterium can use a piece of the viral DNA to recognize the invading virus in subsequent infections—a kind of "record book" called a **CRISPR array**. Using the genetic information in the array to identify invading viral DNA, a **Cas9** protein can cut the viral DNA at specific sequences using a **guide RNA** and destroy it. Together, the **CRISPR/Cas9 system** acts like an adaptive immune system to protect a bacterium from infection.

CART-T cell therapy is a next-generation version of *ex vivo* gene therapy where a person's T cells are modified using CRISPR/Cas9 to make them supercharged cancer-fighting cells that are specific to the person's cancer.

9.4 Genetic Testing

Genotyping determines an individual's genetic makeup by analyzing **single nucleotide polymorphisms (SNPs)**. SNPs are single nucleotide differences and the basis of the variation in human traits.

DNA sequencing determines the order of nucleotide bases. In **Sanger sequencing**, the lack of 3′ —OH on **dideoxynucleosides (ddNTPs)** means they terminate DNA replication as they are added, creating fragments of varying lengths, each ending with a different fluorescently labeled ddNTP. Capillary gel electrophoresis separates the fragments, where a computer can detect the ddNTP, and in doing so, identifies the full sequence, which is displayed as a **chromatograph.** **Next-generation sequencing** allows investigators to sequence millions of different DNA sequences simultaneously, including unknown DNA sequences. Scientists use **bioinformatics** to manage, analyze, and study large data, such as those produced by next-generation sequencing.

Microchimerism describes a person with DNA from more than one individual, which occurs during pregnancy. Due to microchimerism, doctors can use **noninvasive prenatal testing (NIPT)** to test a fetus's genetic makeup using the mother's blood. **Preimplantation genetic diagnosis (PGD)** tests for genetic diseases before implanting a **blastocyst** during ***in vitro* fertilization (IVF)**, an assisted reproductive technique.

Genetic testing can be used as a tool for **personalized medicine**, which uses an individual's genetic profile to guide medical decisions, such as for the treatment of cancer.

Practice It!

Retrieval practice is the process of pulling information from your long-term memory and using it. Here are some questions to help with your retrieval practice.

Self-Quiz

1. ______ is when individuals are identified by variations in their DNA sequences. (LO 9-1)
a. Compulsory newborn genetic testing
b. DNA profiling
c. Gene editing
d. Genetic modification

2. One use of DNA profiling is to ______. (LO 9-2)
a. make genetically modified organisms (GMOs)
b. screen newborns for genetic disorders at birth
c. exonerate wrongfully convicted individuals
d. determine the likelihood of behavioral issues in newborns

3. PCR is often used ______. (LO 9-2, 9-3)
a. in DNA profiling
b. to extract DNA from a sample
c. as a cloning vector
d. to cut DNA at specific sites

4. PCR produces millions of copies of ______ in a few hours. (LO 9-3)
a. DNA
b. carbohydrates
c. protein
d. cells

5. Gel electrophoresis is a technique that separates fragments of DNA by______. (LO 9-4)
a. sequence
b. length
c. species
d. SNPs

6. ______ cut(s) DNA molecules at specific sites. (LO 9-5)
a. Polymerase
b. Molecular probes
c. Restriction enzymes
d. Ligase

7. A transgenic organism ______. (LO 9-7)
a. has a gene from another species
b. has been genetically modified
c. is produced through selective breeding
d. both a and b

8. ______ can correct a genetic defect. (LO 9-9, 9-10)
a. Sequencing an individual's DNA
b. Cloning a gene
c. Gene therapy
d. Electrophoresis

9. ______ is when cells are removed from the body, a faulty gene is repaired, and the cells are put back into the body. (LO 9-9, 9-10)
a. *In vitro* fertilization
b. *Ex vivo* gene therapy
c. DNA profiling
d. *In vivo* gene therapy

10. In gene editing with CRISPR/Cas9, the Cas9 enzyme is directed to cut the DNA by the binding of ______ to the target gene. (LO 9-11)
a. a ribosome
b. a guide RNA
c. DNA polymerase
d. a restriction enzyme

11. Which process can be used to replace specific DNA sequences? (LO 9-11, 9-12)
a. PCR
b. CRISPR/Cas9
c. Next-generation sequencing
d. Sanger sequencing

12. CRISPR/Cas 9 gene editing can be used to inactivate an allele in a technique known as a ______. (LO 9-12)
   a. genetic profile
   b. novel gene
   c. gene knock-in
   d. gene knock-out

13. ______ is a form of personalized medicine that uses CRISPR/Cas9. (LO 9-13)
   a. STR analysis
   b. Sanger sequencing
   c. DNA cloning
   d. CAR-T cell therapy

14. A one-nucleotide difference in an allele is called a ______. (LO 9-15)
   a. harmful mutation
   b. single-nucleotide polymorphism (SNP)
   c. short tandem repeat (STR)
   d. genetically modified organism (GMO)

15. Which of the following are used in both Sanger and next-generation sequencing? (LO 9-16)
   a. ddNTPs
   b. gene-specific primer
   c. ligase
   d. primer pair

Critical Thinking

1. Imagine a child with three possible biological fathers. You want to design a simple experiment using the concept of STRs to find out which of the three is the biological father. Describe the basic steps you would take to do this DNA-based test. (LO 9-1, 9-3, 9-4)

2. A biotech company proposes using CRISPR-Cas9 gene editing to create a transgenic crop plant that can resist a new fungal disease. Another team suggests using CRISPR-Cas9 to deactivate the plant's own susceptibility gene instead, creating a non-transgenic GMO. Explain some potential advantages and risks of choosing the transgenic approach compared to the non-transgenic approach. (LO 9-7, 9-8)

3. Preimplantation genetic diagnosis (PGD) checks embryos for genetic conditions before they are transferred into the uterus of the parent. Explain how PGD helps reduce the chance of genetic disease, and describe one scientific or ethical concern that may arise due to the use of PGD. (LO 9-18)

4. Some healthcare providers offer patients genetic testing. Based on the patient's results, individuals are given guidance for SNPs associated with treatable, but not (currently) untreatable, conditions like Alzheimer's disease. What are some potential pros and cons of this approach? (LO 9-18, 9-19)

5. Suppose you amplified a stretch of human genomic DNA containing gene *X* and wanted to insert it into the cloning vector shown in the figure below.
   a. What method would be best to amplify this specific genetic sequence? (LO 9-6)
   b. Knowing that you don't want to disrupt the ampicillin resistance gene, but you do want to transfer the entire sequence of gene *X*, what restriction enzymes should you select? (LO 9-6)
   c. Supposing you *did* accidently disrupt the ampicillin resistance gene, how might it prevent you from continuing your experiment? (LO 9-6)

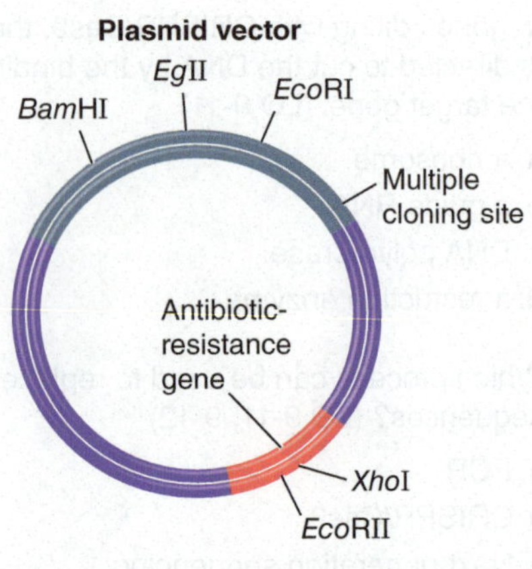

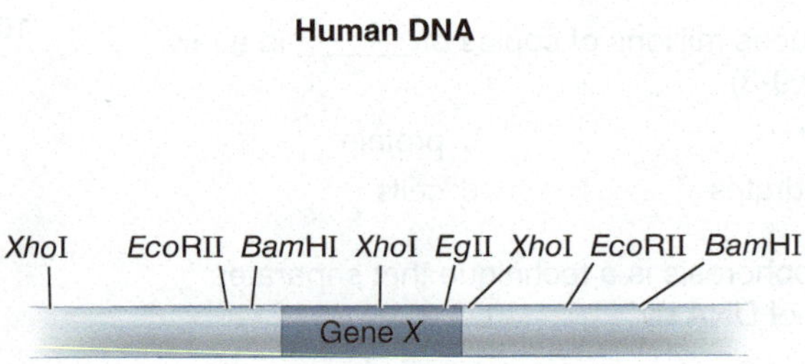

Unit 3 Review

Starting with the fusion of a sperm and egg, genetics touches every aspect of our lives. As we've learned in Unit 3, genetics is a field of study important not only to understand our biology, but also because many of the social issues today are influenced by genetics. From questions of fertility and human development, to the foods we buy at the grocery store, to using genetics as a tool to exonerate those wrongfully convicted, genetics is all around us.

Chapter 7 covered the basics of gene expression in the context of congenital heart defects, the most common type of birth defect. We learned about how genes are expressed, from DNA to RNA to the resulting protein. We also covered the genetic code, how the language of nucleic acids is translated to the language of proteins, and the consequences of mutations.

In Chapter 8, we took a dive into inheritance, including both Mendelian and non-Mendelian inheritance, and the inheritance of gene expression, or epigenetics. We talked about sex, how gametes are formed, and the advantages of sexual reproduction. We also learned about the field of genetic counseling and how traits can be traced across generations.

In Chapter 9, we looked at applications of genetics outside the classroom and how genetic technologies are advancing medicine, biological research, agriculture, and forensics. Although there are some concerns and controversies about issues like genetically modified foods, other uses of genetic engineering, such as in next-generation clothing, will lead to better products for consumers and less waste for the planet overall.

As we leave the human reproduction and genetics unit, we look toward another application of genetics and heredity in Unit 4—evolution.

Credit: Larry Williams/Stone/Getty Images

Unit

4

Tracing Life's Path

Evolution

Have you ever fallen while skating and bruised your tailbone? Perhaps like me, you landed on your tailbone with such force that you required a donut pillow to sit down afterwards. Why do we have tailbones?

During early development, up until about 6 weeks gestation, humans have tails. However, the tail disappears and the vertebrae fuses to form the tailbone (or coccyx, if you're feeling fancy and want to use the anatomical word). Why have a tail? It's an ancestral trait, one shared by all mammals. However, some ancestral primates no longer needed a tail to help with balance and mobility—and so over many generations, having a tail was selected against. Until we reach today, where people have tailbones that are inconvenient after an ice-skating injury, but useful as part of our musculoskeletal system (Chapter 17). For example, muscle groups attach to the tailbone.

Your tailbone, along with other sometimes pesky structures like wisdom teeth, are *vestigial structures*—features that once served a purpose but are no longer functional in the body, left behind over evolutionary time. We aren't the only animals with vestigial structures either—animals like snakes and whales have vestigial hindlimbs.

In this unit, we'll trace life's path and study evolution and how evolutionary processes produced the fantastic array of life on Earth that we know today. We begin with Chapter 10, discussing the evidence for evolution, including vestigial and other important structures, fossils, biogeography, and molecular evidence for evolution.

In Chapter 11, we'll learn about the mechanisms of evolution, such as the different types of selective and non-selective forces that drive evolution. We'll also discuss how speciation and extinction relate to biodiversity on Earth.

In Chapter 12, we'll build on our understanding of speciation to see how phylogenetic trees are used to understand evolutionary history. After learning how species are classified, we'll apply that knowledge to understand human evolution, with a particular focus on how selective and non-selective forces resulted in people with so many different skin tones.

Finally, in Chapter 13, we'll take a field trip across (and near) the United States to explore biodiversity in locations such as a hydrothermal vent, coral reef, estuary, forest, and at your local city park. We'll cover life—starting with the origins of life, moving on to the first cells, and then on to animals and plants. We'll end by discussing the most diverse type of organisms on the planet, land arthropods.

Before we get started with Chapter 10, I want to acknowledge that there are outspoken individuals who contend that biological evolution runs counter to religious beliefs, and that evolution is controversial. Others may say that biologists as a whole reject religion, and one cannot be both a biologist and religious person. The research on how people learn tells us that it is important to talk about these perspectives and emphasize that most people *do not see a controversy*. Rather, the idea of a controversy was propagated by a loud minority, including both scientists and religious leaders.

I and many other biologists (and theologians!) contend that biology and religion are different parts of our lives for seeking answers about the world we live in. Each involves different questions and approaches for answering those questions. Biological inquiry cannot answer questions that religious inquiry can—such as why are we here? What is the purpose of our existence? Vice versa, religion can't tell us about the biological mechanisms that shaped life's path over millennia.

Therefore, to fully embrace and experience the world requires knowing where to direct one's questions. You'll find information about evolution and common misconceptions about evolution in this unit. For information about theological perspectives regarding the lack of a controversy about evolution, see Appendix E.

Preview It!

What is one thing you want to learn in this unit? What are some things you know already about the connection between evolution and biodiversity?

Credit: Ryan Schwark/Wikipedia

10

Darwin, Pokémon, and the Basics of Evolution

Chapter Outline:

A Fish with Arms

In 2004, a team of researchers led by paleontologist Neil Shubin were in northern Canada on Ellesmere Island looking for fossils. Ellesmere Island was once a giant flood plain filled with rivers, swamps, and streams. That season, the scientists made a momentous discovery: three different well-preserved fossil remains of a very special organism with fish-like and land animal-like features.

Like a fish, it had scales and fins. Unlike a fish and more like a land animal, it had a flat head that could rotate on a neck, just like you can turn your head. The most critical discovery was that rather than fish fins, the animal had a forearm, upper arm, and wrist-like structures—the same structures we see in all four-limbed animals today. Further examination of the skull revealed that the animal could breathe air. Since the scientists were working in Nunavut Territory, they deferred to tribe elders to name this special fish. They named this new genus (a biological term used to classify organisms; we'll return to ways of describing lifeforms in Chapter 12) of fish *Tiktaalik*, which means "large freshwater fish." What was so special about *Tiktaalik*?

Shown here is a model of *Tiktaalik*. The discovery of *Tiktaalik* was crucial for understanding the evolutionary processes that led to the major leap from aquatic to terrestrial animals.

Tiktaalik is a transitional organism and demonstrates a step taken on the evolutionary journey between swimming fish and land animals like amphibians, reptiles, birds, and mammals (which include us!). *Tiktaalik* and related animals potentially represent common ancestors for all land animals we see today. What's notable about this discovery is that long before any transitional fossils were found, Charles Darwin hypothesized that they existed. If you think back to Chapter 1, theories have explanatory power, and the best theories can be used to make predictions. In this case, Darwin's prediction was correct, adding more weight to evolutionary theory.

Tiktaalik is only one of many examples of transitional fossils. Fossils are one important line of evidence that supports evolutionary theory. In this chapter, we'll cover evidence from various scientific disciplines that support evolution. We'll explore different lines of evidence for evolution starting with fossil evidence and then moving into biogeography; evidence in form and function; and finally, molecular evidence. Since the Internet is awash in misinformation (Chapter 1) about evolution, look for the trusty avatar in the margin and feature boxes to clear up common misconceptions about the evidence for evolution.

Preview It!

List three things you know about evolution already. Have you heard of any controversies about evolution? What questions do you have about evolution?

10.1 An Introduction to Evolution

Learning Objectives

10-1 Explain what descent with modification means.

10-2 Describe how evolution is descent with modification.

10-3 Describe the role of the environment in evolution.

10-4 Explain how natural selection works.

Note that the criteria that need to be met for Pokémon evolution to occur vary depending on the specific game.

Have you ever played with Pokémon cards? Or maybe played one of the many Pokémon video games? One of my favorite Pokémon is Charmander. Many people can recognize Pikachu, Charmander, Squirtle, and Bulbasaur. Now, Charmander, like many other Pokémon, evolves. We'll call this kind of evolution Pokémon evolution to distinguish it from biological evolution. Collecting 25 candies leads to Charmander becoming Charmeleon. Then if you advance to level 16, Charmeleon evolves to become Charizard (**Figure 10.1**). Although a fun game to play, Pokémon evolution is *not* biological evolution. We start here because Pokémon evolution gives us a good framework for what biological evolution is not.

Figure 10.1 Comparing Pokémon and biological evolution

In Pokémon evolution, individuals evolve into a "super" Pokémon—in this example, from Charmander (left) to Charmeleon (center) to Charizard (right). In reality, biological evolution occurs at the species level, over generations, and the result is a group of organisms that are suited to their particular environment.

In Pokémon evolution, an individual evolves to become a more sophisticated version of themselves. If I could level up and grow wings and fly, I certainly would! But biological evolution doesn't happen to individuals. Rather, it happens across a **species**, a group of individuals that can reproduce with one another to produce offspring that can also reproduce. In biological evolution, populations of organisms change over many generations to produce a new species. Pokémon evolution suggests that there is a final "ideal" powerful organism that is better than all the rest. In reality, there is no ultimate "perfect" organism; instead, there is just a group of organisms that change over generations to be best suited to their environment in that time and place. With that in mind, what is biological evolution?

What about asexually reproducing organisms like bacteria? Bacteria can take up new DNA from their environments or exchange genetic material with one another. This increases genetic diversity and can drive evolution.

What Is Evolution?

As far as basic biological concepts go, I'm not entirely sure any single concept has garnered so much negative press as evolution. Additionally, there is no shortage of misinformation (Chapter 1), found everywhere from billboards to "museums" and of course plastered all over the Internet. Despite all that's been written and said about evolution, the central concept is surprisingly simple. And like most simple things, it is extremely powerful. Evolution, after all, is the driving force we see behind the amazing diversity of life on Earth (Chapter 13). So, what is that central simple idea?

Evolution is all about sex.

Yes. That's it. Species change over time (they evolve) based on how good the individuals of that species are at reproducing (having lots of babies), and how good those offspring also are at surviving and reproducing. That's why **evolution** can be defined as descent with modification. *Descent* refers to passing on your genes—heritability. Like we learned in Chapter 8, parents pass genetic and epigenetic information on to their offspring. The *with modification* piece refers to how genes change over time.

Since evolution happens across a species, this is why it is not true to say humans evolved from monkeys. Rather, our species and chimpanzees, with whom we share 98.8% of our DNA, evolved from a common ancestor who lived 6–7 million years ago.

How good an organism is at reproducing depends on its specific genes and the environment. Different genetic traits arise as the result of mutations (changes in the genetic code; see Chapters 6 and 7). Mutations can lead to phenotypes (Chapter 8) that give an organism an advantage (or disadvantage). If it's an advantage that contributes to an organism's survival such that they successfully reproduce, the beneficial gene variants get passed on to their descendants as well.

But what makes a mutation beneficial? It depends on the environment in which the organism lives. **Natural selection** is the process by which a trait becomes more common if it leads to greater reproductive success, or less common if it leads to decreased reproductive success. In certain environments, a gene variant may provide an advantage, and organisms with that gene variant will survive and reproduce more successfully than others in the population without the gene variant. In other environments, the same gene variant may not be beneficial.

For example, the specific set of gene variants that polar bears have allows them to be successful for surviving life in the cold, harsh conditions in the Arctic (Figure 10.2). Said otherwise, polar bears have **adaptations**, or biological features that arose due to natural selection that contribute to their species' ability to survive in an environment. However, these gene variants (and the adaptations they code for) become a liability if polar bears are in a warmer environment—a concern for the health of polar bear populations as our planet warms (Unit 5).

Figure 10.2 Polar bears contain gene variants that help them survive in extreme cold

Credit: National Geographic Image collection

Although the gene variants that lead to polar bears having adaptations such as an insulating layer of fat, two layers of fur, and small ears and tail are beneficial in the Arctic, these would be considered harmful gene variants in warmer climates.

Natural selection operates on traits at a genetic level, so gene variants that either increase or decrease reproductive success, or **fitness**, tend to become more or less common in a population over time, respectively. Over many, many generations (Section 10.3), new phenotypes or even species can arise as beneficial gene variants become more common in the

Figure 10.3 **A light microscope image of group A *Streptococcus pyogenes***

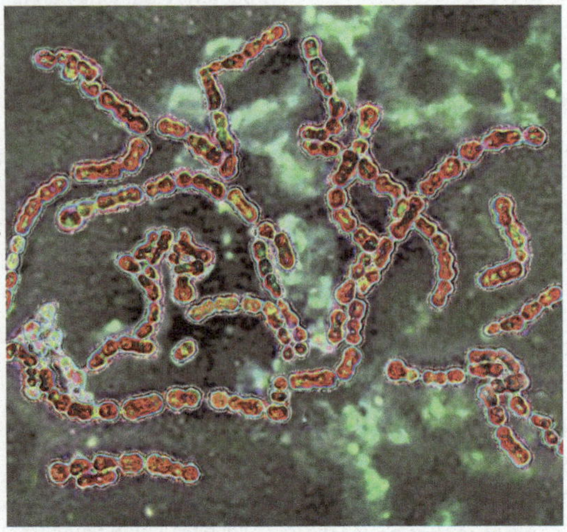

Credit: CAVALLINI JAMES/BSIP/Universal Images Group/Getty Images

S. pyogenes infection can cause everything from strep throat to skin infections.

population. Conversely, if a gene variant occurs that interferes with an organism's ability to survive and pass on its genes, these gene variants will become less common in the **gene pool**, all the genes present in a population, until they're gone entirely. Forces that shape which gene variants are present in the gene pool can be selective, like natural selection, or non-selective, like when a catastrophic natural disaster occurs that randomly removes gene variants from the gene pool. We will discuss these forces in detail in Chapter 11.

Let's look at an example of natural selection in action with a current and significant public health concern: the rise of antibiotic resistance in bacteria.

Natural Selection in Action

It starts as a tickle in the back of your throat. Maybe the air is dry, or you've been talking too much. But then you begin to feel feverish…and before too long, you're running a fever and every time you swallow, it feels like broken glass going down your throat. Then, your friend texts to say they've come down with strep throat—a contagious bacterial infection of the throat caused by group A *Streptococcus* bacteria (**Figure 10.3**).

So you drag yourself to the clinic, where they swab the back of your throat to see if you have group A *Streptococcus* bacteria in your throat. Sure enough, you do—you've got strep throat as well. The doctor sends you home with a prescription for amoxicillin, an antibiotic. You can't help but wonder—why can't we take antibiotics anytime we are sick, instead of just when we have a bacterial infection?

Diseases caused by bacteria, such as scarlet fever and tuberculosis, once caused a quarter of deaths annually in the United States. That is, until 1928, when Sir Alexander Fleming discovered penicillin, a chemical produced by mold that kills bacterial cells. Antibiotics contributed to major advances in medicine and surgery and, alongside other public health interventions like sanitation, were part of the reason why life expectancy in the United States increased from 56.4 years in 1920 to approximately 77 years 100 years later.

Figure 10.4 **The use and development of antibiotics resulted in the evolution of antibiotic-resistant bacteria**

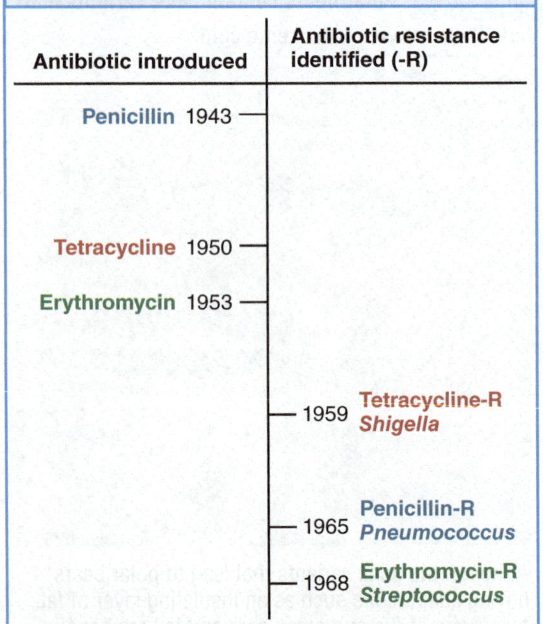

Note that the "-R" following a type of antibiotic indicates the first appearance of an antibiotic-resistant strain.

The benefits of antibiotics are undeniable. However, within a few decades of their development, it became apparent that there was a big problem with using antibiotics—the evolution of antibiotic-resistant bacteria. A natural population of bacteria can evolve astonishingly fast. Consider how each cell division is an opportunity for mutation. Bacteria such as *Salmonella* can divide every 20 minutes. When a natural population of bacteria is exposed to an antibiotic, some cells are likely to survive because they carry an allele that offers resistance. As susceptible cells die and the survivors reproduce, the frequency of the antibiotic-resistant allele in the population increases.

Considering the frequency with which bacteria reproduce, a two-week course of treatment with antibiotics can exert a selective force on hundreds of generations of bacteria. This selective force drives genetic change in bacterial populations so that they become composed mainly of antibiotic-resistant cells.

Thus, using antibiotics on an ongoing basis effectively guarantees the production of antibiotic-resistant bacterial populations. Perhaps not surprisingly, the introduction of new antibiotics over the 20th century also saw the appearance of new antibiotic-resistant strains of bacteria (**Figure 10.4**). As resistant bacteria become more common, the number of antibiotics useful for treating infections in humans dwindles.

Misuse and overuse of antibiotics are a major driver of antibiotic resistance. For example, antibiotics don't help you feel better when you have a viral illness like the flu, since antibiotics kill only bacterial cells, but taking antibiotics creates a selective force for any bacteria living in or on your body, thus promoting the development of antibiotic resistance.

One place of concern for the overuse of antibiotics is on farms. According to the U.S. Food and Drug Administration's Center for Veterinary Medicine, in 2022, farmers used 6.2 million kilograms of medically important antibiotics—more than half of the U.S. supply!—on livestock, causing hot spots for the evolution of antibiotic-resistant bacteria.

People who work with animals on these farms tend to have more antibiotic-resistant bacteria in their bodies, as do neighbors living within a mile radius. The bacteria spread much farther than that, however. For example, bacteria on an animal's skin or in its digestive tract can easily contaminate the meat during slaughter, and contaminated meat ends up in kitchens. It's not just a problem for people who eat meat, either. Antibiotic-resistant bacteria are now found on crops like spinach.

In humans, an infection with multidrug-resistant bacteria tends to be longer, more severe, and more likely to be deadly than an infection caused by non-resistant bacteria. Antibiotic-resistant bacteria cause more than 2 million cases of serious illness each year in the United States alone, and outright kill more than 150,000 people. Many more die because the infection complicates another preexisting illness.

Big Ideas in Biology

Evolution

Evolution is a major unifying theory of biology—all different kinds of biology tie back to evolutionary theory in some way. That's also why it's one of our big ideas introduced in Chapter 1.

Biology for Life

Is Evolution Truly Controversial?

Much of the controversy around evolution doesn't relate to how evolution works—that is, that species change gradually over the course of time in response to a variety of factors. Rather, the controversy involves the question of whether evolution by itself is sufficient to explain the rise of new species and the diversity of life on the planet. In addition, some specifically question whether humans evolved, descending from a common ancestor shared with other species, or if we have existed only in our present form. Furthermore, some people take exception to the evidence for evolution, such as fossil evidence or radiocarbon dating. One consequence of the focus on this controversy is that people get caught up in the conflict and miss out on the importance of timely issues like the evolution of antibiotic-resistant bacteria.

According to the Pew Research Foundation, approximately 81% of U.S. adults say humans evolved over time. Of those people, some believe that a higher power had a role in evolution (which some call theistic evolution), and some do not, believing that evolution occurred solely due to processes like natural selection. Among scientists, the Pew Research Foundation found that 98% accept the evidence for human evolution. Despite this widespread acceptance of evolution and religious beliefs, a few prominent evolutionary biologists and religious leaders have promoted the idea that there is a conflict between religion and evolution (or science in general). This small group of people would have you believe that one cannot be both religious and accepting of the scientific evidence for evolution. However, when one examines the science (as is covered here, in Unit 4) and theological perspectives (which we give a brief overview of in Appendix E), it's clear that you can both hold religious beliefs *and* accept evolution, each with its own role in your life.

For instance, in 2014, Pope Francis, leader of the Catholic Church (which is the religion of approximately 17% of people worldwide), made a statement that there is no conflict between religious belief and scientific theories such as evolution or the Big Bang. The Pew Research Group maintains a list of various religious groups and their stances on evolution—and notably, most religious groups do not see a conflict between their religious teachings and evolutionary theory.

Rather, biology and religious beliefs (or the lack thereof) are both valid and important—and each is concerned with a different aspect of the world. For example, just as you would not consult a biology text for guidance on spiritual matters, a religious text would not be a reliable source of scientific information (see Chapter 1 for a review on what counts as scientific information). Science is a process for exploring and understanding the world we live in, theology is the study of God and spirituality, and they each have their place in our society.

1. Do you feel like there is a conflict between your personal beliefs and evolution?
2. Using the Internet, identify scientists who are religious *and* accepting of evolution. See if you can find someone with a faith background (religious or not) that is similar your own and one who is different.
3. Many people fall on a spectrum between atheistic evolution—that God did not play a role in evolution—and creationism, where humans were created in their current form by a higher power. Using the Internet, research some of these viewpoints, including theistic evolution or agnostic evolution. Where do you fall on the spectrum from atheistic evolution to creationism?

Process of Science

It's Not Just Darwin!

Usually, one can't hear about evolution without also immediately associating it with Charles Darwin, who published *On the Origin of Species* in 1859 and is credited with first describing natural selection. Evolution, like cell theory (Chapter 5), is one of the core theories of biology. Darwin's work wasn't the sole result of just his observations, but also the result of the work of many other scientists, some of whom, like Aristotle, lived thousands of years before Darwin was born. Science is a collaborative effort, and the development of evolutionary theory is not an exception.

The earliest premise for evolution can be traced to 2,300 years ago, when Greek philosopher Aristotle described nature as a continuum of organization, from lifeless matter through complex plants and animals. Aristotle's work greatly influenced later European thinkers, who adopted his view of nature and modified it in light of their own beliefs. By the 14th century, Europeans generally believed that a "great chain of being" extended from the lowest form of life (plants), up through animals, humans, and spiritual beings. Each link in the chain was a species, and each was said to have been forged at the same time, in one place, and in a perfect state. The chain was complete. Because everything that needed to exist already did, there was no room for change.

This was the predominant viewpoint until the 1800s, when European naturalists embarked on globe-spanning survey expeditions and brought back tens of thousands of plants and animals. Each newly discovered species was carefully cataloged. The European explorers began to see patterns in where species live and similarities in body plans. As a result, they started to think about how natural forces might affect life. Some of the key naturalists involved in theorizing on the evolution of species, and descriptions of their work, can be found in Figure 10.5.

Jean-Baptiste Lamarck proposed the idea that a species gradually improved over generations because of an inherent drive toward perfection, up the chain of being. By Lamarck's hypothesis, environmental pressures cause an internal requirement for change in an individual's body, and the resulting change is inherited by offspring. Lamarck's understanding of how inheritance works was incomplete, but he was the first to propose a mechanism for evolution. Not everyone agreed with Lamarck, however. George Cuvier vehemently rejected Lamarck's hypothesis in favor of catastrophism, that Earth's landscape was shaped by periodic, violent geologic events unlike any in recorded history.

Charles Lyell, a geologist, didn't agree with Cuvier, though. Rather, he was a proponent of uniformitarianism, the idea that gradual, everyday geologic processes such as erosion shaped Earth's surface. Lyell realized that ordinary geologic processes that sculpt rock formations in the present could have also sculpted rock formations in the past—if they took place over millions of years. This idea challenged the prevailing belief that Earth was thousands of years old, but Lyell's thorough documentation of rocky evidence in support of geologic continuity led to its general acceptance by other naturalists.

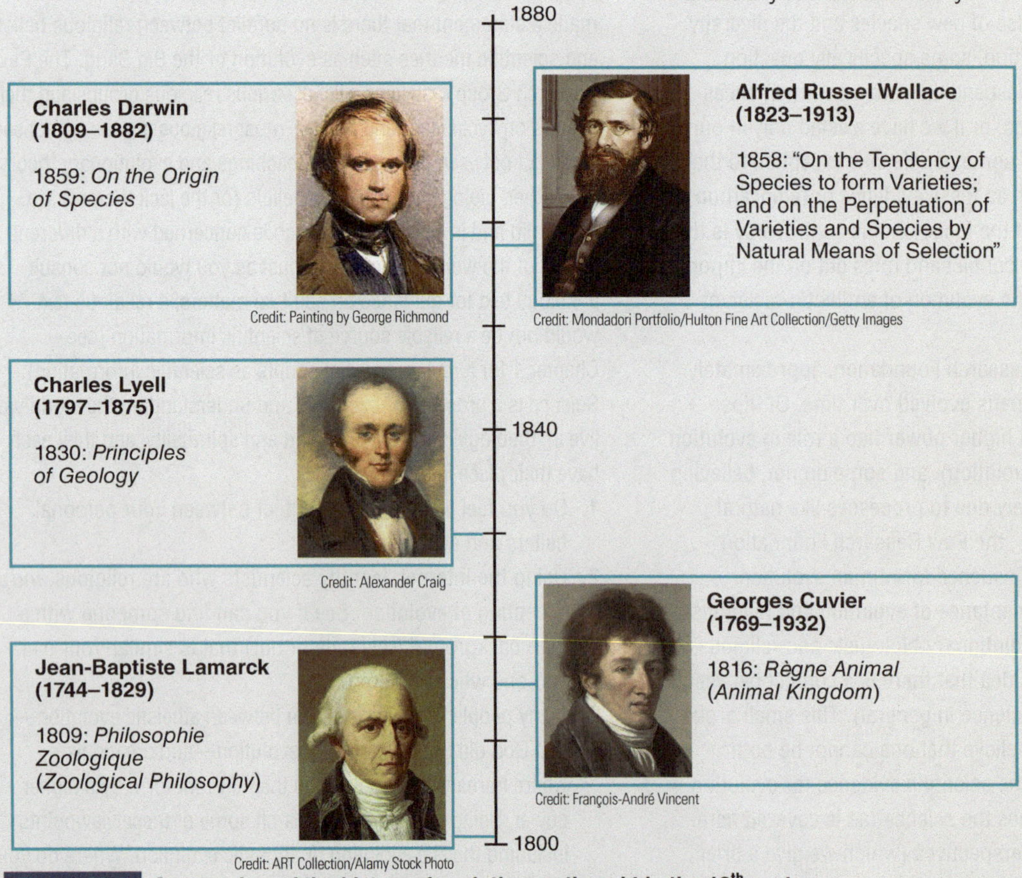

Figure 10.5 An overview of the history of evolutionary thought in the 19th century

Some of the key naturalists of their time are noted, along with their primary work concerning evolution.

Now, we can get to Darwin. Building on his theology degree and the work of Lamarck, Cuvier, and Lyell, his five years spent journeying around the world via sea on the *Beagle*, and his familiarity with selective breeding of domesticated animals, he began to put the pieces together for natural selection. Darwin realized that individuals best adapted to a particular environment would tend to leave more offspring than their less fit rivals, a process he named natural selection. If Earth were millions of years old (as Lyell had proposed), then there had been ample time for natural selection to have driven evolution.

Darwin came up with his hypothesis of evolution by natural selection in the late 1830s, but did not publish it right away. He spent a decade collecting evidence, focusing on other projects, and battling ill health before starting to compile a book on the subject. Meanwhile, Alfred Russel Wallace, who was studying wildlife in the Amazon Basin and the Malay Archipelago, wrote to Darwin about patterns in the geographic distribution of species. When he was ready to publish his own ideas about evolution, he sent them to Darwin for advice. To Darwin's shock, Wallace had come up with the same idea: that evolution can be driven by natural selection.

In 1858, the concept of evolution by natural selection was presented at a scientific meeting. Darwin and Wallace were credited as authors, but neither attended the meeting. The next year, Darwin published *On the Origin of Species*, which laid out detailed evidence in support of evolution. By that time, many people had accepted the idea of descent with modification (evolution). Decades would pass before experimental evidence from the field of genetics led to its widespread acceptance by the scientific community.

1. How does the development and refinement of evolutionary theory over time highlight the collaborative nature of science?
2. Using the Internet, research others (especially from diverse backgrounds) who have contributed to the development of evolutionary theory, both today and historically.

10.1 Summary

- Descent with modification means that parents pass genes on to their offspring. These genes may change over generations as a result of heritable mutations.
- Evolution is the result of cumulative inherited changes. Descent with modification occurs across a species, not in an individual. When individuals of a population survive and reproduce with differing degrees of success, this changes the alleles present in the gene pool and can drive evolution.
- Natural selection occurs among a group of organisms with different genetic traits, where some survive and go on to reproduce more successfully than others in the population. Environmental conditions can make a particular trait an advantage or disadvantage in that environment. A trait that enhances fitness becomes more common in a population over time.
- When a population of bacteria is exposed to antibiotics, most will die. Some bacteria will have genes that allow them to survive long enough to reproduce. The presence of the antibiotic is the selective pressure that leads to antibiotic-resistant bacteria.

The scientists that history remembers as the big contributors to evolutionary thought were also a product of the time in which they lived and worked: It was a time when mainly White male European voices were amplified. In fact, many of the naturalists featured here integrated the ideas of diverse individuals into their work. For example, Charles Darwin regularly communicated with Mary Treat, a fellow naturalist, regarding the biology of carnivorous plants.

10.2 Fossils and the History of Life

Learning Objectives

10-5 Describe the geologic time scale in terms of life's history.

10-6 Describe the kinds of information that fossils provide about past life.

10-7 Relate how fossils form to why fossils are relatively rare.

> *"The two most powerful warriors are patience and time."*
>
> —Leo Tolstoy, *War and Peace*

Time. It's such a strange force. One could argue that time runs our lives. But time doesn't necessarily mean the same thing at any given point in a day. Let's consider 30 minutes. 30 minutes of extra sleep in the morning feels like nothing when you've stayed up too late the night before. But 30 minutes left during your least favorite class? An eternity.

Figure 10.6 Time in layered rocks

Eon	Era	Period	mya*	Major Geologic and Biological Events
Phanerozoic	Cenozoic	Quaternary		Modern humans evolved. The sixth mass extinction is underway.
			2.58	
		Neogene		The tropics and subtropics extend poleward. The climate cools, allowing woodlands and grasslands to emerge. New mammals, insects, and birds evolve.
			23.0	
		Paleogene		Flowering plants diversify and sharks evolve. Primates appear.
			66.0	◀ Mass extinction event
	Mesozoic	Cretaceous		The climate is very warm and dinosaurs flourish. Insects such as bees, butterflies, termites, ants, and grasshoppers appear. The first flowing plants appear.
			145.0	
		Jurassic		Lush vegetation, mainly ferns and coniferous plants. The first birds appear. Pangea breaks apart.
			201.3	◀ Mass extinction event
		Triassic		Turtles, dinosaurs, and mammals appear.
			252	◀ Mass extinction event
	Paleozoic	Permian		Pangea forms. Plants first appear (conifers, ginkgos, and cycads), as well as the first beetles and flies.
			299	
		Carboniferous		High atmospheric oxygen fosters giant arthropods (organisms with a hard exoskeleton, like insects). Spore-releasing plants dominate. Vast coal forests form.
			359	◀ Mass extinction event
		Devonian		The first four-legged land animals appear. An explosion in plant diversity occurs, including the first plants with seeds.
			419	
		Silurian		New marine invertebrates such as corals evolve. The first jawed fish evolve. First evidence of life on land, including fungi, millipedes, and spider-like organisms.
			443.8	◀ Mass extinction event
		Ordovician		The first land plants, fishes, and reef-forming corals appear. Gondwana moves towards the South Pole and becomes frigid.
			485	
		Cambrian		Earth thaws, resulting in an explosion of animal diversity. Most major groups of animals appear in the oceans. The first trilobites and shelled organisms appear.
			541	
Precambrian*	Proterozoic			Oxygen accumulates in the atmosphere. The first aerobic life, eukaryotic cells, appear, followed by protists, fungi, plants, and animals. Earth freezes over in a series of global ice ages between 750–600 million years ago.
			2,500	
	Archean			The origin of bacteria and archaea.
			4,000	
	Hadean			The origin of Earth's crust. The first atmosphere and ocean form.
			~4,600	

The major geologic and biological events in Earth's history are summarized in the timeline shown here and correlated to rock layers in the Grand Canyon in Arizona. Notice that the timeline starts with more recent times at the top, moving back in time as you move down the timeline. Similarly, more recent layers of rock are toward the top of rock formations, with the older rocks toward the bottom.

*mya stands for Millions of Years Ago

Credit: Dates are from the International Commission on Stratigraphy

Now think about a historical event—what was a long time ago? Breakfast? Your fifth birthday? The 9/11 terrorist attacks? The Industrial Revolution? The extinction of the dinosaurs? Time is difficult to grapple with because the way we perceive time is so different, even over the course of a single day, let alone when we consider the age of fossils.

Let's consider geologic time with a pop culture example: Taylor Swift. Swift made history with the Eras Tour—the first concert tour to generate more than $1 billion in revenue. Each performance was divided into 10 acts, or eras. Since then, people have begun to describe phases or times in their lives as eras. One could say that you are in your college era.

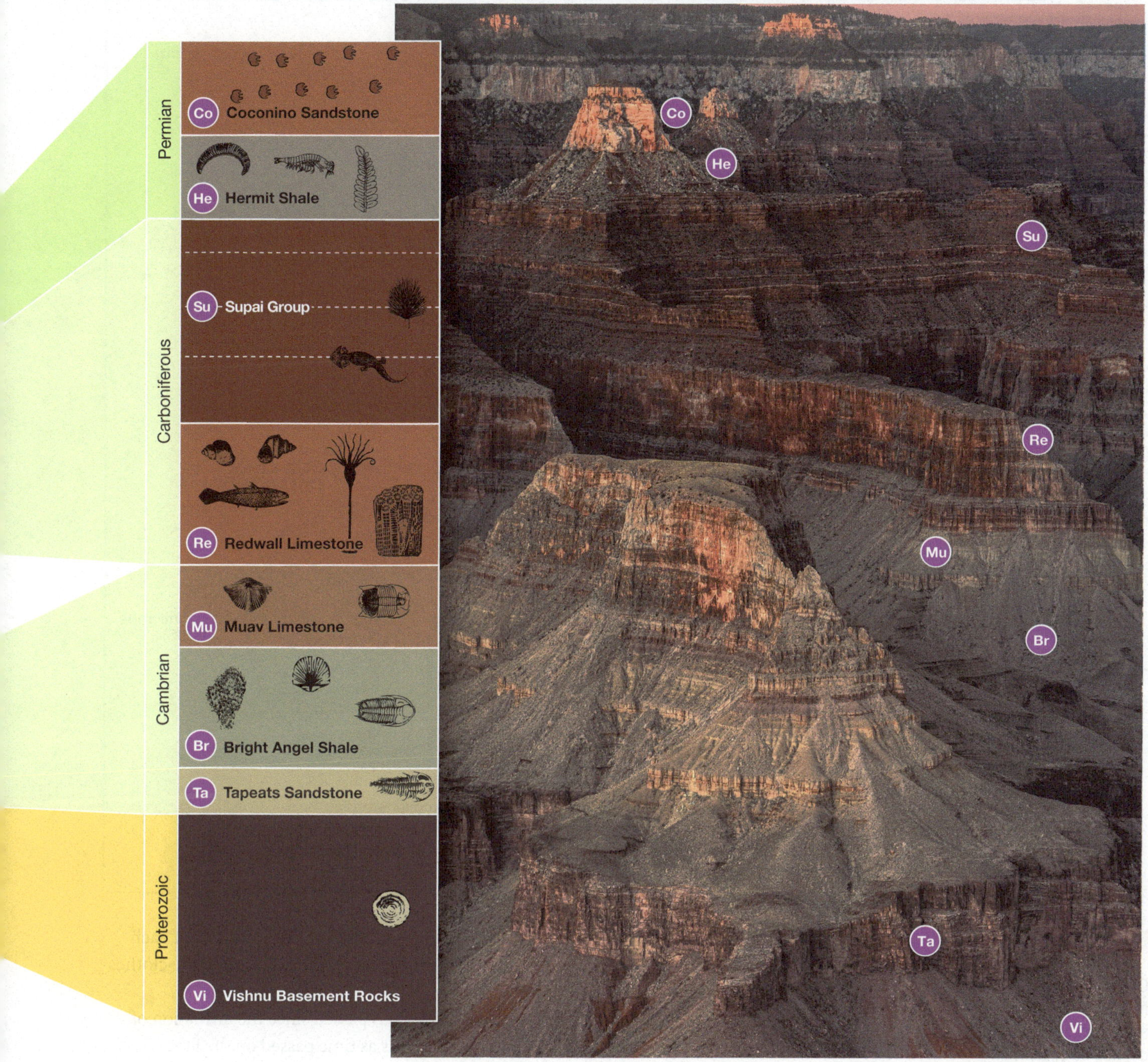

Credit: Cavan Images/Cavan/Getty Images

Considering Earth's Eras: The Geologic Time Scale

Just like you've gone through eras or phases in your life, so too has the Earth—but over a much, much greater span of time. We can summarize the different phases of Earth's history (which can also be parsed into eras, as well as other divisions, such as epochs and periods) into the **geologic time scale**, a chronological history of Earth that correlates layers of rock with great intervals of time (Figure 10.6). The composition of each layer holds information about environmental conditions that prevailed during the time it was deposited. For example, the deepest layers in a formation were the first to form, and those closest to the surface formed most recently. You can draw an analogy with your laundry basket—the clothes you wore most recently are

> **? Did You Know?**
>
> The first penises evolved in reptiles during the carboniferous period. Vaginas evolved much later, and only in mammals.

Figure 10.7 A clock face as a metaphor for the major events in Earth's history

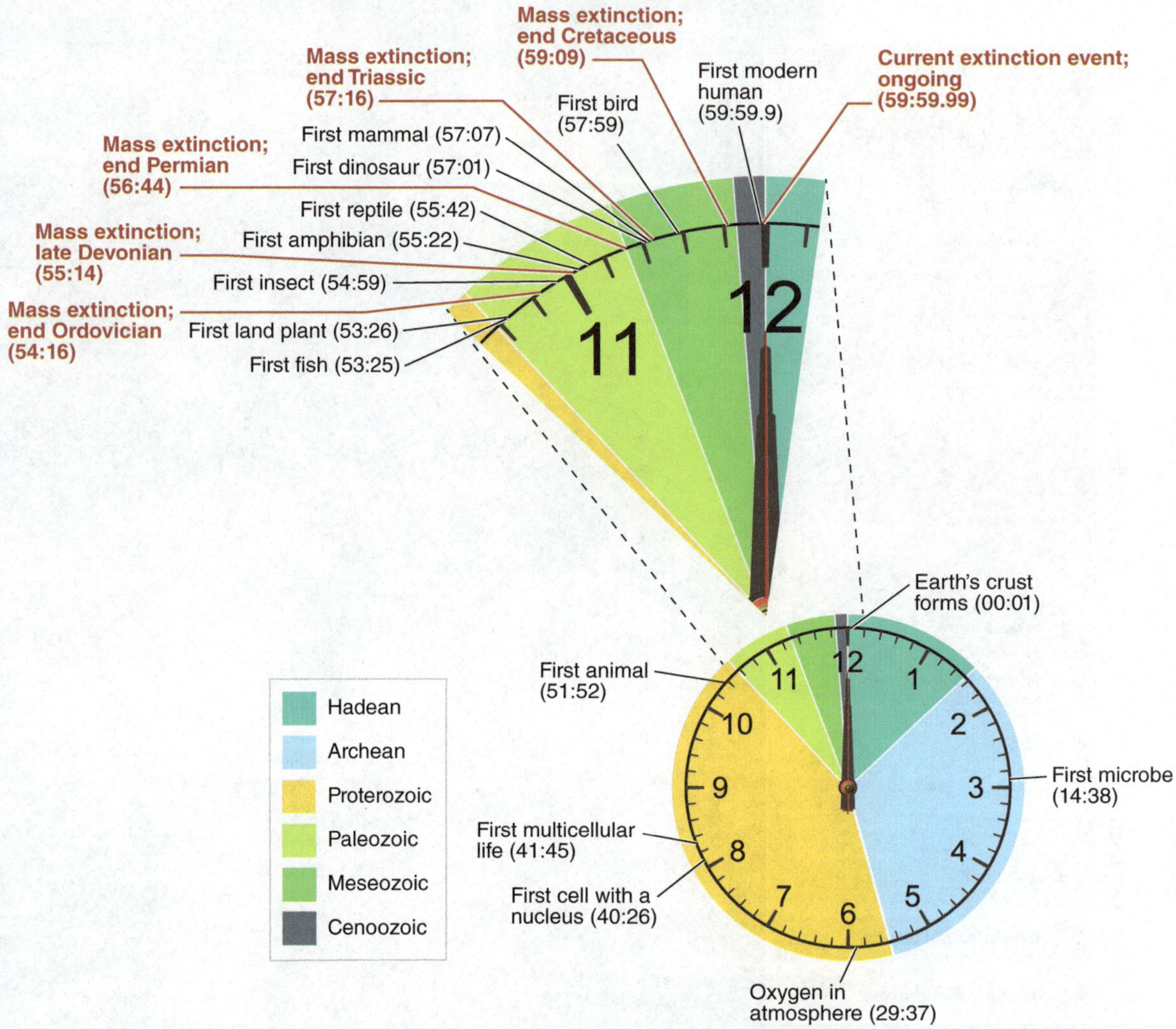

Using the metaphor of a clock, Earth's crust forms just past midnight, and the first eukaryotes appear approximately 40 minutes later. The first humans appear just before the end of the hour. Despite the barely perceptible blip in which humans have existed on Earth, we've had a major impact on life, which we will return to in Unit 5.

? Did You Know?

Over Earth's history, there have been five mass extinction events, periods defined by catastrophic loss of species. The last one was 66 million years ago, when the dinosaurs died out. Some scientists contend that we're currently amid a sixth mass extinction, brought on by human activities.

found on top, and the clothes you wore last week are on the bottom. The composition of each layer also reflects the environmental conditions that prevailed as it formed—just like the clothes you were wearing last week reflect the weather conditions.

One way to help grapple with the vast amount of time that has passed since the formation of the Earth is to visualize the major geological and biological events as time passed on the face of a clock, with the history of Earth to the present compressed into one hour (Figure 10.7). The first bacteria appeared 17 minutes and 37 seconds after the Earth formed, mammals don't appear until the hour has almost passed, and humans appear just as the hour ends. These metaphors stress that the time humans have existed on Earth is a drop in the bucket compared to the age of the Earth, with humans appearing at 59:59.9, in the same second when the ice ages ended. For the tiny amount of time humans have existed in Earth's history, we have had a major impact on the planet—we will return to the current human-induced extinction event in Chapter 11 and climate change in Unit 5.

Fossils: Physical Records of Evolution

Fossils are the remains or traces of organisms that lived long ago (Figure 10.8). We've already seen one example of fossils in this chapter—*Tiktaalik* in the chapter opener. Different layers of rocks contain fossils which serve as a record of life that existed during the same time frame. Most fossils consist of mineralized bones, teeth, shells, seeds, spores, or other durable body parts. Less durable

Figure 10.8 **Fossils reveal not only insight into the history of evolution, but a peek into the lives of ancient animals**

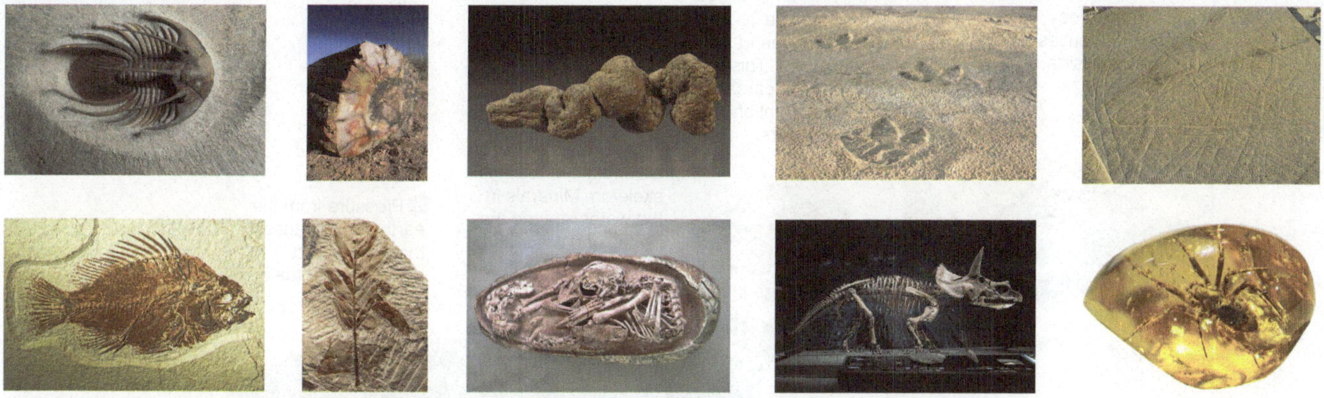

Starting at the top-left and moving clockwise is a trilobite, an ancient aquatic animal; a petrified log in Arizona; a coprolite, which is fossilized dinosaur feces; fossilized dinosaur footprints; fossilized trails from an aquatic, sluglike creature called *Climactichnites*; a spider trapped in amber (tree sap); a triceratops skeleton; a fossilized dinosaur egg containing an embryo; a fossilized leaf; and a fossilized fish.

Credits: National Geographic Image Collection; Yva Momatiuk and John Eascott/Minden Pictures; Breck P. Kent/Shutterstock.com; istock.com/Inner_Vision; Kennethcgass/Wikipedia.org; MILLARD H. SHARP/SCIENCE PHOTO LIBRARY; National Geographic Image Collection; China News Service/Getty Images; Michel Porro/Getty Images; Tomekbudujedomek/Getty Images

body parts, such as tissues and organs, typically degrade before fossilization can occur. Trace fossils—such as footprints and other impressions, nests, burrows, eggs, and even feces!—are all evidence of the organisms that once lived, and how they lived. For some soft-bodied organisms, like jellyfish, the impressions they left in mud are the only historical record we have. For others, like sharks, only their teeth remain—their cartilaginous bodies do not fossilize.

Layers of rock differ in composition and fossil content, and these differences imply transitions in Earth's history. For example, there are two major formations of sedimentary rock that stretch across the western United States. Hermit Shale (see Figure 10.6) consists mainly of compacted mud and silt with abundant fossils of land plants and insects. Therefore, this formation was probably deposited by a river system that stretched across a broad, wet coastal plain.

Coconino Sandstone which was deposited on top of Hermit Shale, consists mainly of compacted, weathered sand. Ripple marks and reptile tracks are the only fossils in it. These and other characteristics indicate that it accumulated as a vast sand desert like the modern Sahara. We can infer that a major climactic shift must have occurred between the deposition of Hermit Shale and Coconino Sandstone that resulted in the river system drying up.

> **? Did You Know?**
>
> Studying pollen or spore fossils in a rock layer can give an indication of the plants that existed at that time and therefore can provide insight about the climate in which the plants lived. For example, a predominance of pine pollen would indicate a cooler, drier climate, whereas plants such palms would be present in a warmer climate.

How Do Fossils Form?

Fossils form only under very specific conditions, a process called **fossilization** (Figure 10.9). This process typically begins when an organism or its traces becomes covered by sediments, mud, or ash. Groundwater then seeps into the remains, filling in the spaces around and inside of them. Minerals dissolved in the water gradually replace the minerals in bones and other hard tissues, which then crystallize inside the cavities to form detailed imprints of internal and external structures. Sediments that slowly accumulate on top of the site exert increasing pressure, and, after a very long time, extreme pressure transforms the mineralized remains into rock.

Most fossils are found in sedimentary rock. Sedimentary rocks form as rivers wash silt, sand, volcanic ash, and other materials from land to sea. Mineral particles in the materials settle on seafloors in horizontal layers that often vary in thickness and composition. After millions of years, the layers of sediments become buried and compacted into layers of rock. Geologic processes can tilt sedimentary rock and lift it far above sea level, where erosion can expose the layers to reveal the fossils within them.

Biologists study sedimentary rock formations to understand the historical context of ancient life. For example, they use the presence of **index fossils**, the remains of organisms that were common during a certain geologic time period, to assign ages to rock samples. One such index fossil is

Figure 10.9 **An overview of fossilization**

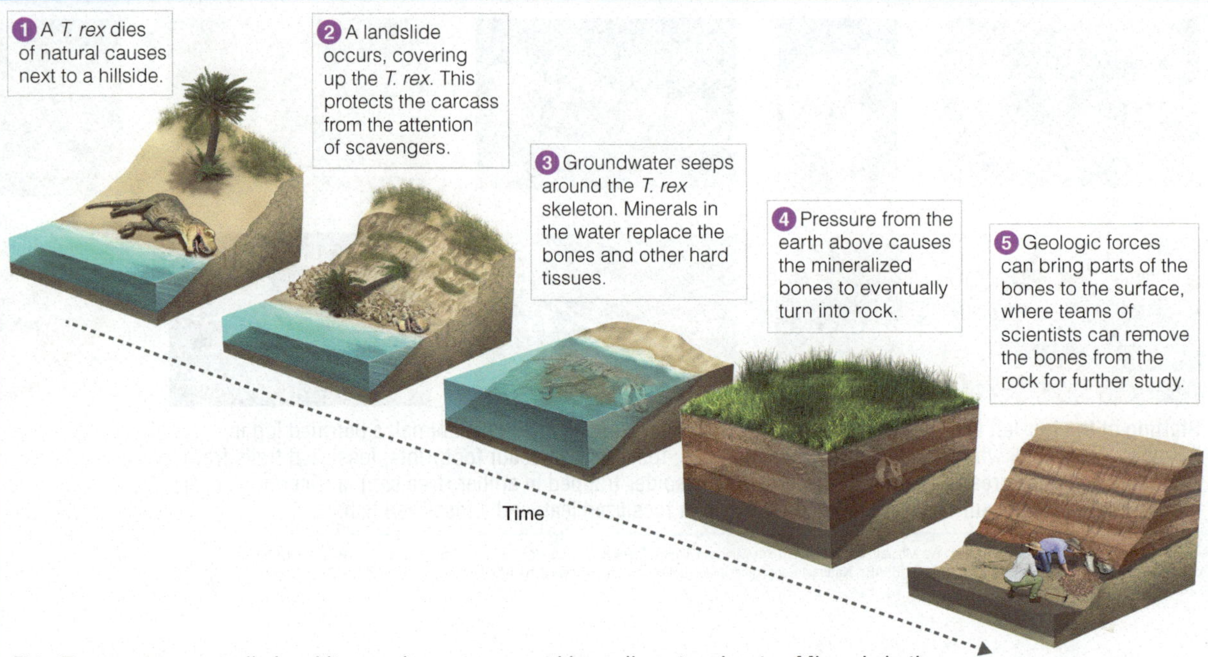

This *Tyrannosaurus rex* died and its remains were covered by sediment and water. Minerals in the water replaced the minerals in bone, where eventually the mineralized remains turned into rock.

foraminifera (Figure 10.10), a single-celled protist (Chapter 13) that lived in ancient oceans and can still be found in marine environments today. Foraminifera have lived in the world's oceans for hundreds of millions of years, and different species of foraminifera have lived at different times. When they die, they leave behind hard calcium carbonate shells, which become fossilized. Scientists can look for specific species of foraminifera to infer the age of both the rock and any other fossils in the same rock.

How Are Fossils Dated?

Using index fossils is one way of determining the approximate age of a layer of rock. Alternately, one can examine the relative positions of different rocks to infer age. Because rock layers form sequentially, one layer on top of another over time, we know generally that a fossil is older than the rocks above it and younger than the rocks below it. This form of dating is called **relative dating**, and it gives us a general sense of the age of a fossil. However, this doesn't tell us the exact age of a particular rock layer.

One way to get a more precise age for a fossil is by determining the exact age of the rocks in the adjacent layers, particularly volcanic rocks. When volcanic rock forms, it contains radioactive

Figure 10.10 **Tracing Earth's history using foraminifera fossils**

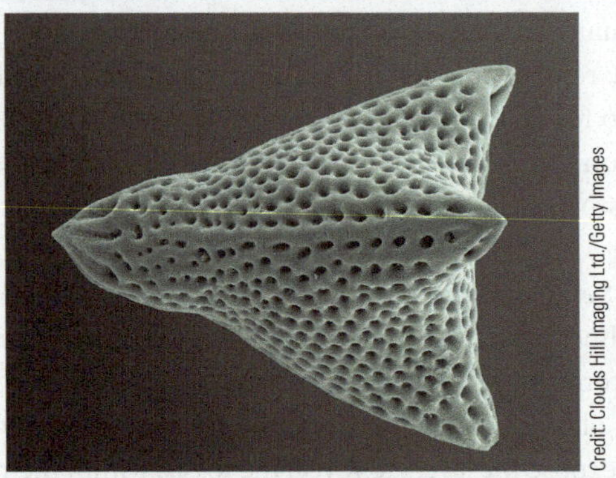

A. Foraminifera are single-celled protists that secrete a double shell of calcium carbonate. This shell is resistant to decay and therefore is more likely to be fossilized.

B. Note the small size of foraminifera—many of these shells can fit through the eye of a needle.

isotopes such as ^{40}K, which has a half-life of 1.27 billion years and decays into ^{40}Ar at a predictable rate (Figure 10.11). By measuring the relative amounts of ^{40}K to ^{40}Ar, scientists are able to determine the age of rock layers with great precision. This process is called **radiometric dating**.

Note that radiocarbon dating (see Chapter 2) is the most well-known form of radiometric dating. However, radiocarbon dating cannot tell us the age of samples older than 60,000 years because the half-life of carbon-14 is too short (see Table 10.1 for a comparison). Since the history of life on Earth goes back billions of years, a different radioisotope is needed. That's why radioisotopes with longer half-lives, such as ^{40}K and ^{238}U, are used instead of ^{14}C to date rocks.

We learned earlier that fossils generally form in sedimentary rocks, but the methods described here are for dating volcanic rocks. So how do we determine the age of a fossil in sedimentary rock using volcanic rock? This is where the principle of relative dating comes in. Scientists can date the volcanic rock layers that flank the sedimentary rock layer containing the fossil, and by doing so, infer the age of the fossil within a certain time frame.

Figure 10.11 Tracking half-lives in rock

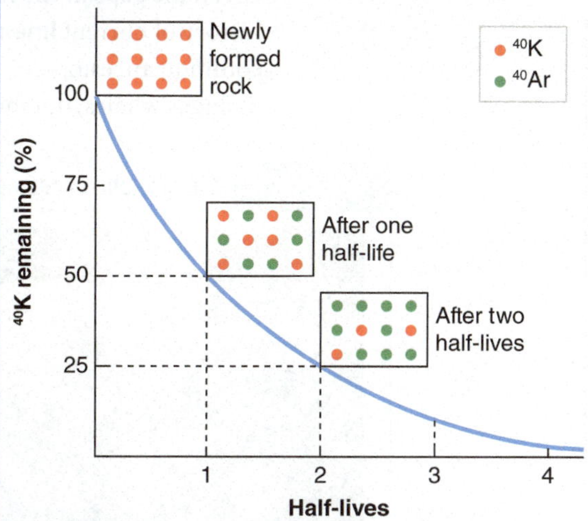

^{40}K–^{40}Ar dating is used to determine the age of rock. Notice that over the course of billions of years, the percentage of ^{40}K remaining in the rock decreases as it decays to ^{40}Ar.

Insights and Holes in the Fossil Record

Despite fossilization occurring only under particular conditions, scientists have described more than 250,000 species using fossils. Considering the current range of biodiversity, these likely represent only a small fraction of life. What about the rest?

For us to know about a species that existed long ago, we must find evidence of it—a fossil. However, only relatively few individuals of a species become fossilized in the first place. Typically, when an organism dies, its remains are eaten or carried away by scavengers. Organic materials decompose in the presence of moisture and oxygen, and so any remains that might escape scavenging can endure only if they dry out, freeze, or become encased in an air-excluding material such as sap, tar, or mud. Additionally, fossils that do form can often be crushed or scattered by erosion and other geologic forces, and of those that remain intact, many are inaccessible, buried deeply in rock.

Most ancient species had no hard parts to fossilize, and so there is little evidence of them in the fossil record. For example, there are many fossils of bony fishes and mollusks with hard shells, but few fossils of the sea jellies and soft worms that were probably much more common in ancient oceans.

Also consider the relative numbers of organisms. Fungal spores and pollen grains are often released by the millions. By contrast, the earliest humans lived in small bands, where few of their offspring survived. As such, the odds of finding even one fossilized human bone are much smaller than the odds of finding a fossilized fungal spore. Finally, imagine two species—one that existed only briefly and another that existed for a hundred million years. Which one is more likely to be represented in the fossil record?

Although the fossil record is not perfect, scientists have discovered enough fossils to make inferences about the evolutionary histories of various organisms. In the next section, we will examine the reconstruction of such evolutionary linkages.

Table 10.1 Forms of Radiometric Dating

Dating method	Material dated	Half-life	Age range dated
$^{14}C \rightarrow ^{14}N$ (radiocarbon)	Organic remains of organisms; archaeological samples	5,730 years	Up to 60,000 years ago
$^{40}K \rightarrow ^{40}Ar$ (potassium–argon)	Volcanic rocks	1.27 billion years	20,000 to 4.5 billion years ago
$^{238}U \rightarrow ^{206}Pb$ (uranium–lead)	Volcanic rocks	4.5 billion years	1 million to 4.5 billion years ago

Constructing Evolutionary Linkages Using Fossils

Given the gaps in the fossil record, how are scientists able to reconstruct the evolutionary history of current **lineages**—a population or group of individuals tracing descent from a common ancestor—of a species? Let's look at the evolution of cetaceans—a type of animal that includes whales, dolphins, and porpoises—as an example (**Figure 10.12**).

Figure 10.12 Comparison of cetacean skeletons

A. ***Pakicetus attocki.*** Considered one of the earliest cetaceans, this wolflike animal lived about 50 million years ago. It was semiaquatic, but its body was specialized for running. *Pakicetus* belonged to a different lineage and it had traits unique to cetaceans.

B. ***Rodhocetus balochistanensis.*** This cetacean lived about 47 million years ago. It was capable of walking (or dragging itself) on land, but its body was built for swimming.

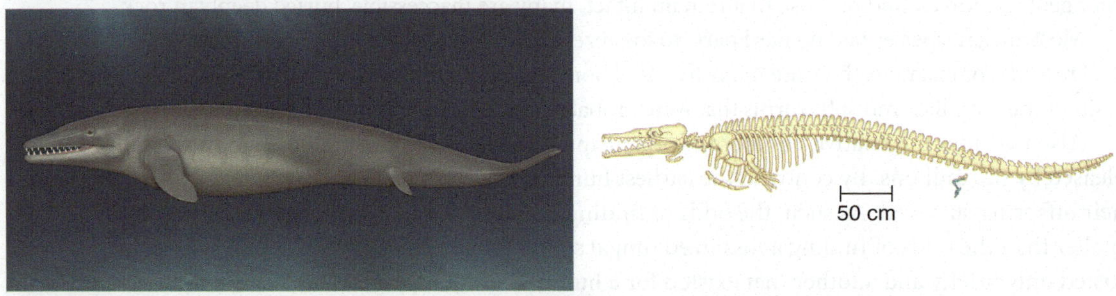

C. ***Dorudon atrox.*** This cetacean lived about 37 million years ago. Its hind limbs worked, but they were tiny and not connected to the backbone, so they would not have supported the weight of the animal's huge body on land. Thus, *Dorudon* had to be fully aquatic.

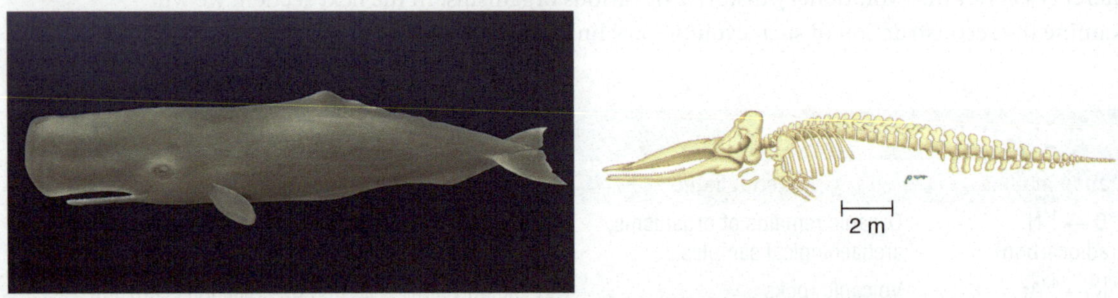

D. Modern whales. Present-day cetaceans, such as the sperm whale, have remnants of a pelvis and leg, and like *Dorundon*, are fully aquatic.

The ancestor of whales was a land animal. Over millions of years, the lineage transitioned from life on land to life in water, and as it did, bones of the hind limbs (highlighted in blue) became smaller.

Skeletons of modern cetaceans have remnants of a pelvis and hind limbs, so evolutionary biologists had long thought that the ancestors of this group walked on land. The hypothesis was supported by the discovery in 1906 of an intermediate form of cetacean from intact fossil skeletons of *Dorudon atrox* (Figure 10.12C), a whalelike animal with hind limbs that lived about 37 million years ago. However, *Dorudon*'s hind limbs were too tiny to have supported its huge body out of water, so it must have been fully aquatic. At that time, no one had found other fossils with intermediate skeletal features that would have accompanied a transition from terrestrial to aquatic life, so the rest of the story remained speculative.

Genome comparisons in the early 1990s suggested that modern cetaceans are related to hooved mammals with an even number of toes (two or four) on each foot. Modern members of the lineage include hippopotamuses, antelopes, and sheep. The discovery was surprising because the common ancestors of these mammals resembled wolves with long tails (Figure 10.12A). How was it possible for whales with gigantic bodies suited to deep ocean swimming to have a much smaller land-based animal as an ancestor?

Then, in 2000, Philip Gingerich and his colleagues discovered intact fossil skeletons of two other ancient cetaceans in a 47-million-year-old rock formation in Pakistan. The skeletons had whalelike skulls and robust hind limbs (Figure 10.12B). Their bodies were built to swim with their feet (not their tails as whales do), and their ankle bones had clearly distinctive features of hooved mammals. Neither was a direct ancestor of modern whales, but they were relatives, offshoots of the family tree connecting modern whales with their terrestrial cousins. The discovery illustrates the many evolutionary changes that occurred as the lineage transitioned from land to water, leading to modern whales (Figure 10.2D).

Big Ideas in Biology

Structure and Function

The evolution of cetacean body structures, such as the changing form of limbs into flippers, shows how the different body forms corresponded to the various functions of those body parts as cetaceans gradually became more aquatic.

Body structures are not the only traits that have evolved—behaviors have evolved as well. For example, different species of birds sing specific songs to attract a mate. Each species of bird has a different song. In humans, our ancestral need to support one another for our own survival may have led to the instinctive desire that many of us have to help one another.

10.2 Summary

- The geologic time scale is the chronological history of Earth. Layers of rock correspond to phases in the Earth's history and contain information about the conditions on Earth when they were deposited.
- Fossils are physical evidence of organisms that lived long ago and yield insight into ancient life. The predictability of radioisotope decay combined with relative dating can be used to determine the age of fossils.
- Fossils are relatively rare because they can form only from durable body parts or traces preserved over time.
- Studying fossils is useful for examining evolutionary relationships and histories.

10.3 Biogeography

Learning Objectives

10-8 Explain how tectonic plate movements influenced the history of life.

10-9 Relate plate tectonics to the unique life found in Australia.

Let's say you won round-trip plane tickets for you and a friend to visit Australia. You have a week to spend down under—what do you go see? Perhaps the Sydney opera house? The outback? Or the Great Barrier Reef? What about the Daintree Rainforest (Figure 10.13)?

If time travel sounds attractive to you, then the Daintree Rainforest should be on your bucket list. The Daintree Rainforest is the oldest rainforest in the world and is full of "living fossils"—living organisms that otherwise exist only in fossil form everywhere else in the world (unless cultivated by humans), like *Gymnostoma australianum*, also known as the Daintree pine (Figure 10.14). The cones of the

Figure 10.13 The Daintree Rainforest

Credit: Kat Clay/Getty Images

The Daintree Rainforest in Queensland, Australia, is a World Heritage site with staggering biodiversity.

Figure 10.14 **The Daintree pine (*Gymnostoma australianum*) is a living fossil**

A. Although this *G. australianum* tree has needlelike leaves and produces cones, it also has both male and female flowers.

B. A modern *G. australianum* cone.

C. A fossil of a *Gymnostoma patagonicum* cone, which is an ancestor of *G. austalianum.*

Daintree pine look very similar to fossils of the same *genus* (a group of related organisms with a common ancestor, Chapter 12). Why did these plants survive here and nowhere else? It comes down, at least in part, to continental drift and is an example of **biogeography**, or how and why different species of animals and plants exist in particular areas of the world.

? Did You Know?

The Daintree Rainforest is home to 12 of the 19 most primitive known flowering plants, and is also home to the oldest fern species.

? Did You Know?

In 2021, the Queensland government returned ownership of Daintree National Park to the Eastern Kuku Yalanji—an Indigenous group with ties to the Daintree dating back at least 50,000 years.

Continental Drift and Pangaea

Wind, water flow, and other natural processes continuously sculpt Earth's surface, but they are only part of a much bigger picture of geologic change. Earth itself also changes. For example, all continents were once part of a bigger supercontinent called **Pangaea**, which split into fragments and drifted apart about 200 million years ago (Figure 10.15). Pangaea was one of at least five supercontinents that formed and split up again since Earth's outer layer of rock solidified 4.55 billion years ago.

Pangea split into two supercontinents, Laurasia and Gondwana (see Figure 10.15). **Laurasia** eventually broke apart to form North America, Europe, and Asia. **Gondwana** eventually split to become Africa, South America, Antarctica, and Australia.

The Daintree Rainforest is estimated to be around 120 million years old, dating back to when Australia was still part of Gondwana. As Australia continued to shift around—breaking off from Antarctica around 30 million years ago—while other parts of Australia moved and changed, the region of Australia where the Daintree Rainforest resides stayed within a consistent latitude, therefore maintaining a consistent climate (see Figure 10.15). The lack of movement, combined with the isolation of Australia from other land masses, explains why animal and plant species that went extinct in other parts of the world are still living there today.

Movement of tectonic plates relative to one another is responsible for phenomena like earthquakes. When a plate moves over a hotspot (a magma plume), this can form volcanoes and island chains, such as Hawai'i. Tectonic plate movement also explains the geologic activity in Yellowstone National Park (Chapter 2).

The Movement of Landmasses and Life

The idea that continents move around, previously referred to as continental drift but now called plate tectonics, was proposed in the early 1900s to explain why the Atlantic coasts of South America and Africa seem to "fit" like jigsaw puzzle pieces. How do continents move?

We need to look to geology for the answer. According to **plate tectonics theory**, the Earth's crust is made up of a series of units called plates, and the movement of these plates creates geological features such as mountains. You can think of the Earth's crust as being like an eggshell, with cracks running through it that create different regions (plates). Magma welling up at an undersea ridge (or continental rift) at one edge of a plate pushes old rock at the opposite edge into a trench. The movement is like that of a giant conveyor belt that slowly transports continents on top of it to new locations. The plates move no more than 10 centimeters

Figure 10.15 **Shifting tectonic plates and moving continents**

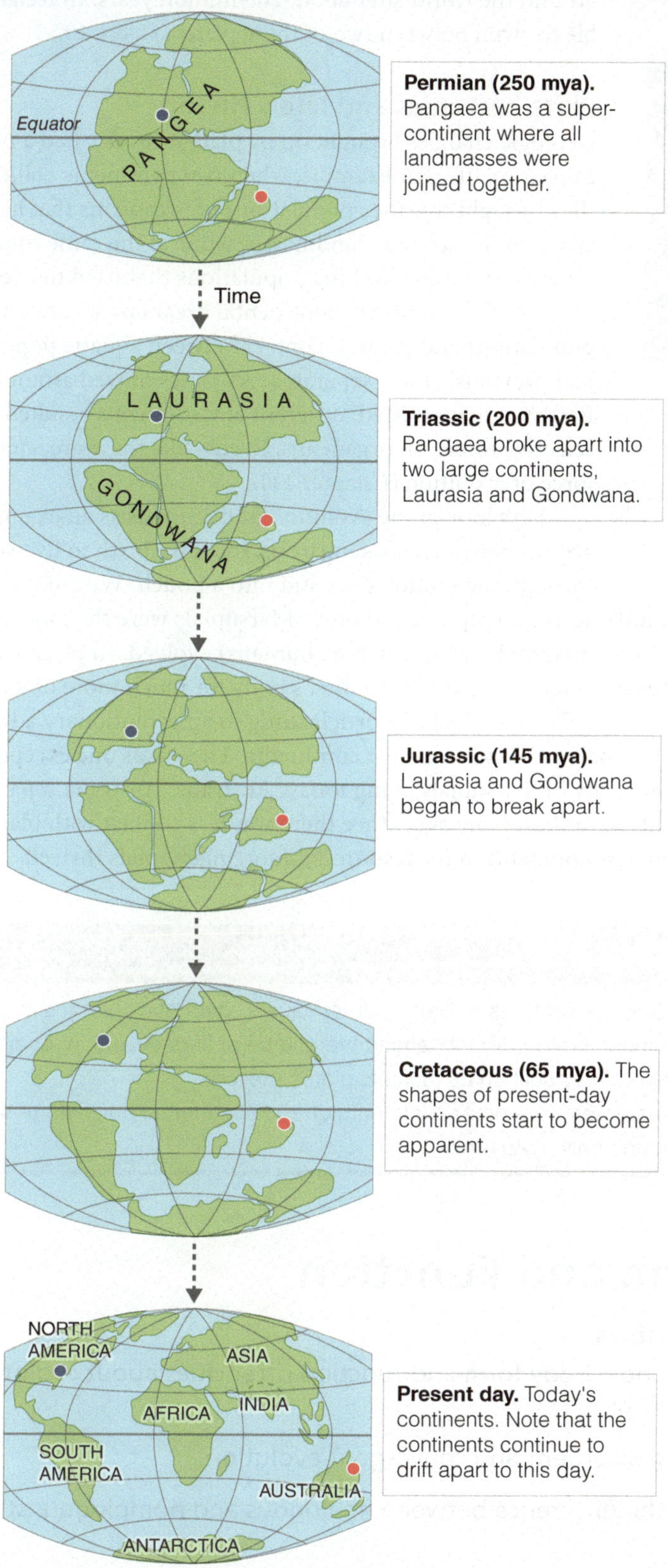

Note how the latitude of the Daintree Rainforest (red dot) remained relatively constant as the continents moved around, but the latitudes of other places, such as the Appalachia region of North America (blue dot), have moved significantly. As a result, the remains of tropical plants that lived near the equator prior to the modern continents forming eventually became the coal deposits found in Appalachia today.

Credit: Historical perspective/USGS, https://pubs.usgs.gov/gip/dynamic/historical.html

Figure 10.16 Exploring the gap between the Eurasian and North American plates

Credit: mauritius images GmbH/Alamy Stock Photo

Located within Thingvellir National Park in Iceland, experienced divers can venture into the frigid water and touch two different tectonic plates at the same time. The gap between the plates forms the Silfra fissure, which is part of the largest mountain range in the world, the Mid-Atlantic Ridge.

(4 inches) per year, but this is enough to carry a continent all the way around the world after about 200 million years. In Iceland, it is possible to swim between two of these plates (Figure 10.16).

Plate Tectonics and Life's History

Geologic changes brought on by plate tectonics had a profound impact on life. For example, when two continents collided into one, they brought together populations of organisms that had been living apart on the separate landmasses while at the same time physically separating ocean-dwelling populations that used to live between the continents. In contrast, continental breakups separated land-based populations and allowed contact between aquatic populations that had previously been separated. As plates moved around and the Earth's climate shifted, different environments resulted in differing selective forces on organisms. Events like these are a major driving force of evolution (Chapter 11).

Let's look at the evolution of mammals as an example. Marsupials are animals, such as kangaroos, that give birth to live young that crawl through the mother's fur and into a pouch. Within the pouch, the young nurse from milk-secreting nipples and grow. Marsupials were the most abundant mammals until the placental mammals (which include humans) evolved. In placental mammals, the developing fetus is protected inside the mother's body for much more of its development, and the offspring is more developed when born, leading to an evolutionary advantage. As a result, placental mammals took over on most continents. There was one exception—and it's where we see the majority of marsupials living today: Australia. Australia separated from the other continents 60–70 million years ago. Since there was less contact with placental mammals (and therefore less competition for resources), more marsupials thrived.

Big Ideas in Biology

Interconnectedness of life

The interactions of animals with each other influences their overall survival. In the case of marsupials, they could not compete with placental mammals for resources and are generally found only in Australia, where there were fewer placental mammals.

10.3 Summary

- Over geologic time, movements of Earth's crust caused dramatic changes in continents and oceans. The changes profoundly influenced the course of life's evolution, as populations were brought together or split up as tectonic plates moved.
- Australia was separated from other continents due to continental drift, where selective forces led to species found only in Australia.

10.4 Form and Function

Learning Objectives

10-10 Describe how body form and function offer clues about evolutionary relationships.

10-11 Compare divergent and convergent evolution.

10-12 Explain the difference between analogous and homologous structures.

What do a bat, butterfly, and hawk have in common? Other than sounding like the beginning of a bad joke, these are all organisms with wings. However, each wing structure is different and represents different evolutionary trajectories. Examining what body structures look like (their form) and what they do (function) provides insights into the evolution of these structures. We will explore these structures and the evolutionary processes behind them next.

Figure 10.17 **Vertebrate forelimbs are an example of homologous structures**

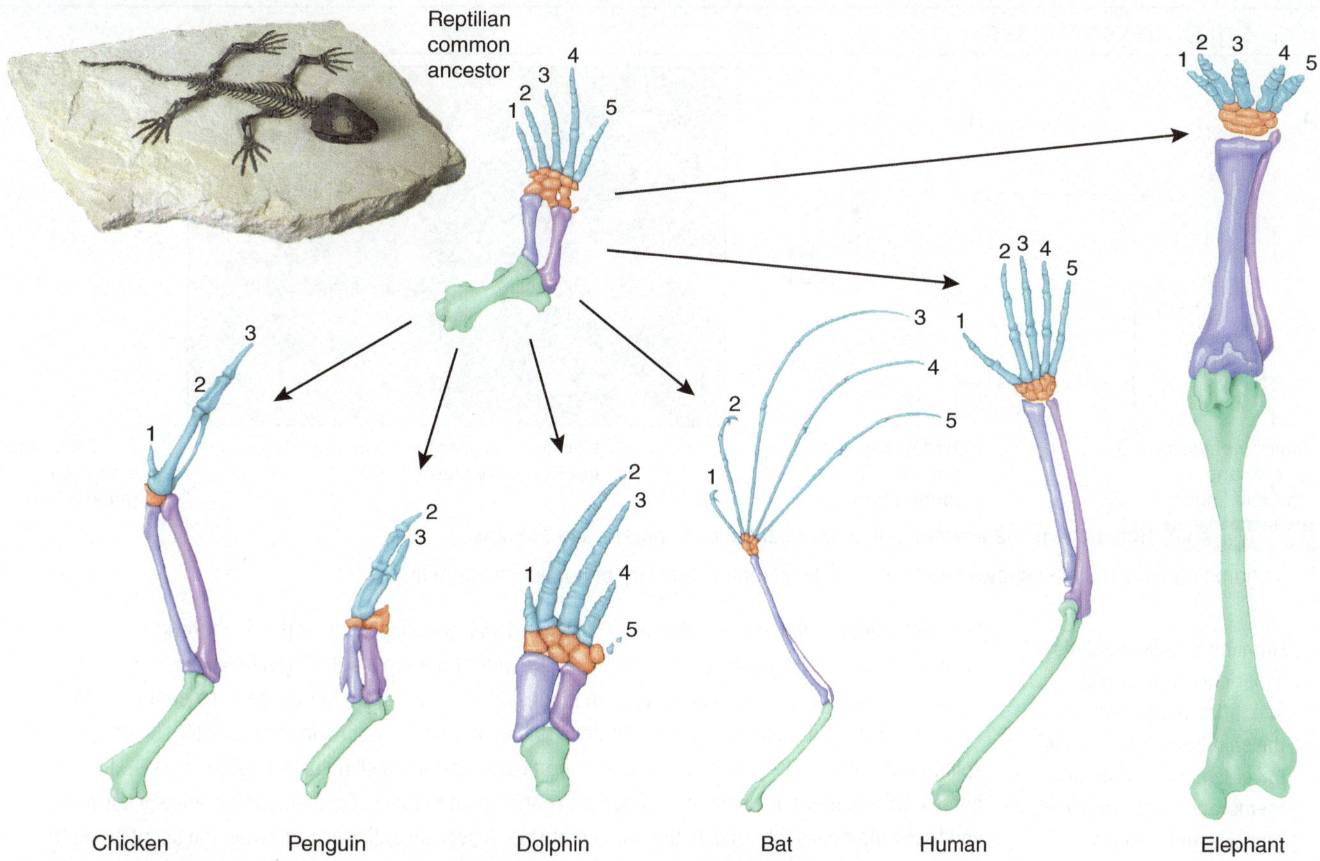

Each of the animals represented here shared a reptilian common ancestor (top) with five-toed limbs. Note how the bones of the forearm and digits appear different and relate to the function (wing, flipper, hand). If you examine the numbering of the digits, you can trace which bones were lost. (Illustrations are not to scale.)

Credit: Richard Paselk, Humboldt State University Natural History Museum

Homologous Structures

Descendants of a common ancestor can evolve in different ways in response to different environmental pressures. The divergence of lineages descended from a common ancestor is called **divergent evolution**. Divergent evolution can give rise to **homologous structures**, which are body parts that may appear unalike in different lineages but are derived from a common ancestral form.

Study Tip

Remember homologous chromosomes (see Chapter 8) that are similar, but not identical? Homologous structures are also similar, but not identical.

Vertebrate forelimbs. The forelimbs of vertebrate animals are examples of homologous structures (Figure 10.17). Although the limbs vary in size, shape, and function, they are alike in the structure and positioning of internal elements such as bones, nerves, blood vessels, and muscles.

Over millions of years, descendants of this ancestral group diversified into different lineages that eventually gave rise to modern reptiles, birds, and mammals. During this process of divergent evolution, five-toed limbs became adapted for specific uses in different environments (water, land, air). For example, human forelimbs are arms with hands that have four fingers and an opposable thumb for grasping objects. Forelimbs of dolphins and other cetaceans are pectoral fins (flippers) that are useful for swimming. In bats and most birds, they are wings that allow flight.

Vestigial structures. Remember human tailbones, which were introduced in the Unit 4 opener? These are **vestigial structures**, or homologous structures that have lost their function. For example, the eyes of cave fish, a species adapted to life in caves and other dark environments, were selected against over many generations, until all that remained were non-functional eyes (Figure 10.18). The remains of hind limbs on whales (see Figure 10.12) are another example of a vestigial structure.

Figure 10.18 **A blind cave fish**

Note the spots where the eyes would have been.

Credit: JOEL SARTORE/NATIONAL GEOGRAPHIC PHOTO ARK

Digging into Data

Vestigial Toes in Horses

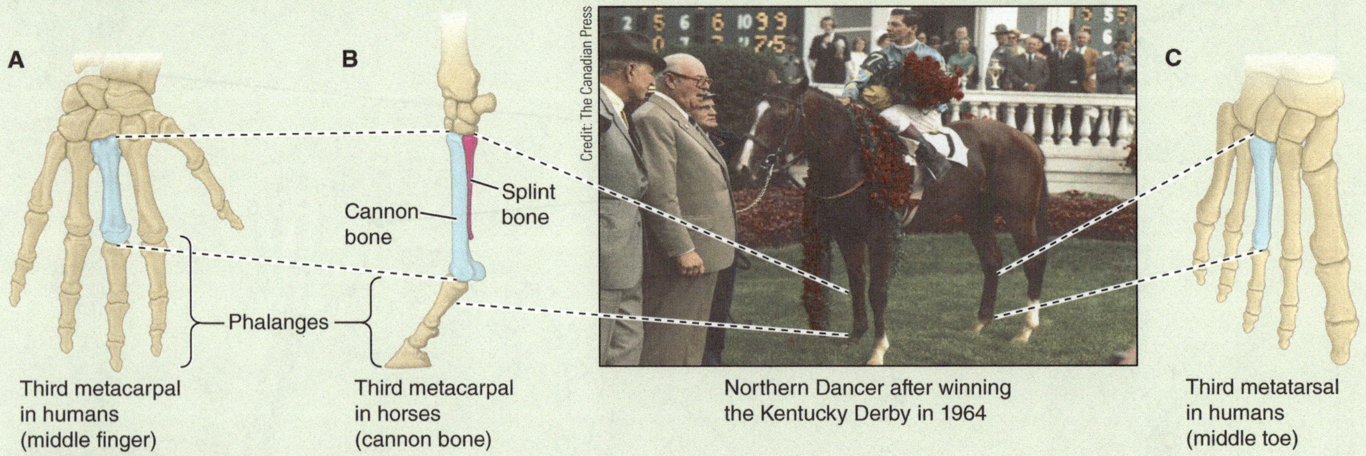

Figure 10.19 **Homologous metacarpal and metatarsal bones in horses and humans**

The homologous structures shown here highlight the common ancestry of horses and humans.

Humans with shin splints experience similar pain and inflammation when the shin bone (tibia) in the lower leg, its surrounding membrane, or the attached muscles and connective tissues become inflamed due to repetitive stress, such as running. Note that unlike horses, humans do not develop a bony bump when the shin split heals.

Northern Dancer, who won the Kentucky Derby in 1964, was one of the top U.S. thoroughbred racehorses of the 20th century (Figure 10.19). Despite "popping a splint" (an inflammation of the splint bone and the surrounding tissue) on the inside of his left front leg at age 2, he continued to race successfully. In horses, the cannon bone is the load-bearing bone in the lower leg, while the splint bones along either side of the cannon bone are supportive structures (Figure 10.19B). In young horses, the ligament between the cannon bone and splint bones is flexible, but it gradually hardens and fuses the bones into one as the horse matures. A popped splint occurs when the splint bone, the ligament, or the surrounding tissue becomes inflamed due to trauma or overwork. This inflammation can be painful until the injury heals into a hard, bony bump.

Functionally, the cannon bone in horses serves a role similar to that of the human tibia, with both being the primary weight-bearing bones of the lower leg. However, the cannon bone in the front leg of horses is homologous to the third metacarpal in humans (the middle finger) (Figure 10.19A and B); the cannon bone in the hind leg of the horse is homologous to the third metatarsal of humans (the middle toe) (Figure 10.19C). The phalanges make up the fingers and toes in humans are homologous to the bones below the cannon bone, closest to the ground.

Understanding homologous structures helps us see the connection between species. The tiny splint bones in horses are vestigial bones resulting from two of the five toes that their distant

? Did You Know?

Horses became extinct in the Americas about 15,000 years ago. How did their descendants get here today? They were brought back on boats by colonizers beginning in 1493, with Christopher Columbus's second voyage to the Americas.

Analogous Structures

Structures that appear similar in different species are not always homologous. Similarities sometimes evolve independently in lineages subject to the same environmental pressures, a pattern called **convergent evolution**. Convergent evolution can give rise to **analogous structures**, which are similar body parts that evolved independently in different lineages. Our opening example in this section with the butterfly, bat, and hawk wings are examples of analogous structures (Figure 10.21). Although similar in function (for flying), the form is different, and these animals do not share a common winged ancestor.

How do we know these aren't homologous structures? All three types of wings have adaptations for the same physical constraints that govern flight, but the adaptations differ. In the case of birds and bats, the limbs themselves are homologous, but other structures that make those limbs useful for flight are not. The surface of a bat wing is a thin, membranous extension of the animal's skin. By contrast, the surface of a bird wing is a sweep of feathers, which are specialized structures derived from skin.

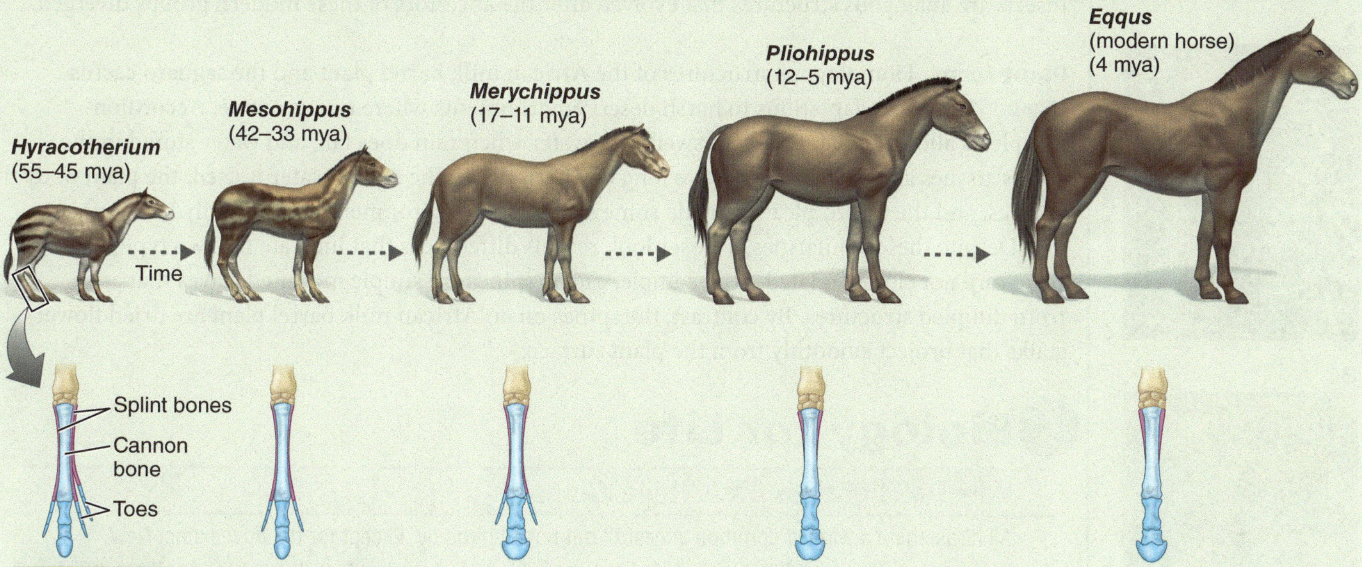

Figure 10.20 Evolution of the horse

Over the past 55 million years, horses gradually transitioned from a four-toed to a single-toed animal.

ancestors once had (Figure 10.20). Millions of years ago, prehistoric horses like *Hyracotherium* (also known as *Eohippus*) had five toes. Over time, as horses adapted for running on open plains—where the extra toes would have been a hindrance to the speed needed to escape predators—these toes were lost. The central toe (the third metacarpal, or metatarsal, and phalanges) became dominant, evolving into the structure that includes the cannon bone and phalanges, plus the single hoof (homologous to a fingernail or toenail) now seen in modern horses.

1. Figure 10.19 shows the homology between horse and human metacarpal and metatarsal bones. How do these homologous bones serve different functions in these two species?
2. Considering that a horse lower leg is like the human third digit, the horse hoof is homologous to a fingernail or toenail. What advantages are provided to horses through these adaptations? What functions did they lose?
3. Horse splint bones can be considered vestigial structures because they are no longer as functional as they were in prehistoric horses. What vestigial structures are found in other vertebrates? How do vestigial structures help us understand evolutionary relationships?

Figure 10.21 Wings are an example of an analogous structure

Although butterflies, bats, and birds all have wings and can fly, each evolved from a different winged ancestor, as shown in the diagram. We will return to diagrams of evolutionary relationships, such as the cladogram shown here, in Chapter 12.

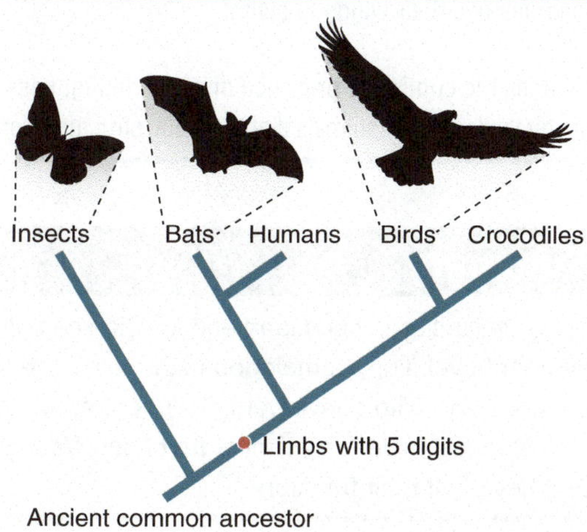

Figure 10.22 Homologous and analogous structures in plants

A

Credit: George Burba/Shutterstock.com

B

Credit: Planet Desert

A. The saguaro cactus and

B. The African milk barrel plant live on different continents, but they share similarities. Their accordion-like pleats are homologous structures, allowing them to store water during rainstorms. However, the spines on the saguaro cactus, which are modified leaves, are analogous to the dried flower stalks of the milk barrel plant, protecting the plant from browsers.

Insect wings differ even more. An insect wing forms as a saclike extension of the body wall. The sac flattens and fuses into a thin membrane around sturdy, forked veins that structurally support the wing. Unique adaptations for flight are evidence that wing surfaces of birds, bats, and insects are analogous structures that evolved after the ancestors of these modern groups diverged.

Plant form. Homologous structures of the African milk barrel plant and the saguaro cactus (Figure 10.22) are adaptations to harsh desert environments where rain is scarce. Accordion-like pleats allow the plant body to swell with water when rain does fall, and water stored in the plants' tissues allows them to survive long dry periods. As the stored water is used, the plant body shrinks, and the folded pleats provide some shade in an environment that typically has none.

Despite these similarities, a closer look reveals differences that indicate the two types of plants are not closely related. For example, cactus spines are simple modified leaves that arise from dimpled structures. By contrast, the spines on an African milk barrel plant are dried flower stalks that project smoothly from the plant surface.

Biology for Life

Why Do Flightless Birds Have Wings?

All birds share a winged common ancestor, but not all birds fly. Except for the now extinct New Zealand moa (Figure 10.23A), all birds have some kind of wing structure. If not used for flight, what is the purpose of these wings? Why would selective forces (Chapter 11) exist to maintain wings?

Ostriches (Figure 10.23B) are the fastest running birds and use their wings both for balance while running and to help slow down (like a parachute). Scientists believe that ostrich wings are also used for wooing a prospective mate (sexual selection, Chapter 11). In the case of penguins, their wings are used to swim (Figure 10.23C). Penguins that could swim better survived better than flying penguins, and so over time, penguins lost the ability to fly.

A

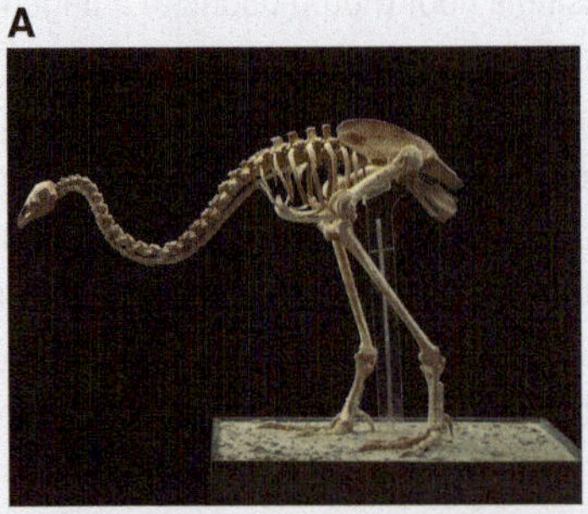

B

C

Figure 10.23 A collage of flightless birds

A. The New Zealand moa, now extinct, lost its wings entirely.

B. Ostriches use their wings for balance and to slow down while running.

C. Penguins use their wings to swim.

Credits: Upland Moa, Megalapteryx didinus, collected March 1987, Honeycomb Hill, Enduro, Map Grid 1385N 720E, New Zealand. Field Collection 1982-1988. CC BY 4.0. Te Papa (S.023700); National Geographic Image Collection; Natural Earth Imagery/Shutterstock.com

1. How does the continued presence of wings in flightless birds provide evidence for evolution?
2. Can you think of other animals that use body structures for reasons other than those previously expected?

10.4 Summary

- Comparing similarities in body form and function can offer clues about evolutionary relationships.
- With divergent evolution, homologous structures are modified differently in lineages from a common ancestor. With convergent evolution, analogous structures are modified in different ancestral lineages to perform similar functions. Vestigial structures are homologous structures that have lost their function.

10.5 Molecular Fingerprints of Evolution

Learning Objectives

10-13 Explain why similarities in patterns of embryonic development can be used as evidence of an evolutionary relationship.

10-14 Describe how similarities between protein or DNA sequences are used as a measure of relative relatedness.

Look at Figure 10.24. Shown here are five animal embryos—and one is a human. Can you figure out which one is the human? How can you tell?

In general, the more closely related animals are, the more similar their development. For example, all vertebrates go through a stage during which a developing embryo has a tail and a series of somites—divisions of the body that give rise to the backbone and associated skin and muscle.

Similar patterns of gene expression in developing embryos from different species is one line of evidence that tells us how complex organs formed. For example, the same genes that control eye development in humans also control eye development in flies.

Evolutionary Developmental Biology

Animals have similar patterns of embryonic development because they have the same **transcription factors**, proteins that regulate gene expression (Chapter 7). Because a mutation in a transcription factor can completely unravel development, these genes tend to be highly **conserved**, meaning that the sequence has remained relatively unchanged in evolutionary history. Even among lineages that diverged a very long time ago, many transcription factors retain similar sequences and functions.

Remember *Hox* genes from Chapter 7? *Hox* genes trigger formation of specific body parts during embryonic development. Insects have a *Hox* gene called antennapedia that causes legs to form wherever it is expressed, which is normally in the midsection of the body (the thorax). Humans and other vertebrates have a version of the antennapedia gene called *Hoxc6*. Expression of the *Hoxc6* gene in a vertebrate embryo causes ribs to develop on a vertebra. Vertebrae of the neck and tail normally develop with no *Hoxc6* expression, and no ribs.

While natural selection leading to evolution is a non-random process, the formation of mutations is largely random. If a mutation leads to an adaptation that helps an organism reproduce, it will be selected for and evolution will occur.

Similarities in DNA and Proteins

Over generations, inevitable mutations change the DNA sequence of a lineage's genome (Chapters 6 and 7). Many of these mutations are neutral, which means they have no effect on an individual's survival or reproduction. Thus, we can assume that neutral mutations accumulate in a lineage's genome at a constant rate.

Mutations accumulate independently in the genomes of separate lineages. When two lineages first start to diverge from a common ancestor, very few mutations differentiate their genomes. Over many generations, the genomes diverge more and more as separate mutations accumulate in them.

The more recently two lineages diverged, the less time there has been for unique mutations to accumulate in their DNA. Thus, **homology**, or similarities in the nucleotide sequence of a shared

Figure 10.24 Similarity in embryonic development between different vertebrates

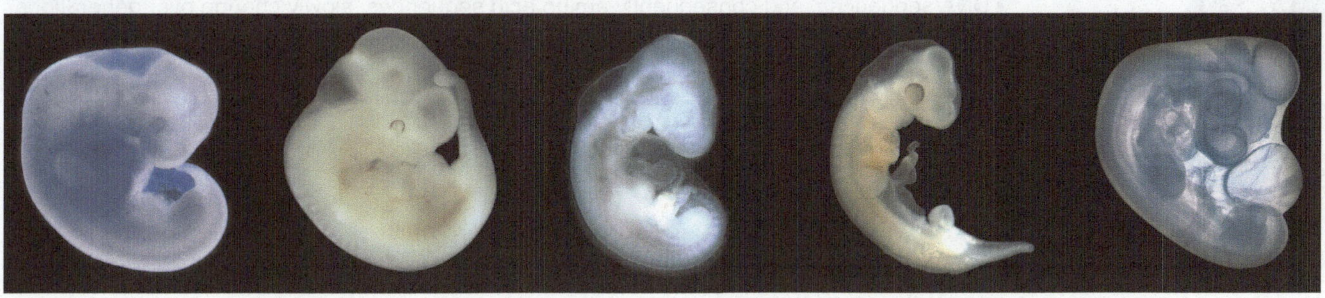

Shown here are the embryos of a mouse, human, chicken, turtle, and lizard. But, who is who? See the end of the Practice It! section for the correct order.

Credits: PROFESSORS P.M. MOTTA and S. MAKABE/Science Source; Natalia Sinjushina & Evgeniy Meyke/Shutterstock.com; Catherine May; Catherine May; Science Stock Photography/Science Source

Figure 10.25 **Comparison of amino acid similarity of hemoglobin between humans and other animals**

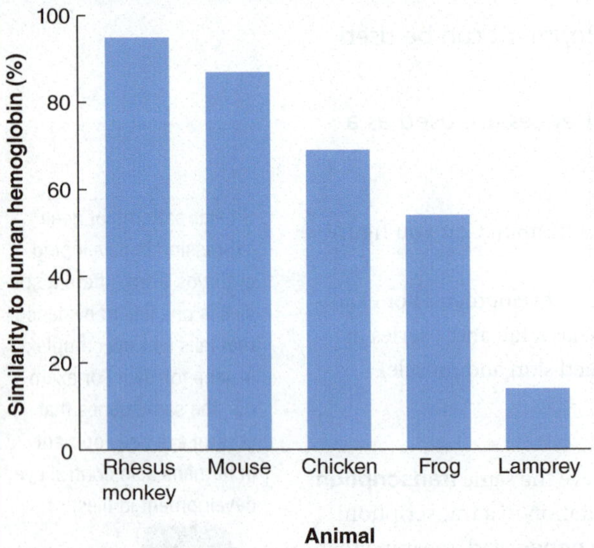

The similarity of the rhesus monkey hemoglobin to that of humans suggests that humans and rhesus monkeys have a more recent common ancestor than, for example, humans and lampreys.

gene (or in the amino acid sequence of a shared protein) are often used as evidence of an evolutionary relationship. Molecular comparisons like these may be combined with morphological comparisons to provide data for hypotheses about shared ancestry.

Comparing proteins. Evolutionary biologists often compare a protein's sequence among species and use the number of amino acid differences as a measure of relative relatedness (Figure 10.25). The amino acids that differ also offer clues. For example, a leucine (Leu) to isoleucine (Ile) change may not affect the function of a protein very much, because both amino acids are nonpolar and are about the same size. By contrast, the substitution of a lysine, which is charged, for an isoleucine, which is nonpolar, will dramatically change the character of a protein, which in turn may affect phenotype. Whether or not a mutation will lead to an advantageous or disadvantageous phenotype depends on the environment the organism lives in.

Comparing DNA. Among species that diverged relatively recently, many proteins have identical amino acid sequences, so comparing proteins may not tell us much about the evolutionary relationship between the species. Nucleotide sequence differences may be instructive in such cases. Even if the amino acid sequence of a protein is identical, the nucleotide sequence of the gene that encodes the protein may be different, due to redundancies in the genetic code (Chapter 7).

Getting useful information from comparing DNA requires more data than protein comparisons. This is because chance matches are statistically more likely to occur with DNA comparisons—there are only 4 nucleotides in DNA versus 20 amino acids in proteins. However, DNA sequencing has become so fast that there are a lot of data available to compare (Chapter 9). In fact, biologists are using changes in DNA sequences to determine the approximate time that two species diverged based on the number of nucleotide differences between the two. Although mutations or genetic changes do result from many other processes, such as DNA polymerase errors (Chapter 6) and crossing over (Chapter 8), comparing DNA sequences provides one valuable line of evidence for understanding evolutionary relationships.

Big Ideas in Biology

Evolution

It's not just structures that represent our shared ancestry. Modern genetic analysis provides additional clues into evolutionary relationships.

Evolution is an ongoing process because our DNA is constantly changing. Accumulation of changes over long periods of time can result in the generation of a new species.

10.5 Summary

- Animal embryos that look similar at early stages of development often share an evolutionary relationship due to conserved transcription factors.
- DNA sequences, and consequently amino acid sequences, slowly change over generations due to cumulative mutations. Species that are more closely related tend to share more similarities in their DNA and proteins.

Understand It! Test Anxiety

Have you ever "blanked" when taking an exam—where you knew the material, but couldn't remember it on the exam? Or you couldn't remember it during the exam, but remembered it later?

Blanking on an exam is a hallmark of test anxiety. Why? It comes down to two different structures in your brain. First is the amygdala. The amygdala is a pair of small structures in the center of your brain. The amygdala controls your emotions, particularly fear, and is part of your limbic system, which controls behavioral and emotional responses. The amygdala and the limbic system are important for survival. In fact, your amygdala can short circuit your frontal cortex—the part of your brain that controls your memory, thinking, learning, problem-solving, and consciousness.

Essentially, the amygdala can take over in high-stress situations and prevent rational thought, and sometimes even cause panic attacks! Think of it this way—you don't need to be aware of a danger in your environment to be able to consciously respond to it. You can run away from a rattlesnake giving you the stink eye and shaking its tail without needing to stop and think, "Oh. That's a venomous snake that's clearly upset with me for nearly stepping on it while it was napping. I'd better move!!" It makes evolutionary sense why our brains evolved this way—getting out of the way quickly without the need for rational thought is clearly an evolutionary advantage!

Now, back to test anxiety. When we are feeling nervous, it's the amygdala's time to shine—not the cerebral cortex. It's difficult to think when we are nervous. Later, when we're relaxed and the cerebral cortex is functioning at full capacity, we often remember what we had forgotten during test time.

Exams are often high-stress situations. Here are some things to think about as you get ready for tests.

1. Are there certain subjects where you are more likely to "blank" on an exam? Which ones?
2. List some strategies you can use to help keep your cerebral cortex in the driver's seat at exam time instead of your amygdala.

Chapter Review

10.1 An Introduction to Evolution

A **species** is a group of organisms that reproduce with each other and produce fertile offspring. **Evolution** is descent with modification across a species, where genes are inherited, along with epigenetic changes and mutations. Inherited changes can be neutral, advantageous, or disadvantageous, depending on the environment.

Natural selection favors traits that enhance the reproductive success, or **fitness**, of an organism in its environment, increasing the alleles for those traits in the population. Alleles that code for traits that decrease reproductive success become less common in the **gene pool**, which is all the alleles in a population.

10.2 Fossils and the History of Life

The **geologic time scale** represents the history of our planet by correlating rock layers with great intervals of time, from the formation of the Earth's crust to the present. **Fossils** are mineralized traces or remains of ancient organisms, typically found in layers of sedimentary rock.

Fossilization relies on rare conditions, but enough fossils have been discovered to reconstruct the evolutionary history of many species.

Index fossils allow us to assign ages to rock samples. **Relative dating** estimates a fossil's general age based on its position in rock layers. **Radiometric dating** uses the characteristic half-life of a radioisotope to determine the age of rocks and fossils. Scientists use fossil evidence to establish the **lineage** of a species.

10.3 Biogeography

Biogeography is the study of the distribution of species around the world. **Plate tectonics theory** describes how Earth's crust is divided into plates that move landmasses. **Pangaea**, the first supercontinent, split into **Laurasia** and **Gondwana** 200 million years ago, with further splits and continental drift forming the current continents. Plate movement influenced species evolution by bringing populations together or separating them during the formation and breakup of supercontinents.

10.4 Form and Function

Comparing similarities in body form and function reveals evolutionary connections among lineages. **Divergent evolution** describes the development of **homologous structures**, body parts that may appear different in different lineages but are derived from a common ancestral form. **Vestigial structures** are homologous structures that have lost their function. **Convergent evolution** describes the development of **analogous structures**, which are similar body parts that evolved separately in different lineages in response to similar environmental pressures.

10.5 Molecular Fingerprints of Evolution

Animals share **transcription factors**, proteins that regulate gene expression. These proteins are highly **conserved** (unchanged through evolution) because changes that disrupt embryonic development are often lethal. **Homology**, or similarities in the DNA or amino acid sequences of a gene shared between different species, provides evidence of evolutionary relationships. These sequences are more similar in lineages that diverged more recently.

Practice It!

Retrieval practice is the process of pulling information from your long-term memory and using it. Here are some questions to help with your retrieval practice.

Self-Quiz

1. Which of the following is a misconception about evolution? (LO 10-1, LO 10-2)
   a. Evolution and religious beliefs cannot coexist.
   b. Evolution says we evolved directly from monkeys.
   c. There is no evidence the species changed over time.
   d. All of the above

2. Natural selection favors traits that increase ______. (LO 10-4)
   a. migration to new regions
   b. reproductive fitness
   c. body size and strength
   d. changes to the gene pool

3. Antibiotic resistance in bacteria developed due to ______. (LO 10-4)
   a. uncontrolled breeding
   b. convergent evolution
   c. selective forces
   d. divergent evolution

4. Layers of rock correspond to conditions on Earth when they were deposited. These relationships are described by ______. (LO 10-5)
   a. a warp in time and space
   b. mass extinctions
   c. plate tectonics
   d. the geologic time scale

5. Fossils of common organisms in a specific geologic time period are called ______ fossils, and they can be used to date rock samples. (LO 10-6)
   a. index
   b. transitional
   c. trace
   d. durable

6. Five times in the past, the loss of up to 50%–95% of all species on Earth dramatically affected the course of evolution. These events are called ______. (LO 10-5, 10-6)
   a. non-selective pressures
   b. mass extinctions
   c. fossilizations
   d. global pandemics

7. Which animal body part would be least likely to produce a fossil? (LO 10-7)
   a. skull
   b. beak
   c. tongue
   d. shell

8. Which supercontinent formed first? (LO 10-8)
   a. Gondwana
   b. Laurasia
   c. Pangaea
   d. Antarctica

9. Unique animals like the kangaroo are only found in Australia due to ______. (LO 10-9)

a. plate tectonics
b. divergent evolution
c. geographic merging
d. convergent evolution

10. Which of the following is an example of a pair of analogous structures? (LO 10-12)

a. human arm and bird wing
b. horse leg and human foot
c. butterfly wing and bird wing
d. dolphin flipper and dog paw

11. Which of the following is an example of a pair of homologous structures? (LO 10-12)

a. horse leg and spider leg
b. horse leg and human pelvis
c. butterfly wing and bird wing
d. bat wing and whale flipper

12. Consider the DNA sequences below. Which two species are most closely related? (LO 10-14)

species A: GATCCGGA	a. species A and B
species B: GCTCCAGC	b. species A and C
species C: CATCAGGA	c. species A and D
species D: CTAGCACC	d. species B and C

13. The fact that the human DNA sequence for a particular transcription factor can be inserted into the genome of a mouse cell in a lab, where the resulting human protein would function as in the human body, is an example of how ______ shows evolutionary relatedness. (LO 10-14)

a. similar form and function
b. radiometric dating
c. chromosomal compatibility
d. sequence homology

Critical Thinking

1. In the late 1800s, a biologist studying animal embryos coined the phrase "ontogeny recapitulates phylogeny," meaning that the physical development of an animal embryo (ontogeny) seemed to retrace the changing form of the species during its evolutionary history (phylogeny). The idea is incorrect, but why would they have thought that development retraces evolution? (LO 10-13)

2. Natural selection makes an adaptive trait more common in a population. After many generations, will all individuals in the population have the same adaptive trait? (LO 10-3, LO 10-4)

3. Why are splint bones considered vestigial structures in horses, but the homologous human metacarpal/metatarsal bones are not? (LO 10-10, LO 10-11, LO 10-12)

4. Current biotechnology allows scientists to insert human genes into the genomes of other organisms, like yeast or mice. Applying concepts from this chapter, explain why this is possible (LO 10-14)

**The correct order of embryos in Figure 10.24 from left to right is human, mouse, lizard, turtle, chicken.*

Credit: Nature Picture Library/Alamy Stock Photo

11

Looking under the Hood: Mechanisms of Evolution

Chapter Outline:

A Tropical Love Nest

Just north of Australia is the world's second-largest island, New Guinea. If you were to hike through West Papua (one of the provinces of New Guinea), you might stumble upon a cone-shaped hut that's around 100 cm (approximately 3 feet) tall and 160 cm (approximately 5 feet) wide. In front of and underneath the structure would be piles of objects sorted by color—flowers, beetles, dung, and perhaps even some human trash. And, nearby, the architect of the structure, a male Vogelkop bowerbird (*Amblyornis inornata*), might be staring at you.

The hut is called a bower—and is created by the male bowerbird to entice a female. Different species of bowerbirds build bowers with different shapes and with different kinds of decorations. For example, in Australia, the satin bowerbird builds circular structures and preferentially decorates with blue objects. Males build and decorate the structures and will even perform dances within it to attract a female. The male birds will even steal high-value objects from each other's bowers to make their own bower more attractive. Females will visit a variety of these bowers before finally choosing a male to mate with.

The male bowerbird invests a significant amount of energy and resources into building a bower. For instance, rather than searching for food, they instead invest effort into maintaining and collecting trinkets for their bowers. The bower may also advertise to any nearby predators that a male bower bird is in the area. With these significant costs in mind, how did these elaborate mating rituals evolve?

The bower of a male Vogelkop bowerbird Male bowerbirds engage in elaborate mating rituals, including building ornate bowers, decorating their bowers, and finally, performing dances for prospective mates near their bowers.

Evolution occurs as the result of a combination of selective and non-selective forces. In the case of the bowerbird and many other bird species, the bright feathers and interesting behaviors are the result of *sexual selection,* or when *fitness* results from certain animals being better than others at securing a mate. Since evolution is descent with modification and fundamentally about reproduction, it makes sense that "sexy" traits get selected for.

In this chapter, we'll explore the selective and non-selective forces that result in evolution. First, we'll discuss alleles and the *gene pool* before talking about selective forces like natural selection, sexual selection, and artificial selection. Then we'll discuss non-selective forces that can cause changes in allele frequencies in the gene pool, thereby resulting in evolution. We'll also discuss how selective and non-selective forces can result in the evolution of new species during *speciation.* We close the chapter with a discussion of the opposite of speciation, *extinction.*

Preview It!

Make a list of three things you know about evolutionary processes already. What are some things you want to learn in this chapter?

11.1 The Gene Pool

Learning Objectives

11-1 Explain variation in shared traits among individuals of a population.

11-2 Describe some sources of allelic variation.

11-3 Relate the concept of alleles to a species' gene pool.

11-4 Explain the relationship between mutations, the environment, and selection.

If you could get on a plane and fly anywhere in the world today, where would you go and why? For me, the answer is...the beach! I love visiting the beach—especially an ocean beach with salty air, crashing waves, and a wide assortment of life present (which we'll return to in Chapter 13). I'm an avid beachcomber and have an extensive collection of treasures I've found on various beaches over the years. Seashells, which are the hard body parts of marine mollusks that wash up on shore after the organism dies, are one of my favorite things to search for. Incidentally, seashells are also a fantastic example of variation in shared traits (Figure 11.1).

Figure 11.1 **Atlantic calico scallops (*Argopecten gibbus*) come in a wide variety of colors**

Scallops are a common find for beachcombers. Although calico scallops are found on Atlantic beaches, other species of scallops can be found on beaches around the world. Note that the colors arise from both genetic differences between individuals and environmental influences, such as the scallop's diet.

Martin Shields/Science Source

Variation in Shared Traits

Individuals of a species, as well as a **population**—a group of individuals that live in the same area and can interbreed with one another—all have similar genetics and therefore share certain features. Common features can include physical characteristics like the presence and type of a shell. These physical features are called **morphological traits** (*morpho-*means "form"). Individuals of a population also share physiological traits, such as how they metabolize food, as well as behavioral traits, such as the ways in which they respond to stimuli. Note that we often use morphological traits as examples of how individuals are different from one another because they are easy to visualize, but there are many examples of biochemical and behavioral traits.

Calico scallops all have a hinged shell and belong to a group of mollusks called bivalves. If you look closely at Figure 11.1, you can see the flattened region where the two halves of the shell used to connect to one another. Although members of a same species have similar features, there are also a wide variety of traits within a single population of scallops. While they're all the same species, there can be various shell colors and patterns, caused in part by genetic variation among the individual scallops. As we learned in Chapter 8, sexual reproduction produces offspring with different combinations of alleles, and so there is variation in almost every shared trait among members of a sexually reproducing population. What it the source of this variation?

Remember, mutations are not purposeful. If a mutation leads to a trait that gives an advantage or disadvantage in a particular environment, it will be selected for or against, respectively.

An Evolutionary View of Mutations

Mutations, or changes to a DNA sequence, have altered genomes for billions of years, and they continue to do so. Cumulatively, they have given rise to Earth's staggering biodiversity (Chapter 13). Think about it: The reason you do not look like an avocado or an earthworm or even your neighbor began with mutations that occurred over vast stretches of evolutionary time, in different lines of descent.

Mutations result in new alleles and are the raw material of evolution. We cannot accurately predict when or in which individual a particular gene will mutate—mutations are largely random. However, which mutations lead to alleles that become more predominant in a population (those that are selected for; Section 11.2) is not a random process.

As discussed in Chapter 10, mutations can result in phenotypes that are beneficial, leading to enhanced reproductive success, or **fitness**. Mutations can likewise be neutral, having no effect on fitness, or can be harmful, leading to decreased fitness. Sometimes there can be evolutionary tradeoffs, resulting in less adaptative mutations (or alleles) persisting in the gene pool. This can occur because the positive impacts of a trait that leads to greater reproductive fitness are greater than the negative impacts. For example, the allele that causes sickle cell disease persists in the population because it also confers resistance to malaria. In addition, if harmful alleles lead to a disease phenotype later in life, after reproduction occurs—such as in the cases of neurodegenerative diseases like Huntington's Disease—the alleles can persist.

Figure 11.2 The cheetah is the fastest land animal

Credit: slowmotiongli/Shutterstock.com

Although its speed is clearly an advantage leading to increased fitness, the smaller stature of a cheetah limits its ability to catch larger prey.

When we consider evolutionary tradeoffs in the context of the cheetah, we see that its speed comes at the expense of the type of prey it can catch. Cheetahs (*Acinonyx jubatus*) are the fastest land animal in the world and can run up to 80 mph as they chase down their prey (Figure 11.2). To be that fast requires cheetahs to be more compact in size. Therefore, although they have an advantage in catching prey, their smaller size means that they're more limited in the prey that they can successfully catch. For example, cheetahs typically prey on smaller animals like antelopes and gazelles, and not on healthy adult zebras or giraffes.

Figure 11.3 The spoonbill has a distinctive bill shape, the result of beneficial mutations

Credit: George Grall/Alamy Stock Photo

The roseate spoonbill shown here hunts shallow waters in Florida using its unique bill to catch prey.

Beneficial mutations. Mutations that lead to beneficial phenotypes tend to become more common in a population over time, even if they bestow only a slight advantage. With **natural selection**, environmental forces result in an increase in the frequency of an adaptive form of a trait in a population over generations.

For example, let's consider a type of wading bird called the spoonbill (Figure 11.3). Spoonbills (genus *Platalea*; you'll learn more about genus and species names in Chapter 12), wade through shallow water and use their unique bill shape (which looks like a flattened spoon) to sweep through the water to find food, like small fish, insects, and shrimp. Once they feel something in their bill, they snap it closed and swallow the prey whole. A random mutation at some point in the spoonbill evolutionary history started the spoonbill on the journey to the unique bill shape we see today, where individuals with the different bill shape were more likely to survive than others without the mutation. As a result, the mutation became more common in the population, and in subsequent generations, eventually led to a new bird species with the spoonbill phenotype.

Figure 11.4 Variation in earlobe attachment

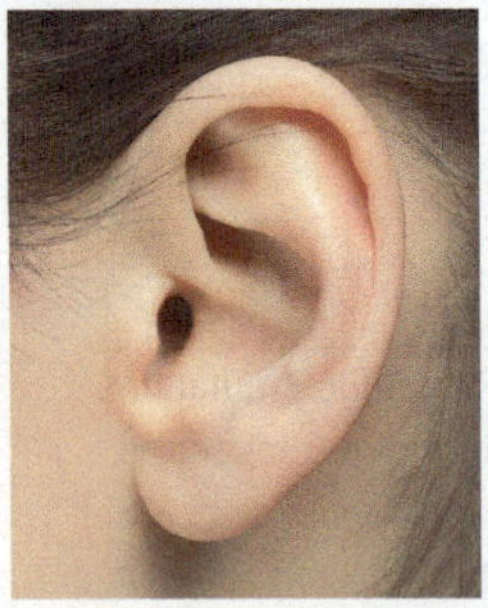

Credit: Tatjana Romanova/Shutterstock.com

Free earlobe

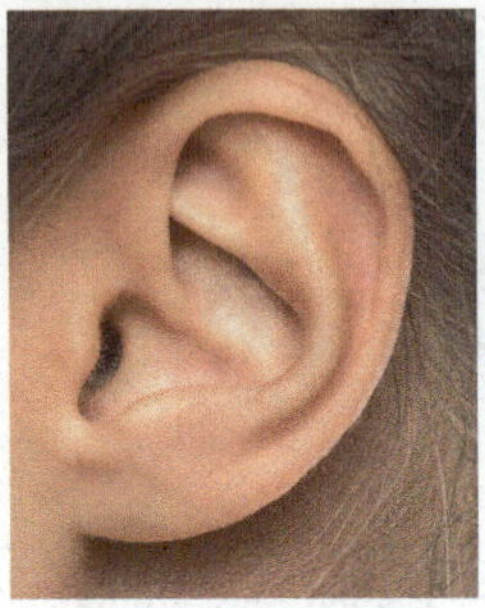

Credit: Blackday/Shutterstock.com

Attached earlobe

Notice that the free earlobe is attached to the head for only part of its length, while the attached earlobe is connected in its entirety.

> **! Point of Confusion** Mutations and natural selection occur in individuals, whereas evolution happens across a species as particular mutations become more (or less) common in a population.

Neutral mutations. Most mutations are neutral, which means they either do not result in a change in phenotype, or they give rise to phenotypes that neither help nor hurt the individual. For example, let's look at the alleles that affect the extent of attachment of your earlobes to your head. Having earlobes that are entirely attached along their length, or that are entirely free, has no effect on survival or reproduction, so natural selection does not shape this trait (Figure 11.4).

Harmful mutations. Some mutations give rise to structural, functional, or behavioral changes that reduce an individual's chance of surviving and reproducing. Some mutations are detrimental to an organism. For example, consider the mutations that lead to leucism, which results in a loss of pigment in an animal's hair or skin. Leucism occurs across the animal kingdom. Since many animals evolved particular coat patterns or coloration to blend in with their surroundings, a predominantly white animal would stick out from its peers and the environment, thus making it a target for predators (Figure 11.5).

Figure 11.5 Leucistic deer lack pigmentation

Credit: iStock.com/gemredding

Caused by a rare genetic mutation, white deer are conspicuous—consider how the white stag shown here stands out from the rest of the herd.

Allele Frequency

Together, all alleles of all genes of a population—including alleles that give rise to beneficial, neutral, and harmful mutations—make up the genetic resources for a population, or its **gene pool** (Figure 11.6). The **allele frequency** in a gene pool is the proportion of one allele relative to all copies of the gene in a population—said differently, it's the percentage of chromosomes in the population that have the allele.

For example, in the population of scallops shown in Figure 11.6, there are individuals that are homozygous dominant (*AA*), homozygous recessive (*aa*), and heterozygous (*Aa*). To determine the allele frequency, you add up the total number of alleles in the population (20) and the number of dominant (*A*; 15) or recessive (*a*; 5) alleles. Keep in mind that diploid organisms have two copies of each chromosome (Chapter 8), which is why the total number of chromosomes in the gene pool is then 20. To calculate the frequency of either allele, you divide the frequency of that allele by the total number of alleles in the population.

Figure 11.6 **A hypothetical gene pool for a population of scallops**

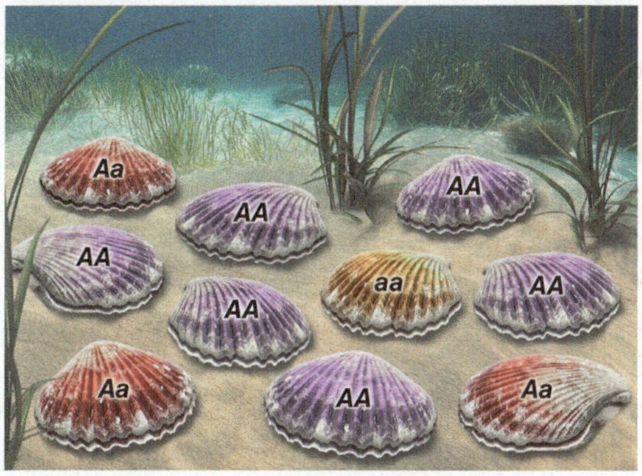

In this hypothetical population of 10 scallops, purple shells have the *AA* allele combination, red shells are *Aa*, and yellow shells are *aa*. Given that the scallops in this population are diploid, there are therefore a total of 20 alleles in the gene pool, 15 alleles are *A* (for a frequency of 15/20 = 75%) and 5 alleles are *a* (for a frequency of 5/20 = 25%).

Note that leucism is not albinism. Animals with albinism have a complete lack of pigmentation, resulting in white fur and pink eye color, whereas animals with leucism have some pigmentation, with white or patchy fur and dark eye color.

11.1 Summary

- A population is a group of interbreeding individuals living in the same area. Members of a population share morphological, physiological, and behavioral traits, but individuals have their own unique alleles. Traits that increase reproductive fitness tend to increase in frequency over time, changing the gene pool.
- Allelic variation occurs due to shuffling of the genome during sexual reproduction, or from mutation. Mutations can be beneficial, neutral, or harmful.
- The sum total of all alleles in a population or species makes up its gene pool. Allele frequency is the proportion of one allele among all alleles of that gene present in a population.
- Natural selection varies based on the alleles in individual organisms and environmental forces. Mutations change genes, and the environment favors individuals with traits leading to higher reproductive fitness. Mutations that increase fitness are maintained, while harmful mutations decrease over time.

11.2 Evolution by Selective Forces

Learning Objectives

11-5 Compare natural, sexual, and artificial selection.

11-6 Describe three patterns of natural selection.

11-7 Explain how sexual selection works and its outcomes.

11-8 Describe artificial selection and its outcomes.

Peppered moths (*Biston betularia*) are commonly found throughout the United Kingdom. These insects feed and mate at night, then rest on trees during the day. In 1850, most peppered moths were light with black speckles, and a very small number were black. At the time, the air was clean, and light-gray lichens grew on the trunks and branches of most trees. Light moths that rested on lichen-covered trees were well camouflaged, but dark moths were not (Figure 11.7A). Therefore, dark moths stood out more against the light background and were consequently easier for predators, like birds, to see and eat.

By 1900, dark moths were more common than light moths. This would suggest that the dark moths were less visible now and therefore had an adaptive advantage over the light-colored moths. What was going on to cause this change?

The change in prevalence of different colored moths was due to the start of the Industrial Revolution and the resulting sharp rise in environmental pollution. With the Industrial

Figure 11.7 The adaptive value of the two color forms of the peppered moth

Credit: Henrik Larsson/Shutterstock.com; Credit: blickwinkel/Alamy Stock Photo; Credit: J. A. Bishop, L. M. Cook; Martin Fowler/Shutterstock.com

A. A light-colored moth (left) on a lichen-covered tree trunk is hidden from predators. A dark moth (right) stands out.

B. On lichen-free tree trunks, the dark color (left) provides more camouflage than the light color (right).

Revolution came coal-burning factories (Figure 11.8). Pollution from these factories killed the lichens and the soot darkened the trees. Under those conditions, the dark moths were better camouflaged from predatory birds on lichen-free, soot-darkened trees than the light-colored moths (Figure 11.7B).

But how did scientists know it was the result of the Industrial Revolution? H. B. Kettlewell set out to test this hypothesis in the 1950s. He bred both kinds of moths in captivity, marked them for easy identification, then released them in several areas. His team recaptured more of the black moths in the polluted areas, and more of the light moths in the less polluted areas. The researchers also observed predatory birds eating more light moths in soot-darkened forests, and more black moths in cleaner, lichen-rich forests. Black moths were clearly at a selective advantage in industrialized areas.

Pollution controls went into effect in 1952. As a result of improved environmental standards, tree trunks gradually became free of soot, and lichens made a comeback. Kettlewell observed that moth phenotypes shifted too: Wherever pollution decreased, the frequency of dark moths decreased as well.

This story illustrates selective forces in action. Selective forces are one mechanism that drives evolution (Table 11.1). A **selective force** is a factor that influences the survival and reproduction of a subset of individuals within a population. For example, in the case of natural selection, animals that have genes that lead to a phenotype that is helpful in a particular environment are more likely to survive and pass on their genes. Moth color is determined by a single gene. Individuals with a dominant allele of the gene are black, and those that are homozygous for a recessive allele are light. Populations of moths living in polluted forests have a higher frequency of the dominant allele; populations in clean forests have a higher frequency of the recessive allele. Over time and under additional selective forces from each environment, it's possible that dark- and light-colored moths could become separate species (Section 11.4).

Figure 11.8 An illustration of smoke rising from steel factories in Sheffield, England, in the 19th century

Credit: North Wind Picture Archives/Alamy Stock Photo

Patterns of Natural Selection

There are three different ways that natural selection works: through directional, stabilizing, or disruptive selection Figure 11.9A. We'll review each of these next.

Directional selection. In **directional selection**, there is one trait that works best in a particular environment. Consider the long necks of giraffes (*Giraffa camelopardalis*). Animals with longer necks would be able to reach food in trees that shorter-necked animals cannot

Table 11.1 Types of Selection

Type of selection	Selective force	Result
Natural selection	Environment, including living (e.g., predators, prey availability) and non-living (e.g., weather) aspects of the environment	Phenotypes that help an organism survive and reproduce persist in the gene pool and become more predominant
Sexual selection	Choosy mates (typically females)	Elaborate coloring or body morphology, or behaviors like singing, dancing, or fighting
Artificial selection	Humans	Organisms with desirable traits

reach, and would therefore have an advantage. Therefore, the environment exerts a selective force leading to one predominant phenotype (Figure 11.9B).

Figure 11.9 Comparing three modes of natural selection

In these examples, blue arrows indicate which alleles of the trait (indicated by differently colored dots) are adaptive and red arrows show alleles that are selected against.

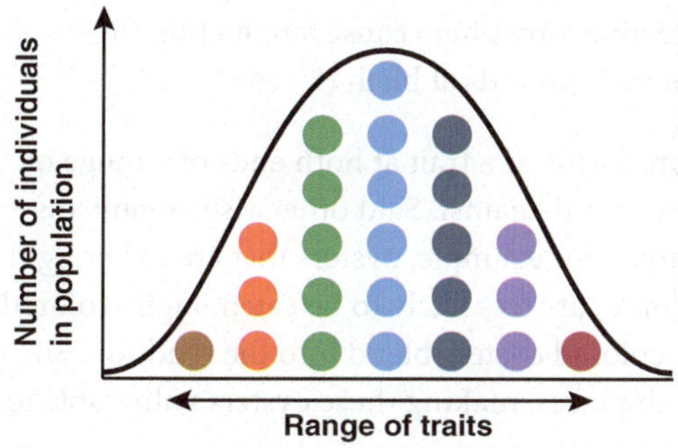

A. A population before selection occurs

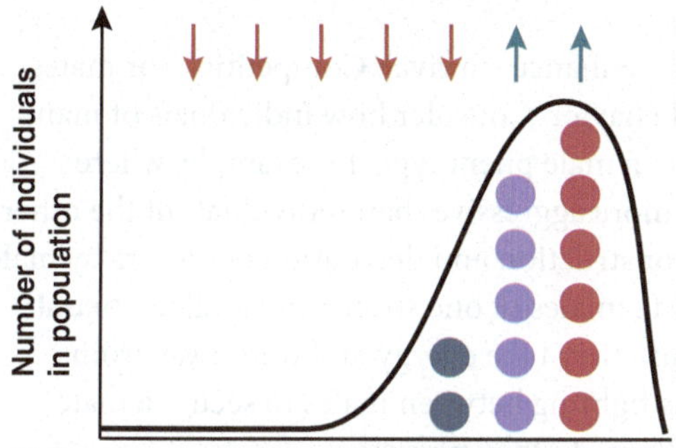

B. Directional selection. A form of a trait at one end of a range of variation is selected.

Giraffes with longer necks can reach leaves higher in trees. Therefore, selective forces resulted in animals with long necks.

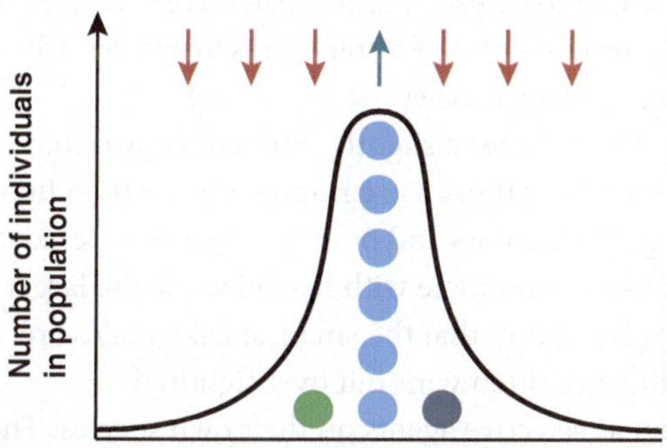

C. Stabilizing selection. Extreme forms of a trait are eliminated, and an intermediate form is maintained.

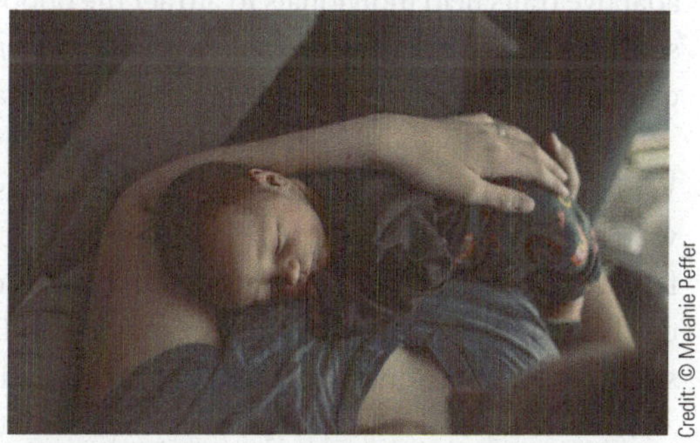

In humans, babies at either extreme in size experience more complications at birth, and so selective forces led to intermediately sized babies.

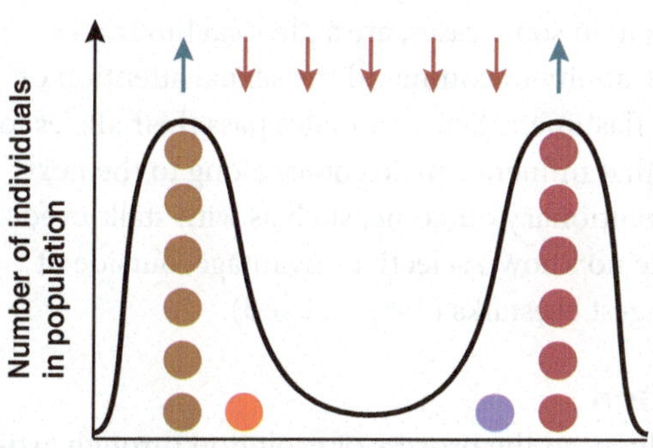

D. Disruptive selection. A midrange form of a trait is eliminated, and extreme forms are maintained.

Medium-colored oysters are more visible to predators than light- or dark-colored ones. Thus, selective forces resulted in more oysters with extreme phenotypes.

? Did You Know?

Evidence from the fossil record suggests that giraffes' long necks also evolved as the result of competition for mates.

Stabilizing selection. With **stabilizing selection**, an intermediate form of a trait is adaptive, and extreme forms are selected against. Since most populations are fairly well adapted to their environments, stabilizing selection is thought to be the most common form of natural selection. Consider human birth weight as an example. Babies born too small often face health challenges outside the womb. In contrast, large babies can get stuck in the birth canal during labor, leading to potentially serious complications. Therefore, selective forces here cause a reduction in variability, where human babies are on average around 7 pounds at birth (Figure 11.9C).

Disruptive selection. With **disruptive selection**, forms of a trait at both ends of a range of variation are adaptive, and intermediate forms are selected against. Said otherwise, organisms with extreme phenotypes, not the average, are favored. For example, oysters that are either light or dark colored, but not those that are medium colored, are less likely to be eaten. Light-colored oysters camouflage nicely into the rocks, and dark-colored oysters blend into the shadows. Since the medium-colored oysters would stand out in both places, making these oysters vulnerable to predators, this phenotype is selected against (Figure 11.9D).

Survival of the Sexiest

Not all evolution is driven by selection for traits that enhance survival. Competition for mates is another selective force that can shape form and behavior. Consider how individuals of many sexually reproducing species have a distinct male or female phenotype; for example, where individuals of one sex are more colorful, larger, or more aggressive than individuals of the other sex. These traits can also include behaviors, such construction and decoration of bowers by male bowerbirds. When a trait differs between males and females of one species, it is called **sexual dimorphism**. These traits can seem puzzling because they take energy and time away from activities that enhance survival, and some—such as fighting between males to secure a mate—can hinder an individual's ability to survive. Why, then, do they persist?

The answer is **sexual selection**, a type of selection in which the evolutionary winners out-reproduce other individuals of a population because they are better at securing mates. With sexual selection, adaptive traits help individuals to attract the opposite sex, or to defeat rivals for mates. For example, in some species, like big-horned sheep (*Ovis canadensis*), males compete for sole access to females, where competition for the ready-made harems favors brawny, combative males (Figure 11.10A).

However, the cuttlefish (*Sepia officinalis*; Figure 11.10B) uses a slightly different approach. While some male cuttlefish will battle for access to females, others use cunning. Rather than fight with large male cuttlefish, small cuttlefish will change their colors and body posture to look like a female, sneak over to where the actual female cuttlefish is, and mate with her, all while the larger cuttlefish watches for potential threats nearby. Biologists found that the small, sneaky males are successful at reproducing 60% of the time—so intelligence often wins out over fighting!

Males or females that are choosy about mates act as selective agents on their own species. The females of some species shop for a mate among males that display species-specific cues, such as a highly specialized appearance or courtship behavior, as is the case with the Goldie's bird-of-paradise (*Paradisaea decora*; Figure 11.10C). The cues often include flashy body parts or movement, which are traits that tend to attract predators and that, in some cases, are a physical hindrance.

However, in terms of reproductive success, the ability to command the sexual attention of females can offset a survival handicap imposed by flashiness. Selected males pass their alleles for their attractive traits while females pass on alleles that influence male choice along to the next generation. Highly exaggerated traits can be an evolutionary outcome, such as with stalk-eyed flies (family Diopsidae). The long eyestalks provide no known selective advantage, outside of females preferring to mate with males with the longest eyestalks (Figure 11.10D).

A Human's Best Friend: Artificial Selection

Perhaps surprisingly, humans have extensively influenced the process of evolution through artificial selection. **Artificial selection** is when evolution occurs due to humans selectively breeding plants or animals for specific traits. Darwin, along with many other Victorians, bred pigeons as

Figure 11.10 **Examples of sexual selection**

A

Credit: iStock.com/mlharing

B

Credit: Nature Picture Library/Alamy Stock Photo

C

Credit: National Geographic Image Collection

D

Credit: Mark Moffett/Minden pictures

A. These male bighorned sheep are fighting each other to mate with females.

B. Two male common cuttlefish display bright colors to impress a female (far right).

C. Two male Goldie's birds-of-paradise engage in a flashy courtship display that has caught the eye (and perhaps, the sexual interest) of a female. Females are choosy, whereas a male will mate with any female that accepts him.

D. Mating in stalk-eyed flies.

Process of Science

Darwin's Distaste for Peacocks

In 1860, Charles Darwin penned a note to a friend stating the sight of a peacock feather "made [him] sick!" Male peacocks, with their large, elaborate tails, make no sense from the perspective of natural selection (Figure 11.11). Having such a tail would seem to be more of a liability than an asset. The male bird wouldn't be easily camouflaged in its environment and the tail would be added weight to drag around—both of which could contribute to male peacocks being more susceptible to predation. So, why the large, elaborate tail if it can't be explained by natural selection?

Now, we must look at the period in which Darwin lived. Much like other men who lived during the Victorian era, he thought that women were inferior to men. That's why it was so difficult for Darwin to make the leap to propose that the evolution of peacock feathers was not due to *natural* selection, but instead was because of *sexual* selection—females preferentially mating with males who have larger, showier tails. How was it possible for something as crucial as evolution to be the result of a *female's choice*?

Darwin managed to look past his cultural biases to propose sexual selection as a driver of evolution in 1871, 12 years after the publication of *On the Origin of Species*. Given the predominant viewpoint, the transformative power of females as drivers of evolution was a radical idea—and one that clearly shows how cultural biases and viewpoints can impact science.

Credit: Shawn Hempel/Adobe Stock Photos

Figure 11.11 **Male peacocks are prized for their beautiful tail feathers**

1. Can you think of other viewpoints that used to be prevalent in science but that are no longer relevant in current times?
2. How can cultural biases unintentionally cause us to miss observations or get in the way of our developing novel ideas?
3. Do some reflection. What cultural biases do you think you might have? How can you address them so that they don't impede your thinking?

Biology for Life

Gift-Giving Spiders and Genetics

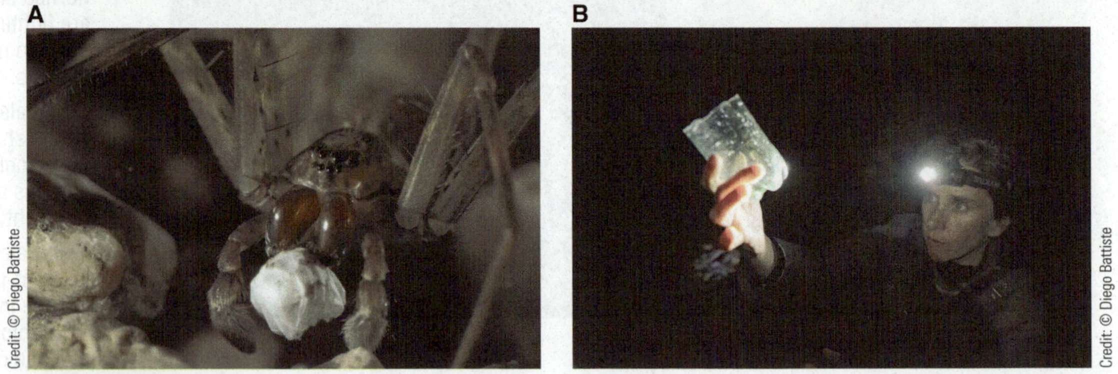

Figure 11.12 **Gift-giving spiders in South America demonstrate the interplay of genes and environment**

A. Is it love? A male *Paratrechalea ornata* spider carrying a nuptial gift to woo his love interest.

B. National Geographic Explorer Dr. Maria Albo examines a trapped spider in a repurposed beverage bottle.

Have you ever received flowers from someone who was trying to win your affection? Or given gifts to try and woo a potential partner? It turns out, humans aren't the only species that offers gifts during courtship. Spiders do it, too (Figure 11.12A). When a male *Paratrechalea ornata* spider wants to mate with a female, he will present her with a nuptial gift. The quality of these gifts can vary and can impact reproduction.

National Geographic Explorer Dr. Maria Albo (Figure 11.12B) studies gift-giving spiders in Uruguay, South America. Females benefit from eating the food gifts that males offer, increasing their number of offspring. However, males can also offer non-nutritive gifts. Albo and her team looked at the quality of spider gifts in two different locations—one with high climatic variation caused by increased rainfall, and the other with low climatic variation.

Surviving in an environment with high climatic variation is difficult for the spiders. Therefore, the ability to present a gift at all is an important mark of survivorship, and an indicator that the male would be a good partner. In this region, male spiders typically brought smaller, less valuable gifts to the female. Rather than a nice insect snack wrapped tidily in spider silk (a quality nutritive gift), they would instead present insect leftovers or plant material (a worthless, non-nutritive gift). Albo and colleagues found that in an environment with high climatic variation, females could produce just as many babies with a male spider presenting a worthless gift as one who presented a quality gift.

In the region with less climatic variation, some male spiders presented worthless gifts, and others gave quality gifts to the females. The females who reproduced with males bearing worthless gifts didn't produce as many offspring. In this case, the worthless gift was an indication of a male who was less able to survive well in an environment that was easier to live in, and ultimately, procured less-desirable gifts. Therefore, the environment likely influenced the quality of the gifts and the overall reproductive success of the spiders—an example of sexual selection in action and how it is influenced by the environment.

1. Why do you think the quality of the gifts was more important for production of offspring in one environment than the other?
2. Over time, how would you predict that the environment might foster or hinder gift-giving behavior? Would the same behaviors be selected for by sexual selection, or would these two selective forces work to oppose each other?

a hobby. They selected specific traits, such as the size of the bird's beak wattles and the arrangement of the tail feathers. There are many examples of artificial selection in our society today that are likely more relatable to you than fancy pigeons. For example, domestication of dogs occurred when humans selectively bred wolves for desirable traits (Figure 11.13).

Humans have also selectively bred many kinds of plants (Chapter 9), by preferentially collecting and planting seeds of plants with the desired trait. For example, familiar cruciferous vegetables such as kale, broccoli, and cauliflower can all trace their ancestry to the wild mustard plant (Figure 11.14).

Figure 11.13 Artificial selection in dogs

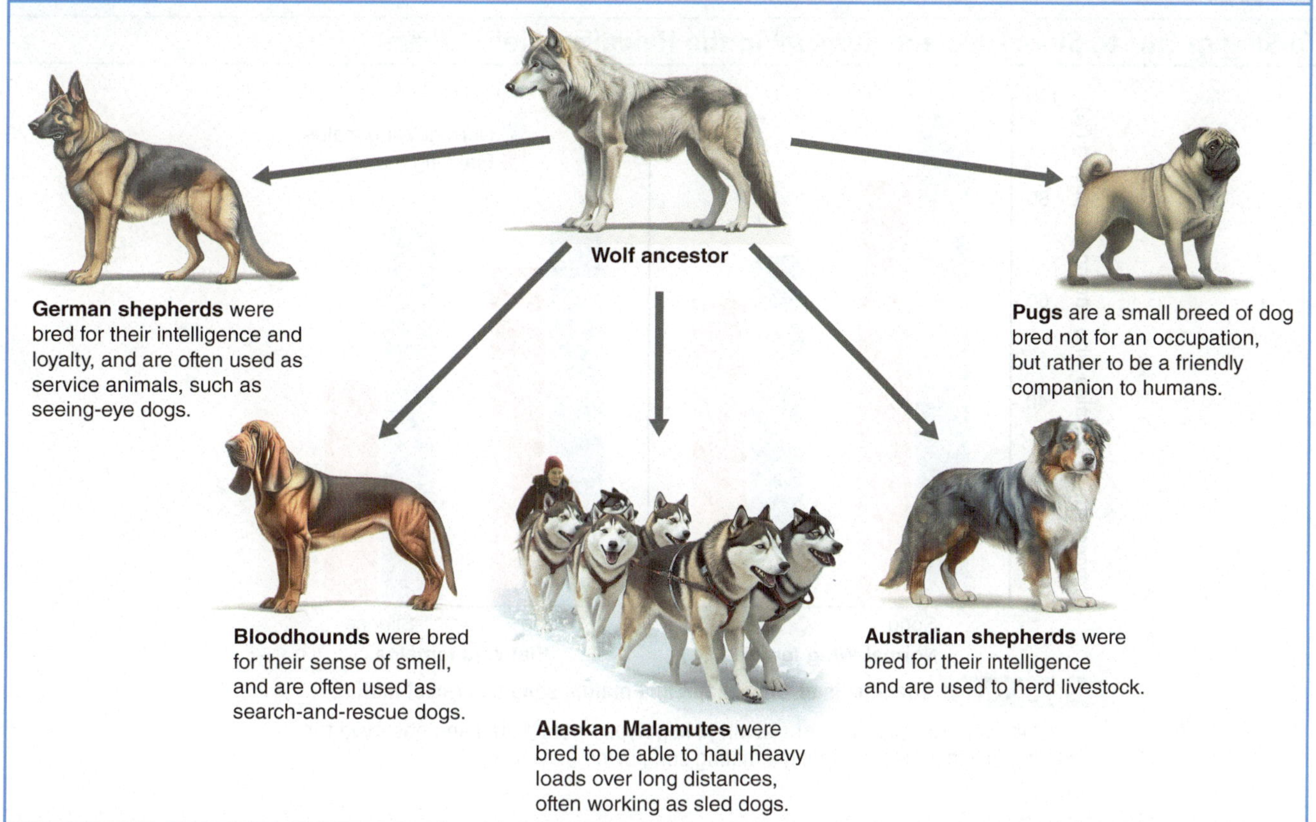

Figure 11.14 Artificial selection in cruciferous vegetables

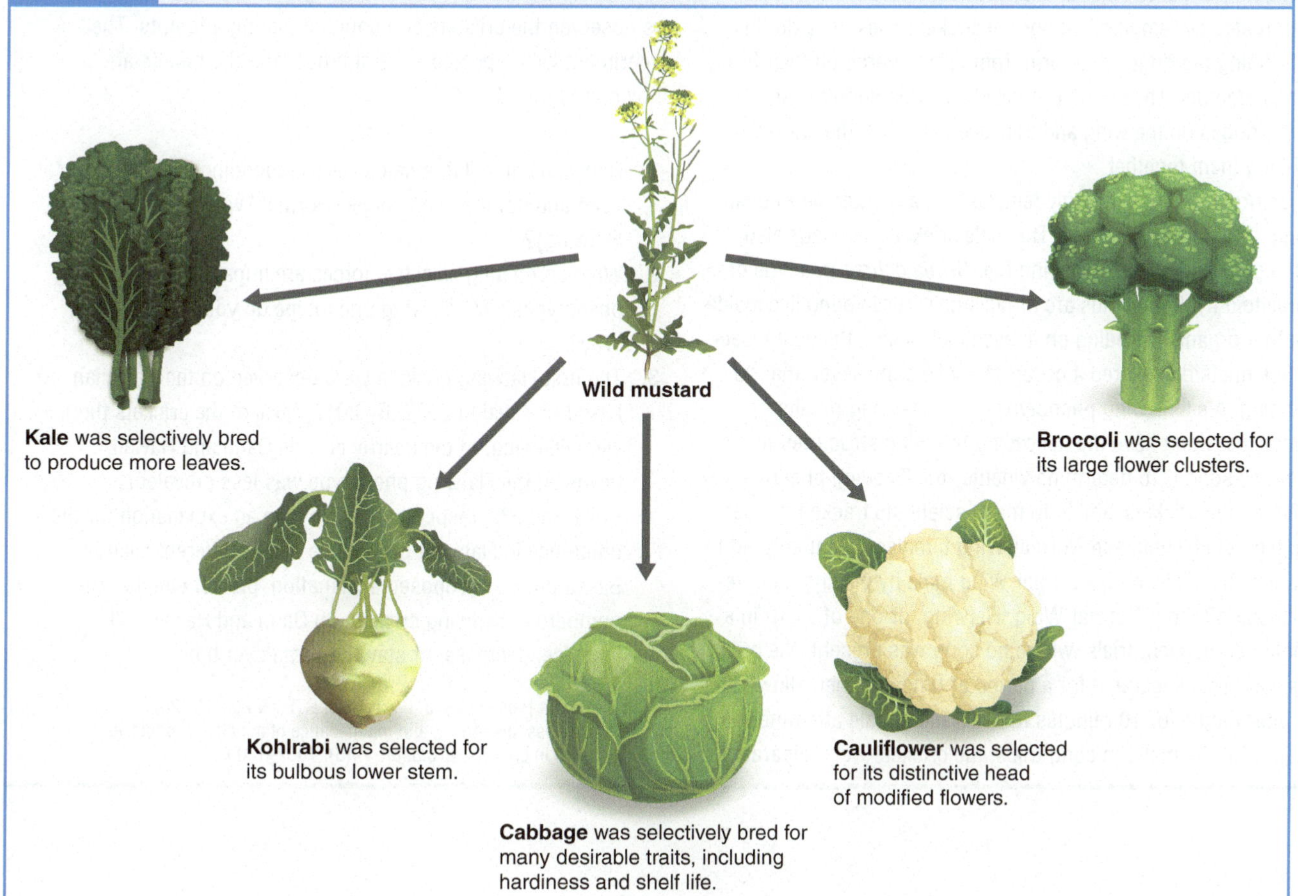

Digging Into Data

To Sing or Not to Sing: Love and Survival in the Hawaiian Field Cricket

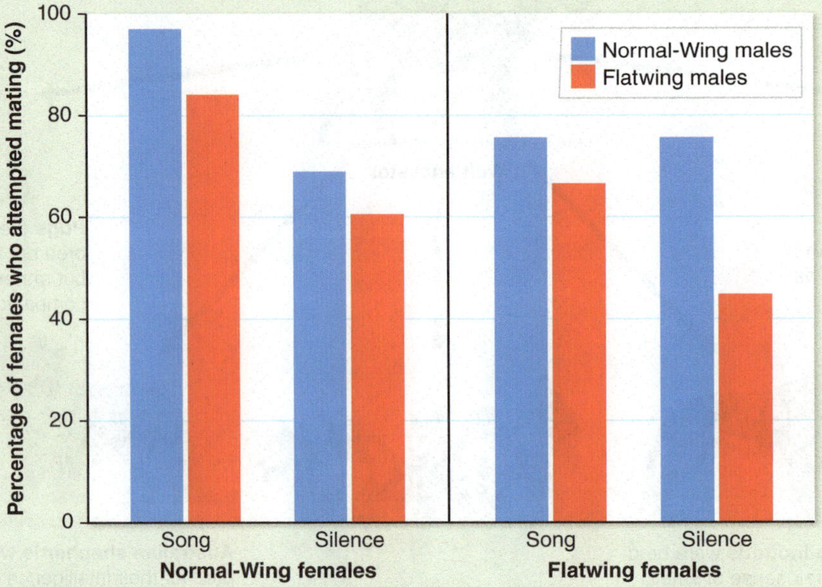

Figure 11.15 The effects of wing shape and mating song on cricket mating

Crickets with Normal-Wing or Flatwing phenotypes were paired and observed for mating in either silence or with a mating song being played.

The Hawaiian field cricket (*Teleogryllus oceanicus*) is a species in which, like some other cricket species, males typically find their mates by "singing." In normal cricket males, they do this by rubbing the ridges that form from vein patterns on their forewings together. The sound produced depends upon the shape of the ridges on the wing and the speed at which the cricket is rubbing them together.

Unfortunately, it's not only females that are attracted to their song, but predators as well. The male cricket's serenade also attracts a deadly eavesdropping foe, *Ormia ochracea*, a type of parasitoid fly. Parasitoids are organisms whose young live inside another organism feeding on it, eventually killing the host insect.

Scientists discovered a group of crickets, however, that do not sing. A silent-wing phenotype called Flatwing results in males that don't have the necessary forewing structures to produce sound. To determine whether the Flatwing phenotype affected the crickets' ability to mate, scientists tracked the percentage of Flatwing and Normal-Wing females that attempted to mate with a Flatwing or Normal-Wing male by placing male-female pairs (either Normal-Wing, Flatwing, or one of each) in a container. In silent trials, where no song was present, the crickets were first separated for a period of time and then allowed to intermingle for 10 minutes to see if the female attempted to mate with the male. In song trials, the crickets were separated as mating song was played on a speaker near the male. Then the crickets were allowed to intermingle while the researchers observed the crickets and counted mating attempts. The experiment was repeated several times, and the results are shown in **Figure 11.15**.

1. How can you tell if females were responding to the mating song and not the wing shape (Normal-Wing versus Flatwing)?
2. What competing selective forces are impacting cricket wing phenotypes? Which wing phenotype do you think is ideal? Why?
3. The first Flatwing crickets were observed on the Hawaiian island of Kauai in 2003. By 2012, 96% of the crickets there were Flatwing. In the nearby islands Oahu and Hawaii, however, the Flatwing phenotype was less prevalent (45% and 2%, respectively). Propose an explanation for the difference in Flatwing prevalence on the different islands. Based on your proposed explanation, predict whether the numbers of Flatwing crickets on Oahu and Hawaii will increase, decrease, or stay the same over time.

Graph based on: Heinen-Kay, J. L., Nichols, R. E., & Zuk, M. (2020). Sexual signal loss, pleiotropy, and maintenance of a male reproductive polymorphism in crickets. *Evolution, 74*(5), 1002–1009.

11.2 Summary

- Selective forces can be from nature (natural selection), from mate choice (sexual selection), or from humans (artificial selection).
- Natural selection occurs in three patterns. With directional selection, one extreme form of a trait is adaptive. Stabilizing selection maintains the intermediate form. Disruptive selection eliminates the intermediate form while maintaining the extreme forms.
- Sexual selection occurs when traits help secure mates. This can result in exaggerated body structures or behaviors that lead to reproductive success.
- In artificial selection, evolution occurs when humans selectively breed plants or animals to produce an organism with a desirable trait.

11.3 Evolution by Non-Selective Forces

Learning Objectives

11-9 Explain how allele frequency can change independently of natural selection.

11-10 Explain why smaller populations are more vulnerable to the loss of genetic diversity.

11-11 Describe the way gene flow stabilizes allele frequency.

In the Straits of Florida is a small island called Key West. Key West is known for its sunsets, close proximity to Cuba...and high number of extra-toed (polydactyly) cats, also called Hemingway cats (**Figure 11.16**). Ernest Hemingway, an American writer, had a home on Key West. Several of his popular works, such as *The Old Man and the Sea*, were written either at his house in Key West or on his boat. In the 1930s, a ship's captain gave Hemingway a white six-toed cat, whom he named Snow White. Now, approximately 60 cats, some of whom are descendants of Snow White, live in the Hemingway House and Museum in Key West. They all carry the polydactyl gene, and more than half of them have the phenotype of polydactyly.

In the general cat population, only about 10% of cats are polydactyl and yet more than half of the cats living at the Hemingway house are polydactyl. Why? And what does this all have to do with evolution? In the last section, we learned about selective drivers of evolution. However, evolution is not always driven by selective forces—anything that causes allele frequencies to change in an area, such as introducing a cat with unique traits to a small island (also called the *founder effect*), can also result in evolution. In the next section, we will learn about non-selective forces of evolution, including genetic drift and gene flow.

Figure 11.16 **A polydactyl cat**

Credit: Rob O'Neal

This resident of the Hemingway House, named Hairy Truman, has polydactyly. Note the presence of additional digits on its front paws.

Genetic Drift

Members of a natural population survive and reproduce with differing success, and the differences are not always an outcome of natural selection. By chance, a perfectly fit and healthy individual may not pass its alleles to offspring—for example, by dying in a random event before the opportunity to reproduce arises. Such events can change a population's allele frequencies. Change in allele frequency brought about by chance alone is called **genetic drift** (Figure 11.17A).

Population bottleneck. A **population bottleneck** is a drastic reduction in population size, and it can reduce the population's genetic diversity (Figure 11.17B). Consider how overhunting during the late 1890s left only about 10 northern elephant seals alive. Since then, hunting restrictions have allowed the population to recover, but genetic diversity among its members remains greatly reduced. Inbreeding, or mating between close relatives, can result from population bottlenecks. Closely related individuals tend to share more harmful recessive alleles than nonrelatives, and inbreeding increases the likelihood that an individual will inherit the same harmful allele from both parents.

The founder effect. A loss of genetic diversity can also occur when a small group of individuals establishes a new population, such as in the case of the Hemingway cats described in our earlier example, or when a group of people leaves the larger population to start a new colony elsewhere. For example, the founder effect can occur in humans when a small group of people leaves to found a new colony. If the people in the new colony do not marry outside of their colony, such as with the Old Order Amish in Lancaster County, Pennsylvania, there can be a higher incidence of genetic disorders due to a limited gene pool. If the founding group is not representative of the original population in terms of allele frequencies, then the new population will not be representative of it either. This outcome is called the **founder effect** (Figure 11.17C).

? Did You Know?

Cheetah populations experienced two different population bottlenecks—one that occurred around 100,000 years ago when cheetahs expanded their range and a second around 10,000 to 12,000 years ago at the end of the last ice age. These historic bottlenecks combined with the loss of cheetahs due to human conflict and habitat loss have led to concerns that despite concerted conservation efforts, there may not be enough genetic diversity left among cheetahs for these efforts to succeed.

Genetic Drift has Different Impacts Depending on Population Size

Genetic drift occurs in all populations, but it makes small populations particularly vulnerable to the loss of genetic diversity. To understand why, imagine a hypothetical gene with two alleles, neither of which confers a selective advantage. These alleles (let's call them *A* and *a*) occur at a frequency of 95% and 5%, respectively. In this example, a population with 10 members has one individual that is heterozygous (*Aa*), and the remaining nine are homozygous (*AA*) (Figure 11.18A). A random event that eliminates the heterozygous individual from the population before it reproduces would therefore also eliminates the *a* allele from the population's gene pool. If this occurs, the other allele (*A*) becomes a **fixed allele**. This means that *all* remaining individuals of the population would be homozygous for the *A* allele. A fixed allele remains that way until a new mutation occurs, or an individual bearing another allele enters the population from elsewhere.

Now, imagine that our population with alleles *A* and *a* consists of 40 members instead of 10 (Figure 11.18B). Given the same frequency of 95% and 5% of the *A* to the *a* allele, respectively, five individuals in this larger population would be heterozygous (*Aa*). For the *a* allele to be lost from the population's gene pool, all five of those individuals would have to be eliminated before they reproduce. Therefore, the chance of random events eliminating all of the heterozygous individuals is smaller in the larger population than in the smaller one. This is a simplified example of a general effect: The loss of genetic diversity is possible in all populations, but it is more likely to occur in small ones.

Figure 11.17 **Genetic drift occurs due to random changes in allele frequency**

A. Genetic drift. Allele frequency changes because of a random event. In this example, the two individuals who have an allele for green coloration are eliminated from the population.

B. Bottleneck. An event that causes a large reduction in population size alters the allele frequency. For example, here, several mice are washed downstream, leaving mainly black mice, one white, and no brown mice at the original location.

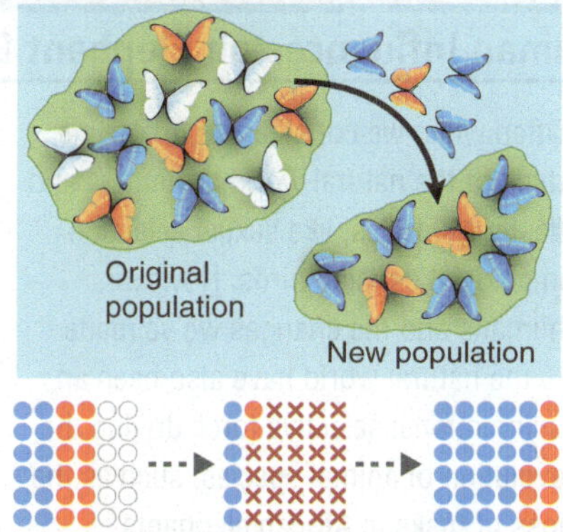

C. The founder effect. A group that founds a new population is not genetically representative of the original population, so allele frequencies differ between the new and the old populations.

Figure 11.18 **A comparison of the impact of genetic drift on a small and large population**

A

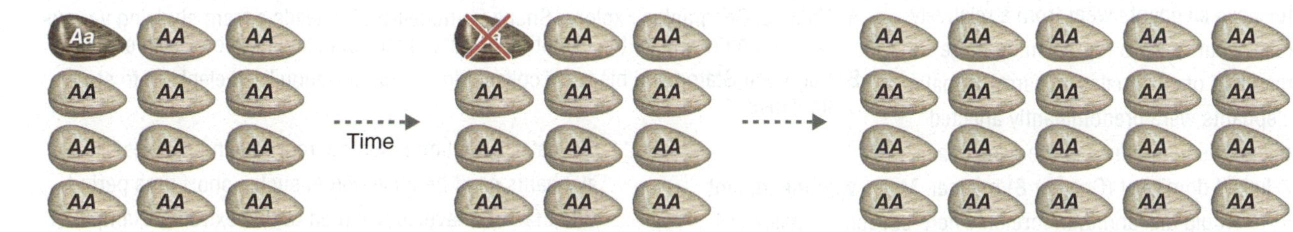

A small population of mollusks, with one heterozygous individual (*Aa*) and the others all being homozygous (*AA*).

A random event removes the *Aa* individual from the population.

The *A* allele becomes fixed in the population in subsequent generations, until new alleles are introduced through gene flow from another population.

B

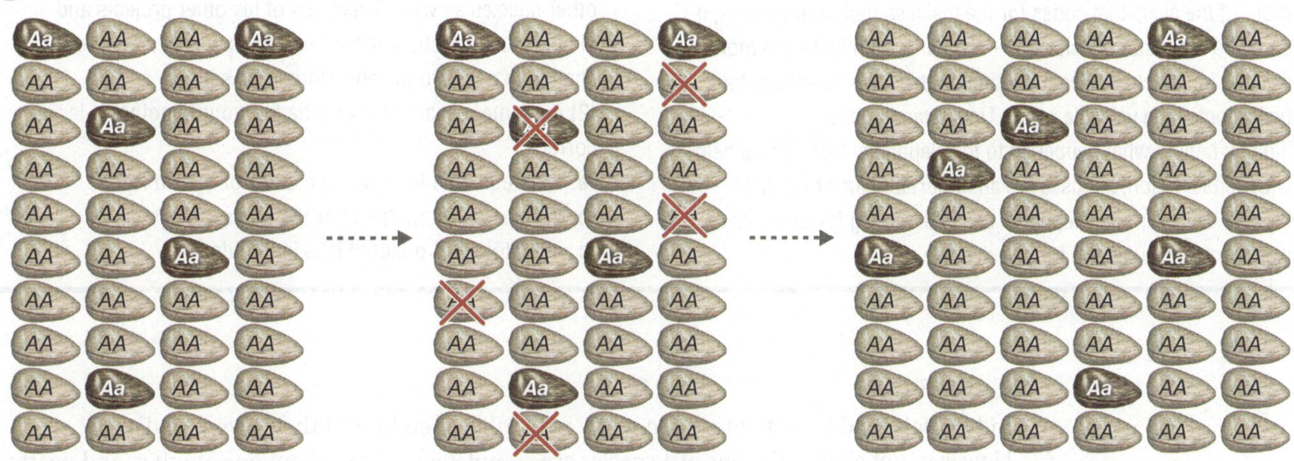

A large population of mollusks, with five heterozygous individuals (*Aa*) and the others all being *AA*.

A random event removes five individuals from the population, only one of which is *Aa*, leaving four individuals with the *a* allele in the population.

The *a* allele persists in subsequent generations.

Biology for Life

Human Influences on Elephant Evolution

Often when we consider evolution, we think of the natural world around us and flashy examples, like sexual selection in colorful dancing birds. However, humans and the changes we've made to the natural world have also been an unintentional selective force driving the evolution of animal species, such as the loss of tusks in African elephants.

National Geographic Explorer Shane Campbell-Staton (Figure 11.19A) and his team are studying the genetic basis of elephant tusk loss in elephants living in Gorongosa National Park (Figure 11.19B). About 90% of the large mammals in Gorongosa National Park were killed as the result of the Mozambican Civil War. Notably, after the war, tuskless elephants went from a relatively rare occurrence to representing more than half of all female elephants. Female elephants were predominantly affected due to the inheritance of the gene in a X-linked dominant (Chapter 8) manner. Males with the mutant form would die, and therefore only heterozygous females will survive. But what happened to the tusks?

There are two possible explanations for this phenomenon. One explanation could be that the loss of genetic diversity led to a population bottleneck. Since there was such a massive loss of African elephants, it could have randomly happened that the survivors had more of the allele that codes for the tuskless trait. Alternatively, it could be the result of selection—elephants with tusks are more likely to be poached (illegal hunting; in this case, to get elephant tusks containing valuable ivory). Understanding the biology behind what happened to the elephants' tusks could help with conservation, as tusks are an important part of elephant biology. For example, elephants use tusks to dig for minerals or water in the ground and for self-defense.

A

B

Figure 11.19 Human-driven selection in African elephants

A. National Geographic Explorer Shane Campbell-Staton leads a team studying various ways humans impact the evolution of other species, including the African elephant.

B. Campbell-Staton and his team collect blood from a tranquilized elephant to study its genes.

1. Typically, evolution of a slow-reproducing species like elephants can't be observed in such a short time period. What factors have contributed to this extreme example of evolution?
2. Do you think poaching-mediated selection should be called natural selection, artificial selection, or a different term entirely? Why?
3. Dr. Campbell-Staton studies human-driven evolution in other species as well. Select one of his other projects and (1) make a prediction about how the species listed might change to adapt to the environment described, and (2) do some research to see whether your hypothesis is correct.
   a. Crested anole lizards in the cities of Puerto Rico
   b. Wolves in Chernobyl after the nuclear disaster
   c. Alligators in a polluted lake in Florida

Gene Flow

Individuals tend to mate most frequently with other members of their own population. However, not all populations of a species are completely isolated from one another, and nearby populations may interbreed. Also, individuals sometimes leave one population to join another.

Gene flow, the movement of alleles between populations, occurs in both cases. In this way, gene flow can keep populations genetically similar by stabilizing allele frequencies, thus countering genetic drift.

Gene flow highlights the importance of wildlife bridges, which serve to connect relatively isolated populations of animals to maintain more diversity in the gene pool (Figure 11.20). Diverse gene pools are important for the health of a species and mitigate the potential impact of population bottleneck events. These bridges also help to keep highways safer by providing a safe passage for wildlife away from cars.

Gene flow is common among populations of animals, but it also occurs in less mobile organisms. Consider acorns dispersed by jays gathering nuts for the winter. Every fall, these birds visit acorn-bearing oak trees repeatedly, and then bury the acorns in the soil of territories as much as a mile away. The jays transfer acorns (and the alleles carried by those seeds) among populations of oak trees that may otherwise be genetically isolated.

Figure 11.20 **Wildlife bridges connect populations of animals isolated by a human made structure, such as a highway**

Credit: Artur Debat/Getty Images

11.3 Summary

- Allele frequency can change in a population due to genetic drift, including events that lead to a population bottleneck or founder effect.
- Smaller populations have fewer alleles in their gene pool than larger populations, leaving them more vulnerable to losing genetic diversity through genetic drift.
- Gene flow stabilizes allele frequencies in a population by replenishing lost alleles through interbreeding with individuals from a different population.

11.4 Speciation and Extinction

Learning Objectives

11-12 Describe how reproductive isolation can lead to speciation.

11-13 Compare different forms of reproductive isolation.

11-14 Describe how different types of separation can lead to allopatric or sympatric speciation.

11-15 Describe how adaptive radiation facilitates speciation.

11-16 Explain the relationship between speciation and extinction.

Between 1940 and 1941 during World War II, there was a German bombing campaign against the United Kingdom called the Blitz. For protection from the German bombs, many people took shelter in London's famous subway system—the Underground (Figure 11.21). As if the constant hazards from bombs falling above and the need to sleep on the hard ground with one's neighbors wasn't enough of a challenge, people faced constant attack by voracious mosquitoes. Although mosquitoes may have initially become trapped in the underground in the 1860s during construction of the subway system, people didn't start complaining about them until World War II.

Figure 11.21 People sheltering in the London Underground during the Blitz in World War II

Credit: Trinity Mirror/Alamy Stock Photo

In the late 1990s, biologist Katharyne Byrne set out to characterize these mosquitoes and concluded that the mosquitoes found in the London underground were a different species from those living above ground—isolation underground separated the two populations of mosquitoes. Mosquitoes living underground learned to prey on humans instead of birds (like their above-ground relatives), since there were plenty of humans to feast upon. Consequently, due to the physical separation, two different species evolved. If the underground and above-ground mosquitoes were put into contact with one another, they could no longer mate and reproduce. This, along with other behavioral lines of evidence, suggests that the mosquitoes trapped underground evolved into a new species of mosquito over a relatively short time frame.

The emergence of a new species—the splitting of one lineage into two—is called **speciation**. Speciation occurs because of the acquisition of genetic differences between the two (or more) separated species due to mutation, natural selection, and genetic drift occurring independently in each gene pool. Over time, the populations may become so different that we consider them to be different species. Although there is some argument about the best criteria used to determine if two organisms are a different species, for our purposes we will define a species as two populations that cannot interbreed.

Remember that speciation occurs in a continuum of geographic, genetic, and ecological contexts. Categories are useful as models, but the relationship between pattern and process is rarely simple. Populations often continue to interbreed even as they are diverging, and populations that have already diverged may come together and interbreed again.

Pre-Reproductive Isolation

Recall from Chapter 10 that evolution is fundamentally about which individuals are best at passing on their genes on to the next generation. Different reproductive success eventually leads to changes in the gene pool of a population and, ultimately, evolution. Perhaps it is not surprising then that speciation typically occurs because of **reproductive isolation**, or the end of gene flow between populations due to an inability to reproduce.

Reproductive isolation can happen in a variety of unique ways, which means that each species is a product of a unique set of evolutionary events. Various mechanisms that prevent successful interbreeding lead to reproductive isolation, reinforcing differences between diverging populations (Figure 11.22).

? Did You Know?

The 13- and 17-year cicadas coemerged in 2024.

Temporal isolation. Differences in the timing of reproduction can prevent interbreeding, which is called **temporal isolation**. Consider periodical cicadas (genus *Magicicada*; Figure 11.22A). Larvae of these insects feed on roots as they mature underground, then emerge as adults to reproduce at the same time, at either 13- or 17-year intervals. Thirteen- and 17-year cicadas are considered different species. They could potentially interbreed, but would only get an opportunity to do so only every 221 years!

Habitat isolation. Adaptation to life in different habitats (where an organism lives) can prevent interbreeding. This is called **habitat isolation**. For example, populations of pea aphids (*Acyrthosiphon pisum*; Figure 11.22B) living on red clover plants do not interbreed with populations living on alfalfa plants, even if the two types of plants are closely intermingled. Individual aphids that migrate to a different plant species have a low survival rate, so these sap-sucking insects reproduce exclusively on the plants where they are born.

Behavioral isolation. Differences in behavior can prevent interbreeding—for example, when specific cues required for sex are not recognized by members of other species. Males and females of many animal species engage in courtship displays, where in a typical pattern,

Figure 11.22 **Processes of speciation can vary, but all involve reproductive isolation**

Shown here are examples of organisms that are reproductively isolated for a variety of reasons, including **A.** periodical cicadas, **B.** pea aphids (arrow), **C.** peacock spiders, and **D.** sage plants.

Credit: MediaNews Group/Reading Eagle via Getty Images/Getty Images; Andy Nowack/Alamy Stock Photo; © Jürgen Otto; piemags/Alamy Stock Photo

the female recognizes the sounds, movements, and appearance of a male of her species as an overture to sex. When those cues aren't recognized, mating will not occur, a phenomenon called **behavioral isolation.** For example, a male peacock spider (genus *Maratus*) signals its intent to mate with a female by raising and waving colorful flaps while also moving its legs in time with abdominal vibrations. If he fails to impress the female, she may eat him (Figure 11.22C)!

Mechanical isolation. Sometimes the morphological features of an organism prevent successful interbreeding, called **mechanical isolation**. Consider two closely related species of sage plants that grow in the same areas, but their flowers are specialized for different pollinators (Chapter 18). Therefore, cross-pollination is unlikely to occur and hybrids rarely form (Figure 11.22D).

Post-Reproductive Isolation

In some cases, different but related species can encounter one another and mate, but either cannot produce offspring or produce offspring that cannot reproduce on their own. We'll look at examples of post-reproductive isolation next.

Gamete incompatibility. Even if gametes of different species do come into proximity with one another, they often have molecular incompatibilities that prevent a zygote from forming. For example, corals often reproduce in mass spawning events, with many different species releasing their gametes into the ocean simultaneously (Figure 11.23A). Although the different gametes are all intermixing into the water at the same time, only the male and female gametes from the same species will result in fertilization.

Figure 11.23 Examples of post-reproductive isolation

A. Corals reproduce in mass spawning events, but fertilization will occur only between gametes from the same species. Note the presence of gametes in the water above the corals.

B. Ligers, a cross between a lion and a tiger, are an example of hybrid inviability.

C. Mules, a cross between a horse and a donkey, are robust animals that cannot reproduce.

Hybrid inviability. Hybrid offspring form in some cases, but most do not reach sexual maturity. For example, hybrid offspring of lions and tigers, called ligers (Figure 11.23B), have more health problems and a shorter life expectancy than individuals of either parent species. If hybrids live long enough to reproduce, their offspring often have lower fitness with each successive generation. Incompatible nuclear and mitochondrial DNA may be the cause (mitochondrial DNA is inherited from the mother only).

? Did You Know?

In the aftermath of Hurricane Helene in 2024, mules were used to carry supplies into communities isolated by washed out roads.

Hybrid sterility. Some interspecies matings produce robust but sterile hybrid offspring. For example, mating between a female horse (64 chromosomes) and a male donkey (62 chromosomes) produces a mule (63 chromosomes—32 from the horse, and 31 from the donkey) (Figure 11.23C). Mules are healthy, but their chromosomes cannot pair up properly for crossing over during meiosis, so this animal makes few viable gametes. Sometimes hybrids can be more vigorous than either of their parents, a phenomenon known as **hybrid vigor**. Mules are prized for their intelligence, strength, and longevity.

Patterns of Speciation

In the opening example for this section, we learned about how a physical barrier between the London Underground and the outside environment was a factor in the evolution of a new mosquito species with a vampiric taste for human blood. **Allopatric speciation** occurs when a physical barrier arises between members of a population and interrupts gene flow between them.

Whether the barrier slows or prevents gene flow depends on the barrier and the species, such as by how it travels (for example, by swimming, walking, or flying), and how it reproduces (for example, by internal fertilization or by pollen dispersal). If gene flow is sufficiently hampered, genetic changes accumulate independently in the two new populations. The divergences may cause the populations to remain reproductively isolated even if the barrier is removed.

? Did You Know?

When the Great Wall of China was built, it interrupted the gene flow of nearby populations of insect-pollinated plants. Today, genetic divergences are occurring between populations of plants living on opposite sides of the wall.

For example, tectonic plate movements resulted in a land bridge between North and South America called the Isthmus of Panama. When this isthmus formed about 3 million years ago, it cut off the flow of water—and gene flow among populations of aquatic organisms—as it separated one large ocean into what are now the Pacific and Atlantic Oceans. Pacific and Atlantic populations of many of these organisms diverged so much that they are now reproductively isolated, separate species (Figure 11.24).

Figure 11.24 **Snapping shrimp (family Alpheidae) separated by the Isthmus of Panama are an example of allopatric speciation**

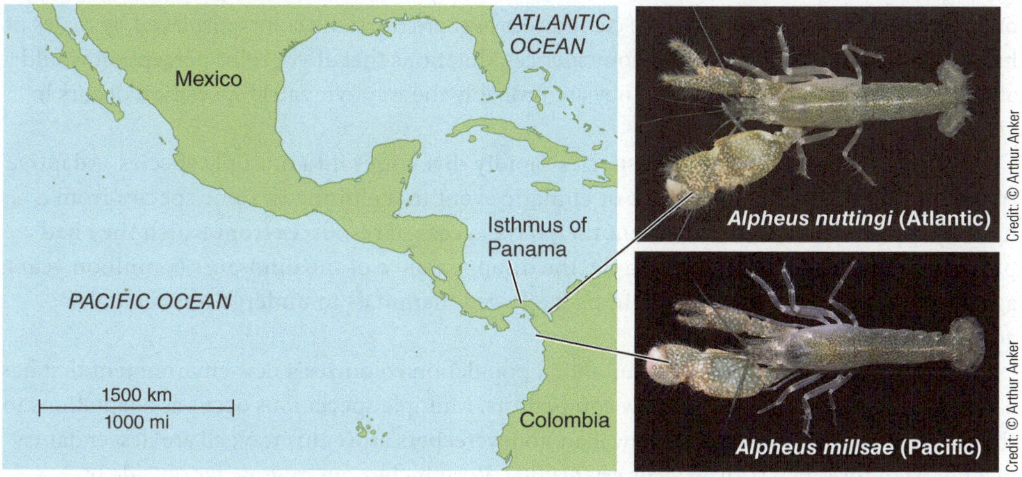

When the Isthmus of Panama formed 3 million years ago, it cut off gene flow between populations of snapping shrimp. If these two species of shrimp are brought together artificially, they aggressively snap their claws at each other and will not mate.

A physical barrier is not always required for speciation to occur. In **sympatric speciation**, genetic changes that lead to reproductive isolation occur within a single population, with no physical barrier to gene flow. For example, more than 500 species of cichlid fishes arose by sympatric speciation in the shallow waters of Lake Victoria. This large freshwater lake sits isolated from river inflow on an elevated plain in Africa's Great Rift Valley. Lake Victoria has dried up completely several times during its 400,000-year history. DNA sequence comparisons indicate that almost all the cichlid species in this lake arose since it was last dry, about 12,400 years ago.

How could 500 species of fish arise so quickly in the same body of water? In this case, the answer involves sexual selection. Consider how the color of ambient light differs in different parts of a lake. The light in the lake's shallower, clear water is mainly blue; light that penetrates the deeper, muddier water is redder. Cichlid species vary in color (Figure 11.25). Female cichlids prefer to mate with brightly colored males. Their preference has a genetic basis. Fish can differ in that they have alleles that lead to an ability to see certain colors of light more easily. Retinal

Figure 11.25 **Red fish, blue fish: Sympatric speciation of Lake Victoria cichlids**

A. Male *Pundamilia nyererei*. This species breeds in deep water, where the ambient light is reddish.

B. Male *Pundamilia pundamilia*. This species breeds in shallow water, where the ambient light is mainly blue.

Today, only 18 Hawaiian honeycreeper species remain, and 16 of them are in immediate danger of extinction. Non-native predators and competitors are an ongoing threat, as is the rise in global temperature. As Hawaii's average temperature rises, mosquitoes carrying diseases, such as avian malaria, are expanding their range into high-altitude environments, decimating populations of high-altitude honeycreepers that had previously avoided these diseases.

pigments made by species that live mainly in shallow areas of the lake are more sensitive to blue light. The males of these species are also the bluest. Retinal pigments made by species that prefer deeper areas of the lake are more sensitive to red light. Males of these species are redder. In other words, the colors that a female cichlid sees best are the same colors displayed by males of her species. This is unlikely to be a coincidence. Mutations that affect color perception would affect a female's choice of mates, so they are probably the way sympatric speciation occurs in cichlids.

With **adaptive radiation**, one species rapidly diversifies into multiple species. Adaptive radiation can occur after a geologic or climatic event that eliminates some species from a habitat; species that survive the event then have access to resources from which they had previously been excluded. For example, the disappearance of the dinosaurs 66 million years ago in the aftermath of an asteroid impact allowed mammals to undergo a spectacular adaptive radiation.

Adaptive radiation can also occur after a population colonizes a new environment that has a variety of different habitats and few competitors. Multiple speciations occur as adaptations to the different habitats evolve. The Hawaiian honeycreepers arose this way; all are descendants of a Eurasian rosefinch (*Carpodacus erythrinus*). Rosefinches can migrate far outside of their normal range, traveling in flocks of thousands or even tens of thousands of individuals. About 7 million years ago, one of these flocks was migrating through southern Asia when it became caught in the winds of a huge storm. The birds were blown at least 6,000 miles (9,600 kilometers) across the open ocean to the Hawaiian archipelago. Enough individuals survived the journey to found a new population.

Isolated from gene flow (Section 11.3) with mainland bird populations, the island colonizers diverged. Local environments on the Hawaiian Islands vary dramatically—from lava beds, rain forests, and grasslands to dry woodlands and snowcapped peaks. Selective forces differ within and between these habitats, and over many generations, populations of birds living in the different environments became hundreds of separate species, each with a beak structure that increases the birds' fitness specific to the area in which they live (**Figure 11.26**).

Big Ideas in Biology

Structure and Function

The beaks fits the purpose! Different kinds of beaks were selected for based upon which type of food was most available, showing the relationship between a biological structure and its function.

The Opposite of Speciation is Extinction

By current estimates, around 98% of all species that ever lived are now **extinct**, which means they no longer have living members. In addition to ongoing extinctions of individual species, called the **background extinction rate**, the fossil record indicates that there have been five **mass extinctions**, events where 75% or more of the world's living species were lost in a "short" period of geological time, less than 2.8 million years. During a mass extinction, species vanish faster than they can evolve through speciation. The last mass extinction was the loss of the dinosaurs, 66 million years ago.

Different lines of evidence suggest that we are amid a **sixth mass extinction**, driven by human activity. These human activities include unsustainable use of land, water, and energy. Climate change is also contributing to loss of biodiversity (Unit 5). According to the World Wildlife Foundation, agriculture alone is responsible for 90% of global deforestation and 70% of freshwater usage. These high numbers, combined with the shocking amount of food wasted in the United States alone highlight both the seriousness of our current usage of natural resources—and opportunities for change.

Are we amid a sixth mass extinction? Although evidence points to "yes" as the answer, it's hard to say for sure since they can take millions of years to occur—or can occur much faster than that. You can think of it like baking a cake...you won't know for sure if the cake turned out

Figure 11.26 **Numerous beneficial mutations led to the diversity of the Hawaiian honeycreeper**

Since environment is key for determining if an allele carries a beneficial, neutral, or harmful mutation, different beak structures were selected for depending on the bird's specific environment. For example, the Kiwikiu (left; *Pseudonestor xanthophrys*) has a large beak used for ripping off tree bark to access and eat the insects within. The Maui ʻalauahio (middle; *Paroreomyza montana*) has a thin bill adapted for eating insects. The Kaua'i amakihi (right; *Chlorodrepanis stejengeri*) has a longer, curved bill used for eating nectar and fruit. Note that only three representative species are shown here, and all are descended from the rosefinch.

well until it is out of the oven, cooled, and served. Right now, we know an extinction event is occurring—but will it reach mass extinction levels? We may not know for many years, possibly beyond the scope of our lifetimes.

Regardless of whether or not we are in the midst of a sixth mass extinction or other extinction event, it is important to remember that species are interconnected with one another (one of the Big Ideas in Biology introduced in Chapter 1, which we'll return to in Unit 5). The loss of one species can cause a ripple effect leading to a catastrophic collapse of ecosystems, ultimately not only harming the animals we share this planet with, but also negatively influencing the ways of life and livelihoods of many people.

Biology for Life

Dire Wolves, Mammoths, and Dodos—Oh My!

Imagine that in the not-so-distant future, your local zoo is advertising an exciting new exhibit. Driving down the road you see billboards advertising the zoo's latest exhibit of…what? A woolly mammoth? A species that has been extinct for around 4,000 years? You check the date—maybe this is an elaborate April fools' prank? Nope. The zoo truly has a mammoth, one brought back to life via *de-extinction*, or the process of bringing back an individual of an extinct species. The mammoth is only one species biotechnology companies are working to de-extinct; others include dire wolves and dodo birds.

A

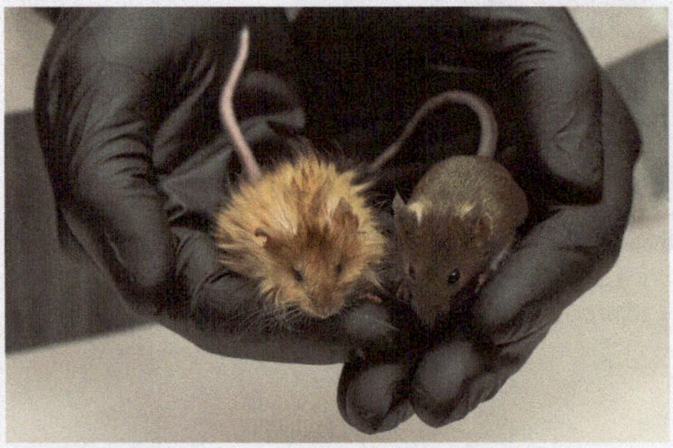

Credit: Colossal Biosciences

B

Credit: Colossal Biosciences

Figure 11.27 **Examples of animals genetically engineered to be like their extinct counterparts**

A. A woolly mouse (left) and normal lab mouse (right). Note the luxurious long hair on the woolly mouse.

B. Two proxy dire wolves.

There are currently two main technologies that could be utilized to bring extinct animals like mammoths back to life. One is **somatic cell nuclear transfer**, or **SCNT** (also colloquially called cloning), where an exact genetic copy of an extinct animal is made. In the case of de-extinction, the nucleus would come from the cell of an extinct animal. The nucleus from the extinct animal would be placed inside the egg of an animal living today with its nucleus removed, and would then be implanted into a surrogate mother. The resulting offspring would be a clone (or exact copy) of the extinct animal.

The second technology is *gene editing*, or changing an organism's DNA, which we learned about in Chapter 9. In the case of de-extinction, biologists are using fragments of mammoth DNA as templates for editing the genes of elephants to look mammoth-like. This will ultimately lead to the creation of a mammoth-elephant hybrid.

In 2025, one of the companies working on de-extinction announced the birth of woolly mice (Figure 11.27A) and dire wolf-like animals (Figure 11.27B). The woolly mice do not contain mammoth DNA, but rather their genes were edited to contain a mix of genes that resemble the profile seen in woolly mammoths. Since shaggy hair is a key mammoth trait, identifying the genetic variants behind these traits will be crucial to eventually editing the DNA of Asian elephants (*Elephas maximus*)—the closest living relative of woolly mammoths—to de-extinct a mammoth.

In the case of the dire wolf-like animal, analysis of extinct dire wolf (*Aenocyon dirus*) genomes, which are largely incomplete, indicates that dire wolves diverged from gray wolves (*Canis lupus*) about 6 million years ago. Starting with a related genome would mean that less gene editing would be required. Scientists focused the edits on physical traits, making 20 changes on 14 genes in the gray wolf genome.

As of the writing of this text, three pups were born with these edits, two males in October 2024 and one female in January 2025. Although these wolves are 99.5% genetically identical to dire wolves, there are still millions of base pair differences between the two—and therefore, it is not accurate to say that dire wolves are truly de-extincted, but rather, that these three individuals are proxy dire wolves.

1. De-extinction raises many potential ethical concerns, from whether bringing back animals to live in an environment they didn't evolve to live in is in their best interest to how social animals, like wolves, will survive without the interactions they rely on as part of their development. Pick an ethical issue involving de-extinction—either one mentioned here or a different one that you can think of or find on the Internet. For the issue you've chosen, identify the primary ethical concerns and whether those concerns can be mitigated, or lessened, by some potential action. What suggestions can you offer to stakeholders to reduce any harmful impacts of making an organism de-extinct?
2. Given the role that humans have in the extinction of many animal species, do we have a moral or ethical obligation to de-extinct animals? Why or why not?
3. Considering the difficulty of sequencing the entire genome of an extinct animal and our inability to truly clone them, do you think we'll ever truly de-extinct a species, or will these proxy species be the best that we can hope for? Defend your answer with scientific principles.

11.4 Summary

- Populations that do not interbreed diverge genetically. This reproductive isolation can lead to speciation as genetic changes accumulate and different selective forces affect each population.
- Pre-reproductive isolation stops gene flow through temporal isolation (different breeding times), habitat isolation (adaption to different habitats), behavioral isolation (failure to recognize mating cues), and mechanical isolation (physical differences prevent mating).
- Post-reproductive isolation occurs after gamete release if gametes are incompatible, hybrids do not mature, or hybrids are sterile, thus preventing viable and fertile offspring.
- Allopatric speciation occurs when a physical separation between populations interrupts gene flow. Sympatric speciation occurs when genetic changes within a population lead to reproductive isolation without physical separation.
- Adaptive radiation leads to speciation by the evolution of multiple species from an ancestral species, as populations adapt to different environmental conditions.
- Extinction is the opposite of speciation and is when there is a loss of all living members of a species.

Understand It: Motivation

What motivated you to sign up for a biology course this semester? Did you sign up because you find biology interesting (intrinsic motivation) or because you needed the credits to graduate (extrinsic motivation)? Both forms of motivation contribute differently to classroom success.

Sometimes our motivation wanes, even if we love the subject at hand or really need the credits to graduate on time. We can use our knowledge of motivation to tweak our study strategies so that we are most likely to be successful. For example, for some people there are times of days when it's easier to work. This is an ideal time of day not only to study, but to study material that might not be our favorite. Give yourself a small motivator—for example, keep a sticky note list with what you need to work on each day and cross off tasks as you accomplish them.

How we think about studying also matters. Saying "I need to study, but I'm tired" makes it really easy to skip out on studying. Slightly changing the wording to "I need to study, *and* I'm tired" acknowledges how we feel and allows us to pivot our studying to work better for us. This is where it can be very advantageous to use the Pomodoro Method (Chapter 6). Try one Pomodoro session, and if it's just too much, come back to it later.

Here's some questions to think about when you're studying this material:

1. Do you find yourself more intrinsically or extrinsically motivated to learn about evolutionary forces? Is your level of motivation to learn evolution different or similar to other material covered this semester?
2. What are strategies you use now to help keep up your motivation while studying? What strategies would you like to try?

Chapter Review

11.1 The Gene Pool

A **population** is a group of interbreeding individuals in one area. Members of a **species** share genes and can interbreed. Members of a species share **morphological traits**, or physical features. **Fitness** in the context of evolution refers to success in passing genes to the next generation. **Natural selection** occurs when environmental forces cause traits to increase or decrease over generations.

Mutations can lead to beneficial, neutral, or detrimental traits. The **allele frequency** is the percentage of a particular allele within the **gene pool**.

11.2 Evolution by Selective Forces

Selective forces can be naturally occurring (natural selection), due to choosy mates (**sexual selection**), or driven by human breeding (**artificial selection**).

There are three patterns of natural selection. In **directional selection**, one extreme version of a trait increases over time. In **stabilizing selection**, the intermediate trait is favored, while in **disruptive selection**, the extreme versions of the trait are favored.

Sexual selection occurs when individuals of a population out-reproduce other individuals because they are better at securing mates. Sexual selection can lead to distinct traits in males and females, which is called **sexual dimorphism**. In **artificial selection**, organisms are selectively bred for traits desirable to humans.

11.3 Evolution by Non-Selective Forces

As a population reproduces, allele proportions can change due to **genetic drift**, a change in allele frequency brought about by chance alone, rather than by natural selection. Genetic drift in small populations can result in a **fixed allele**, where all members of a population are homozygous for one allele.

A **population bottleneck** occurs when a population's size is greatly reduced, decreasing genetic diversity through a loss of alleles, possibly making the population genetically unviable. The **founder effect** occurs when a new population is established by a small number of individuals, reducing the gene pool to the alleles of the founders. **Gene flow** stabilizes allele frequencies through the movement of alleles between interbreeding populations.

11.4 Speciation and Extinction

Speciation occurs when a population splits into lineages that can no longer interbreed successfully. **Reproductive isolation** is the loss of gene flow in sexually reproducing populations. In **temporal isolation**, reproductive timing is out of sync. In **habitat isolation**, species adapt to different habitats and cannot breed. In **behavioral isolation**, changes in behavior prevent individuals from recognizing potential mates. In **mechanical isolation**, physical changes prevent mating.

After fertilization, successful reproduction can be prevented through **gamete incompatibility**, where gametes do not form a viable zygote; **hybrid inviability**, if offspring do not mature; or **hybrid sterility**, if offspring are viable but cannot produce offspring. Some hybrids display **hybrid vigor**, which is when the hybrid animal outperforms either parent in some way.

Allopatric speciation occurs when populations are separated by a physical barrier that interrupts gene flow and leads to speciation. **Sympatric speciation** occurs when genetic changes make some individuals reproductively incompatible with others. Multiple species rapidly arise from a single ancestral species by **adaptive radiation**, as populations adapt to different environmental conditions.

When no individuals of a species remain, the species is considered **extinct**. There have been five **mass extinctions** in the past, where 75% or more of all species went extinct. Evidence suggest we are currently in a **sixth mass extinction** caused by human activity.

Practice It!

Retrieval practice is the process of pulling information from your long-term memory and using it. Here are some questions to help with your retrieval practice.

Self-Quiz

1. Evolution acts at the ______ level. (LO 11-4)
   a. species
   b. regional
   c. individual
   d. global

2. What causes variation in traits within individuals of a population? (LO 11-2)
   a. sexual reproduction
   b. multiple alleles for a gene
   c. mutations
   d. all of these

3. Neutral mutations lead to ______ survival and reproductive fitness. (LO 11-4)
   a. increased
   b. decreased
   c. no change in
   d. unpredictable

4. The total of all alleles found in a population is called ______. (LO 11-3)
   a. the phenotype
   b. natural selection
   c. the gene pool
   d. the population

5. Which of the following statements best describes traits present in a particular *species*? (LO11-1)
   a. Most traits are shared with minor variations.
   b. Most traits are found in only a subset of the species.
   c. A species is defined by having identical traits in all its members.
   d. A species is defined by exhibiting the full range of variation in all its traits.

6. Which of the following statements is *true* when comparing natural selection and sexual selection? (LO11-5)

a. they depend upon variation in the gene pool
b. they typically select for the same types of traits
c. all traits are under both types of selection
d. they drive genetic mutations

7. Which pattern of selection eliminates mid-range forms of a trait? (LO 11-6)

a. stabilizing
b. disruptive
c. directional
d. artificial

8. Which type of selection is due to human influence to choose a specific form of a trait? (LO 11-8)

a. natural
b. disruptive
c. sexual
d. artificial

9. Which of the following examples is due to sexual selection? (LO 11-7)

a. the wide variety of different domestic dog breeds
b. bower bird nests decorated with blue items made by males
c. dairy cows that can produce over 20 gallons of milk per day
d. chickens with coloring that helps them blend in with their surroundings

10. The process by which environmental forces result in the differential survival and reproduction of individuals of a population is called ______. (LO 11-5)

a. catastrophism
b. genetic drift
c. natural selection
d. genetics

11. Which of the following populations would be most vulnerable to loss of genetic diversity? (LO11-10)

a. a small population
b. a large population that has not recently undergone a population bottleneck
c. a population that travels and interbreeds with many other populations
d. a large population with a lot of allelic diversity

12. When a sheep and a goat breed, the resulting "shoat" typically dies before birth. This is an example of ______. (LO 11-13)

a. gamete inviability
b. temporal isolation
c. habitat isolation
d. hybrid inviability

13. When subpopulations of a species spread into different environments, ______ may occur to produce many new species in an evolutionarily short time frame. (LO 11-15)

a. the founder effect
b. adaptive radiation
c. pre-reproductive isolation
d. reduced allele frequency

14. After fire devastates all the trees in a wide swath of forest, populations of a species of tree-dwelling frog on either side of the burned area diverge to become separate species. This is an example of ______. (LO 11-14)

a. allopatric speciation
b. adaptive radiation
c. sympatric speciation
d. genetic drift

Critical Thinking

1. Two species of antelope—one from Africa, the other from Asia—are put into the same enclosure in a zoo. To the zookeeper's surprise, individuals of the different species begin to mate and produce healthy, hybrid baby antelopes. What questions might you ask to determine whether these two species of antelope are in fact one? Why would you choose these questions? (LO 11-13)

2. An island with no sheep was populated with 20 bighorn sheep in 1975. By 1999, this population had expanded to 650 sheep. Suppose you compared the genomes of the 650 island sheep to genomes from a large continental population. Which population would you predict would have less genetic variation? Why is this the case and what force explains this? (LO 11-10)

3. In 2019, the U.S. Environmental Protection Agency (EPA) approved the use of two medically important antibiotics to treat citrus greening, a bacterial disease of plants that has been decimating the citrus industry in Florida. Tiny sucking insects called psyllids harbor and transmit the bacteria that cause the disease. Desperate citrus growers are expected to spray hundreds of thousands of kilograms of these antibiotics per year on their trees. Spraying won't stop citrus greening, but growers hope it may delay the spread of the disease. Ongoing antibiotic treatment may also help infected trees live longer. Explain why environmental and consumer groups are unhappy with the EPA's decision. (LO 11-4)

4. Most Native American people have type O blood, meaning their red blood cells express neither the A or B antigens (sugars) on their surface (see Chapter 8). Propose two possible explanations for this observation using the forces discussed in this chapter. (LO 11-5, 11-9)

Credit: MehmetO/Alamy Stock Photo

12

A Primate Primer

Chapter Outline:

A Shared History of Infant Care

There are three different species of orangutans—all of whom are at risk of extinction. Unfortunately, orangutans continue to face threats from deforestation, illegal hunting, and the illegal pet trade. Accredited zoos play a critical role in the conservation of orangutans by providing refuge and preserving genetic diversity through breeding programs. While helpful, given the small numbers of orangutans living together, there is limited opportunity to learn behaviors like infant care from other orangutans.

Consider Zoe, an orangutan living at the Metro Richmond Zoo in Virginia. She was orphaned at 9 months old, when her own mother died unexpectedly. As a result, she was raised by humans. In 2021, she had her first baby, a boy named Taavi. Zoe didn't know what to make of Taavi, and due to her inability to care for her son, for his own safety, he was raised by humans.

When Zoe became pregnant again, zookeepers played videos for her of orangutans giving birth and caring for their offspring, including demonstrating with a toy orangutan how to care for a baby. Zoe delivered another baby boy named Flynn on December 12, 2022. However, to properly care for Flynn required one more lesson: how to breastfeed. *Primates*—a subset of animals that includes us, apes like Zoe, and monkeys—are *mammals*, meaning they provide milk to their offspring. How would Zoe learn to breastfeed?

Right around the same time that Zoe gave birth, Whitlee Turner, a zookeeper at the zoo, also had a baby. Turner was invited to show Zoe how to breastfeed. During the demonstration, Zoe watched closely to see how Turner fed her baby. Within 24 hours, Zoe began to breastfeed her baby. Zoe went on to raise Flynn herself. What can we learn about our evolutionary history from this story?

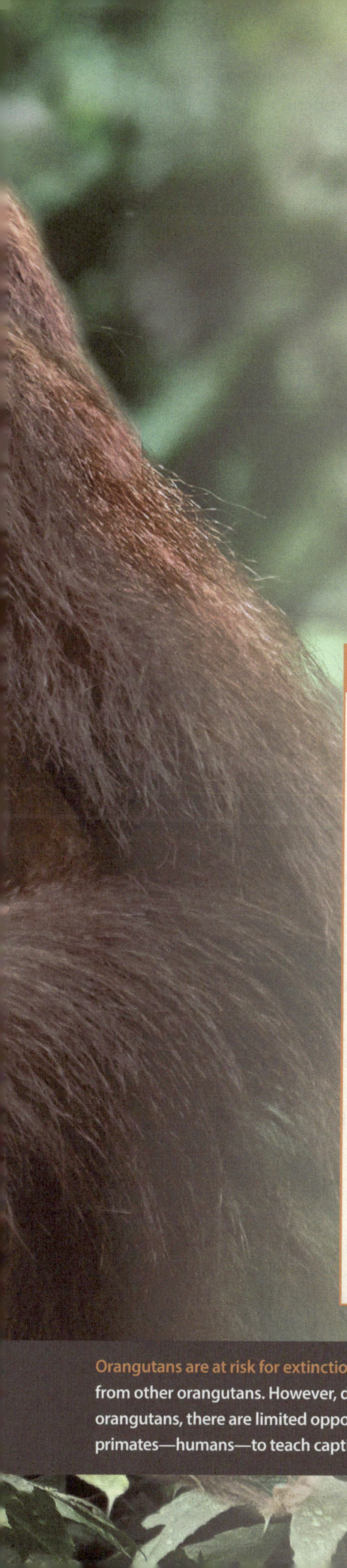

Orangutans are at risk for extinction and find refuge in zoos. Wild orangutans, such as the mother and baby shown here, learn to breastfeed from other orangutans. However, due to declining orangutan numbers and the necessity of using zoos as refuges for the remaining orangutans, there are limited opportunities for them to learn infant care from other orangutans. Therefore, zoos are turning to other primates—humans—to teach captive orangutan mothers.

Did You Know?

The name orangutan means "man of the forest." We share 96.4% of our DNA with orangutans.

Zoe's story is a heartwarming example of the similarities between us and other primates. How do we know about a shared evolutionary history between humans and primates? Where did the lineage of humans come from? In this chapter, we'll explore human evolution. First, we will learn about *taxonomy*, or the scientific study of naming, before moving into how we understand evolutionary relationships with *phylogenies*. Then, in the second half of the chapter, we'll discuss primates, including different kinds of primates (*prosimians*, *tarsiers*, *monkeys*, and *apes*) and evolutionary changes along the primate family tree. We'll end with a discussion of our lineage and the evolution of humans.

Preview It!

What are some things you know already about primates? What questions do you have about human evolution?

12.1 Taxonomy: The Science of Naming

Learning Objectives

12-1 Describe how and why we name species.

12-2 Explain how organisms are arranged in a taxonomic hierarchy.

Study Tip

Think of phylogeny as a kind of genealogy that follows evolutionary relationships.

As we learned in Chapter 1, there's a history of art in science. Arguably, one of the most famous drawings made in the history of biology is the tree of life Darwin sketched in one of his notebooks to explain how species may descend from a common ancestor (Figure 12.1). Today, biologists use *phylogenetic trees* to visually summarize how groups of organisms are related to one another. In fact, there are two related subdisciplines of biology dedicated to understanding relatedness amongst species. **Phylogeny** is the scientific study of evolutionary relationships. Studies of phylogenies can result in the formation of a phylogenetic tree, a diagram that represents species ancestry and can be a useful tool for understanding evolutionary relationships (Section 12.2).

Taxonomy is another subdiscipline of biology that is related to phylogeny. **Taxonomy** is the practice of naming, describing, and classifying species, where a **species** is a group of interbreeding organisms (Chapter 11). By understanding the relatedness of organisms, biologists can understand how different species evolved. In addition, there are practical applications of phylogeny and taxonomy today, such as categorizing the emergence of new viruses (Section 12.2). In this section, we'll discuss taxonomic classifications for organisms before discussing the evolutionary patterns that can be visualized using phylogenies in Section 12.2.

Figure 12.1 An early phylogenetic tree in Darwin's "Notebook B" from 1837

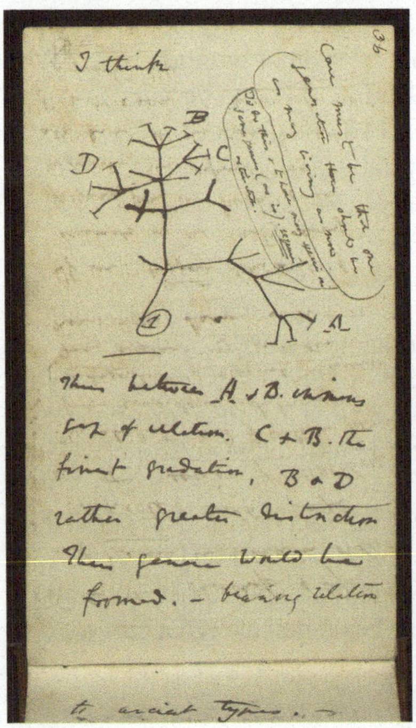

Credit: Photographer: Mark Box/Cambridge University Library

A Rose by Any Other Name...Smells Sweeter with Naming Conventions

Names are important. Names represent many things to people, ranging from capturing their identity to signaling favorite books, such as naming your pet cat Hermione after the character in the *Harry Potter* series. Names can also have significance to expecting parents, such as assigning a name that captures the hopes they may have for their child (such as Felix, meaning "happy" or "lucky"). Learning and using someone's name is how we indicate acceptance of one another. Scientific naming of species is no less important.

When a scientist discovers a new species, they will give it a unique name. Although humans have named life forms in their environments for thousands of years, it wasn't until the 18th century, with the rise of the naturalists (such as those featured in Chapter 10), that a consistent method of naming became a priority. As the European explorers were beginning to understand and catalog the scope of life's diversity, they struggled to communicate with one another because species often had multiple names, depending on who was doing the naming. For example, the dog rose (a plant native to Europe, Africa, and Asia) was alternately known as briar rose, witch's briar, herb patience, sweet briar, wild briar, dog briar, dog berry, briar hip, eglantine gall, hep tree, hip fruit, hip rose, hip tree, hop fruit, and hogseed—and those are only the English names!

The problem of there being too many names for the same species was resolved by Carl Linnaeus, an 18th-century naturalist. Linnaeus introduced the **Linnaean system**, a standardized naming system, used to this day, where every species is given a unique, two-part scientific name. The first part is the **genus** (plural, genera), a group of species that are closely related and share a unique set of inherited characteristics. The second part of a species name is called the **specific epithet**. Together, the genus name and the specific epithet designate one species. Thus, the dog rose now has one official name that is recognized worldwide: *Rosa canina*. *Rosa* is the genus, and *canina* is the specific epithet.

How a species is named is often influenced by its characteristic traits. **Traits** are inherited features that include heritable body shapes (morphology), physiology, or behavioral characteristics that vary between different species. Individuals of the same species have the same set of traits, and that set of traits is unique to the species. For example, let's consider humans, or *Homo sapiens*. The genus (*Homo*) is the Latin word for human. All organisms of the genus *Homo* have large brains, walk on two legs, have opposable thumbs, can perform fine work with their hands, and make tools. This includes modern humans and several extinct species. The specific epithet for our species is *sapiens*, which is Latin for wise, reflecting the ability of our species to, for example, engage in abstract reasoning.

Species and genus are only one way we can categorize an organism. We can also rank a species into ever more inclusive categories based on a subset of traits it shares with other species. Each rank, or **taxon** (plural, taxa), is a group of organisms that share a unique set of traits. Of the major taxa, species is the least inclusive. Each taxon above species—genus, family, order, class, phylum (plural, phyla), kingdom, and domain—consists of all the groups below that level, to the most inclusive taxon, which is the domain. For example, our species is in the genus *Homo*, which is among several genera in the family *Hominidae* and the order of Primates, and so on, to inclusion in the domain Eukarya (**Figure 12.2**).

Recall from Chapter 11 that there is some debate around what exactly differentiates closely related species. Although genetic evidence suggests that *Homo sapiens* interbred with *Homo neanderthalensis* (Neanderthals) and the Denisovans, they are still considered three different species.

As we learned in Chapter 11, it can be difficult to determine if two organisms with similar traits are actually different species. Conversely, observable traits can vary among individuals within a species, such as how eye color can vary among people. How do we decide whether similar-looking organisms belong to the same species? Many pieces of evidence are brought together, such as an examination of anatomy, geographical location, and DNA similarity. Even then, as is the case with all good science (Chapter 1), if new information comes to light, the prevailing scientific understanding can also change. For example, great apes were not originally included in the family *Hominidae*. However, new genetic evidence has revealed how genetically similar we are to the great apes, like the chimpanzee. As a result, the great apes were recategorized to be part of *Hominidae*, with us and ancestral humans.

Figure 12.2 **An overview of taxonomic classifications**

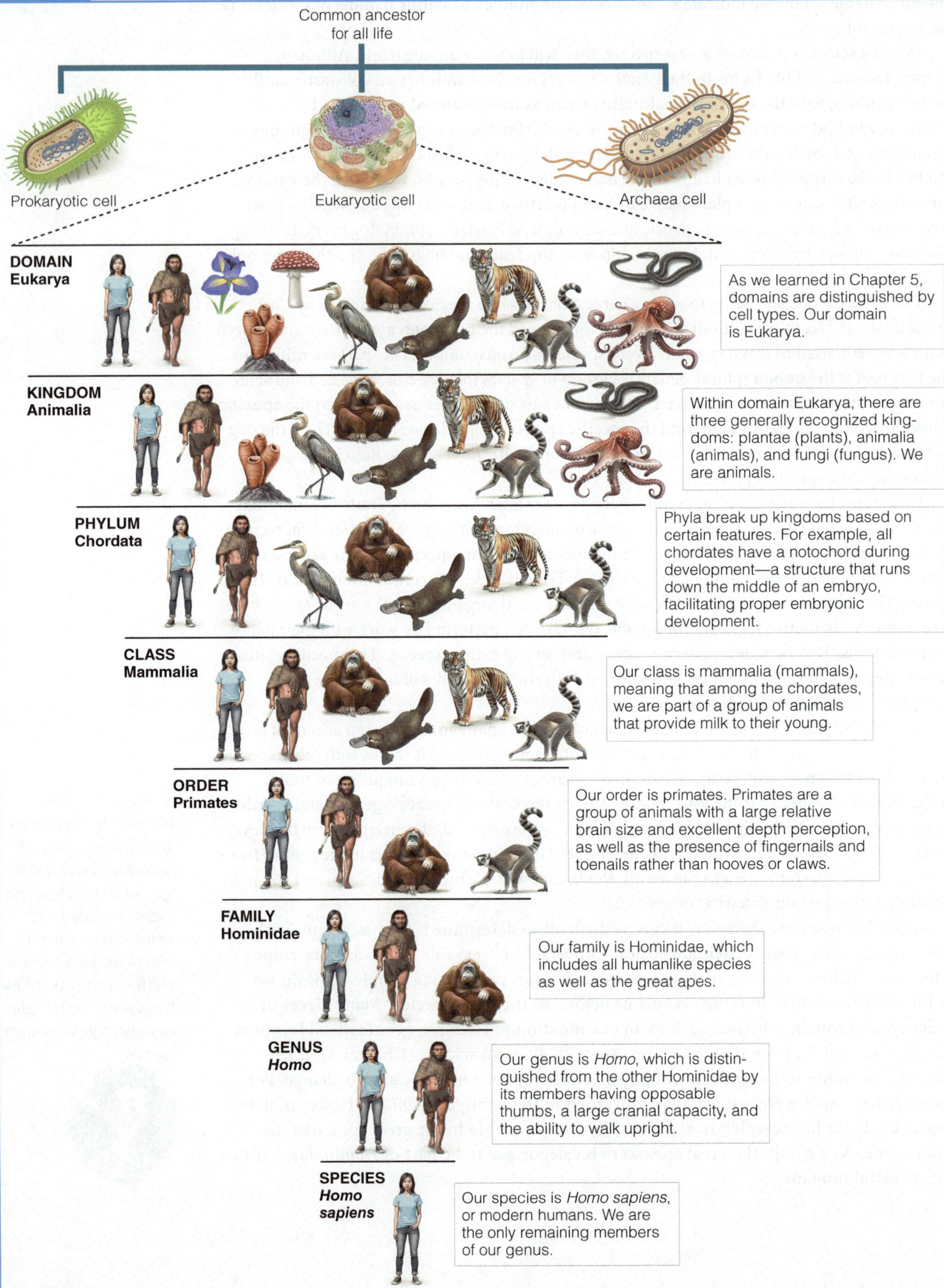

12.1 Summary

- A species is a group of organisms capable of interbreeding. Species are given unique, two-part names that include the genus and specific epithet. The species name allows scientists to distinguish among related species and identify a unique group.
- Organisms are classified into nested groups, starting with a broad group sharing general traits (domain), then narrowing down into successively smaller groups until reaching the genus, which is the smallest taxon and includes closely related species.

12.2 Examining Evolutionary Relationships with Phylogenies

Learning Objectives

12-3 Describe the insights biologists gain from constructing phylogenetic trees.

12-4 Explain the difference between an ancestral and a derived trait.

12-5 Identify evolutionary patterns that reflect how one species influences the evolution of another.

12-6 Explain how phylogenies are used today to trace viral evolution.

Continuing with the theme of overused adages, if a rose by any other name would smell as sweet, then a picture is worth a thousand words. When considering ancestral relationships, **phylogenetic trees** are a useful tool for understanding the evolutionary relationships of different lineages that evolved from a common ancestor (Figure 12.3). A phylogenetic tree is a type of line diagram that contains lines of different lengths, each of which represents a lineage. These lines may branch into two or more lineages. A branch point (node) represents a common ancestor, and the two lineages that emerge from it represent different species. Any complete branch that can be cut from a phylogenetic tree—including the smallest branch possible (a species)—is a **clade**. Two closely related species within a clade are called **sister groups**.

The length of each branch in a phylogenetic tree represents the evolutionary time since the branching point at a node. One way that biologists determine these lengths is through gene sequencing (Chapter 9). Phylogenetic trees may include a scale bar representing how many millions of years have passed between different nodes. Later in this chapter, we will show phylogenetic trees of primates (Figures 12.12) and hominins (Figure 12.14).

Figure 12.3 A phylogenetic tree is a diagram of evolutionary connections

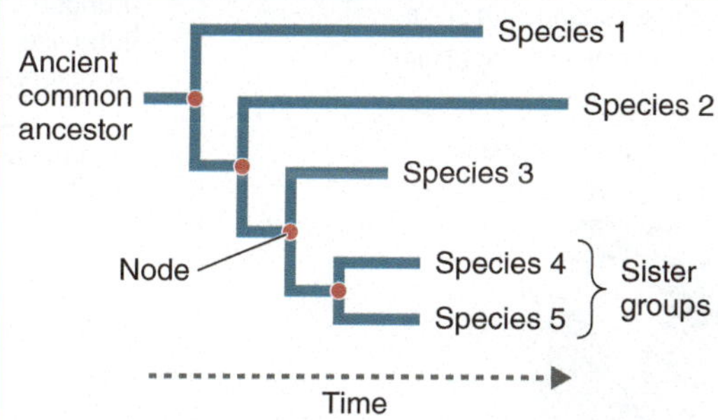

In this diagram, each line represents the evolutionary lineage of a single species. You can see the lineages of related species emerge from a node (highlighted here as red dots), which represents a hypothetical common ancestor for resulting sister groups. Species that are farther apart on the tree are more distantly related. Note that the tree is "rooted" on the left side, meaning that the species to the left are older (ancestral) than the species to the right. The length of each line indicates time since divergence.

! Point of Confusion: Is it a cladogram or a phylogenetic tree? While these diagrams look very similar—compare the cladogram in Figure 12.4 to the phylogenetic tree in Figure 12.12—their usage is slightly different. Cladograms describe historical relationships and are built based on observed characteristics, such as hair. Phylogenetic trees are built based on genetic data, where scientists can therefore determine at which point the species branched from a common ancestor and other species, and scale the length of each branch to represent time. Both types of diagrams are read the same way—with nodes representing common ancestors and the greater passage of time as you read from left to right in the diagram—but the length of the lines between nodes has scientific meaning only in a phylogenetic tree.

Did You Know?

Phylogeny is also used by forensic investigators. Anesthesiologist Juan Maeso was sentenced to life in prison after phylogenetic analysis revealed that he was intentionally infecting his patients with hepatitis C during surgery. Biologists used samples from 311 infected people to construct a phylogenetic tree. The predicted infection dates from the phylogenetic tree correlated with the surgery dates for each of the victims.

Have you heard of macroevolution and microevolution? These terms can be misleading, because they suggest that there are two different kinds of evolution. Rather, evolution is a single, unified process that includes changes in allele frequencies in a population (microevolution) that lead to changes in a species (macroevolution).

Study Tip

Derived traits are *derived* from ancestral traits, just like words are derived from other languages, such as how many English words have roots in Latin (considered a dead language).

Evolutionary Patterns

As we learned in Chapter 11, evolution continues to occur. Millions of years of evolutionary data provide a gold mine of insights that can be gained through examining phylogenies. In this section, we will explore different historical patterns that biologists can visualize using phylogenetic trees and/or cladograms.

Key innovations. A **key innovation** is an adaptive trait that allows its bearer to exploit its environment more efficiently or in a novel way. The evolution of lungs offers an example, because lungs were a key innovation that opened the way for an adaptive radiation (Chapter 11) of vertebrates on land (Figure 12.4).

Derived and ancestral traits. While humans obviously could not directly witness the evolution of most species, there is plenty of evidence available to help us understand ancient evolutionary events. Consider how each species bears traces of its own unique evolutionary history in its traits. Evolutionary biologists determine common ancestry by looking for a **derived trait**—one that is present in a group of species under consideration, but not in any of

Figure 12.4 Key innovations facilitated the evolution of new groups of animals

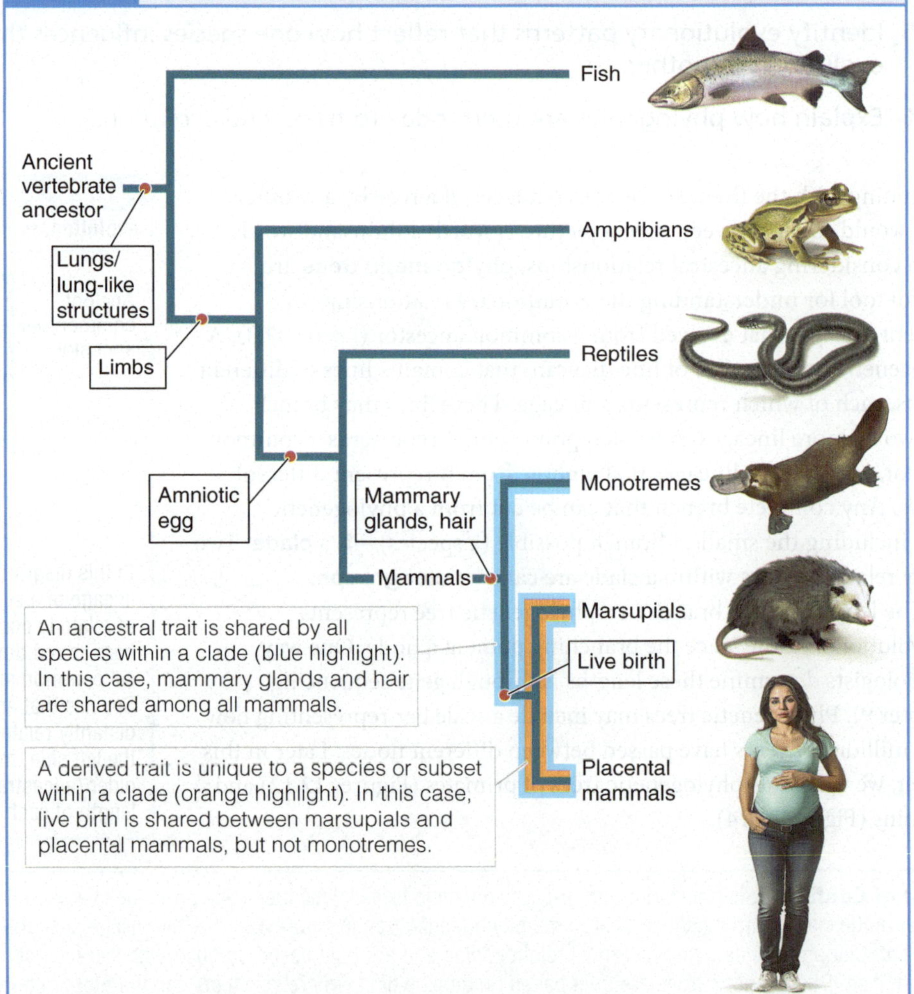

This cladogram calls out key innovations in vertebrates (red dots and labels), beginning with the evolution of the lung. Notice how lungs gave way to the evolution of several new lineages of animals, including amphibians, reptiles, and mammals. The mammal lineage is further broken down to its major groups of monotremes, marsupials, and placental mammals, the latter of which includes us.

the group's ancestors. In contrast, **ancestral traits** are features that are relatively unchanged and inherited from a common ancestor of a clade (see Figure 12.4). Note that ancestral and derived traits can vary, depending on where you look on a cladogram or phylogenetic tree. For example, when considering placental mammals (such as humans) in the context of the entire cladogram, providing milk to young is a derived trait, but in the context of mammals alone, it is an ancestral trait. But if we look at just the mammalian lineage, providing milk to one's young is an ancestral trait (blue highlight), and giving birth to live young is a derived trait (orange highlight).

Repurposing. Major evolutionary novelties stem from the adaptation of an existing structure for a completely new purpose. Said otherwise, the existing structure is repurposed for a different use, a phenomenon called **exaptation**. The feathers that allow modern birds to fly, for example, are derived from dinosaur feathers, which could not have sustained flight when they first evolved. The earliest feathers probably served as insulation and may have been used for courtship displays or camouflage for some dinosaur species.

Coevolution. Two species that have close ecological interactions may evolve jointly, a pattern called **coevolution**. One species acts as an agent of selection on the other, and each adapts to changes in the other. Over evolutionary time, the two species may become so interdependent that they can no longer survive without one another.

Relationships between coevolved species can be quite intricate. Let's look at the Joshua tree and its relationship with the yucca moth as an example (**Figure 12.5**). Joshua trees grow in the desert in the Southwestern United States and Northwestern Mexico, and have a somewhat surprising appearance. A long stem leads to branches that look somewhat like a chandelier covered in short spiny leaves. Since plants are immobile, sexual reproduction usually requires some living organism (for example, a bee) or nonliving intermediary (like the wind) to help move the gametes between plants (we'll return to plant reproduction in Chapter 18). Many plants will produce a sweet substance called nectar to attract organisms such as insects to visit and carry pollen (a plant gamete) from plant to plant. However, Joshua trees do not produce nectar, and their flowers are visited by only two kinds of yucca moths.

Coevolution can also happen on a larger scale. For example, although flowering plants are dependent on pollinators to sexually reproduce, many can be pollinated by a variety of other organisms, such as bees, butterflies, moths, birds, and bats.

The yucca moth needs the Joshua tree to reproduce, as it visits the flowers of the Joshua tree to lay its eggs. The moth has a specialized organ called an *ovipositor* that is the exact size needed to penetrate the part of the flower containing the seeds. As the moths move between Joshua trees to lay their eggs, they transmit pollen between the plants, which facilitates Joshua tree

Figure 12.5 Coevolution of the Joshua tree and yucca moth

A. The Joshua tree (*Yucca brevifolia*) lives mainly in the Mojave Desert in the Southwestern United States.

B. Two yucca moth species (*Tegeticula antithetica* and *Tegeticula synthetica*) are the Joshua tree's only pollinators, and in turn depend on the trees to complete their life cycle.

Big Ideas in Biology

Interconnectedness of Living Systems

The coevolution of the Joshua tree and yucca moth highlights how relationships between organisms can influence their evolution and survival. Interdependent relationships such as this one are less common since if one organism goes extinct, the other will follow. Climate change is impacting this interaction—it's becoming too hot for the yucca moth to survive except in small areas called climate refuges—and could contribute to the loss of both species. It could also harm other organisms that depend on the Joshua tree.

Figure 12.6 **The coelacanth is a living fossil**

Credit: Fossil footprint of the Carboniferous coelacanth (object)/Prehistoric/PATRICE CARTIER/Museum d'histoire naturelle, Toulouse, France/Bridgeman Images

Credit: PETER SCOONES/SCIENCE PHOTO LIBRARY

Until a fisherman caught one in 1938, coelacanths were thought to have gone extinct 70 million years ago. As an example of stasis, compare the similarity between **A.** the fossil and **B.** a live fish.

sexual reproduction. The baby moths develop inside the flower and eat some of the seeds. Consequently, each species requires the other to reproduce.

Stasis. Sometimes, very few changes will occur in some lineages over a very long period of time. **Stasis** refers to a length of time when there is little to no evolutionary change in a species. Consider coelacanths, an ancient order of lobe-finned fish. Modern coelacanth species are similar to fossil specimens hundreds of millions of years old (Figure 12.6). Other examples include dragonflies, horseshoe crabs, and crocodilians.

Human Health and Evolution: Tracking Viral Variants

While discussions of evolution usually result in people thinking about the history of life and time scales in the millions of years, as we learned in Chapter 11, evolution is still happening today. Now that we've discussed how to read a phylogenetic tree and what information can be gleaned from it, we will look at how phylogenetic trees and evolutionary histories are used to understand viral evolution today. First, we will discuss the evolution of coronaviruses. Understanding the pattern of coronavirus evolution can help scientists predict the evolution of future strains to develop targeted medicines *before* a pandemic occurs. Then, we will discuss influenza virus evolution and how viruses can evolve and gain the ability to "jump" between species.

Tracking Coronavirus Evolution

As we learned in Chapter 1, coronaviruses are a class of virus characterized by the spikes on their surface, which give the virus the appearance of wearing a crown (*corona* is the Latin word for crown). Among the different coronaviruses, some, like HCoV-229E, mainly infect the nose and cause a runny nose, akin to having a cold. Others, like SARS-CoV-1 and MERS-CoV, which predominantly impact the lungs, cause severe acute respiratory syndrome (SARS) and Middle East respiratory syndrome (MERS), respectively. From 2002 to 2004, an outbreak of SARS resulted in 774 deaths worldwide, with over 8,000 people from 30 countries becoming infected. MERS emerged on the Arabian Peninsula in 2012 and resulted in approximately 900 deaths.

SARS and MERS outbreaks were contained to the parts of the world where they first emerged. This was not the case with the appearance of SARS-CoV-2, which is the virus that caused the COVID-19 pandemic, where a **pandemic** is the global spread of a new infectious disease. SARS-CoV-2 has features similar to both SARS-CoV-1 and the coronaviruses that cause cold-like symptoms: Like the viruses that cause the common cold, it is highly transmissible, but with much more severe impacts in the lungs, as is the case with SARS-CoV-1. As a result, between when the COVID-19 pandemic was declared in 2020 and the end of 2025, COVID-19 resulted in the deaths of over 7 million people worldwide and is one of the deadliest pandemics in recorded human history.

? Did You Know?

Within 5 years of the first human infections with SARS-CoV-2, thousands of new variants evolved. Scientists follow the variants of concern (those that may be more infectious) closely.

But how are all of these viruses related to one another? That's where phylogeny comes in. Researchers often study the evolution of viruses by grouping them into clades based on biochemical traits. Viruses can mutate every time they infect a host, so their genetic material

Biology for Life

A Shared Evolutionary History between Humans and Sharks

Are you someone who blushes easily? Blushing is a physiological response to a stressful situation—often, one that is embarrassing. Let's say that you are walking through a quiet library carrying a pile of books, trip over your shoelaces, and fall flat on your face, causing your books to fly everywhere. The resulting BOOM causes the rest of the library to look up and see what's going on. The fall and your resulting embarrassment cause a hormone called adrenaline to be released from your adrenal glands, which sit on top of your kidneys. The adrenaline causes the blood vessels in your face and neck to dilate (get bigger), letting more blood into this area—turning the skin red and making you feel hot.

It turns out, hormonally induced changes to skin color in response to one's environment may not be reserved for just humans. In fact, this may be an ancestral response, or it could be an example of convergent evolution (Chapter 10). Preliminary evidence from National Geographic Explorer Gibbs Kuguru (Figure 12.7A) suggests great white sharks have a blush response that is similar to that seen in humans. Great white sharks hunt by coming underneath their prey (seals, for example), launching themselves out of the water, and catching the prey in their mouths. In order to assess shark color, Kuguru and his colleagues observed sharks near the shore and captured images of sharks mid-breach next to a color board attached to a fake seal (Figure 12.7B).

The researchers found that sharks tend to be darker in color mid-attack. However, if a shark is startled (such as if it accidentally hits the boat instead of its prey), it turns pale (like how we might turn pale when afraid) and dives to the bottom of the ocean to hide. Although more study is needed, turning pale when a threat is perceived and swimming to the bottom may allow the shark to camouflage itself—providing an evolutionary reason for the behavior.

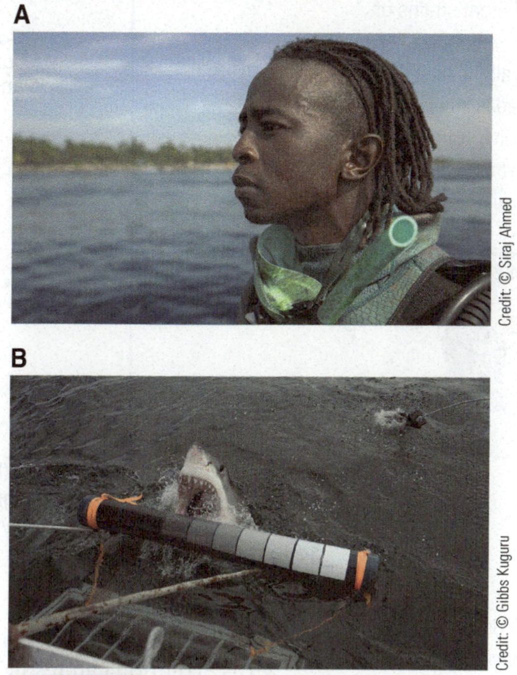

Figure 12.7 National Geographic Explorer Gibbs Kuguru studies shark biology

A. Kuguru in the Maldives getting ready for a dive.

B. A great white shark breaching off the coast of South Africa. Note the color board in the photo.

1. Research other potentially surprising traits that animals share with humans. Which one surprises you the most? Why?
2. Refer back to Figure 12.4, and find fish (sharks are a kind of fish) and humans. Examine the descriptions of ancestral and derived traits there as well. With this in mind, how do you think researchers determine whether a trait is a shared ancestral trait in one clade, or convergent evolution (where similar adaptations evolve independently in different lineages) in more distantly related clades?

changes quickly. Consider the evolution of the coronavirus (Figure 12.8). From a coronavirus common ancestor, we see that coronaviruses that cause cold-like symptoms have diverged over time to form four distinct linages—HCoV-229E, MERS-CoV, and SARS-CoV-1 and -2. We also see that SARS-CoV-1 and MERS-CoV share a more recent common ancestor than SARS-CoV-2. Note that this is a greatly simplified diagram showing only coronaviruses that infect humans, in addition to not including all clades for each type of virus. This same strategy is used by scientists to track viral variants of concern. We will now discuss how new variants arise, particularly variants that can infect different hosts.

Barnyards and Flu Evolution

It seems like every few years, a new influenza (flu) virus with an associated animal name makes the news. Some years it's "bird flu," other years it's "swine flu." What is a bird or a swine flu, and how are they different from the regular flu? And what do different flu viruses have to do with

Why is it possible to get vaccinated for COVID-19 and still get COVID-19? One reason is that there are so many different strains of the virus, with new ones evolving all the time. Getting vaccinated is still important—to give your body a head start on fighting off an infection and decrease the severity of symptoms.

Figure 12.8 Evolution of coronaviruses that infect humans

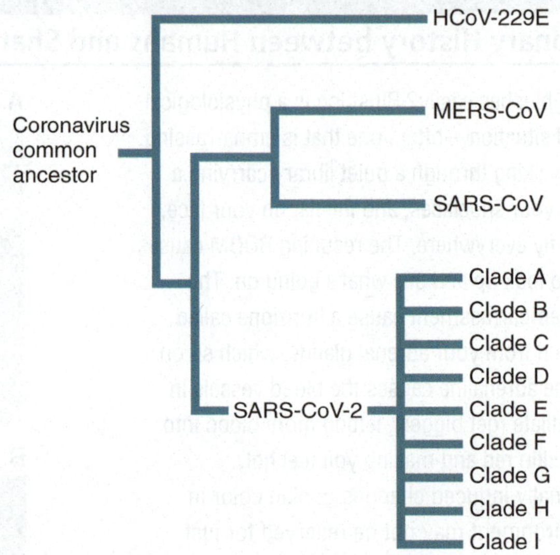

Although other virus lineages can be broken down further into clades, we only highlight one virus here—SARS-CoV-2. Shown is a simplified view of the many different clades within SARS-CoV-2 known as of 2020.

Figure based on: Li, T., Liu, D., Yang, Y., Guo, J., Feng, Y., Zhang, X., ... & Feng, J. (2020). Phylogenetic supertree reveals detailed evolution of SARS-CoV-2. *Scientific reports, 10*(1), 22366

evolution? In this section, we will discuss the evolution of the virus that causes the flu, and how virus evolution can result in these viruses "jumping" between species.

Influenza doesn't cause the "stomach flu," which is caused by other viruses, such as norovirus (Chapter 9).

Odds are that you've had the flu or flu-like symptoms before. The influenza virus predominantly infects the nose, throat, and lungs, and causes fever, cough, headaches, body aches, and chills. There are four different kinds of influenza viruses, and most cases of the flu are caused by two of them: influenza A and influenza B. Here, we will focus on influenza A, which is the only influenza virus known to have caused a pandemic.

Influenza A is further divided into different clades depending on the two types of proteins that it has on its surface: hemagglutinin (H) and neuraminidase (N). These proteins are collectively called *antigens*, structures that can be recognized and destroyed by the immune system (Chapter 17). For example, the 2009 flu pandemic was caused by influenza A (H1N1). The numbers refer to the specific subtypes of H and N proteins found on the virus's surface. If a virus originates from a nonhuman source, its official name will also include the name of the host, such as *avian* influenza A (H1N1) for bird flu.

When you get a flu shot, the shot "teaches" your immune system to recognize the antigens on the virus surface. Therefore, if you are later infected by that virus, your immune system can respond quickly and limit the severity of your illness or prevent it altogether. However, if you are infected with a virus with different antigens than those that you've been vaccinated for, your immune system won't recognize the virus and you could get sick.

The last influenza pandemic to hit the United States was the swine flu pandemic in 2009, which resulted in 60.8 million cases of the flu and over 12,000 deaths in the United States alone. Globally, 80% of deaths were in people under the age of 65, which is unusual for the flu. The virus originated from naturally occurring flu viruses that infect pigs, wild birds, and humans. Since viruses evolve closely with their hosts, they tend to infect only specific hosts…unless an evolutionary leap occurs.

One such evolutionary leap can occur if two different influenza viruses encounter one another—for example, by infecting the same pig—and the viruses swap genetic material in a process called **reassortment**. Note that reassortment is not something that all viruses can do.

In the case of the 2009 swine flu pandemic, the wild pig, bird, and human viruses reassorted to form a new strain called H1N2, which had characteristics of each of the original strains (Figure 12.9). A pig infected with this new H1N2 strain became infected with still another strain of the H1N1 virus from a different pig, forming a new strain of the H1N1 virus through reassortment; this produced the H1N1 strain that infected humans and caused the 2009 swine flu pandemic. Understanding how reassortment occurs is important for developing new vaccines that offer broad protection against many different possible variants.

Figure 12.9 Viruses can swap genetic material, allowing them to infect new hosts

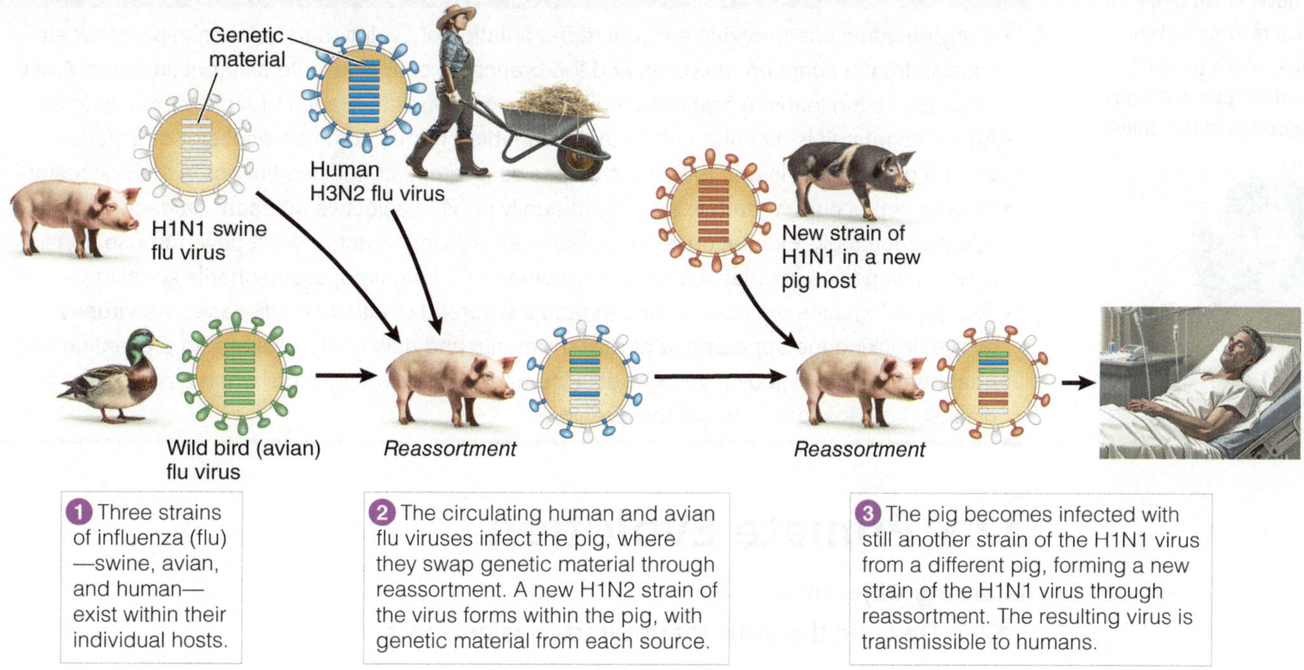

Here, we demonstrate the evolutionary leaps that occurred to generate the H1N1 virus that caused the 2009 swine flu pandemic.

Biology for Life

The Next Pandemic

As the human population grows (Chapter 14), there is increased demand on the Earth's resources. Activities such as deforestation can result in humans having access to unexplored parts of the world. While the devastating impacts of deforestation on biodiversity will be covered in more detail in Unit 5, one potential consequence of human exploration into new areas is that there is greater potential for humans to encounter novel viruses.

One potential source is through the bushmeat trade. Bushmeat is raw or minimally processed wild game meat (including endangered animals) that is harvested for food. The bushmeat can be a source of communicable diseases, such as the Ebola virus. Typically, hunters catch and kill animals and sell them at markets for others to purchase (Figure 12.10). While this is an important part of local communities and can be an important part of Indigenous cultures, the potential for new viruses to infect humans and then spread globally is a real concern.

Credit: Paul Harding 00/Shutterstock.com

Figure 12.10 Bushmeat for sale at a market in Indonesia

For example, Ebola hemorrhagic fever, which causes severe bleeding and can often be fatal, is linked to infection with orthoebolavirus. Humans can become infected with orthoebolavirus and develop Ebola from hunting, butchering, processing, or consuming meat from infected animals. Due to the potential risks from Ebola—and with animal conservation considerations in mind—it is illegal to bring bushmeat into the United States. Other viruses linked to the consumption of bushmeat include human immunodeficiency virus (HIV), monkeypox virus, and coronaviruses. If a novel virus evolves and infects humans, with the ease of global travel, the infected individual(s) could carry the virus around the world. This could ultimately cause another pandemic.

1. The bushmeat trade is an important cultural aspect for some, as well as an important source of protein and economic resource. How do you recommend balancing the needs of the people who depend on bushmeat with the potential for pandemics?
2. Given the impact of the COVID-19 pandemic on our culture, do you think people are more or less receptive to taking steps to prevent another pandemic? Why or why not?

In late 2024, an outbreak of a novel strain of the flu caused record chicken deaths, which (in part) caused the prices of eggs to skyrocket in the United States.

12.2 Summary

- A phylogenetic tree provides a visual representation of evolutionary relationships between species from a common ancestor, and the branching of species into different lineages. A key innovation often leads to branches. A clade is any complete branch of a phylogenetic tree.
- An ancestral trait is a characteristic that is inherited from a common ancestor of a clade, while a derived trait is one present in a species or group of species but not in their ancestors.
- Coevolution occurs when two species mutually provide selective pressure on each other and adapt accordingly. Exaptation is repurposing an existing structure for a new purpose. Stasis is when a species maintains a similar structure for a long time, evolutionarily speaking.
- Today, phylogenies are used to understand the spread of infectious diseases. As viruses evolve, tracking the appearance of reassortments and new strains helps with prevention and treatment efforts, including vaccine development. A pandemic occurs when a new infectious disease spreads to new geographic regions.

12.3 Primate Evolution

Learning Objectives

12-7 Describe the traits that characterize primates.

12-8 Distinguish prosimians, monkeys, and apes.

At the beginning of this chapter, we learned about how biologists organize species by taxonomic classifications. We then discussed how phylogeny is used to understand the relatedness of different species. Now, we will apply these insights to understanding our own evolution. We'll begin with our order (primates) and family (Hominidae), before moving into our genus (*Homo*) and finally species (*Homo sapiens*) in Section 12.4.

An Order of Primates

Many features that distinguish primates should be familiar to you—primates, including us, tend to live in social groups that include adults of both sexes. Females usually give birth to only one or two young at a time and provide care for an extended period after birth. Compared to other mammals, primates have a large brain for their body size. The regions of the brain devoted to vision and to information processing are expanded, and the area devoted to smell is reduced. Said otherwise, we see better than we smell.

Figure 12.11 **Primates, like this chimpanzee, are adapted to climbing**

Credit: iStock.com/MattiaATH

Note the wide range of shoulder motion and grasping ability of the hands and feet. Eyes situated at the front of the animal's face provide excellent depth perception.

Although eye placement can vary across mammals—for example, with prey animals having widely spaced eyes for a wider field of vision and predators having forward-facing eyes for better depth perception—primate eyes tend to be at the front of the head. As a result, both eyes view the same area, each from a slightly different vantage point. The brain integrates the differing signals it receives from the two eyes to produce the three-dimensional view that you see.

A primate's excellent depth perception adapts it to a life spent leaping or swinging from limb to limb. The forest environment shaped many of the key characteristic traits of primates, such as shoulders with an extensive range of motion that facilitates climbing (Figure 12.11). Unlike most mammals, a primate can extend its arms out to its sides, reach above its head, and rotate its forearm at the elbow. Both hands and feet are capable of grasping. In contrast to other mammals, the majority of which have claws, the tips of primate fingers and toes typically have touch-sensitive pads protected by flattened nails.

Primate origins and diversification. Primates most likely arose before the demise of the dinosaurs, sometime between 85 and 66 million years ago. We can track several key innovations over evolutionary time (Figure 12.12).

Big Ideas in Biology

Evolution

The evolution of features such as the moveable upper lip laid the evolutionary groundwork for our own ability to speak.

Monkeys were formerly distinguished from one another by using the terms "old world" and "new world" monkey. These terms reflect a history of colonialism, so scientists are adopting new terms for each of these types of monkeys. Now, we use Monkeys of Africa and Asia instead of "old world" and Monkeys of the Americas instead of "new world."

Figure 12.12 An evolutionary tree for modern primates

A

	Wet nose	Moveable upper lip	Prehensile tail	No tail	Kunckle walking	Bipedalism
Prosimians	✓					
Tarsiers		✓				
Monkeys of the Americas		✓	✓			
Monkeys of Africa/Asia		✓		✓*		
Great apes		✓		✓	✓	
Humans†		✓		✓		✓

*Some monkeys of Africa and Asia have small tails.
†Note that humans are also great apes.

B

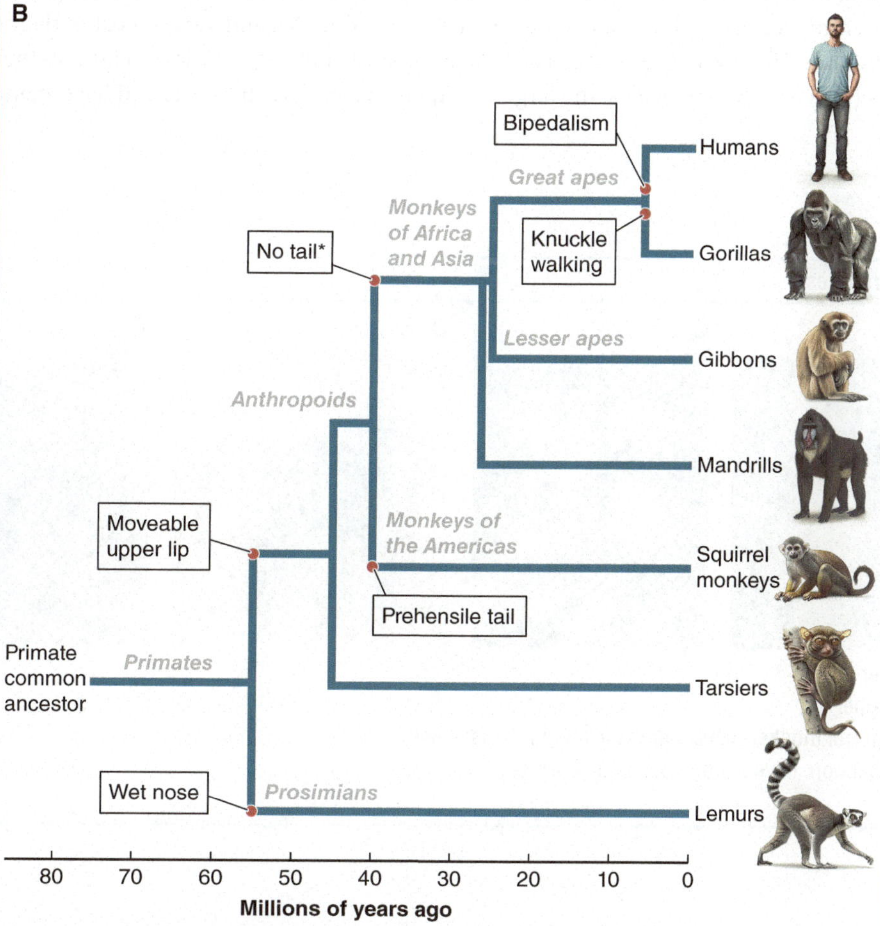

*Some monkeys of Africa and Asia have small tails.

A. Key distinguishing characteristic(s) for different kinds of primates. Note that a check mark indicates if a primate lineage has this feature.
B. Phylogenetic tree of some primate lineages, with key innovations called out.

? Did You Know?

Chimpanzees and bonobos are our closest living relatives. The chimpanzee/bonobo lineage and the lineage leading to humans diverged 6 to 13 million years ago.

Lemurs are members of the oldest existing primate lineage, called the **prosimians**. Like dogs and most other mammals, lemurs are "wet-nosed"; they have a moist nose and an immobile cleft upper lip that attaches tightly to the underlying gum (Figure 12.13A). The second oldest primate lineage is the **tarsiers**. A dry nose and a movable, noncleft upper lip evolved in the common ancestor of tarsiers (Figure 12.13B) and all other modern primates. This innovation gave dry-nosed primates a wider range of facial expressions and allowed more complex vocalizations.

The third lineage is the **anthropoids**, which include monkeys, apes, and, since we are apes, humans. Nearly all anthropoids are diurnal (active during the day) and have good eyesight, including color vision. Within the anthropoid lineage, there are two different types of monkeys. **Monkeys of the Americas** are tree-dwelling and have a flat face and a nose with widely separated nostrils (Figure 12.13C). They have a long prehensile tail that can grasp an object like a branch, acting like an extra appendage. These monkeys also use their tail for balance. **Monkeys of Africa and Asia** also live in trees, but do not have prehensile tails, and in some cases, have no tail at all. They tend to be larger than monkeys of the Americas, with a longer nose and closely set nostrils. Some are tree-climbing forest dwellers. Others, such as baboons (Figure 12.13D), spend most of their time on the ground in grasslands and deserts.

Tailless nonhuman primates are commonly called **apes**. About 15 species of small apes called gibbons inhabit Southeast Asian forests. Gibbons are sometimes referred to as "lesser apes," in comparison with the larger apes, or "great apes." The only surviving Asian great ape is the orangutan, which climbs trees in the forests of Sumatra and Borneo. All other great apes (gorillas, chimpanzees, and bonobos) are native to central Africa and spend most of their time on the ground. When walking, African apes lean forward and support their weight on their knuckles (Figure 12.13E). Gorillas, the largest living primates, live in forests and feed mainly on leaves.

Figure 12.13 A gallery of nonhuman primates

A B C D E

A. A lemur has a wet nose and cleft upper lip.
B. A tarsier has a dry nose and uncleft upper lip.
C. Monkeys of the Americas, like this squirrel monkey, have a flat face and prehensile tail.
D. Monkeys of Africa and Asia, like this baboon, have a long nose and short tail.
E. Gorillas are apes and lack tails.

Credit: ANDY ROUSE/NATURE PICTURE LIBRARY /SCIENCE PHOTO LIBRARY; National Geographic Image collection; Primates.com; iStock.com/JasonRWarren; BellFrost /Shutterstock.com

Biology for Life

Saving Orangutans through Grocery Shopping

As we learned in the chapter opener, orangutans are, like us, great apes, and are genetically very similar to us. Orangutans are also critically endangered. "Critically endangered" is a term used by the International Union for Conservation of Nature (IUCN) to describe a species that has an extremely high risk of going extinct (Chapter 11) in the wild. If they are extinct in the wild, the only remaining members of the species would be found in human care, such as in zoos. We will return to conservation in Chapter 16.

One of the biggest threats to orangutans is the loss of the forests where they live, which are being cut down to create farmland to cultivate African oil palm trees. African oil palm trees produce palm oil, which is found in approximately 50% of household goods and is used in a variety of consumer goods ranging from foods to cosmetics. The majority of palm oil is from trees grown in Borneo and Sumatra—the home of orangutans.

Credit: Roundtable on Sustainable Palm Oil

Figure 12.14 RSPO label

Products with this label contain certified sustainable palm oil or contribute to/support the production of certified sustainable palm oil.

Although healthier to consume than saturated fats like butter (Chapter 3), new oil palm plantations are built at the expense of orangutan habitat—and are consequently decimating the orangutan population. Current practices are causing millions of acres of rainforest to be cut down annually, resulting in the deaths of orangutans and many other species.

What can be done? Given the health benefits of palm oil and its utility—plus the impact on the Indonesian and Malaysian economies—it is difficult to boycott palm oil. African oil palms also produce more oil on less land than comparable plants, such as sunflowers, rapeseed, and soy. In addition, palm oil and its derivatives have over 600 names used on product labels, and so it can be difficult to determine which are the "good" products to buy.

So, what is one to do as a consumer? Grassroots organizations such as Roundtable on Sustainable Palm Oil (RSPO) have created labeling for products (Figure 12.14) that use palm oil harvested using sustainable practices, and companies can choose to join RSPO and commit to using only sustainably generated palm oil in their products.

There's also an app, called PalmOil Scan, created by the World Association of Zoos and Aquariums (WAZA) that you can download to your phone to help you make orangutan-friendly choices when shopping. Some zoos, such as the Cheyenne Mountain Zoo in Colorado Springs, Colorado, have additional resources, such as a candy shopping guide, which is released each year before Halloween. Finally, you can educate those around you on the importance of choosing products with sustainably harvested palm oil, and write to companies that are not currently RSPO members, asking them to consider joining or taking other steps toward using only sustainable palm oil.

1. Considering the environmental impacts palm oil plantations and the potential extinction of the orangutan, do you think farmers have a moral obligation to harvest responsibly? What could each of us do (from individuals to policymakers) to lessen our impact on other species as it pertains to palm oil?
2. Finding ways to use palm oil sustainably balances the competing needs of many different stakeholders, including nonhumans. What wisdom regarding sustainable palm oil might be useful in other situations related to conservation biology?

12.3 Summary

- Primates are defined by traits such as large brains; forward-facing eyes with expanded vision and excellent depth perception; advanced arm dexterity; and grasping hands and feet, with touch-sensitive pads and flattened nails.
- Prosimians include lemurs, which have a long snout and wet nose. Tarsiers have a dry nose and moveable, noncleft upper lip that allows for complex vocalizations.
- Monkeys of the Americas are tree dwellers with prehensile tails. Monkeys of Africa and Asia live in trees or on the ground, and either lack a prehensile tail or have no tail. Apes do not have a tail.

12.4 Human Evolution

Learning Objectives

12-9 Compare different genera of hominins.

12-10 Explain how gene flow influenced human evolution.

12-11 Describe what is known about members of our genus.

12-12 Identify the factors that influenced the evolution of different skin tones.

Imagine that you had a time machine and could go back in time 2 million years. If you were to visit Africa, odds are that you would see a lot of individuals with large heads walking upright. Some might even be using tools. Two million years ago, three out of the four genera (singular: genus) of hominins still lived in Africa. While these species may be somewhat recognizable to us—like the humanoid aliens in Star Trek or Star Wars are recognizable as humanlike—other features would tell you that you were looking at a different species. For example, some, like the australopithecines, had a sagittal crest (a bony ridge at the top of the skull) and large cheeks.

Have you had your wisdom teeth removed? Monkeys, gorillas, chimpanzees, and ancestral humans such as *Australopithecus afarensis* all have wisdom teeth, which they use for extra chewing power. But as human diets have changed over evolutionary time, there has become less of a need for wisdom teeth. As such, approximately 25% of people born today will not develop any wisdom teeth.

We Are Family: Hominidae

In the last section, we learned about our *order*: primates. Now, in this section, we will examine our *family*: **Hominidae** (also called hominins), where modern humans and our closest extinct relatives are grouped together. The defining trait of hominins is **bipedalism**—habitual upright walking. Thus, researchers interested in human origins look for fossil evidence that a species walked upright. For example, to walk upright requires strong thigh bones connected to the hips in a bridge-like shape that can sustain the upper body. A wide area of bone just below the knee, which was first seen in *Australopithecus*, also suggests upright gait, since as one walks, they will intermittently stand on one leg—and that leg must be able to bear the weight of the body. Curved spines also absorb shock during walking—and humans are the only members of the animal kingdom to have this feature.

Scientists have classified four different genera of hominins: *Ardipithecus*, *Australopithecus*, *Paranthropus*, and *Homo* (Figure 12.15A). These classifications are based on both fossil evidence and genetic evidence, which informed the timeline at the bottom of the figure. Note that the **braincase** is the part of the skull that encloses the brain—and the larger the braincase, the larger the brain (Figure 12.15B). Therefore, the size of the braincase is a proxy value for the size of the brain.

Figure 12.15 Hominin phylogenetic tree

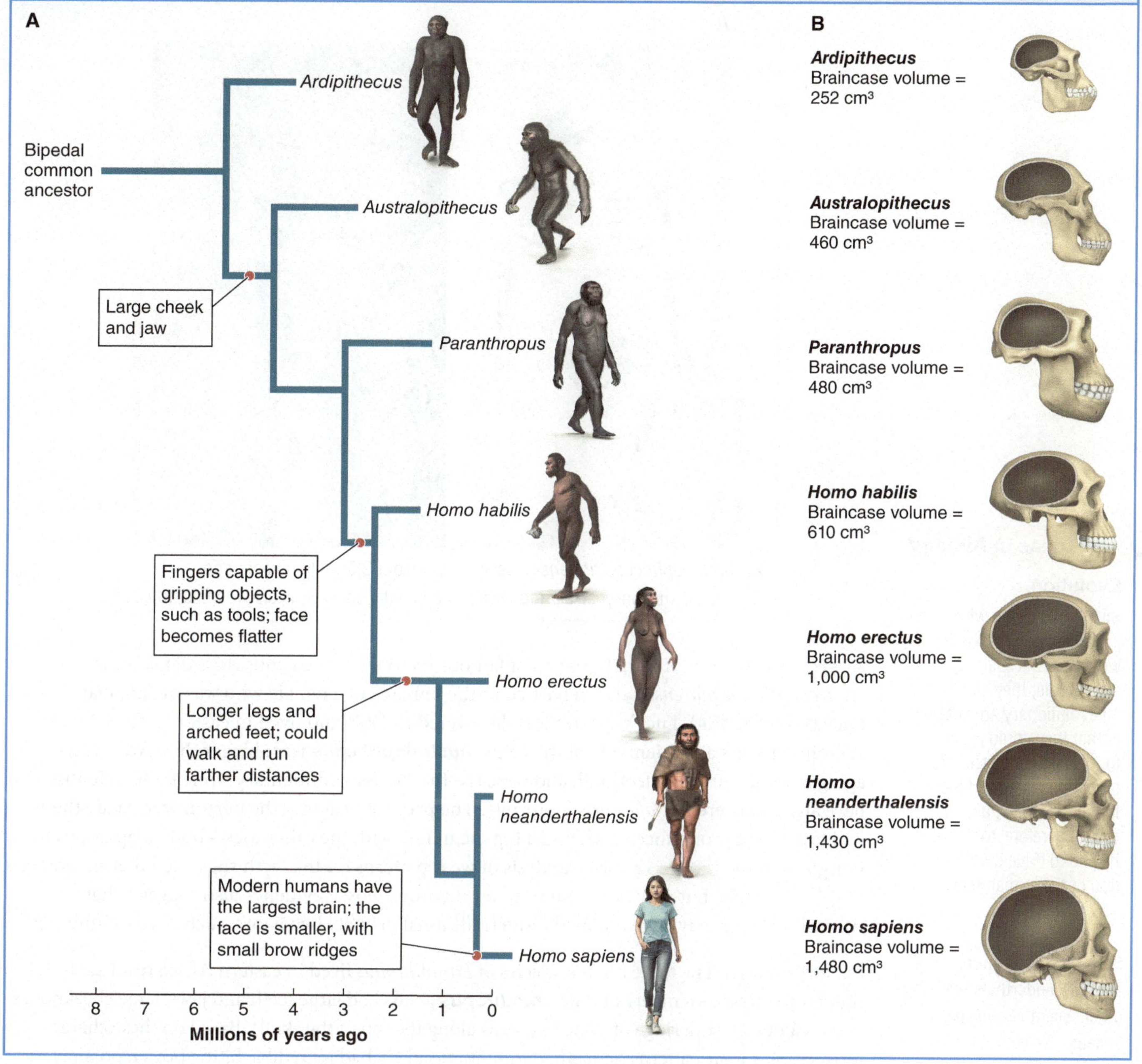

! Point of Confusion: Have you ever seen an illustration of human evolution where there is a line of primates, starting with a chimpanzee and ending with a modern human? This is a misrepresentation, reinforcing the misconception that we "descended from monkeys." Rather, humans, along with the other primates, share a common ancestor. Think of the human common ancestor like a great-great-great—plus many, many more greats—grandmother who had several children. Each child's descendants eventually led to different genera of hominins. So, if we think of our history in terms of a large family tree, these different species of humans were more like very distant cousins rather than direct ancestors or descendants. For example, notice how the horizontal lines overlap somewhat in Figure 12.15, signifying that the species coexisted for some portion of time.

Ardipithecus. The oldest known hominin genus is ***Ardipithecus***, including *Ardipithecus ramidus*, or Ardi for short. Ardi, discovered in 1994, lived in eastern Africa approximately 4.4 million years ago. Ardi was small, about 120 cm (3 feet) tall. While Ardi walked upright, fossil evidence suggests that Ardi also climbed trees. Wear patterns on Ardis' teeth suggest they were omnivorous.

Figure 12.16 **Fossil evidence of *Australopithecus***

A B

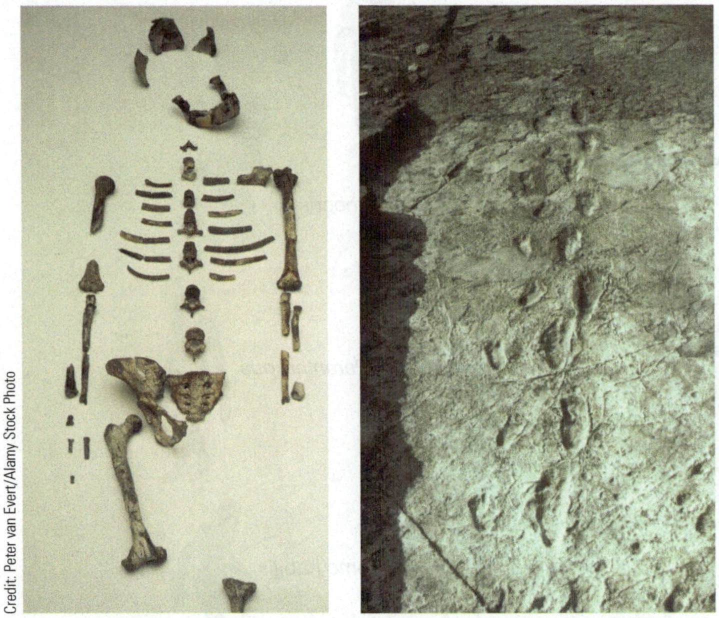

Credit: Peter van Evert/Alamy Stock Photo

Credit: Images of Africa Photobank/Alamy Stock Photo

A. *Australopithecus afarensis* fossil (Lucy) from 3.5 million years ago.

B. The 3.6-million-year-old footprints of a bipedal species, likely *Australopithecus*.

Big Ideas in Biology

Evolution

Since individuals with increased hand dexterity were better at making stone tools, they also had an evolutionary advantage in that they could procure food more easily and protect themselves and their families with the use of tools. Therefore, tools likely helped to drive the evolution of the human hand.

Australopithecus. Arguably, the most famous member of the genus *Australopithecus* is *Australopithecus afarensis*, and in particular the remains of a female who was nicknamed Lucy (**Figure 12.16A**). Lucy's species was discovered in 1974 and lived in eastern Africa 3.9 to 2.9 million years ago. Members of the genus ***Australopithecus*** were bigger than Ardi, averaging around 150 cm (4 feet) tall, and were the first to use rudimentary stone tools. Scientists have also discovered *A. afarensis* footprints. The prints reveal that the person who made them had feet with a pronounced arch and a big toe in line with the other toes—both adaptations to upright walking (**Figure 12.16B**). Analysis of wear patterns on the teeth suggested that *A. afarensis* largely ate plants, but may have also been omnivorous. Fossil evidence also suggests that *Australopithecus* may have retained more traditional primate behaviors, such as tree climbing.

Scientists know more about Neanderthals than other recent *Homo* species because we've found the remains of around 400 individuals. In comparison, the number of known Denisovan bones would fit into a shoe box. More samples are needed before Denisovans can be accurately classified into their own species or as *Homo sapiens*.

Paranthropus. The three known species of *Paranthropus* lived in eastern Africa from 2.7 to 1.2 million years ago. Members of the ***Paranthropus*** genus had large teeth and jaws, large cheekbones, and a sagittal crest (a ridge of bone that runs along the top of the skull). Based on these characteristics, plus wear patterns on teeth, *Paranthropus* likely had incredible biting power, possibly for eating tough grasses. Note that *Paranthropus* is not in the same lineage that eventually led to our own genus, *Homo*.

Early *Homo* species. Early humans were the first species to belong to the genus *Homo*. The genus ***Homo*** is characterized by larger braincases; flatter, narrower faces; and smaller teeth and jaws.

Fossils of the oldest species in the *Homo* genus, ***Homo habilis***, date from 2.4 to 1.4 million years ago. The species was named based on a fossil discovered in Kenya in 1964, where the name *Homo habilis*, which means "handy man," is a reference to the stone tools found near the a *Homo habilis* fossil (**Figure 12.17**). This species was known for regular stone tool use. At the time of the discovery, the ability to make stone tools was considered a distinctly human trait, but we now know that other, nonhuman, species use tools.

The first known hominin with body proportions like modern humans was ***Homo erectus***, which arose in Africa around 1.89 million years ago. The species name means "upright man," and like us, *H. erectus* stood on legs that were longer than its arms. *H. erectus* is the first evidence we have for a hominin outside of Africa, with fossils of this species found in China and the Eurasian country of Georgia.

Recent *Homo* species. The first individuals considered to be part of recent *Homo* species date from a little less than a million years ago, and we are the only subspecies that remains. Recent *Homo* is characterized by flatter faces, smaller teeth, and the largest braincases of any of the human lineages. Recent *Homo* species include us (*Homo sapiens*), Neanderthals (*Homo neanderthalensis*), and the Denisovans, which at the time of this writing did not have a formal species name because there is still insufficient evidence to definitively separate them into their own species, or to classify them as a subpopulation of our species, **Homo sapiens.**

Figure 12.17 **These Oldowan, or pebble tools, were used by *Homo habilis***

The tools shown here are estimated to be 1.8 million years old. Evidence suggests that *Homo erectus* also used stone tools. These stone tools were likely used for chopping up plants and animals, or for scraping hides.

Neanderthals first evolved about 400,000 years ago, and scientists have found Neanderthal fossils in the Middle East, Europe, and central Asia. Reconstructions reveal Neanderthals were shorter than modern humans but stood upright. They lived in regions where winters were cold, where a short, stocky body would minimize the surface area available for heat loss. The Neanderthal braincase was longer and lower than that of modern humans, but their brain was as big as or bigger than ours. Their face had pronounced brow ridges, a large nose with widely spaced nostrils, and no chin.

Fossils of Neanderthal individuals who survived despite disabilities, such as the loss of a limb, suggest they had a compassionate social structure in which individuals cared for others within their group. Neanderthals also created art and buried their dead, suggesting symbolic thought (Figure 12.18). Several lines of evidence suggest Neanderthals were able to speak.

The existence of Denisovans was discovered when researchers sequenced DNA extracted from a fossil pinky (small finger) bone discovered in Denisova Cave in Siberia. The researchers expected to find typical Neanderthal DNA. Instead, the finger's DNA contained unique sequences never seen before. As a result, the fossil was assigned to a group—the Denisovans—named for the site of its discovery.

? Did You Know?

The oldest known *Homo sapiens* remains are from Morocco and date to 315,000 years ago.

? Did You Know?

The idea of Neanderthals as brutish cave dwellers with poor posture arose from an early reconstruction based on a fossil of an individual that was deformed by arthritis.

Figure 12.18 **The oldest known cave art was created by Neanderthals**

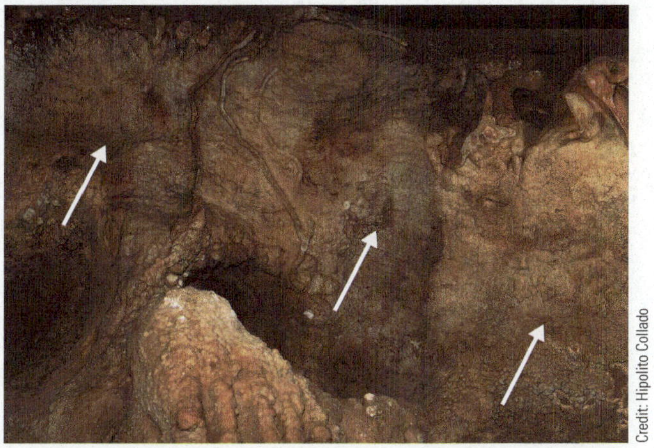

Radiocarbon dating of this art found in a cave in Spain suggests that these handprints (black arrows) were made over 64,000 years ago, when Neanderthals lived in what is now Europe. These paintings were made at least 20,000 years before *Homo sapiens* arrived in Europe from Africa.

You may have heard of discoveries of other recent human ancestors, such as *Homo florensis* (also called the hobbit) and *Homo naledi.* This tells us that yet more *Homo* species may be discovered as we find more fossils and fill in the fossil record (Chapter 10).

Figure 12.19 **A Paleolithic cave painting of a cow**

Credit: Gianpiero/Adobe Stock Photos

This image is one of many found in the Lascaux caves in France. The cave paintings are estimated to be around 20,000 years old.

Why Are *Homo Sapiens* the Only Human Species Remaining?

If there were so many different hominins coexisting millions of years ago, why are we the only species that remains today? It turns out that individuals with larger brains were also better able to plan, communicate, problem solve, and coordinate with others—all of which would give an evolutionary advantage and make *Homo sapiens* more likely to survive. Some lines of evidence suggest that part of the reason there are no Neanderthals today is because *H. sapiens* outcompeted them.

The evolution of our large brain size facilitated better communication between different individuals and the development of culture. **Culture** includes characteristics that define a group, such as beliefs, values, behaviors, and language. Having a shared culture led to stronger group dynamics, facilitating better cooperation among different groups. Culture continues to be an important feature of modern life, where we group ourselves based on similar belief or value systems.

In terms of adaptability, *H. sapiens* has a long history of high adaptability. For example, *H. sapiens* left Africa and went on to migrate around the world (which we will discuss next). This migration happened during climatic shifts, suggesting the ability of these early people to explore despite uncertainty. Today, we still see people adapting to new places and different situations, such as how people adapted during the COVID-19 pandemic.

Our large brain not only allowed us to develop strategies to adapt but also facilitated the pursuit of meaning and use of the imagination. Some of the earliest evidence of human culture is in cave paintings (Figure 12.19), and as a species, we continue to generate art, music, and other cultural artifacts.

Study Tip

Think about how structures inform functions. How have the different changes that have occurred through evolution, like bipedalism and increased braincase size, led to our success as a species?

Did You Know?

The species *Homo sapiens* is the only hominin with a chin, a protruding area of thickened bone in the middle of the lower jawbone. All other primates have a jawbone that does not protrude beyond the teeth.

Out-of-Africa Hypothesis

The oldest *H. sapiens* fossils are all from Africa, indicating that our species originated there. Genetic analyses also support the African origin hypothesis. Modern Africans are more genetically diverse than people of any other region (Figure 12.20). This diversity indicates that African populations have existed for a very long time—long enough to accumulate a very large number

Figure 12.20 **Genetic variation in selected human populations**

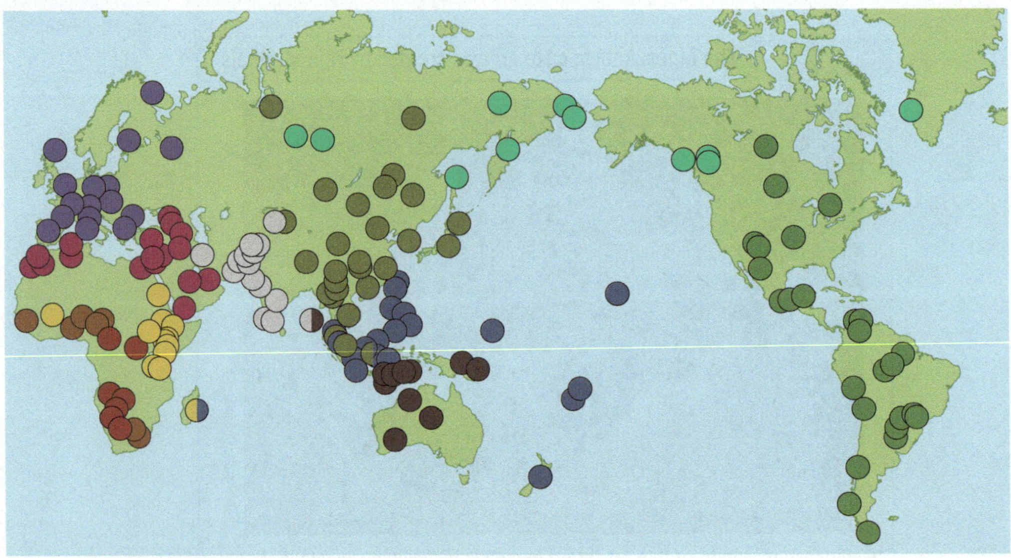

Note that all dots of one color are genetically similar. Thus, looking at the map, there are more genetically diverse populations in Africa than in North and South America.

of random mutations compared with other populations (shown by many differently colored dots in the figure). Furthermore, most genetic variations seen in people native to regions outside of Africa are subsets of ancestral populations in Africa. This pattern is evidence that a series of founder effects (Chapter 11) occurred as small groups of people left their homeland to colonize the rest of the world. For example, if you look at North and South America, you can see less diversity in the color of the dots, indicating less genetic diversity as a result of the founder effect.

Recently discovered *H. sapiens* fossils from Greece indicate that small groups began to venture out of Africa as early as 210,000 years ago (**Figure 12.21**). Other fossils found in China that may be *H. sapiens* date to 100,000 years ago. However, genetic studies of modern humans indicate that most non-African populations are descendants of individuals who left Africa in a second mass migration 60,000 to 70,000 years ago. Our species expanded its range over many generations, as small groups ventured beyond their place of birth. Human pioneers traveled along the coasts of Africa, then to Eurasia and Australia. By about 40,000 years ago, they were creating art on rocks in Europe and in Indonesia. By 13,000 years ago, they were leaving footprints along a beach in British Columbia on the west coast of North America.

Recall from Chapter 11 that whenever one group of organisms leaves and goes to a new place, they take their genes with them, a process called gene flow. At these new locations, different genes—such as those for human skin color—were selected for or against as different populations became adapted to new environments. Therefore, because human skin color will change over generations depending upon the environment, it is not a good trait to use for distinguishing among humans. This is important to note because it highlights that there is no genetic or evolutionary basis for our modern concepts of race, but rather, selective evolutionary forces like natural selection—alongside nonselective forces like gene flow—played a role in shaping skin color.

? Did You Know?

Homo sapiens was not the only species to migrate out of Africa. *Homo erectus* migrated out of Africa around 1.8 million years ago. Descendants of the individuals who migrated out eventually became populations of Neanderthals and Denisovans.

Figure 12.21 Migration of *Homo sapiens*

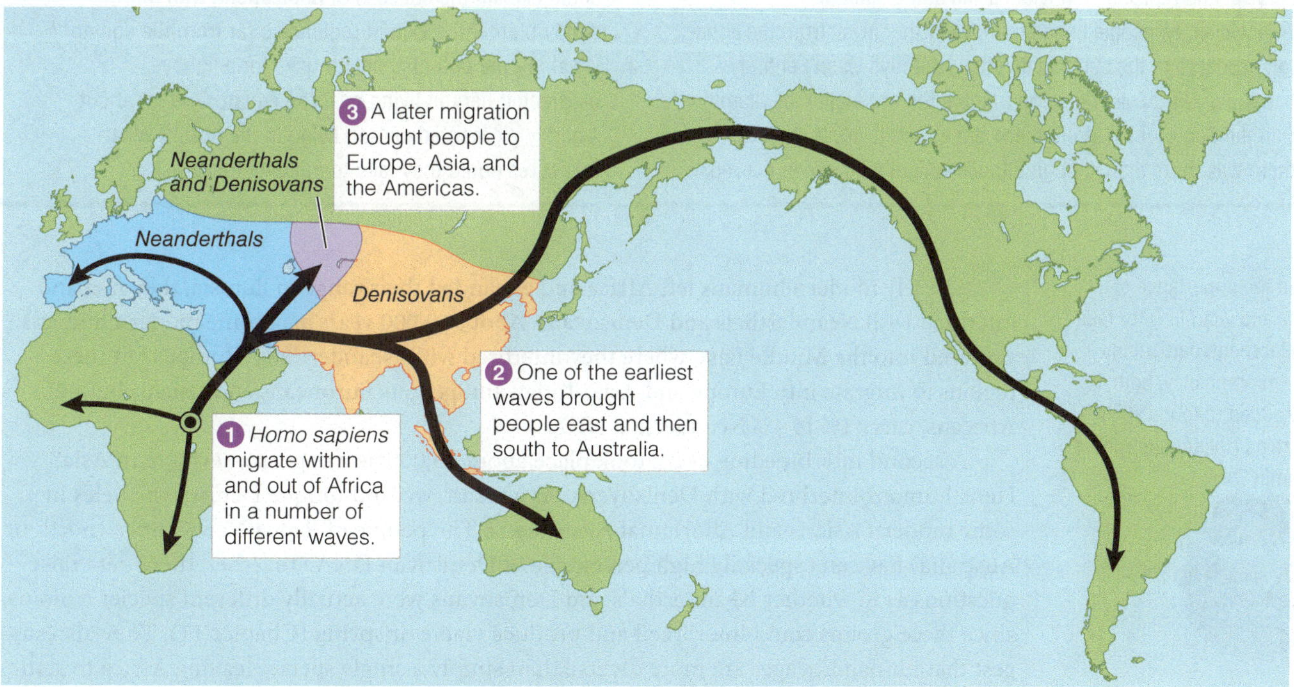

Modern humans first evolved in Africa (black dot) and migrated out of Africa in at least two different waves (shown by the thicker arrows). As *Homo sapiens* migrated, they intermixed and interbred with populations of other hominins, including the Neanderthals (blue) and Denisovans (orange). Note that the purple region was where Neanderthal and Denisovan ranges overlapped.

Digging into Data

Shades of Nature: Why Does Skin Color Vary?

As you go about your day, you may notice that people have a variety of skin tones. The appearance of different skin colors results from wavelengths of light being reflected or absorbed (Chapter 4) by a pigment in the skin called *melanin*, where the higher the melanin content in the skin, the darker the skin tone. Melanin acts like a natural sunscreen, protecting cells from ultraviolet radiation, which can damage our DNA.

Perhaps not surprisingly, early humans living near the equator (where there are higher levels of UV radiation) had darker skin tones. Initially, people thought that this was to prevent cancer. However, there is a problem with this hypothesis: Skin cancer usually develops later in life, after people reproduce. Therefore, it doesn't explain why darker skin color would provide a reproductive advantage (which is essential to evolution!).

So, what explains the prevalence of darker skin—and thus the dominance of alleles for darker skin—in individuals who lived near the equator? It turns out the missing link boils down to two vitamins: vitamin D and a B vitamin that is essential for embryonic development called folate. Without sufficient folate levels during pregnancy, offspring tend to have higher rates of birth defects, meaning that ultimately, the process of passing on one's genes is impaired. High ultraviolet light exposure causes a person's folate levels to go down, while alternatively, melanin protects folate levels. Therefore, people with darker skin tones were better suited for survival and reproduction when living at the equator.

However, as people began to move farther away from the equator, exposure to the sun and its ultraviolet rays wasn't entirely a bad thing. Ultraviolet radiation triggers the production of vitamin D in the body, which strengthens bones and teeth. In the tropics, there was more than enough ultraviolet light from the sun, even with dark skin, for enough vitamin D to be produced. However, this was not the case at high latitudes, especially in the winter. In these places, insufficient vitamin D production in the body of a dark-skinned individual could lead to health issues, and ultimately, death, thus compromising the individual's reproductive success. So, the different environmental conditions selected for those with lighter skin as they migrated away from the equator.

Since a requirement of evolution is the ability to reproduce, a vitamin D or folate deficiency would exert a selective force against reproduction. Note that melanin protects folate levels while interfering with the ability to synthesize vitamin D. How, then, did folate, vitamin D, migration of humans, and UV exposure all work together to influence skin color? Researchers evaluated the varying intensities of ultraviolet radiation (UV) around the world and compared them with skin pigmentation in Indigenous populations, which have existed in the locations shown in **Figure 12.22** since long before European colonization.

1. From what you've learned in the text, where would you expect to see humans with lower amounts of skin pigmentation? With higher skin pigmentation? Do your predictions align with the real-world data shown in the map in Figure 12.22? Explain.
2. Perhaps surprisingly, Indigenous Inuit populations at high northern latitudes have darker skin tones. What could explain the success of populations with more darkly pigmented skin in regions so far from the equator?
3. What are the potential health risks for a lightly pigmented person living near the equator? What about a darkly pigmented person living in the north? What measures could they take to mitigate these risks?

Given the importance of folate and vitamin D for fetal development, particularly early in pregnancy, both are now added to foods such as grains (folate) and milk (vitamin D).

As early modern humans left Africa and expanded their range to Eurasia, they met and interbred with Neanderthals and Denisovans. About 60,000 years ago, some modern humans ventured into the Middle East, where they interbred with Neanderthals. Humans left these regions to migrate into Europe and Asia. Today, Indigenous Europeans and Asians, but not Africans, carry 1% to 4% Neanderthal DNA.

A second interbreeding event took place about 40,000 years ago, somewhere in Asia. Here, humans interbred with Denisovans. As a result, we find unique Denisovan alleles in some modern Asians and Aboriginal Australians. The people of Papua New Guinea (north of Australia) have an especially high percentage of Denisovan DNA. Together, these data raise questions as to whether Neanderthals and Denisovans were actually different species from us, since these groups could interbreed and produce viable offspring (Chapter 11). They also suggest that human lineages are more diverse than simply a single species leaving Africa to settle in new places.

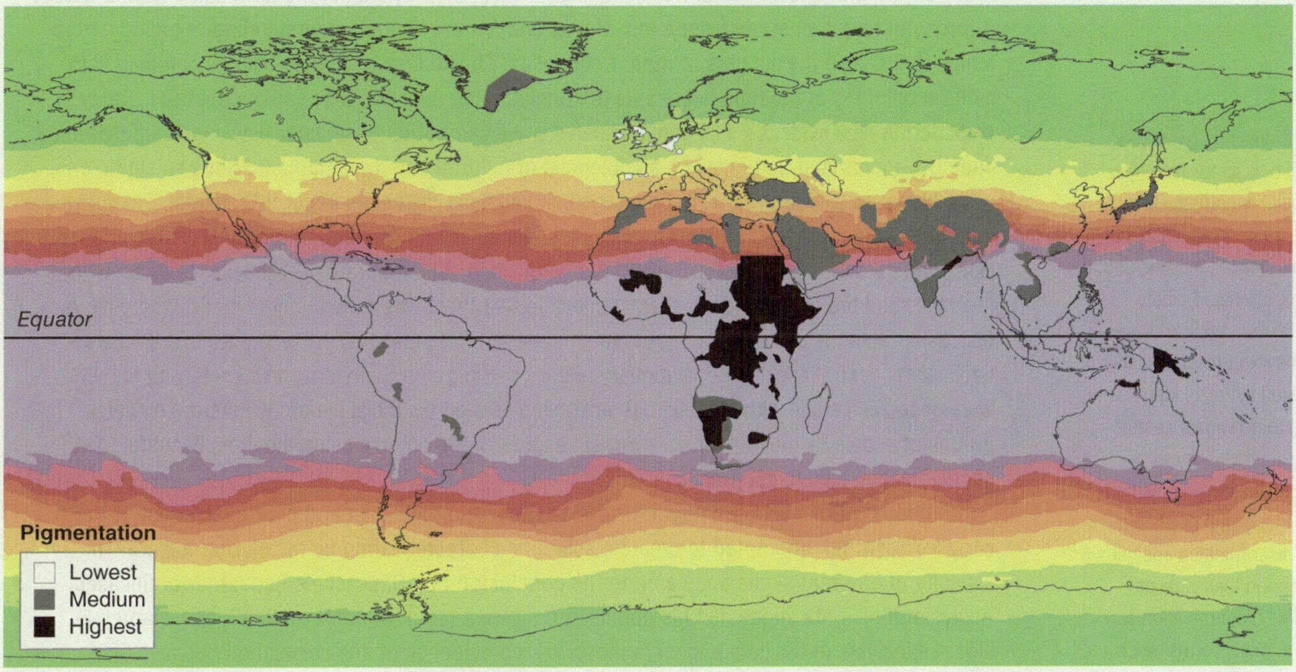

Figure based on: Jablonski, N. G., & Chaplin, G. (2000). The evolution of human skin coloration. *Journal of human evolution*, *39*(1), 57–106.

Exposure	UVI range
Low	<2
Moderate	3 to 5
High	6 to 7
Very high	8 to 10
Extreme	11+

Figure 12.22 The UV light index (UVI) for the world superimposed with predominant skin color for known Indigenous human populations

While there can be seasonal fluctuations, the regions near the equator have the highest levels of UV exposure (purple), with those levels decreasing to their lowest at the poles (green). The locations of assessed Indigenous populations are shown, where skin pigmentation was measured and rated from least to highest amount of pigmentation.

12.4 Summary

- *Ardipithecus*, the oldest hominin genus, climbed trees and walked upright. *Australopithecus* species were the first to use rudimentary stone tools. The genus *Paranthropus* had large, powerful jaws with tremendous biting power.
- *Homo sapiens* originated in Africa and migrated throughout the world, adapting to varied and changing environmental conditions and interbreeding with other groups like *H. neanderthalensis* and Denisovans along the way.
- *Homo habilis* and *Homo erectus* had flat, narrow faces with smaller teeth and jaws as well as a larger braincase, and they were the first to use stone tools. *H. erectus* was bipedal with longer legs than arms.
- *Homo neanderthalensis* were short and stocky, with pronounced brow ridges and no chin. They created art and buried their dead, and likely had a spoken language.
- *Homo sapiens* developed culture, including a complex spoken language and cooperation among social groups, leading to advanced planning and coordination with others, and inevitably, the ability to outcompete other *Homo* species.
- The evolution of human skin color resulted from selective evolutionary forces like natural selection, alongside nonselective forces like gene flow.

In Figure 12.22, you may be wondering why there is no data shown for North America and very little for South America. This is because there are two different methods for collecting data on human skin color. The researchers decided to only include data collected using the more widely used of the two methodologies for their analysis.

Process of Science

Where Did the Idea of Race Come from Anyway?

If there is no biological or scientific evidence for race, then where did the term come from anyway? It turns out that it emerged from a mix of bias and a more limited understanding of biology, from a time that lacked the insights of modern genetics. The studies described below combined with cultural opinions of the time led people to conclude that race was a scientific construct. Today, considering new evidence, like genetic analysis, we can refute these older flawed scientific studies and now understand that race is a sociological or cultural determination, and not a scientific one.

While there is no genetic or scientific basis for race—it is a social construct. Race can be an important aspect of one's identity, culture, and lived experience. Society uses the term race to divide humans into groups based on shared factors, often having to do with ancestral background and social or physical qualities. Racism refers to a group of people being discriminated against based upon their race or ethnicity. Discrimination can come in many forms, such as through social interactions and institutional practices.

Carl Linnaeus—who we met in Chapter 10 and again at the beginning of this chapter—suggested that there were four types of people, basing his characterizations solely on physical characteristics like…you guessed it, skin color. Johann Blumenbach, a contemporary of Linnaeus, came up with five races and took it one step further, suggesting that there was a hierarchy of races, where his race, Caucasian, was the one he listed as the most superior. Note that now we recognize that the name Caucasian doesn't reflect the evolutionary history of the group Blumenbach was referring to, who were generally people with light skin. Blumenbach created his racial hierarchy in terms of which groups of people he thought were the most beautiful (his own race being the most beautiful), with some attention paid to geographic origin.

Continuing in this vein, Samuel Morton, an American physician and natural scientist of the early 19th century, attempted to create a scientific characterization by measuring the cranial capacity of different skulls and came to the conclusion that Caucasians had larger braincases. Analysis of his data suggested his conclusions were a result of confirmation bias (Chapter 1), since reanalysis of his data suggests there was no evidence of any correlation between skin color and braincase size.

Combatting these perceptions was the work of people such as American sociologist and intellectual W. E. B. DuBois and pathologist Julian Lewis (who we met in Chapter 1), who worked to use science to fight racism into the 20th century. For example, DuBois was a Black intellectual who pushed back against the idea that there was a scientific justification for racism. Lewis, as we discussed in Chapter 1, wrote a book titled *The Biology of the Negro* to help counter racial bias in medicine. However, implicit bias in medicine goes on to this day, where data from the National Institutes of Health suggest that 70% of practicing physicians show implicit bias (an unconscious negative attitude or prejudice that colors our understanding, actions, and decisions) against Black people and those who are Hispanic/Latino. These lingering biases likely contribute to poorer health outcomes among Black and Hispanic groups compared to those who are White. Therefore, it is important to reiterate that there is no biological evidence for race, as well as the importance of addressing these implicit biases to create equitable health outcomes.

Since human skin color is a trait that changes in response to one's environment—which can change over evolutionary time—research showing the correlation between skin color and UV exposure is part of a large body of evidence suggesting that there is no genetic or evolutionary basis for human race.

What can be done about this? One of the most important things is to understand the biological basis of skin color and that "race" is a cultural, not a scientific, term. It's also important that we attend to our own biases (Chapter 1), where there can be an inherent distrust of people who are different from us. The important thing is to acknowledge that we all have biases and to be aware of them, but to not make bias-based judgements.

1. As we learned in Chapter 1, we all have biases. Name a bias that you think you might have as it relates to different groups of humans and how you might counteract it.
2. If you were designing training for people working in healthcare, how would you ensure that all people are served fairly, knowing that racial bias exists in medicine? To explore more deeply, you could do some research into medical justice advocacy groups to explore their recommendations.

Understand It! Procrastination

I would procrastinate, but I never get around to it. Procrastination is when we delay or postpone doing something… even if we know we'll regret it later. It's everything from putting off the laundry for another day to staying up late and reading a good book when we know we should be sleeping to delaying completing a homework assignment until the last minute. Procrastination is a normal part of the human experience—we all procrastinate from time to time. What's important is to pay attention to *why* we procrastinate, and if there are patterns in terms of the types of tasks that cause us to put them off.

Procrastination tends to occur when a task makes us uncomfortable, or if we perceive it to be boring, meaningless, frustrating, ambiguous, or stressful. In fact, if you image someone's brain while they are contemplating a task that they don't want to do, the pain centers in the brain become activated. So, even the act of *thinking* about something that we don't want to do can be interpreted by our brains as pain. Procrastination can also feed into itself because anticipation of the task can be painful, making the unpleasant task even harder to take on. Burnout and procrastination are also related to one another—when one is feeling exhausted or overwhelmed, they are also more likely to procrastinate.

At its heart, procrastination is closely tied to our emotions. Sometimes, it can be because we don't feel capable enough to do something—which is a great opportunity to talk to a trusted friend, a counselor, or your professor. At other times, procrastination occurs because it's a simple case of the task being distasteful. In such instances, using strategies to increase motivation (Chapter 11) or applying techniques like the Pomodoro Method (Chapter 6) can be useful. Sometimes it can be helpful to simply acknowledge that, in terms of academics, not every subject is going to be our favorite, and having an awareness of our tendency to procrastinate—and developing strategies for what to do about it—is important for moving on.

1. What tasks and/or subjects tend to make you procrastinate? What do you think is the reason for the procrastination?
2. What are some strategies you've used in the past (or that you'd like to use) to manage procrastination?

Chapter Review

Section 12.1 Taxonomy: The Science of Naming

Phylogeny is the scientific study of evolutionary relationships. **Taxonomy** is a field that names, describes, and classifies species using the two-part **genus** and **specific epithet** of the **Linnaean system**. Organisms are classified in a nested taxonomic hierarchy, from domain to species. A **species** is a group of organisms that can interbreed. A **genus** is a group of closely related species that share **traits**, or inherited characteristics. A **taxon** is a ranking of organisms that share a unique set of traits.

Section 12.2 Examining Evolutionary Relationships with Phylogenies

Phylogenetic trees are diagrams representing evolutionary relationships, including the timeline of divergence between species. A **clade** is a complete branch in a phylogenetic tree, where **sister groups** belong to the same clade. A **key innovation** is a novel trait that increases a species' ability to adapt and better exploit its environment. **Derived traits** are present in a clade, but not in the clade's ancestors, while **ancestral traits** are inherited from a common ancestor.

Exaptation is repurposing of an existing structure for a new purpose. **Coevolution** occurs when two species evolve together, and changes in one species lead to corresponding changes in the other, leading to increased interdependency. **Stasis** occurs when a species remains largely unchanged over a long period of evolutionary time.

Phylogenies are used to understand and track infectious diseases and support vaccine development, and in prevention and treatment efforts. **Reassortment** occurs when multiple viruses swap genetic material when they infect the same organism. A **pandemic** occurs when an infectious disease spreads beyond a single continent.

12.3 Primate Evolution

Primates include **prosimians** like **lemurs** plus **tarsiers**, monkeys, and apes. Monkeys, apes, and humans are **anthropoids**. Tree-dwelling **monkeys of the Americas** have prehensile tails, while **monkeys of Africa and Asia** may or may not have a tail. **Apes** are tailless nonhuman primates.

12.4 Human Evolution

The taxonomic family **Hominidae** includes modern humans and extinct hominins. One key feature of Hominidae is **bipedalism**, or walking upright on two legs. The oldest hominin genus is ***Ardipithecus***, with species that could climb trees and walk upright. ***Australopithecus*** species were larger than *Ardipithecus* and were the first to use rudimentary tools. ***Paranthropus*** species had powerful jaws for biting and chewing. Larger **braincases** led to a complex spoken language and cooperation among social groups.

The genus ***Homo*** includes early and modern humans. ***Homo habilis*** was known for their regular use of stone tools, while ***Homo erectus*** was the first known hominin with body proportions like modern humans, with legs that were longer than the arms. Modern humans (us!) are ***Homo sapiens***.

Culture is the defining collection of a social group's shared beliefs, values, and behaviors, such as language, customs, food, music, and art. Race is a cultural concept not supported by biological evidence.

Practice It!

Retrieval practice is the process of pulling information from your long-term memory and using it. Here are some questions to help with your retrieval practice.

Self-Quiz

1. Which of the following correctly shows a species name? (LO 12-1)

a. striped horse relative
b. mountain zebra
c. *Equus zebra*
d. Perissodactyla Equidae *Equus*

2. Which of the following includes three members of the same genus? (LO 12-2)

a. *Equus caballus*, *Canis familiaris*, *Felis catus*
b. *Equus caballus*, *Equus onager*, *Equus asinus*
c. *Canis familiaris*, *Canis lupus*, *Panthera leo*
d. *Panthera onca*, *Equus zebra*, *Ginkgo biloba*

3. The smallest taxon that includes closely related species is called the ______. (LO 12-2)

a. order
b. family
c. genus
d. phylum

4. ______ show evolutionary relationships over time. (LO 12-3)

a. Genus and species
b. Phylogenetic trees
c. Derived traits
d. Evolutionary charts

5. The evolution of wings helped Pterygota, the winged insect clade, to be successful. In this example, wings are a(n) ______. (LO 12-3)

a. neutral mutation
b. sexually dimorphic trait
c. key innovation
d. all of the above

6. If all members of the taxon that includes reptiles and birds shares the trait of laying eggs, but only birds have feathers, in this group of species, the ancestral trait is ______, while the derived trait is ______. (LO 12-4)

a. egg laying; feathers
b. reptiles; birds
c. feathers; egg laying
d. birds; reptiles

7. In Costa Rica, acacia trees provide food and shelter for *Pseudomyrmex* ants. In turn, the ants have evolved to protect the trees by attacking animals that touch the trees. This is an example of ______. (LO 12-5)

a. exaptation
b. an ancestral trait
c. a key innovation
d. coevolution

8. Primates exhibit all of the following characteristic traits except ______. (LO 12-7)

a. large brains
b. hands and feet with claws
c. excellent depth perception
d. grasping hands and feet

9. A group of nonhuman primates that *all* lack a tail are ______. (LO 12-8)

a. lemurs
b. tarsiers
c. monkeys
d. apes

10. Primates with prehensile tails used to grasp branches when climbing are ______. (LO 12-8)

a. lemurs
b. monkeys of the Americas
c. monkeys of Africa/Asia
d. apes

11. If scientists found a skeleton with bridge-like hip bones and a curved spine, what would be the *most specific* classification they could assign this species without additional information? (LO 12-9)

a. ape
b. primate
c. *Australopithecus*
d. Hominidae

12. Which of the following is evidence of gene flow in early hominids? Select all that are true. (LO 12-10)

a. the presence of Neanderthal DNA in modern human genomes
b. the loss of a tail in this clade
c. the highest genetic diversity is found in Africa
d. the extinction of all non-human hominid species

13. *Homo* ______ first created art and buried their dead. (LO 12-11)

a. *habilis*
b. *erectus*
c. *neanderthalensis*
d. *sapiens*

14. Which of the following contributed to the evolution of different skin tones? (LO 12-12)

a. natural selection
b. gene flow
c. nutrients important for reproduction
d. all of the above

Critical Thinking

1. Suppose a patient is brought to the emergency room with a severe respiratory infection. A patient sample is cultured and a virus is sequenced and identified as influenza A. When a phylogenetic tree is constructed, however, the H gene is found to be closely related to other human influenza strains, while the N gene is closely related to bird influenza strains. What is the most likely explanation for this finding? Explain the process, including the species involved, and how the phylogenetic trees provide evidence for this process. (LO 12-6)

2. According to the phylogenetic tree that follows, members of species E are most closely related to members of which species: A, B, C, D, or F? Explain your reasoning, using correct terminology. (LO 12-3)

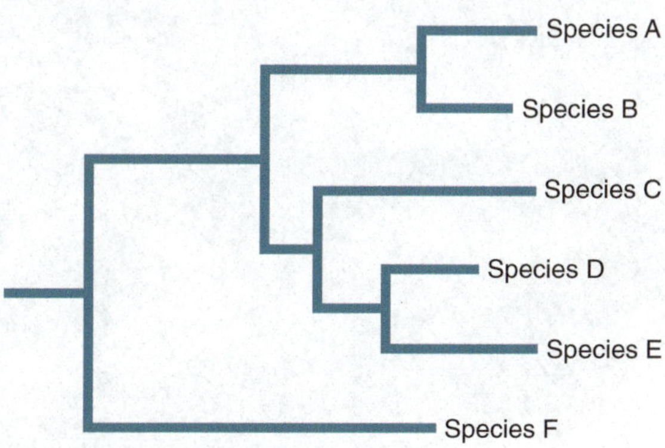

3. Suppose that Species A and Species F on the phylogenetic tree in question 2 had wings, while none of the other species did. Describe two possible explanations for this observation, citing specific evolutionary concepts involved. (LO 12-4)

4. When scientists sequenced two Neanderthal specimens from Spain and Italy, they identified one of the genes that controls the production of melanin, the protein responsible for pigmentation of the hair and skin. Scientists know the function of this gene because modern humans also have this gene. The Neanderthals had a mutation in this gene that likely resulted in a red hair and pale skin phenotype. Modern humans with red hair and pale skin have a similar, but not identical, mutation to this gene.

a. Do you think there was selective pressure on Neanderthals to have light skin? Why or why not? (LO 12-12)
b. Do you think having red hair and pale skin is an ancestral or a derived trait? Explain your answer. (LO 12-4)

Credit: Nature and Science/Alamy Stock Photo

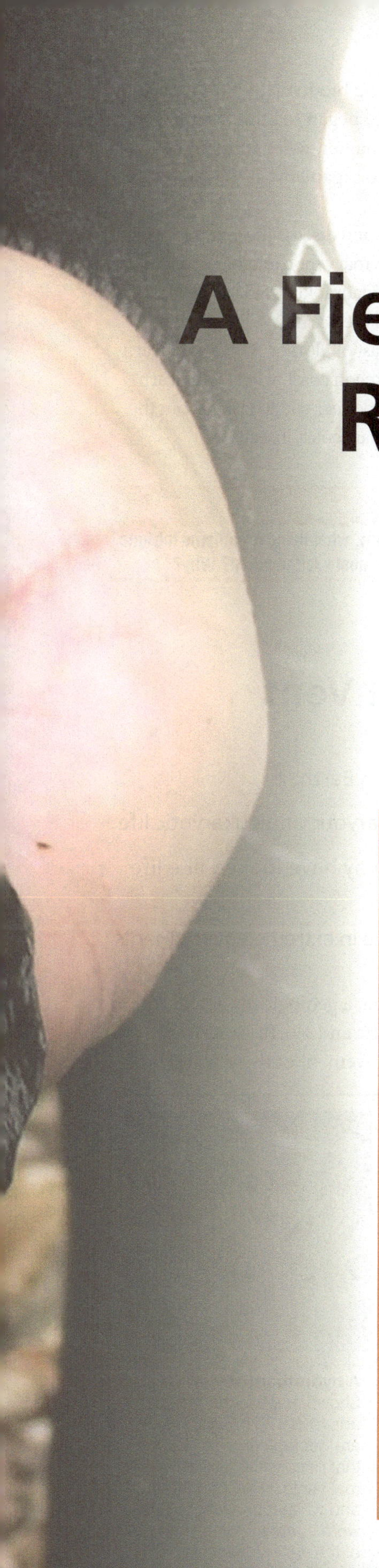

13

A Field Trip to View the Results of Evolution

Chapter Outline:

The Life at Your Fingertips

Imagine that you are tubing with friends down a river somewhere in Appalachia…and you drop your sunglasses. They sink down to the riverbed. You lean down to retrieve them, reach out…and find a salamander the length of your arm staring out at you from among the rocks. Whoa! What is this creature? If you're in Missouri, it could be the Ozark hellbender (see the chapter-opening photo), but more likely you've found the only other species of hellbender, the eastern hellbender (*Cryptobranchus alleganiensis*). It wasn't a matter of bad luck either—these gentle giants won't hurt you, and happenstance sightings like this in the wild are rare.

In fact, hellbenders are special animals. For one, they are *amphibians*—an organism that lives on land but requires water and/or a moist environment to survive. Hellbenders, along with other amphibians, reflect an important evolutionary step between fully aquatic and fully land-based animals. For instance, hellbenders have lungs and can also breathe through their skin. Hellbenders are also considered to be in *stasis* and are *living fossils* (Chapter 12), with relatively few changes since the Carboniferous period (Chapter 10). Hellbenders are also special in that they are an indicator species (Chapter 14), where their populations are at risk from both water pollution and climate change (Chapter 16).

And it turns out hellbenders aren't the only interesting species out there—either from a general interest or ecological perspective. For example, in this chapter, we'll learn about the evolutionary histories of animals ranging from gigantic deep-water sponges to gentle manatees and high-flying, fish-carrying ospreys. Where did the diversity of life we see on Earth today come from?

A person holding a critically endangered Ozark hellbender ***(Cryptobranchus alleganiensis bishopi)***. Hellbenders are North America's largest amphibian and their body style represents an important evolutionary step between aquatic and land animals.

Unit 4 focuses on tracing life's path: how evolution led to the amazing biodiversity that we see today. In Chapters 10 and 11, we looked at how evolution works and began to glimpse into the results of evolution in Chapter 12, with primate evolution. In this chapter, we're going to be taking a field trip across (and near) the United States to learn about the organisms at these special locations, as well as our shared evolutionary history.

On the way, we'll visit a hydrothermal vent in the Gulf of California and discuss the possible origins of life on Earth. We'll look at how conditions such as those at a hydrothermal vent could have created an environment conducive for the first life to evolve. Next, we'll visit Papahānaumokuākea Marine National Monument in Hawai'i to learn about the evolution of the first animals in the ocean. We'll expand our understanding of animals in a trip to the Indian River Lagoon off the coast of Florida, where we'll learn about *amphibians*, *reptiles*, and *mammals*. Our next destination is the Oregon coast, where we'll learn about the evolutionary history of plants and fungi. Finally, we'll visit a park near you to learn about the most diverse type of organisms on the planet—land arthropods, and in particular, insects.

Preview It!

When you visit a zoo, aquarium, or botanical garden (or go for a hike or a swim), what living organisms intrigue you the most? Why? Which stop on our field trip mentioned above are you the most excited about? Why?

13.1 The Beginnings of Life: A Visit to a Hydrothermal Vent

Did You Know?

Some scientists think that life could have originated on another planet, and these lifeforms were carried to Earth on meteorites. Alternatively, meteorites falling to Earth could also have delivered monomers like amino acids, which we know can form in space.

Learning Objectives

13-1 Describe the proposed origins of macromolecules on Earth.

13-2 Identify the proposed steps that led to the first prokaryotic and eukaryotic life.

13-3 Describe how the last common ancestor of all life may have looked like life seen at hydrothermal vents today.

13-4 Explain the adaptations that allow organisms to survive in extreme environments.

The first stop on our tour lies at the bottom of the Gulf of California, a geologically active region located at the northern end of the East Pacific Rise, where the Pacific and North American tectonic plates (Chapter 10) are sliding against one another. The movement of these plates

The first stop in our field trip is the bottom of the ocean, to a hydrothermal vent along the East Pacific Rise in the Gulf of California. Here, we'll learn about the organisms that live there and the evolution of the first lifeforms.

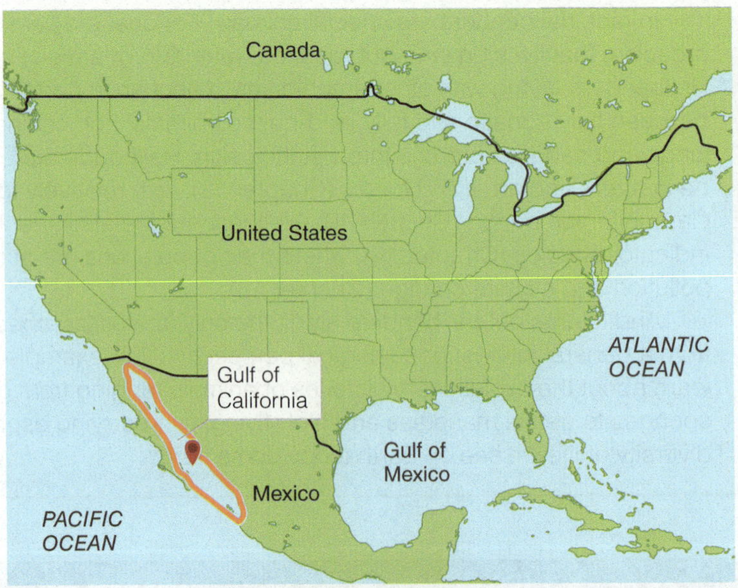

Credit: ROV SuBastian/Schmidt Ocean Institute

A hydrothermal vent. Shown at left in the photo above are chimneys, which are geological formations that occur near hydrothermal vents. Note also the red and white tube worms. Tube worms are only one example of the variety of life found athydrothermal vents.

resulted in the formation of four regions within the Gulf of California with hydrothermal vents—and the life at each site can vary significantly. For example, in two vents 75 kilometers apart, only 7 species (of 61) are the same at both vents.

Hydrothermal vents are of interest to scientists researching the origins of life and how life arose from inorganic chemicals. A **hydrothermal vent** is like an underwater geyser, a place where mineral-rich water heated by geothermal energy streams out through a rocky opening in the seafloor, reaching temperatures of more than 350° C (660° F).

While there are a few hypotheses for where life began, hydrothermal vents are notable candidates. The water from hydrothermal vents contains inorganic compounds that were likely the building blocks for the first macromolecules. Additionally, hydrothermal vents supply abundant heat to the immediate environment, which could provide the energy needed to synthesize the first macromolecules (Chapter 3) in Earth's early seas.

From Inorganic Compounds to Macromolecules and Cells

Earth formed about 4.6 billion years ago (4,600 million years ago, or mya; Figure 13.1). At first, the planet's surface was molten rock, and water existed only as a vapor. Over the next billion years or so, the planet cooled, and by 4.3 billion years ago, liquid water had begun to pool on its surface. From this lifeless beginning, the building blocks for the first life on Earth formed—the first step on the journey to the first cell.

Recall from Chapter 2 that liquid water is considered essential to life because all the metabolic reactions that sustain life take place in water-based solutions. Recall also, from Chapter 3, that all organisms consist of macromolecules: carbohydrates, proteins, nucleic acids, and lipids. Except for lipids, macromolecules have monomer forms that act as the building blocks for more complicated molecules. Therefore, life on Earth likely originated in water—another line of evidence pointing to hydrothermal vents! But where did the original building blocks come from?

Chemical evolution is a theory that describes the process of forming organic molecules from inorganic molecules (Figure 13.2 ①) as theorized to have occurred at hydrothermal vents, leading to the formation of the first macromolecules, and eventually, the first cells. These early inorganic molecules—including carbon dioxide (CO_2), ammonia (NH_3), methane (CH_4), hydrogen gas (H_2), and water (H_2O)—then self-assembled to form organic monomers, such as lipids, monosaccharides, nucleotides, and amino acids ②. The organic monomers then joined to form polymers. These polymers served similar functions as macromolecules today, such as forming vesicles, taking part in metabolism, and becoming the first genome. Protocells formed from polymers and consisted of a fatty acid membrane, an RNA genome, and lysosomes ③.

Figure 13.1 A timeline of major evolutionary events in Earth's history

Era	Period	Major Evolutionary Events
Formation of Earth ~4,600 million years ago (mya)		
Hadean 4,600 to 4,000 mya		First ocean and atmosphere
Archean 4,000 to 2,500 mya		First bacteria and archaea
Proterozoic 2,500 to 541 mya		First multicellular life, fungi, and animals
Paleozoic 541 to 252 mya	Cambrian 541 to 485 mya	Rapid increase in animal diversity in oceans (Cambrian Explosion)
	Ordovician 485 to 443.8 mya	First land plants, fishes, and coral-forming organisms
	Silurian 443.8 to 419 mya	First land fungi, vascular plants, bony fishes, and terrestrial arthropods
	Devonian 419 to 359 mya	First land tetrapods, forests, and seed plants
	Carboniferous 359 to 299 mya	Formation of coal forests
	Permian 299 to 252 mya	First cycads and ginkgos
Mesozoic 252 to 66 mya	Triassic 252 to 201.4 mya	First turtles, dinosaurs, and mammals
	Jurassic 201.4 to 145 mya	First birds
	Cretaceous 145 to 66 mya	First flowering plants and sharks
Cenozoic 66 mya to present	Paleogene 66 to 23 mya	First primates
	Neogene 23 to 2.58 mya	Adaptive radiations of mammals, insects, and birds
	Quaternary 2.58 mya to present	First hominins

Shown here is a simplified geologic timeline noting key evolutionary events in the era or period identified. You may wish to consult this figure as you move through the chapter, to keep an eye on where a group referenced in a given section fits into the context of evolutionary time. For more details on major geologic and biological events, refer to Figure 10.6.

Figure 13.2 A theoretical sequence leading to the first cells at hydrothermal vents

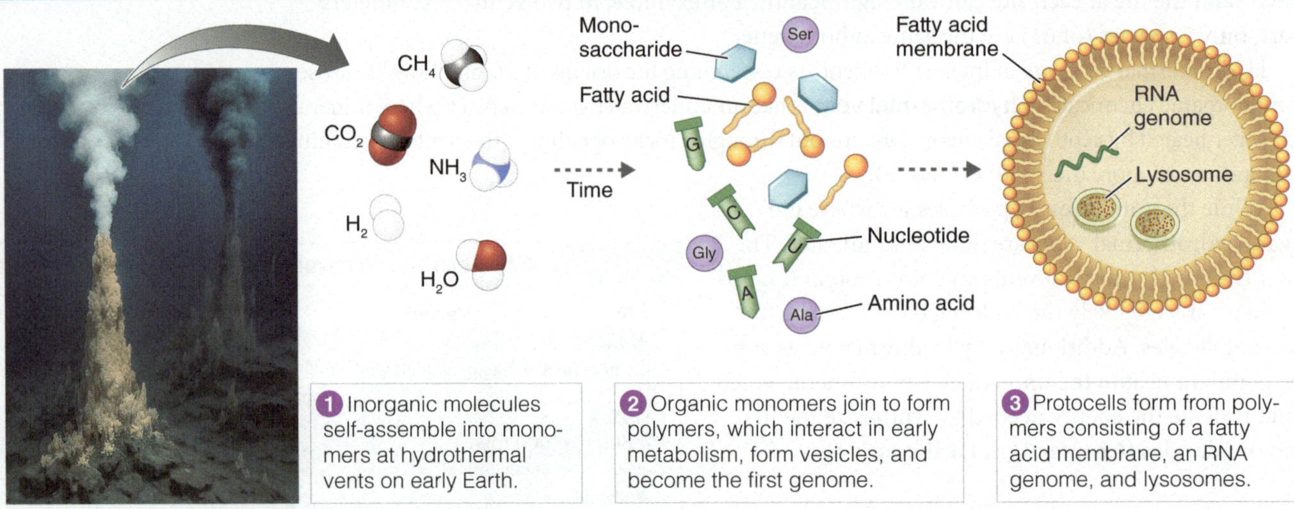

Note that various steps along this pathway have experiments and/or simulations to support their feasibility.

How could this process have happened at hydrothermal vents? The Gulf of California is part of the East Pacific Rise. Along the East Pacific Rise, there are hydrothermal vents that produce either black or white "smoke," called black smokers (Figure 13.3) and white smokers, respectively. The "smoke" is the result of mineral particles in the hot vent fluid contacting the surrounding near-freezing seawater. White smokers get their color because the compounds they contain are lighter in color. Black smokers tend to be hotter than white smokers, with temperatures exceeding 350° C (660° F). White smokers tend to be 50° to 90° C (122° to 194° F). Given the lower temperatures, scientists think that white smokers are the most likely candidate for where the first cells evolved.

? Did You Know?

Modern organisms require iron–sulfur cofactors (Chapter 4), which may be a legacy of life's rocky beginnings.

In addition to minerals, the superheated fluid escaping from white and black smokers contains different organic molecules, like hydrogen gas, methane, and carbon dioxide. Therefore, hydrothermal vents have all the necessary ingredients for organic molecules to form: energy (in the form of heat), water, and chemical compounds.

In addition, the rocks surrounding hydrothermal vents are rich in iron sulfide minerals, and they are porous, with many tiny chambers about the size of cells. Metabolism may have begun when the iron sulfide in the rocks donated electrons to dissolved CO_2, setting in motion reactions that led to formation of larger organic compounds.

Researchers at the National Aeronautics and Space Administration (NASA) modeled conditions at a white smoker hydrothermal vent and were able to create organic molecules from the inorganic molecules found there, thus supporting the hypothesis that life could have originated at hydrothermal vents.

So, it's possible that chemical reactions at hydrothermal vents could have led to the first organic molecules and eventually monomers, but how did we get the first macromolecules? Modern cells selectively take up small organic molecules, concentrate them, and assemble them into larger organic polymers. Before there were cells, a nonbiological process that concentrated organic subunits would have increased the likelihood of polymer formation.

Let's consider nucleic acids and why many scientists think that RNA was the first macromolecule. As we learned in Chapter 3, nucleic acids are information-containing macromolecules. According to the **RNA world hypothesis**, there was a period when RNA served a dual role, functioning both as genetic material (like DNA) and as an enzyme (like a protein).

Figure 13.3 A black smoker hydrothermal vent

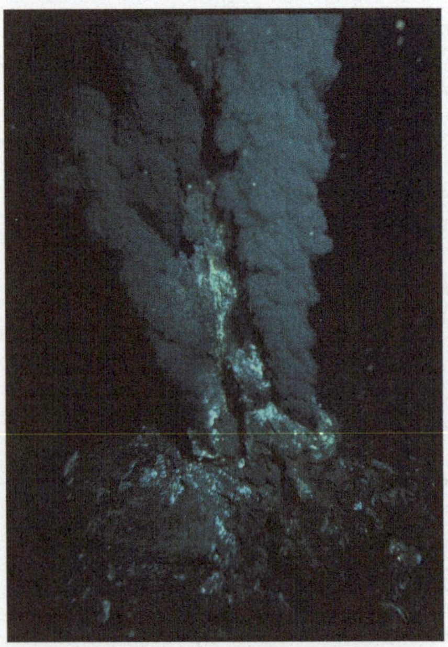

Note the black "smoke" escaping from the chimney-like rock structure.

Process of Science

The Evolving Theory of Chemical Evolution

Chemical evolution was proposed independently by two different scientists: Alexander I. Oparin in 1924 and J. B. S. Haldane in 1929. The Oparin-Haldane proposal wasn't taken seriously until 1953, with the Miller-Urey experiments.

Stanley Miller and Harold Urey proposed that lightning-powered reactions among inorganic gases in Earth's early atmosphere could have produced the organic building blocks of life. To test this hypothesis, they built a glass reaction apparatus containing a flask with water (representing the ocean; Figure 13.4 ①) and filled it with a mixture designed to mimic the early atmosphere: gaseous ammonia (NH_3), methane gas (CH_4), hydrogen gas (H_2), and water vapor (H_2O) ②. As this mixture circulated through the apparatus, sparks from electrodes simulated lightning, thus providing the energy source needed to spur the chemical reactions. Water containing the newly formed simple organic compounds then condensed ③. When the water was sampled, Miller and Urey identified organic compounds like the amino acid glycine, which is found in living organisms today ④.

We now know that the mixture of gases in this first experiment did not accurately represent the atmosphere of the early Earth. Proponents of the hypothesis that life evolved at hydrothermal vents at the bottom of the ocean also question whether life could have evolved in the high-UV conditions present at the surface of early Earth. However, Miller and Urey's work remains notable because it was the first experimental test of a hypothesis about life's origin. Later experimental variations using the same sort of apparatus and mixtures of gases that better simulate Earth's early atmosphere also produced amino acids.

Work in the 2020s at NASA found that organic molecules could be produced when replicating conditions on hydrothermal vents. In addition, genetic analysis suggests that all species on the planet, regardless of domain, have the same set of genes from a common ancestor that lived in a hot, oxygen-free environment and was able to undergo chemosynthesis (Chapter 4). These conditions fit the hypothesis that life evolved at a hydrothermal vent, but aren't consistent with the findings of the Miller-Urey experiment. Taken together, these different experiments show how our understanding of chemical evolution and the origins of life has changed over the last century.

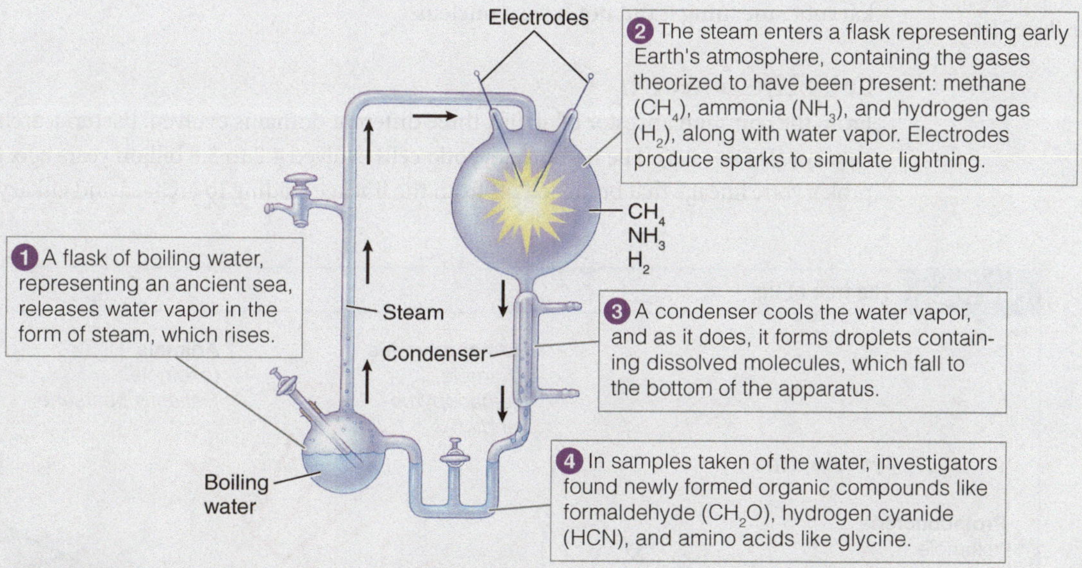

Figure 13.4 Stanley Miller's experimental apparatus

This apparatus was used to test whether lightening-fueled reactions could have formed organic monomers in Earth's early atmosphere. Note that oxygen was not included in the gas mixture—oxygen did not exist on Earth initially.

1. Science is often described as iterative, meaning that we continue circling back to the same topics and refining our understanding rather than making discoveries in a straight line. How does the Miller-Urey experiment show the importance of revisiting experiments of the past?
2. Some scientists believe that life began on a different planet and then arrived on Earth on meteorites. Why is it so hard to either rule out or support this theory?

Scientists think the Lost City Hydrothermal Field in the Atlantic Ocean is the most likely place for the evolution of the first macromolecules. There are similar hydrothermal features in the Gulf of California.

Big Ideas in Biology

Structure and Function

If early genetic systems were RNA-based, why do all cells now have a genome of DNA? It's the difference in stability between the two nucleic acids. DNA's double helix and lack of an additional hydroxyl group make it more stable than RNA.

Therefore, it's likely that RNA predated both protein and DNA. RNA is a simpler molecule than DNA and thus is more likely to have formed by nonbiological processes.

In addition, the fact that **ribozymes**, or catalytic RNAs, can function as enzymes is another line of evidence supporting the RNA world hypothesis. For example, some ribozymes cut introns (Chapter 7) out of newly formed RNAs, and the rRNA in ribosomes catalyzes the formation of peptide bonds (Chapter 7).

The self-replicating molecules that we just described would have floated away from one another unless something enclosed them. In modern cells, the outer membrane serves this function. It is possible that vesicle-like structures may have functioned in a similar manner before there were cells. Such vesicles form spontaneously when amphiphilic lipids mix with water. A membrane-enclosed collection of interacting molecules that can take up material and replicate is called a **protocell** (see Figure 13.2). Protocells are hypothesized to be the ancestors of cellular life.

Membranes of early protocells probably consisted of a variety of fatty acids, rather than the phospholipids that make up the bulk of modern cell membranes (Chapter 5). Like phospholipids, fatty acids are amphiphilic and self-assemble into structures when mixed with water. Since fatty acids are structurally simpler than phospholipids, we can infer that they would have been more abundant on early Earth. Later on in this section, we'll discuss archaea found at hydrothermal vents that have different, more primitive cell membrane compositions.

Ultimately, these processes likely led to the first prokaryotic cells with DNA genomes, where DNA is more stable than RNA. Similarities among modern genomes indicate that all living species descended from a common single-celled ancestor, a species that lived perhaps as early as 4 billion years ago. There was little or no oxygen on the early Earth, so that ancestral species must have been *anaerobic*, capable of living without oxygen. Given fossil evidence and what we know about relationships among modern species, this ancestor was most likely prokaryotic, meaning it did not have a nucleus.

Early Life on Earth

From the common ancestor of all life, three different domains evolved: bacteria, archaea, and eukarya (Figure 13.5). The first **prokaryotic** cells evolved 4.2 to 3.8 billion years ago. **Bacteria** are a prokaryotic lineage that branched off from the lineage leading to archaea and eukaryotes early in

Figure 13.5 **The tree of life**

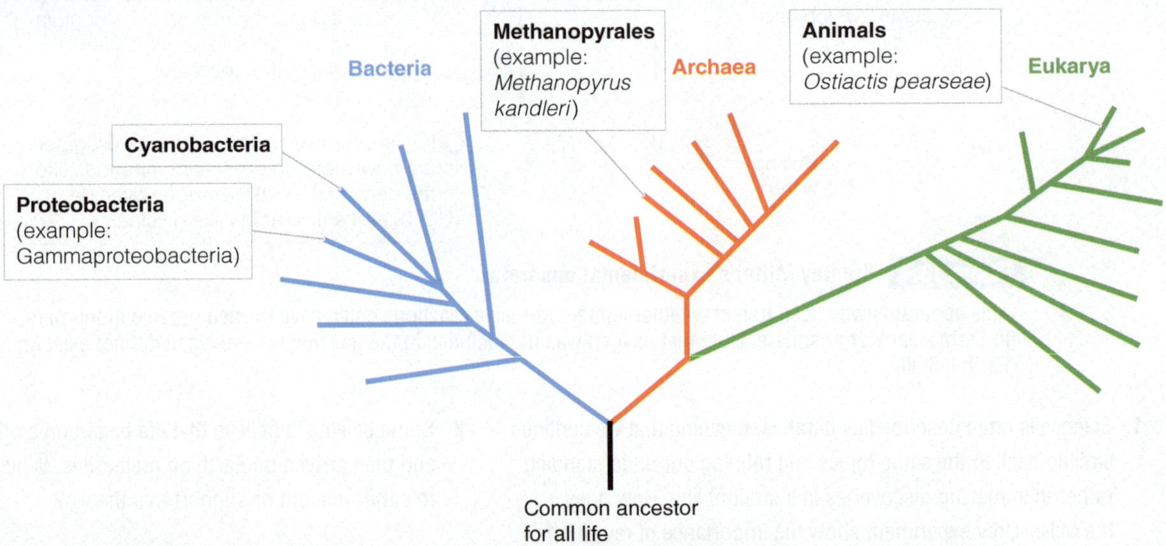

All life evolved from a common ancestor and can be divided up into the three domains based upon cell type: bacteria, archaea, or eukarya. Each branch on the tree indicates a group within its given domain, where some species that we'll encounter in Section 13.1 are noted with white boxes.

the history of life. Notably, **archaea**, which are also prokaryotic, are more closely related to eukaryotes than to bacteria. Archaea differ from bacteria in many respects. For example, their cell wall does not contain peptidoglycan, and their cell membrane contains lipids not found in bacteria.

No eukaryotes can undergo chemosynthesis, so bacteria and archaea are the main *producers* (organisms that capture energy and use it to make their own food from inorganic materials in the environment; Chapter 15) in dark places, such as near hydrothermal vents on the ocean floor.

Bacteria that can obtain energy from chemically processing inorganic materials are called **chemoautotrophs**. As we learned in Chapter 4, the process of making energy from inorganic materials while using a non-solar energy source is **chemosynthesis**. For example, bacteria such as Gammaproteobacteria live on hydrothermal vents in the Gulf of California and are the primary producers, producing organic molecules from inorganic molecules using chemosynthesis (Chapter 4).

Note that bacteria can also be **photoautotrophs**, meaning they capture light energy and carry out photosynthesis. Photosynthesis evolved in many bacterial lineages, so photoautotrophic bacteria like *cyanobacteria* are important producers in well-lit aquatic environments, such as in coral reefs, as we'll see in Section 13.2. Most bacteria carry out photosynthesis by a pathway which does not produce oxygen. The oxygen-producing pathway of photosynthesis evolved only once. As we'll discuss later in this section, chloroplasts are descended from cyanobacteria. Thus, we have cyanobacteria and their chloroplast descendants to thank for nearly all the oxygen in Earth's atmosphere.

So, what kind of microbial life lives at hydrothermal vents? Within the walls of the black smoker chimneys are archaea such as *Methanopyrus kandleri*, which is notable because it is a *methanogen*, meaning it converts hydrogen gas into methane in a process called *methanogenesis* (Chapter 16). The methane then serves as a source of energy for the microbial community on the vent.

When considering the origins of life, the cell membrane of *M. kandleri* contains not phospholipids, but primitive terpenoid lipids (similar to cholesterol)—thought to be one of the most primitive lipid compounds. When we consider eukaryotic cells, for example, cholesterol is a component of the cell membrane, rather than the key structural feature. Therefore, changing lipid composition is one of the possible changes that cells underwent over evolutionary time.

M. kandleri is an **extreme thermophile**, meaning it is an organism that thrives at high temperatures. *M. kandleri* has some specific adaptations that allow it to live on a hydrothermal vent. For example, *M. kandleri* has a special chemical that prevents protein denaturation (Chapter 3) at high temperatures. Its cytosol (the fluid part inside the membrane; Chapter 5) also has a high salt concentration, which helps keep its enzymes functioning at high temperatures.

Note that there are other kinds of organisms that live in extreme environments. For example, an **acidophile** is any organism that can live in highly acidic conditions. Organisms that can live in highly salty environments are called **extreme halophiles**.

? Did You Know?

We share the planet with an estimated 5 nonillion bacterial cells. Bacteria live just about everywhere and play essential roles in all ecosystems as well in your gut, where your gut microbiome (Chapter 5) supports your health.

? Did You Know?

About a third of the human population has significant numbers of methanogens in their intestine, so their flatulence (farts) contains methane, explaining the foul odor.

? Did You Know?

Tardigrades (water bears) are a type of animal that is incredibly resistant to extreme environments—even surviving in space for extended periods of time!

! Point of Confusion: Although we often associate archaea with being found in extreme environments, bacteria and eukaryotes can live in extreme environments as well. As we'll see in this section, all three domains of life can be found in the extreme environment at a hydrothermal vent.

The first **eukaryotic** fossils date to 1.5 billion years ago. The defining feature of eukaryotes is their nucleus. The evolution of a nucleus probably began in the archaeal ancestor of eukaryotes when portions of the cell membrane folded inward (Figure 13.6 ①). Over time, some of this now internal membrane came to surround the DNA, forming the first version of the nuclear membrane. Other portions of internalized membrane gave rise to additional components of the endomembrane system (Figure 13.6 ②; Chapter 5).

When considering their evolutionary history, eukaryotes are similar to archaea in terms of how they store their DNA (wrapped around histone proteins; Chapter 6) and in some aspects of how they transcribe, replicate, and repair their DNA. On the other hand, many eukaryotic genes, especially those involved in energy metabolism, are more like bacterial genes than archaeal ones. The mix of archaeal and bacterial traits in eukaryotes suggests a common ancestor.

Figure 13.6 **The evolution of eukaryotic organelles**

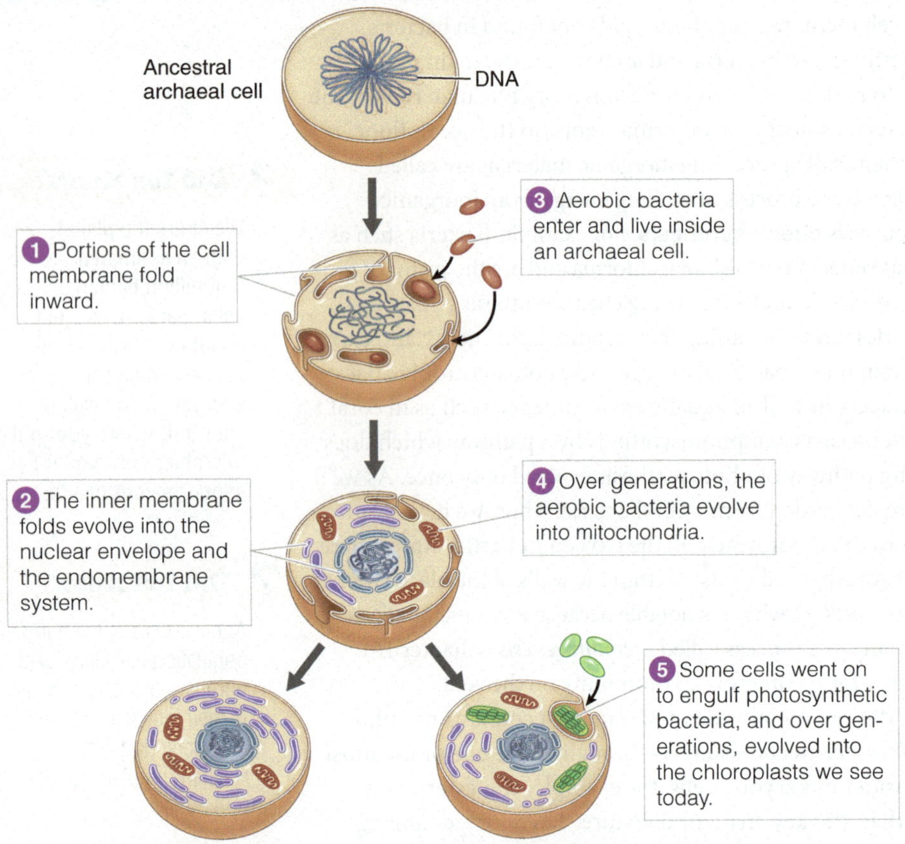

The nuclear envelope and endomembrane components are derived from infoldings of the outer cell membrane. Mitochondria and chloroplasts evolved from bacteria that lived inside a host cell.

The **endosymbiosis hypothesis** states that mitochondria and chloroplasts evolved from bacteria. **Endosymbiosis** refers to a relationship in which one organism lives inside another. Consider that mitochondria and chloroplasts both resemble bacteria. For example, like bacteria, both of these organelles have their own circular piece of DNA. They also have their own protein-building machinery (Chapter 5). The fact that both mitochondria and chloroplasts are surrounded by two membranes also supports the endosymbiosis hypothesis. The innermost membrane in both organelles is similar in structure to a bacterial membrane.

Nearly all eukaryotic lineages have mitochondria or mitochondria-like organelles, but not all have chloroplasts. Thus, biologists think that the two types of organelles were acquired independently in the sequence illustrated in Figure 13.6, with an ancient protocell engulfing

Biology for Life

Why Does It Matter if Mitochondria Have Their Own DNA?

In Chapter 5, we learned that a cell's DNA is stored in the nucleus of the cell. While true, there is a second, independent genome within mitochondria. Not only is this evidence in support of the endosymbiosis hypothesis, it's also important both to understand human evolutionary history and for medical applications today.

Except for a few exceptions that are beyond the scope of this text, mitochondrial DNA is largely inherited from the mother. Therefore, we can trace the ancestral female lines back to one individual who is the matrilineal (the female's) most common recent ancestor. Different branches from the common ancestor describe different mitochondrial DNA signatures. The age of various lineages can therefore be calculated using a molecular clock (Chapter 10), since mutations in mitochondrial DNA occur at a predicted rate. Biologists estimate that the most recent common female ancestor for *Homo sapiens* lived around 200,000 years ago in Africa.

So why does this matter today? Errors or mutations in the mitochondrial DNA can lead to the development of a variety of diseases in humans—many of which are fatal early on in life. For example, Leigh syndrome is a neurological disorder linked to mutations in the mitochondrial genome that typically results in death before age 3. For mothers who carry mitochondria with the defect, there were traditionally very few options—one could either use a donor egg and IVF (Chapter 9) or adopt.

However, in the 2010s, a new technology, *mitochondrial replacement therapy*, allowed people with mitochondrial mutations to have their own biological children. Mitochondrial replacement therapy results in a child who essentially has three biological parents. First, the egg cells are manipulated so that both

? Did You Know?

Eggs have around 100,000 to 600,000 mitochondria, whereas sperm only have about 100, which have no intact mitochondrial DNA!

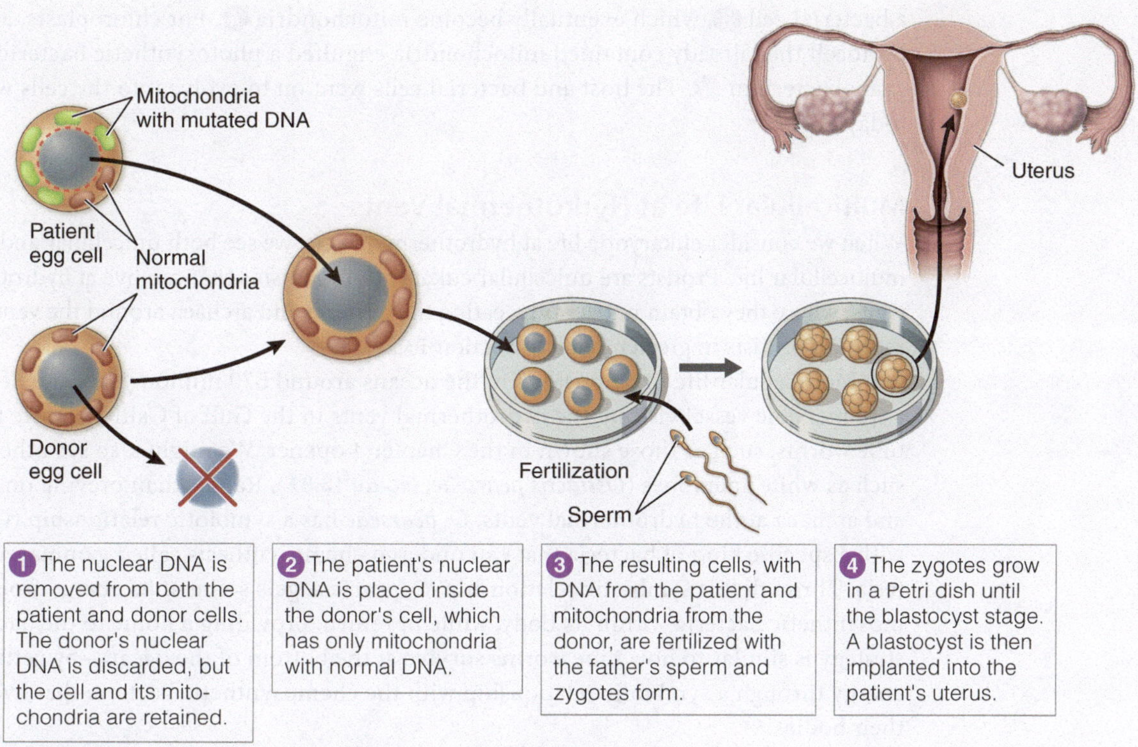

Figure 13.7 An overview of mitochondrial replacement therapy

Mitochondrial replacement therapy involves using the genetic material of three different parents to generate an embryo without the genes that could lead to life-threatening mitochondrial diseases. Note that for simplicity, only one cell is shown in Steps 1 and 2.

the patient (whose cells carry mutated mitochondrial DNA) and donor cells have their nuclei removed, where the donor's nucleus is discarded but the cell and its cytoplasm, along with its normal mitochondrial DNA (mtDNA), are retained (Figure 13.7 1). Next, the patient's nuclear DNA is placed inside the donor's egg cell 2. The resulting cell is then fertilized by a sperm cell containing the father's nuclear DNA. After a few days of growth, a blastocyst forms 3 and is implanted into the patient's uterus 4. In this way, the child has three parents: a donor mother who contributed normal mitochondria, the patient who contributed their nuclear DNA, and the father of the child, who also contributed their nuclear DNA.

The first person with DNA from three parents was born in 2016. The patient (mother 1) was a carrier of Leigh syndrome (some of her mitochondria had the mutation, but not all, which is why she was unaffected). While the child, a baby boy, was healthy at 7 months, the parents chose to not proceed with long-term follow-up with the research team. As of the writing of this text, the boy's current medical status is not publicly known.

1. Mitochondrial replacement therapy raises a number of ethical issues. With a classmate, discuss your thoughts on the following issues.
   a. Recent research suggests that there's a 20% chance the procedure will fail and the offspring would develop the disease that the researchers were trying to prevent. If you were a carrier of a mitochondrial disease, would you risk this treatment or choose a different option, such as adoption or using a donor egg? Why or why not?
   b. Mitochondrial replacement therapy is just one example of new treatments that provide life-saving benefits to a small number of people while costing incredible amounts of money; estimates are at around $50,000 per treatment. For biotechnology companies that create these therapies, is it ethical to provide a treatment that is accessible to only the few that can afford it? Explain your answer.
   c. New, complex treatments like mitochondrial replacement therapy rely upon decades of research advancements, often funded by governments using taxpayer dollars. Findings from government-funded research must be shared with the public because the taxpayers funded it. What do you think about the family of the first three-parent baby opting out of further study? What would you do if it were your family?
   d. Suppose the person who contributed the mitochondria to the child wanted to claim parental rights over the child. Should they have any rights to the child? Should they be required to support the child? Why or why not?

a bacterial cell 3, which eventually become mitochondria 4. For chloroplasts, an ancient protocell that already contained mitochondria engulfed a photosynthetic bacterium, like a cyanobacterium 5. The host and bacterial cells went on to evolve into the cells we know today.

Multicellular Life at Hydrothermal Vents

When we consider eukaryotic life at hydrothermal vents, we see both unicellular and multicellular life. Protists are unicellular eukaryotic organisms that can live at hydrothermal vents, where they obtain nutrients by eating the bacteria and archaea around the vent. We'll return to protists in greater detail in Section 13.2.

Multicellular life first appeared in the oceans around 570 million years ago. If we were in a submersible vessel visiting the hydrothermal vents in the Gulf of California, we might see tube worms, such as those shown in the Chapter 4 opener. We might also see other animals, such as white anemones (*Ostiactis pearseae*; **Figure 13.8A**). Rather than preying on bacteria and archaea at the hydrothermal vents, *O. pearseae* has a symbiotic relationship (Chapter 15) with a specific kind of bacteria that can undergo chemosynthesis called gammaproteobacteria. Through this symbiotic relationship, *O. pearseae* gets some of its energy from the chemosynthetic bacteria within its body, while in return, providing a home to the bacteria. This strategy is similar to how tube worms survive at the bottom of the ocean—by getting their energy through a symbiotic relationship with the chemosynthetic bacteria that live within their bodies.

In contrast, other animals, such as the polychaete worm *Peinaleopolynoe orphanage* (**Figure 13.8B**), or scale worms, are predators that eat bacteria on hydrothermal vents. Like other animals living around hydrothermal vents, *P. orphanage* do not have eyes. They do, however, have very elaborate branching structures that serve a similar purpose as lungs.

Researchers think that the complex branching allows the scale worms to survive in the low-oxygen and high-pressure environment at the bottom of the ocean. Although appealing to humans, the significance of the scale worm's glittery shells (which change colors as the worm ages!) is unknown. We'll return to anemones and polychaete worms in greater detail in Section 13.2.

One subspecies of scale worm is nicknamed the "Elvis worm" because the iridescent shell plates reminded the researchers of Elvis's sequined jumpsuits.

Figure 13.8 **Animals that live at hydrothermal vents in the Gulf of California**

Credit: Goffredi, S.K., Motooka, C., Fike, D.A. et al. (2021). Mixotrophic chemosynthesis in a deep-sea anemone from hydrothermal vents in the Pescadero Basin, Gulf of California. BMC Biol 19, 8. https://doi.org/10.1186/s12915-020-00921-1

Credit: ROV/Subastian Schmidt Ocean Institute

A. White anemones (*Ostiactis pearseae*) obtain some of their energy from chemosynthetic bacteria that live within their bodies.

B. A scale worm, *Peinaleopolynoe orphanage*, with its iridescent shell. The pink mats surrounding it are chemosynthetic bacteria.

13.1 Summary

- Various hypotheses explain the formation of macromolecules on Earth 4.3 billion years ago. During this time, Earth cooled and liquid water started to form, allowing chemical evolution—the formation of organic molecules from inorganic ones—to occur. The energy to drive these reactions may have come from hydrothermal vents.
- The endosymbiosis hypothesis states that an ancient archaeal cell formed a symbiotic relationship with an aerobic bacterium that eventually became the mitochondria. Later, a similar process occurred with photosynthetic bacteria that would eventually form chloroplasts.
- Archaea at hydrothermal vents, such as *M. kandleri*, are thought to be similar to the earliest cells, having a membrane composed of primitive lipids.
- Living in hot water requires adaptations such as unique membrane lipid composition, chemicals that prevent protein denaturation, or elaborate branching structures similar to lungs.

13.2 The First Animals: A Visit to a Coral Reef

Learning Objectives

13-5 Explain the oxygenation of Earth's atmosphere and its biological consequences.

13-6 Describe how colonial protists are similar to sponges.

13-7 Relate animal body structures to evolutionary history.

13-8 Describe the adaptations of protists, mollusks, marine arthropods, and echinoderms found on a coral reef.

Papahānaumokuākea Marine National Monument is not only home to extensive biodiversity, but is also an area of significant importance to native Hawaiians. There are many cultural sites on the islands contained within the sanctuary.

In the last section, we visited hydrothermal vents in the Gulf of California to learn about how life may have first evolved. We also learned about animals, like scale worms. In this section, we'll expand upon our coverage of animals by visiting the coral reefs at Papahānaumokuākea Marine National Monument, near Hawai'i—the largest fully protected marine area within the United States and one of the largest marine sanctuaries in the world.

The next stop on our field trip is to visit Papahānaumokuākea Marine National Monument to learn about the evolution of the first animals, and to explore invertebrate animals on the coral reefs within the monument.

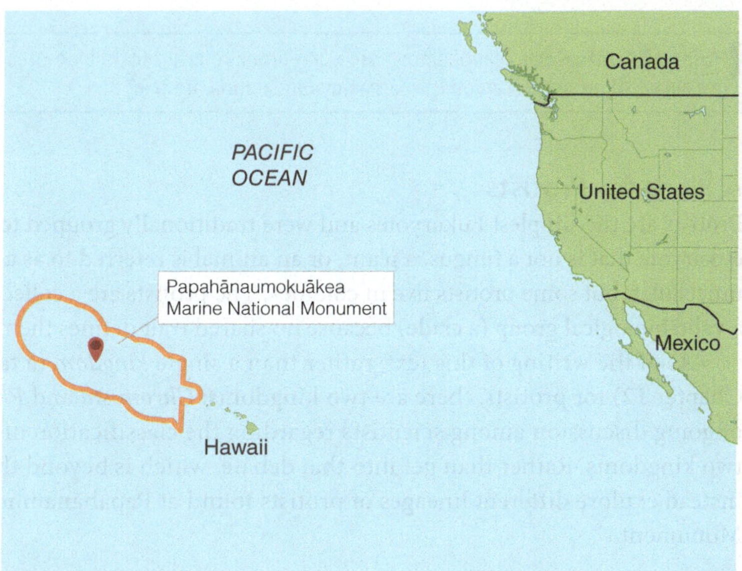

Credit: NOAA

A coral reef. Coral reefs are underwater ecosystems found in oceans that are home to a wide variety of lifeforms, including the first animals: sponges. Coral reefs are characterized by the presence of invertebrate corals, which build reefs.

Papahānaumokuākea is pronounced *Pa-pa-hah-nou-mo-koo-ah-keh-ah.* The name was chosen to represent the traditional story of the creation of the Hawaiian Islands. *Papa* means "Earth Mother," *hānau* means "birth," a *moku* is a small island, and *ākea* means "wide." Taken together, the name represents the birth of islands from the Earth Mother.

While coral reefs account for less than 1% of the total sea floor area, 25% of all marine life is found at coral reefs. Coral reefs are second only to rainforests in terms of biodiversity. Note that coral reefs can appear in both shallow and deep water—we'll look at life from both kinds of coral reefs in this section.

In this section, we'll learn about the evolution of the first animals. We'll begin with cyanobacteria and how they contribute to building coral reefs, in addition to producing oxygen. We'll then discuss protists, the ancestors of animals, before moving on to sponges, the first animals. While at the coral reef, we'll look at each of the major lineages of *invertebrate* animals (animals without a backbone). Note that while *chordates* such as fish and sharks are found at coral reefs, we won't discuss chordate evolutionary lineages until Section 13.3.

The Great Oxygenation Event

Bacteria and archaea arose when Earth had little or no oxygen, so the earliest members of these lineages were anaerobic. Then, about 2.7 billion years ago, a lineage of bacteria called the **cyanobacteria** (Section 13.1) began to carry out photosynthesis, which produces oxygen as a by-product (Chapter 4). As cyanobacteria became increasingly abundant, oxygen began to accumulate in the air and water, leading to the **great oxygenation event**.

During the great oxygenation event, some oxygen atoms combined as ozone gas (O_3) and formed the ozone layer, which reduced the level of UV radiation (a mutagen; Chapter 7) at Earth's surface, making life on land possible, which we'll discuss more in Section 13.3. Water screens out UV light, so aquatic life was possible even without the ozone layer. However, before the ozone layer formed, UV radiation would have destroyed any life that ventured onto land.

The increase in oxygen created a new selective pressure, putting species that could not tolerate oxygen at a disadvantage. Such species were harmed by oxygen because reactions with oxygen damaged their biological molecules, and unlike aerobic cells, they had no enzymes to repair such damage. As time went by, these species became restricted to the limited low-oxygen environments that remained, such as in aquatic sediments.

Cyanobacteria can be found in the oceans today, occupying many different ecological *niches* (the unique ecological role an organism plays in its environment; Chapter 15). They can be found inside corals or sponges, on the ocean floor, or in the water column. In these various roles, they can help build (or break down) a reef, serve as primary producers for the coral reef (Chapter 4 and Chapter 15), and provide a source of nitrogen (Chapter 15). Warming water temperatures, however, promote overgrowth of cyanobacteria. The resulting blooms can produce dangerous toxins that can harm both corals and humans.

! Point of Confusion: Cyanobacteria are sometimes referred to as blue-green algae. However, they are not algae. Algae are eukaryotic, whereas cyanobacteria are bacteria.

A Place for Protists

Protists are the simplest Eukaryotes and were traditionally grouped together by exclusion. A eukaryote that is not a fungus, a plant, or an animal is referred to as a **protist**. Most protists are unicellular, but some protists live in colonies. The protists are a collection of lineages rather than a valid biological group (a clade) because no shared trait defines them.

As of the writing of this text, rather than a single *kingdom* (a taxonomic rank; see Chapter 12) for protists, there are two kingdoms: *Chromista* and *Protozoa*. There is ongoing discussion among scientists regarding the classification of protists within these two kingdoms. Rather than get into that debate, which is beyond the scope of this text, let's instead explore different lineages of protists found at Papahānaumokuākea Marine National Monument.

? Did You Know?

Brown algae can range in size from microscopic filaments to the largest protists—giant kelp (*Macrocystis pyrifera*) that stand 100 feet (30 meters) tall.

Algae. The common term for lineages of photosynthetic protists that live in lakes and seas is **algae.** Here we'll discuss three lineages: brown algae, red algae, and green algae.

Figure 13.9 **A gallery of algae**

A B C

A. *Padina* is a common type of brown algae found throughout the world in both shallow and deep reef habitats, plus the intertidal zone.

B. A newly discovered species of red algae.

C. A recently discovered species of deep-water green algae.

Credit: Papahānaumokuākea Marine National Monument. (n.d.). https://www.papahanaumokuakea.gov/monument_features/creature_padina.html; Greg McFall/NOAA Office of National Marine Sanctuaries, 2012; Credit: Daniel Wagner/NOAA

The brown algae are multicellular photosynthetic protists that live mainly in temperate and cool seas. For example, at Papahānaumokuākea Marine National Monument, one example of a brown algae is brown ear seaweed (*Padina*) (Figure 13.9A).

The red and green algae share a protist ancestor that had chloroplasts. After these two algal lineages diverged, land plants evolved from one lineage of green algae. We'll return to plant evolution in Section 13.4.

Some red algae are single cells, but most are multicellular (Figure 13.9B). Coralline algae (red algae with cell walls hardened by calcium carbonate) are a component of tropical coral reefs. Red algae are tinted red to black by accessory pigments called phycobilins (Chapter 4). These pigments absorb the blue-green light that penetrates deep into water. Phycobilins allow red algae to carry out photosynthesis at greater depths than other algae.

In 2020, scientists discovered a new species of red algae, called *Chondria tumulosa*, in Papahānaumokuākea Marine National Monument. This algae is overgrowing and smothering the corals there, killing them (Figure 13.10). As of the writing of this text, it's not clear if the algae is invasive (Chapter 15) or if it's always been at the monument, but for some reason, only recently started to outcompete other algae.

Green algae includes single-celled, colonial, and multicellular species. Most live in freshwater, but some are marine, and some grow on soil, trees, or other damp surfaces. A few single-celled species of green algae partner with a fungus to form a lichen, which we will discuss in Section 13.4. Within Papahānaumokuākea Marine National Monument, there are also deep-water algae. The specimen of green algae shown in Figure 13.9C was found in 188 feet of water!

A compound extracted from the cell walls of brown algae is used to produce algins, which we use as thickeners, emulsifiers, and suspension agents. Algins are used to manufacture ice cream, pudding, jelly beans, toothpaste, cosmetics, and other products.

Plankton. The community of tiny organisms that swim or drift through Earth's waters are called **plankton**. Plankton can be photosynthetic (phytoplankton) or consumers (zooplankton). In the monument, as with elsewhere in the ocean, plankton form the foundation of the coral reef food web (Chapter 15), providing a food source in the coral reef.

One notable type of phytoplankton are diatoms. A **diatom** is a nonmotile, single-celled photosynthetic protist that secretes a two-part silica shell. The upper and lower parts of the glassy shell fit

Figure 13.10 **Corals covered in an overgrowth of the red algae, *Chondria tumulosa***

Credit: U.S. Fish and Wildlife Service

? Did You Know?

Red algae have many commercial uses, such as nori, the sheets of seaweed used to wrap some sushi, and agar, which helps keep baked goods and cosmetics moist, helps jellies set, and is used to make capsules that hold medicines.

? Did You Know?

Ancient diatom shells are the basis of a silica-rich powder called diatomaceous earth (DE). DE is used as an insecticide because the sharp microscopic edges cut into an insect's body, causing dehydration and, ultimately, death.

Figure 13.11 Diatoms are a type of plankton

A

B

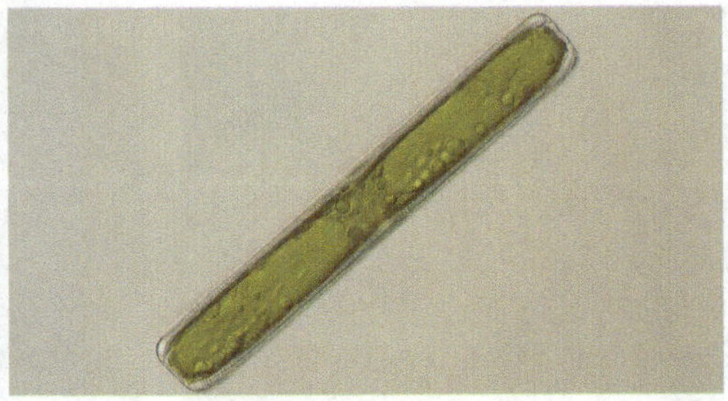

A. Diatoms come in a wide variety of shapes.

B. A living diatom, *Pinnularia*; note the color from their chloroplasts and the small circles within—these are drops of oil.

Credits: Scenics & Science/Alamy Stock Photo; blickwinkel/Alamy Stock Photo

together like a shoe box and its lid and come in a wide variety of shapes (Figure 13.11A). Diatoms are closely related to multicellular brown algae, and like these algae, they have chloroplasts with a brownish accessory pigment that tints them olive green, golden, or dark brown (Figure 13.11B). Diatom cells contain a large amount of oil, which serves as a store of energy and helps them to stay afloat in sunlit waters, as the hydrophobic oil avoids the water (Chapter 2).

? Did You Know?

In tropical seas, dinoflagellates are the most common source of bioluminescence, or light produced by a living organism.

Dinoflagellates, or single-celled protists that typically have two flagella, can be found within the tissues of reef-building corals. Note that dinoflagellates are another kind of phytoplankton. One type of dinoflagellate called zooxanthellae has a mutualistic relationship (one where each organism benefits) with corals (Figure 13.12). The coral provides the zooxanthellae with nutrients, shelter, and the carbon dioxide necessary for photosynthesis. In return, the zooxanthellae supply the coral (a type of animal) with the oxygen it needs for aerobic respiration and with photosynthetically produced sugar. This relationship is crucial to building a reef, as the zooxanthellae are the primary producers. Of note, when corals are stressed, such as by warming waters from climate change, they can expel their resident zooxanthellae. This results in a loss of color, or a bleaching event. We'll learn more about coral bleaching in Chapter 16.

? Did You Know?

The giant blocks of limestone that make up the great pyramids in Egypt consist mainly of the shells of ancient foraminifera.

Another important example of protozoa at coral reefs is foraminifera. **Foraminifera** (singular, *foraminiferan*) are single-celled predators that secrete a shell containing calcium carbonate. We first learned about foraminifera in Chapter 10 when we discussed index fossils (see Figure 10.10). Foraminifera are useful not only for determining the age of fossils, but also for assessing the health of coral reefs, because they occupy similar ecological niches but respond more quickly to environmental stressors than coral because they are single-celled.

Figure 13.12 A healthy colony of *Acropora cythera* coral containing zooxanthelle

The zooxanthelle present here is called *Symbiodinium.*

Credit: M. Stat, E. Morris, & R.D. Gates, (2008). Functional diversity in coral–dinoflagellate symbiosis, Proc. Natl. Acad. Sci. U.S.A. 105(27), 9256–9261. https://doi.org/10.1073/pnas.0801328105.

Foraminifera also play an important role in ocean chemistry. By taking up carbon dioxide from seawater and incorporating it into their shells, foraminifera lower the ocean's carbon dioxide concentration and allow the seawater to absorb more carbon dioxide from the air. Removing carbon dioxide from the air is important to combat climate change.

Colonial protists. The vast majority of protists live as single cells. However, there are also colonial protists. Cells of a **colonial organism** live together and behave in an integrated fashion, but the cells themselves remain self-sufficient, retaining the traits that the cell requires to survive and reproduce on its own. By contrast, a **multicellular organism** has specialized cells, where each cell has a specific function and all cells rely on one another for survival.

When we consider the evolution from protists to the first animals, our closest relatives among the protists are a type of colonial protists called choanoflagellates. **Choanoflagellates** are a small group of aquatic protists that are *heterotrophic* (cannot produce their own food and must get nutrients from other organisms; Chapter 4). While not considered ancestors of animals, they are a group that shared a common single-celled ancestor with animals long ago.

Each choanoflagellate cell has a long flagellum surrounded by a ring (or collar) of tiny filaments reinforced with actin (a protein, Chapter 5; Figure 13.13A). Movement of the flagellum creates a current that draws water through the filaments. After tiny bits of food become entrapped, the cell extends pseudopods to capture the food, which is then digested within the cell body.

Some choanoflagellates form colonies and can work together as a group (Figure 13.13B). The colonies arise when cells divide and the descendant cells stick together with the help of adhesion proteins similar to those that hold cells together in an animal body. Notably, choanoflagellate cells look and function very similarly to the cells of the first animal—the **sponge** (phylum Porifera). For instance, both cell types have flagella for moving water and a collar for catching food (Figure 13.14). The sponge uses the same strategy, but the water is pulled through the entire organism and out through a central opening. As water flows past these cells, the cells capture food in the filaments of their collar and engulf it. The food is then digested in vesicles inside the cell. After the food is digested, the cells share the products with other cells in the sponge body. We'll discuss sponges next, as we delve into the evolution of animals.

The First Animals

An **animal** is a multicellular consumer that takes food into its body, digests it, and then absorbs the released nutrients into its body. An animal body consists of a few to hundreds of types of cells. Animals also share the features of being motile (they can move from place to place) during part or all of their life cycle and generally reproduce sexually.

Figure 13.14 Choanoflagellates look similar to the cells of a sponge

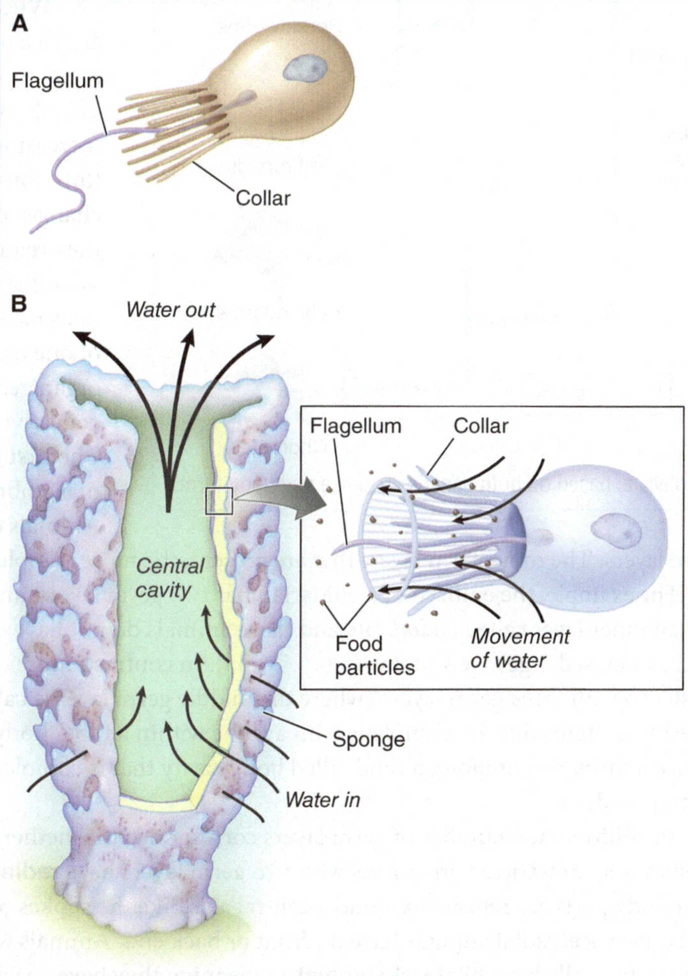

A. A choanoflagellate cell.
B. A sponge consists of choanocyte cells, which are similar in structure to a choanoflagellate.

Figure 13.13 Choanoflagellates can form colonies

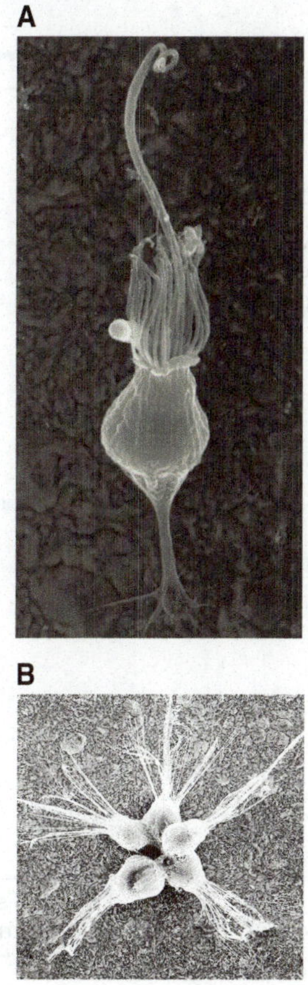

A. A single choanoflagellate, *Salpingoeca rosetta*, and
B. A *S. rosetta* colony.

Credit: López-Escardó, D., Grau-Bové, X., Guillaumet-Adkins, A., Gut, M., Sieracki, M. E., & Ruiz-Trillo, I. (2019). Reconstruction of protein domain evolution using single-cell amplified genomes of uncultured choanoflagellates sheds light on the origin of animals. Philosophical Transactions of the Royal Society B Biological Sciences, 374(1786), 20190088. https://doi.org/10.1098/rstb.2019.0088

Big Ideas in Biology

Structure and Function

Similar structures between the choanoflagellates and the choanocytes in sponges include flagella, which both use to generate water currents to help them eat.

Figure 13.15 The evolutionary history of animals

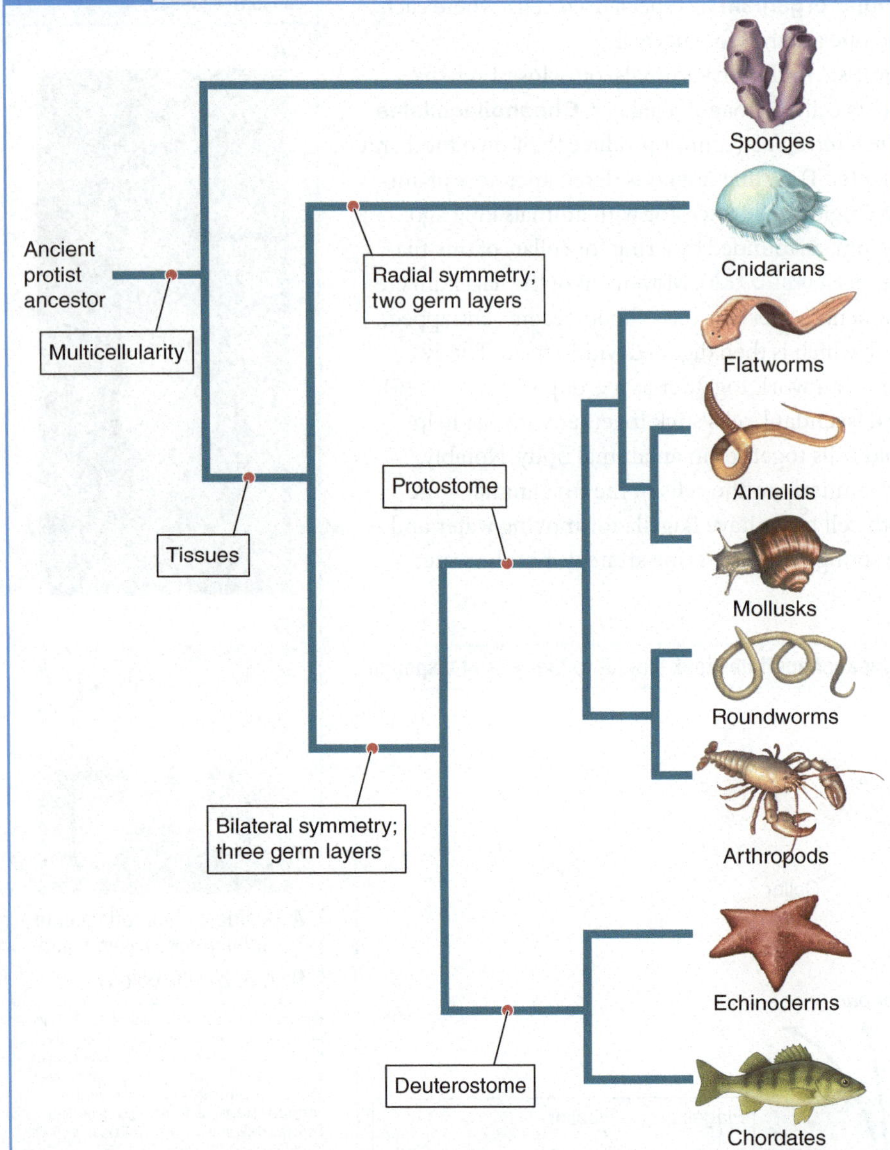

This cladogram for the major animal phyla is based on both genetic evidence and body form.

The **colonial theory of animal origins** states that animals evolved from a heterotrophic protist that formed colonies, similar to choanoflagellates. In fact, genetic analyses indicate that all animals are descendants of the same protist ancestor. At first, all cells in the colony were identical, and each could survive and reproduce on its own. Later, mutations resulted in cells that specialized for a particular task. The division of labor among interdependent cells made the colonies more efficient because it allowed them to collectively obtain more food and produce more offspring. Over time, cells became increasingly interdependent, and more specialized cell types evolved, ultimately leading to the first animals.

When we consider the evolutionary tree of animals (Figure 13.15), the most important feature to note is that as embryonic development changed, increasingly complicated body structures formed. To understand these changes requires knowing a bit about the organization of animal bodies. With the exception of sponges, animals have tissue. A **tissue** consists of one or more types of cells that are organized in a specific pattern and collectively carry out a particular task. The first tissues that form in an animal embryo are called **germ layers**. Animals can have either two or three germ layers. The more germ layers that are present, the more complex body that can form.

For example, the embryo of a cnidarian has two germ layers: an outer layer called *ectoderm* and an inner layer called *endoderm*, and these animals do not have complicated structures, such as a closed digestive system (Figure 13.16A). In contrast, more complex animals, such as lobsters—with three germ layers (where the middle germ layer is called *mesoderm*)—have digestive systems that are complete, with a "tube within a tube" body plan (Figure 13.16B). Their gut runs through a **coelom**, a fluid-filled body cavity that is completely lined with tissue derived from mesoderm.

In addition, the number of germ layers corresponds to whether an animal is radially or bilaterally symmetrical. Organisms with two germ layers have **radial symmetry**, meaning their body parts are repeated around a central axis, like the spokes of a wheel (Figure 13.17A). Said otherwise, radial animals have no front or back end. Animals with a three-layered embryo typically have **bilateral symmetry**, meaning they have a right and left half, with body parts repeated on either side of the body. They also have a distinctive "head end" that has a concentration of nerve cells and sensory structures (Figure 13.17B).

Study Tip

Use the root words to help you remember. *Proto-* means "first," *deutero-* means "second," and *stoma* means "opening."

Did You Know?

Some animals, like sea urchins, are bilaterally symmetrical as juveniles, but not adults.

Figure 13.16 **A body plan comparison of animals with two versus three germ layers**

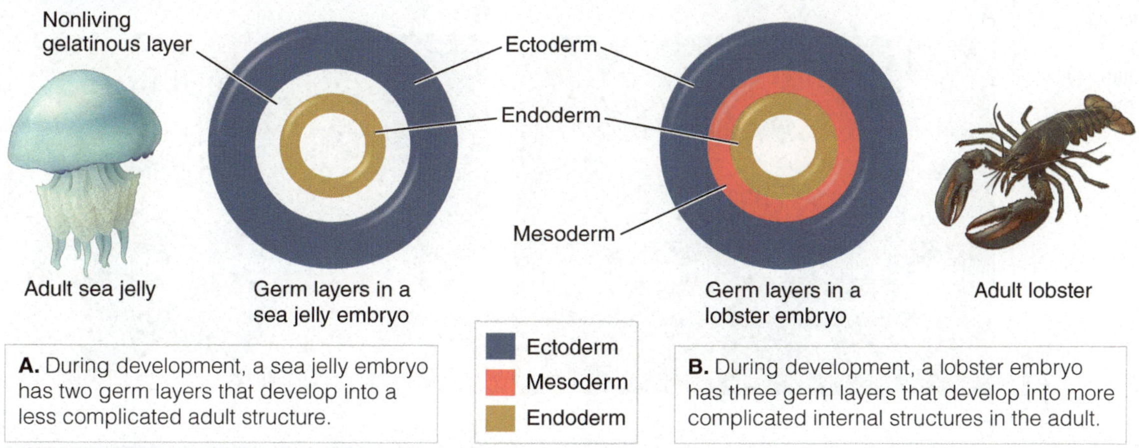

A. During development, a sea jelly embryo has two germ layers that develop into a less complicated adult structure.

B. During development, a lobster embryo has three germ layers that develop into more complicated internal structures in the adult.

Point of Confusion: Recall that complexity doesn't have a value judgement—more complex organisms aren't better than less complex organisms. Rather, the traits we see in an organism are the result of what helps that organism survive and reproduce best in its particular environment.

Animal evolutionary lineages are also distinct based on whether the mouth or anus forms first during embryonic development. In a **protostome**, the first opening that forms on an embryo becomes the mouth. All protostomes are **invertebrates**, or animals without a backbone. Protostomes include flatworms, annelids, mollusks, roundworms, and arthropods. In **deuterostomes**, the mouth develops from the second embryonic opening. Deuterostomes include both invertebrates (echinoderms, invertebrate chordates) and all vertebrates.

Figure 13.17 **A comparison of radial and bilateral symmetry**

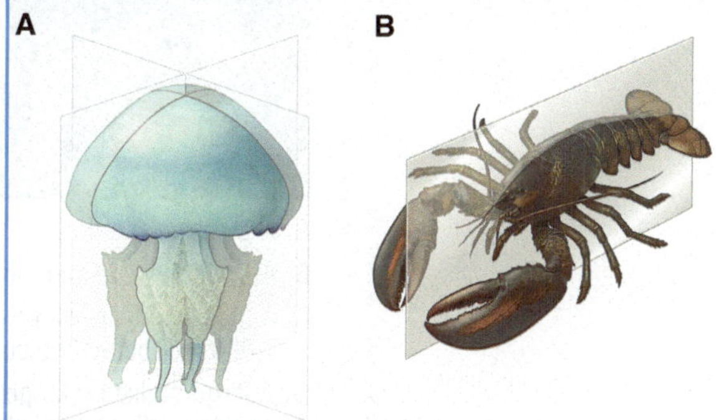

A. A sea jelly is radially symmetrical.
B. A lobster is bilaterally symmetrical—like you!

Diving Down: Looking at Animal Lineages at Papahānaumokuākea Marine National Monument

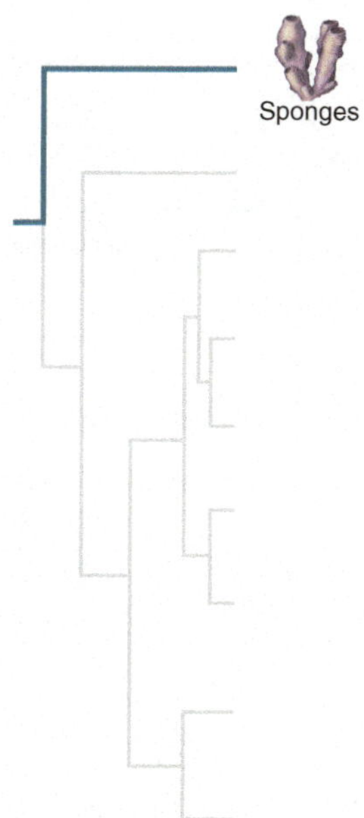

Now that we have a basic understanding of animal organization and evolution, we'll voyage deeper into the Papahānaumokuākea Marine National Monument to explore some different animal lineages. Let's begin with the *invertebrate* (animals without a backbone; Section 13.3) animal lineages using examples found in Papahānaumokuākea Marine National Monument.

Sponges. Have you ever used a natural bath sponge before? If so, you've washed yourself using the remains of a sponge. If we were snorkeling together at a coral reef, it would be easy to spot sponges amongst the corals—although they come in more shapes and colors than your typical bath sponge (Figure 13.18A and B)! If we were to visit a deep-water reef in a submersible, we may even see some examples of gigantic sponges (Figure 13.18C and D).

The earliest animals were aggregations of cells, which did not form into tissues—and likely looked similar to a sponge. Although sponges lack tissues, they are still considered animals based on genetic evidence. In addition, unlike choanoflagellates, which we learned about earlier in this section, sponges are multicellular.

? Did You Know?

Sponges can reassemble into a functional sponge, even after being broken into individual cells by a blender!

Figure 13.18 **A gallery of sponges in Papahānaumokuākea Marine National Park**

A. A *Dysidea* sponge.

B. A *Chondrosiida* sponge.

C. A massive deep-water sponge, approximately the size of a minivan. The sponge is located about 7,000 feet below the surface.

D. A large-stalked sponge (*Bolosoma*). Note the relative sizes of the other animals living on the sponge—if you look closely, you can see red sea stars.

Credits: Scott Godwin/NOAA PMNM, 2012; Photo by: Scott Godwin/NOAA PMNM, 2012; NOAA Office of Ocean Exploration and Research, 2015 Hohonu Moana; NOAA Office of Ocean Exploration and Research, 2015 Hohonu Moana

Molecular evidence suggests that sponges first evolved around 700 million years ago. As of the writing of this text, the oldest known sponge fossil is approximately 550 million years old. Sponges are unique among animals both because they do not have tissues and organs, and because they are the only animals that have intracellular digestion. This is because most sponges are filter feeders, meaning they filter bits of food from the surrounding water (see Figure 13.14).

Adult sponges are sessile, meaning they do not move from place to place, but rather live attached to some surface. A typical sponge is a *hermaphrodite*, an individual that produces both eggs and sperm. Most sponge species release their sperm into the water but retain eggs in their body. After fertilization, the zygote develops into a ciliated larva. A **larva** (plural, *larvae*) is a sexually immature animal that has a different body form than the adult. Sponge larvae exit the parental body, swim briefly, then settle and develop into adults. Sponges can also reproduce asexually when small buds or fragments break away and grow into new sponges.

Cnidarians. The next most complex group of animals are Cnidarians (phylum Cnidaria). Cnidarians first appeared approximately 500 million years ago. **Cnidarians** are aquatic, radially symmetrical animals with stinging tentacles. The two common body plans, medusa and polyp, both consist of two tissue layers (endoderm and ectoderm) with a secreted jellylike matrix between them (see Figure 13.17A). A bell-shaped **medusa** (plural, *medusae*) swims or drifts about. Sea jellies are medusae (Figure 13.19A). A **polyp**, such as a coral (Figure 13.19B) or a sea anemone (Figure 13.19C), is tubular and has one end attached to some surface.

Figure 13.19 **A gallery of cnidarians**

A B C

A. A crown jellyfish (*Coronatae*).

B. Rice coral (*Montipora capitata*); notice the tube-shaped polyps attached to a rocky surface.

C. A sea anemone (*Exocoealactis*); compare the structure of a single organism (the anemone) to the individuals in a colony (the coral in part B).

Credits: NOAA Office of Ocean Exploration and Research, Hohonu Moana 2016; NOAA Office of Ocean Exploration and Research 2015; NOAA Office of Ocean Exploration and Research

In both polyps and medusae, tentacles surround the entrance to the **gastrovascular cavity**, a saclike structure that takes in and digests food, expels wastes, and also helps with gas exchange. An animal with a gastrovascular cavity must fully digest a meal and excrete any resulting wastes before it can ingest the next meal. Nutrients and oxygen taken into the gastrovascular cavity reach body cells by diffusion.

? Did You Know?

The phylum name Cnidaria is derived from *cnidos*, the Greek word for nettle, a kind of stinging plant. The name refers to unique stinging cells (cnidocytes) at the surface of cnidarian tentacles. These cells can sting you, even after the animal is dead!

Flatworms, annelids, and mollusks. Now that we've described the earliest animals, we'll now begin our survey of protostomes, one of the two lineages of bilaterally symmetrical animals. Like the cnidarians, the first flatworms, annelids, and mollusks all appeared around 500 million years ago, during the Cambrian period. All protostomes develop from an embryo that has three germ layers, and all have organs and organ systems. An early evolutionary event produced two protostome lineages. Flatworms, together with annelids and mollusks, are members of one protostome lineage and are what we will discuss first before going on to the second lineage, which includes roundworms and arthropods.

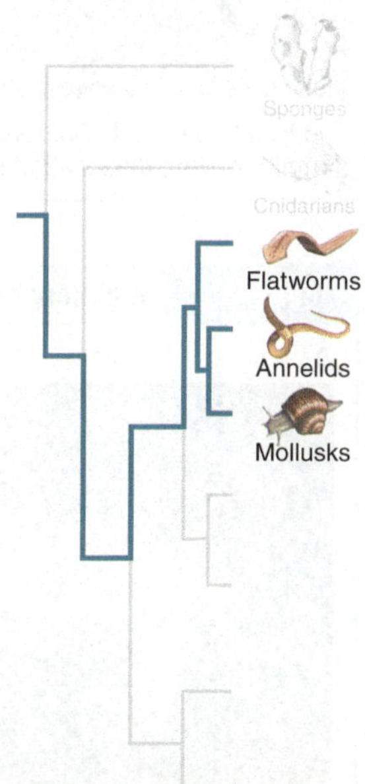

Flatworms (phylum Platyhelminthes) have a flat body with an array of organ systems, but no body cavity other than a gastrovascular cavity. Most free-living flatworms are aquatic, such as the predatory flatworm *Pseudobiceros* (Figure 13.20A). Note that parasitic flatworms can also live inside animals, such as tapeworms.

Annelids are bilateral worms with a coelom and conspicuous segmentation, both inside and out. Segmentation refers to a body plan in which similar units are repeated one after the other along a body's main axis. Most marine annelids are polychaetes, a lineage that typically has many bristles per segment. Although not native to the Hawaiian Islands, the feather duster worm (*Sabellestarte spectabillis*; Figure 13.20B) is an example of a polycheate worm found in Papahānaumokuākea Marine National Monument.

Annelids were one of the first kinds of animals to evolve that had organ systems, including a complete digestive tract and a closed circulatory system. A **complete digestive tract** is a tubular gut with an opening at either end. Different parts of the digestive tract specialize in taking in food, digesting food, absorbing nutrients, or compacting waste. Unlike the gastrovascular cavity of a cnidarian, a tubular gut can carry out all of these tasks simultaneously. A **closed circulatory system** means that all of the organism's blood is contained within vessels (Chapter 17). Note that while annelids have both a complete digestive tract and a closed circulatory system, animals may have one or the other. This variation underscores the fact that evolution varies depending on an organism's environment.

Study Tip

Note the root words: *poly-* means "many," and *chaete* means "bristle."

You may be familiar with annelids like earthworms and leeches. Both have uses for humans—earthworm castings (poo) contribute to healthy soil, and people have used leeches medically for thousands of years. Leech saliva contains a chemical called hirudin, which prevents blood from clotting. Clinically, leeches are useful for preventing blood clots post-surgery, during skin graft surgery, and when a severed body part needs to be reattached.

Figure 13.20 **Flatworms and annelids are two kinds of worms found on coral reefs**

A. A *Pseudobiceros* flatworm.

B. An exotic feather duster polycheate worm (*Sabellidae*). Compare the appearance of these two worms—you can clearly see why flatworms are called flat!

Credits: Scott Godwin/NOAA; Image courtesy of Scott Godwin/NOAA

Mollusks (phylum *Mollusca*) have a soft, unsegmented body, a complete digestive tract, and a reduced coelom. The **mantle**, a skirtlike extension of the upper body wall, drapes over the internal organs (Figure 13.21). In most mollusks, the outermost layer of the mantle secretes a hard shell rich in calcium carbonate. Unlike the invertebrate groups discussed previously, mollusks have respiratory organs. Aquatic mollusks have one or more gills that facilitate gas exchange with the water. Some mollusks that live on land have a lung, which is an organ that facilitates gas exchange with the air.

Figure 13.21 **Body plan of a mollusk**

The triton's trumpet (*Charonia tritonis*) is an example of a mollusk. Note the location of the mantle and its radula.

Credit: David Fleetham/Science Source

Among animals, mollusks are second only to arthropods in diversity. There are three major groups: gastropods, bivalves, and cephalopods. For example, brilliantly colored sea slugs such as *Plakobranchus ocellatus* (Figure 13.22A) are gastropods. **Gastropods** are the largest mollusk subgroup and include around 90,000 species of snails and slugs. *Gastropod* means "belly foot," and members of this group glide about on a broad, muscular foot that makes up most of their lower

Figure 13.22 **A gallery of mollusks**

A. Sea slugs, such as this brilliantly colored *Plakobranchus ocellatus*, are gastropods.

B. Bivalves include animals with a hinged shell, such as this black-lipped pearl oyster (*Pinctada margaritifera*).

C. Cephalopods, such as this dumbo octopus (*Grimpoteuthis*), are predators that move around via jet propulsion.

Credits: Photo by: Scott Godwin/NOAA PMNM, 2012; Photo Credit: Jim Maragos; Photo: Ocean Exploration Trust/NOAA body mass. The gastropod shell, when present, is one piece and usually coiled, such as the shell of the triton's trumpet in Figure 13.21. Most gastropods scrape up algae with a **radula**, a tongue-like organ hardened with a tough polysaccharide called chitin. Note that gastropods include the only land-dwelling mollusks. Glands on the foot of a land snail or slug continually secrete mucus that protects the animal as it moves across dry, abrasive surfaces.

A second type of mollusk is a bivalve. A **bivalve** is a type of mollusk with a hinged, two-part shell, such as the black-lipped pearl oyster, which makes Tahitian black pearls (Figure 13.22B). Bivalves live in both freshwater and the seas. Some, such as mussels, attach to a surface, whereas others, such as clams, burrow into sediments. A bivalve does not have a distinct head, nor does it have a radula. Most feed by drawing water into the mantle cavity and filtering out bits of food.

The third class of mollusks is cephalopods. *Cephalopod* means "head-footed," and in all **cephalopods**, the foot has been modified into tentacles and/or arms that extend from the head. For example, the dumbo octopus (Figure 13.22C) is an example of a cephalopod, along with cuttlefish, squids, nautiluses. Most cephalopods are predators and have beak-like, biting mouthparts in addition to a radula. Cephalopods move by jet propulsion. They draw water into their mantle cavity, then force it out through a funnel-shaped siphon.

Process of Science

Evolutionary History and Model Organisms

In Chapter 1, we learned about clinical trials. Before a clinical trial can begin, though, biological processes are typically explored in model organisms. A *model organism* is a non-human species that is studied in order to understand a biological process or phenomenon. Model organisms are typically a less complex species than humans, where the lack of complexity helps researchers to more easily study a specific process. Since we share an evolutionary history with other animals, biological processes are often similar between different species.

For example, consider the yeast *Saccharomyces cerevisiae*, which grows readily in laboratories and is eukaryotic, like us. Checkpoint genes that regulate the eukaryotic cell cycle (Chapter 6) were first discovered in *S. cerevisiae*. This discovery was the first step toward our current understanding of how mutations of these genes cause human cancers. Genetically engineered *S. cerevisiae* and other yeasts are also now used to produce proteins that serve as vaccines or other medicines.

Yeast isn't the only species of model organism—multicellular animals are also used. For example, we looked at *Drosophila* (fruit fly) mutations in Chapter 7. Another example is *Caenorhabditis elegans*, a soil roundworm that is easily grown in the lab. *C. elegans* has the same tissue types as more complex organisms, but it is transparent and has fewer than 1,000 body cells (Figure 13.23). Such traits make it easy for scientists to monitor each cell's fate during development. Several scientists have made Nobel Prize-winning discoveries while studying *C. elegans*.

Zebrafish (*Danio rerio*) are also common model organisms used for genetic studies. Since they are fertilized externally, they are particularly useful for studying embryonic development, since the embryo can be easily observed in a Petri dish without being harmed. Zebrafish are cheaper to maintain than mammals like mice or rats. However, mice are still a valuable model organism because they are the most genetically similar to humans.

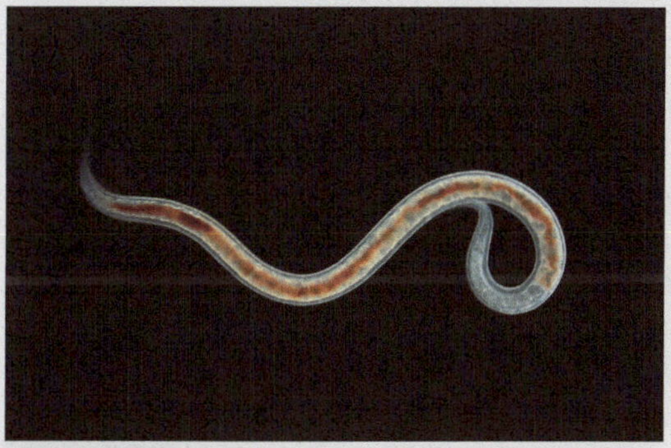
Credit: Ruben Duro/Science Source

Figure 13.23 Roundworms like *C. elegans* are used as model organisms

1. While our closest genetic relative, the chimpanzee, would respond most similarly to humans during experimental testing, the National Institutes of Health (NIH) stopped supporting chimpanzee research in 2015. What concerns do you think led to this move away from chimpanzees as a model system?
2. Often, scientists will begin in a simple model system, like *C. elegans*, and move to a more complex system, like a mouse, when their research is shown to be promising. Why not start in mice? Why not only use *C. elegans*? How does using both species help to balance the pros and cons of the use of each organism?

Biology for Life

Medicines from the Sea

Figure 13.24 A cone snail (*Conus*) in captivity, with its fish prey

After administering a toxin that puts the fish into a stupor, the snail engulfs and devours it.

Many marine invertebrates produce compounds that are toxic to other organisms. Such toxins can protect an animal from predators, help it fend off pathogens, or assist in the capture of prey. Some invertebrate toxins also have effects in the human body—and some of these chemicals can be useful as medicines.

Consider the venom that some fish-eating cone snails use to subdue their prey (Figure 13.24). The snail's venom anesthetizes and paralyzes a fish, thus preventing the fish from struggling and possibly harming the snail. Human and fish nerves use some of the same chemical communication signals. Thus, a toxin that acts on fish nerves can also affect the function of human nerves. A person stung by a fish-eating cone snail may become numb at the site of venom injection, suffer from temporary paralysis, or even die.

The painkiller ziconotide (Prialt®) is a synthetic version of a peptide in the venom of one cone snail. Ziconotide is injected into the spinal cord to suppress pain that cannot be controlled by other means. Additional peptides isolated from cone snail venom are being tested as treatments for epilepsy, diabetes, and cancer.

Compounds derived from other marine invertebrates are also in use. Azidothymidine, or AZT, was the first drug successfully used to treat HIV. AZT is a synthetic version of a molecule discovered in a sponge. Other compounds discovered in sponges can be used to treat infections caused by the herpes viruses. Sea whips, which are relatives of sea anemones, produce a variety of anti-inflammatory compounds, one of which is used in face creams.

Finding a compound that might have medicinal value is only the first step in developing a new drug. For a compound to be used in clinical tests, researchers must obtain sufficient amounts. This can be difficult because many compounds of interest occur only at very low concentrations in animals.

Consider eribulin, a synthetic drug based on a molecule first extracted from a Japanese sea sponge (*Halichondria okadai*). Eribulin is now used to treat some types of advanced-stage breast cancers. We use a synthetic version for two reasons. One, sponges make eribulin in tiny amounts. Obtaining 300 milligrams of the sponge compound—enough to test it for anticancer activity—required processing more than a metric ton of sponge tissue! Second, the use of synthetically produced compounds can prevent overharvesting of the species in which the compound was discovered.

1. The World Health Organization estimates that about 40% of pharmaceutical products today draw from nature and traditional knowledge. Think about natural remedies you have heard about or tried. What do they treat? How might you go about testing whether they could be developed into a pharmaceutical?
2. The painkiller ziconotide is a peptide (a short chain of amino acids). Peptides can be produced synthetically, so researchers could make peptides with random amino acid sequences, then test those peptides for medically useful activity. Why is testing peptides isolated from animal venoms a more promising way to find new drugs?
3. Searching for economically valuable compounds in natural organisms is called bioprospecting (Chapter 16). One criticism of bioprospecting is that it often allows companies in developed nations to profit from the biodiversity of less-developed regions. For example, venom from a cone snail native to the Philippines was the basis for ziconotide, a drug currently manufactured in Ireland. Do people in less-developed regions have a right to share in jobs and profits made possible by the exploration of their country's biological diversity? How would such profit sharing encourage people to maintain biodiversity?

Roundworms and arthropods. Our next protostome lineage features another type of worm—this time roundworms. Roundworms and arthropods also likely first appeared during the Cambrian era. Roundworms and arthropods are distinguished from other protostomes in that molting is part of their life history. **Roundworms**, or nematodes (phylum *Nematoda*), are cylindrical, unsegmented worms with a pseudocoelom. They do not have circulatory or respiratory organs. Roundworms secrete a flexible, protein-rich cuticle that is molted and replaced as the worm grows. Roundworms live in the seas, in freshwater, and in soil, and the majority are free-living decomposers less than a millimeter long. Roundworms like *Caenorhabditis elegans* are also an important *model organism* used in laboratories to study biological processes.

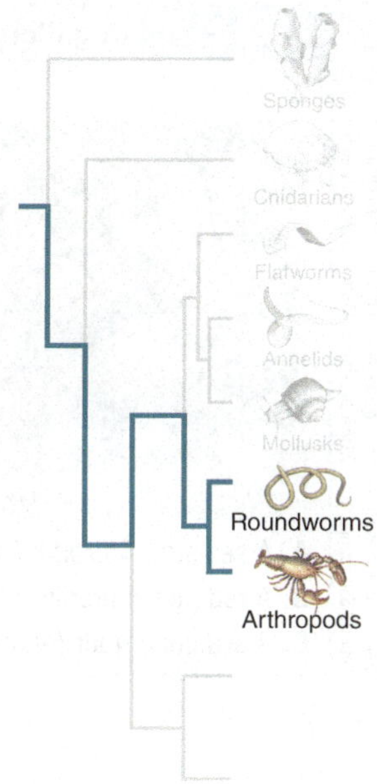

Arthropods (phylum *Arthropoda*) are invertebrates with jointed legs. Unlike roundworms, they have a coelom, a circulatory system, and respiratory organs. There are an estimated 5 to 10 million arthropod species, making arthropods the largest and most diverse animal phylum. Note that arthropods include marine animals as well as what we colloquially call "bugs," which includes chelicerates (spiders and other eight-legged friends), myriapods (centipedes and millipedes), and the true insects. Note that crustaceans live predominantly in the seas (the focus of this section), but can also live on land. Given the massive biodiversity of the arthropods, we'll introduce arthropods here, as we talk about marine crustaceans, and will then return to "bugs" in Section 13.5.

A variety of traits have contributed to the evolutionary success of arthropods. For example, arthropods have an external skeleton or **exoskeleton** made up of chitin (Chapter 3). The exoskeleton helps the animal fend off predators while also serving as an attachment point for muscles. In land arthropods, like insects, the exoskeleton also helps conserve water and supports the animal's weight. A hard exoskeleton does not restrict growth because, like roundworms, arthropods periodically molt their exoskeleton.

? Did You Know?

Parasitic roundworms also infect humans.

Most arthropods have paired eyes, such as the banded spiny lobster (Figure 13.25A). Insects and crustaceans have compound eyes that consist of many units, each with a lens. Compound eyes excel at detecting movement. Many arthropods also have one or two pairs of **antennae** (singular, *antenna*), which are sensory structures on the head that detect touch, odor, and vibrations (Figure 13.25B).

The crab body form is so successful that it independently evolved five different times and is an example of convergent evolution (Chapter 10).

The **crustaceans** are a lineage of mostly marine arthropods that have two pairs of antennae and at least five pairs of legs. Many, such as shrimps, crabs, and lobsters, are harvested for food. Other relevant examples of crustaceans include krill, a shrimplike crustacean found in cool ocean waters. While krill are only a few centimeters long, they are so abundant and nutritious that a 100-ton blue whale can subsist almost entirely on the krill that it filters from seawater.

Figure 13.25 A gallery of crustaceans

A B C

A. A banded spiny lobster (*Panulirus marginatus*); note the presence of paired eyes and jointed legs.
B. Two banded coral shrimp (*Stenopus hispidus*); note their prominent white antennae.
C. An acorn barnacle (*Balanomorpha*). Unlike most arthropods, barnacles are sessile and extend their feathery feet (which are jointed, like the other arthropods) to feed when underwater.

Credits: Photo by: James Watt/NOAA Office of National Marine Sanctuaries, 2002; Photo by: Greg McFall/NOAA Office of National Marine Sanctuaries, 2011; E.R. Degginger/Alamy Stock Photo

Figure 13.26 **A gallery of echinoderms**

A. A sea star, *Mitrodia fisheri.*
B. A red pencil urchin (*Heterocentrotus mamillatus*) surrounded by corals.
C. A sea cucumber (genus *Stichopus*).

Credits: Toni Parras/NOAA; Photo by: James Watt/NOAA Office of Marine National Sanctuaries, 2002; NOAA and Robert Whitton/Bishop Museum

? Did You Know?

Some barnacles are notable for the length of their penis, which can be eight times that of their body length. If we draw a parallel with humans, men are on average 5.6 feet tall, which means that if a man's penis were similarly sized, it would be approximately 45 feet long!

Barnacles are marine crustaceans that secrete a calcium-rich external shell (Figure 13.25C). Barnacle larvae swim, then settle and develop into adults that attach to a surface—you can often see them attached to rocks in the intertidal zone. Adults filter food from seawater with their feathery legs.

Echinoderms. With this section, we begin our survey of deuterostomes, another lineage of bilaterally symmetrical animals, and will begin with the echinoderms. In Section 13.3, we'll discuss the other lineage of deuterostomes, the chordates.

Echinoderms (phylum *Echinodermata*) include over 7,000 marine invertebrates such as sea stars, sea urchins, and sea cucumbers. Like the other invertebrate animals, they also appeared around 500 million years ago. Their phylum name means spiny-skinned and refers to interlocking spines and plates of calcium carbonate embedded in their skin. The plates form an internal skeleton called an **endoskeleton**.

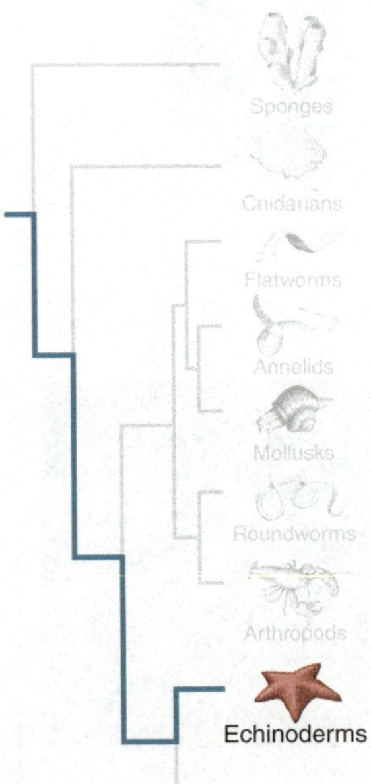

Figure 13.27 **A sea star (family *Keratosididae*) feeding on bamboo coral (*Evoplosoma forcipifera*)**

Note the tiny tube feet and how they hold onto the coral.

Credit: NOAA Office of Ocean Exploration and Research, 2015 Hohonu Moana

Sea stars are the most familiar echinoderms (Figure 13.26A). They do not have a brain, but they do have a decentralized nervous system. Eyespots at the tips of their arms detect light and movement. A typical sea star moves about on tiny, fluid-filled tube feet. You may also be familiar with echinoderms such as urchins, like this red pencil urchin (Figure 13.26B). Sea cucumbers (Figure 13.26C) are another example of an echinoderm.

To feed, a sea star slides its stomach out through its mouth (Figure 13.27). The stomach secretes acid and enzymes that kill the prey and begin to digest it. Partially digested food enters the stomach, and digestion is completed with the aid of enzyme-secreting digestive glands in the arms.

? Did You Know?

A sea star's reproductive organs are in its arms.

13.2 Summary

- Photosynthesizing cyanobacteria evolved around 2.7 billion years ago and began the great oxygenation event, creating a selective pressure that resulted in new types of organisms.
- Eukaryotes that are not fungi, plants, or animals are called protists. Colonial protists are groups of single cells that cooperate in an integrated manner, while still functioning individually. The earliest multicellular organisms are thought to have evolved from colonial protists like choanoflagellates, whose cells resemble those of the first animals (sponges).
- Animal tissues form from germ layers during embryonic development. Animals with two germ layers have radial symmetry and evolved first. Animals with three germ layers are bilaterally symmetrical with a distinct "head end."
- Organisms have many unique adaptations that allow them to thrive in coral reefs. For example, protists like dinoflagellates supply coral with oxygen while the coral provides protection and nutrients. Animals like mollusks have adaptations that include respiratory organs and a calcium carbonate shell.

Genetic comparisons indicate that the animal lineage may have arisen as early as 850 million years ago. The oldest known fossil animal is Dickonsonia, which was an oval, pancake-like bottom dweller that lived in the sea 570 million years ago. We know that Dickonsonia was an animal because a chemical analysis of fossilized remains showed that the organism contained a large amount of cholesterol—a hallmark of animals.

13.3 The Diversity of Vertebrates: A Visit to an Estuary

Learning Objectives

13-9 Describe the evolutionary history of chordates.

13-10 Compare invertebrate and vertebrate chordates.

13-11 Describe the characteristics and/or adaptations that distinguish fish, amphibians, reptiles, and mammals, allowing them to live in their respective environments.

13-12 Describe the anatomical changes that occurred as vertebrates moved onto land.

13-13 Describe structural and physiological traits that facilitate flight in birds.

Life first evolved in the seas—so perhaps it's not surprising that our third stop also happens to be another watery area. However, rather than a purely marine environment, we're moving into brackish (somewhat salty) waters to begin to look at animals who live on land and in the air. Our next tour stop is the Indian River Lagoon, which covers 40% of Florida's eastern coast!

Credit: iStock.com/Ryan Tishken

An estuary. Estuaries are largely enclosed coastal regions where fresh water and salt water intermingle. Since these are areas where multiple environments intersect (land, freshwater and saltwater), they tend to be areas of high biodiversity.

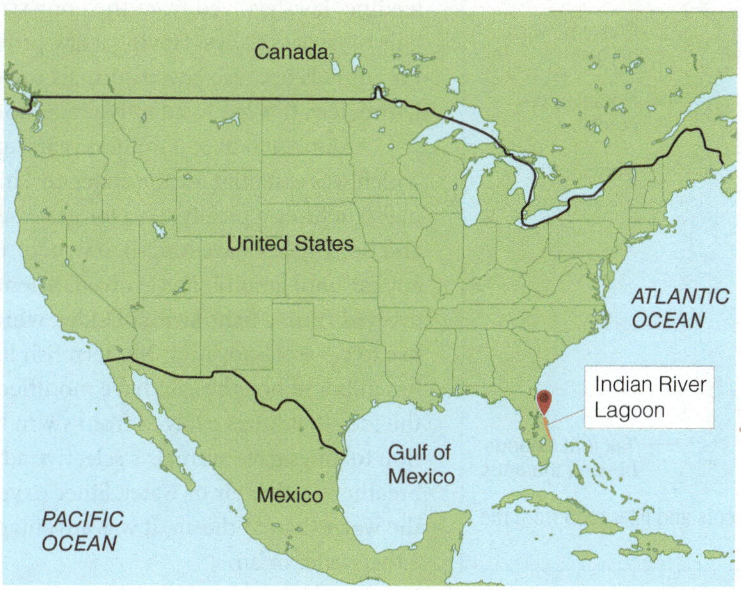

The next stop on our field trip is to the Indian River Lagoon, on the eastern coast of Florida, to learn about the evolution of different chordates/vertebrates, including the transition of animals from the sea to land.

The Indian River Lagoon is special because it includes a wide range of habitats (Chapter 14) and spans both temperate and subtropical zones (Chapter 15). In addition to its location in two different climatic zones, estuaries are areas where saltwater and freshwater mix. All of these different factors together mean that the Indian River Lagoon is a hotbed of biodiversity. In fact, as of the writing of this text, researchers have documented over 11,000 individual species in the Indian River Lagoon.

Although the Indian River Lagoon is home to many kinds of animals we've already discussed in Section 13.2, in this section, we'll be focusing on chordate evolution, starting with invertebrates like lancelets and ending with a discussion of mammals.

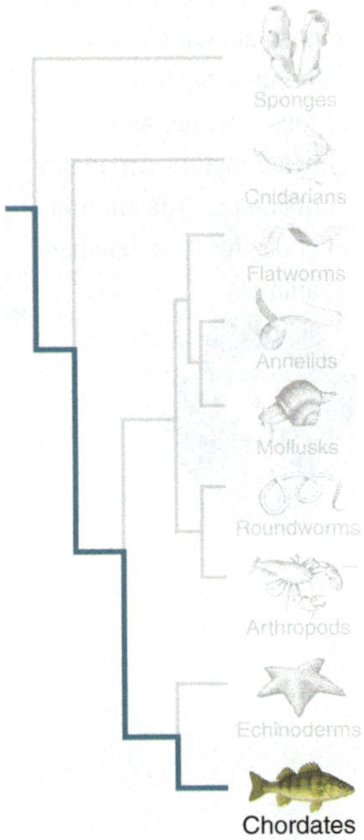

A Chorus of Chordates

Like the rest of the animals, chordates first appeared during the Cambrian explosion some 500 million years ago. Similar to the rest of the animals, chordates are distinguished based on embryonic features. As embryos, **chordates** (phylum *Chordata*) have: (1) A **notochord**, a rod of stiff but flexible connective tissue that extends the length of the body and provides support; (2) a dorsal hollow nerve cord that parallels the notochord; (3) gill slits open across the wall of the pharynx (throat region); and (4) a muscular tail that extends beyond the anus. **Figure 13.28** shows all of these structures in a lancelet, which is a chordate that has all these traits present as an adult.

Depending on the chordate lineage (**Figure 13.29**), some, all, or none of these traits persist in the adult. Let's consider humans as an example, since we are chordates. During embryonic development, we have all of the chordate traits. However, by birth, we no longer have a notochord, gills slits, or a tail.

Like the animal family tree (see Figure 13.15), different chordate lineages are characterized by key innovations (Chapter 12). The older chordate lineages are invertebrates. Invertebrates are animals that do not have backbones. All of the evolutionary lineages that we learned about in Section 13.2 are invertebrates, as are lancelets and tunicates. The development of a **backbone**, a bony structure that extends from the pelvis to the skull, was a key innovation that led to many different animal lineages. Animals with a backbone are called **vertebrates**, and they first appeared during the Cambrian explosion some 500 million years ago. While all chordates have a notochord, in vertebrates, the backbone replaces the notochord as the embryo develops.

Figure 13.28 **Lancelets are chordates that have all defining traits as adults**

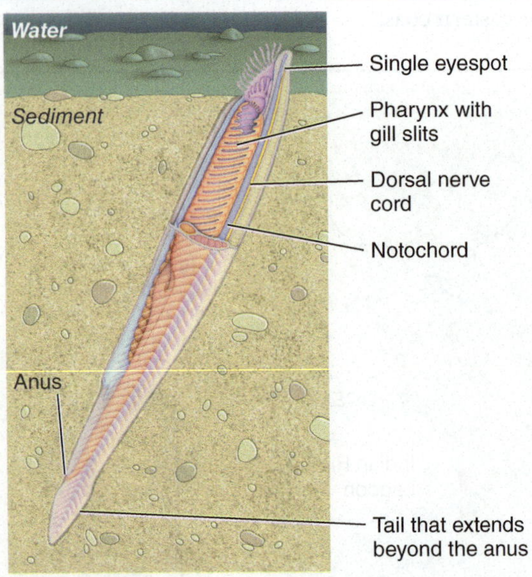

Lancelets live in sandy sediments and filter food from the water with their pharynx.

The next key innovation was the development of jaws, around 450 million years ago. A **jaw** is a hinged skeletal element used for feeding. Jaws evolved from the bony structures that supported the gills of jawless fishes. Having a jaw provided a selective advantage for animals because jawed animals can eat a wider range of foods. Ultimately, this led to adaptive radiation (Chapter 11).

Around 400 to 350 million years ago, the first lungs appeared—which was essential for the water-to-land transition. **Lungs** are internal structures typically used for gas exchange. Note that cartilaginous and bony fishes have lung-like structures, but the first true lungs didn't appear until amphibians evolved. Scientists think that lungs likely evolved from a fish's swim bladder, which is an internal air sac that fish use to control buoyancy. Modern fish like the African lungfish, which has gills and breathes air, have modified swim bladders, supporting the idea that lungs evolved from swim bladders. Having gills in addition to lungs gave animals a selective advantage in that they could breathe in either air or water. Since oxygen levels are always lower in the water than in the air, it was advantageous to be able to breathe in either water or air.

Figure 13.29 The evolutionary history of chordates

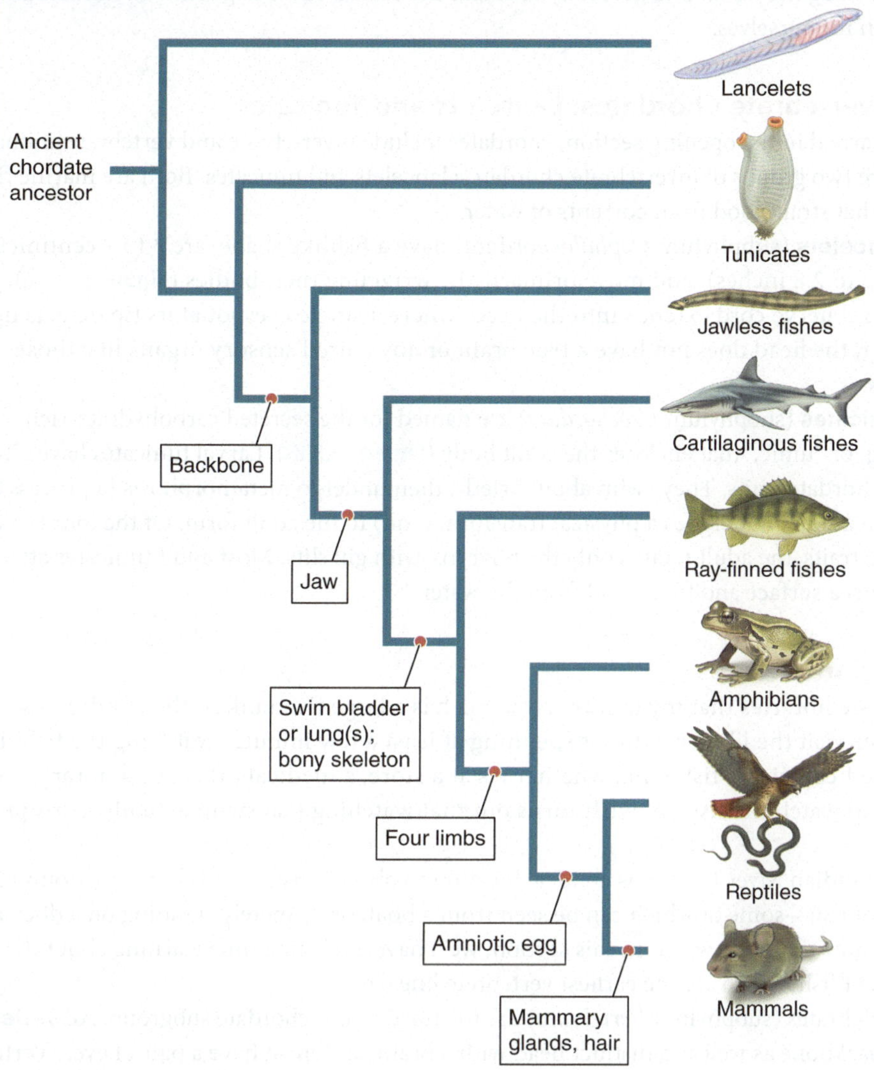

Note that the majority of chordates are vertebrates. Key innovations are indicated on the cladogram with boxed text.

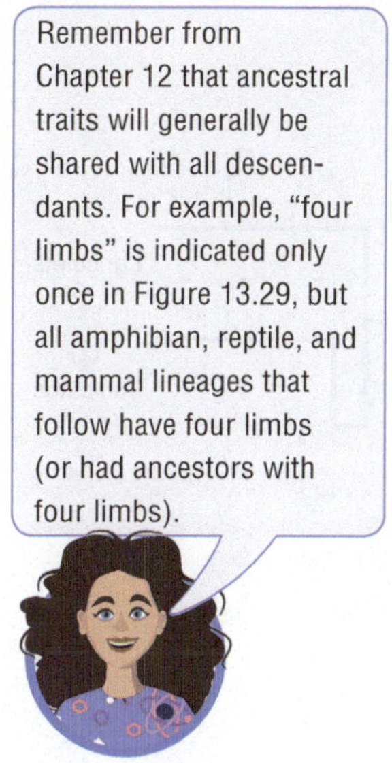

The next key innovation was the appearance of a **bony skeleton**, a trait essential for the transition from water to land, allowing an animal to support its body weight on land. Animals such as Tiktaalik (Chapter 10) had a bony skeleton, as well as the ability to breathe in water or in air. Tiktaalik is thought to be one of the first **tetrapods**, animals with four bony limbs. Tetrapods first evolved around 400 million years ago.

In addition to needing the structures to move and breathe on land, animals also needed to be able to reproduce successfully out of the water. Successful land-based reproduction required the evolution of the **amniotic egg**, which provided a protective environment for the developing offspring. Ultimately, this meant that animals were not dependent on water and could access new habitats, giving them an evolutionary advantage. The evolution of the amniotic egg ultimately led to the evolution of reptiles (which includes birds) and eventually mammals. As we learned in Chapter 12, mammals are characterized by having **mammary glands** that are used to provide milk to their offspring, as well as having hair. Except for the monotremes, mammals also give birth to live young rather than laying eggs. Growing within an adult's body provides additional protection to the offspring.

Now that we've highlighted some key innovations in chordate evolution, it's time to hop onto our imaginary boat and get ready to cruise the Indian River Lagoon to see the results of evolution for ourselves.

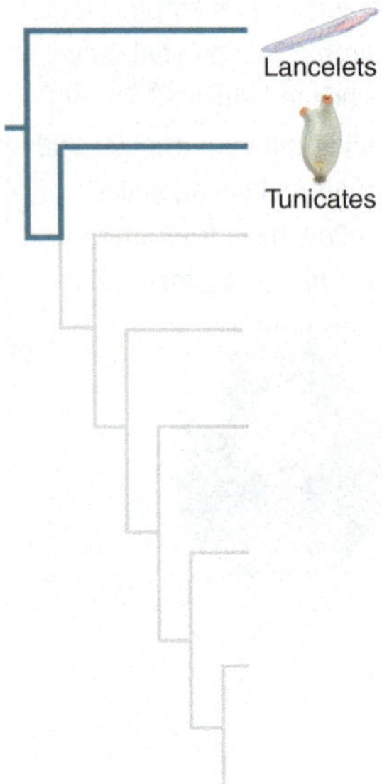

The Invertebrate Chordates: Lancelets and Tunicates

As we learned in the opening section, chordates include invertebrate and vertebrate animals. There are two groups of invertebrate chordates: lancelets and tunicates. Both are marine filter feeders that strain food from currents of water.

Lancelets (subphylum *Cephalochordata*) have a fishlike shape, are 3 to 7 centimeters long (1.2 to 2.8 inches), and move primarily by wriggling their bodies (**Figure 13.30A**). The dorsal nerve cord extends into the head, where a single eyespot at its tip detects light. However, the head does not have a true brain or any paired sensory organs like those of fishes.

Tunicates (subphylum *Urochordata*) are named for the secreted carbohydrate-rich covering, or "tunic," that encloses the adult body (**Figure 13.30B**). Larval tunicates have all the typical chordate traits. They swim about briefly, then undergo metamorphosis (a process by which an animal undergoes a physical transformation) to the adult form. Of the four typical chordate traits, the adult retains only the pharynx with gill slits. Most adult tunicates attach to an undersea surface and filter food from the water.

A Foray into Fishes

One of the libraries that my family frequents has a large fish tank in the children's section. We never visit the library without spending at least a few minutes watching the fish. In fact, any time I can visit a fish tank, whether it's at a store, a medical office, or a library, I always stop by to watch the fish. Why? It turns out that watching fish swim actually helps people to relax.

The Indian River Lagoon is an ideal location to observe fish, as it is home to around 700 species of fish—some of which can be seen from a boat, or by merely standing on a dock and looking down into the water. In this section, we'll have a relaxing time learning about the four lineages of fish, which are the earliest vertebrate lineage.

Vertebrates (subphylum *Vertebrata*) are the third major chordate subgroup. All **vertebrates** have a backbone as well as a distinct head with a brain, and most have a pair of eyes. Vertebrates also have a *circulatory system*, with a heart and blood that is transported in vessels, as well as a complete digestive tract (Chapter 17). We'll begin our survey of vertebrates with jawless fish, before moving onto cartilaginous and bony fishes.

Neither lancelets nor tunicates are ancestors of vertebrates. All three groups evolved from a now extinct invertebrate chordate ancestor; each has unique traits that put it on a separate branch of the animal family tree.

Figure 13.30 Representative examples of lancelets and tunicates

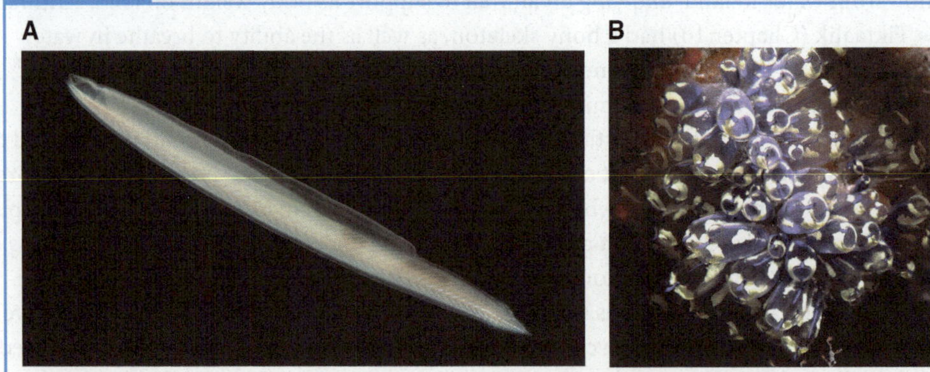

A. A Florida lancelet (*Branchiostoma floridae*). Note that lancelets are not actually found in the Indian River Lagoon, but they are found in estuaries along the western coast of Florida, such as in Tampa Bay.

B. A grouping of bulb tunicates (*Clavelina*).

Credits: National Geographic; Amar and Isabelle Guillen - Guillen Photo LLC/ Alamy Stock Photo

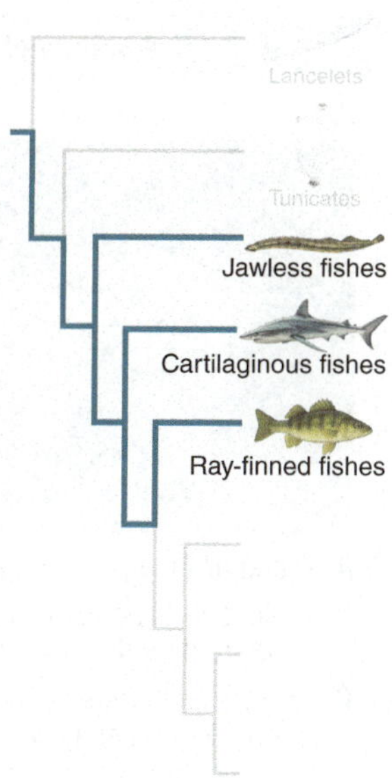

Jawless fish. Modern **jawless fishes** have a smooth, elongated body without paired fins. Like lancelets, they move by wriggling their body. The skeleton of a jawless fish is made of cartilage, the same type of connective tissue that makes up your nose and ears. There are two kinds of jawless fish living today: lampreys and hagfish.

Despite the incredible biodiversity in the Indian River Lagoon, there are no jawless fish present. However, the southern brook lamprey (Figure 13.31A), with its distinctive mouth (Figure 13.31B), can be found in nearby central Florida. As adults, most lampreys feed on other fish. Lacking jaws, a lamprey cannot bite. Instead, it attaches to a fish using an oral disk ringed by horn-like teeth made of the protein keratin. Once attached, the lamprey secretes enzymes and scrapes up bits of its host's flesh with a tooth-covered tongue.

Figure 13.31 **The southern brook lamprey (*Ichthyomyzon gagei*)**

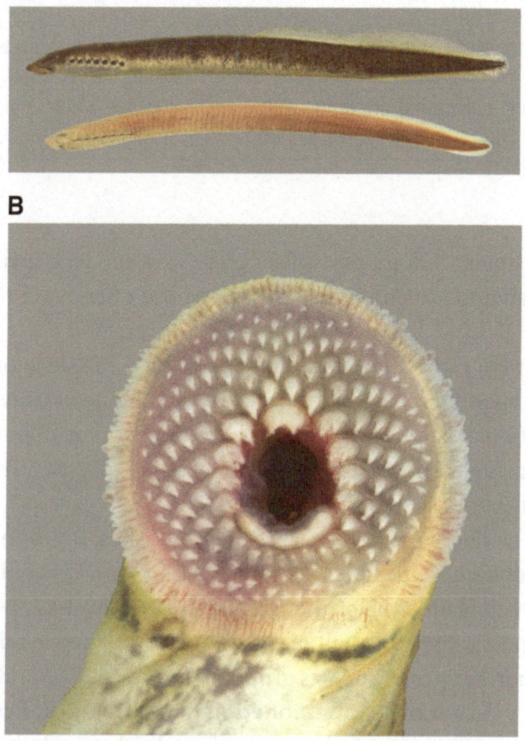

A. Shown here are two examples of southern brook lampreys. Note the lack of paired fins and visible gill slits.

B. The oral disk of a lamprey, which it uses like a suction cup to attach itself to other fish. It then uses its oral disk to scrape off the host's flesh.

Source: Florida Museum photo by Zachary Randall

Cartilaginous fish. Remember Bruce from *Finding Nemo*? Sharks are one of the most well-known examples of cartilaginous fishes—and one you might spot from a boat in the Indian River Lagoon. As their name suggests, **cartilaginous fish** are jawed fishes with a skeleton made of mineral-hardened cartilage. They include 850 species of sharks, skates, and rays. In the Indian River Lagoon, there are many kinds of cartilaginous fish, including blacktip reef sharks (Figure 13.32A), the longtail pygmy skate (Figure 13.32B), and rays like the Atlantic stingray (Figure 13.32C). Note that while we commonly think of sharks as predators that swim in upper ocean waters, others strain plankton from the water or suck up food from the seafloor.

Note that cartilaginous fishes were also the first animals to have the key innovation of a jaw. **Jawed fish** have movable jaws and well-developed teeth. Jaws evolved from gill arches, which are skeletal

Shark attacks on humans receive a lot of attention, but they are rare. Worldwide, sharks kill about five people a year. For comparison, dogs kill about 30 people each year in the United States alone.

? Did You Know?

Skates lay eggs, and rays give birth to live young. Skates and sharks also produce egg cases, informally known as mermaid's purses, which you can occasionally find as you walk on the beach.

Figure 13.32 **A gallery of cartilaginous fish**

A. A blacktip reef shark, *Carcharhinus limbatus*, swimming through the water.

B. A longtail pygmy skate, *Fenestraja plutonia*. Note that it has a bumpy back and shorter tail than a ray.

C. An Atlantic stingray, *Dasyatis sabina*, resting on the seafloor. Note the long tail, which has a venomous barb at the end.

Credits: Cdascher/E+/Getty Images; NOAA; © Andy Murch/Oceanwidelmages.com

Credits: Picture Partners/Alamy Stock Photo; Gary Meszaros/Science Source; Paulo de Oliveira/Biosphoto/Minden Pictures

Figure 13.33 **A gallery of bony fishes**

A. A gag grouper, *Myceriperca microlepis.*

B. A bluegill, *Lepomis macrochirus*. Note in both part A and part B how the fins have a web of skin supported by thin spines.

C. An African lungfish (*Protoperus annectens*) breathing at the surface of the water. Note that the fins are thick and fleshy with bony supports inside. (Note: This is a composite photo).

elements that support a fish's gills. Most modern jawed fishes have paired fins and **scales**, or hard, flattened structures that grow from and often cover the skin.

Bony fish. In **bony fishes**, an embryonic skeleton of cartilage develops into an adult skeleton consisting mainly of bone. Both cartilaginous fishes and bony fishes have paired fins, but those of bony fish have a much greater range of motion. Bony fishes also differ from other fishes in having their gill slits hidden beneath a gill cover. In jawless and cartilaginous fish, gill slits are visible at the body surface.

There are two lineages of bony fishes: ray-finned fishes and lobe-finned fishes. **Ray-finned fishes** have flexible fins supported by thin rays of bony spines derived from skin. A gas-filled swim bladder helps them adjust their buoyancy. With about 30,000 species, ray-finned fishes constitute nearly half of the vertebrates. They include most familiar freshwater fish as well as marine species such as grouper (Figure 13.33A) and bluegills (Figure 13.33B).

In contrast, **lobe-finned fishes** have thick, fleshy pelvic and pectoral fins with internal bony supports. Modern lobe-finned fishes include coelacanths (Chapter 12) and lungfishes (Figure 13.33C). As their name suggests, lungfishes have one or two lungs in addition to their gills. The fish inflates its lungs by gulping air, then air in the lungs exchanges gases with blood. Having lungs in addition to gills allows lungfishes to survive in low-oxygen waters. Note that neither coelacanths nor lungfish are found in Florida.

Land Ho! The First Land Animals: Amphibians

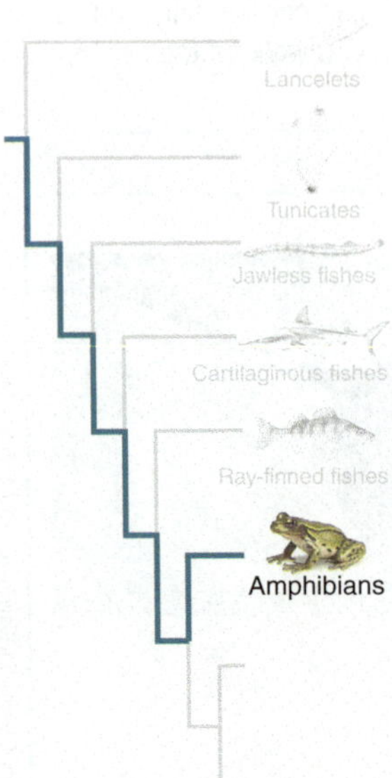

We've now reached a critical point in animal evolution—the emergence of the first animals that could live on land. Tetrapods, or vertebrates with four limbs, descended from an ancestral lobe-finned fish. Genetic comparisons indicate that lungfishes are closely related to the first tetrapods, the amphibians. **Amphibians** are scaleless vertebrates that generally spend time on land but require water to breed. Amphibian larvae are aquatic and have gills. Most species lose their gills and develop lungs during the transition to adulthood.

> **! Point of Confusion:** Amphibians are a specific kind of animal. Other animals, like birds and reptiles, can be *amphibious*, meaning they are found in water and on land, but they are not amphibians.

Looking at fossils, we can see the skeletal changes that occurred as fishes, with bodies adapted to swimming, evolved into the four-legged walkers (Figure 13.34). Amphibian limb bones and the bones of a lobe-finned fish's pelvic and pectoral fins are homologous structures

Figure 13.34 Fossil species from the late Devonian show how vertebrates made the transition from water to land

A. *Eusthenopteron.* This fish lived around 385 million years ago. Its skeleton suggests that it had features common to tetrapods, including limb-like bones within its fins that it may have used to drag itself across land.

B. *Tiktaalik.* This fish lived 375 million years ago. Evidence from this animal's skeleton indicates that it had bony appendages for dragging itself and a structure in the skull associated with primitive lungs in addition to gills.

C. *Ichthyostega.* *Ichtyostega* was one of the first tetrapods and lived around 365 million years ago. Despite being a land dweller, *Ichthyostega* shared many features with fish, such as its spine. Based on its skeleton, *Ichthyostega* likely moved like a mudskipper on land by dragging itself through crutching, which is a movement similar to a person walking using crutches.

Notice how the body structure changed over time. In the illustrations of fossil imprints at the right, ribs are highlighted in purple, digits in red, and forelimb and pelvic structures in orange.

(Chapter 10). During the transition to land, these bones became larger and better able to bear weight. Ribs also enlarged, and a distinct neck emerged, allowing the head to move independently of the rest of the body.

In addition to skeletal changes, the transition to land resulted in changes to other organ systems. For example, lungs became larger and more complex, and a more complex heart was needed to divert blood into the lungs. In addition, changes to the inner ear improved detection of airborne sounds, while the evolution of eyelids prevented delicate eye tissues from drying out.

Within the Indian River Lagoon, particularly in freshwater or brackish areas, one can find examples of amphibians, such as salamanders like the eastern newt (Figure 13.35A). You can also find the American green tree frog (Figure 13.35B), and while there are no toads that are native to the Indian River Lagoon, the non-native cane toad is established there (Figure 13.35C). As adults, nearly all amphibians are carnivores, feeding on mainly insects and worms.

Why would living on land be evolutionarily favorable? Individuals on land were safe from aquatic predators and had access to a new food—insects—which evolved around the same time.

Figure 13.35 **A gallery of amphibians**

A B C

A. The eastern newt, *Notophthalmus viridescens*; note how the front and back limbs are the same size. Salamanders and newts most closely resemble early tetrapods in body form.

B. An American green tree frog, *Dryophytes cinereus*); note how the frog has long muscular legs and short forelimbs

C. A cane toad, *Rhinella marina*; note the stubbier body and shorter hind legs than a frog.

Digging into Data

A Frog Swan Song

In Chapter 11, we learned about how new species form through speciation and are lost through extinction. We also learned that we are likely in the midst of a mass extinction event. Who warns us when a species is at risk of extinction? A consortium of scientists called The International Union for Conservation of Nature (IUCN). The IUCN tracks species and documents their extinction risk in their Red List Index (RLI; Chapter 16). You may have heard of their rankings before: Species can be extinct (entirely, or just in the wild), critically endangered, endangered, vulnerable, near-threatened, or of least concern. Based on their RLI, amphibians are the most threatened vertebrates (Figure 13.36).

To help learn more about threatened amphibians, the IUCN collaborated with a group of international scientists that included National Geographic Explorer Jodi Rowley (Figure 13.37) and her team, to understand where most threatened amphibians were located (Figure 13.38). The IUCN also wanted to identify major threats to their survival.

When the research team considered the threats affecting amphibians in all regions, the most commonly documented threats by rank were: (1) agriculture, (2) timber and plant harvesting, (3) infrastructure development, (4) climate change effects, and (5) disease.

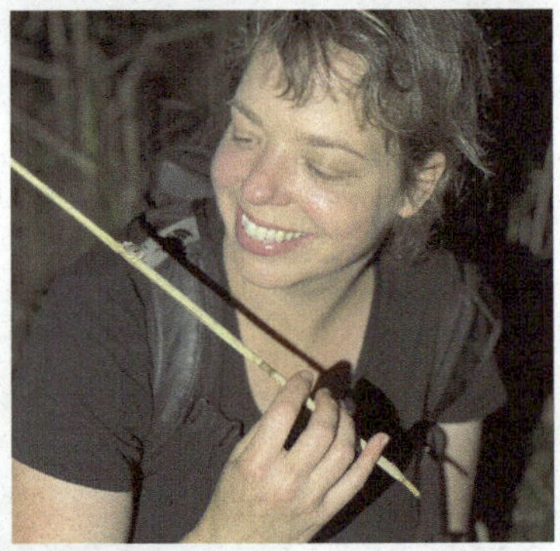

Figure 13.37 **National Geographic Explorer Jodi Rowley**

Dr. Rowley is an Australian conservation biologist who collaborated with the IUCN to study amphibian species under threat.

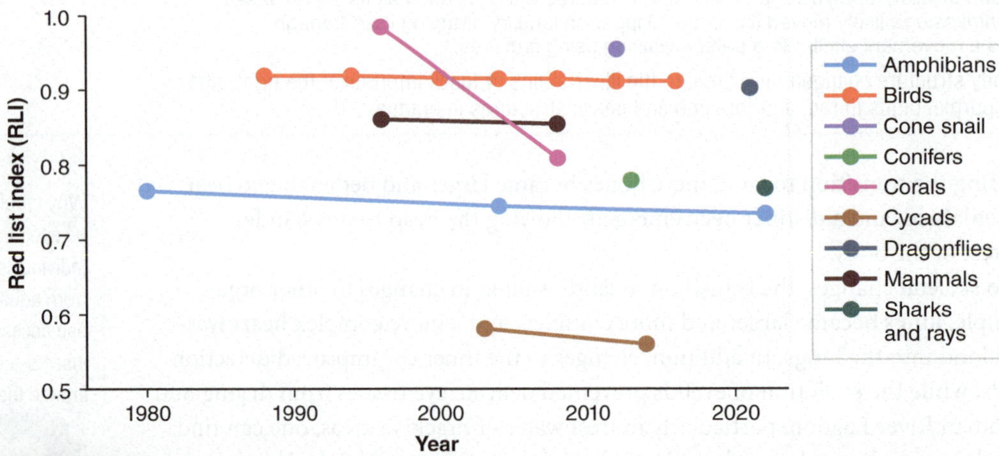

Figure based on: Luedtke, J.A., Chanson, J., Neam, K. *et al.* (2023). Ongoing declines for the world's amphibians in the face of emerging threats. *Nature* 622, 308–314. https://doi.org/10.1038/s41586-023-06578-4

Figure 13.36 **Extinction risk**

The RLI of all comprehensively assessed taxonomic groups on the IUCN Red List. Note that data points are only shown for years in which that group was studied.

The Arrival of Amniotes

During the Carboniferous period, amphibians were the dominant animals on land. Late in that period, around 340 million years ago, the first amniotes—which include reptiles (including birds) and mammals—evolved from an amphibian ancestor.

> **Point of Confusion:** Birds are considered reptiles because of their evolutionary history. Birds, dinosaurs, and crocodiles are all descended from Archosaurs. Therefore, it's correct to say that birds are dinosaurs! Note that this is a different lineage from other reptiles, such as snakes, lizards, and turtles.

The defining trait of an **amniote** is the unique type of egg that they lay, which allows them to reproduce on land. Inside an amniote egg, the embryo develops enclosed within protective membranes and surrounded by fluid. In most lineages, females lay eggs that have a hard or leathery shell (Figure 13.39), but in some mammals, development takes place inside the mother's body.

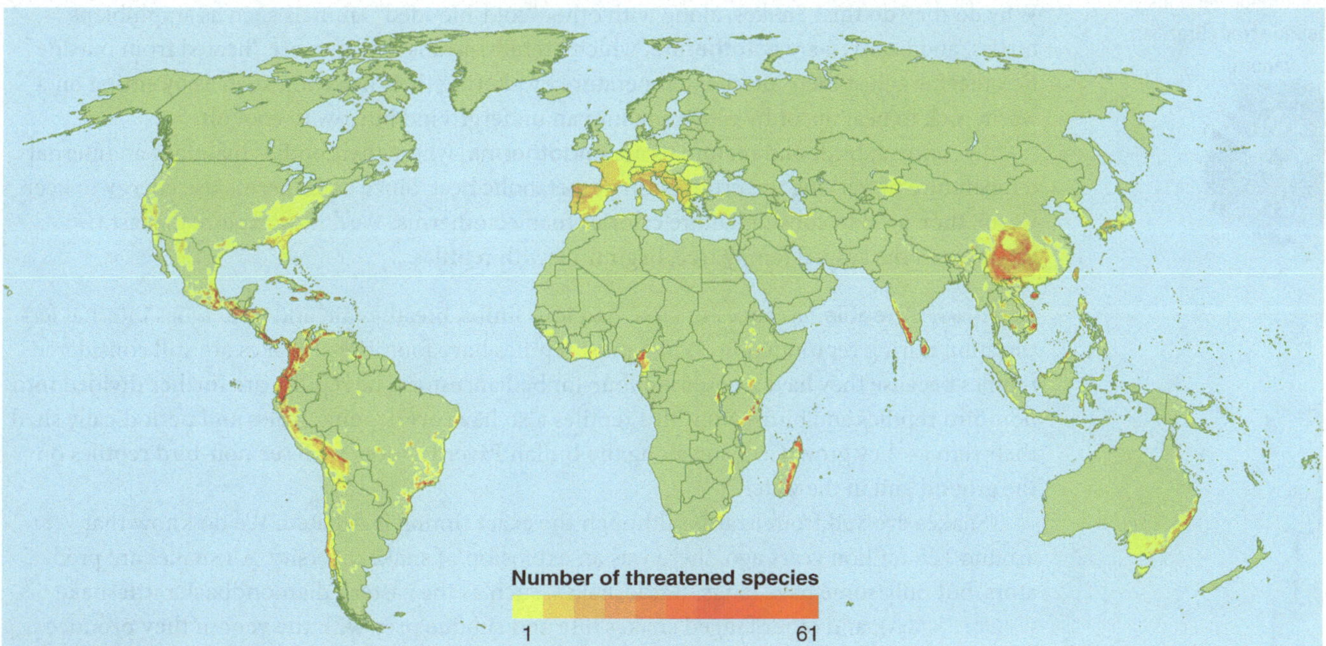

Figure based on: Luedtke, J.A., Chanson, J., Neam, K. *et al.* (2023). Ongoing declines for the world's amphibians in the face of emerging threats. *Nature* 622, 308–314. https://doi.org/10.1038/s41586-023-06578-4

Figure 13.38 The distribution of globally threatened amphibian species

The colors correspond to the locations of threatened species worldwide, on a color scale from yellow (fewer species threatened) to red (more species threatened), with a maximum number of 61 species assessed.

1. Examine Figure 13.36. Which group shows the greatest RLI change? Is that group becoming more or less threatened? Do some research (for example, see Chapter 16) and identify possible explanations for the trend.
2. Examine the map in Figure 13.38. Can you identify a region where amphibian species are threatened? Based on the color key, about how many species are threatened in this area? Do some research. Name a specific species that is threatened in this area and the factors that are putting it most at risk.
3. Consider the ecosystems explored in this chapter and cross reference Figure 13.38. Which ecosystem has the highest number of threatened amphibians? What features of this ecosystem do you think explain the diversity of amphibians that live in this region?
4. Some taxonomic groups have only one data point. What factors might affect whether a study can be performed over multiple time frames? What conclusions can be made with only one data point?

Figure 13.39 Amniote eggs protect reptile embryos

A. Birds such great egrets (*Ardea alba*) lay hard-shell eggs.
B. American alligators (*Alligator mississippiensis*) lay eggs with soft shells. Note how the egg is crumpling around the baby alligator's body as it emerges.

Credits: iStock.com/Jrubacha; NASA/Kim Shiflett

Venom from lizards and snakes, like the venom from marine invertebrates, contains compounds that can have medical benefits. ACE inhibitors, a class of drugs used to treat high blood pressure, are synthetic versions of a peptide from Brazilian viper venom.

In addition, other traits that allow amniotes to live away from water include an abundance of keratin in the skin, which allows the skin to be waterproof. Amniotes also have a well-developed pair of kidneys to help conserve water.

As we journey through the Indian River Lagoon, perhaps we'll see a snake sitting on a rock. Why do they do this? Snakes, along with other "cold-blooded" animals such as amphibians, turtles, and lizards—are **ectotherms**, which literally means that they are "heated from outside." Ectotherms adjust their internal temperature by altering their behavior, such as by sitting on a warm rock to heat up, or by retreating into an underground burrow to cool off.

In contrast, birds and mammals are **endotherms**, where their bodies maintain an internal temperature by varying the production of metabolic heat. Since endotherms use energy to keep warm, they need to consume more calories than ectotherms. We'll now explore the last two branches in the animal family tree, beginning with reptiles.

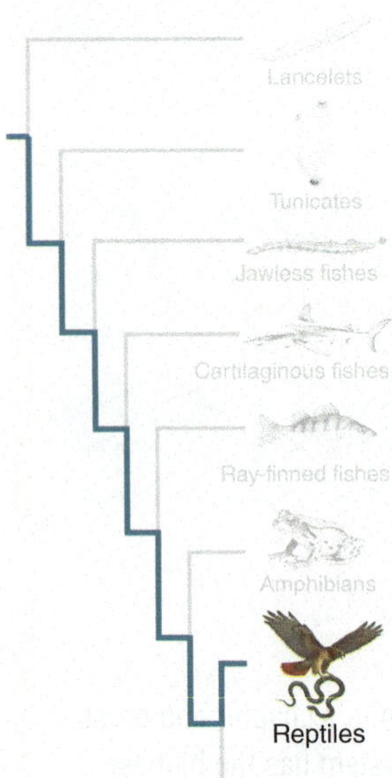

Reptiles. A **reptile** has ancestors that had four limbs, breathes air, and undergoes internal fertilization during reproduction. While most reptiles have four limbs, snakes are still considered reptiles because they have a common four-limbed ancestor. The reptiles are further divided into non-bird reptiles and birds. Non-bird reptiles also have overlapping scales and periodically shed their skin as they grow. Cruising along the Indian River Lagoon, we'll see non-bird reptiles on the ground and in the water.

Snakes evolved from lizards, although the exact timing is debated. We do know that around 128 million years ago, there was an explosion of snake diversity. All snakes are predators, but only some have fangs. Rattlesnakes, such as the eastern diamondback rattlesnake (Figure 13.40A), and other fanged snakes bite and subdue prey with the venom they produce in modified salivary (saliva-making) glands. Other snakes are constrictors that wrap themselves around a prey animal, suffocating it. Lizards, such as the endemic island glass lizard (which doesn't have legs; Figure 13.40B) and exotic green iguanas (Figure 13.40C) can also be found in the Indian River Lagoon. Legless lizards share a slightly different evolutionary history than snakes. Legless lizards are descended from lizards with legs. Lizards differ from snakes in that they have movable eyelids, an inflexible jaw, and ears that open to the outside. In contrast, the earliest snakes had legs. That is why some snakes, like pythons, have vestigial (Chapter 10) legs.

Turtles, like the loggerhead sea turtle, have a bony, protective shell attached to their backbone (Figure 13.40D). Modern turtles do not have teeth. Instead, a thick layer of the protein keratin covers their jaws and forms a horn-like beak. Most turtles live either in the sea or in freshwater. Those that live entirely on land, like the gopher tortoise (Figure 13.40E), are commonly called tortoises.

Crocodilians live in or near water and include crocodiles, alligators, and caimans—all three of which can be found in the Indian River Lagoon. They are predators with powerful jaws, a long snout, and sharp teeth (Figure 13.40F).

Figure 13.40 **A gallery of non-bird reptiles**

A. An Eastern diamondback snake (*Crotalus adamanteus*); note how the scales create a diamond pattern.
B. The island glass lizard (*Ophisaurus compressus*) is a type of legless lizard.
C. A green iguana, *Iguana iguana*, which can be found on land and in water.
D. A hawksbill sea turtle, *Eretmocheyls imbricata*; note how the shell is part of the animal's body.
E. Gopher tortoises (*Gopherus polyphemus*) live entirely on the islands within the Indian River Lagoon.
F. An American alligator, *Alligator mississippiensis*; note its sharp teeth and elongated snout.

Credits: David A. Northcott/Corbis Documentary/Getty Images; Photos by J.D. Willson; iStock.com/Sandi Smolker; Source: Caroline S. Rogers/NOAA/U.S. Fish and Wildlife Service; iStock.com/BrianLasenby; iStock.com/Zanthra

Birds are the only modern animals with feathers, which are modified scales. Feathers and other traits adapt the bird body to flight. Bird wings are homologous to our arms. Contraction of one set of muscles produces a powerful downstroke that lifts the bird. A less powerful set of muscles contracts to raise the wing.

Flying also requires good eyesight and a great deal of coordination. Compared to a lizard of a similar body mass, a bird has much larger eyes and a larger brain. Several birds that live in the Indian River Lagoon have specialized adaptations that allow them to spot fish from the air and dive to catch their prey. For example, ospreys (Figure 13.41A) have large talons for grasping fish, and can toss the fish in the air to move it into a more aerodynamic position. Pelicans dive into the water (Figure 13.41B) and capture fish in the large fleshy pouches that hang below their bills.

Most birds are surprisingly lightweight. Air cavities inside a bird's bones keep its body weight low, as does the lack of a bladder, an organ that stores urinary waste in many other vertebrates. Rather than heavy, bony teeth, bird jaws are covered with a lightweight bill made of keratin. Note how a sanderling (a type of sandpiper) uses its bill to sift through the sand looking for food (Figure 13.41C).

Birds, such as these indigo buntings (Figure 13.41D), also use song and visual cues (compare the brightly colored male to the dun-colored female) to communicate with one another—an example of sexual selection (Chapter 11). In the case of indigo buntings, only the males sing, and only during breeding season to attract a mate. Indigo bunting males may also dance as well as part of breeding. In addition, many species of birds, including indigo buntings, mate for life.

In birds, as in other reptiles, fertilization is internal. Unlike most male non-bird reptiles, male birds do not typically have a penis. Thus, to inseminate a female, a male bird must press a structure called a *cloaca*—an opening or cavity that serves as the sole exit for urine, feces, and gametes in birds and other reptiles—against hers, in a maneuver poetically described as a cloacal kiss. A female bird lays an egg with membranes, which is characteristic of amniote eggs. Nutrients from the egg's yolk and water from the egg white (albumen) sustain the developing embryo. In nearly all bird species, one or both parents incubate the eggs until they are ready to

? Did You Know?

Hawksbill turtles are endangered due to the use of their shells to make "tortoise shell" jewelry.

Alligator or crocodile? Alligators have a rounded snout, whereas crocodiles have more of a V-shaped snout.

? Did You Know?

Birds, like the red-winged blackbird, also sing to denote their territory!

Figure 13.41 **A gallery of birds**

A. An osprey (*Pandion haliaetus*) in flight, carrying its lunch. Note the talons that are holding onto the fish, and the position that the osprey holds the fish so as to minimize drag on its flight.

B. A brown pelican (*Pelecanus occidentalis*) in mid-dive.

C. A sanderling, *Calidris alba,* forages for food in the surf.

D. A pair of indigo buntings (*Passerina cyanea*); the male (left) is a bright blue, whereas the female (right) is brown.

Credits: Zoonar GmbH/Alamy Stock Photo; iStock.com/CarolinaBirdman; ImageBROKER.com/Alamy Stock Photo; Bonnie Taylor Barry/Shutterstock.com

hatch. Many birds hatch in a relatively undeveloped state and require extensive parental care before they can live on their own.

Mammals. We now arrive at the last class of animals we'll be considering in this section: the mammals. **Mammals**, as we learned in Chapter 12, are amniotes in which females nourish their offspring with milk secreted from mammary glands. The class name is derived from the Latin *mamma*, meaning "breast." Mammals are the only animals that have hair or fur. Like feathers, hair and fur are modifications of scales. Mammals are endotherms, and their hair or fur helps them to maintain their core temperature.

Mammals have distinctive skeletal and dental traits. Compared to other vertebrates, they have a larger skull and brain for their body size, and they are the only vertebrates with three bones in their middle ear. Most mammals have multiple types of teeth, whereas in other jawed vertebrates, all of the teeth are similarly shaped. Having a variety of different kinds of teeth allows mammals to eat a wider variety of foods than most other vertebrates.

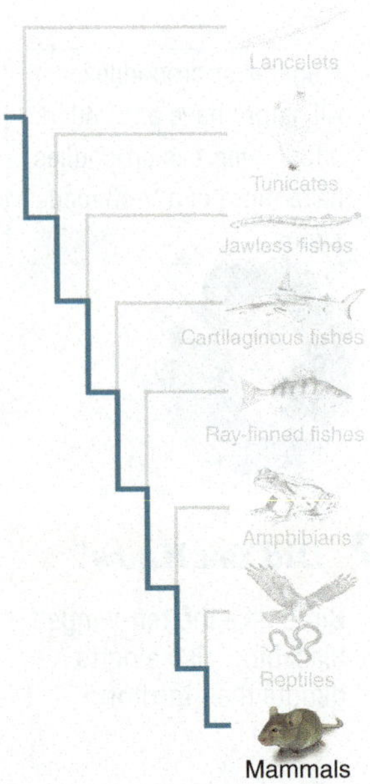

Mammals evolved early in the Jurassic period, around 225 million years ago, when early, mouselike species coexisted with dinosaurs. By 130 million years ago, three lineages had evolved: **monotremes**, which are egg-laying mammals that are only found in Australia or on islands near Australia; **marsupials**, which are pouched mammals, like the opossum (Figure 13.42A); and **placental mammals**, mammals in which an organ called the placenta provides nutrients to developing offspring.

Placental mammals are the most diverse and abundant mammalian lineage. Placental mammals have a competitive edge over monotremes and marsupials because they have a higher metabolic rate, better body temperature control, and a more efficient way to nourish embryos. Compared to other mammals, placental mammals develop to a far more advanced stage inside their mother's body.

In the Indian River Lagoon, there are both aquatic and terrestrial mammals. Aquatic mammals include charismatic species such as manatees (Figure 13.42B) and dolphins (Figure 13.42C). It's quite common to spot both manatees and dolphins while boating in the Indian River Lagoon. In the vicinity of the Indian River Lagoon, one can also find several species of the only flying mammals—bats (Figure 13.42D). Although bats may look like flying mice, bats are more closely related to carnivores such as wolves and foxes than to rodents.

We've now concluded our tour of animals, from the first animals (Section 13.2) to the chordates (this section). Remember, information about primates and human evolution was covered in Chapter 12. From here, our journey will head from the East Coast back to the West Coast, where our focus will switch from the evolutionary history of animals and protists to plants and fungi.

Figure 13.42 **A gallery of mammals**

A. Opossums (*Didelphis virginiana*) are the only marsupials in North America.

B. Manatees (*Trichenchus manatus*), like all other mammals, provide milk to their young. Here, a calf nurses from a nipple below its mother's flipper.

C. Bottlenose dolphins (*Tursiops truncatus*) are often spotted from the shore or a boat.

D. A hoary bat (*Lasiurus cinereus*) is the second-largest bat species in Florida. These bats eat only insects.

Credits: Ivan Kuzmin/Alamy Stock Photo; Gregory Sweeney/Getty Images; iStock.com/SBphotos; MerlinTuttle.org/Science Source

13.3 Summary

- Chordates are a group of bilateral deuterostomes that include invertebrate and vertebrate animals that have (or had, during development) a notochord, a hollow nerve cord, gill slits, and a muscular tail.
- The older chordate lineages are invertebrates, those that do not have backbones, like lancelets and tunicates. More recently evolved chordates with a backbone are called vertebrates, which include fish, amphibians, reptiles, and mammals.
- Fish have no limbs. Jawless fish have skeletons made of cartilage and have no jaws, while jawed fish have jaws and their skeletons can be made of either cartilage or bone.
- Amphibians were the first vertebrates with limbs (tetrapods); they are scaleless and spend time on land but require water to breed.
- Reptiles—such as lizards, snakes, turtles, and birds—have four limbs (or ancestors with four limbs), and they breathe air and undergo internal fertilization.
- Mammals have hair or fur, and they feed their offspring with milk.
- Fish move through water either by wiggling (jawless fish) or using fins. Amphibians live in water as larvae, at which point they have gills, but as they mature into adults and move onto land, they develop lungs and breathe air.
- Reptiles lay eggs with protective membranes that help them to survive on land. Their skin is rich in keratin, making them waterproof, and they have kidneys that help them to conserve water.
- Mammals have large brains and multiple types of teeth that allow them to eat a variety of foods. Placental mammals have competitive advantages due to their higher metabolic rate, body temperature control, and more effective method of supporting their embryos.
- As animals adapted to life on land, limb bones evolved from fins to become larger and better able to bear weight. A distinct neck evolved, allowing the head to move independently, and lungs and hearts became larger and more complex.
- Birds have additional adaptations for flight, including feathers, a larger brain and eyes than lizards, and a lightweight body.

13.4 The Diversity of Plants: A Visit to a Temperate Rainforest

Learning Objectives

13-14 Describe the key traits that led to evolutionary advantages for each plant lineage.

13-15 Explain how plant structure relates to its growth and reproduction.

13-16 Compare seedless and seeded plants.

13-17 Differentiate between plants and fungi.

13-18 Describe the key characteristics of kingdom fungi.

> **? Did You Know?**
>
> Several movies were filmed in Ecola State Park, including *The Goonies* and *Twilight*.

> **Study Tip**
>
> Use the names of the different lineages to help you remember the key features of each evolutionary lineage.

The next stop on our field trip is different from the last three. Not only are we completely on land for the first time, but rather than focusing on the evolution of the animal or protist kingdoms, we'll be focusing on the evolution of two other eukaryotic kingdoms: plants and fungi. For this tour stop, we're visiting coastal Oregon, in the vicinity of Ecola State Park. As with our other tour stops, the Oregon coast is home to a variety of interesting animals similar to those we've already discussed, but our focus here will be on plants and fungi.

For this section, we'll be taking a hike through a temperate rainforest along the Pacific coast. On our hike, we'll view the four major evolutionary lineages of plants: nonvascular plants, seedless vascular plants, gymnosperms, and angiosperms. We'll learn how the different plant lineages reproduce and the evolutionary advantages of each. As our hike continues, we'll switch our focus to examine fungi and how they can partner with protists to form lichens. Grab your hiking boots—it's time to begin our journey!

A Portrait of Plants

Before we start on our hike, let's review the basics of plant evolution. A **plant** is a eukaryotic organism that is land dwelling and multicellular. Plants evolved from freshwater green algae and share many traits with this group. The defining trait that sets plants apart from algae is that plants have

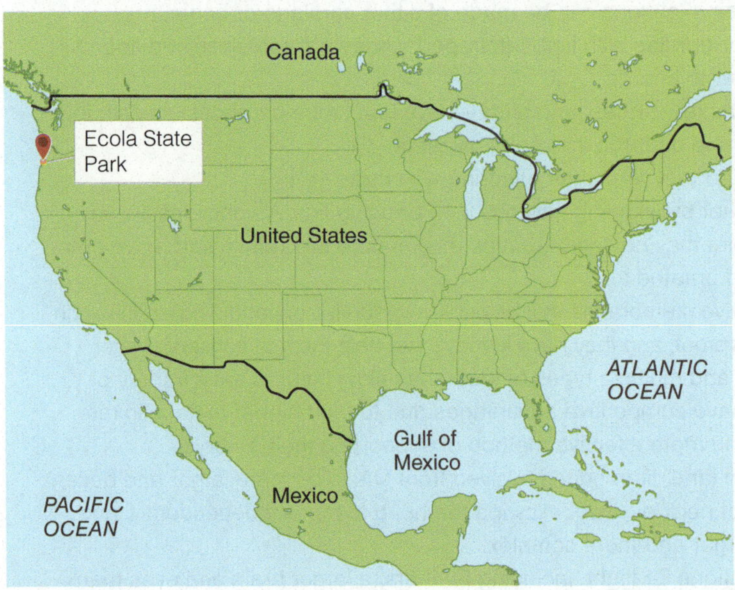

Credit: Halbergman/E+/Getty Images

A temperate rain forest. Temperate rain forests are generally in coastal regions, such as the Pacific Northwest, and receive high annual rainfall, experience mild temperatures, and are home to many species of plants.

a multicellular embryo that forms and develops within the parent plant and is nourished by it. Plants are also defined by having a waxy covering called a **cuticle**, as well as pores for gas exchange called **stomata** (Chapter 4). Both of these structures help to minimize water loss in land plants, which first evolved 475 million years ago.

The story of plant evolution is one of adaptation to increasingly drier environments. An aquatic green alga lives surrounded by water, so it can absorb water and dissolved nutrients across its entire surface. Water buoys the alga's parts, thus helping it stand upright. On land, plants face the constant threat of drying out, and they often must hold themselves upright. Adapting to these new challenges involved changes in life cycle, body structure, and altered mechanisms of reproduction and dispersal. When we consider evolutionary lineages of plants, the two main features we examine are (1) if a plant has vascular tissue and (2) how the plant reproduces. The four lineages of modern plants are shown in Figure 13.43.

Figure 13.43 The evolutionary history of plants

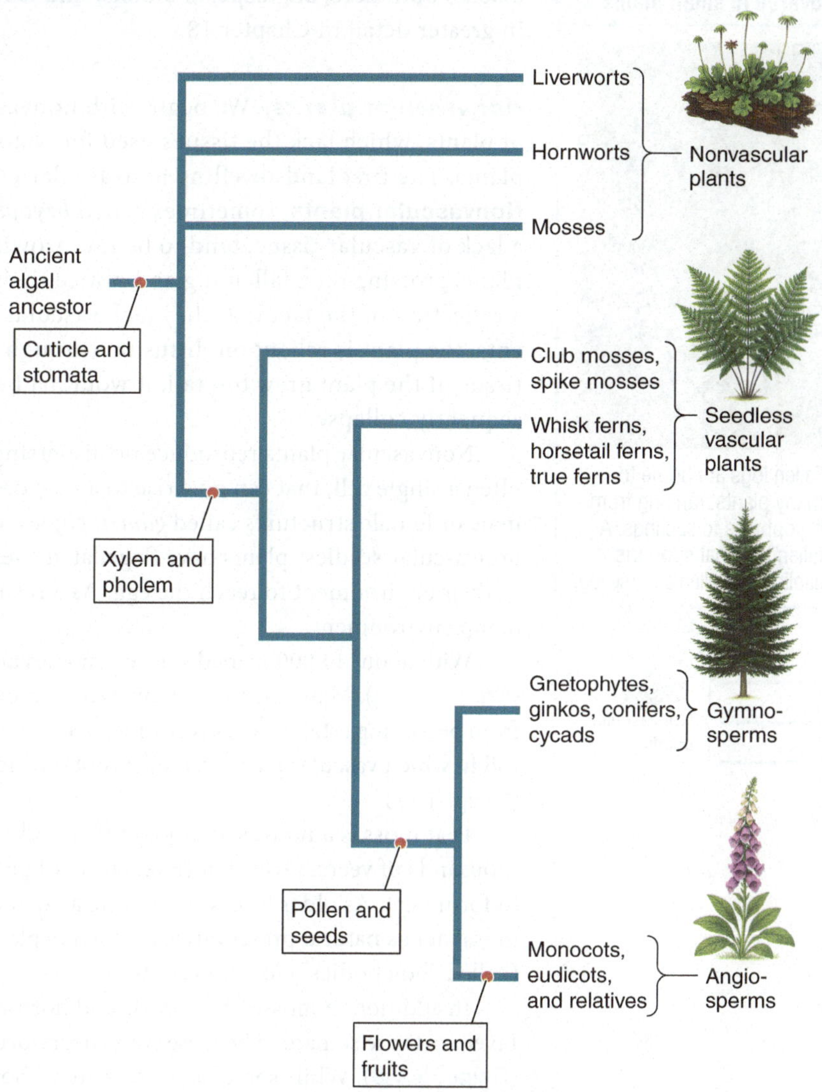

Key innovations are indicated on the cladogram with boxed text.

For a plant to stand up against gravity on land requires vascular tissue. Plants with vascular tissue first evolved around 420 million years ago. Vascular tissue serves dual purposes in that it provides structural support while also supplying water and nutrients to the plant. There are two types of vascular tissue. **Xylem** is the vascular tissue that distributes water and mineral ions. **Phloem** is the vascular tissue that distributes sugars produced by photosynthetic cells. More than 90% of modern plants have xylem and phloem and thus are classified as vascular plants. We'll return to xylem and phloem and discuss plant physiology in greater detail in Chapter 18.

Now, let's consider the evolution of reproductive traits. Nonvascular plants and seedless vascular plants disperse by releasing spores, but **seed plants** disperse by releasing seeds. A **seed** consists of an embryo sporophyte and food to support it, enclosed within a protective coat. A seed is formed following fertilization of an egg by a sperm. A **pollen grain** is an immature male *gametophyte,* which produces sperm cells once the pollen grain is germinated (Chapter 18). Wind or animals can convey pollen grains between plants, so seed plants can reproduce even in dry environments.

Plants that produce pollen and seeds, the *gymnosperms*, evolved around 350 million years ago. Flowering plants, the *angiosperms*, evolved around 160 million years ago. Angiosperms evolved to produce **flowers**, a specialized reproductive structure that consists of modified leaves arranged in concentric whorls (see Figure 18.16). A **fruit** is the structure

Figure 13.44 **A fallen log covered in small plants**

Fallen logs are home to many plants, ranging from bryophytes to saplings. A fallen tree that supports saplings is called a nurse log.

Credit: Melanie Peffer

that contains the seeds that form after the plant produces flowers. Note that while we'll discuss both here, angiosperm biology and its relationship to our food supply are discussed in greater detail in Chapter 18.

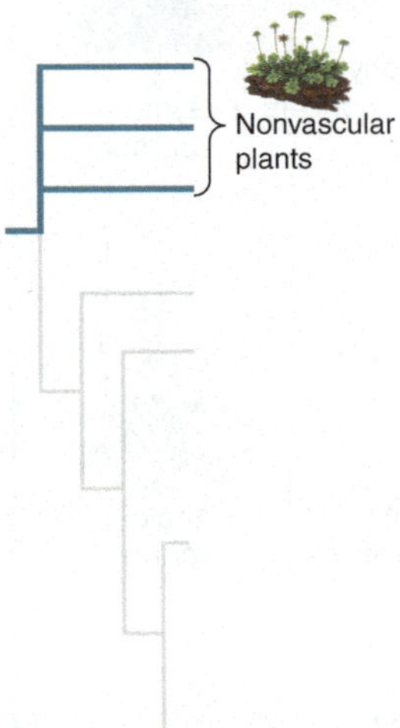

Nonvascular plants. We begin with nonvascular plants, the oldest known lineage of plants, which lack the tissues used for water and mineral transport seen in vascular plants. The first land-dwelling nonvascular plants evolved around 470 million years ago. **Nonvascular plants**, sometimes called *bryophytes*, are simple, seedless plants that, due to a lack of vascular tissue, tend to be low growing. While hiking you may see nonvascular plants growing on a fallen log and notice their tiny size (Figure 13.44). This small size is a reflection of the fact that they lack vascular tissue. In order to obtain water and nutrients, the plant is reliant on diffusion through its surface. In addition, without vascular tissue, if the plant grew too tall, it wouldn't be able to support its weight and would consequently collapse.

Nonvascular plants reproduce by dispersing spores. A **spore** is a reproductive unit, often a single cell, that can give rise to a new organism. These spores then develop into male or female structures called *gametophytes*, which produce male or female gametes. All nonvascular seedless plants have flagellated sperm, which swim through droplets of water in their environment to reach the egg. As a result, these plants can reproduce only in a damp environment.

With about 14,000 named species, mosses are the most diverse nonvascular plants (Figure 13.45A). Many mosses colonize rocky areas where the lack of soil prevents other plants from becoming established. Over time, decomposition of dead moss helps to create a layer of soil in which vascular plants can take root—which is an important part of ecological succession (Chapter 15).

Peat moss is a nonvascular plant that makes up peat bogs, many of which have existed for thousands of years. Over that time, layers of partially decayed plant remains have compressed to form peat. As older layers rot, humic acid and sphagnum (produced by dying sphagnum moss) act as natural preservatives, which explains why bogs can preserve human bodies (called "bog bodies") for thousands of years.

In addition to mosses, liverworts and hornworts are also seedless nonvascular plants. Liverworts are so named because some liverworts are flat and multilobed like a human liver (Figure 13.45B). While some hornworts are ribbonlike, their common name refers to the hornlike spore capsule, which is shown in (Figure 13.45C).

Figure 13.45 **A gallery of nonvascular plants**

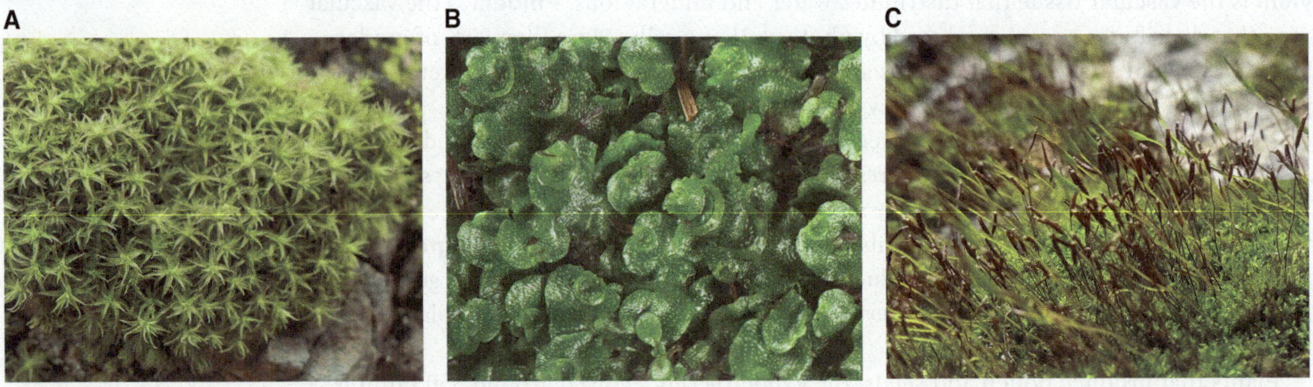

A. A *Dicranum* moss.

B. An exotic crescent-cup liverwort *Lunularia cruciate,* with its moon-like shape.

C. Hornwort, *Phaeoceros carolinianus*; note the slender "horns" that give the plant its name. The horns or capsule of the plant will split to release spores.

Credit: najeebbajauri/Shutterstock.com; Nigel Cattlin/Alamy Stock Photo; Sergey Shakuro/Shutterstock.com

Seedless vascular plants. Like nonvascular plants, **seedless vascular plants** produce spores and require water to reproduce, as their flagellated sperm must swim to reach an egg. However, the distinguishing characteristic of seedless vascular plants is in their name—they have vascular tissue. Recall that vascular tissue is a series of tubes inside the plant that distribute water, sugars, and minerals. This key innovation in support and plumbing allowed seedless vascular plants to be larger and more structurally complex than nonvascular plants, with roots, stems, and leaves.

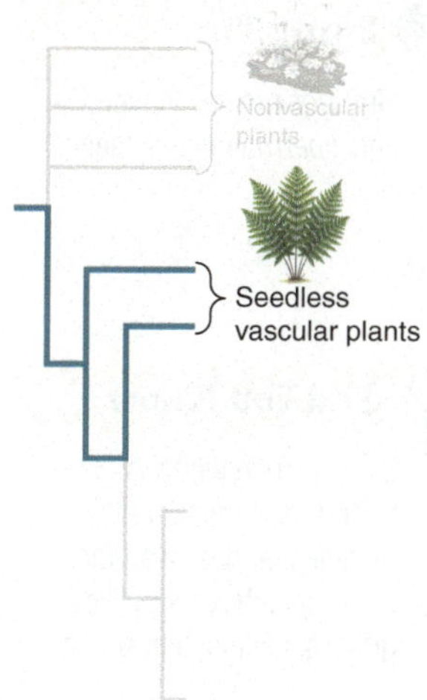

As we continue on our hike, we may see **lycophytes**, which are the oldest living group of vascular plants. Lycophytes include club and spike mosses, although these are not true mosses, since they have vascular tissue (**Figure 13.46A**). Club mosses, the most common type of lycophyte, are common on the floor of temperate forests in places like Ecola State Park. Roots and upright stems are covered by tiny, scalelike leaves, where each leaf has a single, unbranched vein.

While hiking, it's hard to miss ferns along hiking trails (**Figure 13.46B**). You don't even need to go on a hike to see them—people often have ferns in their homes or offices. Along a hiking trail, you may also see fiddleheads, or immature fern leaves (fronds), that are wound up like the scroll at the end of the neck of a fiddle. Some people enjoy eating fiddleheads in salads. Finally, along our hike we might see a whisk fern, which doesn't have leaves or roots, and instead gets its water and nutrients through *mycorrhizae*, which we'll discuss later in this section.

Horsetails (**Figure 13.46C**) are also relatively easy to spot close to hiking trails. Horsetails thrive along streams and roadsides, and in areas disturbed by humans. A horsetail has hollow stems with tiny nonphotosynthetic leaves at the joints, where photosynthesis instead occurs in stems and leaflike branches. The stems of horsetails contain silica, a gritty mineral that helps the plant fend off insects and snails. Before the invention of modern abrasive cleansers, the silica-rich stems of some horsetail species were used to scrub pots and polish metals.

Gymnosperms. Seed plants evolved from a lineage of seedless vascular plants during the Carboniferous period. They survived alongside nonvascular plants and seedless vascular plants until the late Carboniferous, then rose to dominance as the climate became drier. Unique traits of seed plants give them a competitive advantage over seedless plants in places where water can be scarce.

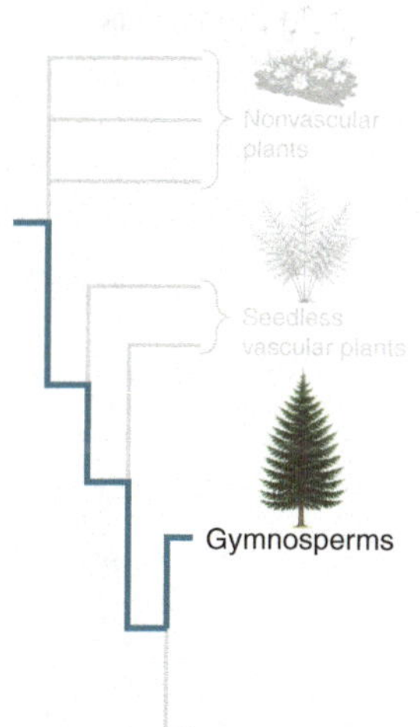

Gymnosperms are seed plants that produce seeds on the surface of an *ovule,* which is a part of the female reproductive structures. Their seeds are said to be "naked," because unlike seeds of angiosperms, they are not inside a fruit. However, many gymnosperms enclose their seeds in a fleshy or papery covering.

Figure 13.46 **A gallery of seedless vascular plants**

A. Common clubmoss, or stag's horn clubmoss (*Lycopodium clavatum*); notice the pine tree-like features—these plants are sometimes called ground pines.

B. A western sword fern (*Polystichum munitum*) has long fern leaves or fronds and veins of vascular issue.

C. Great horsetails (*Equisetum telmateia*) look like a horse's tail.

Credit: Manfred Ruckszio/Shutterstock.com; Donald E. Hall/Moment/Getty Images; Federica Grassi/Moment/Getty Images

Study Tip

Gymnos means "naked," and *sperma* means "seed."

? Did You Know?

Sitka spruce wood is prized for its resonant qualities, and is used for making instruments like guitars, violins, and pianos.

? Did You Know?

Conifers include the world's tallest trees—redwoods—and the longest-lived plants—bristlecone pines.

Figure 13.47 Sitka spruce (*Picea sitchensis*) trees

The Sitka spruce is a conifer that can grow to be over 100 m (330 feet) tall.

Credit: iStock.com/Wirestock

As we continue on our hike, it's difficult to miss the massive Sitka spruce trees (Figure 13.47). Sitka spruce trees are a type of conifer. **Conifers** are the most diverse gymnosperms, with more than 600 species. All are woody trees or shrubs with needle-like or scalelike leaves. Most conifers are **evergreen plants**, meaning they keep their leaves throughout the year.

Evergreen conifers are the main plants in cool northern hemisphere forests. A conical shape helps many conifers shed snow easily. Needlelike leaves with a heavy wax coating help a conifer to minimize water loss during a long winter, when soil is frozen, or during a hot, dry season. Conifers are of great economic importance. We use trees for building homes and furniture; flavor gin with juniper "berries" (which are actually female seed cones); use oils from cedar in cleaning products; and eat the seeds, or "pine nuts," of some pines.

As part of their life cycle, conifers produce **cones**, which are reproductive structures that contain either pollen or ovules. Let's consider the cones of another tree we may see hiking, which you might have had in your house during the holidays—the Douglas fir (*Pseudotsuga mensiesii*). Conifers have two different types of cones: a small one that produces pollen (male; Figure 13.48A) and large seed-containing cones (female; Figure 13.48B). Both types of cones have scales arranged around a central axis. Depending on the type of cone, each scale will contain either male pollen sacs or female ovules. Wind causes pollen to be released from male pollen-producing cones, where it is then carried to a different tree. The pollen can then become trapped on the scales of a female seed cone, where the pollen grain germinates, producing sperm. The sperm fertilize an egg, and an embryo will then form inside a conifer seed.

Note that conifers are not the only kind of gymnosperms—there are also cycads and ginkgos. We talked about cycads in Chapter 10, which were common when dinosaurs walked the Earth, but now species like sago palms are commonly used in landscaping. Cycads and ginkgos are the only plants with seeds and flagellated sperm, similar to the seedless vascular plants. Ginkgos are an ancient lineage of gymnosperms—one with leaves instead of needles. Ginkgos are also unique in that they are a gymnosperm whose female trees produce fruit—which have an unmistakable bad odor. Male ginkgo trees are often planted in cities because they are resistant to air pollution, people like their pretty fan-shaped leaves, and they don't produce stinky fruit.

Figure 13.48 A comparison of Douglas fir cones

A. These are pollen-bearing male cones.
B. These are seed-bearing female cones.

Credits: Danita Delimont/Alamy Stock Photo; Milan Sommer/Shutterstock.com

Figure 13.49 **A gallery of angiosperms**

A. A foxglove, *Digitalis purpurea*. Note the tube-like flowers, which bees will climb into to obtain nectar.
B. Pink purslane, *Claytonia sibirica*, also called the candy flower for its stripes
C. A Pacific trillium, *Trillium ovatum*. The *tri* in its name is for the three petals in the flower.

Credits: iStock.com/Nikamata; Martin Fowler/Shutterstock.com; iStock.com/Joshua Moore

Angiosperms. A highlight of any hike is seeing flowers—and while hiking in coastal Oregon, there are many beautiful flowers to see (Figure 13.49). **Angiosperms** are the only plants that make flowers and disperse their seeds inside of fruits. We will cover fruits and flowers in greater detail in Chapter 18.

Flowering plants (which include flowering trees) constitute 90% of all modern plant species. Their success comes from how fast they can grow (think of the weeds in your yard!) and from the coevolution of flowering plants with animals, such as insects and birds (Chapter 12), which makes plant reproduction more efficient. After pollen-producing plants evolved, some insects began feeding on the plants' highly nutritious pollen (Figure 13.50A). Plants give up some pollen (and a sugary substance called *nectar*) but gain a reproductive edge when insects unknowingly facilitate **pollination**, where male pollen is moved to a different plant, coming into contact with the female part of a flower. This results in fertilization, where an ovule matures into a seed and the ovary becomes the fruit (Figure 13.50B).

Plants evolved many strategies for helping to disperse their seeds, such as by flying through the air or sticking to animal fur. We will discuss such seed dispersal strategies—as well as flower anatomy and the process of plant fertilization—in greater detail in Chapter 18.

? Did You Know?

Foxglove produces digitalis. Now sold under the brand name Lanoxin®, digitalis is used to treat heart conditions like atrial fibrillation.

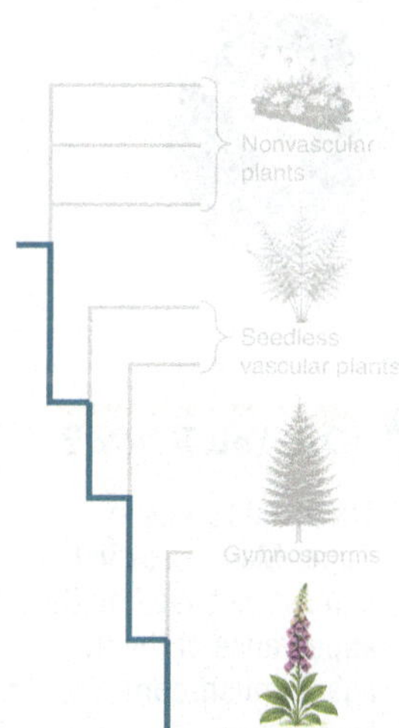

Not a Plant, but Still Fun: Kingdom Fungi

Now that we've covered the evolutionary history of plants, we'll move onto another major lineage of eukaryotes: the fungi. Note that fungi are a completely different evolutionary lineage than plants (Figure 13.51). A **fungus** (plural, fungi) is a eukaryotic heterotroph, meaning it eats

Figure 13.50 **Coevolution with pollinators was critical for angiosperm success**

A. A pollen-covered bee on a pink purslane flower. Note the bulbous female part of the flower below the bee.
B. The resulting pink purslane fruit forms after pollination and successful fertilization.

Credits: Peter Pearsall/USFWS; Janet Horton/Alamy Stock Photo other organisms. They do so by secreting digestive enzymes onto their food, then absorbing the resulting breakdown products.

Most fungi are decomposers that feed on organic wastes and remains, but some live on or in other living organisms—including us! Fungi typically grow on our body surface and eat the outer layers of skin, causing raised, red, and itchy skin. For example, athlete's foot and ringworm are both fungal infections. The ring-shaped lesion associated with ringworm is from the growth of *hyphae* outward from the initial infection.

Figure 13.51 Fungi are a distinct lineage from plants in the tree of life

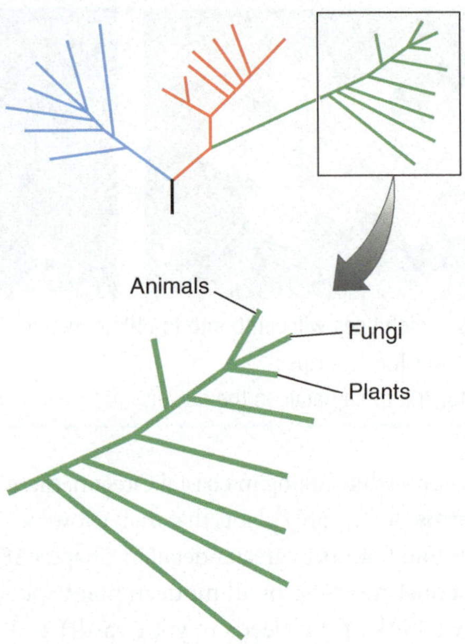

Shown here is a view of the eukaryotic branch of the tree of life. Note that animals, plants, and fungi are three distinct lineages. The remaining unlabeled branches are all different lineages of protists.

! Point of Confusion: Fungi may seem like plants because, like plants, they spend their lives fixed in place. In fact, scientists used to think that fungi were plants for this reason, but reclassified based on new evidence (Chapter 1). Fungi are more closely related to animals than to plants. For example, fungi and animals are both heterotrophs, whereas plants are autotrophs, producing their own food.

If you are considering mushroom hunting, keep in mind that many edible species have poisonous look-alikes. Mushrooms taken from the wild should always be identified by an experienced mushroom forager before they are eaten.

Fungi come in a variety of sizes. For instance, **yeast** is a single-celled fungus used for making bread and beer (Chapter 4). The largest known living organism, a fungus called *Armillaria ostoyae* in Oregon, weighs over 35,000 tons and covers over 2.5 square miles, mostly underground as part of its *mycellium.* All fungi disperse by producing spores. Spores can form either asexually (by mitosis) or sexually (by meiosis). Fungi also require water in order to reproduce. Most fungi are multicellular—and these fungi are relatively easy to spot while hiking.

Mushrooms are the most familiar multicellular fungi. While you are hiking, you might see mushrooms with the familiar cap and stalk (Figure 13.52A) or shelf-like fungi (Figure 13.52B). Mushrooms can even take on surprising shapes, looking like animals, such as coral (Figure 13.52C).

The body of a multicellular fungus is composed of many tiny filaments called **hyphae** (singular, *hypha*). Each fungal hypha is thinner than a human hair. A fungus grows by adding cells to the tips of its hyphae. As fungal hyphae grow through soil or over organic material, they form a branching network called a **mycelium** (plural, *mycelia*; Figure 13.53).

? Did You Know?

The mushrooms in Super Mario Bros video games are based on the appearance of the fly agaric mushroom.

Figure 13.52 A gallery of mushrooms

A B C

A. A fly agaric mushroom (*Amanita muscaria*) is easily recognizable, with its red cap and white spots.
B. A turkey tail mushroom, *Trametes versicolor*; notice the different colors that make up the mushroom.
C. A coral mushroom, *Ramaria ariaiospora*, with its coral-like shape.

Fungi do not have vascular tissue; however, material can move within their hyphae. Any walls that exist between hyphal cells are porous, so nutrients or water taken up in one part of the mycelium can be shared with cells in other regions. Note that mushrooms often appear seasonally, although the mycelium is constantly present in the soil. This is because the mycelium is a feeding structure, whereas a mushroom is a reproductive structure.

Figure 13.53 **Mycelium spreading on a log**

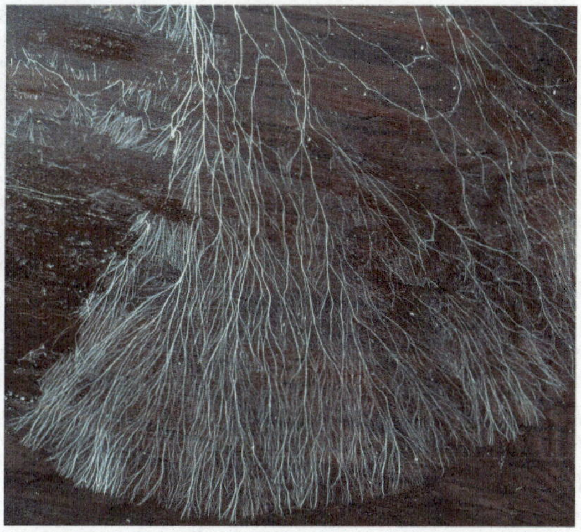

Note the fine filaments that make up the individual hyphae.

Credit: Gilbert S. Grant/Science Source

White-nose syndrome is a fungal disease decimating North American bats. The fungus kills bats during the winter by causing the bats to wake up from hibernation due to skin irritation. Awakening and fidgeting during winter, when no food or water is available, causes infected bats to waste away.

Most fungi provide an important ecological service by breaking down complex compounds in organic wastes and remains. Fungal hyphae can play an important part in this process. Hyphae can extend deep into a bulky food source, such as a dead log, and break it down from the inside out. When a fungus secretes digestive enzymes onto organic materials, some soluble nutrients escape into the nearby soil and water. Plants and other producers can then take up these nutrients to meet their own nutritional needs. We will return to fungi and their role as decomposers in Chapter 15.

Many mushrooms are cultivated or gathered for human consumption. Mushroom farms produce different varieties of mushrooms by inoculating sterilized compost with spores of the desired crop species. *Mycorrhizal* fungi, including chanterelles, morels, and truffles, cannot be easily cultivated, so they are typically gathered from the wild.

Mold is a multicellular structure that some fungi can form. You may see it in your kitchen, growing on old fruit or bread. **Mildew**, which you might find in your bathroom, is a parasitic multicellular fungus. You may also spot mildew while hiking (**Figure 13.54**). Mildew removes nutrients from the host plant, and the resulting loss of nutrients stunts the plant, reducing its ability to form seeds, and may eventually kill it. However, a plant usually does not die before the fungus has produced spores on the surface of its infected parts, which allows the fungus to continue to spread to other plants.

? Did You Know?

The initial source of the antibiotic penicillin was the mold *Penicillium*. Fungi have also yielded drugs, such as ones that lower blood pressure, reduce cholesterol levels, or prevent rejection of transplanted organs.

Fungi Form Mutually Beneficial Relationships

Finally, we'll end our hike by having a look at lichen, which you can find growing on many exposed surfaces, such as rocks, logs, and wooden structures. **Lichens** are an example of a mutually beneficial relationships in nature, being a composite organism consisting of a fungus and a single-celled photosynthetic species, either a green alga or a cyanobacterium, or sometimes all three (**Figure 13.55A**). The fungus makes up most of the lichen's mass and shelters the photosynthetic species, which shares nutrients with the fungus. Lichens are ecologically important as colonizers in places that are too hostile for other organisms, such as on bare rock (**Figure 13.55B**). By releasing acids and retaining water that freezes and thaws, lichens help to break down rocks and form soil, which is also an important part of ecological succession (Chapter 15).

Figure 13.54 **Powdery mildew**

The white "dust" shown growing on a Norway maple leaf (*Acer platanoides*) is powdery mildew (order: *Erysiphales*).

Credit: Kazakov Maksim/Shutterstock.com

Figure 13.55 A gallery of lichens

A. Lung lichen (*Lobaria pulmonaria*) is a composite organism made up of individuals from three different evolutionary lineages: fungi, protist, and bacteria.

B. An elegant sunburst lichen (*Rusavski elegans*) growing on a rock.

Credits: Ed Reschke/Stone/Getty Images; iStock.com/Gerald Corsi

? Did You Know?

Fungal partners also enhance the nutrition of some animals, such as termites, where the fungus helps in the digestion of compounds like cellulose, which would otherwise be indigestible.

Nearly all plants form mutually beneficial relationships, or **mutualisms**, with fungi. For example, many soil fungi take part in a **mycorrhiza** (plural, mycorrhizae), a mutually beneficial relationship with plant root cells (Figure 13.56). This relationship may involve hyphae either penetrating the root cell or growing in between the root cells.

Hyphae of all mycorrhizal fungi provide a benefit to their plant partner by functionally increasing the absorptive surface area of the roots. Hyphae are thinner than even the smallest roots and can grow more easily between soil particles. In this mutually beneficial relationship, the fungus shares water and nutrients taken up by its hyphae with root cells, and in return, the plant supplies the fungus with photosynthetically produced sugar.

Figure 13.56 Mycorrhizal filaments surrounding a plant's root tips

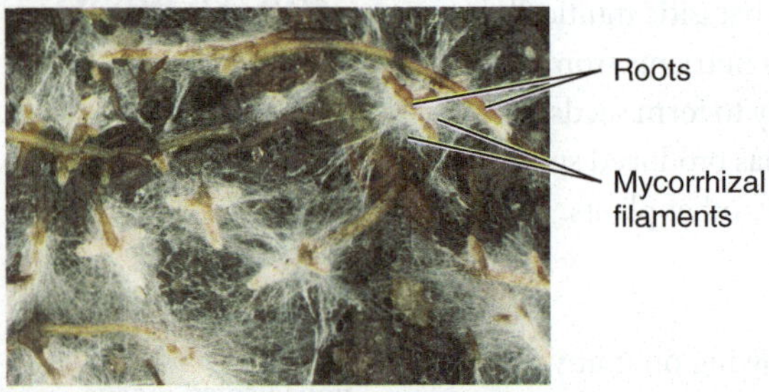

Credit: Dr. Jeremy Burgess/Science Source

In this section, we hiked through Ecola State Park on the Oregon coast and learned about the evolutionary histories of plants and fungi. In the upcoming final section, we'll bring our field trip close to home to consider the biodiversity immediately around and sometimes on us, with the lineage that makes up more than half of all known species worldwide—the land arthropods, or "bugs!"

While hiking, you may also see protists, too! Slime molds are the only protists that live on land. If you see a yellow splotch on a tree while hiking in Ecola State Park, it may be the aptly named scrambled egg slime (also known as the dog vomit slime mold), *Fuligo septica*, a type of plasmodial slime mold.

13.4 Summary

- The four lineages of plants have varying levels of adaptations to life on land. They are classified by reproductive strategies and the presence of vascular tissue.
- Lacking vascular tissues, nonvascular plants tend to live low to the ground and need wet environments to reproduce. Vascular plants have vascular tissue to hold them upright.
- The nonvascular plants and the seedless vascular plants produce sperm that must swim through droplets of water to reach an egg. Seed plants, gymnosperms, and angiosperms, however, reproduce using pollen grains that are transmitted through air or by animals, making them better able to reproduce in dry environments.
- While plants and fungi can be found growing in similar places, they acquire energy in different ways: Plants are autotrophs and fungi are heterotrophs.
- Fungi are often decomposers. They secrete digestive enzymes onto their food and absorb the resulting breakdown products. Fungi can be single-celled, like yeast, or multicellular, like mold, mildew, and mushrooms.

13.5 The Diversity of Land Arthropods: A Visit to a City Park

Learning Objectives

13-19 Compare modern arthropod lineages.

13-20 Describe the beneficial and negative interactions between humans and insects.

Consider for a moment—where is the closest public park to you? Are you sitting in a park right now? What do you see in your park? Probably a variety of lifeforms—mammals like squirrels or dogs, flowering plants, and trees, depending on the climate where you live. Wherever you are reading this text, whether it's a park or not, there's almost definitely a land arthropod, or "bug," nearby. What we colloquially call "bugs" represent multiple evolutionary lineages: Chelicerates, myriapods, crustaceans, and insects. Why do "bugs" get their own section? It's because they're the most diverse organisms on the planet. Consider that 1 in every 4 animals on the planet is a single type of insect, a beetle!

? Did You Know?

Horseshoe crabs are chelicerates, not crustaceans. They've lived in Earth's oceans for at least 470 million years.

In this section, we're going to take a walk in a city park, turning over some logs and rocks and keeping our eye alert to the air for sightings of land arthropods as we learn about their evolutionary history. We'll also explore how insects are relevant to our lives—good, bad, or indifferent—from their role in the production of goods like honey to how some of them transmit disease. Here, we'll be visiting City Park in Denver, Colorado, but many of the arthropod groups covered here can be found in any city park.

Creepy-Crawlies and the Life All Around Us

Imagine that you are in sunny Denver laying on your back on a picnic blanket. Maybe you're reading a book, or perhaps you're just taking in the scenery when suddenly, a spider crawls on you! How would you respond? Are you a bug-loving sort who catches spiders and put them outside when they get in your house? Or do you have arachnophobia, a condition of extreme fear in encounters with spiders?

"Bugs" are land arthropods, which we were first introduced to in Section 13.2 when we learned about the crustaceans that live at coral reefs. Recall from Section 13.2 that arthropods are invertebrates with jointed legs, an exoskeleton, paired compound eyes, and antennae.

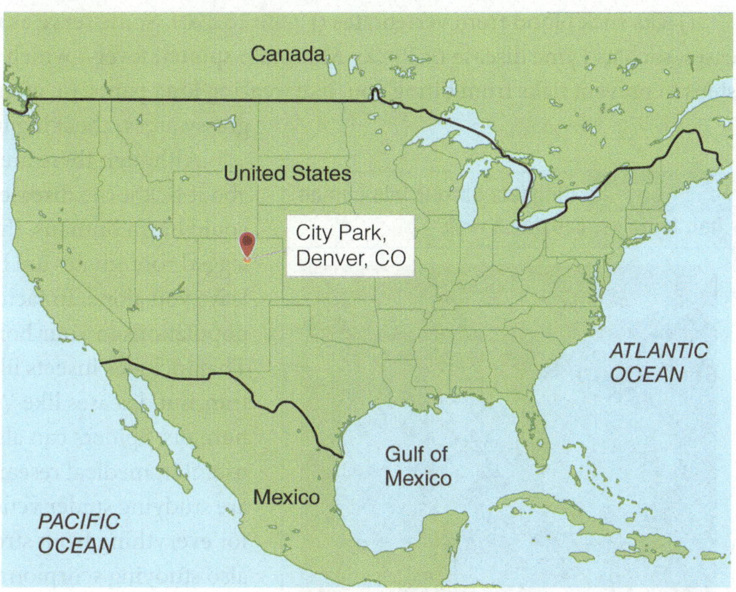

Credit: Lightvision, LLC/Moment/Getty Images

A city park. Parks are great locations to see land arthropods and for people to connect with the natural world. The "bugs" you see will depend upon where you live, but there are general types that you will see, such as spiders, centipedes, and beetles.

Figure 13.57 **The evolutionary history of modern arthropods**

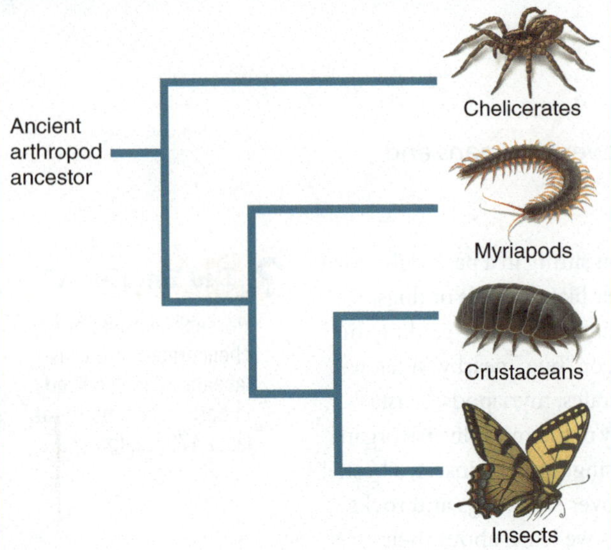

Note that crustaceans include sea-dwelling animals like lobsters, crabs, shrimp, and barnacles. Our focus in this section, however, is on land-dwelling arthropods.

Within a city park, you're likely to run into examples from all four different evolutionary lineages of arthropods that still live today (Figure 13.57). The oldest lineage is the **chelicerates**, which includes spiders, scorpions, mites, ticks, and horseshoe crabs. Chelicerates evolved over 500 million years ago, during the Cambrian period. The crustaceans, which include pill bugs and sowbugs, evolved around the same time. Around 425 million years ago, the first myriapods appeared. The **myriapods** are a lineage that includes segmented animals with many legs, such as centipedes. Then around 400 million years ago, the first **insects** evolved, which include many familiar examples like ants, cockroaches, and moths. In this section, we'll learn about each of these evolutionary lineages.

Chelicerates. In the *Metamorphoses*, Ovid, a Roman poet, tells the story of Arachne, who boasts of her skill as a weaver, proclaiming herself a better weaver than the goddess Minerva. Of course, challenging a goddess never works out well, and so after Arachne's death, she was turned into a spider. Arachnids are named for Arachne in the myth.

While the chelicerates include marine animals like horseshoe crabs and sea spiders, in a city park (or just in your house!), the chelicerates you're most likely to encounter are arachnids.

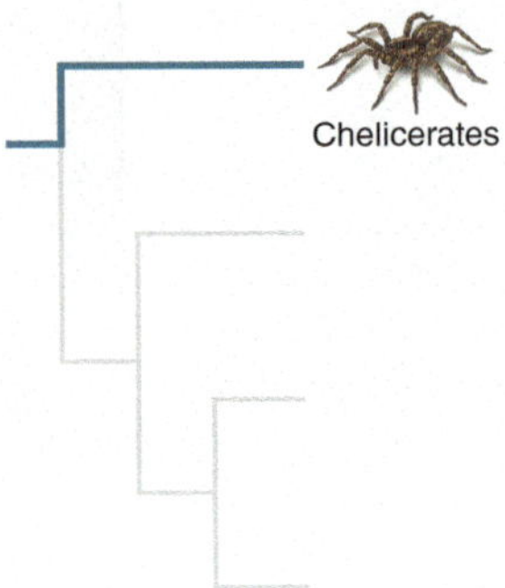

Arachnids have four pairs of walking legs, a pair of touch-sensitive appendages called palps, and no antennae. Arachnids include mites, spiders, scorpions, and ticks.

Mites are the smallest arachnids, and are typically less than a millimeter long. Mites that burrow beneath the skin are responsible for everything from mange in animals to scabies in humans. Chiggers are larval mites that crawl into a human hair follicle (Figure 13.58), causing an itchy rash.

Spiders and scorpions are carnivorous, venomous predators. Spiders deliver venom through fanglike mouthparts. They have a two-part body, with the cephalothorax (fused head and thorax) separated from the abdomen by a narrow "waist." The abdomen contains silk-making glands, which spiders can use to spin webs (Figure 13.59A). Spider silk is incredibly strong—stronger even than steel. It's also stretchy. Therefore, researchers are studying how to make bulletproof clothing based on spider silk. Scorpions catch their prey with clawlike palps and subdue it with a venom-producing stinger on the last abdominal segment (Figure 13.59B). They eat insects, spiders, and even small lizards.

Ticks suck blood from vertebrates (Figure 13.59C). Some serve as vectors for bacterial diseases, such as Lyme disease or Rocky Mountain spotted fever—which is why you'll want to take steps to prevent ticks from biting you, like wearing long pants, or, after hiking or walking through grassy areas, checking for ticks on your body.

Fear of spiders is a relic of our coevolution with spiders. One study found that even when distracted, people are hypersensitive to the movement of spiders in their immediate environment. This makes sense as some species of spiders, like black widows, can harm humans.

Figure 13.58 **Chiggers burrow into human hair follicles, causing a rash**

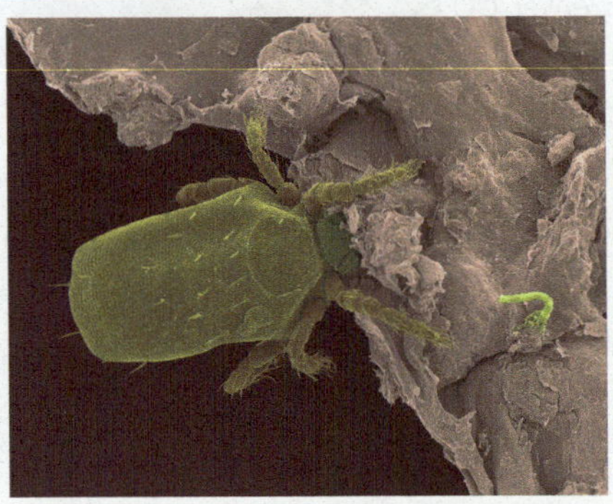

Credit: Dennis Kunkel Microscopy/Science Source

Although there are some legitimate concerns about tickborne diseases and spider venom that could harm humans, the vast majority of our eight-legged roommates are harmless and will generally leave you alone. In fact, spiders help keep insect populations in your home or garden under control. That includes insects like mosquitoes, which can transmit diseases like West Nile or Zika virus to humans. Spiders can also be pollinators and are useful in medical research. For example, biologists are studying spider venom to develop treatments for everything from stroke to cancer. Scientists are also studying scorpion venom and finding that it can kill antibiotic-resistant tuberculosis.

Figure 13.59 **A gallery of arachnids**

A. A type of orb weaver spider (*Araneida*) sits on its web. Note the narrow waist that separates the large abdomen from the smaller cephalothorax.

B. Scorpions, such as this striped bark scorpion (*Centrutoides vittatus*), are predatory arachnids. Note its venomous stinger and the clawlike palps it uses for grabbing onto its prey. The striped bark scorpion is found in Colorado, but not in the Denver area.

C. A Rocky Mountain wood tick (*Dermacentor andersoni*) at the tip of a leaf, waiting for its next meal.

Credits: Nenad Nedomacki/Shutterstock.com; Wolfgang Kaehler/Alamy Stock Photo; Center for Disease Control (CDC) - PHIL

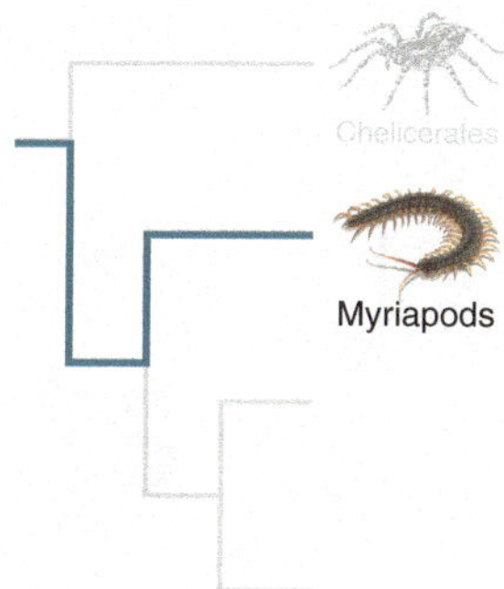

Myriapods. Early in this chapter we talked about the transition of life in the water to land. One of the reasons animals could survive so well outside of water was because of the abundance of food. What was the food? Well, tetrapods weren't the first animals to appear on land—the arthropods arrived first, starting with the myriapods around 490 to 443 million years ago.

The myriapod lineage includes only terrestrial arthropods, all of whom have mandibles and antennae followed by repeating units containing the legs. There are two common subtypes: centipedes and millipedes. Centipedes and millipedes are nocturnal (active at night) ground dwellers. **Centipedes** are venomous predators with a flat, low-slung body and one pair of legs per segment. Within City Park, you may find stone centipedes under a rock or log (Figure 13.60A). Depending on where you live, you may be familiar with house centipedes (Scutigera coleoptrata), or "thousand leggers." While surprising to encounter at night, they are useful house guests as they eat cockroaches, spiders, silverfish, carpet beetles, and other pests.

Most **millipedes** (Figure 13.60B) eat plant material and are harmless to humans. Their cylindrical body has two legs per segment, and their cuticle is hardened with calcium carbonate. Millipedes are important detritovores (organisms that eat organic waste; Chapter 15) and help with cycling nutrients and creating healthy soil. So, they are great neighbors to have in your garden!

? Did You Know?

In 2018, geologists discovered the largest fossil of a land-based arthropod ever found, that of an *Arthropleura*, likely a hybrid of a millipede and centipede that lived over 300 million years ago and was 2.6 meters (8.5 feet) long—roughly the size of a grizzly bear!

Figure 13.60 **Comparing centipedes and millipedes**

A. Stone centipedes (genus *Lithobius*) are speedy predators that can be found under rocks and logs.

B. Millipedes, like this spotted snake millipede (*Blaniulus guttulatus*), are scavengers that eat decaying plant material.

Credit: The University of Georgia; Tomasz Klejdysz/Alamy Stock Photo

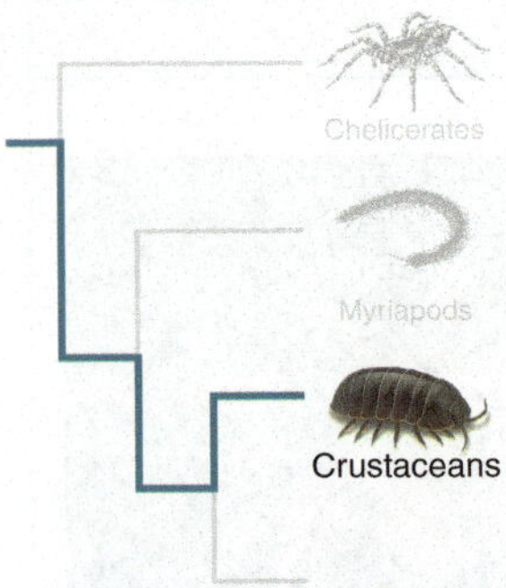

Land crustaceans. Have you ever played with a roly-poly, or pill-bug, before (Figure 13.61)? These critters are common to see in parks and are endearing to interact with, because if you poke at one, it will roll into a ball to protect itself. While the majority of crustaceans are found in (or in the case of crabs, near) the seas, one notable exception is woodlice, which includes pill-bugs (*Armadillidium vulgare*) and sowbugs (*Porcellio scaber*). Crustaceans like pill-bugs have two pairs of antenna and at least five pairs of legs, and are an important part of the ecosystem, serving as decomposers, helping to break down and recycle organic matter.

Figure 13.61 A pill-bug

Credit: FrankRamspott/Getty Images

Insects. It's impossible to go about your day and not experience insects, the most diverse group of arthropods. Approximately 900,000 insect species exist today, comprising about 80% of all species living worldwide. Insects are descended from freshwater crustaceans, which is supported by morphologic and genetic data. Insects have three pairs of legs and one pair of antennae, and make up around 90% of arthropods

Early insects were wingless, ground-dwelling scavengers that did not undergo **metamorphosis**, a change in the insect body form into an adult. Insect metamorphosis first evolved between 300 and 280 million years ago. The ability to undergo metamorphosis gave insects a selective advantage. Larvae, or immature forms of insects, often live in different environments and eat different foods than adult insects.

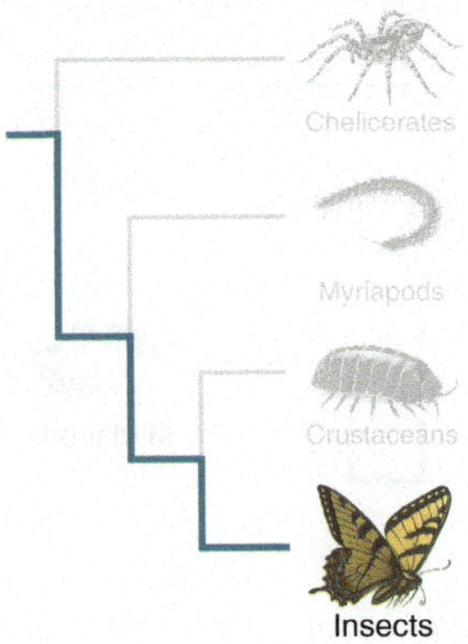

While you are walking through the park, you might encounter one of the most well-known examples of an organism that undergoes metamorphosis—butterflies—although flies and beetles undergo this process as well. After an egg is laid, the larva—in the butterfly's case, a caterpillar—emerges and grows, eating plant matter like leaves and molting without altering its form. It then undergoes pupation. A pupa is a nonfeeding body in which larval tissues are remodeled into the adult form (Figure 13.62). In the case of a butterfly, the pupa is called a chrysalis. The adult butterfly that emerges from the chrysalis looks and act completely differently from the caterpillar, flying, eating nectar, and serving as a pollinator.

Did you know that approximately 75% of our crops depend on insects (Chapter 18). Or that the economic value of honeybees is over $15 billion in the United States? Insects are essential, not only for humans but for other animals as well.

In addition to being pollinators, insects serve a variety of other important ecological roles. For example, insects are a primary food source for a variety of organisms, including birds, fish, amphibians, various other animals, and even other insects! The aquatic larvae of dragonflies and mayflies serve as food for fish and in turn are food for amphibians and reptiles like birds (Figure 13.63A). Insects, like other land arthropods, prey on one another to help manage arthropod numbers (Figure 13.63B). Flies and beetles are essential for disposing wastes and

? Did You Know?

Insect populations worldwide are in decline. Since the 1980s, insect populations have dropped by 45%.

? Did You Know?

Monarch butterflies get their name because of the gold line that appears on their chrysalis.

Figure 13.62 Metamorphosis of a monarch butterfly (*Danaus plexippus*)

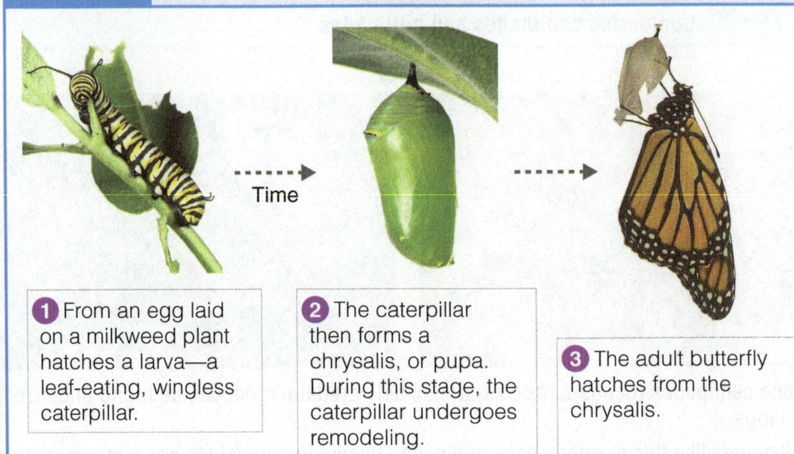

Credits: Jacob Hamblin/Shutterstock.com; Jacob Hamblin/Shutterstock.com; Laurie Barr/Shutterstock.com

Figure 13.63 **Insects you might see in a park**

A. A sparrow eats a dragonfly.

B. An ambush bug (*Phymatinae*) on a sunflower eating a meal. Ambush bugs sit on yellow flowers, where they blend in and lay in wait to capture and devour other unsuspecting insects visiting the same flower.

C. A fly (*Sarcophaga carnaria*) eating a rotting apple.

Credit: Wirestock Creators/Shutterstock.com; Melanie Peffer; iStock.com/Ralf Menache

organic remains, like animal corpses, piles of feces, or rotten food (Figure 13.63C). Flies lay eggs in or on organic material, and the larvae (in the case of flies, maggots) hatch and devour it. By their actions, these insects help to distribute nutrients through the ecosystem. In addition, without the action of flies and beetles recycling nutrients, there would be a layer of unrecycled poo everywhere we went.

As you might have noticed as you've walked around parks or just about anywhere, insects can take many forms. The great diversity among insects is a reflection of how well the insect body plan can adapt to a variety of environments. The specific genes that control body plan (which we learned about in Chapter 7) can be mixed and matched together to produce the many different body forms we see, ranging from beetles to flies to butterflies. Insects also saw such great evolutionary success because they are able to reproduce frequently and have many offspring—which anyone whose had an insect infestation at home is all too familiar with! Insects were also the very first animals to fly, long before reptiles and birds took to the sky, which gave flying insects a selective advantage.

> **? Did You Know?**
>
> Insects are a highly nutritious and environmentally friendly food source. Grasshoppers, for example, are high in protein and have anti-inflammatory properties. They also contain antioxidants, which help to protect your DNA from mutations (Chapter 7).

Insect pests. But now that we've talked about all the friendly aspects of insects, we'd be remiss if we didn't also talk about the negative aspects, too. While insects are important crop pollinators, others can damage crops. This can be done directly, such as by an insect eating the crop, or indirectly, by transmitting a disease to the plant. This is of serious concern, as the U.S. Department of Agriculture estimates that 20% to 40% of global crop production is lost due to pests.

Insects can likewise damage structures like homes. For example, termites are insects that eat wood, causing billions of dollars in structural damage each year. Carpenter bees bore holes into wooden structures to build their nests (Figure 13.64A). While this can ultimately weaken the structure, carpenter bees are not nearly as destructive as termites.

And as we carry on with our walk, if there's a group of people around, like at a city park, odds are there are also insects around that evolved to feed on humans. This includes everything from mosquitoes (which can transmit malaria, Zika, and West Nile virus) to fleas, lice, and bedbugs. Fleas (Figure 13.64B), incidentally, are one of the culprits behind the Bubonic plague, which killed as many as 50 million people, approximately 50% of

> **? Did You Know?**
>
> Bedbugs are experiencing a population boom due to developing resistance to pesticides and ease of travel.

Figure 13.64 **A gallery of insect pests**

A B C

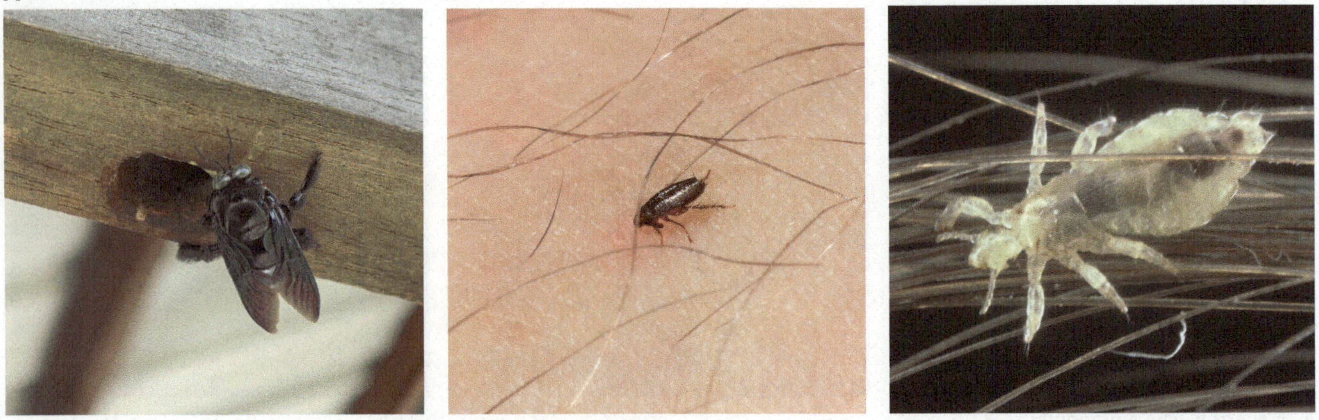

A. A carpenter bee (*Xylocopa*) next to damage in an outdoor structure.

B. A flea biting a human. There are many different species of fleas, including dog, cat, and human fleas.

C. A head louse crawling through strands of human hair.

Europe's population in the 14th century and is still around today. Note that cases of the plague are rarer today, and when they do occur, they can be treated with antibiotics. The plague is not caused by fleas, but by a bacteria they carry called *Yersinia pestis*. If a flea with the bacteria bites a human, they can transmit the bacteria to the human. It turns out fleas may not have been alone in transmitting the plague-causing bacteria either—recent research suggests that lice may also have transmitted the bacteria between people during feeding.

Lice remain a problem today. Head lice (**Figure 13.64C**) spread easily, particularly amongst children, and can be a nuisance to get rid of. Head lice are typically killed using a 1% permethrin solution, but some lice are now resistant to treatment. Just like antibiotic-resistant bacteria evolved, so did lice that are resistant to commonly used pesticides. Now, people are using alternative treatments like hand removal of lice and their eggs through combing.

From the bottom of the ocean to a typical city park, arthropods are everywhere. They are the most diverse type of organism living on this planet, and odds are there is one ambling (or flying!) nearby (but hopefully not on) you as you read this text. The next time you see a "bug," remember all that they represent in terms of evolution—a highly successful body structure—and their existence is part of why life was able to survive outside of water.

? Did You Know?

There are three kinds of lice that infect humans, named based on where you can find them: head lice, body lice, and pubic lice. Each of these lice is a different species, showing how lice have coevolved with humans.

Feeling itchy reading this section? You're experiencing psychosomatic itching, where anxiety about being infected with something like lice makes you feel itchy or like something is crawling on you.

13.5 Summary

- Arthropods are invertebrates with jointed legs, an exoskeleton, paired compound eyes, and antennae. There are four arthropod lineages still alive today.
- Chelicerates, like spiders, have eight legs. Myriapods, including centipedes and millipedes, have mandibles, antennae, and repeating body segments with legs on each one. Crustaceans are largely marine arthropods, although they can exist on land. Insects—including bees, termites, butterflies, and more—have three pairs of legs and one pair of antennae.
- Insects benefit humans in many ways, including serving as pollinators and food sources, and disposing waste. Negative impacts of insects include damaging crops, structures, and by spreading disease, both among crops and among humans.

Biology for Life

Our Relationship with Insects

A B C

Credits: Universal History Archive/Getty Images; Greatstock/Alamy Stock Photo; Melanie Peffer

Figure 13.65 A gallery of insects and their significance

A. A beautifully embellished winged scarab found in the tomb of King Tutankhamun.
B. The dung beetle (*Scarabaeidae*) is the basis for scarabs. Here, a dung beetle rolls up a ball of dung to use as shelter and/or food.
C. Silk dyed with crushed cochineal beetles. Note the dried beetles within the vial sitting on top of the silk.

When I was growing up, I remember being mesmerized by fireflies. Maybe you also grew up catching fireflies, or spent time watching the fireflies come out in the evenings. While it might be true that many of us can be grossed out by insects, it's also true that insects hold significant religious, personal, and symbolic meaning.

Take the scarab beetle (Figure 13.65A), which is based upon the dung beetle (Figure 13.65B). To the ancient Egyptians, the scarab represented the cycle of life, rebirth, and regeneration. Butterflies are symbols of hope, renewal, and change—one that we use in Chapter 16 to call out our reasons to hope. For example, a "butterfly baby" is a term used in the pregnancy loss community used to describe a child lost pre- or postnatally—the idea that butterfly represents that nothing is lost, only changed. When we consider this in the context of metamorphosis, the caterpillar is not lost, merely changed into a butterfly. Therefore, the butterfly has symbolic meaning that although a baby has died, they are not lost.

In addition to symbolic meanings, insects also have many practical purposes. Insects like ladybugs are a natural and effective way to control plant pests like aphids or mealybugs. As we'll discuss in Chapter 18, insects, and particularly bees, are critically important to our food supply. Insects are important for clothing production—silk (Figure 13.65C) is made by silkworms, which are the larval stage of silk moths. Silk (and other fibers) can be dyed with insect-based dyes. For example, cochineal beetles (in the bottle in Figure 13.65C) are ground into a powder and used to create pink and red dyes, such as those used to dye the silk shown in Figure 13.65C.

1. Do insects have personal significance to you? What insects are they, and what is their significance?
2. Has your opinion of insects changed after reading this section? What do you see as their greatest benefit?

Understand It! Memory Networks

Learning unequivocally requires memory—which is why we first learned about short- and long-term memory networks in Chapter 1, and have returned to the topic in various Understand It! features throughout the text. Much like how organisms within nature are interconnected with each other, the physical structures within our brain that hold onto memory are the result of connections between neurons (brain cells). Whenever you learn something, new connections form between neurons. Within your brain, there are tens of thousands of connections.

These connections (or synapses) get stronger the more a memory is reinforced, such as occurs during the practice phase of the study cycle (Chapter 1). The less that a connection is reinforced, the weaker that connection is. Weakening connections explain why we forget information.

When learning new information, it helps to connect the new information to something that we already know. Things we already know are called prior knowledge. The prior knowledge that we have about a subject is a starting place for new memories to connect. This is also why it's important to learn and remember quality information. For example, if someone remembers something false, new information will be colored by that false memory, which can cause a person's memory to become problematic.

1. Visually organizing the material can help to create organized memory networks in your brain. Working with a partner, draw a concept map outlining this chapter.
2. How does the knowledge in this chapter connect to previous chapters?
3. How do you think the knowledge obtained in this chapter might connect to future chapters?

Chapter Review

13.1 The Beginnings of Life: A Visit to a Hydrothermal Vent

Life at **hydrothermal vents** provides hints about how life originated on Earth. Current thinking suggests that organic molecules formed from inorganic molecules through **chemical evolution**, eventually forming a self-replicating RNA. The **RNA world hypothesis** is supported by the presence of RNAs, called **ribozymes**, that catalyze reactions in our cells today. A membrane formed around this self-replicating RNA to form a **protocell**, which increased in complexity over time, eventually forming ancestral bacterial and archaeal cells.

Prokaryotes, including **bacteria** and **archaea**, are the producers at hydrothermal vents making energy by using inorganic substances in a process called **chemosynthesis**. **Chemoautotrophs** perform chemosynthesis, while **phototrophs** obtain their energy from sunlight. Organisms that survive in extreme heat are called **extreme thermophiles**, while those in extreme acidity are called **acidophiles**, and in high salt concentrations, **extreme halophiles**. **Eukaryotes** at the vents are heterotrophs and dependent on chemosynthetic producers.

The endosymbiosis hypothesis proposes that a symbiotic relationship formed between an aerobic bacterium, and later, a photosynthetic bacterium, in an ancestral cell. These bacteria eventually became the mitochondria and chloroplasts in modern cells through a process called **endosymbiosis**.

13.2 The First Animals: A Visit to a Coral Reef

Coral reefs are home to many prokaryotes, like oxygen-producing **cyanobacteria**, as well as eukaryotes such as **protists** (any eukaryote that is not a fungus, plant, or animal) and **animals**. Cyanobacteria are thought to have increased the oxygen on Earth enough to support terrestrial organisms during the **great oxygenation event**.

Protists like photosynthetic **algae** (brown, green, and red) are found on coral reefs. Protists are also a component of **plankton**, a community of small organisms that swim or drift through water. Some plankton is photosynthetic (phytoplankton) and includes protists like **diatoms** and **dinoflagellates**, while other plankton contains consumers (zooplankton), like **foraminifera**. While most protists live as single cells, some are **colonial organisms** that can both cooperate with each other and work independently. According to the **colonial theory of animal origins**, colonial protists like **choanoflagellates** share an ancestor with the first **multicellular organisms**.

Most **animals** are made up of groups of specialized cells that work together, called **tissues**. The first tissues that form in an embryo are called **germ layers**. Animals with two germ layers have **radial symmetry**, while animals with three germ layers have **bilateral symmetry**, as well as more complex systems, like a **coelom**, a fluid-filled body cavity, and organ systems, like a **complete digestive system** or **closed circulatory system**.

Animals found in coral reefs include **sponges**, which don't have tissues or organs. They have intracellular digestion and are hermaphrodites, making both sperm and eggs. Sponges develop from **larva**, which have a different body plan and can swim briefly before settling and becoming a sessile adult.

Reef animals with radial symmetry include **cnidarians**, which have stinging tentacles and a **gastrovascular cavity** that digests food, expels waste, and performs gas exchange. Cnidarians have two common body plans: a **medusa**, which swims or drifts about, and a **polyp**, which has one end attached to a surface. Reef animals with bilateral symmetry can be either **protostomes**, where the first opening that forms in an embryo becomes a mouth, or **deuterostomes**, where the mouth forms from the second embryonic opening.

Protostomes are all **invertebrates**, including flatworms, annelids, mollusks, roundworms, arthropods, and crustaceans. **Flatworms** have a flat body with organ systems but no coelom, while **annelids** are bilateral worms with a coelom and a **complete digestive tract**. **Mollusks** have a complete digestive track, a reduced coelom, and a **mantle** that drapes over the organs. There are three types of mollusks: (1) **gastropods**, like snails; (2) **bivalves**, like oysters; and (3) **cephalopods**, like squid. Gastropods and cephalopods also have **radula** made of chitin used for scraping.

Roundworms are small decomposers that molt as they grow. They have a pseudocoelom, and no respiratory or circulatory organs. **Arthropods** like **crustaceans** have a coelom and respiratory organs. They have **exoskeletons** that molt as they grow, compound eyes, and often one or more sets of sensory structures called **antennae**.

Some invertebrates and all vertebrates are deuterostomes, such as **echinoderms** like sea stars, which are invertebrates with spines and an **endoskeleton**.

13.3 The Diversity of Vertebrates: A Visit to an Estuary

Chordates include invertebrates like **lancelets** and **tunicates**, plus all **vertebrates**. Vertebrates have a **backbone**, a distinct head with a brain, a complete digestive tract, **lung(s)** typically used for gas exchange, and paired kidneys for filtering blood. As embryos, they have a stiff rod of connective tissue called a **notochord**, a hollow nerve cord, gill slits, and a muscular tail.

Fish are divided into two major groups: the **jawless fishes**, such as lamprey and hagfish, and the jawed fish. **Jawed fish** have **scales**, and their skeletons can be made of either cartilage or bone. **Cartilaginous fish** include sharks, rays, and skates. **Bony fish** can be either **ray-finned fish** like grouper and blue gill, or **lobe-finned fish**, such as coelacanths and lungfish.

Tetrapods are vertebrates with **bony skeletons** that descended from lobe-finned fish. They include amphibians and amniotes. **Amphibians**—such as newts, salamanders, frogs, and toads—typically spend their adult lives on land and have lungs but have offspring with gills that live in the water.

Amniotes evolved from an amphibian ancestor and lay **amniotic eggs** with protective membranes that help them survive on land. Reptiles and mammals are amniotes. **Reptiles** include lizards, snakes, turtles, and **birds**. Non-bird reptiles have overlapping scales and periodically shed their skin as they grow. Both kinds of reptiles lay eggs that are fertilized internally. **Mammals** nourish their young with milk from **mammary glands** and have hair or fur on their bodies. Mammals can be divided into three groups: **monotremes**, which include a few species found in and near Australia; **marsupials**, like opossums; and **placental mammals**, which are the most diverse and abundant type of mammal.

Snakes, amphibians, turtles, and lizards are all **ecotherms** (cold-blooded), while birds and mammals are warm-blooded **endotherms**.

13.4 The Diversity of Plants: A Visit to a Temperate Rainforest

Plants are distinguished from algae by their multicellular embryo and the presence of a waxy **cuticle** and pores for gas exchange called **stomata**. Plants are classified by the presence or absence of vascular tissues—**xylem** and **phloem**—and whether they use spores or seeds. **Spores** develop into gametes that require water to reach each other. **Seeds** house the embryo and the food to support it in a protective coat. **Seed plants** also produce **pollen grains**, which carry the immature male gametophytes.

There are four lineages of plants. **Nonvascular plants** like moss, liverwort, and hornwort have no xylem or phloem and reproduce with **spores**. **Seedless vascular plants**—like **lycophytes**, ferns, and horsetails—have xylem and phloem, but reproduce with spores, not seeds.

Gymnosperms are seed plants that do not use fruit to disperse their seeds. **Conifers** are the most diverse gymnosperms. They produce **cones** that contain pollen or ovules, and are usually **evergreen plants** that have leaves throughout the year. **Angiosperms** are seed plants that have **flowers** that require **pollination** and disperse **seeds** in **fruit**.

Fungi are commonly mistaken for plants, but they are eukaryotic heterotrophs and often decomposers. Fungi can be single-celled like **yeast**, or multicellular, with thin filaments called **hyphae** that form branching networks called **mycelia**. Examples of multicellular fungi include **mold**, **mildew**, and **mushrooms**. Fungi often form **mutualistic** relationships. For example, soil fungi form beneficial relationships with plant root cells in **mycorrhiza**, and with single-celled photosynthetic organisms in **lichens**.

13.5 The Diversity of Land Arthropods: A Visit to a City Park

Arthropods are invertebrates with jointed legs, an exoskeleton, paired compound eyes, and antennae. There are four lineages of arthropods: chelicerates, myriapods, crustaceans, and insects. **Chelicerates** include **arachnids**, such as spiders, scorpions, ticks, and mites. Myriapods include **centipedes** and **millipedes**, which have repeating body segments with legs. Crustaceans like pillbugs have two pairs of antenna and at least five pairs of legs. **Insects**, like bees, three pairs of legs and one pair of antennae.

Practice It!

Retrieval practice is the process of pulling information from your long-term memory and using it. Here are some questions to help with your retrieval practice.

Self-Quiz

1. The conditions on Earth 4.3 billion years ago and the conditions at current-day hydrothermal vents are similar in terms of having very little of which of the following? (LO 13-1)

a. oxygen
b. methane
c. salt
d. water

2. The endosymbiosis hypothesis states that early aerobic bacteria and cyanobacteria became which cell structures? (LO 13-2)

a. mitochondria and chloroplasts
b. nucleus and lysosomes
c. cell membrane and nucleus
d. lipids and proteins

3. The first prokaryotes are thought to have survived on inorganic molecules, like the producers found in hydrothermal vents today. Which of the following terms best describes these organisms? (LO 13-3)

a. phototrophs
b. chemoautotrophs
c. endosymbiotes
d. ribozymes

4. Which of the following adaptations could best help a prokaryote to survive high heat, like that found at hydrothermal vents? (LO 13-4)

a. a complex branching lung system
b. a chemical that prevents proteins from denaturing
c. symbiotic relationships with other prokaryotes
d. many chloroplasts and mitochondria

5. Which of the following organisms are thought to be responsible for the great oxygenation event? (LO 13-5)

a. protists
b. archaea
c. cyanobacteria
d. chonanoflagellates

6. Which of the following observations suggests that choanoflagellates and other colonial protists are the ancestors to the first animals? (LO 13-6)

a. When choanoflagellates form a colony, it looks similar in structure to a sponge.
b. Colonial protists have distinct tissues that perform specific functions.
c. Colonial protists have organs and organ systems for respiration and digestion.
d. Colonial protists have a fluid-filled body cavity called a coelom.

7. In which of the following ways do sponges differ from typical animals? (LO 13-7)

a. Sponges have bilateral symmetry.
b. Sponges don't have specialized tissues.
c. Sponges are deuterostomes.
d. Sponges are unicellular.

8. Which of the following adaptations are shared by mollusks, arthropods, and echinoderms? (LO 13-8)

a. They are all protostomes.
b. They all have a full coelom.
c. They all have specialized tissues.
d. They are all vertebrates.

9. Which of the following statements correctly relates the terms "chordates" and "vertebrates"? (LO 13-9)

a. All vertebrates are chordates, but not all chordates are vertebrates.
b. Animals can be either chordates or vertebrates.
c. Chordate and vertebrate are different names for the same group of organisms.
d. All chordates are vertebrates, but not all vertebrates are chordates.

10. Suppose you found an egg from an unknown animal on the ground by a pond. The egg is a little smaller than a chicken egg and has a thick, leathery shell. Which of the following animals would you guess laid this egg? (LO 13-11)
    a. a frog
    b. a salamander
    c. a turtle
    d. a bird
11. On land, without water to help carry sperm to fertilize eggs, how do amniotes typically fertilize eggs? (LO 13-12)
    a. by depositing sperm directly on the eggs
    b. by depositing sperm into the female's body
    c. by individual organisms making both sperm and egg
    d. by making more sperm than water animals
12. Which of the following adaptations help to decrease a bird's weight, allowing it to fly? Choose all that are correct. (LO 13-13)
    a. having feathers instead of fur
    b. having large brains and eyes
    c. having bones with air pockets
    d. having skeletons made of cartilage
13. Which of the following statements correctly relates the terms "angiosperm" and "vascular plant"? (LO 13-14)
    a. Angiosperms and vascular plants are two ways to describe the same group of plants.
    b. All angiosperms are vascular plants, but not all vascular plants are angiosperms.
    c. All vascular plants are angiosperms, but not all angiosperms are vascular plants.
    d. Plants can either be angiosperms or vascular plants, but no species is both.
14. A dry alpine forest would be more likely than a humid rainforest to have which of the following groups of organisms? (LO 13-16, LO 13-17)
    a. nonvascular plants
    b. seed plants
    c. fungi
    d. seedless vascular plants
15. Bamboo has the following structures: xylem, phloem, and flowers. How would you predict bamboo would reproduce? (LO 13-15)
    a. asexually
    b. using water
    c. with seeds and pollen
    d. similar to fungi
16. Which of the following characteristics are shared between most plants and fungi? (LO 13-17)
    a. the ability to perform photosynthesis
    b. the presence of vascular tissue
    c. the presence of hyphae and mycelium
    d. They are multicellular eukaryotes.
17. How do most fungi get their nutrients? (LO 13-18)
    a. by secreting enzymes and decomposing organic remains
    b. by absorbing sunlight and photosynthesizing
    c. by parasitizing other fungi
    d. from inorganic materials like rocks
18. How would you characterize a solifuge, a small animal a few centimeters in length with an exoskeleton, four pairs of legs, no antennae, and one pair of palps? (LO 13-19)
    a. an arthropod of the chelicerate lineage
    b. an arthropod of the insect lineage
    c. an insect of the arthropod lineage
    d. an insect of the chelicerate lineage
19. Suppose your neighbors were arguing about whether to build an enclosure around their garden to completely eliminate the presence of all insects. Neighbor A argues that the enclosure would protect their garden from pests, making their plants healthier and their yield better. Neighbor B disagrees, saying that eliminating insects will prevent pollinators from entering, making their yield worse. Which neighbor is right? (LO 13-20)
    a. Neighbor A; insects are pests that feed on and kill plants.
    b. Neighbor B; insects are pollinators that are necessary for plant reproduction.
    c. Both neighbors are correct; some insects are harmful to plants while others are important pollinators.
    d. Both neighbors are incorrect; insects have neither positive nor negative effects on plants.

Critical Thinking

1. While no ribozyme currently exists in nature that is able to self-replicate, scientists have successfully made a ribozyme in the lab that is able to replicate itself without any other enzymes. How does this support and/or refute the RNA world hypothesis? (LO 13-1)
2. Some scientists argue that protists should be further classified into more specific groups. What currently determines whether an organism is a protist? What attributes might be used to further classify these organisms into more specific groups? Do you think the

current classification makes sense, or would the more specific grouping you describe be better? Why? (LO 13-8)

3. The platypus, a mammal found in Australia, is one of only five living species of monotremes. The platypus has fur and a duck-like bill, and lays eggs. They can sense prey in cloudy water using electrolocation, a method of detecting electrical signals that works only in the water. Males have spurs that can deliver a painful venom. They have a cloaca as the sole exit for waste and reproductive cells (gametes). Considering these individual traits as well as what you've learned about the various animal groups in this chapter, how would you classify the platypus, looking only at the single traits? How might DNA evidence be used to help decide the correct way to classify the platypus? (LO 13-11)

4. Suppose your classmate made the claim that evolution is working to make organisms more complicated—that more recently evolved organisms are more complicated, and because they are more complicated, they are better adapted to their environment. Can you describe an example of a species that is very well adapted to its environment but that has a simple body plan? How are they able to survive? Can you describe a species that has not changed in a very long time that is still very well adapted to its environment? (LO 13-7)

5. In this chapter, you have learned about many adaptations that have led to evolutionary success. But what will happen in the future? Select one of the kingdoms covered in this chapter and think about how their members might have to adapt to a climate that is getting warmer and drier, with less oxygen and more carbon dioxide. (LO 13-4, 13-7, 13-11, 13-12, 13-14)

Unit 4 Review

In this Unit, we learned about evolution—one of the major theories of biology, and one of the Big Ideas that we first learned about in Chapter 1. While it may be tempting to think of evolution as merely history, it's an active process that is still occurring, and we can see the fingerprints of evolution all around us.

In Chapter 10, we learned the basics of evolution. We discussed the history of evolutionary thought while also covering the evidence for evolution. We examined a variety of different lines of evidence in support of evolution, including an examination of anatomical structures such as vestigial and analogous structures, as well as fossil evidence, biogeography, and molecular evidence.

In Chapter 11 we looked at the forces that drive evolution. We examined selective forces like natural selection and non-selective forces like gene flow. We also talked about these principles in the context of extinction and speciation.

Chapter 12 built on the concepts of speciation to describe how we use phylogenetics to understand evolutionary history. We learned how species are classified and applied that knowledge to understand human evolution. We also applied evolutionary principles from Chapter 11 to understand the evolution of human skin color and why race doesn't have a scientific basis, only a cultural one.

Finally, in Chapter 13 we took a tour across (and near) the United States and visited five different locations to learn about the results of evolution: the diversity of life on the planet. At the bottom of the ocean, we learned about the first life forms. In the ocean, at a coral reef, and then at an estuary we learned about the evolution of the first animals and the evolution of different animal lineages. We also visited coastal Oregon to learn about the evolution of plants as well as fungi, and then considered the most diverse group of animals, the land arthropods, within a city park.

Now that we've finished our survey of how life evolved, we'll next turn to a discussion of how lifeforms interact with one another, with a focus on how Earth and its lifeforms are impacted by climate change.

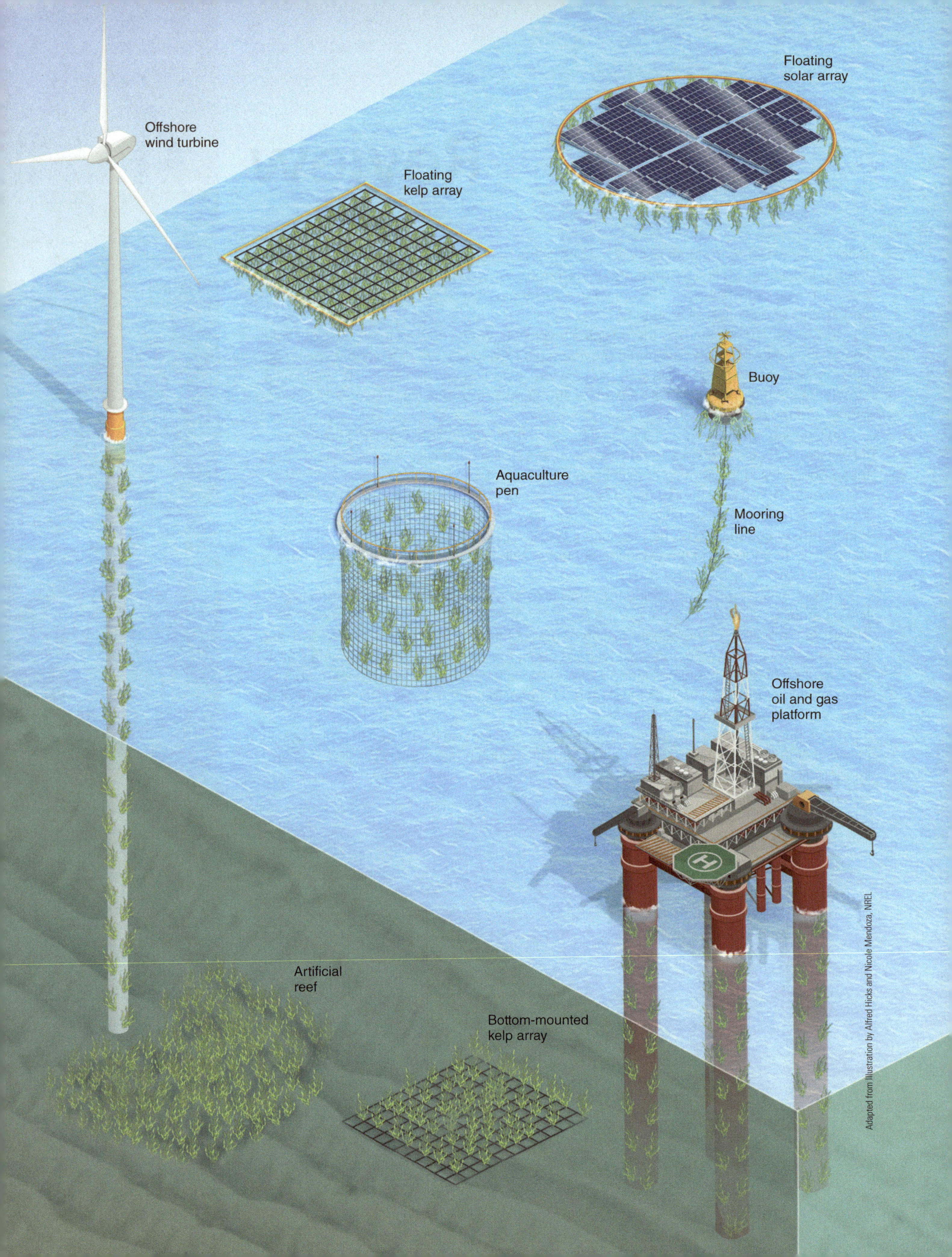

Adapted from Illustration by Alfred Hicks and Nicole Mendoza, NREL

Unit

5

Climate Change

Ecology

What do you think of when you hear about *climate change*? Is it doom and gloom scenarios with death and destruction? Is it changing weather patterns leading to more intense hurricanes and increasingly severe wildfires? Maybe it's political figures arguing about the climate crisis or climate alarmism.

All in all, climate change tends to bring up negative images. As I've done throughout this textbook, I endeavor to present the scientific consensus (Chapter 1) so that you can make informed choices either at the polls or as part of your daily routine. Although human-driven climate change is happening and is supported by scientific data and the consequences are serious, in this unit, I want to offer you hope for an eco-friendly and sustainable future as we discuss *ecology*, the study of interactions between living organisms and their environment.

Throughout the next few chapters, I'll highlight not only how the study of ecology can highlight the dangers of climate change and its impact on life, but also simple yet powerful changes you can make as part of your daily life to mitigate the effects of climate change. I'll also cover exciting new technologies in development that will help to minimize our environmental impact.

Upcoming technologies such as pKelp (highlighted in the image that opens the unit) take inspiration from nature—in this case, the movement of kelp (a type of algae) in water. As pKelp fronds move in water currents, electricity is generated. Since this is a renewable form of energy that doesn't emit greenhouse gases, it can help fight climate change—and provide a cheaper, locally based energy source. In addition, as shown in the illustration, you can see how pKelp could be used in combination with other existing marine structures.

So, there is reason to hope—if policy makers and other leaders prioritize investing in these technologies. But there are also simple yet powerful changes you can make as part of your daily life to help fight climate change. I want you to leave this unit feeling energized and hopeful about the possibility of a greener future for all life on this planet.

In Chapter 14, we will learn about *population ecology*, or how populations change over time, such as in response to a warming planet. Chapter 15 looks at how populations interact with one another in a *community*, how communities change over time, and the interactions of living and nonliving factors that make up an *ecosystem.* We'll also talk about how substances like phosphorus and water move through the environment. In Chapter 16, we'll discuss conservation, which is a subdiscipline of ecology that is focused on preserving and managing biodiversity and natural resources. We'll discuss pollution, biodiversity, and the impacts of climate change, with an eye on practical actions you can take and reasons for hope.

Preview It!

What questions do you have about climate change? Write down three things you want to learn in this unit.

Credit: National Geographic Image Collection

14

Penguins, Pikas, and Population Ecology

Chapter Outline:

A Potato-Sized Sentinel for Climate Change

Have you ever climbed a mountain? In Colorado, one popular activity for locals and visitors alike is to climb a "14er," one of the 58 mountains in Colorado that rises more than 14,000 feet above sea level. Climbing or driving into an alpine region—an area where conditions are too harsh to allow trees to grow, which starts at around 11,500 feet in Colorado—is an incredible voyage. As you climb the mountain, the types of plants and animals present slowly start to change. Smaller and smaller trees dot the landscape before disappearing altogether.

Once you cross 8,000 feet above sea level, you may hear the characteristic *eep! eep!* of the American pika. These potato-sized mammals are an *indicator species*, a species whose abundance (or lack thereof) provides an indication of the state of the environment it lives in. An indicator species is sensitive to changes in its environment, and changes in its population will be the first sign (or indicator) of a changing environment.

Pikas thrive in cold, alpine environments and are very sensitive to temperature changes, where exposure to temperatures just a few degrees above their body temperature can result in death. One of the impacts of climate change is steadily warming temperatures, and monitoring pika *populations* (a group of organisms of the same species living in a location) is one way of detecting the effect that warming temperatures have on the species. Unfortunately, due to climate change, pikas have disappeared entirely from several regions in Nevada, Oregon, and Utah, but populations currently remain stable in Colorado.

How do biologists measure populations? What else can we learn by studying populations?

The American pika (*Ochotona princeps*) is a heat-sensitive mammal that lives at high elevation (>8,000 feet above sea level) in places like the Rocky Mountain Range, which stretches from New Mexico to British Columbia.

Understanding the impact of climate change requires knowing not only how many different species live in a certain area, but also the traits that characterize the different populations of a species (like pikas), and how those populations interact with one another and with other species. Questions about population characteristics fall under the science of *ecology*, the study of interactions among organisms and between organisms and their environment.

We start this chapter with an overview of ecology and how ecologists can track and understand the impact of a changing climate on ecosystems. Next, we'll examine characteristics of populations and how biologists capture population data. Then, we'll discuss how populations grow, before getting into how studies of populations give insight into an organism's life history.

Preview It!

List three things you know about populations already. What questions do you have about population ecology?

14.1 Introduction to Ecology

Learning Objectives:

14-1 Describe ecology and the insights we can gain from its study.

14-2 Describe the different levels of ecological interactions.

14-3 Explain how ecologists study the impact of climate change.

How do you interact with your environment? In regard to the nonliving parts of your environment, you are breathing air and have likely consumed water at some point today. Did you take a form of transportation to get somewhere? If so, was it powered by you—such as by walking or riding a bike—or by electricity or fossil fuels? Have you seen any animals as you've gone about your day, such as birds outside your window or someone out walking their dog?

How organisms interact with one another and their environments is at the core of ecology. **Ecology** is a subdiscipline of biology concerned with studying the interactions among organisms, as well as how organisms interact with their environment. There are many levels at which these interactions can occur (Figure 14.1), starting from the most restrictive and getting more inclusive.

The first level is the organism level. An **organism** is one individual. Studies at this level are concerned with the interactions that an individual, like one pika, has with its environment. Next is the population level. A **population** is a group of organisms of the same species living at the same location, such as all pikas that live in Rocky Mountain National Park. The study of populations is the subject of this chapter.

A **community**, or all the populations in a region, is the next level of interaction. If we continue with our Rocky Mountain National Park example, this could be the different species that live in the alpine region, such as elk, marmot, four o'clock flowers, and grasses. An **ecosystem** includes the community—all of the living organisms in an area, or the **biotic** components—plus non-living, or **abiotic** parts of the environment. For example, in the alpine ecosystem, rocks, snow, and air temperature are all important abiotic factors. We will explore communities and ecosystems in more depth in Chapter 15.

The largest and most inclusive level of interaction is the biosphere. The **biosphere** is where life is found on Earth. We will discuss the biosphere in Chapter 15.

Big Ideas in Biology

Interconnectedness of Living Systems

Different organisms are connected to and influence each other, which is one of the foundational principles of ecology.

Point of Confusion: Ecology is not the same thing as environmentalism, which is advocacy for protection of the natural environment. Environmentalists often cite the results of ecological studies when drawing attention to environmental concerns.

Figure 14.1 A hierarchy of ecological interactions

Individual. A single organism, such as a pika.

Population. A group of the same species, such as pika, in a given area.

Community. The living and non-living aspects of a location. Note the presence of animals like elk, marmot, and pika, in addition to plants such as grasses and flowers. Abiotic factors include water, snow, rock, and cool temperatures.

Biosphere. The inhabitable part of Earth.

Note that there are four levels of interactions, and each successive level builds on the previous levels.

Process of Science

Pink River Dolphins and Policy Change

The pink river dolphin (*Inia geoffrensis*), or boto, is a species of dolphin that lives in the rivers of the Amazon rainforest in South America. These charismatic mammals have a distinctive pink color when engaging in intense physical activity, such as chasing fish; at rest, they are gray in color. Due to their diminishing populations, botos are classified as endangered by the International Union for Conservation of Nature. Their populations are threatened by various human impacts on their ecosystem, including fishing practices, mining, and deforestation.

Credit: National Geographic Image Collection

Figure 14.2 **National Geographic Explorer Fernando Trujillo with a pink river dolphin**

National Geographic Explorer Fernando Trujillo (Figure 14.2) is working to protect pink river dolphins through his work with local and Indigenous communities in the Amazon to come together to create policy changes that work for all stakeholders. One example of Trujillo's work is collaborating with local communities to develop sustainable fishing practices that protect the need for subsistence fishing while also protecting the dolphins. In recognition of his work, the Tikuan Indigenous community named Trujillo *omacha*, which means the dolphin who can turn into a person.

Credit: Joy Asico-Smith/National Geographic Society

Figure 14.3 **Trujillo speaking on the importance of scientific research and policy in Washington, D.C.**

Trujillo's work highlights an important aspect of science: that science is interconnected with society. What people need influences research priorities, and people in turn benefit from the insights provided through scientific study. Therefore, an essential outcome of many scientific investigations, including the work of Trujillo, is informing policy decisions related to conservation while also addressing societal issues.

For example, in 2023, Trujillo helped the Colombian government organize a meeting to create protections for pink river dolphins and the rivers they inhabit, both in South America and Asia. Ten different countries signed a declaration indicating their commitment to developing research-based solutions to protect river dolphins worldwide. Trujillo also participated in a panel discussion in Washington D.C. in 2024 (Figure 14.3) about science and policy.

Trujillo says, "We need to combine scientific research with actions…[and]…look for solutions now." What's at stake is not just the population of the pink river dolphins, but the overall health of the biosphere, which includes us!

1. Why is it important that scientists work with policy makers? What is at stake if policy makers do not use scientific and evidence-based research?
2. How does Trujillo's work impact both the pink river dolphins and Indigenous people in the Amazon? How might his work impact you?
3. Why is it important to protect endangered species? What unintended consequences may arise if we do not?

What Ecological Data Tell Us about Climate Change

We learned in the last section about the different levels of interaction that an ecologist would be interested in studying. Since we are focusing on the population level in this chapter, how might an ecologist look at a population, say of pikas, and assess its health and the potential impact of a changing climate on that population?

Such research might begin by examining the geographic range of pikas. **Geographic range** is an area where a species can be found. As mentioned in the chapter opener, pikas are very sensitive to temperature changes and are an **indicator species**, a species whose presence or absence, or relative abundance, allows scientists to gauge an ecosystem's health. Although pikas can move "up" the mountain to where temperatures are cooler, eventually, pikas would reach the top of the mountain and could not go any farther.

To understand the impact of climate change on pika populations, volunteers from the Colorado Pika Project have collected data since 2010 on pika numbers and their *habitats* (where they live). These "pika patrollers" are trained volunteers who collect data such the size of boulders pikas live near and note any pika scat (poo) and hay piles (which pikas use to make nests to keep them warm in the winter), along with temperature at 145 sites. Data from the 2023 season suggest that pikas are leaving areas where they have been spotted before, likely due to rising temperatures. A study of pika populations specifically in the Southern Rockies suggests that at 89% of locations surveyed, pikas have moved up the mountain by about 1,160 feet.

These data are used to inform computer models that predict what would happen to pika populations in the future given the current rates of temperature change and pika population behavior. If climatic changes were to continue at the current rate, pikas are expected to disappear from Rocky Mountain National Park by the end of the century.

Although the pika's geographic range is shrinking, other organisms, like the *Aedes aegypti* mosquito—which transmits dangerous viruses like the Zika virus—have expanded their range. As the climate has warmed, their range has steadily expanded northward, leaving people in places like the United States at risk for contracting tropical diseases.

14.1 Summary

- Ecology is a subdiscipline of biology concerned with the interactions between living organisms, and between organisms and their environment.
- A population is all individuals of one species in a specific area. Communities include all populations in a region (the biotic components), while ecosystems consist of communities plus the abiotic, or non-living, components. The biosphere consists of all ecosystems on Earth.
- Ecologists study the effects of climate change by monitoring measurable characteristics of populations, such as the number of individuals and where they are found.

14.2 Characteristics of Populations

Learning Objectives

14-4 Identify quantifiable characteristics of populations.

14-5 Describe two methods of estimating population size.

14-6 Explain carrying capacity and how it limits population size.

14-7 Identify factors that contributed to a rapid increase in the human population.

14-8 Explain how population growth is influenced by age structure.

14-9 Relate human population growth to impacts on ecology.

As we learned in the last section, ecologists study quantifiable characteristics of a population, such as geographic range, to understand how a population is impacted by an outside force, such as climate change. But geographic range is not the only quantifiable characteristic of a population. In this section, we will learn about other quantifiable characteristics, such as population size, density, and distribution, as well as age structure and fertility rate.

For example, Colorado Pika Project volunteers assess pika populations by using *direct counts* to determine the **population size**, the number of individuals at the specified location, as well as **population density**, or the number of individuals per unit of land area and its range. Each population occupies some subset of the species' geographic range and is influenced by and adapted to the unique conditions in that area. As we'll see throughout this chapter, population size, density, and range are all influenced by a changing climate (as well as other factors) and are therefore often critical datapoints in ecological studies.

Population Distribution

One important feature of a population is the physical distance between individuals in the population, which can provide insight into how individuals can be affected by biotic and abiotic factors in the environment. **Population distribution** describes where individuals are relative to one another. For example, members of a population may clump together, be distributed in a uniform manner, or be randomly distributed (Figure 14.4).

Pando, a grove of quaking aspens in Utah, is considered the largest organism on the plant. The clones, spread over 106 acres, include over 40,000 individual trees and weigh nearly 13 million pounds!

Clumped distribution. Most populations have a **clumped distribution**, meaning members of the population are closer to one another than would be predicted by chance alone. Often, a patchy distribution of an essential resource draws individuals together, as when hippopotamuses (*Hippopotamus amphibius*) gather in muddy river shallows (Figure 14.4A). Within the river shallows, hippopotamuses clump further into social groups. Different social behaviors result in animals living in closer proximity to one another. For example, animals may choose to live in social groups because it increases the likelihood of detecting predators, permits cooperate defense, and may allow the group to more readily find food resources.

Different factors can cause a clumped distribution. For example, some plants, like poppies, have limited ability to spread their seeds. Therefore, seeds fall near the parent plant, where a clump of poppies will grow. Asexual reproduction (Chapter 8) can also result in clumping. For example, a female aphid that reproduces asexually during the summer will produce many female offspring, which will live beside her as they mature. Similarly, asexual reproduction produces large stands of trees such as aspens, which appear in a clumped distribution pattern.

Figure 14.4 Examples of different types of population distribution

A. Clumped distribution. Note that each grouping of hippos represents a clump.

B. Uniform distribution. Seabirds nest at a consistent distance away from one another, creating a largely uniform distribution.

C. Random distribution. Dandelions are common plants found in random places around one's yard.

Uniform distribution. Competition for resources among members of a population can produce a nearly **uniform distribution**, with individuals more evenly spaced than would be expected by chance. Creosote bushes (*Larrea tridentata*) in deserts of the American Southwest grow in this pattern. Competition for water among the plants' root systems prevents them from growing near one another. Seabirds in breeding colonies often show a near-uniform distribution, too. Each bird repels others that get within reach of its beak as it sits on its nest (Figure 14.4B).

Random distribution. A random population distribution is rare in nature. **Random distribution** arises when resources are uniformly available, and proximity to others neither benefits nor harms individuals. For example, when wind-dispersed dandelion seeds land on the uniform environment of a suburban lawn, dandelions (*Taraxacum officinale*) grow in a random pattern (Figure 14.4C).

? Did You Know?

Red winged blackbirds (*Agelaius phoeniceus*) sing to identify their territorial boundaries. Within a marshy area, you can witness males perched at evenly spaced intervals, making their characteristic trilling *conk-la-ree* call.

The observed pattern of distribution can be influenced by the timing of the observations. For example, nesting sea birds crowd together in the breeding season, but disperse at other times.

Collecting Population Data

All the metrics of populations that we've discussed to this point rely on knowing the size of a population. But how do scientists determine the size of a population? And how do they gather data regarding the characteristics of that population? We'll now explore direct and indirect methods scientists use to capture population data.

Direct counts. For small, well-defined populations such as waterfowl on a lake, scientists can do **direct counts**, a method for quantifying individuals by counting all the members of the population, just like you might count the number of cans of soup in your pantry before going shopping. Aerial photography, such as using drones, can be used to measure the number of antelope in a grassland area or the number of whales in a bay. Drones are particularly useful for photographic surveys and can minimize the likelihood of disturbing animals (Figure 14.5).

What about for populations that are too large or dispersed for scientists to directly collect data on all the members of a population? Under these circumstances, scientists sample a population—that is, they take a subset of the larger population, then use data from that subset to estimate characteristics of the larger population. One approach is called plot sampling. In **plot sampling**, scientists directly count the number of individuals in a marked area, such as a 1-meter-by-1-meter square (Figure 14.6). This process is repeated across several areas of the same size, where ecologists can then calculate the average number of individuals across all spaces by the total area that the population inhabits, providing an estimate of the total population size. Although a useful technique, it is limited to organisms, like coral, that tend to stay in one place and do not move around.

National Geographic Explorer **Dr. Isla Meyers-Smith**

Credit: Gergana Daskalova /National Geographic Society

Figure 14.5 National Geographic Explorer Dr. Isla Meyers-Smith preparing to launch a drone Dr. Meyers-Smith's research focuses on the Canadian tundra ecosystem—an ecosystem consisting of vast, flat, and treeless regions of the Arctic—and changes to it in response to climate change.

Indirect counts. For mobile animals that cannot be counted directly, scientists use **indirect counts**, which rely on measuring signs left by animals or counting only a subset of the population. **Mark–recapture sampling** is an indirect count often used to estimate population size. Using this method, researchers capture a sample of animals from the population, tag each of them with a mark that identifies them (such as by giving each a unique number), and then release the animals back into the wild (Figure 14.7). Later, the researchers repeat the capture procedure, where the number of marked animals in the second sample is divided by the total number of animals captured in that sample. The proportion of marked animals in the sample is then taken to be representative of their proportion in the whole population. For example, suppose that scientists capture, mark, and release 100 deer in the first sample. Later, they return and once again capture 100 deer. Of these, 50 are marked.

Any study that draws conclusions based on only a sample of a population is susceptible to sampling error. Thus, the larger the sample size, the more likely it is that conclusions drawn from that sample will be accurate.

The proportion of marked deer in the second sample (50 percent) indicates that half of the deer in the population have been marked. Thus, the 100 deer marked in the first sample must be members of a population of 200 deer.

In addition to establishing population numbers, the information gained from using sampling techniques can be used to infer other characteristics of the population, such as its age structure or sex ratio. For example, if a third of the deer recaptured in a mark–recapture study are of reproductive age, then a third of the population is assumed to share this characteristic.

Figure 14.6 An example of plot sampling

Credit: Leo Francini/Alamy Stock Photo

Here, a scuba diver is photographing a defined region to count the number of corals. The sample can then be used to estimate the total number of corals at the location.

Figure 14.7 The mark-recapture method with eastern carpenter bees (*Xylocopa virginica*)

A B

A. Top view of a bee squeezer, a device used for marking live bees without causing injury

B. A marked eastern carpenter bee. Note the presence of two different colors—a green dot on top of a yellow splotch. In this study, bees were marked with different colors depending on the day of capture; the presence of yellow and green suggests that this bee was caught twice.

Source: BRIGGS, E. L., BARANSKI, C., MÜNZER SCHAETZ, O., GARRISON, G., COLLAZO, J. A., & YOUNGSTEADT, E. (2022). Estimating bee abundance: can mark-recapture methods validate common sampling protocols? Apidologie, 53(1). https://doi.org/10.1007/s13592-022-00919-4

Digging Into Data

Penguin Tales: The Effects of Changes in Sea Ice

Antarctica is one of the most remote places in the world, with one of the most forbidding climates. In the past, scientists who wanted to study the distribution and abundance of biodiversity in Antarctica would have to carefully track species by foot. Today, new technologies allow researchers greater access to the species being tracked—for example, by using satellite and drone imagery (**Figure 14.8**). Using computers, scientists can then analyze massive amounts of imagery data to assess populations.

One scientist using this technology is National Geographic Explorer Heather Lynch (**Figure 14.9**). She's using satellite and drone imagery to study and monitor the impact of climate change on hundreds of penguin populations in the genus *Pygoscelis*—including populations of the Adélie (*P. adeliae*), chinstrap (*P. antarctica*), and gentoo penguin (*P. papua*). Lynch developed a mapping application for monitoring penguin populations that includes almost 50 years of satellite imagery. Satellites can detect

Credit: National Geographic Society

Figure 14.8 **A new colony of gentoo penguins on Galindez Island**

Notice that penguins, which you can see as black dots on the rocks (one of them is circled), are nesting only where the snow has already melted. As a result of climate change, snow melts earlier in this region each year, giving penguins like these greater opportunity to settle down and raise chicks.

Credit: National Geographic Society

Figure 14.9 **Studying penguins in Antarctica**

Here, National Geographic Explorer Heather Lynch is counting gentoo penguins on the Antarctic Peninsula.

the pinkish guano (seabird excrement; see Figure 14.8) on rocks inhabited with penguin populations, allowing researchers to identify and analyze even previously unknown penguin populations.

In recent years, the Antarctic and its Southern Ocean ecosystem have experienced significant warming and changes in sea ice distribution. Snow and ice are important to Antarctic penguins for several reasons. Adélie, chinstrap, and gentoo penguins need exposed rock to nest, and increased air temperatures melt the snow earlier and in more places. For gentoo penguins, which are known to colonize newly exposed rock habitat quickly, this means more places where they can nest as their populations expand (Figure 14.8). At the same time, reduced winter sea ice allows gentoo penguins—which remain near the colony all winter long—to settle in areas that were previously too icy for penguins to find food through the winter.

While ice and snow free areas can make it easier for penguins to access the ocean for food and bare rock of nesting, these changes to the ice environment are having detrimental impacts of the marine food web. For example, penguins, particularly Adelie and chinstrap penguins, rely overwhelmingly on krill (*Euphausiacea*) for food. The krill life cycle is tied to the availability of sea ice, and the loss of sea ice is one of the reasons why krill populations have been declining in many of the areas where penguins are trying to forage.

As part of her research, Lynch developed the Penguindex, a Living Planet Index, to more easily estimate trends in the relative abundance of Antarctic penguins (Figure 14.10).

Examine the graph in Figure 14.10. The black line indicates a penguin population that is stable through time, signified as a constant value of 1.0. Populations that are growing relative to this baseline lie above the line and populations that are shrinking are below the line. Each species is indicated by its own color.

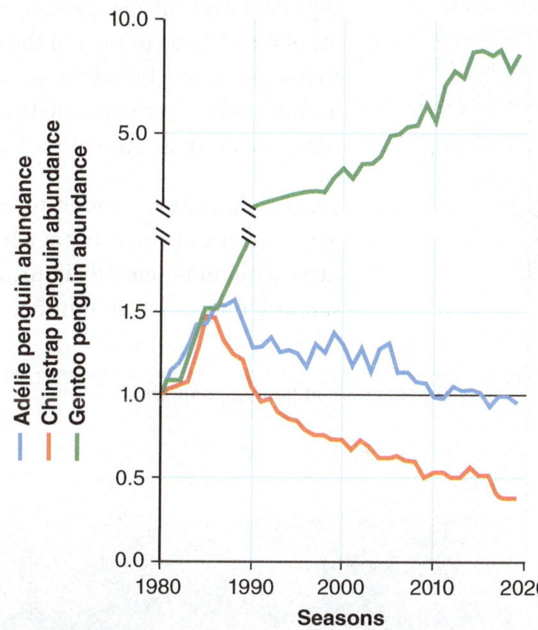

Graph based on: Talis, E.J., Che-Castaldo, C., Hart, T., McRae, L., and Lynch, H.J. Penguindex: a Living Planet Index for Pygoscelis species penguins identifies key eras of population change. *Polar Biology.* 2023; 46: 707-718. https://doi.org/10.1007/s00300-023-03148-2.

Figure 14.10 **Population data for Adélie, chinstrap, and gentoo penguin populations in Antarctica from 1980 to 2020**

The breaks in the *y*-axis and the green data line signify a gap in the data, where several values are included between where the line concludes and where it picks up again.

1. What trends do you see in the population data for Adélie, chinstrap, and gentoo penguins? How could warming temperatures explain these data?
2. Do you think the trend observed for gentoo penguins will continue as the climate continues to warm, or will there be limiting factors?
3. How does examining the data on three species of penguins in Antarctica affect your views on the complexity of studying the impacts of climate change?

Examining Human Populations

For most of human history, the human population grew very slowly (Figure 14.11). The growth rate began to pick up around 10,000 years ago, and during the past two centuries, it has soared. A worldwide decline in death rates, without an equivalent drop in birthrates, is responsible for the ongoing explosion in human population size. It took more than 100,000 years for the human population to reach a billion in number. Since then, the rate of increase has risen steadily.

The human population is now about 8.2 billion, and it is expected to reach 9.7 billion by 2050—and the impact on the planet cannot be understated (Figure 14.12). As we learned in Chapter 10, humans (both modern and ancestral) have been on Earth for a relatively small amount of time. In that time, we've changed the climate and are contributing to a potential sixth mass extinction. While we are also capable of reversing this damage, in this section, we'll examine some of the factors that have contributed to human population growth.

Although the carrying capacity for humans may have increased due to things like the use of synthetic fertilizers, the use of these fertilizers is also having a negative impact on the environment, lowering the carrying capacity for other organisms.

Three trends have contributed to human population growth. First, humans migrated into new habitats and expanded into new climate zones. Second, humans developed technologies that increased the *carrying capacity* of existing habitats. Third, humans sidestepped factors that typically restrain population growth.

Expansion into new habitats. As we learned in Chapter 12, modern humans evolved in Africa approximately 200,000 years ago, and about 60,000 years ago they began to spread out across the globe. A large brain and the capacity to master a variety of skills gave humans an unmatched ability to live in a broad range of habitats. Humans learned how to start fires, build shelters, make clothing, manufacture tools, and cooperate in hunts. With the advent of language, knowledge of such skills did not die with the individual but could be recorded and passed onto others.

Increased carrying capacity. When you come home from grocery shopping, how many bags of groceries can you carry at a time? Just like there is a limited amount of groceries you can carry without potentially injuring yourself or your food, the environment can also "carry," or support, only a certain number of individuals. **Carrying capacity** is the maximum population

Figure 14.11 A growth curve of human population

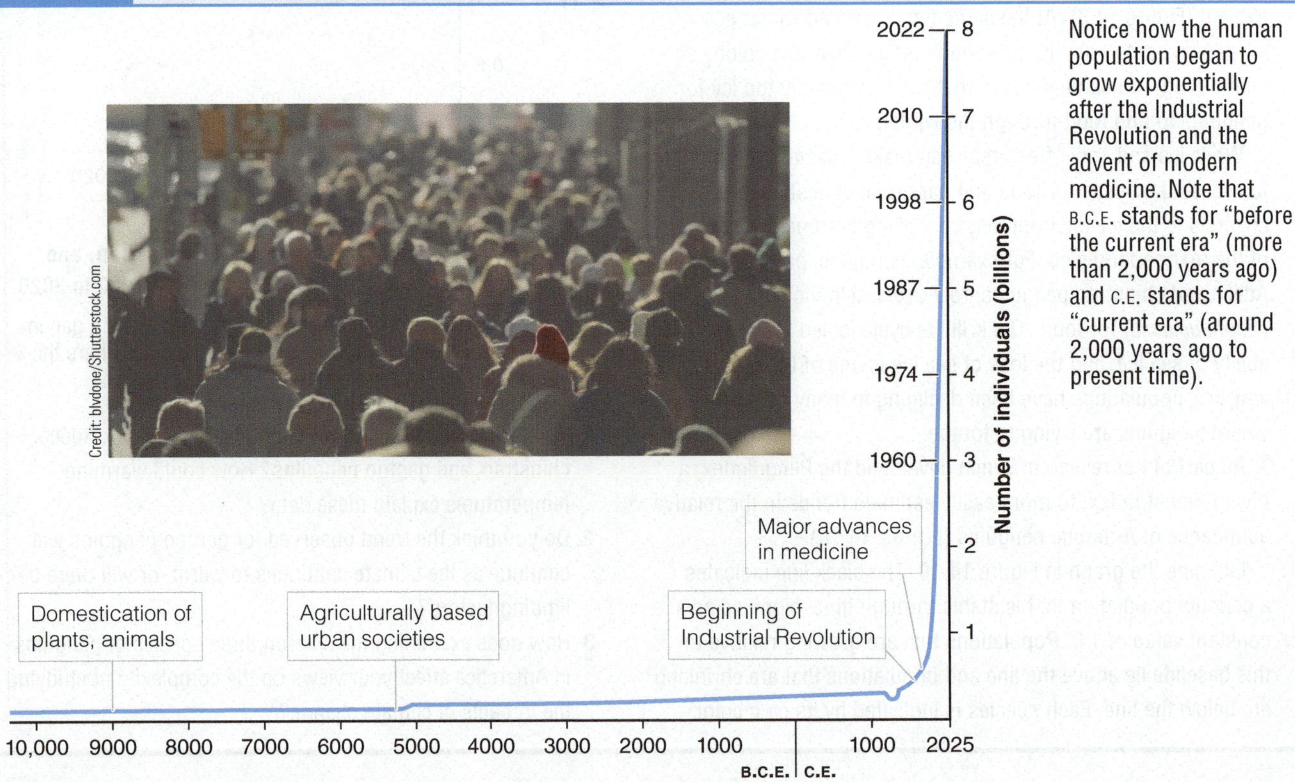

an environment can sustain before irreversible damage to the environment occurs (Figure 14.13). Exceeding carrying capacity eventually leads to a population crash as resources are depleted. For example, in areas where deer populations surged after wolves were eliminated, deer populations continued to grow until the food source was significantly damaged, resulting in little food for the deer and ultimately leading to starvation.

Note that carrying capacity is both environment-specific and species-specific. An acre of land with regular rains can support more grass plants than an acre of desert can. The carrying capacity of an environment for a particular species can also change over time. For example, the carrying capacity for a plant species decreases when nutrients in the soil become depleted.

When we think about the history of humans, the invention of agriculture and the domestication of wild grasses as food crops provided a more dependable food supply than traditional hunting and gathering, resulting in an increased carrying capacity. In addition, the human carrying capacity was increased with the Industrial Revolution and the arrival of synthetic fertilizers. Both of these innovations allowed people to increase food production.

Figure 14.12 **A humorous view on the impact of humans on Earth**

"I'm afraid you have humans."

Credit: Eric Lewis/cartoonstock.com

Figure 14.13 **A graphical depiction of carrying capacity**

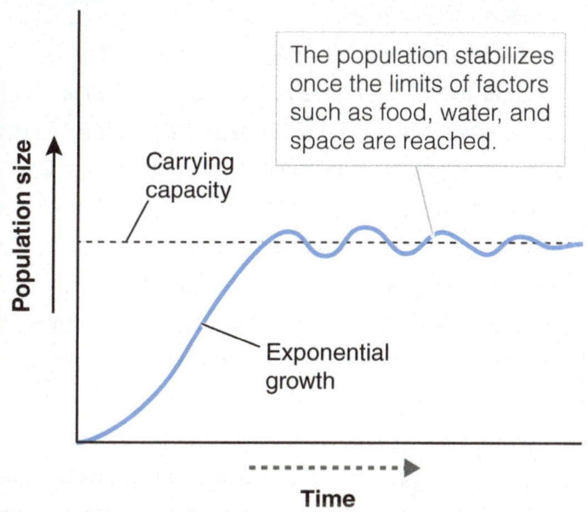

Notice that exponential growth occurs until carrying capacity is reached, where the population stabilizes.

Removal of population-limiting factors. Disease has historically dampened human population growth and is one factor that influences carrying capacity. For example, the more individuals there are in a location, the more potential there is for disease to spread among that population. Beginning in the mid-1800s, an increased understanding of the link between microorganisms and illness led to improvements in food safety, sanitation, and medicine. For example, people began to pasteurize or heat foods and drinks to reduce the number of harmful bacteria. In addition, there was a greater emphasis on hand washing and water supply protection. Vaccines and antibiotics also helped lower death rates. Vaccinations became widespread in Western countries during the 1800s, with the first antibiotics arriving in the 1940s (Chapter 10). As such, the advent of modern medicine contributed to the rapid increase in the human population (see Figure 14.11).

Reproduction Influences Population Growth, Fertility Rates, and Future Growth

Populations of organisms are always in flux as individuals are born and die. For example, in 2023, there were 134 million babies born around the world and 61 million people died. This means that the global population of humans increased by 73 million people in 2023. Statistical models suggest that the human birth rate is going to stay the same, but the number of deaths is expected to increase as the population ages. Once the death rate surpasses the birth rate, the total population will fall.

The **total fertility rate** is the average number of children a person is expected to give birth to during their childbearing years. For a population to remain stable, the total fertility rate must remain above the replacement level, or the number of children a couple must have to replace themselves. The United Nations estimates that a couple must produce 2.1 children to maintain a stable population. This rate can vary, however, since not everyone chooses to or is able to reproduce. The rate may also be higher in countries with higher infant or childhood mortality rates.

The total fertility rate also influences the age structure of a population. **Age structure** refers to the distribution of individuals among various age categories within a population. Age structure also accounts for the number of males and females in each age category.

? Did You Know?

The first modern sewer system was constructed in London, England, in the late 1800s. By diverting wastewater downstream of the city's water source, the system lowered the incidence of waterborne diseases such as cholera and typhoid fever.

Biology for Life

Assessing Risk to Humans from Climate Change

Many people think that the growing human population is the cause of climate change. This is a classic example of mixing up correlation and causation (Chapter 1). It's not that there are more people causing climate change, it's that the resource-intensive lifestyles people adopt cause the greenhouses gas emissions that lead to a warming planet. In fact, the wealthiest 10% of the global population is responsible for half of all greenhouse gas emissions. By 2070, as many as 3 billion people will be living in areas considered too hot for human habitation—which includes heavily populated cities like Los Angeles and Paris. The impacts of climate change on heavily populated areas is something that we've already witnessed—for example, with the major destruction caused by wildfires in the Los Angeles area in 2025.

And it's not just warming temperature, but also extreme weather events that have impacts beyond those strictly limited to the environment. Take, for example, insurance costs. A consortium of actuaries from the United States and Canada generated the Actuaries Climate Index. Actuaries are people who analyze the financial costs of risk and uncertainty. The data they generate are useful to many stakeholders, such as policy makers, the general public, and in particular, insurance companies.

The Actuaries Climate Index is a measure of extreme weather changes based on temperature, precipitation, wind, and sea level that represent the risk of climate impacts on people and the economy (**Figure 14.16**). The same consortium also generates the Actuaries Climate Risk Index, which uses historical data regarding economic losses, deaths, and injuries from extreme weather events to assess their financial impact. Insurance companies can use this index to adjust rates of property or life insurance costs accordingly based on risk.

Unfortunately for many of us, if insurance companies are to remain in business, they must accurately gauge risk so that they can offer valid quotes to consumers, while also spreading the risk across a large pool. This means that even those of us not living in areas frequently at risk for extreme weather events will likely see rising homeowners' insurance costs, or even unavailability of coverage.

1. Research the Actuaries Climate Index website and examine their regional graphs and maps, which show average risk since 1961.
   a. Look at the graph for the United States and Canada in the Actuaries Climate Index, also shown in Figure 14.16. How has the index changed from 1961 to today? What does it say about the risk of extreme weather events?

You may have heard news reports that the United States population is declining. Overall, the population decreased by 2% per year between 2015 and 2020, and has fluctuated since then. More data are needed to definitively say that the population is in decline and these aren't normal fluctuations.

Ecologists use age structure to make inferences and predictions about how a population may change in the future. For example, a population with a large proportion of young individuals who have not begun to reproduce has a greater potential for growth than one composed mainly of older individuals that are beyond the optimal reproductive years. Considerations of age structure are also important when determining if a population is at risk for extinction, where a population with fewer breeding individuals is at greater risk of extinction.

Let's look at total fertility rates and age structure in human populations. When we consider three human populations—in Nigeria, the United States, and Japan (**Figure 14.14**)—we see that there are more young people in Nigeria than in Japan, meaning that the population in Nigeria is expected to grow while the population in Japan will shrink. The United States has a stable level of growth, where the population should remain at stable numbers. Note that for all graphs, the number of people in their post-reproductive years shrinks as people reach the maximum life span.

Ecological Footprints

Just like walking on the beach leaves a trail of footprints imprinted in the sand, consumption of natural resources also leaves an imprint on the environment. An **ecological footprint** is the amount of Earth's surface required to support a particular level of development and resource consumption in a sustainable fashion. It includes the area needed to secure food and manufacture products, as well as the natural area needed to take up any excess carbon dioxide produced by human activities.

The United States accounts for about 4.6 percent of the world's population, yet it uses about 25 percent of the world's minerals and energy supply, and so has a large ecological footprint (**Figure 14.15**). Billions of people living in places like Nigeria and other emerging economies would like to own the same kinds of goods that people in highly industrialized countries enjoy. However, given current technology, Earth does not have the resources to make that possible. Finding ways to meet the wants and needs of expanding populations on a planet with finite

b. Go to the map portion of the website and look up the region where you live. Do you notice any changes that have happened over time? How does it compare to the United States and Canada combined average?

c. Go to the map portion of the website and explore how the map changes over time when you choose different components, such as the warm temperature index or extreme precipitation index. What observations did you make?

2. Some argue that the planet is not warming, since they have experienced extremely *cold* temperatures and blizzards where they live. Is what they are experiencing evidence for or against a warming planet? Hint: Consider the definition of climate as part of your answer.

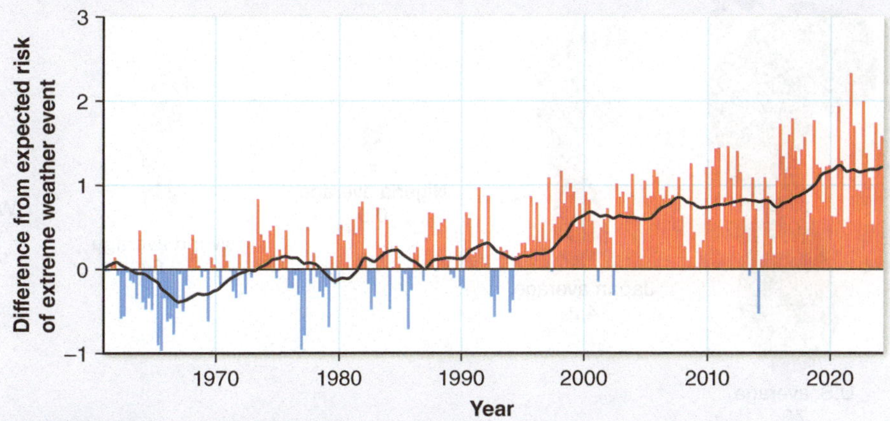

Figure 14.16 The Actuaries Climate Index (ACI)

This index shows the risk of climate impacts on people and the economy in the United States and Canada combined. The value of 0 on the *y*-axis represents baseline, and the height of the bars represents the risk for climate-related incidents. Blue indicates lower than normal risk, and orange indicates higher than normal risk. Note how high the index has climbed above baseline since the late 1990s.

Figure based on: Component Graphs - Actuaries Climate Index. (2016, November 8). Actuaries Climate Index. https://actuariesclimateindex.org/explore/component-graphs/

Figure 14.14 Age structure diagrams for three countries

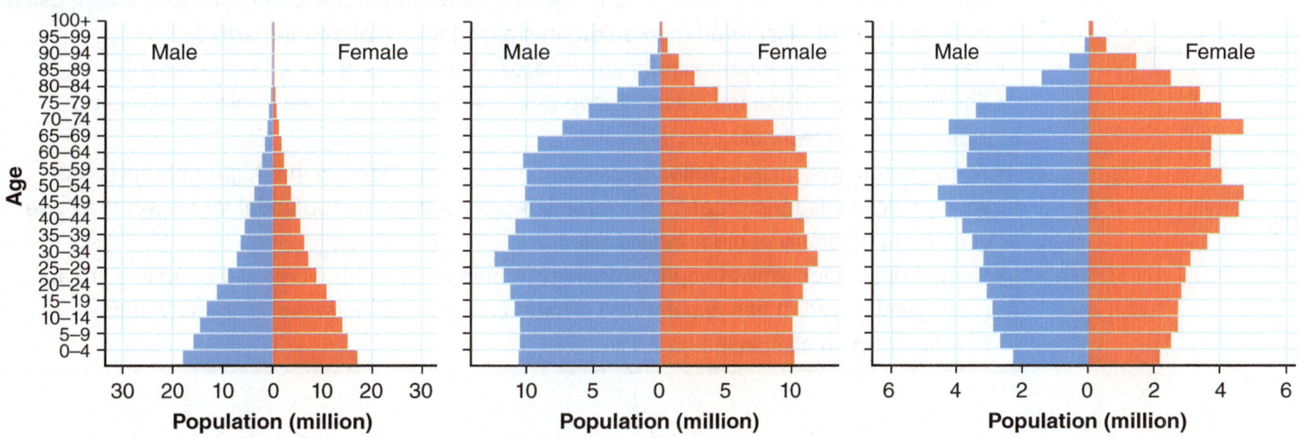

A. Increasing growth. Nigeria has many young people who are still within their peak fertility. Therefore, we would expect the population of Nigeria to grow.

B. Stable growth. The United States has a relatively even distribution of people across different ages. We would therefore expect the population to remain stable.

C. Decreasing growth. Japan has fewer young people, and most of its population is beyond its reproductive years. Therefore, we would expect that the population will decline.

The width of each bar represents the number of individuals in a 5-year age group. Ages 0–15 are pre-reproductive, 16–45 are reproductive, and over 45 are post-reproductive. The left side of each chart shows the number of males, and the right side shows the number of females, in millions.

Figure 14.15 **Ecological footprints**

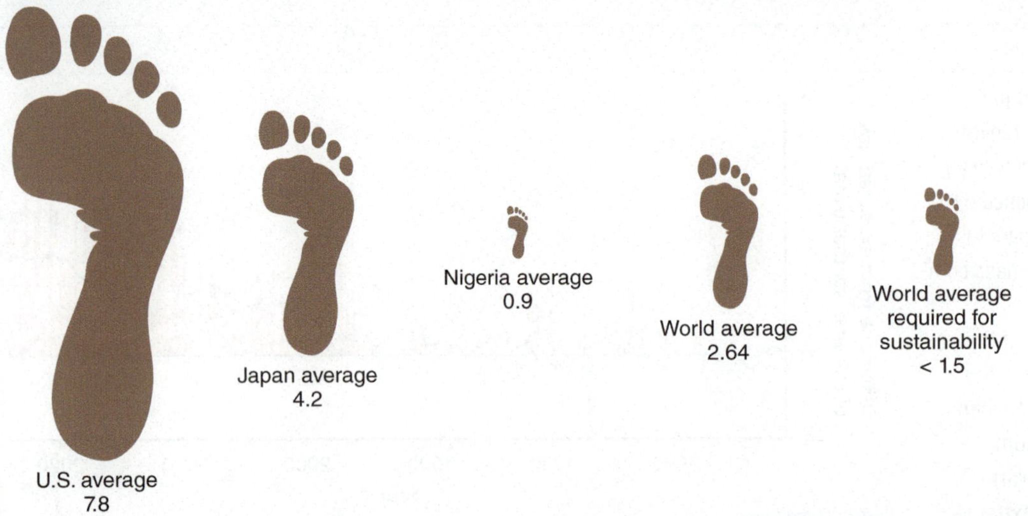

This figure symbolically shows per capita global footprint data for the United States, Japan, and Nigeria by scaling the ecological footprints to each other. Notice that the average person in the United States has an ecological footprint nearly three times that of an average world citizen, which on its own is above the world average required for sustainability.

resources will be a challenge. For example, if we were to estimate the number of planet Earths needed to sustain various populations at current levels of consumption, the United States would require 5.02 Earths, China would need 2.26 Earths, and Nigeria would need 0.55 Earths.

14.2 Summary

- Populations can be quantified by their size, density, geographic range, distribution (clumped, uniform, random), age structure, and total fertility rate. A species' geographical range includes the total area occupied by all of its populations.
- Using direct counts, researchers can observe and count individuals using plot sampling, where direct counts in multiple plots estimate population size. Indirect counts can estimate population size by using mark–recapture sampling, where a sample of organisms is captured and tagged, and the population size is then estimated when a second sample is taken, using the proportion of recaptured marked organisms to those that are unmarked.
- Carrying capacity is the maximum sustainable population size for an environment.
- The human population rapidly increased due to expansion into new regions, increased agriculture, and technological innovations.
- Age structure affects population growth because a population with a higher proportion of young individuals (and greater total fertility rate) has a greater potential for growth than one composed of mainly of older individuals.
- Human population growth impacts ecology in many ways. Our increased collective impact (ecological footprint) has accelerated climate change and increased extreme weather events, affecting all species on Earth.

14.3 Models of Population Growth

Learning Objectives

14-10 Compare exponential and logistic growth.

14-11 Describe the factors affecting a population's growth.

As we learned in the last section, population sizes grow and shrink depending on the overall birth and death rates. Ecologists track growth using the **per capita growth rate**, which is the per capita birth rate minus the per capita death rate. Per capita means for each individual.

Let's look at an example. Say that among a population of 2,000 mice, 1,000 new mice are born every month. This means that the per capita birth rate is 1,000/2,000, or 0.5 birth per mouse per month. Now, let's do the same with the death rate. If the death rate in the population of 2,000 mice is 200 per month, we would calculate a rate of 200/2,000 (0.1 per mouse per month). Our per capita growth rate is then 0.5 – 0.1, or 0.4.

So why do we care? Different per capita growth rates influence how fast a population grows. Knowing the dynamics of population growth is important for tracking the efficacy of, for example, conservation measures (Chapter 16) or the growth of a population of an invasive species (Chapter 15) and its impact on the local flora and fauna. Let's explore two types of population growth—exponential and logistic.

Figure 14.17 **Exponential growth**

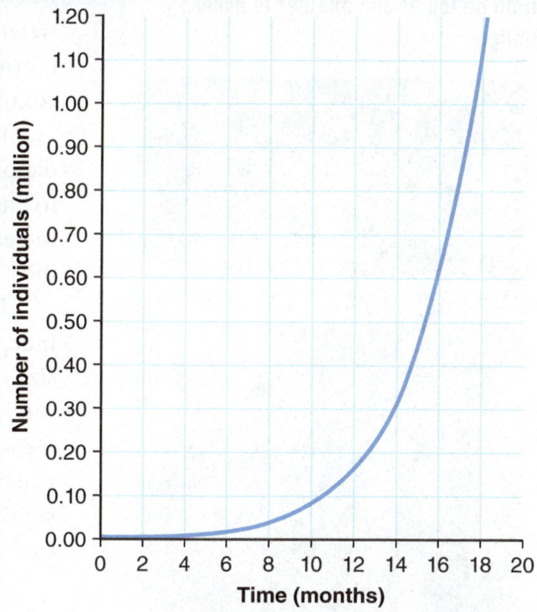

In this fictious population of mice with a per capita growth rate of 0.4 per mouse per month and an initial population size of 2,000, there would be 1.1 million mice after 18 months.

Exponential Growth

The **exponential model of population growth** describes how a population would change in size over time if its per capita growth rate were constant and its resources were unlimited. Under these conditions, the population growth in any interval (G) can be calculated as follows:

$$\underset{\text{(number of individuals)}}{N} \times \underset{\text{(per capita growth rate)}}{r} = \underset{\text{(population growth per unit time)}}{G}$$

Now that you've learned about exponential population growth, look at Figure 14.11 again. Note that the human population is growing exponentially and has been for many years.

Suppose we apply this to our population of 2,000 mice with their per capita growth rate of 0.4 per month. In the first month, the population grows by 2,000 mice × 0.4. This brings the size to 2,800. In the next month, 2,800 × 0.4, or 1,120 mice, are added, and so on. At this growth rate, the number of mice would rise from 2,000 to more than a million in less than two years! Plotting the size of this population against time produces a J-shaped, or "hockey stick," curve, which is characteristic of exponential growth (Figure 14.17).

Exponential population growth is analogous to compound interest on a bank account that pays a fixed rate of return and from which there are no withdrawals, where you earn interest on what you've saved *plus* the interest that you've earned along the way added to the overall balance. Although the interest rate does not change, each year the *amount* of interest paid increases with the higher and higher balance.

Density-Dependent Limiting Factors

The exponential model of population growth cannot accurately predict the long-term growth of real populations because it assumes that resources are unlimited—which isn't true 100% of the time. However, it does provide insight into the expected short-term growth of a population with plentiful resources. For example, when a few individuals of a species colonize a new habitat, the resulting population often grows exponentially for a period. Once there are too many individuals in one place, growth will begin to taper off. Density-based factors that impact birthrates and/or death rates are called **density-dependent limiting factors**. Note that these factors are typically biotic in nature, as they result from the interactions of living organisms. Let's look at some examples.

Have you ever competed before? As an athlete, musician, or something else? When we compete, we are pitted against, for example, another sports team, to see who will win. The same phenomenon occurs throughout the animal kingdom. **Competition** is when organisms

Figure 14.18 **In the rainforest, plants climb on top of one another to access sunlight**

Credit: Genevieve Vallee/Alamy Stock Photo

The more plants there are in an area (higher density), the harder it is to access sunlight (greater competition).

challenge one another for the same resources. Said otherwise, organisms, whether plants or animals, will compete for essential resources like food, water, hiding places, and nesting sites. For example, plants compete for nutrients, water, and access to sunlight (**Figure 14.18**). We will return to competition in Chapter 15.

Disease can also impact carrying capacity. Parasitism, disease, and predation are also density-dependent limiting factors. The closer individuals are to one another, the more easily parasites and pathogens can spread. Predation increases with density, because predators often concentrate their efforts on the most abundant prey species.

In contrast to density-dependent factors, there are also density-independent factors. **Density-independent limiting factors** impact population size regardless of the density of the population. Density-independent factors can include natural or human-made disasters that impact population size. Density-independent limiting factors are typically abiotic. For example, when the *Exxon Valdez* oil tanker ran aground in 1989, it killed thousands of seabirds, sea otters, seals, bald eagles, fish, and whales irrespective of the size of each population.

? Did You Know?

Two pods (groups) of orcas were impacted by the *Exxon Valdez* oil spill. Neither pod fully rebounded to its original numbers.

Logistic Growth

The **logistic model of population growth** describes how population growth changes as the size of the population approaches carrying capacity (Section 14.2). Unlike the exponential growth model, this model incorporates the effects of density-dependent limiting factors. With logistic growth, the rate of population growth does not remain constant, but rather declines as population density increases.

When there are few individuals relative to the amount of resources, the population grows exponentially (**Figure 14.19**). As the number of individuals rises, density-dependent limiting factors begin to put the brakes on growth. As a result of these factors, population growth begins to slow. Population growth continues until population size levels off at the environment's carrying capacity.

Density-dependent and density-independent factors interact to determine the fate of a population. Ideal environments can result in an increased carrying capacity. However, changes to those environments, such as a particularly harsh winter, can cause populations to crash.

Figure 14.19 **A hypothetical population experiencing logistic growth**

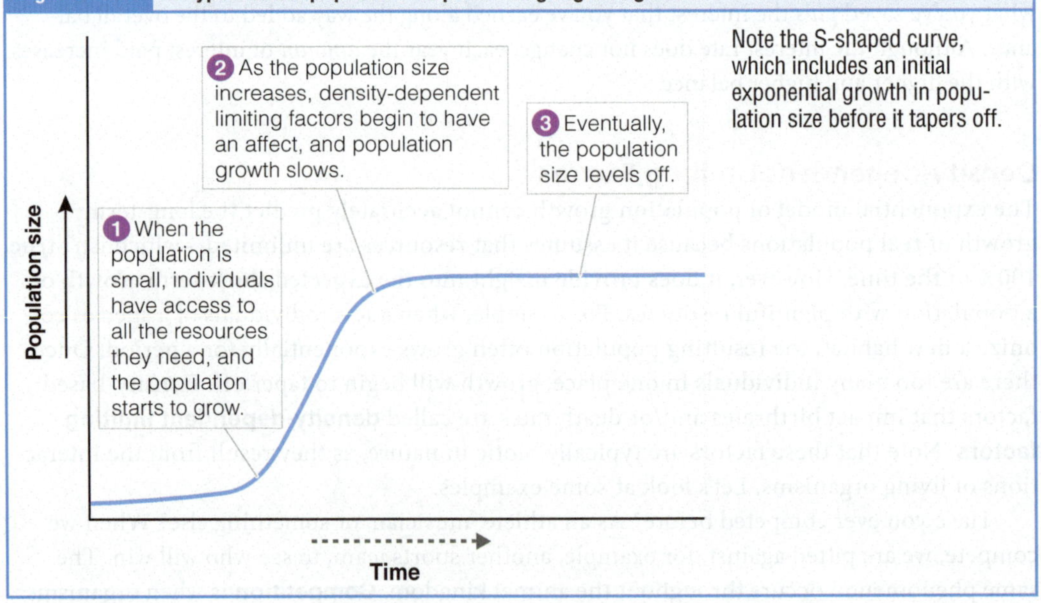

Biology for Life

Car Accidents Regulating Population Sizes

Credit: istock.com/Pascal-L-Marius

Figure 14.20 **Have you ever been involved in a collision with a deer?**

According to Pew Research, there are between 1 million and 2 million crashes between motor vehicles and large animals, such as deer, each year, resulting in hundreds of human deaths and thousands of human injuries, as well as approximately $8 billion in costs related to property damage and medical treatment (Figure 14.20).

Although eventually deer populations will decrease as competition for food increases, that point is above and beyond what is safe for deer and humans to live near one another. Additionally, large populations of deer are decimating forests due to their tendency to eat new growth, which negatively impacts the forest ecosystem (Figure 14.21).

Deer populations began to spiral out of control when humans hunted *predators* (animals that eat other animals) of deer, such as mountain lions and wolves. Now, deer populations are held in check by human hunters, who harvest deer after securing a hunting license. Culling the deer population also protects the forests and minimizes the chances of collisions with cars. While hunting is certainly helpful, as of 2022, half as many Americans hunted than in the 1980s. Solutions to this problem include re-introduction of apex predators (Chapter 15), expanding hunting-accessible public lands, and programs geared toward teaching people how to hunt for subsistence or to close gender gaps in hunting, such as the Becoming an Outdoors-Woman program.

1. What type of growth curve would you expect if humans stopped hunting deer altogether? What would the curve look like if the optimal amount of hunting was occurring?
2. Do humans have an obligation to manage wild populations of deer? Does your answer change when considering instances where humans encroach into the territories of wild animals?

Credit: Nature Picture Library/Alamy Stock Photo

Credit: Nature Picture Library/Alamy Stock Photo

Figure 14.21 **A comparison of a healthy forest and an overgrazed forest**

A. A healthy forest. Note the presence of a variety of short and tall plants. This variety of plant life provides habitat for other organisms, such as places for birds to nest.

B. An overharvested forest. Note the lack of small plants and bushes.

What about apex predator populations? Wolves and mountain lion populations are dependent on the populations of their food sources. If a lot of food (like deer) is available, apex predator populations grow. If the deer population declines, predator populations will also decline. Therefore, the two populations regulate one another.

14.3 Summary

- Exponential growth occurs when resources are unlimited and the per capita growth rate is constant; this displays in a graph as a J-shaped curve. Logistical growth starts exponentially but levels off when a population is affected by density-dependent limiting factors, such as competition, and reaches carrying capacity, forming an S-shaped curve.
- Both forms of growth are affected by density-independent limiting factors, which include natural or human-made disasters that impact population size.

14.4 Life History Patterns

Learning Objectives

14-12 Compare opportunistic and equilibrial life histories.

14-13 Describe the three types of survivorship curves.

Throughout this chapter, you've learned about factors that influence a population, and the ways of analyzing and tracking those changes. Together, different factors of a population can be combined to define a life history. A **life history trait** is a characteristic of an organism that impacts its ability to survive and reproduce. Life history traits include factors like average rate of growth, age at first reproduction, number of breeding events, and life span—all of which influence the population. In this section, we'll look at how life history traits vary.

Describing Life Histories

One way to gather information about the life history traits of a population is to focus on a **cohort**, or a group of individuals that are all born at about the same time. Information about age-specific death rates can also be illustrated by a **survivorship curve**, a plot that shows how many members of a cohort remain alive over time. Survivorship curves are based on life tables. A **life table** is a chart that represents how many individuals are expected to survive to at each age (Figure 14.22).

Ecologists describe three generalized types of survivorship curves. A **type I curve** is convex (bulges outward), indicating survivorship is high until late in life (Figure 14.23A). This pattern is characteristic of humans and other large mammals that produce one or two young and care for them.

A **type II curve** is diagonal, indicating that the death rate of the population does not vary much with age (Figure 14.23B). A type II curve is characteristic of lizards, small mammals, and birds. In these groups, old individuals are as likely to die of disease or predation as young ones are.

? Did You Know?

Life tables are used by insurance companies to calculate life insurance premiums.

Figure 14.22 A life table and survivorship curve for people living in the United States

A

Age	Male	Female
0	100,000	100,000
5	99,304	99,400
10	99,238	99,348
15	99,143	99,279
20	98,715	99,105
25	97,926	98,802
30	96,850	98,361
35	95,512	97,730
40	93,937	96,887
45	92,003	95,756
50	89,523	94,215
55	86,089	92,040
60	81,272	88,915
65	74,786	84,502
70	66,732	78,758
75	56,849	70,872
80	44,198	59,469
85	29,314	44,170
90	14,571	26,043
95	4,386	10,192
100	698	2,208

B

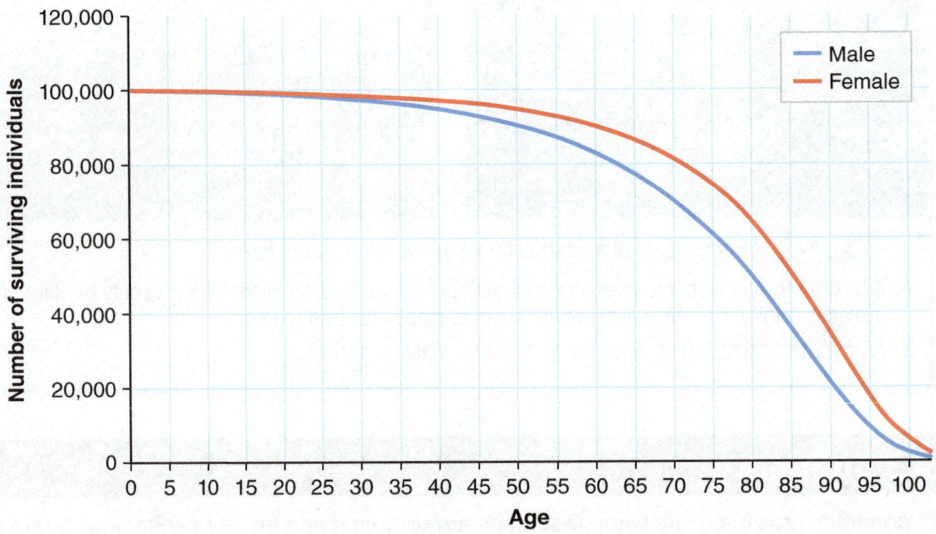

A. A life table representing survivorship amongst males and females living in the United States.
B. A survivorship curve, based on the life table in part A. These data are from the U.S. Social Security Administration and show the number of individuals who are expected to survive to each age.

Figure 14.23 **Types of survivorship curves**

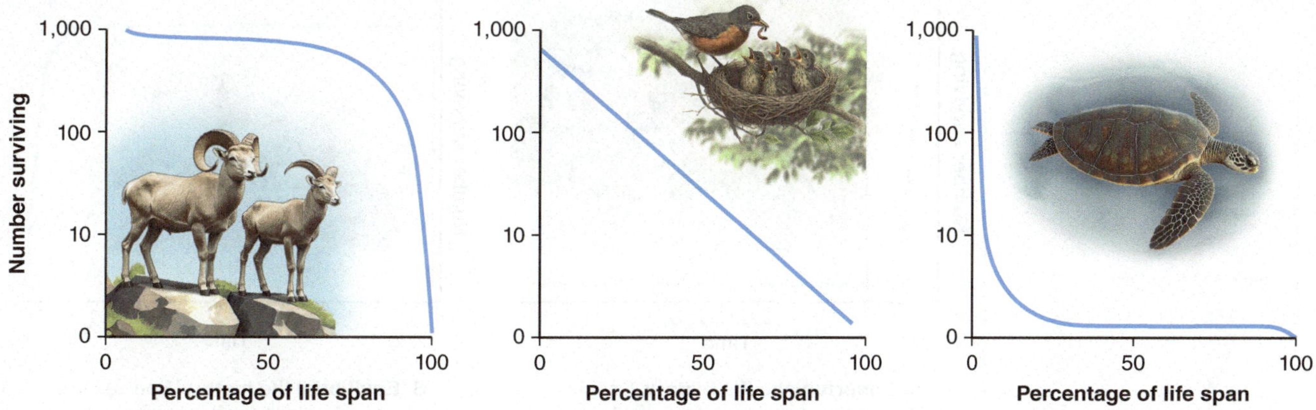

A. Type I curve. Bighorn sheep are an example of an organism with a type I curve, where mortality is highest late in life.

B. Type II curve. Birds, such as robins, have a type II curve. For these organisms, mortality does not vary much with age.

C. Type III curve. Organisms like sea turtles have a type III curve, where mortality is highest early in life.

A **type III curve** is concave (bulges inward), indicating that the death rate of a population is highest early in life (Figure 14.23C). Marine animals that lay many eggs at once and experience high mortality early in life have this type of curve, as do plants that release enormous numbers of tiny seeds.

Adaptive Value of Life History Traits

When producing offspring, an individual uses resources that it could otherwise use to grow and maintain itself, called the parental investment. Note that parental investment occurs even in species that do not provide parental care. For example, the parental investment includes resources used to produce gametes, as well as resources used to nurture developing young in or on a parental body.

Species differ in the way in which they distribute their parental investment among offspring and over the course of a lifetime. Some invest little in each of many offspring; others invest a lot in only a few offspring. Some reproduce once, others many times. The sequence of growth, development, and reproduction-related events that occurs over an individual's lifetime is called a **life history**. Life histories vary continuously, but ecologists have described two theoretical extremes at either end of this continuum (Figure 14.24). Both allow an individual to maximize the number and survival of its offspring, but each is adaptive under different environmental conditions.

Opportunistic life history. When a species lives where conditions vary in an unpredictable manner, its populations seldom reach their carrying capacity. As a result, there is usually little competition for resources among members of the same species. Such conditions favor an opportunistic life history, in which individuals produce as many offspring as possible, as quickly as possible. Because parental investment is spread across many offspring, each offspring receives a relatively small share. Opportunistic species tend to have a small body and to reach the age of reproduction quickly. They usually have a type III survivorship curve, with mortality highest early in life.

Equilibrium life history. When a species lives in a stable environment, its populations are often near their carrying capacity for that environment. Under these circumstances, competition for resources can be fierce, and an **equilibrium life history**, in which parents produce a few high-quality offspring, is more common.

Species with an equilibrium life history tend to have a large body and a long generation time (the typical time between birth and becoming a parent). Large mammals such as elephants and whales also have this kind of life history. They take years to reach adult body size and begin reproducing. A mature female whale typically produces only one large calf at a time, and she continues to invest in the calf by nursing it after its birth.

Figure 14.24 **Comparing life histories**

A. Opportunistic life history. Species like beetles and flies grow quickly and reproduce frequently, with many offspring. This is characteristic of a type III survivorship curve, representing an opportunistic life history.

B. Equilibrial life history. Species like orcas grow slowly and reproduce later in life, with fewer offspring and greater parental investment. This is characteristic of a type I survivorship curve and an equilibrial life history.

14.4 Summary

- When survivorship of a cohort is plotted against the percentage of life span, the convex type I curve shows the greatest mortality late in life due to high parental care for few offspring; the type II curve is diagonal, indicating a species where age does not affect mortality rates; and the concave type III curve shows the highest mortality early in life, for species producing many offspring with little or no parental input.
- In unpredictable environments, species that quickly and robustly reproduce have opportunistic life histories. In stable environments, species that raise few offspring with high parental investment have equilibrium life histories.

Understand It! Forgetting

Have you ever walked into a room to get something, only to arrive and find out that you forgot what you were doing there in the first place? In Chapter 13, we learned about how we remember and why memory is important to learning. The opposite of remembering something is forgetting it—and if our goal is to learn something, we need to make sure we don't forget it!

It's easy to think of forgetting as something negative. For example, we sit down to take an exam but no matter how hard we tried couldn't remember something we had recently studied. Forgetting is a normal process—what we remember tends to decline overtime, unless that learning is reinforced, such as during retrieval practice (Chapter 1). Each additional time you study the material, you forget a little less after each study session.

Sometimes it helps to forget, such as an unpleasant memory like getting a cavity filled at the dentist. In the context of learning, a little bit of forgetting might lead to better long-term retention. Some psychologists think that allowing for time to forget is why spaced practice (Chapter 15) is an effective study strategy. This is because as we repeatedly look at information, we forget the less important parts while the important parts are reinforced in our memory. Since there is less information clogging up our brain, it's more efficient to remember the important information, because we've forgotten the less important stuff.

1. Reflecting on your learning, how has forgetting helped you? Hurt you?
2. Retrieval practice helps us remember, while spacing out that retrieval practice to allow for forgetting is helpful. What do you think is the optimum interval for your study to learn? Do some research—how does your answer compare to the research?

Chapter Review

14.1 Introduction to Ecology

Ecology is the scientific study of interactions among organisms, and between organisms and their environment. One individual is an **organism**. A **population** is a group of organisms of one species in a specific area. A **community** consists of all populations in a region. An **ecosystem** includes the **biotic** (living) components of a community plus the **abiotic** (non-living) components. The **biosphere** includes all ecosystems on Earth. The total area in which all populations of a species is found is its **geographical range**.

Ecologists study the impact of climate change by comparing quantifiable characteristics of populations over time.

14.2 Characteristics of Populations

The **geographic range** is the total area occupied by all populations of a species. **Population size** is the number of individuals in a population, while **population density** is the number of individuals in a specified area or volume.

Population distribution describes the way members of a population are arranged within their environment. A **clumped distribution** means members of the population are closer to one another than predicted by chance alone, while **uniform distribution** occurs when individuals are evenly spaced. A **random distribution** is rare in nature because it relies on resources being uniformly available, and proximity to others is neither beneficial nor harmful.

When possible, scientists do **direct counts** to determine the number of individuals in a population. In **plot sampling**, scientists directly count the number of individuals in a marked area, such as a 1-meter-by-1-meter square. For mobile animals that cannot be counted directly, scientists do **indirect counts**, like **mark–recapture sampling** to estimate population size.

Age structure is the distribution of individuals among various age groups. Populations with greater proportions of younger individuals have a higher growth potential than those with higher levels of older individuals.

The **total fertility rate** is the average number of children born to individuals in a population over the course of their lifetimes. **Carrying capacity** is the maximum number of individuals of a species supported indefinitely by their environment. The **ecological footprint** is the area of Earth's surface required to sustainably support a particular level of development and consumption.

14.3 Models of Population Growth

The **per capita growth rate** provides the rate of change in a population per individual. The **exponential model of population growth** describes population growth with a constant per capita growth rate and unlimited resources. **Density-dependent limiting factors** slow growth in dense populations. Competition among members of the same species is called **competition.**

The **logistic model of population growth** is when numbers initially increase exponentially, then the growth rate slows and population size levels off at carrying capacity. **Density-independent limiting factors** limit growth in populations of any density.

14.4 Life History Patterns

Life history traits are heritable characteristics that directly affect an organism's ability to survive and reproduce. A **cohort** is a group of individuals all born at about the same time. A **life table** shows the probability of a member of a particular population living or dying at a certain age.

Three types of **survivorship curves** plot how many members of a cohort remain alive over time. **Type I curves** are convex, indicating the highest mortality late in life; **type II curves** are diagonal, where the death rate does not vary by age; **type III** curves are concave, indicating the highest death rate early in life.

Species in unpredictable environments favor an **opportunistic life history**, producing as many offspring as fast as possible with little or no parental input. Species in stable environments favor an **equilibrium life history**, producing few offspring with high parental investment.

Practice It!

Retrieval practice is the process of pulling information from your long-term memory and using it. Here are some questions to help with your retrieval practice.

Self-Quiz

1. The study of interactions among organisms and between organisms and their environment is called ______. (LO 14-1)

a. environmentalism
b. conservation
c. ecology
d. population studies

2. A fellow student describes their weekend camping trip. They mention the beautiful trees, the singing birds, and the babbling brook at their campsite. Which level of the hierarchy of interactions did they describe? (LO 14-2)

a. population
b. community
c. ecosystem
d. biosphere

3. Why are randomly distributed populations rare in nature? (LO 14-4)

a. Because resources are not typically evenly distributed in the environment
b. Because all organisms have some sort of community organization
c. Because it would make them more susceptible to predation
d. Because members of any population must cooperate to survive

4. When sampling methods are used to estimate population size, ______ counts are done. (LO 14-5)

a. direct
b. indirect
c. cohort
d. fictional

5. Which of the following is NOT an example of how humans have increased the carrying capacity of their environment? (LO 14-6)

a. By developing genetically engineered crops
b. By moving to new areas of the world
c. By developing medications that treat disease
d. By eating more meat and fewer plants

6. An increase in the population of a prey species would most likely ______ the carrying capacity for that species' predators. (LO 14-6)

a. increase
b. decrease
c. not affect
d. stabilize

7. Human population growth occurred due to ______. (LO 14-7)

a. the invention of agriculture
b. medical advances
c. improved sanitation
d. all of the above

8. The total fertility rate of a population is ______. (LO 14-8)

a. the number of fertile years divided by the length of gestation
b. the total number of children born to a population per year
c. the average number of children one has in their reproductive years
d. the maximum number of children one could have if resources were unlimited

9. Why does age structure affect the total fertility rate? (LO 14-8)

a. Because reproductive rates differ by age
b. Because individuals of all ages are needed to care for offspring
c. Because reproductive rates remain constant, while members in each age group vary
d. Because fertility increases steadily as individuals age

10. Compared to countries with emerging economies, highly industrialized nations have a higher ______. (LO 14-9)

a. death rate
b. birthrate
c. total fertility rate
d. ecological footprint

11. The exponential model of population growth assumes ______. (LO 14-10)

a. the death rate declines as population density increases
b. industrialization causes a fall in birthrates
c. resources are limited
d. per capita growth rate does not change

12. The logistical model of population growth starts with ______ as carrying capacity is approached. (LO 14-10)

a. exponential growth then speeds up
b. steady growth then decreases
c. exponential growth then slows
d. steady growth then increases

13. Disease and competition for resources are ______ controls on population growth rates. (LO 14-11)

a. age structure
b. fertility rate
c. density-dependent
d. density-independent

14. A population exhibiting a type III survivorship curve is likely living in ______ environment that favors ______ life histories. (LO 14-13)

a. a stable; opportunistic
b. an unpredictable; opportunistic
c. a stable; equilibrium
d. an unpredictable; equilibrium

15. Species that invest a lot of parental care over a long time are likely to show a ______ survivorship curve with the highest mortality ______ in life. (LO 14-13)

a. type I; late
b. type I; early
c. type III; late
d. type III; early

Critical Thinking

1. How might you determine whether climate change is having any effects where you live? If you had unlimited time and money, describe an experiment that could determine the effects of climate change in your area. Be sure to explain which level of ecological interaction you would study and how you might determine whether climate change has affected it. (LO 14-2, 14-3)

2. A biologist is hired to assess the population of Canada geese at a city park. She captures 50 geese in August, marks them, and then releases them. She returns in November and again catches 50 geese. Ten of them are marked. What would she estimate the local Canada goose population to be, based on these data? What might she do to get a more accurate estimate? (LO 14-5)

3. In 1989, Martin Wikelski started a long-term study of two marine iguana populations in the Galapagos Islands, on the islands of Genovesa and Santa Fe, located about 80 miles apart. The iguanas have no predators, so deaths are usually the result of food shortages, disease, old age, or numbers can decline during El Nino events, when water surrounding the islands heats up. In January 2001, an oil tanker ran aground and leaked a small amount of oil near Santa Fe Island. The graph below shows the number of iguanas on each island, just before the oil spill and again a year later.

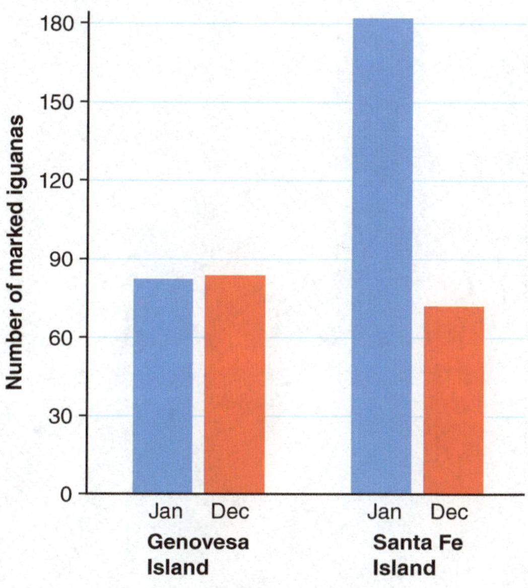

a. Which island had more marked iguanas at the time of the first census? What about during the second? (LO 14-5)

b. How much did the population size on each island change between the first and second census? (LO 14-5)

c. Wikelski concluded that changes on Santa Fe Island were the result of the oil spill, rather than climatic factors common to both islands. Do you agree with his conclusions? What, if any, additional information is required to be confident in why iguana population numbers differed? (LO 14-6)

4. Terrariums are self-sustaining ecosystems that can be built in jars or bottles. Water, oxygen, and carbon dioxide levels need to be considered to sustain the ecosystem, but by combining carefully selected plants, a terrarium can be closed and survive indefinitely. Suppose you built a closed terrarium with three types of plants (A, B, and C). You notice after a few months that plant A is beginning to die, while plant B seems to be taking over the terrarium. Plant C is healthy, but has not spread to the degree that plant B has. Make a list of the (1) density-dependent and (2) density-independent factors that can explain this phenomenon. Then, describe an experiment that could help explain why plant A is dying and plant B is thriving. (LO 14-6)

Credit: Nature Picture Library/Alamy Stock Photo

15

Biological Neighborhoods

Chapter Outline:

15.1 Community Structure
15.2 Species Interactions
15.3 Changing Communities
15.4 The Nature of Ecosystems
15.5 Climate and the Distribution of Biomes
15.6 Biogeochemical Cycles

The Power of Poo

Whether you call it poo, poop, dookie, or something more vulgar, defecation is part of both our and other animals' daily routines. Giggling aside, poop, and especially bird poop (also called guano), was essential for life to form on Surtsey.

Named for the Norse god Surtur, Surtsey is an island off the coast of Iceland that appeared suddenly in November 1963, with ongoing volcanic eruptions adding to its size until 1967. Since its growth in the 1960s, Surtsey has shrunk somewhat due to erosion and is currently 1.4 square kilometers in size. During the first decade that Surtsey existed, there was little in the way of plant or animal life. Today, however, the island is home to many plants and animals, ranging from birds to insects and seals. Humans have generally been barred from the island to allow the island to develop biologically on its own. In 2008, Surtsey was declared a World Heritage Site by UNESCO, as a testament to its scientific value.

Poop, and in particular bird guano, was key to the development of a thriving ecosystem on Surtsey. But why?

A black-legged kittiwake with its chick on a cliff face in Iceland. Shore birds such as black-legged kittiwakes (*Larus tridactyla*) were among the first animals to arrive at Surtsey. Their guano (excrement, or poo) is full of nutrients and is left behind in white streaks.

In Chapter 3, we learned about macromolecules and the atoms that make them up. Nitrogen is required to make DNA and proteins, and without proper amounts of nitrogen, these macromolecules that sustain life cannot form. Therefore, without nitrogen, the majority of plants and animals on Earth could not survive. Volcanic rock is full of phosphorus (which is also needed for making DNA), but devoid of nitrogen, and so, not surprisingly, Surtsey initially lacked life—during its first 20 years, only 10 species of plants grew in its nutrient-poor soil. At least until birds began to visit the island, leaving behind their nitrogen-rich guano. They also left behind eggshells, seeds, and carcasses, all of which provided nutrients to Surtsey.

The story of Surtsey highlights many of the concepts we will learn about in this chapter. We will begin by discussing *communities*, the populations of organisms that live together, such as those on Surtsey. Then we will discuss how species interact with one another, such as through *competition* and *symbiosis*. We will then discuss how communities change over time through *succession*, and how the addition or loss of a single organism can substantially impact a community. Then we will turn to *ecosystems*, which are specific communities and their surrounding environment. Finally, we'll close with a discussion of *biomes*, which are large regions that share a similar climate and lifeforms. We'll end the chapter with a discussion of how nutrients like nitrogen and phosphorus move through the environment.

Preview It!

What do you know already about communities? What do you want to learn in this chapter?

15.1 Community Structure

Learning Objectives

15-1 Describe what a community is and the factors that impact community structure.

15-2 Compare species richness and species evenness.

15-3 Explain the difference between a species' habitat and its niche.

Think about where you live. What does your neighborhood look like? Do you live in a dorm? An apartment? A house? What about others (humans or not) who live close to you? If you look out your window, what do you see? For example, how many trees can you see?

Rather than using the term "neighborhood," biologists instead refer to all the populations of organisms that live in a particular area as a **community**. Note that although communities technically include the different populations of living organisms in an area and not the abiotic factors (nonliving parts of an ecosystem; Chapter 14), these abiotic factors still influence the community. Biological communities vary in size and often are nested one inside another. Let's look at tide pools as an example of a community (Figure 15.1). Within any given tide pool, you may find a variety of different organisms. In this example, let's say we find populations of sea stars, giant green anemones, mussels, and plants such as sea cabbage. These distinct populations make up the community.

If you ever visit a tide pool, be careful not to pick anything up—not only could you hurt the tide pool inhabitants, but you could also be stung or injured!

When we consider a community, we can describe it in terms of its **species diversity**, or the variety of species at a specific location, which includes two components. The first component is **species richness**, which refers to the number of different species present at the location. The second is **species evenness**, or the relative abundance of each species at the same location.

There are varying degrees of diversity when considering the species richness and evenness of a community. A tidepool could have many different species present, which means it has high species richness (Compare Figure 15.2A and 15.2B). In terms of evenness, if there is high species evenness, then there is a similar abundance of each species present. Said otherwise, the numbers

Figure 15.1 **A tide pool**

Tide pools form when water becomes trapped in rocky bowls near the ocean. At high tide, the tide pool is completely underwater. At low tide, water (and the various organisms inside of it) become trapped until the next high tide. Tide pools offer visitors a chance to peek into the underwater lives of animals such as sea stars, anemones, and mussels (inset). These species interact with one another as part of a community.

Note that tide pools are transient communities—after the next high tide comes in, mobile species like fish and crabs will move to different tide pools or stay farther out at sea. However, the concepts of species diversity, richness, and evenness can apply to any community, such as the relative numbers of different types of trees in a forest.

of no one population is substantially greater than any other. If there is low species evenness, then the abundance of one species is much higher than the other species.

For example, Figure 15.2C, top left, shows high species richness and low evenness, where there are many different species present, but in relation to their numbers, the mussel population is substantially larger than any other species' population. This is a situation where a community has intermediate species diversity. In intermediate species diversity, there is either high species richness or high species evenness, but not both at the same time (see Figure 15.2C).

Figure 15.2 **Species diversity is a factor of species richness and evenness**

A B C

A. A tide pool with high diversity. The most diverse communities have high species richness, where a number of different species are present, and high species evenness, where there is a relatively even abundance of each species.

B. A tide pool with low diversity. Low diversity is characterized by low species richness and evenness. This means that there are not many different species present, nor are there many individuals of each species.

C. Tide pools with intermediate diversity. Intermediate diversity can mean high species richness/low species evenness (upper left) or low species richness/high species evenness (lower right). For example, at the upper left, notice how the mussels dominate (low evenness) but there are many different species present (high richness).

Community Structure

Much like your neighborhood is shaped by who lives there and nonliving features of the environment—like the presence of a river or the amount of rain you receive—biological communities are also influenced by these factors. Together, **community structure** refers to the types of species within a community, as well as their abundance and how different species interact with one another. Community structure is influenced by a combination of abiotic and biotic factors, and it can change over time.

Study Tip

Use root words to help you remember. The prefix "a-" in Latin means away from or without, and "bio" comes from the Greek word for life. So "abiotic" means without life.

Abiotic factors. Nonliving components in an ecosystem, or **abiotic** factors such as geography and *climate* (the average weather over many years; Section 15.5) affect community structure. These factors include soil quality, sunlight intensity, rainfall, and temperature. Nonbiological factors vary with latitude (distance from the equator) and elevation (distance above sea level). In aquatic environments, they also vary with depth. In our tide pool example, the organisms living there are impacted by the intensity of sunlight and the depth of the tide pool—which influences whether they are at risk of drying out—as well as the temperature and salinity of the water. Tide pool organisms must also be able to withstand pounding waves as the tides go in and out, and have adaptations to cope with such conditions, such as the ability to cling tightly to rocks.

Biotic factors. The living components, or **biotic** factors, such as the evolutionary history and adaptations of the species in a community, also influence community structure. Each species evolved in and is adapted to a specific **habitat**, or the natural environment where it typically occurs. Although many species in a community may share the same habitat, such as a tide pool, each species has a unique ecological role that sets it apart. This role is the species' **niche**, which is described in terms of the conditions, resources, and interactions necessary for the survival and reproduction of that species. For example, within a tide pool, species are located within the tide pool based upon how well they can tolerate exposure to sunlight, air, and waves. Aspects of a species' niche include the temperature range it can tolerate, the kinds of foods it can eat, and the types of places where it can breed or hide. A description of a plant's niche would include its requirements for soil, water, and sunlight.

Did You Know?

Between 2013 and 2014, around 20 species of sea stars died in large numbers, likely due to a disease. The exact nature of this disease is not known.

Species interactions also affect community structure. The presence of one species often enhances or inhibits the population growth of another species. For example, sea stars are top predators (Section 15.2) in tide pools and are also *keystone species* (a species that many other organisms depend on; Section 15.3). Sea stars eat the mussels in a tide pool, which allows room for other species to exist. Therefore, the health of the community is strongly impacted by the presence (or absence) of sea stars.

15.1 Summary

- A community refers to all the populations of organisms that live in a particular area. Community structure is affected by its abiotic (nonliving) and biotic components, and can change over time as these interactions change.
- Communities are described in terms of species diversity, which is made up of species richness and species evenness. Species richness refers to the numbers of different species present in an area, while species evenness refers to the relative abundance of each species present.
- A habitat is the natural environment where a species typically lives. A niche is the ecological role a species plays in its habitat, such as how it interacts with other organisms.

15.2 Species Interactions

Learning Objectives

15-4 Compare the different kinds of species interactions.

15-5 Compare different types of symbiosis.

Just like you have a variety of interactions with your neighbors—ranging from a friendly hello to perhaps a less friendly gesture given to someone while driving—organisms within a community can interact with one another in a variety of ways. These interactions can be between members of the same species or between different species. The interactions can have a positive (+) or negative (−) effect on the species involved, or can have no effect, depending on the type of interaction. The types of species interactions are summarized in Table 15.1 and are explained in this section.

Table 15.1 Summary of Species Interactions

Type of interaction	Effect on first party	Effect on second party
Competition (intraspecific or interspecific)*	−	−
Predation	+	−
Herbivory	+	−
Parasitism	+	−
Mutualism	+	+
Commensalism	+	None

*Note that the "winner" in a competition may have a positive benefit, such as being able to mate with a female, but the resources and risks associated with competition (fighting other males) are considered negative.

Competition

Have you ever competed before, perhaps as an athlete or a musician? For example, maybe you've run track and have competed with others to have the fastest time, or auditioned for a seat in a musical ensemble. Competition involves pitting ourselves against others. In biological terms, **competition** is when species interact with each other for access to limited resources.

In some cases, like running a race, we are competing against others similar to us—for example, in the same age group. This is akin to **intraspecific competition**, which is competition among members of the same species. Because competition has a negative effect on both parties, intraspecific competition is a density-dependent factor that limits population growth (Chapter 14). Intraspecific competition takes many forms and can drive evolution—such as when males compete for mates (sexual selection; Chapter 11). Individuals of the same species may compete for a limited resource, such as food (Figure 15.3).

Interspecific competition is competition between members of different species. Like intraspecific competition, it has a negative effect on both competitors. Competition can take many forms. For example, members of one species can actively prevent members of another species from using a resource, such as when scavengers like an eagle and a fox fight over a food source, such as a carcass (Figure 15.4). Plants can also compete with their neighbors. Some plant species can actively interfere with other species for the space they occupy, such as when a sagebrush plant secretes chemicals that prevent other plant species from growing nearby.

Species compete most intensely when the supply of a shared resource is the main limiting factor for both. The **principle of competitive exclusion** states that species that require the same limited resources and access them in the same way cannot coexist indefinitely. One will eventually "win" and outcompete the other. As you will learn about in Section 15.3, *invasive*

Figure 15.3 Two greater flamingo (*Phoenicopterus roseus*) males compete by twisting their necks together

Credit: ROGER POWELL/NATURE PICTURE LIBRARY/SCIENCE PHOTO LIBRARY

While flamingos may compete for a variety of reasons, food is a common resource that flamingos fight over.

Figure 15.4 **Active competition among different species of scavengers**

Credit: WENN Rights Ltd/Alamy Stock Photo

Credit: WENN Rights Ltd/Alamy Stock Photo

After facing off over a carcass (left), an eagle attacks a fox with its talons (right). The fox then retreated, leaving the eagle to exploit the carcass.

species, which are species that were introduced to a habitat that they aren't originally native to, provide a classic example of the principle of competitive exclusion, outcompeting and harming native species.

When the resource needs of two species overlap somewhat, the presence of each may reduce the carrying capacity (Chapter 14) of the habitat for the other. Over time, this may result in **resource partitioning**, where different species within a community have different niches for resources, like food, that minimizes competition. For example, consider three species of bats that live in one location. The bats all compete for the same food source—insects. However, each species of bat hunts for food in a different part of the habitat (Figure 15.5)—in the more densely forested spaces between trees, at the edge of the forest, or in an open area with no trees.

Figure 15.5 **Resource partitioning in bats**

Since this hypothetical population of bats all eat the same prey (insects), to coexist in the same place, different species of bats hunt in different parts of the environment.

Antagonistic Interactions

We don't always get along with our neighbors…but unlike in many natural communities, we generally don't have to be concerned about our neighbors eating us. In this section, we will discuss three types of **antagonistic interactions**, where an individual from one species benefits by extracting resources from another, which is harmed as a result. These interactions include predation, herbivory, and parasitism.

Figure 15.6 **An example of predation**

Ed Cesar/Science Source

Here, a lynx (*Lynx canadensis*; the predator) is attempting to catch its lunch, a snowshoe hare (*Lepus americanus*; the prey).

Predation. An interaction where an individual from one species captures, kills, and eats an organism from another species is called **predation** (Figure 15.6). Predators exert selective pressure on prey, so prey individuals with the best anti-predator defenses will have an advantage. Alternately, predatory individuals that are best able to overcome prey defenses will have an advantage. As a result, predators and prey may engage in an evolutionary arms race that continues over many generations.

For example, some species evolved to contain chemicals that taste bad or sicken predators. Consider monarch butterfly caterpillars (*Danaus plexippus*), which feed exclusively on milkweed (Figure 15.7A) and store toxins from the milkweed in their bodies. Monarch caterpillars are not sickened by the toxin because they have an evolved resistance to its effects, but if a bird were to eat a monarch caterpillar or butterfly, the plant-derived chemicals would sicken the bird. As such, the bright orange color of monarch (Figure 15.7B) acts as a type of **warning coloration**, fending off potential predators with a conspicuous pattern or color that predators learn to avoid.

To avoid predation, some species have evolved to have a coloration pattern that is similar to that of a poisonous species, even if they themselves are not poisonous. Consider the viceroy butterfly (*Limenitis archippus*; Figure 15.7C). Like the monarch, it has orange and black coloration, but because the viceroy caterpillar does not feed on milkweed, its body does not contain the milkweed toxins. This is an example of **mimicry**, an evolutionary pattern in which a species evolves to resemble a different species to gain a selective advantage—in the case of the viceroy butterfly, to avoid predation. The viceroy benefits from their similar appearance to monarchs, because a predator that became ill after eating a monarch is likely to avoid other butterflies with similar coloration.

Mimicry also explains why flies that cannot sting can have black and yellow stripes, mimicking a wasp or bee that can indeed sting.

Figure 15.7 **An example of mimicry**

A B C

A. Monarch caterpillars exclusively eat milkweed, accumulating high amounts of distasteful milkweed toxins in their bodies.

B. An adult monarch butterfly, with characteristic orange and black warning coloration.

C. The viceroy butterfly mimics the monarch's coloration. Can you spot the difference between the two?

Credits: Danita Delimont/Alamy Stock Photo; National Geographic Image Collection; iStock.com/bookguy

Figure 15.8 **The giant owl butterfly (*Caligo eurilochus*) has eye spots that deter predators**

Credit: Wild Horizon/Getty Images

Figure 15.9 **Camouflage in prey and predator**

A B

Credit: Menno Schaefer/Shutterstock.com

Credit: Auscape International Pty Ltd/Alamy Stock Photo

A. The bird in this photo, a least bittern (*Ixobrychus exilis*), points its bill upward and sways with the wind to blend in with marsh vegetation.

B. This Bengal tiger (*Panthera tigris tigris*) blends in with the dry grass and leaves that it relaxes in.

? Did You Know?

Porcupine quills require very little force to pierce the skin. To develop vaccine needles that will result in less pain when you get your shots, scientists are studying the ease with which they break the skin.

In addition, many prey species have hard or sharp parts that make them difficult to eat, such as a snail's shell or a porcupine's quills. Other species may engage in behaviors designed to scare a predator away, such as butterflies with eyespots that can trick a predator into thinking a larger, scarier predator was looking out at them (Figure 15.8).

Another strategy used by animals to avoid detection is camouflage. **Camouflage** is a body shape, color pattern, or behavior that allows an individual to blend in with its surroundings and avoid detection. Prey benefit when they have camouflage that allows them to hide from predators (Figure 15.9A), and predators benefit from camouflage that makes them less visible to prey, such as a tiger blending into the grass (Figure 15.9B).

Figure 15.10 **Scottish thistle (*Cirsium vulgare*) contains leaves with sharp spikes to deter grazing**

Credit: Ian Kenny/Alamy Stock Photo

Herbivory. An antagonistic interaction in which an organism feeds on a plant is called **herbivory**. Note that plants may or may not die as a result. Some plants can withstand the loss of their parts and quickly grow replacements. For example, grasses are seldom killed by herbivores. They have a fast growth rate and store enough resources in their roots to replace the shoots lost to grazers. Other plants have traits to fend off herbivores, including physical deterrents such as spines, thorns, and fibrous, difficult-to-chew leaves (Figure 15.10).

Plants can also produce compounds that taste bad to herbivores or can sicken them. Capsaicin, a compound that makes some peppers hot, helps the plant defend its seeds against seed-eating mammals. For example, rodents find capsaicin-rich pepper fruits unpalatable, leaving them to be eaten by birds, who cannot detect capsaicin. A pepper benefits because rodents chew up and kill the seeds, whereas birds eat the seeds whole, and excrete the alive, intact, and with their own source of fertilizer.

? Did You Know?

Scottish thistle is the national flower of Scotland. Legend has it that in the 13th century, a Norse army invaded Scotland. They landed at night and removed their shoes to be extra sneaky…but upon stepping on the thistle, the accompanying yelling caused the Scots to wake up and defeat the invaders.

Is It Love? Symbiotic Relationships

As we've learned through this section, organisms can have a variety of interactions with one another. While relationships like predation or herbivory are more transient interactions, sometimes organisms have a longer relationship with one another. **Symbiosis** refers to any

Figure 15.11 Parasites can be insects or plants

A

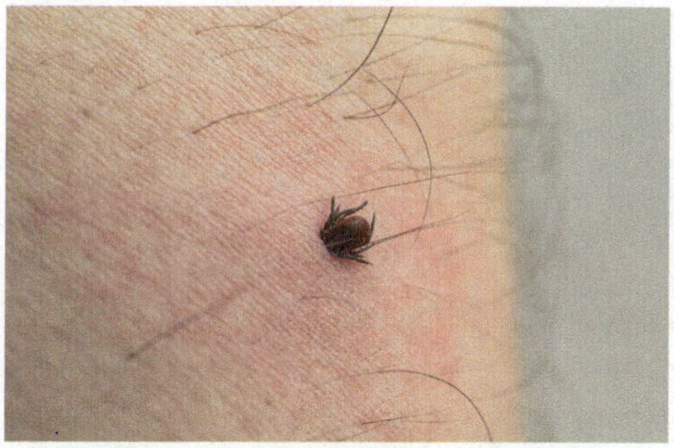

Credit: iStock.com/kmatija

B

Credit: Arterra/Getty Images

A. This deer tick (*Ixodes scapularis*) is biting a human. Note how the head is below the skin, and the associated redness that forms on the person's skin.

B. The green balls in these trees are mistletoe (*Viscum album*), a type of parasitic plant that takes water and nutrients from host trees.

interspecific interaction in which two species have a permanent or long-lasting, physically close association. In this section, we will review three types of symbiotic relationships: parasitism, commensalism, and mutualism.

! Point of Confusion: In nonbiological contexts, symbiosis is often used to explain loving relationships, where people have a mutually beneficial relationship. In biological terms, it refers to any prolonged relationship between two different species that may or may not be mutually beneficial, and could in fact harm one of the organisms.

Parasitism. With **parasitism**, a parasite withdraws nutrients from a living individual of another species, its host, often by living on or within the host. In parasitism, the host does not gain any benefit and is in fact harmed by the interaction. For example, the presence of a parasite can weaken a host, making it more vulnerable to predation or less attractive to potential mates.

There are many kinds of parasites, including bacteria, protists, and fungi, and invertebrates such as tapeworms, fleas, and ticks (Figure 15.11A). Ticks, like other parasites, evolved adaptations that allow the parasite to locate hosts and feed on them undetected. For example, ticks locate mammals and birds by moving toward sources of heat and exhaled carbon dioxide. When the tick bites, a chemical in its saliva acts as a local anesthetic, preventing the host from noticing it. Plants can also be parasitic. For example, mistletoe grows on trees or shrubs and uses its roots to penetrate the host plant, where it steals water and nutrients from the host (Figure 15.11B).

Some parasites have unique ways of obtaining nutrients from their host. For example, **brood parasites** deplete their hosts' resources by stealing parental care, tricking an animal of a different species into raising its young. One such example is the North American cowbird (*Molothrus ater*), which lays its eggs in the nests of other birds. When a cowbird egg hatches, the host parents care for the young cowbird as if it were their own, where the parasitic chick can outcompete the host birds' actual chicks for food and resources (Figure 15.12).

? Did You Know?

Capsaicin is the main ingredient in pepper spray, a personal defense tool.

? Did You Know?

Stripes have long been thought to protect zebras against parasitic flies. When researchers outfitted horses in black, white, or zebra-striped coats, biting flies landed on the single-color coats significantly more often than the striped coats.

? Did You Know?

Some insects, called parasitoids, lay their eggs in other insects. The eggs hatch into larvae that devour their host from the inside out, eventually killing it.

Figure 15.12 A wren's nest with a cowbird chick

The larger cowbird chick (right) pushes itself above the wren chick to get more food.

Credit: Rolf Nussbaumer Photography/Alamy Stock Photo

Figure 15.13 Examples of commensalism

A

B

A. In this commensal relationship, the moon orchid benefits from its position in the tree's canopy to capture sunlight, but unlike mistletoe, there is no negative impact on the tree.

B. Here, remora are attached to a lemon shark in the Bahamas.

Commensalism. With **commensalism**, one species benefits and the other is unaffected. For example, some orchids, like the moon orchid (*Puspa pesona*), live attached to a tree trunk or branch and benefit by having a perch in the sun, while the tree that provides this support is unaffected (Figure 15.13A). In an interesting example from the animal kingdom, remora fish (*Echeneis naucrates*) hitch a ride on other large marine animals, such as lemon sharks (*Negaprion breviorstiris*), to help them move throughout the ocean (Figure 15.13B). The remora uses less energy, while the shark is neither harmed nor helped by their presence.

Big Ideas in Biology

Evolution

From an evolutionary standpoint, mutualism is best considered as reciprocal exploitation. Each participant increases its own fitness by extracting a resource from its partner, such as protection or food. If taking part in the mutualism has a cost, individuals who minimize that cost will be at a selective advantage. For example, nectar production by a plant is energetically costly, but nectar also makes the plant more likely to be pollinated, thereby allowing it to successfully reproduce.

Mutualism. An interspecific interaction that benefits both species is called **mutualism.** Flowering plants and their pollinators (Chapter 18) are an example of mutualism, since the plant gets assistance in reproducing while the pollinator gets food in the form of nectar or pollen. Another example is closer to home, where the digestive tract of humans, and most animals, contains mutualistic bacteria that assist their host by aiding in digestion and synthesizing vitamins. In return, the bacteria benefit by having a warm, nutrient-rich place to live.

Mutualistic relationships can also involve defense. For example, a clownfish uses a sea anemone to avoid predation, hiding within the anemone's stinging tentacles (Figure 15.14). The clownfish has a protective coating of mucous on its body, leaving it unaffected by the anemone's toxins that can paralyze other fish. In return, the clownfish helps the anemone by removing parasites and protecting it from predators.

Figure 15.14 A clownfish nestled among the tentacles of a sea anemone

In this mutually beneficial relationship, each species protects the other.

15.2 Summary

- Competition is when species compete for access to limited resources, either within the same species (intraspecific) or with members of another species (interspecific). Due to the principle of competitive exclusion, one species may outcompete another. To minimize competition, different species in a community may use resource partitioning, with each having a different niche for resources, like food.
- Antagonistic interactions occur when one species benefits by extracting resources from another, which is harmed as a result. Predation occurs when an individual of one species (the predator) captures, kills, and eats another (their prey); herbivory occurs when an organism feeds on plants. Prey species use defenses such as warning coloration and mimicry to avoid predation. Camouflage is used by both predators and prey to hide from each other.
- Symbiosis refers to any interspecific interaction in which two species have a permanent or long-lasting, physically close association, and includes parasitism (when one species benefits and the other is harmed), commensalism (when one species benefits but the other is unaffected), and mutualism (when both species benefit).

15.3 Changing Communities

Learning Objectives

15-6 Distinguish between primary and secondary succession.

15-7 Explain what a keystone species is and its role in community structure.

15-8 Differentiate exotic and invasive species and their impacts on community structure.

Succession can occur in aquatic communities as well. For example, when a lake or pond is gradually filled in with sediment.

In the chapter opener, we learned about how after the island of Surtsey formed, it eventually became colonized by plants and animals. Any time a major disturbance occurs to a habitat, from a major geological event like a volcanic eruption to events like a forest fire, there are a series of changes to the environment that allow new organisms to live in that environment. In this section, we'll look at how communities change, due to either ecological succession or the addition or subtraction of species.

Ecological Succession

The process by which the array of species shifts as organisms alter their own habitat is called **ecological succession**. One array of species is replaced by another, beginning with **pioneer species**, the first organisms to colonize new habitats, which are replaced by other species.

Primary succession occurs in habitats that lack soil and thus have few or no existing species, such as right after the formation of Surtsey. These first species often include organisms like lichens and plants that require little soil. Once the soil builds up, other plants can move in (Figure 15.15). In contrast, **secondary succession** occurs after a natural or human disturbance removes the natural array of species but leaves the soil in place (see Figure 15.15). We observe this kind of succession after a forest or grass fire. In such cases, an array of different species will move in and take over. In both types of succession, there are initial pioneer species like grasses and flowers, followed by intermediate species like shrubs and trees, until finally there is a mature forest present. When a mature forest is present and a community enters the final stage of succession, it is referred to as a **climax community**.

Fireweed is a plant with showy pink flowers and has a wide geographic range in North America. Fireweed seeds will grow only after being exposed to fire, which is an event that can precede secondary succession. They grow quickly in the absence of competition from other plants, giving them a selective advantage.

Scientists now recognize that three types of factors affect succession: (1) physical factors, such as climate; (2) chance events, such as the order in which pioneer species arrive; and (3) the frequency and extent of disturbances. Because the sequence of species arrivals and the

Figure 15.15 **Overview of succession**

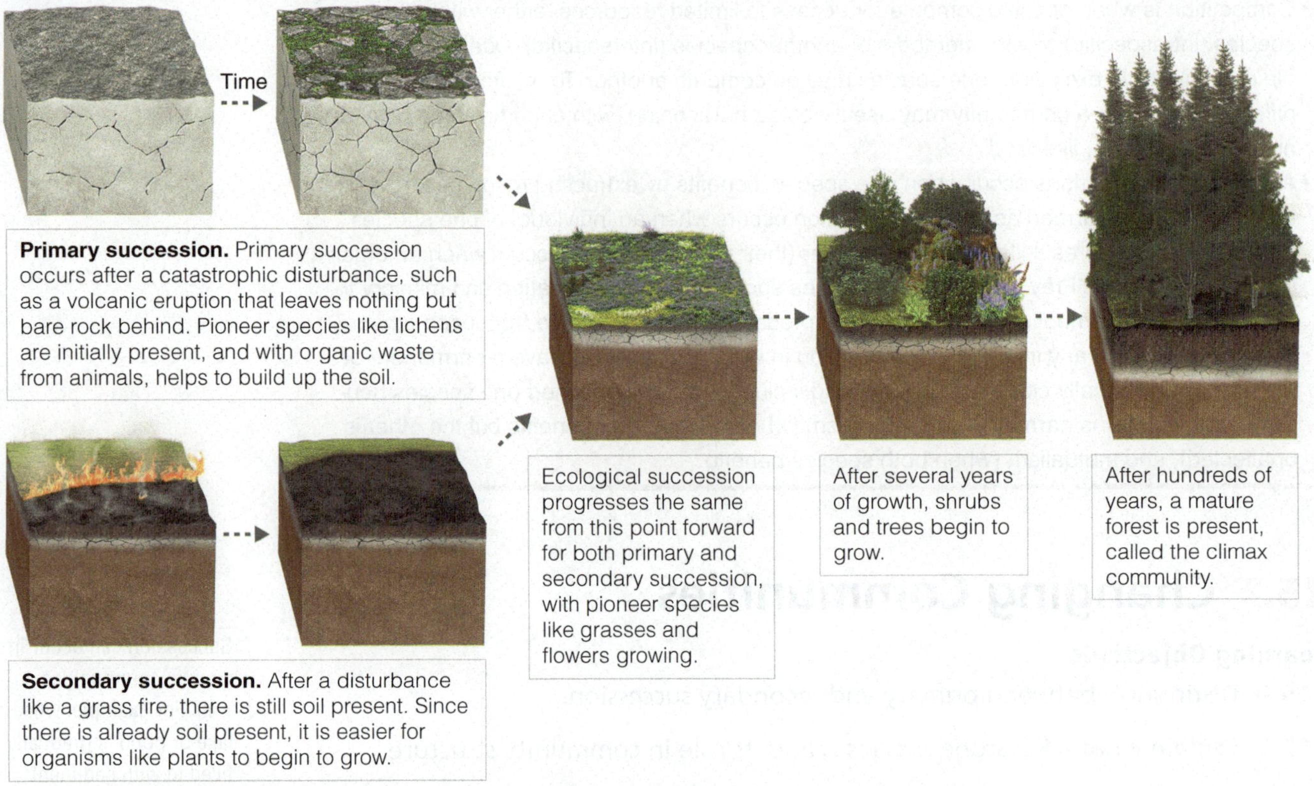

Here are two examples of succession—the top shows primary succession after a volcanic eruption, such as when Surtsey formed, and the bottom shows secondary succession after a grass fire. Note that at any given location on Earth, you would see different amounts of succession, depending on how much time has passed. Surtsey, for example, has not reached the climax community stage.

Have you ever seen signs prohibiting the transport of firewood? This is with the intent of trying to stop the spread of invasive species like the emerald ash borer, which decimates ash tree populations.

frequency and extent of disturbances vary in unpredictable ways, it is difficult to predict exactly how the composition of any particular community will change.

Species Losses or Additions

Interactions between species are an essential part of an ecosystem. Some species, called **keystone species**, can have a disproportionally large impact on a community relative to their abundance. Said differently, a few individuals can have a major impact on the community. For example, the sea stars in Section 15.1 are keystone species along rocky coastlines in the Pacific Northwest.

Another example of a keystone species is beavers. Beavers are large rodents that cut down trees, then use the branches and logs to build a dam across a stream. Construction of a beaver dam creates a deep pool where none previously existed. By altering the physical conditions in a section of the stream, the beaver affects the types of fish and aquatic invertebrates that can live there. Therefore, when beavers are present, there tends to be more species diversity. Due to their role in physically altering the environment that impacts the entire community, beavers are also termed **ecosystem engineers**.

Exotic and invasive species. The United States is home to more than 15,000 exotic species. An **exotic species** is a species introduced to a habitat that it isn't native to, and that has become established at the new location. Many exotic species do no harm, but some are invasive. An **invasive species** is an exotic species whose introduction disrupts the community structure in its new home, sometimes outcompeting native species. Of the over 15,000 exotic species in the United States, 6,500 are considered invasive. Table 15.2 summarizes some invasive species in the United States. We will return to the impact of exotic and invasive species on the environments—and creative mitigation efforts—in Chapter 16.

? Did You Know?

One of the most beloved *Star Trek* episodes from the original series, *The Trouble with Tribbles*, was inspired by the biology of invasive species.

Table 15.2 Examples of Invasive Species and Their Ecological Impacts

Invasive species	Range in the United States	Ecological impacts
Nutria (*Myocastor coypus*)  Credit: Monika Hodanova/ Shutterstock.com	In Gulf Coast and Southeastern states, Chesapeake Bay, and Pacific Northwest	Damage wetland and marsh areas; overconsume aquatic plants and eat crops
Spotted lantern fly (*Lycorma delicatula*)  Credit: Amy Lee/Alamy Stock Photo	Established in states along the East Coast and Midwest	Remove nutrients and sugars from plants; threaten over 70 species of plants, including crops like grapes and fruit trees
Burmese python (*Python bivittatus*)  Credit: Wayne Holt/ 500px/Getty Images	Established largely in southern Florida and southern Texas, with some reported sightings in Georgia, Louisiana, Ohio, and Puerto Rico	Decimates mammal and bird populations; competes with native predators for food and habitat
Lionfish (genus *Pterois*)  Credit: iStock.com/ TeamDAF	In coastal waters off Louisiana, Florida, Georgia, and the Carolinas	Voracious and indiscriminate eaters that outcompete native fish; nicknamed "vacuum cleaner fish"
Carp (*Cyprinus carpio*)  Credit: USGS	In rivers throughout the midwestern United States	Outcompetes native fish for food and space; some species can jump out of the water to invade new bodies of water by jumping over physical barriers, like dams
Purple loosestrife (*Lythrum salicaria*)  Credit: iStock.com/Oleg Marchak	In habitats throughout the United States	Outcompetes native plants
Kudzu (*Pueraria montana*)  Credit: iStock.com/7000	Established in the Southeastern United States, Midwest, and East Coast, into New England	Quickly overgrows native plants, interfering with their ability to undergo photosynthesis and killing the covered plant; can also cover structures, such as buildings

Biology for Life

How Wolves Shape Ecosystems

And I'll huff, and I'll puff, and I'll blow your house down!
—The Big Bad Wolf, *The Three Little Pigs*

Wolves get a bad reputation. From the Big Bad Wolf in the classic children's story *The Three Little Pigs* to Sergei Prokofiev's musical children's story, *Peter and the Wolf*, we grow up hearing cautionary stories about wolves. Humans have hunted wolves for approximately 10,000 years, when early humans began to settle down into established communities and raise livestock. Do wolves deserve all the bad press?

Let's look at the story of wolves in Yellowstone National Park, who, along with beavers, are keystone species. Wolves were hunted to extinction in the park in 1926. With the wolves gone, their prey species, which include elk, became more populous in the park. More elk meant more grazing of young trees, which eventually became overgrazed so that there were fewer trees available for beavers to build their dams.

Overall, the quality of the communities in Yellowstone dramatically suffered without the wolves. After 20 years of witnessing the decline in the health of the communities in Yellowstone, people began to campaign for the reintroduction of wolves. Finally, after decades of advocacy work and policy changes, such as the Endangered Species Act of 1973, the first wolves were reintroduced to Yellowstone in 1995. As of 2023, there were 11 different packs living in Yellowstone, encompassing some 124 individual wolves.

The ecological response to wolf reintroduction to the park was nothing short of remarkable. With declining elk populations due to wolf predation (and changes in elk behavior) (Figure 15.16A), there are now healthier tree stands that reach maturity…and as a result, more beavers building dams (Figure 15.16B). The new beaver dams and subsequent pools of water led to greater numbers of moose, otter, mink, and water birds.

Given the benefits, wolf reintroduction campaigns are occurring in California and Colorado. Wolves are also beginning to move into their former territories in Michigan, Minnesota, and Wisconsin. Healthier ecosystems mean not only a boon for animals, but also for humans as well. Ecotourism is a major industry. For example, studies done in 2023 suggest that ecotourism in Yellowstone National Park has a benefit to the local economy of $828 million and supports jobs for 8,560 people. Many of the 4.5 million people who visit Yellowstone come to see the wildlife—and in particular, the wolves.

Despite the immense benefits wolves have on the ecosystem, wolves are still hunted by humans—for sport, for their skins, or to protect livestock. However, there are many nonlethal ways to deter wolves from property, such as by blasting air horns or using animal deterrence lights on livestock fences, which produce a flashing light that scares wolves away. It's also important to remember that wolves pose little danger to humans—on average, fewer than two people per year *worldwide* are fatally attacked by wolves. The majority of fatalities are caused not from wounds, but rather from rabies infection (which can be treated).

1. How has the reintroduction of wolves impacted the populations of other animals in Yellowstone National Park? Does this affect your impression of wolves? Explain.
2. Do some research into the wolf population nearest to you. What efforts are being taken on their behalf, or to control them? What would you recommend for helping wolves while keeping human interests in mind?

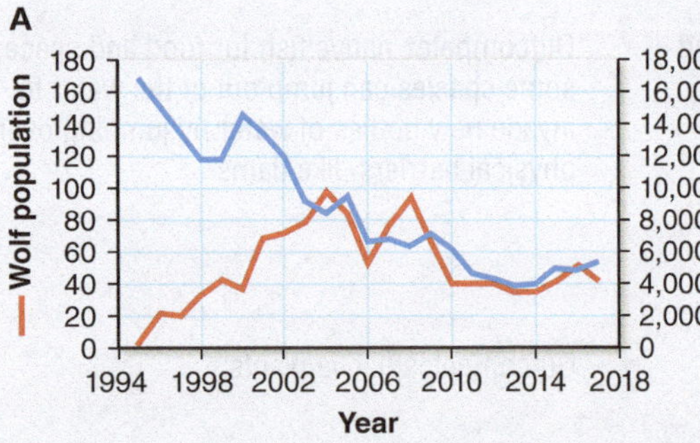

Credit: Donald M. Jones/ Minden Pictures

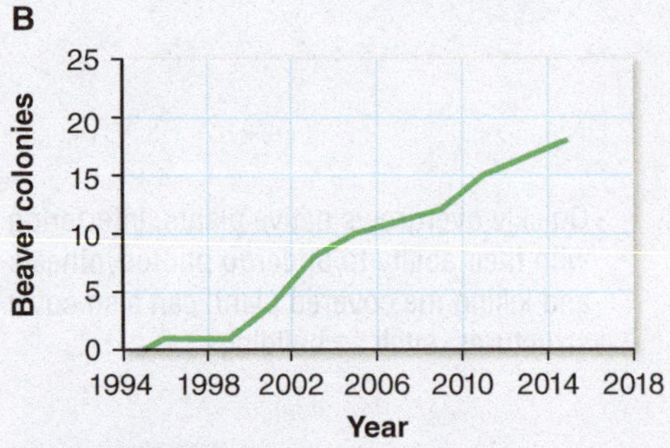

Credit: Troy Harrison/Getty Images

Figure 15.16 Populations of wolves, elk, and beaver in Yellowstone National Park

A. Populations of wolves and elks, and

B. Numbers of beaver colonies from 1995 to 2017. Notice how the populations of each species change relative to one another.

Digging into Data

Zebra Mussels: A Tiny Mussel Packs a Big Punch

Credit: Jonathan Larsen/Alamy Stock Photo

Figure 15.17 A sign on the shore of Lake Mille Lacs in Minnesota warning of invasive species, including zebra mussels

Credit: ZUMA Press, Inc./Alamy Stock Photo

Figure 15.18 A native mussel infested with zebra mussels

Zebra mussels (*Dreissena polymorpha*) are a species of mussels native to Eastern Europe. These tiny mussels, named for their characteristic striped pattern, are about the size of a fingernail. But however diminutive they might be, they have a big impact when introduced into aquatic habitats far from their native waters (Figure 15.17).

Zebra mussels were first introduced into North America when a transatlantic cargo ship released infested ballast water into the Great Lakes. The first established colony was discovered in Lake St. Clair (a lake connecting Lake Erie and Lake Huron) in 1988, and then the mussels spread throughout all five Great Lakes within a decade. Currently, zebra mussels are distributed as widely as the Mississippi River basin to bodies of water in the eastern and western United States, including as far away as Texas and California.

One such body of water is Lake Mille Lacs in Minnesota. Zebra mussels were first detected on the northeast side of the lake in 2005, where they found their ideal habitat of rocky and gravel areas with high water volumes. Researchers analyzed the zebra mussel population growth in Lake Mille Lacs from 2006 to 2016 and found that the population started at very low density but then experienced a period of exponential growth (Chapter 14). The fast spread of the zebra mussel caused significant impacts on human infrastructure, such as damaging boats and clogging water intake pipes.

? Did You Know?

As of the writing of this text, 28 of 51 native mussel species in Minnesota are listed as endangered, threatened, or of special concern, with five species believed to be extinct and two in imminent danger of extinction.

But their impacts aren't reserved to the human realm. Zebra mussels are filter feeders that compete with the lake's native fish and invertebrates for phytoplankton, therefore disrupting native food webs. Additionally, because colonizing zebra mussels cannot distinguish between abiotic and biotic surfaces, they colonize on the shells of native mussels (Figure 15.18). The attachment of zebra mussels to the affected native mussels renders them unable to move or feed, ultimately leading to their death.

Researchers began to study the impact that zebra mussels were having on native mussel species in the *Unionidae* family in Lake Mille Lacs. The graph in Figure 15.19 summarizes counts of native *Unionidae* mussels made using plot sampling (Chapter 14) from 2009 to 2012. Researchers counted the average number of native mussels, and of those, the number infested with zebra mussels.

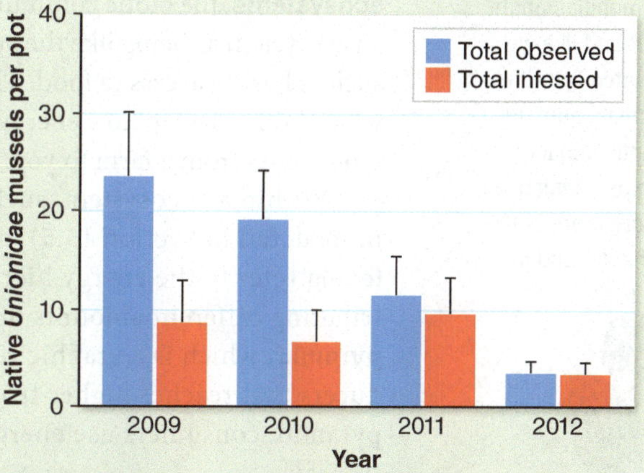

Graph based on: Jones, T. S., & Montz, G. R. (2020). Population increase and associated effects of zebra mussels Dreissena polymorpha in Lake Mille Lacs, Minnesota, USA. *BioInvasions Record, 9*(4).

Figure 15.19 A graph showing *Unionidae* abundance per plot in Lake Mille Lacs from 2009 to 2012, and of those, the number infested with zebra mussels

Error bars (the lines above the bars) represent uncertainty in the results, indicating how far the actual numbers might vary from the numbers reported.

1. What does the graph in Figure 15.19 indicate about the effect of zebra mussels on native unionid mussels? What could these results mean for the native mussel species?
2. The state of Minnesota has taken steps to minimize further zebra mussel invasions by requiring that visitors use high-pressure water cleaning stations to clean their boats as they enter and leave the lake, and to drain all water and dispose of all bait. Do some research. Are these mussel control and eradication strategies working in Minnesota? How about in your home state, if it's affected by zebra mussels?

15.3 Summary

- Primary succession occurs after a disturbance creates a soilless habitat. Secondary succession is when a disruption occurs, but soil is still present after the disruption.
- A keystone species impacts a community more than expected based on its abundance, such as when ecosystem engineers physically alter the environment.
- Exotic species that become established in a new habitat may become invasive and alter community structure by outcompeting or otherwise negatively impacting native species.

15.4 The Nature of Ecosystems

Learning Objectives

15-9 Describe what an ecosystem is and how energy moves through it.

15-10 Explain the relationship between producers, consumers, and decomposers.

15-11 Compare a food chain and a food web.

15-12 Explain the process of bioaccumulation.

> Although the analogy between ecosystems and our communities might make it sound like we are separated from the natural areas, humans significantly impact ecosystems—which we will explore more both in this chapter and in Chapter 16.

Throughout this chapter, we've considered the parallels between the places in which we live and the biological communities in natural areas. If we continue with this analogy, when we consider **ecosystems**, the biotic communities and abiotic factors at a particular location, we can think of an ecosystem as being like the systems that make it possible for us to live in our communities, such as having access to food. The parallels to natural systems are *trophic levels* and *food webs*, which both relate to how energy (food) is produced and moves through the ecosystem, like how food comes from a farm to your table to nourish you.

Within an ecosystem, such as in a grassy area like a prairie (we'll discuss prairies in more detail in Section 15.5), different organisms occupy different trophic levels. **Trophic levels** refer to the energy hierarchy within an ecosystem, with different organisms contributing different amounts of energy to the environment. **Figure 15.20** shows an **energy pyramid**, which is a graphic representation of the proportion of energy captured by producers that reaches higher trophic levels. With each higher trophic level in an energy pyramid, consumers use energy captured by producers, losing efficiency with each energy transfer.

As we learned in Chapter 4, the flow of energy is critical for life—and the flow of energy begins with **producers**, which capture energy and use it to make their own food from nonbiological (inorganic) materials in the environment. In Chapter 4, we focused on how plants can capture energy from sunlight and carbon dioxide from the atmosphere and produce glucose (photosynthesis), which we (or plants) can break down to use for energy (cellular respiration). In other ecosystems, like at the bottom of the ocean, producers use chemosynthesis to convert hydrogen sulfides into useable organic materials. Thus, producers can be diverse and are responsible for the most energy production in an ecosystem.

An ecosystem's **consumers** obtain energy and other nutrients by feeding on tissues, wastes, and remains of producers and one another. Consumers that eat plants (herbivores) are **primary consumers** (see Figure 15.20). Primary consumers are eaten by **secondary consumers**, which are in turn consumed by the top predators in the ecosystem, or **tertiary consumers**.

Figure 15.20 **A sample food pyramid using animals common in an American prairie**

Tertiary consumer. Apex predators like hawks eat primary and secondary consumers.

Secondary consumer. Black-footed ferrets eat primary consumers, like prarire dogs.

Primary consumer. Prairie dogs eat prairie grasses.

Producers. In a prairie ecosystem, plants like grasses are the producers.

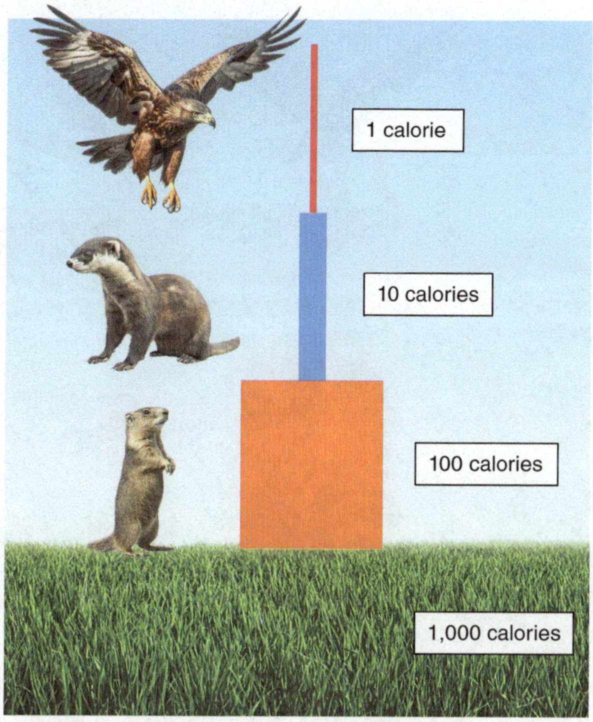

Notice that at successively higher trophic levels, the number of calories available decreases tenfold.

All organisms—whether producers or consumers—produce waste. Waste comes in many forms, from excrement to dead leaves to deceased organisms; this leftover material is called **detritus**. Waste not, want not—there is still energy in detritus, and that energy can be captured and recycled. **Decomposers** are organisms that recycle leftover organic matter into inorganic building blocks that can be used by primary producers to make energy. Decomposers include microbes, like bacteria, as well as fungi, such as mushrooms.

When people promote a vegetarian or vegan diet by touting the benefits of "eating lower on the food chain," they are referring to the inefficiency of energy transfers. For example, eating plant material involves only a single energy transfer, and there is more energy available to the consumer.

Food Chains and Food Webs

The trophic level that an organism occupies relates directly to its position in a food chain. A **food chain** is a sequence of steps by which the energy initially captured by producers moves to higher trophic levels. For example, if you consider Figure 15.20, the main producers in this ecosystem—grasses and other plants—are at the first trophic level. Energy flows from the plants to the prairie dog to the black-footed ferret, and finally, to the hawk. Note that as you move up the food chain, less energy is available at each consumer level. This is because each organism needs energy to sustain itself and produces heat, which is lost to the environment as a byproduct.

Food chains cross-connect with one another as a **food web**. Figure 15.21 shows some participants in a prairie food web. Nearly all food webs include two types of food chains: a grazing and a detrital food chain. In a grazing food chain, energy stored by producers flows next to herbivores, which can be relatively large animals like bison or small animals like prairie dogs. An herbivore can also be an insect, like a grasshopper. In a detrital food chain, energy production starts with decomposers, such as bacteria and fungi. The decomposers

Figure 15.21 **A typical American prairie food web**

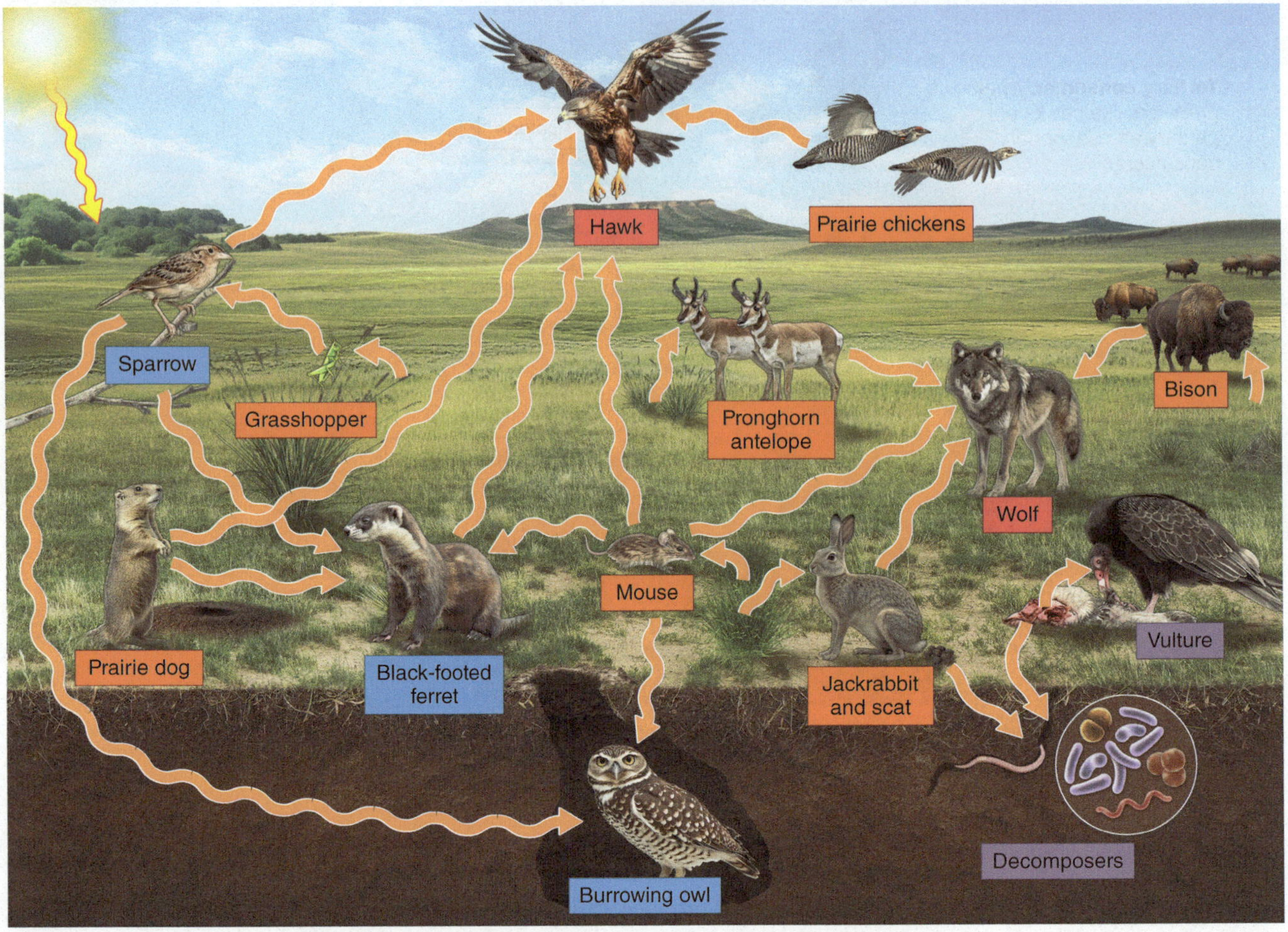

Animal names are shown in the figure, where the colors used represent the trophic levels shown in Figure 15.20; decomposers are shown in purple. Energy enters the ecosystem in the form of sunlight, and producers, such as grasses, capture the sunlight. From there, the orange arrows signify nutrients moving through the ecosystem, where the arrow points to the consumer (for example, the wolf) that eats the organism at the other end of the arrow (for example, the jackrabbit). Decomposers break down animal remains and their leavings, such as scat (feces), fur, feathers, and so on. Note that as we move up trophic levels (shown by the colored boxes), there are fewer individuals supported by the food web.

To learn more about the healthiest seafood choices (including low-mercury seafood options), check out Seafood Watch, a program by the Monterey Bay Aquarium.

free up nutrients in the soil for organisms like producers to use. Detrital food chains and grazing food chains interconnect as an ecosystem's overall food web.

Bioaccumulation

Have you ever heard that pregnant women should limit their consumption of tuna or avoid eating large predatory fish like shark or swordfish entirely? In fact, consumption of shark is not recommended for anyone. Why? It comes down to mercury—a toxic metal that can cause severe health problems if consumed, either by being eaten, inhaled, or absorbed through the skin. And predatory fish like sharks have dangerously high levels of mercury in their bodies. Where does the mercury come from? And why the concern over tuna and shark, and not other seafood, like shrimp?

Like nutrients and energy, some toxic substances move through food webs. Over the course of a predator's life, the predator consumes all the toxic substances previously taken up and accumulated by its many prey, and the prey of their prey, and so on. Thus, by a process called biological accumulation, or **bioaccumulation**, a toxin becomes increasingly concentrated as it moves up a food chain (Figure 15.22).

In animals, fat-soluble toxins, such as methylmercury—which forms when inorganic mercury from coal-burning power plants dissolves in water—are ingested or absorbed across the skin in marine life and can accumulate in fatty tissues. The concentration of fat-soluble toxins in an animal's body increases over time, so long-lived species tend to have a higher concentration of these substances in their bodies than shorter-lived species do. Within a species, older individuals have a higher toxin load than younger ones. While all sea life has some level of methylmercury present, animals like shrimp, which are lower on the food chain, have less in their bodies than do predatory fish like tuna. This is why there are no specific guidelines for eating shrimp, but there are for tuna.

And it's not just toxic chemicals like mercury that can bioaccumulate. Plastic pollution in the oceans is a major problem (Chapter 16). Microscopic plastic, or **microplastic**, can also move up the food chain. Analyses of deceased humans indicate the presence of microplastics in the kidneys, liver, and brain. Scientists have also noted a correlation where people with dementia have higher amounts of microplastics in their brains than people without dementia.

Figure 15.22 Mercury bioaccumulation

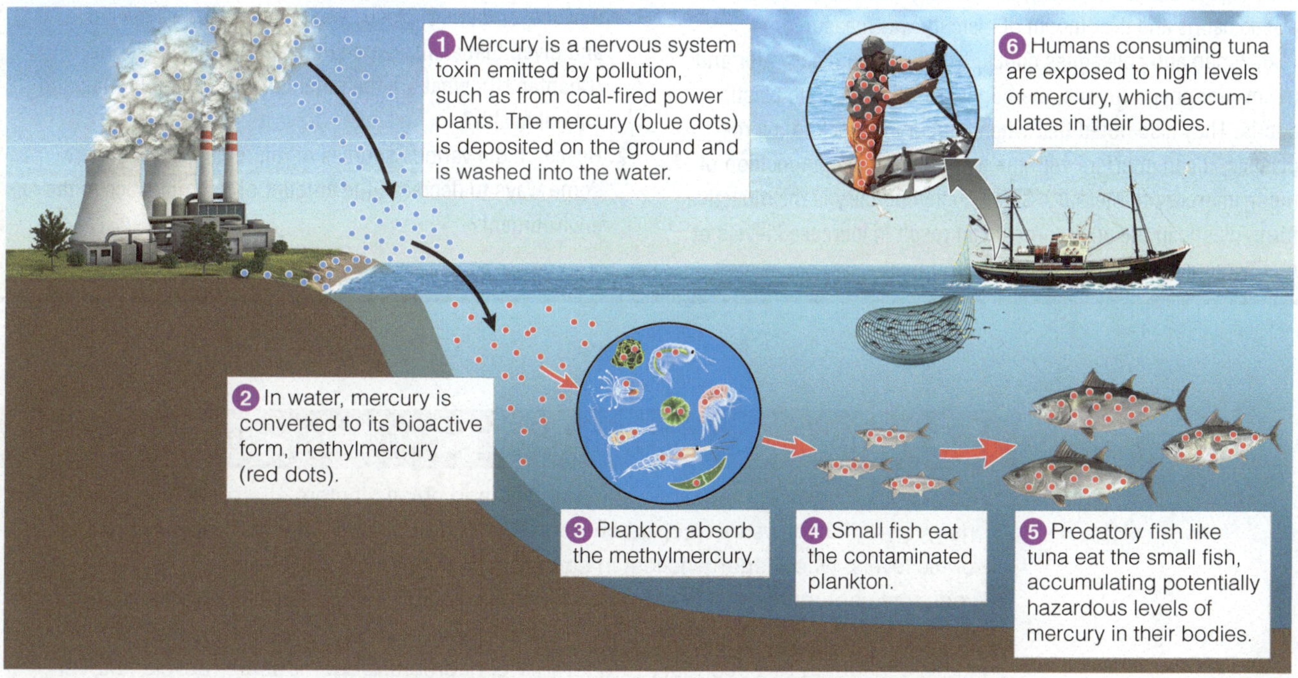

Blue dots represent mercury, and red dots represent methylmercury, the form of mercury that affects the body. Arrow sizes represent the relative amount of accumulated methylmercury in the aquatic food web.

Biology for Life

The Impact of Microplastic Pollution on Mercury Bioaccumulation

How long can you go during your day without touching something that is plastic? If you're sitting at a computer, you're likely touching plastic computer keys. You might be wearing plastic, too. Clothing often contains polyester and nylon, both of which are plastic-based. The last snack you ate was likely packaged in plastic, where you either took a wrapper off to eat it or it was packaged in a bag—both of which often contain plastic.

Bioaccumulation of microplastics and mercury unfortunately go hand in hand. Plastic tends to have a negative charge, whereas mercury has a positive charge, so the two tend to stick together (Chapter 2). Therefore, tiny pieces of plastic floating in the ocean (microplastics) tend to be highly concentrated in mercury.

National Geographic Explorer Katlin Bowman (Figure 15.23) studies the relationship between plastic and mercury pollution in Lake Erie and San Francisco Bay. In 2021, Bowman and colleagues published a paper describing the amount of mercury that was stuck to plastic debris found near urban areas. First, Bowman and her team collected plastic debris with a large net. Then, the plastic debris was analyzed to see what it was composed of (for example, foam, plastic pellets, and the plastic tag fasteners that attach price tags to clothing). Finally, the mercury was removed from the plastic debris and the amount of mercury was measured.

Bowman and colleagues concluded that in both saltwater and freshwater environments, mercury was found stuck to plastic debris. They also noted that when mercury is stuck to microplastics, it can interfere with the natural photodecomposition of methylmercury to mercury. Since methylmercury is the bioactive form, plastic in the environment can result in increased levels of methylmercury lingering in the environment.

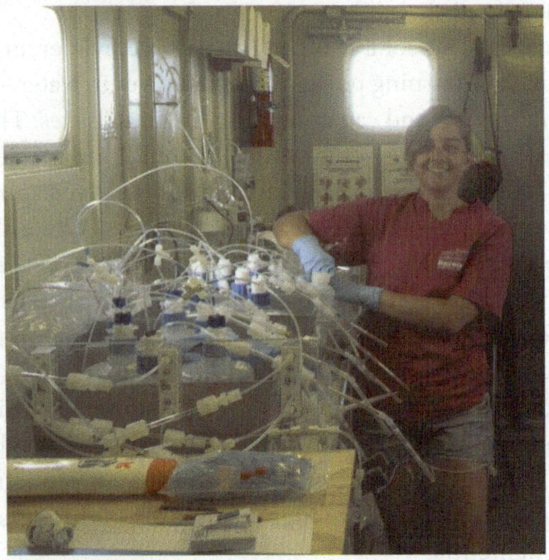
©Gretchen Swarr

Figure 15.23 National Geographic Explorer Katlin Bowman in a ship-based mercury analysis lab

1. Bowman's work demonstrated that methymercury can adhere to plastic debris in the ocean. Since there is a massive amount of plastic waste in the world's oceans, and given that animals are known to eat plastic debris (Chapter 16), what are the potential impacts on human and animal health?
2. Research the various sources of microplastics. What are some ways to decrease the amount of microplastics in the environment?

15.4 Summary

- An ecosystem includes the biotic communities and abiotic factors at a particular location. An energy pyramid shows the proportion of energy captured by producers that reaches higher trophic levels, where energy is lost in each transfer.
- Producers capture energy from sunlight and make their own food using inorganic materials from the environment. They are consumed by primary consumers (herbivores), which are in turn consumed by secondary and tertiary consumers. Decomposers recycle leftover organic matter into inorganic building blocks that can be used by primary producers to make energy.
- A food chain is a sequence of steps by which energy is captured and moves from producers to consumers into progressively higher trophic levels. A food web includes more than one food chain that cross-connects with one another.
- Bioaccumulation occurs when a toxin, such as mercury or microplastics, becomes increasingly concentrated as it moves up a food chain.

15.5 Climate and the Distribution of Biomes

Learning Objectives

15-13 Describe climate and the factors that impact climate.

15-14 Indicate factors that determine biome classification.

15-15 Compare terrestrial biomes.

15-16 Compare aquatic biomes.

Recall from Chapter 14 that ecology is all about relationships, starting from the individual, to its population, to the interaction of different populations together in a community. Communities are also influenced by abiotic factors, such as the transfer of nutrients and energy between the trophic levels of an ecosystem. Now, we've arrived at the biggest level of interaction—the biosphere. The **biosphere** includes all regions of Earth where life can exist. What lives where in the biosphere depends largely on differences in regional climate. **Climate** refers to the average weather conditions, such as temperature and precipitation, in a region over a long period of time, usually decades. This is different from "climate change," which refers to long-term changes in the weather patterns, rather than one unusual week, winter, or year. We will return to the global impacts of climate change in Chapter 16.

Solar Energy and Latitude

Light energy from the sun, or **solar energy**, varies by **latitude**, the distance from the equator. Solar energy tends to be most intense at the equator and least intense at the poles of the Earth. Why? One reason is the amount of atmosphere that sunlight travels through. The more atmosphere that the sunlight travels through, the less solar energy that reaches Earth's surface (Figure 15.24). Another reason is the angle at which sunlight hits the Earth. Near the equator, the sunlight is direct, whereas at the poles, the sunlight is at an angle.

Latitudinal differences in solar energy impact the air and water. As a result, we see global patterns of air circulation and rainfall (Figure 15.25). Note that elevation (distance above sea level) also affects temperature. At any given latitude, the average temperature decreases as elevation increases. This temperature decrease occurs because air at high altitudes is thinner and less able to retain heat than the air at lower altitudes. You will see the effect that elevation—plus latitude, air circulation, and rainfall—has on *biomes*, ecosystems resulting from different climates, as we explore Earth's biomes later in this section.

Figure 15.24 Solar intensity varies depending on latitude, affecting regional temperature

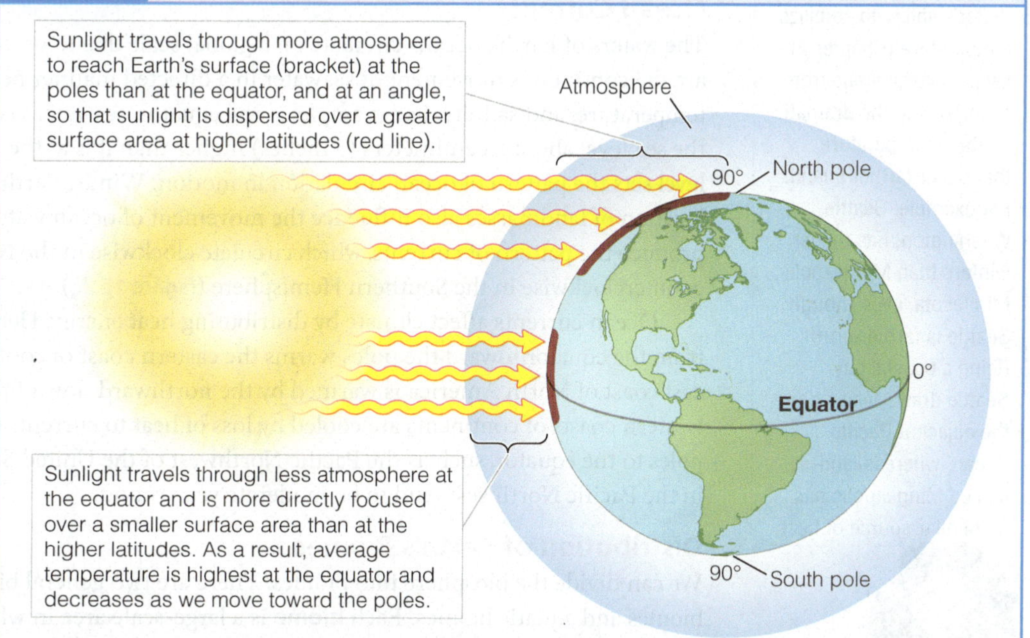

The view shown here is Earth on either the spring or fall equinox, which is when Earth's axis is tilted neither toward nor away from the sun and where equal amounts of incoming solar radiation arrive perpendicular to Earth's axis. For simplicity, we show light energy from the sun falling on the north pole and equator only. Note that the equator is a latitude of 0° and the poles are at 90°.

? Did You Know?

Deserts can form due to mountains. The presence of the mountain forces air to move upward. As the air rises, it cools and releases the moisture as rain or snow. By the time the air descends on the downwind side of a mountain (the side facing away from the wind), the air is dry, forming a desert.

Figure 15.25 Air circulation patterns and rainfall result in different climatic regions on Earth

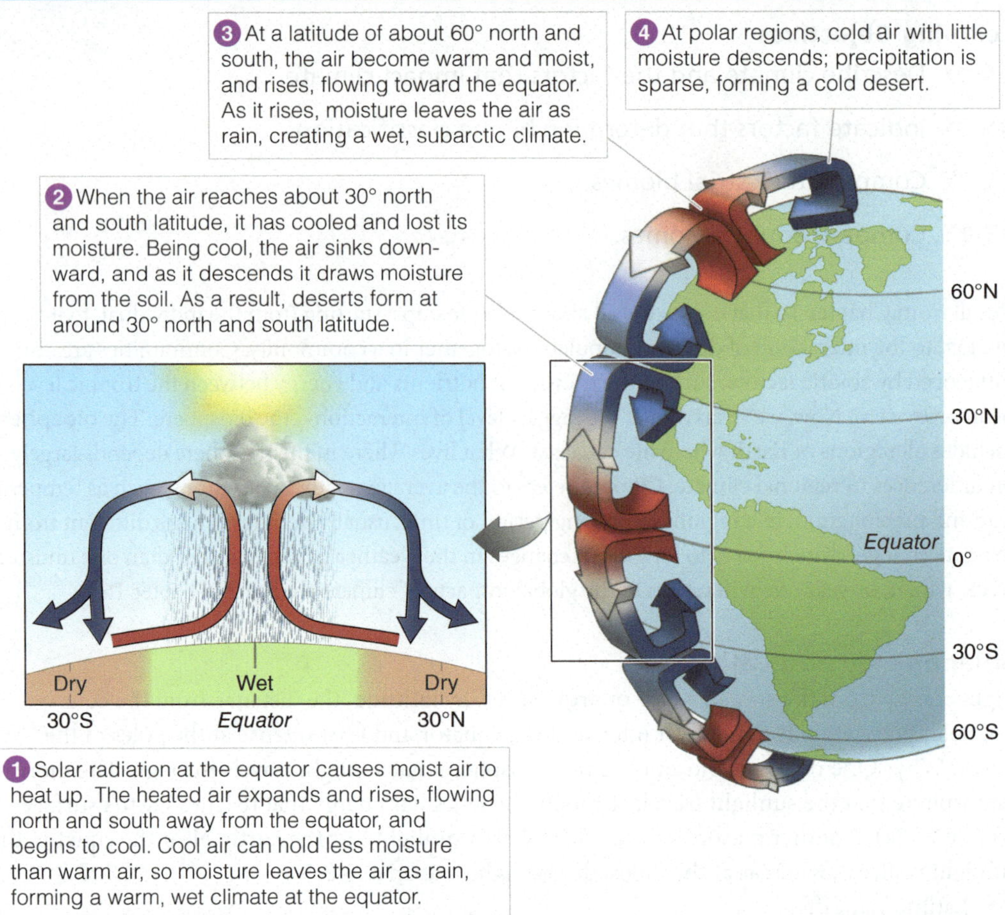

Water's ability to stabilize temperature (Chapter 2) keeps coastal areas from experiencing the dramatic shifts in temperature that occur farther inland. For example, Seattle, Washington, has milder winters than Minneapolis, Minnesota, even though Seattle is farther north. Being a coastal city, Seattle draws heat from the adjacent Pacific Ocean, whereas landlocked Minneapolis has no similar source of heat.

Ocean Currents

The waters of Earth's oceans circulate on a global scale due to ocean currents. **Ocean currents** are the continuous movement of seawater in a directed manner because of forces like wind, temperature, and salinity. For example, at the equator, seawater warms and expands, making the sea level about 8 centimeters (3 inches) higher than it is at the poles. This difference in sea level sets the pattern of ocean circulation in motion. Winds, Earth's rotation, and the shape and location of landmasses also influence the movement of ocean water. Collectively, these factors produce the pattern of currents, which circulate clockwise in the Northern Hemisphere and counterclockwise in the Southern Hemisphere (Figure 15.26).

Ocean currents affect climate by distributing heat energy. Heat given off by water flowing from the equator toward the poles warms the eastern coast of continents. For example, the east coast of North America is warmed by the northward flow of the Gulf Stream. Conversely, western coasts of continents are cooled by loss of heat to currents of cold water flowing from the poles to the equator, such as the Pacific Northwest of the United States. Therefore, temperatures in the Pacific Northwest tend to be moderate year-round.

Distribution of Earth's Biomes

We can divide the biosphere into biomes. There are two general biome categories: terrestrial biomes and aquatic biomes. Each **biome** is a large-scale area in which specific physical conditions support a distinctive biological community of plants and animals adapted to those conditions. We'll first cover terrestrial biomes before moving on to aquatic biomes.

! Point of Confusion: Ecosystem or biome? While both involve the living and nonliving components of an environment, biomes represent a much larger area than an ecosystem.

Figure 15.26 Global ocean circulation

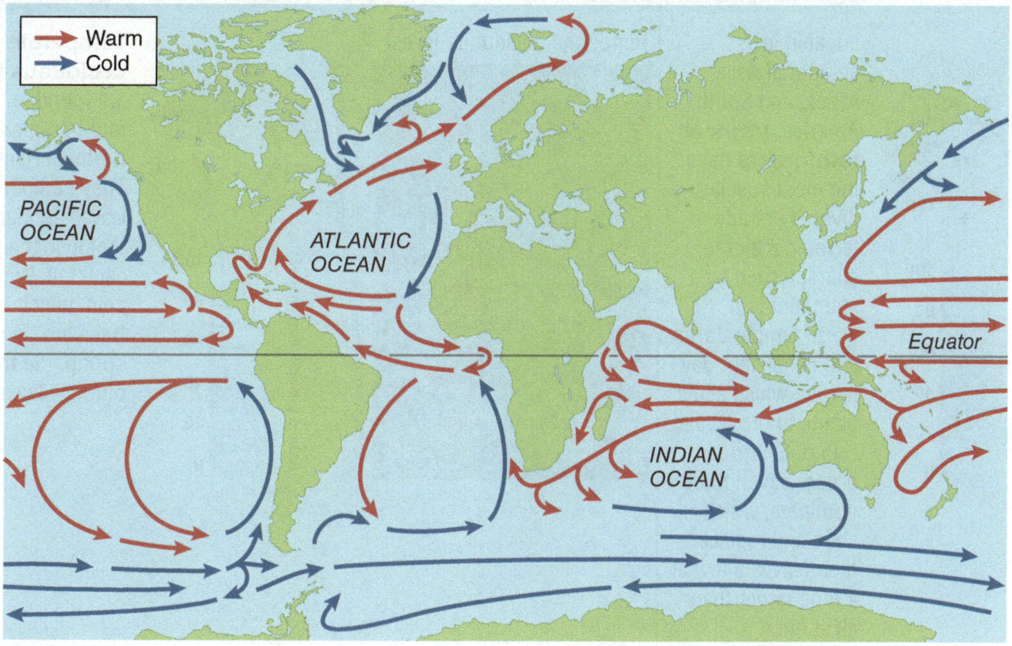

Warm surface currents (red arrows; blue arrows represent cold currents) start moving from the equator toward the poles, but prevailing winds, Earth's rotation, gravity, and the shape of ocean basins or landmasses influence the direction of flow. Water temperatures, which differ at different latitudes and different depths, contribute to regional differences in air temperature and rainfall.

? Did You Know?

Curtis Ebbesmeyer, an oceanographer, studied the movements of 28,800 toys—which included ducks, beavers, turtles, and frogs—that accidentally spilled into the Pacific Ocean in 1992. He and his team were able to predict where and when the toys would wash up on shore based upon ocean currents.

Terrestrial biomes. As we learned in the previous section, solar energy, wind currents, rainfall, and ocean currents all influence the climate in different regions. These factors combined contribute to the formation of several distinct terrestrial biomes. **Terrestrial biomes** are large areas of land with similar climate and lifeforms. Most terrestrial biomes are geographically discontinuous, meaning they consist of widely separated regions on different continents. Figure 15.27 shows the distribution of nine major terrestrial biomes, and Table 15.3 provides a brief description of the

Figure 15.27 Geographic distribution of Earth's terrestrial biomes

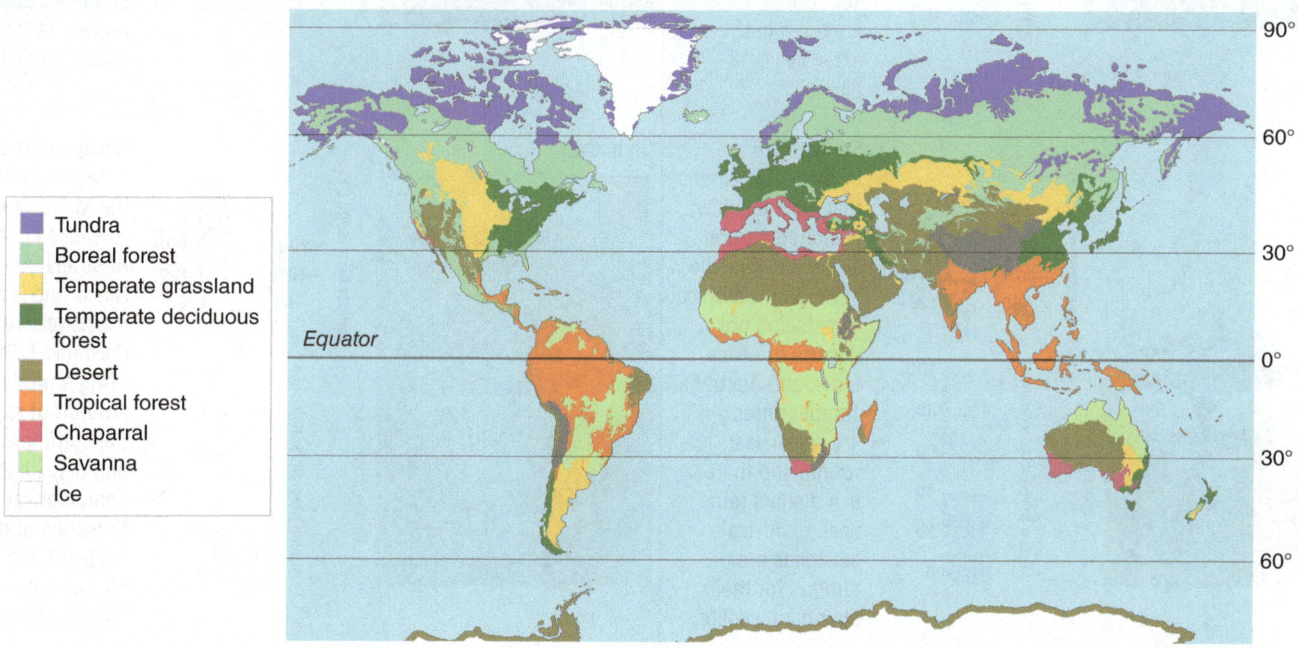

Table 15.3 Key Features of Earth's Terrestrial Biomes

Biome	Description	Biome	Description
Tropical rainforest  Credit: Maris Maskalans/Shutterstock.com 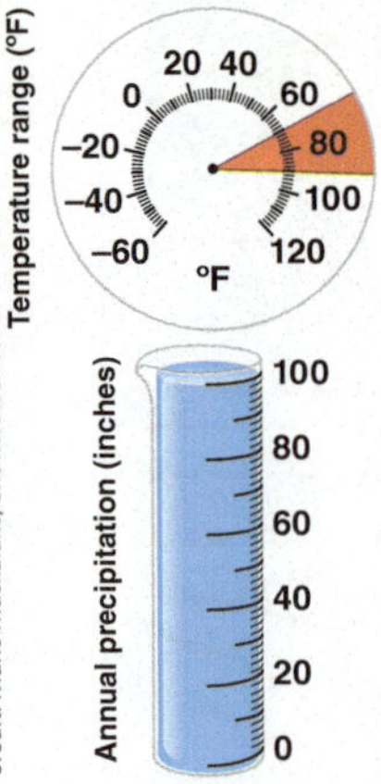 	Located in equatorial Asia, Africa, and South America, **tropical rainforests** have the most species diversity of any terrestrial biome, characterized by abundant rain, warm temperatures, and consistent day length, which allows plants to grow year-round. Evergreen broad-leaved trees dominate, which shed and replace their leaves a few at a time rather than all at once.	Temperate deciduous forest  Credit: Dean Pennala/Shutterstock.com 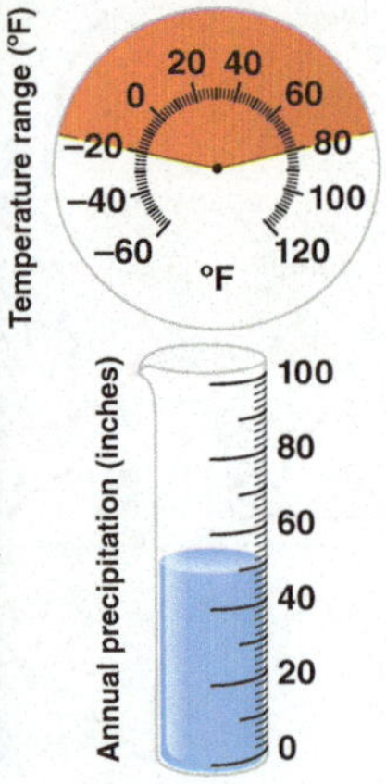 	**Temperate deciduous forests** are found in the mid-latitudes of the Northern Hemisphere. They contain broad-leaved deciduous trees that lose their leaves each fall. Winters are cold, which is when trees are dormant. In spring, the trees put out new leaves.
Boreal forest  Credit: ssawpics/Shutterstock.com 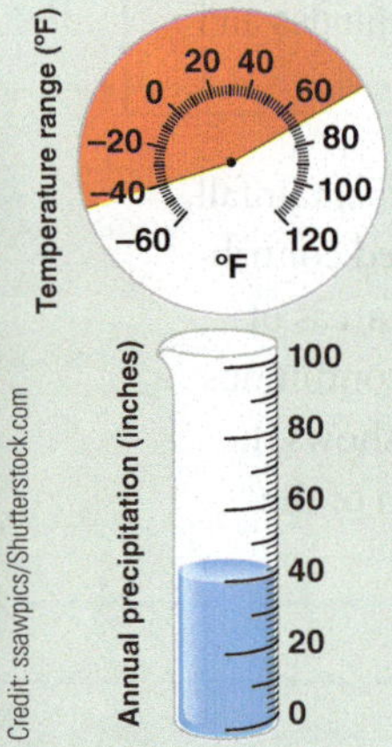 	Located at high latitudes in broad bands in northern Asia, Europe, and North America, **boreal forests** experience cool, dry conditions year-round. Spruce, fir, and pine trees dominate, and they are tolerant of drought, cold, and nutrient-poor soil. Boreal forests are the most extensive land biome.	Temperate grassland  Credit: Tetra Images/Getty Images 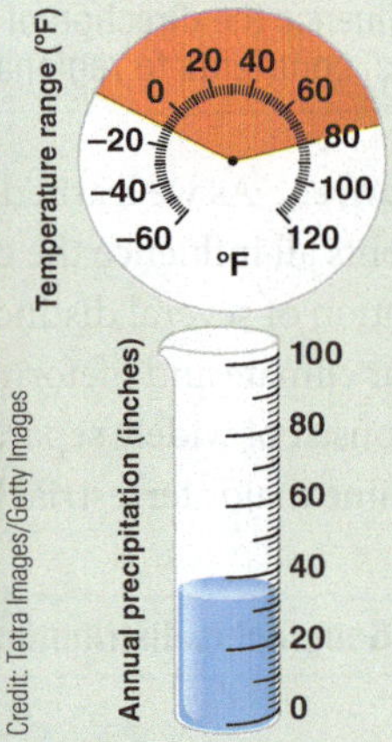 	Covering about 25% of Earth's surface, **temperate grasslands** contain perennial grasses and other nonwoody plants, and typically occur in the middle of continents, bordered by shrublands or deserts. North American grasslands, called **prairies**, experience hot summers and cold winters, with moderate to low precipitation.
Savanna  Credit: iStock.com/czekma13 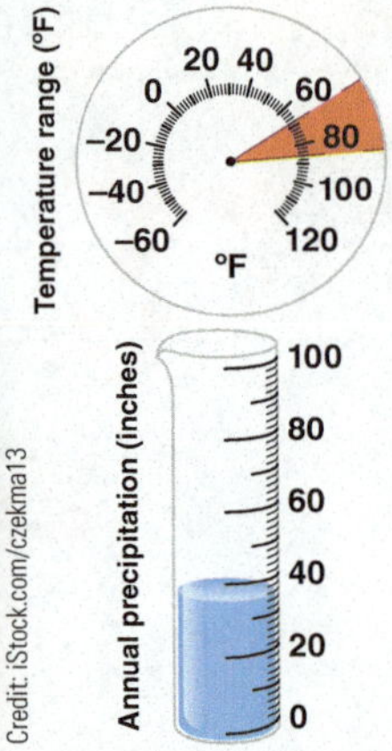 	**Savannas** are tropical grasslands that include a few scattered shrubs and trees. They lie between the tropical forests and hot deserts of Africa, India, and Australia. Temperatures are warm year-round, and there is a distinct rainy season. Africa's savannas are famous for their abundant wildlife.	Chaparral  Credit: Tim Gray/Shutterstock.com 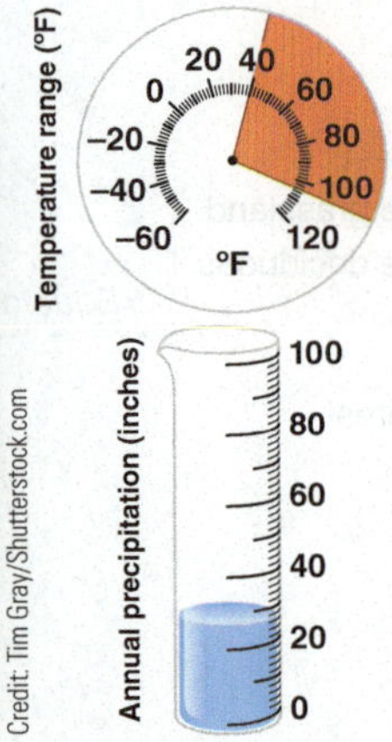 	**Chaparral** occurs in regions bordering the Mediterranean, as well as in Chile, Australia, and South Africa, and is the most extensive ecosystem in California. Mild winters bring a moderate amount of rain, and the summers are hot and dry. Chaparral is characterized by the presence of drought-resistant, fire-adapted shrubs with small, leathery leaves.

Biome	Description
Desert  Credit: National Geographic Image Collection 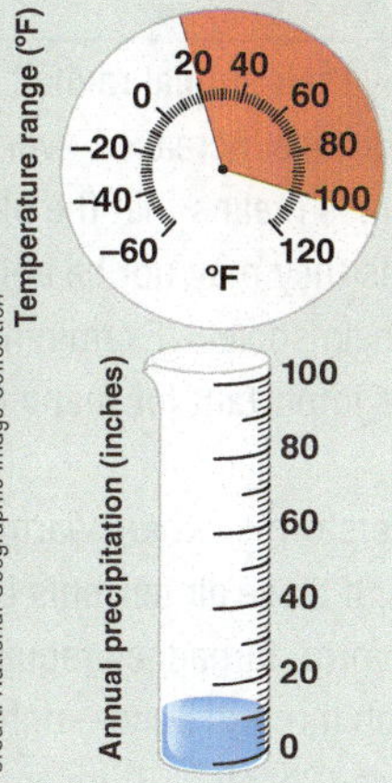 	Defined by low annual precipitation, **deserts** cover about one-fifth of Earth's land surface, in regions where global air circulation causes dry air to sink. Deserts contain many heat-adapted plants and animals. Lack of rainfall keeps the humidity in a desert low. As a result, deserts tend to have larger daily temperature shifts than other biomes.
Tundra  Credit: Darrell Gulin/Corbis Documentary/Getty Images 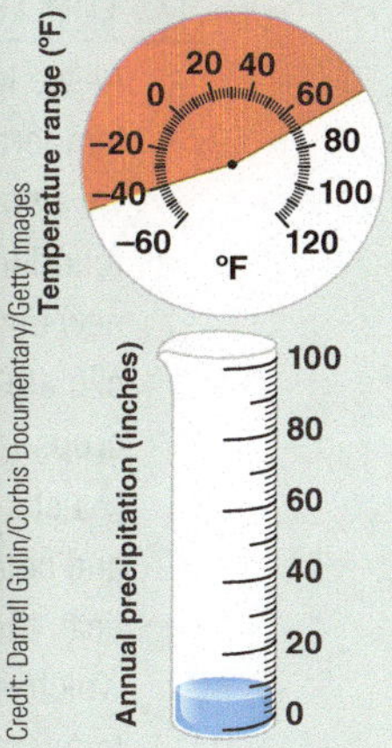 	**Tundra** lies between the polar ice cap and the belts of boreal forests in the Northern Hemisphere, mainly in Russia and Canada. Snow blankets the Arctic tundra for up to nine months of the year. The tundra also has a layer of **permafrost**, soil that is frozen for two or more consecutive years. During the brief growing season, nearly continuous sunlight fuels the growth of shallow-rooted low-growing plants and lichens.
Ice  Credit: iStock.com/Straystone 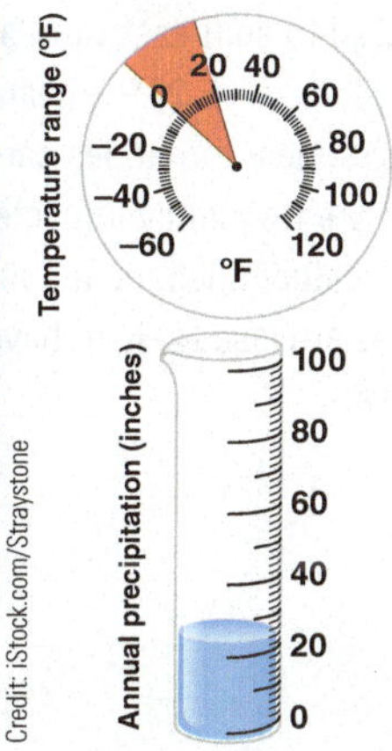 	The **ice** biome includes areas that are largely covered in ice year-round. The ice biome includes much of Greenland, Antarctica, and the Arctic. This is the coldest of all biomes. The ice biome encompasses 10% of all land surface on Earth. Despite the frigid environment, a variety of microorganisms—including bacteria, archaea, and fungi—live there.

biomes, along with icons showing each biome's temperature range (in degrees Fahrenheit) and annual precipitation (in inches).

Aquatic biomes. Although the term "biome" was coined to describe land areas, there are also aquatic biomes. **Aquatic biomes** are water-based environments that vary depending on their temperature, salinity, sunlight, dissolved oxygen, and nutrient concentration. Like terrestrial biomes, there are also characteristic communities of species they support. Table 15.4 describes different aquatic biomes.

Process of Science

The Impact of Changing Biomes at Home

As we learned in this section, terrestrial biomes are defined by their annual rainfall and average temperature. Perhaps not surprisingly, as global temperatures have increased over the past century, we also see a shifting distribution of biomes. The shift of biomes means that the plants and animals found in particular regions are potentially in trouble as well, as they may not be able to adapt to the new environment. Recall that science is a process serves to help society. Examining how a changing climate alters aspects of biomes, such as weather patterns, is important for many reasons, from our crops (Chapter 18) to keeping people safe.

Let's consider tornadoes. Tornadoes form from large thunderstorms, where warm air rising and cold air falling can eventually lead to spinning air currents. If these air currents become vertical, they can drop out of a cloud in the form of a tornado. To form, tornadoes require warm air, which is why peak tornado season in the United States is March through June—right when the temperatures are beginning to warm up. Warming temperatures (Chapter 16) mean there are more tornadoes.

Tornado Alley is the region of the Great Plains where most of these devastating storms occur. It is also a key region where crops are grown in the United States. As we'll see in Chapter 18, the plants that produce the food we depend on are threatened by climate change; changes in tornado activity add an additional threat to our food supply.

However, in the past few decades, Tornado Alley has appeared to shift eastward, away from the Great Plains and into the more populous Southeastern states (Figure 15.28)—meaning that more people are at risk of the damaging effects of tornadoes. The strongest tornadoes can cause massive property damage, injury to people, and even death. Areas that are frequently impacted by tornadoes are often prepared and have, for example, designated severe weather shelters in public places. However, areas that are not traditionally impacted by tornadoes are less likely to have these safeguards in place and are therefore more vulnerable to tornadoes.

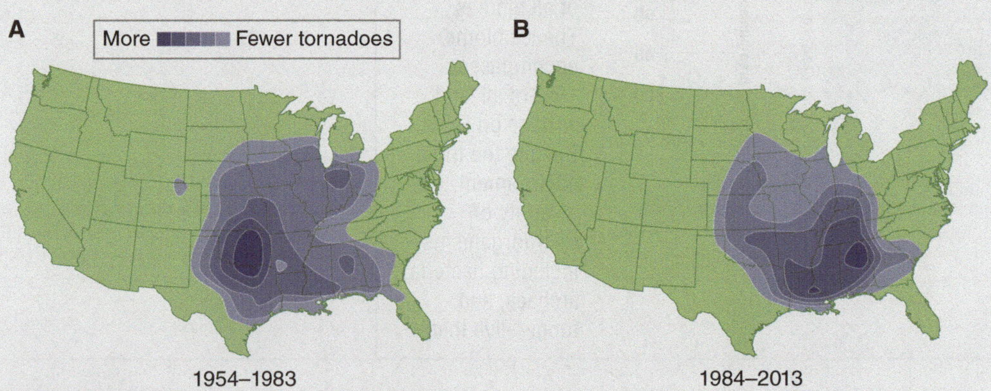

Figure 15.28 The number and locations of tornadoes in the United States

A. Between 1954 and 1983, most tornadoes formed in the central United States

B. In the past few decades, more tornadoes are forming eastward.

1. Using the map in Figure 15.27, find the biome where you live. How could your biome be impacted by climate change? What new risks are possible in your area?
2. What can we do as a society to help people in areas traditionally not impacted tornadoes—or other extreme weather conditions, like floods—to be prepared?

Table 15.4 Key Features of Earth's Aquatic Biomes

Biome	Description	Biome	Description
Streams and rivers  Credit: Apolinariy/Shutterstock.com	Moving freshwater includes **streams**, which typically originate from runoff or melting snow or ice, and then grow in size and merge to form **rivers** as they flow downhill. Properties of a stream or river vary in terms of solute concentration, oxygen content, and temperature, so they support different species along their length.	Lakes  Credit: Michael Piepgras/Dreamstime	A **lake** is a body of standing freshwater. Near the shore, where sunlight penetrates all the way to the lake bottom, rooted aquatic plants and algae that attach to the bottom are primary producers. All but the shallowest lakes have zones that differ in their physical characteristics and species composition.
Estuaries  Credit: imageBROKER.com/Alamy Stock Photo	An **estuary** is a mostly enclosed coastal region where seawater mixes with nutrient-rich freshwater from a river or rivers. Plants adapted to withstand high salinity and changes in water level serve as producers. Due to the meeting of freshwater and seawater, estuaries have significant biodiversity.	Intertidal zone  Credit: NOAA	Organisms that live along rocky and sandy shores, called the **intertidal zone**, are adapted to the mechanical force of the waves and to tidal changes. Many species are underwater during high tide, then are exposed to the air when the tide is low. Algae that cling to rocks are the producers.
Coral reefs  Credit: Andrea Izzotti/Shutterstock.com	**Coral reefs** are found in saltwater, typically in warm, shallow, well-lit tropical seas. These reefs are formations made primarily of calcium carbonate secreted by generations of corals, which are invertebrate animals. Tropical coral reefs are home to an extraordinary assortment of species. The main producers in a coral reef community are photosynthetic dinoflagellates (Chapter 13) that live inside the reef-building corals. Note that in addition to shallow reefs, there are deep-sea coral habitats that thrive in dark, cold water.	Open ocean  Credit: John Foxx/Getty Images	In the brightly lit saltwater of the **open ocean**, photosynthetic protists and bacteria are the primary producers. On the seafloor, most species live along the edges of continents. Under most conditions, light penetrates only about 650 feet (200 meters). Below that, organisms live in darkness, sustained by organic material that drifts down from above.

15.5 Summary

- Climate is the average weather conditions in a region over a long period of time. Factors affecting climate include solar energy, latitude, air circulation, and ocean currents.
- A biome is a large-scale area with similar physical conditions, supporting distinctive biological communities. A terrestrial biome is determined by the annual temperature and rainfall, while an aquatic biome is determined by sunlight and salinity.
- Terrestrial biomes are land-based biomes and include tropical, temperate deciduous, and boreal forest biomes, as well as temperate grasslands, savannas, chaparral, deserts, tundra, and ice.
- Aquatic biomes are water-based biomes and include streams, rivers, lakes, estuaries, the intertidal zone, coral reefs, and the open ocean.

15.6 Biogeochemical Cycles

Learning Objectives

15-17 Describe the water, phosphorous, nitrogen, and carbon cycles.

15-18 Explain the consequences if there is too much phosphorus, nitrogen, or carbon in an environment.

Earlier in this chapter, we described how nutrients and energy move throughout the environment. Essential elements (Chapter 2), such as carbon and nitrogen, also move throughout ecosystems. This movement of essential elements—and molecules, like water—from one or more environmental (abiotic) reservoirs, through the living (biotic) components of an ecosystem, and back again is called a **biogeochemical cycle**. Depending on the element, environmental reservoirs may include Earth's rocks and sediments, bodies of water, and/or the atmosphere.

Chemical and geologic processes move elements to, from, and among environmental reservoirs. Some elements locked in rocks become part of the atmosphere due to volcanic activity. Movement of Earth's tectonic plates (Chapter 10) can uplift rocks, so an area that was once an ancient seafloor becomes part of a landmass. On land, erosion breaks down rocks, allowing the elements in them to enter rivers and flow to seas. Compared to the movement of nutrients within a community, the movement of elements among nonbiological (abiotic) reservoirs is far slower. Processes such as erosion and geologic uplift operate over thousands or millions of years.

In this section, we will learn about four biogeochemical cycles: the water cycle, phosphorus cycle, nitrogen cycle, and carbon cycle. Each of these cycles is represented by a detailed figure that includes reservoirs and how they move through the environment (shown by orange arrows). As you read this section and review the figures, start with a reservoir and trace the movement of water or the element as it moves through the environment until you return to the starting point.

The Water Cycle

From Chapter 2, we know that water is essential to life. As we learned in the previous section, aquatic biomes are the most predominant biomes, as water covers the majority of Earth's surface. Of the water on Earth, 97% is saltwater. The 3% of remaining water is freshwater, where the majority of freshwater is locked away in ice caps and glaciers. Only about 1% of all freshwater on the planet is available to sustain life. We can find freshwater on the surface as lakes and ponds, or as **groundwater** buried beneath our feet. Pockets of groundwater are held in rock formations called **aquifers**. We can extract water from aquifers for our usage.

Water moves between saltwater and freshwater reservoirs as part of the water cycle. The **water cycle** moves water from oceans to the atmosphere, onto land, and into freshwater ecosystems, and back to the oceans (Figure 15.29). Much of the process is driven by phase changes of water, where it's converted between its liquid, gas, and solid forms. For example, water turns from a liquid to a gas (evaporation) and then condenses into liquid and falls from the sky as precipitation. Note that as part of this cycle, water also can pass through living organisms, such as us or plants. Plants can also release water back to the environment in a process called *transpiration* (Chapter 18).

Note that the movement of water results in the movement of soluble nutrients. For example, water plays a critical role in moving the soluble forms of nitrogen and phosphorus throughout

Figure 15.29 The water cycle

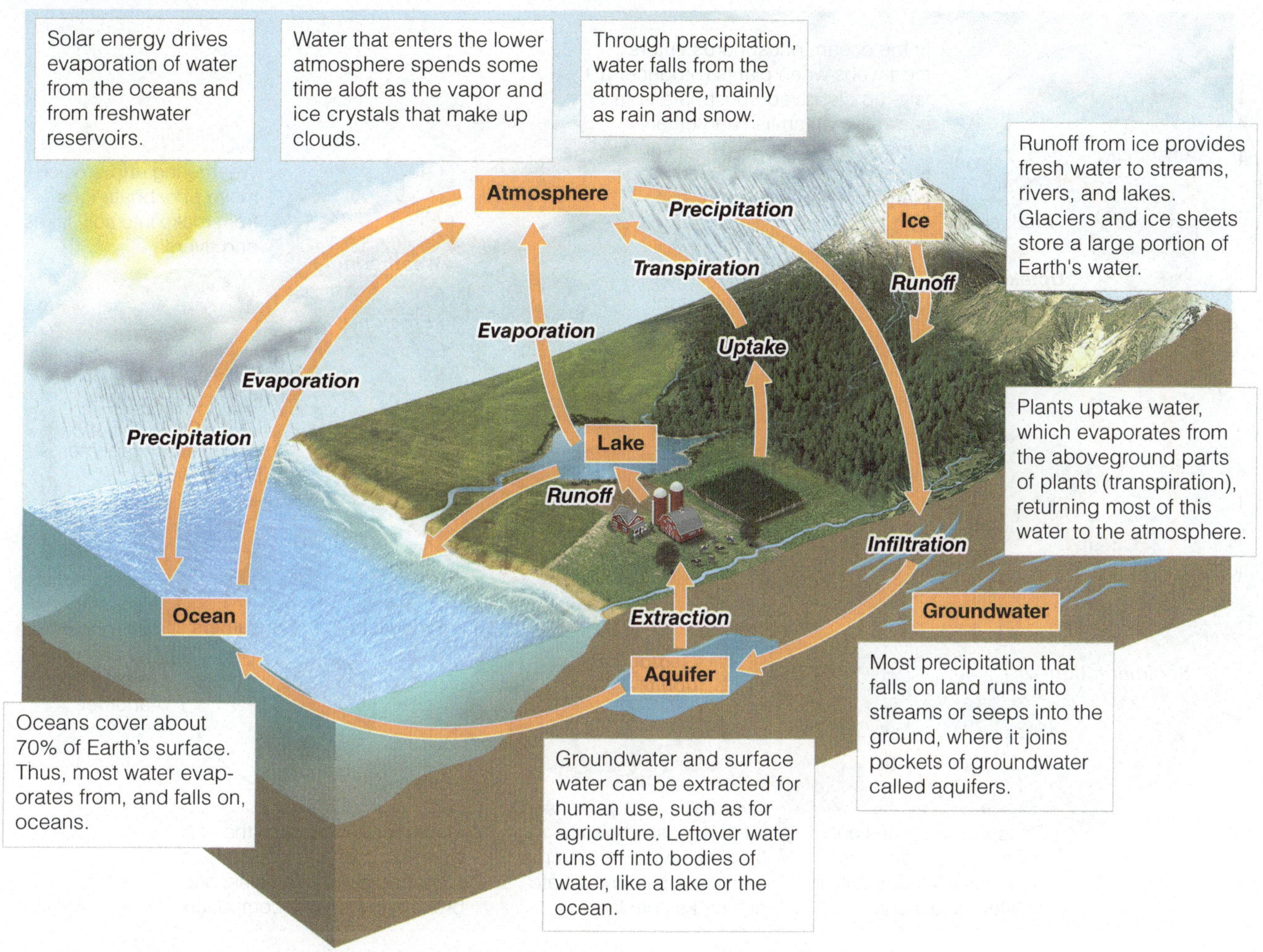

Terms in orange boxes represent water reservoirs, and the orange arrows represent the movement of water throughout the ecosystem. The labels overlaid on the arrows indicate the various processes involved.

the environment, which we'll explore next. Flowing water can transport pollutants too; runoff from heavily fertilized lawns and agricultural fields carries dissolved phosphates and nitrates into streams and lakes.

The Phosphorus Cycle

In the **phosphorus cycle**, phosphate moves among Earth's rocks, soil, and water, and into and out of food webs (Figure 15.30). All organisms require phosphorus to build ATP, nucleic acids, and phospholipids. The biological portion of the phosphorus cycle begins with the uptake of phosphate by producers, through consumers that eat producers, and then to decomposers in the soil that free phosphorus for uptake by plants. Wastes and remains are continually broken down by decomposers on land and in the ocean, replenishing the supply.

In the chemical portion of the cycle, most of Earth's phosphorus is bonded to oxygen as phosphate (PO_4^{3-}), an ion abundant in rocks and sediments. This is due to the high reactivity

? Did You Know?

Phosphorus-rich droppings from seabird and bat colonies can be harvested as a natural fertilizer.

Figure 15.30 The phosphorus cycle

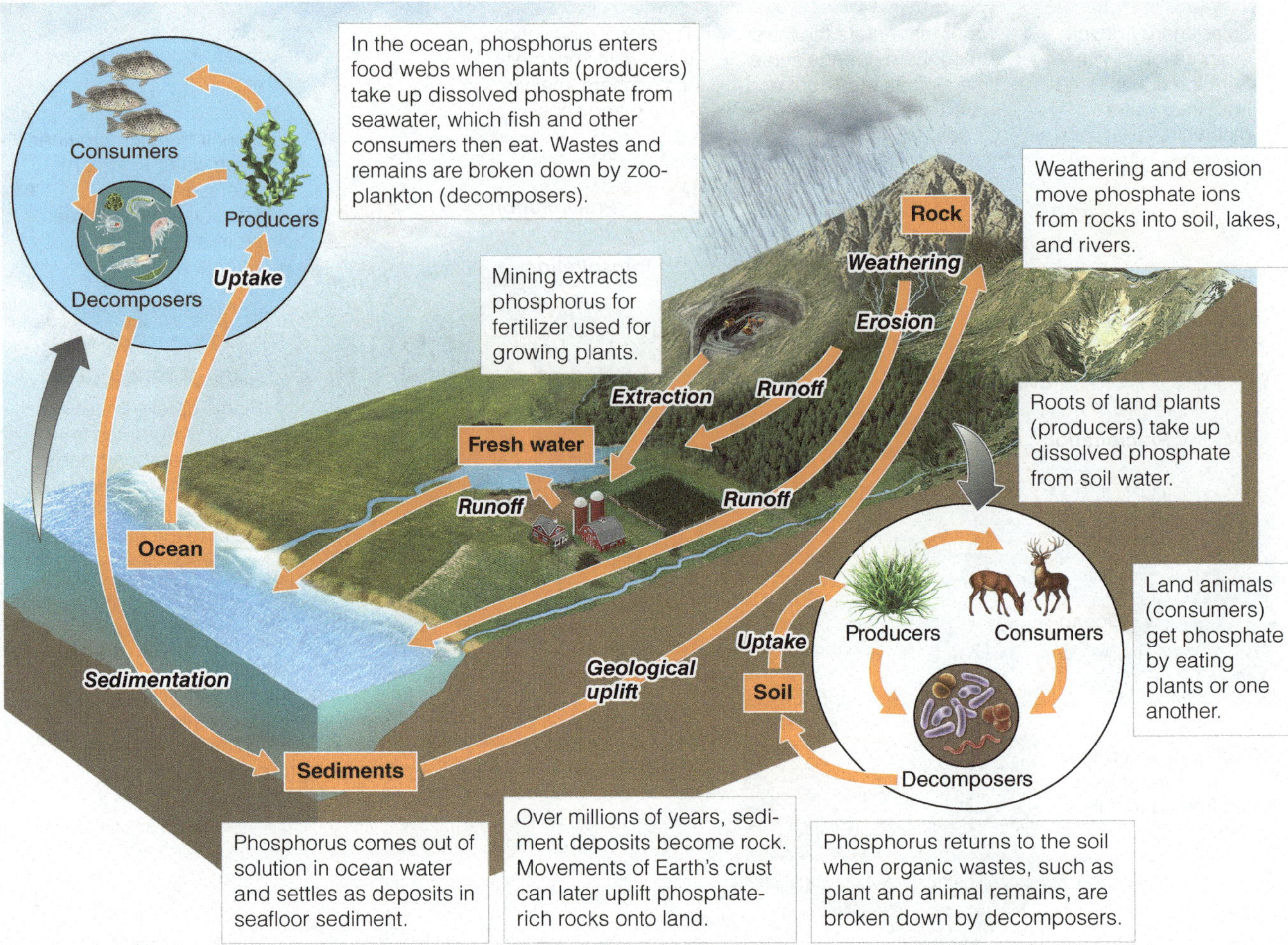

Terms in orange boxes represent phosphorus reservoirs, and the orange arrows represent the movement of phosphorus throughout the ecosystem. The labels overlaid on the arrows indicate the various processes involved.

Crop rotation, or growing plants with nitrogen-fixing bacteria on fields in alternating years with plants without nitrogen-fixing bacteria, is a way of maintaining crop health without over-reliance on commercial fertilizers.

of elemental phosphorus. Therefore, phosphorus does not occur naturally in its elemental form. Phosphate ions move through the cycle by the weathering of phosphorus-containing rocks and runoff into bodies of water. Phosphorus can become a solid in ocean water and settle on the seafloor. Over millions of years, these sediments can be compressed to form rock and be uplifted to form new mountains, beginning the process again.

The Nitrogen Cycle

As is the case with phosphorus, all organisms require nitrogen. For example, nitrogen is an essential component of nucleic acids and proteins, and the lack of nitrogen on Surtsey (from the chapter-opening story)—until birds arrived with their nitrogen-rich excrement—was a major limiting factor for primary succession. In the **nitrogen cycle**, nitrogen moves among the atmosphere (the largest nitrogen reserve), soil water, and food webs (**Figure 15.31**).

Although the greatest source of nitrogen on the planet is nitrogen gas (N_2), plants cannot use gaseous nitrogen. This is because plants do not have an enzyme that can break the triple covalent bond between its two nitrogen atoms. However, some bacteria have such an enzyme that breaks apart nitrogen gas and produces ammonia in a process called **nitrogen fixation**. Plants and bacteria can form a symbiotic relationship with one another so that the bacteria get sugars produced by

Figure 15.31 **The nitrogen cycle**

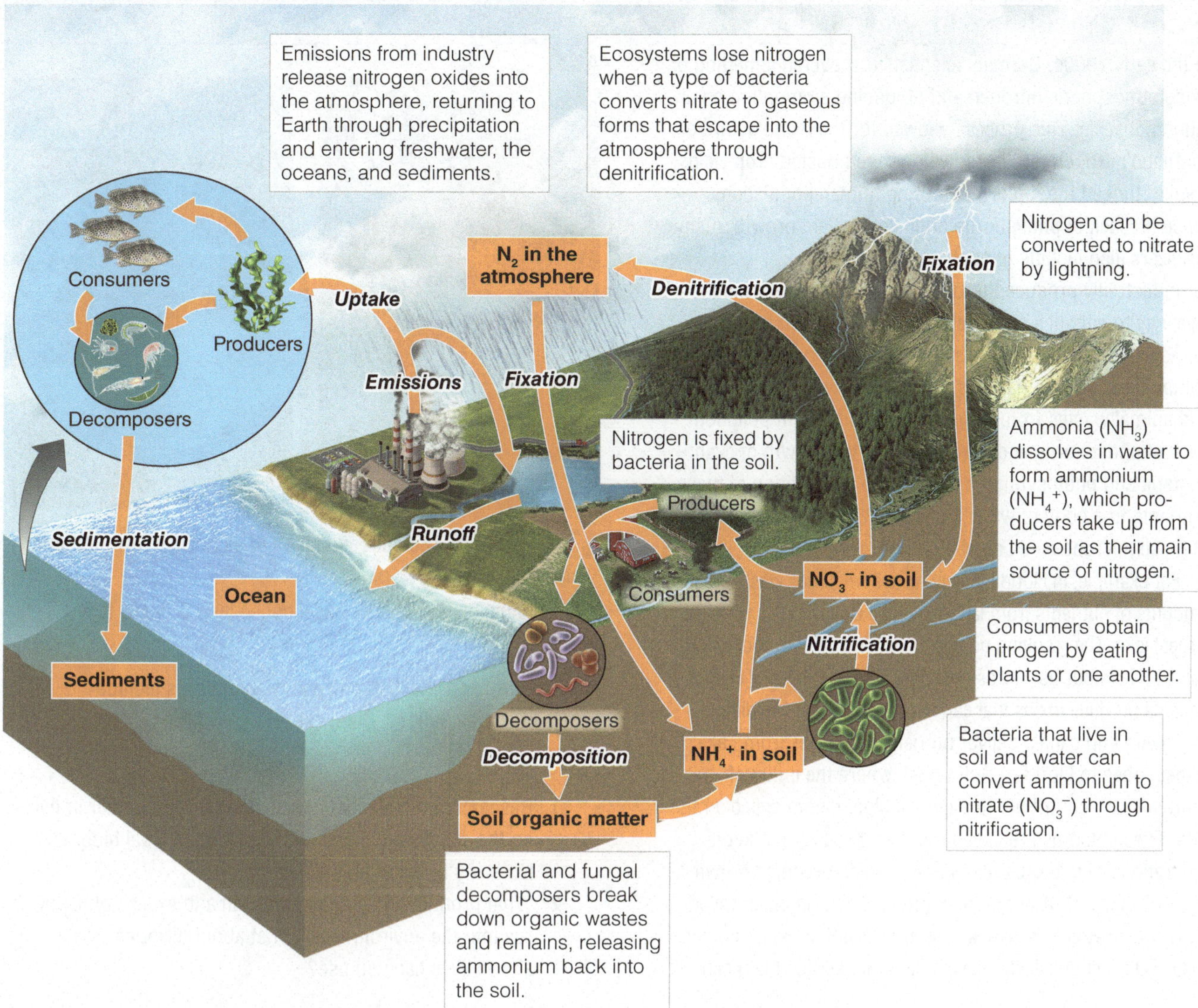

Terms in orange boxes represent nitrogen reservoirs, and the orange arrows represent the movement of nitrogen throughout the ecosystem. The labels overlaid on the arrows indicate the various processes involved.

the plants, and the microorganisms produce a useable chemical form of nitrogen called ammonia (NH_3), which can be converted to ammonium (NH_4^+) in the soil or water. In plants with these bacteria, the bacteria live on the plants' roots in small bumps called nodules (Figure 15.32).

Ammonia can also be converted to another useable form of nitrogen, nitrate (NO_3^-), during **nitrification**. Bacteria and archaea in the soil and ocean can perform nitrification. Microbes also help close out the cycle through the process of **denitrification**, which is the conversion of nitrate back into nitrogen gas.

Notice that the cycles we covered here require balance—too much phosphorus and nitrogen can wreak havoc on the environment. For example, the burning of fossil fuels such as coal releases nitrogen into the atmosphere, contributing to acid rain (Chapter 16). In addition, overuse of phosphorus or nitrogen-containing commercial products like fertilizers can lead to run off and cause **eutrophication**, or water having too many nutrients. Eutrophication can cause harmful algal blooms. These algal blooms can cause a depletion of dissolved oxygen in the water, causing mass die-offs of fish.

Figure 15.32

Nodules on the roots of a legume

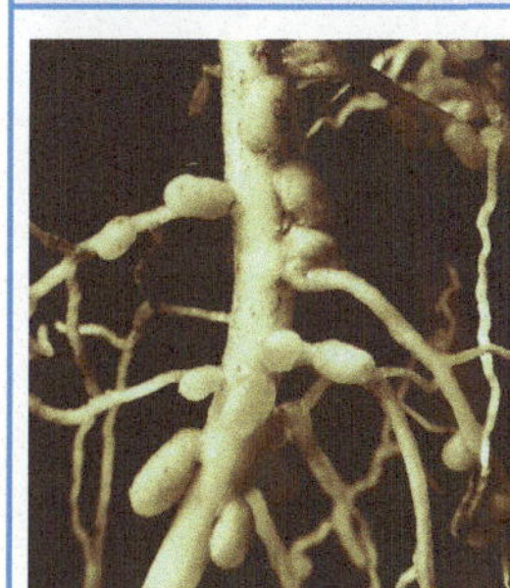

The bumps on the plant's roots contain colonies of nitrogen-fixing bacteria.

Credit: Hugh Spencer/Science Source

Biology for Life

Nutrient Pollution and Algal Blooms

In the early 1900s, German scientists discovered a method of fixing atmospheric nitrogen and producing ammonium on an industrial scale. This process allowed for the manufacture of synthetic nitrogen fertilizers, which have boosted crop yields, helping to feed the rapidly increasing human population. Since the lack of phosphorus often limits plant growth, many fertilizers also include phosphate.

Industrially produced fertilizers are a relatively inexpensive way to enhance crop yield and maintain lush lawns. However, nutrients from these fertilizers can run off and contaminate aquatic habitats. Similarly, sewage from cities and animal wastes from farms can cause nutrient pollution. Lack of phosphorus and nitrogen limits the growth of cyanobacteria and photosynthetic protists, so the addition of these nutrients to a body of water can cause a population explosion of these microbes—an algal bloom (Figure 15.33A).

For example, in Florida, heavy rains sometimes wash large amounts of nutrients from land into Lake Okeechobee, the state's largest lake. The resulting population explosion of cyanobacteria can be hazardous as well as unsightly. Some species of cyanobacteria produce toxins that can harm humans and wildlife.

Rivers and canals deliver nutrient-rich water from Lake Okeechobee to Florida's Gulf Coast, where the influx of nutrients can contribute to a type of algal bloom called a red tide. These blooms are called red tides because high concentrations of this organism color seawater reddish brown (Figure 15.33B). A nerve toxin produced by an algae called *Karenia brevis* kills fishes, sea turtles, dolphins, and manatees. Eating shellfish that have accumulated this toxin can cause neurotoxic shellfish poisoning. Symptoms include gastrointestinal distress, dizziness, and loss of motor control.

Figure 15.33 The consequences of excessive nitrogen and phosphorus

A. An algal bloom in Lake Okeechobee, and **B.** a red tide.

1. Research the impacts of algal blooms. What other impacts on the ecosystem occur as a result of an algal bloom? Note any biotic or abiotic factors.
2. What products do you use that can add excess phosphorus into the environment? What about nitrogen? What alternatives can you use?

The Carbon Cycle

Carbon occurs abundantly in the atmosphere, combined with oxygen as carbon dioxide (CO_2). Additionally, all molecules of life have a carbon backbone (Chapter 3). In the **carbon cycle**, carbon moves among rocks, water, and the atmosphere, and into and out of food webs (Figure 15.34).

Recall from Chapter 4 that plants can take in carbon dioxide, and along with energy from the sun, build carbon-containing high-energy molecules like glucose. Therefore, plants remove carbon dioxide from the air and chemically convert it into a form that consumers can use. Decomposers can liberate stored carbon as carbon dioxide. These processes occur on land and in the ocean.

Like the phosphorus and nitrogen cycles, a balance of carbon is important for the well-being of the biosphere. Consider the burning of fossil fuels. Deposits of fossil fuels formed over hundreds of millions of years from carbon-rich remains, where high pressure and temperature transformed the remains of ancient land plants to coal. A similar process transformed the remains of marine diatoms into deposits of oil and natural gas. Until we began burning these fossil fuels, the carbon in them, like the carbon in rocks, had little impact on ecosystems. Today, burning this fuel releases billions of tons of CO_2 into the atmosphere every year. We will explore the consequences of too much carbon dioxide in the atmosphere in Chapter 16.

Figure 15.34 The carbon cycle

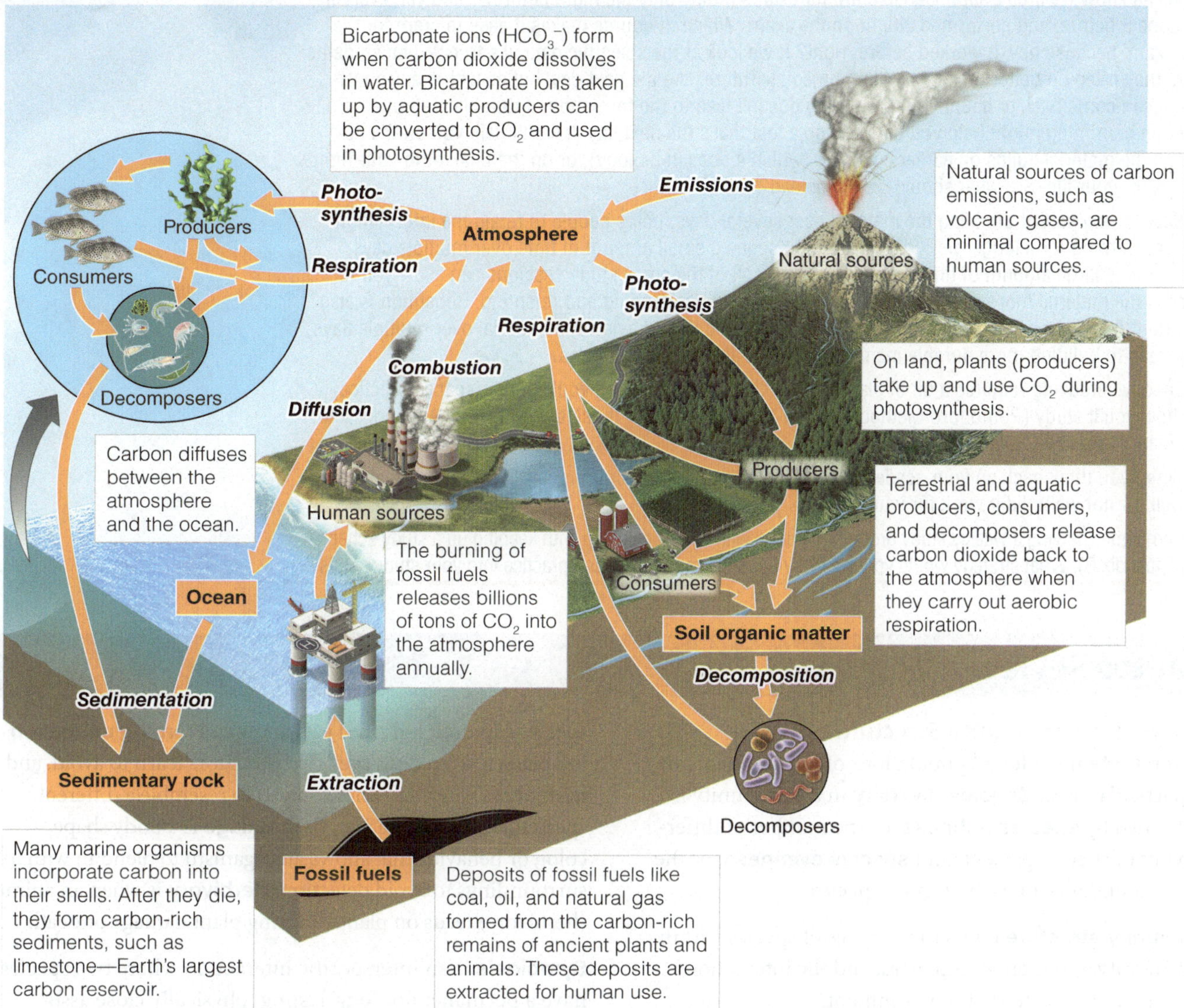

Terms in orange boxes represent carbon reservoirs, and the orange arrows represent the movement of carbon throughout the ecosystem. The labels overlaid on the arrows indicate the various processes involved.

15.6 Summary

- Biogeochemical cycles show the movement of water and essential elements (phosphorus, nitrogen, and carbon) through ecosystems.
- The water cycle describes the movement of water into the atmosphere from the ocean via evaporation, to terrestrial ecosystems as precipitation, and then back to the ocean.
- The phosphorus cycle moves phosphate ions from rocks, soil, and water, and into and out of food webs. The nitrogen cycle moves nitrogen from the atmosphere, and through soil and water reservoirs and food webs. Too much phosphorus or nitrogen in the environment can lead to algal blooms in water, which can pose a threat to life.
- The carbon cycle moves carbon between rocks, water, the atmosphere, and into and out of food webs when carbon fixation occurs. The burning of fossil fuels releases billions of tons of CO_2 into the atmosphere every year.

Understand It! Spaced Practice

So, we all know we aren't *supposed* to cram for exams (study only the night before)...and yet, everyone has done it before, and performed alright on the exam. Which of course makes it easy to cram for the next exam, because hey, it worked before, right? If we look at the scientific data, it's true. When students cram, they often do better on a test in the immediate future, like the next day, but as with many short-cuts, it can come back to haunt you. Cramming doesn't lead to the formation of durable memories. So, although cramming might help you get through a test that's the next day, it won't help you on the next exam if the material builds on earlier coverage (which it does in biology), or on the final, or in your future classes, or your life. So, what should we do instead?

Spaced practice, or studying the material over several days, helps people to retain the information they are learning for a longer period. The key is to spend a small amount of time over multiple days rather than a large amount of time on a single day. Each of the small study sessions can allow you to process the material more effectively, leading to greater understanding and retention. Repetition is also important for memory (part of *practicing*; Chapter 1)—and if you review the material over multiple days, it will lead to building more durable memories that are also easier to access.

Effective studying requires a bit of planning—both in terms of when you'll study (spaced practice) and how you'll study (Pomodoro Method; Chapter 6) during each of those sessions. See Appendix B for information on how to plan out your study.

1. When you think back on your academic career, did you do more cramming or spaced practice? How well (or not well) did it work for you?
2. Sometimes, having a weekly study group or meeting with a tutor can help with keeping up a study schedule (Appendix B). What strategy would you like to try to integrate more spaced practice into your study habits?

Chapter Review

Section 15.1 Community Structure

A **community** includes all populations of living organisms in a particular area. **Species diversity** in a community is determined by **species richness**, or the number of different types of species present, and **species evenness**, or the relative abundance of each of those species.

Community structure refers to the types of species within a community and their abundances, and the interactions among them and with their environment.

The **abiotic** factors in an environment refer to the nonliving factors affecting community structure, while the **biotic** factors are the living components. The natural environment where a species typically lives is its **habitat**, while its **niche** is the combination of conditions, resources, and interactions that it needs for survival and reproduction.

Section 15.2 Species Interactions

Competition is when species interact with each other for access to limited resources. **Intraspecific competition** is competition between members of the same species. **Interspecific competition** is competition between members of different species. The **principle of competitive exclusion** states that species that compete over the same limited resources cannot coexist indefinitely. **Resource partitioning** is when species use the same resources differently to minimize competition.

Antagonistic interactions occur when one species benefits by extracting resources from another. **Predation** is when one species captures, kills, and eats another. **Warning coloration** is a pattern or color in prey that predators learn to avoid, and **mimicry** is when a species evolved to resemble a different species to avoid predation. **Camouflage** is a body shape, color, or behavior that allows an organism to blend in with its surroundings to avoid detection. **Herbivory** is when an animal (herbivore) feeds on plants, causing plant damage or death.

Symbiosis is an interspecific interaction where two species have a permanent or long-lasting, physically close association. **Parasitism** occurs when one species withdraws nutrients from a living member of another species (the host), by living in or on the host, who is in turn harmed from the interaction. A **brood parasite** steals parental care and depletes the resources of its host by tricking another species into raising its young.

In **commensalism**, one species benefits and another is unaffected, while **mutualism** is an interspecific interaction that benefits both species.

Section 15.3 Changing Communities

Ecological succession is the process by which the species present in an area change as organisms alter their habitat. **Pioneer species** are the first to colonize new habitats. **Primary succession** starts after a disturbance creates a soilless habitat. **Secondary succession** occurs after a disturbance where species are removed, but the soil remains, and leads to a mature forest, or **climax community**.

Keystone species have a disproportionately large impact on a community relative to their abundance. **Ecosystem engineers**

are species that physically alter the environment, impacting the entire community. **Exotic species** become established after being introduced into a non-native habitat. **Invasive species** are exotic species that disrupt the community structure in their new habitat, sometimes outcompeting native species.

Section 15.4 The Nature of Ecosystems

Ecosystems are communities plus the abiotic factors at a particular location. **Trophic levels** refer to the energy hierarchy within an ecosystem. An **energy pyramid** graphically represents the proportion of energy at each trophic level, starting with the **producers**, which capture and store energy from inorganic materials. **Consumers** obtain energy and nutrients by feeding on tissues, wastes, and remains of producers and other consumers. **Primary consumers** eat algae or plants, **secondary consumers** eat primary consumers, and **tertiary consumers** are the top predators in an ecosystem.

Some animals consume **detritus**, the organic waste from organisms. **Decomposers** are organisms that recycle leftover organic matter into inorganic building blocks that can be used by primary producers to make energy.

A **food chain** is a sequence of steps by which the energy initially captured by producers moves to higher trophic levels. A **food web** consists of cross-connected food chains. **Bioaccumulation** occurs when a toxin or **microplastic** (microplastics) become increasingly concentrated as it moves up a food chain.

Section 15.5 Climate and the Distribution of Biomes

The **biosphere** includes all regions on Earth where life can exist. **Climate** refers to the average weather conditions in a region over a long period of time.

Solar energy is the light energy from the sun, and it varies by **latitude** (distance from the equator), being most intense at the equator and least intense at the poles of the Earth. **Ocean currents** are the continuous movement of seawater in a directed manner due to wind, temperature, salinity, landmass location, and the Earth's rotation.

A **biome** is a large-scale area with similar physical conditions that support distinctive biological communities adapted to those conditions. **Terrestrial biomes** are large areas of land defined by their average rainfall and temperature.

A **tropical rain forest** has abundant rain, warm temperatures, and consistent day length, and is dominated by evergreen broad-leaved trees. A **temperate deciduous forest** is dominated by broad-leaved trees that shed their leaves and become dormant in winter. **Boreal forests** experience cold, dry winters and are characterized by evergreen conifers that tolerate dry soil and cold.

The **temperate grassland** biome, called **prairie** in North America, supports perennial grasses and nonwoody plants adapted to hot summers, cold winters, and moderate to low precipitation. Warm temperatures and a distinct rainy season define **savannas**, tropical grasslands with scattered shrubs and trees, located between tropical forests and deserts.

Chaparral is characterized by drought-resistant, fire-adapted shrubs, with mild winters; moderate rain; and hot, dry summers. **Deserts** have large daily temperature ranges and the lowest levels of annual precipitation of any biome, supporting heat-adapted plants. **Tundra** lies between the polar ice cap and the belts of boreal forests in the Northern Hemisphere, and is covered by a layer of **permafrost**. The **ice** biome is the coldest biome, covered in ice year-round, but nonetheless supports a variety of organisms.

Aquatic biomes are water-based environments determined by temperature, salinity, sunlight, dissolved oxygen, and nutrient concentration, and they support different communities at different depths and distances from shore. A **stream** is moving freshwater that originates from runoff or melting snow or ice, growing and merging into a **river** as it flows downhill. A **lake** is a body of standing freshwater.

An **estuary** is a coastal region where seawater mixes with the freshwater from a river, supporting plants that can withstand high salinity and changing water levels. The **intertidal zone** along rocky or sandy shores can be above or below the water's surface, depending on the tides. It supports organisms adapted to tidal changes and the force of waves. A **coral reef** is a species-rich ecosystem often found in warm, shallow, tropical seas with abundant sunlight. The **open ocean** supports photosynthetic and microscopic primary producers where sunlight is prevalent. Below that, organisms live in darkness, sustained by organic material that drifts down from above.

Section 15.6 Biogeochemical Cycles

A **biogeochemical cycle** is a cycle in which water or an essential element—such as phosphorus, nitrogen, or carbon—moves from environmental reservoirs, through the living components of an ecosystem, and back to environmental reservoirs.

Groundwater is freshwater buried under Earth's surface. **Aquifers** are pockets of groundwater held within rock formations. The **water cycle** moves water from oceans and into the atmosphere through evaporation, then into land and freshwater ecosystems, and back into the oceans.

The **phosphorus cycle** moves phosphate ions among rocks, soil, and water, and into and out of food webs.

The **nitrogen cycle** moves nitrogen from the atmosphere, through soil and water reservoirs, and through food webs. **Nitrogen fixation** is the process by which certain bacteria break down nitrogen gas to form ammonia, a form of nitrogen that producers can use. **Nitrification** is the process by which bacteria and archaea in the soil and ocean convert ammonia into nitrate, another useable form of nitrogen, while **denitrification** is the conversion of nitrate back into nitrogen gas. **Eutrophication** occurs when water contains too much nitrogen and/or phosphorus, leading to algal blooms that deplete the water of dissolved oxygen.

The **carbon cycle** moves carbon among rocks, water, and the atmosphere, and into and out of food webs.

Practice It!

Retrieval practice is the process of pulling information from your long-term memory and using it. Here are some questions to help with your retrieval practice.

Self-Quiz

1. A community consists of species ______. (LO 15-1)
   a. that share a habitat
   b. that do not compete
   c. that share genes
   d. with the same niche

2. Imagine two ponds. Pond A is home to five species of fish and three species of frogs, and Pond B is home to three species of fish and one species of frog. Which pond has a higher species richness? (LO 15-2)
   a. Pond A
   b. Pond B
   c. Pond A and B have the same species richness.
   d. There is not enough information given to answer this question.

3. The role a species plays in its environment is called its ______. (LO 15-3)
   a. habitat
   b. community
   c. niche
   d. competition style

4. When individuals from different species eat the same food in a habitat, they exhibit ______. (LO 15-4)
   a. interspecific competition
   b. intraspecific competition
   c. predation
   d. mutualism

5. One species of snake gradually disappears from an area due to competition with another species of snake. This is an example of ______. (LO 15-4)
   a. parasitism
   b. a niche
   c. the principle of competitive exclusion
   d. community structure

6. A harmless fly has the yellow and black stripes of a stinging wasp. This is an example of ______. (LO 15-4)
   a. herbivory
   b. mimicry
   c. camouflage
   d. predation

7. A mosquito sucks your blood. This is an example of ______. (LO 15-5)
   a. mutualism
   b. competition
   c. predation
   d. parasitism

8. A bee pollinates a flower while eating floral nectar. This is an example of ______. (LO 15-5)
   a. mutualism
   b. competition
   c. predation
   d. parasitism

9. The establishment of a biological community on a newly formed volcanic island is an example of ______. (LO 15-6)
   a. competitive exclusion
   b. resource partitioning
   c. primary succession
   d. secondary succession

10. A species that has a disproportionally large impact on its community relative to its abundance is a(n) ______ species. (LO 15-7)
    a. exotic
    b. keystone
    c. exotic
    d. mutualistic

11. Zebra mussels were accidentally introduced into the Great Lakes when a ship discharged ballast water. They have now spread throughout the Great Lakes, becoming the predominant species in many areas. Zebra mussels are an example of a(n) ______ species. (LO 15-8)
    a. predatory
    b. keystone
    c. invasive
    d. mutualistic

12. The trophic level with the most energy, found at the bottom of the energy pyramid, consists of ______. (LO 15-9)
    a. producers
    b. consumers
    c. decomposers
    d. predators

13. An organism that degrades organic wastes and remains into inorganic forms that producers can use is a ______. (LO 15-10)

a. producer
b. consumer
c. decomposer
d. predator

14. A food web is defined as a system made up of multiple, cross-connected ______. (LO 15-11)

a. energy pyramids
b. food chains
c. carbon cycles
d. detritivores

15. ______ occurs over the course of a predator's lifetime as it consumes toxic substances that were in its prey. (LO 15-12)

a. Herbivory
b. Resource partitioning
c. Bioaccumulation
d. Brood parasitism

16. The amount of solar radiation reaching Earth is ______. (LO 15-13)

a. greatest at the equator
b. greatest at the north pole
c. the same at all points on the globe
d. greatest at the south pole

17. Climate is affected by all of the following **except** ______. (LO 15-13)

a. solar energy
b. latitude
c. precipitation
d. community structure

18. ______ refers to the average weather conditions in a region over a long period of time, usually decades. (LO 15-13)

a. Biome
b. Precipitation
c. Climate
d. Microbiome

19. Which of the following is NOT used to classify a biome? (LO 15-14)

a. rainfall
b. temperature
c. weather
d. salinity

20. Which biome is a tropical grassland? (LO 15-15)

a. chaparral
b. tundra
c. savanna
d. desert

21. Plants in a ______, also called a prairie, are adapted to grazing. (LO 15-15)

a. desert
b. boreal forest
c. tropical rain forest
d. grassland

22. Which of the following terrestrial biomes has the lowest annual precipitation? (LO 15-15)

a. temperate forest
b. temperature grassland
c. chaparral
d. desert

23. The aquatic biome where freshwater and saltwater meet is called a(n)______. (LO 15-16)

a. coral reef
b. intertidal zone
c. open ocean
d. estuary

24. In the ______ cycle, this resource is moved from the ocean and into the atmosphere by evaporation, returning to it by precipitation. (LO 15-17)

a. water
b. phosphorus
c. nitrogen
d. carbon

25. Addition of excessive ______ to water increases algal blooms. (LO 15-18)

a. carbon
b. mercury
c. lead
d. phosphate

Critical Thinking

1. The viceroy butterfly has a coloration pattern that mimics that of the monarch butterfly, whose toxins deter predators. Researchers have found that in such mimicry systems, mimics benefit most when they are rarer in the habitat than the toxic species that they mimic. Can you explain why? (LO 15-4)

2. With antibiotic resistance on the rise, researchers are looking for ways to reduce the use of antibiotics, where instead of giving cattle antibiotics, the cattle are given feed containing beneficial bacteria. Explain why giving cows beneficial bacteria can help to fight antibiotic resistance. (LO 15-4)

3. In a seaside community in New Jersey, the U.S. Fish and Wildlife Service suggested trapping and euthanizing feral cats, which are domestic cats living in the wild. The goal was to protect piping plovers (*Charadrius melodus*), which are endangered shorebirds that nest on the town's beaches. Many residents were angered by the proposal, arguing that the cats have as much right to be there as any bird. Do you agree? Why or why not? What is another possible solution? (LO 15-8)

4. Two different seals of the same species have different levels of a toxin in their bodies. If they are the same species, what could explain the difference? (LO 15-12)

5. Scientists study air bubbles trapped in ancient glacial ice to determine the concentrations of nitrogen and carbon dioxide gas in the atmosphere at the time the air bubble formed and how those concentrations are different from current levels. However, air bubbles in glacial ice cannot provide information about changes to phosphorus concentrations. Explain why, and propose an alternative method to study how the amount of phosphorus in a region has changed over time. (LO 15-17)

Credit: iStock.com/WendyOlsenPhotography

16

Reasons to Hope: Conservation Biology

Chapter Outline:

16.1 The Impacts of Pollution

16.2 The Impacts of Climate Change

16.3 Biodiversity Protection and Restoration

16.4 Allies for the Future of Conservation

The Rebirth of the Elwha River

Olympic National Park in Washington is a unique place to visit. One can visit the mountains, a rainforest, or the coast—all in one park! The Elwha River runs from the mountains to the Pacific Ocean and serves as an important "highway" for bringing nutrients from the mountains to the ocean and allowing passage for trout and salmon, which are fish that live in the ocean but swim inland to breed.

However, in the early 1900s, two hydroelectric dams were built along the river to generate electricity for local communities. With the arrival of the dams, nutrient flow stopped, fish could no longer breed, and areas of cultural significance to the Lower Elwha Klallam Tribe were now underwater. As we've learned throughout this unit, species and the components of their environments are all interwoven with one another. The building of the hydroelectric dams had a catastrophic impact on the ecosystem of the Olympic Peninsula. Riverbanks eroded. The fishing industry was decimated. The nutrients that the fish carried up the river from the ocean were lost, impacting the overall health of the river and the surrounding ecosystem. Estuary and intertidal habitat along the coast disappeared.

Then, in 1992, the Elwha River Ecosystem and Fisheries Restoration Act passed. After 20 years of planning, including "before" surveys of the life around the Elwha River, the Elwha Dam and Glines Canyon Dam were removed. Although full restoration of the ecosystem around the Elwha dam will likely take a generation to complete, in the first decade after dam removal, the river demonstrated a remarkable rebound. Several species of fish populations are increasing, and natural sediment flows have facilitated the rebound of the near-shore ecosystem at the mouth of the Elwha River. What can we learn from this story about the impacts that humans have on the environment…and critically, how to plan and implement future restoration efforts?

The Elwha River flows through Olympic National Park in Washington. The removal of dams along the Elwha River in 2012 and 2014 led to a resurgence of wildlife in and around the river and its watershed—and is an excellent example of how we can work together to conserve natural resources, for the benefit of all.

In this chapter, we'll be learning about *conservation biology*, with a focus not only on the environmental issues we are facing as a society, but also on the paths we can take to address those issues. We'll begin with a discussion of *pollution*, including a discussion of water, plastic, and air pollution. We'll also explore how greenhouse gas pollution is driving *climate change*, or long-term shifts in temperatures and weather patterns. We'll discuss *biodiversity* and its value, not only to the health of the planet, but also the direct benefits it provides to us. Then, we'll discuss the relationship between biodiversity and *habitats*, with stories about biodiversity and habitat recovery. Finally, we'll bring the entire chapter together with a discussion of what you can do to help conserve our natural resources.

Reason for Hope While many stories about pollution, habitat destruction, and loss of biodiversity are sobering, there are reasons to hope. Like the Lorax tells us in the classic Dr. Seuss book *The Lorax*, caring about conservation and *stewardship* of natural resources is the first step to creating a better world for all. A major theme of this chapter is looking at past examples of conservation and environmental issues, and how these problems were addressed through the actions and choices of individual people, just like you. Watch for the butterfly icon and Reason for Hope headings, such as shown at the beginning of this paragraph, throughout the chapter to see examples of how people have solved environmental issues before. Look for themes such as the relationship between science and policy, the impact of our everyday choices, and community science efforts. As you read this chapter, remember that *you can make a difference*…if you choose to do so.

Preview It!

What is a natural area that you'd like to save or preserve for either yourself or future generations? What do you think are the first steps that you can take to learn more about preserving that natural area? What do you want to learn about the conservation challenges we face as a society, as well as their potential solutions?

Figure 16.1 Characters from the Captain Planet cartoon series encouraged children to care about environmental issues

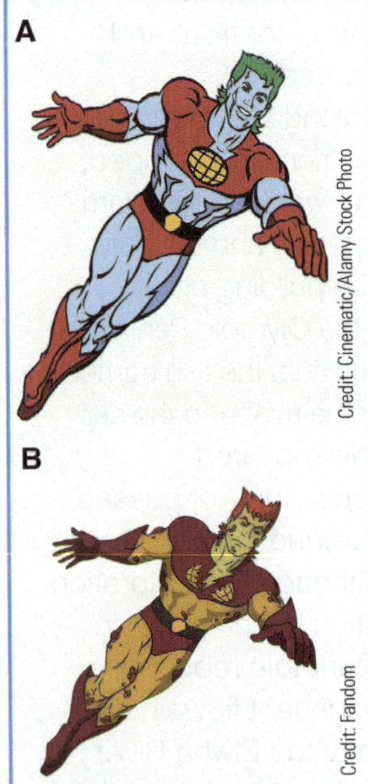

A. Captain Planet and his nemesis,
B. Captain Pollution.

16.1 The Impacts of Pollution

Learning Objectives

16-1 Identify what a pollutant is and how it impacts the environment.

16-2 Describe the types and sources of water pollution.

16-3 Identify concerns surrounding plastic pollution.

16-4 Describe the types of air pollution and their sources.

16-5 Explain the greenhouse effect and its relationship to global climate change.

16-6 Describe the evidence that atmospheric carbon dioxide is increasing due to human activity.

16-7 Identify examples of success stories for mitigating pollution.

In the early 1990s, a new cartoon superhero appeared to entertain children: Captain Planet. The general premise is that there are five Planeteers chosen by Gaia (the goddess of Earth and Mother of Life in Greek mythology, or the personification of Earth) who are tasked with protecting the planet from environmental disasters while also educating others about conservation. The villains in the series are termed Eco-villains, and each represents a threat to the environment, like greed, pollution, and destruction. When the Planeteers can't resolve a situation on their own, they can combine their individual superpowers to summon Captain Planet for help (**Figure 16.1A**). In the guise of "edutainment," each episode of Captain Planet ended with a connection of the episode's content to an environmental, social, or political issue, and how the viewer could be part of the "solution" and not the "pollution."

Now, like other superheroes, Captain Planet has a weakness: pollution. In fact, Captain Planet's nemesis is Captain Pollution (Figure 16.1B)! Unlike the usual tagline of being part of the solution and not the pollution, Captain Pollution's motto is instead "The polluting power is yours!!!"

Although Captain Planet may be a vintage cartoon series, the issues highlighted by the series are just as relevant today as they were decades ago. In fact, the threats of pollution—particularly carbon dioxide and plastic pollution—are worse today than they were in 1990, when the cartoon first aired.

? Did You Know?

The Captain Planet Foundation, which directed revenues from the show's merchandise to environmental projects, still exists today.

In this section, we'll explore **pollution**, which occurs when harmful materials enter the environment. These materials are collectively called pollutants. A **pollutant** is a natural or artificial substance that is released into soil, air, or water in greater than natural amounts. Pollutants disrupt the physiological processes of organisms that evolved in their absence, or that are adapted to lower levels of them. In each section, we will focus on different kinds of pollutants, including those that cause water, plastic, and air pollution. Keep an eye out for examples of how to be part of the *solution instead of the pollution*!

? Did You Know?

London's Thames River was so polluted by raw sewage in the 1800s that many people were dying from cholera, a disease associated with drinking contaminated water. Thames water pollution came to a head in the summer of 1858 with the Great Stink, when the hot weather exacerbated the smell of the sewage and spurred the development of a new sewer system.

Water Pollution

Water is essential for life—one of the earliest concepts we learned about back in Chapter 2. While water itself isn't a pollutant, various pollutants can contaminate water and negatively impact life. For example, in Chapter 15, we discussed how agricultural runoff can lead to **eutrophication**, where bodies of water with too much nitrogen or phosphorus runoff can lead to excessive plant and/or algae growth, depleting oxygen content in the water and harming aquatic life.

Water pollutants can be grouped into two general categories: inorganic and organic. **Inorganic water pollutants** do not contain carbon atoms and include a variety of different elements, including nitrogen, phosphorus, mercury, lead, and copper, and/or poisons like cyanide. In contrast, **organic water pollutants** include carbon-based compounds, such as sewage, oil, pharmaceutical drugs, and persistent organic pollutants, or *forever chemicals*. These can have many different impacts on humans and the environment. For example, we will explore the impact of forever chemicals like per- and polyfluoroalkyl substances (PFAS) as well as bisphenol A (BPA) on human fertility in Chapter 17.

Human pharmaceuticals excreted in urine can enter the environment through wastewater treatment plants, where they may not be fully removed. Once in aquatic ecosystems, these substances can have serious ecological impacts. For example, the presence of antibiotics can promote the evolution of antibiotic-resistant bacteria (Chapter 10), while hormones from birth control pills can cause the feminization of male fish, impairing their ability to reproduce and potentially leading to population declines.

Reason for Hope Although the sources and types of water pollution and their impacts are staggaring to consider—especially when we consider the necessity of water for life—there are examples where water pollution was successfully addressed. Let's take the Cuyahoga River in Ohio as an example. For over a century, the Cuyahoga River was a dumping ground for various manufacturing companies. As a result, the river was a bubbling cesspool covered with oil slicks. The water was so polluted that it *caught fire* not once, but *multiple times* (Figure 16.2A).

The tipping point for change came in 1969, when a fire received national attention. The burning river became a symbol of rampant pollution and was one of the examples used to support the passage of the National Environmental Policy Act in 1969, which is the precursor to today's Environmental Protection Agency, and the Clean Water Act in 1972. Today, the Cuyahoga River is a tourist destination (Figure 16.2B). In 2019, people in northeast Ohio celebrated the 50th anniversary of the fire in an event called the Xtinguish Torch Fest, which recognized the cleanup of the river and revitalization of communities along the Cuyahoga.

The Flint Water Crisis began in 2014 when a decision was made to use water from the Flint River as the city's water source, instead of getting it from the Detroit water system, to save money. This resulted in widespread lead poisoning of the citizens of Flint because of lead leaching from old pipes when the heavily polluted water from the Flint River was piped through.

Figure 16.2 **The Cuyahoga River before and after restoration**

A

Credit: "Photo by Frank Aleksandrov from the Cleveland Press Collections, courtesy of the Michael Schwartz Library Special Collections, Cleveland State University"

B

Credit: Kenneth Sponsler/Alamy Stock Photo

A. Firefighters battle a burning Cuyahoga River in 1952.
B. Paddlers enjoying the Cuyahoga River in 2022.

Plastic Pollution

In the Pacific Ocean, about halfway between Hawaii and California, is a **garbage patch**, an area full of marine debris created by ocean currents (Chapter 15), called the Great Pacific Garbage Patch. Over 99% of the garbage patch is plastic. This garbage patch covers an area twice the size of Texas and extends deep into the ocean. To put that into context, if it were possible to drive across the Great Pacific Garbage Patch—which is approximately the distance between Denver, Colorado, and Dallas, Texas—assuming travel at typical highway speeds, *it would take 12 hours.*

> **? Did You Know?**
>
> Plastic straws are one of the largest components of plastic pollution in the oceans. During a 5-year period, people picked up nearly 7.5 million straws during beach cleanups in the United States.

What's perhaps more sobering is that this is only one of the five main garbage patches in the world's oceans (Figure 16.3). Historically, humans buried unwanted material in the ground or dumped it out at sea. The United States no longer dumps its trash at sea, but plastic and other

Figure 16.3 **The world's five major ocean gyres, which create garbage patches at those locations**

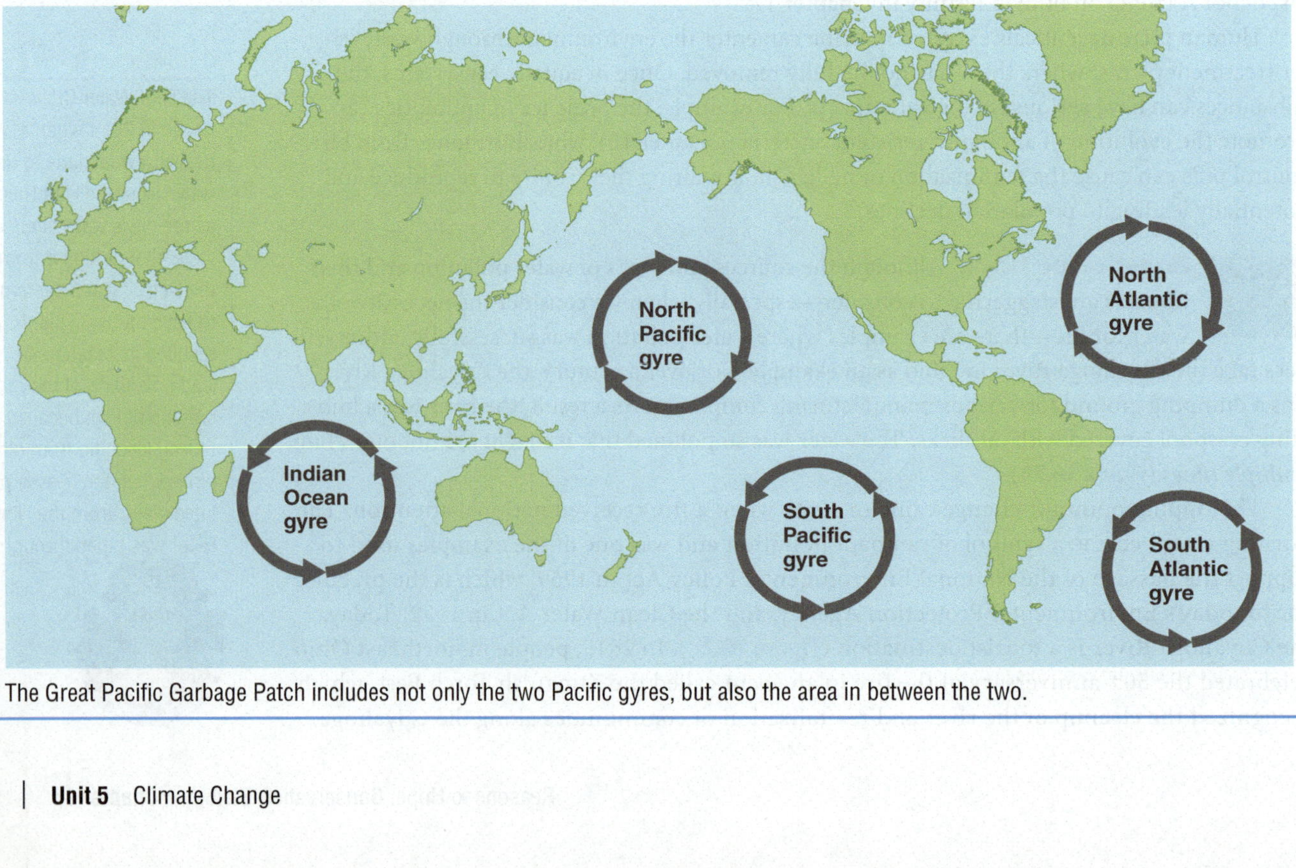

The Great Pacific Garbage Patch includes not only the two Pacific gyres, but also the area in between the two.

Biology for Life

The Quandary of Plastics

Look around you. Plastic is everywhere, from the case on your phone to the seat you might be sitting on to your sunglasses. So unfortunately, it's not so simple to say, "Well, let's just get rid of plastic!" Plastic is an important material that is used in a wide variety of applications, from medical and industrial applications to consumer products. These include single-use plastics, like those used in disposable diapers or food packaging.

Figure 16.4 Mr. Trash wheel intercepting trash

Mr. Trash Wheel was the first trash interceptor developed. The googly eyes and catchy name were created to spark community interest—and turn a messy problem into a must-see attraction in Baltimore.

So, we're faced with two quandaries: (1) What can we use as alternatives to plastics, and (2) what do we do with the plastic waste we already have? One possible alternative to plastics is using *bioplastics*, which are made using biomass such as plants, as opposed to fossil fuels that are used for conventional plastics. You might be familiar with bioplastics from everything from to-go containers to plant-based trash bags. And while bioplastics are better for the environment than conventional plastics because they are made from renewable resources, they do not break down effectively in landfills, as they need aerobic (oxygen) conditions that are lacking in anaerobic landfills. We will return to bioplastics and circular economies at the end of this chapter.

Which brings us to the question of what to do with the plastic waste that we already have. Unfortunately, at the writing of this text, there is no silver bullet for how to deal with plastic waste, but there are some promising developments. For example, in 2001, Japanese scientists discovered plastic-eating bacteria. To combat plastic pollution in the ocean, the scientists isolated the genes that these bacteria use to digest plastic and inserted them into bacteria that live in the ocean. Although this technology is still in development, both naturally occurring and transgenic bacteria (Chapter 9) could be possible sources of bioremediation to break down plastic.

Innovative solutions like Mr. Trash Wheel (Figure 16.4) can also help specifically with mitigating plastic pollution in the ocean. Mr. Trash Wheel was created by environmental scientist John Kellet in partnership with the Waterfront Partnership of Baltimore and is at turns a social media darling, a tourist attraction, and a means of keeping trash from flowing into the ocean. Mr. Trash Wheel and his friends, Professor Trash Wheel and Captain Trash Wheel, live in various locations around Baltimore, Maryland, catching floating trash in the harbor using containment booms. A solar- and water-powered rake lifts the trash out of the water and onto a conveyer belt, which moves the trash into a dumpster.

Unfortunately, this solution isn't perfect. Current sorting technologies are unable to pull plastic waste from the other miscellaneous debris (such as mattresses, tires, and trees) that the trash wheels pick up. For now, the waste the trash wheels eat up is incinerated to generate electricity.

While we have yet to find a meaningful way to deal with plastic pollution, there are still some things that we can do, regardless of whether we live near the ocean. Each of us can help by consuming less and reusing more, thus reducing demand for manufacturers to produce new plastic.

1. Can you think of places in your daily life where you can substitute single-use plastics for reusable items? What about non-plastic alternatives to single-use items?
2. Is it worth keeping the oceans clean if it means that the trash is instead added to landfills? What are some potential things that can be done to manage the amount of trash in the waste stream?

Did You Know?

Disposable diapers take around 550 years to decompose in a landfill… and the United States alone throws away more than 18 billion disposable diapers each year. Cloth diapers can be used repeatedly over many years and for multiple children, and are often made of biodegradable materials. The problem is accessibility—many families can't afford to spend $500 upfront on reusable diapers, but will instead buy a package of diapers each week for $30, ultimately spending thousands of dollars per child.

Big Ideas in Biology

Evolution

Where did plastic-eating bacteria come from? A single bacterium likely had mutations in genes that allowed it to consume plastic, giving it a selective advantage that was then passed on to its offspring, eventually leading to a new strain of bacteria.

? Did You Know?

Video taken from the bottom of the ocean's deepest trench, the Mariana Trench in the western Pacific Ocean—which records a maximum depth of about 36,000 feet below sea level, or roughly equivalent to the height at which commercial airplanes fly above sea level—showed a plastic grocery bag and a candy wrapper, highlighting the extent of plastic pollution in the world's oceans.

Figure 16.5 Death by plastic

Credit: NOAA ONMS

When a scientist dissected this recently deceased Laysan albatross (*Phoebastria immutabilis*) chick, they found more than 300 pieces of plastic in its body. The chick died when one piece punctured its gut wall.

garbage still enters coastal waters worldwide…and this has led to a massive plastic pollution problem in the oceans.

And plastic isn't just in the oceans…there are massive amounts of plastic waste on land as well. The United Nations estimates that only about 9% of all plastic produced worldwide is recycled…and 350 million tons of new plastic is created each year. The issue compounds itself further when we look at recycling. Unlike glass, plastic cannot be recycled indefinitely, and the quality of the plastic decreases with each new use. Plastic recycling is also not straightforward—there are often labels, food remains, or other materials that complicate the recycling process. There are also many kinds of plastics, which cannot be recycled together. Ultimately, recycling plastic is a short-term solution to the problem—eventually, plastic cannot be recycled further and will ultimately pollute land or oceans, or will end up in a landfill. We return to solutions for single-use plastics in Section 16.4.

Plastic waste has an impact on our health and the health of other organisms. Plastics in landfills can leech dangerous chemicals into our water supplies. Waste plastic that enters the ocean poses a threat to marine life. Many marine animals mistake pieces of plastic for food, sometimes with deadly results (**Figure 16.5**). Marine species that filter food from the water take in microscopic bits of plastic along with plankton. This microplastic can then pass up the food chain (bioaccumulation; see Chapter 15) to other organisms and then on to us.

Remember microplastics from Chapter 15? The plastic wrappings on laundry and dishwasher pods also contribute to plastic pollution, as do plastic fibers shedding from your clothes when they are washed.

Figure 16.6 An advertisement raising awareness about radon pollution

A poster advertising National Radon Awareness Month in Colorado.

Credit: 2025 State of Colorado (https://cdphe.colorado.gov/marketing-campaigns)

Air Pollution

Harmful contamination of air, either indoors or outdoors, is called air pollution. There are many kinds of air pollution. For example, some specific gases are pollutants, such as greenhouse gases like carbon dioxide and methane. We will discuss *greenhouse gases* and their relationship to climate change later, in Section 16.2. Other gases, like radon, occur naturally. **Radon** is produced when radioactive (Chapter 2) metals in the soil break down. This can result in high levels of radon in basements, where prolonged exposure can lead to lung cancer. States and cities often have campaigns to increase awareness about radon to encourage home radon testing and remediation when high levels are detected, to protect the health of the household (**Figure 16.6**).

Tiny particles in the air, or **particulates**, are the source of particulate pollution, either inside or outside of the home. For example, inhaling lead-containing dust or fumes (such as in old buildings, which may contain lead-based paints) is the most common cause of lead poisoning, which can significantly impact mental and physical development in children. Particulate pollution can also result from smoke. Exposure to wildfire smoke can have a significant impact on lung function.

Sometimes, chemical reactions occur in the atmosphere to generate dangerous pollutants. For example, **ground-level ozone** (O_3) forms when gases like carbon monoxide and nitrogen oxides (which are both part of car exhaust) react with sunlight. Ground-level ozone is a component of smog. **Smog**, or "smoke fog" (Figure 16.7), is a mixture of ground-level ozone and particulate matter. In addition to the potential health impacts of smog, it can reduce visibility and cause traffic accidents and flight cancellations.

Figure 16.7 Smog over Los Angeles, California

Credit: iStock.com/steinphoto

Point of Confusion: Although all ozone is O_3, only ground-level ozone is considered a pollutant. Ozone gas in the atmosphere is important for protecting us from harmful ultraviolet radiation. We'll return to the ozone layer later in this section.

Air pollutants released by burning coal and other fossil fuels can combine with atmospheric water vapor to form **acid rain**. Acid rain is 10× more acidic than normal rain (Chapter 2) and forms when sulfur dioxide and nitrogen oxides react with water and oxygen to form sulfuric and nitric acids. Since pollutants are distributed by the winds, acid rain can affect regions hundreds of miles from the source of the pollution. Acid rain can damage buildings (Figure 16.8A) as well as forests (Figure 16.8B), where it can burn tree leaves. As a result, trees become malnourished and are more susceptible to disease.

? Did You Know?

The World Health Organization (WHO) estimates that air pollution is responsible for 6.7 million premature deaths each year.

Reason for Hope In the United States, the incidence of acid rain peaked during the 1970s. In the 1990s, new legislation was introduced to alter the Clean Air Act to include a section on acid rain. The legislation sets a cap on the amount of sulfur dioxide that can be emitted throughout the United States. Since the new regulations were put into place in 1995, reduced sulfur dioxide emissions have greatly reduced the acidity of precipitation (Figure 16.9).

Does your community have ozone action days in the summer? These are days that are expected to have ground-level ozone levels that exceed the safety limits outlined by the Environmental Protection Agency. People are encouraged to either stay home during the hottest parts of the day, when the most ozone is likely to be produced, and/or limit driving, which produces the raw materials needed for ozone to form.

The Greenhouse Effect and Greenhouse Gas Pollution

Have you ever sat in a car on a sunny day and noticed how warm the interior of the car can become, even if the air temperature outside is cold and the car's heater is not on? The inside of your car warms up because of the solar energy coming in through the windows, which becomes trapped inside the car and warms the air, much like how a greenhouse works to cultivate plants.

Figure 16.8 Impacts of acid rain

A. A stone carving eroded by acid rain.
B. Acid rain-damaged trees in North Carolina.

Figure 16.9 **Changing policies lead to less acid rain in the United States**

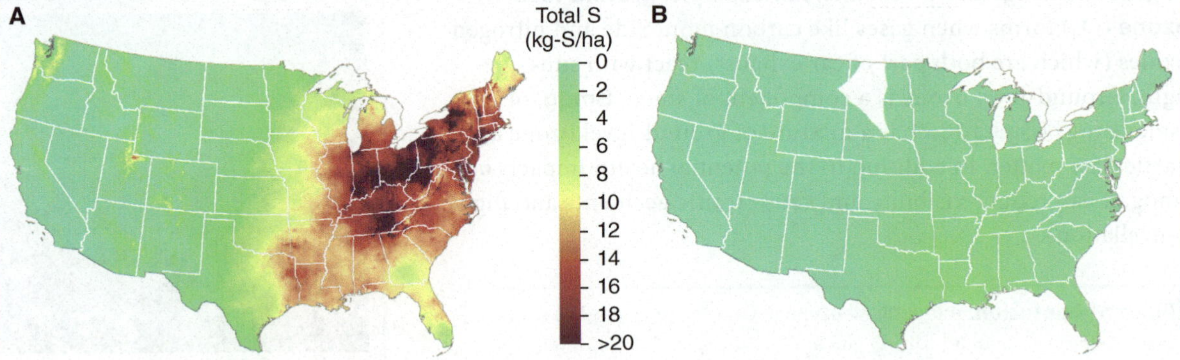

The maps show annual amounts of acid rainfall, where green indicates rainfall with a normal pH, and yellow and red indicate decreasing pH and therefore increasing acidity.

A. Levels of acid rain in the early 1990s and

B. Approximately 30 years later.

Figure based on: EPA. (2020, March 27). Acid Rain Program Results. US EPA. https://www.epa.gov/acidrain/acid-rain-program-results

Note that the white spot over North and South Dakota in Figure 16.9B represents locations that did not have enough data to calculate a representative annual value. Missing data can happen due to drought, sample collection concerns, or equipment malfunction. Given the logistics involved in collecting data from across the country, the white spot highlights the vigilance that scientists use in presenting accurate data, rather than falsifying it.

When we think about the greenhouse effect and Earth, the **greenhouse effect** occurs when heat energy is trapped by gases in our atmosphere (Figure 16.10). **Greenhouse gases** are atmospheric gases, including methane and carbon dioxide (CO_2), that keep Earth warm by absorbing and re-radiating heat. When we consider greenhouse gases, we can make an analogy to the blankets on your bed on a cold winter night. If you didn't have any blankets, heat would escape away from your body as you tried to sleep, and you would be cold. Add a blanket or two, and you could sleep comfortably through the night.

Now think of the layers of blankets in terms of greenhouse gases. We need some greenhouse gases in the atmosphere to sustain life on Earth. The problem is, now there is too much greenhouse gas in the atmosphere (see Figure 16.12), which means the surface of the planet is warming. It's as if you have 20 blankets on your bed and are adding more. Each additional blanket (more greenhouse gas) would trap more heat as you slept, reflecting it back to you and making it too warm to sleep.

Reflected heat is ultimately what's driving up global temperatures and fueling **climate change**, or lasting changes to average temperature and weather patterns. Recall from Chapter 15 that *climate* refers to the average weather conditions in a region over a long period of time. Before we get into the impacts of climate change on the biosphere in Section 16.2, we will first look more closely at greenhouse gas pollution.

Methane pollution. Have you ever heard that cow farts are contributing to climate change? Or that people decide to adopt either a vegetarian or a vegan diet to mitigate the impact of climate change? What does cow flatulence and someone's diet have to do with greenhouse gases anyway? If you recall the definition of a pollutant from earlier in this section, where it is defined as an excess amount of something—naturally occurring or not—that is harmful to the environment, **methane pollution** occurs when excessive amounts of methane gas enter the atmosphere due to human activities.

While most greenhouse gas emissions come from the burning of fossil fuels and carbon dioxide emissions (around 64%), methane (CH_4; a component of natural gas) from sources like livestock manure and landfills is responsible for about 11% of all greenhouse gas emissions. Now, that may seem like a drop in the bucket, until you consider that methane is a 28 × more potent greenhouse gas than carbon dioxide. If we go back to our blanket analogy, if carbon dioxide is like a lightweight quilt, methane is like a heavy wool blanket.

So, what do landfills have to do with methane? Trash is typically placed into plastic trash bags. Plastic blocks the flow of oxygen. These plastic bags then pile up in landfills, creating an oxygen-free (anaerobic) environment (Figure 16.11A). Without oxygen, the bacteria that usually

Figure 16.10 **The greenhouse effect**

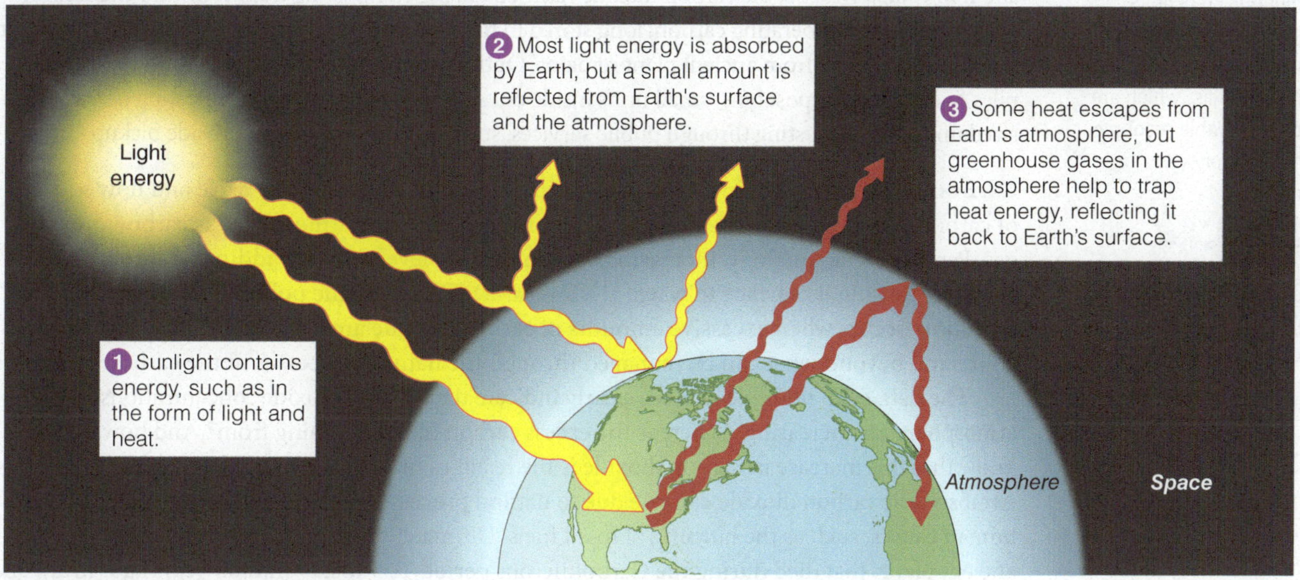

The yellow arrows represent light energy, and the red arrows represent heat. Note that the thickness of an arrow represents its relative abundance.

break down organic food waste cannot survive. Instead, a group of archaea (Chapter 5) called **methanogens** break down the organic waste using **methanogenesis**. In methanogenesis, hydrogen gas (H_2) and carbon dioxide from decaying organic matter are used to produce methane and water. The methane is sometimes captured for use in other applications, but much of it goes into the atmosphere, contributing to climate change.

Reason for Hope The good news is there's a relatively easy and accessible way to reduce methane emissions: composting. **Composting** is the process of breaking down organic material like food scraps in the presence of oxygen by decomposers such as bacteria and fungi (Figure 16.11B). You may be wondering what's so special about composting, since doesn't organic waste break down in landfills? The key is that composting uses oxygen. The result of composting is a type of rich soil called **compost**, which one can add to their garden to create healthy soil that will nourish their plants (Chapter 18).

Figure 16.11 **Comparison of greenhouse gases released in different settings**

Credit: EyeEm Mobile GmbH/getty images

Credit: Evan Lorne/Shutterstock.com

A. Under anaerobic conditions, such as in landfills, methanogens, a type of archaea, break down organic material and produce methane.

B. Under aerobic conditions, bacteria and fungi break down organic waste and produce carbon dioxide.

In addition to reducing greenhouse gas emissions, composting also minimizes the burden on public landfills, which are not a renewable resource and cost money to maintain.

Composting is essentially how nature recycles and is part of the biogeochemical cycles we learned about in Chapter 15. So, the carbon dioxide released during composting doesn't disturb the carbon cycle—unlike how liberating carbon, long stored in the ground in the form of fossil fuels, adds carbon into the atmosphere from ancient plant remains. Options for composting include backyard compost piles or rotating compost bins; small units inside one's home, such as worm composting or trash can-sized units; or composting through public services, such as drop-off sites or curbside pickup.

Carbon dioxide pollution. In Chapter 15, we discussed the carbon cycle and how sources of carbon (fossil fuels) are being burned and released into the atmosphere as carbon dioxide gas. Per our earlier analogy, increasing carbon dioxide levels are like adding additional blankets to Earth, causing its surface to warm (Figure 16.12). Like methane pollution, **carbon dioxide pollution** occurs when excessive amounts of carbon dioxide are released into the atmosphere, above and beyond what can be managed through the natural processes of the carbon cycle.

The data from multiple sources clearly indicate that carbon dioxide concentrations in the atmosphere are increasing. Where is the excess carbon dioxide coming from? And how do we know that this increase in carbon dioxide is the result of human activities like burning fossil fuels? Atmospheric carbon dioxide can rise due to natural processes, like volcanic eruptions, or due to human causes, such as the burning of fossil fuels. Ultimately, fossil fuels are derived from plants—ancient plants that died during the Carboniferous period, 359 to 299 million years ago (mya), and are buried deep in the ground (Figure 16.13A). Recall from Chapter 4 that plants take carbon dioxide from the atmosphere and use it to make glucose, which they then store in their tissues or use as energy. Then consumers like ourselves can, in turn, use the produced glucose for energy.

But plants are picky in the kind of CO_2 they prefer to use during photosynthesis. Recall from Chapter 2 that there are different isotopes of carbon. Plants prefer to use "light" carbon-12 (^{12}C) during photosynthesis, rather than "heavy" carbon-13 (^{13}C), which contains one extra neutron compared to ^{12}C. Volcanoes, on the other hand, spew both ^{12}C and ^{13}C into the atmosphere. So natural sources of carbon dioxide have a higher ratio of ^{13}C to ^{12}C (Figure 16.13B) than plant-based sources of CO_2 do.

Therefore, one can examine the ratio of CO_2 containing ^{13}C to ^{12}C to determine if the CO_2 in the atmosphere is derived from fossil fuels (which are a plant-based source) or volcanoes (a non-plant-based source). Evidence strongly suggests that the increased level of CO_2 in the atmosphere is due to humans burning fossil fuels, and is therefore a result of human activities rather than natural sources. Scientists have tracked the ratio of ^{13}C to ^{12}C since the 1990s, determining that while the amount of CO_2 in the atmosphere has increased over time (see the black data line in Figure 16.13), the amount of ^{13}C-containing CO_2 has steadily declined (Figure 16.14).

? Did You Know?

While most people think about road transportation as a major source of carbon dioxide emissions, the largest culprit is residential buildings, due to the use of coal-burning power plants for energy.

Figure 16.12 **Atmospheric carbon dioxide levels and average temperature changes**

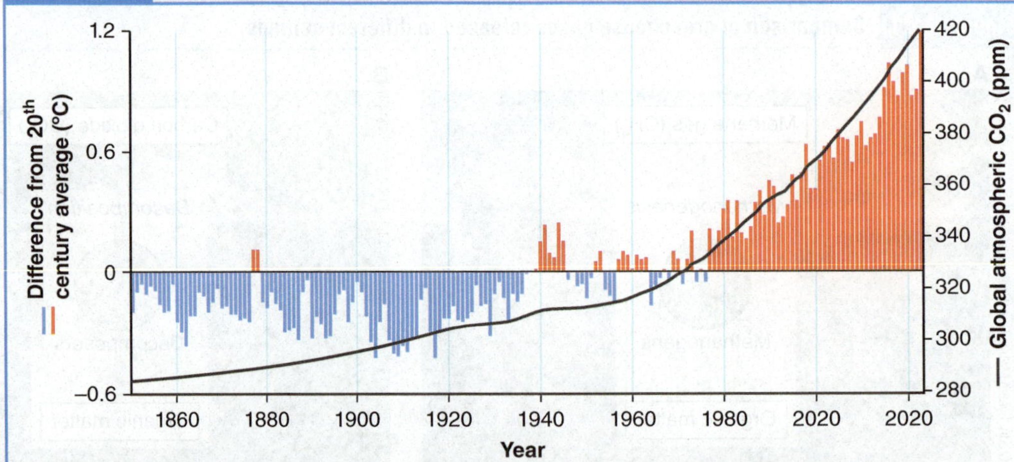

The black line represents average global carbon dioxide concentration in parts per million (ppm; right axis), whereas the blue and orange bars indicate whether yearly average temperatures are above or below the average temperature in the 20th century (in degrees Celsius; left axis). Note that a 1.2-degree Celsius increase in temperature corresponds to a 2.1-degree increase in Fahrenheit.

Graph based on: NOAA Climate.gov. (2024, May 16). www.climate.gov. https://www.climate.gov/media/13840

Figure 16.13 The type of CO_2 emitted varies depending on its source

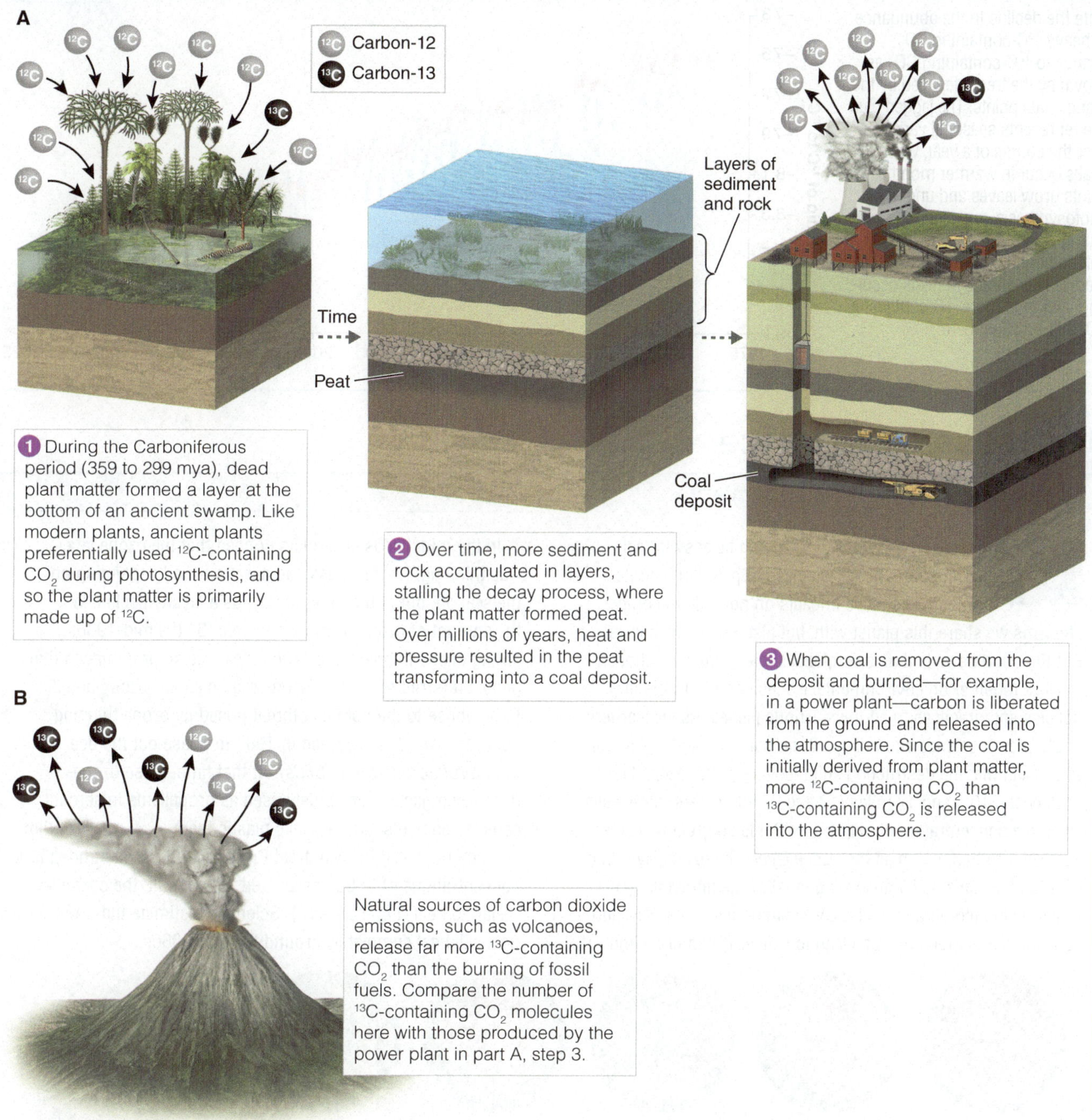

A. Since plants preferentially use ^{12}C-containing CO_2 over ^{13}C-containing CO_2, burning of fossil fuels releases more ^{12}C-containing CO_2.

B. Natural sources of CO_2, such as volcanoes, release more ^{13}C-containing CO_2 than ^{12}C-containing CO_2.

Source: Based on Pathfinder Merit Fleet LLP https://www.pmfias.com/coal-formation-of-coal-types-of-coal-peat-lignite-bituminous-coal-anthracite-coal/

Reason for Hope Carbon offsets are financial instruments intended to reduce global emissions of carbon-based gases that are causing climate change. Companies and individuals can buy carbon offsets to counterbalance activities that release these gases, where the funds are then used to support projects aimed at reducing emissions or that remove the gases from the atmosphere. Many of these projects involve forests, including replanting deforested areas or protecting forests from logging. This is because forests remove a lot of carbon dioxide from the atmosphere—about 3 billion tons per year!

Figure 16.14 **The average ratio of heavy versus light carbon in the atmosphere**

Note the decline in the abundance of heavy ^{13}C-containing CO_2 relative to ^{12}C-containing CO_2 as shown by the trend depicted by the orange data points. The jagged blue line represents seasonal variation over the course of a year, where peaks occur in warmer months, as plants grow leaves and undergo photosynthesis, and troughs occur during colder months, as trees enter a period of dormancy, when there is less photosynthesis occurring, and thus more ^{12}C-containing CO_2 in the atmosphere.

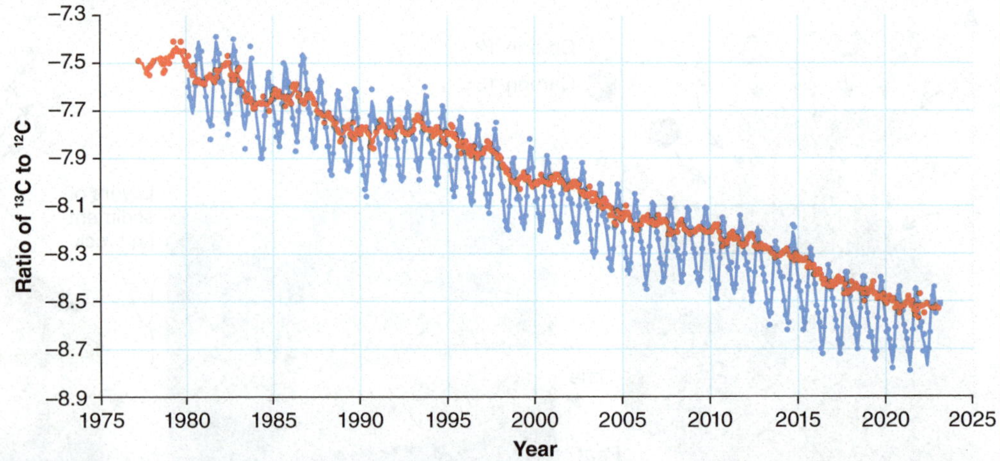

Graph based on: Scripps CO2 Program. (2022). Ucsd.edu. https://scrippsco2.ucsd.edu/graphics_gallery/isotopic_data/mauna_loa_and_south_pole_isotopic_c13_ratio.html

Process of Science

Ozone Layer Damage and Recovery

Reason for Hope It can be easy to feel glum about the seriousness of pollution and its potential negative impacts on not only the other lifeforms we share this planet with, but also on our own lives. But the good news is that, as a global society, we have chosen to work together on environmental issues before, developing policies and practices to address various issues. As we learned in Chapter 1, science and policy can go hand in hand. Let's look at an example: the destruction and recovery of the ozone layer.

Between 10.5 and 17 miles (17 and 27 kilometers) above sea level, the concentration of ozone gas (O_3) is so great that scientists refer to this region as the *ozone layer*. The ozone layer benefits land organisms by absorbing most of the ultraviolet (UV) radiation in incoming sunlight. UV radiation damages DNA and can cause mutations that can lead to cancer (Chapters 6 and 7).

In the mid-1970s, scientists noticed that the ozone layer was thinning. Its thickness had always varied depending on the season, but at the time, there was a steady decline in the amount of ozone from year to year. By the mid-1980s, spring ozone thinning over Antarctica was so pronounced that people referred to the low-ozone region as an "ozone hole." In response to the potential threat posed by ozone thinning, countries worldwide agreed in 1987 to phase out the use of *chlorofluorocarbons (CFCs)*. At that time, these ozone-destroying gases were widely used as propellants in aerosol cans, as coolants, and to make plastic foam. As a result of that agreement, called the Montreal Protocol, the rise in atmospheric concentrations of CFCs has slowed. As a result, the ozone layer began to heal (**Figure 16.15**). Scientists estimate that the ozone hole will heal completely around the mid-2060s.

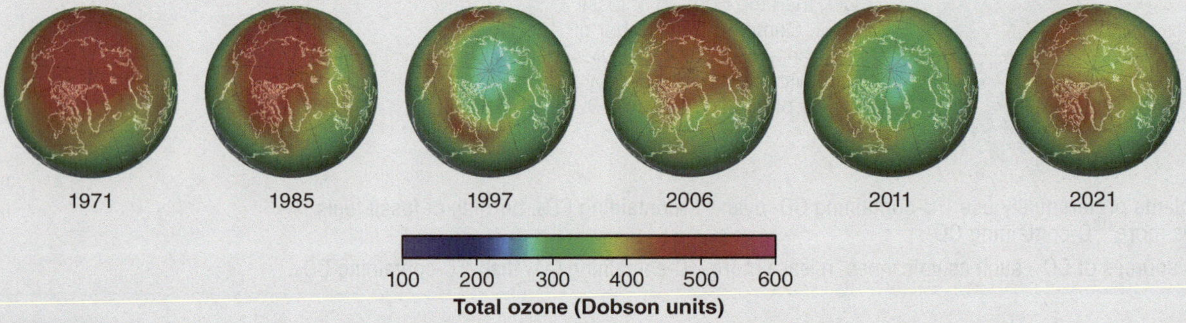

Source: Twenty Questions and Answers About the Ozone Layer. Csl.noaa.gov. https://csl.noaa.gov/assessments/ozone/2022/twentyquestions/

Figure 16.15 **Progress in restoring the ozone hole**

Here we show data for ozone levels in the atmosphere for select years, collected in March for each year indicated. Note that red indicates high levels of ozone, as described in Dobson units, and blue indicates thinning. Notice how dark red is prevalent in the 1970s and 1980s, but mostly disappears after that point. By 2021, yellow and red are more prevalent than blue, suggestive of ozone layer recovery.

1. Research the Montreal Protocol. Who agreed to participate, and what specifically did they agree to do?
2. Now, research the Paris Agreement. How are the Montreal Protocol and Paris Agreement like one another?
3. What can you conclude about the role of international cooperation between policymakers and scientists in reversing the impacts of pollution on the biosphere?

16.1 Summary

- Pollution occurs when harmful substances (pollutants) are introduced into air, water, or soil, disrupting ecosystems and health.
- Water pollution includes inorganic contaminants like lead and nitrogen, and organic contaminants like oil and forever chemicals. Water pollution causes issues like eutrophication. Historic water crises highlight long-standing risks, but cleanup successes like the Cuyahoga River offer hope.
- Plastic pollution, especially in oceans, is widespread and persistent. Very little plastic is recycled, and microplastics affect the health of humans and other organisms, as well as the environment.
- Air pollution comes in various forms, such as harmful gases like carbon dioxide and radon, as well as particulates. Chemical reactions involving pollutants can lead to ground-level ozone and smog. Air pollutants released by burning fossil fuels can combine with atmospheric water vapor to form acid rain. New regulations put into place in 1995 have greatly reduced the acidity of precipitation.
- Greenhouse gases, including carbon dioxide and methane, trap heat in the atmosphere. Excess greenhouse gas emissions from human activity, as shown by the changing ratio of ^{12}C-containing carbon dioxide to ^{13}C-containing carbon dioxide in the atmosphere, led to an overabundance of carbon dioxide and methane in the atmosphere, causing climate change. Methane generated by methanogens, though less abundant, is far more potent than carbon dioxide as a harmful greenhouse gas. Greenhouse gas production can be mitigated through actions such as composting and legislation.

16.2 The Impacts of Climate Change

Learning Objectives

16-8 Explain the impacts of increasing atmospheric carbon dioxide on the climate.

16-9 Identify strategies for mitigating climate change.

In the last section, we learned about the greenhouse effect and how excessive levels of greenhouse gases in the atmosphere are resulting in a warming planet. Warming temperatures are causing changes in weather patterns, and ultimately, the climate. Earth's climate has varied greatly over its history. Scientists can correlate some historical large-scale temperature changes with shifts in Earth's orbit, which varies in a regular fashion over 100,000 years, and Earth's tilt, which varies over 40,000 years. Changes in solar output, the arrangement of landmasses, and the frequency of volcanic eruptions also influence Earth's temperature. As a result, Earth has gone through shifts where average temperatures were warmer or cooler.

However, as we've seen throughout Unit 5, scientific evidence suggests that natural factors alone cannot explain the extent of climate change that we are currently seeing. The scientific consensus (Chapter 1) is that the recent change in Earth's climate is the result of rising greenhouse gases from human activities. In this section, we'll review some of the consequences of climate change on the biosphere.

> **? Did You Know?**
>
> The first scientist to predict warming as a result of greenhouse gas emissions was Svante Arrhenius, a Nobel Prize–winning chemist from Sweden, in 1896. Arrhenius was not concerned by this prospect; he expected the resulting increase in temperature to make Sweden's climate more pleasant.

Consequences of Climate Change

In North Carolina, people are observing a decline in trout. Trout require cold water as part of their life history, and as waters warm, people who fish for trout are coming up empty-handed. Throughout North America, warming temperatures and a lack of rainfall are contributing to dangerous wildfires. Farmers in Georgia are observing flooding and droughts that have a direct negative impact on crop yields. Powerful hurricanes, which get their energy from warm ocean waters, are causing major damage in the southeast. These are only snippets of the dozens of personal stories about the impacts of climate change from people *just like you*.

As we've learned throughout Unit 5, ecosystems are interconnected with one another. We've looked at many examples of what happens when either abiotic or biotic factors get out of balance—and the potentially catastrophic impacts on ecosystems. Many people feel like "ecosystems" are nature, and that the natural world is different from the human world. In truth, we are part of the natural world. The reality is that the negative consequences of climate change aren't limited to faraway polar bears. Climate change is impacting each of us personally—and presents a danger to us and future generations.

Note that "global warming" is a misnomer. While average temperatures on the planet are increasing, some areas are cooler than normal. This is why "climate change" is the better term.

Although I could devote an entire chapter of this book to the varied impacts of climate change on the biosphere, instead, I will give a brief overview of the impacts of climate change on terrestrial and aquatic biomes, closing with action steps that you can take to address climate change. I encourage you to visit one of the many repositories of climate stories available online, which include videos and podcasts from people around the world discussing the personal impact of climate change on their lives—and to ponder how your experience with climate change is similar to or different from theirs.

Impacts on Terrestrial Biomes

Recall from Chapter 15 that terrestrial biomes are large areas of land with similar climate and lifeforms, and include a variety of forests, grasslands, and deserts. While we could step through the impact of climate change on each of the biomes individually, here we'll cover only a few examples.

Melting permafrost. In the tundra biome (Chapter 15), the soil contains **permafrost**, a component of the soil that stays frozen for two or more years. Permafrost is typically found below the layer of soil that freezes and thaws each year as part of normal seasonal variation. However, as global temperatures increase, permafrost is beginning to disappear. The loss of permafrost means that structures built on soil containing permafrost, such as homes and roads, no longer have a strong foundation—and some are beginning to sink into the earth (Figure 16.16). In fact, people in Fairbanks, Alaska, are faced with either expensive repairs to their property or abandoning their property and heading southward.

Thawing permafrost is also, unfortunately, a problem that compounds itself. As the water contained in permafrost melts, the organic material stored within, like plant and animal remains, begins to decompose. As this organic matter decomposes, more stored carbon is released into the atmosphere, thus resulting in continually increasing temperatures, which drives more melting, and so on.

Figure 16.16 **This Fairbanks, Alaska, home is sinking due to compromised permafrost below**

Credit: Vladimir Romanovsky

As the ice within permafrost melts, damage can occur to properties, leaving homeowners with the difficult decision of either spending thousands on repairs or abandoning their homes.

Changing weather patterns. The temperature of the land and seas affects evaporation, winds, and currents. As a result, weather patterns change as the temperature rises. Altered precipitation patterns can result in reduced water availability in some regions and overabundance in others. For example, warmer temperatures are correlated with extremes in rainfall patterns, with periods of drought interrupted by unusually heavy rains. Warmer seas also increase the intensity of hurricanes. As we discussed in Chapter 14, the increasing severity of weather is also causing insurance rates for homeowners to increase, with some people losing their homeowner's insurance altogether.

Increased drought and fire risk. Warmer temperatures combined with less rain mean that terrestrial biomes will be more susceptible to fire (Figure 16.17). In addition, dry conditions cause stress on trees and increase their vulnerability to pest damage. Since trees are important carbon sinks—and provide shade, and therefore protection from increasing temperatures—the loss of trees negatively impacts our quality of life while also making climate change worse.

In addition, droughts degrade soil conditions, which impacts the amount and quality of food that farmers can produce, meaning that climate change poses a direct threat to our food supply. Elevated temperatures will result in a higher incidence of heat stroke and an increased reliance on air conditioning—and without newer, greener technologies, more air pollution.

Figure 16.17 A scorched road sign in Colorado after the Cameron Peak Fire in 2020, the largest wildfire in Colorado history

In addition to burning over 208,000 acres, 469 structures were destroyed by the fire, including the primary homes of 42 families.

Credit: Miles Blumhardt/Fort Collins Coloradoan

Changing wildlife habitats. In Chapter 14, we learned about the geographic range of animals, such as the pika, and how changing temperatures mean that there is less habitable space for animals. Wildfires don't just translate into lost homes for humans—animals are losing their habitats as well, and as a result, are encroaching into areas where people live. Warming temperatures also mean a change in the geographic range of pests, such as mosquitoes and ticks. For example, as mosquitoes move northward, scientists predict that 1.3 billion new people will be at risk from the Zika virus, a virus that causes birth defects, by 2050. Overall, climate change is causing both loss and expansion of habitats, which is causing some animal populations to soar, and others to decline.

Impacts on Aquatic Biomes

As we learned in Chapter 15, aquatic biomes are water-based environments and make up most of the biosphere. Aquatic biomes can be comprised of salt or freshwater, or a mixture of the two. Although the impacts of climate change are felt by both, since most of the water on the planet is saltwater, here we'll focus on the impacts of climate change on the oceans.

Ocean acidification. As carbon dioxide in the atmosphere increases, more carbon dioxide dissolves in the oceans. When dissolved carbon dioxide reacts with water to form bicarbonate, hydrogen ions are released, making the ocean more acidic (Chapter 2). This process is called **ocean acidification**. Increased acidity interferes with calcium carbonate formation, so ocean acidification poses a threat to animals that have a calcium carbonate shell (including bivalves, such as oysters) or a skeleton (such as corals). In a more acidic ocean, these animals must expend extra energy to build their hard body parts, and the parts that they do make are thinner and weaker than normal. The stress of doing so can cause coral bleaching events (Figure 16.18), and ultimately, the death of coral reefs and the animals sustained by the reefs (Chapter 13).

Figure 16.18 Coral bleaching in a coral reef in American Samoa

A B C

Healthy coral, December 2014 | Bleached coral, February 2015 | Dead coral, August 2015

Credit: National Oceanic and Atmospheric Administration (NOAA)

Figure 16.19 **Glaciers are melting due to climate change**

A

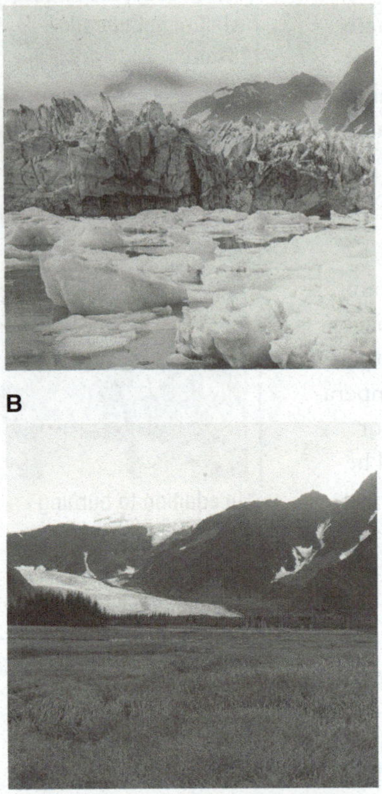

B

Shown here is Pedersen Glacier in Alaska in **A.** 1917 and **B.** 2005. Note the presence of grass and lack of ice when comparing B to A. As glaciers melt, more water enters the oceans, and sea levels rise.

Credit: NOAA

Rising sea levels. Another consequence of climate change on the world's oceans is rising sea levels. An increase in average annual temperature elevates sea levels by two mechanisms. First, it causes the volume of the ocean to increase directly, because water expands as it absorbs heat. Second, warming temperatures melt glaciers (Figure 16.19) and ice sheets, so water that has been locked on land enters the ocean.

Loss of ice accelerates warming because ice reflects sunlight, whereas water and soil absorb it, re-emitting the energy as heat. Thus, melting of ice from water or land causes increased warming, which in turn causes more melting. As a result of this feedback process, the temperature is rising faster in the Arctic than at lower latitudes. And much like the melting of permafrost, as sea ice melts, trapped methane gas is released (Section 16.1), further exacerbating climate change.

According to the National Aeronautics and Space Administration (NASA), between 1993 and 2024, sea levels rose by 10.3 cm (4 in.). The average sea level rise each year is 0.43 cm (0.17 in.)—and this rise in sea level is expected to continue. Consequently, wetlands are beginning to disappear all along the coastal United States as they are covered up by rising seas. Recall from Chapter 5 that saltwater kills plant cells by creating a hypertonic environment outside the cells. As a result, the water rushes out of the plant's cells towards the hypertonic environment. The loss of water ultimately kills the plant. Therefore, as saltwater levels rise, plants that are not adapted to growing in brackish (salty) water will die off.

But not only habitats are affected by rising sea levels. The infrastructure of coastal communities is also at risk, with water supplies threatened by saltwater intrusion and homes at risk for property damage. According to the United Nations, millions of people are under direct threat of rising sea levels. As sea levels rise, people may be forced to abandon their homes and migrate inland, causing strain on the availability of resources at the new locations.

Solutions to Climate Change

Reason for Hope In Chapter 12, we learned about how *Homo sapiens* out-survived other early humans, like the Neanderthals, due to our ability to adapt and shape our environment. Ironically, what helped our species survive may also be our undoing. One could argue that climate change is the greatest threat that humanity has ever faced. But it turns out that despite the monumental problem of climate change, there are many things we can do in our daily lives that can help, again learning to adapt to our environment.

One of the best things we can do to combat climate change is to minimize our use of fossil fuels and reduce our carbon footprint. Some strategies for doing so are outlined in Table 16.1. Additionally, we can support research into more efficient uses of existing solar and wind technology and/or new forms of renewable energy, such as the pKelp technology we learned about in the Unit 5 opener. Renewable energy not only helps combat climate change, but it also promotes energy security and independence.

Finally, as we've seen several times already throughout this chapter, policy change is critical to protecting our natural resources. Therefore, paying attention to elections and candidate positions on climate change and environmental legislation—and actively voting in

The impacts of climate change will be disproportionally felt by those who are poor and/or minoritized. The recognition of how people will be differentially impacted—with some being disproportionately negatively affected—plus work to address any disparity, is called *climate justice*.

Table 16.1 Things You Can Do to Reduce Your Carbon Footprint	
Action	**Example**
Conserve energy	Use energy-efficient appliances and lightbulbs at home; insulate your home
Reduce carbon emissions	Reduce your use of gas-powered vehicles and drive a hybrid or electric car instead; where possible, walk, bike, or use public transportation
Minimize waste	Recycle and reuse, thus reducing your purchases, and make repairs where possible; buy secondhand clothes and goods
Eat lower on the food chain	Adopt a predominantly plant-based diet to reduce the resources needed to produce your food
Compost	Compost food scraps to reduce the amount of food waste in landfills, and consequently, to reduce methane pollution

all elections, local and national—is critical to making your voice heard and affecting policy. You can also speak up by contacting your representatives and asking them to support common-sense climate change legislation. Speaking up can also involve creating artwork that raises awareness about climate change (and other environmental issues; Figure 16.20), or educating others.

? Did You Know?

As part of the Climate Stories Project, people are sharing their experiences with climate change through music.

Figure 16.20 A gallery of artworks created to raise awareness about climate change

A. Priscilla the Parrot fish is one of several sculptures made of marine debris by the Washed Ashore Project to raise awareness about plastic pollution.
B. An image from the Changing Balance/Balancing Change dance performance to teach climate change and highlight impacts in Montana.
C. A series of three quilts showing the impact of rising sea levels on coastal cities. Note how the progression of quilts shows ocean water steadily overtaking the land in each of the successive images.

Credits: Michael Siluk/Alamy Stock Photo; NPS/Melissa Sladek; Source: Courtesy of Don Tuttle

16.2 Summary

- The ongoing changes to climate that we now see are driven by excess greenhouse gases in the atmosphere due to human activity, and are not due to natural factors alone. The consequences of climate change are already visible through shifting weather patterns, rising sea levels, and impacts on ecosystems globally.
- Solutions that individuals can take to reduce their carbon footprint include conserving energy, reducing carbon emissions, minimizing waste, eating lower on the food chain, and composting.

16.3 Biodiversity Protection and Restoration

Learning Objectives

16-10 Describe aspects of biodiversity.

16-11 Present arguments for conserving biodiversity.

16-12 Compare extinct, endangered, and vulnerable species.

16-13 Relate habitat loss to species extinction.

16-14 Describe strategies for preserving habitats and biodiversity.

16-15 Identify success stories for recovery of biodiversity.

High in the mountains of central Mexico is a conifer forest so special that the United Nations has designated it a World Heritage site. Most of North America's monarch butterflies (*Danaus plexippus*) spend the winter here, at the Monarch Butterfly Biosphere Reserve. The large orange and black butterflies cluster on fir trees, where the densely packed evergreens keep the local humidity high and provide protection against strong winds (**Figure 16.21**).

A monarch butterfly that spends the winter in Mexico is a long-distance migrant. The butterfly's life began a few months earlier, somewhere in the United States or southern Canada. After hatching from an egg as a caterpillar, the monarch fed for a few weeks on milkweed, then underwent metamorphosis to become a butterfly. Over the next month or so, it flew south to Mexico, sustaining itself along the way by sipping nectar from flowers.

If this butterfly survives until spring, its reproductive system will become active, and the butterfly will head north, once again feeding on flower nectar as it travels. The butterfly will not fly as far north as its birthplace. Rather, it will travel partway, find a patch of milkweed, lay its eggs, and die. Successive generations will travel still farther north, mate, and die—until a sexually dormant generation arises in the late summer. This generation (the great-grandchildren or great-great-grandchildren of the previous long-distance migrants) will fly back to Mexico for the winter.

The extraordinary, long-distance, multigenerational migration of the monarch butterfly makes this species vulnerable to environmental changes in any of the ecosystems it inhabits. Logging of conifer forests in Mexico threatens the butterflies' winter home. Conversion of prairies to agricultural fields or housing developments in the United States reduces the availability of the monarch caterpillars' sole food source—milkweed. Due to these impacts, the monarch butterfly is now classified as *endangered*.

Figure 16.21 Monarch butterflies on a tree in Mexico's Monarch Butterfly Reserve

Credit: Sylvain Cordier/Getty Images

Credit: Catherine Avilez/Shutterstock.com

The inset photo shows a monarch on milkweed, the only plant monarch caterpillars eat.

Declining populations mean that **biodiversity**—the variety of life on Earth, including living organisms, their genetics, and ecosystems—is also declining. In this section, we'll examine the value of preserving biodiversity and habitats. Recall from Chapter 11 that species form through speciation and are lost through extinction. As we learned in that chapter, many scientists interpret the current rapid loss of biodiversity to mean that we are amid a sixth mass extinction. The byproducts of human activities are causing declines to biodiversity at every level.

The need to address declining biodiversity is so pressing that a new subdiscipline of biology emerged in the late 1970s called conservation biology. **Conservation biology** is a field that aims to determine the best ways to preserve biodiversity through protecting species, habitats, and ecosystems. In addition, conservation biologists also work to educate others on the value of biodiversity, bringing light to issues concerning conservation biology and how people can help.

? Did You Know?

Biofuels help to combat climate change by reducing greenhouse gas emissions, being part of the natural carbon cycle rather than liberating stored carbon from the ground.

The Value of Biodiversity

Why should we care if wild species disappear? Although one can argue for the importance of environmental **stewardship**—the practice of responsibly managing resources—at the end of the day, protecting biodiversity is an investment in our future. Healthy ecosystems are essential to our survival. Other organisms produce the oxygen we breathe and the food we eat. They also remove waste carbon dioxide from the air and decompose and detoxify other wastes. Plants take up rain and hold soil in place, preventing erosion and reducing the risk of flooding. Insects, like butterflies, are crucial to pollination, and thus our food supply (Chapter 18).

Biology for Life

Shipworms and Biofuels

What do you think of when considering the value of biodiversity? Is it a verdant green forest teeming with plant and animal life? An African savannah? A coral reef? What about...worms? Shipworms may look like worms but they are not worms. Rather, they are clams that use their shells to carve into wood (Figure 16.22A). The name reflects a history of when people traveled the globe on wooden ships...which were being eaten from under their feet! While shipworms are often still considered pests today because they eat through underwater wooden structures (including culturally significant shipwrecks), they are also an important part of the carbon cycle (Chapter 15), and understanding how they digest wood could lead to the discovery of new enzymes (Chapter 4) for producing biofuels, which can help fight climate change.

A

Credit: Gary G. Gibson/Science Source

B

Credit: Marco Carmignan/National Geographic

Figure 16.22 Shipworms are the subject of study by National Geographic Explorer Reuben Shipway

Biofuels are a source of energy that is derived from living matter. Unlike fossil fuels, biofuels can be produced quickly and in a renewable manner. Lignocellulose, which is a major structural component of woody plants, is the most abundant source of carbon on Earth. Finding a way to convert lignocellulose into a biofuel would help us transition from reliance on fossil fuels to a pathway to source fuel renewably. Since shipworms are capable of processing lignocellulose, what could we learn from them?

That is the question that National Geographic Explorer Reuben Shipway (Figure 16.22B) and colleagues are working to answer. For example, in 2018, they characterized the digestive system of a particular shipworm, *Lyrodus pedicellatus*, to determine the enzymes it used to digest lignocellulose. Comparing these enzymes to the enzymes of other organisms that can digest lignocellulose, such as termites and fungi, will help biorefineries determine the most beneficial enzymatic cocktail to use to convert plant biomass into biofuel.

1. How are shipworms important for the global carbon cycle? How does this mesh with the fact that they cause trouble for humans? Can you think of ways to balance both?
2. Research such organisms that have been helpful to humans, but that some might consider "gross" or unimportant. Why is it important to consider all organisms when bioprospecting?

Bioprospecting—or the search for natural resources such as genes, chemicals, and macromolecules—has resulted in scientists identifying useful compounds. For example, two widely used chemotherapy drugs, vincristine and vinblastine, were extracted from the rosy periwinkle, a low-growing plant native to Madagascar's rainforests. In Chapter 6, we learned about Taxol, another chemotherapeutic drug, derived from tree bark. Antibiotics in the penicillin family are compounds that are produced by fungus. In Chapter 13, we also learned about pharmaceutical applications of snail venom.

Additionally, animal poisons and venoms (or synthetic versions of them) have different clinical applications. For example, bee venom has anti-inflammatory properties and is used to treat osteoarthritis pain. Gila monster lizard venom was used by scientists in the development of medicines to treat diabetes, like Ozempic®. Derivatives of snake venom are used to treat everything from high blood pressure to cardiac failure.

The various compounds discovered and synthesized for medical applications previously described (and many others) would not have been possible if not for bioprospecting. However, for bioprospecting to continue, it requires the protection and restoration of natural areas. Next, we will examine the conservation of habitats and species to preserve biodiversity.

As of the writing of this text, the IUCN estimates that 46,300 species are threatened with extinction—which is about 28% of all species assessed by the IUCN.

Preserving Species and their Genes

Have you ever visited a zoo and, looking at signage outside an animal's enclosure, noticed whether it was labeled as critically endangered, threatened, or of least concern? Or perhaps it had the designation of extinct in the wild? The International Union for Conservation of Nature (IUCN) categorizes species based on their extinction risk (Figure 16.23). Biologists with IUCN assess species using criteria such as population size and geographic range (Chapter 14) to determine their designation. These designations are used to inform conservation strategies, to guide research and policy changes, and for international agreements.

A species is considered extinct if repeated, extensive surveys of its known range repeatedly fail to turn up signs of any individuals. A species is considered "extinct in the wild" if the only known members of the species are in captivity. **Endangered species** are those facing a high risk of extinction in the wild. **Vulnerable species** are those that are likely to become endangered.

Figure 16.23 IUCN species designations

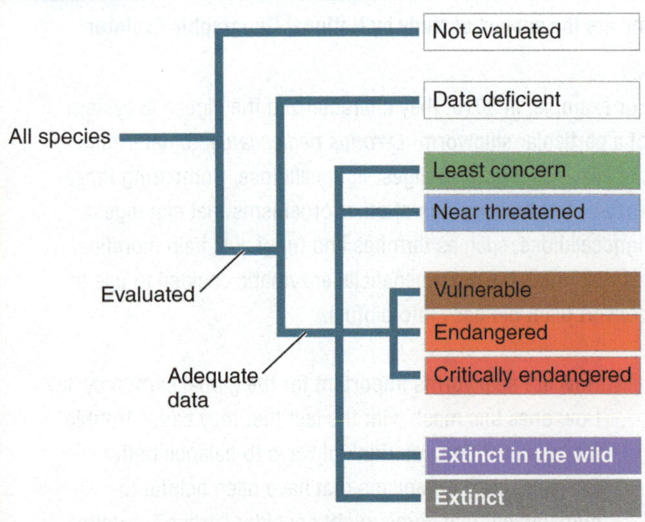

Species are first classified by whether biologists have adequately assessed their populations. For species that have been evaluated and have sufficient data available to make a determination, biologists classify them on a spectrum ranging from least concern for extinction to extinct. Notice that as we move down the figure, the risk of extinction increases.

Source: Based on the IUCN red list

Note that not all rare species are threatened or endangered. Some species have always been uncommon, such as organisms that are highly adapted to a particular environment, like the saguaro cactus, which is only found in the Sonoran Desert in the United States (Figure 16.24A). Species like the saguaro, which are confined to the areas in which they first evolved, are called **endemic species**. Endemic species are more likely to go extinct than a species with a more widespread distribution, because of the particular requirements they need to survive in an environment. In addition, a species with highly specific needs is more likely to go extinct than one that thrives under a wider range of conditions. Consider giant pandas (Figure 16.24B), which are endemic to China's bamboo forests and subsist almost entirely on bamboo. As humans cut down bamboo forests, panda numbers declined. Their population, which may once have been as high as 100,000 animals, has now been reduced to about 1,600 animals in the wild.

Threats to Biodiversity

As the plight of pandas illustrates, humans can increase the risk of species extinction through actions that lead to habitat destruction. But as we learned earlier in this chapter, there are other major threats to biodiversity, including climate change and pollution. Here, we'll specifically discuss the impacts of habitat loss, overexploitation of animals, and invasive species on biodiversity.

Habitat loss. Each species requires a specific type of environment, and any loss, degradation, or fragmentation of its habitat can pose a threat. These problems can compound one another as well. For example, as Asian forests are disappearing due to human activity, elephants overgraze the land that is left and destroy crops to find food—causing friction between elephant and human populations.

When all trees are removed from a forested area, called **deforestation**, it often has broad implications beyond just the loss of the trees. For example, deforestation encourages flooding because water that would otherwise be taken up by tree roots instead runs off into streams. The water runoff removes mineral ions from the soil, leaving behind nutrient-poor soil. For hilly areas, the loss of trees raises the risk of landslides, since there are no longer trees to hold the soil in place.

Figure 16.24 Endemic species are more at risk for extinction

A

B

A. Saguaro cacti are found only in the Sonoran Desert in the southwestern United States.

B. Giant pandas are at risk not only because they are endemic to the mountains of China, but also because they can eat only bamboo.

Credits: iStock.com/Brent_1; iStock.com/powerofforever

Loss of trees can also alter the local weather. When a forest is cut down, lack of shade and reduced transpiration (the release of water vapor into the atmosphere from a plant's leaves; Chapter 18) cause daytime temperatures to rise. The loss of transpiration also means there is less water in the atmosphere, and therefore, less rainfall, resulting in drier conditions. Drier conditions may lead to a rapid change in the habitat, so species that once lived there can no longer survive.

Deforestation is a major contributor to climate change. Forests act as a **carbon sink**, or a natural reservoir for carbon, removing more carbon dioxide from the atmosphere than they release. When trees are clearcut, there is a twofold negative impact on carbon levels in the atmosphere. First, the removed trees are no longer alive and undergoing photosynthesis, thereby removing carbon dioxide from the atmosphere and storing it in their tissues. Second, clearcut trees are often burned, and so the stored carbon is released back into the atmosphere. A major source of deforestation is the clearing of forest for agriculture. While the clearcut trees are replaced with crops, crop plants are much less efficient than trees at removing and sequestering carbon from the atmosphere.

Another form of habitat loss is desertification. Rapidly changing habitats can result in **desertification**, a rapid shift from a grassland or woodland to a desert, where the soil becomes unsuitable for cultivation. One famous historical example of desertification is the Dustbowl in the United States in the 1930s. The Dustbowl resulted from farmers plowing large areas of prairie (Chapter 15), exposing the soil to wind which carried away more than a billion tons of topsoil in great plumes (Figure 16.25). Coupled with drought, the result was an economic and ecological disaster.

Desertification remains a problem today. Scientists estimate that as of 2020, 6% of the world's drylands (the area largely used for crop cultivation) underwent desertification because of climate change and unsustainable farming practices.

Overexploitation of species. Do you remember during the early days of the COVID-19 pandemic when there was a scramble to buy toilet paper and hand sanitizer? Everyone was in such a panic about not having any toilet paper that people bought more than they needed, creating a shortage. This story illustrates an important ecological and economics principle: the tragedy of the commons.

? Did You Know?

Saguaro cacti can live to be 200 years old and grow to over 50 feet tall. They serve as habitat for many different bird species, like the Gila woodpecker, which builds its home inside the cactus. To avoid drying out, the saguaro forms a hard lining around the space left by the bird. When the cactus eventually dies and decomposes, these "Saguaro boots" are left behind. Indigenous tribes such as the Tohono O'odham people collected these boots and used them to store water.

Figure 16.25 A dust storm near Rolla, Kansas, in 1935

Credit: Franklin D. Roosevelt Presidential Library & Museum

In the **tragedy of the commons**, when people have unrestricted access to a finite resource, the resource can be depleted and destroyed for all. So, in the case of toilet paper, rather than buying only what each person needed, overbuying led to a shortage. We see the same thing happening in the environment, where populations of animals are overharvested for food, traditional medicines, the pet trade, or for making jewelry.

For example, overfishing—or the depletion of a stock of fish due to their being caught in numbers that are unsustainable for the survival of the species—can cause a collapse of the fisheries for everyone. However, if everyone takes a sustainable amount, there will be enough fish for everyone. To protect the world's fisheries, policy changes such as the United Nations Convention on the Law of the Sea were put in place to protect the world's fisheries from overexploitation. Organizations such as the Monterey Bay Aquarium also work to educate people on sustainable seafood choices. For example, you can visit the Seafood Watch website or download the app to your phone to identify if the seafood you are purchasing at the store was sustainably caught.

Invasive species. In Chapter 15, we learned about invasive species. Recall that invasive species are not only exotic species—a species introduced from another region—but are considered "invasive" if they cause significant disruptions to the ecosystems in which they are introduced, for example, by outcompeting native species. While we discussed invasive species and their impacts in depth in Chapter 15, here we'll discuss some creative solutions that various communities are using to control invasive species.

? Did You Know?

In the 2024 Florida Python Challenge, the grand champion caught and culled 20 Burmese pythons.

Reason for Hope What do fine dining, hunting competitions, and invasive species have in common? Lionfish! Recall from Chapter 15 that these beautiful fish are invasive. Lionfish indiscriminately eat any fish or crustaceans they can catch. Due to how fast they can reproduce, biologists estimate that 35% to 65% of all lionfish in the waters around Florida need to be caught each year to prevent damage to the ecosystem.

To manage lionfish numbers, prizes are available for catching lionfish through the Florida Fish and Wildlife Conservation Commission's Reef Rangers program (Figure 16.26A). Taking advantage of the program, chefs are now devising lionfish dinners (Figure 16.26B), where diners across the country can try an exotic new fish dish while also helping to manage the number of lionfish in the environment. Similar community-based approaches are also taking place in Louisiana to manage nutria populations, and in Florida to manage the number of Burmese pythons in the wild.

Preserving Species and their Genes

Preserving biodiversity isn't limited to the number of individual animals of a species remaining, but also to preserving their genes. For example, wild relatives of crop plants are reservoirs of genetic diversity that plant breeders can use to protect and improve crops. Wild plants often have genes that make them more resistant to disease or adverse conditions than their domesticated relatives are. Plant breeders use traditional crossbreeding methods or biotechnology to introduce genes from wild species into domesticated ones, thus creating improved varieties.

As we've learned in Chapters 8 and 11, sufficient genetic diversity is critical for a species' survival. Therefore, assessing the genetic diversity of a species is crucial for understanding its susceptibility to extinction. For example, National Geographic Explorer Gibbs Kuguru (who we first met in Chapter 12) studies shark genetics (Figure 16.27).

Figure 16.26 **Managing invasive species through dinner and fishing competitions**

A B

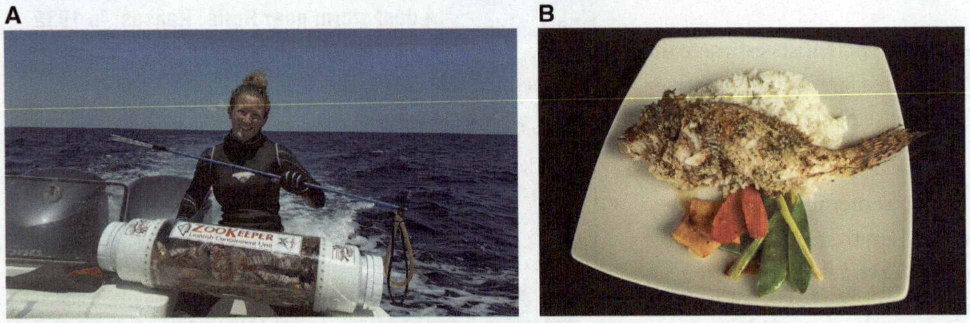

A. People can win prizes for the number of lionfish caught. Lionfish are caught by divers using spears, such as the one shown here.
B. Adventurous eaters can now dine on lionfish.

Credit: Rachel Lynn Bowman/USA Today; Nature Picture Library/Alamy Stock Photo

Understanding shark genetics can guide wildlife management organizations on how best to conserve populations. As Gibbs says, "Once we destroy the quilted mosaic of nature, repairing the diversity in the animal kingdom becomes an unrealistic proposition."

In fact, **conservation genomics** is an emerging field dedicated to studying the genes of species and how genetics can help conservation efforts. For example, biologists are saving tissues of animals on the brink of extinction that could be used in future "de-extinction" (Chapter 11) efforts.

National Geographic Explorer
Gibbs Kuguru

Credit: Siraj Ahmed/National Geographic Image Collection

Figure 16.27 National Geographic Explorer Gibbs Kuguru captures a photo of a tiger shark

What can be done to preserve biodiversity? Let's return to our earlier example of a trip to the local zoo. Have you ever noticed if an animal is participating in a Species Survival Plan? The **Species Survival Plan (SSP)** program was developed in 1981 by the Association of Zoos and Aquariums (AZA) to breed animals that are threatened, vulnerable, or endangered. SSPs help maintain genetic diversity through careful breeding. In addition, if a species goes extinct in the wild, SSPs ensure there is enough genetic diversity in captive animals to maintain the species. SSPs are a cooperative endeavor between AZA-accredited facilities, where animals are often moved between zoos for the purposes of mating.

Biology for Life

Zoos, Conservation, and the Importance of AZA Accreditation

The care of animals in captivity is a concern for many people—and rightfully so. Over human history, wild animals were often captured for human spectacle or given as elaborate gifts to royalty. There are many instances where the care of the animals was questionable at best…and neglectful at worst. Unfortunately, we still see examples of animal mistreatment today, from organizations offering selfies with endangered animals to the illegal pet trade.

When considering animals in captivity—whether at a zoo, a wildlife sanctuary, or an aquarium—look to see if the organization is accredited by the Association of Zoos and Aquariums (AZA). As of late 2025, there were 240 AZA-accredited institutions and 15 AZA-accredited facilities. All of these organizations must agree to meet the AZA's stringent animal welfare guidelines and are subject to inspection by the AZA as well as the U.S. Department of Agriculture (USDA). The AZA accreditation process is based on scientific evidence and best practices in the profession.

As such, as new evidence comes to light on best practices for animal welfare, animal care approaches at zoos have also changed, allowing for more animal choice in how and where they spend their time. For example, the Philadelphia Zoo in Pennsylvania opened its first Zoo360 trail in 2011. Zoo360 is a network of mesh trails that traverse the zoo, allowing animals to leave their primary habitat to explore other locations in the zoo. This benefits the animals in allowing them to roam freely throughout these pathways—and it's exciting for visitors to see animals up close and from a different perspective (Figure 16.28).

Credit: Tom Mihalek/EPA/Shutterstock.com

Figure 16.28 **Visitors at the Philadelphia Zoo experience a tiger walking overhead**

Changes in how we view animal care, particularly in captivity, are also resulting in a decline in animal-based entertainment, such as circus performances and dolphin and whale shows. For example, Ringling Bros and Barnum & Bailey closed in 2017 and are planning to re-open without animal acts. The National Aquarium in Baltimore stopped offering dolphin shows in 2012 and is planning to move its dolphins to a protected coastal habitat.

AZA-accredited zoos are important resources for understanding animals, raising awareness about conservation, and advocating for animals that live far from us. Be sure to visit and support your local AZA-accredited facility—you never know what you'll learn about biodiversity!

1. Research the history of animals in captivity, as well as current practices in an AZA-accredited facility near you. What differences in animal care do you notice between modern AZA-accredited facilities and historical facilities?
2. Do you think it's important to sacrifice the freedom of a small number of animals in zoos to preserve the genetic diversity of threatened and endangered species? Explain.

Digging into Data

A Tiny Weevil Saves the Manatees

A

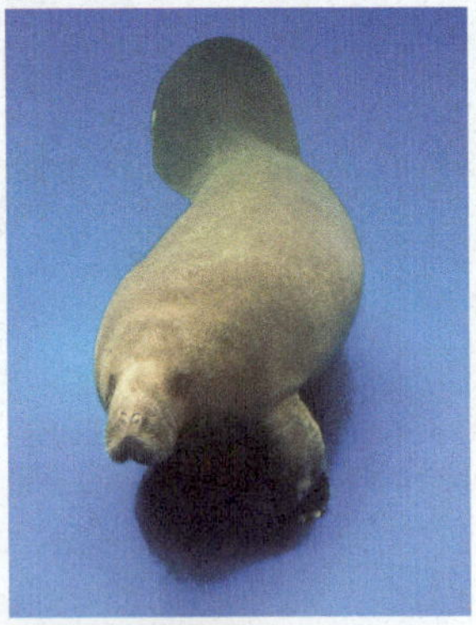

Credit: WaterFrame/Alamy Stock Photo

B

Credit: Jahawi Bertolli

Figure 16.29 National Geographic Explorer Dr. Aristide Takoukam Kamla studies African manatees

A. An African manatee; these peaceful mammals are smaller than their North American cousins.

B. Kamla on Batoke Beach in his native Cameroon.

Bodies of water like Lake Ossa in Cameroon are essential to people in West Africa. They provide drinking water, support agriculture, and serve as fisheries. They are also important sources of biodiversity as habitats for birds, amphibians, and endangered species such as the African manatee (Figure 16.29A).

Unfortunately, habitat destruction and runoff from agriculture have allowed salvinia—an invasive weed that can double its biomass in only three days—to take over Lake Ossa. Salvinia forms floating mats that lower the oxygen content of the water and block out sunlight to the underwater plants and animals. By 2021, salvinia covered nearly half of Lake Ossa's 4,000-hectare surface, an area the size of a large city like Scottsdale, Arizona. As a result, there was little space for fishing, where fishers reported that they lost about 80% of their income. And without underwater plants, the lake could no longer support the manatees, which rely on the plants as a food source.

To save the manatees from this ecological crisis, Cameroonian marine biologist and National Geographic Explorer Dr. Aristide Takoukam Kamla (Figure 16.29B) worked to address the salvinia issue using an unexpected ally: a small insect called the salvinia weevil.

The salvinia weevil destroys plants by laying hundreds of eggs in the leaves. The weevil's larvae burrow through the stems, killing the plants and causing them to sink. Dr. Kamla and his team released salvinia weevils into Lake Ossa. To monitor their success at mitigating the salvinia, and other invasive plant species like sedge, Kamla and team used satellite and drone imagery to quantify salvinia coverage on Lake Ossa (Figure 16.30).

1. Based on the graph and satellite imagery, would you say the salvinia weevils were effective at controlling salvinia specifically? Use data from the graph to support your answer.
2. Recall from Chapter 15 how a non-native species can have major ecological impacts on the habitat into which they are introduced. With this in mind, would you advise using an exotic species like salvinia weevils to mediate an invasive species like salvinia? What are the potential benefits and consequences of this approach?

Preserving and Restoring Habitats

Preservation of critical habitats is crucial to conservation efforts. Mexico's Monarch Butterfly Biosphere Reserve, described earlier in this section, is one such area. Prohibition of logging in the reserve has helped protect the fir trees essential to the butterfly's survival. Marine areas are also protected, which include a variety of habitats from marshes to seamounts (underwater mountains). For example, Biscayne National Park, which is located near Miami, Florida, is comprised of about 95% water and preserves coral reefs, mangrove forests, and shipwrecks.

Many zoos donate the proceeds of ticket sales to conservation efforts. Some allow people to vote at the entrance on which conservation efforts they would like the zoo to fund.

Travel to natural areas like Biscayne National Park, in ways that conserve the environment and well-being of the ecosystem and the people living there, is called **ecotourism**. Ecotourism can be an effective (and fun!) way of preserving biodiversity and habitats.

With limited resources available, how do people choose which areas to protect? One way is for conservation biologists to identify **biodiversity hotspots**, critical habitats of high biodiversity that are under the greatest threat. Biodiversity hotspots often occur in areas where two ecosystems meet, such as along coasts (Chapter 13). By identifying these biodiversity hotspots, high-priority areas can be targeted for conservation.

Figure 16.31 shows the location of 36 global biodiversity hotspots. Although the combined area of these biodiversity hotspots constitutes about 2% of the world's landmass, they collectively contain more than half of the world's endemic plant species and 40% of its endemic vertebrates.

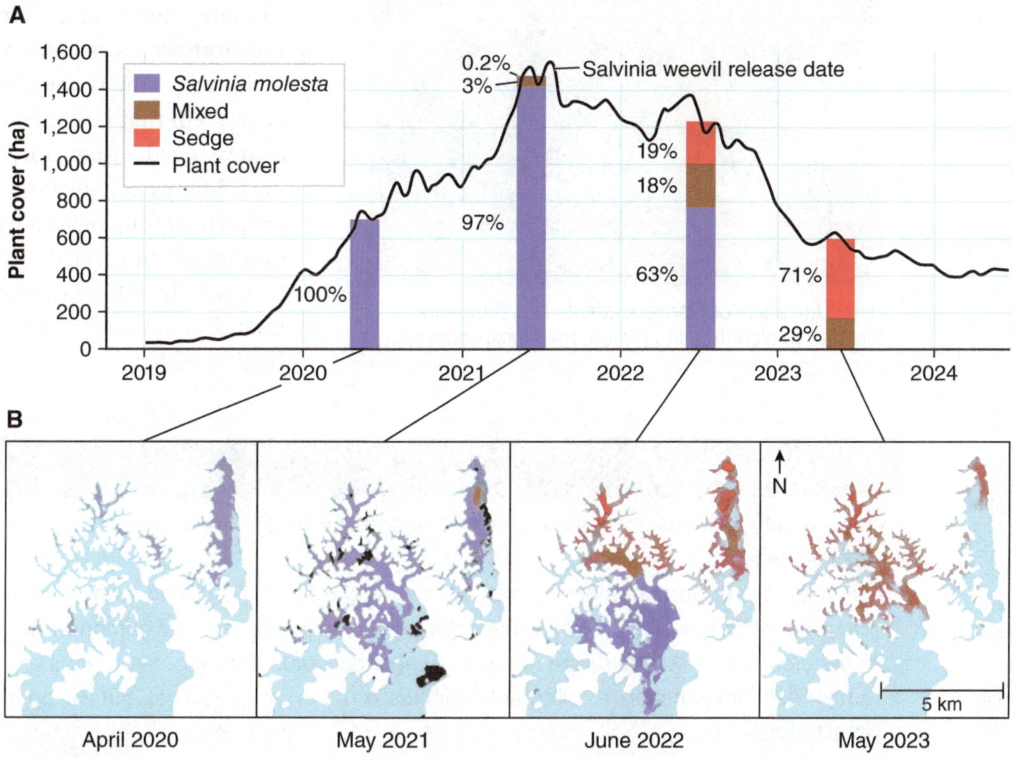

Figure 16.30 **Changes in the percent salvinia and sedge plant cover in Lake Ossa from 2019 to 2024**

A. The colored bars show the percent plant cover in hectares (ha) for each plant species, with "mixed" signifying mats that contained a mixture of both salvinia and sedge. The solid line represents the percent plant cover on the lake, determined by imagery, like those in part **B**.

Figure based on: Herbert, I., Zanga, A., Kinsler, D., Ayala, V., Takoukam Kamla, A., & Diaz, R. (2024). Optimizing aquatic weed management in Lake Ossa, Cameroon: Harnessing the power of biological control and real-time satellite monitoring. Biological Control, 199, 105650. https://doi.org/10.1016/j.biocontrol.2024.105650

Figure 16.31 **Biodiversity hotspots**

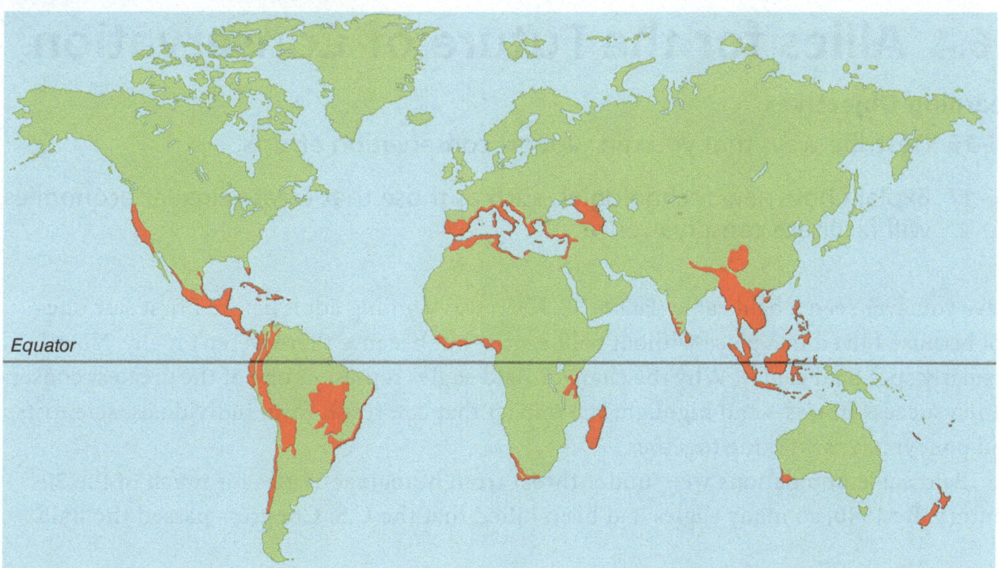

These areas were designed as hotspots by the Critical Ecosystems Partnership Fund because they are home to a large number of endemic species and have lost much of their native vegetation.

Figure based on: Myers, N., Mittermeier, R. A., Mittermeier, C. G., Da Fonseca, G. A., & Kent, J. (2000). Biodiversity hotspots for conservation priorities. Nature, 403(6772), 853-858.

? Did You Know?

Biscayne National Park is home to over 600 species of fish, several bird species, and 29 threatened and endangered species.

? Did You Know?

There are two biodiversity hotspots in the United States. One is in Northern California, where there are many endemic conifers, such as the giant sequoia, and the other is in the Southeast, around the Gulf and Atlantic Coasts.

Figure 16.32 Restoration projects are often staffed by volunteers—people like you

Credit: Nature and Science/Alamy Stock Photo

Here, University of Florida students and staff work together to plant trees to restore their local ecosystem.

Reason for Hope Sometimes, an ecosystem is so damaged, or there is so little of it left, that protection alone is not enough to sustain biodiversity. **Ecological restoration** is work designed to bring about the renewal of a natural ecosystem that has been degraded or destroyed, either fully or in part. Many ecological restoration projects are supervised by trained biologists but carried out primarily through the efforts of volunteers, including undergraduate students (Figure 16.32).

Worldwide, about 17% of Earth's land and 8% of oceans are designated as protected, meaning they are managed with the goal of conserving biodiversity. Various conservation organizations recommend that we work together to protect at least 30% of land, freshwater, and oceans by 2030, an initiative called 30×30.

16.3 Summary

- Biodiversity includes ecosystems, as well as the variety of living organisms on Earth and their genetics.
- Biodiversity supports human survival by providing oxygen, food, and medicines, among other benefits. For example, bioprospecting yields valuable medical compounds.
- A species is extinct if there are no more living individuals. Endangered means a species has a high risk for extinction. A vulnerable species is not endangered yet, but will be if changes aren't made.
- Habitat loss through deforestation and desertification can result in species extinction. Protecting habitats includes the removal of invasive species, habitat restoration, and preserving biodiversity hotspots.
- Success stories for the recovery of biodiversity include ecological restoration projects, which bring about the renewal of a natural ecosystem that has been degraded or destroyed, and breeding programs in AZA-accredited zoos.

16.4 Allies for the Future of Conservation

Learning Objectives

16-16 Describe ways that you can support conservation efforts.

16-17 Explain how new technologies, such as those that create circular economies, will facilitate conservation.

? Did You Know?

When bald eagles in captivity lose their feathers, zoos collect the feathers and give them to Native Americans for use in religious ceremonies.

Have you ever seen a bald eagle (Figure 16.33)? I was a young adult before I first saw one—not because I lived in a place without bald eagles, but because there weren't many. Now, I regularly spot bald eagles. Why the change? Bald eagles represent one of the greatest conservation success stories—and highlight the impact that can result from individuals, scientists, and policymakers working together.

Bald eagle populations were under threat from human activities for much of the 20th century. By 1940, so many eagles had been killed that the U.S. Congress passed the Bald

Eagle Protection Act, which outlawed killing or disturbing bald eagles. The act also made it illegal for people to possess eagle eggs, feathers, or nests.

However, bald eagles were still to face their greatest threat yet, not from guns...but from a pesticide. Dichloro-diphenyl-trichloroethane, or DDT, was one of the most widely used pesticides in the United States from the 1940s to the 1970s. While great at killing insects, DDT also bioaccumulates in organisms (Chapter 15), and eventually worked its way up the food chain to bald eagles. The chemical interfered with the eagles' ability to produce sufficiently strong eggshells. During nesting, the eggs broke, and the chicks died. And the bald eagle population plummeted. By 1963, there were only 417 nesting pairs remaining in the wild.

Figure 16.33 A bald eagle prepares to grab a fish

Bald eagles are the U.S. national bird and have served as the national emblem since 1782. They can have a wingspan of up to 7.5 feet and can reach nearly 100 miles per hour when diving to catch their prey. Their size, distinct flapping patterns, and white heads make them easy to identify.

Credit: iStock.com/Wirestock

In 1962, Rachel Carson (Figure 16.34), a nature author and marine biologist, published *Silent Spring*, which detailed her research on how DDT entered the food chain, bioaccumulated in animals (including humans), and ultimately caused massive harm to the ecosystem. Carson included 55 pages of notes and extensive references, plus had her data peer-reviewed (Chapter 1), which helped to dispel the storm of attacks on her work by people in the chemical pesticide industry. Her work eventually caught the notice of President John F. Kennedy. Kennedy ordered the President's Science Advisory Committee to investigate, which found her data to be sound. Ultimately, DDT was banned in 1972.

Figure 16.34 Rachel Carson examines a specimen on a beach in Maryland

Credit: Alfred Eisenstaedt/The LIFE Picture Collection/Shutterstock.com

Now, let's return to the eagles. Once DDT was banned, bald eagle populations steadily began to recover. By 2007, the population was large enough that bald eagles were no longer listed as endangered. As of 2020, there were estimated to be 316,700 individuals (71,400 nesting pairs) in the United States. Between a bald eagle survey in 2009 and the most recent survey in 2020, the population had quadrupled!

Reason for Hope The story of bald eagles touches upon many themes of this chapter, from pollution to biodiversity, and highlights the collaboration necessary between different people, entities, and government bodies to enact change that benefits us all. It isn't too late to avert the sixth mass extinction (Chapter 11) or to stop climate change—if we work together toward that goal. Rather than being consumers of the planet, we can instead choose to be stewards, or people who take responsibility for the care and management of the planet. In this section, we'll explore strategies that we can work toward together to be good stewards of planet Earth—or as Captain Planet would put it, to be part of the solution.

Community Science Programs

As we've seen throughout the book, science is a process that exists as part of our daily lives and in service to all. With the rise in **community science**, programs that allow the public to collaborate with professional scientists, more people can directly participate in scientific research endeavors. This is beneficial because it increases public awareness and ownership surrounding local issues, builds beneficial relationships between scientists and the public, and can make a big difference in achieving conservation goals. Participating in community science also allows the public to have firsthand experience with data collection, analysis, and reporting. One way people participate in

Figure 16.35 **Monarch Watch is a community science program**

A B

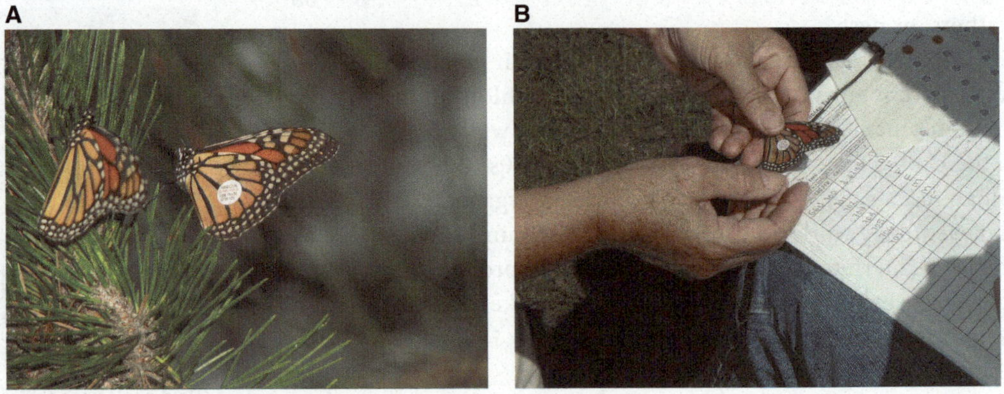

A. These monarch butterflies bear tags placed by a volunteer.
B. A volunteer checks the tag on a monarch and records the data.

Credits: Mark Garland/USDA; Anne Okonek/USDA

Generally speaking, touching butterflies shortens their lives. Monarchs are an exception—their bodies need to be tougher to survive migration and so can survive gentle human manipulation.

community science endeavors is through sharing bird sightings using the eBird mobile app, a program run by Cornell University. Birder reports on eagle numbers from eBird were used in recent bald eagle population surveys.

In another example, Monarch Watch is a community science program, based at the University of Kansas, that monitors monarch butterfly numbers. Working with biologists in the program, volunteers in the United States and Canada learn how to tag these migrating butterflies (Figure 16.35). Using the information on these tags, volunteers can help monitor butterfly populations and their geographic range, as well as the course, timing, and pace of their migration.

Community science makes a difference—for example, allowing for better sampling (Chapter 14) of populations to determine the success of conservation efforts (as in the bald eagle example) or in identifying species of concern, such as through butterfly counts. For example, in a study published in 2025, scientists analyzed over 76,000 surveys conducted by 35 different community science programs in the United States to determine that across 554 species of butterfly, abundance fell by 22% between 2000 and 2020. Most populations were found to be in decline.

The collaborative efforts of professional and community scientists translated to the most comprehensive analysis of butterflies in the United States ever undertaken. As we learned in Chapter 1, obtaining a sufficient sample size is important to obtain generalizable results. Therefore, efforts within your community—be it community science, habitat restoration, or competitions to remove invasive species—are all ways that you can get involved to become a better steward, and part of the solution.

Technology and the Future of Conservation

Reason for Hope In 1976, Michigan was the first state to enact legislation that added a 10-cent surcharge onto each individual bottle purchased. If consumers returned the bottle at a redemption center, the bottle deposit was refundable. Within the first few years, bottle litter alongside highways dropped by 84%. Since then, several other states have followed suit, witnessing a similar drop in litter and an increase in recycling.

So, what does this have to do with technology and the future? Recall from Section 16.1 that plastic cannot be indefinitely recycled. Eventually, it will end up in a landfill.

This represents a linear lifecycle for plastics—they are produced, are possibly recycled, and then end up in a landfill (**Figure 16.36A**). What if we could instead change the lifecycle of plastics? In a **circular economy**, materials or products are regenerated after they are used (**Figure 16.36B**), where the net amount of waste produced is either minimal or non-existent.

For example, instead of using traditional fossil fuel–based plastics, which can be recycled only a few times, a new bioplastic (see Section 16.1) material made of polymers (Chapter 3) that can be indefinitely recycled could be developed. When the empty bottles made with the new bioplastics are returned to collection facilities, they could be broken

Figure 16.36 **Comparing the lifecycles of a conventional plastic bottle and a bioplastic bottle**

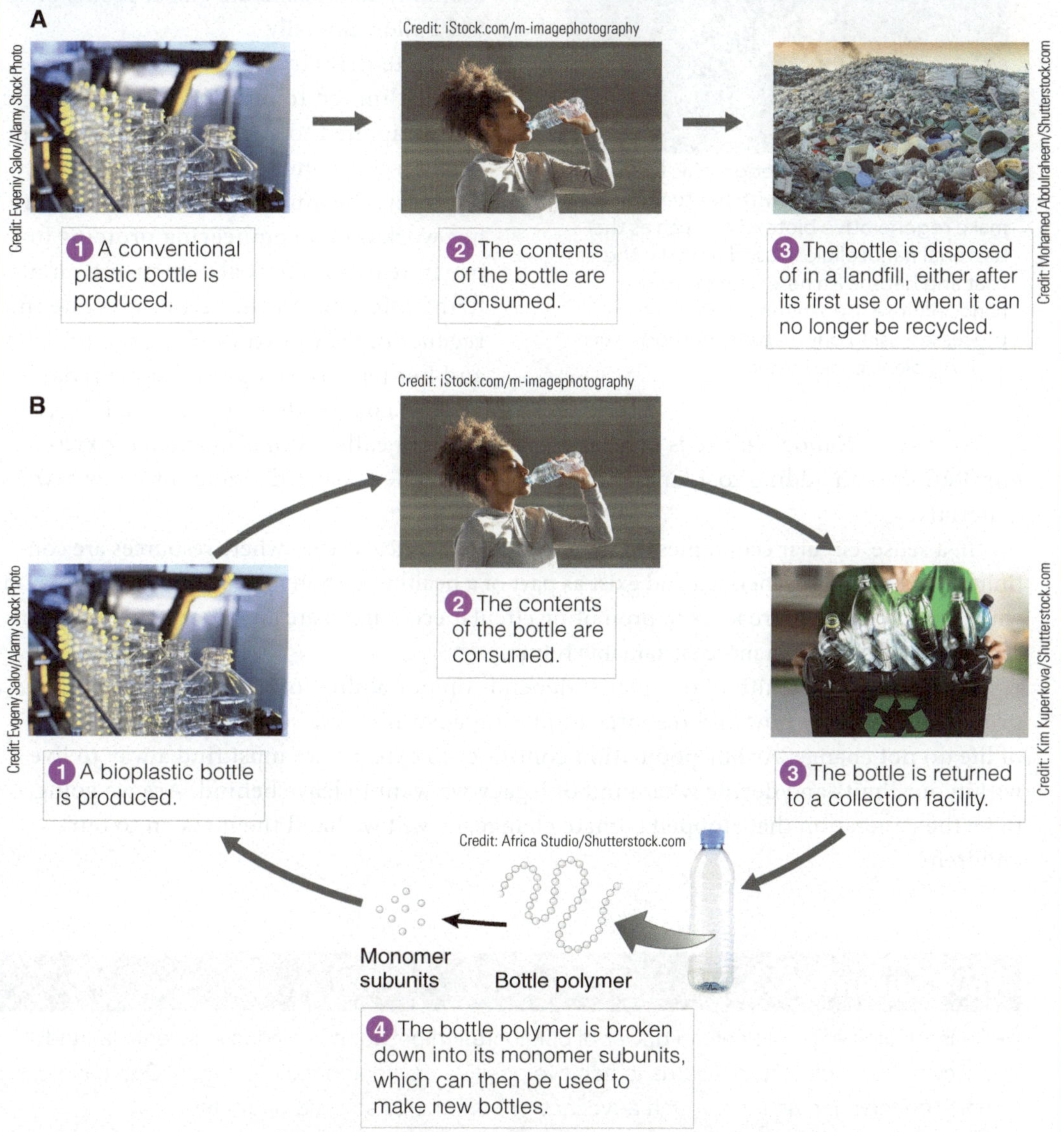

A. In a linear economy, plastics are produced, used, and discarded, requiring the continual production of new plastic and contributing to an abundance of plastic in landfills.

B. In a circular economy, bioplastic bottles can be broken down into their monomer subunits (Chapter 3) and made into new bioplastic bottles indefinitely, without the need for additional resources.

down under the required conditions, allowing manufacturers to recapture the monomers needed to build new bioplastics.

The good news is that the groundwork for the infrastructure required for a circular economy already exists, with current bottle bills, centralized drop-off locations, and curbside recycling programs. The payoff in terms of both environmental and human health is well worth it. In addition, such a model can provide the potential for profit—scientists estimate that a plastic circular economy could generate global profits of $55 billion annually.

Note that circular economies aren't strictly limited to plastics; rather, they can be applied to any consumer good. In Chapter 9, we met National Geographic Explorer Theanne Schiros. In addition to her work on bioengineering proteins for use in textiles, she is also focusing on utilizing microbes to convert food waste into regenerative biotextiles. For example, she and her team developed a biodegradable hat and coat made from nanocellulose (**Figure 16.37**). Nanocellulose is created by a bacterium called *Komagataeibacter xylinus*. Rather than adding to a landfill, the biodegradable textiles degrade into new raw materials.

National Geographic Explorer
Theanne Schiros

Credit: Jon Brown

Figure 16.37 National Geographic Explorer Theanne Schiros works with her team to make regenerative biotextiles, such as the bacterial nanocellulose used to make the coat and hat shown here. Note that the nanocellulose used in the clothing was processed using only natural methods, such as using plant-based dyes.

In a sense, circular economies mimic the biogeochemical cycles, where resources are continuously used and regenerated and exist as part of a healthy ecosystem. New technologies, along with legislation and approaches to promoting circular economies, are an important part of conservation and building a more sustainable future.

Ultimately, the health of our planet depends on our ability to recognize that the same principles of energy flow and resource limitation governing the survival of all systems of life do not change. As our population continues to expand, we must find a way to live within our limits and decide what kind of legacy we want to leave behind. Are we going to be the generation that stopped climate change, or will we hand the mess on to our children?

16.4 Summary

- Conservation efforts show the power of collaboration between individuals, scientists, and policymakers acting as stewards to achieve positive environmental outcomes. One can support conservation efforts through advocacy or community science programs.
- Using existing infrastructure in concert with new technologies that are focused on creating a circular economy is crucial for addressing climate change and building a more sustainable future.

Understand It! Wellness

"Self-care" and "wellness" are common buzzwords in society, and their use tends to pop up with increasing frequency toward the end of the semester. Wellness, or the state of being in good health, has several different aspects, including physical, mental, and emotional wellness. There is even environmental wellness, where we build and foster healthy interactions between ourselves and the Earth—one of the benefits of stewardship! But what does wellness have to do with learning?

Let's consider Maslow's Hierarchy of Needs, a psychological theory that describes what we need as humans, from our basic self-care needs to self-actualization. To reach the higher levels, we must first meet the needs at the lower levels. So, while we can sacrifice our wellness for a while, it will ultimately lead to poor health, unhappiness, and burnout. When we think about Maslow's theory in terms of learning, the indication is that we need to first meet our basic physical and psychological needs before we can reach the higher levels, where learning occurs.

But checking in on our mental health is also important. Penn Medicine recommends that we do a mental health check each day to see how we're feeling—to identify sources of worry as well as sources of support, and to assess our success at meeting our basic physiological needs, like eating nutritious foods and sleeping. Mental health checks are even more important when we are learning material that can be stressful to grapple with, such as the serious implications and threats of climate change.

Note that there is a difference between feeling stressed and feeling anxious. Stress is normal for students to experience. In fact, research tells us that a bit of stress helps you do better academically! Too much stress, however, can lead to anxiety. Anxiety can be a debilitating medical condition that interferes with your ability to perform well in school and to live life to the fullest. How do you know if it's stress or anxiety? Stress is a temporary response to an external stressor, whereas anxiety doesn't go away, even when a stressor is removed.

Remember, wellness isn't just about not being sick—it's about meeting our other needs as well. And meeting the demands of busy class and/or work schedules can often mean that wellness activities are the first things to get cut. I encourage you to plan wellness time as part of your study plan (Appendix B) each week. Even if it feels counterintuitive to take time away from a task, you'll be more productive if you take breaks.

1. What are some wellness activities that you already do? How could you plan more time for wellness into your schedule?
2. Meditation, yoga, and other mindfulness-based stress reduction activities are all evidence-based ways of decreasing stress that are relatively simple to try. Look around at your school's health page and/or on YouTube. Can you find any meditations or visualization activities that you would like to try?

Chapter Review

Section 16.1 The Impacts of Pollution

Pollution is the introduction of natural or artificial substances called **pollutants** into soil, air, or water in greater than natural amounts. **Eutrophication** occurs when a body of water is depleted of dissolved oxygen due to excessive plant or algae growth from too much nitrogen or phosphorus runoff. **Inorganic water pollutants** include elements and molecules without carbon atoms, while **organic water pollutants** contain carbon atoms. An area of marine debris created by ocean currents is a **garbage patch**.

Air pollution includes gases like **radon**, a naturally occurring radioactive gas produced in the soil; particulate pollution, which occurs when there are **particulates**, or tiny particles, in the air; and **ground level ozone (O_3)**, which forms when gases like carbon monoxide and nitrogen oxides react with sunlight. **Smog** consists of ground-level ozone and particulates, and **acid rain** is formed when fossil fuels combine with atmospheric water vapor.

The **greenhouse effect** occurs when heat energy is trapped by **greenhouse gases** in Earth's atmosphere. **Climate change** refers to lasting changes to average temperature and weather patterns in a region. Climate change is driven by **methane pollution** and **carbon dioxide pollution**.

Methanogens are a group of archaea that break down organic waste in anaerobic conditions, releasing methane gas in a process called **methanogenesis**. **Composting** occurs when aerobic microbes break down food scraps into a rich, organic soil called **compost**.

Section 16.2 The Impacts of Climate Change

Warming temperatures affect terrestrial biomes in many ways, such as the loss of **permafrost**, a soil layer that stays frozen for at least two years, as well as changing weather patterns, increased drought and fire risk, and changing wildlife habitats. Impacts on aquatic biomes include increasing **ocean acidification** and rising sea levels.

Mitigation strategies include reducing fossil fuel usage and supporting research in renewable energy.

Section 16.3 Biodiversity Protection and Restoration

Biodiversity is the variety of life on Earth. **Conservation biology** is a field dedicated to preserving biodiversity. **Stewardship** is the practice of responsibly managing resources, such as through conservation. **Bioprospecting** is the search for useful natural resources such as genes, chemicals, and macromolecules for use as medicines.

An **endangered species** is a species at high risk of extinction in the wild, and a **vulnerable species** is one that is likely to soon become endangered. **Endemic species** are confined to areas where they first evolved and are more likely to go extinct than a species with a more widespread distribution.

Habitat loss is a major threat to biodiversity, where **deforestation**, or the removal of all trees in a forested area, can cause runoff and flooding, and alterations to local weather. Deforestation also contributes to climate change, altering forests' role as a **carbon sink** or a natural reservoir for carbon. **Desertification** is a rapid shift from a grassland or woodland to a desert where the soil becomes no longer suitable for cultivation.

Overexploitation of species is an example of the **tragedy of the commons**, where when people have unrestricted access to a finite resource, the resource can be depleted and destroyed for all.

Conservation genomics is a field that studies the genes of species to support conservation efforts. **Species Survival Plans** (**SSP**) were developed to breed animals that are typically endangered. **Biodiversity hotspots** are threatened habitats with high biodiversity. **Ecological restoration** works to renew a degraded or destroyed natural ecosystem. Travel in ways that conserve the environment, and the well-being of an ecosystem and the people living there, is called **ecotourism**.

Section 16.4 Allies for the Future of Conservation

Everyone can act as stewards, people who take responsibility for the care and management of the planet. These efforts can include participating in **community science**, projects in which the public collaborates with scientists to accomplish research endeavors.

New technologies can support a **circular economy**, where materials or products are regenerated after they are used.

Practice It!

Retrieval practice is the process of pulling information from your long-term memory and using it. Here are some questions to help with your retrieval practice.

Self-Quiz

1. What characteristic is shared by all pollutants? (LO 16-1)
   a. They are all synthetic.
   b. They are all released into the water system by humans.
   c. They are all harmful to the environment.
   d. They all impact the quality of the air we breathe.

2. Which of the following examples of environmental pollution commonly impacts water directly? (LO 16-2)
   a. oil spills
   b. carbon dioxide emissions from vehicles
   c. noise from construction sites
   d. methane emissions from farms

3. Which of the following is NOT a reason why plastic is a significant environmental concern? (LO 16-3)
   a. Plastics are biodegradable.
   b. Plastics cannot be recycled indefinitely.
   c. Plastics breakdown into microplastics that can bioaccumulate.
   d. Plastics take up room in landfills.

4. Which of the following is a major contributor to particulate matter, a type of air pollution? (LO 16-4)
   a. smoke
   b. UV radiation
   c. acid rain
   d. radon

5. What role do greenhouse gases play in climate change? (LO 16-5)
   a. They block sunlight, leading to deforestation.
   b. They prevent precipitation, causing droughts.
   c. They increase the temperature on Earth by trapping heat in the atmosphere.
   d. They create ozone, which warms the planet.

6. Which of the following gases is the most potent greenhouse gas? (LO 16-5)
   a. carbon dioxide
   b. methane
   c. nitrogen
   d. oxygen

7. Atmospheric carbon dioxide from natural sources like volcanoes differs from that produced by burning fossil fuels in the ratio of ______. (LO 16-6)
   a. methane to oxygen
   b. carbon-13 to carbon-12
   c. particulate to clean air
   d. latitude to longitude

8. Since the Clean Air Act was amended in the 1990s, reduced sulfur dioxide emissions have greatly reduced the prevalence of ______. (LO 16-7)
   a. methane
   b. greenhouse gases
   c. plastic pollution
   d. acid rain

9. Which of the following strategies is most likely to mitigate climate change? (LO 16-9)
   a. reducing plastic waste
   b. removing acid from the oceans
   c. participating in community science
   d. decreasing fossil fuel consumption

10. ______ describes the variety of life on Earth. (LO 16-10)
    a. Bioprospecting
    b. Biodiversity
    c. Biological sciences
    d. Extinction

11. Which of the following statements is true regarding endangered species as compared to vulnerable species? (LO 16-12)
    a. Endangered species have no members left, while vulnerable species have a stable population.
    b. Endangered species are abundant in their environment, while vulnerable species are rare.
    c. An endangered species is at a high risk of extinction in the wild, while a vulnerable species is likely to soon become endangered.
    d. Endangered species will definitely go extinct, while vulnerable species are likely to survive.

12. Which of the following factors directly contribute to habitat loss? (LO 16-13)
    a. climate change
    b. desertification
    c. deforestation
    d. All of the above contribute to habitat loss.

13. Which of the following activities can be categorized as ecotourism? (LO 16-14)
    a. hiking and kayaking
    b. birdwatching
    c. community-led tours
    d. All of the above are correct.

14. Which of the following activities is a way you can personally contribute to conservation efforts? (LO 16-14)
    a. participating in community science
    b. buying an exotic pet from an illegal vendor
    c. increasing your personal fossil fuel emissions
    d. removing vulnerable species from your yard

15. Which of the following statements best describes the concept of a circular economy? (LO 16-17)
    a. an economic system focused on commerce that circles the globe
    b. an economic system based on the regeneration of materials in a sustainable manner
    c. an economic system based on recycling conventional plastics
    d. an economic system specifically for reducing greenhouse gas emissions

Critical Thinking

1. Scientists in Mexico have been transplanting oyamel fir (*Abies religiosa*) saplings in the Monarch Butterfly Biosphere Reserve at higher elevations than where they currently grow. Why is this change in elevation necessary? (LO 16-8, 16-9)

2. A large industrial company is proposing to build a factory in a region known for its rich biodiversity. The proposed location is home to several ecosystems that are vital for the local climate and water cycle. The company argues that the factory will bring significant economic benefits, such as job creation. Present arguments for conserving biodiversity at the proposed factory site. (LO 16-11)

3. Provide examples of success stories in habitat preservation and biodiversity recovery at three levels: the individual level, the community level, and the legislative level. Which of these strategies has the largest impact? (LO 16-14, 16-15, 16-16)

4. Using the example of the rise of large community science projects like Monarch Watch, discuss how emotional appeal, species visibility, and public engagement influence the way biodiversity conservation priorities are defined. Critique the effectiveness of these approaches in addressing broader biodiversity goals, including less visible or charismatic species. (LO 16-16)

5. When trying to combat plastic pollution, should the companies that produce the plastic packaging be held responsible for the pollution they cause? Who should be responsible for shifting us away from disposable, single-use plastics to a circular economy, where items can be recycled and reused? (LO 16-3, 16-17)

Unit 5 Review

In this unit, we discussed ecology and climate change. Throughout the unit, we looked specifically at the different lines of evidence that indicate that climate change is an indisputable and potentially devastating force. We also looked forward to future technologies to address climate change, while also consulting examples from the past, where people and governments were able to work together to address the environmental challenges of their day…and make positive changes for the future.

In Chapter 14, we learned about population ecology. We first discussed ecology, and the insights that can be gained from its study. We then moved on to examine the characteristics of populations, and how biologists monitor population growth and life history patterns. We also examined the changing human population, and the overall global footprint among nations.

In Chapter 15, we learned about communities and ecosystems. We started with discussing community structure before going on to how different species interact with one another, and how communities change over time. We then discussed the nature of ecosystems, including an overview of Earth's biomes and the factors that influence climate in a given area. We closed with a discussion of Earth's biogeochemical cycles, and how different nutrients or chemicals move through the environment.

In Chapter 16, we looked at conservation biology through a hopeful lens, examining the environmental issues of today as well as some examples from earlier generations. When we look at historical environmental issues, we can see the value of people coming together purposely to enact change. We examined the impact of pollution on the environment, from water and plastic pollution to the role of air pollution in contributing to climate change. We also discussed the impacts of climate change, and examined biodiversity, threats to biodiversity, and ways to help preserve biodiversity. Finally, we examined conservation in the context of our everyday lives and things that we can do to spur positive environmental change, as well as future technologies like circular economies that will help society and the biosphere in the future.

Credit: FangXiaNuo/Getty Images

Unit

6

Biology and You

Public Health

Do you garden? Or maybe you've participated in gardening as part of a class or community project? Growing up, my grandfather always had an extensive garden, which he tended to until he was no longer able to physically do so—well into his 80s. Perhaps it isn't surprising that he was able to garden for so long. Scientists have found evidence suggesting the many benefits of gardening, including to one's health and overall sense of well-being. Gardeners, incidentally, also tend to live longer, so my grandfather isn't an exception! Gardening is an example of an activity that encompasses a primary theme of this final unit—the topic of public health.

Public health is a discipline concerned with improving the health of individuals and the communities that they call home. Some people might think of those who work in public health, or public health officials, as people who are primarily concerned with infectious disease prevention. However, as we'll learn in this unit, disease prevention goes well beyond fighting infectious disease. In this final unit, we'll bring together insights from the earlier chapters—ranging from molecules to cells to the diversity of life and how living things exist as part of a bigger system—to see how all of these seemingly disparate concepts connect to paint a picture of public health.

But what does gardening have to do with public health? Well, spending time moving one's body outdoors is good for both physical and mental wellness—something that we'll explore in greater detail in Chapter 17. We'll also learn the basics of our anatomy and physiology, exploring select organ systems that make up the body—including the integumentary, nervous, endocrine, reproductive, muscular, skeletal, and immune systems—specifically as they relate to a public health issue. For example, we'll learn about endocrine disruptors, which can be found in the products that we use every day, and how they can impact our reproductive and endocrine systems and contribute to infertility. Other topics include mental health and our nervous system, vaccines and the immune system, stem cells for wound healing, as well as an exploration of whether sitting really is the new smoking!

How we grow food, and the economic importance of plants, are topics we'll explore in Chapter 18, where we'll learn about plant anatomy and physiology. We'll start by learning about the structures that make up a plant, as well as the myriad ways that plants reproduce and how those reproductive styles reflect plant and pollinator evolutionary history. Throughout, we'll apply the concepts of plant structure and function to agriculture and our food supply. We'll also discuss the value of forests and how plants are important for our future.

Preview It!

What questions do you have about public health? What do you think animal and plant physiology have to do with public health? Write down three things you want to learn in this unit.

Credit: MoMo Productions/Getty Images

17

A User's Guide to the Human Body

Chapter Outline:

Living in Our Bodies

Have you ever attended a yoga class? Yoga is an ancient practice, dating back some 5,000 years. Collectively, yoga is a physical, mental, and sometimes spiritual exercise that exists in many forms. Yoga classes typically involve movement, stretching, and breathing, in different combinations and with different focuses. Some yoga classes are more meditative, while others are more physically intense. Although there's a stereotypical "yoga body," anyone can practice yoga, from young to old, regardless of body composition or physical limitations. Numerous scientific studies point to the health benefits of yoga. Benefits to mental health and the nervous system, the cardiovascular system, the musculoskeletal system...and more! But what are all these systems? How do they work together? And what does it mean to be *healthy* anyway?

Oftentimes, we don't think about the organ systems in our bodies unless something goes wrong. We breathe without thinking about it...until we get bronchitis. We move around and use muscles in our bodies, but don't notice our musculoskeletal system until we have muscle aches or pull a tendon. We're also constantly receiving messages about "good health," but what does that even mean? And how does all of this relate to public health, the subject of this unit?

Yoga is an activity suitable for all that promotes wellness across body systems. Regular yoga practice yields positive benefits to mental and physical health.

In this chapter, we are going to integrate the various topics you studied in many of the previous chapters of the book to collectively tell a story about our bodies and relate this knowledge to our lifestyle choices. Drawing on insights from Unit 1 (nutrition), the cells that make up our bodies (Chapters 5 and 6), our genetics and epigenetics (Chapter 7 and 8), the evolution of the human body (Unit 3), and the passage of nutrients and pollutants through the environment (Chapter 15 and 16), we will learn about how our bodies are put together, and the relationship between public health and our anatomy and physiology.

We'll begin with an overview of the human body. We'll discuss the levels of organization of the human body and give an overview of the organ systems. From there, the rest of the chapter will have a slightly different feel from your typical textbook chapter. Each section after our overview will contain a vignette about a public health issue highlighting the anatomical and physiological processes behind the story. We'll learn about chronic wounds and the integumentary system before getting into the physiology of mental health and addiction. We'll also learn about how chemicals in the environment impact our reproductive and endocrine systems and how environmental toxins are impacting both systems and contributing to rising rates of infertility. We'll also explore the relationship between muscle mass and bone density and our overall health. We'll close with a discussion of how vaccines mimic our immune system to protect us from disease.

Preview It!

How many body systems can you name? What is one question you have about how your body works? Which public health issue are you most excited to dive into?

Study Tip

Look for common themes! As you read through the following sections, find the different levels of organization in each system. What are the cells? Tissues? Organs?

17.1 An Introduction to the Human Body

Learning Objectives

17-1 Describe the levels of organization in the human body.

17-2 Compare the studies of anatomy and physiology.

17-3 Identify and describe the functions of each human organ system.

Big Ideas in Biology

Structure and Function

In earlier chapters, we learned about the relationship between molecular structure and function, such as how the active site of an enzyme determines the substrate that it is specific to. This same idea applies to anatomy and physiology, such as how the hairlike cilia (Chapter 5) on the cell surface in Figure 17.1 (3) rhythmically move to clear debris from the lungs.

Before we can dive into public health issues as they relate to each system, we'll first need to "eat our Wheaties" and cover some foundational content. In Chapter 1, we learned that one of the characteristics of life is that there is an underlying structure. While proceeding through different chapters, we learned more about the structure of living things, starting with atoms (Chapter 2; Figure 17.1 (1)); the molecules of life, or macromolecules (Chapter 3; (2)); and cellular structure (Chapter 5; (3)).

All animals are multicellular, and nearly all have cells organized as tissues (4). A **tissue** consists of one or more cell types, plus there is often an extracellular matrix, arranged in a way that allows cells to collectively perform a specific task or tasks. Humans have four tissue types: **epithelial tissue**, which covers body surfaces and lines internal cavities; **connective tissue**, which holds body parts together and provides structural support; **muscle tissue**, which moves the body; and **nervous tissue**, which detects stimuli and relays information. Note that each tissue type has unique cell types that are not present in other tissues. For example, contractile cells are present in muscle tissue, but not in nervous or epithelial tissue.

Tissues make up organs. An **organ** is a structural unit of two or more tissues organized in a specific way to carry out specific tasks. For example, the lungs (5) are organs that facilitate gas exchange between the body and its environment. Lungs are part of an organ system, in this case, the respiratory system (6). In **organ systems**, two or more organs and other components interact physically, chemically, or both, to perform a common task. For example, in the respiratory system, air enters through your mouth and nose, goes through a passageway called the trachea, and then arrives in your lungs. The gas travels into the smallest parts of the lung, the *alveoli*,

Figure 17.1 Levels of organization in the human body

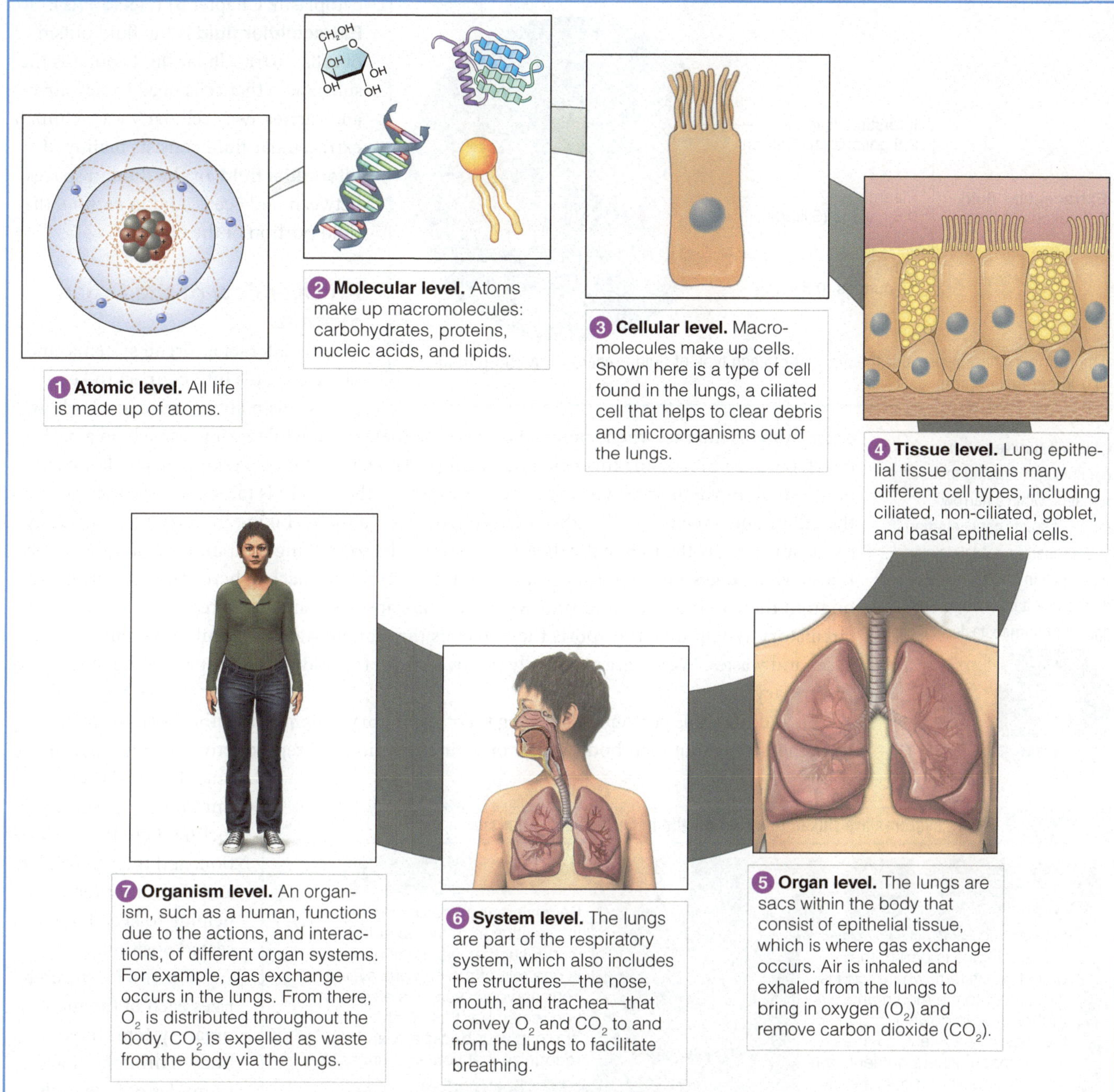

where gas diffuses (Chapter 5) between the lungs and the blood vessels of the circulatory system. Humans have 11 body systems, which come together to support the **organism**, or any individual lifeform 7. Collectively, all organ systems contribute to **homeostasis**, the process by which the body maintains conditions within a range that is required by the body's cells.

The study of the structure of the body is called **anatomy.** Anatomy is closely related to another discipline, physiology. While anatomy refers to the form of the body, **physiology** is concerned with how the body functions. Since form (or structure, per the Big Idea in Biology) influences function, and vice versa, examining both gives us a well-rounded understanding of how the body works.

In addition to the cells, tissues, and organs that make up our bodies, there is also quite a lot of watery fluid. In Chapter 2, we learned how life relies on water to survive. When we consider the human body, we see that in an average human adult, body fluids (which are mainly water) account for about 60% of body weight. Most body fluid is **intracellular fluid**, located in the

Figure 17.2 **Distribution of human body fluids**

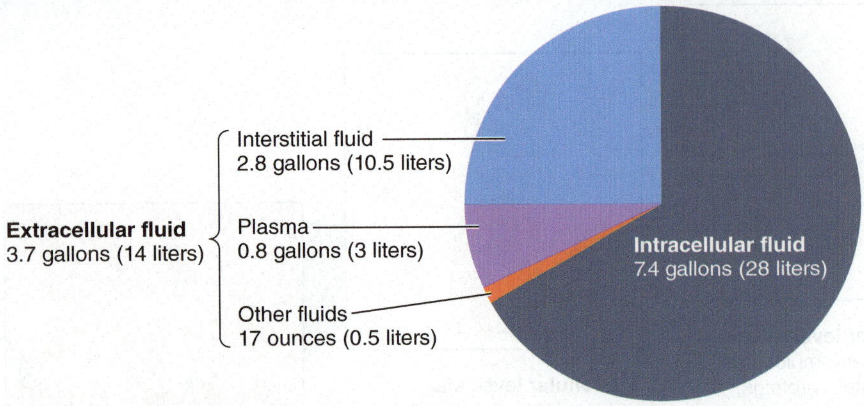

The numbers here represent the amount of each type of fluid present for a 150-pound (70-kilogram) person.

cytosol of cells (the fluid portion of the cytoplasm; Chapter 5) (Figure 17.2). **Extracellular fluid** is the fluid outside of cells. Extracellular fluid contains the substances that cells need to stay alive and carries away cellular waste. Human extracellular fluid consists mainly of **interstitial fluid**, or the fluid in spaces between body cells, and in *plasma*, the fluid portion of the blood.

An Overview of the Organ Systems

Organs interact as organ systems, and organ systems interconnect to keep an organism alive. Let's consider how organ systems interact when someone is checking their email while eating a snack on a park bench (Figure 17.3). While this person is sitting outside, their body is taking in a wide variety of information via their nervous system. For example, their brain is taking in information from the cell phone screen as well as the environment around them. Taking in nutrients, such as by eating an apple, is the role of the digestive system. The food is ingested into the mouth, which is also where digestion begins. Digestion continues in the stomach, where the nutrients move on from there to the small intestine, where they are absorbed and delivered to the blood. The circulatory system then transports the nutrients throughout the body while also transporting gases and wastes. Wastes are eventually removed from the body as either part of the digestive or the urinary system.

Cells will survive only if the fluid surrounding them remains within a narrow range of solute concentration and temperature. Maintaining this range in the cellular environment is an important aspect of homeostasis.

In addition, this person is also using their respiratory system to breathe; hormones are traveling throughout their body as part of their endocrine and reproductive systems; their integumentary (skin) and immune systems are protecting them from infection; and their muscular and skeletal systems are helping to keep them in a seated position.

Figure 17.4 summarizes each of the main organ systems in your body. Throughout are examples of how each given organ system interacts with others. Note that we'll discuss the muscular and skeletal systems together as the *musculoskeletal system* later in the chapter. We'll also return to some of the body systems described here in order to understand our physiology in the context of a public health issue.

Figure 17.3 **Organ systems are linked together for the benefit of an organism**

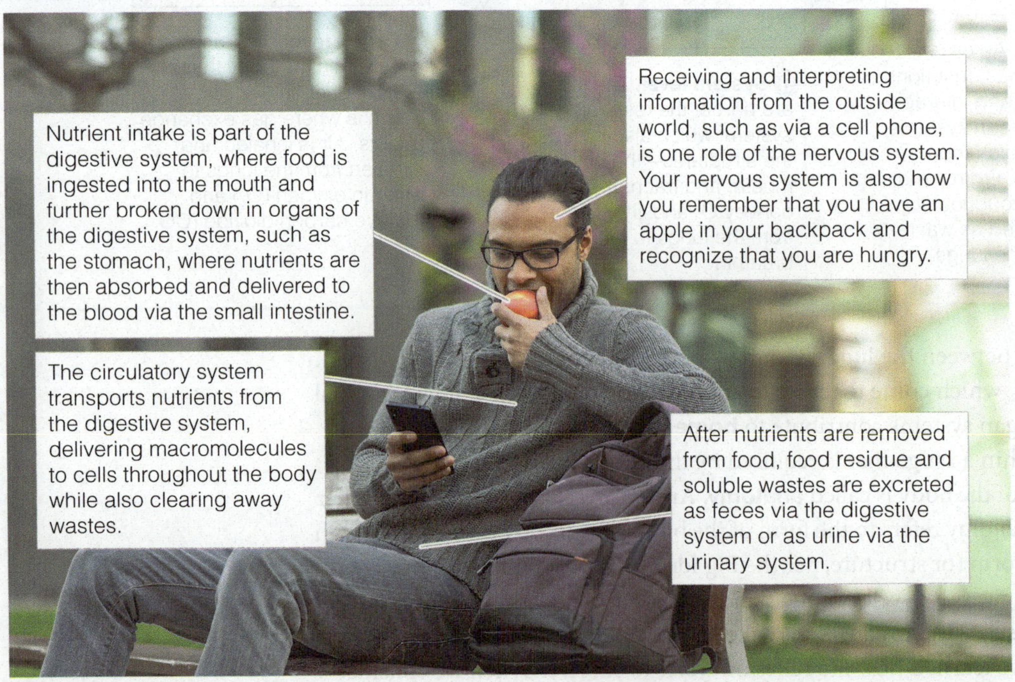

As you review this figure, identify some ways that organ systems interact to keep the body supplied with essential substances and free of wastes. Note that not all organ systems are represented here.

Credit: Mimagephotography/Dreamstime LLC

Figure 17.4 An overview of human organ systems

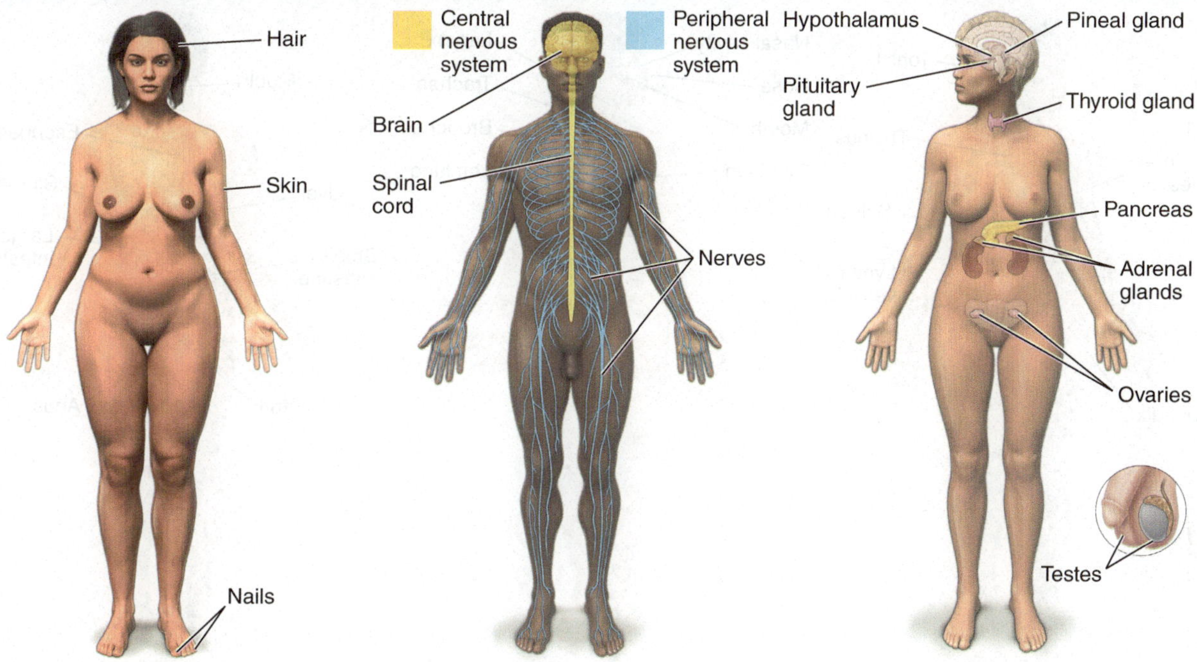

The **integumentary system** protects the body from injury and infection by pathogens, prevents fluid loss, and helps to regulate body temperature. The skin interfaces with the nervous system.

The **nervous system** monitors the internal and external environment, integrates information, and controls muscles and glands. The brain and spinal cord are part of the central nervous system, and the peripheral nervous system includes the body's nerves.

The **endocrine system** consists of hormone-secreting endocrine glands. It ties in with the reproductive system, which include reproductive glands like the ovaries and testes. It links with the nervous system as well, and like that system, controls other organ systems.

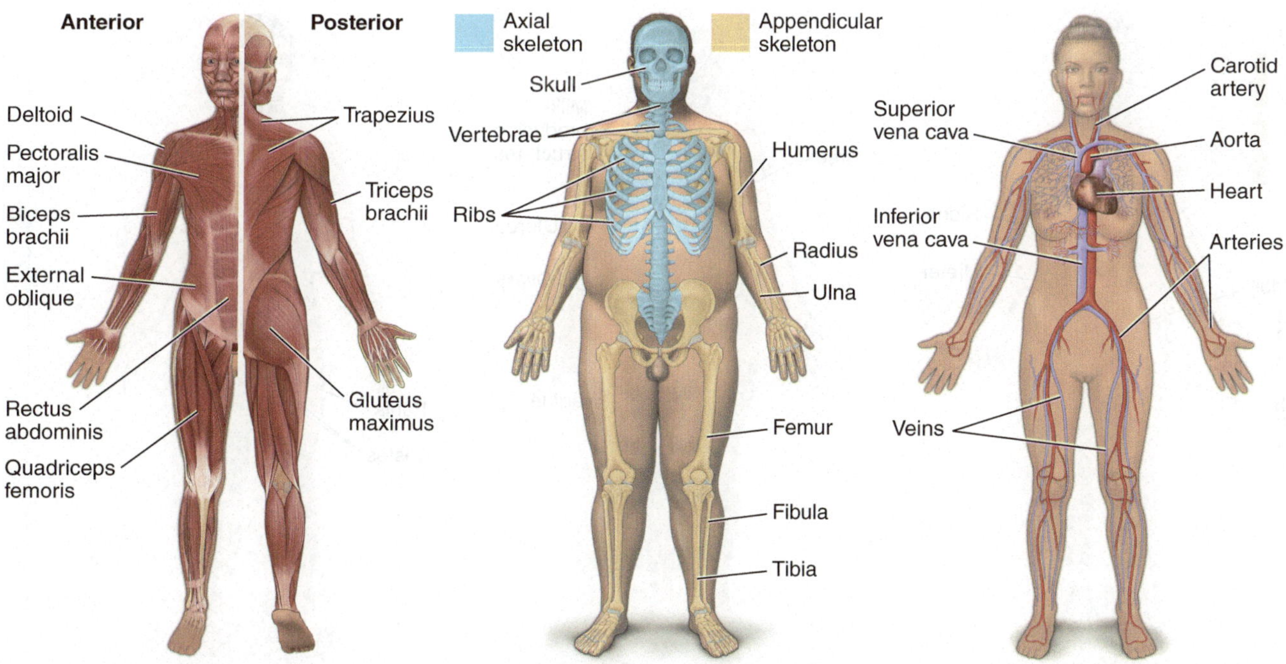

The **muscular system** is comprised of muscles that move the body and its parts, such as the skeleton. This system plays a role in regulating body temperature by generating heat. In this image, only a handful of muscle groups that you might be familiar with are labeled.

The **skeletal system** is comprised of bones, which protect internal organs, serve as a point of attachment for skeletal muscles, store minerals, and produce red and white blood cells. The muscular and skeletal systems work together to move the body.

The **circulatory system** consists of the heart and blood vessels. It works in coordination with the respiratory, digestive, and urinary systems in delivering O_2 and nutrients to cells and clearing away CO_2 and other wastes.

Figure 17.4 An overview of human organ systems, *continued*

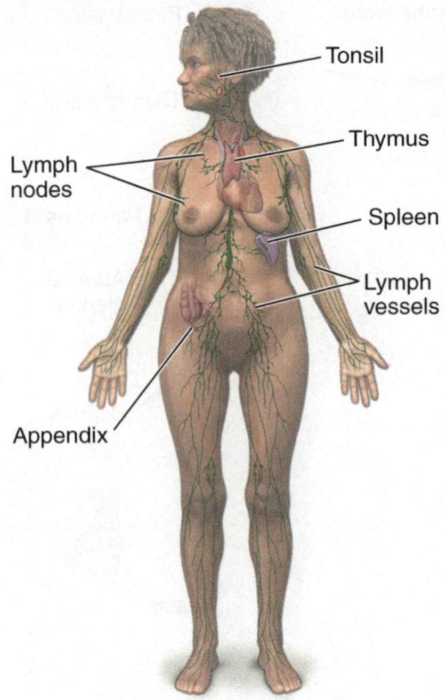

The **lymphatic system** has vessels that move fluid (lymph) from tissues to the blood. Lymphatic vessels and organs, such as the lymph nodes, contain **immune system** cells, thereby working with that body system to protect and rid the body from infection.

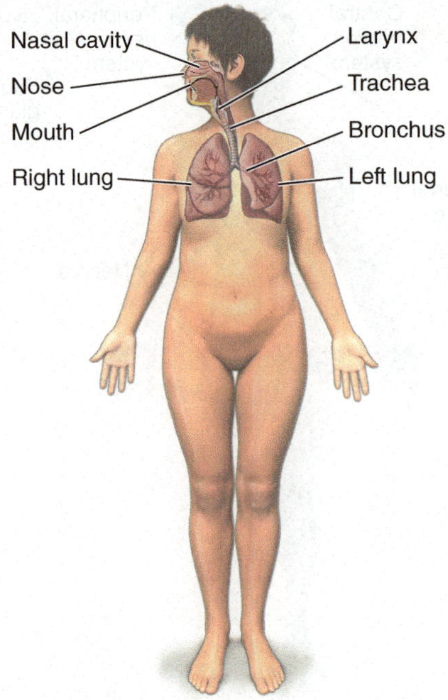

The **respiratory system** includes two lungs, the airways leading to them, and the structures within the lungs where gas exchange occurs with the circulatory system. Gas exchange results in bringing O_2 to the body's tissues and removing CO_2.

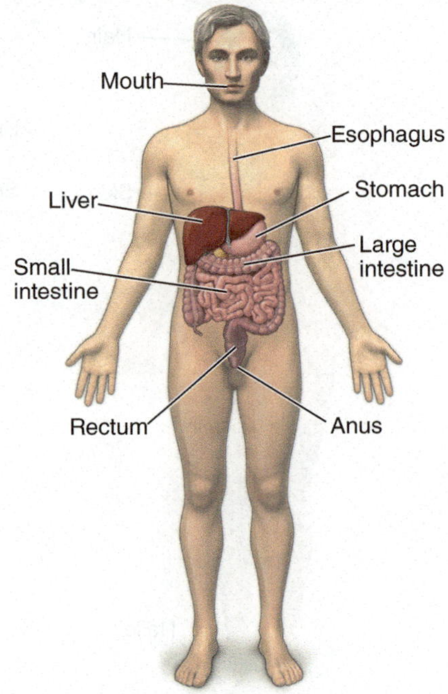

The **digestive system** takes in food, breaks it down, delivers nutrients to the blood, and eliminates undigested wastes. It includes organs of the digestive tract, and glands such as the salivary glands, liver, and pancreas, which supply substances that break down food.

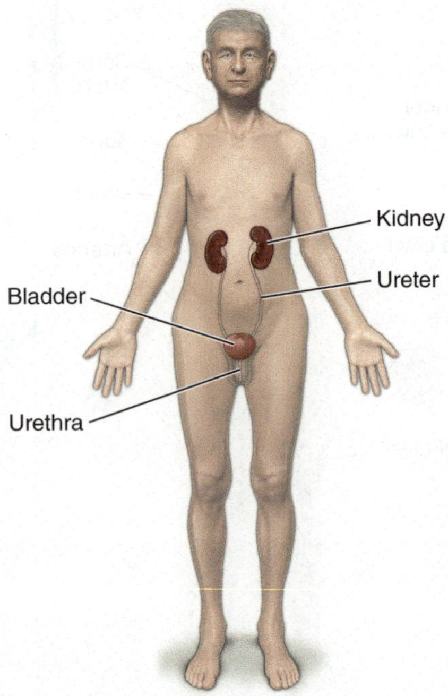

The **urinary system** consists of the kidneys (organs that filter blood and create urine), the bladder, and ducts that convey urine to be excreted from the body. It removes wastes from blood and adjusts blood volume and solute composition.

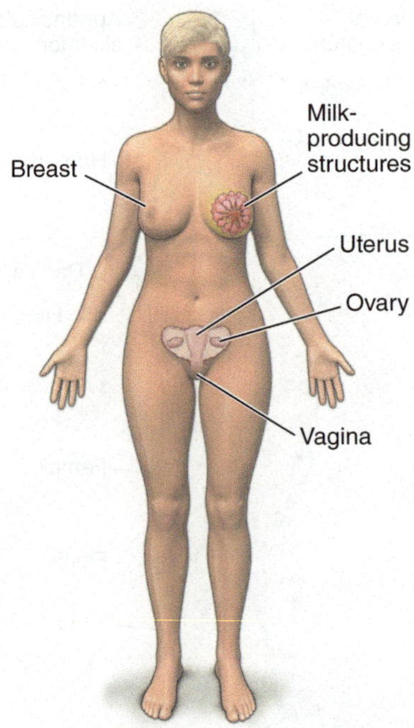

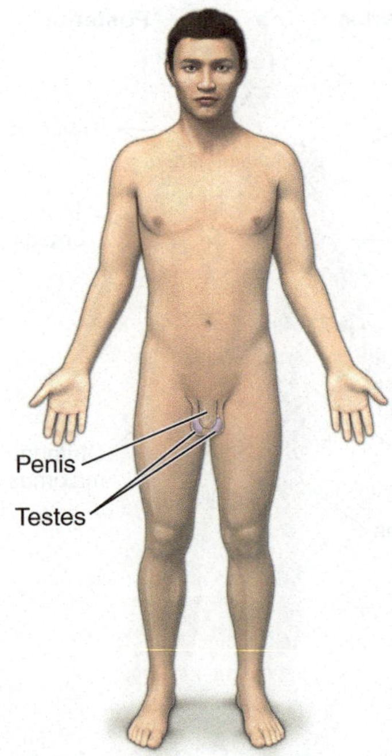

The **reproductive system** of both sexes include glands that are involved in gamete production (ovaries or testes), and structures like the penis and vagina, which are used during sex to put gametes in contact with one another. A female typically has structures for housing the developing fetus during pregnancy, and for milk production, to sustain the offspring once born. Note that intersex individuals may have a mix of these structures.

17.1 Summary

- The human body is organized into levels from smallest to largest, beginning with atoms, then macromolecules, cells, tissues, organs, organ systems, and finally the organism.
- The study of the structure of the body relates to anatomy, while physiology is concerned with how the body functions.
- Organs interact as organ systems, and organ systems interact to keep an organism alive. These systems function individually and cooperatively to respond to our changing needs.

17.2 Healing Chronic Skin Wounds and the Integumentary System

Learning Objectives

17-4 Explain what chronic skin wounds are and why they are a public health burden.

17-5 Describe the structure and function of the skin.

17-6 Identify the steps of wound healing.

17-7 Discuss how stem cells can be used to heal chronic skin wounds.

Our skin acts as an important barrier between our bodies and the outside world. Perhaps it's not surprising then that our skin is continually at risk for damage. Therefore, it's important for our bodies to have processes to heal wounds when they occur. Minor injuries are common occurrences, whether from papercuts, scrapes, or splinters. These small, superficial wounds usually heal within a few days.

However, some wounds fail to heal in a reasonable amount of time. A non-healing wound, or **chronic wound**, is an injury that fails to heal after several weeks, even with proper treatment. Chronic wounds typically occur because there is a physiological process that either prevents healing or causes healing to slow—which we'll look at in more detail later in this section. For example, **diabetes** is a condition where one's blood sugar (Chapter 3) is difficult to regulate. If diabetes isn't treated, ongoing high levels of blood sugar can cause poor circulation. Without good blood flow, it is difficult for wounds to heal because the cells that are needed to heal the wound travel in the bloodstream.

Therefore, people with uncontrolled diabetes are at risk for developing chronic wounds like ulcers in areas of the body with poor circulation, such as the feet. Uncontrolled diabetes isn't the only condition that can cause chronic wounds. Lack of mobility can cause pressure ulcers, which are also known as bedsores. Chronic wounds can also occur due to a person having a compromised immune system or poor blood circulation. Finally, as the body ages, its ability to heal wounds declines due to fewer growth factors and stem cells in the skin.

Chronic wounds are a big problem—and not just for the 6.5 million people annually who are affected by a chronic wound in the United States. The economic burden of medically managing chronic wounds is staggering, ranging from \$28.1 to \$96.8 billion *per year*. These numbers don't include the negative impacts on someone's life—chronic wounds can be debilitating. In severe cases, chronic wounds can lead to gangrene (tissue death), and ultimately, amputation. Currently, approximately 11% of the U.S. population—or about 38 million Americans—have diabetes, and each year, 154,000 individuals will have a body part amputated as a result of diabetes complications.

What can be done to relieve the burden of chronic wounds? Before we get there, we need to understand our skin and how damage to our skin (wounds) heals. Then, we'll learn about how stem cells are used to help treat chronic wounds.

Skin: Your Largest Organ

Your skin is the largest organ in your body, and since it interfaces with many different bodily systems, it also performs myriad functions. For example, it coordinates with the nervous system through sensory reception, allowing you to detect temperature and touch. Skin is also an important part of your immune system, where it helps to prevent infection by keeping pathogens out of your body. Skin prevents water loss, and with the circulatory system, helps to maintain your internal temperature. Skin, along with your hair and nails, makes up the integumentary system.

The outermost region of skin, the **epidermis**, is made of a stratified squamous epithelium (Figure 17.5). Human epidermis consists mainly of **keratinocytes**, epithelial cells that synthesize the waterproofing protein keratin. Keratinocytes in deep epidermal layers continually divide by mitosis, so new cells displace the older ones toward the skin's surface. As cells move outward, they become flattened, lose their nuclei, and die. Dead keratinocytes accumulate at the skin surface, forming a tough layer that blocks the entry of many pathogens and helps the body to conserve water.

Did You Know?

A tan is the result of increased delivery of melanin to keratinocytes. It's your body's response to protect your cells as a result of increased UV exposure.

Melanocytes, another type of epidermal cell, make pigments called melanins. Melanocytes have cellular extensions that look like the tentacles of an octopus that the cell uses to transfer the melanins to keratinocytes. Variations in human skin color arise from differences in the distribution and activity of melanocytes in the skin, and in the type of melanin they produce. As we learned in Chapter 12, melanin functions as a sunscreen, absorbing ultraviolet (UV) radiation that can damage DNA and other biological molecules. When skin is exposed to sunlight, melanocytes produce more of the brownish-black melanin, which in part accounts for the variation we see in skin color.

Beneath the epidermis is the **dermis**, which consists mostly of dense connective tissue. Within the dermis are small blood vessels, sweat glands, and nervous tissue. Additionally, the dermis contains the base of your hair follicles. Within hair follicles are the fastest-dividing cells of the human body. As cells at the base of each follicle divide, they push the cells above them upward, lengthening the hair fiber. The part of a hair that extends beyond the skin surface is composed of the keratin-rich remains of dead cells. Secretions from oil glands adjacent to each hair follicle help to keep the hair smooth and shiny. The smooth muscle that attaches to each hair can reflexively pull the hair upright in response to cold or fright—causing goosebumps.

Skin sits atop a layer of connective tissue, nerves, and **adipose tissue**, which stores fat. Together, they constitute the **hypodermis**. This layer contains larger blood vessels that branch to

Study Tip

Use the root words to help you remember the different tissue layers. *Epi* means "on," and it's the layer on top. *Hypo* means "under," and *dermis* means "skin." Consider that a hypodermic needle is one that goes under (*hypo*) the skin (*dermis*).

Figure 17.5 A cross-section showing the structure of human skin

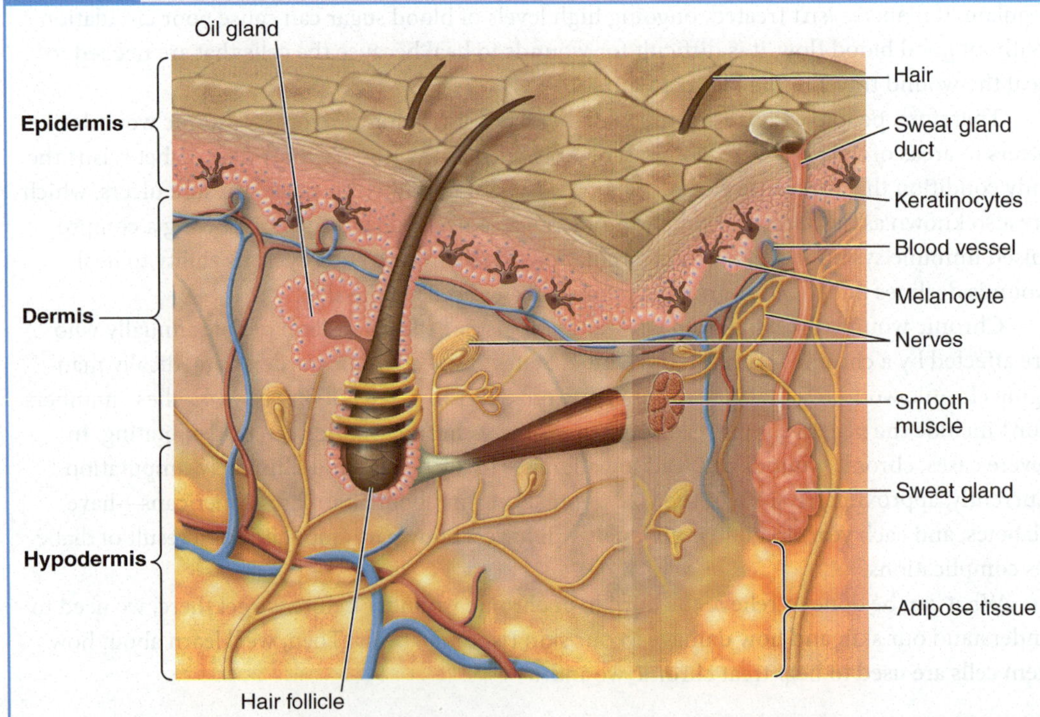

the small vessels that run through the dermis. The depth of the hypodermis varies among body regions. For example, the hypodermis beneath the skin of the eyelids is thin, with only a few adipose cells, while the hypodermis of the buttocks is thickened by many adipose cells.

How Skin Wounds Heal

Let's consider what happens in normal and chronic wounds (**Figure 17.6**). Say you are walking outside and cut your foot. One of the first things that happens is the release of *histamine,* which causes inflammation and signals for immune system cells, such as *white blood cells* (Section 17.6), to migrate from the bloodstream into the wound site. If the wound is deep enough that bleeding occurs, **platelets**, which are clotting factors, enter the wound via the blood vessels to help stop

Figure 17.6 A comparison of a post-injury response in normal and chronic wounds

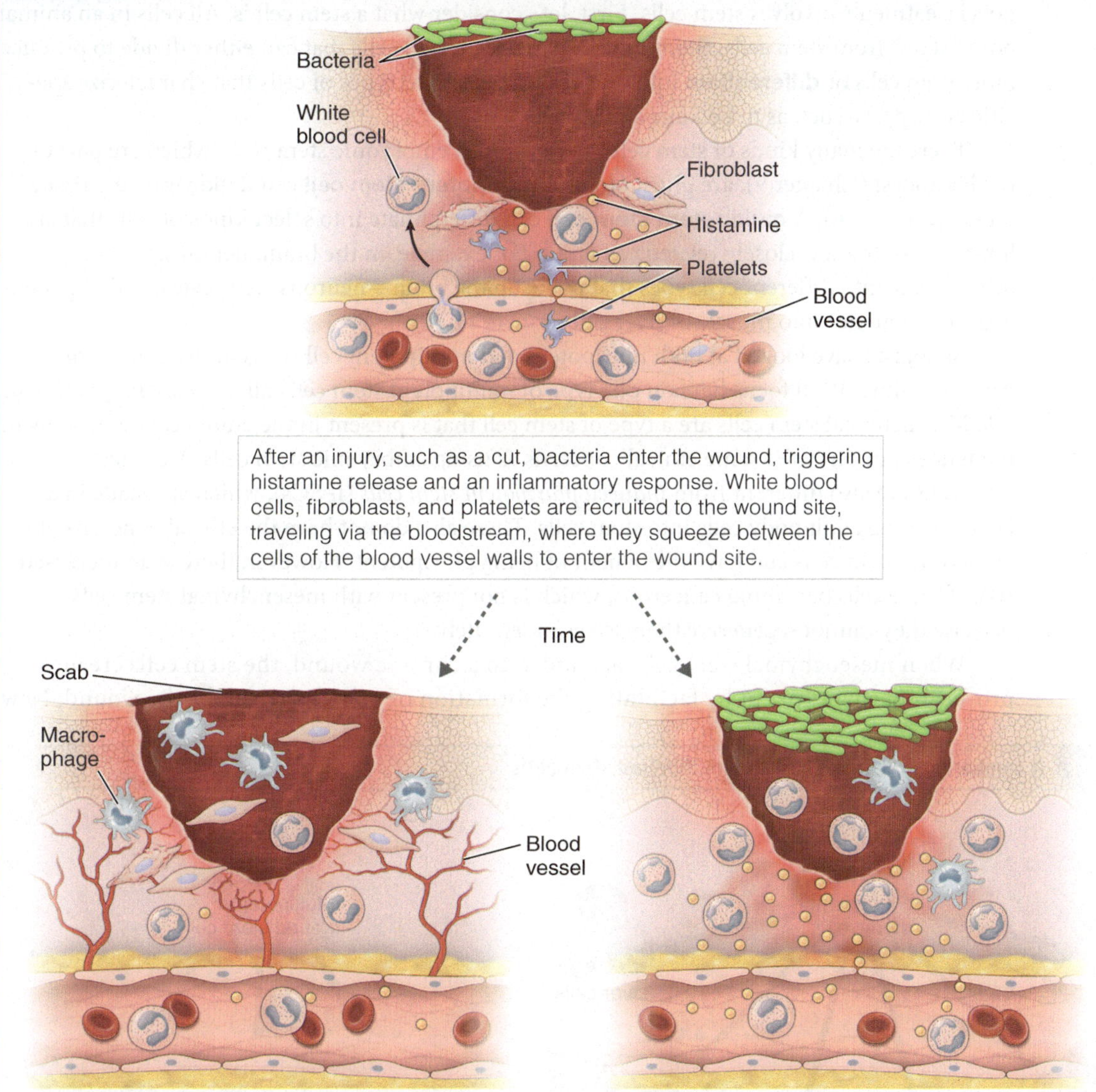

After an injury, such as a cut, bacteria enter the wound, triggering histamine release and an inflammatory response. White blood cells, fibroblasts, and platelets are recruited to the wound site, traveling via the bloodstream, where they squeeze between the cells of the blood vessel walls to enter the wound site.

Normal wound repair. A scab forms over the wound and the bacteria are removed by white blood cells, such as macrophages. Blood vessels grow into the wound site to promote healing, and fibroblasts add collagen fibers upon which cells will grow. The scab will eventually go away and the wound will heal.

A chronic wound. No scab forms and bacteria overgrow the wound. White blood cells persist, and fibroblasts have reduced or no function. The wound stays in the inflammatory phase, and no healing occurs. Note how blood vessels do not grow into the wound site.

The top image shows what happens during an injury. The bottom-left represents normal wound healing, and the bottom-right shows a chronic wound.

the bleeding. A scab forms due to the action of **fibroblasts**, which also migrate into the wound. During wound healing, the fibroblasts create a collagen matrix (Chapter 5), which acts like a base layer for the new epithelial cells to form. Once the new base layer is formed, epithelial cells migrate over the wound, eventually closing it (Figure 17.6, bottom left).

In contrast, in a chronic wound, there is decreased blood flow, meaning that there is a lack of immune cells, platelets, and fibroblasts entering the wound site. As a result, no scab forms and bacteria overgrow in the wound. The wound stays stuck in the inflammatory phase of the immune response, and no healing occurs (Figure 17.6, bottom right), leading to a chronic wound.

Can Stem Cells Help Treat Chronic Wounds?

In this section, we've learned about the public health burden of chronic wounds, the physiology of wound healing, and the anatomy of the skin. Given the public health burden of chronic wounds, it's important for scientists to develop new ways of treating them—and one of those novel treatments involves stem cells. First, let's consider what a stem cell is. All cells in an animal body "stem" from stem cells. **Stem cells** are unspecialized cells that can either divide to produce more stem cells or **differentiate** into one of the specialized types of cells that characterize specific body parts, such as to become a liver cell.

Embryonic stem cells are more attractive options for use in healing chronic wounds, but are less studied due to the ethical, religious, and legal implications of their isolation and use.

There are many kinds of stem cells. For example, embryonic stem cells, which are part of the blastocyst (Chapter 9), are pluripotent. A **pluripotent stem cell** can divide into any tissue type (Figure 17.7A). A **multipotent stem cell** can differentiate into select kinds of cells that are limited to only a few closely related cell types. For example, in the brain, neural stem cells can differentiate into different cell types that make up the brain—neurons, astrocytes, and oligodendrocytes—but not into muscle or liver cells (Figure 17.7B).

Biologists have looked at both pluripotent and multipotent cell types in the context of wound healing. We'll focus here on one type of multipotent stem cell called a *mesenchymal stem cell*. Mesenchymal stem cells are a type of stem cell that is present in the adult body, and so using them does not carry the same ethical concerns as using embryonic stem cells. Mesenchymal stem cells are also different from *induced pluripotent stem cells* (iPSCs), which are made in a lab by turning adult body cells into stem cells. These also do not have the ethical concerns of embryonic stem cells and have the benefit of being pluripotent. However, there is an increased risk of these cells becoming cancerous, which is not present with mesenchymal stem cells because they cannot regenerate themselves indefinitely.

When mesenchymal stem cells are added to a chronic wound, the stem cells create a pro-healing environment by facilitating the formation of new tissue within the wound. New

Figure 17.7 **A comparison of pluripotent and multipotent stem cells**

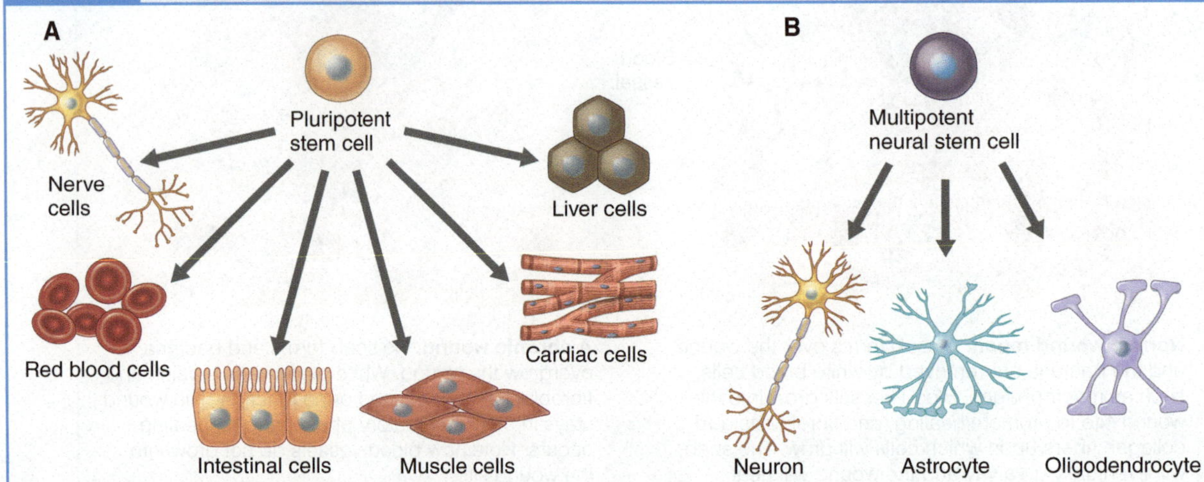

A. A pluripotent stem cell can differentiate into any cell type in the body.

B. A multipotent stem cell can differentiate into only limited cell types, such as the different cells found in one organ: the brain.

tissue is formed when the mesenchymal stem cells differentiate. These cells also give off signals, a form of paracrine signaling (Chapter 6), in the wound environment. These signals recruit immune system cells and other cells to the wound to facilitate the healing process.

Although research in mesenchymal stem cells is promising, these cells are limited in their utility due to their finite ability to regenerate themselves. Overall, the use of stem cells in the treatment of chronic wounds could potentially lead to improved quality of life for people suffering from chronic wounds, speeding up the healing of chronic wounds and ultimately reducing healthcare costs.

17.2 Summary

- Chronic skin wounds are wounds that fail to heal after several weeks and are often due to underlying health conditions, such as diabetes. Chronic wounds affect millions of people in the United States and require prolonged treatment, have high associated healthcare costs, and can have a significant negative impact on a person's quality of life.
- The skin consists of three layers: the epidermis, dermis, and hypodermis. The skin coordinates with the nervous system to provide sensory perception, protects from infection, prevents water loss, and helps to regulate body temperature.
- In normal wound healing, first a scab forms to stop the bleeding, then white blood cells and macrophages clear away the bacteria. Fibroblasts create a collagen matrix, where epithelial cells will eventually close the wound.
- Stem cells can be used to promote tissue regeneration and wound healing by creating a pro-healing environment and differentiating into the damaged tissues.

17.3 The Physiology of Anxiety and Depression: The Nervous System

Learning Objectives

17-8 Describe the public health burden of untreated anxiety and depression.

17-9 Explain the process of neurotransmission.

17-10 Describe the organization of the nervous system and how the different subsystems function.

17-11 Identify various treatment options for anxiety and depression, and explain how they work.

We often hear about mental health. At least for me, on any given day I'll see advertisements for counselors in my area, or will read a slogan somewhere about self-care or living with a mental health condition. Perhaps you have had a similar experience. There's a good reason for all of the campaigns to raise awareness around mental health—about 1 in 5 adults in the U.S. deal with mental illness, most commonly anxiety.

Whether it's boarding an airplane, having a medical procedure done, or giving a public talk, we've all experienced anxiety at some point in our lives. From an evolutionary standpoint, anxiety makes sense—attentiveness to one's environment and the ability to sense and react to danger led to increased fitness and survival. Depression, which goes hand-in-hand with anxiety, has its roots in our evolutionary history as well. One of the hallmarks of depression is rumination, or persistent thoughts around a problem. People who ruminate tend to be highly analytical and good problem solvers, which, from an evolutionary perspective, could increase one's ability to adapt and survive. So, one could argue that anxiety and depression are not disorders, but rather important evolutionary adaptations.

Statistics show that people in Generation Alpha, those born between 2010 and 2024, have higher rates of anxiety and depression than other generations. There are myriad reasons why this could be the case, including having Generation X and Millennial parents—who discuss mental health more openly than earlier generations—as well as the effect of social media driving maladaptive behaviors, such as doomscrolling or trying to live up to unrealistic standards.

The problem is, in our fast-paced society, our brains can't keep up—and anxiety and depression become less adaptive and more maladaptive. According to the Centers for Disease Control and Prevention's 2024 National Health Statistics Report, the number of adults in the United States with significant anxious or depressive symptoms increased between 2019 and 2022. This is particularly concerning given the link between mental and physical health, and the potential negative impact that poor mental health can have on finances as well as performance at work or school.

In addition, untreated mental illness is associated with higher incidences of chronic disease, homelessness, incarceration, and disability. In fact, the American Psychiatric Association estimates that the economic burden of major depressive disorder was around $236 billion in 2018. This estimate is based on costs associated with medical services and/or medications, missed work or low productivity at work, and costs related to suicide.

Aspects of mental health, such as anxiety and depression, are the result of nervous system physiology. Understanding that physiology can lead to both treatments for clinically significant levels of anxiety and depression, as well as strategies that anyone can use to manage their mental health. In this section, we'll focus on examples of nervous system physiology at the molecular and systems level, specifically as it relates to the physiological basis of mental health.

Cellular Function of the Nervous System

The star of the show in the nervous system is a specialized cell called a **neuron** (Figure 17.8). Neuron structure varies, but all neurons have a central **cell body** (or soma) that contains the nucleus. Connected to the cell body are dendrites and an axon. **Dendrites** are short, branched extensions that receive electrical signals from other neurons. An **axon** is a cytoplasmic extension that conveys the *action potential* along the neuron length, releasing chemical signals from its endings, called **axon terminals**.

The human body has three different types of neurons, including *sensory neurons,* which help you to detect touch; *interneurons*, which respond to and transmit signals from other neurons; and *motor neurons*, which control movements, like muscle contraction. Most interneurons reside entirely within the central nervous system, whereas axons of sensory neurons and motor neurons extend throughout all regions of your body.

Note that neurons aren't the only important cells in your nervous system. There are many other cell types that take on a supporting role, such as *glia*. Glial cells are responsible for making **myelin**, a fatty substance that surrounds the axon and functions like insulation on an electrical wire.

? Did You Know?

In patients with multiple sclerosis, the immune system mistakenly attacks the myelin-making cells of the central nervous system. Damage to the myelin sheath results in progressive weakness, impaired balance, and vision problems.

Neuronal Chatter: How Neurons Signal to One Another

In Chapter 6, we learned about the importance of cell signaling, which is the process of how cells communicate with one another. Neurons have an end that receives a signal (the dendrites on the cell body) and an end that sends a signal (the axon terminals). But how does a signal move from one end of a neuron to the other?

Figure 17.8 **The anatomy of a neuron**

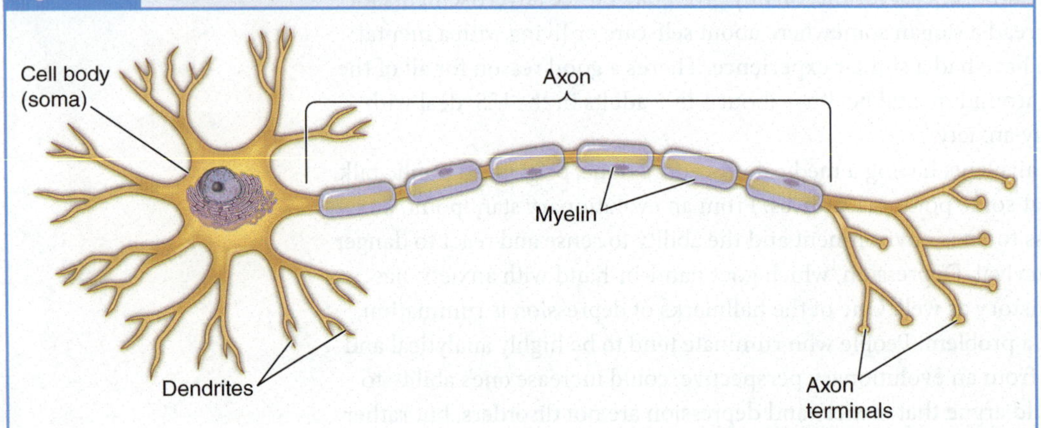

The signal moves from one end of a neuron to the other by changing the charge across the membrane, creating an electrical signal. This electrical signal is called an **action potential**. Action potentials occur in a wavelike fashion, moving down the axon of the neuron (Figure 17.9). Once an action potential reaches the axon terminals, the signal-sending neuron releases a chemical-signaling molecule called a **neurotransmitter**. The neurotransmitter communicates between two neurons in a region called a **synapse**. At the synapse, a narrow space called the **synaptic cleft** separates the axon terminal of a signal-sending neuron from the dendrites of an adjacent signal-receiving cell. This process of neurons sending and receiving chemical messages is similar to the cell signaling processes that we talked about in Chapter 6.

Neurotransmitters are stored in vesicles inside the signal-sending neuron's axon terminals. When an action potential arrives at these endings, the vesicles move to the cell membrane and fuse with it, releasing neurotransmitter via exocytosis (Chapter 5) into the synaptic cleft. Neurotransmitter molecules diffuse across the synaptic cleft and bind to receptor proteins, such as ion channels, on the adjacent signal-receiving neuron or effector cell. Movement of these ions can alter the membrane potential of the signal-receiving cell, which may result in an action potential—or the inhibition of an action potential—in the receiving cell. Any neurotransmitter that is not taken up by the signal-receiving cell can be broken down or recycled by the body through reuptake.

Figure 17.9 **An overview of an action potential**

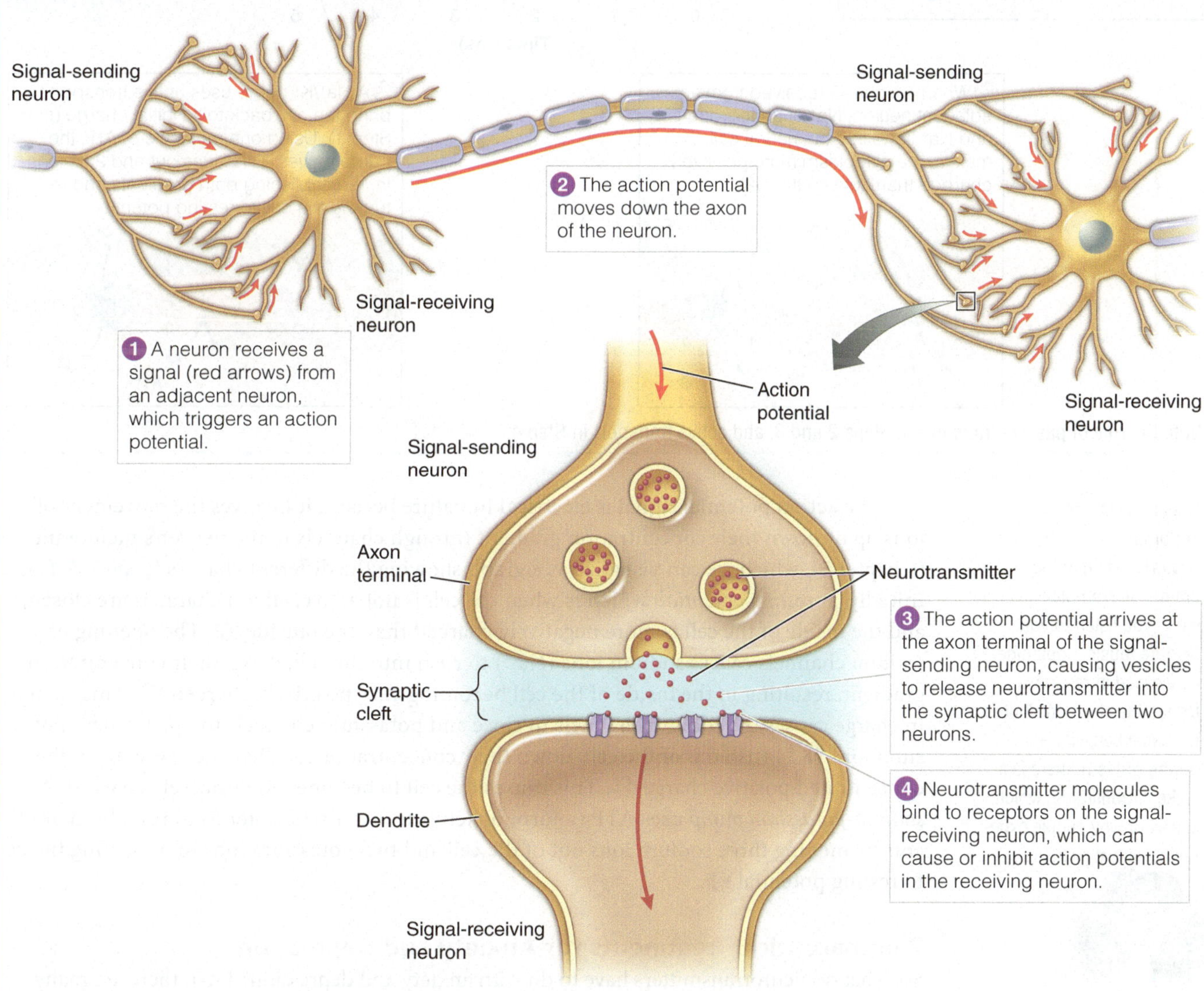

Note neurons both receive and send signals. For example, the central neuron shown here is receiving a signal from the neuron to the left while sending a signal to the neuron shown at right.

Figure 17.10 Action potentials are characterized by a series of chemical changes

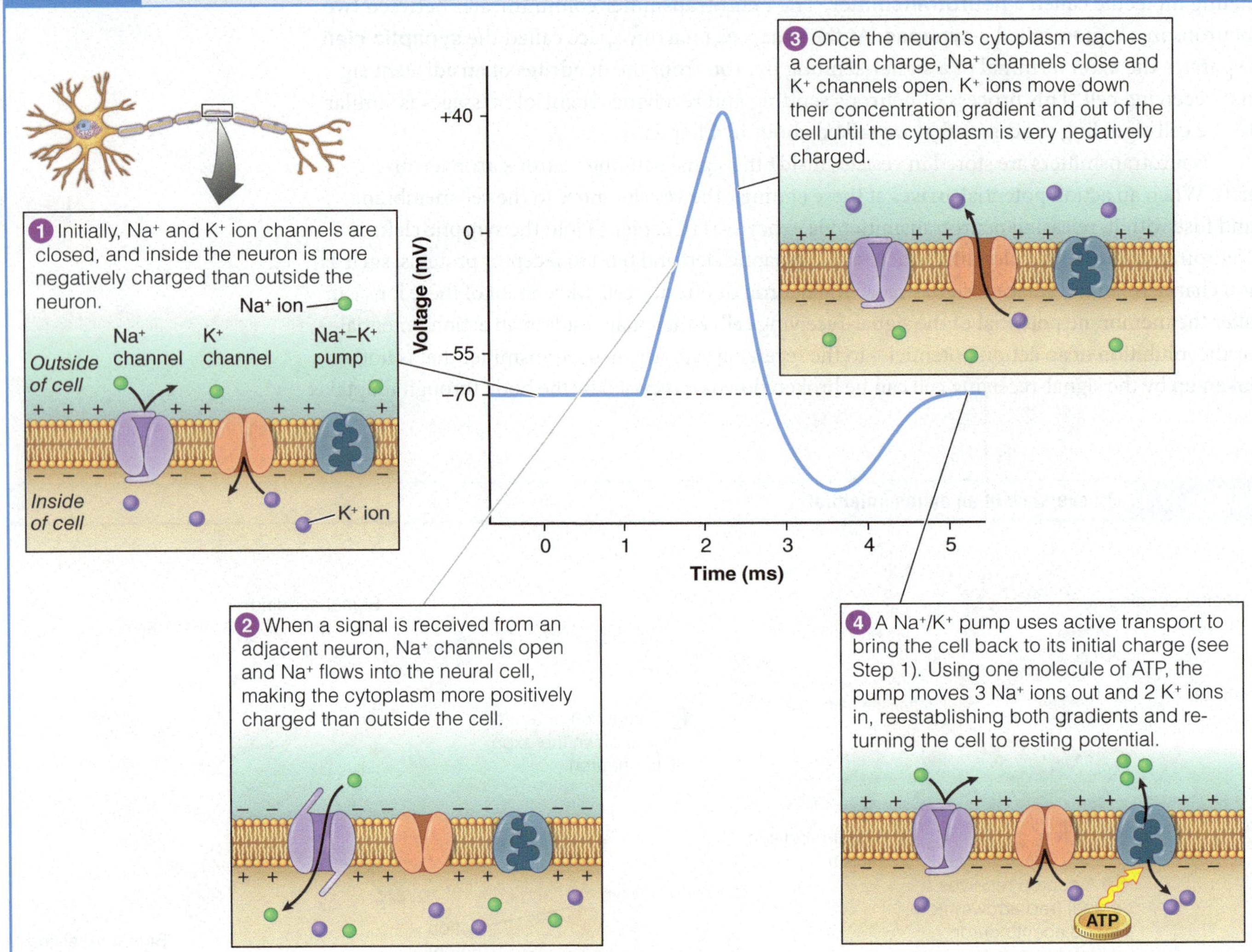

Note the role of passive transport in Steps 2 and 3, and active transport in Step 4.

Dopamine is also an important part of the reward system in the brain. For example, when you listen to your favorite music, dopamine is released in response. Dopamine also contributes to sexual behavior—during orgasm, there is a spike of dopamine, which is part of why people experience sexual pleasure.

The action potential signal is electrical in nature because it involves the movement of ions up or down their concentration gradient through channels in the neuron's membrane (Chapter 5), which we can visualize as a graph showing the different charges (**Figure 17.10**). Initially, at *resting potential*, which is when the cell is not active, all ion channels are closed, and the inside of the cell is more negatively charged than the outside ①. The opening of sodium channels causes sodium ions (Na^+) to rush into the cell, down their concentration gradient, resulting in the inside of the cell becoming very positively charged ②. This change in charge causes the sodium channels to close and potassium channels to open, where potassium ions (K^+) rush out of the cell, down their concentration gradient and away from the concentrated positive charge ③. This causes the cell to become too negatively charged. A *sodium-potassium pump* uses ATP (a form of active transport; Chapter 5) to reset both gradients by moving three sodium ions out of the cell and two potassium ions in, returning the cell to resting potential ④.

Pharmaceutical Treatments for Anxiety and Depression

So, what do neurotransmitters have to do with anxiety and depression? First, there are many different types of neurotransmitters, including serotonin, dopamine, norepinephrine, and gamma-aminobutyric acid (GABA). Any imbalance in the amount of neurotransmitter released—or in the number of receptors that can take up the neurotransmitter—can result in mood changes associated with anxiety and depression.

For example, common medications such as Zoloft®, Prozac®, and Paxil® work by increasing the effect that serotonin has at synapses by blocking the proteins involved in its reuptake, meaning there is more serotonin present at the synaptic cleft to be taken up by receptors on the signal-receiving neuron. In the case of anxiety, there is less GABA neurotransmission. GABA is an inhibitory neurotransmitter. When an inhibitory neurotransmitter is present, the action potential on a signal-receiving neuron is prevented. In the case of GABA, this occurs because the receptor that GABA binds to is also a chloride channel (a *gated channel*, which will only open when its "key," or signal—in this case, GABA—is present). When GABA binds to its receptor, the channel opens and chloride ions rush into the cell, causing it to become too negative on the inside, therefore blocking the action potential from firing (Figure 17.11A). In a person with anxiety, the lack of GABA neurotransmission (either because of less GABA or too few GABA receptors) means that the neuron stays active too often, and the person can experience anxiety and/or have trouble sleeping. (Figure 17.11B).

So how do anti-anxiety medicines work? Let's look at an example related to anxiety more closely, focusing on what happens in the brain when a specific kind of antianxiety medication called a benzodiazepine is used. Benzodiazepines such as Valium®, Xanex®, and Ativan® have an antianxiety effect on the body by increasing the activity of GABA. Specifically, benzodiazepines bind to the same receptor on chloride channels as GABA does (Figure 17.11C). When the chloride channel opens, chloride ions rush into the signal-receiving neuron, which then becomes very negatively charged and therefore cannot undergo an action potential. Since neuronal activity goes down, the result is a calming effect.

? Did You Know?

Linalool, one of the chemicals in lavender essential oil, acts in a similar manner in your brain as Zoloft.

Figure 17.11 A comparison of GABA neurotransmission in a non-anxious and an anxious person, and the effects of benzodiazepines

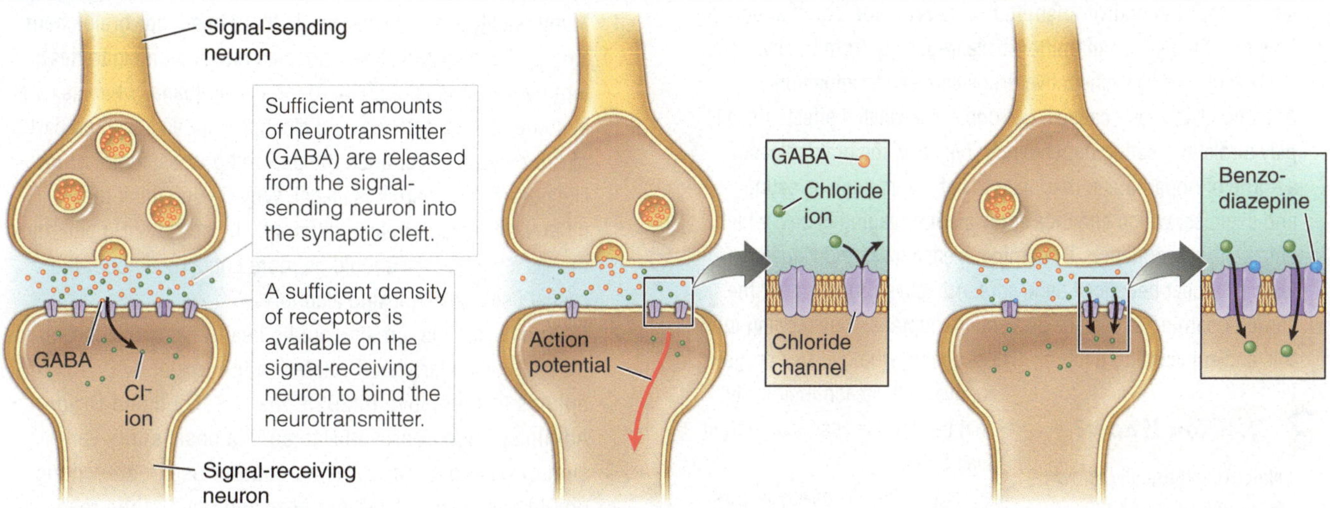

A. In the neural cells of a non-anxious person, GABA is released by a signal-sending neuron and binds to receptors on chloride channels on a signal-receiving neuron, which open, allowing chloride ions (Cl^-) in. The incoming chloride ions make the cell interior negative, preventing an action potential from occurring. Therefore, GABA acts as an inhibitory neurotransmitter, producing a calming effect in the body.

B. In the neural cells of an anxious person, there is less GABA released by the sending neuron and/or fewer chloride channels—and thus GABA receptors—on the receiving neuron. Therefore, channels on the receiving neuron stay closed and the neuron remains active. The result is that a person may feel anxious or have difficulty sleeping.

C. Treatment with benzodiazepine causes the chloride channels to stay open, allowing chloride ions to pass into the receiving cell. Then, even if there are fewer receptors and/or there is less GABA around, the person can experience a calming effect.

Biology for Life

Psychoactive Drugs and Addiction

Psychoactive drugs are chemicals that enter the brain and alter mood or perception by acting at synapses. Psychoactive drugs include prescription medications, street drugs, and readily available chemicals such as caffeine, which acts as a stimulant by binding to and inactivating receptors in the brain for adenosine, a neurotransmitter that makes people tired. In addition, caffeine consumption increases dopamine production, which boosts mood and focus.

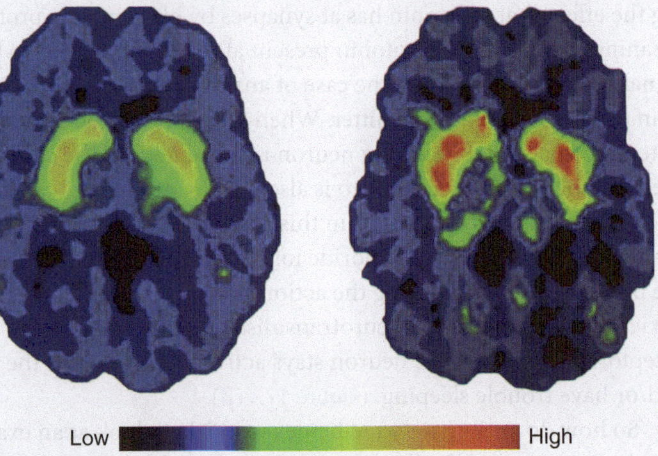

Figure 17.12 PET brain scans showing dopamine receptors in a person who uses drugs (left) and after recovery (right)
Warmer colors indicate higher levels of dopamine receptors, whereas cooler colors indicate fewer dopamine receptors.

Other psychoactive drugs include opioids such as morphine, codeine, oxycodone, and the street drug heroin, which bind to and activate receptors for endorphins. These drugs dull pain and, when administered in larger doses, produce a rush of euphoria. But these drugs can have a darker side, leading to dependence on greater quantities of the drug to achieve the desired effect. Continual use of a drug can result in tolerance, meaning that the effectiveness of a given dose of the drug decreases over time. Tolerance occurs because, by repeatedly taking a drug, one can alter their body function and chemistry. For example, tolerance to a drug that causes release of a neurotransmitter may arise due to decreased synthesis of either the neurotransmitter or its receptors, or changes in how fast the neurotransmitter is cleared away from the synapse.

Use of some psychoactive drugs can lead to addiction. Addiction has many causes, but dopamine-related effects usually play a role in creating dependency. Typically, the brain releases a surge of dopamine in response to behaviors such as eating and exercise, which enhance survival, or engaging in sex, which enhances reproduction. Dopamine release helps individuals learn to repeat beneficial behaviors. Addictive drugs trigger the same dopamine release or prevent its reuptake, thus tapping into this ancient learning pathway. Thus, people who overuse drugs inadvertently teach their body that the drug is essential to their well-being.

However, recovery is possible. Consider Figure 17.12, which shows the number of dopamine receptors in a person who uses drugs (left) and after recovery (right). Medications such as methadone and Suboxone partially block opioid receptors in the body, thus helping to decrease cravings. In fact, scientists find that using these medications can help to decrease cycles of dependence while also cutting overdose rates in half.

1. Many readily available chemicals that impact our brain chemistry and can be addictive—such as nicotine (in cigarettes), tetrahydrocannabinol (THC; found in marijuana, which is legal in some states), caffeine, and alcohol—are an accepted part of our culture. In contrast, the use and possession of drugs like cocaine and heroin is illegal. Why do you think these different drugs are treated differently by society? Do you think that as a society, we should be more concerned about the casual use of caffeine and alcohol?
2. Drug overdose is a major public health issue in the United States, particularly as it relates to the use of synthetic opioids like fentanyl. In 2023, the U.S. Food and Drug Administration approved Narcan®, a nasal spray version of naloxone, for over-the-counter use in preventing opioid overdose. Weigh the pros and cons of the ready availability of Narcan. Do you think its use will help to address the problem of drug overdose, or will it potentially make matters worse, by helping to normalize drug overdose?
3. Narcotics Anonymous lays out a 12-step plan to help people recover from drug addiction. The twelfth step is Sustainable Sobriety, which helps establish life-long healthy habits that can help to prevent relapse. Based on your knowledge of dopamine, what types of healthy habits might help to prevent relapse?

? Did You Know?

Narcan, generically known as naloxone, works to prevent overdose by binding to the same receptors in the brain that opioids do. Essentially, naloxone has a stronger affinity for the receptors than the opioids do, binding to the receptors and preventing the opioid molecules from doing so, thus reversing the effects of an overdose.

Systems within a System

As we learned in the last section, neurons are the cellular unit of the nervous system. Neurons make up the nervous system, and within the nervous system, there are two anatomical divisions—the central and peripheral nervous systems (Figure 17.13). The brain and spinal cord are the organs of the **central nervous system** (**CNS**). The brain is divided into different functional regions, many of which we've learned about in various Understand It! features. The spinal cord is the portion of the central nervous system that conveys signals between peripheral nerves and the brain.

The human **peripheral nervous system** includes 31 pairs of spinal nerves that connect to the spinal cord, and 12 pairs of cranial nerves that connect to the brain. The peripheral nervous system is divided further into the somatic and autonomic nervous systems. The **somatic nervous system**, in part, controls skeletal muscles and informs the central nervous system about external conditions and the body's position. The **autonomic nervous system** controls smooth muscle, cardiac muscle, and glands, and it informs the central nervous system regarding conditions inside the body.

There are two divisions of the autonomic nervous system: sympathetic and parasympathetic. The **parasympathetic nervous system** is active in times of relaxation. Signals from these neurons promote routine maintenance tasks such as digestion and urine formation. They are responsible for the "resting-and-digesting" response. Part of the parasympathetic nervous system is the *default-mode network*, a system within the brain that works to promote a state of wakeful rest where the mind can wander, which has a restorative function on the body.

The **sympathetic nervous system** is most active in times of excitement or danger. The sympathetic nervous system, in conjunction with the endocrine system, is responsible for initiating the fight-or-flight response in the body. Signals received by the brain cause a surge of the hormones *adrenaline* and *cortisol* to be released, which prepares the body for escape by triggering physiological changes, such as increasing breathing and heart rates. The problem is that with chronic stress, the fight-or-flight response becomes overactive, leading to health issues such

? Did You Know?

An injury that damages the spinal cord can cause a permanent loss of sensation and paralysis. The effects can be permanent because axons in the central nervous system—unlike those elsewhere in the body—do not regrow after injury.

? Did You Know?

Dysfunction of the default mode network is thought to occur in people with anxiety disorders.

Figure 17.13 A schematic representation showing the anatomical and functional divisions of the nervous system

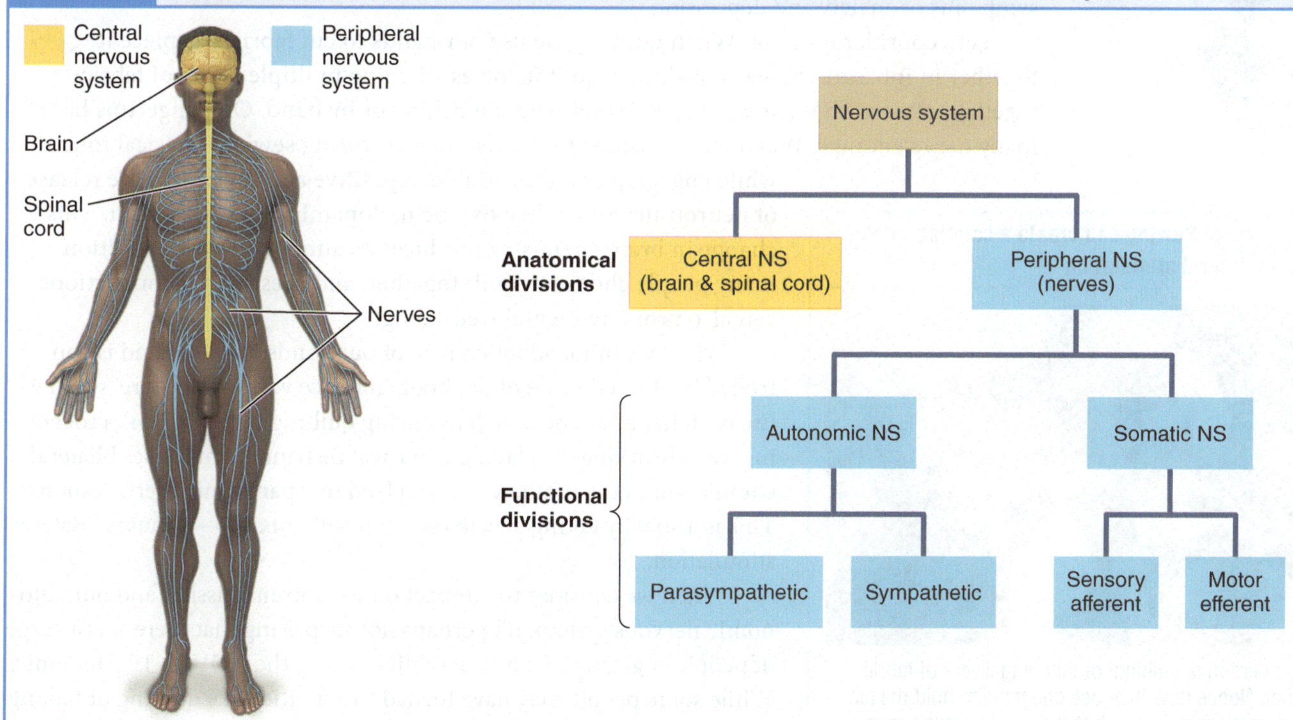

Note that components of the central nervous system are in gold and those of the peripheral nervous system are in blue.

? Did You Know?

Deep breathing has a calming effect because it stimulates the vagus nerve, which is part of the parasympathetic nervous system. Stimulating the vagus nerve through belly breathing is thought to help deactivate the sympathetic nervous system and activate the parasympathetic nervous system.

as anxiety, depression, headaches, or sleep problems, among others. Symptoms such as hypervigilance, or difficulty relaxing—which is common in people who've experienced traumatic events—are due to the person being in a perpetual state of fight or flight.

! Point of Confusion: One is rarely in the extreme states described here, but throughout the day, our sympathetic and parasympathetic nervous systems go through different levels of activity.

Managing Anxiety and Depression

In this section, we learned about the high incidence and economic burden of anxiety and depression, and the physiological basis for each. Here, we'll discuss some effective strategies for managing the symptoms of anxiety and depression, where investing in one's mental health can provide a huge benefit in terms of improved outlook as well as productivity.

One effective strategy is to seek out the help of a mental health professional. Ten percent of American adults have received counseling or therapy. People can choose to see a therapist for short periods of time, for example, when recovering from trauma, grief, or a stressful life event. Others may choose to work with a therapist for longer periods of time, such as those experiencing post-traumatic stress disorder (PTSD).

Therapy works because the brain is *plastic*, or capable of changing. Learning to change behaviors with the help of a therapist can literally change brain structure, strengthening or weakening neural circuits and resulting in a lessening of symptoms. And as we learned earlier in this section, medications can also be an important and useful treatment option to alter neurotransmission—one used by many individuals. The National Health Interview Survey conducted by the Centers for Disease Control and Prevention in 2020 suggest that 16.5% of adults in the United States take some form of prescription medication for a mental health concern.

Planning, executing, and completing a craft also give a person a sense of agency and reward, which can help to reduce anxiety and increase one's sense of well-being. Interestingly, working on crafts can affect reward circuits in a way similar to the effects of using heroine or smoking cigarettes.

Hobbies are also a great option for reducing the symptoms of anxiety or depression. Exercise, for example, is excellent not only for your musculoskeletal system (Section 17.4), but also for reducing stress hormone levels by releasing endorphins, which improves one's mood. Playing a wind instrument can stimulate the vagus nerve, resulting in feelings of relaxation. Traditional crafts such as knitting, quilting, or woodworking can also be used to manage the symptoms of anxiety and depression.

Let's consider quilting. When quilting, we use our hands to cut fabric and piece it together in intricate shapes. Finishing a quilt involves stitching multiple layers of fabric together, often with intricate shapes, either with a machine or by hand. Our fingertips have many nerve endings. When we are using our hands, such as during sewing, physical touch while engaging in a pleasurable, repetitive activity causes the release of neurotransmitters like oxytocin, dopamine, and serotonin. This change in brain chemistry produces a calming effect. In addition, many people choose to quilt together, and these social connections can also promote mental well-being.

Figure 17.14 **Sewing can create a calming effect through bilateral stimulation**

Here, a person is quilting, or stitching layers of fabric together. Notice how they use one hand to hold the fabric and another to manipulate the needle, which causes bilateral stimulation in the brain and has a calming effect.

Credit: Diane Labombarbe/Getty Images

When we think about control of our hands, the left hand is controlled by the right side of the brain and vice versa. Therefore, when we use both hands at once, such as during quilting (Figure 17.14), crocheting, woodworking, or playing a musical instrument, it causes bilateral stimulation in the brain, which can lead to a parasympathetic response. This is also why taking a walk can help with anxiety—it causes bilateral stimulation.

When we consider the impact on neurotransmission and our autonomic nervous system, it's perhaps not surprising that there was a surge of people engaging in crafty activities during the COVID-19 pandemic. While some people may have turned to activities like quilting or baking bread for something to do, many people may have stuck with it to foster a sense of control and relieve anxiety.

17.3 Summary

- Untreated mental illness and addiction are associated with higher incidences of chronic disease, homelessness, incarceration, disability, and mortality, all of which can strain public resources.
- During neurotransmission, neurotransmitters are released from the axon terminal of a signal-sending neuron into the synaptic cleft, where they bind to receptors on the dendrite of a signal-receiving neuron. This transmits an electrical signal, or action potential, which moves through the receiving neuron.
- The central nervous system consists of the brain and spinal cord, which process and integrate information. The peripheral nervous system includes the rest of the nerves in the body, which gather and deliver information to the CNS and control voluntary and involuntary bodily functions.
- Effective strategies for managing anxiety and depression include seeing a therapist, taking medication, and pursuing a hobby. Hobbies like exercise and those that involve bilateral stimulation or social support help to relieve symptoms of anxiety and depression.

17.4 The Endocrine and Reproductive Systems, Endocrine Disruptors, and Fertility

Learning Objectives

17-12 Summarize the public health impacts of endocrine disruptors, particularly as they pertain to fertility.

17-13 Describe how the endocrine system regulates the reproductive system.

17-14 Explain how endocrine disruptors can interfere with normal cell function.

17-15 Identify endocrine disruptors, their environmental sources, and ways to protect yourself from them.

If your teenage years were anything like mine, you grew up hearing from the adults around you that "It takes just one time to get pregnant!" This fear-based narrative suggests that it's "easy" to get pregnant and that if one wants to have a baby, it's as simple as having unprotected sex.

While that may be the case for some, for millions of people—around 1 in 6 couples worldwide—it isn't quite so straightforward. **Infertility** is the inability for a couple to conceive after a year of regular unprotected sex, or for people over the age of 35, not conceiving within 6 months. There are many potential causes for infertility, including physical abnormalities, obesity, sexually transmitted infections, and polycystic ovary syndrome (PCOS), a hormonal imbalance that causes cysts in the ovaries, irregular menstrual cycles, and infertility.

> **? Did You Know?**
> The most common cause of female infertility is polycystic ovary syndrome...and rates of PCOS are increasing. PCOS also increases one's risk for heart disease, type 2 diabetes, and endometrial cancer.

Infertility can have a significant mental health toll. For instance, anxiety and depression related to an inability to get pregnant are common. Infertility problems can also lead to significant relationship problems. Failed infertility treatments and pregnancy loss can result in grief, with about a third of people who have experienced pregnancy loss also experiencing post-traumatic stress, with some people going on to develop post-traumatic stress disorder as a result of their loss. One study estimates that the economic burden of infertility treatments alone (not including mental health treatments) is approximately \$81 million per 10,000 women between the ages of 18 and 50.

In addition to infertility, some couples alternately experience **impaired fecundity**, a broader term that refers to having difficulty getting pregnant or carrying a pregnancy to term. We learned about one kind of impaired fecundity in Chapter 8—recurrent miscarriage. While infertility rates have remained fairly stable over the years, rates of impaired fecundity are on the rise. What can be the reason for this?

While discussions around infertility and miscarriage typically focus on age at the time of conception and egg quality, sperm quality likewise declines with age and can be impacted by one's health. Therefore, sperm quality likely plays a greater role in fertility and impaired fecundity than previously thought.

In this section, we will learn about the relationship between the endocrine and reproductive systems. We'll also discuss *endocrine disruptors*—chemicals in the environment that can interfere with the function of those body systems. We'll specifically focus on the relationship between endocrine disruptors and impaired fecundity. At the end of this section, we'll discuss how to limit your exposure to endocrine disruptors.

An Overview of Reproductive Endocrinology

We'll begin our discussion by talking about the interrelationship between the endocrine and reproductive systems. In a typical textbook that includes coverage of human anatomy and physiology, a whole chapter would be devoted to each topic. But for the purposes of our discussion, we're going to focus on the interrelationship between the endocrine and reproductive systems as it pertains to **reproductive endocrinology**, which is a medical field dedicated to the study of the hormonal problems that impact fertility, as well as the basis of infertility and the treatment of it.

Like the nervous system, the endocrine system facilitates long-distance communication among the cells of an animal body. **Hormones**, the intercellular signaling molecules of the endocrine system, are secreted by an endocrine gland or cell, and typically travel in the bloodstream to their target cells (Figure 17.15). The target cell of a hormone is any cell with receptors for that specific hormone—similar to how enzymes bind only to specific substrates (Chapter 4). Like neurotransmitters, hormones affect cells by binding to receptor proteins. However, hormones can often travel farther than neurotransmitters, exert their effects on more cells, and persist longer in the body.

Many organs of the endocrine system might already be familiar to you. Figure 17.16 summarizes the glands of the endocrine system and their functions. For the purposes of our

Figure 17.15 Endocrine signaling

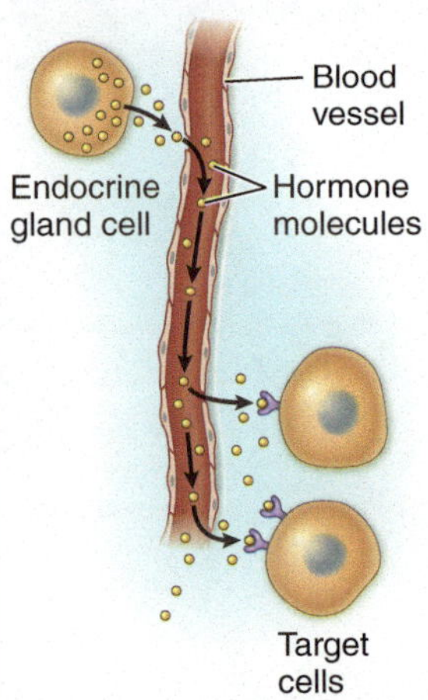

While there are exceptions, endocrine glands release hormones that travel in the bloodstream to their target cells.

Adipose tissue (fat) also has endocrine functions, secreting hormones that regulate metabolism and that can also contribute to obesity. Obesity is a complex metabolic disease resulting from hormonal problems as well as poor diet and lack of exercise.

Figure 17.16 An overview of the glands of the endocrine system

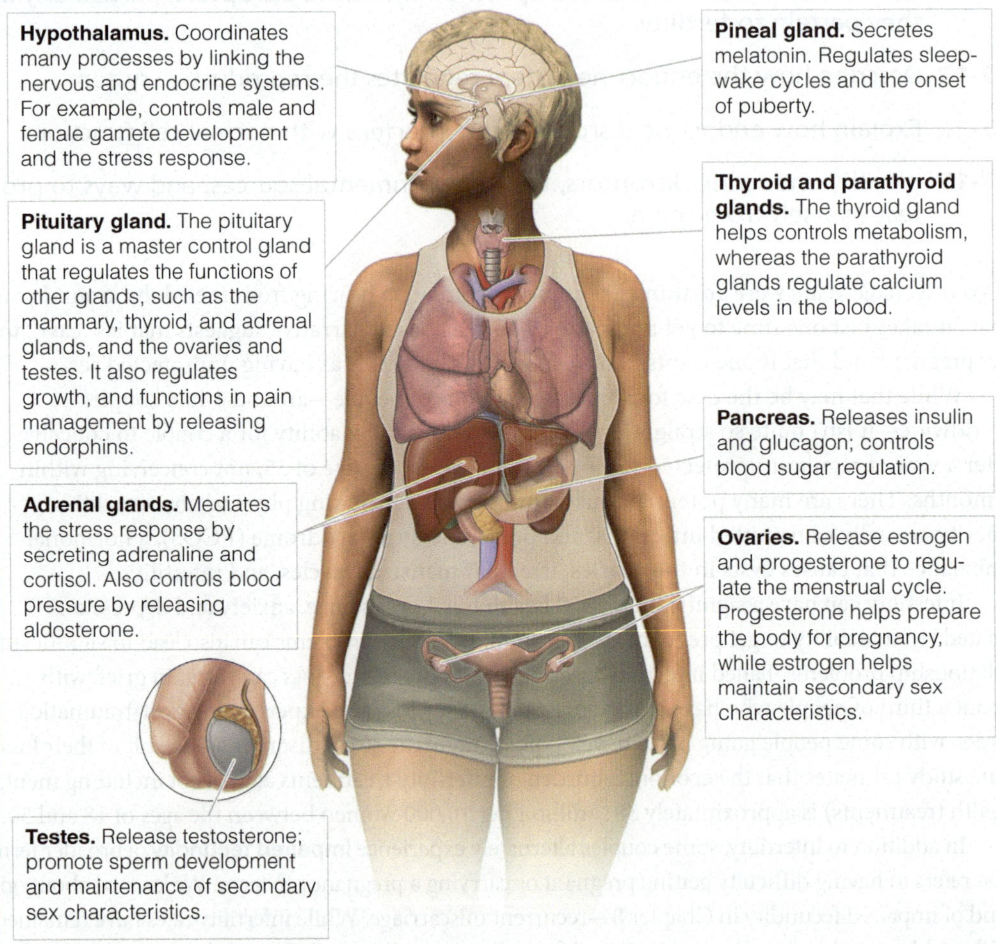

discussion, we will focus on the structures involved with human reproduction, which include the *hypothalamus*—a part of the brain that controls both the nervous and endocrine systems—the pituitary gland, and testes or ovaries.

Hormonal Control of the Gonads

Gonads are an animal's primary reproductive organs and include **testes** and **ovaries**. In addition to producing gametes (sperm or eggs, respectively; Chapter 8), gonads secrete sex hormones—steroid hormones such as **testosterone**, **estrogen**, and **progesterone**, which are essential for reproductive function. Note that males and females both have a mix of testosterone and estrogen, although higher testosterone levels can be associated with males and higher estrogen levels with females. While the levels may vary, the normal ranges of each hormone overlap between males and females, meaning that one cannot tell a male from a female based on hormones alone. Sex hormones also affect secondary sexual traits, such as facial hair and prominent breasts, which are traits that differ between the sexes but do not play a direct role in reproduction.

Note that the testes also produce estrogen and progesterone, and ovaries can produce testosterone as well.

In both males and females, a hormonal pathway called the **hypothalamic-pituitary-gonadal axis** controls reproduction and sexual development (Figure 17.17). The **hypothalamus**, which is a structure that connects the nervous and endocrine systems, sends a signal to the pituitary gland. The **pituitary gland** is a pea-sized gland that is directly beneath the hypothalamus and controls other glands in the body, including the testes and ovaries. The secretion of luteinizing hormone (LH) and follicle-stimulating hormone (FSH) causes the testes and ovaries to release testosterone and estrogen/progesterone.

Figure 17.17 The hypothalamic-pituitary-gonadal axis

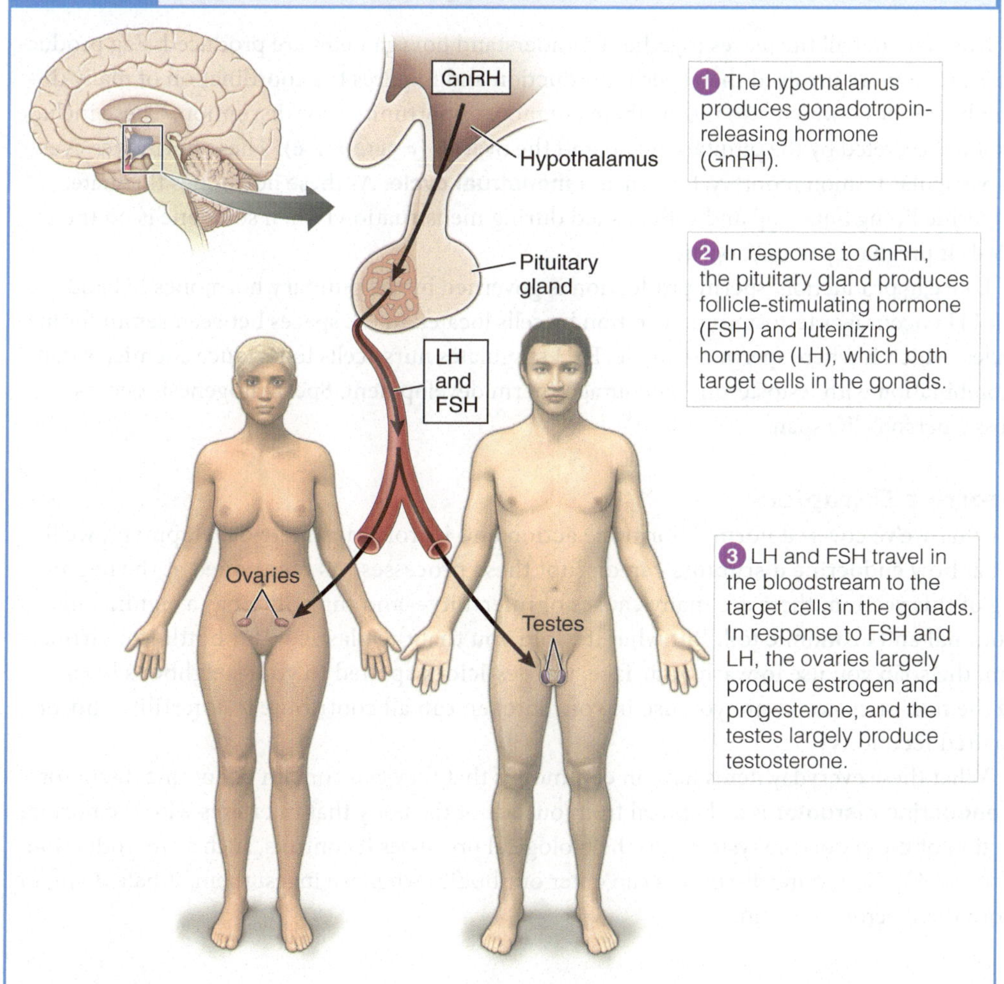

Recall from Unit 3 that there's a range of "normal" regarding biological sex and that this is different from one's gender identity. For our purposes, we define "male" and "female" as they pertain to which gametes one's body produces or produced, since eggs, unlike sperm, are not produced throughout one's lifetime, and which organ structures are typically found in a body (e.g., people with ovaries usually have a uterus).

Figure 17.18 The correlation of hormones to the menstrual cycle

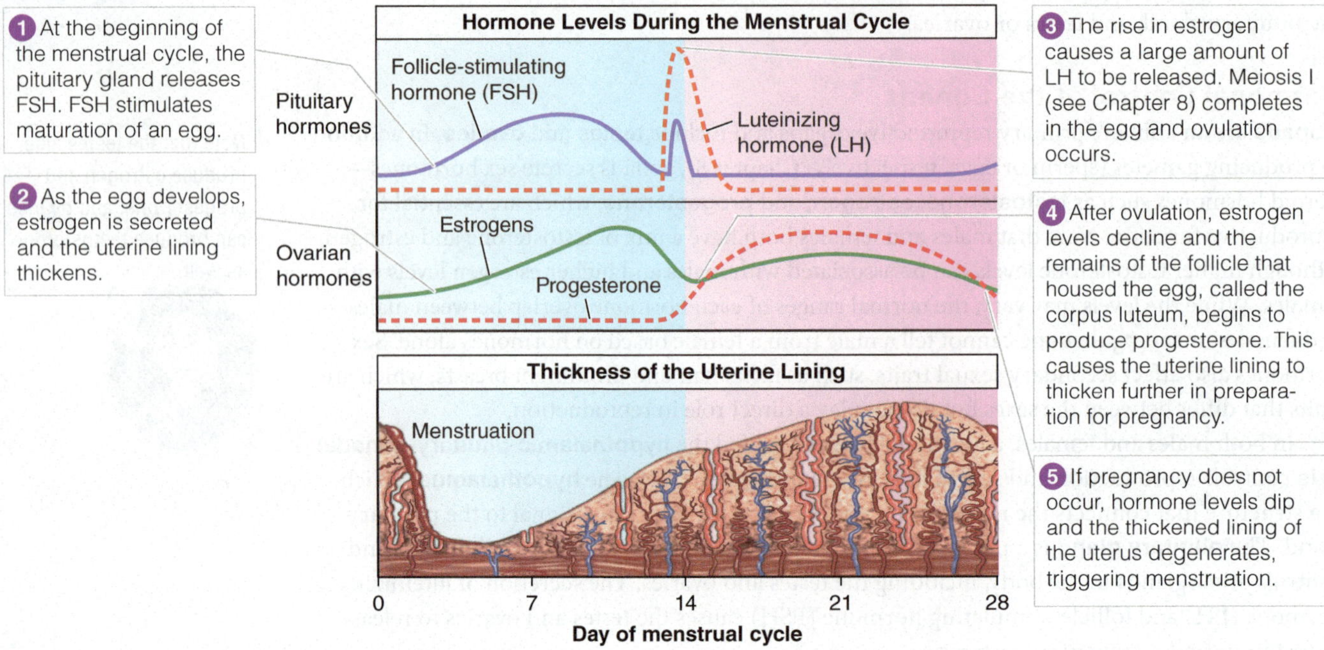

Note that the boxes guide you through the hormonal changes, while the diagram at the bottom indicates what is occurring in the uterus.

A woman enters menopause when all the oocytes in her ovaries have either been released during menstrual cycles or have disintegrated due to aging. Menopause is the end of fertility. With no oocytes left to mature, production of estrogen and progesterone diminishes dramatically, and there are no more menstrual cycles.

Now, let's put all the pieces together to understand how gametes are produced. Egg production is much more involved than sperm production and requires the coordination of many different hormones—which also means there are more opportunities for disruption. These include hormones secreted by the pituitary gland and the ovaries (Figure 17.18). These hormones cycle in a particular fashion monthly, known as a **menstrual cycle**. As these hormones fluctuate, the uterine lining builds up and is then shed during menstruation (when someone is on their period) if pregnancy doesn't occur.

Like egg production, sperm production is governed by the pituitary hormones LH and FSH. LH encourages testosterone secretion by cells located in the spaces between seminiferous tubules, which is where sperm are made. FSH stimulates nurse cells to produce chemicals that, in combination with testosterone, encourage sperm development. Spermatogenesis occurs across a person's life span.

Endocrine Disrupters

Now that we've covered normal hormone action and its role in gamete development, we'll look at how endocrine disruptors can disrupt these processes. As mentioned at the beginning of this section, there are many causes of infertility—and infertility has a significant emotional and economic toll. But what if I told you that the plastic water bottle you drink from, the soap you use to wash your face, the pesticides applied to your neighbor's lawn, and the non-stick cookware you use in your kitchen can all contribute to infertility and/or impaired fecundity?

What these everyday items have in common is that they can contain endocrine disruptors. An **endocrine disruptor** is a chemical from outside of the body that interferes with the normal function of the endocrine system and the biological processes it controls, such as reproduction (Figure 17.19). Endocrine disruptors can enter our bodies when we ingest them, inhale them, or absorb them across our skin.

Figure 17.19 A simplified view of endocrine disruptors

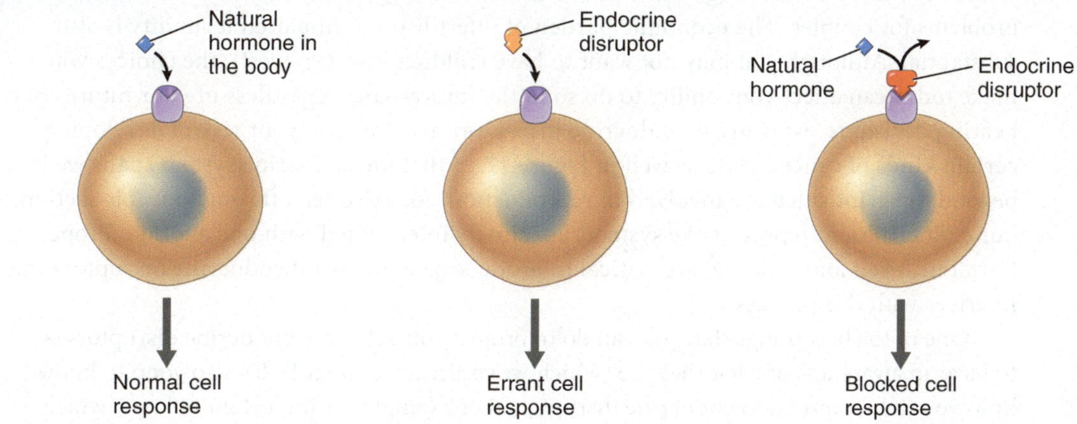

Under normal conditions, a hormone is released from a gland and travels in the bloodstream to a target cell, where it binds to a receptor, eliciting a response in the target cell.

Some endocrine disruptors interfere with normal function by binding to cell receptors and changing cell behavior, absent from any signal from a hormone produced by the body.

Other endocrine disruptors bind to a cell's receptor, thereby preventing the natural body hormone from accessing the receptor, interfering with normal cell function.

More than 1,000 synthetic chemicals are now listed as endocrine disruptors by the Endocrine Society, an international organization for physicians and scientists who study endocrine function. Note that sometimes endocrine disruptors are referred to as forever chemicals. **Forever chemicals** are synthetic (human-made) chemicals that persist in the environment as well as in the human body, and can also bioaccumulate (Chapter 15). For example, per- and polyfluoroalkyl substances (PFAS) are chemicals used to repel oil or water in cookware, carpets, and textiles. PFAS are a form of both air and water pollution (Chapter 16), and have known negative effects on fertility because they block the hypothalamic-pituitary-gonadal axis (see Figure 17.17). Consequently, FSH and LH are not produced at sufficient levels. Since both are required for egg and sperm production, overall fecundity suffers.

You may have heard about the endocrine disruptors bisphenol A (BPA) and phthalates during your routine grocery shopping. As awareness of the hazards of BPA and other endocrine disruptors such as phthalates increases, manufacturers are beginning to phase out their use and actively label products as BPA- or phthalate-free. BPA can also be found in hard plastics and the thermal paper used to print cash register receipts, and both BPA and phthalates can be found in food and beverage containers.

But how do BPA and phthalates affect the body? According to the Endocrine Society, BPA is associated with reduced egg quality, meaning that one's eggs are more likely to have meiotic defects, and consequently, the wrong number of chromosomes (Chapter 8). BPA exerts its endocrine disrupting function by directly binding to estrogen-specific receptors in cells and causing activation, even when there is no estrogen. Since the estrogen receptor is also a transcription factor, BPA can cause gene expression at inappropriate times, and as we learned through the story of Levi's heart in Chapter 7, can wreak biological havoc.

Phthalates negatively impact fertility by interfering with the production of testosterone and oocyte maturation. Phthalate exposure is also linked to premature ovarian failure by increasing FSH levels. This causes people to undergo menopause early, limiting their fertility.

In addition to their effects on fertility, endocrine disruptors are also implicated in the development of cancer in hormone-sensitive organs—such as the breast, ovary, and prostate—by interfering with normal hormone function.

? Did You Know?

The endocrine-disrupting effects of BPA were discovered by accident. The researchers detected an estrogen-like chemical in their yeast cultures. At first, they thought they had discovered that yeast produced estrogen. However, additional studies showed the estrogen-like chemical was BPA—it was leaching out of the plastic bottles in which the yeast was grown.

? Did You Know?

DDT, which we learned about in Chapter 16, is an estrogen mimic and therefore an endocrine disruptor.

Endocrine Disruptors and Your Health

Infertility can be devastating, causing mental health struggles for individuals and relationship problems for couples. The economic burden of infertility and impaired fecundity is also substantial. Although you may not want to have children now (or ever!), the choices you make today can affect your ability to do so in the future—and regardless of your future child-bearing decisions, exposure to endocrine disruptors also impacts your risk of developing certain kinds of cancer. Note as well in Figure 17.15 that the endocrine system spans well beyond the glands that are involved in reproduction. As we've seen throughout this section, our endocrine and reproductive systems are closely intertwined with one another. Proper hormone levels and function are critical to produce gametes—and endocrine disruptors can interfere with this process.

One of the best things that you can do to protect yourself from endocrine disruptors is to have an awareness of what they are (which we've already covered!). It's also good to know how we can be exposed to endocrine disruptors. For example, it's important to know which household products contain endocrine disruptors and what activities can increase your exposure to them.

Let's start with per- and polyfluoroalkyl substances (PFAS). Common ways that people are exposed to PFAS in their households include through foods contaminated with PFAS from food-packaging materials, and from cooking with non-stick cookware. People are also exposed by wearing fabrics that contain PFAS, such as water-resistant materials. When shopping, you can protect yourself by looking for foods packaged in PFAS-free packaging, and by purchasing and using non-stick cookware that does not contain PFAS, such as ceramic (**Figure 17.20A**) or stainless-steel cookware.

BPA is often a component in plastic, and so one of the simplest ways to avoid BPA is to buy products packaged in glass, paper, or metal instead of plastic. Since this isn't always feasible, you can also look at the recycling codes of plastic objects in your house to see if they contain BPA. Typically, #7 plastics contain BPA, whereas #1, #2, #4, and #5 plastics are usually BPA-free. Manufacturers are increasingly labeling products as BPA free (**Figure 17.20B**).

It's also important to note that BPA can be released from plastic during heating. Therefore, one may choose to avoid heating or storing hot food in plastic containers. As you shop, look for beauty and healthcare products that are listed as BPA-free. You may also choose to write to companies and/or your congressional representatives about removing known endocrine disruptors from common household goods.

Figure 17.20 Examples of ways to avoid endocrine disruptors

A B

Credit: Pixel-Shot/Shutterstock.com

Credit: Avector/Shutterstock.com

A. Choosing to cook with ceramic or stainless steel cookware decreases your exposure to endocrine disruptors like PFAS.

B. When grocery shopping, look for labels like this one on food and household goods to avoid incidental exposure to endocrine disruptors.

Biology for Life

Ozempic® and Why Obesity Involves More Than Diet and Exercise

Have you heard of Ozempic? Or maybe you've heard of it by its other names, like semiglutide or Wegovy®? Ozempic received U.S. Food and Drug Administration approval to treat type 2 diabetes in 2017...and to treat obesity under the name Wegovy in 2021. These drugs work by mimicking a naturally occurring hormone in your body called GLP-1. GLP-1 causes insulin, another hormone, to be released from the pancreas, resulting in a consequent drop in blood sugar.

So, in the context of type 2 diabetes, which occurs when people do not use and/or produce enough insulin, it can help to manage blood sugar levels. Since GLP-1 slows down digestion and helps people to feel fuller after eating, it can thus cause weight loss and be used to treat obesity—another major public health concern. Its success at promoting weight loss, particularly as the result of social media influencers highlighting its positive benefits, is part of the skyrocketing interest in the drug.

It turns out there may be other benefits to Ozempic, and its use might also be fueling what some doctors are calling an "Ozempic baby boom." As of the writing of this text, scientists were just beginning to describe this observation and correlation (Chapter 1), where taking Ozempic and losing weight appear to cause an increase in fertility. Obesity is implicated in increased estrogen levels and decreased fertility, but with weight loss, estrogen levels can return to normal, where one's fertility could likewise increase. Since uncontrolled diabetes can also impact fertility, it could also be that since the medication is effective at treating diabetes, then a boost in fertility could be expected when one's diabetes is under control.

Regardless of the reason why, Ozempic highlights the connection between the endocrine and reproductive systems, as well as the influence that a drug can have on a body system that is unrelated to its intended purpose. It also highlights how obesity isn't simply a result of diet, but is directly influenced by the endocrine system.

1. Weight loss is an "off label" use of Ozempic, which was approved to treat type 2 diabetes. Do you think it is important for drugs to go through additional studies, FDA approval, and/or clinical trials when they are used for a condition they weren't initially intended to treat? Why or why not?
2. Ozempic is a peptide (a short chain of amino acids). Often, peptide drugs need to be injected, rather than taken orally. What might happen to the drug if it passed through the digestive system?

17.4 Summary

- Infertility affects millions of people, causing mental health issues, relationship strain, and economic burden.
- Reproductive function, including the production of gametes, is regulated by the hormones of the endocrine system.
- Endocrine disruptors alter hormone function, either by activating receptors inappropriately or by blocking receptors, which ultimately causes changes in cell behavior.
- Endocrine disruptors include per- and polyfluoroalkyl substances (PFAS), which are found in non-stick cookware and certain types of fabric, and bisphenol A (BPA), which is found in plastics. You can avoid exposure by reading the labels on the products you use.

17.5 Is Sitting the New Smoking? The Musculoskeletal System

Learning Objectives

17-16 Describe the public health impact of a sedentary lifestyle as it relates to the musculoskeletal system.

17-17 Identify and describe the components of the musculoskeletal system.

17-18 Explain how different forms of exercise contribute to a healthy musculoskeletal system.

Have you ever heard someone say that sitting is the new smoking? The implication is that sitting for eight hours a day, whether in a vehicle or at a desk, means that your risk of an early death is equivalent to if you smoked cigarettes, which are known to cause numerable health

Note that skeletal muscle contraction occurs due to neurotransmitter release, much like the process described between neurons in Section 17.3, but with the key difference being that rather than the receiving cell being another neuron, it is instead a muscle fiber.

problems. It turns out there's reason to be concerned—a meta-analysis (Chapter 1) of 13 different studies and over a million people suggested a connection between a sedentary lifestyle and health problems.

Why does this occur? When you sit, your leg muscles relax and make less lipoprotein lipase (LPL). LPL is an enzyme that contributes to health by facilitating the uptake of lipids from the blood and production of HDL, the "good" cholesterol. When LPL declines, fewer lipids are removed from the blood and production of HDL decreases, causing an increased risk of heart attack or stroke. Resting muscles also use less glucose, meaning that, combined with poor diet, glucose levels can remain elevated, which increases the risk of diabetes. A sedentary lifestyle can also impact mental health, potentially leading to depression.

Considering all of this, a sedentary lifestyle can profoundly impact society and public health due to the myriad diseases that can result from it—from obesity to diabetes, depression, and accelerated aging that occur simply because of not moving the body more. According to the Centers for Disease Control and Prevention, physical inactivity contributes to 1 in 10 premature deaths while costing $117 billion yearly in healthcare costs alone.

So, is it time to sign up for a gym membership? Not necessarily. Research suggests that approximately an hour of moderate physical activity per day can offset the negative effects of a lot of time spent sitting. How is that possible? Our bodies move because of the muscular and skeletal systems working together, collectively called the **musculoskeletal system**, under the control of the nervous system. In this section, we will learn about the musculoskeletal system and how exercise can promote a healthy life.

Study Tip

Focus on the roots of vocabulary words. "Myo" refers to muscle, and "osteo" refers to bone.

? Did You Know?

Having good muscle mass is associated with a better prognosis in cancer patients.

An Overview of Muscle Structure and Function

Let's begin by looking at the muscular system. Muscles generate force to move the skeleton, which is the framework for the body. **Skeletal muscles** are voluntary muscles, meaning that they function in intentional movement. In contrast, our bodies also have involuntary muscles, such as *smooth muscle* that helps with digestion and *cardiac muscle* in our hearts. Skeletal muscle acts to maintain posture and participate in reflexes. For example, the shivering reflex helps maintain core body temperature when the external temperature plunges. When someone shivers, their muscles shake, which creates warmth.

A sheath of dense, regular connective tissue encloses each muscle and extends beyond it as a straplike **tendon**. Most often, one or more tendons attach each end of the muscle to a bone. A skeletal muscle consists of connective tissue-wrapped bundles of individual muscle cells called **muscle fibers** (Figure 17.21). Each fiber runs the length of a muscle, so some fibers, such as those in your thigh, can be quite long. The fibers also contain many mitochondria that supply the ATP (Chapter 4) needed for muscle contraction.

The bulk of a muscle fiber is taken up by thousands of threadlike structures called **myofibrils**. A myofibril consists of many contractile units, or **sarcomeres**, attached end to end. Sarcomeres contain **actin,** which makes up the thin filaments, and **myosin**, which makes up the thick filaments. The **sliding-filament model** describes how interactions

Figure 17.21 Muscle fiber structure

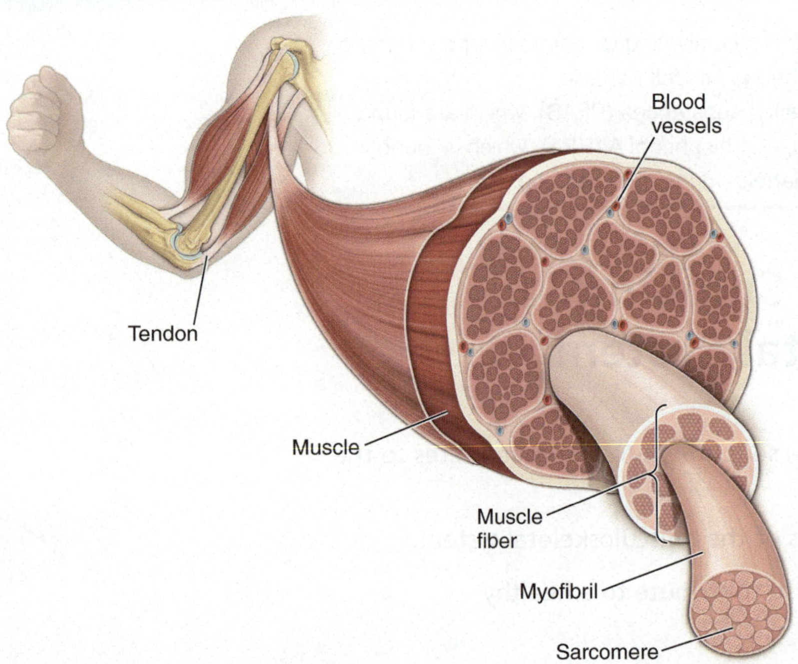

Note how the muscle is made up of many muscle fibers all bound together, and that these muscle fibers are made up of sarcomeres.

Figure 17.22 The sliding-filament model of muscle contraction

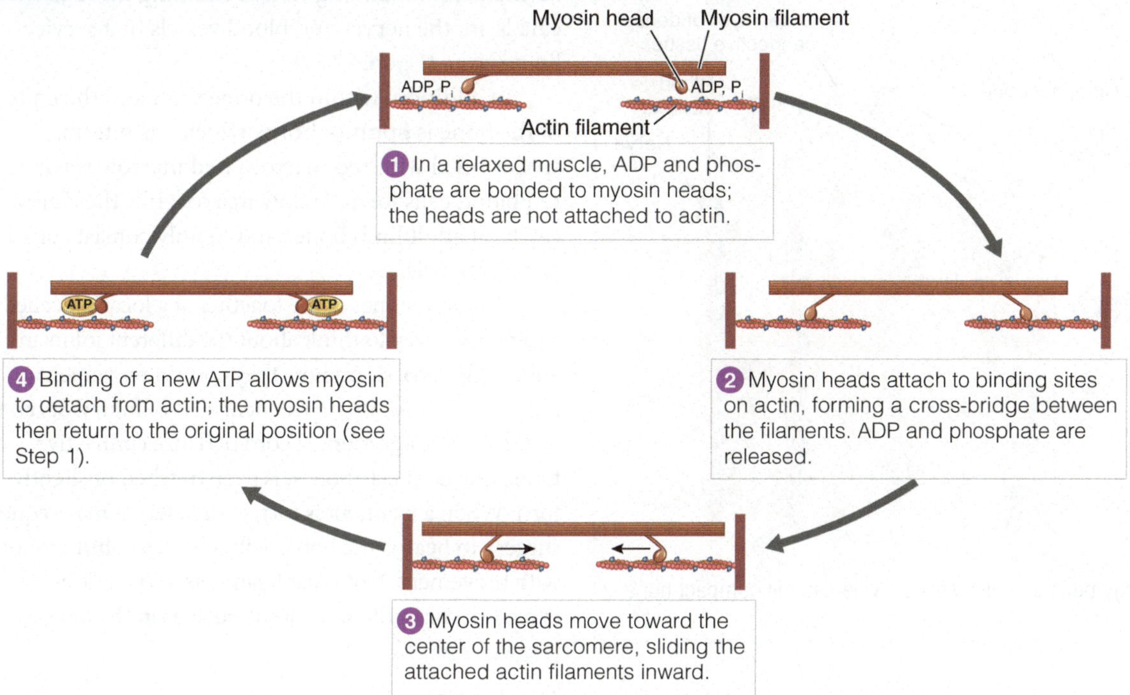

Although there are many sarcomeres in a muscle fiber, for simplicity, only one is shown here. Notice the role of ATP and how the actin and myosin interact with one another during muscle contraction.

between actin and myosin bring about muscle contraction. Note that neither the actin nor the myosin filaments change length, and that the myosin filaments do not change position. Instead, myosin heads bind to actin filaments and slide them toward the center of a sarcomere using ATP for energy (Figure 17.22). As actin filaments are pulled inward, the ends of the sarcomere are drawn closer and the sarcomere shortens so that in a fully contracted muscle, the thin filaments overlap. Muscle bundles, muscle fibers, myofibrils, and thick and thin filaments of a sarcomere all have the same parallel orientation. As a result, they all pull in the same direction when a muscle contracts.

Note that within a muscle bundle, there are different kinds of muscle fibers that move at different rates. For example, **fast-twitch muscle fibers** allow you to make rapid movements. These muscle fibers are most useful when catching yourself from falling or while sprinting. In contrast, **slow-twitch muscle fibers** are useful for endurance, like while on a long walk or taking a jog. Each of these muscle fibers uses ATP (Chapter 4) at different rates, which is why one can only sprint for a short amount of time: The ATP is used up quickly in a fast-twitch muscle.

Study Tip

If you interlace your fingers and slide them against one another, closer together (contracted) and farther apart (relaxed), this is similar to the movement that muscle fibers make when they slide next to one another.

Not all muscle moves bones. Some muscle moves other tissues, such as the muscles in your face that allow you to make facial expressions, and the muscles that move your tongue, enabling you to speak.

An Overview of Bone Structure

When muscles contract, not only are the muscles moving, but bone can move as well. Now that we've discussed muscles, we'll turn to the skeletal system, which is comprised of bones. **Bone** is a dynamic, living organ with many different biological functions. For example, new blood cells are produced in the bone marrow. Bones are also continually being broken down and built up again, or remodeled, which we will discuss shortly. An outer sheath of connective tissue covers the bone, and nerves and blood vessels extend through tiny passageways in the bone's interior. Bone tissue itself consists of living bone cells in a secreted extracellular matrix that contains collagen and is hardened with calcium and phosphorus.

Within your body are two types of bone: compact bone and spongy bone. **Compact bone** makes up the outermost, weight-bearing part of a limb bone, such as a femur. Compact bone

Figure 17.23 **The structure of compact bone and spongy bone tissue**

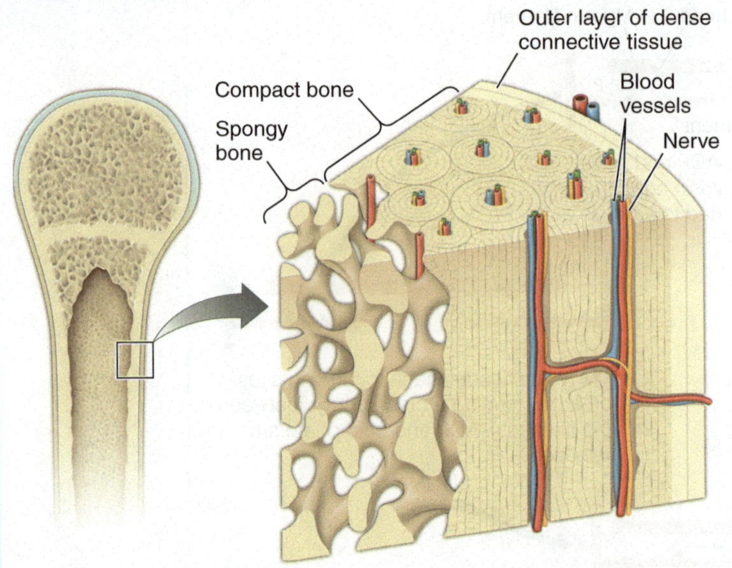

Notice how the spongy bone contains spaces, whereas the compact bone does not.

consists of many thin, concentric layers of extracellular matrix surrounding canals. Running through these canals are the nerves and blood vessels that service living bone (Figure 17.23).

In addition, within the bone shaft and the ends of the bone is **spongy bone**, which has internal spaces filled with red marrow. Red marrow is where red blood cells form. Yellow marrow fills the central cavity of adult limb bones and mainly consists of adipose (fat) cells.

Different bones come together at a location called a **joint**. One way to think about the different joints in your body is based on how they control movement (Figure 17.24). Joints are frequent sites of injury. Sprains occur when a **ligament**, a cord of dense connective tissue that connects bones, is overstretched or slightly torn. When a ligament is torn completely, it may require surgery to heal as the bones will otherwise shift around with movement. Note that ligaments and tendons can co-occur in the same joint, such as in the knee (Figure 17.25).

? Did You Know?

Bone is remodeled to meet the strain it is put under. When astronauts are in space, their bones become about 1% less dense for every month they spend in microgravity conditions.

Bone Remodeling

Bones are dynamic organs that undergo continual remodeling. Bone matrix is your main reservoir of calcium, and your body adds to and draws upon this reserve as needed by breaking down and building up bone via the actions of two types of cells: osteoclasts and osteoblasts, respectively. Within your bones, **osteoblasts** are cells that form new bone while also contributing to bone growth and healing after a break. Osteoblasts are found in the circle of compact bone around the internal cavity of a bone, where they lay down new bone matrix, thickening the bone. In contrast, **osteoclasts** break down bone that lines the medullary cavity. This is

Figure 17.24 **A comparison of different joint movements**

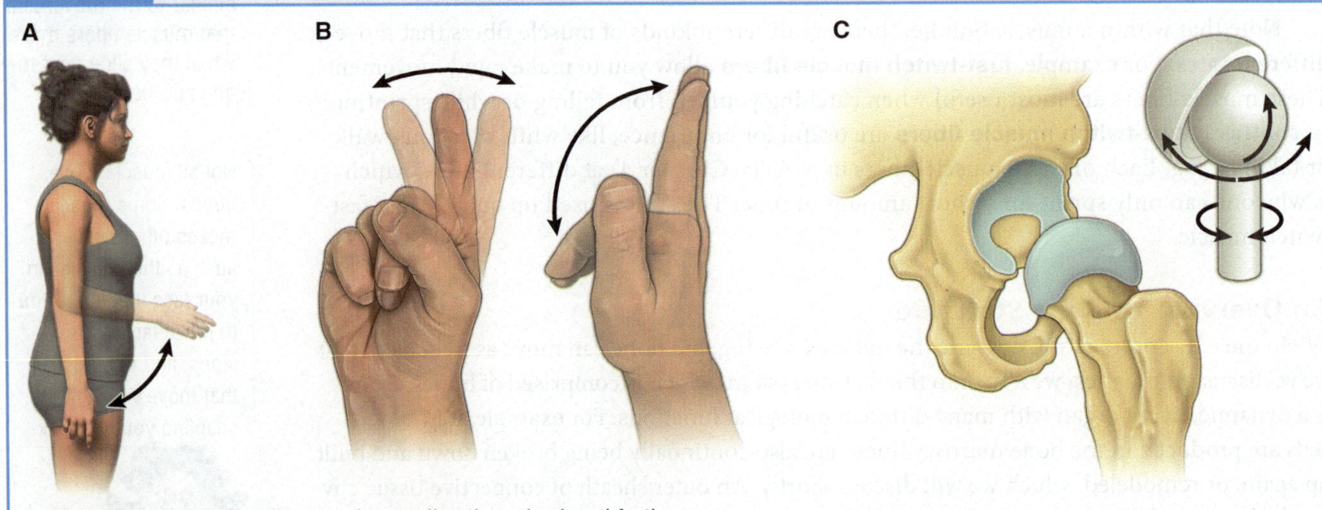

A. Your elbow joint can only move in one direction—back and forth.
B. The joints between your fingers allow movement in two directions, so you can move your finger back and forth and side to side.
C. The hip joint allows circular motion.

important because it helps to "refresh" the bone by removing old cells so that osteoblasts can add new bone tissue.

Until people reach their mid-20s, the rate of matrix formation exceeds the rate of matrix breakdown. After that, matrix production slows, and bone density can decline. This can lead to osteopenia and eventually, osteoporosis. Both of these are conditions where bone loss greatly outpaces bone formation, and so bones become weaker and more likely to break (Figure 17.26). Rates of bone deposition are impacted by lifestyle, medical conditions, and medications. Note that although osteopenia and osteoporosis are most common in postmenopausal women—because the ovaries no longer produce estrogen, which encourages bone deposition—these conditions can occur in women and in men of any age. We'll return to how exercise can mitigate these effects at the end of this section.

Figure 17.25 **Ligaments work in conjunction with tendons at the knee joint**

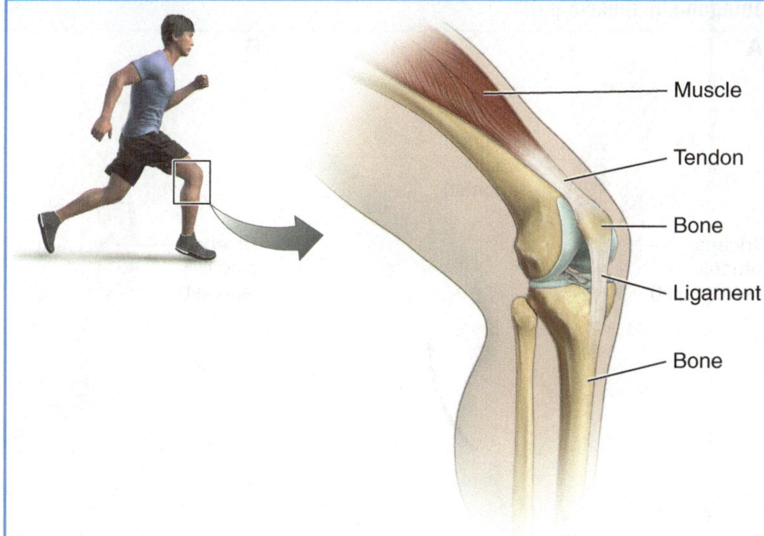

Note how the muscle connects to the bone via a tendon, and the bones are connected to one another via a ligament.

How Muscles Move Bones

As noted earlier, muscles move the skeleton. Muscles can pull on bones, but they cannot push on them because contraction involves the shortening of sarcomeres, thus creating a pulling force. Therefore, the muscles associated with an action often work in opposition: Motion generated by contraction of one muscle is reversed by the contraction of another.

For example, the biceps brachii muscle on the anterior (front) side of the upper arm opposes the triceps brachii on the dorsal (back) side. Together, they form an **antagonistic muscle pair**, where a pair of muscles pulls a bone in opposite directions, each at a different time. In the context of the arm, when the biceps contracts, the triceps relaxes, and the forearm is pulled toward the shoulder (Figure 17.27A). Contraction of the triceps coupled with relaxation of the biceps reverses this movement, extending the arm (Figure 17.27B). The force that muscles exert on bone during movement promotes bone deposition—so when one strengthens their muscles, they are concurrently strengthening their bones as well. We'll continue to address the positive benefits of exercise on both systems next.

? Did You Know?

Arthritis is the chronic inflammation of a joint. The most common type, osteoarthritis, usually appears in old age, after cartilage has thinned at a frequently jarred joint. Knees and hips are most often affected. Rheumatoid arthritis can affect people of any age. It occurs when the immune system mistakenly attacks synovial joints throughout the body.

Figure 17.26 **The progression of healthy bone through osteopenia to osteoporosis**

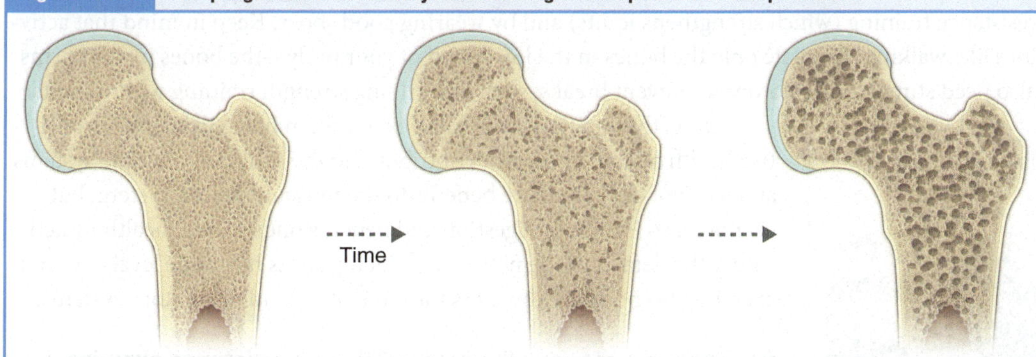

Shown here is the head of a femur (which is part of the hip joint). Osteoporosis in this area increases the risk of a hip fracture after falling.

? Did You Know?

When a bone breaks, cartilage, a flexible connective tissue, forms in the gap created by the break. Matrix-producing bone cells move in and replace the cartilage with spongy bone and then compact bone. This is similar to how your skeleton formed during prenatal development. A skeleton of cartilage formed first, then most of that skeleton was converted to bone.

Figure 17.27 The biceps brachii and triceps brachii, muscles of the upper arm, form an antagonistic muscle pair

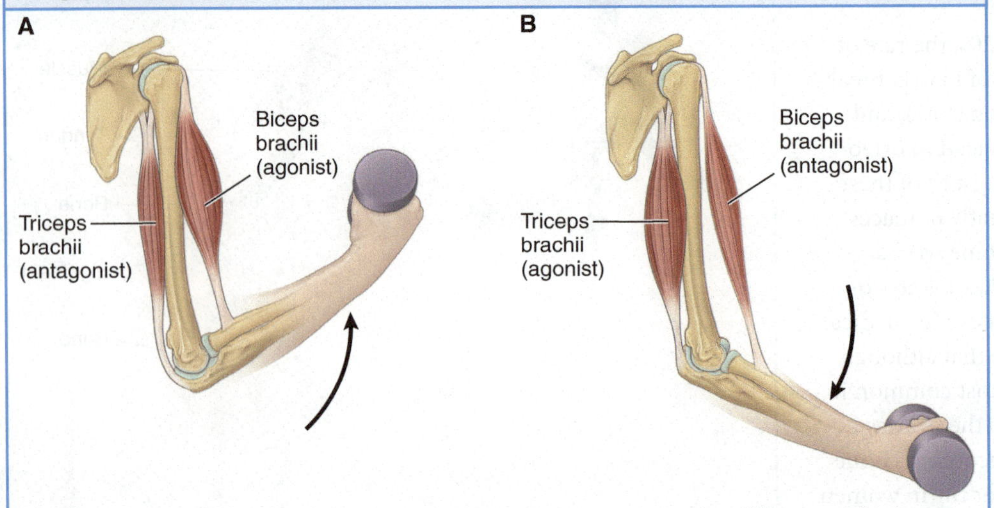

A. When someone lifts their forearm towards the upper arm, such as when lifting a weight, the biceps brachii contracts and the triceps brachii relaxes.

B. When the triceps brachii is activated, the biceps brachii relaxes, the forearm straightens, and the weight lowers.

The Health Benefits of Exercise to the Musculoskeletal System

In this section, we have learned about muscle fibers and how they contract, as well as how bone remodels itself. Now it's time to bring it all together to look at how exercise benefits musculoskeletal health and why an hour of exercise daily can have such a positive benefit on the body. We'll specifically focus on how aerobic exercise and strength training benefit the musculoskeletal system, and how each of these types of exercise can protect against disorders like osteoporosis.

Aerobic exercise. When unrelenting stimulation keeps a skeletal muscle excited, such as during exercise, muscle fatigue occurs. Muscle fatigue means that the muscle's capacity to generate force declines despite ongoing stimulation. Regularly engaging in **aerobic exercise**—exercise that is of low intensity but long duration—changes the number of fast-twitch or slow-twitch muscle fibers in your muscles.

You can feel the biceps brachii contract if you extend one arm out, palm up, then place your other hand over your biceps and slowly bend your arm at the elbow. Although the muscle shortens only about a centimeter during contraction, your hand moves through a much greater distance.

Exercise is important for preventing age-related decline, as people slowly lose their fast-twitch muscle fibers as they age. Aerobic exercise, such as jogging, biking, swimming, and walking all help preserve your slow-twitch muscles, whereas jumping, high-intensity interval training, and sports like hockey, which require quick bursts of energy, help to maintain and form fast-twitch muscle fiber. Strength training is also an excellent way of building fast-twitch muscle fibers, which we will return to shortly.

Weight-bearing aerobic exercise benefits not only the muscles, but also your skeletal system. When we are on our feet, aerobic exercise stimulates the osteoblasts to deposit bone. Ultimately, this can help to slow age-related bone loss and may even help with building bone. Higher-impact activities can have an even greater effect, but can also put additional strain on the body's joints. Therefore, one needs to balance the benefit to the bones while protecting joints, such as through resistance training (which strengthens joints) and by wearing good shoes. Keep in mind that activities like walking stimulate only the bones in the lower half of your body—the bones in your arms also need stimulation in order to prevent breaks, such as by doing strength training with weights.

Altogether, an hour of aerobic exercise can have such a major positive health benefit because it contributes to the health of multiple systems at once. Not only are there benefits to the musculoskeletal system, but exercise also improves digestion and benefits one's mental health by activating the default mode network, reducing stress hormone levels. Aerobic exercise also positively benefits the circulatory and respiratory systems.

Figure 17.28 Resistance training builds strong muscles and bones

Credit: Hirurg/Getty Images

Resistance exercise. Strength training, or **resistance exercise**, is any exercise that causes force on muscles, such as during weightlifting (Figure 17.28). Research suggests that resistance exercise is one of the best ways to slow, or in some cases, reverse, age-related muscle loss. During resistance exercise, the force on a muscle causes it to increase in size. For example, when lifting heavy weights, damage can occur to the sarcomeres. When this damage occurs, the sarcomeres respond by growing and/or forming more sarcomeres, which causes the muscle to ultimately increase in size. As the muscle fibers lengthen, the muscle can bear more weight.

And it's not just muscles that receive a benefit from exercise. Any activity that puts strain on your muscles also causes your bones to become stronger, thus helping to maintain bone health. This is particularly important for those at risk of osteoporosis. Within the bone, resistance exercise stimulates osteoblasts, promoting bone formation so that remodeling occurs, where bone is rebuilt faster than it is broken down, causing the bone to get stronger.

So, by engaging in exercise that stimulates the musculoskeletal system, we are building up the valuable muscle and bone tissue that we need to enjoy a healthy life now, while increasing the likelihood of a future of greater mobility with a decreased risk for injury.

Let's consider this in the context of osteoporosis and hip fractures, which are relatively common injuries that occur when people fall, especially with older individuals. Building up fast-twitch muscles through either aerobic activity or resistance training can not only strengthen muscle and bone, but can also help to prevent someone from falling in the first place. Building up slow-twitch muscles through activities like walking not only helps to promote muscle formation, but also bone deposition, which means that if the person does fall, their bones are much more likely to be stronger and less likely to break.

You don't need a gym membership or expensive equipment to engage in resistance training. Body weight exercises such as lunges, squats, and push-ups are all good options to try at home. You can also buy resistance bands at various retailers for under $20, opening up the potential for even more resistance exercises at home.

17.5 Summary

- A sedentary lifestyle increases risks of heart attack, stroke, and diabetes. It also can affect mental health and cause premature death, costing billions in healthcare costs each year.
- Muscles generate force to move the body, whereas the skeleton is the rigid framework for the body. Tendons connect muscles to bones, and bones are connected to one another by ligaments. A joint is where two bones connect.
- Aerobic exercises help strengthen muscle and bone while positively supporting other body systems. Resistance training can also help mitigate age-related decline and prevent osteoporosis.

As we age, we naturally lose muscle mass and bone density. Getting into an exercise routine and sticking to it while you're still young can help to mitigate some of the physical impacts of aging, keeping you mobile and healthier throughout your life.

17.6 Your Lymphatic and Immune Systems and Vaccines

Learning Objectives

17-19 Describe the economic, personal, and societal impacts of communicable disease.

17-20 Explain the connection between the lymphatic and immune systems.

17-21 Summarize how innate and adaptive immunity work.

17-22 Explain how vaccines elicit immunity.

17-23 Identify positive benefits of vaccines on public health.

Have you ever visited an old cemetery? Walking around the old stones, did you ever notice a stone with a lamb on top of it, typically on a child's headstone (Figure 17.29)? And did you notice that most children's headstones are old? That isn't a fluke—before the advent of modern medicine, it wasn't uncommon for children to die in their youth. The rates of childhood mortality (death before 5 years of age) have dropped significantly in the last 200+ years (Figure 17.30). Why?

Part of the decline was due to changes in public sanitation. For example, cholera, a major cause of death among young children in the early 20th century, is transmitted through consumption of water or food contaminated with the bacterium *Vibrio cholerae*. The other reason for the decline in childhood death rates? The advent of widespread **vaccination**, which is the process of giving someone a weakened, killed, or portion of an infection agent (a vaccine) for the purposes of protecting them from that disease.

Figure 17.29 **Lambs are common atop children's graves**

Some people see the lambs as religious symbols, while others feel they represent a life gone too soon.

Credit: Daniel Borzynski/Alamy Stock Photo

Figure 17.30 **Declining rates of childhood mortality**

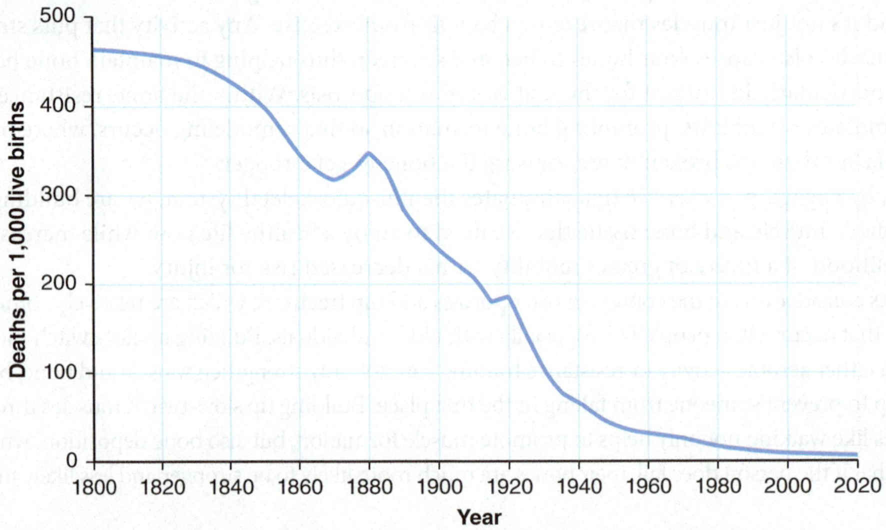

In 1800, 46% of children born did not survive to their fifth birthday. In 2020, the rate had dropped to 0.7%. Spikes in childhood mortality in the 1870s were the result of cholera, smallpox, and yellow fever outbreaks. The increase in 1910 was the result of the Spanish flu pandemic.

Graph based on: Aaron O'Neill, Child mortality in the United States 1800-2020, Aug 9, 2024,Statista,https://www.statista.com/statistics/1041693/united-states-all-time-child-mortality-rate/

Although Jenner is credited as developing the first vaccine, variolation (transferring small amounts of material from a sick person, such as a smallpox sore, to a healthy person) may have occurred as early as 200 B.C.E.

Edward Jenner was the first to scientifically test vaccines. In 1796, he took pus from a cowpox blister and put it into a cut on a person's skin for the purpose of preventing that person from getting smallpox. While rudimentary, the method was successful at preventing the spread of smallpox, a highly contagious disease that can cause scarring and death. Childhood vaccination did not become commonplace until 1948, when the diphtheria, pertussis, and tetanus shots became available.

Childhood vaccines became popular in the 1950s with the development of the polio vaccine. Polio, also called infantile paralysis, first came into the public eye in 1894, when it killed 18 children and left 123 permanently paralyzed. Each summer thereafter, outbreaks of polio killed or paralyzed millions of people, mainly children.

Working at the University of Pittsburgh School of Medicine, Jonas Salk, a virologist and medical researcher, and his team went on to develop a vaccine containing inactivated viral material from three different strains of polio (Figure 17.31A). In the early 1950s, they conducted the largest clinical trial to date, with over 1.8 million children volunteering to participate. In 1955, the data were in: The vaccine was safe and effective—and it received approval for licensing and distribution, to the relief of many who waited in long lines to receive the vaccine (Figure 17.31B).

Figure 17.31 **The development of the polio vaccine was a milestone for modern public health**

A B

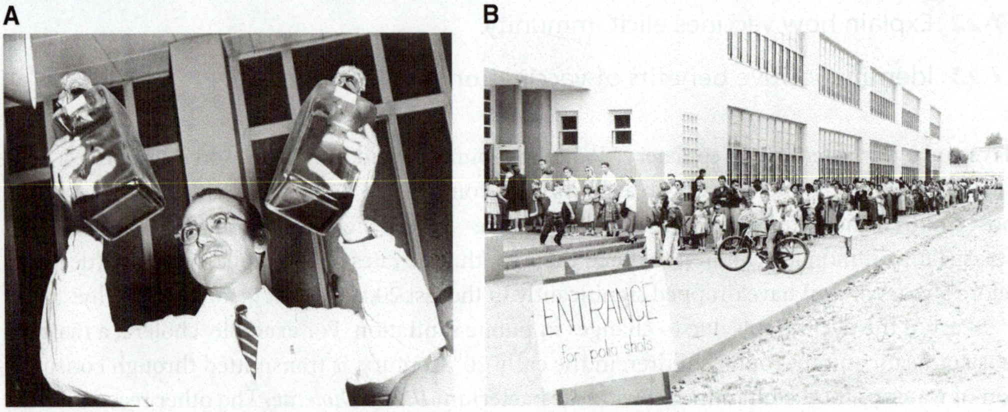

A. Virologist Jonas Salk holds up bottles containing the polio vaccine.
B. First and second grade students wait in line at the Kit Carson School in San Diego, CA, to receive their vaccines in 1955.

Credits: Bettmann/Getty Images; Bettmann/Getty Images

? Did You Know?

Jonas Salk didn't make any money from developing the polio vaccine. He decided to not patent it, stating that the vaccine belonged to the people and should be available to all.

Process of Science

Evidence-Based Medicine

When we consider how science works in the real world, one of the benefits and outcomes of science revolves around building knowledge not only for the sake of increasing our knowledge, but also for addressing societal issues; for example, using quality evidence (Chapter 1) in deciding treatment options.

Let's consider measles, a highly contagious disease caused by a virus. Typical symptoms include high fever, cough, runny nose, and body-wide rash (Figure 17.32). Serious complications are not uncommon. The virus outright kills 1 to 2 of every 1,000 children it infects. Optic nerve infection results in blindness, and encephalitis (dangerous swelling of the brain) can cause permanent intellectual disability or death. Rarely, a fatal central nervous system disease can appear decades after a measles infection resolved.

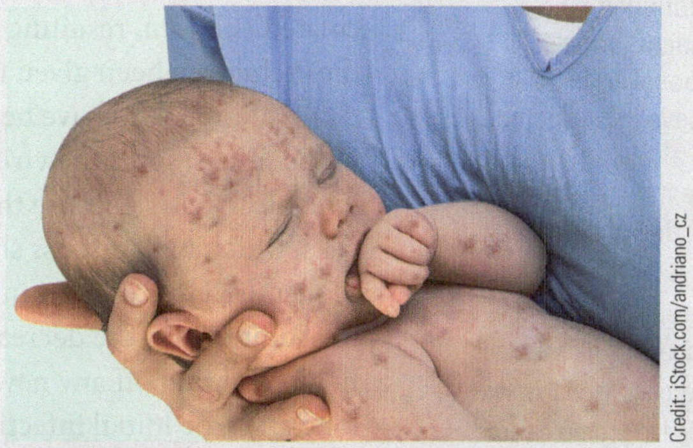

Figure 17.32 A baby infected with measles, a dangerous and vaccine-preventable disease

Measles is dangerous, particularly for children who are too young to receive vaccinations. Outbreaks are increasingly common as people decline to vaccinate their children.

The measles virus spreads easily from person to person via the air, through droplets expelled when a sick person coughs or sneezes. The virus can stay infectious for hours in the air. An infected person is contagious for four days before the telltale rash appears. An unvaccinated person who encounters someone who is contagious has a 90% chance of contracting measles!

As such, an effective vaccination program requires comprehensive community participation, since measles can spread quickly through a community in which fewer than 92% of people are vaccinated. Some people have no choice but to rely on *community immunity*, where people with immune deficiencies, for example, or children who are too young to be vaccinated, are provided with some degree of protection due to enough people in the community being immune to the disease.

Unfortunately, due to concerns about vaccination, some people turned away from an evidence-based practice (vaccinating to prevent measles) during the 2025 measles outbreak in the southwestern United States. People instead relied on a nonevidence-based treatment: giving their children vitamin A to treat the disease. Like many misinformation campaigns (Chapter 1), there is a nugget of truth here: Vitamin A can reduce the deadliness of a measles infection in people who are deficient for vitamin A (largely not an issue in the United States). So, the treatment would not be expected to work here. In addition, vitamin A does not prevent someone from getting the measles or spreading it to others.

Sadly, during the 2025 measles outbreaks, parents attempting to treat their children with vitamin A resulted in their hospitalization due to vitamin A toxicity. Overall, this story highlights the consequences of not following evidence-based guidelines, and the tragic outcomes that can result from children not being vaccinated against this highly contagious and avoidable disease.

1. States differ in the types of vaccine exemptions they allow. All states allow medical exemptions for those who have conditions that prevent them being vaccinated or who have had reactions to vaccines in the past. Where states differ is regarding other exemptions, such as those related to religious and personal beliefs. What type of evidence might you expect a person to provide for a personal belief exemption? What do you think they should know from a community perspective before being granted such an exemption?
2. Some people oppose vaccinations because they believe that they "baby" the immune system and think it's important for our health to fight infections naturally.
   a. How might measles specifically *reduce* someone's ability to fight off infection when one becomes infected with a different infectious agent?
   b. Consider that we are exposed to billions and billions of pathogens in our lifetime. Why do you suppose we don't receive even close to that number of vaccines? What does this say about the few (around 15) diseases that we are vaccinated against?

Research suggests that the polio vaccine alone prevented approximately 24 million cases of polio worldwide between 1988 and 2021.

Today, polio and many other dangerous childhood illnesses are relegated to the past. Or are they? Many have argued that vaccines have become victims of their own success. We don't remember visiting the park and seeing children in braces. We don't remember family members dying of the measles. We don't remember the fear of *getting* the measles and wondering if it was going to kill you next.

Unfortunately, the lack of personal experience with these deadly diseases combined with rampant mis/disinformation campaigns around vaccination (Chapter 1) have resulted in a

Did You Know?

The National Foundation for Infantile Paralysis was founded by President Franklin D. Roosevelt in 1938. Roosevelt contracted polio at the age of 39 and was paralyzed from the waist down as a result.

decline in vaccination rates—which has fueled outbreaks of diseases such as the measles in 2019 and in 2025—and resulted in the preventable deaths of children. What's especially heartrending is that measles was declared eliminated from the United States in 2000. However, in 2025, there were over 1,300 cases of the measles in the United States alone, almost exclusively among unvaccinated children, resulting in two deaths—deaths that could have been entirely prevented had these children been given the measles vaccine.

Children who have been infected with the measles virus will go on to face many unknowns in the future, since the virus can cause a loss of immunity. The loss of immunity, called *immune amnesia*, occurs because the measles virus infects memory B and T cells (key immune system cells, which we'll discuss shortly), causing a significant reduction in this pool of cells, and consequently, a loss of immunity to any pathogens that the person had been previously exposed to. Additionally, the decrease in memory B and T cells means that the individual may have difficulty fighting off any new infection, meaning that they can suffer from secondary infections long after the initial infection with the measles virus.

Grandparents for Vaccines is an organization that was founded to promote vaccines by encouraging people who are old enough to remember life before childhood vaccination to share their personal experiences with diseases like polio.

How does vaccination work? And what does it have to do with our immune and lymphatic systems? In this section, we'll explore how our immune system works, with a specific focus on how vaccines work with our immune systems to keep us safe from deadly diseases.

Defending the Body Involves the Immune and Lymphatic Systems Working Together

Every day, you continually cross paths with a tremendous array of bacteria (Figure 17.33), viruses, fungi, parasites, and other potential agents of disease. The good news is, humans coevolved (Chapter 11) with these pathogens, so your body has defenses against them. As multicellularity evolved, so did the mechanisms of distinguishing self (cells belonging to one's own body) from nonself (foreign matter or microbial invaders).

The body can protect itself through biological, physical, and chemical barriers that prevent microorganisms from entering. In addition, the body can also protect itself through the recognition of foreign cells and viruses once they enter the body. This recognition is possible through the action of cells that are part of the **immune system**, which help the body to recognize itself while also removing foreign invaders. **White blood cells**, the cells of the immune system, are housed within the **lymphatic system**, which includes organs like the thymus, where T cells mature, and the lymph nodes and spleen, where white blood cells are stored and then activated and released as needed. We'll learn more about different types of white blood cells later in this section.

The lymphatic system collects the **lymph**, a type of interstitial fluid that leaks from blood vessels, filters it, and then returns it to the circulatory system. In doing so, it helps to remove pathogens, cancerous cells, foreign substances, and other waste products and cellular debris that might be circulating in the body through interactions with cells of the immune system, which we'll now examine in greater detail.

Figure 17.33 **Bacteria found on a cell phone**

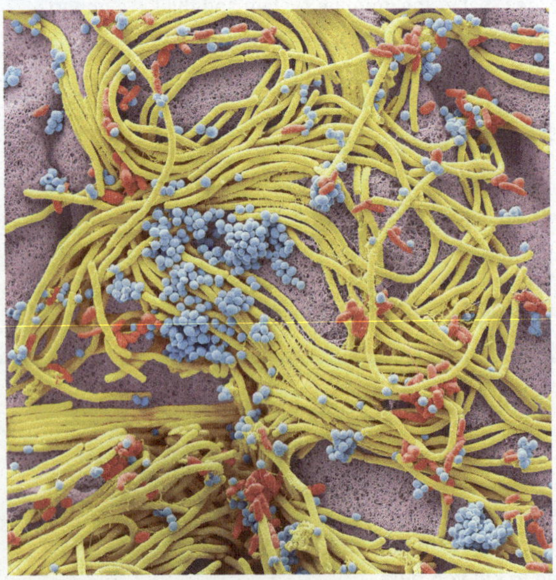

The average cell phone has about 18× more bacteria on it than a toilet handle in a public bathroom.

Credit: STEVE GSCHMEISSNER/Science source

The Defenders: The Cells of the Immune System

White blood cells of vertebrate animals participate in both innate and adaptive immune responses. **Innate immunity** refers to a set of immediate, general immune defenses that help protect organisms from infection and includes barriers, such as the skin, plus fever and the involvement of various types of white blood cells, like phagocytes. In addition, innate immunity includes **inflammation**—a fast, local response that destroys infected or damaged tissue and jump-starts the healing process. Components of the innate immune system are turned on quickly and in a nonspecific manner. For example, you may get a fever regardless of whether you're suffering from a viral or a bacterial infection.

Adaptive immunity refers to a set of defenses that can be tailored in billions of different ways to fight specific pathogens that an individual may encounter during their lifetime. Cells of the adaptive immune system include lymphocytes, such as T cells. Innate and adaptive immunity have different features, but they work together to protect the organism from disease.

Almost all cells of the immune system spend at least part of their life circulating in the blood and lymph, and many populate the lymph nodes, spleen, and other solid tissues. White blood cells communicate by secreting and responding to chemical-signaling molecules called **cytokines**. Cytokines act as a chemical messenger between immune system cells, allowing them to coordinate their activities throughout the body during an immune response.

If you've ever wondered why you get a fever when you're sick, it's because of cytokines. Some cytokines stimulate brain cells to raise the body's internal (core) temperature during an infection, which causes a fever. Fever enhances almost every step in the immune response—for example, by speeding up the production and activity of white blood cells while also greatly reducing the rate of viral replication and bacterial division. However, dangerously high fevers (>105°F, or 40.6°C) can cause brain damage or death.

There are three different kinds of white blood cells (Table 17.1). We'll explore each type of white blood cell in more depth next.

Table 17.1 Types of White Blood Cells

White blood cell type	Examples	General function
Phagocytes 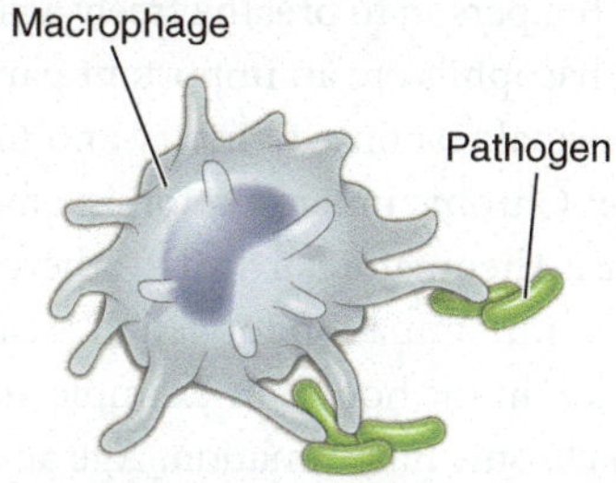	Macrophages, neutrophils, dendritic cells, and basophils	Clear out bacteria or cellular debris via engulfment
Granulocytes 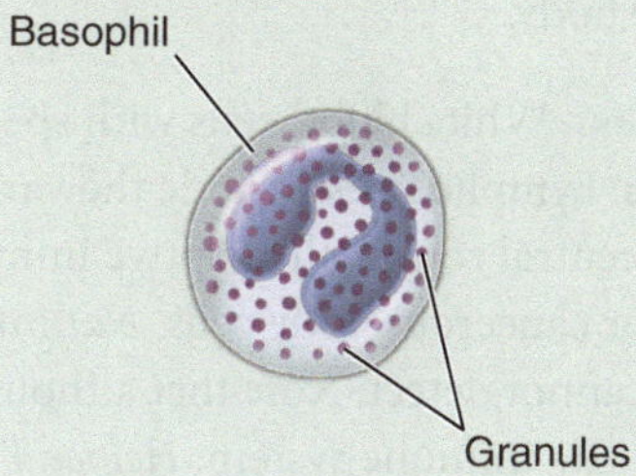	Basophils and neutrophils	Cause inflammation
Lymphocytes 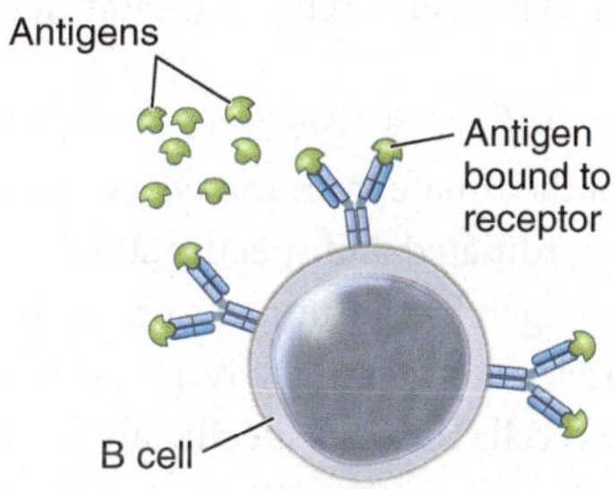	B cells and T cells	Remove infected or cancerous cells; integral to adaptive immunity

Phagocytes. While many blood cells can undergo *phagocytosis* (when a foreign invader is surrounded, engulfed, and ultimately destroyed; Chapter 5), some cells, like neutrophils, macrophages, and dendritic cells, are highly specialized phagocytes. A **phagocyte** is a cell that specializes in finding and removing bacterial or cellular debris. **Neutrophils** are the most abundant circulating white blood cells, so they are often the first responders at a site of infection or tissue damage, where they engulf invading pathogens or clean up cellular debris such as dead cells.

Macrophages are large phagocytes that patrol tissues and tissue fluid, engulfing essentially everything except undamaged body cells. A **dendritic cell** specializes in antigen presentation, where an **antigen**—a molecule that evokes an immune response—is recognized by a dendritic cell, which captures it and, after processing, presents it on its surface, alerting cells of the adaptive immune system of the threat.

Granulocytes. A **granulocyte** is a cell with vesicles called granules, which contain cytokines, local signaling molecules, destructive enzymes, and toxins such as hydrogen peroxide. These cells degranulate (release the contents of their granules) in response to a trigger such as antigen binding, thereby killing the invader. Note that many neutrophils are also granulocytes.

Basophils are white blood cells and a type of granulocyte that release the chemical signals that cause inflammation. Despite the fact that basophils are the least abundant type of white blood cell, they are probably the immune cell type that is most on your radar—especially if you have allergies! Basophils are the culprit behind allergies. For example, basophils in your nose release *histamine*, which causes swelling and a runny nose. Antihistamines like Benadryl® counter the release of histamine to keep you comfortable. In the case of *anaphylaxis*, which is a life-threatening allergic reaction that among other symptoms causes respiratory distress, epinephrine is used to reduce the swelling brought on by histamine release, relaxing smooth muscle in the airways and allowing the person to breathe freely again.

Notably, basophils are an important part of the inflammatory processes. Short-term inflammation is a normal response to injury and infection. But what about long-term, or *chronic, inflammation*? Chronic inflammation is a factor in several serious diseases, including asthma, Crohn's disease, rheumatoid arthritis, atherosclerosis, diabetes, and cancer. Some people have higher levels of inflammation for genetic reasons. Body composition can also impact the degree of inflammation in the body. For example, *adipokines* are chemicals produced by fat cells that contribute to chronic inflammation. Age and diet can also contribute to chronic inflammation. The good news is that a healthy diet and regular exercise help to decrease many types of inflammation in the body.

? Did You Know?

Humans can make billions of unique B and T cell receptors because the genes that make them can be alternatively spliced (Chapter 7).

Lymphocytes. White blood cells with special roles in the adaptive immune response are called **lymphocytes**. Lymphocytes include the various kinds of T and B cells. **T cells** (or T lymphocytes) play a central role in all adaptive immune responses. *Cytotoxic T cells* are specialized to kill infected or cancerous body cells. *Natural killer cells* can kill cancerous body cells that cytotoxic T cells cannot detect. Note that although natural killer cells are lymphocytes, they act as part of the innate immune system. *Helper T cells* help to activate **B cells** (or B lymphocytes), which make antibodies that bind to and help inactive microscopic invaders. We'll discuss the function of T cells and B cells in greater detail in the next section.

? Did You Know?

Since antibodies tightly bind to their target antigen and not much else, they are useful in research. For example, an antibody may be labeled with a fluorescent dye and used to identify the location of a protein of interest in cells.

A Specific Defense System: Adaptive Immunity

If we think of the innate immune system as a general response to infection, the adaptive immune system is a coordinated and specific attack on an invading pathogen. The immune system can identify a pathogen by the antigens on its surface—carbohydrates or proteins (Chapter 3) that mark it as foreign—triggering a response from B and T cells. An **antibody** is a Y-shaped protein produced by B cells that can specifically recognize antigens. The antigen-binding sites of each antibody bind only to an antigen that has a complementary distribution of bumps, grooves, and charge (Figure 17.34). Antibodies produced by a new B cell stay attached to its surface as **B cell receptors**.

B cells that become activated in an adaptive immune response secrete antibodies into the bloodstream. Within the bloodstream, if an antibody encounters its antigen (such as the surface of a bacterium), it will attach to it. Once the antibody attaches, it can directly inhibit a bacterium by preventing it from attaching to body cells, or it can signal other immune cells,

Figure 17.34 Antibody structure

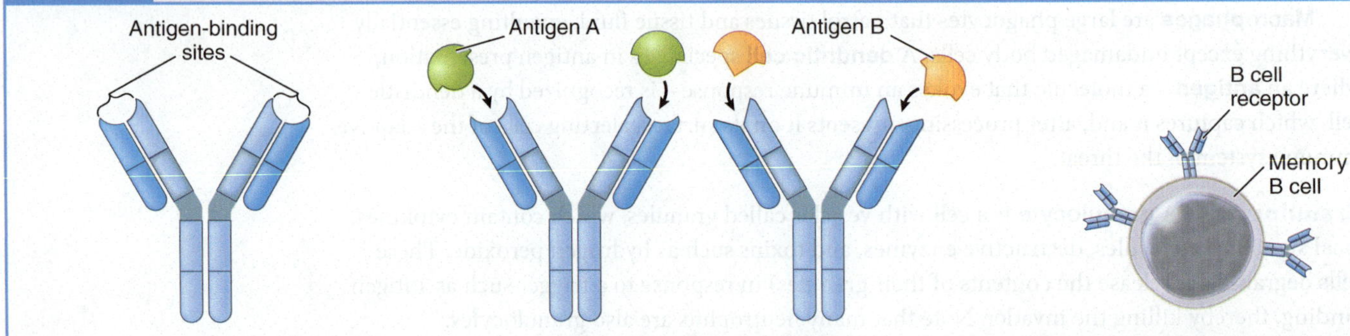

A. An antibody molecule consists of four polypeptide chains joined in a Y-shaped configuration. The chains fold so that the antigen-binding sites form at the tips of the Y.

B. The antigen-binding sites of each antibody bind to a specific antigen—generically labeled as antigens A and B above—based on its shape and charge.

C. Antibodies produced by a new memory B cell stay attached to its membrane as B cell receptors.

like macrophages, to the bacterium's presence, marking it for destruction by phagocytosis. Note that T cells are also able to recognize antigens through the **T cell receptors** on their surface.

Teaching the Immune System to Identify Invaders: Antigen Processing and Display

Suppose that you accidentally nick your finger while cutting vegetables. Being opportunists, some bacterial cells on your skin immediately enter the cut, invading your internal environment (Figure 17.35 1). Dendritic cells engulf, or phagocytose, the invading bacteria, which begins the process of **antigen processing**, where a foreign invader (such as a bacterium) is digested by an immune system cell, which then displays the antigen on its surface, thus "teaching" other cells of the immune system to recognize the antigen 2. Think of it like the dendritic cell holding up a sign showing a photo of the invader. Note that for simplicity, only a single antigen is displayed on the surface of the dendritic cell in Figure 17.35, but, in fact, thousands of antigens could be displayed.

The dendritic cell bearing its antigen-presenting "sign" then moves via lymph into a **lymph node**, which is a small, bean-shaped organ that filters the lymph while also removing pathogens. Once in the lymph node, the dendritic cell presents the antigen to a **helper T cell** by binding to a T-cell receptor 3. This helps to coordinate the immune response by making more copies of the T cells, called **memory T cells**—which can then readily identify the invader upon a subsequent infection—as well as activating effector T cells 4. **Effector T cells** can circulate in the blood, directly killing infected cells, but can also move to the *spleen*, a secondary lymphoid organ, where they can recognize antigens displayed on the surface of B cells, causing the T cell to release cytokines 5.

Cytokine release stimulates the production of **memory B cells** 5, which, like memory T cells, help with recognition of the same pathogen if it infects again. Cytokine release also stimulates plasma cells. A **plasma cell** is a cell that makes and releases the antibodies that circulate in the blood and lymph. Once in the

Figure 17.35 Antibody-mediated response

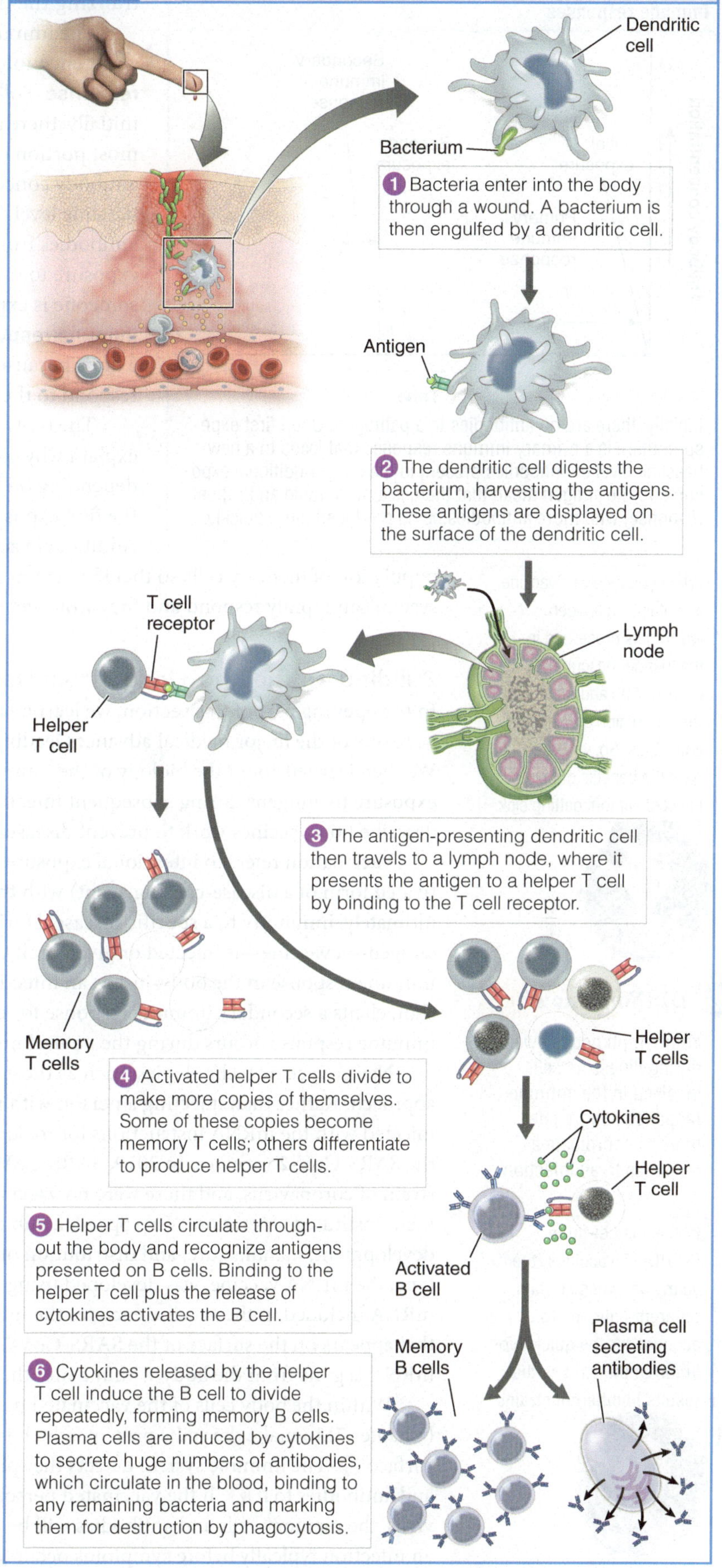

Figure 17.36 **A comparison of the primary and secondary immune responses**

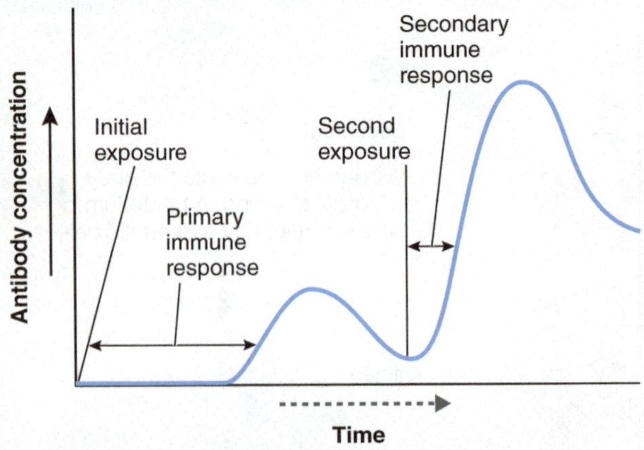

Initially, there are no antibodies to a pathogen. Upon first exposure, there is a primary immune response that leads to a new baseline level of antibodies present in the body. Additional exposures to the antigen mean that there is a more rapid and robust response, with more antibodies being produced more quickly.

blood and lymph, they bind to any remaining bacteria, neutralizing the infection.

The immune system's initial reaction to a new foreign agent encountered in the body is called the **primary immune response**. Following the data line in Figure 17.36, note how initially, there are no antibodies to the antigen (the flat, leftmost portion of the data line), but after the initial exposure, the antibody concentration increases, then drops to a new, higher baseline level. This new baseline level represents circulating antibodies from the memory B cells generated after the initial exposure to the antigen. Due to this immunological memory, if someone is exposed to the same antigen again, the **secondary immune response** is quicker and much greater than the first. This is because the immune system can rapidly mobilize and respond to the threat, making more antibodies more quickly.

The body's primary response and then secondary response explain why some vaccines require boosters or multiple doses, depending on how many memory B cells are produced after the first exposure and how long those memory B cells persist. Additional vaccines for the same antigen can create a robust population of memory cells so that if a person encounters, say, the polio virus, their immune system can rapidly respond and they won't get sick.

When you receive a vaccine, it contains an antigen, which gets processed in the manner described in Figure 17.35 and results in the production of specific antibodies. So, you can have the benefits of being infected without getting sick.

? Did You Know?

Your lymph nodes swell during illness as cells involved in the immune response congregate there to coordinate a response to an infection.

Building on the Knowledge from Nature: Vaccination

In the opening part of this section, we learned about how vaccinations, especially for children, were one of the major medical advances leading to the sharp decline in childhood mortality. We then learned about the biology of the immune system, and how the body can remember exposure to antigens during subsequent infection. Now, we're going to bring it all full circle to describe how vaccines work to prevent diseases—not only in children, but also in adults!

Vaccination refers to intentional exposure to antigens (such as part of a virus or an inactivated form of a disease-causing agent) with the goal of inducing an immune response, and ultimately, immunity to a specific disease. During vaccination, a preparation that contains antigen—a vaccine—is injected or given orally. The vaccine elicits an antibody-mediated immune response in the body, just as an infection would, and a second vaccination, or booster shot, elicits a secondary immune response for enhanced protection, similar to how a secondary immune response occurs during the natural immune response.

Newer vaccine technologies, such as those that use mRNA, take a slightly different approach. Rather than injecting a person with part of a virus or bacterium, the person is injected with the mRNA instructions for making a part of the virus (Chapter 7). Let's consider the SARS-CoV-2 vaccines. In 2019, SARS-CoV-2 (the virus that causes COVID-19) was a new strain of coronavirus, and there were no vaccines available for it. As a result, the U.S. government instituted Operation Warp Speed, a public–private partnership designed to accelerate the development, manufacture, and distribution of vaccines for SARS-CoV-2. One such vaccine type, the mRNA vaccine, was developed using knowledge of the central dogma (Chapter 7). The mRNA included in the vaccine encodes the information to make the spike protein (Chapter 1) that appears on the surface of the SARS-CoV-2 virus (Figure 17.37 ①). Note that all the injected mRNA is gone from the body within a month.

Within the body cells of the vaccinated person, the mRNA is then translated on a ribosome (Chapter 7) to make the spike protein, which is then displayed as a harmless protein on the cell surface ②. The immune system detects the spike protein as foreign, generating memory cells and antibodies to it ③. If the vaccinated person has a subsequent infection with SARS-CoV-2 virus, the memory cells and antibodies will be able to identify and neutralize the virus, stopping an infection typically before symptoms occur ④.

In the case of flu or COVID-19 vaccines, the yearly vaccines contain different antigens to account for the quick evolution of the viruses that results in different strains.

Figure 17.37 **An overview of SARS-CoV-2 vaccination with mRNA technology**

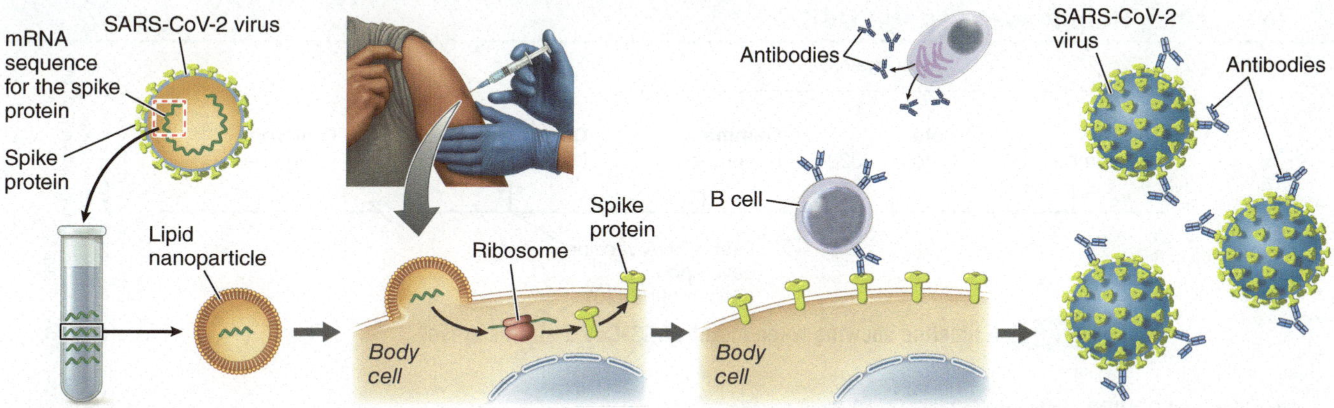

① The mRNA sequence for the spike protein on the SARS-CoV-2 virus is synthesized in the lab and packaged in a lipid nanoparticle.

② When a person is vaccinated, lipid nanoparticles with the mRNA sequence for the spike protein enter the body cells and are translated by ribosomes. The spike proteins produced then travel to and appear on the cell surface.

③ An immune system cell, such as a B cell, recognizes the spike as foreign, binding to it and triggering an immune response. Antibodies, as well as memory B cells, are produced against the spike protein.

④ If the immunized person is later infected with the SARS-CoV-2 virus, the body's antibodies recognize the spike protein and bind to it, which stops the virus from replicating.

Biology for Life

Autoimmunity and Women

Our immune systems are critically important for keeping our bodies safe from invading pathogens. The body has ways of examining the antigens on the surface of a cell to determine if any given cell is "self" and should be left alone, or whether it is "non-self" and should be attacked and destroyed. But what happens when the body gets it wrong and begins to attack its own healthy tissues, which is called *autoimmunity*? The resulting internal civil war can be devastating, causing autoimmune disorders ranging from rheumatoid arthritis to ulcerative colitis. Autoimmunity is also implicated in recurrent pregnancy loss, when one's immune system mistakenly attacks the developing fetus.

According to the Autoimmune Association, over 50 million Americans live with an autoimmune disorder—and the number is increasing. The number of Americans with new autoimmune disorder diagnoses has increased from 3% to 12% each year, with many people taking on average 5 years to receive a diagnosis. In 2025 alone, the projected cost of autoimmune disorders—including treatment, disability rates, and lost productivity—is expected to be around $180 billion.

Notably, women are 4× more likely than men to develop an autoimmune disease. The reason for this stark difference isn't entirely clear, but it could be that individuals with more than one X chromosome (Chapter 8) have naturally higher levels of antibodies, which suggests a genetic component. Additionally, antibody levels, and consequently, autoimmunity, tend to increase post-pregnancy, which suggests that hormones may play a role. Evolutionarily speaking, this makes sense. During breastfeeding, the baby receives a temporary benefit by ingesting the mother's antibodies. The mother also has additional immune protection during the post-birth recovery period. However, it comes at a cost to the mother's own cells—overactivation of the immune system can damage cells. Again, in evolutionary terms, this makes sense—evolution fundamentally requires successful reproduction (Chapter 10), so if it helps the offspring survive, even if it's at the expense of the mother, it would be favorable in future generations.

1. Describe the negative effects of having too few antibodies or plasma cells. Why is it also detrimental to have *too many* antibodies or antibody-producing cells?
2. Specific antibodies can trigger release of histamine from basophils. What effects would this have in the body? Would antihistamines potentially help with these types of autoimmune disorders?
3. When histamines are released from basophils, it is typically a response to an infection. In the case of autoimmune disorders, however, there is no infection. Why would you expect the effects of histamines to be long-term (chronic) and therefore detrimental in the case of autoimmune disorders?

? Did You Know?

Massive immunological changes happen during pregnancy that allow foreign cells (the fetus) to remain and grow inside the mother's body. The presence of semen in the body starts the immunological changes necessary for pregnancy to occur.

Digging into Data

Do Vaccines Help Prevent Long COVID?

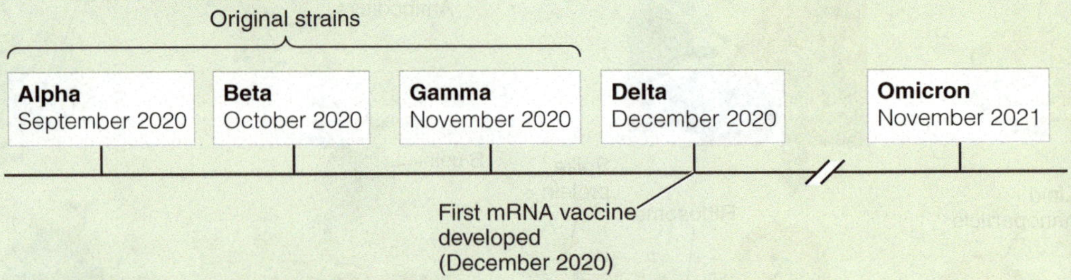

Figure 17.38 **A timeline showing when major SARS-CoV-2 strains arose**

Graph based on: Taha BA, Al-Juboori Q, Al Mashhadany Y, Zan MSDB, Bakar AAA, Fadhel MM, Arsad N. Photonics enabled intelligence system to identify SARS-CoV 2 mutations. Appl Microbiol Biotechnol. 2022 May; 106(9-10):3321-3336.

Have you had COVID? Maybe more than once? Most people who are infected by the severe acute respiratory syndrome coronavirus 2, or SARS-CoV-2 for short, have acute (sudden and short-term) symptoms of coronavirus disease (COVID). These symptoms include fever, cough, tiredness, loss of sense of taste, nausea, and many more. These symptoms usually last for about a week and then fully resolve.

About 10% of people infected with COVID, however, continue to experience symptoms long after the infection has resolved. This condition is called *long COVID*. The most common symptoms are debilitating fatigue, changes in sense of smell and taste, insomnia, and headaches. In addition, people can experience complex long-term effects on many organ systems, including the nervous, digestive, urinary, and cardiovascular systems. While doctors can provide options to manage symptoms, there is currently no cure for long COVID.

Since the pandemic began, SARS-CoV-2 has gradually evolved (Chapter 12). Random genetic mutations have caused changes in how well the virus spreads, as well as the types and severity of the symptoms it causes. When enough mutations occur, scientists designate a new strain (Figure 17.38). No vaccines existed when the original pre-delta strains (alpha, beta, and gamma) circulated.

In research published in the *New England Journal of Medicine* in July 2024, Ziyad Al-Aly and his team analyzed data from the U.S. Department of Veterans Affairs. Using the medical records of over 4 million veterans, the researchers sorted them by (1) the strain of SARS-CoV-2 they were infected with; (2) whether and when they were diagnosed with long COVID (defined by symptoms persisting one year or more after acute infection); and (3) whether or not they had received a SARS-CoV-2 vaccine. The results of the research are provided in the graph, which shows the incidence of long COVID for each group per 100 people (Figure 17.39).

1. Examine the data in the graph provided. How does the risk of long COVID differ by SARS-CoV-2 strain? What might explain the difference?

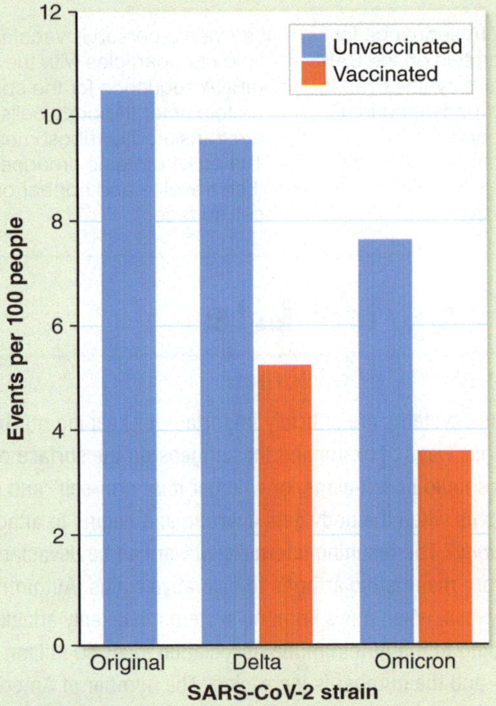

Graph based on: Xie, Y., Choi, T., & Al-Aly, Z. (2024). Postacute sequelae of SARS-COV-2 infection in the Pre-Delta, Delta, and Omicron eras. New England Journal of Medicine, 391(6), 515–525

Figure 17.39 **Long COVID by strain of infection and vaccination status**

Incidence of long COVID one year after SARS-CoV-2 infection among unvaccinated persons during the original (pre-delta), delta, and omicron eras; and among vaccinated persons during the delta and omicron eras. Values are shown in events per 100 people.

2. For each strain, compare the percentage of vaccinated versus unvaccinated individuals who developed long COVID. How does vaccination affect one's risk of acquiring long COVID? What explains these results?
3. Do some research on a reputable medical website, like the Mayo Clinic. Weigh the potential side effects of mRNA vaccines against the symptoms of long COVID. How might the risks associated with each affect one's decision to be vaccinated? What other considerations might affect a person's decision?

Scientists estimate that COVID-19 vaccinations prevented 14.4 to 19.8 million deaths during the first year the vaccines were available (**Figure 17.40**). From COVID-19 vaccines to the polio vaccine, vaccination campaigns have undoubtedly saved lives. But even with easy access to vaccines, some parents are choosing not to vaccinate their children, citing fears about vaccine safety. Is this concern justified? Like all medical treatments, vaccines are not risk-free...but neither are the diseases they intend to prevent. Consider COVID-19. Many people (your author included!) had a sore arm after getting their vaccine. When I *had* COVID-19, I was in bed for a week. But I was lucky, since I didn't end up in the hospital or die, or develop long COVID, like 20 million Americans have—a number that includes young adults.

Figure 17.40 **A person displaying their COVID-19 vaccine status**

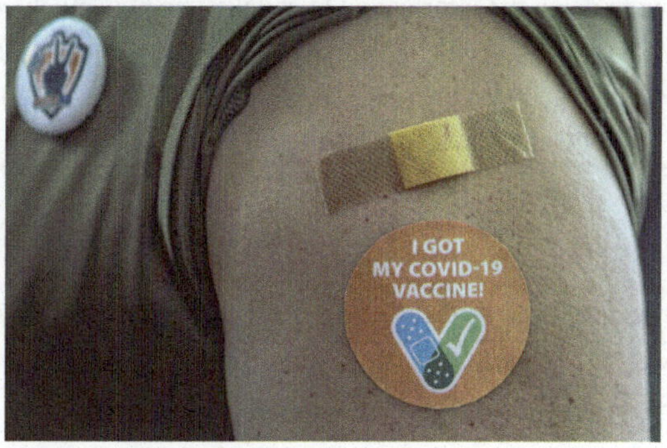

Credit: Ezra Acayan/Stringer/Getty Images

We remember the COVID pandemic and the relief experienced once vaccines became available—just like the greatest generation remembers the relief that the polio vaccine brought to them and how they no longer needed to worry about their children becoming paralyzed or dying during the summer waves of polio infection. So, the next time you're visiting an old cemetery and notice how the number of children's tombstones declines after the 1950s, remember the story of vaccination.

As with all vaccines, SARS-CoV-2 vaccines come with certain risks and often have side effects, such as a sore arm.

17.6 Summary

- Communicable diseases have high personal, economic, and societal costs. For example, while vaccines have greatly reduced childhood mortality, recent vaccine hesitancy has led to a resurgence of vaccine-preventable diseases like measles.
- The organs and tissues of the lymphatic system house the immune system.
- The first line of immune defense includes physical barriers to pathogens, like the skin and mucous membranes. The second line of defense, innate immunity, involves inflammation and the activation of white blood cells like phagocytes to defend the body. The third line of immunity, the adaptive immune response, involves lymphocytes, which recognize specific pathogens that have been encountered before.
- Vaccines elicit immunity by introducing a harmless form of an antigen into the body, triggering a primary immune response and prompting the immune system to produce antibodies and memory cells to the antigen. If a person is exposed to the same antigen again, the body can respond more quickly, in a stronger manner, and more efficiently to prevent illness.
- Vaccines protect entire communities by preventing harmful pathogens from spreading. When you get vaccinated, you are protecting not only yourself but also the people in your community.

? Did You Know?

Vaccine hesitancy can begin in childhood. Lying to children about the pain experienced during vaccination or exaggerating the pain from shots can result in children building negative associations with vaccination.

Understand It! Chunking

In Chapter 1, we learned about short- and long-term memory, and in the Chapter 13 Understand It!, we learned about memory networks. Here, we're going to learn about another way to utilize our memory to help us remember lots of content—a strategy called *chunking*.

The human brain can remember large chunks of related information better than many smaller pieces of information. This relates to the limits of our short-term memory—we can store only so much information at one time before we begin to forget it. So, when presented with a lot of information at once, which is common when studying anatomy and physiology, chunking as a study strategy can be used to make connections between different pieces of information.

You likely already use chunking as part of your daily routine, such as to help you remember tasks or items. For example, you know what you need to bring to class, even if you don't have a checklist for each item to put in your bag. Additionally, if you've ever used a mnemonic device to remember something—such as ROY G. BIV for the colors of the electromagnetic spectrum—that's another way that you've chunked content. Essentially, when we chunk information together, we take small pieces of information and organize them into larger chunks that can be more easily remembered.

So, when you're studying the information in this chapter, how can you connect the different systems and their components? You could do this by looking for the functional relationship between structures. For example, how is the brain part of both the nervous system and the endocrine system? Once you write down the information in chunks, see how much you can remember!

1. Have you used chunking before in another class? What did you do and what was helpful about it?
2. Pick two to three body systems. What chunks can you come up with? What about when considering all the body systems together?

Chapter Review

17.1 An Introduction to the Human Body

Organs are a combination of the four **tissue** types: **epithelial**, **connective**, **muscle**, and **nervous tissue**. Each tissue has a population of cells that are made up of molecules, which in turn are made up of atoms.

Homeostasis is maintained by the **organ systems**, in which two or more organs (for example, the lungs) work together to perform a common task (like breathing). These systems combine to form an **organism**, such as a human.

Anatomy is the study of the form of the body, while **physiology** studies the function of the body.

Fluids in the body are found both inside cells (in the cytosol) as **intracellular fluid**, and outside of cells as **extracellular fluid**. Much of the extracellular fluid is either surrounding cells in tissues as **interstitial fluid** or circulating throughout the body as plasma.

Organs are organized into 11 systems. The **integumentary system** includes the skin and protects the body from the external environment. The **nervous system** monitors both inside and outside the body and responds to changes. The **endocrine system** is made up of the glands that make hormones, which control other systems. The **musculoskeletal system** combines the **muscular system** (the muscles) and the **skeletal system** (the bones), which protect the organs and allow the body to move. The **circulatory system**, the heart and blood vessels, delivers nutrients to the tissues of the body and cycles away wastes. The **lymphatic system** provides fluid balance and houses the cells of the **immune system**. The **respiratory system** cycles oxygen in and carbon dioxide out. The **digestive system** breaks down and absorbs nutrients from food. The **urinary system** removes wastes through the urine. The **reproductive system** includes the structures necessary for sexual reproduction, including glands and the gonads, which make gametes.

17.2 Healing Chronic Skin Wounds and the Integumentary System

Many types of injuries and diseases, for example, **diabetes**, can lead to **chronic wounds**.

The skin has three layers: the **epidermis**, which is epithelial tissue made up of **keratinocytes** and **melanocytes**; the **dermis**, which is connective tissue; and the **hypodermis**, which is connective tissue, specifically, **adipose tissue**. Skin and accessory structures like hair make up the integumentary system.

During wound healing, **fibroblasts** make collagen to help scaffold new cell growth, and **platelets** help the wound to clot.

The **stem cells** found in adults are **multipotent**, or able to **differentiate** into a few closely related cell types, while those found in embryos are **pluripotent**, or able to form any cell type. Stem cells can be used as a treatment to help chronic wounds heal.

17.3 The Physiology of Anxiety and Depression: The Nervous System

The nervous system is made up of electrical signal-conducting **neurons** and supporting glia. Neurons have a distinct structure, with a central nucleus, or **cell body**; branching **dendrites** that receive information; and a long **axon** that transmits an electrical signal called an **action potential**. The axon is often wrapped in **myelin**.

Neurotransmitters are chemicals released from a neuron's **axon terminals** that communicate information through a **synapse**, a junction between two neurons. The neurotransmitters from one neuron are released into the **synaptic cleft** and then bind receptors on the next neuron.

The brain and spinal cord are called the **central nervous system**, while the nerves that branch from them are called the **peripheral nervous system.** The peripheral nervous system has two branches: the **somatic nervous system**, which relays information about the outside world to the central nervous system and controls the muscles, and the **autonomic nervous system**, which relays information about the body to the central nervous system and controls the organs.

The **autonomic nervous system** has two branches: the **parasympathetic nervous system**, which controls rest-and-digest functions, and the **sympathetic nervous system**, which controls fight-or-flight functions.

17.4 The Endocrine and Reproductive Systems, Endocrine Disruptors, and Fertility

Millions of people struggle to become pregnant, suffering from **infertility** and **impaired fecundity**.

Hormones are the signaling molecules of the endocrine system that travel through the blood to affect target cells all over the body. **Reproductive endocrinology** is a medical field that studies the hormones that regulate the reproductive system.

Gonads, including **ovaries** and **testes**, are the body's primary reproductive organs. They make gametes and the sex hormones **testosterone, estrogen**, and **progesterone**. The sex hormones are regulated by the hypothalamic-pituitary-gonadal axis, which involves the **hypothalamus** and **pituitary gland** of the brain communicating with the gonads through hormones.

During the **menstrual cycle**, a person's body has predictable hormonal changes that contribute to egg maturation and prepare the uterus for implantation. If the egg does not become fertilized and implant, the uterine lining is shed during menstruation.

Endocrine disruptors interfere with the endocrine system and the processes it controls. Some endocrine disruptors are also **forever chemicals**, which persist in the environment and can bioaccumulate.

17.5 Is Sitting the New Smoking? The Musculoskeletal System

The **musculoskeletal system** is the collective name used to describe how the skeletal and muscular systems work together under the control of the nervous system to control movement.

Skeletal muscles are made up of **muscle fibers** that contain **myofibrils**. Myofibrils contain contractile units called **sarcomeres**, in which the fibers **actin** and **myosin** overlap during muscle contraction. This process is called the **sliding-filament model**.

Fast-twitch muscle fibers can give us fast, explosive movements, while **slow-twitch muscle fibers** allow slow, extended contractions.

Tendons connect muscles to **bones**. The junction between two bones is called a **joint**. Joints can be stabilized by **ligaments**.

The bones in our limbs have a combination of strong **compact bone** and **spongy bone** that houses our blood stem cells. Bone is a living tissue, in which **osteoclasts** are continuously breaking down bone and **osteoblasts** are laying down new bone.

Muscles often function in **antagonistic muscle pairs**, where when one muscle contracts, the opposing muscle relaxes.

There are benefits to many types of exercise. **Aerobic exercise**, which is low intensity exercise done for long amounts of time, makes muscles more resistant to fatigue and has cardiovascular benefits, while **resistance exercise** makes muscle fibers larger and helps to make bones stronger.

17.6 Your Lymphatic and Immune Systems and Vaccines

As the **lymphatic system** circulates fluid or **lymph**, it screens the body for toxins and pathogens. Many **white blood cells** of the immune system are found in the lymphatic system, often in the **lymph nodes**.

The immune response has two branches. **Innate immunity** targets general threats, while **adaptive immunity** provides protection from specific pathogens one has previously encountered.

Immune system cells fight off threats in different ways. **Phagocytes**—including **neutrophils, macrophages**, and **dendritic cells**—engulf pathogens and perform **antigen processing** so they can display them to **lymphocytes** to mount the adaptive immune response. **Basophils** release histamines, leading to **inflammation**, which promotes tissue healing but can be damaging when chronic. B cells and T cells are lymphocytes that must recognize specific **antigens** (pieces of carbohydrates or proteins) with the **B-cell receptors** and **T-cell receptors** on their surface, respectively, to become activated. **T cells** can either be **helper T cells**, which help activate **B cells**, or **cytotoxic T cells**, which can directly destroy pathogens or infected cells. When **B cells** are activated, they make **antibodies** that bind the **antigens** of pathogens, targeting them for destruction. Cells of the immune system communicate with chemical messengers called **cytokines**.

During the first exposure to a particular pathogen, or the **primary immune response**, the body depends more on the innate immune response for protection. After the primary response, however, the body stores **memory B cells** so that **antibodies** can be made more quickly and at higher levels during a subsequent exposure, initiating a **secondary immune response** to that pathogen. **Vaccination** depends upon this secondary immune response; the vaccine primes the primary immune response to an antigen, so if a person encounters the real pathogen, the secondary immune response occurs.

Practice It!

Retrieval practice is the process of pulling information from your long-term memory and using it. Here are some questions to help with your retrieval practice.

Self-Quiz

1. Which of the following correctly represents the sequence of increasing complexity in the levels of organization in the human body? (LO 17-1)
   a. organ system → organ → tissue → cell
   b. cell → organ → tissue → organ system
   c. tissue → cell → organ → organ system
   d. cell → tissue → organ → organ system

2. The study of body structures is called ______, and how they work is called ______. (LO 17-2)
   a. physiology; anatomy
   b. anatomy; physiology
   c. biology; anatomy
   d. biology; physiology

3. Which organ system is primarily responsible for transporting oxygen, nutrients, and wastes throughout the body? (LO 17-3)
   a. the respiratory system
   b. the endocrine system
   c. the circulatory system
   d. the lymphatic system

4. Chronic skin wounds differ from typical skin wounds in which of the following ways? (LO 17-4)
   a. the amount of time they take to heal
   b. the layers of the skin affected
   c. the cell types involved
   d. the types of tissue involve

5. Which of the following best describes the primary function of the skin? (LO 17-5)
   a. to produce hormones for regulating body functions
   b. to break down nutrients for absorption
   c. to cover the body and separate the internal and external environments
   d. to filter blood and remove waste products

6. In wound healing, what is the role of the fibroblasts? (LO 17-6)
   a. to divide and fill the hole in the tissues
   b. to make collagen to form a scaffold
   c. to differentiate into keratinocytes
   d. to alert immune cells to fight infection

7. When multipotent stem cells are used to treat chronic wounds, what types of cells can they differentiate into? (LO 17-7)
   a. a few closely related cell types
   b. any adult cell type
   c. any epithelial cell type
   d. any connective tissue type

8. Neurons communicate using chemicals called ______. (LO 17-9)
   a. synapses
   b. myelin
   c. axons
   d. neurotransmitters

9. Which structures are part of the central nervous system? (LO 17-10)
   a. the brain and spinal cord
   b. ligaments and tendons
   c. sensory receptors and muscles
   d. reflex arcs and skin

10. How do medications that treat anxiety and depression impact neurotransmission? (LO 17-11)
    a. by causing permanent changes to neuron structure
    b. by altering DNA in nerve cells
    c. by enhancing or blocking the action of neurotransmitters
    d. by stopping all nerve signals entirely in the spinal cord

11. How does the endocrine system interact with the reproductive system? (LO 17-13)
    a. by producing antibodies that fight infection
    b. by controlling movement through muscle contraction
    c. by releasing hormones that regulate reproductive processes
    d. by digesting reproductive cells in the ovaries and testes

12. Which of the following is a common source of exposure to BPA? (LO 17-15)
    a. fresh fruits and vegetables
    b. natural spring water
    c. food packaging
    d. organic cotton clothing

13. Which structure in the musculoskeletal system is primarily responsible for generating movement through contractions? (LO 17-17)
    a. bones
    b. muscles
    c. ligaments
    d. joints

14. The lymphatic system helps regulate the body's fluids and contains many cells of which of the following systems? (LO 17-20)
    a. immune
    b. endocrine
    c. nervous
    d. musculoskeletal

15. How do lymphocytes recognize specific pathogens? (LO 17-21)
    a. by randomly attacking all foreign materials
    b. by detecting body temperature changes
    c. by consuming nutrients from invaders
    d. by expressing receptors that bind to specific antigens

16. How do vaccines provide immunity against disease? (LO 17-22)
    a. They kill pathogens directly in the bloodstream.
    b. They cause illness in a controlled way to build resistance.
    c. They introduce harmless antigens that trigger the formation of memory cells.
    d. They block pathogens from entering the body.

Critical Thinking

1. Image you are an epidemiologist, a scientist who studies patterns of diseases and their social impacts. Your local government asks for your recommendation on how to spend taxpayer money to improve health in your community. Which health concern discussed in this chapter—chronic skin wounds, anxiety and depression, endocrine disruptors and fertility, sedentary lifestyle, or communicable diseases—would you recommend as the top priority for government spending? Explain your answer using what you've learned in this chapter and what you see as important to your specific community. (LO 17-4, 17-8, 17-12, 17-16, 17-19)

2. Imagine a patient has a diabetes-related chronic wound that has not responded well to traditional treatments. Their doctor suggests stem cell-based therapy to help promote healing. The patient is concerned that these stem cells will need to come from embryos and is morally against these types of treatments. Based on your understanding of stem cell treatments, explain how stem cells might be used while avoiding these ethical issues. (LO 17-7)

3. Cocaine inhibits serotonin reuptake in neurons. What effect do you think that has on serotonin levels? How would that affect mood? How might cocaine use be an example of self-medicating? (LO 17-11)

4. Suppose you were studying the effects of PFAS, a class of forever chemicals, on male fertility. How would you predict PFAS would affect the endocrine system? What effects to the reproductive system might you test for to determine whether PFAS indeed affects male fertility? (LO 17-14)

5. You are designing a workout plan for yourself. You are primarily interested in making your muscles more resistant to fatigue. Explain what type of exercise (aerobic exercise or resistance exercise) you would focus on to achieve this goal and why, and provide a few sample exercises that you might engage in. (LO 17-18)

6. Some vaccines use only a piece of a pathogen, like a protein from the surface, to elicit immunity. How can this approach still produce immunity, and why might it be a better solution than using an entire virus? Why might it be worse? Propose a way to improve vaccines based on these concerns. (LO 17-21, 17-22, 17-23)

Credit: Inga Spence/Alamy Stock Photo

18

Fields, Forests, and Your Future

Chapter Outline:

The Importance of Plants

It may be impossible to overstate the importance of plants, and particularly flowering plants. As the most numerous and diverse plants in nearly all land habitats, flowering plants provide food and shelter for land animals. They also supply many products that meet human needs, such as the clothing we wear and the various household items we have around the home, such as furniture.

Flowering plants provide nearly all human food, either directly or as feed for livestock. Name a part of a plant, and humans eat it. In addition to the seeds of grains and legumes, we dine on leaves of lettuce and spinach, stems of asparagus, developing flowers of broccoli, modified roots of carrots and beets, and fruits of cherry trees and tomato plants. The male parts of crocus flowers provide the spice saffron, and cinnamon comes from the bark of a tropical tree.

We extract medicines and psychoactive drugs from flowering plants as well. Aspirin is derived from a compound in willow trees. Digitalis, from the foxglove plant, strengthens a weak heartbeat. Coffee, tea, and tobacco are widely used plant-derived stimulants. Worldwide, cultivation of opium poppies (the source of heroin) and coca (the source of cocaine) have wide-reaching health, economic, and political effects.

Of any organisms on the planet, plants have by far the most diverse and widespread impact on other organisms. What is it about plants that makes them so special?

Workers harvesting watermelons. Plants are a critically important parts of our biosphere. For example, we rely on plants for their ability to use sunlight to convert inorganic molecules into organic molecules that we can consume, such as those within these watermelons.

As we learned in Chapter 4, plants are autotrophs, meaning they use the energy from sunlight to convert inorganic molecules into organic ones through photosynthesis, serving as an energy source for both the plant and other animals. A plant's ability to undergo photosynthesis is essential for most life on Earth. In addition, a plant's ability to capture carbon dioxide from the atmosphere and convert it into its body structures is an important part of the carbon cycle (Chapter 15) and critical for mitigating climate change (Chapter 16). What is special about plant physiology that allows them to do this? What does fruit production have to do with plant reproduction?

In this chapter, we'll start with an exploration of plant form and function. We'll look at the tissues that make up a plant, its anatomy and how anatomy relates to function, and how a plant moves nutrients throughout its body. Next, we'll explore flowering plant or *angiosperm* reproduction and how and why some plants produce fruit. We'll also discuss pollinators and their importance not only to angiosperms but also to agriculture in general. We'll relate plant form, function, and reproduction to agriculture and forestry. In closing, we'll discuss our relationship with plants, with an eye toward how cultivating a healthy relationship with plants will lead to a brighter future.

Preview It!

List three ways you've interacted with plants already today. What do you want to learn about plants in this unit?

18.1 Plant Form and Function

Indigo dye is a traditional plant-based dye used to produce blue jeans. The blue pigment, found in indigo leaves, is hydrophobic ("water fearing"; Chapter 2). When dyeing, acid-base reactions change the dye molecules from hydrophobic to hydrophilic, allowing the dissolved molecules to penetrate the clothing. When the clothes are removed from the solution, exposure to air causes the dye molecules to become hydrophobic agains and consequently, become trapped within the clothing.

Did You Know?

Some plants, like watermelons, have specialized stems that allow them to climb.

Learning Objectives

18-1 Describe the basic anatomy of a plant.

18-2 Relate plant structures to their functions.

18-3 Compare the types and functions of plant tissues.

18-4 Explain how water and nutrients are transported throughout a plant.

18-5 Explain how plants grow, both above and below ground.

Imagine you are making a salad. What do you like to put into a salad? For the purposes of this exercise, let's assume it's a vegetarian salad. Would you put spinach in your salad? How about carrots? Maybe some celery? Or, my favorite part of a salad, bell peppers? Each of these different salad ingredients represents a different plant structure. We typically eat spinach *leaves* in salad, carrots are the *root* of the plant, celery is the *leaf stalk*, and bell peppers are *fruit*. So even in something as homey as a salad, we see an overview of the anatomy of a plant—which we'll explore more in this section.

The Anatomy of a Plant

Like other multicellular organisms, plants are organized into tissues, organs, and organ systems. A vascular plant (Chapter 13) consists of two organ systems: shoots and roots (Figure 18.1). **Shoots** include stems, leaves, and reproductive organs, such as flowers. In a typical plant, shoots are above the ground and roots are below it. In the following sections, we'll review each part of the plant and its function.

Stems. A plant's **stem** forms the framework of most flowering plants, providing support and keeping leaves best positioned for photosynthesis. Stems characteristically have **nodes**, which are regions where leaves originate. New shoots (and, in some cases, roots) can form at nodes. Note that the area between nodes is called the **internode**.

Running lengthwise within stems are a series of tubes called vascular bundles, which include *xylem* and *phloem*. **Vascular bundles** conduct water, ions, and nutrients between different parts of the plant, which we will return to later in this section. These bundles also provide mechanical support for upright stems in vascular plants.

Leaves. As we learned in Chapter 4, **leaves** are the main organs where photosynthesis occurs in most flowering plant species. They also function in gas exchange. For instance, during photosynthesis, carbon dioxide (CO_2) needed for photosynthesis enters a leaf through pores on the underside of the leaf called **stomata**, then diffuses through air spaces to **mesophyll** cells, which are the cells where photosynthesis occurs. Oxygen released by photosynthesis diffuses in the opposite direction through the stomata, from the inside to outside of the leaf—yet another service provided by plants, providing us with the oxygen we breathe.

Stomata are also the major site of evaporative water loss, where water taken up by roots travels up the plant and evaporates from leaves and other plant structures. This water loss is one reason you feel a cooling effect under trees. We'll discuss this process, called *transpiration*, later in this section.

Leaf structure varies widely among species, but a typical leaf is thin, with a high surface-to-volume ratio (Chapter 5). Vascular bundles in leaves are called leaf veins. **Leaf veins** carry water and nutrients to leaf cells. Leaves can vary between different species, where a leaf surface can be smooth, sticky, or slimy. Leaves can also have specialized structures such as hairs, scales, spikes, or hooks. For example, the spines on a cactus are modified leaves (Figure 18.2).

Figure 18.1 Anatomy of a tomato plant

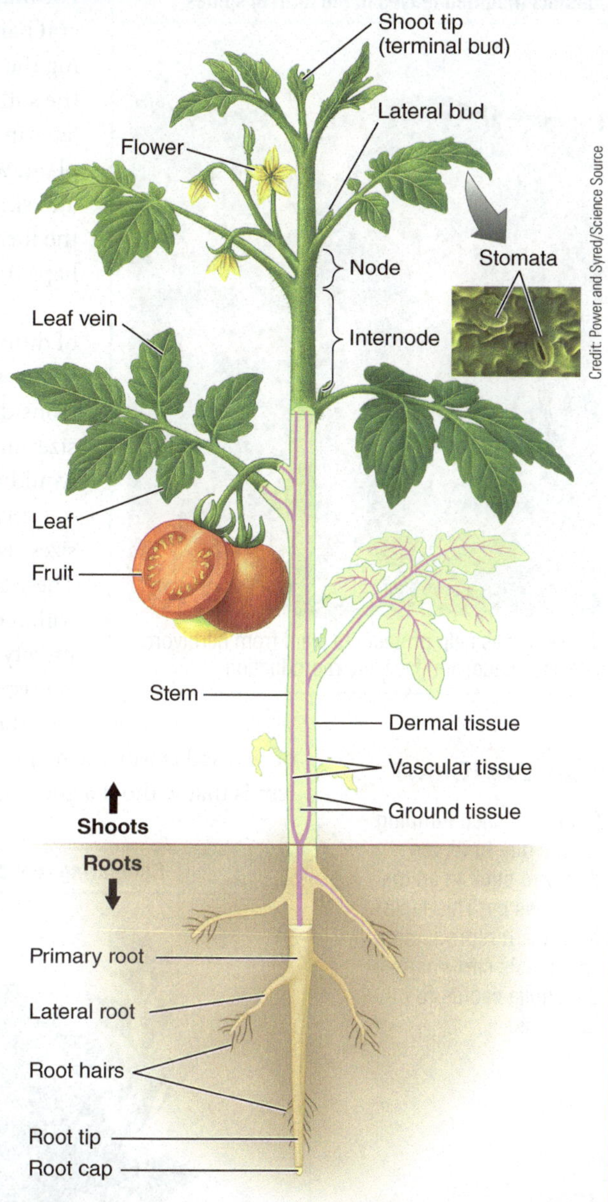

The aboveground (shoots) and belowground (roots) parts of a plant are shown. Vascular tissues (purple) conduct water and solutes throughout the plant, threading through the ground tissue that makes up most of the plant body. Dermal tissue covers the plant's surface.

Roots. The **roots** of a typical plant are as extensive as its shoots, and often more so. The main function of roots is the uptake of water and mineral ions from soil. They also anchor the plant and (in some species) store nutrients. In addition, root systems can help stabilize soil and in connection with microorganisms, such as mycorrhizae (Chapter 13) and/or nitrogen fixing bacteria (Chapter 15), promote the creation of rich soil.

When a seed begins to grow, the first structure to emerge is a **primary root** (see Figure 18.1). Over time, branches called **lateral roots** will form from the primary root. At the tip of each root is the **root cap**, which protects the root as it grows. We'll return to the root cap later in this section.

Tissue on the surface of roots is the plant's absorptive interface with soil. Epidermal cells of young roots have hairlike extensions on their surface called **root hairs**. Root hairs are tiny, but a huge number of them form. Collectively, they greatly increase the root's surface area, thus maximizing its ability to absorb soil water.

Depending on the type of plant, there are two types of root structures that can form. If there is a dense network of shallow roots, this is called a **fibrous root** (Figure 18.3A). In other kinds of plants, the primary root that emerges from a seed thickens and lengthens to become a **taproot**. A taproot can give rise to other lateral-branching roots, but even so, it remains the largest, dominant root (Figure 18.3B).

Figure 18.2 **The cholla cactus (genus: *Cylindropuntia*) contains modified leaves in the form of spines**

These spines help protect the plant from herbivores, provide shade, and facilitate reproduction.

Credit: Mattia Cioni/Shutterstock.com

The Dirt on Soil

For many plants, roots ground the plant into the soil. **Soil** consists of mineral particles derived from broken-down rocks, the water and air surrounding the particles, and the organic materials bound to the particles. From the soil, the plant can take up water and nutrients, which we'll discuss more later in this section. Before we talk about transport from the soil into the plant, we'll dig into the soil a bit more. It's easy to think of soil as just dirt, and since we spend a non-trivial amount of our time dealing with dirt—in the forms of dirty hands, clothes, dishes, floors, and so on—it's easy to perhaps think that dirt or soil in a negative light.

However, soil is more than simply dirt. Healthy soil contains a variety of different particle sizes that stay connected regardless of whether the soil is wet or dry. For example, most plants tend to not grow well in sand. Consider what dry sand feels like—the particles are generally similar in size, and dry sand doesn't stick together. Therefore, sand alone or sandy soil wouldn't be considered healthy.

In addition to impacting whether soil stays together, the different particle sizes also cause different amounts of space to occur in between the particles. These spaces between particles allow for water to infiltrate and to also be stored within the soil. In addition, healthy soil is nutrient rich and is home to a community (Chapter 15) of organisms, including insects, worms, bacteria, and fungi. A teaspoon of healthy soil has between 100 million and 1 billion bacteria alone!

Rich soil means fertile soil—which is ideal for growing crops. The legacy of rich soil is why the majority of America's tallgrass prairie was converted to farmland. The problem is that without a good management strategy, large-scale farming operations can contribute

? Did You Know?

Cholla is called jumping cactus due to its tendency to stick to animals that pass by. This facilitates plant reproduction as animals carry parts of the cholla cactus to other locations.

? Did You Know?

Some plants store nutrients in swollen, modified lateral roots called root tubers. A sweet potato is an example of a root tuber.

Figure 18.3 **Comparing root systems**

Credit: Casther/Alamy Stock Photo

Credit: Richard Griffin/Alamy Stock Photo

A. In a fibrous root system, similarly sized roots branch from the stem, such as from the bottom of a celery plant.

B. A tap root system consists of a large main root together with any lateral roots that branch from it, as is the case with carrots.

to soil erosion. We first learned about soil erosion in Chapter 16, where we discussed how the Dust Bowl in the United States in the 1930s occurred due to poor soil management, leading to severe soil erosion and significant ecological, economic, and personal impact.

To prevent erosion, some farmers are choosing to integrate strips of native prairie into their fields in areas called prairie strips (Figure 18.4). These prairie strips are full of native plants, which have much more extensive root systems than agricultural plants. These extensive root systems reduce soil erosion while also sequestering carbon from the atmosphere and providing habitat for pollinators and birds.

Figure 18.4 **A prairie strip among crops in Iowa**

Credit: Flickr

The Soil Your Undies Challenge is a way of testing the health of your soil. To participate, you plant a pair of cotton underwear 3 inches deep and then dig the pair back up after 60 days. The more degraded the underwear, the healthier your soil.

Study Tip

The next time you eat a plant, try to identify the three main parts: stem, leaf, and root. What parts did you eat?

Plant Tissues

Now that we've discussed some basic plant anatomy with familiar structures, we're going to dig a little deeper into the tissues that make up the different parts of a plant. Both roots and shoots are composed of dermal, vascular, and ground tissues (Figure 18.5).

The **dermal tissue** system consists of tissues that cover and protect the plant's exposed surfaces. The first dermal tissue to form on a plant is **epidermis**, which in most species consists of a single layer of epidermal cells on the plant's outer surface. Epidermis of leaves and young stems includes specialized cells, such as those that form stomata. Recall from Chapter 4 that stomata (singular: *stoma*) are pores or openings typically found on the undersides of leaves that help with gas exchange and water regulation. A pair of specialized epidermal cells called **guard cells** border each stoma. When these two guard cells swell with water, they bend slightly so that a gap (the stoma) forms between them. When the guard cells lose water, they collapse against one another, so that the stoma between them closes (Figure 18.6A). We'll return to how plants manage water loss through stomata via *transpiration* later in this section.

Figure 18.5 **Plants contain dermal, ground, and vascular tissue**

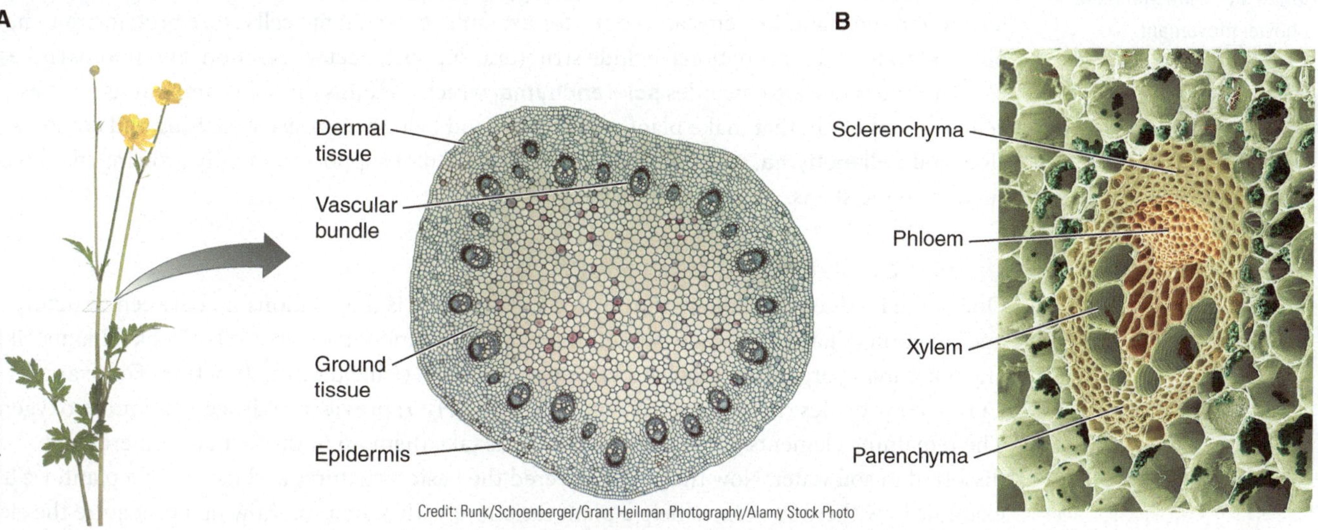

Credit: Dabjola/Shutterstock.com

Credit: Runk/Schoenberger/Grant Heilman Photography/Alamy Stock Photo

A. A buttercup and a cross-section of its stem. Note the presence of dermal tissue on the outside, and vascular tissue surrounding by ground tissue on the inside.

B. A magnified view of a vascular bundle. Note the different types of vascular tissue present (xylem and pholem) and the surrounding ground tissue (parenchyma and sclerenchyma).

Plants can sense CO_2 levels, opening stomata in low-CO_2 environments and closing them in high-CO_2 environments. Given increasing levels of atmospheric CO_2 (Chapter 16), understanding how stomata work can yield insights into genetic engineering (Chapter 9) approaches to sequester more CO_2 from the atmosphere.

Figure 18.6 Stomata are sources of water loss

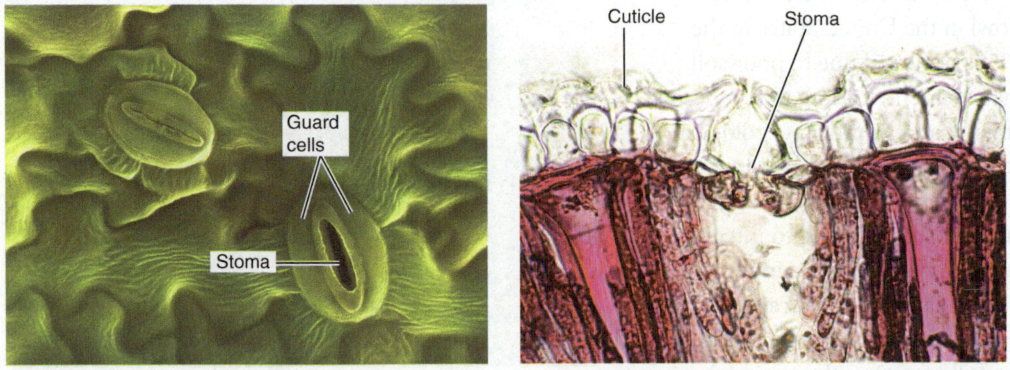

A. Whether a stoma is closed (top-left) or open (bottom-right) depends on how much water is pumping through the guard cells bordering it.

B. In this plant leaf, the stoma is recessed beneath a ledge of cuticle. The ledge minimizes evaporative water loss because it creates a small cup that traps moist air above the stoma.

Credits: Power and Syred/Science Source; Dr. Keith Wheeler/Science Source

There are also specialized cells with the dermal tissue that secrete a waterproof, protective **cuticle** on their outward-facing cell walls, which helps the plant to conserve water and repel pathogens (Figure 18.6B).

The **vascular tissue** system includes the tissues that carry water and nutrients from one part of the plant body to another through a series of tubes, or vascular bundles, that consist of elongated conducting tubes bundled with sclerenchyma fibers and parenchyma (see Figure 18.5). **Xylem** is the vascular tissue responsible for conducting water, and the substances dissolved in it, from the roots to the shoots. In addition to conducting water, xylem tubes contain **lignin**, which stiffens the walls of xylem and provides structural support to plant parts. **Phloem** is the vascular tissue that conducts sugars and other organic solutes to all parts of the plant.

Since humans cannot digest cellulose, people take cellulose supplements such as Metamucil to help with regularity. Since carbohydrates are water soluble (Chapter 3), the cellulose absorbs water and forms a gel-like substance that can help soften stool and stimulate a bowel movement.

The **ground tissue** system includes all tissues that are not part of the dermal or vascular tissue systems, while also making up the bulk of the plant. Most ground tissue is made up of parenchyma cells (see Figure 18.5). **Parenchyma** cells vary in shape, but all have a thin, flexible cell wall and remain alive at maturity. Parenchyma has specialized roles that vary with its location in the plant. In stems and roots, for example, parenchyma cells store proteins, starch, oils, and water. Other functions include structural support, nectar secretion, and photosynthesis.

Ground tissue also includes **sclerenchyma**, which contains durable compounds such as cellulose and lignin that make plant parts sturdy and help them resist stretching and compression, and **collenchyma**, which provides pliable structural support to rapidly growing plant parts such as young stems.

Uptake of Water and Nutrients

One of the Big Ideas in Biology introduced in Chapter 1 is the relationship between structure and function (Chapter 1)—how something works is a function of what it looks like. Plants, like any other living organism, require a variety of elements (Chapter 2) to function. For example, CO_2 in air provides carbon and oxygen, and water (H_2O) provides hydrogen and more oxygen. The remaining elements occur in soil, and plants take them up in the form of mineral ions dissolved in soil water. Now that we've covered the basic structures and tissues of a plant, we'll examine how these structures help a plant function, with a focus on how plants acquire the elements they need to survive.

Turgor pressure. Water moves from soil, through a root's epidermis and ground tissue, and into the vascular cylinder. Osmosis (Chapter 5) drives this movement, because fluid in the plant typically contains more solutes than soil water, drawing water into the plant through the roots. After water enters the vascular cylinder, xylem distributes it to the rest of the plant. The stiff cell walls of plant cells keep them from expanding very much, so an inflow of water causes pressure to build up inside of the cells. Pressure that a fluid exerts against a structure that contains it is called **turgor pressure**. When enough pressure builds up inside a plant cell, water stops diffusing into it. The amount of turgor pressure that is enough to stop osmosis is called **osmotic pressure**.

Osmotic pressure keeps plant cells plump, just as high air pressure inside a tire keeps it inflated. A young land plant can resist gravity to stay erect because the turgor pressure inside of its cells is high (**Figure 18.7A**). If the plant does not get enough water to replace what it uses, the cytoplasm of its cells shrinks. As turgor pressure inside the cells decreases, the plant wilts (**Figure 18.7B**).

Figure 18.7 **Plants stand upright or wilt due to turgor pressure**

A. This plant is standing upright due to turgor pressure.

B. This plant is wilted because there is not enough water available to maintain turgor pressure.

Credit: izzy71/Shutterstock.com

Transpiration. Recall that plant dermal tissue contains stomata. Open stomata allow water—a lot of it—to exit the plant. Around 95% of the water taken up from soil is lost through open stomata. Even under conditions of high humidity, the interior of a leaf or stem contains more water vapor than air. Thus, water vapor diffuses out of a stoma whenever it is open. The evaporation of water from aboveground plant parts is called **transpiration**.

Water is important for land plants, so why do they let almost all of it evaporate away? Recall from Chapter 4 that the stomata must remain open for gases to exchange and photosynthesis to occur. Without photosynthesis, the plant would die. Therefore, it's important for the plant to balance water loss with gas exchange. We'll return to regulation of stomata opening and closing in Section 18.3.

Let's go back to transpiration. Transpiration requires that water be pulled from the roots up the stems and to the leaves, where it exits the plant. How is this possible?

Recall from Chapter 2 that water can stick to itself (cohesion) and other surfaces (adhesion). These properties of water explain how water can enter in through a root and travel through the rest of the plant through tubes of xylem. Water is pulled up through a plant by evaporation though the stomata, which creates negative pressure (**Figure 18.8**). Water is pulled upward in continuous columns because xylem tubes are narrow. The water moving through a xylem tube (akin to a straw) resists breaking into droplets because the water molecules are attracted to each other through cohesion and to the hydrophilic materials that make up the xylem tube through adhesion.

? Did You Know?

During the day, each open stoma loses about 400 molecules of water for every molecule of carbon dioxide it takes in.

Have you heard of corn sweat? Corn sweat occurs because of transpiration from the plant and evaporation from the soil. While not the only factor impacting the humidity in a region, the U.S. Department of Agriculture estimates than corn can add thousands of gallons of water to the atmosphere in just a day or two.

Translocation. When it comes to breakfast condiments, the one I can't live without is maple syrup. Maple syrup is made from the sap of maple trees, and one can obtain it only in early spring. Why? It comes down to **translocation**, or the movement of sugars and other organic molecules through phloem, which is what we'll discuss next. We'll return to making maple syrup at the end of the section.

Recall from Chapter 4 that a plant produces sugar molecules during photosynthesis. Phloem is the vascular tissue that conveys the products of photosynthesis—sugars—throughout the plant. Some of these molecules are used by or are stored in the cells that make them; the rest of the molecules are conducted to other parts of the plant via phloem pipelines called sieve tubes. A **sieve tube** is a stack of living cells, called *sieve elements*, that are connected end to end at perforated sieve plates. An associated **companion cell** provides each sieve element with metabolic support, and also transfers sugars into it. The sugars move through the sieve tubes to all parts of the plant.

Study Tip

Sieves are used to keep larger items in while letting smaller items out, like when you use a flour sifter to sift flour.

Figure 18.8 **How water moves through a plant**

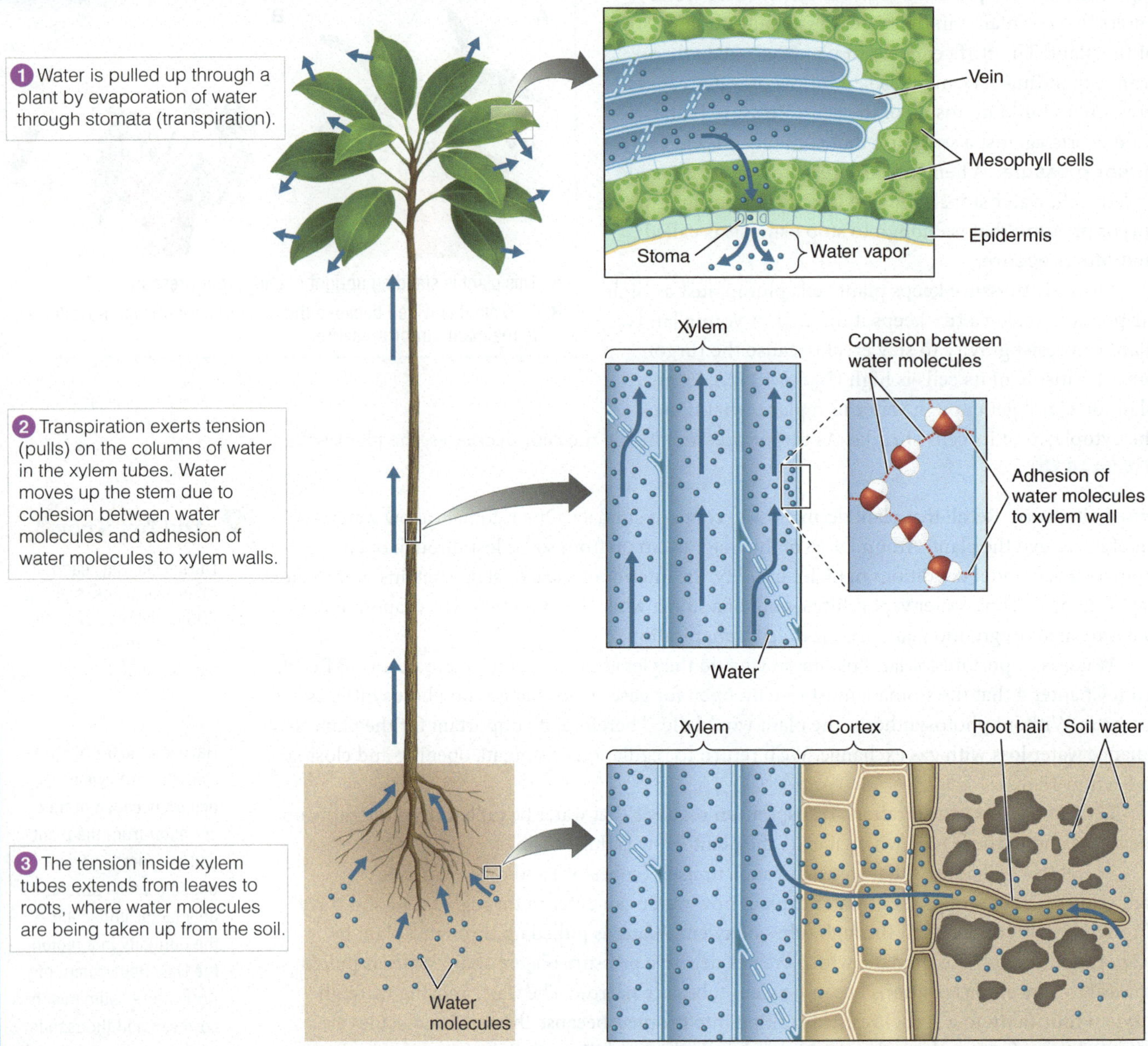

Inside sieve tubes, fluid rich in sugars—mainly sucrose—flows from a **source**, or region of the plant where sugars are being produced or released from storage, to a **sink**, a region where sugars are being broken down or stored for later use. Photosynthetic tissues are typical source regions; tissues of developing shoots and fruits are often sink regions. As sugar moves into phloem, water follows via osmosis. This creates high pressure inside the phloem at the source. Once the sugar reaches a sink, it moves into the fruit, followed by water, which creates a low-pressure spot. Together, a pressure gradient forms between the source and the sink, which ultimately causes the sugars to move throughout the plant (Figure 18.9).

This pressure gradient explains why maple sap (the raw material for making maple syrup) will flow outside the tree if tapped (Figure 18.10). In early spring, as temperatures vacillate between above and below freezing, ice inside the xylem begins to melt while the sucrose becomes concentrated. This creates positive pressure inside the tree. If the tree has a

Study Tip

To remember the functions of xylem and phloem, you can associate "food" and "phloem" with the same starting sound, since phloem transports food. Because a plant also needs water, by default, xylem transports water.

Figure 18.9 How sugars move through a plant

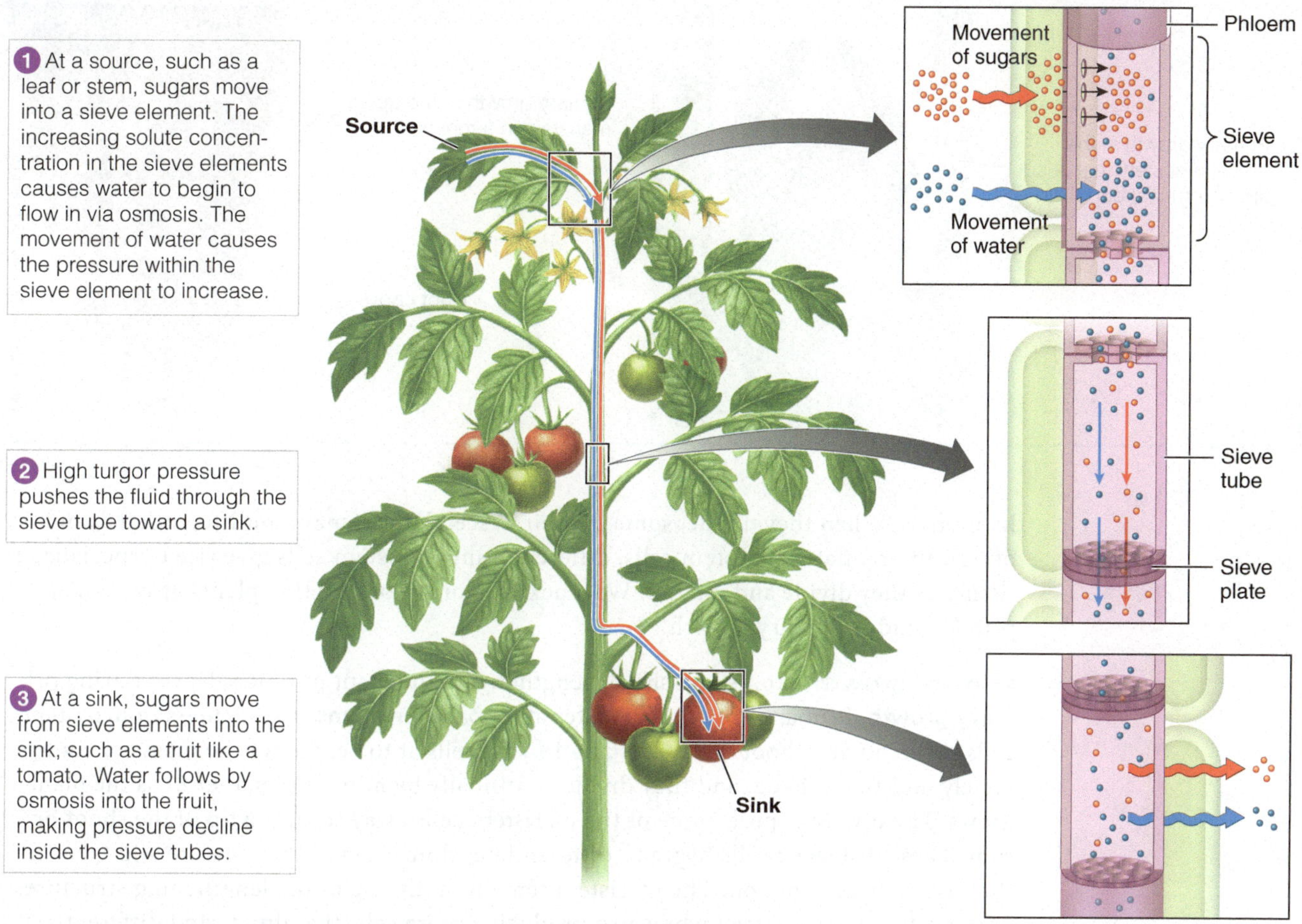

Note that the blue arrow represents movement of water, and the orange arrow the movement of sugars. For clarity's sake, only one source and one sink are labeled.

wound—or is tapped by a person seeking the sweet sap—the pressure inside the tree pushes the sap outwards. In late spring, once the temperatures are consistently above freezing and new buds and leaves form on trees, the maple sap stops flowing. This is because once there are leaves, water is pulled up and out of the tree via transpiration. This means that the positive pressure inside the tree is released, and so the sap will no longer flow outwards.

Figure 18.10 Sap dripping out of a maple tree and into a collection bucket

A B

A. Maple tree sap is the raw material used for producing **B.** maple syrup.

Patterns of Growth

Consider the weeds that inevitably sprout in a yard. Initially, young plants sprout and get taller. Now, let's say that one of the unwanted plants was actually a tree sapling. As the sapling grows, it not only gets taller, but thicker as well, until eventually you can't easily pull the sapling out. Whether growing taller or thicker, new plant tissues arise from the activity of **meristems**, which are regions of undifferentiated cells that can divide continually during a growing season. Meristem cells are analogous to *stem cells* (cells that can turn into other cells; Chapter 17)

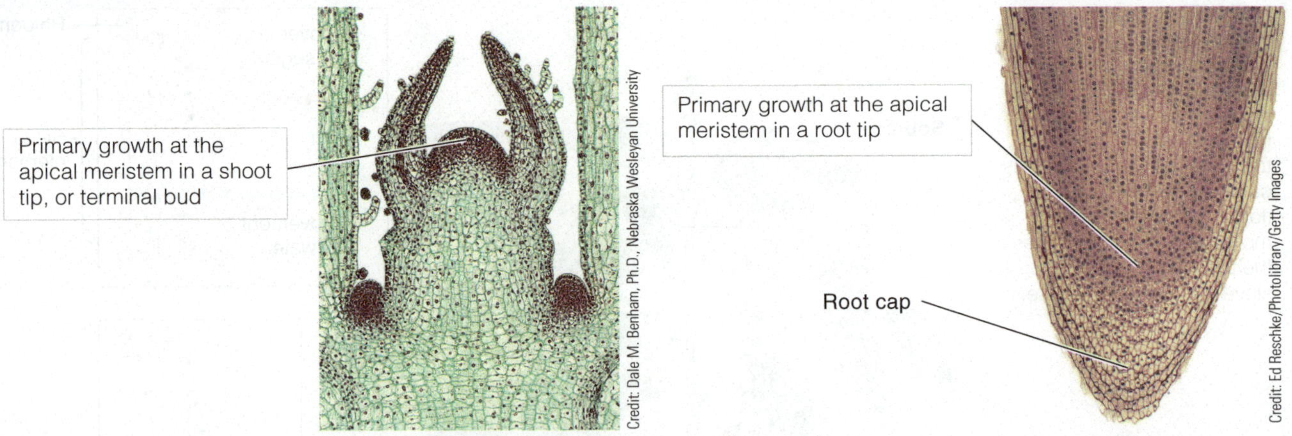

Figure 18.11 Primary growth in the shoots and roots

in animals: When they divide, some of their descendants remain undifferentiated, and others differentiate. Like stem cells, differentiating meristem cells give rise to specialized tissues as they divide and mature. We'll next explore two ways that plants grow, called primary and secondary growth.

Primary growth. Roots and shoots lengthen and the plant produces leaves during **primary growth**. Primary growth originates at **apical meristems**, each a mass of meristem cells at the tip of a shoot or root (Figure 18.11). Cells at the center of the mass are completely undifferentiated, and they divide continually by mitosis (Chapter 8) as the plant grows. The divisions push some of the meristem cells away from the tip of the shoot or root. These displaced cells begin to differentiate, elongating as they mature and take on their specialized function. The meristem remains at the tip of the lengthening structure, continually renewing itself while also producing more cells that divide and differentiate behind it.

Primary growth in a shoot originates at its tip—its **terminal bud**. As a typical shoot lengthens, masses of tissue near the terminal bud bulge out and then develop into leaves or new shoots. These structures form in orderly tiers along the lengthening stem. As they do, lateral buds form in places where the leaves attach to the stem. Lateral buds also contain an apical meristem, but this tissue may stay dormant until signals trigger the cells to divide. Depending on these signals, an apical meristem in a lateral bud can give rise to a leaf, flower, or new stem that branches from the main stem.

Root **primary growth** begins with dividing apical meristem cells in root tips (see Figure 18.11). The dome-shaped root cap (see Figure 18.1) protects the tip of the root by secreting a slimy substance that provides lubrication as the new root grows and pushes through the soil. Cells of the root cap are shed continually as the root lengthens. Lateral roots that form on an existing root originate with division of cells just inside the root's vascular cylinder.

> **? Did You Know?**
>
> Removing the bark in a circular fashion from the base of a tree, or "girdling" it, will kill the tree. When the bark is removed, the phloem and vascular cambium are removed as well. The result is that sugars and other nutrients cannot be transported from the roots to the rest of the tree and vice versa, ultimately killing the tree.

Secondary growth. Many plant species undergo **secondary growth**, during which their shoots and roots thicken and become woody, such as in trees. Secondary growth involves two lateral meristems, vascular cambium and cork cambium. **Vascular cambium** produces secondary phloem on its outer surface, and secondary xylem on its inner surface (Figure 18.12, top). Secondary xylem, commonly called **wood**, accumulates during secondary growth. The accumulation displaces the vascular cambium toward the outer surface of the stem or root (Figure 18.12, bottom). Continued divisions of cells in the vascular cambium maintain its cylindrical shape.

With continued secondary growth, a second lateral meristem forms just under the dermal tissue of the stem or root. This lateral meristem—**cork cambium**—produces a tissue called cork

Figure 18.12 Comparison of primary and secondary growth in a stem

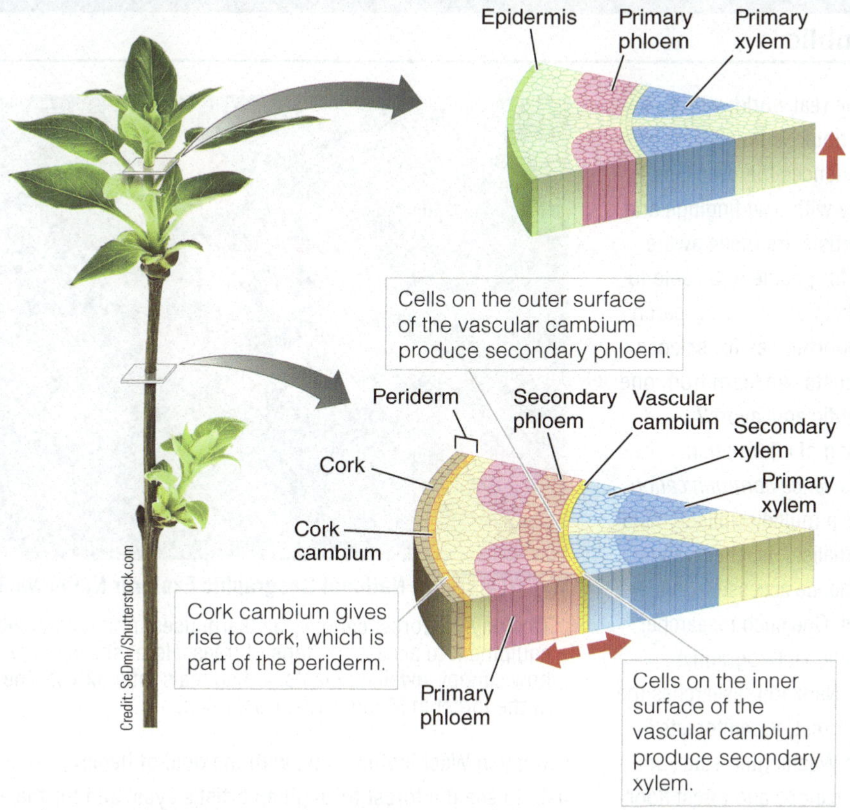

During primary growth (top), the stem or roots lengthen at apical meristems, and the tree grows taller (lengthens; red arrow). During secondary growth in trees (bottom), the stem begins to thicken (red arrows) at lateral meristems and wood is produced.

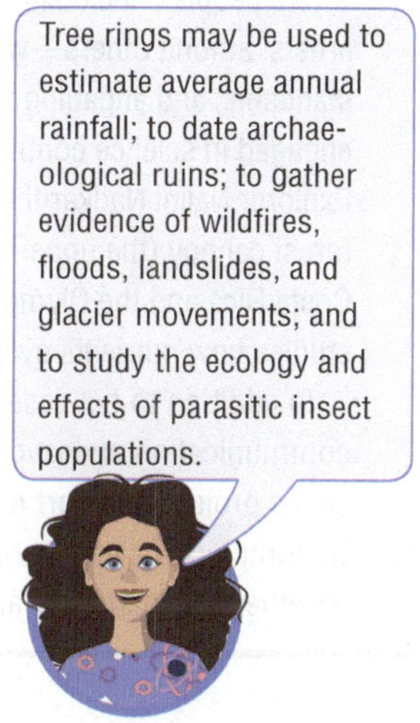

on its outer surface. **Cork** consists of densely packed dead cells with waxy walls, and it protects and insulates the surface of the stem or root. Cork is part of periderm, the dermal tissue that replaces epidermis on the surfaces of older stems and roots. "Bark" is an informal term for periderm and all other tissues that lie outside the cylinder of vascular cambium.

A tree adjusts its growth in response to seasonal changes in its environment, and these adjustments are reflected in the structure of its wood. For example, many trees native to temperate zones become dormant during winter. No growth occurs during this period. Dormancy ends in spring, when environmental conditions favor rapid growth. Wood produced during rapid growth—earlywood—consists of large-diameter, thin-walled xylem. Growth slows in summer, and wood produced during this time—latewood—consists of small-diameter, thick-walled xylem. Early wood is more porous, less dense, and lighter in color than late wood.

These differences are often visible in a cross-section of an older tree, as bands called growth rings or tree rings. Each ring is composed of a layer of early- and latewood. In most temperate-zone trees, one ring forms each year (Figure 18.13). In tropical regions where seasonal variation in weather is minimal, many trees grow at about the same rate all year and have inconspicuous rings or none at all.

Figure 18.13 Cross-section of a tree trunk showing its rings

Note that the earlywood, which forms in the spring, is lighter in color than latewood.

Process of Science

Connecting Science and the Public

A key theme of how science works in the real world, which we've visited several times through the book, is how science is intertwined with society—society influences the process of science, which in turn influences society with new findings and discoveries. It's important that nonscientists are made aware of important scientific discoveries, and for people to be able to interact with science and scientists, to help to inform research priorities. What are ways of creating opportunities for science engagement where the public and scientists can learn from one another and collaboratively shape scientific endeavors?

The practice of engagement and sharing of information between scientists and society is called *science communication.* Science communication, or SciComm, is a multidisciplinary field where people—including scientists, journalists, teachers, and artists, among others—work to make science accessible, understandable, and engaging to nonscientists. One such researcher engaged in science communication is National Geographic Explorer Nalini Nadkarni (**Figure 18.14**). Nadkarni researches the forest canopy (the uppermost layer in a forest) in rainforests in Costa Rica and the Olympic Peninsula in Washington state, and studies how nutrients cycle between the canopy and forest floor.

Figure 18.14 National Geographic Explorer Nalini Nadkarni

To study the forest canopy, Nadkarni uses mountain-climbing equipment to access the tops of trees. Here, she is studying how canopy-dwelling plants and animals interact with the rest of the forest in Monteverde, Costa Rica.

In addition to her research, Nadkarni has led many science communication endeavors. For example, the Canopy Confluences project was part of Nadkarni's work with the National Geographic Society. During this project, Nadkarni brought together scientists and artists to spend a week in the forest canopy in Washington state, with the goal of helping the scientists to see the forest through an artist's eyes, and for the artists to learn about the biology of the forests. Each party would gain a different understanding of the other's work through their experiences in the canopy and their conversations with each other. Sitting on platforms that were suspended on branches 50 feet above the forest floor (**Figure 18.15A**), people created art,

18.1 Summary

- Vascular plants have two organ systems: (1) shoots, including stems, leaves, and reproductive organs like flowers, and (2) roots.
- Stems provide scaffolding for the other structures and have vascular bundles that contain xylem and phloem. Leaves perform photosynthesis and gas exchange. Roots are branching structures that anchor the plant and take up water and minerals from the soil.
- Both roots and shoots are composed of dermal, vascular, and ground tissue. Dermal tissue covers and protects the plant, vascular tissue carries water and nutrients around the plant, and ground tissue makes up the bulk of the plant.
- The vascular tissues, xylem and phloem, transport water and nutrients throughout the plant. Water moves into the xylem from the soil by osmosis, where it exerts turgor pressure inside of the plant cells, keeping the plant erect. Water can leave the plant through stomata, evaporating via transpiration. Sugars and other organic molecules are transported in the phloem via translocation.
- Plants grow from seeds. When the seed germinates, a primary root emerges and then becomes either a fibrous root or a taproot. The plant grows by lengthening and thickening as meristems grow and differentiate. Primary growth occurs at apical meristems at the tips of shoots (terminal buds) and roots (root caps). Secondary growth involves lateral meristems forming either wood or cork.

A B

Credit: National Geographic Image Collection

Credit: Benj Drummond

Figure 18.15 **Examples of science communication and outreach work**

A. Inspired by the canopy, a person plays music on their guitar.

B. Inmates participating in the INSPIRE program raise rare plants for prairie restoration.

dance, or music about what they saw or heard while in the canopy, and then shared their experiences in art exhibits or theater performances.

Nadkarni also finds ways to bring science to people who can't access nature easily on their own, such as individuals who are incarcerated. Many people who are incarcerated have limited access to nature and the outdoors, let alone the ability to engage in community science (Chapter 16). Nadkarni found that providing environmental lectures to inmates resulted in their seeking more information about science, and helped the participants see themselves as science learners. These lectures developed to include inmates in Utah participating in sustainability projects through INSPIRE (Initiative to bring Science Programs to the Incarcerated). For example, inmates could participate in activities such as gardening and conservation programs, including rearing rare plants and butterflies for prairie restoration (Figure 18.15B). In addition to the obvious benefit of exposure to nature, inmates learn scientific skills, as well as how to work as a team and, most importantly, to gain the satisfaction of positively contributing to society and the Earth.

Whatever your major is, or whether you do or do not identify as a science person, everyone can be an ambassador for biology and the natural world. Science communication and engagement activities don't necessarily need to be big programs—they can be as simple as participating in a community science project (Chapter 16) or building bridges between different communities.

1. When you consider your career aspirations and/or hobbies, what are ways that you can be a science communicator and ambassador for nature?
2. Why do you think it is important for science communication to go both ways, with scientists sharing their work with the public, and people sharing their thoughts with scientists?

18.2 The Lifecycle of a Flowering Plant

Learning Objectives

18-6 Describe the components of a flower, their functions, and the factors that influence flowering.

18-7 Explain the relationship between floral traits and pollinators.

18-8 Compare sexual and asexual reproduction in a flowering plant.

18-9 Explain the evolutionary purpose of seeds and fruits.

18-10 Describe agricultural applications of plants.

One of my favorite things to do with my kids in the spring is to walk around the neighborhood and spot flowers. It's always exciting to see the first pops of color appear after a long winter. As spring goes on, different flowers appear and disappear, and by the end of summer, we begin to see abundance in our neighbors' vegetable gardens.

Humans have an interesting relationship with flowering plants, or **angiosperms**. We enjoy looking at and smelling flowers. Flowers are often used in celebrations, such as weddings, and can be given as a gift. But as we'll learn in this section, flowers are more

Figure 18.16 **Typical flower anatomy**

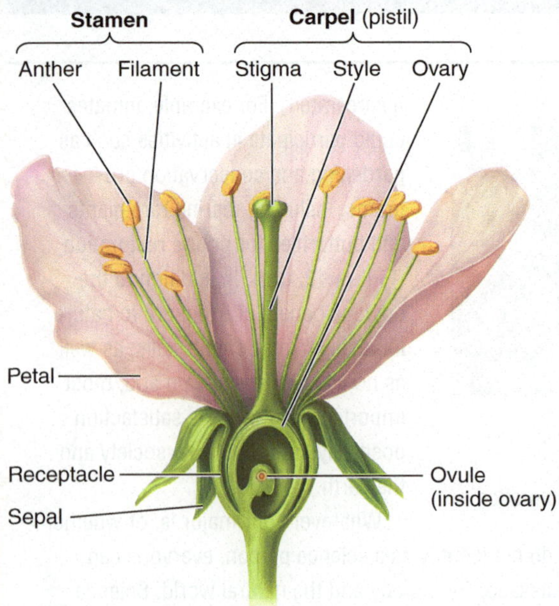

Male gametophytes are produced by stamens; female gametophytes are produced by carpels. A stamen consists of a pollen-bearing anther on a slender filament. A carpel consists of an ovary, a stigma, and a style. Note in this illustration that the ovary is cut away to reveal the ovule within.

than just attractive adornments that smell good: Flowers are an important part of a plant's lifecycle and are vital to the production of crops.

Did you realize that when you smell a flower, you're appreciating the plant's reproductive structures? Or that the reason we and so many other animals are attracted to flowers is due to evolution, where plants evolved flowers to help aid in successful reproduction? In this section, we'll learn about reproduction in flowering plants, with a focus on the agricultural applications of flowering plants.

Flower Structure

The reproductive structures of angiosperms are collectively referred to as a **flower** (Figure 18.16). The parts of a flower are modified leaves that develop from a thickened region of stem. Floral form and composition vary among species, but a typical flower has four concentric rings (whorls) of modified leaves. The outermost ring consists of leaflike sepals. **Sepals**, which are often photosynthetic and inconspicuous, enclose and protect the flower's internal tissues before it opens. Just inside the sepals is a ring of petals. **Petals** are often the largest and most brightly colored parts of a typical flower.

Inside the petals is a ring of **stamens**, the reproductive organs that produce the plant's male **gametophytes**, or gamete-making structures. A stamen is a thin filament with an anther at the tip. The **anther** is a structure with four pouches called pollen sacs. **Pollen grains**, which are immature male gametophytes, form inside pollen sacs.

Male trees are preferentially planted in cities to minimize the cleanup from petals and fruits. However, all the pollen produced by the male trees makes seasonal allergies much worse!

The innermost ring of modified leaves folds and fuses into one or more **carpels**, or reproductive organs that produce female gametophytes. A flower may have one carpel or several, depending on the species. The upper region of a carpel is a sticky or hairy **stigma** that is specialized to receive pollen grains. Typically, the stigma sits on top of a slender stalk called a **style**. The lower, swollen region of a carpel is the **ovary**, which contains one or more ovules. In seed plants, an **ovule** is the structure in an ovary that gives rise to the female gametophyte. Seeds are mature ovules. The stigma, style, and ovary are collectively referred to as the carpel. One or more carpels fused together are called a **pistil**.

Flowers vary widely in form. Some flowers are single blossoms (Figure 18.17A); others occur in clusters called *inflorescences* (Figure 18.17B). A composite flower is an inflorescence of many flowers that cluster in a single head (Figure 18.17C). A "regular" flower is radially symmetric, which means if it were cut like a pie, the pieces would be roughly identical. "Irregular" flowers are not radially symmetric, like the flower in 18.17A.

A flower with all four sets of modified leaves (sepals, petals, stamens, and carpels) is a complete flower. Flowers of many species are incomplete, which means they lack one or more of these structures (Figure 18.17D).

Flowering

The process of producing flowers is called **flowering**. Flowering in many plant species is a response mediated by the length of sunlight exposure over the course of a day. Such species are termed long-day or short-day plants depending on the season in which they flower.

Figure 18.17 **Examples of structural variation in flowers**

A. The solitary flower of a lady slipper's orchid (*Cypripedioideae*).
B. A hyacinth flower (*Hyacinthus orientalis*) is an elongated inflorescence.
C. A daisy (*Bellis perennis*) is a composite flower with many individual florets.
D. The flower from an elm tree (*Ulmus*) is incomplete because it does not have petals.

! Point of Confusion: The names long-day and short-day plant are somewhat misleading, as the main trigger for flowering is the length of *night*, not the length of *day*.

Irises and other long-day plants flower only when the hours of darkness fall below a critical value, typically in summer (**Figure 18.18A**). Christmas cactuses and other short-day plants flower only when the hours of darkness are greater than a critical value, which is why you can often find them flowering in stores around the holidays, when nights are longer (**Figure 18.18B**).

Researchers suspect that plant cells perceive temperature via their cell membrane, because the lipid composition and consequently membrane permeability vary with temperature.

Figure 18.18 **The impact of night length on the flowering of long-day and short-day plants**

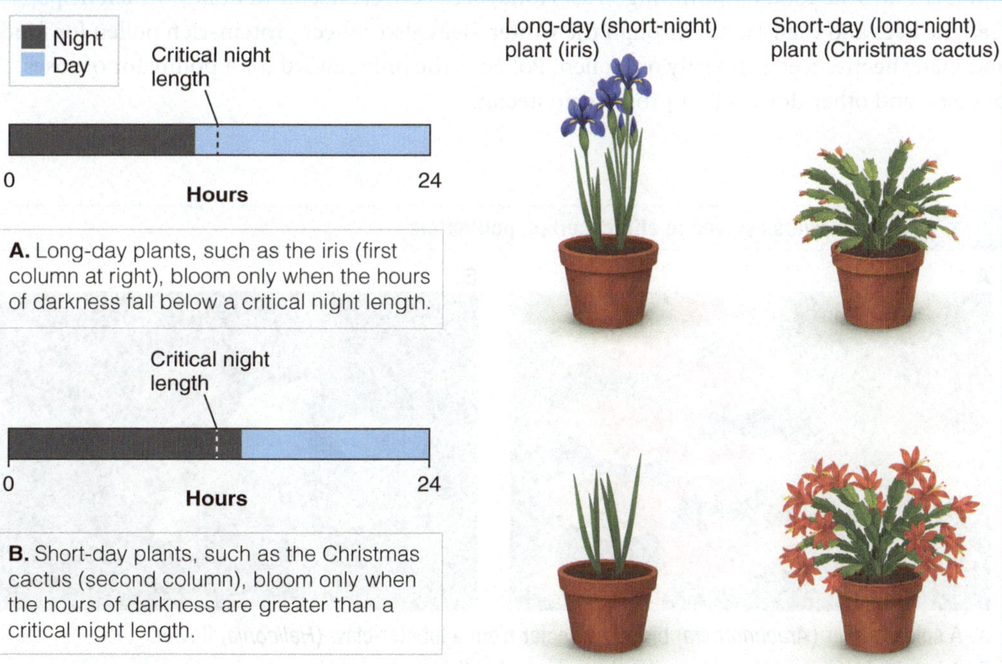

A. Long-day plants, such as the iris (first column at right), bloom only when the hours of darkness fall below a critical night length.

B. Short-day plants, such as the Christmas cactus (second column), bloom only when the hours of darkness are greater than a critical night length.

? Did You Know?

Cabbage, spinach, and other edible long-day plants that are grown for their leaves are said to "bolt" when they switch from producing leaves (desirable) to producing flowers (undesirable). Farmers minimize bolting in these plants by growing them in short-day seasons.

Figure 18.19

UV-reflecting patterns in petals of flowers pollinated by bees

A

B

A. Silverweed flowers as our eyes see them.

B. The same flowers photographed with a special filter that allows us to see reflected UV light, here represented by the color red. The pattern guides bees, which can see UV light, directly to the flower's reproductive parts.

Credits: Bjorn Rorslett/Science Source; Bjorn Rorslett/Science Source

Wind-pollinated flowers are typically small, nonfragrant, and green, with insignificant petals. Large stamens release pollen grains by the billions to reach the stigma of another plant of the same species.

Big Ideas in Biology

Evolution

Flowers having similar structures to attract the same kinds of pollinators is an example of convergent evolution (Chapter 12).

Pollination

In angiosperms, **pollination** is the arrival of pollen from the male part of a plant to a receptive stigma, which is one of the female parts of a flower. In many species, stigmas are receptive to pollen from either the same flower or another flower on the same plant. Individuals of these species can pollinate themselves.

Self-pollination can be adaptive because it does not require a partner. However, self-pollination is a form of inbreeding, so it can lower a population's genetic diversity and reduce the quality of offspring. Accordingly, many plants have adaptations that encourage or even require cross-pollination. For example, some flowers, like those of an avocado plant (*Persea americana*), release pollen only after their stigma is no longer receptive to pollen.

Pollinators. Pollination is essential to sexual reproduction in flowering plants. The diversity of flower form reflects a dependence on **pollination vectors**, which are animals or environmental agents that transfer pollen from anther to stigma. Animal pollination vectors are called **pollinators**. An animal attracted to a particular flower often picks up pollen on a visit, then inadvertently transfers it to another flower on a later visit. The more specific the attraction, the more efficient the transfer of pollen among plants of the same species. Thus, floral traits tend to reflect the sensory abilities and preferences of particular pollinators.

The flowers of angiosperm species that depend on the same pollinator (for example, all plants pollinated by bats) often share a certain set of traits. Bee-pollinated flowers tend to be fragrant, with a central pattern of UV-reflecting pigments in their petals, which attract bees (Figure 18.19). Bees can also sense distinct patterns of electric charge on the surfaces of flowers. Among other purposes, these electric fields help bees to discriminate between flowers of different species.

Birds have excellent color vision and a relatively poor sense of smell, and so many bird-pollinated flowers are bright and unscented (Figure 18.20A). Flowers pollinated by night-flying bats, which have a good sense of smell but poor vision, tend to be large and light-colored, with a strong nighttime fragrance (Figure 18.20B). Not all flowers smell sweet; flowers that produce odors similar to dung or rotting flesh beckon beetles and flies.

Rewards offered by a flower reinforce a pollinator's memory of a visit and encourage the animal to seek out other flowers of the same species. Many flowers exude a sweet fluid called **nectar** that is prized by many pollinators. Nectar is the only food for most adult butterflies, and it is a favorite food of hummingbirds. Honeybees convert nectar to honey, which helps feed the bees and their larvae through the winter. Bees also collect protein-rich pollen for food, and many beetles feed primarily on pollen. Pollen is the only reward for a pollinator of roses, poppies, and other flowers that produce no nectar.

Figure 18.20 **Flowers evolved to attract certain pollinators**

A

B

A. A spiderhunter (*Arachnothera*) bird sips nectar from a lobster-claw (*Heliconia*) flower.

B. A bat drinks from a night-blooming Saguaro cactus flower.

Credits: iStockphoto.com/mormoi; Johner Images/Getty Images

Reproduction in Flowering Plants

In Chapter 8, we learned how organisms, including plants, can reproduce either sexually or asexually. Each of these ways of reproducing has particular applications for agriculture, which we will explore next.

Sexual reproduction. Angiosperms, like all other plants, have a lifecycle in which a diploid generation alternates with a haploid generation. Flowers form on the **sporophyte**, a diploid, spore-producing plant body that grows by mitosis from a fertilized egg, or zygote. Spores that form by meiosis in flowers develop into haploid male or female gametophytes, which produce sperm or eggs, respectively. Male and female gametes meet at fertilization, and a new diploid generation begins (Figure 18.21).

Figure 18.21 **Lifecycle of a typical angiosperm**

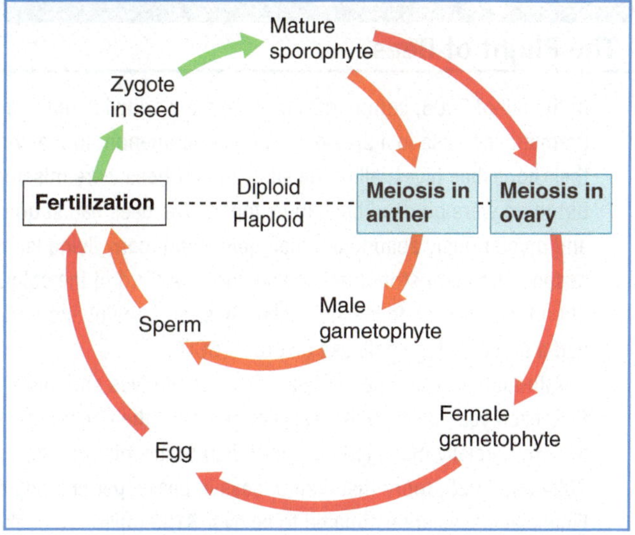

Plant sexual reproduction produces two valuable commodities for humans: seeds and fruits. Seeds form after fertilization occurs. A **seed** is a self-contained package that consists of a developing plant, or embryo, and its reserves of food, the *cotyledon*. The seed is surrounded by a protective layer called the **seed coat** (Figure 18.22).

Figure 18.22 **The anatomy of a peanut**

Peanuts are the seeds of peanut plants. The papery covering is the seed coat, and the cotyledon provides nutrition to the embryo. Note the small leaves on the embryo.

Credit: Mahirart/Shutterstock.com

Seeds form inside the ovule (see Figure 18.16). While inside the ovule, the seed receives nutrients from the parent plant. These nutrients accumulate in **endosperm** mainly as starch with some lipids and proteins. When a seed sprouts, its stored nutrients will sustain rapid growth of the seedling until new leaves form and begin photosynthesis.

We consume the seeds of various plants valued for their high protein and starch content, such as beans, lentils, and peas. We cultivate many cereals as well—grasses such as rice, wheat, rye, oats, and barley—for their nutritious seeds. A cereal seed's embryo (the germ) contains most of its protein and vitamins, and the seed coat (the bran) contains most of its minerals and fiber. Milling removes the bran and germ, leaving only the starch-packed endosperm. This is why whole wheat bread (which includes the bran and germ) is healthier for you than white bread, which is made from milled flour.

A **fruit** is a seed-containing mature ovary, often with fleshy tissues that developed from the ovary wall as the seed formed. Apples, oranges, and grapes are familiar fruits, but so are many "vegetables" such as beans, peas, tomatoes, grains, eggplant, and squash. Fruits may be categorized by the composition of their tissues, how they originate, and whether they are fleshy (juicy) or dry. Fleshy true fruits such as oranges consist of tissues derived from the ovary wall (Figure 18.23A). Accessory fruits have tissues derived from floral parts other than the ovary—such as petals, sepals, stamens, or the receptacle—which expand along with the developing ovary. Most of the flesh of an apple is an enlarged receptacle, so apples are accessory fruits (Figure 18.23B). Apples are also *pomes*, or fruits with fleshy tissues derived from the receptacle enclosing a core derived from the ovary wall.

? Did You Know?

The technical definition of a berry is a fruit that is produced from one ovary, as is the case with grapes, tomatoes, citrus fruits, and watermelons.

Simple fruits are derived from one ovary, or a few fused ovaries, of a single flower. Cherries, pea pods, and acorns are examples. By contrast, aggregate fruits such as blackberries are derived from multiple unfused ovaries of one flower that matured as a cluster. A multiple fruit such as a pineapple or a fig develops as a unit from several individually pollinated flowers.

Figure 18.23 **Fruits develop from different parts of the flower**

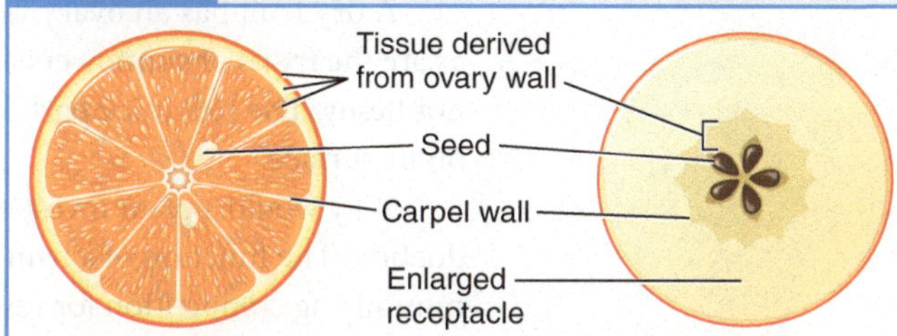

A. The tissues of an orange develop from the ovary wall.
B. The flesh of an apple is an enlarged receptacle.

Biology for Life

The Plight of Bees

In the fall of 2006, commercial beekeepers worldwide noticed a dramatic increase in a previously rare phenomenon. In many of their honeybee hives, all of the adult worker bees were missing, but there were no dead bees to be found. The bees had suddenly and mysteriously abandoned their queen and many living larvae, despite abundant stored pollen and honey. A third of the colonies did not survive the following winter. By spring, the phenomenon had a name: colony collapse disorder (CCD).

Although a certain percentage of managed hives suffered from CCD each year, in 2025, beekeepers observed the largest die-off of commercially managed bees in United States history. Between 2024 and 2025, more than half of bees in beekeeper custody died. Financial losses are estimated to be over $139 million to beekeepers, not counting the potential harm to honeybee-pollinated crops.

Nearly all of our food crops are the fruits of flowering plants. Like other seed-bearing plants, angiosperms produce pollen, and their reproductive success requires the transfer of pollen from one flower to another. Fruits do not form from unpollinated flowers, and in some species, they form only when flowers receive pollen from a different plant of the same species. Even plants with flowers that can self-pollinate tend to make bigger fruits, and more of them, when they receive pollen from another plant (**Figure 18.24A**).

Why does CCD occur? CCD is likely the result of many different stressors on honeybees. Stressors to honeybees can include exposure to pesticides, habitat changes, or diseases. One considerable threat is the varroa mite, which is a honeybee parasite (Chapter 15). *Varroa* coevolved with Asian honeybees but spread rapidly around the world after it began parasitizing Western honeybees in the 1950s. Since it began parasitizing Western honeybees, the *Varroa* mite was able to invade bees on all continents with bee colonies (**Figure 18.24B**).

A *Varroa* mite attaches itself to a honeybee and sucks out tissue from an organ called a fat body. Damage to the fat body does not directly kill bees or cause them to behave abnormally. However, it weakens their immune system and can lead to the transmission of viruses, such as the viral infection that causes deformed wings. This ultimately leads to bee mortality and the hive's demise. If the mite population in a hive rises above about 1,000 individuals, bees with deformed wings appear at a high rate and the colony is likely to fail. Thus, today's honeybee hive management practices now include

Credit: Courtesy of James H. Cane, USDA-ARS Bee Biology and Systematics Lab, Utah State University, Logan, UT

Credit: iStock.com/Bee-individual

Figure 18.24 ***Varroa*** **mites and the importance of honeybees as pollinators**

A. Raspberry flowers can pollinate themselves, but the fruit that forms from a self-fertilized flower is of lower quality than fruit from a cross-pollinated flower. The two raspberries on the left formed from self-pollinated flowers. The one on the right formed from an insect-pollinated flower.

B. A Varroa mite (arrow) crawling on the back of a honeybee.

A dry fruit has an ovary wall that dries out as it matures. Acorns and grains are dry fruits, as are the fruits of sunflowers and strawberries. Strawberries are not berries, and their fruits are not fleshy. The red, juicy part of a strawberry is an enlarged receptacle with individual dry fruits on its surface.

Why would a plant invest energy into producing fruit? It comes back to enhancing reproduction: The function of a fruit is to disperse seeds. Dispersal increases reproductive success by minimizing competition for resources between parent and offspring, as the offspring would be transported away from the parent plant. Just as flower structure reflects adaptation to certain pollination vectors, fruit structure reflects adaptation to certain dispersal vectors: environmental agents of movement such as wind or water, or mobile organisms such as animals, birds, or insects.

ongoing pesticide treatments to keep mite populations below this threshold.

Foraging behavior makes honeybees particularly efficient pollinators, but it also spreads *Varroa* and viruses. About one-third of each hive's population—up to 20,000 bees—are foragers, which means their job is to bring pollen and nectar to the hive. Each forager visits hundreds or thousands of flowers per day, which can be as much as 8 kilometers (5 miles) from the hive. A hitchhiking mite can climb on to a flower during one of these visits, then attach itself to another honeybee that visits the same flower later. Deformed wing virus can also spread this way—to other honeybees as well as to wild bees, wasps, ants, and other insects.

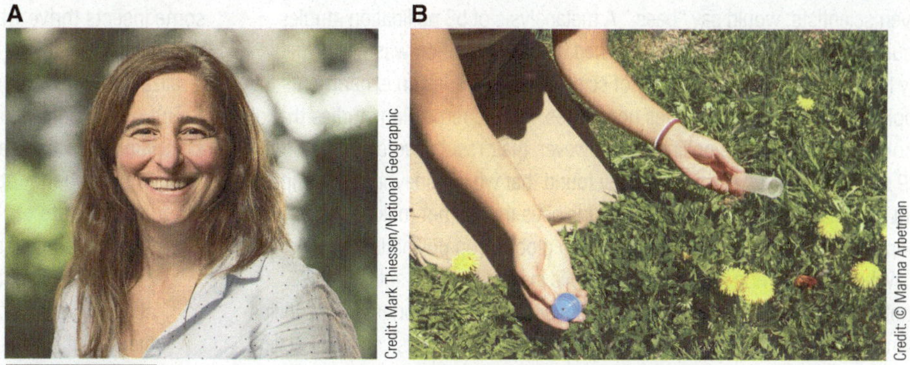

Figure 18.25 Protecting native bumblebees from parasitic infection

A. National Geographic Explorer Marina Arbetman studies how to save the Patagonian bumblebee.

B. Arbetman's team collects bee poop to track if native bumblebees are infected with a parasite. Here a member of the research team releases a bee after collection.

One important group of bees is the furry and charismatic bumblebee. National Geographic Explorer Marina Arbetman (**Figure 18.25A**) at the National Research Council in Argentina has been working to save the Patagonian bumblebee (*Bombus dahlbomii*) from extinction. Patagonian bumblebees, like other organisms, already face stressors from climate change—and the introduction of exotic (Chapter 15) bees for crop pollination adds additional stressors. For example, with more bees, there is increased competition (Chapter 15) for nectar. The commercial bees brought parasites with them as well.

One of the ways that Arbetman and her team are working to protect Patagonian bumblebees is by tracking the extent of parasite infection in the bees. To screen for parasites without harming the native bees, her team catches bees in plastic tubes (**Figure 18.25B**). The bees poop inside the tube and are then released. Arbetman and her team collect the poop and use polymerase chain reaction (PCR; Chapter 9) to analyze whether the bee is infected with a parasite. They also use genomic analysis to understand why some bees are more at risk for parasitic infection than others, which could reveal possible strategies for protecting the native bees.

1. Populations of wild bees have been declining in California's Central Valley and other areas of the United States where large swaths of land are used for intensive agriculture. What factors would you predict account for this decline? Do some research about wild bees in California's Central Valley. Were your predictions correct? Do you think wild bees in your local area would have the same risk factors?
2. Are *Varroa* mites a problem where you live? Are any measures being taken to control them?
3. What do you think you can do to help the bees? Are there any bee-friendly programs for non-experts that you can take part in?

For example, fruits dispersed by wind tend to be lightweight with breeze-catching specializations (**Figure 18.26A**). Fruits dispersed by water have water-repellent outer layers, and they float (**Figure 18.26B**). Many fruits have specializations that facilitate dispersal by animals. Some have hooks or spines that stick to the feathers, feet, fur, or clothing of more mobile species—like the cocklebururs that can get stuck on your shoes! Colorful, fleshy, or fragrant fruits attract birds and mammals that can scatter seeds (**Figure 18.26C**). The animal may eat the fruit and discard the seeds, or eat the seeds along with the fruit. Abrasion of the seed coat by teeth or by digestive enzymes in an animal's gut can help the seed germinate after it departs in feces, which in turn provides fertilizer for the seed.

? Did You Know?

Tomato plants tend to grow spontaneously in sewage treatment plants because their seeds are particularly difficult to digest. In fact, tomatoes were found growing on Surtsey (see the opening story for Chapter 15) for the same reason! Now, visiting scientists are more careful about the organic matter that they leave behind.

Digging into Data

Bees Get All the Credit

When you hear the word "pollinator," what comes to mind? Most people, even scientists, would say "bees." A metanalysis of 58 pollination studies found that 33% of the studies assumed that all pollination was performed by honeybees. The vast majority of scientists did not design experiments to look for any pollinators besides honeybees. For the remaining studies that did look for non-honeybee pollinators, however, their results were surprising. In 39 studies from five continents, they found that while honeybees are important pollinators, wild bees (which include bees that do not make honey) and non-bees were significant contributors to global crop pollination.

Who are these non-bee pollinators? Wild insects like flies, ants, beetles, moths, and butterflies are equally and sometimes more important to the pollination of some crops than bees are (Figure 18.27). While bees can carry pollen more efficiently, non-bee pollinators tend to visit plants more frequently, resulting in overall comparable levels of pollination.

Wild pollinators also provide some additional benefits. Different pollinators tend to visit varying parts of a flower or the plant overall. These wild pollinators are also active at different times of day or different times of year. Wild insects are also more likely to carry pollen long distances, especially if they don't need to return to a hive or nest each day, as is the case with bees. As land use changes, becoming more agricultural or urban, some insects thrive while others do not. Overall, these diverse pollinators improve the seed set (the production of viable seeds after pollination and fertilization), quality, shelf life, and commercial value for a variety of crops.

But which pollinators are most important and for which crops? By compiling data from all over the world, researchers identified different pollination-dependent plant species and the insect species that were found to pollinate each plant (Figure 18.28).

1. Consider what you know about ants. From the graph above, what plant is most dependent on ants for its pollination? How might their pollination abilities differ from those of honeybees? What part of the plant might they reach? What environments do they thrive in?
2. Oilseed rape, a yellow flowering plant used to make vegetable oil, is shown in the graph multiple times. This plant thrives in many different environments, all over the world. Summarize the findings that the authors made about oilseed rape pollinators, then consider Table 18.1, which shows the locations where the individual oilseed rape data were collected. Propose an explanation for the trends you observe for oilseed rape pollinators in the graph, using the information from the table. Table 18.1.

Credit: Gerrit Lammers/Shutterstock.com
Credit: iStock.com/FrankyDeMeyer
Credit: Mlraj madurawala/Shutterstock.com

Figure 18.27 **Non-bee pollinators and the plants they pollinate**

A. A black hoverfly (*Episyrphus balteatus*) on an avocado (*Persea americana*) flower in New Zealand.
B. A cabbage butterfly (*Pieris rapae*, in the order Lepidoptera) on an oilseed rape (*Brassica napus*) flower.
C. An ant (family Formicidae) on a mango flower (*Mangifera indica*).

? Did You Know?

The squirting cucumber spreads its seeds by shooting them outwards. The ripe fruit falls, causing seeds to spurt out up to 10 meters (33 feet) away!

Figure 18.26 **Fruit specializations that aid dispersal**

Credit: PRILL/Shutterstock.com
Credit: David Fleetham/Alamy Stock Photo
Credit: Vishnevskiy Vasily/Shutterstock.com

A. Wind-dispersed fruits. Wind that lifts the hairy modified sepals of a dandelion (genus *Taraxacum*) may carry the attached seed for miles.

B. Water-dispersed fruits. Buoyant fruits of the coconut palm have tough, waterproof tusks. Coconuts (*Cocos nucifera*) can float for thousands of miles in seawater.

C. Animal-dispersed fruits. The fruits of hawthorn plants (genus *Crataegus*) are an important food source for birds, such as this cedar waxwing (*Bombycilla cedrorum*), which disperse seeds in feces.

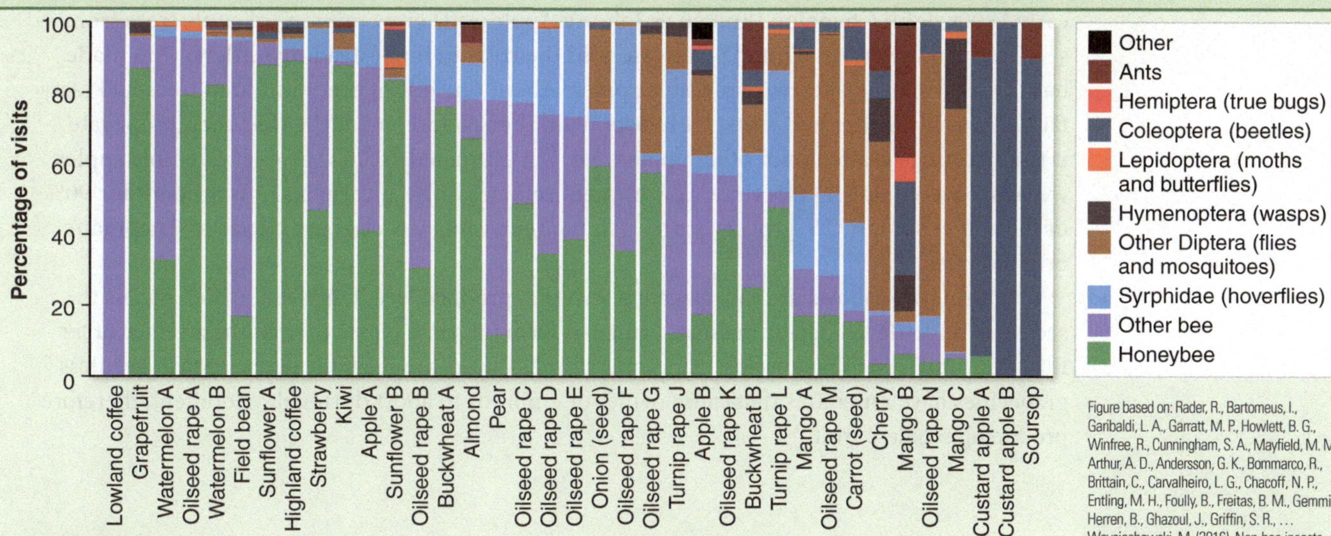

Figure based on: Rader, R., Bartomeus, I., Garibaldi, L. A., Garratt, M. P., Howlett, B. G., Winfree, R., Cunningham, S. A., Mayfield, M. M., Arthur, A. D., Andersson, G. K., Bommarco, R., Brittain, C., Carvalheiro, L. G., Chacoff, N. P., Entling, M. H., Foully, B., Freitas, B. M., Gemmill-Herren, B., Ghazoul, J., Griffin, S. R., … Woyciechowski, M. (2016). Non-bee insects are important contributors to global crop pollination. Proceedings of the National Academy of Sciences of the United States of America, 113(1), 146–151. https://doi.org/10.1073/pnas.1517092112

Figure 18.28 The contribution of different insect groups to flower visitation

The 37 crop studies for which visitation data were available are ordered, left to right, from mostly bee-dominated to mostly non-bee-dominated. When different studies investigated the same plant, they are labeled with letters after the plant name. Pollinator categories are listed in the key on the right.

Table 18.1 Data Collection Sites for Oilseed Rape Pollinators

Oilseed Rape Data Set	Data Collection Site
A, E, G, M	Sweden
B	Germany
C	Netherlands
D, F	Ireland
K	Australia
N	United Kingdom

3. Consider that 33% of studies that were analyzed were omitted because pollination was entirely attributed to honeybees. Describe an experiment that *would not* be able to distinguish honeybee pollinators from wild pollinators, and then describe an experiment that *would* be able to distinguish between the different types of pollinators.
4. Suppose a company developed a genetically modified honeybee that was resistant to an insecticide that could kill all other insects. With this insecticide, the company reasons, farmers can kill harmful insect pests without hurting beneficial pollinators. Based on what you've learned from this study on non-bee pollinators, is it possible this product might *decrease* pollination? In what ways?

Asexual reproduction. Most flowering plants can reproduce asexually. During asexual reproduction, new roots and shoots grow from extensions or pieces of a parent plant (**Figure 18.29**). Each new plant is a clone, or a genetic replica of the parent. New plants can form via roots and shoots that sprout from nodes on stems, or in some cases from cells in the vascular cylinder of roots.

A flowering plant's ability to reproduce asexually has many agricultural advantages. For example, orchard trees are clones grown from stem fragments (cuttings) of a parent plant. Propagating a plant from cuttings may be as simple as jamming a broken stem into the soil, a method that harnesses the root- and shoot-forming ability of nodes on a stem. Plants grown from cuttings may mature more quickly when they are **grafted**, or induced to fuse with and become supported by another established plant for optimal growth.

Figure 18.29 Asexual reproduction in a potato

Credit: iStock.com/huasui

New roots and shoots sprout from the nodes or "eyes" of the potato.

Propagating a plant from cuttings ensures that offspring will have the same desirable traits as the parent. Consider familiar orchard apples, which are descendants of a wild species native to central Asia. Most apple trees grown from seed produce unpalatable fruit. In the early 1800s, John Chapman (known as Johnny Appleseed) grew millions of apple trees from seed in the midwestern United States. He sold the trees to homesteading settlers, who made hard cider from the otherwise inedible apples. About one out of every hundred trees had tasty fruit. If a lucky owner had a tree with tasty fruit, they would patent the tree, then propagate it from cuttings grafted onto other apple trees. An estimated 16,000 varieties of edible apples were discovered during this time. Few of them remain; only 15 varieties now account for 90% of the apples sold in U.S. grocery stores. All the remaining varieties are clones of Chapman's original trees.

Grafting also allows us to propagate plants that produce "seedless" fruit, which means the seeds are underdeveloped or missing. Fruit can form on some plants even in the absence of fertilization, and fruits that develop from unfertilized flowers are seedless. Plants such as seedless grapevines have mutations that cause ovules or embryos to abort during development, therefore producing seedless fruit.

Biology for Life

The Importance of a Diverse Gene Pool

By definition, a fruit is a plant reproductive structure that contains the plant's seed or seeds. Now, if bananas are fruit, then where are the seeds? It turns out that the cheap, readily available Cavendish bananas that we know so well don't have seeds, because—as with other fruits manipulated by humans, such as seedless grapes and watermelons—they've been bred to not have them.

If there are no seeds, then how does a banana plant reproduce? The only way to cultivate more banana plants is by taking a shoot from a parent plant and replanting it elsewhere. Therefore, each banana plant is genetically identical to all other banana plants, with no genetic diversity between them.

A diverse gene pool facilitates evolution. If, say, a new disease appears, where some individuals in the population have a beneficial mutation that allows them to be resistant to that disease, then these individuals will survive and multiply. As was the case with the antibiotic-resistant bacteria we learned about in Chapter 10, this is an example of natural selection in action. However, the lack of genetic diversity in banana plants means that if a disease or some other threat to the banana crop occurs, without a beneficial mutation in the banana plant gene pool to counteract the threat, then bananas as we know them could go extinct.

Unfortunately, this is not a thought exercise. Panama disease, a fungal disease that wiped out another strain of banana in the 1950s, evolved to be able to infect the previously infection-resistant Cavendish banana. In 2019, banana farmers in South America discovered the disease infecting their banana plants. While we could find an alternative for a breakfast fruit, many people in South America and the Caribbean depend on banana farming for their livelihoods. According to botanists (plant biologists) at Kew Royal Botanic Gardens in London, we may need to look towards other banana relatives, like the pink hairy banana instead (Figure 18.30).

Credit: Marcophotos/E+/Getty Images

Figure 18.30 **The pink banana (*Musa velutina*) grows in the wild in Southeast Asia and the South Pacific**

1. Considering the current threat to our banana crop, what are the pros and cons of humans using asexual reproduction at a large scale for crop production?
2. Are human preferences for seedless fruit worth the potential risk to our banana supply? Why or why not?

Fruit Ripening

As you likely know from personal experience, ripe fruit just tastes better. **Ripening** is the process by which fruit matures. Ripe fruit is less green, and consequently sweeter and softer. Consequently, ripe fruit is more palatable than unripe fruit. How does ripening occur? It comes down to the action of cell signaling (Chapter 6) and two plant hormones: *ABA* and *ethylene.*

A **hormone** is an extracellular signaling molecule that exerts its effects at very low concentrations. Hormones work by binding to receptor proteins on (or in) target cells (Chapters 5 and 6). When a hormone binds to a receptor protein on the surface of a cell, it triggers a change in cellular activities within the cell. The change that is triggered may differ depending on cell types, allowing the same hormone to have different effects.

Hormones operate somewhat differently in plants and animals. Animals have organs specialized for hormone secretion, and plants do not. Instead, all of a plant's cells can make and release hormones. Animal hormones and plant hormones are similar in their ability to elicit an effect in a distant tissue. For example, a plant hormone released from cells in an actively growing root keeps the above-ground shoot tips growing as well. Plant hormone release can also impact ripening on nearby fruits, when a cell in a ripening fruit releases a hormone that affects itself as well as its neighbors.

The two plant hormones that are involved in ripening are abscisic acid and ethylene. **Abscisic acid (ABA)** is a hormone that helps the plant respond to environmental stress. As levels of this hormone rise, there are changes in gene expression (Chapter 7) that help the plant survive the adverse conditions. The other hormone, **ethylene**, is a gas produced by mature fruit. Since ethylene is a gas, it can diffuse from one fruit to initiate ripening in another. Humans use this property to artificially ripen several types of fruit. Hard, unripe fruit can be transported long distances with less damage than if one were transporting soft, ripe fruit. Upon arrival at its final destination, the unripe fruit is exposed to synthetic ethylene, which jump-starts ripening.

As a fruit forms and enlarges, ABA accumulates in its tissues. When the hormone reaches a certain level, it triggers production of ethylene. The resulting ethylene surge enhances transcription of genes whose products have several effects associated with ripening, while also triggering the production of even more ethylene (**Figure 18.31**). The fruit changes color as chlorophyll (which is green) in chloroplasts break down, and red, orange, yellow, or purple accessory pigments are produced. During ripening, the firm cell walls break down, making the fruit softer. Starch and acids are converted to sugars, and aromatic molecules are produced—so the fruit is now sweeter and smells good. The resulting color change, softening, and increased palatability attract animals that can disperse the fruit's seeds.

Figure 18.31 **Ethylene production during strawberry formation and ripening**

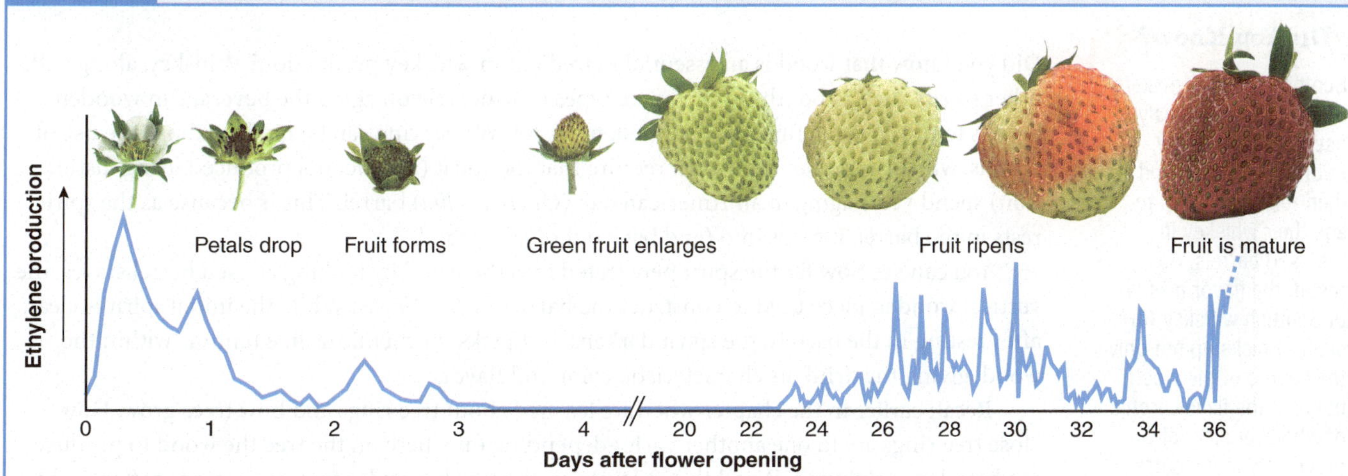

Oscillations are normal daily cycles in the plant. Note the dashed line under the mature fruit represents values that extend beyond the graph.

Credits: Madlen/Shutterstock.com; Westend61/Superstock; Alena Brozova/Shutterstock.com; Anest/Shutterstock.com

The flowering heads of the teasel plant contain hooks that are used to help with seed dispersal. Another use for these hooks is in the production of woven wool blankets, where the blankets are brushed with the heads to help raise the nap on the cloth, making it ultra-soft. Johnstons of Elgin in Scotland uses this traditional method to this day when producing luxury cashmere blankets.

18.2 Summary

- Flowers are the reproductive structures of angiosperms and typically have four concentric rings: a ring of sepals, petals, stamens, and carpels.
- Male reproductive structures include the stamen, which has an anther full of pollen sacs, where pollen forms. The female reproductive organs are called the carpel or pistil, which have the stigma that receives pollen grains, and an ovary with ovules (seeds). Flowers with all these structures are called complete, while those missing one or more of those structures are called incomplete.
- Flowering is regulated by night length. Plants that flower in the winter, spring, or fall and are called short-day plants; plants that flower in the summer days are called long-day plants.
- Pollination is the arrival of pollen from the male part of a plant to a receptive stigma. Floral traits vary by their pollinators; flowers with bright colors tend to attract pollinators that can perceive bright colors, while flowers with strong smells attract pollinators that are able to smell them. The flowers offer rewards like nectar or pollen to encourage the animal to pollinate them.
- Sexual reproduction in flowering plants occurs when spores that form by meiosis in flowers develop into gametes, which meet at fertilization to form a diploid offspring called seeds. Asexual reproduction occurs when new roots and shoots grow from extensions or pieces of a parent plant.
- Producing fruits and seeds helps to minimize competition between parent and offspring while also facilitating dispersal.
- Sexual reproduction produces seeds, which can be used as food (like grains) or planted. Asexual reproduction can be harnessed to clone or graft a plant and make seedless fruit.

18.3 Plants and Your Future

Learning Objectives

18-11 Identify ways in which you encounter trees and wood products in your life.

18-12 Describe the challenges with feeding a growing human population.

18-13 Identify sustainable solutions for feeding a growing population.

18-14 Describe ways in which plants can improve living conditions for humans.

? Did You Know?

Scottish whiskey makers buy barrels previously used to age whiskey, bourbon, or wine, and then refurbish them to age their whiskey in the used barrels. As a result, the flavor profile of Scottish whiskey (or scotch) picks up not only the flavors of the wood, but also the flavor profile of whatever alcohol was previously in the barrel.

Did you know that wood is an essential ingredient in whiskey production? Whiskey, along with other spirits such as bourbon, and some types of wine, rely on aging the beverage in wooden barrels to produce the final product (Figure 18.32). While wine can be made without the use of barrels, whiskey and bourbon both require that the spirit (any alcohol produced using distillation) spend years aging in an American oak (*Quercus alba*) barrel. This is because as the spirit rests in the barrel, it seeps into (and back out of) the wood.

You can see how far the spirit penetrated into the wood by looking at used barrel staves, the vertical wooden piece used to construct the barrel (Figure 18.33). While the initial spirit is clear, after resting in the barrels, the spirit darkens as it picks up chemicals like tannins within the wood, giving the drink its characteristic color and flavor.

Recall earlier in the chapter when we learned about tree rings and how trees grow. How close tree rings are to one another varies depending on where on the tree the wood to produce the barrel was obtained. Wood that is lower on the tree has wider tree rings and is softer—and easier for the spirit to penetrate during aging. Therefore, barrels produced from wood found lower on the tree tend to have a more robust flavor profile. In contrast, wood that is higher up on the tree has closer tree rings and is harder, and therefore, the spirit doesn't penetrate as easily, leading to a whiskey with a more limited flavor profile. The flavor from the wood is also influenced by the climatic conditions where the tree grew. Due to the variability that wood can introduce to the process of producing aged spirits, many commercially available whiskeys are aged in many different barrels and blended together to produce a characteristic flavor.

Which brings us to a very important point. Wood products are essential in many industries, not just in the production of aged spirits. And obtaining that wood requires allowing trees to mature for decades, often beyond our own lifetimes. Consider the lumber used to build our houses. Depending on the species of tree that produces the wood, it can take anywhere from 20 to 70 years or longer from planting until a tree is ready to be harvested for lumber. The toilet paper in your bathroom was produced from a tree that was 10 to 20 years old. Trees used to produce paper have a similar age span. Think of it this way—the toilet paper you used in the bathroom earlier today may have been from a tree as old as you are! In the next section, we'll explore the importance of forests, with a focus on *stewardship* (Chapter 16).

Figure 18.32 **Wooden barrels are required to produce spirits such as whiskey and bourbon**

Shown here are barrels containing aging bourbon in a warehouse in Kentucky.

Credit: iStock.com/Try Media

Figure 18.33 **Used barrel stave showing spirit penetration line**

The arrow points to the remaining spirit in the barrel stave.

The Importance of Forests

We need trees—not only for the consumer goods made from trees, but also because of the critical roles that trees play in ecosystems and the biosphere. For example, trees are crucial habitats for other plants and animals, provide wind breaks and shade, and, as we learned about in Chapter 16, are important carbon sinks. A **carbon sink** is a natural or artificial carbon reservoir that absorbs more carbon than it releases. Trees, and especially mature trees, are carbon sinks and therefore valuable in the fight against climate change because they remove carbon dioxide from the atmosphere. Like other plants, trees capture CO_2 from the air and fix it into sugars during photosynthesis. As a plant grows, the sugars are continually remodeled into other carbon-containing compounds that compose the plant's tissues. Growth does not slow with age in long-lived trees; rather, more wood is produced and consequently more carbon is captured and incorporated into the tree's tissues.

But forests are under threat. In the United States, for example, there are 754 million acres of forest. At the start of European settlement, there were estimated to be 1,023 million acres of forest. So, over the last 300 years, around 256 million acres of forests have been converted for human purposes, mainly for agriculture.

The forests in the United States, while still vast, are shrinking. In 2024 alone, the United States lost 3.8 million acres of forests. The United Nations estimates that globally, around 24.71 million acres of forests are lost each year. While there are some efforts to replant trees, worldwide we are losing trees faster than we are planting them. Much of this loss is related to climate change. For example, the increasing severity of wildfires means that more forests are lost to fires.

? Did You Know?

The trees used to produce barrels are typically 100+ years old.

Given the long time periods needed to generate a climax community (Chapter 15), it's critical to be good stewards of trees and forests. How can we be good stewards? The simplest way is to plant trees and promote the planting of trees in your community. Trees, and particularly

Figure 18.34 **Logos used to indicate that tree-based products are sustainably harvested**

A

B

A. Forest Stewardship Council labeling, and
B. The Rainforest Alliance's little green frog seal represents a commitment to better farmer livelihoods, human rights, environmentally friendly practices, and climate resilience.

Credits: Forest Stewardship Council; Rainforest Alliance

mature trees, raise real estate values while also providing benefits to the local ecology and biosphere. You can also contact your local and/or state representatives and ask them to support policies that protect forests.

Finally, we can also be good stewards of trees by paying attention to whether the wood and/or paper goods we are using are made from sustainably managed forests. For example, you can check to see if the goods you are buying are labeled with a Forest Stewardship Council (FSC) label (**Figure 18.34A**). FSC labels appear on goods that are derived from trees that are harvested responsibly (meaning *deforestation*, or the loss of forest (Chapter 16), is not occurring), from recycled materials, or from a mix of recycled materials and new wood from sustainably managed forests. The Rainforest Alliance (**Figure 18.34B**) specifically focuses on conservation and stewardship of tropical forests. The latter is particularly important when you consider that well-loved products like chocolate and coffee depend on trees grown in the tropics!

Up to this point, we've mainly focused on using trees to produce wood products. However, trees are not just useful for producing wood—they are also the source of many delicious foods, from fruits like cherries to nuts like almonds…and even spices like cinnamon! We'll now turn our discussion to the use of plants to feed the ever-growing human population (Chapter 14).

Climate Change and Feeding a Growing Human Population

Did you know that 47 million Americans (and 1 in 5 children) struggle with food insecurity, or the inability to access sufficient food and/or good quality food? The far-reaching impact of climate change (Chapter 16) is likely to make matters worse. Climate change threatens crop production, livestock and fisheries, and our ability to distribute food.

Let's consider the threats to agriculture, which is sensitive to changes in weather and/or climate. Changes to the growing season, like the timing of first frost, and alterations in temperature can cause crops to have reduced yields. Heavy rains and flooding can also cause soil erosion, taking away the valuable soil needed for growing healthy plants. Plants that can adapt to changing climate conditions will be the "winners" as the planet warms, where other species, less able to adapt, face the threat of diminishing numbers and ultimately extinction. We'll now examine the relationship between the different types of photosynthesis and resistance to climate change, and how genetic engineering (Chapter 9) might be used to help preserve our food supply.

C3 and C4 plants. Depending on the exact processes that a plant uses when undergoing photosynthesis, some plants will be better able to adapt to climate change. Recall from Chapter 4 that photosynthesis is the process of using the sun's energy to convert inorganic molecules like CO_2 into glucose ($C_6H_{12}O_6$), which the plant and other organisms can use for energy. Most plants are **C3 plants**, meaning that when carbon is fixed by rubisco, the result is two 3-carbon molecules of PGA (Chapter 4, Figure 4.16), hence the name C3.

While most plants—including crucial commercial crops such as rice, wheat, and barley and many species of tree—are C3 plants, C3 plants are less efficient at higher temperatures. When temperatures are high, C3 plants close their stomata to reduce water loss. When the stomata are closed, the concentration of oxygen increases and the concentration of CO_2 decreases, and the plant undergoes a metabolic process called **photorespiration**. During this time, the plant fixes less carbon and uses energy (ATP and NADPH) rather than producing it. You can think of photorespiration as the opposite of photosynthesis—while photosynthesis produces energy, photorespiration uses it. Photorespiration is necessary to eliminate the toxic products that are produced when oxygen concentrations are high and CO_2 concentrations are low. Because more photorespiration occurs at increased temperatures, prolonged periods of high heat and closed stomata can ultimately put the plant in jeopardy.

How is it then that plants can live in hot, dry areas? C4 and crassulacean acid metabolism (CAM) plants are better adapted to hot, dry environments because their process of performing photosynthesis is slightly different from typical photosynthesis: C4 plants separate the steps of photosynthesis in different cells, whereas the steps of CAM photosynthesis happen at different

Biology for Life

Leafy Cleanup Crews

From 1940 until the 1970s, the U.S. Army tested and disposed of weapons at J-Field, Aberdeen Proving Ground, in Maryland. Obsolete chemical weapons and explosives were burned in open pits, together with plastics and other wastes (**Figure 18.35A**). Lead, arsenic, mercury, and other metals heavily polluted the soil and groundwater. So did hazardous organic solvents, including one called trichloroethylene (TCE). TCE is toxic to humans and other animals: It damages the nervous system, lungs, and liver, and exposure to large amounts can be fatal.

A polluted site like J-Field can be cleaned up by removing contaminated soil, but this approach is expensive and risky. A huge amount of soil must be excavated and moved to another location for decontamination or permanent storage, and spillage or release of toxins may occur during any part of the process. Instead of this traditional approach, the Army and the Environmental Protection Agency (EPA) turned to *phytoremediation*, or the use of living plants to remove and/or degrade environmental contaminants.

Phytoremediation of J-Field began in 1996, when a large grove of poplar trees was planted to remove organic solvents in groundwater (**Figure 18.35B**). Like other vascular plants, poplar trees take up water through their roots. Dissolved substances, including TCE, are taken up along with the water. TCE does not harm poplars. The trees break down some of the solvent, but release most into the atmosphere. Airborne TCE is the lesser of two evils: It breaks down more quickly in air than it does in water. Within a few years, the trees had removed 60 pounds of TCE from the soil. Genetically engineered poplar trees (Chapter 9) are currently used at other sites to break down not only TCE but also other dangerous organic compounds such as carbon tetrachloride and formaldehyde, and pesticides like chlorpyrifos.

Unlike organic solvents, metal pollutants cannot be broken down. Phytoremediation of sites contaminated with toxic metals uses plants that take up metal ions from soil water and store the ions in aboveground tissues. The metal-laden plant parts can then be harvested for safe disposal. This strategy is being used in the Fukushima region of Japan, where a nuclear accident in 2011 released huge amounts of radioactive material—including radioactive cesium ions that have a very long half-life. Sunflowers planted in a 30-mile radius around the damaged Fukushima nuclear reactors are removing these metal ions from contaminated soil. Sunflowers were also used in phytoremediation of radioactive soil around Hiroshima and Chernobyl, so it is perhaps not surprising that the sunflower is an international symbol of nuclear disarmament.

A

Credit: Billy Wrobel, 2004

B

Credit: © OPSEC Control Number #4 077-A-4

Figure 18.35 Phytoremediation of J-Field

A. Decades of burning chemical weapons, explosives, plastics, and other wastes at J-Field Aberdeen Proving Ground left the soil heavily contaminated with dangerous organic compounds.

B. Today, poplar trees are removing TCE and other contaminates from the soil at J-Field.

1. Compared with traditional methods of toxic site cleanup, phytoremediation is less costly and less risky. What other benefits might poplar trees, planted for phytoremediation, provide to the environment that other methods of remediation would not?
2. In what tissue type of poplar trees would you expect to find high levels of contaminants stored? Justify your answer.
3. Is phytoremediation happening near you? Do a web search for your area and phytoremediation efforts and summarize a project you find.

times of day. **C4 plants**, which include corn, sugarcane, and sorghum (a type of cereal crop), synthesize a 4-carbon molecule during carbon fixation (**Figure 18.36A**). In C4 plants, within the mesophyll cells, incoming CO_2 is first fixed into a 4-carbon intermediate. This molecule travels into neighboring *bundle-sheath cells*, where the Calvin cycle is occurring. There, it is converted back to CO_2 and enters the photosynthetic pathway. The goal of the first step is to create a high concentration of CO_2 in the bundle sheath cells, reducing photorespiration.

Figure 18.36 **Schematic representation of C4 and CAM photosynthesis**

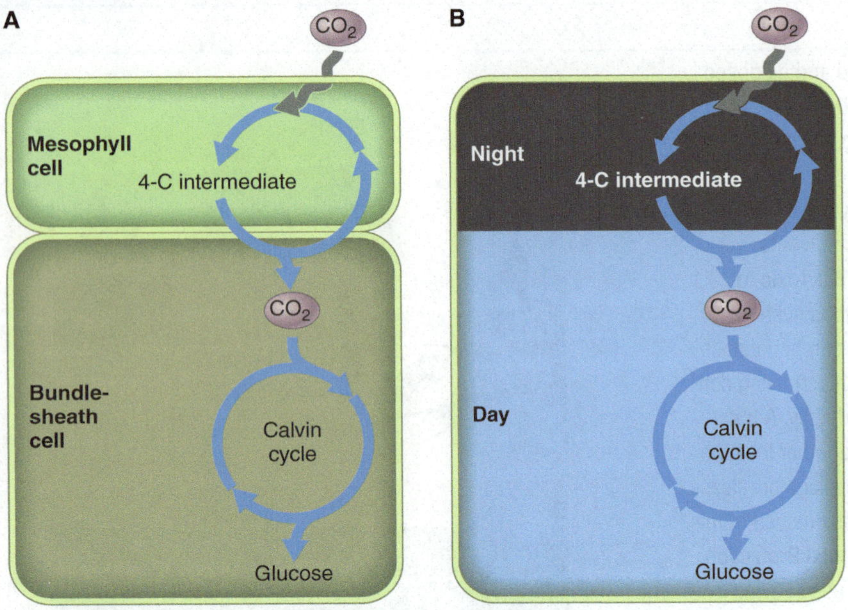

A. C4 photosynthesis occurs in different cells. In C4 plants, incoming carbon dioxide is converted to a four-carbon (4-C) intermediate in the mesophyll cells. This molecule is moved into the bundle sheath cells, converted back to carbon dioxide, and fed into the Calvin cycle.

B. CAM photosynthesis occurs at different times of day. CAM plants fix carbon dioxide into a four-carbon intermediate only at night, when it is cooler and the plant's stomata are open. During the day, the four-carbon intermediate is converted to carbon dioxide and then fed into the Calvin cycle.

CAM plants survive in hot temperatures by opening their stomata only at night, when temperatures are cooler. So instead of two reactions occurring at the same time in different places, which is what C4 plants do, CAM plants separate carbon fixation and energy products by time of day (Figure 18.36B). During the night, the stomata are open, and the plant is fixing CO_2 into a 4-carbon intermediate. Then during the day, the stomata are closed, and the 4-carbon intermediate is converted to CO_2 and enters the Calvin cycle and energy is produced. Converting the 4-carbon intermediate into CO_2 during the day keeps the concentration high enough to avoid photorespiration, even when the stomata are closed. CAM plants are limited to desert dwellers. Note that agriculturally significant plants such as pineapples and agave are CAM plants. Therefore, while some agriculturally significant crops like pineapples and corn will be more resilient to climate change, other staples such as wheat are at greater risk.

Engineering climate change–resistant crops. What can we do to protect our food supply and prevent increasing food insecurity? One solution comes from biotechnology. Recall from Chapter 9 that genetically modified (GM) organisms either contain the DNA of other organisms (transgenic) or had their DNA altered using a process such as gene editing with the CRISPR/Cas9 system. When referring to agriculture specifically, most people refer to genetically altered plants as genetically modified organisms or GMOs.

Remember, GMOs do not include organisms bred for specific traits using artificial selection (Chapter 11).

While there are some legitimate concerns about GMOs (Chapter 9), there are also many real and recognized benefits to their use, particularly related to climate change. For example, if a plant is engineered to have greater yield, then the same amount of food could be produced on less land. This means that more land can be retained as native vegetation—which is good for the ecosystem, especially if carbon sinks like trees are retained. A smaller field would also mean less acreage to tend to, translating into fewer greenhouse gas emissions (Chapter 16) from farm equipment.

To make plants more resilient, biologists are engineering plants and soil microbes to be more efficient at capturing and storing carbon. They are also developing strains of crops, such as corn and soybeans, to be more resilient to extreme weather conditions, such as very dry or very wet conditions. As we learned throughout Unit 5, extreme weather conditions are increasingly common due to climate change. Therefore, engineering more resilient plants will help to protect our food supply. Finally, researchers are engineering plants to be more efficient at photosynthesis. For example, engineering C3 plants to use alternative metabolic pathways to increase photosynthesis and crop yield.

Taken together and looking toward the future, GM plants will be essential for feeding a growing human population. With the threat of climate change to our food supply, using genetically modified plants may be the best way to protect agriculture and livelihoods in affected regions, and by extension, protect our own food supply. However, as discussed in Chapter 9, the use of genetically modified organisms in agriculture requires transparency and oversight to ensure that the modifications aren't having unintended consequences to humans or the environment.

Biology for Life

What's the Big Deal with Genetically Modified versus Organic Crops?

It is difficult to answer the question of which is better, non-genetically modified (GM) organic foods (which, by definition, cannot be GM) or GM foods. The prevalence of biases (Chapter 1) complicates the issue further.

Let's start by considering organically grown produce. Many people think of organic crops as a healthier and more environmentally friendly option. Organic farming produces fewer greenhouse gases and has less of a negative impact on soil and water. There is also no use of synthetic fertilizers or pesticides. However, organic farms have a smaller yield than conventional farms, which may result in greater net greenhouse gas emissions, because the crops need to be transported long distances to meet demand. The lower yields also contribute to a higher price tag at the store, limiting those who can afford to buy organic products. Additionally, pesticides can still be applied to organic produce, provided they are naturally occurring. Note that naturally occurring does not necessarily mean safer for human and environmental health.

When we look at GM foods, we see plants that have been genetically engineered, which can give some shoppers hesitancy. As we discussed in Chapter 9, there are concerns about the risk of engineered genes escaping into the environment and potentially disrupting the local ecology. There are also concerns regarding overreliance of the herbicide RoundUp® on RoundUp-ready crops, fueling the evolution of herbicide-resistant super weeds. In addition, there are potential health concerns regarding RoundUp residues on food.

Regarding overall nutrition, the scientific consensus is that there's no nutritional difference between GM and organic crops. Additionally, plants engineered to require less water or fertilizer—particularly given drier conditions with climate change—could be a boon for the environment.

So, some people will make consumer decisions based upon concerns about their health or the health of the planet, while others might focus more on the best method for feeding the most people. The debate will only continue as the planet warms—and it is necessary as a society to consider the best way to feed future generations.

1. When considering the potential impacts of organic and GM crops on the environment, what is most important to you as a consumer and a citizen?
2. Often, plants are genetically engineered to be pest resistant, so they can be farmed organically without pesticides. Considering this, would you change the definition of "organic" to include GM crops? What would be the pluses and minuses of creating a third category of crops that are both organic and GM?

Plants and Our Future

As we've learned both in this chapter and throughout this entire text, plants are crucial to our lives. It's perhaps surprising that we don't often think about plants unless we're weeding or taking allergy medicine to deal with pollen....and yet, we wouldn't exist without plants and their ability to use sunlight and carbon dioxide to generate glucose. Plants provide us not only with food, but also with goods that are important to our survival, such as wood and medicines. When we think about allies for fighting climate change, plants are one of our best choices.

Figure 18.37 Example of a green roof

Here an urban farmer tends a rooftop garden in Denmark.

Credit: Pixdeluxe/E+/Getty Images

Consider green roofs (Figure 18.37). A green roof is a gardening technique where the roof of a building is either partially or totally covered with vegetation. Green roofs can be atop structures, such as in the urban garden in Figure 18.37. They can be park-like regions of grass and trees, or simply consist of native grass. In addition to green roofs, scientists in the Netherlands are developing bricks that can grow moss. The bricks are designed to foster plant growth, therefore allowing entire buildings, not just the roof, to be covered with plants.

Green roofs and bricks that facilitate plant growth can help to reduce heat islands. **Heat islands** occur when structures like buildings and roads absorb and re-emit the sun's heat energy and are particularly a problem in cities. The amount of heat reflected is much higher than a similar space containing plants. Therefore, in highly developed areas like cities, temperatures can be decreased by several degrees by using plants to create a natural space.

The Environmental Protection Agency (EPA) estimates that temperatures in heat islands can be 1 to 7 degrees Fahrenheit warmer during the day, and 2 to 5 degrees Fahrenheit warmer at night. Because green roofs absorb more heat energy and reflect less of it, this can serve to cool the

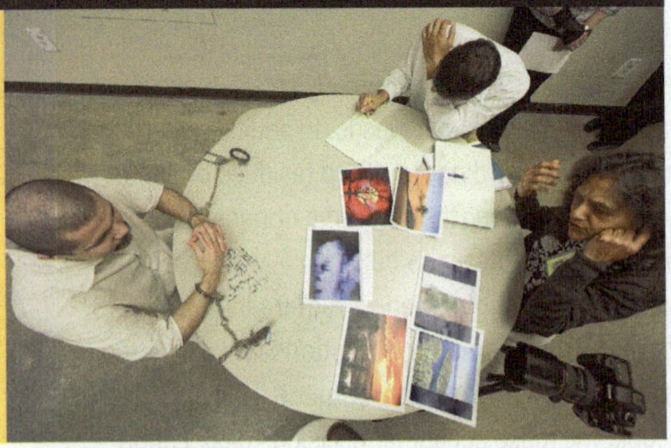

Figure 18.38 National Geographic Explorer Nalini Nadkarni interviews an inmate about their experience watching nature videos

air temperature in the immediate vicinity by as much as 11 degrees Celsius (about 20 degrees Fahrenheit)—resulting in lower cooling bills for the building! Additionally, since cities can often be food deserts, that is, areas with limited access to fresh food, rooftop farming can provide a resource for targeting food insecurity by increasing the availability of fresh produce within cities.

Finally, above and beyond their role in addressing climate change and food accessibility concerns, let's not forget that plants also provide positive mental health benefits. Recall Nalini Nadkarni, the National Geographic Explorer featured earlier in this chapter. One of her research projects found that showing nature videos and playing nature sounds to inmates in maximum security prisons—a nature-starved location—resulted in 26% fewer violent infractions by the inmates (Figure 18.38). Inmates also reported feeling calmer after watching the videos.

The mental health benefits of exposure to nature aren't limited to inmates, of course. Many studies show the positive benefit of nature on one's mental health. Time in nature is associated with less anxiety and better mood overall, plus better memory and cognition (the ability to think). Time in nature can also help people to cope with conditions such as depression, anxiety, post-traumatic stress disorder, and attention-deficit/hyperactivity disorder.

Ultimately, having a healthy relationship with plants is essential to our physical and emotional well-being, as well as the well-being of the planet. As we face the serious realities of climate change, plants may be one of our greatest assets as we work to remove excess carbon dioxide from the atmosphere. So, the next time you see a tree or a flower, pause for a minute and think about not only their form and function, but why plants are important to your life.

? Did You Know?

In 1976, Mort Garson recorded an album called *Plantasia*, which was given for free to anyone who bought a houseplant at a store called Mother Earth in Los Angeles. The claim was that the music was for plants to listen to, which would help them grow. Today, we remember the album as an early example of electronic music, and, at least in your author's opinion, an excellent album to listen to while working on the computer!

18.3 Summary

- Trees are important for producing products ranging from whiskey barrels to toilet paper, while also providing us with food and fighting climate change.
- Climate change threatens many of our food sources, including plants. However, some plants are more tolerant to heat; C4 and CAM plants have adaptations to help reduce the negative effects that C3 plants experience during photorespiration.
- Genetically modifying plants, so they require less water or are more efficient at photosynthesis, can help to make them more resilient to climate change, ultimately helping to address the issue of feeding a growing population amidst climate change.
- Plants absorb heat, so planting green roofs can help to counteract temperatures in heat islands such as cities. Being in natural spaces can also have a calming effect.

Understand It! Neuroplasticity

It's easy to think about the various structures in our bodies as being steady and constant. In fact, we often learn about structures that way in an introductory biology course. For example, one may learn the different organs and their associated systems, and how those systems contribute to the overall function of the body. However, organs can change over time; for example, as one ages, their heart and lung function decreases. In addition, did you know that your brain is dynamic and changes throughout your life? And that learning causes physical changes in the brain?

In the Chapter 13 Understand It! we learned about how memory networks form within the brain. The ability of the brain to change because of learning or being exposed to new experiences is called *neuroplasticity*. Everyone can change their brain through learning. So how easy one learns isn't simply a matter of intelligence, but rather about the connections that one makes in their brain while studying (see also the growth mindset Understand It! in Chapter 4). When we deliberately study in such a manner that we are making and strengthening neuronal connections, we are more likely to be successful at changing our brains and learning new material.

While most of our neurons form prior to birth, the formation of new neurons, or neurogenesis, occurs throughout one's lifetime in the hippocampus. The hippocampus is the region of the brain involved in memory formation. When we learn something new, new neurons form in the hippocampus and/or connections between neurons are strengthened. For example, biologists looked at the area of the brain involved in the perception of moving objects between people who practiced juggling over a period of three months to those who did not. This area of the brain was larger in the jugglers than in the non-jugglers…and returned to the normal size when the juggling practice ceased. A later study found that the changes in the brain structure were due to increased neuronal connections when juggling was practiced.

We see a similar phenomenon in how people study—which is why practice (Chapter 1) and drawing connections between concepts is so important to our learning. In addition, building good habits can help people who are neurodivergent change their brains in such a manner that helps them to succeed in their educational and professional pursuits. For example, people with attention-deficit/hyperactivity disorder (ADHD) have different brain structures than people who do not have ADHD. These different brain structures can lead to, for example, more spontaneity—which can be a positive attribute, but a liability while learning in a traditional classroom. Practicing techniques to help one concentrate can change the brain in such a way that these skills can be drawn upon to help one study.

1. Reflect on your experiences in the biology course you took this semester. How do you think your brain has changed because of studying biology?
2. After Albert Einstein's death in 1955, his brain was removed and studied. Compared to other people who were the same age as Einstein when he died, Einstein had more neuronal connections between the left and ride sides of his brain. What can you infer about the relationship between intelligence and learning from these findings? How does this impact your studying and/or learning strategies?

Chapter Review

Section 18.1 Plant Form and Function

Flowering plants provide nearly all human food, either directly or as feed for livestock. Above ground, vascular plants have **shoots** with **stems**, **leaves**, and reproductive structures. New shoots form at **nodes**, and the spaces between nodes are called **internodes**. **Vascular bundles** run through the stems, providing water, nutrients, and structural support. Leaves have **mesophyll cells**, where photosynthesis occurs, and **stomata** through which gas exchange occurs. Vascular bundles that run through the leaves are called **leaf veins**.

Below ground, vascular plants have **roots** that acquire water and nutrients from the **soil**. **Primary roots** grow from a seed and then branch to form **lateral roots**. At the tip of each root is a **root cap** that protects the root. Young roots have many **root hairs** for absorption. As they grow, roots will eventually form either **fibrous roots** or **tap roots**.

Plants are made up of three tissue types: dermal, vascular, and ground. **Dermal tissue** covers and protects exposed surfaces, especially the outermost **epidermis** layer that makes a waterproof protective **cuticle**. Xylem and phloem are made up of **vascular tissue**. **Xylem** tubes conduct water through the plant and contain **lignin**, which provides structural support. **Phloem** transports sugars and other organic solutes. **Ground tissue** makes up the bulk of the plant and contains parenchyma, sclerenchyma, and collenchyma. **Parenchyma** can be a storage tissue, provide structural support, or be a site of photosynthesis. **Collenchyma** provides pliable support at rapidly growing portions of the plant, like the **sclerenchyma**.

Water is drawn into plants through osmosis, creating osmotic pressure. Because plant cell walls are rigid, the water pushes on the plant walls, creating **turgor pressure**. Water evaporates from the plant shoots in a process called **transpiration**. Water loss can be reduced by **guard cells** closing stomata.

Movement of sugar and organic molecules through phloem pipelines called sieve tubes is called **translocation**. A **sieve tube** is supported by a **companion cell** that transfers sugars into it. The sugars move through the sieve tubes to all parts of the plant. Flow occurs from a **source** where sugars are produced or released from stores, such as leaves, to a **sink**, where they are broken down or stored, such as in a fruit.

Plant growth occurs at **meristems** in two phases: primary and secondary growth. All plants undergo **primary growth**, which involves **apical meristems** at the **terminal buds** and **root caps**. Some plants continue onto **secondary growth**, where lateral meristems form **vascular cambium** that makes **wood**, and **cork cambium** that produces **cork**.

Section 18.2 The Lifecycle of a Flowering Plant

Flowers are plant reproductive structures in **angiosperms** that form from **sporophytes** (diploid, spore-producing plants) that typically have four concentric rings made up of **sepals**, **petals, stamens**, and **carpels**. The male part of the plant, the stamen, has an **anther** with **pollen grains**, which are immature **gametophytes**. The female part of the plant, the **pistil**, has a **stigma** that receives pollen, and **ovaries** with ovules.

Plants go through the process of **flowering** so that they can reproduce, a process that requires **pollination**, where the pollen from the male part of a plant arrives at a receptive stigma. Some plants self-pollinate, while others use **pollination vectors**. Anything, including animals or environmental agents, that transfers pollen is called a pollination vector, while specifically, animal pollination vectors are called **pollinators**. Some plants provide pollinators a reward in the form of **nectar**.

Sexual reproduction occurs when sporophytes produce spores that develop into haploid gametophytes. When fertilization occurs, plants produce **seeds** that are protected by a **seed coat**. Seeds contain an **endosperm** to provide nutrients from the sprouting seed. **Fruit** helps disperse seeds away from the parent plant. Most flowering plants can also reproduce asexually when new roots and shoots grow from pieces of a parent plant or if cuttings are **grafted** to another plant.

The process of **ripening** makes fruit softer and sweeter. It involves two hormones, **abscisic acid (ABA)** and **ethylene**.

Section 18.3 Plants and Your Future

Trees, especially mature trees, are important because they provide habitats, wind breaks, and shade. In addition, they are **carbon sinks**. We depend upon trees for many products, and, therefore, forests are shrinking. Caring for our forests is essential for our future survival.

Climate change threatens crop production, livestock and fisheries, and food distribution. Warming temperatures will further threaten food security and will likely lead to the die-off of certain food sources.

When temperatures increase, **C3 plants** (those that fix carbon dioxide into a 3-carbon intermediate) close their stomata and undergo **photorespiration**, a process that wastes energy and doesn't fix CO_2. **C4 plants** and **CAM plants**, however, use different mechanisms to concentrate CO_2, making them more resilient to hot, dry conditions. Genetic modifications can be used to make plants more resilient to climate change stressors like heightened temperature and drought.

Plants can help combat climate change by providing carbon sinks, preventing **heat islands** by cooling the environment, such as by using green roofs. They also benefit our mental health.

Practice It!

Retrieval practice is the process of pulling information from your long-term memory and using it. Here are some questions to help with your retrieval practice.

Self-Quiz

1. Stems are part of which of the two major organ systems of plants? (LO 18-1)
   a. shoots
   b. leaves
   c. roots
   d. reproductive organs

2. Young roots have root hairs that increase surface area so that there is additional absorption of which of the following substances? (LO 18-2)
   a. water and minerals
   b. CO_2 and sugar
   c. sunlight and nitrogen
   d. salts and oxygen

3. Xylem and phloem are composed of which type of plant tissue? (LO 18-3)
   a. dermal tissue
   b. epidermal tissue
   c. vascular tissue
   d. ground tissue

4. Sugars are carried to the cells of the plant by ______. (LO 18-4)
   a. xylem
   b. phloem
   c. taproots
   d. lignin

5. Primary growth occurs ______. (LO 18-5)
   a. at apical meristems
   b. at lateral meristems
   c. in the seed
   d. in cork and wood

6. What is the function of a flower's pistil? (LO18-6)
   a. It receives pollen and houses ovules.
   b. It displays pollen sacs for visiting pollinators.
   c. It increases the structural stability of the stamen.
   d. It attaches to the base of the flower and grows upward.

7. How is nectar involved in pollination? (LO 18-7)
   a. It is a nutrient-rich reward for the pollinator.
   b. It carries the gametes of the flower.
   c. It is sticky and distracts the pollinator.
   d. It encourages the pollinator to visit a different flower instead.

8. Succulents can be propagated by cutting off a leaf and setting it in soil. The cutting will develop roots and then can be planted. What type of reproduction does this describe? (LO 18-8)
   a. sexual reproduction
   b. asexual reproduction
   c. both sexual and asexual reproduction
   d. neither sexual nor asexual reproduction

9. A single squirrel can bury thousands of nuts (a type of seed) each season, creating food caches to survive the winter when food is scarce. They often, however, forget where these caches are located. Would it be

advantageous for a tree to produce nuts that squirrels like to eat? (LO 18-9)

a. yes, because nuts that the squirrel didn't find would be "planted" far from their parent tree
b. yes, because the nuts could then fertilize other nuts and make more diverse trees
c. no, because the nuts would be eaten, and the seeds inside cannot survive digestion
d. no, because the nuts would not have access to sunlight, which all plants need to germinate

10. Which of the following is an example of an agricultural application of sexual reproduction in plants? (LO 18-10)

a. grafting a high-producing grapevine onto a disease-resistant grapevine
b. taking a cutting of an herb and placing it in water until it grows roots
c. growing a new green onion from the bulb and roots that have been cut off
d. harvesting the seeds from a tomato to grow a new tomato plant

11. Most products from trees are harvested from which of the following types of tree? (LO 18-11)

a. healthy, mature trees that are many years old
b. trees that are less than a year old
c. trees that have been salvaged from forest fires
d. trees that have died from disease

12. Suppose you planted a crop of spinach and were wondering if it was a C3 or C4 plant. Which of the following observations might lead you to believe it was a C3 plant? (LO 18-13)

a. Its leaves are pale and low in chlorophyll.
b. Its stomata do not close in the night.
c. Its growth slows noticeably when the weather is hot and dry.
d. Its nutrition content is low in rubisco.

13. Which of the following plant types would be most susceptible to reduced yield caused by climate change? (LO 18-13)

a. C3 plants
b. C4 plants
c. CAM plants
d. both C4 and CAM plants

14. What process would make a C3 plant better able to thrive in a hot, dry climate? (LO 18-13)

a. moving it from ground level to roof top gardens
b. genetic modification to be more like C4 plants
c. genetic modification to remove its stomata
d. selective breeding to increase its glucose levels

15. Which of the following is NOT an expected benefit of adding a green roof to a household? (LO 18-14)

a. decreased cooling costs
b. increased carbon dioxide
c. a calming effect
d. a potential source of food

Critical Thinking

1. While plants can't move to escape damage, like being eaten by an insect, they can do things to protect themselves. When bitten by an insect, there is a stress response that triggers changes throughout the entire plant. These changes include shifts in turgor pressure, tissue deformation, and stomata closure.

   a. Explain why plants might wilt or close their stomata in response to being bitten by an insect. (LO 18-2)
   b. Scientists at Cornell University are studying the stress response in plants by fluorescently tagging stress response genes. Using this genetic modification, plants could signal that they were in stress within seconds of the stress occurring, and the source of the stress could be removed. How is this similar to other types of GM plants? How is it different? Do you think this is a better or more ethical use of the technology? Why or why not? (LO 18-10)

2. Venus fly traps are carnivorous plants that supplement their energy from photosynthesis by trapping and digesting insects. They reproduce both sexually and asexually. Make a list of structures from this chapter and predict whether Venus fly traps would or would not have each structure. Explain your reasoning. (LO 18-1, 18-2)

3. When freshly cut celery sticks are placed in water with food coloring, the water is drawn up into the stalk and leaves. The plant does not swell or change size, but the water level goes down and the dye stays in the celery. Describe the processes and structures that are involved in this phenomenon. (LO 18-4)

4. Trees are considered carbon sinks because they take carbon from the atmosphere and use it to make body structures. Consider these two statements. (LO 18-11)

   Student 1: "Young trees are growing and therefore more actively metabolizing, so they will take in more carbon to make more structures for their rapidly dividing cells."

   Student 2: "Old trees are larger, so they have more cells, whose structures are made up of carbon, and more leaves performing photosynthesis, so they fix more carbon."

   Which student do you agree with and why? Based on your answer, decide whether you would invest more money in planting new trees or preserving old forests.

Unit 6 Review

Public health impacts every aspect of our lives. In this unit, we connected themes discussed over the course of the text to bring an understanding of how anatomy and physiology relate to public health. We also explored the importance of plants to our health, and how the structure and function of plants influences that relationship.

In Chapter 17, we learned about the overall structure of the human body and looked at a subset of systems in the body while specifically connecting physiology to public health. The topics covered ranged from mental health to vaccination. Throughout the chapter, we connected foundational concepts such as the structure and function of cells and cellular behavior to how organ systems function, and the public health–related repercussions associated with a breakdown in the organ system.

In Chapter 18, we explored the structure and function of plants, as well as our relationship with plants. We learned about how flowering plants, which are critically important to our food supply, reproduce. We examined the importance of plants and plant-derived products, and their importance as a source of food and materials for creating goods, as well as their utility in fighting climate change. Throughout Chapter 18, we related foundational concepts—such as the principles of reproduction, evolution, and osmosis—to plant form and function.

One of the central messages of this book is that biology is a relevant part of your life—and in this unit, we went full circle with applying what could seemingly be abstract biological concepts to your personal health and the health of society. Where will you experience biology next?

Appendix A

Arrow Styles

While working on *Biology for Life*, I worked closely with the development team to implement a consistent arrow style, so that every time you see an arrow, you know what it means. Here is a guide to the arrow styles used throughout the text, what they signify, and examples in which a specific arrow type might be used.

Type	Example	Description and use
Expansion arrow		An expansion arrow is a gray arrow with a gradient from lighter at the base to darker at the arrowhead, and it is used to indicate expanding to a more close-up view of something.
Cycle arrow (cellular)	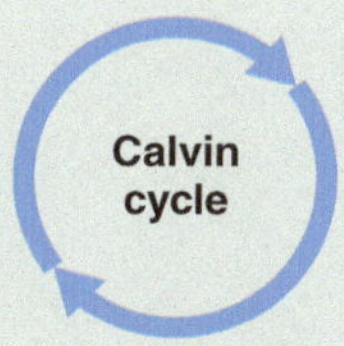	Cycle arrows appear in a circular pattern, with one arrow following another, and can appear in any color depending on the process. These are used to represent the steps that occur as part of any cyclic chemical process, such as the Calvin cycle.
Cycle arrow (in the environment)		Orange arrows—the same size and thickness as cycle arrows, but not in a strictly circular pattern—are used to represent the cycling of a nutrient within an environment, such as carbon within the carbon cycle.
Process step arrow		A process step arrow is a straight dark gray arrow that appears between the steps of a process.
Time passing arrow		A dashed black arrow indicates the passage of time.
Movement arrow	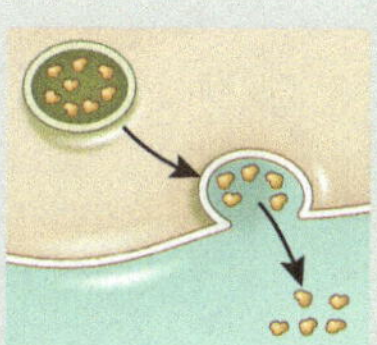	Different from the thin, straight step arrow, movement arrows are thin and curved, and are used to signify the movement of an object from one location to another.
Chemical conversion arrow	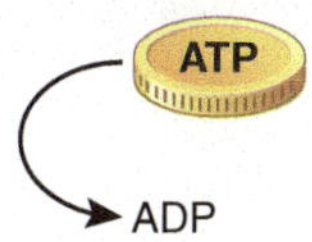	A direct chemical conversion is shown by a curved, thin black arrow, with the starting compound at the base of the arrow and the ending compound at the arrowhead. Here, ATP is shown being converted to ADP.
Energy arrow	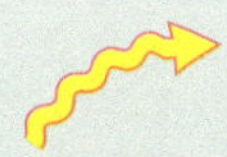	A wavy gold arrow with an orange outline and a pointed base represents energy loss or gain during a chemical reaction.
Light energy arrow		Wavy yellow arrows with an orange outline emanating from the sun signify the input of light energy into a process or system.
Heat energy arrow		A wavy red arrow represents the movement of heat energy.
Nutrient energy arrow		A wavy orange arrow represents the movement of nutrient energy in an ecosystem.
Crossing a membrane arrow	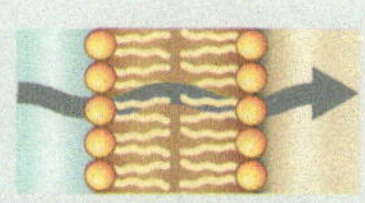	The passage of something across a membrane, such as a molecule, is shown by a wavy arrow passing through the membrane. These arrows can appear in various colors, to track the movement of different molecules as they pass through the same membrane.

Appendix B

Planning Your Study

In Chapter 1, we learned about the phases of the study cycle—preview, understand, and practice. In each chapter, a chance to preview, advice for building understanding, and a chance to practice were included. Some of the Understand It! features include details on how to manage a study session (Pomodoro Method, Chapter 6) and the importance of spaced practice (Chapter 15). Here, I'll give specific advice on making a study plan based on the study cycle, and how to assess your study plan so that you can adjust as needed. Please know that it's normal and okay to make mistakes—adjusting one's plan is an important part of the process.

First, a piece of advice I learned the hard way: If you don't manage your time, it will manage you. Said otherwise, without a good study plan in place, you are likely to find yourself stressed out and scrambling at the end of the semester…and ultimately, not getting the grades you want. So, it's important to have a plan in place. Here, I'll describe how to do this by hand, but some generative AI technologies can also help you to devise a study plan. Note that in the following figures, the colors correspond to the phases of the study cycle as introduced in Chapter 1.

When you're considering your study plan, think of each task or responsibility you have as a ball that you are juggling. Some balls are glass—you can't drop them, or they'll break. These are the non-negotiables you'll consider in Step 1. Other balls are rubber—you can drop these, and they'll bounce back. And some balls are made of paper. It doesn't hurt to either let them fall or set them down for a while—you'll have a chance to pick them up again later. When going through this exercise, consider what responsibilities are akin to a glass, rubber, or paper ball.

Step 1. Draw out what a typical week would look like. What are the non-negotiable time commitments, like when you have class or are scheduled to work?

	Monday	Tuesday	Wednesday	Thursday	Friday	Saturday	Sunday
9:00		English		English			
10:00	Work		Work		Work		
11:00							
12:00		History		History			
1:00							
2:00	Bio		Bio		Bio		
3:00							
4:00	Spanish		Spanish		Spanish		
5:00							
6:00							
7:00		Band rehearsal		Band rehearsal			
8:00							

Step 2. Block off time for self-care. This is an important part of wellness (Chapter 16), and if you don't make time for it, it's usually the first thing to get cut. For example, when will you pursue hobbies or exercise each week?

	Monday	Tuesday	Wednesday	Thursday	Friday	Saturday	Sunday
9:00		English		English		Long run	
10:00	Work		Work		Work		Sleep in!
11:00							
12:00		History		History			
1:00							
2:00	Bio		Bio		Bio		
3:00							
4:00	Spanish		Spanish		Spanish		
5:00							
6:00							
7:00	Short run	Band rehearsal	Short run	Band rehearsal			
8:00							

Step 3. Think about when you'll study. Add to your calendar when you'll preview for class and what you plan to do. This can include answering the Preview It! questions at the beginning of the chapter that will be covered in lecture that week, or reading through the Chapter Review or your instructor's PowerPoint slides and taking notes. Decide as well if you want to read the chapter before or after class—this is a matter of personal preference.

	Monday	Tuesday	Wednesday	Thursday	Friday	Saturday	Sunday
9:00		English		English		Long run	
10:00	Work		Work		Work		Sleep in!
11:00							
12:00		History		History			
1:00							
2:00	Bio		Bio		Bio		
3:00							
4:00	Spanish	Go over slides for bio	Spanish	Go over slides for bio	Spanish		
5:00							
6:00							
7:00	Short run	Band rehearsal	Short run	Band rehearsal			Review Preview It! questions and skim text. Go over slides for Monday
8:00							

Step 4. Write down when you will work on the understanding piece of the study cycle. I recommend students do this within 24 to 48 hours of class, before forgetting too much of the information covered in class. This phase will vary depending on the class, but may involve using strategies like elaboration (Chapter 3), dual coding (Chapter 5), or self-explanation (Chapter 7) to move the information into your long-term memory. This may be the point at which you'd want to read the text while cross-referencing your notes from class to see if you've missed anything. You may or may not want to meet with a tutor or your professor at this stage, in the event there is content that you don't understand.

	Monday	Tuesday	Wednesday	Thursday	Friday	Saturday	Sunday
9:00		English		English		Long run	
10:00	Work		Work		Work		Sleep in!
11:00							
12:00		History		History			
1:00							
2:00	Bio		Bio		Bio		
3:00		Review yesterday's bio notes		Review yesterday's bio notes		Review yesterday's bio notes	
4:00	Spanish	Go over slides for bio	Spanish	Go over slides for bio	Spanish		
5:00							
6:00							
7:00	Short run	Band rehearsal	Short run	Band rehearsal			Review Preview It! questions and skim text. Go over slides for Monday
8:00							

Step 5. Determine when you plan to do homework and/or meet with a study group. For the practicing phase, I suggest having dedicated time each week to work on practice. In the spirit of spaced practice, you may do your homework on your own but then meet with a study group to do additional practice problems and/or to quiz one another. Remember, if you are trying to work through problems and must continually go back to look things up, that means that you haven't yet successfully completed the understanding phase of the study cycle and the information isn't available for you to retrieve. If this is the case, you'll want to use strategies like self-explanation, dual coding, and concrete ideas to encode the information in your memory.

	Monday	Tuesday	Wednesday	Thursday	Friday	Saturday	Sunday
9:00		English		English		Long run	
10:00	Work		Work		Work		Sleep in!
11:00							
12:00		History		History			
1:00							
2:00	Bio		Bio		Bio		
3:00		Review yesterday's bio notes		Review yesterday's bio notes		Review yesterday's bio notes	
4:00	Spanish	Go over slides for bio	Spanish	Go over slides for bio	Spanish	Bio study group	
5:00		Bio Homework		Bio Homework			
6:00							
7:00	Short run	Band rehearsal	Short run	Band rehearsal			Review Preview It! questions and skim text. Go over slides for Monday
8:00							

Step 6. Assess your study plan and adjust, if necessary. After you build a study plan, how do you know if it's working? Assessments like quizzes and exams are a great way to determine how well your study plan is working. After every assessment, take a moment to reflect on your study plan (Chapter 8) to see if you need to change it in some way, such as to improve your notetaking. The Study Plan Assessment is an activity that I do with my students to help them assess their study plan after an exam and identify necessary changes before the next exam.

Study Plan Assessment

After you've completed an exam, assess each question and put a check mark next to the reason why you missed it.

- I didn't have this information in my notes.
- I didn't think this information was important.
- I misunderstood the concept.
- I studied this for recognition.
- I misread the question.
- I "blanked" on the question and remembered the material later.
- I changed the right answer to a wrong one.
- Other:______________________________

Now that you've gone through your exam or quiz, identify the top three reasons why you missed questions and read the targeted advice for each given below.

1. ____________________
2. ____________________
3. ____________________

I didn't have this information in my notes. *or* I didn't think the information was important. *or* I misunderstood this concept

These categories go together and indicate a problem at either the Preview or Understanding phase. Throughout this text, key information is called out with learning objectives at the beginning of each section and the bolding of key terms in the sections, where concepts are then summarized in section summaries and the chapter review. I also suggest looking over your professor's materials for their own learning objectives and the concepts that they emphasize. Going back through your notes with the text after class can also help you to identify if there were any possible gaps in your notes or if you had some misunderstanding. Trading notes with a study partner can also help you to identify knowledge gaps.

I studied for recognition.

This is a very common reason for getting questions wrong and is often a sign of when one is trying to memorize information rather than trying to learn it. One sign that you might be doing this is if you can visualize the information from your notes during the quiz or exam, but don't know how to answer a question. This can indicate problems at the Understanding and Practicing phases of the study cycle. Especially in cases where you're assessed on your ability to apply class material, you'll want to make sure that you have a lot of practice working with the material ahead of an exam. For example, you can draw concept maps, write your own test questions and trade with a partner, or quiz each other in study groups.

I misread the question.

This is likely due more to test-taking skills than your study habits—which you can work on during the practice phase. Some test-taking habits to consider include using a different colored pen or highlighter to call out important words. If you are taking a virtual test, write down on scratch paper what the question is asking you to do. You can also notate at this point if there's any information that you are missing and might need to calculate. If you are answering multiple-choice answers, it can help to cross out answers that you know are wrong.

I "blanked" on the question and remembered the material later.

Blanking on questions during an exam is a hallmark of test anxiety (Chapter 10), and therefore, it may be helpful to seek help from the academic resource center or a counselor. It may also be helpful to take practice exams, especially if you're able to do so in a test-like environment. For example, it can be helpful to take a practice exam in the same seat and classroom where your exam will be. Consider also simple things that you can do during an exam to help you stay present, such as to sip cold water or suck on a mint or sour candy. I once had a student who wore a suit to all exams, because it helped get him in the right mindset to succeed!

I changed the right answer to a wrong one.

Changing the right answer to a wrong one indicates a lack of confidence with the material, which could be the result of problems in any of the phases of the study cycle. One of the best things you can do here is to ensure that you're spending adequate time at each phase of the study cycle, especially during the Understanding and Practicing phases. The more you work with the material and draw connections between concepts and/or do practice problems, the more confidence you will build with the material. Teaching concepts to a study partner during a study group session can help to build your confidence. Teaching a friend can also help you in identifying any misunderstandings that you might have with the material.

Other

This means there are likely multiple things going on, or something entirely unrelated. See above for advice.

Appendix C

Periodic Table

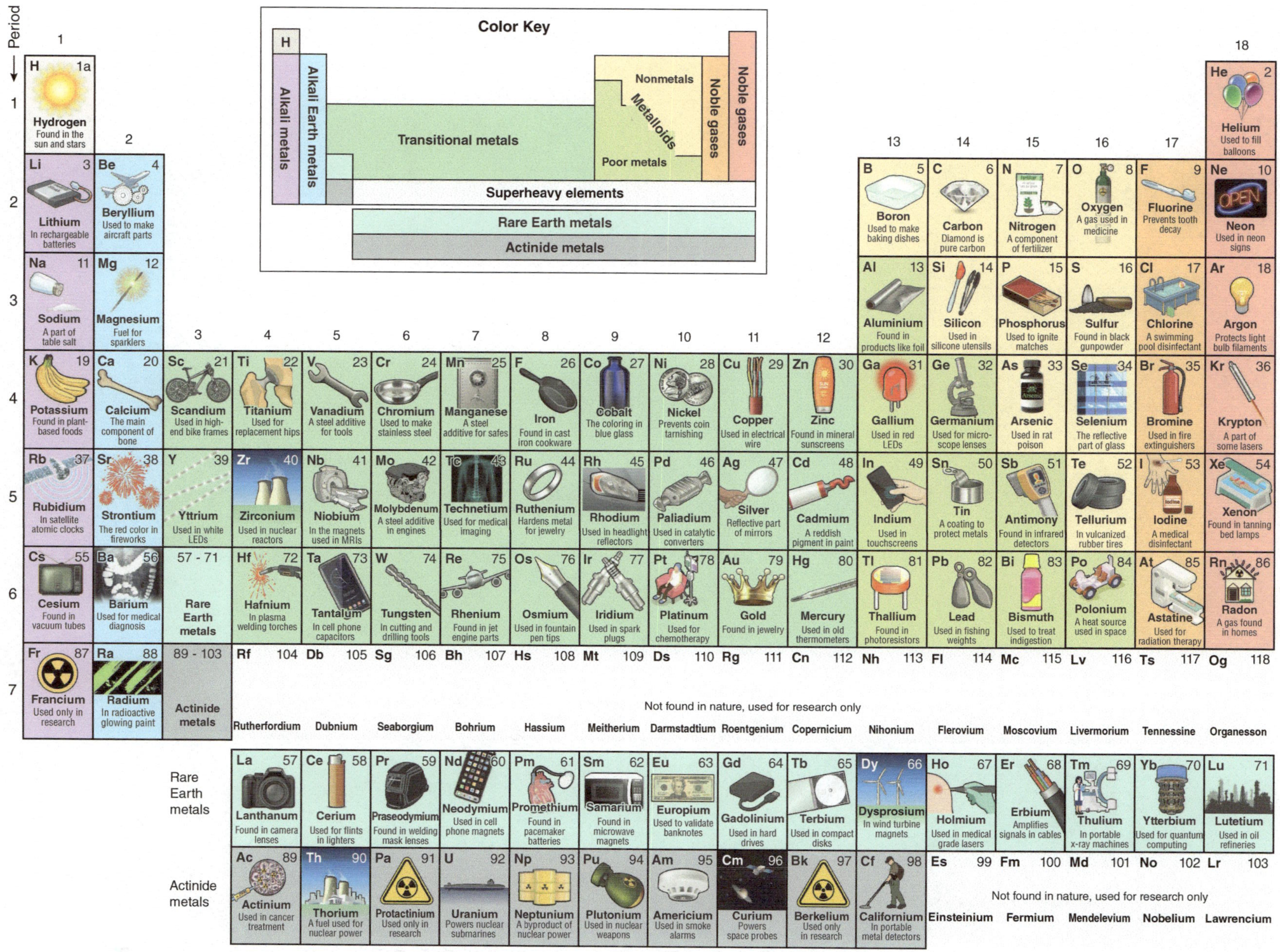

Appendix D

Graph Interpretation

Throughout the text, there are a variety of Digging into Data exercises and graphical representations of data. Visual representations of data help to display large amounts of information in an understandable way, and can help to reveal relationships between *variables*—quantities that can be manipulated or measured in an experiment—or trends in the data. Practicing interpreting graphs and charts is an important skill for you to acquire, not only for this class, but for making informed decisions based on data as part of your life.

When interpreting a graph, there are a few principles you can use to ensure that you get the best understanding of the data possible. First, read the graph's title (in this text, titles are in bold and part of the figure legend) and examine its labeling before looking at the specific data. For example, if a table has a key, view the key to make sure that you know what orange versus blue means. This will help to ensure that you're understanding things correctly, to prevent any room for misinterpretation.

Then, note that graphs have an *x*-axis, which is the horizontal axis and represents the *independent variable* (**Figure D.1**); the independent variable is not changed by the other variables in the experiment. The *y*-axis is the vertical part of a graph and shows the *dependent variable*, which can be changed by other variables. The dependent variable is the variable measured in the experiment; in the case of Figure D.1, this would be pH levels. The dependent variable *depends* on the independent variable—time, in this case, where the pH changes as a factor of time passing.

Figure D.1 **A sample graph**

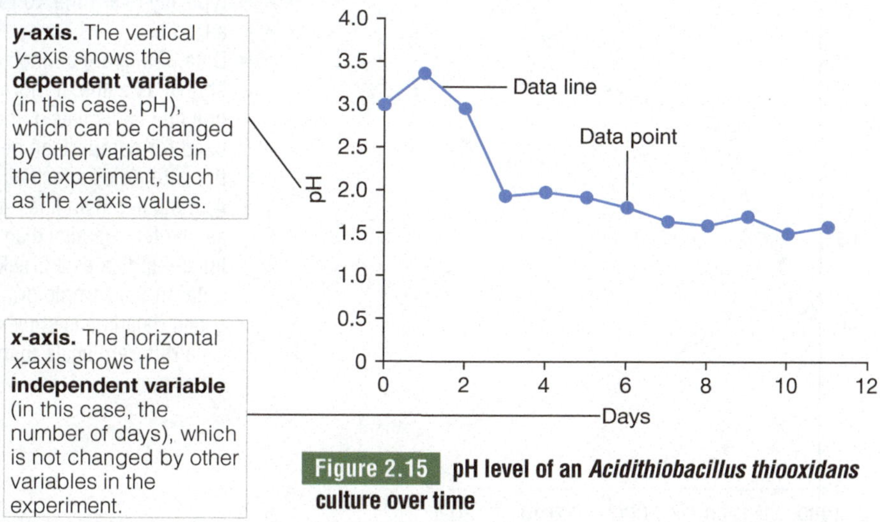

Figure 2.15 **pH level of an *Acidithiobacillus thiooxidans* culture over time**

Here is a line graph that is showing change over time. The data line shows a downward trend and then a leveling off of pH levels over time. Note that data points are shown on this graph; in other examples, such as those shown in Table D.1, the values are there but are instead shown as a smooth line.

There are some common ways that you'll see biological data visually displayed, with some examples from the text provided in **Table D.1**. The table indicates when each type of graph is typically used, along with an explanation of how to interpret it. You may want to refer back to this table as you work through the data provided in the text to be sure that you're interpreting the graph correctly.

Table D.1 Common Types of Graphs

<table>
<tr><th>Graph Type</th><th>Sample Graph</th><th>Details</th></tr>
<tr><td>Bar graph</td><td>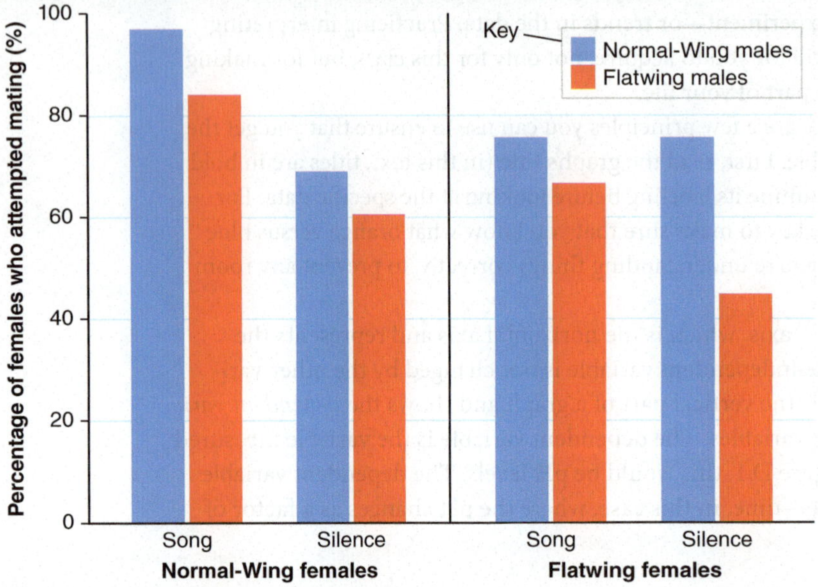
</td><td>• A bar graph is used to compare quantities between different categories.
• Numerical values are displayed as bars, where the height of the bar correlates to the value on the y-axis.
• Different colors are used for different categories as specified in a key and/or on the x-axis.</td></tr>
<tr><td>Line graph</td><td>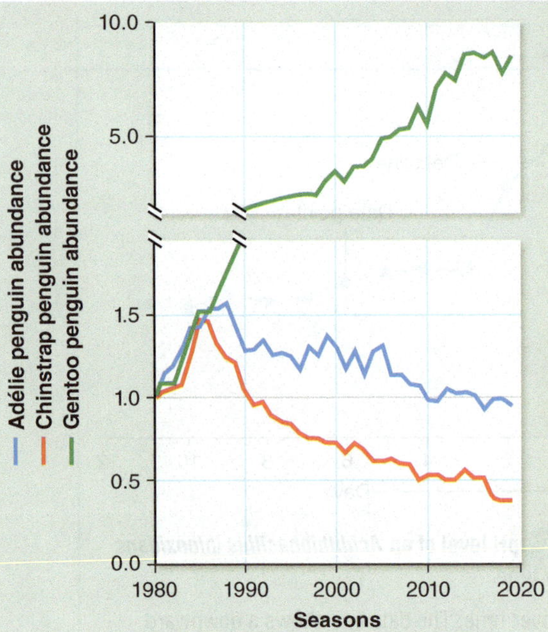
</td><td>• A line graph is used to show how a variable changes, typically over time, to reveal a trend.
• Data points are connected and shown as a line; if more than one line is displayed, it can be used for comparison between the categories specified.
• A break in a data line—shown as slanted, parallel diagonal lines—indicates a break in the data. In the sample graph, the green data line is shown to be on a different scale than the other two.</td></tr>
</table>

<table>
<tr><th>Graph Type</th><th>Sample Image</th><th>Details</th></tr>
<tr><td>Pie chart</td><td>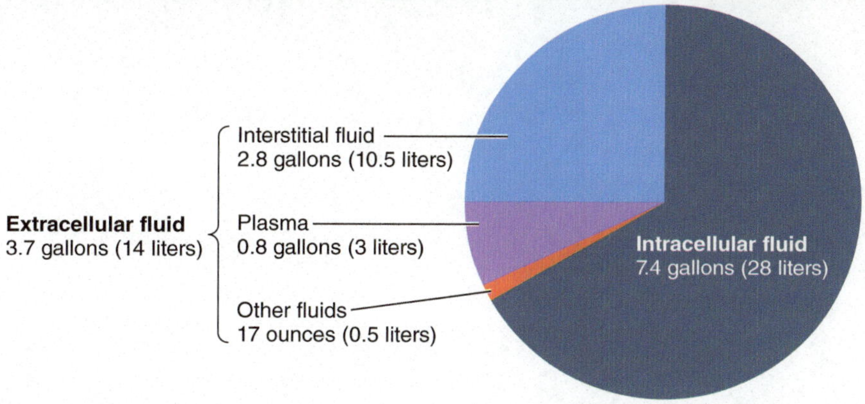
</td><td>• A pie chart shows data where each slice represents a portion of some whole.</td></tr>
<tr><td>Histogram</td><td>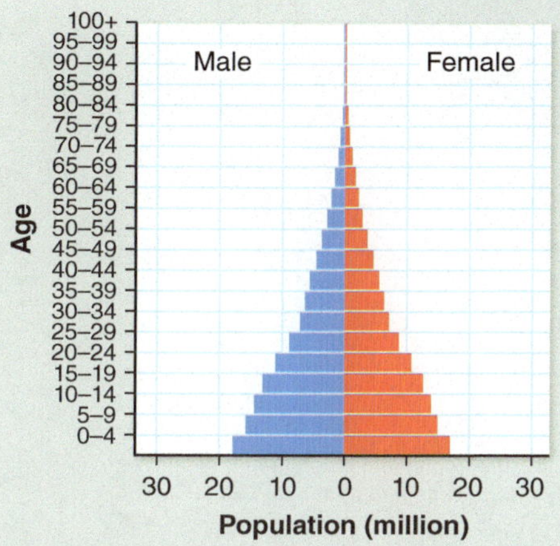
</td><td>• A histogram is a type of bar graph that is used to display the distribution of data.
• Each bar (called a bin) represents a range of values, with the height of the bar indicating how many data points fall into that range. For example, in the sample graph, approximately 18 million males fall into the category for those aged 0 to 4.</td></tr>
</table>

Appendix E

Theology and Science

The process of science is a major theme of this text—one that comes up in all chapters. One important aspect of science is how it intertwines with our daily lives. Sometimes this can lead to perceived conflicts. One area of relevance to the study of biology revolves around evolution. Evolution is a foundational concept to biology, which is why it's one of the big ideas identified in Chapter 1. Perhaps it is difficult to see, at first, how evolution is relevant to life outside the classroom. However, understanding the evolution of antibiotic-resistant bacteria, why tax dollars go to funding animal models to learn about human diseases, and, in general, knowing what is or is not scientific information are all important ways that evolution pops up in our daily lives.

Research in the United States suggests that among Americans, the main factor that predicts whether someone accepts evolution is if they identify as a religious person[1,2,3]—and, particularly, if people *perceive* a conflict between their religion and evolution. For example, a 2020 survey of undergraduate students indicated that more than half of students surveyed thought that one needed to be an atheist to accept the science of evolution[4]. When instructors remain mute about a potential conflict to remain neutral, it can feel to students like the instructor is not supportive of their religious beliefs, further flaming the fires of a conflict[5,6].

As your textbook author and someone who has spent decades teaching biology—in both academic and religious settings—I want to acknowledge that your religious beliefs (or lack thereof!) are important and valid. I also care that you understand evolution, since it's a critically important concept to understanding life and biology. I want to emphasize that there is room for both religion and acceptance of evolution (and other scientific principles)—and that the widespread perception of conflict may be due more to our cultural tendency to amplify louder voices on the extremes than the actual lived experiences of a majority of people. Research by the Pew Foundation[7] found that the greatest perception of conflict was not in highly religious people, but rather among people who are less religious.

While it's up to you to make sense of this in your own life, what makes sense to me is that we as humans are curious beings and want to understand the world around us. There are different ways of making sense of the world around us—and one can use different tools to do so. Science is a useful process for telling us how the world works. Evolution, for example, explains how it is possible for so many different lifeforms to exist. However, evolution can't tell us about any deeper meaning behind why it happened. Science in general, let alone evolution, can't tell us if there is any supernatural being such as a God or other deities. Some people of faith may tell

[1] Glaze, A. L., & Goldston, M. J. (2015). US science teaching and learning of evolution: A critical review of the literature 2000–2014. Science Education, 99(3), 500–518.

[2] Dunk, R. D., Petto, A. J., Wiles, J. R., & Campbell, B. C. (2017). A multifactorial analysis of acceptance of evolution. Evolution: Education and Outreach, 10(1), 4.

[3] Jensen, J. L., Manwaring, K. F., Gill, R. A., Sudweeks, R. S., Davies, R. S., Olsen, J. A., ... & Bybee, S. M. (2019). Religious affiliation and religiosity and their impact on scientific beliefs in the United States. BioScience, 69(4), 292–304.

[4] Barnes, M. E., Dunlop, H. M., Sinatra, G. M., Hendrix, T. M., Zheng, Y., & Brownell, S. E. (2020). "Accepting evolution means you can't believe in god": Atheistic perceptions of evolution among college biology students. CBE—Life Sciences Education, 19(2), ar21.

[5] Barnes, M. E., & Brownell, S. E. (2016). Practices and perspectives of college instructors on addressing religious beliefs when teaching evolution. CBE—Life Sciences Education, 15(2), ar18.

[6] Stahi-Hitin, R., & Yarden, A. (2022). Scientists' and teachers' attitudes toward relating to religion when teaching evolution. Evolution: Education and Outreach, 15(1), 18.

[7] Smith, Gregory, Alan Cooperman, Becka Alper, Besheer Mohamed, Chip Rotolo, Patricia Tevington, Justin Nortey, Asta Kallo, Jeff Diamant and Dalia Fahmy. (2025). "Decline of Christianity in the U.S. Has Slowed, May Have Leveled Off." Pew Research Center. doi: 10.58094/4kqq-3112.

you that you don't need scientific evidence to believe, because the basis of faith is having confidence without requiring proof.

I also want to point out that theology is a scholarly discipline in its own right—and not only has biology been taken out of context and warped to perpetuate the idea of a conflict, but theological scholarship has as well. Take the work of John Walton, an Old Testament scholar. In his book, *The Lost World of Genesis One*, he points out that people get so hung up on what it means to "create" that they forget that the people who originally wrote religious texts existed in a different culture...and used words differently. It's the same phenomenon we see today where different groups of people use different words to describe the same thing: While my grandparents had a davenport in their living room, I have a couch. It's the same kind of furniture, just described differently. Today, we understand creating as making something new. So when we consider the book of Genesis, Walton's argument is that when "God [created] humankind in His own image," in modern society, we would interpret that to mean a supernatural being made humans. However, as Walton points out, the original author could have been using the word "create" in the sense of instilling purpose. So, for instance, we can create a desk out of a table by putting a computer on it and using it for the purposes of working. So Walton's argument is that when God "created," they assigned a purpose to humanity, rather than creating humans from nothing—which would not suggest a conflict with evolution. Note that I've highlighted only one of many examples of theological scholarship that suggests that evolution need not be in conflict with religious beliefs—if you're interested, I encourage you to research more on your own.

There are many different religious and non-religious perspectives to consider, even within any single tradition. Within these perspectives and traditions, there are many resources addressing views of religion and evolution from a variety of viewpoints, and that literature is always growing. If you are interested in learning more about this from your own religious tradition, here are a few places you might start:

- The Pew Research Center report called Religious Group's Views on Evolution[8]
- The American Association for the Advancement of Science (AAAS) has a program dedicated to fostering engagement among scientific and religious communities on all kinds of science topics. Visit the Dialogue on Science, Ethics, and Religion's resource website (scienceandreligiondialouge.org) for more information.
- There are many organizations that address scientific issues from a variety of particular religious perspectives, including BioLogos (evangelical Christian—biologos.org), the Society of Catholic Scientists (Catholic—catholicscientists.org), and Sinai and Synapses (Jewish—sinaiandsynapses.org), as well as many others.
- Many religious denominations also have their own doctrinal or policy statements about science and evolution. You can look for "[religious tradition] statement on evolution" in the search engine of your choice.
- Consider talking to religious leaders or others in your community. There may be student groups on your campus where you can connect with others. It's okay to be nervous to ask about these issues because of the perception of conflict, but as we saw above, most people's personal experience is one of compatibility. Even if you don't agree, having conversations about these issues helps to encourage dialogue and break down barriers while also opening new avenues for inquiry.

In closing, I hope this appendix helped you realize that there are many people, including religious leaders and scientists, who do *not* believe there is any conflict between the religious beliefs and evolution. While these resources are by no means an exhaustive list, I hope that they will be a starting point for you as you delve into these ideas on your own.

---

[8] Religious Group's Views on Evolution. Pew Research Center, Washington, D.C. (2014). https://www.pewresearch.org/religion/2009/02/04/religious-groups-views-on-evolution/).

Glossary

A

abiotic Non-living components of an ecosystem.
abscisic acid (ABA) A plant hormone that helps plants respond to stress and helps trigger fruit ripening.
acid A substance that releases hydrogen ions into water.
acid rain Rain that forms when sulfur dioxide and nitrogen oxides react with water and oxygen to form sulfuric and nitric acids that can damage human-made structures or ecosystems.
acidophile An organism that can live in highly acidic conditions.
actin The protein that makes up microfilaments and the contractile filaments in muscle.
action potential An electrical signal that moves in a wave-like fashion down the axon of a neuron.
activation energy Minimum amount of energy required to start a chemical reaction.
active site Location on an enzyme where substrates bind and a reaction occurs.
active transport Energy-requiring mechanism in which a transport protein pumps a solute or molecule across a cell membrane against the concentration gradient.
adaptation A biological feature that arises due to natural selection.
adaptive immunity A set of defenses tailored to fight specific pathogens that an individual encounters.
adaptive radiation When one species rapidly diversifies into multiple species.
adhesion proteins Membrane proteins that help cells stick together in animal tissues.
adhesion The ability of water to stick to other surfaces due to the attraction between water molecules and another substance.
adipose tissue A type of connective tissue that stores fat.
ADP (adenosine diphosphate) Nucleotide containing one less phosphate group than ATP; used for transferring and providing cells with energy.
aerobic exercise An activity that increases heart rate and respiration for a long duration; improves cardiovascular fitness and muscle fatigue resistance.
aerobic respiration When cells break down molecules like glucose for energy in the presence of oxygen.
age structure The distribution of individuals among various age groups in a population.
alcoholic fermentation When cells break down molecules like glucose for energy in the absence of oxygen, producing ATP, ethanol, and CO_2.
algae Lineages of photosynthetic protists that live in lakes and seas.
allele frequency The proportion of one allele among all alleles of that gene present in a population.
alleles Different forms of the same gene.
allopatric speciation Speciation that occurs when a physical barrier separates populations and interrupts gene flow.
alternative splicing When different portions of the same gene are cut out to produce different proteins.
amino acid A monomer of protein with a central carbon attached to an amino group, a carboxylic acid group, a hydrogen atom, and an R group.
amniote A type of animal whose embryo develops within an amnion and chorion.
amniotic egg A self-contained egg that provides a protective environment for developing offspring.
amphibians Scaleless vertebrates that generally spend time on land but require water to breed.
amphiphilic A molecule that has both hydrophilic and hydrophobic characteristics.
analogous structures Body parts that are comparable in function and appearance but that evolved independently in different species due to similar environmental pressures.
anaphase The stage of mitosis during which sister chromatids detach from one another and move toward opposite ends of the cell.
anatomy The study of the structure of the body.
ancestral trait A characteristic trait inherited from a common ancestor.
aneuploidy When there are too many or too few copies of a particular chromosome, which can lead to miscarriage.
angiosperms Flowering plants that produce seeds enclosed within fruits.
animal A multicellular consumer that takes food into its body, digests it, and then absorbs the released nutrients into its body.
annelids Bilateral worms with a coelom and clear segmentation.
antagonistic interactions An interaction where an individual from one species benefits by extracting resources from another, which is harmed as a result.
antagonistic muscle pair A pair of muscles that work in opposite directions, where one contracts while the other relaxes.
antennae Sensory structures on the head that detect touch, odor, and vibrations.
anther The part of a stamen that makes pollen.
anthropoids A primate group that includes monkeys, apes, and humans.
antibody A y-shaped protein produced by B cells that can specifically recognize and bind to antigens.
anticodon a triplet of nucleotides on a tRNA that can base-pair with a particular codon on mRNA.
antigen processing The process in which an immune cell digests a foreign invader and displays the antigen on its surface.
antigen A molecule that evokes an immune response.
antiparallel Occurs due to strands of DNA that are hydrogen-bonded together running in opposite directions (one $5' \rightarrow 3'$ and the other $3' \rightarrow 5'$).
apes A group of primates including humans, gorillas, and chimpanzees, that are intelligent and do not have tails.
apical meristems Growing tips of roots and shoots that cause lengthening.
apoptosis The self-destruction of a cell or cells through programmed cell death.
aquatic biomes Water-based environments.
aquifers Rock formations that hold pockets of groundwater.
arachnids Chelicerates with four pairs of walking legs, a pair of palps, and no antennae.
Archaea One of the three domains of life. Archaea are prokaryotic organisms that are more closely related to eukaryotes than to bacteria.
arthropods Invertebrates with jointed legs.
artificial selection Evolution that occurs due to humans selectively breeding plants or animals for specific traits.
asexual reproduction The process by which offspring are produced by a single parent.
atom The smallest unit of matter.
atomic number Number of protons in an element, which uniquely defines that element.
ATP (adenosine triphosphate) A nucleotide used for energy storage and as the main energy source for cells.
ATP synthase Enzyme that makes ATP.
autonomic nervous system The part of the peripheral nervous system that controls smooth muscle, cardiac muscle, and glands; informs the central nervous system about internal conditions.
autophagy A mechanism by which a cell's lysosomes ingest cellular debris such as old or damaged organelles and recycle the components for reuse; "self-eating."

autosomal dominant pattern An inheritance pattern where a trait is associated with a dominant allele on an autosome.
autosomal recessive pattern An inheritance pattern where a trait is associated with a recessive allele on an autosome.
autosomes Chromosomes that make up the majority of an organism's chromosomal pairs, typically numbered 1 through 22 in humans.
autotroph An organism that makes its own food by taking non-nutritional inorganic substances such as CO_2 and converting them into nutritious organic molecules, such as sugars.
axon terminals The endings of an axon where chemical signals are released.
axon A cytoplasmic extension of a neuron that conveys the action potential along the neuron's length.

B

B-cell receptors Receptors on a B cell that are specific for a particular antigen.
B cells Lymphocytes that, when activated, produce antibodies.
backbone A bony structure that extends from the pelvis to the skull.
background extinction rate The normal rate at which species go extinct.
Bacteria One of the three domains of life. This bacterial lineage contains unicellular prokaryotes.
bacteriophage A type of virus that infects bacteria.
balanced translocation A condition that occurs when part of one chromosome completely breaks off and reattaches to another chromosome.
ball-and-stick model Three-dimensional representation of a molecule using colored balls for atoms and sticks to represent bonds.
Barr body A structure formed by the condensation of a deactivated X chromosome.
base A substance that picks up (accepts) hydrogen ions in water.
basement membrane A type of extracellular matrix consisting of a sheet of fibrous material that structurally supports and organizes tissues.
basophils A type of granulocyte that releases chemical signals, such as histamine, that cause inflammation.
behavioral isolation Differences in behavior that prevent interbreeding, such as a failure to recognize mating cues.
benign An abnormal growth of cells that is slow growing and unlikely to spread.
bias A predisposition to a belief, often in a way considered to be unfair or distorted, that results in systematic errors in the collection or interpretation of scientific data or information.
bilateral symmetry Having a right and left half, with body parts repeated on either side of the body.
bioaccumulation When a toxin, such as mercury or microplastic, becomes increasingly concentrated as it moves up a food chain.
biodiversity hot spots Ares with high biodiversity, such as rain forests and coral reefs, that are home to an abundance of species and often prioritized for conservation.
biodiversity The variety of life on Earth, including living organisms, their genetics, and ecosystems.
biogeochemical cycle The movement of essential elements—and molecules, like water—from one or more environmental (abiotic) reservoirs, through the living (biotic) components of an ecosystem, and back again.
biogeography The study of how and why different species of animals and plants exist in particular areas of the world, influenced by factors such as continental drift.
bioinformatics A discipline concerned with the management, analysis, and understanding of large genomic data sets.
biology the scientific study of life, including what it means to be alive, where life came from, how life forms are organized, and how living things interact with each other and their environment.
biome A large-scale area with the same physical conditions, supporting distinctive biological communities.
bioprospecting The search for natural resources such as genes, chemicals, and macromolecules.
biosphere The total area in which life is found, including all ecosystems on Earth.
biotic Living components of an ecosystem.
bipedalism Walking upright on two legs.
birds An endothermic reptile with feathers; can typically fly.
bivalve A type of mollusk with a hinged, two-part shell.
blastocyst An early embryonic structure shaped like a ball that develops 5–6 days after fertilization.
bone A dynamic, living organ that provides structural support and produces blood cells.
bony fishes Fishes with a skeleton consisting mainly of bone.
bony structures Limbs with internal bony supports.
boreal forests Forests that experience cool, dry conditions year-round.
braincase The part of the skull that encloses the brain.
brood parasites When one organism tricks an animal of a different species into raising its young, thereby deplete its host's resources by stealing parental care.
buffer Set of chemicals that can keep the pH of a solution stable by reacting with substances that affect pH.

C

C3 plants Plants that use the Calvin cycle directly, forming a 3-carbon compound; most common type of plants.
C4 plants Plants that first fix carbon into a 4-carbon compound to minimize photorespiration.
Calvin cycle Second stage of photosynthesis in which the energy in ATP and NADPH is converted and ultimately stored as glucose.
CAM plants Plants that open stomata at night to conserve water and use stored CO_2 for photosynthesis during the day.
camouflage A body shape, color pattern, or behavior that allows an individual to blend in with its surroundings and avoid detection.
carbohydrate Macromolecule that consists of carbon, hydrogen, and oxygen in a ratio of approximately 1:2:1.
carbon cycle Cycle in which carbon moves through the atmosphere, rocks, living organisms, and the oceans.
carbon dioxide pollution A type of pollution that occurs when excessive amounts of carbon dioxide are released into the atmosphere, above and beyond what can be managed through the natural processes of the carbon cycle.
carbon fixation Taking carbon atoms from an inorganic source (such as CO_2) and incorporating them into an organic molecule.
carbon sink A natural reservoir for carbon where more carbon dioxide is removed from the atmosphere than is released.
carbon-14 (^{14}C) dating A method for determining the age of various carbon-containing materials. ^{14}C dating relies on the fact that carbon exists in nature as stable and unstable (radioactive) isotopes, and that the decay rate of the radioactive isotope of carbon is constant.
carcinogens Cancer-causing chemicals.
carpel The female reproductive organ of a flower.
carrying capacity The maximum number of individuals of a species that can be supported indefinitely by their environment.
CAR-T cell therapy A cancer treatment in which a person's own immune cells are edited to be cancer-fighting cells.
cartilaginous fish Jawed fish with a skeleton made of mineral-hardened cartilage.
Cas9 A bacterial enzyme that molecular biologists have adapted for gene editing applications in eukaryotes.
catalyze To speed up a chemical reaction.
cell body The central part of a neuron that contains the nucleus.
cell cycle checkpoints A regulatory mechanism that interrupts the progression of cell division to allow a cell to "decide" whether or not to continue with division.

cell division limit The maximum number of times cells can divide before they senesce.
cell signaling The process by which cells talk to themselves and to each other.
cell The smallest functional unit of life.
cell theory A foundation of modern biology. It has three parts: first, all living organisms are made up of one or more cells; second, the cell is the basic structural and functional unit of life; and finally, cells come from preexisting cells.
cellular respiration A metabolic pathway used to break down high-energy molecules like glucose and produce ATP.
cellulose A tough, insoluble polysaccharide that is the major structural material in plants.
centipedes Venomous predators with a flat body and one pair of legs per segment.
central dogma of molecular biology How genetic information flows in cells from DNA → RNA → protein.
central nervous system The division of the nervous system that includes the brain and spinal cord.
centromere A constricted region on a chromosome where sister chromatids attach to each other.
cephalopods Mollusks where the foot has been modified into tentacles and/or arms that extend from the head.
chaparral A shrub-dominated ecosystem with wet winters and dry summers.
chelicerates A lineage of arthropods that includes spiders, scorpions, mites, ticks, and horseshoe crabs.
chemical bond A strong attractive force connecting two atoms that arises from the loss, gain, or sharing of electrons.
chemical evolution A theory that describes the process of forming organic molecules from inorganic molecules.
chemoautotrophs An organism that can obtain energy from chemically processing inorganic materials.
chemosynthesis The process of making energy from inorganic materials while using a non-solar chemical energy source.
chemotherapy A cancer treatment that uses chemicals to kill or inhibit the growth of fast-growing cells, such as cancer cells.
chlorophyll *a* Main photosynthetic pigment in plants, algae, and photosynthetic microorganisms.
chloroplast Organelle of photosynthesis in the cells of plants and photosynthetic protists. Outer membranes of the chloroplast enclose stroma and a highly folded thylakoid membrane.
choanoflagellates A small group of aquatic protists that are heterotrophic and the closest living relatives to animals.
cholesterol The most common steroid in animal tissue; used to make other essential molecules and found in the cell membrane.
chordates Animals that at some point in their life have a notochord, a dorsal nerve cord, gill slits, and a tail.
chromatin Complex of DNA and protein that make up chromosomes.
chromatin remodeling A way of controlling gene expression by winding up or loosening chromatin to make DNA more or less accessible.
chromatograph The output or record produced during Sanger sequencing showing different base pairs.
chromosome Structure found inside the nucleus of a cell consisting of DNA with associated proteins.
chromosome number The number of chromosomes in a cell of a given species.
chronic wound An injury that fails to heal after several weeks, even with proper treatment.
cilia Short, hairlike motile structures that project from the membrane of some eukaryotic cells.
circular economy An economic model focused on eliminating waste by reusing old materials to regenerate new products in a continuous cycle.
citric acid cycle Cyclic pathway that dismantles acetyl-CoA to produce CO_2, NADH, $FADH_2$, and ATP. Also called the Krebs cycle.
clade A complete branch in a phylogenetic tree representing organisms with a common ancestor.
claims Data-informed statements that can be challenged through experimentation.
climate change Lasting changes to average temperatures and weather patterns.
climate The average weather conditions, such as temperature and precipitation, in a region over a long period of time.
climax community When a mature forest is present and a community enters the final stage of succession.
cloning vector A piece of DNA, typically a plasmid, that can be modified to contain foreign DNA for the purposes of genetic engineering.
clumped distribution A distribution in which members of a population are closer to one another than would be predicted by chance alone.
cnidarians Aquatic, radially symmetrical animals with stinging tentacles.
codominance When two alleles equally contribute to a phenotype.
codons Three-nucleotide units that code for either a specific amino acid or signal protein synthesis to stop.
coelom A fluid-filled body cavity that is completely lined with tissue derived from mesoderm.
coenzymes Organic molecules that are required for an enzyme to function.
coevolution When two species evolve together due to reciprocal interactions.
cofactors Inorganic ions that bind to enzymes to enable or enhance their biological activity.
cohesion The ability of water to stick to itself due to the attraction between individual water molecules.
cohort A group of individuals all born at about the same time.
collenchyma Flexible cells that give support to young, growing parts of a plant.
colonial organism Cells that live together and behave in an integrated fashion but remain self-sufficient.
colonial theory of animal origins States that animals evolved from a heterotrophic protist that formed colonies.
commensalism When one species benefits and the other is unaffected.
community science Scientific research and environmental monitoring conducted by nonprofessionals, such as local community members or students, who contribute data and observations to support conservation efforts.
community structure The types of species within a community, as well as their abundance and the interactions among them.
community All populations of different species in a region.
compact bone The dense, weight-bearing tissue that makes up the outer part of a bone.
companion cell Cells next to sieve tubes that help load sugars into them.
competent cells Cells that can take up new DNA via transformation; cells can be competent naturally or can be chemically treated to become competent in a biology lab.
competition When species challenge one another for access to limited resources.
competitive inhibitor A compound that binds to the active site of an enzyme and competes with the substrate for the active site.
complementary bases Specific pairs of nucleotides that match up and form hydrogen bonds connecting strands of DNA or RNA.
complete digestive tract A tubular gut with an opening at either end.
compost A type of rich organic material that can be added to gardens to create healthy soil that nourishes plants.
composting The process of breaking down organic material (e.g., food scraps) by decomposers such as bacteria and fungi as part of the natural carbon cycle.
compound Molecule that has atoms of more than one element.
concentration amount of solute per unit volume of a solution.
concentration gradient A gradual change in the concentration of a substance over a distance.
concrete ideas Part of a learning strategy in which abstract scientific concepts are illustrated with real-world examples.
cones Reproductive structures that contain either pollen or ovules.

confirmation bias The tendency to seek evidence that aligns with one's personal beliefs or values while denying, dismissing, or ignoring evidence that is contrary to it, even unconsciously.
conflict of interest In the context of scientific research, when a person's interests (financial or personal) or beliefs could compromise their ability to conduct an investigation and/or the objective interpretation of data.
conifers Woody trees or shrubs with needlelike or scalelike leaves.
connective tissue Tissue that holds body parts together and provides structural support.
conservation biology A field that aims to determine the best ways to preserve biodiversity through protecting species, habitats, and ecosystems.
conservation genomics An emerging field dedicated to studying the genes of species and how genetics can help conservation efforts.
conserved Refers to transcription factors that remain relatively unchanged in evolutionary history, retaining similar sequences and functions over a very long period of time.
consumers Obtain energy and other nutrients by feeding on tissues, wastes, and remains of producers and one another.
control group A group of individuals that does not receive the intervention that researchers are studying. The control group is identical to the treatment group in all other ways.
convergent evolution When organisms evolve body parts of similar appearance and function along different lines of descent.
coral reef Formations made primarily of calcium carbonate secreted by generations of corals.
cork Dead cells with waxy walls that protect and insulate woody plants.
cork cambium A ring of dividing cells that produces cork.
correlation Type of statistic that describes the relationship between two variables.
covalent bond Type of chemical bond in which two atoms share electrons.
CRISPR array A portion of a bacterial or archaeal genome that contains repeated sequences interspersed with spacer sequences containing segments of DNA from previous viral infections.
CRISPR/Cas9 system A bacterial defense mechanism that protects against viral infections; used by biologists to target gene sequences for gene editing.
crossing over The process by which a chromosome and its homologous partner exchange corresponding pieces of DNA during meiosis.
crustaceans Mostly marine arthropods that have two pairs of antennae and at least five pairs of legs.
culture The shared beliefs, values, and behaviors of a social group.
cuticle A waxy, waterproof layer on the outer surface of stems and leaves that reduces water loss and blocks pathogens.
cyanobacteria A lineage of bacteria that carry out photosynthesis. Formerly known as blue-green algae.
cytokines Chemical-signaling molecules that allow immune system cells to coordinate their activities.
cytokinesis The final part of cell division during which the cell's cytoplasm is divided between two daughter cells.
cytoplasm In a eukaryotic cell, collective term for everything between the cell's membrane and its nucleus. In a prokaryotic cell, everything enclosed by the membrane.
cytoskeleton Network of protein filaments that support, organize, and move eukaryotic cells and their internal structures.
cytosol A jellylike mixture of water, sugars, ions, small molecules, and proteins that surrounds the cell's organelles and is enclosed by the cell's membrane.

D

data Factual information generated by both deductive and inductive methods.
decomposers Organisms that recycle leftover organic matter into inorganic building blocks.
deductive methods A method of scientific inquiry that starts with a hypothesis which is then tested by running experiments.
deforestation Occurs when many or all trees are removed from a forested area.
dehydration synthesis The process by which monomers join to form a polymer, during which the formation of a covalent bond requires the removal of a water molecule.
dendrites Short, branched extensions of a neuron that receive electrical signals from other neurons.
dendritic cell A phagocyte that specializes in capturing antigens and presenting them to cells of the adaptive immune system.
denitrification The conversion of nitrates into nitrogen gas.
density-dependent limiting factors Factors that slow growth in dense populations, such as competition and disease.
density-independent limiting factors Factors that limit population growth regardless of population density, such as natural disasters.
deoxyribonucleic acid (DNA) Nucleic acid that has sugars made of deoxyribose. Deoxyribose sugar does *not* have an oxygen atom at the second carbon.
derived trait A modified ancestral trait found in descendants, but not the ancestor.
dermal tissue The outer protective covering of a plant.
dermis The layer of skin beneath the epidermis, which consists mostly of dense connective tissue.
desertification A rapid shift from a grassland or woodland to a desert, where the soil becomes unsuitable for cultivation.
deserts Areas defined by low annual precipitation.
desmosomes Elements that join cells together, consisting of anchor proteins that work like a rivet or button to connect two cells.
detritus Leftover material, such as excrement, dead leaves, and deceased organisms.
deuterostomes Animals where the mouth develops from the second embryonic opening.
diabetes A condition where there is difficulty regulating blood sugar.
diatom A nonmotile, single-celled photosynthetic protist that secretes a two-part silica shell.
dideoxynucleosides (ddNTPs) A DNA base with its 3′ —OH group removed.
differentiation The process by which a cell becomes a specialized cell type.
diffusion The spontaneous spreading of molecules or atoms through a fluid or gas.
dinoflagellates Single-celled protists that typically have two flagella.
diploid (*2n*) Having two complete sets of chromosomes, one from each parent.
directional selection A form of natural selection where one extreme form of a trait is adaptive.
disaccharide Sugar formed from two monosaccharides.
disinformation The intentional and malicious spread of false information, often with an ulterior motive.
disruptive selection A form of natural selection where traits at both ends of a range of variation are adaptive, and intermediate forms are selected against.
divergent evolution The divergence of lineages descended from a common ancestor.
DNA cloning The process of making multiple, identical copies of a piece of DNA.
DNA methylation The addition of methyl groups to cytosines in DNA.
DNA polymerase The enzyme that carries out DNA synthesis by adding complementary nucleotides to the new strand of DNA.
DNA profiling The process of identifying an individual by analyzing specific regions of their genome that are highly variable. Also called DNA fingerprinting.
DNA replication The process in which a molecule of DNA is copied in its entirety, producing two identical DNA molecules.
DNA sequencing The process of determining the exact order of nucleotide bases in a DNA molecule; also known as its sequence.
dominant A type of allele that masks a recessive allele.
dual coding Learning strategy in which the learner uses two different formats for representing information. For example, one could draw pictures *and* make written notes.

E

echinoderms Marine invertebrates with spiny skin.
ecological footprint The area of Earth's surface required to sustainably support a particular level of development and consumption.
ecological restoration The process of repairing and revitalizing damaged ecosystems, such as when removing dams or reintroducing species, to restore natural ecological functions and biodiversity.
ecological succession The process by which the array of species shifts as organisms alter their own habitat.
ecology The scientific study of interactions among organisms, and between organisms and their environment.
ecosystem Includes the biotic (living) components of an area plus the abiotic (non-living) components.
ecosystem engineer A species that physically alters the environment, and thus impacts the entire community.
ecotourism Responsible travel to natural areas that supports conservation efforts, raises awareness about biodiversity, and helps local communities by promoting sustainable tourism practices.
ectotherms Animals that obtain heat from outside the body.
effector T cells T cells that circulate and directly kill infected cells or help activate other immune cells.
elaboration Learning strategy in which one connects new information that they are learning to things they already know.
electron microscopy A type of microscopy where a beam of electrons in a magnetic field is used to visualize cellular structures that are too small to see with light microscopy.
electrons Subatomic particles surrounding the nucleus of an atom that have a negative charge.
electron transport chain A series of coupled redox reactions leading to the production of ATP, in which various proteins pass electrons and energy to one another.
elements Substances that cannot be broken down into anything else.
endangered species A species currently at a high risk of extinction in the wild.
endemic species Species that are confined to the areas in which they first evolved.
endocrine disruptor A chemical from outside the body that interferes with the normal function of the endocrine system.
endocrine signaling A cell signal that impacts the body's cells over long distances.
endocytosis Process by which a cell takes extracellular molecules inside itself.
endomembrane system A multifunctional network of membrane-enclosed organelles that occur throughout a cell's cytoplasm.
endoplasmic reticulum (ER) Membrane-enclosed organelle that consists of a continuous system of sacs and tubes extending from the nuclear envelope. Rough ER makes and modifies proteins; smooth ER makes phospholipids, stores calcium, and has additional functions in some cells.
endoskeleton An internal skeleton.
endosperm Nutrient-rich tissue inside seeds that feeds the embryo.
endosymbiosis A relationship in which one organism lives inside an organism of a different species.
endosymbiosis theory A theory stating that that mitochondria and chloroplasts evolved from prokaryotes that were engulfed by host cells.
endotherms Animals whose bodies maintain an internal temperature by varying the production of metabolic heat.
energy pyramid A graphic representation of the proportion of energy captured by producers that reaches higher trophic levels.
energy The capacity to do work.
enzyme An organic molecule that can speed up a chemical reaction but is not consumed or changed during that reaction. Enzymes are typically proteins, but some RNAs can act as enzymes.
epidermis (animals) The outermost layer of skin.
epidermis (plants) The first dermal layer; usually one layer of cells covering stems, leaves, and roots.
epigenetic inheritance When DNA modifications that do not change the DNA sequence itself are passed on to offspring, altering gene expression patterns in the offspring.
epithelial tissue Tissue that covers body surfaces and lines internal cavities.
equilibrium life history A life history in which parents produce a few high-quality offspring; common in stable environments.
estrogen A steroid sex hormone essential for reproductive function and secondary sex characteristics like breast development.
estuary A mostly enclosed coastal region where seawater mixes with freshwater.
ethylene A gaseous plant hormone that promotes fruit ripening.
eukaryote Organisms whose cells characteristically have a nucleus, membrane-bound organelles, and a comprehensive cytoskeleton.
eutrophication When an excess of nutrients in a body of water, often due to runoff, leads to dense plant growth and the death of animal life.
evaporation Process in which molecules escape from the surface of a liquid and become vapor.
evergreen plants Plants that keep their leaves throughout the year.
evolution The change in the gene pool of a population from generation to generation through processes such as mutation, natural selection, and genetic drift; also known as descent with modification.
***ex vivo* gene therapy** A therapy in which cells are removed from the body, modified to replace a faulty gene, and reintroduced into the body.
exaptation The repurposing of an existing structure for a new function.
exocytosis Process by which a cell expels a vesicle's contents to extracellular fluid.
exon A DNA sequence that contains protein-coding information.
exoskeleton An external skeleton made up of chitin.
exotic species A species that is introduced to a habitat that it isn't native to and becomes established at that new location.
experiment A procedure designed to evaluate a prediction, utilizing variables to generate data that will show whether a hypothesis is supported or refuted.
exponential model of population growth Describes population growth when resources are unlimited and the per capita growth rate is constant, forming a J-shaped curve.
extinct When no individuals of a species remain.
extracellular fluid The fluid outside of cells.
extracellular matrix (ECM) Complex mixture of substances secreted by a cell onto its surface; composition and function vary by cell type.
extreme halophile Organism that can live in highly salty environments.
extreme thermophile Organism that thrives at high temperatures.

F

F_1 generation The first generation of offspring resulting from a cross between two individuals.
facilitated diffusion Passive transport mechanism in which a solute follows its concentration gradient across a membrane by moving through a transport protein.
falsifiable Able to be proven wrong by experimentation or other means of collecting data.
fast-twitch muscle fibers Muscle fibers that allow for rapid, explosive movements but fatigue quickly.
fat A type of lipid, consisting of a molecule of glycerol and three molecules of fatty acids; also referred to as a triglyceride.
fatty acid Molecule with a long hydrocarbon "tail" with a carboxylic acid functional group "head." The tail is hydrophobic (fatty) and the carboxyl group makes the head hydrophilic and acidic, hence the name fatty acids.
feedback inhibition Regulatory mechanism in which the product of a reaction slows or stops the metabolic pathway that produces it.

fermentation Any of several energy-producing pathways that occur in the absence of oxygen, break down organic molecules (typically glucose), and produce ATP.
fertilization The fusion of an egg and a sperm cell.
fibroblasts Cells in connective tissue that make collagen and fibers.
fibrous root A dense network of many shallow roots branching from the stem.
first law of thermodynamics During a chemical reaction, energy is always conserved, meaning it can be transferred or converted, but not created or destroyed.
firsthand bias The tendency to believe something is more common than it is because it happened to us or someone close to us.
fitness In evolutionary terms, one's success in passing genes to the next generation.
fixed allele When all members of a population are homozygous for an allele, and other alleles disappear.
flatworms Animals with a flat body and an array of organ systems, but no body cavity other than a gastrovascular cavity.
flowering The process of producing flowers; often triggered by night length.
flower A specialized reproductive structure that consists of modified leaves arranged in concentric whorls.
fluid mosaic model Characterization of a cell membrane as a two-dimensional structure of proteins contained in a dynamic phospholipid bilayer.
fluorescence microscopy A type of microscopy where fluorescent dyes are used to visualize cellular structures.
food chain A sequence of steps by which energy is captured and moves from producers to consumers into progressively higher trophic levels.
food web Interconnection of multiple food chains in an ecosystem.
foraminifera Single-celled protists that secrete a shell containing calcium carbonate.
forever chemicals Synthetic chemicals that persist in the environment and in the human body; can bioaccumulate.
fossil The remains or traces of an organism that lived long ago.
fossilization The process by which fossils form under specific conditions, typically beginning when an organism or its remnants are covered by sediment, mud, or ash.
founder effect When a new population is established by a small number of individuals, reducing the gene pool to the alleles of only the founders.
frameshift mutation A change in DNA that occurs when bases are added or deleted in an interval other than three.
free radicals Atoms with unpaired electrons.
fruit The structure that contains the seeds that form after a plant produces flowers.
functional groups Special groups of atoms that give specific properties to molecules.
fungus A eukaryotic heterotroph that eats other organisms.

G

G_0 A resting phase in which cells are neither dividing nor preparing to divide.
G_1 The first phase in the cell cycle, in which the cell grows and produces the molecules needed for DNA replication.
G_1 checkpoint The point in the cell cycle between the G_1 and S phases, in which various proteins assess the ability of the cell to divide.
G_2 The phase in the cell cycle when cells make the proteins and other cellular components needed for cell division.
G_2 checkpoint The point in the cell cycle prior to mitosis in which a cell's DNA is checked to ensure the DNA was properly replicated.
gametes Haploid reproductive cells formed by through meiosis.
gametophyte Haploid structure that produce gametes (sperm or eggs).
gap junctions Channels that open, close, and connect the cytoplasm of adjoining animal cells.
garbage patch An area full of marine debris created by ocean currents.
gastropods Mollusks that include snails and slugs.
gastrovascular cavity A saclike structure that takes in and digests food, expels wastes, and helps with gas exchange.
gel electrophoresis The process by which nucleic acids are separated by size using an electric field to pull the fragments through a semisolid gel.
gene A unit of DNA that codes for a specific protein or RNA molecule.
gene dosage The proper number of genes needed for optimum health.
gene editing A form of genetic engineering in which the sequence of an organism's DNA is changed.
gene expression Conversion of information from DNA into a product, usually protein, in a multistep process.
gene flow The movement of alleles between populations, which stabilizes allele frequencies.
gene knock-in A genetic engineering process of replacing a gene with a different version.
gene knock-out A genetic engineering process where a gene is inactivated.
gene pool The sum total of all alleles in a population or species.
gene silencing A post-transcriptional regulatory process where mRNA is not translated, either due to degradation or physical blockage from interacting with the ribosome.
gene therapy An approach to curing a genetic disease by replacing a faulty gene with a functioning one.
generalizable Can be applied to larger populations or has wider application.
genetic code The set of 64 codons used by all cells to determine the sequence of amino acids in a protein.
genetic counselor A medical professional who supports and educates prospective parents about the risks of passing on disease-causing genes to their offspring.
genetic drift Change in allele frequency brought about by chance alone.
genetically modified organism (GMO) Any organism that has been genetically engineered.
genotype An organism's set of alleles.
genotyping The process of determining the genetic makeup of an organism at specific locations that are related to large changes in phenotype.
genus A group of closely related species that share inherited characteristics.
geographic range The total area occupied by all populations of a species.
geologic time scale A chronological history of Earth that correlates layers of rock with great intervals of time.
germ layers The first tissues that form in an animal embryo.
glycogen A highly branched polysaccharide produced by animals for energy storage.
glycolysis Set of reactions that collectively convert one molecule of glucose to two molecules of pyruvate, two ATP (net), and two NADH. First part of fermentation and aerobic respiration.
Golgi apparatus Organelle that modifies polypeptides and lipids, then sorts and packages the finished products into vesicles. Also called the Golgi body.
gonads An animal's primary reproductive organs (testes and ovaries).
Gondwana A supercontinent formed from the southern part of Pangaea, which eventually split to become Africa, South America, Antarctica, and Australia.
graft An area where a person joined one piece of one plant to another so they grow together.
granulocyte A white blood cell with vesicles (granules) that contain materials used to fight infection.
great oxygenation event When oxygen began to accumulate in the air and water due to cyanobacteria.
greenhouse effect A warming phenomenon that occurs when heat energy is trapped by gases in the atmosphere.
greenhouse gases Atmospheric gases, including methane and carbon dioxide (CO_2), that trap heat and therefore contribute to the greenhouse effect.
ground tissue Tissue that makes up most of the plant body; used for support, storage, and photosynthesis.
ground-level ozone (O_3) A hazardous gas that forms when carbon monoxide and nitrogen oxides (which are both part of car exhaust) react with sunlight.

groundwater Freshwater found beneath Earth's surface.
guard cells Specialized cells around stomata that open and close them.
gymnosperms Seed plants that produce seeds on the surface of an ovule.

H

habitat isolation When two closely related species live in different habitats, but are unable to interbreed, which may ultimately lead to speciation.
habitat The natural environment where a species typically lives.
half-life The length of time it takes for half of the atoms in a radioactive element to decay.
haploid (*n*) A cell that contains one copy of each chromosome.
helicase An enzyme that unwinds and separates DNA into single strands, allowing it to be copied.
helper T cell A type of T cell that helps to activate B cells and coordinate the immune response.
herbivory An antagonistic interaction in which an organism feeds on a plant.
heterotroph An organism that gets its energy entirely from consuming organic molecules produced by other organisms.
heterozygous An individual with different alleles of a gene on each homologous chromosome.
histones Type of protein that DNA winds around; one of many proteins that structurally organize eukaryotic chromosomes.
homeostasis The process by which the body maintains internal conditions that differ from the outside environment within a range that is required by the body's cells.
Hominidae The family that includes modern humans and extinct hominins.
homologous chromosomes Chromosomes that are the same type and have the same genes but are not genetically identical (they have different alleles or gene variants).
homologous structure A body part that may vary in different lineages but that is derived from a common ancestral form.
homology Similarities in the nucleotide sequence of a shared gene or in the amino acid sequence of a shared protein.
homozygous An individual with the same allele on both homologous chromosomes.
hormones The intercellular signaling molecules of the endocrine system, secreted by a gland or cell and typically traveling in the bloodstream to target cells.
hybrid vigor When a hybrid animal outperforms either parent in some way.
hydrogen bond Weak electrostatic (charged-based) attractions between a slightly positively charged hydrogen on one molecule and a slightly negatively charged atom (oxygen, nitrogen, or fluorine) on another molecule.
hydrolysis The breaking of a covalent bond by the addition of a water molecule, which results in a molecule splitting into two smaller molecules.
hydrophilic Describes a substance that dissolves in water.
hydrophobic Describes a substance that does not dissolve in water.
hydrothermal vent A place where mineral-rich water heated by geothermal energy streams out through a rocky opening in the seafloor.
hypertonic Describes a fluid that has a high overall solute concentration relative to another fluid.
hyphae Tiny filaments that make up the body of a multicellular fungus.
hypodermis The layer beneath the skin, consisting of connective tissue, nerves, and adipose tissue.
hypothalamic-pituitary-gonadal axis A hormonal pathway that controls reproduction and sexual development, involving the hypothalamus, pituitary gland, and gonads.
hypothalamus A structure in the brain that connects the nervous and endocrine systems.
hypothesis A testable explanation for a natural phenomenon.
hypotonic Describes a fluid that has a low overall solute concentration relative to another fluid.

I

ice A biome that is largely covered in ice year-round.
immune system The body system that helps recognize itself or foreign invaders and assists in the removal of foreign invaders.
impaired fecundity A broader term than "infertility" that describes difficulty getting pregnant or carrying a pregnancy to term.
***in vitro* fertilization (IVF)** An assisted reproductive technique that combines male and female gametes outside the body.
***in vivo* gene therapy** A type of gene therapy in which a vector transports a functional gene into the patient's cells with the non-functional gene with the purpose of disease treatment.
incomplete dominance When one allele is not fully dominant over the other.
index fossils The remains of organisms that were common during a certain geologic time period.
inductive method A method of scientific inquiry that starts with data, often in the form of observations about the natural world, that are then used to generate a hypothesis.
infertility The inability for a couple to conceive after a year of regular unprotected sex (or 6 months for people older than age 35).
inflammation A fast, local immune response that jump-starts healing.
infodemic The rapid, worldwide dissemination of unreliable information.
initiation The first phase of either transcription or translation when the process is getting started.
innate immunity A set of immediate, general immune defenses that protect against infection, including barriers, fever, and inflammation.
inorganic water pollutants Pollutants that do not contain carbon atoms but do include a variety of different elements, such as nitrogen, phosphorus, mercury, lead, and copper, and/or poisons like cyanide.
insects A lineage of arthropods with three pairs of legs and one pair of antennae.
interdisciplinary research When a scientific pursuit involves perspectives and/or methods from multiple scientific disciplines.
intermediate filaments Cytoskeletal elements made of fibrous proteins that form a stable framework lending structure and resilience to cells and tissues.
internode The area between nodes.
interphase The time between cell divisions when the cell carries out basic cellular functions, such as growth, copying DNA, and making proteins.
interspecific competition Competition between members of different species.
interstitial fluid The extracellular fluid in the spaces between body cells.
intertidal zone The area along shorelines adapted to wave forces and tidal changes.
intracellular fluid Body fluid located in the cytosol of cells.
intraspecific competition Competition among members of the same species.
intron Typically the noncoding regions of DNA found between exons.
invasive species An exotic species whose introduction disrupts the community structure in its new home.
invertebrates Animals without a backbone.
ionic bond Type of chemical bond in which a strong mutual attraction links ions of opposite charges.
ions Atoms or molecules with an unequal number of protons and electrons; thus, they carry a net charge.
isotonic Describes two fluids that have the same overall solute concentration.
isotopes Atoms of the same element that have different numbers of neutrons.

J

jaw A hinged skeletal element used for feeding.
jawed fish Fish with movable jaws and well-developed teeth.
jawless fish A type of fish with no jaw, a smooth, elongated body, and no paired fins.
joint The location where different bones come together.

K

karyotype An image that displays chromosomes, allowing someone to view the chromosome number and chromosome structure of an individual cell.
keratinocytes Epithelial cells in the epidermis that synthesize keratin.
key innovation A novel trait that increases a species' adaptability.
keystone species A species that can have a disproportionally large impact on a community relative to its abundance.
kinetic energy The energy of motion.

L

lactic acid fermentation Fermentation pathway that produces ATP and lactic acid.
lagging strand The portion of DNA that is synthesized in short pieces by DNA polymerase during replication; the strand where helicase and DNA polymerase are moving in opposite directions.
lake A body of standing freshwater.
lancelets Fishlike chordates that retain all chordate characteristics into adulthood.
larva A sexually immature animal that has a different body form than the adult.
lateral roots Branches that grow out from the primary root.
latitude The distance from the equator.
Laurasia A supercontinent formed from the northern part of Pangaea, which eventually broke apart to form North America, Europe, and Asia.
law A well-tested statement, often expressed in mathematical terms, that describes a fundamental aspect of the natural world.
law of independent assortment One of Mendel's laws that states the movement of alleles during meiosis does not influence the movement of other alleles.
law of segregation One of Mendel's laws that states alleles on different chromosomes move separately from one another during meiosis.
leading strand The portion of DNA that is synthesized continuously by DNA polymerase during replication; the strand where helicase and DNA polymerase are moving in the same direction.
leaf veins Vascular bundles in leaves that carry water and nutrients to leaf cells.
leaves The main organs where photosynthesis occurs in most flowering plants. They also function in gas exchange.
lichens A composite organism consisting typically of a fungus and either a type of algae or cyanobacteria.
life history trait A heritable characteristic that directly affects an organism's ability to survive and reproduce.
life history The sequence of growth, development, and reproduction-related events that occurs over an individual's lifetime.
life table A chart representing how many individuals are expected to survive to a given age.
ligament A cord of dense connective tissue that connects bones at a joint.
ligase An enzyme that joins the short strands of DNA created by DNA polymerase mainly on the lagging strand of DNA during replication.
light microscopy A type of microscopy where visible light is used to illuminate a sample.
light-dependent reactions Reactions in the initial stage of photosynthesis that capture light energy and convert it to chemical energy in the form of ATP and NADPH.
lignin A compound in xylem that strengthens and stiffens cell walls.
Linnaean system A standardized naming system where each species is given a unique two-part name.
lipid A broad group of hydrophobic and amphiphilic molecules that do not dissolve in water.
lipid bilayer A two-layered sheet of phospholipids that forms the membranes around cells. The hydrophilic heads of one layer face the cell's watery interior and exterior, and the hydrophobic tails make up the interior of the bilayer.
lobe-finned fishes Bony fishes with thick, fleshy pelvic and pectoral fins with internal bony supports.
logistic model of population growth A model that describes how population growth changes as the size of the population approaches carrying capacity, forming an S-shaped curve.
lungs Internal air sacs most often used for gas exchange.
lycophytes One of the oldest types of vascular plants.
lymph node A small, bean-shaped organ that filters lymph and houses immune cells.
lymph A type of interstitial fluid that is collected, filtered, and returned to the circulatory system by the lymphatic system.
lymphatic system The system that houses immune cells and includes organs like the thymus, lymph nodes, and spleen.
lymphocytes White blood cells with special roles in the adaptive immune response, including T cells and B cells.
lysosome Enzyme-filled vesicle that breaks down particles such as cellular debris.

M

macromolecules Large molecules that are necessary for life. Carbohydrates, proteins, nucleic acids, and lipids are the four classes of macromolecules, each with their own distinguishing characteristics.
macrophages Large phagocytes that patrol tissues and engulf foreign material and cellular debris.
malignant An abnormal growth of cells that is cancerous and capable of spreading to other parts of the body.
mammals Amniotes in which females nourish their offspring with milk secreted from mammary glands.
mammary glands Glands used to produce milk.
mantle A skirtlike extension of the upper body wall that drapes over the internal organs.
manuscript A written report describing the methods, results, and conclusions of a scientific study submitted for publication in an academic journal.
mark–recapture sampling A method to estimate population size by capturing and marking individuals, and then recapturing them later to estimate the total population based on the proportion of recaptured marked individuals.
marsupial A mammal that gives birth to underdeveloped young that complete their development within a pouch on the mother's body.
mass extinctions Events where 75% or more of the world's living species are lost in a short period of geological time.
mass number Total number of protons plus neutrons in an atom.
matter Anything that has mass and takes up space.
M checkpoint The point in the cell cycle between metaphase and anaphase during which chromosomes are inspected to ensure they are properly attached to the microtubules making up the spindle.
mechanical isolation Morphological features of an organism that prevent successful interbreeding.
medusa A cnidarian with a bell-shaped body form that swims or drifts about.
meiosis A nuclear division mechanism that halves the chromosome number and is used to form haploid reproductive cells.
melanocytes A type of epidermal cell that makes pigments called melanins.
membrane Structure separating a cell's contents from the external environment.
memory B cells Long-lived B cells that help with rapid recognition of a pathogen upon reinfection.
memory T cells Long-lived T cells that can readily identify an invader upon a subsequent infection.
menstrual cycle The monthly cycle in which hormones fluctuate, causing the uterine lining to build up in preparation for pregnancy and be shed if fertilization does not occur.
meristems Regions of undifferentiated cells where new plant growth occurs.
mesophyll The cells inside leaves where photosynthesis occurs.
messenger RNA (mRNA) An RNA copy of a specific gene that carries its information from the nucleus to the cytoplasm.
meta-analysis A method for examining the combined results of multiple separate scientific studies.

metabolic pathway Process that takes a starting material and ends with a product in a series of enzymatic reactions.
metamorphosis A change in body form a juvenile to adult form.
metaphase The stage of mitosis in which the chromosomes line up in the center of the cell.
metastasis The spread of cancer cells to other parts of the body.
methane pollution A form of pollution that occurs when excessive amounts of methane gas enter the atmosphere due to human activities.
methanogens A group of archaea that break down organic waste using methanogenesis, producing methane.
microchimerism The presence of a small number of cells in an individual that originated from another genetically distinct individual.
microfilaments Fine fibers composed primarily of actin subunits found in eukaryotes that reinforce cell membranes and function in cell movement.
microplastic Microscopic plastic debris that result from the breakdown of plastic.
microtubules Cytoskeletal elements of eukaryotes that consist of a hollow filament of tubulin subunits and support cellular processes involving movement.
mildew A parasitic multicellular fungus.
millipedes Animals that eat plant material and have two legs per segment.
mimicry An evolutionary pattern in which a species evolves to resemble a different species to gain a selective advantage.
miscarriage Fetal death prior to 20 weeks of pregnancy.
misinformation The unintentional or accidental spread of false information.
mismatch repair The process by which enzymes correct replication errors by removing the section of DNA containing mismatched bases and using the remaining parental strand as a template to replace them with the correct nucleotides.
missense mutation A mutation that occurs when a base-pair change in DNA leads to translation of a different amino acid.
mitochondrion (plural *mitochondria*) Organelle where cellular respiration occurs and most of the cell's ATP is produced.
mitosis The process of generating two new, genetically identical daughter cells from a single parent cell.
mold A type of fungus that grows in dark, damp places and breaks down organic matter.
molecule Two or more atoms joined by a chemical bond.
mollusks Animals with a soft, unsegmented body; a complete digestive tract; and a reduced coelom.
monkeys of Africa and Asia Monkeys that may or may not have tails, typically found in Africa and Asia; formerly called Old World monkeys.
monkeys of the Americas Tree-dwelling primates with prehensile tails, typically found in the Americas; formerly called New World monkeys.
monomer A small molecule that forms a single subunit of a larger molecule.
monosaccharide A molecule consisting of a single sugar molecule. Also called simple sugars. Monosaccharides are the monomers from which all carbohydrates are built.
monosomy When an individual has one copy of a particular chromosome instead of two.
monotremes Egg-laying mammals; typically found in Australia and New Guinea.
morphological traits Physical features shared by members of a species.
motor proteins Type of energy-using protein that interacts with cytoskeletal elements to move the cell's parts or the whole cell.
multicellular organism An organism with specialized cells where each cell has a specific function and all cells rely on one another for survival.
multiple cloning site A sequence in a plasmid that contains many restriction enzyme sites.
multipotent stem cell A stem cell that can differentiate into different cells but is limited to only a few closely related cell types.
muscle fibers The individual muscle cells that make up muscle tissue.
muscle tissue A tissue type made up of fibers that is used to produce movement.
musculoskeletal system The collective name for the muscular and skeletal systems, which work together under nervous system control to facilitate movement.
mushrooms The fruiting body of a fungus that produces spores.
mutagens Agents that cause DNA damage.
mutations Changes to a DNA sequence that can be beneficial, neutral, or harmful.
mutualism An interspecific interaction that benefits both species.
mycelium A branching network of fungal hyphae.
mycorrhiza A mutually beneficial relationship between fungi and plant root cells.
myelin A fatty substance made by glial cells; surrounds the axon and functions like insulation.
myofibrils The contractile portion of muscle fibers.
myosin The protein that makes up the thick filaments in a sarcomere.
myriapods A lineage of arthropods that includes segmented animals with many legs.

N

natural selection The process in which environmental pressures result in the differential survival and reproduction of individuals in a population.
necrosis The death of cells due to injury or disease.
nectar A sugary liquid produced by flowers to attract pollinators.
nervous tissue A tissue type that detects stimuli and relays information.
neuron A specialized cell of the nervous system that conducts electrical signals.
neurotransmitter A chemical released by a neuron that communicates with another neuron across a synapse.
neutrons Subatomic particles within the nucleus of an atom that have no charge.
neutrophils The most abundant circulating white blood cells; often the first responders to an infection.
next-generation sequencing A process that allows investigators to sequence millions of different DNA sequences simultaneously, then combine the sequence data to determine a longer DNA sequence.
niche The ecological role a species plays in its habitat, such as how it interacts with other organisms and uses resources to survive and reproduce.
nitrification The conversion of ammonia to nitrites and nitrates.
nitrogen cycle Cycle in which nitrogen moves through the atmosphere, soil, water, and food webs.
nitrogen fixation The process of converting nitrogen gas into ammonia.
nodes The regions where leaves originate.
nondisjunction The failure of chromosomes to separate properly during meiosis.
noninvasive prenatal testing (NIPT) Prenatal testing where the mother's blood is sampled to test the fetus's DNA for potential chromosomal abnormalities.
nonpolar covalent bonds A type of covalent bond in which electrons are shared equally between atoms.
nonsense mutation A mutation that occurs when a base-pair change in DNA leads to a premature stop codon.
nonspontaneous chemical reaction A chemical reaction that requires an input of energy to proceed.
nonvascular plants Small seedless plants that lack vascular tissue and are dependent on water to complete their lifecycle.
notochord A rod of stiff but flexible connective tissue that extends the length of the body and provides support.
nuclear envelope A double membrane that constitutes the boundary of the nucleus.
nuclear pores Small "holes" in the nuclear envelope that control the entry and exit of molecules.
nucleic acids Macromolecules formed from nucleotide monomers that can transmit information.
nucleotides Molecules that contain a sugar, a nitrogenous base, and phosphate functional groups. Nucleotides are the monomers that form nucleic acids, and there are five different kinds of nucleotides: guanine, adenine, cytosine, thymine, and uracil. Note that thymine is found only in DNA, and uracil is found only in RNA.

nucleus (chemistry) In chemistry, the central core of an atom, consisting of protons and neutrons. The nucleus of a hydrogen atom is an exception and consists of a proton only.
nucleus (biology) Organelle with a double membrane that holds, protects, and controls access to the cell's DNA.

O

ocean acidification A process that occurs when carbon dioxide dissolves in the oceans and reacts with water to form bicarbonate, which releases hydrogen ions, making the ocean more acidic.
ocean currents The continuous movement of seawater in a directed manner because of forces like wind, temperature, and salinity.
oligosaccharide A short-chain carbohydrate consisting of approximately three to six sugar monomers.
oncogene Mutated form of a proto-oncogene; results in cells that divide continuously.
oogenesis The process of making egg cells that results in a single large egg cell.
open ocean Brightly lit salt water away from the coast and seafloor where photosynthetic protists and bacteria are the primary producers.
organ A structural unit of a system that carries out specific tasks.
organ systems Systems in which two or more organs and other components interact physically, chemically, or both to perform a common task.
organelles Structures, such as the nucleus, that carry out specialized functions inside a cell.
organic water pollutants Pollutants made up of carbon-based compounds, such as sewage, oil, pharmaceutical drugs, and forever chemicals.
organism Any individual lifeform.
organization In biology, the recognition that there is a hierarchical structure inherent to living organisms, from the smallest molecules to planet-wide interacting biological systems.
osmosis Diffusion of water through a semipermeable membrane; occurs when there is a difference in solute concentration between the fluids on either side of the membrane.
osmotic pressure The pressure that builds up when water moves into cells by osmosis.
osteoblasts Cells that form new bone and contribute to bone growth and healing.
osteoclasts Cells that break down bone tissue.
ovaries (animal) Gonads that produce eggs and secrete predominantly estrogen and progesterone.
ovary (plant) The swollen base of a carpel that contains ovules.
ovule The structure inside a plant ovary that develops into a seed after fertilization.

P

pandemic When an infectious disease spreads worldwide.
Pangaea The largest supercontinent, which all continents were once part of.
paracrine signaling A cell signal that impacts cells in the immediate vicinity of the signaling cell.
parasitism When an organism (a parasite) withdraws nutrients from a living individual of another species (a host).
parasympathetic nervous system The division of the autonomic nervous system that is active in times of relaxation, promoting "rest and digest" functions.
parenchyma Thin-walled living plant cells with flexible walls; involved in storage, photosynthesis, and secretion.
parental (P) generation The first pair of organisms that are crossed in a genetic experiment to produce offspring.
particulates Tiny liquid or solid particles in the air.
passive transport Movement of molecules requiring no energy input.
pedigrees Charts that mark the appearance of a trait among generations of family members.
peer review The process in which other experts in the same field as a scientist (that is, a scientist's peers) review and critique the scientist's work.
peptide bond A covalent bond between the amino group of one amino acid and the carboxylic acid group of another; connects amino acids to form a polypeptide
per capita growth rate The rate of change in a population per individual.
periodic table A graphic display that organizes all elements based on their atomic number and chemical properties.
peripheral nervous system The division of the nervous system that includes all the nerves outside the central nervous system.
permafrost A component of some soils that stays frozen for two or more years.
personal anecdote A short story about a person's experience; is not a source of scientific evidence.
personalized medicine Medical care that is designed based on the patient's genetic profile, whether for cancer treatment or the prevention of diseases.
petal Often brightly colored part of a flower; attracts pollinators.
pH A measure of the number of protons (hydrogen ions) in a solution.
phagocyte A cell that specializes in finding and removing bacterial or cellular debris via engulfment.
phagocytosis "Cell eating"; a process by which a cell engulfs a large particle such as another cell.
phenotype An individual's observable traits.
phloem The vascular tissue that distributes sugars and other nutrients through a plant.
phospholipid A type of lipid with two long hydrocarbon tails and a head with a phosphate group. The tails are hydrophobic, and the polar phosphate group makes the head hydrophilic
phosphorus cycle Cycle in which phosphorus moves soil and water, as well as living and dead organic material.
phosphorylation Reaction in which an enzyme facilitates the attachment of a phosphate to a molecule.
photoautotrophs Organisms that capture light energy and carry out photosynthesis.
photorespiration A process where oxygen, instead of carbon dioxide, is fixed; reduces photosynthesis efficiency.
photosynthesis Process used by most autotrophs; utilizes the energy in sunlight to make sugars from carbon dioxide and water.
photosystem I Protein-pigment complex in the thylakoid membrane of photosynthetic organisms that captures the energy of the sun to move electrons from one side of the membrane to the other.
photosystem II Protein-pigment complex in the thylakoid membrane of photosynthetic organisms that captures the energy of the sun to split water molecules and produce oxygen.
phylogenetic trees Diagrams representing evolutionary relationships between species.
phylogeny The scientific study of evolutionary relationships.
physiology The study of how the body functions.
pigments Organic molecules that selectively absorb light of specific wavelengths.
pioneer species The first organisms to colonize new habitats.
pistil A single carpel or several fused carpels.
pituitary gland A pea-sized gland beneath the hypothalamus that controls other glands in the body by secreting hormones.
placebo A harmless inert substance like saline or a procedure that has no therapeutic effect.
placental mammals Mammals in which an organ called the placenta provides nutrients to developing offspring.
plankton The community of tiny organisms that swim or drift through Earth's waters.
plant A eukaryotic organism that is a land-dwelling autotroph and generally stationary.
plasma cell An activated B cell that makes and releases large amounts of antibodies.
plasmid A small extrachromosomal DNA molecule that can replicate independently from the chromosome; often used in the process of molecular cloning as a vector.
plate tectonics theory A scientific theory that describes the Earth's crust as made up of a series of units called plates.
platelets Clotting factors that enter a wound via blood vessels to help stop bleeding.
pleiotropy When a single gene influences more than one trait.

plot sampling A method where direct counts in multiple plots estimate population size.
pluripotent stem cell A stem cell that can divide into any cell type in the body.
polar bodies Cells produced during oogenesis that cannot be fertilized and undergo apoptosis.
polar covalent bonds A type of covalent bond in which electrons are shared unequally between atoms, resulting in positively and negatively charged regions in a molecule.
pollen grain An immature male gametophyte.
pollination The transfer of pollen to a receptive stigma.
pollination vectors Animals or environmental factors that move pollen.
pollinators Organisms that carry pollen between flowers.
pollutant A natural or human-made substance that is released into soil, air, or water in greater than natural amounts, harming the environment.
pollution The introduction of harmful materials to the environment that can make the environment unsafe for living organisms.
poly-A tail String of adenines at the 3′ end of an mRNA.
polygenic inheritance When multiple genes influence a phenotype.
polymer A chain of monomers linked together.
polymerase chain reaction (PCR) A laboratory technique for rapidly amplifying DNA.
polyp A tubular body form that has one end attached to some surface.
polypeptide A chain of amino acids linked by peptide bonds.
polysaccharide A macromolecule consisting of a long chain of hundreds or thousands of sugar monomers.
population A group of interbreeding individuals in one area.
population bottleneck A drastic reduction in population size that reduces genetic diversity.
population density The number of individuals in a specified area or volume.
population distribution The way members of a population are arranged within their environment.
population size The number of individuals in a population.
potential energy A form of energy that arises from an object's position or other property of matter that has the potential to become another form of energy. The chemical energy stored in chemical bonds is a form of potential energy.
practicing Phase of the learning cycle that involves pulling information out of long-term memory and using it, such as when working on practice problems.
prairie A type of temperate grassland, with mainly grasses and shrubs.
predation An interaction where an individual from one species captures, kills, and eats an organism from another species.
preimplantation genetic diagnosis (PGD) A type of prenatal genetic testing used alongside IVF to identify genetic disorders before the blastocyst is transferred to the uterus.
previewing Learning strategy in which a student gains some familiarity with the material prior to class that activates prior knowledge around the subject and piques curiosity.
primary consumers Organisms that eat plants (herbivores).
primary growth Growth that makes roots and shoots longer.
primary immune response The immune system's initial reaction to a new foreign agent.
primary root The first root structure to emerge when a seed germinates.
primary structure The sequence of amino acids in a protein.
primary succession Occurs in habitats that lack soil and thus have few or no existing species.
primase An enzyme that synthesizes primers during replication.
primer A short single strand of RNA or DNA that serves as an attachment point for DNA polymerase during DNA replication, PCR, or sequencing.
principle of competitive exclusion States that species that require the same limited resources and access them in the same way cannot coexist indefinitely.
producers Organisms that capture energy and use it to make their own food from nonbiological (inorganic) materials in the environment.
prokaryote A single-celled organism with no nucleus; includes the bacteria and archaea domains.
promoter A region of DNA containing special sequences that control when and where the gene is expressed; where transcription factors bind.
prophase The first part of mitosis, during which chromosomes pack into their most compact forms.
prosimians An ancient group of non-anthropoid primates; includes lemurs.
protein Macromolecule that contains amino acids linked together in a chain that folds into a particular shape to achieve a particular function.
protist A eukaryote that is not a fungus, a plant, or an animal.
protocell A membrane-enclosed collection of interacting molecules that can take up material and replicate; precursor to modern cells.
protons Subatomic particles within the nucleus of an atom that have a positive charge.
proto-oncogenes Genes that encode proteins that tell cells when to divide.
protostome An animal in which the first opening that forms on an embryo becomes the mouth.
pseudopodia "False feet"; temporary arm-like extensions of a cell's membrane and cytoplasm that enable a cell to move and capture food.
pseudoscience Claims, arguments, or methods that are presented as scientific, but do not follow scientific principles.
Punnett squares Graphical representations of the potential gametes that a parent can produce and the potential genotypes of the offspring.
pyruvate Three-carbon molecule produced by the splitting of glucose during glycolysis.

Q

quaternary structure The association of two or more proteins into a larger functional unit through bonding and other attractive forces.

R

radial symmetry Body parts repeated around a central axis, like the spokes of a wheel.
radioactive decay The spontaneous transformation of a subatomic particle in the nucleus of an atom into a different subatomic particle. Radioactive decay occurs at predictable rates and forms predictable products.
radiometric dating A method used to determine the precise age of rock layers by measuring the relative amounts of radioactive isotopes.
radon A gas produced when radioactive metals in the soil break down.
radula A tongue-like organ hardened with chitin used for scraping.
random distribution A pattern of distribution where resources are readily available and there is no order in which one finds organisms.
ray-finned fishes Bony fishes with flexible fins supported by thin rays of bony spines.
reassortment When multiple viruses swap genetic material within the same host.
receptor proteins Membrane proteins that trigger a change in cell activity in response to a stimulus such as hormone binding.
recessive A type of allele whose traits are seen only if a person has two copies of that allele.
recombinant chromosome A completely new chromosome formed as a result of crossing over.
recombinant DNA The resulting DNA molecule formed when genetic material from multiple sources is combined in a laboratory setting.
reduction-oxidation (redox) reaction A chemical reaction in which an atom loses or gains an electron.

relative dating A method used to determine the general age of a fossil by comparing its position within rock layers.
release factor A protein that binds to the ribosome where a tRNA would typically bind, ending translation.
replication bubble The structure created by the separation of two DNA strands by helicase enzymes during replication.
reproduce To produce offspring.
reproductive endocrinology A medical field that specializes in diagnosing and treating infertility.
reproductive isolation The end of gene flow between populations due to an inability to reproduce.
reptile An amniote that breathes air and undergoes internal fertilization.
resistance exercise Exercise that causes muscles to work against an outside force like a weight with the goal of increasing muscle size and strength.
resource partitioning When different species within a community have their own niche for resources, like food, that minimizes competition.
restriction enzyme An enzyme that cuts DNA at specific nucleotide sequences.
retrieval practice A type of study strategy when an individual actively pulls information from their long-term memory, rather than passively reviewing notes.
ribonucleic acid (RNA) Nucleic acid that has sugars made of ribose. Ribose has two hydroxyl groups at the second and third carbons.
ribosomal RNA (rRNA) A type of RNA found in the ribosome that catalyzes the reaction between amino acids to form peptide bonds.
ribosome The organelle where protein synthesis occurs. An intact ribosome has two subunits, each composed of rRNA and proteins.
ribozymes Catalytic RNAs that can function as enzymes.
ripening The process where fruits mature, involving changes in texture, taste, and color.
river A waterway that forms as streams merge and flow downhill.
RNA polymerase An enzyme that makes mRNA.
RNA world hypothesis A hypothesis that RNA existed before DNA and proteins during chemical evolution. The hypothesis also states that RNA once stored genetic information and catalyzed its own replication.
ribonucleic acid (RNA) A type of nucleic acid that is typically single stranded and made up of ribonucleotides that have diverse functions.
root cap The tip of the root that protects it as it grows through soil.
root hairs Tiny extensions on root epidermal cells that increase surface area for absorption.
roots Parts of the plant that take up water and minerals from soil, anchor the plant, and sometimes store nutrients.
roundworms Cylindrical, unsegmented worms with a pseudocoelom.
rubisco Key carbon-fixing enzyme in photosynthesis; the most abundant enzyme on Earth.

S

S phase The point in the cell cycle when DNA replication occurs.
Sanger sequencing A sequencing process, also known as chain termination sequencing, used to determine the sequence of a DNA molecule.
sarcomeres Repeating unit making up myofibrils.
saturated fatty acids Fatty acids that have only single bonds linking the carbons in their tails.
savanna A tropical grassland that includes a few scattered shrubs and trees.
scales Hard, flattened structures that grow from and often cover the skin.
scientific methodologies Collection of scientific approaches and techniques that enable researchers to answer scientific questions.
sclerenchyma Thick-walled cells containing cellulose and lignin that provide strength and rigidity.
second law of thermodynamics A physical principle stating that over time, the universe tends to become disordered.
secondary consumers Organisms that eat primary consumers.
secondary growth Growth that makes stems and roots thicker.
secondary immune response The quicker and stronger immune response upon a second exposure to an antigen; due to memory cells.
secondary structure In a protein, the local arrangement of amino acids in space. The most common secondary structures are alpha helices and beta sheets.
secondary succession Occurs after a natural or human disturbance removes the natural array of species but leaves the soil in place.
seed Consists of an embryo sporophyte and food to support it enclosed within a protective coat.
seed coat The protective outer covering of a seed.
seed plants Plants that reproduce by releasing seeds.
seedless vascular plants Plants that have vascular tissue but produce spores rather than seeds and are dependent on water to complete their lifecycle.
selection bias When subjects in a study are intentionally chosen so that the study's results will support a claim. Selection bias can also be unintentional and occurs when the experiment or analysis of the data is flawed.
selective force A factor that influences the survival and reproduction of a subset of individuals within a population.
selectively permeable A membrane characteristic where some substances can pass through a membrane, but others cannot.
self-explanation Learning strategy that checks one's understanding by explaining the material to someone else.
senescence A state in which a cell is permanently in a nondividing state, cell cycle phase G_0.
sepal Leaflike part that encloses and protects a flower before it opens.
sex chromosomes The 23rd pair of chromosomes; one way of characterizing sex in humans.
sexual dimorphism When a trait, such as a behavior or morphological characteristics, varies within a species based on sex, often as a result of sexual selection.
sexual reproduction The process in which offspring arise from two parents and inherit genes from both.
sexual selection A type of selection in which the evolutionary winners out-reproduce other individuals of a population because they are better at securing mates.
shells (chemistry) Specific regions surrounding the nucleus of an atom in which electrons can be found.
shoots In most plants, the parts that grow above the ground like the stems, leaves, and reproductive organs such as flowers.
short tandem repeats (STRs) Sequences of DNA in which a short (usually two- to six-nucleotide) sequence is repeated many times in a row. Also called microsatellites.
sieve tube Long tubes in phloem made of stacked cells that transport sugars.
silent mutation A mutation that changes the DNA, but not the amino acid.
single nucleotide polymorphisms (SNPs) An area where genes differ by a single different nucleotide. SNPs are the most common genetic variation among humans.
single-trait cross A way of tracking the inheritance of one gene.
sink A region where sugars are used or stored, such as fruits or roots.
sister chromatids Identical chromosomes created during DNA replication.
sister groups Two closely related species within a clade.
sixth mass extinction A current mass extinction event driven by human activity, including land use, water use, and climate change.
skeletal muscles Voluntary muscles that function in intentional movement and maintain posture.
skepticism When someone approaches any new scientific claim with an attitude of doubt.
sliding-filament model The model describing how interactions between actin and myosin filaments bring about muscle contraction.

slow-twitch muscle fibers Muscle fibers that allow for slow, extended contractions and are resistant to fatigue.

smog A mixture of ground-level ozone and particulate matter.

soil A mix of mineral particles, water, air, and organic matter that provides plants with nutrients and support.

solar energy Energy from the sun.

solutes Substances dissolved in a liquid such as water.

solvent A substance, usually a liquid, in which other substances dissolve.

somatic nervous system The part of the peripheral nervous system that controls skeletal muscles and informs the central nervous system about external conditions.

source A region where sugars are made or released, such as leaves.

spaced practice When multiple study sessions are spread over time.

space-filling model A representation of a molecule used to show a molecule's overall three-dimensional shape.

speciation The emergence of a new species due to genetic differences accumulating between populations.

species A group of individuals that can reproduce with one another to produce offspring that can also reproduce.

species diversity Assortment of species in a given area; described in terms of species richness and species evenness.

species evenness The relative abundance of each species present in an area.

species richness The numbers of different species present in a given area.

Species Survival Plan (SSP) A program developed to breed animals that are threatened, vulnerable, or endangered.

specific epithet The second part of a species name, distinguishing it within its genus.

spermatogenesis The process of making sperm; results in four equally sized sperm cells.

spindle Array of microtubules that move chromosomes during mitosis and meiosis.

spindle microtubules A temporary structure that forms at the spindle poles and attaches to chromosomes at their centromeres and moves chromosomes during nuclear division.

spindle poles The areas on opposite sides of the cell where the spindle originates.

sponge The first animal to evolve. Sponges are multicellular but lack tissues.

spongy bone Bone tissue with internal spaces filled with marrow; found in places such as the bone shaft and ends.

spontaneous chemical reaction A chemical reaction that releases energy and will proceed on its own without the input of energy.

spore A reproductive unit, often a single cell, that can give rise to a new organism.

sporophyte The diploid, spore-producing stage of a plant's life cycle.

stabilizing selection A form of natural selection where an intermediate form of a trait is adaptive; extreme forms are selected against.

stamen The male reproductive part of a flower.

starch A coiled polysaccharide produced by plants for energy storage.

start codon The first AUG in a typical mRNA that acts as a signal to start translation.

stasis A long period during which a species experiences little or no evolutionary change.

statistics A branch of mathematics that includes analytical tools (or tests) for organizing and interpreting data. Statistics enable a researcher to assess whether an observed effect is the consequence of the variable being measured or is due to random chance.

stem The part that forms the framework of most flowering plants, providing support and keeping leaves best positioned for photosynthesis.

stem cells Unspecialized cells that can either divide to produce more stem cells or differentiate into specialized cell types.

steroids Lipids characterized by a rigid backbone containing four carbon rings.

stewardship In the context of the environment, people taking responsibility for managing and protecting natural resources, such as thorough daily efforts toward conservation and long-term commitment to environmental health.

sticky ends The complementary single strand ends created when some restriction enzymes cut at a restriction site.

stigma The sticky top part of a carpel that catches pollen.

stillbirth Fetal death after 20 weeks of pregnancy.

stomata Pores or openings, usually on the underside of leaves, that help with gas exchange and water regulation.

stop codon A codon that signals for release factors to bind, signaling the end of translation.

stream A waterway that typically originates from runoff or melting snow or ice.

stroma Cytosol-like fluid between the thylakoid membrane and the two outer membranes of a chloroplast; contains enzymes and the chloroplast's own DNA.

structural formula The structural formula of a molecule lists the atoms present using their shorthand letter abbreviations.

study cycle Framework for learning that has three parts: previewing/reviewing, understanding, and practicing.

substrates Starting materials (reactants) in an enzyme-catalyzed reaction.

surface-to-volume ratio The relationship between the surface area of an idealized cell to its volume. As a cell increases in size, its volume typically increases faster than its surface area, which sets limits on how large the cell can grow.

survivorship curve A plot showing how many members of a cohort remain alive over time.

symbiosis Any interspecific interaction in which two species have a permanent or long-lasting physically close association.

sympathetic nervous system The division of the autonomic nervous system that is most active in times of excitement or danger, initiating the "fight-or-flight" response.

sympatric speciation Speciation that occurs without a physical barrier, due to genetic changes leading to reproductive isolation.

synapse A space between two neurons where chemical signals pass.

synaptic cleft The narrow space that separates the axon terminal of a signal-sending neuron from the dendrites of an adjacent signal-receiving cell.

system The group of organs, including the brain, spinal cord, and a wide network of nerves, that facilitates our responses to stimuli such as heat, cold, and pain.

T

T-cell receptors Receptors on the surface of T cells that allow them to recognize antigens.

T cells Lymphocytes that play a central role in adaptive immunity, including helper T cells and cytotoxic T cells.

taproot A large main root that grows deep, sometimes with smaller lateral roots.

tarsiers Small primates with dry noses and noncleft upper lips.

taxon A group of organisms that share a unique set of traits within the classification hierarchy.

taxonomy The practice of naming, describing, and classifying species.

telomerase An enzyme that maintains the length of telomeres by adding new nucleotides to their ends.

telomeres Short DNA sequences that are repeated hundreds or thousands of times that bind to proteins, shielding chromosome ends from damaging cellular processes.

telophase Stage of mitosis during which chromosomes arrive at the spindle poles and become enclosed by a new nuclear envelope.

temperate deciduous forest Contains broad-leaved deciduous trees that lose their leaves each fall.

temperate grassland Contains perennial grasses and other nonwoody plants.

temperature Measure of molecular motion.

template strand The strand that RNA polymerase reads when making mRNA.

temporal isolation Differences in the timing of reproduction that prevent interbreeding.

tendon A straplike extension of connective tissue that attaches muscle to bone.

terminal bud The bud at the tip of a stem where the plant grows longer.

terrestrial biomes Land-based biomes.

tertiary consumers Organisms that consume primary and secondary consumers. Often the top predators in the ecosystem.
tertiary structure The three-dimensional shape of a protein that results from amino acid side chain interactions.
testes Gonads that produce sperm and secrete predominantly testosterone.
testosterone A steroid sex hormone essential for reproductive function and secondary sex characteristics like body hair.
tetrad A structure formed during prophase I of meiosis when homologous chromosomes pair up closely so that their corresponding chromatids align.
tetrapods Animals with four bony limbs.
theory A well-tested and widely accepted explanation that is useful for making predictions about a wide range of natural phenomena.
thermocycler A machine that has various temperatures in a cyclical manner; commonly used for polymerase chain reaction (PCR).
thylakoids Disk-shaped structures found in stacks inside the stroma of a chloroplast; they contain the photosynthetic pigments used for capturing light and are the location of photosynthesis.
tight junctions Proteins that join the membranes of adjacent cells.
tissue One or more types of cells that are organized in a specific pattern and collectively carry out a particular task.
topoisomerase The enzyme that keeps knots from forming in front of helicase as the DNA strands separate during replication.
total fertility The average number of children born to individuals in a population over their lifetimes.
tragedy of the commons An economic theory that states when people have unrestricted access to a finite resource, if they act in their own short-term self-interest, then the resource can be depleted and destroyed for all.
traits Inherited features such as body shape, physiology, or behavior.
trans fatty acids Fats created industrially by the hydrogenation of vegetable oils, resulting in a product that has longer shelf life but imposes health risks on consumers.
transcription bubble An opening in the DNA helix caused by polymerase as it transcribes the gene.
transcription factors Proteins that regulate gene expression by binding to specific DNA sequences.
transfer RNA (tRNA) An RNA molecule used during translation to carry the amino acid to the ribosome.
transgenic Any organism whose DNA has been altered to include DNA from a different species.
translocation The movement of sugars and other nutrients through phloem.
transpiration The evaporation of water from leaves and stems.
transport proteins Membrane protein that passively or actively helps specific ions or molecules move across the membrane.
treatment group A group of individuals that receives an intervention such as a medication that researchers are studying. The treatment group is identical to the control group in all other ways. Also called the experimental group.
triglyceride A type of lipid consisting of a molecule of glycerol and three molecules of fatty acids. Also referred to as a fat.
trisomy An aneuploid condition in which a zygote has three copies of one type of chromosome.
trophic levels The energy hierarchy within an ecosystem.
tropical rain forest Areas having the most species diversity of any terrestrial biome, characterized by abundant rain, warm temperatures, and consistent day length.
tumor A mass of abnormally dividing cells in a tissue.
tumor suppressor A protein that prevents tumors from forming.
tundra An area with cold temperatures and a short growing season.
tunicates Chordates named for the secreted "tunic" that encloses the adult body.
turgor pressure The pressure of water pushing against plant cell walls, keeping the plant upright.
type I curve A convex curve where survivorship is high until late in life.
type II curve A diagonal curve where the death rate does not vary much with age.
type III curve A concave curve where mortality is highest early in life.

U

understanding Phase of the learning cycle in which newly learned material is moved to long-term memory.
uniform distribution A distribution wherein individuals are more evenly spaced than would be expected based on chance alone.
unsaturated fatty acids Fatty acids that have at least one double bond linking the carbons in their tails.

V

vaccination The intentional exposure to an antigen (via a vaccine) to induce an immune response and provide immunity to a specific disease.
valence electrons Electrons found in the outermost shell of an atom, which most readily participate in chemical reactions.
valence shell Outermost shell of an atom, containing the electrons that most readily participate in chemical reactions.
vascular bundles Tubes inside stems that conduct water, ions, and nutrients between different parts of the plant and also provide mechanical support.
vascular cambium A ring of dividing cells that produces secondary xylem (wood) inside and secondary phloem outside.
vascular tissue The system that transports water, minerals, and sugars throughout the plant.
vertebrates Animals with a backbone.
vesicles Membrane-enclosed "containers" that make it possible for the cell to move materials in bulk, including inside the cell as well as into and out of the cell.
vestigial structure A body part that has lost its function as a species has evolved.
vulnerable species A species likely to soon become endangered.

W

warning coloration A conspicuous pattern or color that predators learn to avoid.
water cycle Cycle in which water moves from oceans to the atmosphere, onto land and into freshwater ecosystems, and back to the oceans.
wavelength Distance between the crests (or troughs) of two successive waves.
wax A water-repellent substance that consists of a complex, varying mixture of lipids.
white blood cells Cells of the immune system that help prevent infection.
wood Layers of secondary xylem that give trees their trunks and branches.

X

X-inactivation The process by which an X chromosome can be turned off.
X-linked disorders Conditions related to genes on the X chromosome.
X-linked pattern A pattern of inheritance in which traits arise from genes located on the X chromosome and are typically seen in people who have a single X chromosome.
xylem The vascular tissue that distributes water and mineral ions.

Y

yeast A single-celled fungus with many uses for humans, such as in bread baking.

Z

zygote A diploid cell produced from the fusion of two haploid gametes at fertilization; the first cell of a new individual.

Index

Page numbers followed by "*f*" denote figures and "*t*" denote tables.

A

B

C

D

E

F

G

H

I

J

K

L

M

N

O

P

Q

R

W

X

Y

Z